International SAMPE Conference and Exhibition 2020

Online
1 June 2020

Volume 1 of 2

ISBN: 978-1-7138-2963-8

Printed from e-media with permission by:

Curran Associates, Inc.
57 Morehouse Lane
Red Hook, NY 12571

CURRAN ASSOCIATES INC. proceedings.com

Some format issues inherent in the e-media version may also appear in this print version.

Printed with permission by Curran Associates, Inc. (2021)

SAMPE
21680 Gateway Center Drive, Suite 300
Diamond Bar, California, USA
91765-2454

Phone: +1-626-521-9460

www.sampe.org

Additional copies of this publication are available from:

Curran Associates, Inc.
57 Morehouse Lane
Red Hook, NY 12571 USA
Phone: 845-758-0400
Fax: 845-758-2633
Email: curran@proceedings.com
Web: www.proceedings.com

TABLE OF CONTENTS

VOLUME 1

ADDITIVE MANUFACTURING

ADVANCED DESIGN AND ANALYSIS

AEROSPACE APPLICATIONS

EMERGING MATERIALS AND PROCESSES

INSPECTION & REPAIR

VOLUME 2

MANUFACTURING TECHNOLOGY

MARKET APPLICATIONS

MULTIFUNCTIONAL MATERIALS AND STRUCTURES

WORKFORCE DEVELOPMENT

HYBRID ADDITIVELY MANUFACTURED TOOLING FOR LARGE COMPOSITE AEROSTRUCTURES

Scott R. Huelskamp[1], Vern Benson[2], and Maria Pilar Muñoz-Lopez[3]

[1]University of Dayton Research Institute, 300 College Park, Dayton, OH 45469
[2]Northrop Grumman Innovation Systems, S Freeport Industrial Pkwy, Clearfield, UT 84015
[3]Airbus Spain, Madrid, Spain

ABSTRACT

Additively manufactured (AM) tooling has been shown to reduce cost and time in composite fabrication. However, AM tooling suffers from high, anisotropic thermal expansion that makes molding high accuracy composite parts at high temperatures very difficult. This is especially true for large structures with high aspect ratios and complex surface features. This paper discusses a new approach to overcome this challenge and allows AM tooling to be used in previously infeasible applications. A sandwich tool approach, made with a combination of a polymeric AM core and conventional tooling prepreg facesheets, was used to take advantage of the inherent strengths of each material while offsetting for the weaknesses. This approach was ultimately demonstrated on a multi-piece C-spar tool in support of an Airbus tooling development program. The tool exhibited isotropic thermal expansion that was nearly equivalent to that of invar at 180 degrees Celsius in both the x and y direction. The tooling was shown to be durable and dimensionally accurate through more than ten autoclave cycles.

1. INTRODUCTION

The commercial aviation industry has relied on composite materials for over 30 years. Although their use was limited upon their inception, there are aircraft in the skies today that have over half of their structure made using composite materials. Airbus' A350 XWB is a prime example. These materials are also being used in larger, more critically loaded components as composite material and processing technology continue to evolve. The wings of the A350 XWB are made from carbon fiber composite laminates. Measuring 32 meters long and 6 meters wide, these are among the largest composite structures ever made for commercial aviation [1].

While there are many advantages to composite aerostructures, they come with significant investment, particularly in regard to tooling. Traditional carbon/epoxy prepreg requires tooling able to survive 180 °C through hundreds of cure cycles while maintaining dimensional stability. Matching the thermal expansion of the tooling to the composite material, which in the case of carbon composites is almost zero, is also critical for hitting dimensional tolerances. Invar or composite tooling is mostly used to meet these requirements, but both are extremely expensive and can have very significant lead times. Steel may be used to help lower cost, but its higher thermal expansion limits its application.

Additive manufacturing (AM) is a relative newcomer to composite tooling, but it is proving to be a viable option for lowering tooling costs and lead times. The United States Air Force, for example, has sponsored numerous studies proving the effectiveness of AM tooling and continues to support such efforts [2] [3]. The commercial sector has also publicized multiple AM tooling efforts, showing significantly reduced costs and lead times [4] [5]; however, to date, no AM tool has fully addressed the need for low, isotropic coefficient of thermal expansion (CTE) at elevated temperatures. Without this trait, tool sizes and geometries will always be limited. Larger tool sizes will suffer from higher dimensional error at cure temperatures. Using a compensation value to design out the error in tools addresses some of the concern, but this approach is ineffective in female cavity tools or with parts containing trapping features down the length. Tool expansion and contraction will damage parts with these geometries during thermal cycling. Also, highly anisotropic CTE values, with are the norm for AM tools due to feedstock filler alignment during extrusion, make determining a correct compensation value very difficult and prone to error.

Much work is being done in the industry to address the CTE concern. Feedstock development, hardware changes to the AM equipment, and modeling efforts to better predict complex tool CTE are ongoing, but none of these have fully addressed the issue. The study presented in this paper takes a different, more brute force approach. A hybrid tool construction is used that begins with an AM shell structure printed net shape to the final part. The AM shell is laminated with a conventional tooling prepreg, which is then cured and machined to final shape. The AM shell allows for a "tool-less" lay-up of the tooling prepreg. The prepreg provides a more durable and nonporous surface as compared to the AM shell, and it also physically restrains the AM core from expanding at temperature. The end result is a tool shown to exhibit invar-like CTE values isotropically across the tool face at a significant cost reduction as compared to traditional tooling.

2. EXPERIMENTATION

2.1 Geometry

2.1.1 Tool Geometry

For this study, a c-spar from an Airbus tooling development program was selected as an ideal candidate to investigate this tooling approach. The c-spar is over 10 meters in length. A representative 3 meter long demonstrator of the c-spar was used to better match facility capabilities.

A female tool was selected so that the critical part surfaces were hard-tooled surfaces. The spar caps had a slight inward angle that would create a negative draft and tool lock condition in a single piece tool. This necessitated a split tool design.

2.2 Materials

2.2.1 AM Feedstock

SABIC provided the AM feedstock used in this project. A 20 % carbon fiber-filled polyetherimide (PEI) (Thermocomp™ EX004EXAR1 Ultem®) was selected due to the researchers' familiarity with the material and its proven ability to withstand typical autoclave cure cycles when used in composite tooling.

2.2.2 Tooling Prepreg

Hexcel's Hextool™ M81 epoxy tooling prepreg was selected for lamination of the AM shell. The material is a 2000 grams per square meter mat made of chopped carbon fiber impregnated with Hexcel's 8552 toughened epoxy system. The material boasts nearly isotropic properties that allows for post-machining without tool movement due to residual stress relief. It is a proven material for composite tooling. A bismaleimide matrix version of Hextool is also available, but with the limited expected lifespan of this tooling, the extra cost for this material was deemed unwarranted.

2.3 Equipment

2.3.1 Additive Manufacturing Equipment

Cincinnati Incorporated's (CI) Big Area Additive Manufacturing (BAAM) 1000 equipment was used in this study to print the thermoplastic composite tooling (Figure 1). BAAM was the first among the large format additive manufacturing (LFAM) platforms now on the market, and it has been used for countless aerospace tooling demonstrations. BAAM printers at CI (Harrison, Ohio) and Additive Engineering Solutions (AES) (Akron, Ohio) were used during this program.

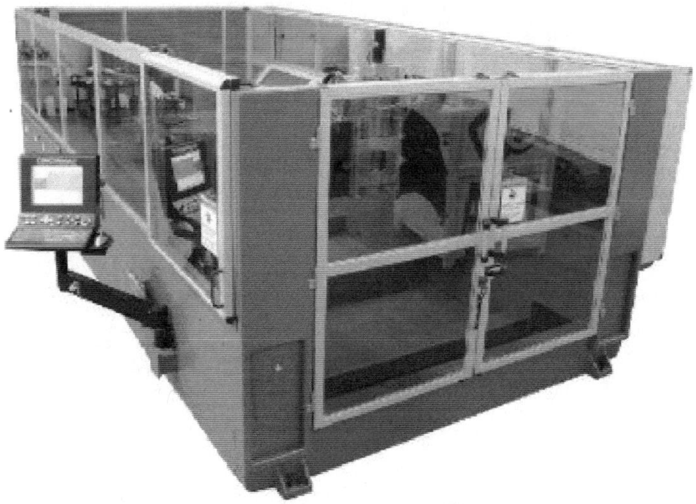

Figure 1. Cincinnati Incorporated's Big Area Additive Manufacturing printer

2.4 Coupon Fabrication and Testing

2.4.1 Fabrication

Initial modeling and hand calculations predicted that five plies of Hextool on each side of the AM shell would be sufficient to restrain the thermal expansion of the tool to a target range of less than 6 ppm/ °C. Two coupon panels measuring 150 mm by 150 mm were cut from a larger BAAM print. After preparing the panel surfaces for bonding, both panels were laminated with Hextool.

One panel was laminated with five plies and the other was laminated with 6 plies. Both panels were cured according the Hextool M81 datasheet [6].

2.4.2 Testing

After curing, holes were drilled into the four corners of each coupon panel (Figure 2). Precision metal dowels, measuring 4.5 mm in diameter, were installed in the holes. A handheld caliper was used to measure the distance between the dowels in the x and y direction. After recording these values, the panels were soaked in a 180 °C oven until the panels had reached steady state, as determined by an attached thermocouple. The panels were removed individually from the oven, and the distances between the dowels were quickly measured again while the panel was hot. This process was repeated with a 150 mm by 150 mm BAAM print without tooling prepreg as well. Once cool, the panels were inspected for signs of delamination, but none were found (Figure 3).

Figure 2. Test panel with Hextool lamination and metal dowels installed. Y-direction is the printed in-flow direction, while the x-direction is the printed cross-flow direction.

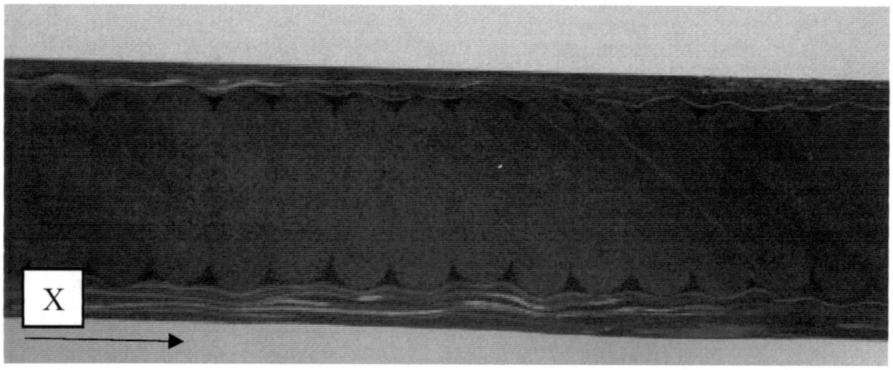

Figure 3. Cross section of laminated test panel after thermal cycling. Y-direction (in-flow) is into the page. X-direction (cross-flow) is labeled.

Using Equation 1, the CTE in the x and y direction of each panel was calculated. These are shown in Table 1.

$$\propto_L = \frac{\Delta L}{L\Delta T} \quad [1]$$

Table 1. CTE values for test panels

		PEI Tool (as-printed)	PEI Tool (with 5 plies HexTool)	PEI Tool (with 6 plies HexTool)
$CLTE_{177C}$	flow	5.8	2.6	2.4
ppm/C	x-flow	57.1	6.3	5.4

As can be seen, the CTE difference between the x (cross-flow) and y (flow) direction is an order of magnitude different in the as-printed panel. This confirms the high anisotropy expected in the AM shell that results from fiber alignment during printing. The CTE in the cross-flow direction is almost equivalent to the base polymer since there is almost no carbon aligned to that direction. After laminating with Hextool, the CTE in both directions drop, but the cross-flow direction drops significantly more at almost 90 % since it had such a high starting value as compared to the Hextool. The panel with six plies of Hextool provides a CTE in the flow direction of 2.4 ppm/ °C and in the cross-flow direction of 5.4 ppm/ °C. By comparison, invar tooling exhibits a CTE of approximately 2 ppm/ °C.

2.5 Sub-scale Fabrication and Testing

2.5.1 Design

A 0.6 meter tool was designed using a subsection of the C-spar geometry. The tool, shown in Figure 4, was split to allow for demold of the negatively drafted spar caps. An o-ring groove was added at the split line to interface with a surface bag to seal the tool during use.

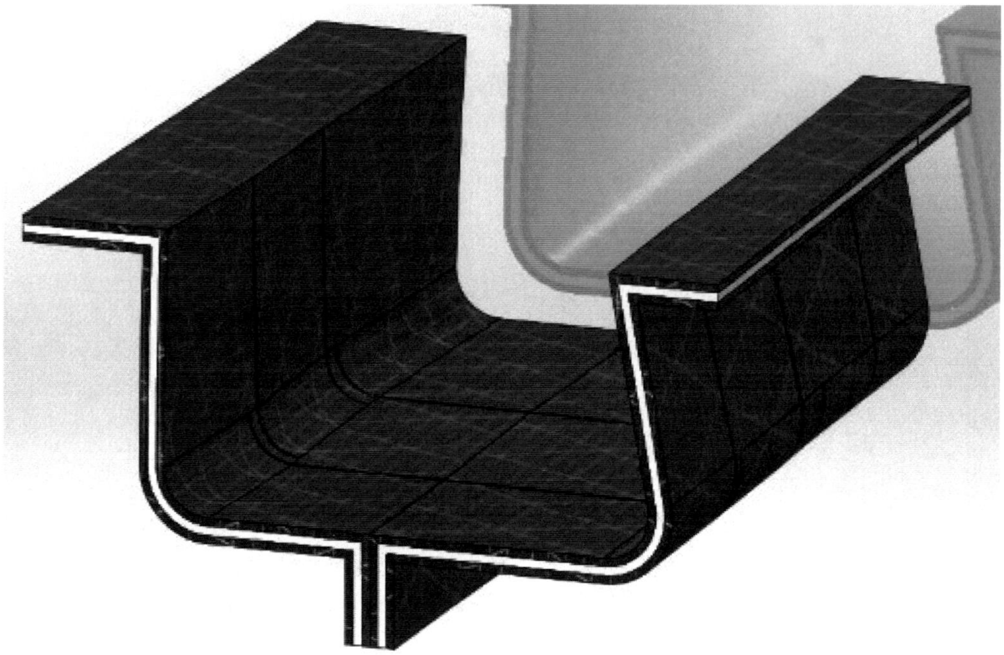

Figure 4. Subscale tool model

2.5.2 Fabrication

After validating at the coupon level that the tooling approach would successfully lower the CTE of the tool and survive thermal cycling, fabrication efforts moved to a subscale tool measuring 0.6 meters. For ease of manufacture, the preferred printing direction for the AM shell was a vertical build. At the subscale level, this was not a concern. However, a full-scale tool could not be printed to its full length in this orientation. The research team elected to split the tool down the length so that a method of bonding the tool could be established at the subscale level. The tool was also split into left and right sections to allow for the undercut issue mentioned in Section 2.1.1.

The tool sections were printed back-to-back to be more efficient during the print (Figure 5). After printing, the pieces were cut and the two right sections were bonded together (Figure 6), followed by the two left sections. Loctite EA9394 (Henkel) epoxy adhesive was used for the bonding.

Figure 5. Both sections of right half of the subscale tool BAAM print. Red lines denote cut lines after printing.

Figure 6. Right half of BAAM print after bonding

After bonding, each tool half was laid up with Hextool without any machining of the BAAM print. Avoiding this machining operation was deemed critical for achieving the tool cost and lead times needed to make this tooling approach an attractive alternative to conventional tooling. The backside of the tool, which would remain unmachined, was laid up with 5 plies of Hextool. The top side was laid up with 9 plies to allow for extra material to be removed during machining. Hot debulks (one hour soak at 93 °C) were preformed after ply 1, ply 4, and ply 7. The pregreg was cured using the "High Temperature/Standard" cure cycle provided by Hexcel (ramp to 120 °C, hold for 60 minutes, ramp to 177 °C, hold for 120 minutes with 690 kPa pressure).

After cure, the tool was sent for machining (Visioneering in Auburn Hills, Michigan). Pre-machining inspection indicated a risk of cutting into the core if a full surface clean-up was done due to thin areas in the prepreg facesheet. These areas were primarily isolated to the male radii and were a result of pressure intensification that thinned out the prepreg during cure. The decision was made to machine as deep as possible without risking a punch-through of the laminate. After machining, the areas that were not fully cleaned up were filled in with a high temperature tool filler and reworked to get the desired finish.

2.5.3 Testing

The finished tool was probed with a FARO arm at room temperature to determine overall length and width at four discrete pick-up points. The tool was then placed in a 180 °C oven and allowed to soak until reaching steady-state. It was removed from the oven and the four pick-up points were quickly probed to determine the tool's length and width at elevated temperature. The tool was also inspected for physical damage, most notably delamination between the face sheets and core, of which, none was found. The two sets of measurements allowed for the change in width and length, and subsequently, the linear coefficients of thermal expansion, to be calculated.

A leak check was performed on the tool. Initially, a surface bag was applied that tied into the silicone o-ring cordstock seal at the mating surfaces between the two tool halves. The tool exhibited excessive vacuum loss, and it was determined, through the use of dye penetrant, that the leak was across the cordstock. Given time constraints, the decision was made to envelop bag the entire tool and improve the cordstock seal at a later date.

A section of the C-spar lay-up was applied to the tool surface. Unidirectional carbon fiber/epoxy prepreg (M21E/34%UD194 IMA12k, Hexcel) was laid up across the tool surface with a ply drop at the midpoint of the tool. The root section of the lay-up was 54 plies while the tip end reduced to 38 plies. The ply distribution was as follows: 0 deg = 26%, +/-45 deg = 48%, 90 deg = 26%. The part was debulked and cured according to Hexcel's M21 datasheet [7]. It demolded easily from the tooling (Figure 7).

Figure 7. C-spar laminate laid up on tool (left) and after demold (right)

After cure, the part was demolded and the tool was subjected to further cycling to test durability. The tool was vacuum bagged (without prepreg) and cycled in an autoclave so that it experienced the same ramps and high temperature soaks that would be present during actual part cures. A pressure of 690 kPa (100 psi) was maintained on the tool during the cycling. Ten total cycles were ran. A second C-spar lay-up, identical to the first, was fabricated on the tool at the conclusion of the "dry" cycles.

A vacuum leak check was redone on the tool at this point, with the goal of improving the mating flange seal so that a surface bag could be used. This was especially important in advance of fabricating a full-scale tool. The o-ring cordstock was increased in size from the designed-for 5.33 mm diameter stock to the next bigger stock size of 5.7 mm. The material was also switched to Viton® to improve the adherence of the vacuum tape to the cordstock. These changes resulted in a leak rate of less than 3.4 kPa (1.0" Hg) in five minutes and gave assurance that with these modifications, a full-scale tool would also seal well.

2.6 Full-scale Fabrication

2.6.1 Design

A full-scale, 3.3 meter tool was designed using the entire geometry of the C-spar (Figure 8). The same design methodology used for the subscale tool was repeated, although a few additions were included. First, scribe lines were added to denote the end-of-part (EOP) of the cured C-spar. A double o-ring groove was added to give additional confidence in achieving a good seal in the finished tool. Alignment pins were added on the mating flange to drive repeatable alignment between both tool halves. Lastly, 6.35mm (0.25in) bushings were added on the top flange of the tool halves (four total). During fabrication, plates stamped with the actual center locations of these bushings would be adhered to the tool to aid in tool alignment during part lay-up, as well as serve as pick-up points for confirming tool CTE.

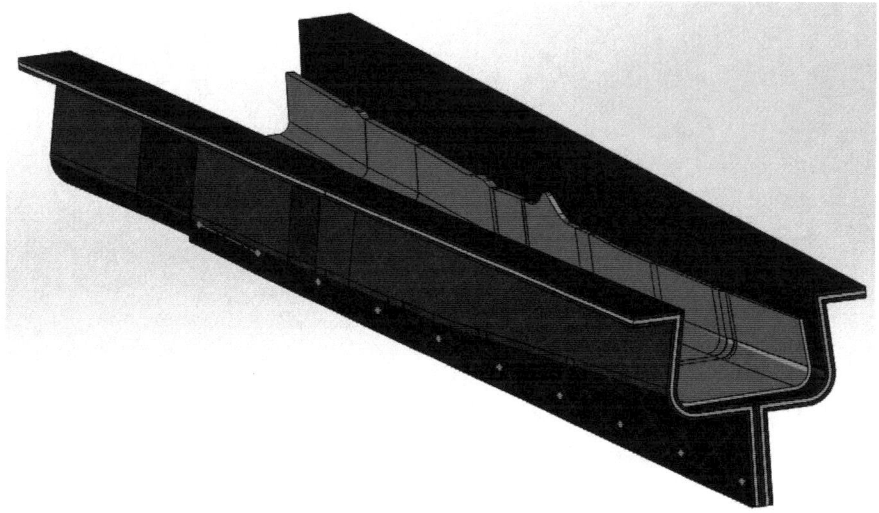

Figure 8. Full-scale C-spar tool design

2.6.2 Fabrication

The full-scale tool core was printed in two pieces, split at roughly the midspan, to accommodate print height restrictions on the BAAM printer. The left and right half was combined into a single print for each of the two pieces to allow for more efficient printing. The two pieces were bonded together with Loctite EA9394 and clamped together with pipe clamps prior to cutting out the left and right halves (Figure 9). After the adhesive had set at room temperature, the left and right halves were cut apart to accommodate for the parting line on the final tool.

Figure 9. Full-scale tool, clamped for bonding. Red line denotes split line, cut after bonding.

The two tool halves were scanned and the surfaces were compared to CAD to confirm that the 3D structures hadn't sprung out of shape when cut out of the main structure. Both tool halves showed isolated areas of excessive movement. These areas were locally heated with a heat gun and manipulated so that they better conformed to the original design intent.

The two halves were laid up with Hextool prepreg, with a slight modification from the subscale tool. The prepreg was redistributed so that four plies were added to the backside and 10 plies were added to the topside (layup side). In addition, an extra strip of prepreg was added on all male radii. These two modifications allowed for more face sheet thickness on the tool surface for machining. As before, hot debulks were done after plies 1, 4, and 7, and the tool was cured using the "High Temperature/Standard" cure cycle.

After cure, the tooling surface was machined to meet geometric tolerance and surface finish requirements.

3. RESULTS

3.1 Sub-scale Test Results

The completed tool was inspected with a CMM probe to confirm that the required profile tolerance was achieved. A machined surface of + / - 0.25 mm (0.010") from nominal was confirmed on the layup surface (Figure 10).

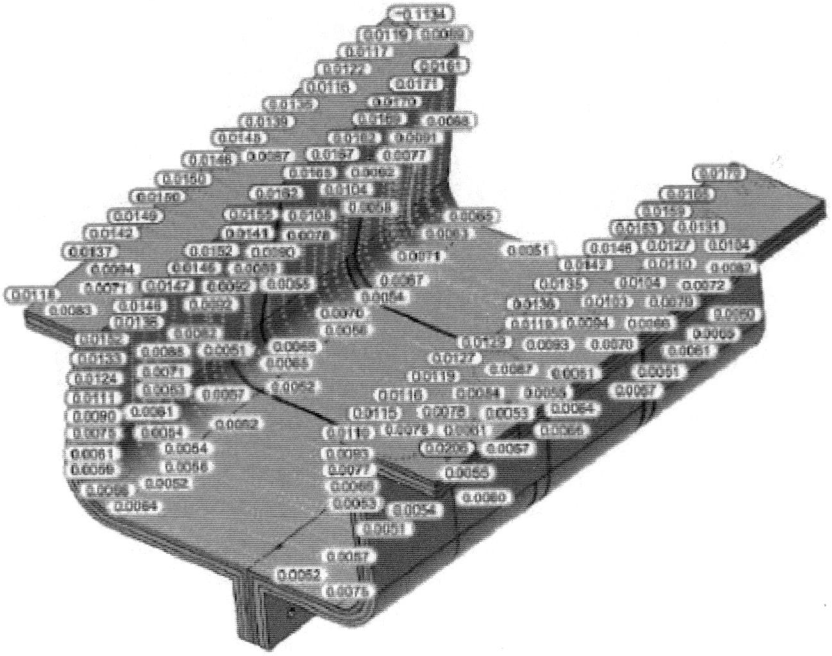

Figure 10. CMM scan map of subscale tool showing deviation from CAD model (shown in inches)

In addition to a profile scan, the location of four pick-up points located in each corner of the top flanges were measured at room temperature. After heating the tool to 180 °C, these pick-up point locations were measured again while the tool remained hot. These measurements were used to determine the tool expansion in the x and y direction as a result of temperature (Figure 11), and subsequently, the linear CTE of the tool in both principle directions. Down the length of the tool, the CTE was calculated to be 2.9 ppm/ °C, while the calculated CTE in the transverse direction was 6.0 ppm/ °C.

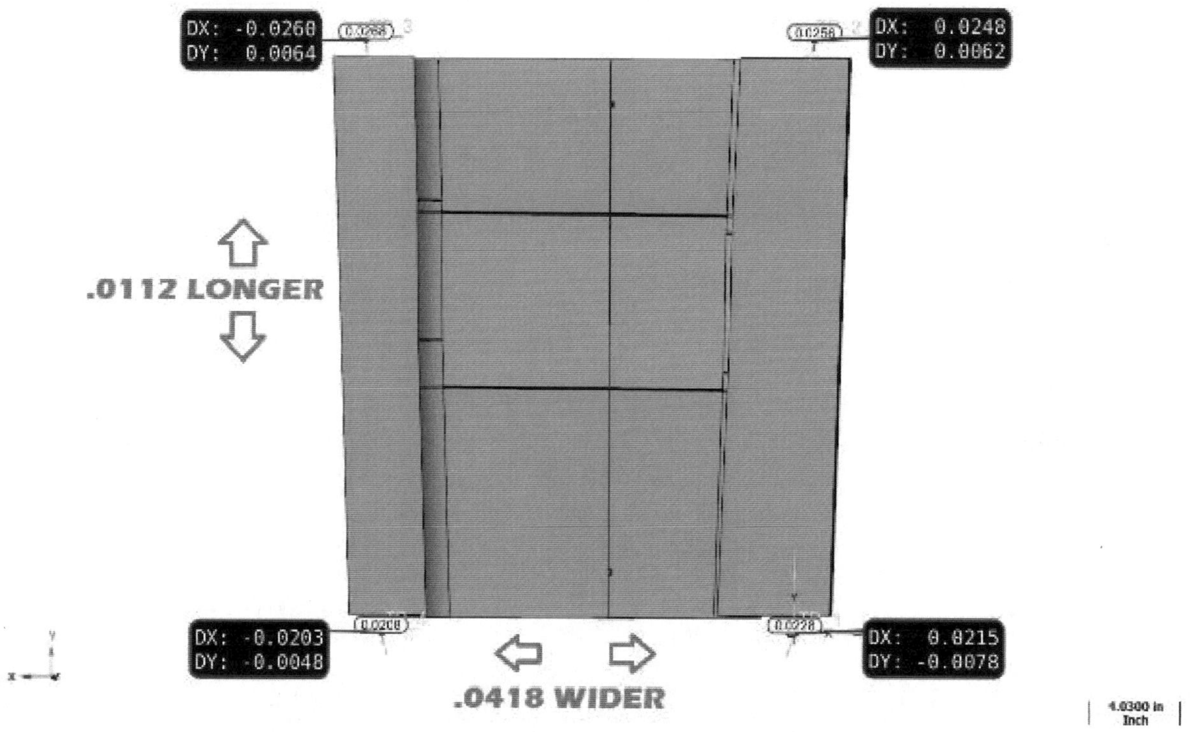

Figure 11. Tool expansion as a result of heating from room temperature to 180 °C (shown in inches).

The transverse direction of the tool was also the flow direction of the underlying BAAM print. The unreinforced BAAM print was shown to have a CTE of 5.8 ppm/ °C, already less than what was being measured on the actual tool. It is suspected that residual stresses in the tool were causing a springback condition, causing the top flanges to move further away from each other. This artificially inflated the calculated CTE value in this direction. The actual transverse CTE is expected to be 3 ppm/ °C or lower, although this has yet to be tested.

No delamination or other physical damage was identified at any point during the thermal cycling of the tool. A surface scan of the tool prior to fabricating Part 1 was compared to a scan of the tool after fabricating Part 2 (Figure 12). After 12 cure cycles, the tool was shown to have moved less than + / - 0.125 mm (0.005").

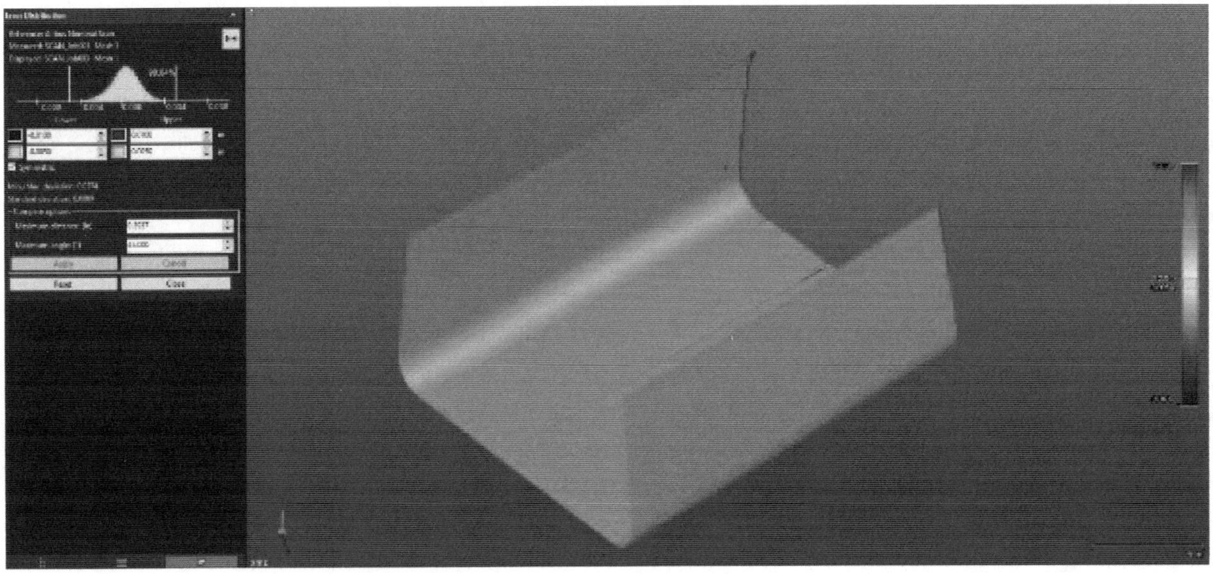

Figure 12. Comparison scan of subscale tool surface before and after 12 thermal cycles. Green shading indicates permanent movement between cycles of less than 0.125 mm.

4. CONCLUSIONS

4.1 Economics

An important aspect in this research was showing a significant cost reduction in the fabrication of this tooling when compared to traditionally machined invar tooling. Costs of the hybrid tool were recorded as the tool was being manufactured. Actual quotes from AES and Visioneering were used to capture the printing and machining costs, respectively. Industry-average labor rates were used to approximate labor costs so that university research labor rates did not skew the data. After compiling, the total cost of the tool was $24,136. Costs were roughly split equally between machining costs and printing/lamination costs. By comparison, Visioneering provided a quote to reproduce an identical tool using invar. This came to $46,775, for a cost savings of almost 50 %.

4.2 Ancillary Benefits

In addition to controlling the CTE of the tool, the prepreg used in the hybrid tooling approach was shown to provide a pit-free surface able to maintain vacuum integrity without additional sealing or coating. In fact, this approach was so successful that it allowed for a mulit-piece tool design sealed with an o-ring, something that to the best knowledge of the authors, has not been previously demonstrated in AM tooling at high temperatures.

AM tooling alone often suffers from significant vacuum loss at cure temperatures and pressures, stemming from internal porosity exposed during machining. Additional filling rework and surface coatings are required to improve this, although coating failure through thermal cycling is common and often results in the premature scrapping of the tool. Using the tooling prepreg surfaces negated these concerns.

A second ancillary benefit is the reduced weight of the tool. The 0.3 m hybrid tooling was able to be picked up and moved by hand. An invar tool would weigh in the range of 50 to 100 kg. This weight savings makes moving a tool significantly easier, allowing for much lower rated lifting provisions. This becomes increasingly important for 10 meter long tooling that can exceed the lifting capacity of a gantry in some manufacturing spaces. The reduced mass also allows easier storage on racks and shorter heat up and cool down rates in an autoclave.

4.3 Final Conclusion

The hybrid tooling approach taken in this program yielded a tool that met the operational performance required by Airbus and Northrop Grumman to be considered a viable replacement to invar tooling. A cost savings of almost 50 % was shown on a subscale tool. The approach was also shown to be appropriate for limited production applications, such as prototyping or one-off replacement part that require production-quality tooling but cannot afford high tool costs due to low amortization numbers. The end-of-life of the hybrid tool was not determined in this program, and could extend considerably beyond the 12 cycles that were already demonstrated. The Hextool material has been shown to produce tools capable of hundreds of cycles, so the possibility of extended tool life using the hybrid approach is not far-fetched [8].

The manufacture of a larger 3.3 m tool was demonstrated. Further work is needed to verify that the tool performance matches that of an invar tool. Cost comparisons will also need to be done to ensure the savings shown in the smaller tool scales to larger tooling.

5. ACKNOWLEDGMENTS

The authors would like to thank the following organizations for their financial support of this effort: Department of Energy (DOE), Institute for Advancement of Composite Manufacturing Innovation (IACMI), JobsOhio, and Airbus Americas. Also, the program would not have been possible without the technical contributions of the following partners: Northrop Grumman, Cincinnati Incorporated, Additive Engineering Solutions, SABIC, and Hexcel.

6. REFERENCES

[1] "Composites: Airbus continues to shape the future," Airbus, 1 August 2017. [Online]. Available: https://www.airbus.com/newsroom/news/en/2017/08/composites--airbus-continues-to-shape-the-future.html. [Accessed 19 November 2019].

[2] "Air Force Research Laboratory, Boeing, Thermwood Partner on Low Cost Responsive Tooling Program," Additive Manufacturing, 22 August 2019. [Online]. Available: https://www.additivemanufacturing.media/news/air-force-research-laboratory-boeing-and-thermwood-partner-on-low-cost-responsive-tooling-program. [Accessed 27 January 2020].

[3] S. Huelskamp and T. Osborn, "Large Format Additively Manufactured Tooling for Out-of-Autoclave Aerospace Composites," in *SAMPE*, Seattle, WA, 2017.

[4] "SABIC Explores Effects of Autoclave Tooling on Large Format Additive Manufacturing Parts," 3DPrint.com, 28 February 2019. [Online]. Available: https://3dprint.com/237358/sabic-explores-effects-of-autoclave-tooling-on-large-format-additive-manufacturing-parts/. [Accessed 27 January 2020].

[5] E. Milberg, "ORNL Tool for Boeing 777X Sets World Record for Largest 3-D Printed Item," Composites Manufacturing Magazine, 27 September 2016. [Online]. Available: http://compositesmanufacturingmagazine.com/2016/09/oak-ridge-national-lab-sets-guinness-world-record-largest-3-d-printed-item/. [Accessed 27 January 2020].

[6] "Hextool M81 Product Datasheet," Hexcel, 2016. [Online]. Available: https://www.hexcel.com/user_area/content_media/raw/HexTool_M81_DataSheet_eu.pdf. [Accessed 19 November 2019].

[7] "HexPly M21 Product Datasheet," Hexcel, [Online]. Available: https://www.hexcel.com/user_area/content_media/raw/HexPly_M21_global_DataSheet.pdf. [Accessed 31 January 2020].

[8] "Tooling," Composites World, 23 March 2016. [Online]. Available: https://www.compositesworld.com/articles/tooling. [Accessed 30 12 2019].

ADDITIVE MANUFACTURING OF MULTI-FUNCTIONAL LATTICE STRUCTURE-BASED CONCEPTS FOR THERMAL PROTECTION SYSTEM

Yu-Chuen Chang[1], Hao Wu[2], Colin Yee[1], Roderick Reber[3], Jon Langston[1], Steven Kim[2], Joseph H. Koo[1,2]

[1]Department of Mechanical Engineering, The University of Texas at Austin
Austin, TX/USA

[2]KAI, LLC.
Austin, TX/USA

[3]Arkema Inc.
King of Prussia, PA/USA

ABSTRACT

The potential of lattice structure as a dual-layer meta-material for ablative thermal protection system (A-TPS) was investigated. The geometry of the dual-layer sample comprises a solid Polyether-ketone-ketone (PEKK) regression layer and a lattice-structure insulating layer. The lattice structure was printed by Fused Filament Fabrication (FFF) process and designed with three different topologies (3D honeycomb, Gyroid, Schwarz D) with different relative densities (50% and 70%). The dual-layer lattice structures were tested in Oxy-Acetylene Test Bed (OTB) with 100 W/cm^2 heat flux for 30 seconds. The temperature at the end of regression layer and insulating layer was compared and the feasibility of using lattice structure as light weight heat insulation material was validated. The results can integrate with impact absorption properties of lattice structure and optimize the performance of TPS for various energy mission requirements.

1. INTRODUCTION

The advancement of Additive Manufacturing (AM) allows for producing complex geometry and printing material with multiple properties in a single process. Therefore, it could potentially be applied to optimize the design of the ablative thermal protection system (A-TPS) for various entry missions. The A-TPS is often a polymetric ablatives, which is a combination of regression layer (RL) with high thermal capacity and an underlying thermal insulating layer (IL) to protect space vehicle from the aerodynamic heating. Recently, Milos et. al. [1] developed a new dual-layer woven carbon phenolic for Heatshield for Extreme Entry Environment Technology (HEEET) project and changed the thickness of each layer to reduce the A-TPS mass. In this study, the potential of the AM process (*e.g.*, Fused Filament Fabrication, FFF) for the dual-layer A-TPS application was investigated, since the AM technology might provide greater flexibility to optimize the mechanical-thermal properties of materials.

Mechanical-thermal behavior is the primary consideration for designing A-TPS. The A-TPS needs to be lightweight and barrier of the aerodynamic heating loads for the space application. The

RL serves as flame resistant purpose and requires high ablation resistance, high decomposition temperature, and moderate thermal conductivity [2]. The IL maintains the interior temperature of the space vehicle and is typically low density and low thermal conductivity [1-2]. The design of lattice structure by FFF potentially enable to use the same material but vary the material properties for RL and IL. Catchpole-Smith et. al. [4] studied the change of thermal conductivity of three different lattice structures (gyroid, Schwarz D, and Schwarz P) made by Hastelloy-X and Ti6Al4V for the heat exchanger application. They interpreted the effective thermal conductivity of lattice structure as the function of solid material properties and volume fraction. Additionally, under the same volume frication, gyroid and Schwarz D structure provides lower thermal properties than Schwarz P. Habib et. al. [5] investigated bending, buckling, and stretch dominated structure and found the bending-based structures have lower stiffness and compression yield strength but better energy absorption. The stretch and buckling-based structures perform stiffer and stronger but have less capability to absorb energy. Maskery et. al. [6] also found Schwarz P (stretch and buckling-dominated architecture) has higher elastic modulus than Schwarz D and gyroid (bending architecture). In general, the increment of volume fraction will increase mechanical strength but also the thermal conductivity [3]. It is necessary to determine the trade-off between mechanical properties and thermal conductivity for A-TPS application.

To evaluate the feasibility of applying lattice structure on A-TPS designs, this study examined the heat insulation performance of dual-layer lattice structure by the OTB test. The RL of OTB specimen was solid for enhanced ablation resistance, and the IL was investigated with different lattice structures. The heat insulation capacity of lattice structure with various cell topologies and densities was experimentally determined by collecting temperature gradients conforming to ASTM E285-08. Additionally, the energy absorption performance of the lattice structures was characterized to better understand thermo-mechanical properties. The data collected enable us to demonstrate the efficiency of the lattice structure-based concept for A-TPS in the future research.

2. EXPERIMENTATION

2.1 Lattice Designs

Gyroid, Schwarz D, and 3D honeycomb were the three selected lattice structures in this study. Gyroid and Schwarz D are minimal surfaces and have good energy absorption properties [6]. Additionally, they have lower thermal conductivity in the family of minimal surfaces [4]. The lattice structure of 3D honeycomb represented the closed cell type, which is widely used for heat insulation in the industry [7]. The honeycomb structure is also bending-dominated and provides an energy absorption behavior.

The lattice structures were generated using infill function in ANSYS SpaceClaim and exported the files as STL (STereoLithography). The parameters to generate lattice structures are length l, wall thickness t, and relative density ρ. It should be noticed that the definition of length setup in SpaceClaim is "the spacing between the walls" instead of "length of unit cell". The term of l in this study followed the definition in SpaceClaim. The wall thickness of unit cells could vary the effective thermal conductivity of a lattice structure even at the equivalent level of density [4], hence the wall thickness kept constant for most of compression and OTB tests. All specimens were built using FFF process. Due to the precision of the FFF machine used in this study, the geometric precision was constrained to 0.1 mm.

2.2 Compression Test

2.2.1 Specimen designs and manufacturing

The lattice-structure specimens for compression tests were 5.08- × 5.08- × 2.54-cm (2- × 2- × 1-inch) cuboids conforming to ASTM D1621. All specimens were designed with 1 mm wall thickness and 25% of relative density. The range of relative density is recommended by Gibson et. al. [7]. Due to the precision limitation, the actual relative densities are slightly different between each unit cell type. Table 1 summarizes the geometric parameters of compression test samples.

Table 1. Specimen geometries of compression test

Unit Cell Type		Cell Design Parameters			Overall Specimen
		l (mm)	t (mm)	ρ^* (SpaceClaim)	
3D Honeycomb	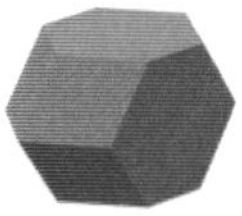	5	1	24.5%	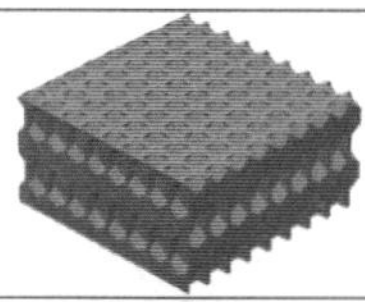
Gyroid	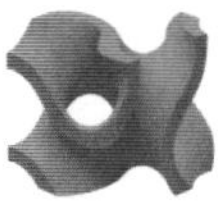	2.5	1	24.7%	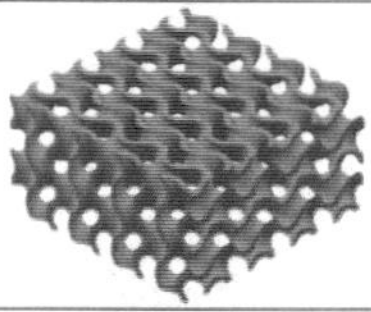
Schwarz D		3.1	1	24.6%	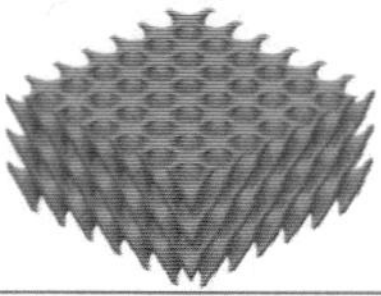

*: The density ρ is the modeling density shown in ANSYS SpaceClaim.

The specimens were produced using Craftbot XL 3D printers from polylactide (PLA) filament (procured form H&H 3D Plastics, https://www.hh3dplastics.com/). The print bed was parallel to the x-y plane. The FFF setting for PLA in this study was 215°C for the printing nozzle and 60°C for the print bed. The maximum speed of the printing was 120 mm/s. The printing infill is 100%. Figure 1 shows the printed specimens with three different lattice structures.

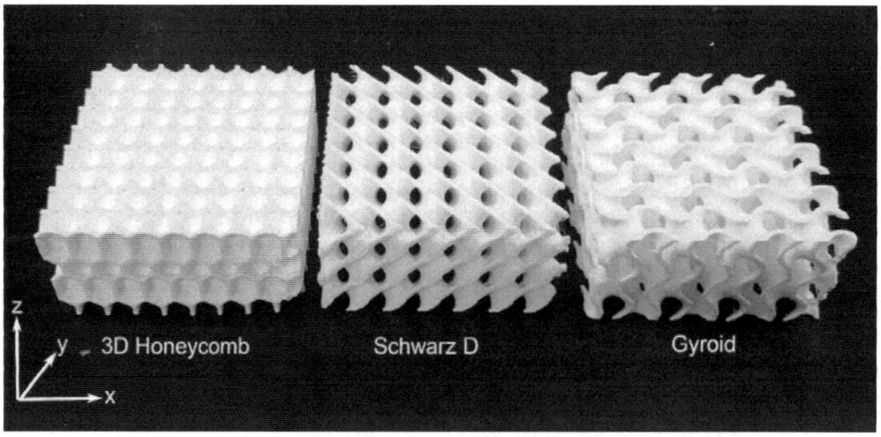

Figure 1. Compression test specimens produced by Craftbox XL 3D printer.

2.2.2 *Compression apparatus*

The compression tests were conducted with Instron 1125 of capability 20,000 lbf (88.96 kN) and followed the guidance of ASTM D1621. The specimens lied on the compression plate in x-y plane and were compressed in z direction. The testing speed was set to 0.2 in/min (5.08 mm/min). A compression test was continued until reached the densification regions. Three replicates of lattice structures were tested in this study. The initial collected data were displacement versus force and postprocessed to stress and strain curve with theoretical dimensions of samples in Microsoft Excel. Figure 2 shows the example of setup and orientation of the compression test.

Figure 2. Compression test setup and specimen orientation.

2.3 OTB Test

2.3.1 *Dual-layer lattice specimens*

The specimens of OTB tests were 30-mm diameter × 30-mm thickness dome-shapes. The OTB specimens had a 21-mm extruded hole with 2.54-mm of diameter to insert the thermocouples. The thickness ratio of the RL and IL was 3:7, which is close to the thickness setup in NASA's Heatshield for Extreme Entry Environment Technology (HEEET) project [1]. The ablative RL layer was 100% infill, and the IL layer was designed as lattice structures with 1 mm of outside wall thickness. Figures 3 and 4 illustrate the drawing and photography of the OTB specimens.

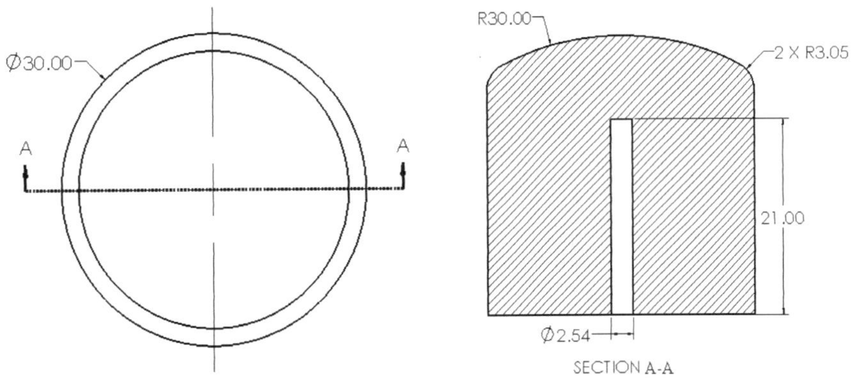

Figure 3. Overall dimensions of OTB specimens (*Left*) Top view (*Right*) Side view.

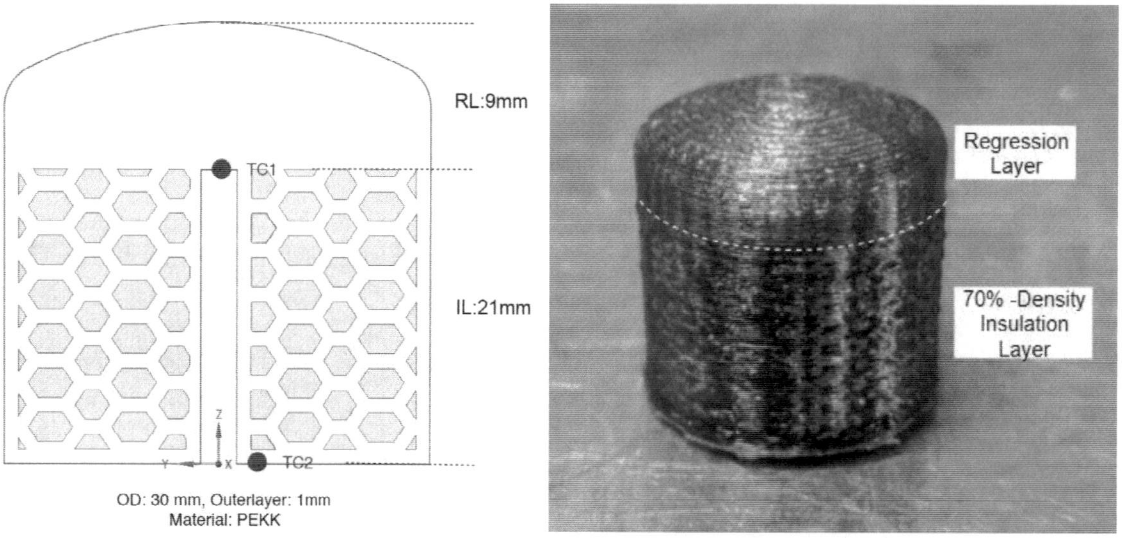

Figure 4. (*Left*) Illustration of OTB specimen designs and (*Right*) Sample D1 of dual-layer OTB test (the white dashed line was added to highlight boundary of RI and IL).

Cell topology and relative density have critical impact on the heat insulation capacity [4,7] and were examined in this study. Table 2 summarizes the designs of specimens for OTB tests. Sample R1 represented the reference material, which is solid for both RL and IL. The purpose of Sample D was to test material with the same cell topology and wall thickness but different densities. Cell D1 and D2 were 3D honeycomb with 70% and 50% of density. Sample T aimed to investigate the impact of different cell topologies with the same density and wall thickness. Cell T1 and T2 were gyroid and Schwarz D with 50% of density. Sample C was to study the influence from the wall thickness/cell size. In our study, C1 and D2 had the same cell topology (3D honeycomb) and density (48.8%) but different wall thickness/cell size.

Table 2. Design parameters of the lattice-structure insulating layers

Name	Cell structure	Spacing between the walls l (mm)	Min. wall thickness t (mm)	Ideal relative density	Modeling relative density ρ^*
R1	Solid	-	-	100%	100%
D1	3D honeycomb	1	1	70%	70.4%
D2	3D honeycomb	2	1	50%	48.8%
T1	Gyroid	1.27	1	50%	48.6%
T2	Schwarz D	1.57	1	50%	48.6%
C1	3D honeycomb	4	2	50%	48.8%

*Due to constraints of minimum printable features, the density ρ would vary slightly than ideal density.

2.3.2 Material and manufacturing

Polyether-ketone-ketone (PEKK) is the high-temperature thermoplastic with high strength-to-weight ratio and low flammability [8]. As a member of the polyaryletherketone (PAEK) family, PEKK has high temperature stability [9]. For example, Tadini et. al. [9] studied the chemical degradation behavior of carbon/PEKK and carbon/phenolic composites in 30°C to 1000°C and found that carbon/PEKK degraded at the higher temperature. Wu et. al. [8] compared the ablation performance of PEKK, PEEK (Polyether-ether-ketone), and PEI (Polyetherimide). The results of Thermogravimetric Analysis (TGA) and Micro-scale Combustion Calorimetry (MCC) tests indicate that PEKK has highest char yield, lower heat release capacity, and so better flame-resistance property. The specimens of OTB tests were produced by Arkema Inc. with Ultimaker 3S printer and Kepstan ® 7002 PEKK filament. Table 3 presents the material properties of PEKK properties at room temperature.

Table 3. Thermal properties of PEKK at 23 °C

Property	PEKK	Ref.
Density, g/cm^3	1.29	[10]
Thermal conductivity, W/(m·k)	0.23	[11]*
Heat capacity, J/(g·K)	1.02	[10]
Melting temperature, °C	331~334	[10]
Glass transient temperature, °C	162	[10]

*Thermal conductivity of PEKK is obtained by the regression result in [11].

2.3.3 OTB apparatus

The oxyacetylene test bed (OTB) utilizes a welding torch nozzle which mixes compressed oxygen and acetylene to generate a high heat flux, high speed flowfield for aerothermal testing of ablative materials. The OTB used for these tests is pictured below in Figure 5. A combination of radiative and convective heat flux is generated by the torch and applied to a test model held in place with a vice in order to simulate an extreme oxidative environment, such as a solid rocket exhaust plumes [12]. The heat flux applied to the test model is varied by adjusting the oxygen to acetylene flow rate ratio and the test model's distance from the nozzle and is calibrated using a Gardon gage.

Figure 5. Unshielded glass/phenolic test model undergoing OTB testing.

OTB testing is advantageous and cost effective tool as it provides a large amount of data from a single test, including surface temperature, back-side/heat soak temperature, in-depth temperature, mass change, volume change, depth of ablation, ablation rate, and pyrolysis zone depth. Morphological and compositional changes in the charred and pyrolyzing material can be examined after testing.

For this series of experiments, a carbon/carbon heatshield was glued to the PEKK test models using Aremco's GraphiBond high temperature carbon adhesive. A 31-mm diameter hole in the center of the heat shield permits heating to be applied to the solid dome portion of the test model while minimizing sidewall heating by deflecting flow out and away from the rest of the test model. The modified OTB test setup is shown in Figure 6. The insulative nature of the GraphiBond minimized heat transfer between the heat shield and the test model in order to prevent additional heating affects from the larger surface area or cooling affects due to heat sinking during the heat-soaked portion of the tests.

Figure 6. Modified OTB test sample and experimental setup.

One K-type thermocouple (TC1) was positioned at the back side of the solid dome while a second TC (TC2) was positioned at the back side of the metamaterial portion in order to determine the temperature difference across the metamaterial backing section and determine the peak heat-soaked time for each section of the PEKK test model (see Figure 3). The OTB applied a 100 W/cm^2 heat flux for 30 seconds to the solid dome of the 30 mm diameter, 30 mm thick PEKK test models. After 30 seconds, the test model was removed from the flowfield and TC temperature measurements continued until the peak heat-soaked time was detected.

3. RESULTS AND DISCUSSION

3.1 Compression Test

The stress-strain curve (σ-ε curve) of lattice structures shown in Figure 7 has three regions: an elastic region (stage I), a plateau region (stage II), and densification region (stage III). When the compressive load is applied, the cell wall of lattice structures bends at lower stress initially and provides a linear elastic deformation. When the compressive force is higher than a certain level, the cell wall begins collapsing due to elastic buckling (*e.g.,* elastomer), plastic yielding (*e.g.,* metal), or brittle crushing (*e.g.,* ceramic). The collapsing process of lattice structure causes a significant increase in compressive strain, but the stress maintains nearly constant, which is so-called plateau region. The plateau region allows greater absorption of impact energy without noticeable increment of the load on lattice structures. The cell walls will keep collapsing until they contact each other. In subsequent densification region, the lattice structure performs like a parent material and the stress increases rapidly [5,7].

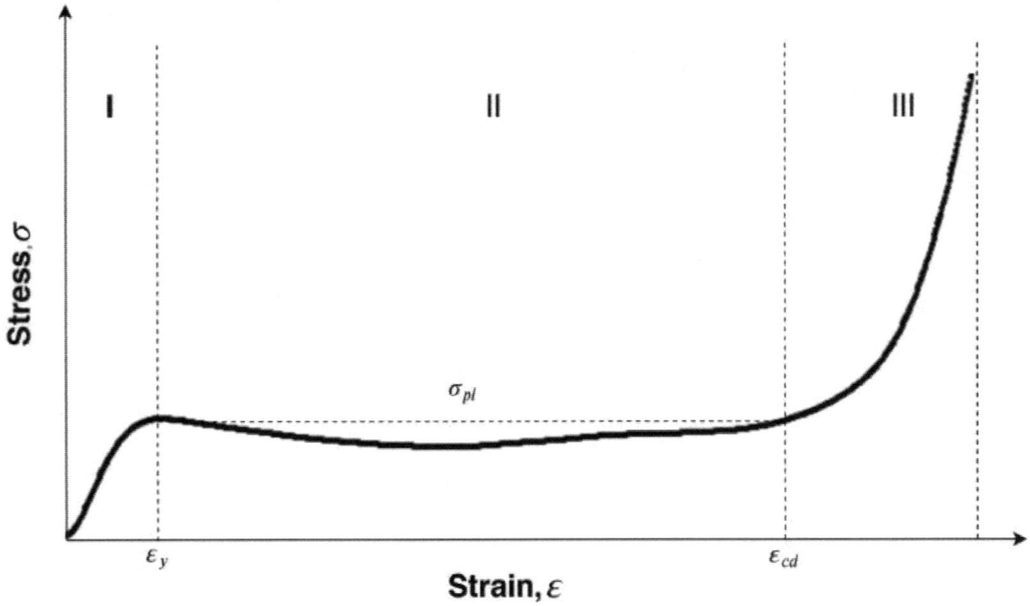

Figure 7. Schematic of compressive stress-strain curve for cellular materials (I), elastic region (II), and plateau region (III) densification region (modified from [5]).

The energy absorption diagram [7] is commonly used to determine the boundary of the plateau and densification region. It describes the relation between the peak stress and absorbed energy *W* in the specific range of strain ε. The absorbed energy W is equivalent to the area of σ-ε curve [13] and defined as

$$W = \int_0^{\varepsilon} \sigma(\varepsilon)d\varepsilon \tag{1}$$

The plateau stress σ_{pl} is another important parameter to observe the energy absorption property of lattice structure. The plateau stress defined the average of stress in the plateau region [14] (between yield strain ε_y and critical strain at the end of plateau region ε_{cd}) as shown in Figure 7. The plateau stress is given by

$$\sigma_{pl} = \frac{\int_{\varepsilon_y}^{\varepsilon_{cd}} \sigma(\varepsilon)d\varepsilon}{\varepsilon_{cd} - \varepsilon_y} \tag{2}$$

where the yield strain ε_y is obtained by the peak stress of the elastic region. The energy absorption efficiency η represents the ratio of actual energy absorption to the critical stress σ_{cd} at the end of plateau region [5,13] and described as

$$\eta(\varepsilon) = \frac{1}{\sigma_{cd}} \int_0^{\varepsilon_{cd}} \sigma(\varepsilon)d\varepsilon \tag{3}$$

Figure 8 shows the compression test results for the selected lattice structures and examples of energy-absorption diagram (*right* column in Figure 8).

A. 3D Honeycomb at 25% of the density

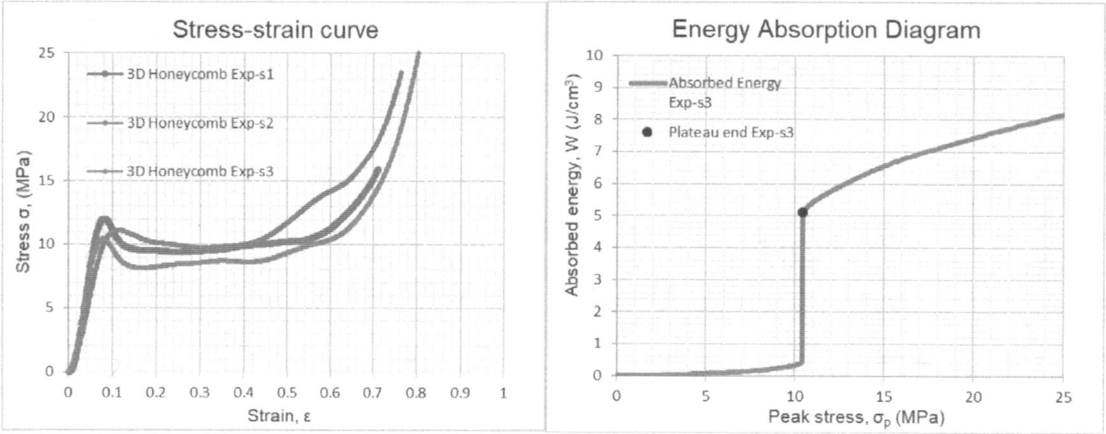

B. Schwarz D at 25% of the density

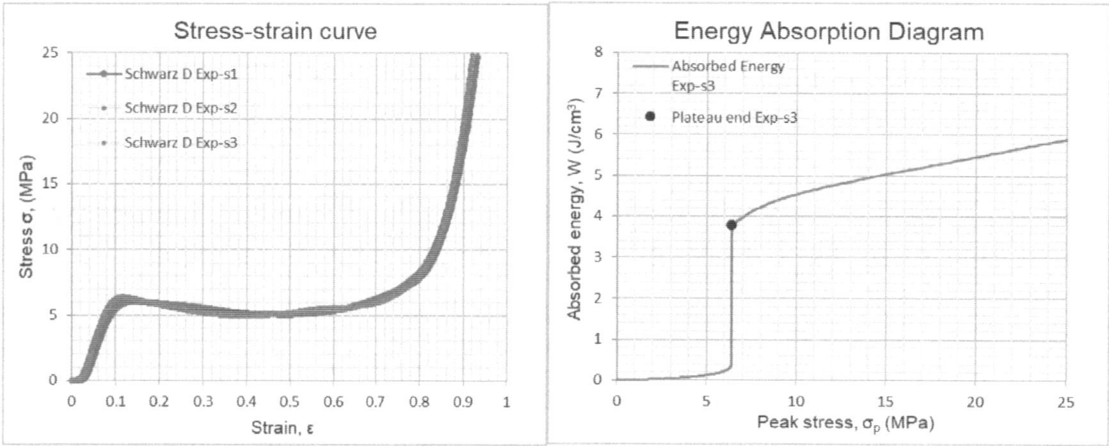

C. Gyroid at 25% of the density

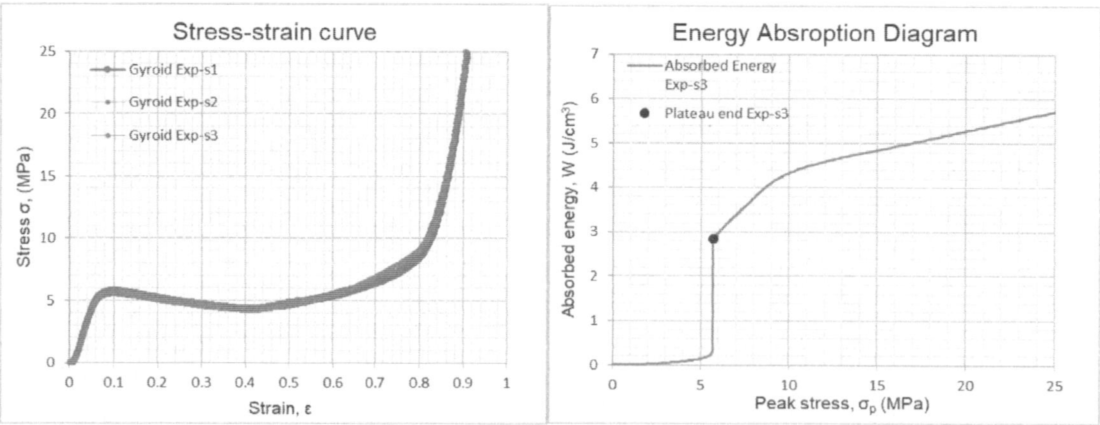

Figure 8. Compression stress-strain curves and representative energy absorption diagrams for the 3D printed PLA lattice structures with different cell topologies (a. 3D Honeycomb, b. Schwarz, and c. Gyroid).

Table 4 summarizes the average compressive properties of three different lattice structures. The stress-strain curves of gyroid and Schwarz D are similar and have a longer plateau region than 3D honeycomb. The strain at the end of the plateau region is 0.71 for Schwarz D and 0.64 for gyroid. However, both of them also has lower plateau stress than 3D honeycomb. The average plateau stress for Schwarz D and gyroid is 5.43 MPa and 4.95 MPa, respectively. Additionally, the range of plateau region in our test generally agrees with the experimental data in [6]. Overall, the 3D honeycomb structure has the best compressive properties. The 3D honeycomb has the highest plateau stress (9.69 MPa) and absorbed the largest amount of energy (5.15 J/cm^3).

Table 4. Average energy absorption properties of 3D-printed lattice structures

Lattice structure	Plateau stress σ_{pl} (MPa)	Critical strain ε_{cd}	Critical stress σ_{cd} (MPa)	Absorbed Energy W (J/cm^3)	η (%)
3D Honeycomb	9.69	0.57	11.18	5.15	47.82
Schwarz D	5.43	0.71	6.21	3.57	58.02
Gyroid	4.95	0.64	5.72	2.94	51.81

3.2 OTB Test

Figure 9 shows the temperature difference between TC1 and TC2. During OTB test, the temperature of TC2 for most of the sample rapidly increased in the first 30 sec. The thermal shock was due to the heat conduction from the carbon/carbon heat shield. The solid reference, R1, experienced a thermal shock as high as 80°C. The thermal shock for sample D1 (3D honeycomb with 70% of density) was around 55°C. Sample D2 (3D honeycomb with 50% of density) did not have significant thermal shock within the operating time. Similar behavior can be obtained from sample T1 (gyroid with 50% of density). Sample T1 also showed a smooth temperature curve of the TC2. This characteristic explains that a lattice structure with high porosity can provide a better thermal shock resistance. Furthermore, wall thickness/cell size has impact on the thermal shock resistance. Sample D1 and C1 have the same density and cell topology, but the wall thickness of sample C1 was doubled and so did cell size. Although sample C1 only have 50% of density, it still experienced a thermal shock of 67°C. The temperature curves of TC2 suggest that the ideal designs for thermal shock resistance prefer lower density and smaller wall thickness/cell sizes.

Table 5 summarizes the collected OTB data for the five different IL designs. The initial mass in Table 5 refers to the gross mass of the OTB specimen, heat shield, and high temperature carbon adhesive, and they were compared with the residual mass after heating. An IR camera was used to record the average surface temperature. To exclude the impact of thermal shock, The TC2 data in Table 5 only considered the temperature change after the first 100 sec. The residual mass of all five specimens are similar. The temperature difference between TC1 and TC2 demonstrates the insulation capability of lattice structure: the larger temperature difference is preferable. All sample with lattice structure has higher temperature difference than solid OTB sample. The preliminary tests indicate that adding lattice structure might improve the capability of thermal insulation. However, before getting into a concrete conclusion, a primary improvement of experimental setups is required: during the experiments, the dome of sample R1 and C1 melted and separated from the rest of the specimens. This explains why the maximal temperature of TC1 for sample R1 and C1 is much lower than D1, D2, and T1. To prevent this melt and drip behavior, all printed OTB specimens will be crystalized to improve their thermal property. Additionally, the data recording in sample T2 was incomplete, and additional tests are needed.

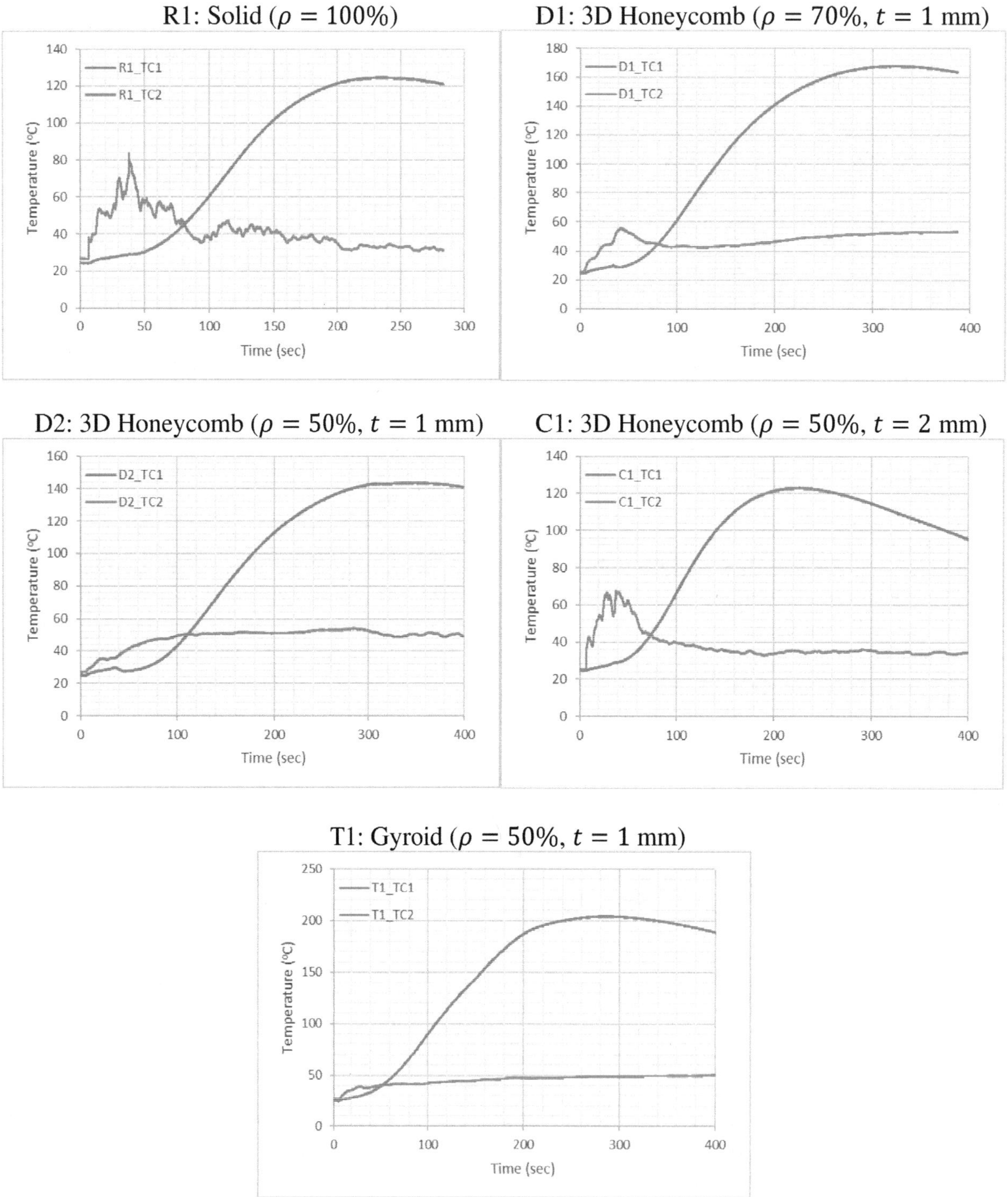

Figure 9. The measured temperature for the back of solid dome (TC1) and heat soak temperature (TC2) with different IL designs.

Table 5. Summary of the OTB data with five lattice structure designs

Sample	Initial mass (g)	Final mass (g)	Residual mass (%)	Average surface temperature (°C)	Max. TC1 (°C)	Max. TC2 (°C)	TC1-TC2 (°C)
R1	54.4	53.2	97.8	1588	124.6	47.5	77.1
D1	49.3	48.1	97.6		167.8	53.4	114.4
D2	50.8	49.5	97.5		143.9	54.1	89.8
C1	51.8	50.7	98.0		123.1	36.3	86.8
T1	48.8	47.4	97.2		204.6	50.6	154.0

4. CONCLUSIONS

This study evaluated the feasibility of using lattice structures for the insulating layer (IL) design of thermal protection systems. For mechanical properties, the 3D honeycomb can absorb more energy than gyroid and Schwarz D and has higher plateau stress. The results of OTB tests show that a lower density and smaller cell size of lattice structure provide a better heat insulation property. In addition, the lattice structured IL has better thermal shock resistance. The OTB data indicate that the lattice structure could be a possible solution for the IL. However, more data are still required to have a more solid conclusion and a more comprehensive characterization is currently in progress and the results will be presented in the coming conference.

5. REFERENCES

1. Milos, F. S., Chen, Y. K., & Mahzari, M. (2018). Arcjet Tests and Thermal Response Analysis for Dual-Layer Woven Carbon Phenolic. Journal of Spacecraft and Rockets, 55(3), 712-722. doi: https://doi.org/10.2514/1.A34142
2. Kumar, S., & Mahulikar, S. P. (2016). Selection of materials and design of multilayer lightweight passive thermal protection system. Journal of Thermal Science and Engineering Applications, 8(2). doi: https://doi.org/10.1115/1.4031737
3. Natali, M., Kenny, J. M., & Torre, L. (2016). Science and technology of polymeric ablative materials for thermal protection systems and propulsion devices: A review. Progress in Materials Science, 84, 192-275. doi: https://doi.org/10.1016/j.pmatsci.2016.08.003
4. Catchpole-Smith, S., Sélo, R. R. J., Davis, A. W., Ashcroft, I. A., Tuck, C. J., & Clare, A. (2019). Thermal conductivity of TPMS lattice structures manufactured via laser powder bed fusion. Additive Manufacturing, 30. doi: https://doi.org/10.1016/j.addma.2019.100846
5. Habib, F. N., Iovenitti, P., Masood, S. H., & Nikzad, M. (2018). Fabrication of polymeric lattice structures for optimum energy absorption using Multi Jet Fusion technology. Materials & Design, 155, 86-98. doi: https://doi.org/10.1016/j.matdes.2018.05.059
6. Maskery, I., Sturm, L., Aremu, A. O., Panesar, A., Williams, C. B., Tuck, C. J., ... & Hague, R. J. (2018). Insights into the mechanical properties of several triply periodic minimal surface lattice structures made by polymer additive manufacturing. Polymer, 152, 62-71. doi: https://doi.org/10.1016/j.polymer.2017.11.049

7. Gibson, L. J., & Ashby, M. F. (1999). Cellular solids: structure and properties. Cambridge University Press, Cambridge, UK. doi: https://doi.org/10.1017/CBO9781139878326

8. Wu, H., Kafi, A., Yee, C., Atak, O., Langston, J. H., Reber, R., ... & Koo, J. H. (2020). Ablation Performances of Additively Manufactured High-Temperature Thermoplastic Polymers. AIAA-2020-1125, AIAA SciTech 2020 Forum, Orlando, FL. doi: https://doi.org/10.2514/6.2020-1125

9. Tadini, P., Grange, N., Chetehouna, K., Gascoin, N., Senave, S., & Reynaud, I. (2017). Thermal degradation analysis of innovative PEKK-based carbon composites for high-temperature aeronautical components. Aerospace Science and Technology, 65, 106-116. doi: https://doi.org/10.1016/j.ast.2017.02.011

10. Arkema Inc. (2012). Kepstan by Arkema: Technical Data – 7000 Series. Retrieved January 26, 2020 from https://www.arkema.com/export/shared/.content/media/downloads/products-documentations/incubator/arkema-kepstan-7000-tds.pdf

11. Coulson, M., Dantras, E., Olivier, P., Gleizes, N., & Lacabanne, C. (2019). Thermal conductivity and diffusivity of carbon-reinforced polyetherketoneketone composites. Journal of Applied Polymer Science, 136(38), 47975. doi: https://doi.org/10.1002/app.47975

12. Yee, C., Ray, M., Tang, F., Wan, J., Tatuaca, R., Natali, M., & Koo, J. H. (2014). In situ ablation recession sensor for ablative materials based on ultraminiature thermocouples. Journal of Spacecraft and Rockets, 51(6), 1789-1796. doi: https://doi.org/10.2514/1.A32926

13. Avalle, M., Belingardi, G., & Montanini, R. (2001). Characterization of polymeric structural foams under compressive impact loading by means of energy-absorption diagram. International journal of Impact Engineering, 25(5), 455-472. doi: https://doi.org/10.1016/S0734-743X(00)00060-9

14. Li, Q. M., Magkiriadis, I., & Harrigan, J. J. (2006). Compressive strain at the onset of densification of cellular solids. Journal of Cellular Plastics, 42(5), 371-392. doi: https://doi.org/10.1177/0021955X06063519

HIGH-PERFORMANCE MATERIALS FOR MASK-PROJECTION STEREOLITHOGRAPHY VIA IN-SITU SEQUENTIAL INTERPENETRATING NETWORK

Anh N. Huynh, Nicolas J. Alvarez, Giuseppe R. Palmese
Drexel University
Philadelphia, PA/USA

ABSTRACT

Stereolithography, a common technique used in additive manufacturing, offers high dimensional accuracy. However, there are limited choices of materials for this process. Acrylates and epoxies are commonly used, but their thermal properties are not suitable for applications in which high temperature performance is needed. In contrast, high-performance thermosets such as bismaleimide and cyanate ester are cured using high temperature processing techniques, and their use for stereolithography is limited. In this work, a photocurable formulation of bismaleimide, a reactive diluent, and cyanate ester was developed to improve thermal and mechanical properties as well as cure shrinkage of stereolithography resins. *In-situ* sequential interpenetrating polymer network (IPN) were investigated wherein the copolymerization reaction between bismaleimide and diluent occurs during printing, resulting in a cyanate ester swollen network with sub room temperature glass transition temperature (T_g). The polymerization of cyanate ester takes place during post processing. The resulting material possesses a T_g well above 200°C (loss modulus peak), significant decrease in cure shrinkage, and improved toughness.

Keywords: 3D printing, interpenetrating polymer network, high temperature, shrinkage
Corresponding author: Giuseppe R. Palmese, grp27@drexel.edu

1. INTRODUCTION

Additive manufacturing has gained great popularity in recent years due to its advantages such as design freedom, customization, waste reduction, and low cost.[1] Stereolithography processes (SLA) and derivatives, like mask-projection stereolithography, are popular techniques in additive manufacturing due to their ability to provide prints with high resolution. The materials used for this process are constrained by photo-curing chemistry and are typically acrylates, methacrylates, and epoxies.[2] Therefore, the use of materials with high thermal properties are limited. Recently, direct ink writing (DIW) of cyanate ester with fillers was reported by Lewicki et al.[3] Other attempts in designing materials with high temperature stability using aromatic polyimides have demonstrated printability but have high dimensional shrinkage and poor mechanical properties.[4] There are few reports in the literature regarding the T_g of SLA materials, and the reported T_g's range from 87-160°C (tan delta).[5, 6] Parts processed by SLA can exhibit significant residual stress [7] because polymerization of thermosetting monomers results in cure shrinkage due to the formation of a covalent bonds.[8] During printing, residual stress is caused by material shrinkage while being restrained between substrates. Residual stress can result in warpage or delamination, which can affect mechanical properties of printed parts.[9] The degree of shrinkage depends on chemical structure and varies generally from 5–10% for common acrylates and methacrylates.[2] Another limitation of SLA is low extent of reaction during room temperature printing, particularly for high T_g systems. The reactions of thermosets are limited by vitrification so that the T_g attained during isothermal cure is close to the reaction temperature.[10] Lange et al. demonstrated that the photocurable materials can generate high residual stress because of significant shrinkage close to vitrification when the modulus increases significantly.[11] Our work focuses on interpenetrating polymer network (IPN) which utilize high-performance materials to improve thermal and mechanical properties, conversion during print, and shrinkage.

The use of IPNs for SLA has been reported. A common combination of chemistries used for the formation of sequential IPNs entails the free radical photopolymerization of acrylates followed by cationic polymerization of epoxies.[12,13] Another type of IPN used in SLA is the two-stage curing of acrylates and epoxies, where acrylates are photopolymerized, and the epoxies are thermally cured with a curing agent.[6] Our study utilizes two-stage curing but allows the processing of high-performance material that are traditionally thermally cured. Bismaleimide (BMI) and cyanate ester (CE) are not favorable for SLA as they require high curing temperature.[14-16] In addition, bismaleimide is a brittle material and generally requires toughening for more practical use. It is reported in literature that the formation of IPN between bismaleimide and cyanate ester provides a toughening effect.[17-19] A reactive diluent that can copolymerize with bismaleimide is needed in order to formulate a BMI-CE for SLA applications. Previous works showed that bismaleimide copolymerizes with acryloylmorpholine through electron donor acceptor mechanism under electron beam irradiation or UV irradiation in the presence of appropriate photoinitiators.[18,20] In this work, we investigate sequential IPNs wherein a first network of bismaleimide and acryloylmorpholine is formed during printing, and the second network of cyanate ester is obtained subsequently by thermal cure. The concept is demonstrated in Figure 1. This scheme is expected to result in delayed vitrification due to the presence of unreacted second network allowing a greater conversion of double bonds in the rubbery state and overall lower cure shrinkage because cyanate ester is known to have a low degree of shrinkage.[16]

Figure 1: The reaction sequence in IPN.

2. EXPERIMENTATION

2.1 Resin preparation

Poly(phenylmethane) maleimide (kindly supplied by Miki Sangyo), 4-acryloylmorpholine (TCI Chemicals), bisphenol E cyanate ester AroCy L-10 (Huntsman), Copper (II) napthenate (Strem Chemicals), and phenyl bis(2,4,6trimethylbenzoyl) phosphine oxide (Sigma Aldrich) were used as received. Poly(phenylmethane) maleimide (BMI) was dissolved in 4-acrcyloylmorpholine (AMP) at 70 °C at a concentration of 30 wt% BMI in AMP. Bisphenol-E cyanate ester (CE) was well mixed with copper (II) napthenate at concentration of 200 ppm Cu^{2+}. The photoinitiator, phenyl bis(2,4,6 trimethylbenzoyl) phosphine oxide (PPO), was dissolved in the resins at a concentration of 0.5 wt.% with respect to the total mass of the resin. Two resin systems were investigated: (1) 30 wt% BMI in AMP and (2) the IPN resin composed of 50 wt% of system (1) and 50 wt% of cyanate ester. Throughout the paper, system (1) is denoted as BMI-AMP and system (2) is denoted as IPN 50:50 (BMI-AMP):CE.

2.2 Printing and post-curing procedures

The resins were printed using Anycubic Photon printer with the LED light of 405 nm. The power density of the light measured at LCD surface was 0.51 mW/cm^2 (measured by ILT 2400 with sensor range of 275–450 nm). Both resins were printed using 100-micron layer thickness with 80 s exposure per layer. The printed samples were first post-cured in a UV oven (FormLabs) at 75 °C for 1 hr. The BMI -AMP system was then thermally post-cured at 120 °C for 3 hrs, 180 °C for 3 hrs, 220 °C for 4hrs, and 280 °C 1 hr. The IPN system was post-cured at 150 °C for 2 hrs, 180 °C overnight, 220 °C for 3 hrs, and 280 °C for 1 hr. The thermal postcuring procedures were divided into multiple temperature stages in order to prevent the evaporation of AMP and the highly exothermic reaction of cyanate ester.

2.3 Polymer characterizations

The T_g was measured by Dynamic Mechanical Analysis using a TA Instruments DMA 2980. typical size was 35 x 12 x 3.2 mm. The tests were carried out using single cantilever mode with amplitude of 10 μm, frequency 1 Hz, and a temperature ramp rate of 2 °C/min. Sample dimensions were 35 x 12 x 3.2 mm.

The density of printed and postcured parts were measured using a density gradient column following ASTM 1505-03. Samples were rinsed with water before being placed in the column, and the position of the sample was recorded after 2 hours of submersion. The volumetric shrinkage was calculated using the following equation:

$$volumetric\ shrinkage = 100 * \frac{(density\ at\ cure\ stage - resin\ density)}{resin\ density} \quad (1)$$

Mechanical tests were performed at room temperature using an Instron Model 8872. The compact tension samples, followed ASTM 5045-99, had the dimension of 15.5 x 14.88 x 5.5 mm and the notch length of 8.7 mm. A crack was initiated at the bottom of the notch by a sharp razor blade. For each formulation, five to ten compact tension samples were tested at crosshead speed of 0.5 mm/min. The flexural tests were conducted following ASTM D790 in 3-point bending configuration to characterize the flexural strength and modulus using five samples per formulation with a span to depth ratio of 16:1. The flexural specimens had the dimension of 120 x 12.7 x 3.2 mm.

Fourier transform infrared (FTIR) spectroscopy in the mid infrared region (mid-IR) was used to evaluate the extent of cure of printed and post-cured parts. The C=C of AMP and BMI were characterized by peaks at 790 cm^{-1} and 690 cm^{-1}, respectively. The fractional conversions of AMP and BMI double bonds were quantified using the equations below:

$$Conv\ AMP = \frac{I_{790,printed\ or\ postcure}/I_{2850,printed\ or\ postcure}}{I_{790,resin}/I_{2850,resin}} \quad (2)$$

$$Conv\ BMI = \frac{I_{690,printed\ or\ postcure}/I_{2850,printed\ or\ postcure}}{I_{690,resin}/I_{2850,resin}} \quad (3)$$

The OCN (cyanate) group of the CE resin was characterized by a group of peaks at 2200-2300 cm^{-1}. The reaction of CE was confirmed by the disappearance of OCN group and presence of triazine ring represented by peaks 1360 cm^{-1} and 1558 cm^{-1}.

3. RESULTS

Poly(phenylmethane) dissolved well in AMP without recrystallizing after cooling to room temperature from 70°C. Although 50 wt% of this BMI can be dissolved in AMP, a mixture containing 30 % BMI in AMP was selected for the study as it possesses sufficiently low viscosity for printing. The selected cyanate ester also possesses low viscosity so that the IPN formulation could also be printed. This IPN formulation has the viscosity of 125 c.Ps, which is easy for DLP processing.

3.1 Resin characterization

Working curves were obtained in order to assess the depth of penetration (D_p), critical energy (E_c) for the systems investigated, and appropriate print times. Figure 2A shows cure depth vs energy on a semi-log plot for both resin systems investigated. The slope of the best-fit line to these data is D_p, and the intercept is the product of $-D_p*\ln(E_c)$ which allows for the determination of E_c. Values of D_p and E_c for the two systems are also shown in Figure 2A. The D_p of the IPN system is higher due to lower concentration of BMI, and a higher E_c is observed because of the dilution CE effect. The UV-VIS spectra of these two systems (Figure 2B) shows that BMI absorbs light at 405 nm, while AMP and n-CE do not. Therefore, BMI acts as a reactant as well as a light absorber. With a lower concentration of BMI, the IPN has higher D_p. The resins were printed using the same exposure time of 80 s and layer thickness of 100 microns. The same exposure time was used for the purpose of comparing conversion even though the BMI-AMP system could adequately be printed at lower exposure time.

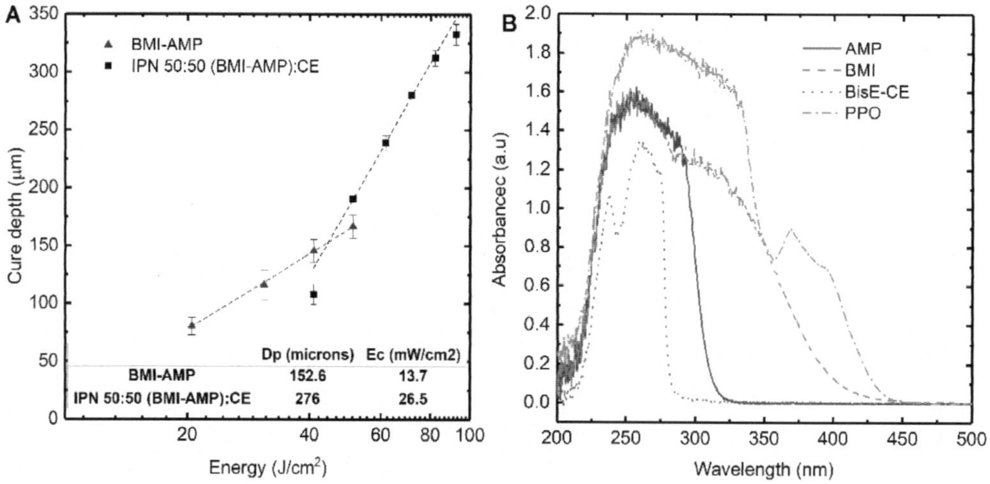

Figure 2: (A) Working curve of the resins, (B) UV-VIS spectra of individual components in the resins.

3.2 Dynamic mechanical analysis and volumetric shrinkage

Figure 3A and 3B show the storage modulus and loss modulus vs. temperature of the samples at different curing stages for the BMI-AMP and IPN systems, respectively. The corresponding conversions at various stages of cure are also given in these figures. The as-printed BMI-AMP system shows a T_g close to room temperature based on storage modulus. Therefore, this system reaches vitrification at room-temperature reaction conditions. The loss modulus curve shows a second peak at higher temperature which can be attributed to continued reaction during DMA testing. The as-printed IPN, on the other hand, has a T_g below 0 °C. This is due to the plasticizing effect of unreacted cyanate ester in the network. The conversion of BMI and AMP was calculated according to equation (2) and (3). It is observed that the conversion of BMI and AMP of the as printed IPN is significantly higher than that of the BMI-AMP system thus indicating vitrification is delayed for the IPN system, providing a less diffusion limited environment for the reaction between AMP and BMI to reach higher conversion. The T_g of the IPN system following the print stage is significantly lower than the cure temperature. This indicates that further reaction is possible if higher energy doses (exposure time) were applied. The UV-oven curing step for the IPN resulted in complete conversion of AMP and 52% conversion of BMI. Although curing was conducted at 75 °C, the T_g of the material remained below room temperature. For both systems, conversion and T_g increase significantly after thermal post-curing.

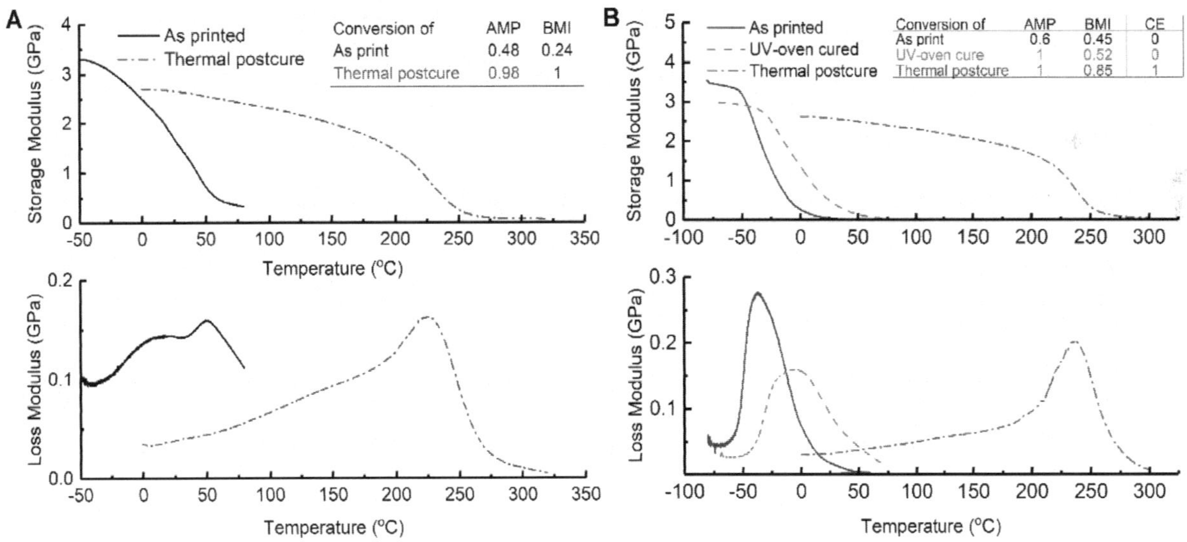

Figure 3: The storage and loss modulus vs. temperature at different stages of (A) BMI-AMP system, (B) IPN 50:50 (BMI-AMP):CE system

The final post-cure T_g of the IPN system is 236 °C, as defined by the peak position of the loss modulus curve. The T_g by tan delta is 260 °C, which is higher than current values reported in literatures.[5-6] The T_g by loss modulus for IPN system is compared with that estimated using the Fox equation:

$$\frac{1}{T_{g,f}} = \frac{w_1}{T_{g,1}} + \frac{w_2}{T_{g,2}} \quad (4)$$

Where $T_{g,f}$ is the final T_g of the IPN, $T_{g,1}$ is the T_g of BMI-AMP system, and $T_{g,2}$ is the T_g of pure CE system, w_1 is the weight fraction of BMI-AMP, and w_2 is weight fraction of CE. The T_g values defined above and the estimated T_g of the IPN are summarized in Table 1. The estimated T_g of the IPN is higher than the measured value by 15 °C. This could be due to a small amount of unreacted BMI in the system.

Table 1: The T_g of individual network and estimated T_g of IPN.

Material	**T_g (°C)**
BMI-AMP	225
IPN 50:50 (BMI-AMP):CE	236
Pure CE	280
Estimated T_g for IPN through Fox Equation	251

The volumetric shrinkage was calculated according to equation (1). Table 2 shows the density of materials at each stage and the accompanying volumetric shrinkage after post-cure relative to the unreacted resin. The volumetric shrinkage of the IPN after printing is roughly half that of the BMI-AMP system. The IPN also exhibits lower overall shrinkage after post-cure. Interestingly, the change in density between print and post-cure of the BMI-AMP sample is lower than the change in conversion would suggest. Why this is the case is an open question that merits further investigation. Possible reasons include mass loss during post-cure and physical constraints on shrinking imposed by vitrification at lower temperature. The observed lower shrinkage of the IPN as well as having a low modulus after printing could reduce residual stresses of the printed part.

Table 2: Density and % volumetric shrinkage of BMI-AMP and IPN system.

Density (g/cm^3)	**BMI-AMP**	**IPN 50:50 (BMI-AMP):CE**
Resin	1.1748	1.1780
As printed	1.2757	1.2340
Thermal post-cure	1.2817	1.2571
% Volumetric shrinkage as print vs. unreacted resin	8.58	4.58
% Volumetric shrinkage post-cure vs. unreacted resin	9.1	6.71

3.3 Mechanical properties

Compact tension and 3-point bending tests were carried out to examine fracture toughness and flexural properties, respectively. Figure 4A shows the critical strain energy release rate (G_{Ic}) and critical stress intensity factor (K_{Ic}) of the two systems. Bismaleimide is known to be brittle because of its high crosslinking density.[14] The low fracture toughness of BMI-AMP is expected. The IPN has significantly higher fracture toughness compared to the BMI-AMP system. This observation is in agreement with reports in the literature that IPN network architectures result in toughened systems.[22-24] The reported G_{Ic} of pure bisphenol E cyanate ester is 190 J/m^2.[15] The IPN increases the toughness of BMI-AMP system but to a level lower than that of the pure cyanate ester system. Figure 4B shows the flexural strength and modulus of the BMI-AMP and IPN materials. Both systems possess good flexural properties. The flexural moduli are similar, but the IPN system has significantly higher flexural strength. The resulting IPN has good mechanical properties in addition to thermal properties.

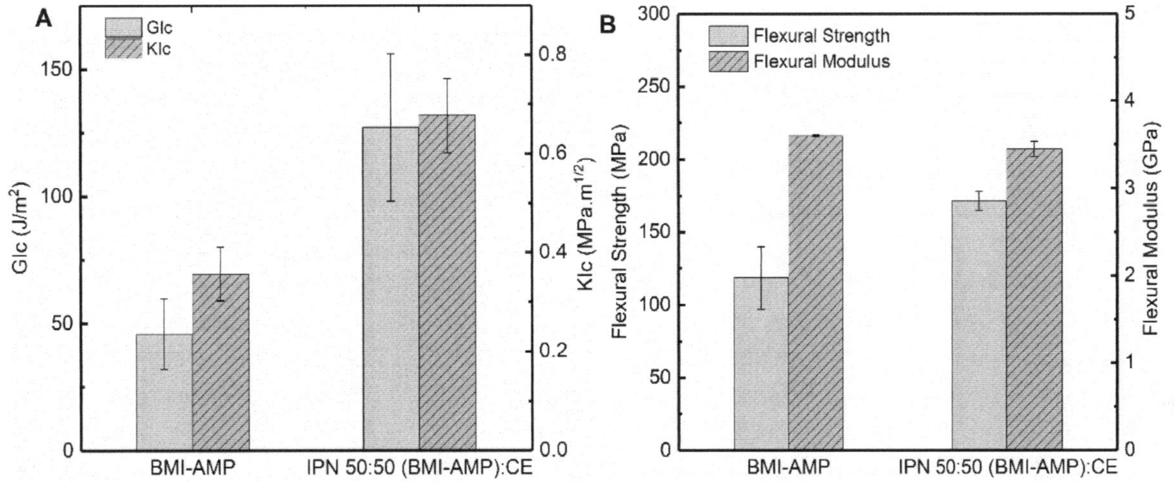

Figure 4: (A) Fracture toughness, (B) Flexural properties of BMI-AMP and IPN system.

4. CONCLUSIONS

This investigation demonstrated that high-performance thermosetting materials can be printed using stereolithography by formulating *in-situ* sequential IPN networks of BMI and CE. A copolymerization reaction between bismaleimide and a diluent occurs during printing, resulting in a cyanate ester swollen network with sub room temperature glass transition temperature (T_g). The polymerization of cyanate ester takes place during post processing. The resulting material possesses a T_g close to 240 °C (loss modulus peak), significant decrease in cure shrinkage, high strength, and improved toughness.

5. ACKNOWLEDGEMENTS

Research was sponsored by the Army Research Laboratory and was accomplished under Cooperative Agreement Number W911NF-14-2-0227. The views and conclusions contained in this document are those of the authors and should not be interpreted as representing the official policies, either expressed or implied, of the Army Research Laboratory or the U.S. Government. The U.S. Government is authorized to reproduce and distribute reprints for Government purposes notwithstanding any copyright notation herein.

6. REFERENCES

[1] Guo N. and Leu M. C., "Additive manufacturing: technology, applications and research needs," *Frontiers of Mechanical Engineering,* vol. 8, no. 3, pp. 215-243, 2013. DOI: 10.1007/s11465-013-0248-8

[2] Ligon S. C., Liska R., Stampfl J., Gurr M., and Mülhaupt R., "Polymers for 3D printing and customized additive manufacturing," 2017. DOI: 10.1021/acs.chemrev.7b00074

[3] Chandrasekaran S., Duoss E. B., Worsley M. A., and Lewicki J. P., "3D printing of high performance cyanate ester thermoset polymers," *Journal of Materials Chemistry A,* vol. 6, no. 3, pp. 853-858, 2018. DOI: 10.1039/C7TA09466C

[4] Hegde M. *et al.*, "3D printing all-aromatic polyimides using mask-projection stereolithography: processing the nonprocessable," *Advanced Materials,* vol. 29, no. 31, p. 1701240, 2017. DOI: 10.1002/adma.201701240

[5] Zhang X., Xu Y., Li L., Yan B., Bao J., and A. Zhang, "Acrylate-based photosensitive resin for stereolithographic three-dimensional printing," *Journal of Applied Polymer Science,* vol. 136, no. 21, p. 47487, 2019. DOI: 10.1002/app.47487

[6] Kuang X., Zhao Z., Chen K., Fang D., Kang G., and Qi H. J., "High-Speed 3D Printing of High-Performance Thermosetting Polymers via Two-Stage Curing," *Macromolecular rapid communications,* vol. 39, no. 7, p. 1700809, 2018. DOI: 10.1002/marc.201700809

[7] Bychkov P., A. Chentsov, Kozintsev V., and Popov A., "Determination of residual stresses in objects at their additive manufacturing by layer-by-layer photopolymerization method," in *Journal of Physics: Conference Series*, 2018, vol. 991, no. 1, p. 012016: IOP Publishing.

[8] Karalekas D., Rapti D., Gdoutos E., and Aggelopoulos A., "Investigation of shrinkage-induced stresses in stereolithography photo-curable resins," *Experimental mechanics,* vol. 42, no. 4, pp. 439-444, 2002. DOI: 10.1007/BF02412150

[9] Shah D. U. and Schubel P. J., "Evaluation of cure shrinkage measurement techniques for thermosetting resins," *Polymer Testing,* vol. 29, no. 6, pp. 629-639, 2010. DOI: 10.1016/j.polymertesting.2010.05.001

[10] Palmese G. and Gillham J., "Time–temperature–transformation (TTT) cure diagrams: Relationship between Tg and the temperature and time of cure for a polyamic acid/polyimide system," *Journal of applied polymer science,* vol. 34, no. 5, pp. 1925-1939, 1987. DOI: 10.1002/app.1987.070340511

[11] Lange J., Toll S., Månson J.-A. E., and Hult A., "Residual stress build-up in thermoset films cured above their ultimate glass transition temperature," *Polymer,* vol. 36, no. 16, pp. 3135-3141, 1995. DOI: 10.1016/0032-3861(95)97876-H

[12] Li W., Noodeh M. B., Delpouve N., Saiter J.-M., Tan L., and Negahban M., "Printing continuously graded interpenetrating polymer networks of acrylate/epoxy by manipulating cationic network formation during stereolithography," *Express Polymer Letters,* vol. 10, no. 12, 2016.

[13] Sitzmann E. V., Anderson R. F., Koljack M. P., Cruz J. G., and Srivastava C. M., "High temperature performance polymers for stereolithography," ed: Google Patents, 2000.

[14] Iredale R. J., Ward C., and Hamerton I., "Modern advances in bismaleimide resin technology: A 21st century perspective on the chemistry of addition polyimides," *Progress in Polymer Science,* vol. 69, pp. 1-21, 2017. DOI: 10.1016/j.progpolymsci.2016.12.002

[15] Hamerton I. and Hay J. N., "Recent developments in the chemistry of cyanate esters," *Polymer international,* vol. 47, no. 4, pp. 465-473, 1998. DOI: 10.1002/(SICI)1097-0126(199812)47:4<465::AID-PI88>3.0.CO;2-S

[16] Nair C. R., Mathew D., and Ninan K., "Cyanate ester resins, recent developments," in *New Polymerization Techniques and Synthetic Methodologies*: Springer, 2001, pp. 1-99.

[17] Gu A., "High performance bismaleimide/cyanate ester hybrid polymer networks with excellent dielectric properties," *Composites Science and Technology,* vol. 66, no. 11-12, pp. 1749-1755, 2006. DOI: 10.1016/j.compscitech.2005.11.001

[18] Geng X., "Bismaleimide and cyanate ester based sequential interpenetrating polymer networks for high temperature application," 2005.

[19] Hamerton I., "High-performance thermoset-thermoset polymer blends: A review of the chemistry of cyanate ester-bismaleimide blends," *High performance polymers,* vol. 8, no. 1, pp. 83-95, 1996. DOI: 10.1088/0954-0083/8/1/006

[20] Fan S., Boey F., and Abadie M., "UV curing of a liquid based bismaleimide-containing polymer system," *Exp Polym Lett,* vol. 1, pp. 397-405, 2007. DOI: 10.3144/expresspolymlett.2007.56

[21] Francis L., McCormick A., Vaessen D., and Payne J., "Development and measurement of stress in polymer coatings," *Journal of Materials Science,* vol. 37, no. 22, pp. 4717-4731, 2002. DOI: 10.1023/A:1020886802632

[22] Ligon-Auer S. C., Schwentenwein M., Gorsche C., Stampfl J., and Liska R., "Toughening of photo-curable polymer networks: a review," *Polymer Chemistry,* vol. 7, no. 2, pp. 257-286, 2016. DOI: 10.1039/C5PY01631B

[23] Sperling L. H., "Interpenetrating polymer networks: an overview," ACS Publications, 1994. DOI: 10.1021/ba-1994-0239.ch001

[24] Lin M. S., Liu C. C., and Lee C. T., "Toughened interpenetrating polymer network materials based on unsaturated polyester and epoxy," *Journal of Applied Polymer Science,* vol. 72, no. 4, pp. 585-592, 1999. DOI: 10.1002/(SICI)1097-4628(19990425)72:4<585::AID-APP15>3.0.CO;2-M

Additive Manufacturing of Polyamide 6/66 Copolymer Nanocomposites: Thermal and Flammability Properties

Steven D. Kim[1,2], Abdullah Kafi[3], Hao Wu[2], Stuart Bateman[3], Joseph H. Koo[1,2]*

[1]Department of Mechanical Engineering, The University of Texas at Austin, Austin, TX 78712, USA

[3]KAI, LLC, Austin, TX 78739, USA

[2]School of Engineering, RMIT University, Melbourne, VIC 3000, Australia

*jkoo@mail.utexas.edu

ABSTRACT

As one of the most commonly used methods of additive manufacturing (AM), fused filament fabrication (FFF) has drawn a great amount of interest in both industry and academia. The purpose of this study is to develop a flame retardant (FR) polyamide 6/polyamide 66 (PA6/66) nanocomposite with reduced flammability and enhanced thermal and mechanical properties, while simultaneously maintaining the printability for commercially available FFF machines. In this study, varying amounts of FR, thermoplastic elastomer (K), nanoclay (NC), and multiwall carbon nanotubes (MWNT) were melt compounded via twin-screw extrusion to prepare PA6/66 nanocomposite pellets. The flame retardant and thermal properties of these formulations were characterized using microscale combustion calorimeter (MCC) and thermogravimetric analysis (TGA). Based on MCC, it was found that formulations 2, 4, and 5 yielded best results, while formulation 3 and 2 gave best results in TGA testing. Finally, three FR PA6/66 formulations (2, 4, and 5) were down-selected, 3D printed via FFF, and used for flammability, thermal, and mechanical properties characterizations. Due to difficulties in 3D printing, only formulation 5 was successfully printed and tested. However, despite not being the best performing formulation, formulation 5 yielded promising results in UL 94 and tensile tests. Lastly, Micro-CT analysis displayed that formulation 5 was much more porous in comparison to the neat PA6/66 system.

1. INTRODUCTION

Additive manufacturing is a method of production in which products are produced by building the object layer by layer. While many unique AM technologies, such as fused filament fabrication (FFF), stereolithography apparatus (SLA), and selective laser sintering (SLS) all offer methods for rapid prototyping, FFF is more widely used for its ease of use and low cost. Due to this rapid rise in AM technologies, it is important to develop new and improved 3D printable high-performance materials. However, there are not many papers published in the literature studying FFF compatible flame retardant materials.

Although FFF was developed in the 1980s, it is more widely known today as fused deposition modeling or FDM. FFF creates parts by feeding a filament that is often 1.75 mm in diameter through a heated extruder head which moves in the X and Y axes. This extruder head then deposits the molten material onto a heated build platform in a predetermined pattern. Once a layer is complete, the build platform is lowered by a specific layer height distance, and the extruder begins to deposit the material for the following layer. This process is repeated until the final product is complete. Due to its greatly improved user interface, low cost, and versatility, FFF is the preferred method for use in this experiment.

Polyamide 6/polyamide 66 copolymer is a thermoplastic polymer developed to improve the existing properties found in polyamide 6 and polyamide 66. PA6/66 offers varying desirable properties, such as good chemical resistance, great impact strength, and low density. Despite the various advantages, PA6/66, like other polyamides, readily absorbs moisture which in turn can hinder the processability and printability of the copolymer. Since PA6/66 is derived from polyamide 6 and polyamide 66, its physical properties do not vastly differ; the density and melting temperature of PA6/66 are slightly lower than those of PA66.

Neat PA6/66 is inherently not flame retardant as per UL 94 test rating. When compounded with intumescent FR additives, the flammability properties are expected to drastically improve. Although adding greater amounts of flame retardant (FR) additive results in greater flammability results, this also results in weaker mechanical properties as there is a lesser amount of PA6/66 remaining. Aside from the FR additive, some other additives were used in compounding PA6/66 formulations; these include, thermoplastic elastomer, nanoclay, and multiwall carbon nanotubes [1-5]. Adding specific amounts of nanomaterials is often attributed to enhancing various material properties, such as thermal, electrical, mechanical, or thermal. However, adding too much of a single additive can have adverse effects. Although polyamides are compatible with varying nanomaterial additives, the compatibility between the nanomaterials and between nanomaterial and polymer must be considered before compounding. Incompatibility between any of the materials or polymer and poor dispersion of nanomaterials can results in poor material properties.

In our previous study, it was found that FR will decompose when exposed to a flame, creating a protective char, thus protecting the material below it. However, as mentioned above, adding excessive FR may deteriorate mechanical properties. NC and MWNT are expected to improve the heat deflection temperature (HDT) of the polymer and help form a more solid char layer around the material. With the addition of just small amounts, such as 2.5 wt.% to 5 wt.% in the composite, NC and MWNT both exhibited significant increase in mechanical and FR performance [2, 6, 11-12]. Although both NC and MWNT show improved properties, increased loading exceeding 5 wt.% often leads to deterioration in mechanical properties, causing the polymer to become very brittle [1, 6-10]. Therefore, to combat the onset of brittleness, a thermoplastic elastomer K, which improves the elongation at break, will be added to restore the mechanical properties that are lost from FR, NC, and MWNT.

The objective of this research is to prepare an FFF compatible polyamide 6/polyamide 66 copolymer nanocomposite that exhibits improved flammability and mechanical properties since there is currently a lack of commercially available FR PA6/66 filaments. Although flammability properties will be improved with the addition of multiple components, it is also vital to maintain adequate levels of mechanical properties by adding an elastomer (K) component. To achieve this purpose, PA6/66 were blended with multiple additives consisting of FR, K, NC, and MWNT.

2. EXPERIMENTATION

2.1 Material and Processing

The pelletized PA6/66 was acquired from Zig Sheng Industrial Co. Ltd. and was used as the base material in the composite matrix. The FR and elastomer were provided by Clariant and Kraton, respectively. The NC was received from BYK and the MWNT was provided by Arkema as the last two additives.

PA6/66 was dried at 70°C for 24 hours before the compounding and extrusion process. As shown in Table 1, the FR varies from 15 wt.% to 10 wt.%. Although 15 wt.% FR loading is very likely to result in UL 94 V-0 rating, achieving this rating with just 10 wt.% FR loading would be preferred in order to retain more of the PA6/66 mechanical properties. Varying the wt.% of K from 10wt.% to 15wt.% will allow analysis for any significant changes to the elongation at break for the varying formulations. Lastly, the slight variation in the wt.% of NC and MWNT will allow analysis for any significant differences in mechanical properties. The following nine formulations (also denoted as F#) have been melt-compounded with the twin-screw extruder. Initial thermal and flammability properties on all nine formulations were tested via MCC and TGA. After initial analysis, three best performing formulations were down-selected for 3D printing and further analyzed for flammability and mechanical properties.

Table 1. PA6/66 Formulation Matrix

Formulation #	Formulation	PA 6/66 (wt.%)	FR (wt.%)	K (wt.%)	NC (wt.%)	MWNT (wt.%)
0	PA 6/66	100	-	-	-	-
1	15FR_10K_2.5NC_2.5MWNT	70	15	10	2.5	2.5
2	15FR_15K_2.5NC_2.5MWNT	65	15	15	2.5	2.5
3	15FR_10K_3.5NC_3.5MWNT	68	15	10	3.5	3.5
4	15FR_15K_3.5NC_3.5MWNT	63	15	15	3.5	3.5
5	10FR_10K_2.5NC_2.5MWNT	75	10	10	2.5	2.5
6	10FR_15K_2.5NC_2.5MWNT	70	10	15	2.5	2.5
7	10FR_10K_3.5NC_3.5MWNT	73	10	10	3.5	3.5
8	10FR_15K_3.5NC_3.5MWNT	68	10	15	3.5	3.5

In order to compound and process the PA6/66 formulations, a twin-screw extruder (Figure 1) was used. The Thermo Scientific Inc. Process 11 parallel twin-screw extruder serves to melt, mix, and extrude FFF compatible filament. The entire twin-screw extruder setup consists of formulation feeder, heating chamber, motorized roller, and motorized spool. The mixed compound is first poured into the formulation feeder and the heating chamber, which consists of eight heating zones with temperatures varying from 230°C to 270°C, melts the compound and thoroughly mixes the polymer and additives into a filament. The twin screws that run along the length of the heating chamber is rotated at 150 rpm for optimal speed and quality of the filament. The extruded filament is then pulled by the motorized roller at a constant speed to ensure constant filament diameter. The motorized spool which rotates at approximately the same speed as the motorized roller then collects the filament into a spool for ease of transport. The extruded filaments are then used to conduct thermal characterization.

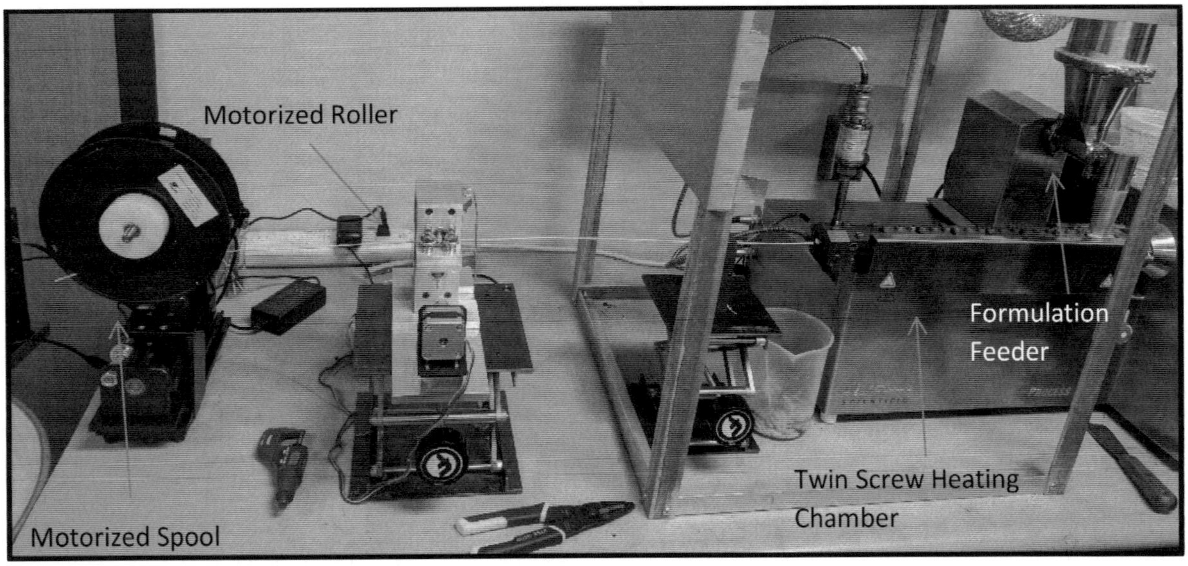

Figure 1. Image of twin-screw extruder setup.

2.2 Thermal Characterization

2.2.1 Microscale Combustion Calorimetry

Microscale Combustion Calorimetry (MCC) manufactured by Govmark Inc. was used to analyze heat release properties to predict flammability properties of each formulation. MCC utilizes approximately 2 to 3 mg of the formulation sample to conduct the analysis. The sample is loaded onto the ceramic rod showing in Figure 2 which is raised into the combustion chamber. The combustion chamber is held at constant 900°C while the pyrolyzer chamber is gradually heated from 100°C to 750°C at a rate of 1°C/s. The testing environment is controlled to be 80% nitrogen and 20% oxygen. This nitrogen and oxygen gas mixture is pumped through a tube containing desiccant to remove all moisture from the gas to ensure accurate results. MCC analyzes the oxygen levels in the testing chamber to provide heat release (HR) capacity, peak heat release rate (HRR), and total HR as a function of temperature. Each sample was tested in three non-repeated trials in accordance to ASTM D7309 standards.

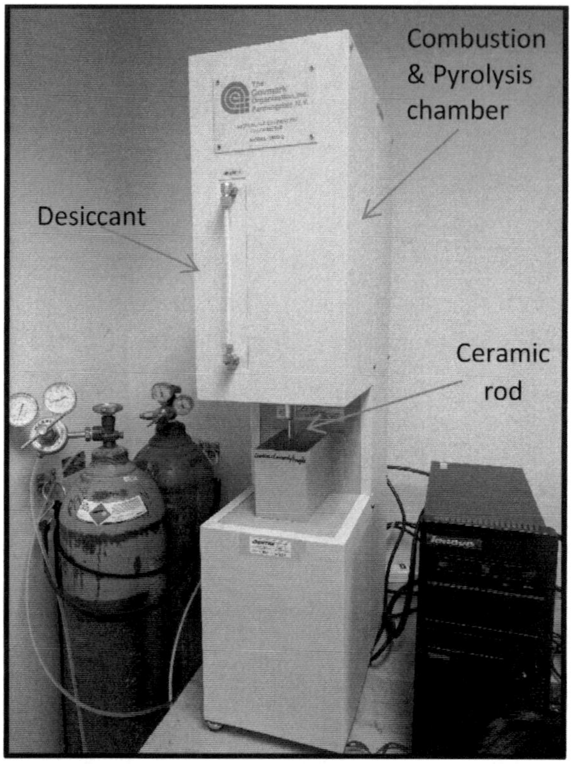

Figure 2. Image of Microscale Combustion Calorimetry instrument.

2.2.2 Thermogravimetric Analysis

After the completion of MCC analysis, thermogravimetric analysis (TGA) of all formulations was performed using TGA-50 developed by Shimadzu Scientific Instruments (Figure 3). Approximately 17 to 18 mg of the sample will be loaded into the heating chamber and heated to 1,000°C at a rate of 20°C/min in a nitrogen testing environment to measure the char yield as a function of temperature. Each sample will be analyzed once, then the six best performing formulations will be down-selected to be analyzed once again in the second trial. Then lastly, the top four best performing formulations will be analyzed for the third time in the third trial. After conducting the third TGA trial, data collected from MCC and TGA will be analyzed to select top three best performing formulations which will then be extruded using the twin-screw extruder at constant filament diameter of 1.75 ± 0.09 mm.

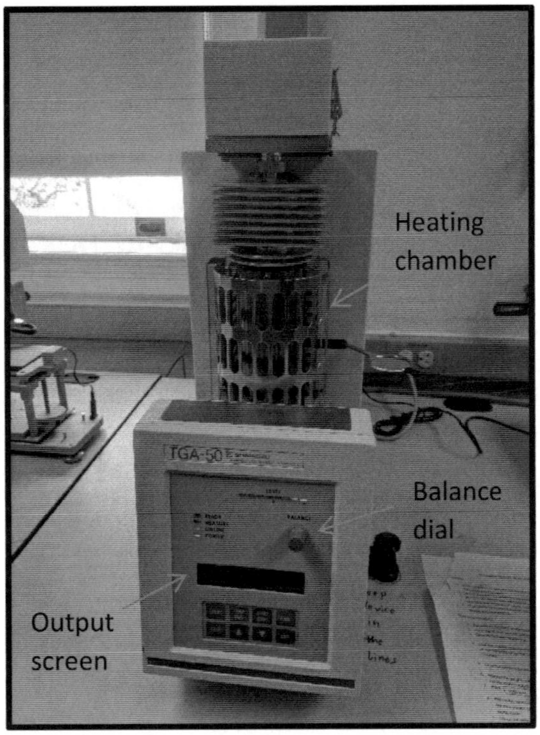

Figure 3. Image of Thermogravimetric Analysis instrument.

1.1 FFF Additive Manufacturing

After thorough data analysis of MCC and TGA results, a Craftunique CraftBot XL 3D printer was used to print five UL 94 and five tensile test bars (Figure 4) for each of the top three down-selected formulations. Same number of neat PA6/66 test bars will also be printed for control. The UL 94 test samples were 125mm long, 13mm wide, and 3mm thick. The tensile test specimens were 115mm long, 19mm wide, and 3mm thick. Additionally, the narrow sections of the tensile specimen were 33mm long and 6mm wide. The build platform will be maintained at 60°C and extruder head temperature of 260°C. The samples will be printed in a +45°/-45° raster infill pattern at 100% infill and a layer height of 0.1 mm. The printing speed will be limited to 60 mm/s.

Figure 4. Image of FFF printed neat PA 6/66 test bars.

1.2 UL 94 Flammability Test

UL 94, which is a standardized thermoplastics flammability test used to analyze the safety of ablated plastics, was used after 3D printing UL 94 test models of the top three best performing formulations. The vertical UL 94 test ablates the test bar from the bottom using a flame at a 45-degree angle from the vertical test bar. The gas supply of the flame is timed such that after ten seconds of open flame, the gas supply shuts off automatically. Then, the first time (t1) is recorded until the test bar self-extinguishes. As soon as the test bar is fully extinguished, the flame is activated again for another ten seconds; the self-extinguishing time is once again measured as t2 and time for the after-glow to dissipate as t3. During the ablating and timing process, the test bar is also observed to determine if the test article melts and any dripping occurs. Using the three times and observation for dripping, the formulations are then rated as V-0, V-1, V-2, from best to worst. The specific rating criteria for UL 94 is shown in Table 2.

Table 2. UL 94 Rating Criteria

Criteria	V-0	V-1	V-2
Afterflame for each individual specimen (t1 and t2)	≤ 10s	≤ 30s	≤ 30s
Total afterflame for any condition set (t1+t2 for the 5 specimens)	≤ 50s	≤ 250s	≤ 250s
Afterflame plus afterglow time for each individual specimen after the second flame application	≤ 30s	≤ 60s	≤ 60s
Afterflame or afterglow of any specimen up to the holding clamp	No	No	No
Cotton indicator ignition	No	No	Yes

1.3 Tensile Test

In accordance with ASTM D638 type IV, an Instron Tension Tester 5966 was used to analyze the mechanical properties, such as tensile strength, modulus, and elongation at break, of each 3D printed tensile test bar. A crosshead type extensometer will be used to grip the test bars and then tested at room temperature with crosshead speed of 5 mm/min. With five tensile test bars for each of the top three down-selected formulations, each formulation will be tensile tested five times for better accuracy.

1.4 Micro-CT

First, the samples were scanned on a Bruker Skyscan 1275 operating at 40 kV 100 μA with no filter in place. The sample was placed on a brass stage and positioned to achieve an image pixel size. The height was adjusted to capture the thickest part of the sample. Samples were scanned with the following settings: 0.4° rotation step, 360° scanning, 4 frame averaging, total of 901 shadow projections (raw images). Camera settings were: 1944x1536 resolution, 67 ms exposure.

Second, a program, NRecon, is used to calculate cross sections from the raw images. With the software, post processing corrections were adjusted; corrections included smoothing, misalignment compensation, ring artifacts reduction, and beam-hardening correction (constant at

30% across all samples). The dynamic image range was set from 0.0 to 0.0170 for all samples. In order to analyze the volume inside the object, a rectangular shape ROI was set. The ROI (red rectangle) was created over a 2D slice and applied to all slices. Each slice was checked to make sure the ROI stays within the object. Adjustments to the ROI was made when necessary. Figure 5 below shows an image from the NRecon software window.

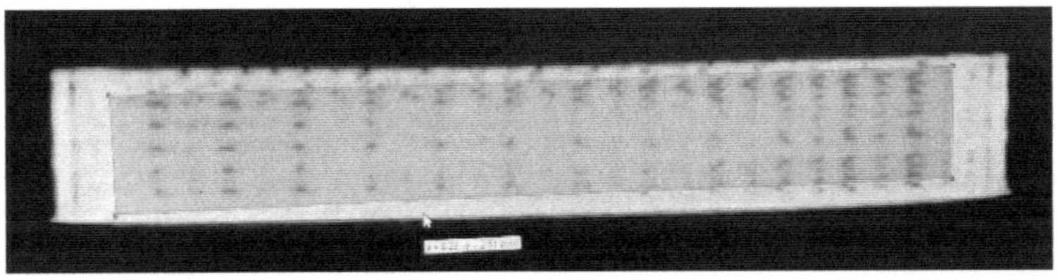

Figure 5. Snapshot from NRecon software showing raw data and reconstructed slice.

Analysis of the images were done using a CTan software. This software is used to binarize images in order to segregate areas of different density. Cross sections are represented as a binary image of material (1) and air (0). Porosity is calculated from the binarized images. Binarized images were exported as a model file and made into videos in CTVol.

2. RESULTS

2.1 MCC Results

MCC results shown in Table 3 display the average values from the three trials, the standard deviation, and the percentage improvement in comparison to the neat system (Formulation 0). The collected data indicate that formulations 2, 3, 4, and 5 had the best results.

Table 3. PA 6/66 MCC Data

	HR Capacity (J/g-K)		Peak HRR (W/g)		Total HR (kJ/g)	
Formulation	Avg	StDev	Avg	StDev	Avg	StDev
0	1301	39.6	1295	39.6	63.07	2.3
1	1022	181.9	1017	180.9	60.57	4.4
2	819	42.5	814	42.7	59.50	4.3
3	851	62.6	840	58.7	59.00	4.4
4	829	18.8	823	17.1	57.67	0.7
5	830	80.6	826	80.5	53.93	5.2
6	1060	102.8	1054	101.7	64.50	6.6
7	870	107.6	865	108.3	54.20	6.6
8	900	7.9	895	7.4	56.93	1.5

Figure 6 displays the average HR capacity of the formulation data acquired from MCC. When compared to the neat system, all formulations showed enhanced properties; of those formulations, F2 had the best results, closely followed by F4 and F5.

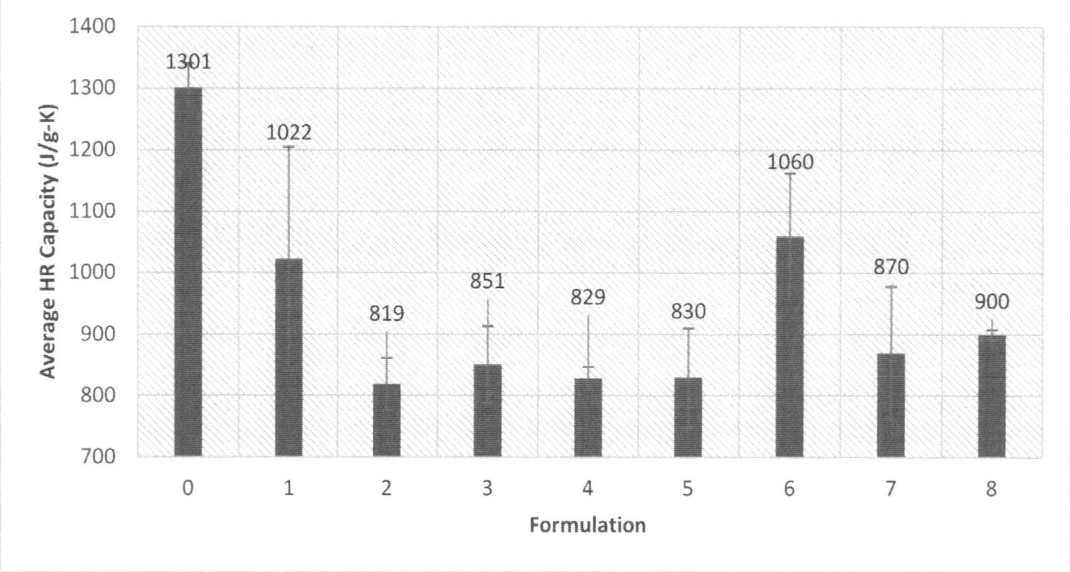

Figure 6. Graph of average HR Capacity of PA6/66 nanocomposites.

2.2 TGA Results

The averaged results from TGA are shown in Table 4, which include the temperatures at 10% mass loss, 50% mass loss, as well as temperatures at greatest mass loss (Max DrTGA). The main differences are shown in the temperatures at 10% mass loss. F1 through F4 all experienced 10% mass loss, on average, at 424°C, whereas F5 through F8 experienced it, on average, at a slightly higher temperature of 427°C. This factor can largely be attributed to F1~F4 having 15 wt.% FR while the subsequent formulations have only 12.5 wt.% FR. This result shows that having greater amounts of FR in a formulation causes the FR to decompose at a lower temperature, which forms the protective char layer earlier.

Table 4. PA6/66 TGA Data

	10% Mass Loss (°C)		50% Mass Loss (°C)		Max DrTGA (°C)	
Formulation	Avg	StDev	Avg	StDev	Avg	StDev
0	432	6.99	468	9.94	476	16.12
1	424	1.74	470	1.07	477	0.94
2	424	0.71	466	0.26	462	5.79
3	424	1.30	464	0.85	458	4.97
4	424	0.85	467	0.72	470	1.29
5	426	0.40	466	0.36	469	1.06
6	428	0.39	468	0.50	471	2.07
7	427	0.53	468	0.51	473	0.50
8	427	0.50	469	0.77	473	3.25

Figure 7 shows the TGA degradation curves of all nine formulations. All formulations begin at 100% weight but begins to deteriorate at a certain temperature. In the middle of the curve, the neat PA6/66 system lost its weight percentage noticeably faster than other formulations. This indicates that the additives used in all other formulations successfully improved the properties of PA6/66 by enhancing its thermal stability.

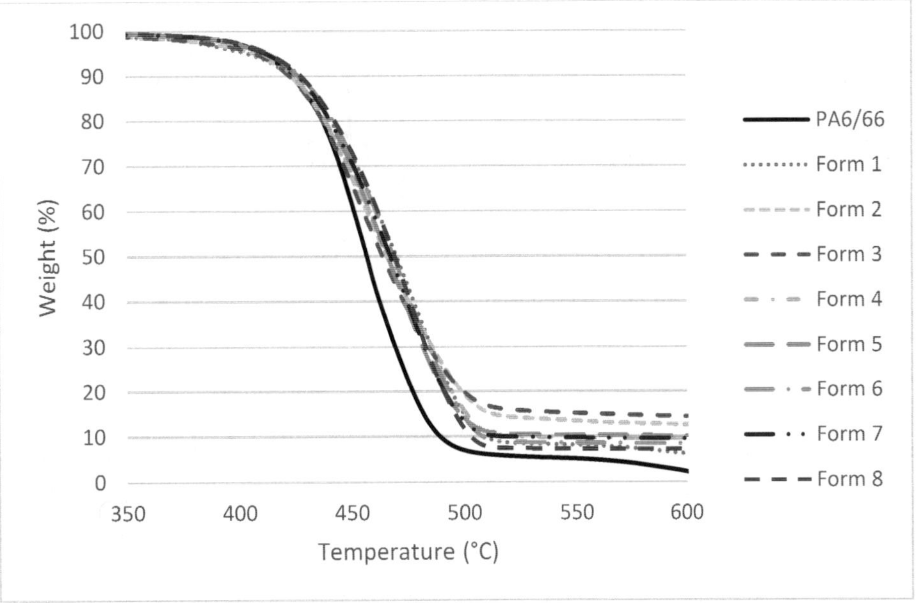

Figure 7. Degradation curves for all formulations (TGA, scan rate 20°C/min in N_2).

In observing the ends of the degradation curves, it is seen that F3 yielded the highest weight percentage, closely followed by F2. It is also important to note that all formulations performed better than the neat PA6/66 system by retaining more of its weight percentage after degradation.

Figure 8 shows the DrTGA curves for all formulations. As previously mentioned, DrTGA indicates at which temperature the greatest amount of mass loss occurred as well as the greatest amount of mass loss. Figure 8 indicates that all formulations either had a peak DrTGA value at a higher temperature or had lower peak mass loss values than the neat PA6/66 system. Furthermore, despite reaching the peak mass loss temperature earlier than other formulations, F1 through F4 display noticeably lower peak DrTGA than F5 through F8. This indicates that a lower peak mass loss temperature was caused by the additional FR, but this FR also assisted in lowering the peak mass loss value.

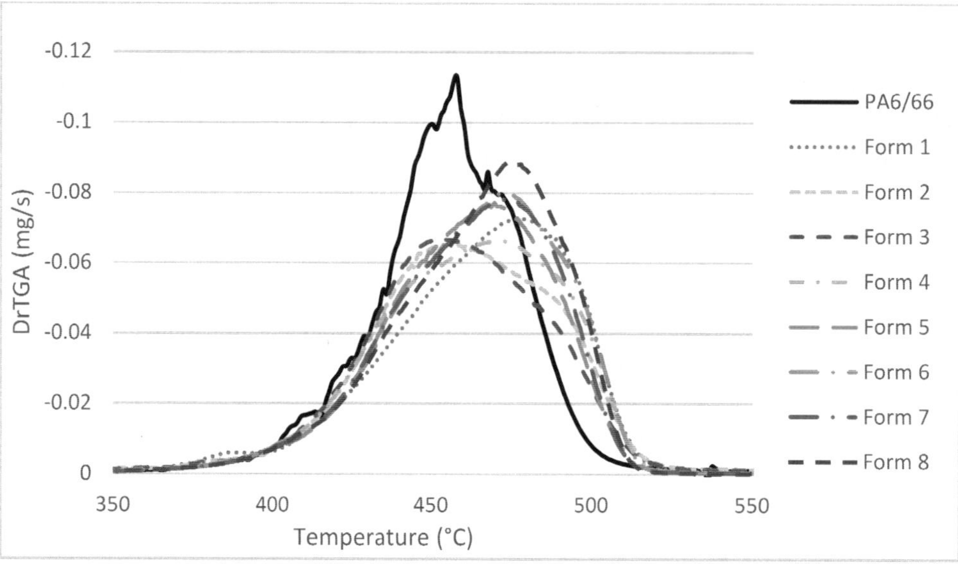

Figure 8. Peak derivative TGA curves for PA6/66 nanocomposites.

Upon analyzing the overall TGA data, it is seen that formulations F1 through F4, which had a higher weight percentage of FR at 15 wt.%, performed better than the subsequent formulations. However, the values between the formulations containing 15 wt.% and 12.5 wt.% FR were very similar. Also, the testing environment of TGA was 100% nitrogen gas, whereas the testing environment of MCC was 80% nitrogen and 20% oxygen gas, which more closely resembles real life atmospheric conditions. Therefore, the MCC data are weighted more heavily in the final decision-making to move forward with 3D printing and further testing. Taking these factors into consideration, the experiment moved forward with formulations 2, 4, and 5 to 3D print and perform further testing since these three formulations were the top performers in MCC and showed promising values in TGA.

2.3 FFF Printing

After close inspection of both MCC and TGA data, formulations 2, 4, and 5 were chosen as the top three down-selected formulations to attempt 3D printing with. These three were chosen because TGA results did not greatly vary between the formulations and F2, F4, and F5 displayed the best MCC results. In preparation for 3D printing, the top three formulations were extruded using the previously mentioned twin-screw extruder (Figure 1). Each of the three formulations were twice extruded to ensure good dispersion of the additives. When attempting to 3D print, however, it was found that F2 and F4 could not be printed on the CraftBot XL. It is speculated that this is due to the high concentration of additives that may be clogging the 0.4mm extruder nozzle. Although F2 and F4 could not be printed, F5 fortunately printed without any complications. After printing five test samples of tensile and UL 94 each for F5, the same amount was printed using the neat PA6/66 system as control. An image of the 3D printed UL 94 samples is shown in Figure 9.

Figure 9. Image of FFF printed F5 UL 94 test samples.

2.4 UL 94 Flammability Test Results

Due to previously mentioned difficulties in printing F2 and F4, UL 94 and tensile testing was only performed on F5. Table 5 shows the results from five UL 94 trials conducted on F5. The obtained results were not very consistent, which could be caused by a multitude of factors, from exact placement of the test specimen to accuracy of the timer used by the operator and the dispersion of FR within the polymer matrix. However, based on the retrieved data, it was concluded that this formulation did not achieve the UL 94 V-0 rating due to t2 being greater than the V-0 criteria of 10s.

Table 5. UL 94 Test Results of F5

PA 6/66 Formulation 5			
Sample #	t1 (s)	t2 (s)	Drip
1	1.8	57.6	No
2	0.5	1.7	No
3	1.7	113.8	No
4	0.6	58.9	No
5	0.0	15.1	No

Although the initial UL 94 test did not provide consistent t1 and t2 results, the test sample did not have any drips on all five trials. Furthermore, none of the five trials showed any afterglow after initial ablation. Therefore, it is likely that if the printing difficulties of F2 and F4 are resolved, these two formulations will show better results than F5 since F2 and F4 contain higher weight percentage of FR than F5. Figure 7 below shows an image of the UL 94 test samples after testing. Although two out of 5 samples warped due to the heat from the flame, none of the five samples experienced any heavy deformities, such as dripping.

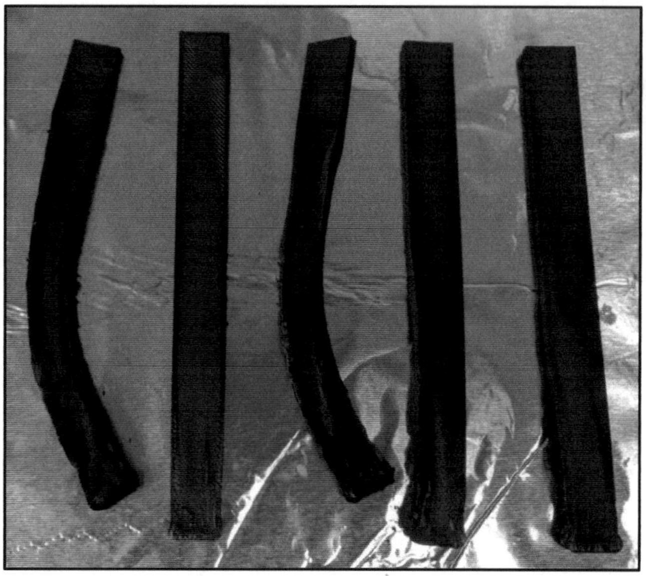

Figure 10. Image of post-test F5 UL 94 test samples.

2.5 Tensile Testing

Tensile testing was only performed on the neat PA6/66 and F5. Table 6 shows the overall data collected from the neat PA6/66 tensile test (F0). These data are immediately followed by Table 7, which showcases tensile test data of F5 with comparison to F0.

Table 6. Tensile test results of Neat PA6/66 (F0)

Neat PA6/66			
Sample ID	Ultimate Tensile Strength (MPa)	Modulus (MPa)	Elongation at Break (%)
1	32.50	225.67	116.66
2	40.89	186.75	196.46
3	31.99	161.71	134.46
4	48.15	212.20	215.28
5	37.67	189.60	153.82
Std. Dev.	6.66	24.71	26.99
Avg.	38.24	195.19	163.33

Table 7. Tensile test results of PA6/66 F5 with % improvement from F0

Sample ID	Ultimate Tensile Strength (MPa)	Modulus (MPa)	Elongation at Break (%)
1	24.75	178.74	62.17
2	26.45	193.77	63.09
3	26.04	99.45	92.71
4	25.00	93.14	81.44
5	24.72	81.70	96.32
Std. Dev.	0.80	52.60	16.05
Avg.	25.39	129.36	79.15
% Improvement	-32%	-34%	-49%

As shown in Table 7, the average tensile strength of F5 was 25.39 MPa, which is 32% lower than F0. Although the tensile strength degraded, it was to be expected since multiple additives were added to each formulation. The modulus and elongation at break both show similar trends; modulus degraded by 34% while elongation at break degraded by 49%. Despite the deterioration of these values, it was as expected. Figure 11 below shows the stress vs. strain curve of both neat PA6/66 and F5. This figure further showcases the degraded mechanical properties of F5 when compared to the neat system. The enhanced flammability properties of F5 outweigh the decline of mechanical properties.

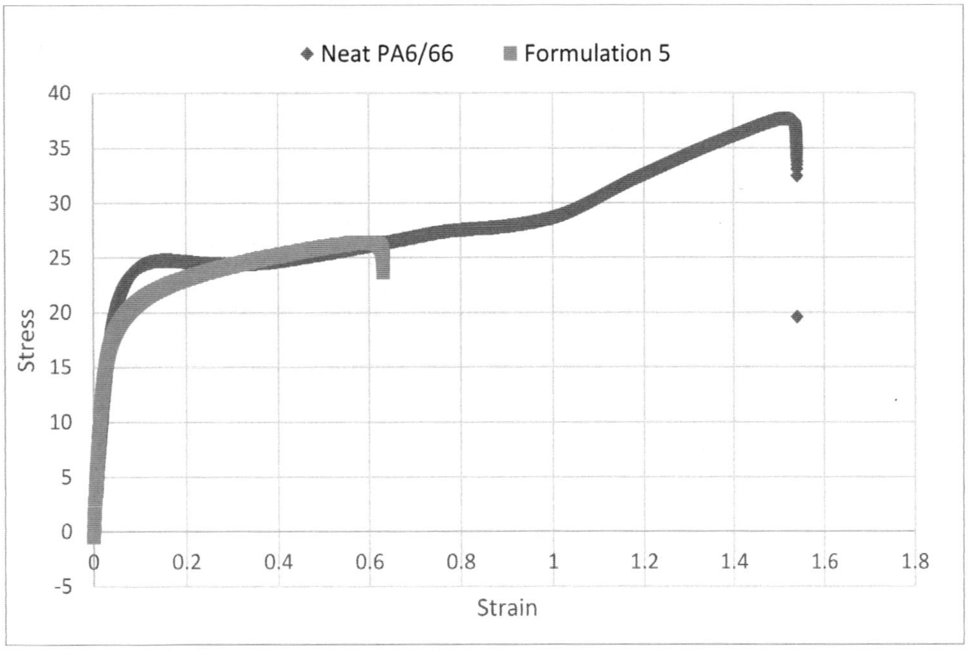

Figure 11. Stress vs. Strain curve of neat PA6/66 (F0) and F5.

2.6 Micro-CT Analysis

The total porosity is calculated as the volume of open and closed pores. Open pores are classified as the property of the volume of interest (VOI) as they intersect the boundary of VOI. A 'closed pore' in contrast a group of black pixels that is fully surrounded by a border of white pixels. Subsequently, closed pores are classified as a property of the material itself. Table 8 shows all porosity values are significantly lower for PA666 control (F0) than F5.

Table 8. Porosity results of PA6/66 samples (F0 and F5)

Properties	PA666 F0 (Control)	PA6/66 F5
Open Porosity, %	0.015	7.80
Closed Porosity, %	0.107	0.20
Total Pore Volume, mm^3	1.286	75.15
Total Porosity, %	0.121	7.99

F0 contains less pores and has a uniform distribution of pores localized around the center of tensile bar as shown in Figure 12. On the other hand, 3D models obtained from F5 showed porosity distributed throughout the surface of the sample. Results are in conformity with tensile properties as reported in Table 7.

Figure 12. 3D models obtained from F0 (left) and F5 (right).

Micro-CT images of F0 (Figure 13) and F5 (Figure 14) further show differences in porosity between the two samples. Row b of both figures show the reconstructed model of F0 and F5. In comparing the two samples, F5 has visibly larger and greater number of pores within its cross-section. It is plausible that the greater porosity of F5 is due to the additives within the nanocomposite. These additives may have decreased the flow rate of the filament in the 3D printing process, resulting in less material extruding out of the printer nozzle. This decrease in material extruding from the nozzle may have resulted in the creation of pores and cavities in the 3D printed samples.

Figure 13. Micro-CT images of F0 (a) Binarized (1st Row) and (b) Reconstructed (2nd Row).

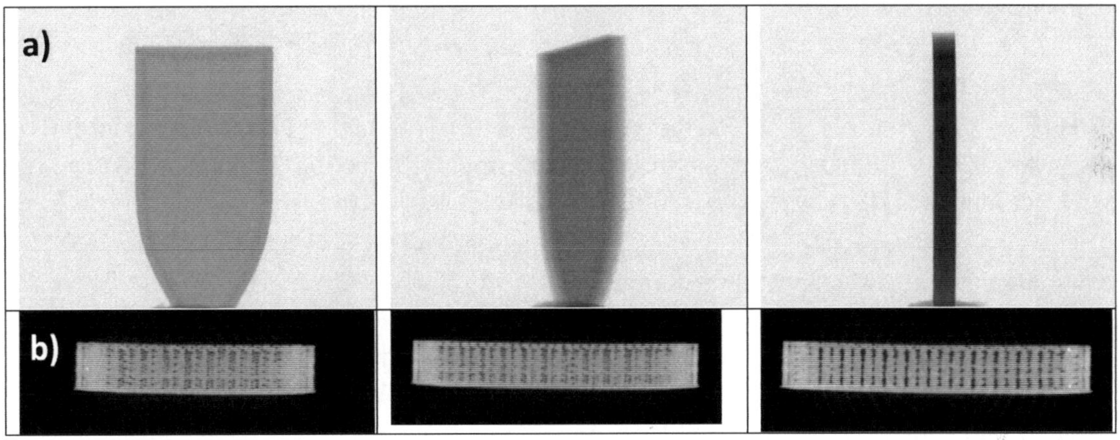

Figure 14. Micro-CT images of F5 (a) Binarized (1st Row) and (b) Reconstructed (2nd Row).

3. CONCLUSIONS

In this study, PA6/66 and its derivative formulations were successfully processed, printed, and characterized. After analysis of all tests, it was found from MCC that all formulations outperformed the neat PA6/66 system; among the results, F2, F4, and F5 yielded the best results. From TGA testing, all formulations outperformed the neat system. Furthermore, TGA displayed clear differences between the formulations containing 15 wt.% FR and 12.5 wt.% FR. UL 94 flammability testing also yielded promising results as F5 showed no drip formation, therefore, successfully 3D printing F2 and F4 may yield even better flammability properties. Although mechanical properties of F5 degraded in comparison to the neat system, it is believed that the other additives (NC, MWNT, and K) assisted in preventing the mechanical properties from deteriorating even further. Micro-CT analysis proved that F5 has much higher porosity in comparison to the control system, F0. This increase in porosity was most likely caused by the additives in the nanocomposite which decreased the material flow rate during the printing process.

4. FUTURE WORKS

Although SEM and TEM images are not taken yet, they are expected to show good dispersion of the additives. However, the SEM images will be closely observed to ensure that flame retardant is homogenously dispersed within the PA6/66 matrix because uneven dispersion of FR may have caused inconsistent UL 94 date. Furthermore, SEM can also be used to observe the surface morphology of the samples before and after UL 94 tests. Moving forward, more research will be conducted to adjust the 3D printing parameters to ensure successful prints with more PA6/66 formulations as well as printing test samples with lower porosity. Doing so will allow further UL 94 testing, tensile testing, and micro-CT analysis of more formulations.

Acknowledgments

This research was partially funded by KAI, LLC, Austin, Texas, USA. The authors acknowledge the facilities, and the technical assistance of the RMIT Advanced Manufacturing Precinct and RMMF Facility.

REFERENCES

[1] J. Koo, *Fundamentals, Properties, and Applications of Polymer Nanocomposites*, Cambridge University Press, Cambridge, UK, 2016, pp 389-424.

[2] H. Wu, M. Krifa, J.H. Koo, Rubber (SEBS-G-MA) Toughened Flame Retardant Polyamide 6: Microstructure, Combustion, Extension, and Izod Impact Behavior, *Polymer-Plastics Technology and Engineering* 57(8) (2017), 727-739.

[3] H. Wu, R. Ortiz and J. H. Koo, Self-Extinguishing and Non-Drip Flame Retardant Polyamide 6 Nanocomposite: Mechanical, Thermal, and Combustion Behavior, *Flame Retardancy and Thermal Stability of Materials* 1(1) (2018) 1-13.

[4] H. Wu, R. Ortiz, J.H. Koo, Rubber toughened flame retardant (FR) polyamide 11 nanocomposites Part 1: the effect of SEBS-g-MA elastomer and nanoclay, *Flame Retardancy and Thermal Stability of Materials* (2018), 1(1) (2018) 25-38.

[5] H. Wu, R. Ortiz, J.H. Koo, Rubber toughened flame retardant (FR) polyamide 11 nanocomposites Part 2: synergy between multi-walled carbon nanotube (MWNT) and MMT nanoclay, *Flame Retardancy and Thermal Stability of Materials* 2(1) (2018) 19-29.

[6] S.C. Lao, C. Wu, T.J. Moon, J.H. Koo, A. Morgan, L. Pilato, G. Wissler, Flame-Retardant Polyamide 11 and 12 Nanocomposites: Thermal and Flammability Properties, *Journal of Composite Materials* 43(17) (2009) 1803-1818.

[7] S.C. Lao, W. Yong, K. Nguyen, T.J. Moon, J.H. Koo, L. Pilato, G. Wissler, Flame-retardant Polyamide 11 and 12 Nanocomposites: Processing, Morphology, and Mechanical Properties, *Journal of Composite Materials* 44 (25) (2010) 2933-2951.

[8] S.C. Lao, J.H. Koo, T.J. Moon, M. Londa, C.C. Ibeh, G.E. Wissler, L.A. Pilato, Flame-retardant Polyamide 11 Nanocomposites: Further Thermal and Flammability Studies, *Journal of Fire Sciences* 29(6) (2011) 479-498.

[9] S.C. Lao, J.H. Koo, T.J. Moon, W. Yong, C. Lam, J. Zhou, B. Haisujoto, G. Wissler, L. Pilato, Z.P. Luo, Flame Retardant Intumescent Polyamide 11 Nanocomposites-Further Study, *Journal of Composite Materials* 47(923) (2013) 2973-2986.

[10] A. Hao, I. Wong, H. Wu, B. Lisco, B. Ong, A. Sallean, S. Butler, M. Londa, J.H. Koo, Mechanical, Thermal, and Flame-Retardant Performance of Polyamide 11-Halloysite Nanotube Nanocomposites, *Journal of Materials Science* 50 (1) (2015) 157-167.

[11] J.W. Gilman, T. Kashiwagi, J.D. Lichtenhan, Nanocomposites: a revolutionary new flame retardant approach, *SAMPE Journal* 33(4) (1997) 40-46.

[12] J.W. Gilman, Flammability and thermal stability studies of polymer layered-silicate (clay) nanocomposites, *Applied Clay Science* 15(1–2) (1999) 31-49.

COLD SPRAY ADDITIVE MANUFACTURING OF POLYMERS: TEMPERATURE PROFILES DICTATE PROPERTIES UPON DEPOSITION

Tristan W. Bacha,[1†] Elizabeth M. Henning,[2] Luke Elwell,[2] Francis M. Haas,[2] Joseph F. Stanzione III[1]
[1]Department of Chemical Engineering, Rowan University
[2]Department of Mechanical Engineering, Rowan University
Glassboro, NJ, 08028
[†]Corresponding Author: bachat9@students.rowan.edu

ABSTRACT

The cold spray additive manufacturing process enables deposition of coatings and structures by impact bonding microparticles to surfaces through preheated, nozzle-accelerated gas flow. Most established models of gas-particle heat transfer in cold spray assume lumped thermal capacitance, which ignores particle temperature distributions and is typically valid for metals. However, recent advances in cold spray consider polymeric systems that differ categorically and considerably in thermophysical properties compared to metals. Particularly, the low Biot number (*Bi*) assumption may be invalid, impacting efficiency and quality of polymeric particle deposition. We describe herein a 1D transient gas-particle heat transfer model that simulates temperature profile evolution in a representative particle traveling through a cold spray nozzle. With this model, we simulated thermal histories of polymeric materials at realistic processing conditions. Polymer sprays with Bi consistently near 10^{-1} or greater demonstrated thermal inhomogeneities, in some cases straddling the glass transition temperature. Thermal stratification provides insight into physical conditions of a particle before deposition, which can guide processing parameters to improve deposition efficiency and quality of polymer cold sprays.

1. INTRODUCTION

1.1 Cold Spray

Cold spray was first discovered by Russian scientists in the 1980s, and has been widely studied to deposit a wide variety of metal coatings on surfaces at high deposition rates [1]. The cold spray process uses high-velocity gas flow to accelerate micron-sized particles, subsequently impacting them on a surface. The impact causes plastic deformation between the particle and surface, and with some probability termed the deposition efficiency (η_D), effects a bond between the particle and substrate for particle velocities exceeding a material-specific critical velocity (V_c). Gas conditions are tuned to avoid the melting phase change of the material in the process; accordingly, the process is termed "cold" spray and relies (macroscopically) on kinetic energy instead of thermal energy to achieve interlayer bonding. To impart this kinetic energy, an aerosolized stream of particles is combined with hot gas and accelerated through a nozzle. Converging nozzles are used to achieve (sub)sonic gas conditions, and converging-diverging nozzles are used to obtain supersonic conditions [2]. Nozzle choice depends on the critical velocity requirements.

To date, research in cold spray has focused on metal coatings [3]. The cold spray of polymers has received less attention, with a few research groups demonstrating polymeric deposition in recent years [4-9]. Reported deposition efficiencies for polymers rarely exceed a few percent, while metals have reached nearly 100 % [4-7, 10]. Bush et al. demonstrated and characterized the deposition of semicrystalline high-density polyethylene (HDPE) [5], and subsequent work by Khalkhali and Rothstein reports extremal η_D approaching ~10 % for HDPE sprayed on to LDPE [8]. The deposition efficiency improved with increasing particle preheat and substrate temperatures; however, the particle temperatures were limited by the cold spray apparatus used. Thus, there are still unexplored process conditions that could be optimal for deposition. For instance, changing the nozzle geometry results in different temperature profiles as discussed herein. This temperature profile may have substantial impacts on deposition, especially above the glass transition temperature (T_g) of a given polymer.

Below T_g polymers are glass-like, with little motion of individual polymer chains. As the temperature rises above T_g, polymer chains move more freely and the material becomes less elastic. Amorphous polymers typically experience a decrease in elastic modulus by a factor of ~1000 in the glass transition range [11]. From this change in properties around the T_g, we hypothesize that the deposition efficiency can be greatly improved in the cold spray of polymers by carefully manipulating the spray conditions to bring portions of a polymer particle well above T_g. Therefore, modeling particle temperature as a function of spray conditions is an important prerequisite to achieving desired particle temperatures upon impact.

1D modeling has been used to rapidly simulate the temperature evolution of a representative particle as it traverses a nozzle and predict its temperature upon impact [5, 12-14]. Champagne and Helfritch coupled the 1D equations to a deposition efficiency model to determine optimal particle size [14]. Each of these models assumes lumped thermal capacitance, which implies essentially uniform temperature throughout the particle. However, this assumption is only valid when the particle Biot number (Bi), described by Equation 1, is less than 0.1.

$$Bi = \frac{h_c L_c}{k} \tag{1}$$

L_c is a characteristic length defined as the ratio of the object volume to surface area, h_c is the convective heat transfer coefficient, and k is thermal conductivity of the particle. The Biot number is the ratio of conduction resistance L_c/k to convective resistance $1/h_c$. For a sphere, L_c is equal to one third of the radius R [15]. The rate of heat transfer inside a particle is markedly slower than the heat transfer to or from the edge in contact with fluid at values greater than ~0.1 [15]. Thus, a temperature gradient may exist depending on the timescale the object is subject to heat transfer. In micro-scale metallic particles assuming lumped capacitance is typically valid, meaning the entire particle can be treated as one temperature. However, this is not always the case in thermal spray processes. For polymers and ceramics, the Biot number tends to be close to or greater than the 0.1 threshold. Katanoda showed that a substantial temperature difference exists in cold sprayed Al_2O_3 and Ti6Al4V particles [16]. Ivosevic and colleagues showed that polymers in an High-Velocity Oxy-Fuel process (HVOF) exhibit extremely non-uniform temperature profiles [17]. Previous works studying polymers in real cold spray systems have utilized 1D Isentropic models assuming lumped capacitance to estimate average particle temperatures upon impact. While proving to be a useful tool, these models ignore the expected temperature inhomogeneities in polymer particles. Bush et al. assumed a lumped capacitance model in making predictions for their experiments, but acknowledged that 3D temperature profiles may be crucial for accurate modeling of impact simulations [5].

Herein, we developed a model to simulate the temperature profile inside a representative spherical particle as it travels through high-speed nozzles characteristic of cold spray processing. We demonstrate with our results that the lumped capacitance assumption is not appropriate for the cold spray of polymeric systems by simulating temperature stratification in an experimentally demonstrated spray. A second hypothetical nozzle designed to spray polystyrene is used to demonstrate the substantial effects of a temperature distribution in an amorphous polymer. The range of temperatures observed in a particle may fall within the glass transition range of a given polymer, resulting in heterogeneous mechanical properties through a particle. Around the T_g, the mechanical properties change considerably in a 20 °C to 30 °C window [11]. Therefore, we expect temperature distributions to have profound implications on spray quality and deposition efficiency. The results herein can be used in conjunction with impact simulations to better understand the effect of temperature stratification on deposition physics, with the goal of maximizing deposition efficiency and deposition quality in polymer cold spray systems for additive manufacturing.

2. SIMULATION

2.1 Stratified Temperature Modeling

For transient heat transfer processes, Geankoplis and Bird et al. outline the shell balances for several geometries [15, 18]. When applied to a sphere, the unsteady-state energy transport equation as a function of radius r and time t becomes Equation 2.

$$\frac{\partial T}{\partial t} = \frac{-1}{4\pi\rho C_p r^2}\frac{\partial(Q_r)}{\partial r} \qquad [2]$$

Where T is the temperature, ρ is the particle density, C_p is the particle heat capacity, and Q_r is the total heat flow at r. Q_r is described by Equation 3, where k is the thermal conductivity of the particle.

$$Q_r = -4\pi r^2 k\frac{\partial(T)}{\partial r} \qquad [3]$$

To solve these equations, we employed the numerical method of lines. Equation 4 gives the boundary condition used to solve for the center of a sphere.

$$\left.\frac{\partial T}{\partial r}\right|_{r=0} = 0 \qquad [4]$$

Similarly, boundary conditions are needed for the surface of the sphere at the outer radius R. The heat transfer at the edge of the sphere is modeled using the convective heat transfer coefficient h_c. Equation 5 gives the condition for the surface of a sphere.

$$-k\left.\frac{\partial T}{\partial r}\right|_{r=R} = h_c(T_R - T_\infty) \quad [5]$$

The initial condition for time ($t = 0$) for particle temperature assumes a uniform temperature through the particle.

2.2 Particle Velocity and Gas Modeling

Particle motion was modeled by integrating the acceleration function derived from Newton's second law in Equation 6 [19].

$$\frac{dV_p}{dt} = \frac{\pi C_D \rho_g D_p (V_g - V_p)|V_g - V_p|}{8m} \quad [6]$$

Where V_p is the particle velocity, V_g is the gas velocity, m is the mass of the particle, D_p is the diameter of the particle, C_D is the drag coefficient, and ρ_g is the gas density.

C_D was calculated using correlations from Henderson that account compressibility, and span the subsonic, supersonic, and transition flow regimes [20]. The convective heat transfer coefficient was calculated using the Ranz and Marshall model [21]. The isentropic flow equations described by Shapiro were used to model the gas [22]. Champagne and Helfritch outline the 1D method used herein in detail [14, 19]. It was assumed that the gas properties remained constant from the end of the nozzle to the bow shockwave before the substrate. At the shockwave, the Mach number was assumed to decrease linearly to zero at the substrate [19]. The shockwave location was estimated using Billig's approximation given by Equation 7 [19, 23].

$$\Delta = 0.143Dl\left[\exp\left(\frac{3.24}{{M_e}^2}\right)\right] \quad [7]$$

Where Δ the distance from the bow shockwave to the substrate, Dl is the nozzle diameter at the exit, and M_e is the Mach number at the nozzle exit.

The gas properties were calculated as a function of the linearly decreasing Mach number with the isentropic gas equations and reservoir conditions [19].

2.3 Numerical Integration

MATLAB version R2019a was used to program the simulation. Equation 6 was integrated using the Runge-Kutta 4 algorithm. A time step of 10^{-6} s was used from the entrance of the nozzle to the bow shockwave, and a time step of 10^{-9} s was used after the shockwave. The particle temperature partial differential equation was solved using the numerical method of lines. One hundred temperature nodes were used for polymer particles, and 20 were used for the metal particles. The temperature equations were integrated using the ode45 function in MATLAB.

2.4 Physical Properties and Modeling

When available, data from previously simulated results or real experimental setups were used. The thermal conductivity, heat capacity, and density of each particle were assumed to be constant. Gas thermal conductivities and viscosities were assumed to be a function of only temperature except for the helium case, where the constant viscosity of 1.86×10^{-5} kg m^{-1} s^{-1} from Marrocco et al. was assumed [13]. Sutherland's law was used to model the viscosity for other cases [24]. The heat transfer properties for air and nitrogen were taken from data compiled in Incropera [15]. Table 1 outlines the model setup for the conditions studied.

Table 1. Nozzle dimensions and gas conditions

Nozzle	***Tr*, (°C)**	***Pr*, (kPa)**	***Lc*, (mm)**	***Lt*, (mm)**	***Ld*, (mm)**	***De*, (mm)**	***Dt*, (mm)**	***Dl*, (mm)**	***Ls*, (mm)**	**Gas**
HDPE (1)	70, 100	250	23.9	26	0	9.5	1.6	1.6	15	Air
PS (2)	127	400	29.9	0	2.97	9.5	1.6	1.7	15	Air
Ti (3)	25	2900	-	-	100	-	2.9	8	-	He
Al (4)	300	3000	30	0	180	20	2	6	-	N_2

Tr and *Pr* are the reservoir temperature and pressure, respectively. *Lc*, *Ld*, and *Lt* are the converging, diverging, and throat lengths. *De*, *Dt*, and *Dl* are the entrance, throat, and exit diameters. The standoff distance is denoted as *Ls*. Figure 1 represents the simple conical nozzle geometry used in the simulation. The distance denoted in later figures is the distance from the nozzle inlet.

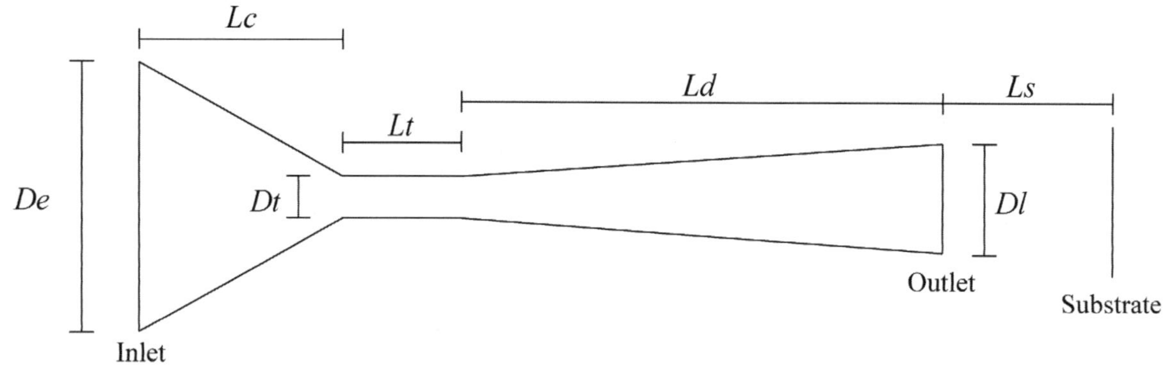

Figure 1. Representation of the simple conical nozzle geometry

Nozzle 1 is a geometry designed by Bush et al., who used the nozzle to cold spray HDPE. The spray conditions simulated included a maximum temperature condition limited by the experimental setup [5]. Several theoretical conditions with independent initial particle temperature and gas reservoir temperature were also modeled. Nozzle 2 is similar to a design used by Bush; however, the experimental conditions used for modeling were not from actual experimental conditions [25]. The inlet pressure and temperature to achieve reasonable target particle temperatures and impact velocities from the results of Khalkali et al., who used a polystyrene particle size of 44 μm in single impact studies [6]. The larger 90 μm particle size was simulated to study the effect of changing particle size on the temperature distributions. Two metal particles were simulated from experimentally demonstrated conditions. The titanium conditions were replicated from a report by Marrocco et al., and the aluminum conditions were from Jenkins et al. [13, 26]. The material properties of the particles are given in Table 2.

Table 2. Material properties of simulated particles

Material	Thermal Conductivity (W/mK)	Heat Capacity (J/kgK)	Particle Density (kg/m^3)	Particle Diameter (μm)
HDPE (Nozzle 1)	0.49	2250	990	48
Polystyrene (Nozzle 2)	0.163	1460	1000	44, 90
Titanium (Nozzle 3)	21.6	528	4500	47
Aluminum (Nozzle 4)	234	1005	2700	40

The properties for HDPE are the same as Bush used for their simulations [25]. The density and thermal conductivity of polystyrene at 350 K were from Algaer and assumed to be constant [27]. The heat capacity value for polystyrene is from Incropera [15]. Titanium properties were those used by Marrocco et al. in their simulations [13]. The thermal conductivity and heat capacity values for aluminum were taken from Incropera at 250 °C, and the density at 27 °C [15].

2.4.1 Assumptions

The following assumptions apply to the nozzle model:

1. The gas flow is isentropic and frictionless
2. The particle has zero initial velocity
3. The particle does not affect gas temperature or velocity (one-way interaction)
4. The initial particle temperature distribution is uniform
5. The gas returns to reservoir conditions at the substrate

3. RESULTS

3.1 Polymeric Temperature Profiles

3.1.1 HDPE in a sonic nozzle

Four cases for the HDPE particle were modeled with varying initial particle temperature *Tpo* and reservoir temperature *Tr*. The gas velocity, pressure, and temperature for the sonic nozzle spraying HDPE are reported in Figure 2. It was assumed that the Billigs approximation applies to this case, and the Mach number decreases linearly from 1 to 0 at the predicted location. Because the gas is sonic, there is no step change in properties at the estimated shock location. Although more complex phenomena contribute to the gas property profiles, this assumption lets us model the nozzle up to the point of impact in a computationally inexpensive way and gain valuable insight into the heat transfer and temperature-dependent property changes experienced by the particles.

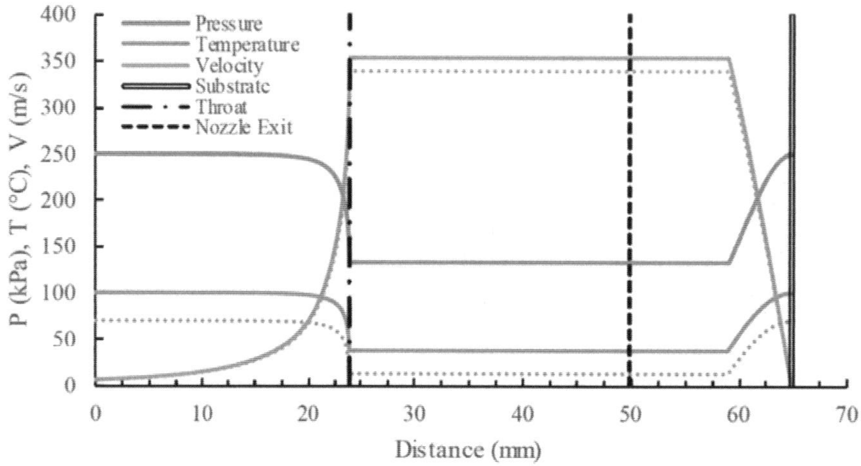

Figure 2. Gas pressure (*P*), temperature (*T*), and velocity (*V*) profiles through the sonic nozzle with a *Tr* of 100 °C (solid line) and *Tr* of 70 °C (dotted line).

Figure 3 presents the particle velocity profiles corresponding to these gas conditions. The higher reservoir temperature condition resulted in slightly faster particles.

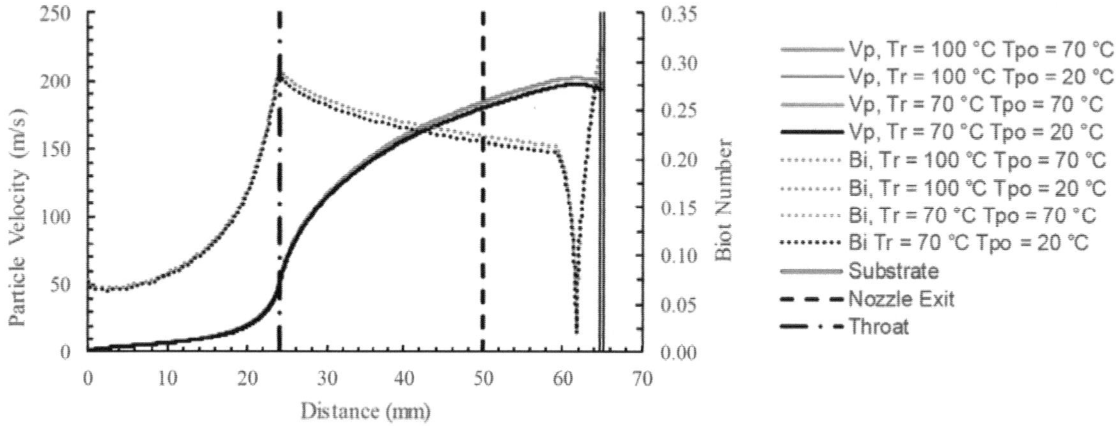

Figure 3. Particle velocity (*Vp*) and Biot number profiles of HDPE particles subject to the gas conditions in Figure 2.

Particles with the same reservoir temperature condition have velocity profiles and Biot numbers that overlap. The orange and blue lines overlap, and the black and grey lines overlap. The Biot number changes along the nozzle because it is a function of the velocity difference between the particle and gas. The low Biot number observed at 62 mm is a result of the gas velocity matching the particle velocity. This matching velocity results in a minimum convective heat transfer coefficient, hence the Biot number achieves a minimum. The particle spends a considerable time at relatively large Biot numbers, indicating a potential temperature inhomogeneity within the particle. Figure 4 shows the resulting internal temperature profiles at several discrete radii.

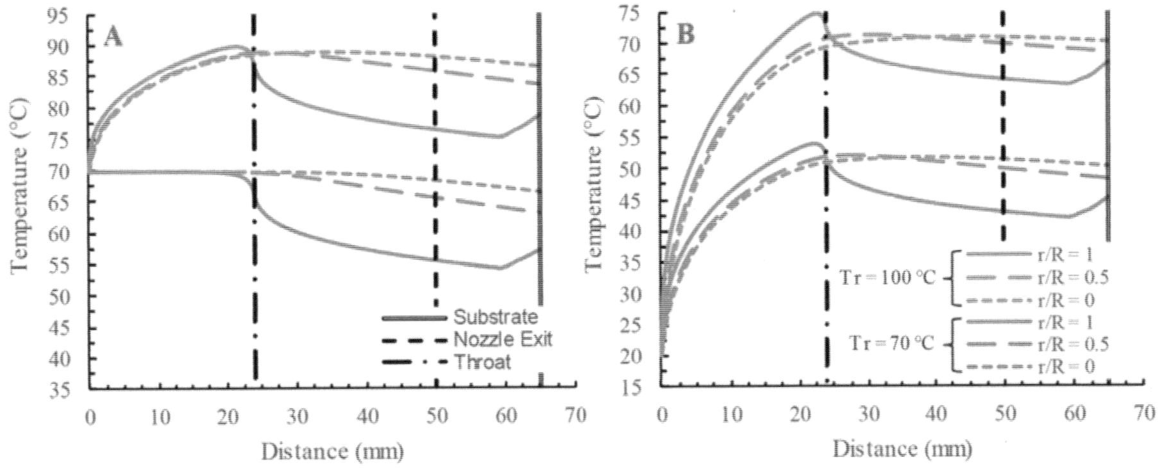

Figure 4. Particle temperature as a function of nozzle distance and dimensionless radius *r/R* at three discrete locations. A) 48 μm HDPE particle at *Tpo* = 70 °C and B) at *Tpo* = 20 °C.

The particle at equal initial particle temperature and reservoir temperature (*Tr* = *Tpo* = 70 °C) is the highest temperature condition achievable by the Bush et al. cold spray apparatus [5]. In their experiments, the particle hopper is surrounded by the reservoir gas, linking the reservoir temperature and initial particle temperature. Clogging was experienced at temperatures over 70 °C as particles became tacky in the hopper. The *Tr* = 100 °C and the *Tr* = 70 °C, *Tpo* = 20 °C conditions are outside of the operating range as they would require a different particle feeding mechanism than possible experimentally. The particles at lower temperatures than the reservoir temperature undergo heating until reaching the nozzle throat, where the gas then begins to cool them. The particle with an initial temperature equal to the reservoir only experiences cooling. For all particles, the internal temperatures appear to converge as they approach the substrate. To further investigate this, the temperature profiles just before substrate impact are discussed in Section 3.1.3.

3.1.2 Polystyrene in a converging-diverging nozzle

The polystyrene case considers a polymer particle with a much higher glass transition temperature than the HDPE case (T_g of 100 °C vs. -90 °C) [11, 25]. In the HDPE case, the particles were always well above the T_g. For the polystyrene simulations, two particle sizes and two initial particle temperatures were studied in a converging-diverging nozzle, unlike the purely converging nozzle for the HDPE case. The gas reservoir conditions were chosen to set the final temperature and velocity of the 44 μm polystyrene particle to values within the deposition window reported by Khalkhli et al. [6]. Figure 5 depicts the gas property profiles the polystyrene particles were simulated with.

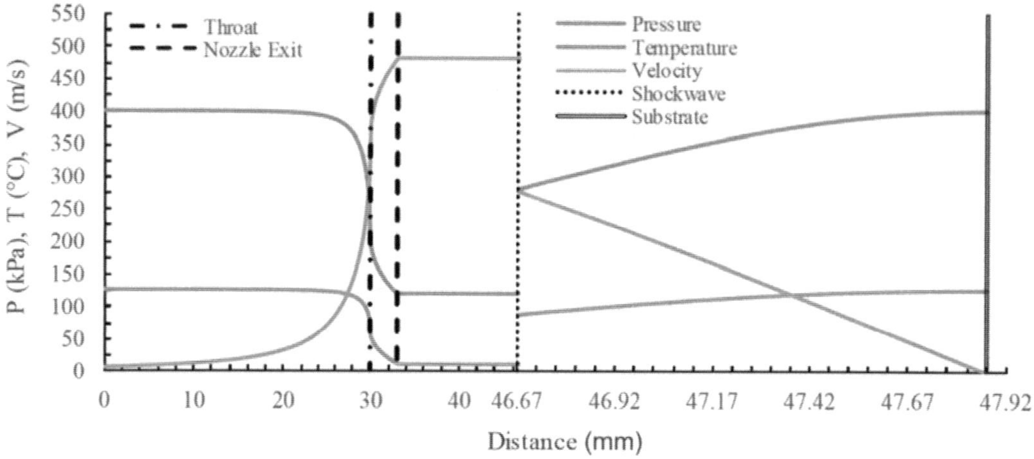

Figure 5. Pressure (P), temperature (T), and velocity (V) of the gas in the supersonic nozzle.

The Biot numbers were calculated for both particle sizes studied. The same size particles have velocities and Biot numbers independent of the initial particle temperature, exemplified by the HDPE case in Figure 3. Figure 6 shows the corresponding Biot and velocity profiles encompassing both initial particle temperatures (20 and 100 °C).

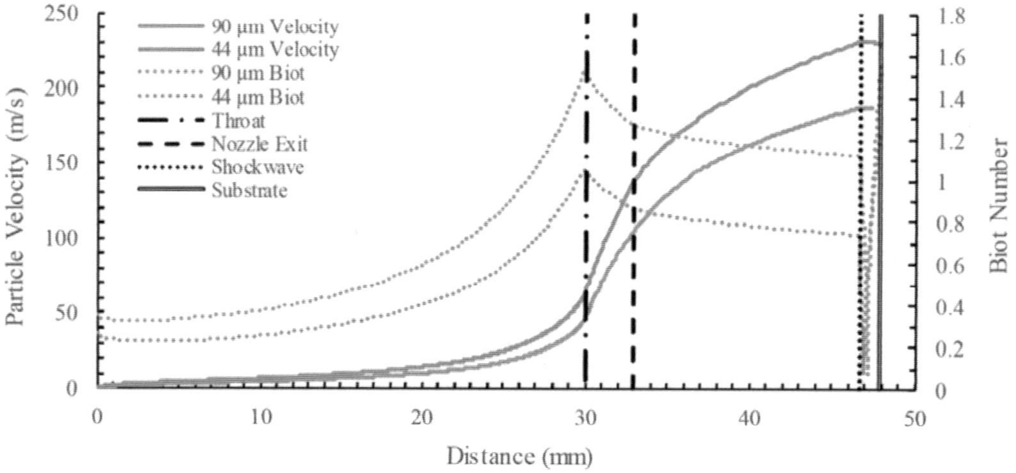

Figure 6. Polystyrene particle velocity and Biot number profiles in the supersonic nozzle.

The Biot numbers for the polystyrene system are higher than the HDPE system, and consistently above the 0.1 threshold. This is attributed to higher gas velocity and lower thermal conductivity of the particles. Because of the high Biot number, the polystyrene system cannot be adequately represented by the lumped capacitance model, and thus a temperature distribution is observed. Figure 7 reports the resulting temperature profiles as a function of distance from the particle inlet to the substrate.

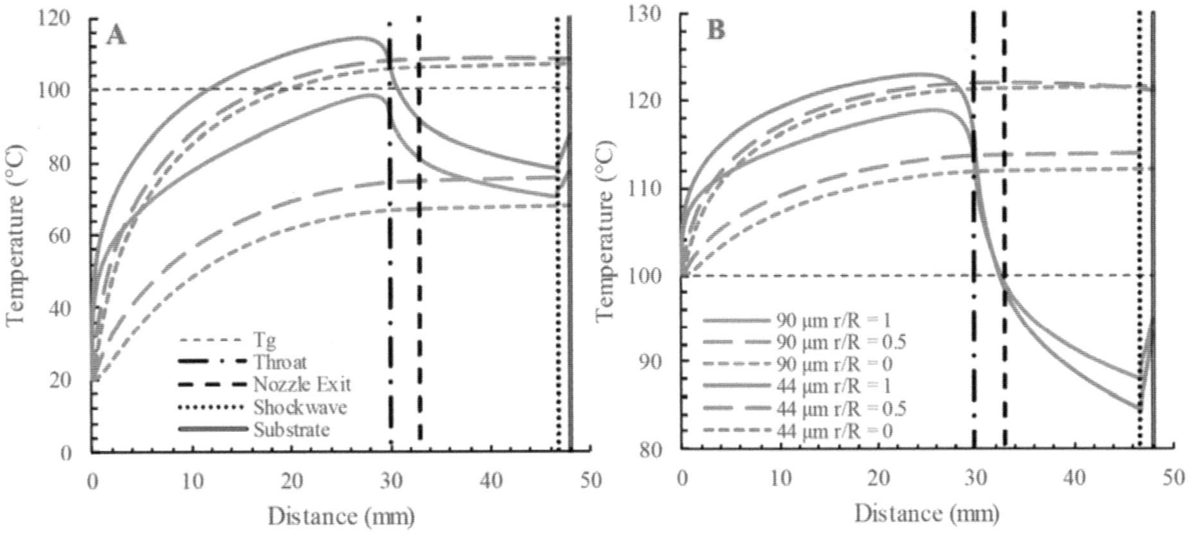

Figure 7. Temperatures of 90 and 44 μm polystyrene particles as a function of distance and dimensionless radius *r/R*, and at initial particle temperatures A) 20 °C and B) 100 °C. The horizontal dashed line at 100 °C represents the T_g of polystyrene. *Tr* and *Pr* are fixed at 127 °C and 400 kPa respectively.

Heating and cooling in the particles occur more rapidly at the surface than at the core and midpoint. The particles cool rapidly through the diverging section of the nozzle and subsequently heat again after passing through the bow shockwave. The internal temperatures however barely see this heating effect in the small region between the shockwave and the substrate, resulting in a polymer sphere with a cool shell and warm inside. The temperature profiles of the polystyrene particles upon impact with the substrate are presented in the following section.

3.1.3 Polymer particle impact temperature

The temperature profiles upon impact from the conditions in Figure 4 and Figure 7 were plotted as a function of particle radius. Figure 8 summarizes these temperature profiles. The polystyrene and HDPE particles exhibit distinct differences in the internal temperature distributions. Notably, the polystyrene particles cross the glass transition region in several cases. A 30 °C temperature differential in the region between the surface and the core implies a significant deviation in mechanical properties near the T_g. If the maximum temperature is close to the surface, we expect this deviation to cause localized deformation at the surface of the particle. The localized deformation would improve bonding, as bonding in cold spray is attributed to plastic deformation between the particle and substrate. We expect this shift in deformation localization to occur regardless of whether the bonding mechanism is shear instability or hydrodynamic deformation [28, 29].

The shear modulus profile as a function of radius for two polystyrene particles was predicted using data from Mott et al. to quantify expected mechanical property deviations in the polystyrene case [30]. Figure 9 shows the internal shear modulus profile for the denoted cases from Figure 8A, and two additional cases where all points in the respective temperature profiles are shifted a specified temperature. This shift could be achieved by adjusting intial particle temperature and nozzle geometry. The 90 μm particle at *Tpo* = 20 °C impacts in the glassy state, and shows relatively little change in shear modulus (0.09 MPa) in a ~20 °C window. The 44 μm *Tpo* = 100 °C particle straddles the glass transition upon impact, resulting in maximum 0.43 MPa difference in shear modulus (ΔG_{max}) inside the particle. Both differences are considerably increased with the specified temperature shifts, increasing to $\Delta G_{max} =$ 0.34 and 0.62 for the 90 and 44 μm particles respectively. This highlights that a broad range of mechanical properties are achievable by tuning the particle temperature.

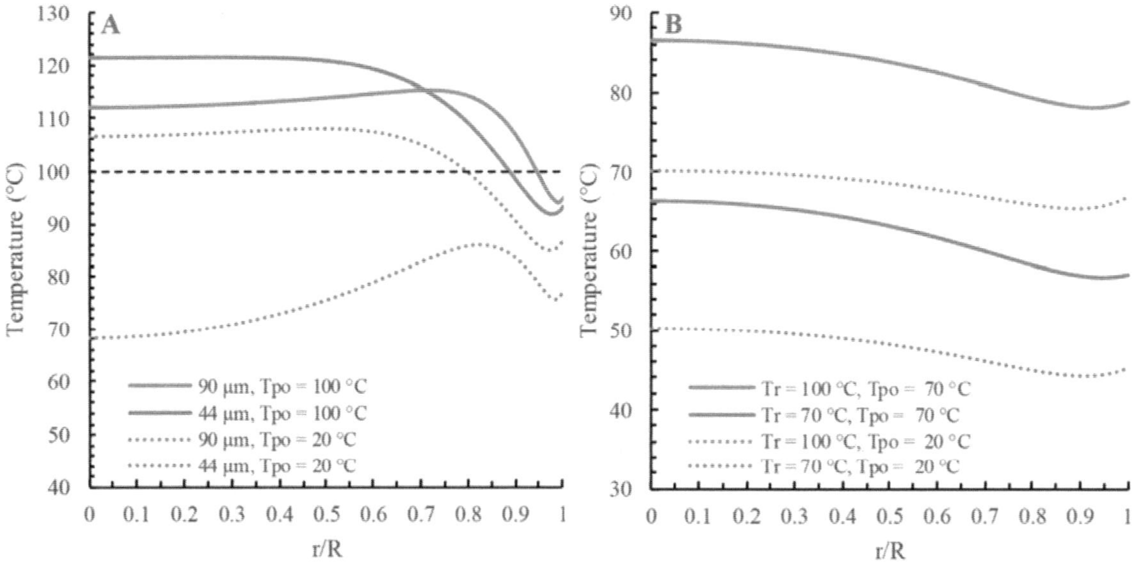

Figure 8. Temperature profiles upon impact as a function of dimensionless radius *r/R* in the A) polystyrene and B) HDPE particles.

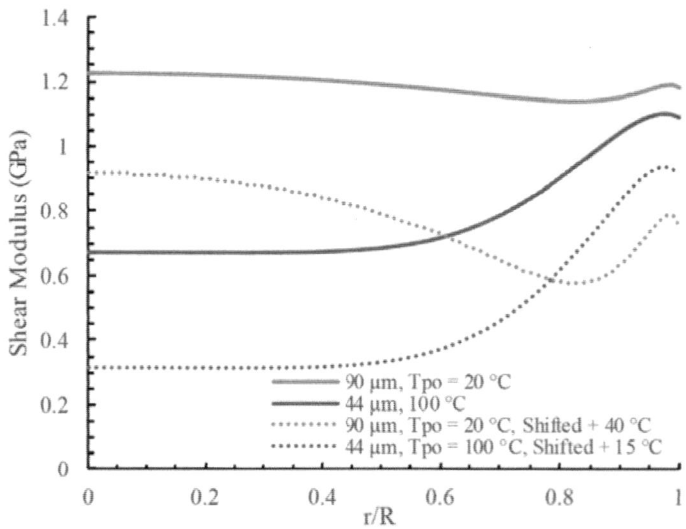

Figure 9. Shear modulus vs. dimensionless radius for two polystyrene particles. The dotted lines represent the modulus curves if the corresponding temperature profiles in Figure 8 are shifted the specified temperature increase. Shear modulus data is from Mott et al. [30].

The HDPE particles are expected to behave similarly when initially crossing the T_g. However, the HDPE particles are highly crystalline compared to polystyrene, an amorphous polymer. Amorphous polymers experience a rapid decrease in modulus with increasing temperature in the region of T_g. Semicrystalline polymers experience this decrease to a lesser degree. Additionally, a broader rubbery plateau region is observed in semicrystalline polymers all the way up to the crystal melting temperature [11]. In this region, the modulus decreases only slightly. For HDPE, this range extends from the -90 °C T_g to the 135 °C melt temperature. Thus, only slight decrease in modulus is expected for the HDPE particles studied. The representative curves for HDPE would be analogous to the shear modulus profile for the glassy 90 µm *Tpo* = 20 °C polystyrene particle, with minor deviations in the shear modulus.

3.2 Metal Particle Simulations

To demonstrate the substantial differences in thermal behavior between metals and polymers in cold spray, we simulated two cases for metal particles inside a cold spray nozzle from experimental spray conditions. Titanium and aluminum were chosen as representative cases for low and high conductivity metal, respectively. Figure 10A shows the predicted temperatures for the surface and core of the two particles plotted as a function of dimensionless nozzle distance (distance/total nozzle length) from the particle injection point.

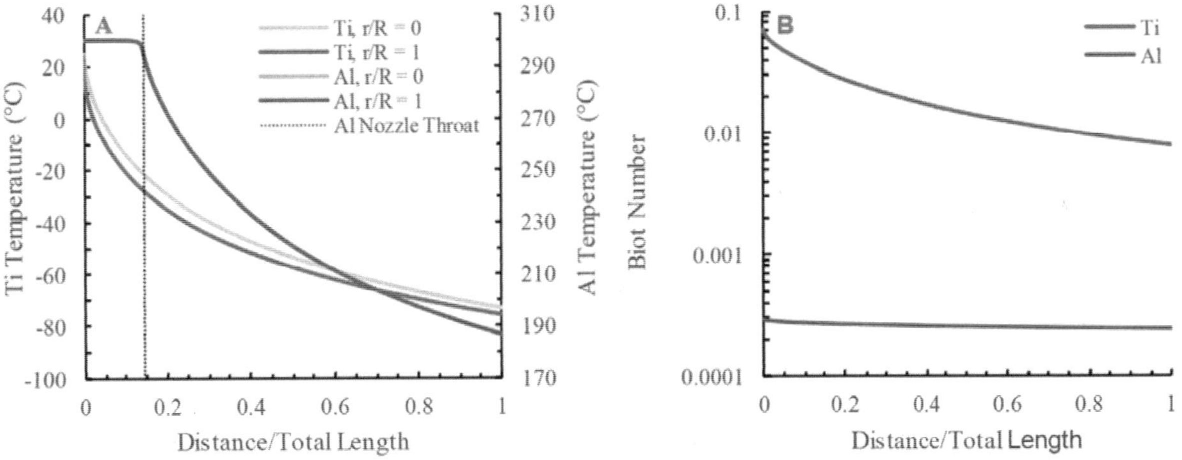

Figure 10. A) Temperature profiles in converging-diverging nozzles for the titanium and aluminum spray conditions. The values at the center and edge of the particles are plotted as a function of dimensionless length. B) Biot number profiles along the nozzles.

The aluminum particle was injected before the nozzle throat, while the titanium particle was injected after the nozzle throat according to the experimental conditions [13, 26, 31]. Both particles show Biot numbers lower than the 0.1 lumped thermal capacitance threshold along the length of the nozzle. The aluminum case shows little temperature difference from the surface to the core of the particle at the end of the nozzle. However, the titanium case shows a greater difference, which is expected from its lower thermal conductivity, and subsequently higher Biot number. At the nozzle exit, the maximum temperature difference inside the aluminum and titanium particles is

0.53 and 4.9 °C, respectively. Unlike the polymer particles discussed in the previous section, this small difference is expected to have negligible effects on particle properties upon impact.

4. CONCLUSIONS

4.1 Discussion

The results herein emphasize that polymer particles have internal temperatures that vary considerably with location in the particle. Metal particles show relatively little temperature distribution upon impact comparted to polymers. This temperature variation in polymers results in a substantial mechanical property variation within the same particle at the substrate. Not only are the temperature profiles evident in the particles, but their shape depends on the polymer properties, gas conditions, and nozzle geometry. If an amorphous polymer particle is around or above its T_g, drastic changes in mechanical properties are observed.

The various temperature distributions observed indicate that not only can the average temperature and velocity of a particle be controlled, but the temperature distribution and resulting mechanical properties are variable as well. The nozzle type, geometry, and gas conditions govern the resulting temperature profile upon impact. We plan to use this model to predict deposition temperature profiles and velocities of polymer particles upon substrate impact and target specific temperature distributions. Additionally, the heat transfer principles used herein could be applied to more complex CFD simulations that can better estimate flow properties in later studies. The resulting profiles will be analyzed in impact studies like Sagar et al. and Yildirim et al. performed to predict optimal processing conditions, and subsequently used as experimental conditions [32, 33]. Future investigations into gas properties and particle temperatures simulated with the 1D model will be validated by comparing to computational fluid dynamics (CFD) simulations and experiments.

4.2 Conclusion

The thermal histories of polymer particles inside cold spray nozzles were simulated using the 1D isentropic gas relations. Compared to metal particles, polymers are not adequately represented by the lumped capacitance assumption. The simulated radial temperature profiles at the substrate showed substantial deviation in polymer particles and were found to potentially straddle the material-dependent glass transition temperature. We expect temperature profiles and resulting mechanical property inhomogeneities to have substantial effects on deposition rate, efficiency, and bond quality. This temperature profile can be altered by changing the initial particle temperature, nozzle geometry, and gas conditions. The simulated temperature profiles on impact will be used for initial conditions in impact simulations to analyze deformation patterns. These simulations will guide experiments to optimize deposition efficiency and quality for polymer cold spray additive manufacturing.

5. REFERENCES

1. Fauchais, P., Heberlein, J., and Boulos, M., *Thermal Spray Fundamentals*. New York: Springer, 2014
2. Champagne, Victor K., *The Cold Spray Materials Deposition Process*. Cambridge England: Woodhead, 2007
3. Moridi, A., Hassani-Gangaraj, S. M., Guagliano, M., and Dao, M., "Cold spray coating: review of material systems and future perspectives," *Surface Engineering* 30 (2014): 369-395. doi: 10.1179/1743294414y.0000000270.
4. Xu, Y. and Hutchings, I. M., "Cold spray deposition of thermoplastic powder," *Surface and Coatings Technology* 201 (2006): 3044-3050. doi: 10.1016/j.surfcoat.2006.06.016.
5. Bush, T. B., Khalkhali, Z., Champagne, V., Schmidt, D. P., and Rothstein, J. P., "Optimization of Cold Spray Deposition of High-Density Polyethylene Powders," *Journal of Thermal Spray Technology* 26 (2017): 1548-1564. doi: 10.1007/s11666-017-0627-5.
6. Khalkhali, Z., Xie, W., Champagne, V. K., Lee, J.-H., and Rothstein, J. P., "A comparison of cold spray technique to single particle micro-ballistic impacts for the deposition of polymer particles on polymer substrates," *Surface and Coatings Technology* 351 (2018): 99-107. doi: 10.1016/j.surfcoat.2018.07.053.
7. Ravi, K., Ichikawa, Y., Deplancke, T., Ogawa, K., Lame, O., and Cavaille, J.-Y., "Development of Ultra-High Molecular Weight Polyethylene (UHMWPE) Coating by Cold Spray Technique," *Journal of Thermal Spray Technology* 24 (2015): 1015-1025. doi: 10.1007/s11666-015-0276-5.
8. Khalkhali, Z. and Rothstein, J. P., "Characterization of the cold spray deposition of a wide variety of polymeric powders," *Surface and Coatings Technology* 383 (2020): doi: 10.1016/j.surfcoat.2019.125251.
9. Ravi, K., Sulen, W. L., Bernard, C., Ichikawa, Y., and Ogawa, K., "Fabrication of micro-/nano-structured super-hydrophobic fluorinated polymer coatings by cold-spray," *Surface and Coatings Technology* 373 (2019): 17-24. doi: 10.1016/j.surfcoat.2019.05.078.
10. Ravi, K., Deplancke, T., Ogawa, K., Cavaillé, J.-Y., and Lame, O., "Understanding deposition mechanism in cold sprayed ultra high molecular weight polyethylene coatings on metals by isolated particle deposition method," *Additive Manufacturing* 21 (2018): 191-200. doi: 10.1016/j.addma.2018.02.022.
11. Sperling, Leslie H., *Introduction to Physical Polymer Science*. Hoboken, New Jersey: Wiley, 2006
12. Champagne, V. K., Helfritch, D. J., Dinavahi, S. P. G., and Leyman, P. F., "Theoretical and Experimental Particle Velocity in Cold Spray," *Journal of Thermal Spray Technology* (2010): doi: 10.1007/s11666-010-9530-z.
13. Marrocco, T., McCartney, D. G., Shipway, P. H., and Sturgeon, A. J., "Production of titanium deposits by cold-gas dynamic spray: Numerical modeling and experimental characterization," *Journal of Thermal Spray Technology* 15 (2006): 263-272. doi: 10.1361/105996306X108219.
14. Helfritch, D. and Champagne, V. K., "Optimal Particle Size for the Cold Spray Process," in *Thermal Spray 2006*, Seattle, Washington, USA, 2006.
15. Incropera, F. P., DeWitt, D. P., Bergman, T. L., and Lavine, A. S., *Fundamentals of Heat and Mass Transfer 5th Edition*. Wiley, 2001

16. Katanoda, Hiroshi., "Numerical Simulation of Temperature Uniformity within Solid Particles in Cold Spray," *Journal of Solid Mechanics and Materials Engineering* 2 (2008): 58-69. doi: 10.1299/jmmp.2.58.
17. Ivosevic, M., Cairncross, R. A., and Knight, R., "3D predictions of thermally sprayed polymer splats: Modeling particle acceleration, heating and deformation on impact with a flat substrate," *International Journal of Heat and Mass Transfer* 49 (2006): 3285-3297. doi: 10.1016/j.ijheatmasstransfer.2006.03.028.
18. Bird, R. B., Stewart, W. E., and Lightfoot, E. N., *Transport Phenomena*. New York: Wiley, 2006
19. Helfritch, D. and Champagne Jr, V., "A Model Study of Powder Particle Size Effects in Cold Spray Deposition," in *Army Science Conference*, Florida, 2008.
20. Henderson, Charles B., "Drag Coefficients of Spheres in Continuum and Rarefied Flows," *AIAA Journal* 14 (1976): 707-708. doi: 10.2514/3.61409.
21. Ranz, W. and Marshall, W. R., "Evaporation from drops," *Chem. eng. prog* 48 (1952): 141-146.
22. Shapiro, Ascher. H., *The Dynamics and Thermodynamics of Compressible Fluid Flow*. New York: Ronald Press Company, 1953
23. Billig, Frederick S., "Shock-wave shapes around spherical-and cylindrical-nosed bodies," *Journal of Spacecraft and Rockets* 4 (1967): 822-823. doi: 10.2514/3.28969.
24. Tan, Zhongchao., *Air Pollution and Greenhouse Gases: From Basic Concepts to Engineering Applications for Air Emission Control*. Springer Singapore, 2014
25. Bush, Trenton., "Cold Gas Dynamic Spray - Polymer Deposition," Master of Science M.S., Mechanical Engineering, University of Massachussets Amherst, 2016.
26. Jenkins, R., Aldwell, B., Yin, S., Chandra, S., Morgan, G., and Lupoi, R., "Solid state additive manufacture of highly-reflective Al coatings using cold spray," *Optics & Laser Technology* 115 (2019): 251-256. doi: 10.1016/j.optlastec.2019.02.011.
27. Algaer, Elena., "Thermal Conductivity of Polymer Materials - Reverse Nonequilibrium Molecular Dynamics Simulations," Ph.D, Physical Chemistry, Technische Universität, Darmstadt, 2010.
28. Hassani-Gangaraj, M., Veysset, D., Champagne, V. K., Nelson, K. A., and Schuh, C. A., "Adiabatic shear instability is not necessary for adhesion in cold spray," *Acta Materialia* 158 (2018): 430-439. doi: 10.1016/j.actamat.2018.07.065.
29. Grujicic, M., Zhao, C. L., DeRosset, W. S., and Helfritch, D., "Adiabatic shear instability based mechanism for particles/substrate bonding in the cold-gas dynamic-spray process," *Materials & Design* 25 (2004): 681-688. doi: 10.1016/j.matdes.2004.03.008.
30. Mott, P. H., Dorgan, J. R., and Roland, C. M., "The bulk modulus and Poisson's ratio of "incompressible" materials," *Journal of Sound and Vibration* 312 (2008): 572-575. doi: 10.1016/j.jsv.2008.01.026.
31. Meyer, M. C., Yin, S., McDonnell, K. A., Stier, O., and Lupoi, R., "Feed rate effect on particulate acceleration in Cold Spray under low stagnation pressure conditions," *Surface and Coatings Technology* 304 (2016): 237-245. doi: 10.1016/j.surfcoat.2016.07.017.
32. Shah, S., Lee, J., and Rothstein, J. P., "Numerical Simulations of the High-Velocity Impact of a Single Polymer Particle During Cold-Spray Deposition," *Journal of Thermal Spray Technology* 26 (2017): 970-984. doi: 10.1007/s11666-017-0557-2.
33. Yildirim, B., Muftu, S., and Gouldstone, A., "Modeling of high velocity impact of spherical particles," *Wear* 270 (2011): 703-713. doi: 10.1016/j.wear.2011.02.003.

EVALUATION OF FIBER PLACEMENT STRATEGIES FOR THE IMPLEMENTATION OF CONTINUOUS REINFORCEMENT FIBERS IN SELECTIVE LASER SINTERING PROCESS

Patrick Moll, Felix Pirrung, Michael Baranowski, Sven Coutandin, Jürgen Fleischer
wbk Institute of Production Science, Karlsruhe Institute of Technology
Karlsruhe, Germany

ABSTRACT

Among engineering materials today continuous fiber reinforced polymers (FRP) show some of the highest stiffness and strength to weight ratios. To rival the traditional manufacturing methods of continuous FRP many investigations have sought to combine the outstanding mechanical performances of these materials with the freedom in design and the economic benefits of additive manufacturing (AM). This paper focuses on the fiber placement strategies and their interaction with Selective Laser Sintering (SLS) specific machine features. The goal is to develop and conduct test series to gain a deeper understanding of how the process, the polymer, and the reinforcement fibers interact. For this investigation different patterns of glass fiber rovings are embedded into specimens made from PA 12 on a Sintratec Kit printer. The rovings are put up onto a frame in varying patterns to be able to relate fiber tension and curvature as well as the stack height of intersecting rovings to the quality of embedding. Additionally the time of placement, the clamping and the interaction of the fibers with the recoater have been investigated. Based on these results an SLS printer with automated continuous fiber implementation will be developed in the future.

Corresponding author: Patrick Moll, patrick.moll@kit.edu, +4915239502600

1. INTRODUCTION

The material group of fiber reinforced polymers is relatively young and still subject to many investigations. But it has already opened up possibilities and applications that could not be realized with other materials. Especially the ratios of stiffness and strength to density are unrivaled and combined with the ability to tailor the mechanical properties along different orientations it is easy to see why composite structures are used in high end applications in many industries. However these outstanding performances come at a cost. Common to many manufacturing methods of continuous reinforced polymer parts are high costs due to a large amount of manual labor, a poor freedom of design and long lead times due to the necessity of molds. The field of polymer additive manufacturing shows its strength in exactly those three areas. On the other hand the mechanical properties of an additively manufactured part often lack behind its injection-molded or extruded counterpart [1]. By combining continuous fiber reinforcement with polymer additive manufacturing it is possible to cancel out the drawbacks and create a new method of manufacturing composite parts. As will be shown in chapter 2 a lot of research has been done to implement continuous fibers in extrusion based additive manufacturing processes and systems are already commercially available [2, 3]. For industrial applications the selective laser sintering process is considered a promising process due to the good mechanical properties, thermal stability, precision and surface quality [4].

In the research project FiberAdd conducted by the wbk Institute of Production Science at Karlsruhe Institute of Technology and funded by the Vector foundation (Stuttgart, Germany) the implementation of endless fibers in the selective laser sintering (SLS) process is examined. The overall project goal is the development of an SLS printer with an additional unit to automatically insert continuous fibers in the process to reinforce SLS printed polymer parts. This paper presents the evaluation of fiber path strategies with the aim to better understand the interaction between the continuous fibers, the SLS printer and the process. It addresses researchers in the field of continuous fiber reinforced AM and development engineers for AM equipment alike.

In this paper first the current state of the art in continuous fiber reinforced additive manufacturing is presented. Then the experimental set-up for the evaluation of the fiber paths will be explained. Afterwards the results for the test series are presented and discussed. Finally a conclusion and an outlook of the next steps is given.

2. STATE OF THE ART

In the following the current state of the art for the additive manufacturing with continuous fibers will be presented and analyzed. Previous investigations have mainly focused on three additive manufacturing methods that are either widespread and therefore well understood or show characteristics which makes it particularly easy to include continuous fiber reinforcement. Material extrusion is the most commonly used AM technique and makes up most of the research as well. The next common technique is stereolithography were also some work was done to introduce continuous fiber reinforcements to the process. Laminated Object Manufacturing (LOM) is not widely spread in neat polymer applications but due to its sheet feedstock continuous reinforcement fibers can easily be included in the process.

The most investigations have been made in the area of material extrusion and here especially in Fused Deposition Modelling (FDM). Baumann et al. suggested to cluster these methods by the point of fiber implementation [5]. The fibers can be brought together with the polymer matrix prior, inside, or after the extruding nozzle. Since the formation of the fiber-matrix interface is a crucial part of the print process a lot of work went into the development of the print nozzle itself. Fischer et al. argued that the fiber feeding could be done by the extruded polymer eliminating the need for a separate fiber-speed control system. Also, the interface formation would work best inside the hot nozzle where the polymer is at its lowest viscosity. They designed and manufactured a FDM nozzle via a selective laser melting process out of which the fibers and matrix material are extruded together [6]. Other works that were focusing on print nozzle development are Bettini et al., Li et al. and Yang et al. All of them combined the fibers and the matrix material inside the nozzle to form a fully reinforced part. Tests parts with a PLA matrix and aramid [7] or carbon [8] fiber reinforcement or an ABS/carbon combination [9] were printed and then the samples were tested mechanically to compare them to the neat matrix polymer. Li et al. went one step further and prepared a set of samples where the carbon fibers were coated in PLA prior to printing with the aim to investigate the influence of the interface formation. Several ways of placing the fibers inside the polymeric matrix have been tested and evaluated by Baumann et al. They found that only a fiber introduction after the nozzle would enable the machine to print a neat polymer and a composite with just one nozzle [10]. They investigated several ways of introducing a roving of carbon and glass fibers into a just printed bead of ABS using a modified Arburg freeformer. The fibers were laid down and overprinted, injected with a hot hypodermic needle, and sunken into the polymer with a solvent. The US American company Markforged offers with its Mark Two printer a machine capable of locally reinforce the build parts with continuous fibers. Unlike the approach of Baumann et al. the Mark Two uses two nozzles with one reserved for the neat polymer and the other to lay down a fiber rowing that is already impregnated with the matrix polymer. Using this commercial system or its predecessor Mark One, Dickson et al., Zhu et al. and van der Klift et al. investigated the mechanical reinforcement achievable with glass, carbon and aramid fibers [11–13].

Using stereolithography Karalekas and Antoniou fabricated dog-bone shaped tensile specimen from a polyurethane and epoxy based photo-polymer and embedded a nonwoven glass fiber mat into the middle layer. The specimens out of the epoxy based photo-polymer were reinforced with mats of four different basis weights while the polyurethane specimens only had two different mats. Generally the reinforcement effect increased with increasing basis weight except for the highest weight in the epoxy based photopolymer [14]. The second work of Karalekas used the same technique and test set-up but altered the materials investigated. An acrylic based photo-polymer was reinforced and tested with two glass and carbon nonwoven mats and an aramid mat [15].

Since 2D fiber mats and prepregs are the most used semi-finished products in composite manufacturing anyway it is the logical step to utilize them in LOM to additively manufacture composite parts. Klostermann et al. investigated the interface formation of glass fiber / epoxy prepreg. A post LOM consolidation cycle was necessary to reduce the void content and strengthen the layer interface [16]. To avoid the post treatment Parandoush et al. used a uni- and bidirectional glass fiber / PP prepreg composite sheet to manufacture their specimens. The individual layers were bonded by heating the surfaces of both strips with a CO2 laser and pressing them together with the consolidation roller [17]. In a later work Paradoush et al. used the same technique to manufacture multidirectional carbon fiber / PA 6 samples and did further investigations about the heat transfer inside the already build stack [18].

Table 1 summarizes the presented state of the art and shows which properties of continuous fiber reinforced additive manufactured parts have already been investigated. It can be seen that the mechanical properties of differently manufactured specimen have been investigated thoroughly. Also the interface between the fibers and the matrix polymer has been studied. However the combination of the SLS process and continuous fiber reinforcement is uninvestigated so far.

Table 1. Overview of the current state of the art in continuous fiber reinforced additive manufacturing

Process	References	Reinforcement material	Tested properties								
			Young' s modulus	Tensile strength	Strain at break	Stiffening	Flexural modulus	Flexural strength	Storage modulus	Interface	Porosity and voids
Material extrusion	Baumann, Scholtz et al. 2017	C, G	+	+	-	+	-	-	-	0	-
	Baumann, Sielaff et al. 2017	C	+	+	-	+	-	-	-	-	-
	Dickson et al. 2017	C, G, Ar	+	+	0	-	+	+	-	+	+
	Li et al. 2016	C	-	+	-	-	-	+	+	+	0
	Tian et al. 2016	C	-	-	-	-	+	+	-	+	-
	Fischer et al. 2017	C	-	-	-	-	-	-	-	0	-
	Bettini et al. 2017	Ar	+	+	+	+	+	+	-	0	-
	Yang et al. 2017	C	+	+	-	-	+	+	-	+	+
	Zhu et al. 2017	C	-	-	-	-	+	+	-	-	+
	Van der Klift et al. 2016	C	+	+	+	-	-	-	-	-	0
LOM	Klostermann et al. 1998	G	-	-	-	-	-	-	-	0	0
	Parandoush et al. 2017	G	+	+	+	-	+	+	-	0	+
	Parandoush et al. 2019	C	+	+	+	-	+	-	-	-	+
SL	Karalekas 2003	C, G, Ar	+	+	+	-	-	-	-	0	-
	Karalekas, Antoniou 2004	G	+	+	+	-	-	-	-	0	0

Symbols:
C – Carbon fiber, G – Glass fiber, Ar – Aramid fiber
+ fully examined, 0 partly examined, - not examined

3. EXPERIMENTAL SET-UP

This work wants to lay the base on which further research can develop a process to additively manufacture composite parts with the SLS process. To be able to automatically introduce continuous reinforcement fibers the interaction between the fibers and the process must be well understood. In contrast to the processes presented in chapter 2 the SLS process poses some unique challenges to the introduction of continuous reinforcement fibers. Unlike other common AM machines, a SLS printer shows many individually moving parts. This relative movement of polymer grains, machine parts and reinforcement fibers poses a challenge on process reliability. Hence the placement strategies of the reinforcement fibers need to take into account all the moving parts inside the build chamber. This interaction of the reinforcement fibers with parts like the recoater blade is crucial to maintain a stable and repeatable process. The main questions that are tried to be answered within the scope of this paper are the placement time of the continuous fibers, the achievable complexity of the fiber patterns, how well the fibers can be embedded in one layer height and whether fixation of the fibers after placement is needed.

3.1 Development of test series

The conducted experiments are split up in two test series differing in the timing of the fiber placement during the SLS process. In test series one the fibers are placed manually on top of an already sintered layer before the recoater applies a new powder layer. In the second test series a new layer of powder is applied first and then the fiber is placed manually, followed by the pass of the laser to sinter the material.

The objective of the first test series is to see how the recoater as well as the fresh powder moved by it would alter the position of a fiber roving placed on the build chamber. Different placements and fixation points make the fibers successively more vulnerable to the recoater movement. It is recorded how, and at which step in complexity the initial placement fails to be transferred to the final part. To evaluate if the location of the roving was changed a simple specimen is designed with a cutout in the middle and a notch on either side. The roving can be placed exactly over the notches and the cut-out. So after printing the movement is easy to determine by its position visible at the cut-out. Figure 1 shows the steps in fiber-pattern complexity and the shape of the specimen. To secure the rovings in place two sets of fiber brackets are printed on the machine itself. With these the roving can be secured either parallel or perpendicular to the recoater movement.

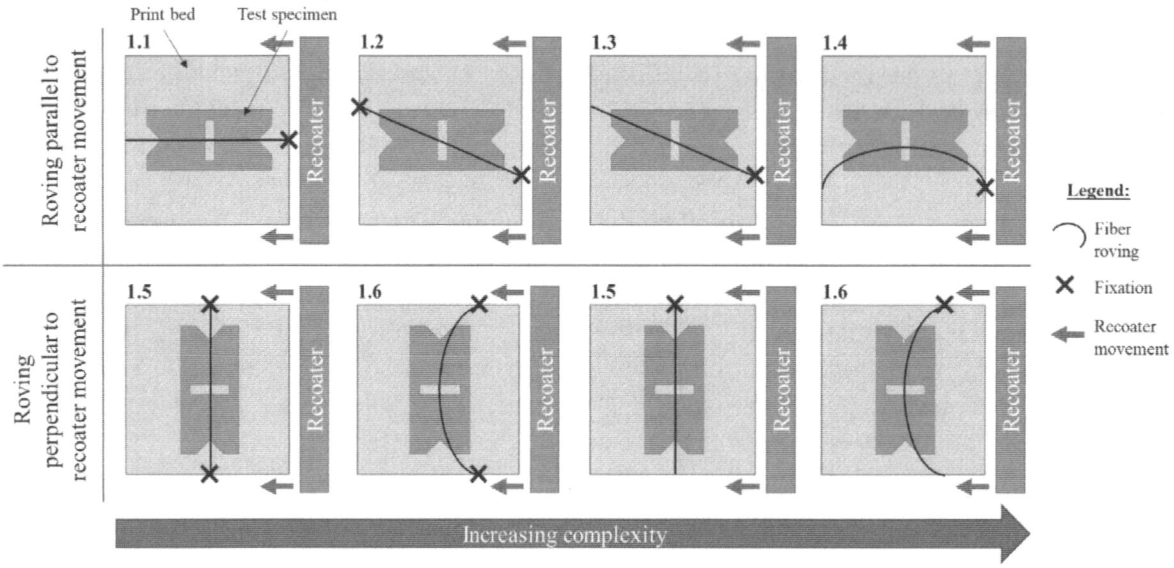

Figure 1. Test set-up for test series one with increasing complexity to the right

The main goal of the second series of tests is to investigate how well a roving can be embedded into the part when it is introduced to the powder bed after the recoating step. The target parameters here are the stability of a roving's fixation inside one layer and the maximum stack height of a fiber pattern before it cannot be securely embedded in one layer anymore. Additionally the effect of curvature and tension of the fibers on embedding and interface formation was analyzed. In order to be able to place the rovings in the powder bed a frame was printed onto which they could be stretched and placed on top of the powder surface. Four different fiber patterns were designed to investigate the above mentioned research goals. The first two tests 2.1 and 2.2 are focused on how well the roving can be embedded within one layer height and how strong this connecting is. Test 2.3 investigates the maximum fiber stack height that can be securely embedded, and 2.4 investigates the effect of curvature and fiber tension on the way the roving is embedded. For this second test series the specimen shape is a simple rectangular plate. Figure 2 shows the specimen shape, the frame and the different fiber patterns for the test scenarios of the second test series.

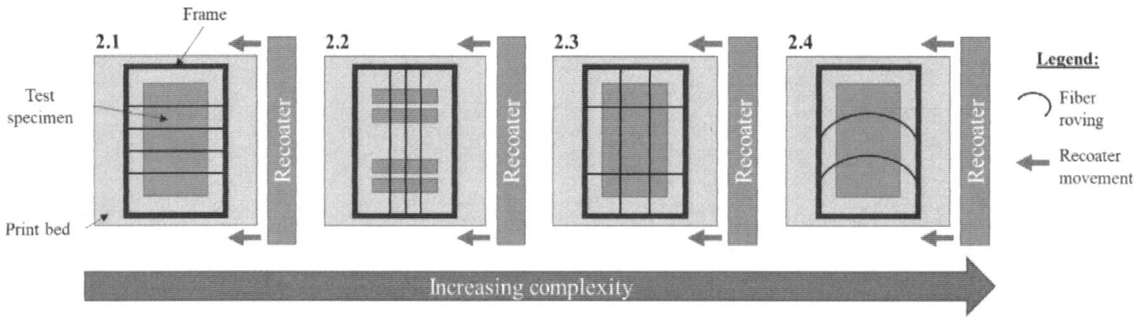

Figure 2. Test set-up for test series two with increasing complexity to the right

3.2 Experimental procedure

The previously described test series were conducted with a Sintratec Kit desktop SLS printer from the company Sintratec (Brugg, Switzerland) shown in Figure 4. For the experiments the materials and process parameters shown in Table 2 were used.

Table 2. Material and process parameters for conducted test series

Material	
Fiber	Glass fiber roving D450 60 tex (Fulltech Fiber Glass Corp.)
Powder	PA 12 Black Powder (Sintratec)
Process parameters	
Heat up temperature powder surface	150 °C
Heat up temperature build chamber	140 °C
Print temperature powder surface	170 °C
Print temperature build chamber	140 °C
Laser speed	500 mm/s
Layer height	0.1 mm

The Sintratec Kit is a low-cost desktop SLS printer widely used in academia and research. It comes with an own closed control system and a software called Sintratec Central to preprocess the proprietary control code for the machine. The adjustment of process parameters and process steps is very limited and restricted to basic settings like the print temperature, laser speed and layer height. After the start of the print job no further control of the printer is possible. After slow heating of the build chamber (approx. 1.75 hours) and powder bed preparation, the sintering of the part is done in a cyclic manner. This cyclic procedure consists of recoating fresh powder from the reservoir, infrared heating of the new layer to print temperature and sintering with the laser. After the last layer the printer cools down for another 1.5 hours. In order to be able to manufacture the planned composite specimen on the machine a specific experimental procedure has been developed as a workaround solution.

For the first test series the specimen is created as a .stl file with half of the plate thickness and then imported to the Sintratec Central software. At the start of each new experiment the machine is always loaded with fresh powder and heated up. After the heat-up the lower half of the specimen is printed classically. Right after the last layer of the part is scanned by the laser and molten, the machine is stopped via the emergency stop button. The printer is opened, and the roving placed manually on the last layer. This positions the roving onto the consolidated polymer, ready to spread the next powder layer over it. The time during which the machine was open is kept as short as possible since the process chamber would lose a lot of heat which in turn could alter the results. After preparing the build chamber in one of the ways shown in Figure 1 the machine is closed, and the same print job is started again. Because every new print job starts with the preparation of a hot powder base this step has to be skipped to not burry the first half of the specimen and the roving under the powder. By starting the new print the recoater spreads a layer of powder over the already build specimen and the suspended roving. It is recorded how the fibers behave under the influence of the recoater and how many layers it took to completely embed the roving into the part. After the specimen is printed for the second time, the printer performs its normal cooldown program. Figure 3 shows the step-by-step procedure of this first test series.

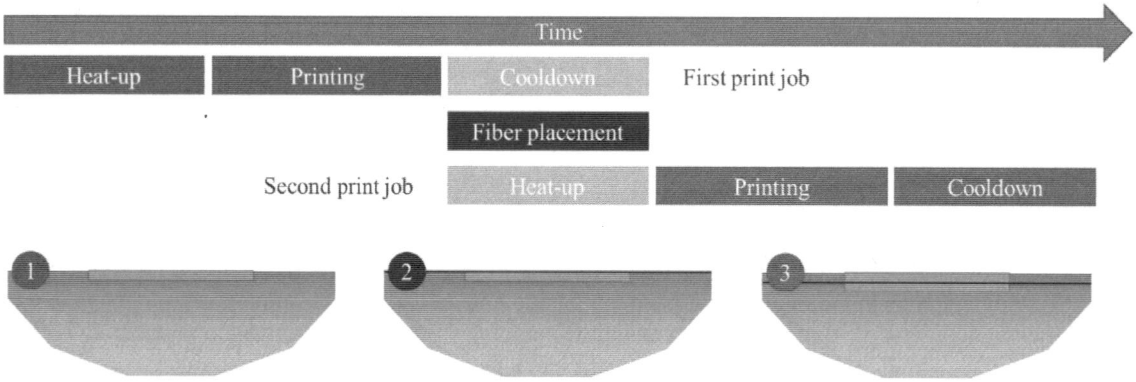

Figure 3. Specific step-by-step test procedure for the first test series

The control problem for the second test series is different. The Sintratec Kit always executes the previously described three basic build steps successively. After the initial powder bed recoat the infrared lamps heat it up to the pre-set sinter temperature and then the laser scans the cross-section of the part to be printed. In order to only heat and scan a previously applied powder layer the blade of the recoater must be detachable. This still allows the printer to perform the recoat operation but the frame or the recoater does not come in contact with the powder bed or any device sitting on top of it. Similar to the first test series also the specimens for the second test series are printed in two steps. The first print is performed in the known way but this time the printer is not stopped after but before the sintering of the last layer. This leaves the last coat of the powder bed untouched. The machine is opened and the recoater blade is removed. Then the frame with the fiber rovings is placed onto the powder bed according to the set-up shown in Figure 2 and the machine is closed again. Just like in the previous test series, the time when the machine was open is kept at a minimum to reduce the temperature drop inside the process chamber. With the machine closed a new print job is started according to the test set-up. Again, the powder bed preparation is skipped manually. The printing of one layer always starts by moving the recoater and spreading new powder. Due to the detached recoater blade only the recoater frame moves without applying fresh powder. Hence the powder bed with the fiber frame is not disturbed (cf right side of Figure 4). Then the infrared lamps bring the powder surface back to sintering temperature and the laser scans the cross-section. After that the print is manually aborted and the machine is given time to cool down.

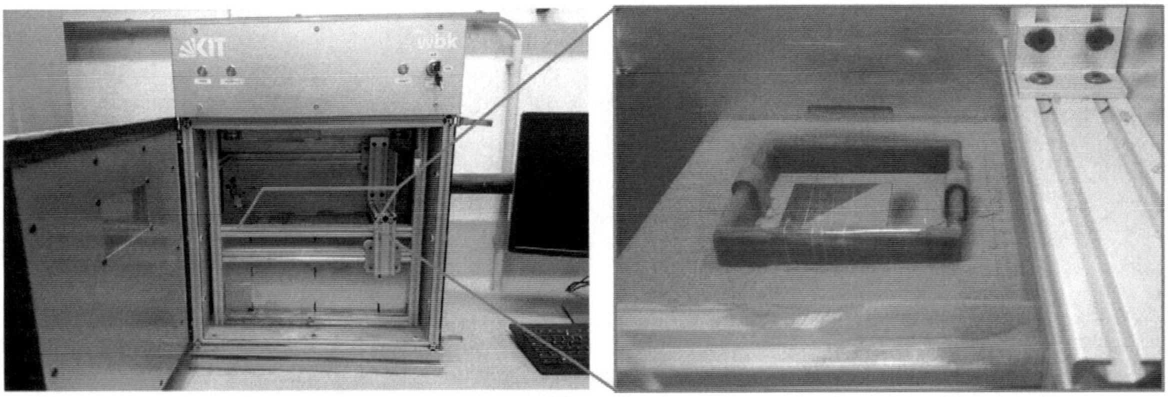

Figure 4. Sintratec Kit desktop SLS printer used for test series (left side) and view of build chamber during print of second test series after insertion of the frame (right side)

4. RESULTS

4.1 First test series

In the first test series the roving was placed manually prior to the application of fresh powder for a layer. For the two roving orientations the complexity of the set-up was raised until the fibers could not be embedded in the part anymore and their positions were altered by the movement of the recoater. As the complexity in the tests increases no more complex test has been carried out after a failed test scenario.

For test 1.1 the intended location of the fiber and in the actual part differ only a little bit (cf. Figure 5). This small displacement was not caused by the recoater blade but during the manual placement of the roving in the first place. Due to the alignment of the recoater movement and reinforcement fibers the roving is tensioned by the recoater and therefore keeps its position parallel to the recoater movement. During the recoating the blade and the fresh powder stretches the roving out and covers it in powder. This is enough to prevent it to be touched and moved again by the retracting recoater blade. After the first laser scanning the roving is secured on the part and none of the following steps could alter its position. In test 1.2 we can also observe a good placement of the fiber inside the test specimen. In the center of Figure 5 we can see that the fiber is well tensioned inside the test specimen. The small position error can also be attributed to the manual placement of the fiber.

For both test 1.1 and 1.2 we can observe a good integration of the fiber into the polymer. The embedded roving is completely embedded in the sintered powder which has been applied on it. The bonding is strong enough to withstand a slight manual pull on the fiber. However it is yet to find out if the bonding mechanism is purely mechanical or if it relies on adhesion forces as well.

In test 1.3 the roving was only fixed on one side but orientated at an angle to the recoater movement. In this test scenario the recoater blade significantly changed the position of the roving during the print process as can be seen on the right of Figure 5.

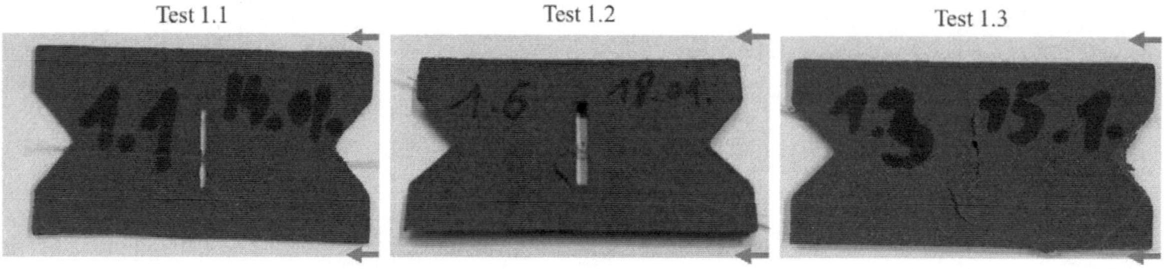

Figure 5. Red highlighted roving position of test specimen with roving parallel to the recoater movement (direction of recoater movement shown by arrow)

The roving in test 1.5 the positioning of the roving is at the desired place in the center of the test specimen without a negative effect of the recoater movement. Also, here the small deviation was due to an error in the manual placement of the roving.

The position of the roving without tension in test 1.6 on the other hand was highly altered by the recoater movement. In Figure 6 on the right side we can see the position of the roving in the finished specimen is curved to the opposite direction of its initial manual placement. This is caused by a movement of the roving by the retracting recoater blade. It can also be seen, that the two halves of the specimen did not align perfectly, which is caused by the recoater moving the lower half of the specimen.

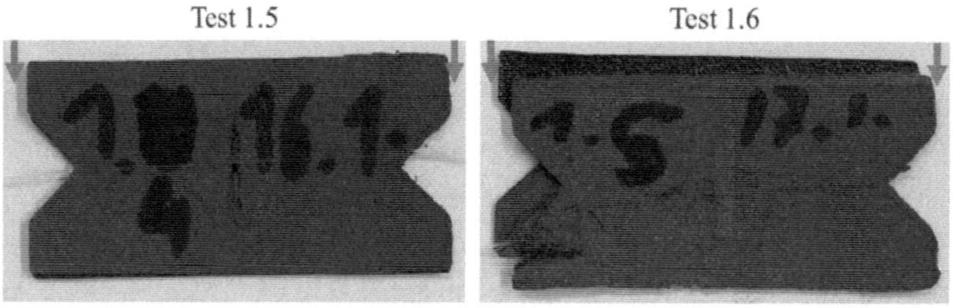

Figure 6. Red highlighted roving position of test specimen with roving perpendicular to the recoater movement (direction of recoater movement shown by arrow)

We can conclude from this first test series that the implementation of fibers before applying new powder for a layer fails if the fiber alignment is not straight and without tension. Without the fixation of the fiber the recoater moves the fiber in an uncontrolled manner. On the backstroke the recoater blade picks the rowing up and pulls it out of the powder again. This leads to the roving loosely lying on top of the powder bed during the laser scanning. Due to the fibers low weight and inherent spring only a few points actually contact the powder bed so no real interface can be established.

4.2 Second test series

In the second test series the rovings were placed manually after the application of fresh powder for the layer. The roving was fixed at both ends by the frame and kept under tension (except test 2.4).

Figure 7 shows that in test 2.1 the roving is embedded into the polymer layer after one sintering step. The fibers are not completely covered by the matrix, but bridges of molten polymer particles have formed over the roving can be observed. The fibers are not bonded to the matrix by the formation of a strong interface but rather by mechanically locking them in place through powder particles forming bridges over the roving.

Test 2.2 is a similar scenario than 2.1 but with the fibers placed perpendicular to the recoater movement. Here the roving is only fixed in specimens of 5 mm width. We can observe that enough polymer bridges have formed over this short distance, so the fibers are secured in place as in test 2.1.

In test 2.3 fibers were placed in two directions to observe the behavior at fiber intersections. It can be seen that the lower roving is attached to the specimen through the polymer bridges like in the previous scenarios, but the upper roving is not attached at all (cf. right side of Figure 7).When the mechanism of bonding is not changed the thickness of the complete fiber stack up cannot exceed the height of one layer or issues in fiber fixation can occur. Because no real interface can be formed the roving must be fully embedded to secure it to the specimen plate.

The final test 2.4 without tension in the rovings shows no embedding or fixation of the rovings at all. The specimen plate is not attached to the rovings, but lines of unmolten powder were the fibers blocked the light from the laser can be seen.

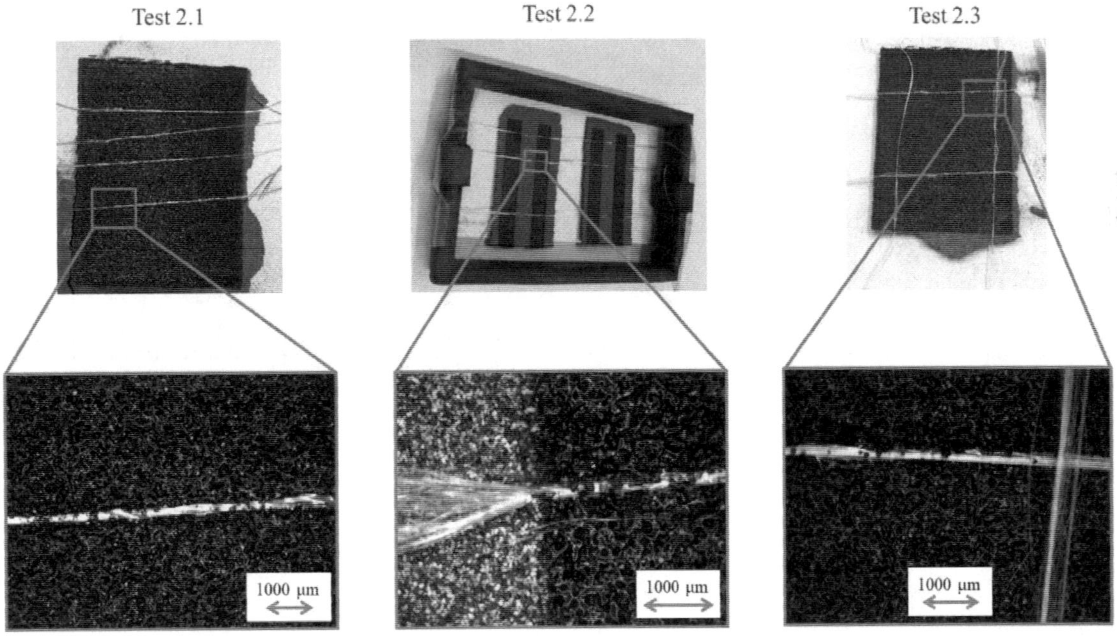

Figure 7. Results of second test series with fiber placement after powder application

For test series two it can be summarized that a placement of a single layer of fibers without intersection is possible after the application of powder for a new layer. The fixation of the fibers is achieved by small randomly formed polymer bridges rather than a real bonding of the roving to the polymer. An intersection of fibers is not possible with the used layer height due to the increased height and the resulting contact with the recoater.

5. CONCLUSION AND OUTLOOK

From the results presented in this paper conclusions can be drawn for fiber placement strategies. First, it can be stated that the fibers need to be under tension during interface formation. Regardless whether the reinforcement is introduced before or after the application of fresh powder for a layer the fibers will not remain in the intended place during the sintering step without tension. This also means that no curved fiber path can be realized with the tested approach. Therefore additional trials and development have to be made in order to overcome this issue. For instance a curve could be approximated by short straight segments under tension instead.

It was shown that fibers can be sufficiently embedded in one layer height to act as a fixation point in a fiber placement strategy. Even without a dedicated thermoplastic sizing and adapted laser paths the roving was securely and repeatably fixed to the specimen. This proves that a fiber reinforcement within the SLS process is possible. However with the mechanical fixation mode seen here the fiber stack height is limited to the layer thickness of the process.

As mentioned before, this work contributes to the development of a SLS printer with an additional unit to automatically insert continuous fibers in the process to reinforce SLS printed polymer parts. The results of the presented work have shown that an automated fiber placement is needed to allow further process development and an improvement of the achievable fiber paths. The following steps in this field are the investigation of interface formation, the development of more complex fiber paths for a load-dependent part design and the design and implementation of a fiber placement module inside an SLS printer. After the technical implementation the complete system needs to be characterized and an appropriate selection of materials and process parameters needs to be found.

6. REFERENCES

[1]. P. Pinter; S. Baumann; C. Lohr; A. Heuer; L. Englert & K.A. Weidenmann. "Mechanical Properties of Additively Manufactured Polymer Samples Using a Piezo Controlled Injection Molding Unit and Fused Filament Fabrication Compared with a Conventional Injection Molding Process" *Solid Freeform Fabrication Symposium*. Austin, Texas, USA. August 13-15, 2018.

[2] *Markforged Carbon Fiber Filament*. Markforged. 03 Jan 2020 <https://markforged.com/materials/carbon-fiber>

[3] *Arburg: Freeformer 300-4X für faserverstärkte Bauteile - Zusätzliche Achse ermöglicht lokale Bauteilverstärkung - Praxisbeispiel: Individuelle Handorthesen für die Medizintechnik.* 21 Oct 2019. Plasticker – the home of plastics. 03 Jan 2020 < https://plasticker.de/Kunststoff_News_36078_Special_k19_Arburg_Freeformer_300_4X_fuer_faserverstaerkte_Bauteile___Zusaetzliche_Achse_ermoeglicht_lokale_Bauteilverstaerkung___Praxisbeispiel_Individuelle_Handorthesen_fuer_die_Medizintechnik?special=k19>

[4] Wendel, B.; Rietzel, D.; Kühnlein, F.; Feulner, R.; Hülder, G.; Schmachtenberg, E. „Additive Processing of Polymers“, *Macromolecular Materials Engineering* 293 (10) (2008): 799–809. DOI: 10.1002/mame.200800121.

[5]. Baumann, F.; Sielaff, L. & Fleischer, J. “Process Analysis and Development of a Module for Implementing Continuous Fibres in an Additive Manufacturing Process”, *SAMPE Europe Symposium*. Stuttgart, Germany. November 14 – 16, 2017. Society for the Advancement of Material and Process Engineering.

[6]. Fischer, A.; Rommel, S. & Bauernhansl, T. “New Fibre Matrix Process with 3D Fibre Printer - A Stragic In-Process Integration of Endless Fibres Using Fused Depostion Modeling”, *IFIP TC 5 International Conference, NEW PROLAMAT 2013*. Dresden, Germany. October 10-11, 2013.

[7]. Bettini, P.; Alitta, G.; Sala, G. & Di Landro, L. “Fused Deposition Technique for Continuous Fiber Reinforced Thermoplastic”, *Journal of Materials Engineering and Performance* 26(2) (2017): 843–848. DOI: 10.1007/s11665-016-2459-8.

[8]. Li, N.; Li, Y. & Liu, S. “Rapid prototyping of continuous carbon fiber reinforced polylactic acid composites by 3D printing”, *Journal of Materials Processing Technology* 238 (2016): 218–225. DOI: 10.1016/j.jmatprotec.2016.07.025.

[9]. Yang, C.; Tian, X.; Liu, T.; Cao, Y. & Li, D. “3D printing for continuous fiber reinforced thermoplastic composites. Mechanism and performance“, *Rapid Prototyping Journal* 23(1) (2017): 209–215. DOI: 10.1108/RPJ-08-2015-0098.

[10]. Baumann, F.; Scholz, J. & Fleischer, J. “Investigation of a New Approach for Additively Manufactured Continuous Fiber-reinforced Polymers“, *Procedia CIRP 66* (2017): 323–328. DOI: 10.1016/j.procir.2017.03.276.

[11]. Dickson, A. N.; Barry, J. N.; McDonnell, K. A. & Dowling, D. P. “Fabrication of continuous carbon, glass and Kevlar fibre reinforced polymer composites using additive manufacturing“, *Additive Manufacturing* 16 (2017): 146–152. DOI: 10.1016/j.addma.2017.06.004.

[12]. Zhu, P.; Li, S.; Ashcroft, I.; Jones, A. & Pu, J., “3D Printing of continuous fibre reinforced thermoplastic composites“, *21st International Conference on Composite Materials*, Xi’an, China. August 20-25, 2017.

[13]. van Der Klift, F.; Koga, Y.; Todoroki, A.; Ueda, M.; Hirano, Y. & Matsuzaki, R. “3D Printing of Continuous Carbon Fibre Reinforced Thermo-Plastic (CFRTP) Tensile Test Specimens“, *Open Journal of Composite Materials* 6(1) (2016): 18–27. DOI: 10.4236/ojcm.2016.61003.

[14]. Karalekas, D. & Antoniou, K. “Composite rapid prototyping. Overcoming the drawback of poor mechanical properties“, *Journal of Materials Processing Technology* 153-154 (2004): 526–530. DOI: 10.1016/j.jmatprotec.2004.04.019.

[15]. Karalekas, D. E. “Study of the mechanical properties of nonwoven fibre mat reinforced photopolymers used in rapid prototyping“, *Materials & Design* 24(8) (2003): 665–670. DOI: 10.1016/S0261-3069(03)00153-5.

[16]. Klostermann, D.; Charthoff, R.; Graves, G.; Osborne, N. & Priore, B. “Interfacial characteristics of composites fabricated by laminated object manufacturing“, *Composites Part*

A: Applied Science and Manufacturing 29(9-10) (1998): 1165-1174. DOI10.1016/S1359-835X(98)00088-8.

[17]. Parandoush, P.; Tucker, L.; Zhou, C. & Lin, D. “Laser assisted additive manufacturing of continuous fiber reinforced thermoplastic composites“, *Materials & Design* 131 (2017): 186–195. DOI: 10.1016/j.matdes.2017.06.013.

[18]. Parandoush, P.; Zhou, C. & Lin, D. “3D Printing of Ultrahigh Strength Continuous Carbon Fiber Composites“, *Advanced Engineering Materials* 21 (2019). DOI: 10.1002/adem.201800622.

GEOMETRIC PARAMETER ANALYSIS OF VERTICALLY EXTRUDED PINS FOR STRENGTH IMPROVEMENT IN ADDITIVE MANUFACTURING WITH FIBER-REINFORCED THERMOPLASTIC

Seokpum Kim[1], Tyler Smith[2,3], Justin Condon[2,4], Alexander Lambert[1], Vlastimil Kunc[1,3], Chad Duty[2,3,*]

[1] Energy and Transportation Science Division, Oak Ridge National Laboratory, Oak Ridge, TN 37831

[2] Materials Science and Technology Division, Oak Ridge National Laboratory, Oak Ridge, TN 37831

[3] Mechanical, Aerospace and Biomedical Engineering, University of Tennessee at Knoxville, Knoxville, TN 37996

[4] Civil and Environmental Engineering, University of Tennessee at Knoxville, Knoxville, TN 37996

*Corresponding author: cduty@utk.edu

ABSTRACT

Vertical directional strength is significantly weaker than planar directional strength in polymer extrusion additive manufacturing due to its layer-wise deposition method. The disparity between the planar strength and vertical strength is even larger for a printed part with a fiber-reinforced thermoplastic. A printing technique has been proposed in which continuous material is extruded across layers throughout the body of the part in order to improve the strength in the vertical direction. The mechanical engagement between the material extruded in vertically aligned holes (called "z-pins") and the surrounding layers is primarily influenced by the geometry of the hole and the extrusion volume of the z-pin. Previously, the z-pinning parameters that provide good penetration in the hole has been evaluated for neat polylactic acid (PLA). The current study investigates z-pinning parameters for carbon fiber-reinforced PLA as a function of the quality of mechanical engagement of the pins with the surrounding structure. The results of this parametric study will provide guidance for general printing applications across multiple platforms with various materials.

1. INTRODUCTION

Additive manufacturing (AM) is "the process of joining materials to make objects from three-dimensional model data, usually layer upon layer, as opposed to subtractive manufacturing methodologies" as in Wohlers report [1]. Extrusion-based polymer AM is one of the common AM technologies which involves a viscous polymer extruded from a nozzle and deposited layer by layer. There are mainly three types of material systems used in extrusion-based polymer AM: thermoplastic, thermoset, and metals or ceramics with a polymer as a binding material. For metals or ceramics with a polymer as a binding material, polymer is removed in the printed part via a debinding process, and the part undergoes a sintering process, which increases the layer-wise bonding strength. For thermoset system, the layer-wise bonding strength increases during the curing process. For a thermoplastic system (thermoplastic with or without fillers), a molten plastic is extruded through a heated nozzle and deposited on a previously-deposited layer. The extruded material starts to cool after the exit of the nozzle. If the layer is cooled too much before the next layer is deposited, then due to not enough cross-linking of polymer chains between the two layers, the printed part often suffers from a weak bonding strength between layers.

To increase the bonding strength between the layers, we have introduced the z-pinning approach [2-5] which is currently patent pending [6]. In Ref. [2], z-pinning was used for neat polylactic acid (PLA) and showed 20% higher strength and 100% higher toughness as compared to a non-pinned sample. In Ref. [4], three parameters were varied, including hole depth, hole width, and the amount of pin extrusion (= pin volume) for neat PLA to determine the optimal parameters for z-pins. In Ref. [5], a z-pinned sample of carbon fiber reinforced PLA demonstrated 3 times higher strength and 8 times higher toughness as compared to a non-pinned sample. These studies show that the increase in mechanical performance by z-pinning can be more significant with carbon-fiber reinforced polymer than that with neat polymer. The mechanical interlocking between the z-pins and the surrounding base structure determines the effectiveness of the z-pin approach in terms of vertical strength. The mechanical interlocking is influenced by geometric z-pin parameters and the material. In Ref. [5], it was evident that a further study was needed to optimize the geometry of pin and hole pairs and their relative spacing using carbon fiber-reinforced PLA. In the current study, the z-pin parameters are varied to investigate the z-pin engagement with the surrounding structure using carbon fiber-reinforced PLA.

2. Z-PINNING

The z-pinning approach involves vertical direction extrusion (pinning) into aligned holes across multiple layers of a pre-deposited structure. Specifically, in the z-pinning approach, the material from the nozzle is deposited to create a three-dimensional structure with pre-determined holes in a layer-by-layer fashion. The holes are not made by machining or drilling after the deposition. The holes are vertically aligned locations where materials are not deposited. After a given number of layers are deposited, the nozzle visits the holes and extrudes a material to fill the holes. The term "z-pin" is used to indicate the direction of the vertical fill. The material for the z-pins can be, but may not have to be, the same material as the base structure. In Ref. [2], the term "void" is used to indicate the hole intentionally designed in the structure. In this paper, we used "hole" instead of "void" to indicate a vertically aligned, intentionally non-deposited location. The term "void" will be used to indicate a sub-millimeter scale (or millimeter scale) gap between a z-pin and the surrounding base structure. The z-pinning process is illustrated in Figure 1. The details of the z-pinning process are provided in Ref. [3].

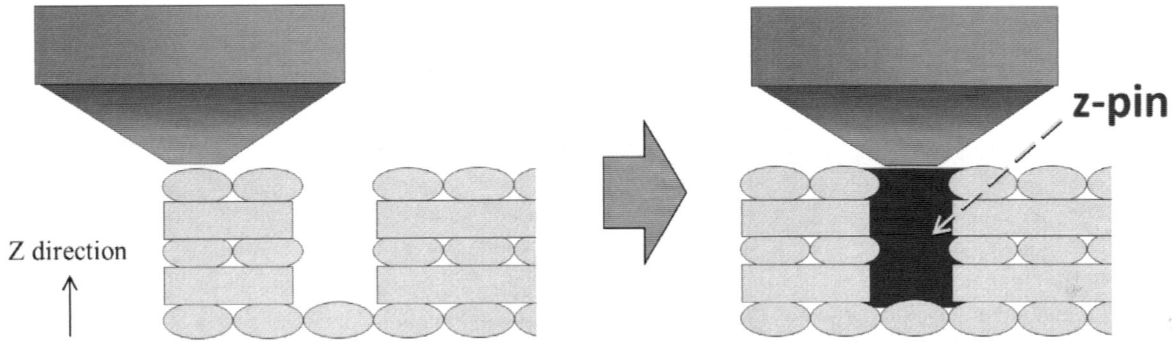

Figure 1: Illustration of z-pin approach

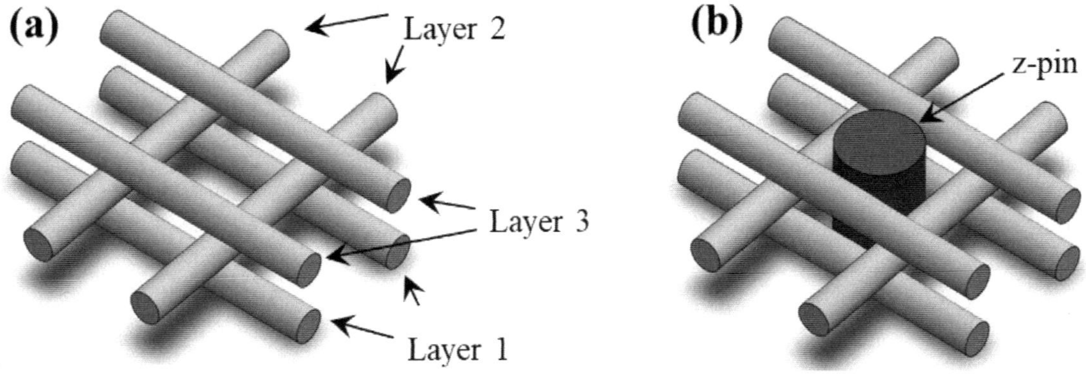

Figure 2: (a) Base structure layup configuration used in this study and (b) z-pin extruded in the hole of the base structure.

In this study, the base structure consists of multiple layers of [0/90/0/90] cross-ply stacking, each of which consists of unidirectionally printed lines in parallel from one line to another within the layer but perpendicular to the printed lines in the previous layer, which is similar to a log cabin layup as shown in Figure 2(a). This structure is the same as an infill structure generated by Slic3r [7] with a rectilinear pattern. The infill lines have a regular spacing between them, and therefore, the base structure has periodic holes in the x- and y-directions from the top view. The z-pins are extruded into the holes as shown in Figure 2(b).

3. EXPERIMENTS

A standard MakerGear M3 printer was used for printing. The filament was carbon fiber-reinforced PLA (model name: CarbonX™ PLA+CF) with the diameter of 1.75 mm purchased from 3DXTech. The bed temperature was 65°C and the extruder temperature was 220°C. The nozzle diameter was 0.5 mm, the layer height was 0.25 mm, and the bead width was 0.55 mm.

Figure 3: A sample with z-pins. The dimensions vary depending on the spacing of the infill lines.

The sample geometry was a box shape as shown in Figure 3. The length in x- and y-directions of the sample varied from 25 mm to 38 mm, and the height of the sample varied from 16 mm to 38 mm depending on the spacing of the infill lines. As the spacing between the infill lines was increased, the sample dimensions became longer and higher to maintain a consistent geometric ratio. The spacing between the infill lines determined the hole width.

The effect of three z-pin parameters were investigated in this study: hole width, hole depth, and the amount of pin extrusion (= pin volume). The hole width varied from 1.5 mm to 3 mm, the hole depth varied from 8 layers to 20 layers (= 2 mm – 5 mm), and the pin volume varied from 80% to 120% of the hole volume. The hole width was the same as the spacing between the infill lines subtracted by the bead width (0.55 mm). In Ref. [4], the hole volume was calculated by a square of the line spacing multiplied by the hole depth subtracted by the infill line volume as illustrated in Figure 4(a). Therefore, although practically not possible, if every hole was filled with a z-pin with 100% of the hole volume, the sample would have 100% solid block of material. In this study, the method of calculating the hole volume was changed to equal the square of the hole width multiplied by the hole depth as illustrated in Figure 4(b). Therefore, if every hole in the sample were filled with a z-pin with 100% of the hole volume, the sample would still have empty spaces between the infill lines as marked by a red circle in Figure 4(b). This modified method of calculating the hole volume provides a more intuitive relation between the z-pin volume ratio to the visually observable hole volume from the top view. The hole width and depth have been normalized by the nozzle diameter (0.5 mm). Therefore, a hole with of 1.5 mm – 3 mm corresponds to a normalized hole width (W_h) of 3 – 6 and a hole depth of 8 – 20 layers corresponds to a normalized depth (D_h) of 4 – 10. The z-pin volume was also normalized by the hole volume, and was varied from 80% to 120%.

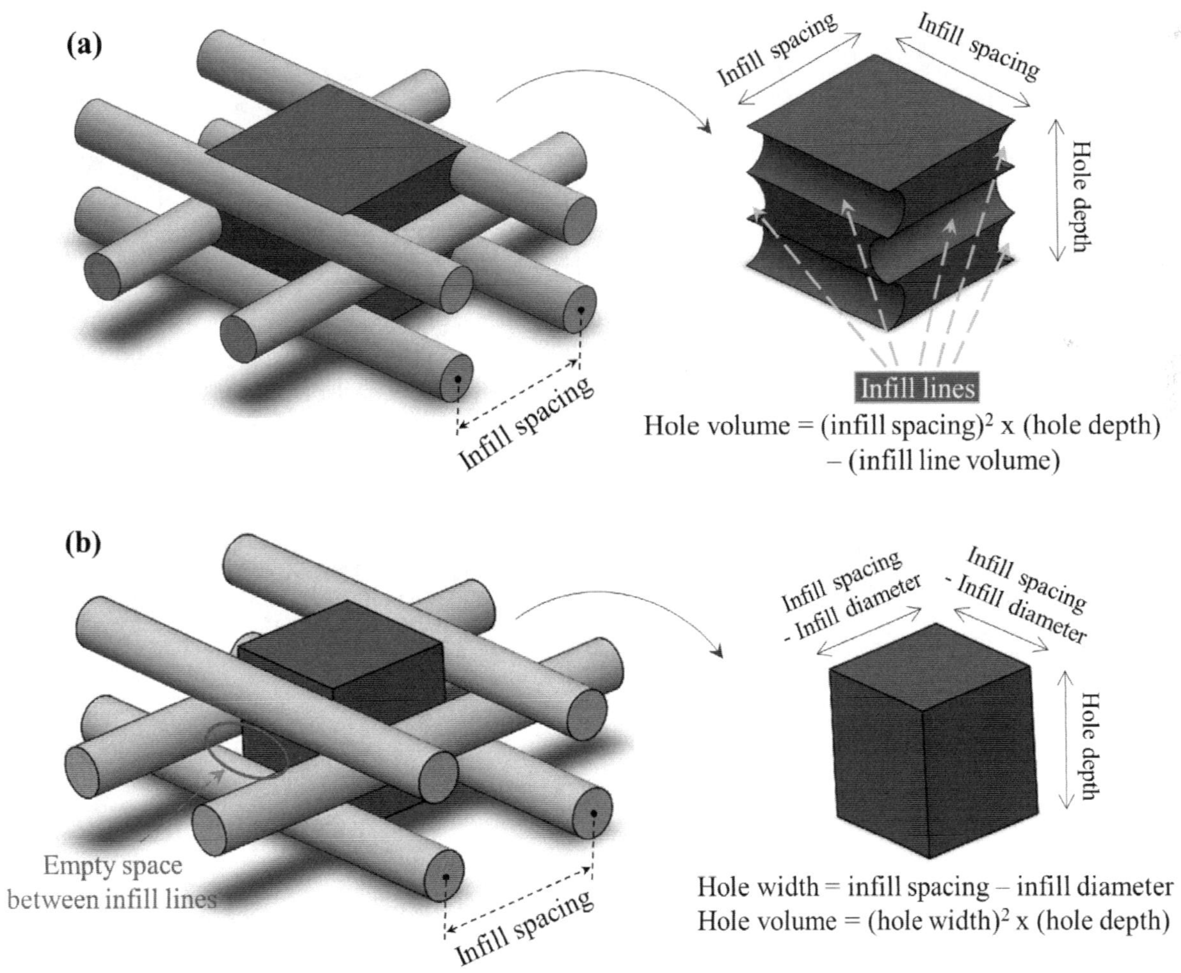

Figure 4: Method of calculating hole volume (a) in Ref. [4] and (b) in this study

3.1 Effect of pin volume

The pin volume was varied from 80% to 120% of the hole volume. Figure 5 shows the cross-section view of z-pins with a hole depth of $D_h = 8$, a width of $W_h = 4$, and a varying pin volume of 80%, 100%, and 120%. The sample with 80% pin volume shows voids between the pins and the base grid structure, especially near the interfaces between upper pins and lower pins. It is presumably because the amount of material extruded as z-pins was insufficient to fill the holes. On the other hand, the sample with 120% pin volume showed distorted boundaries adjacent to the z-pins. This was presumably because the amount of material extruded as z-pins was too much for the hole, and the excessive material flowed into the adjacent holes.

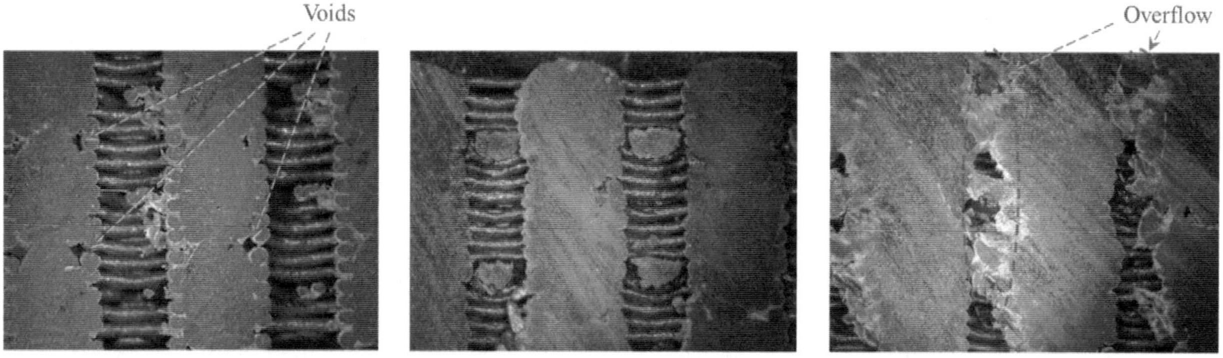

(a) Vol **80%**, $D_h = 8$, $W_h = 4$ (b) Vol **100%**, $D_h = 8$, $W_h = 4$ (c) Vol **120%**, $D_h = 8$, $W_h = 4$

Figure 5: Cross-section views of z-pins with consistent hole depth ($D_h = 8$) and width ($W_h = 4$), and varying pin volume (80% - 120%). Voids were seen for small pin volume (80%), and overflow materials were seen for large pin volume (120%).

(a) Vol **80%**, $D_h = 4$, $W_h = 5$ (b) Vol **100%**, $D_h = 4$, $W_h = 5$ (c) Vol **120%**, $D_h = 4$, $W_h = 5$

Figure 6: Cross-section views of z-pins with varying z-pin volume. The quality of z-pins was good for all volume cases (80%, 100%, and 120%). The red dotted lines show that the shape of the interface between stacked z-pins (= upper pin and lower pin) changed from upward convex to downward convex as the pin volume increased from 80% to 120%.

It was, however, interesting to see that some samples with different hole dimensions (depth and width) showed a good quality of z-pins even for small pin volume (80%) and large pin volume (120%). Figure 6 shows the cross-sections of z-pins with a hole depth of $D_h = 4$, a width of $W_h = 5$, and a varying pin volume of 80%, 100%, and 120%. All three figures (Figure 6 (a),(b), and (c)) with the pin volume of 80%, 100%, 120% showed a good quality of z-pins. The disparity of z-pin quality between Figure 5 and Figure 6 indicates that it is not sufficient to study a single volume fraction (e.g., 100% volume fraction) to determine the z-pin quality for different hole dimensions. A quality analysis required samples with 80%, 100%, and 120% pin volume to be made with all combinations of hole widths and depths.

Another interesting feature shown in Figure 6 is the interface shape (= top surface shape of z-pin) between two vertically stacked pins (lower pin and upper pin) as indicated by the red dotted lines. For a low pin volume (80%), the material bulged upward in the z-pin which created an upward convex shape at the interface between two stacked pins. This shape was consistantly seen at all interfaces and the top surface of the top pins in Figure 6(a). As the pin volume increased from 80% to 100% and to 120%, the interface shape changed to a downward convex shape as shown in Figure 6 (b-c). The difference of the shape was presumably because of the material flow near the nozzle tip. It was assumed that the pressure from the material at the nozzle tip pushed the existing material as illustrated in Figure 7(a). If the volume was too much for given hole dimensions, the extra material flows upward as illustrated in Figure 7(b). The top surface shape of z-pin was also influenced by the aspect ratio of the z-pin, which is presented in the next two sections.

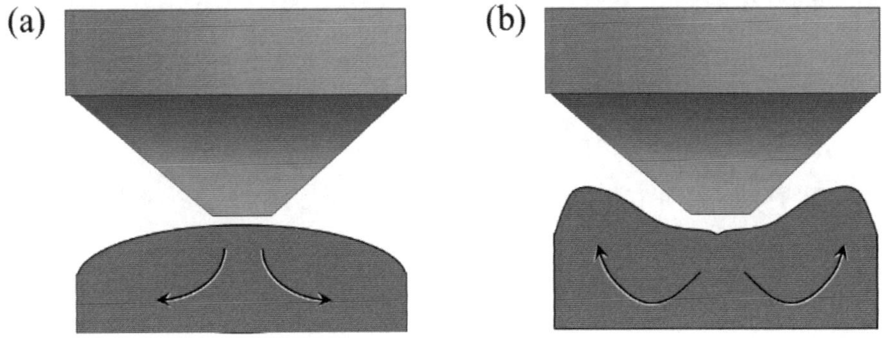

Figure 7: Estimated flow mechanisms of the two shapes of top surface of z-pin (a) Upward convex: material flows down and toward sides (b) Downward convex: material flows down and sides and pushes the existing material to the upward direction.

3.2 Effect of hole width

The hole width was varied from $W_h = 3$ to $W_h = 6$. Figure 8 shows the cross-sections of z-pins with the pin volume 100% (= same as the hole volume) and the hole depth of $D_h = 4$. As the hole width became larger, the aspect ratio reduced, and the top surface of z-pin changed from an upward convex shape to a downward convex shape.

(a) Vol 100%, $D_h = 4, \boldsymbol{W_h} = \mathbf{3}$ (b) Vol 100%, $D_h = 4, \boldsymbol{W_h} = \mathbf{5}$ (c) Vol 100%, $D_h = 4, \boldsymbol{W_h} = \mathbf{6}$

Figure 8: Cross-section views of z-pins with varying hole width

3.3 Effect of hole depth

The hole depth was varied from $D_h = 4$ to $D_h = 10$. Figure 9 shows the cross-sections of z-pins with a pin volume 100% (= same as the hole volume) and a hole width of $W_h = 6$. As the hole depth became deeper, the aspect ratio increased, and the top surface of z-pin changed from a downward convex shape to an upward convex shape.

(a) Vol 100%, $\boldsymbol{D_h} = \mathbf{4}, W_h = 6$ (b) Vol 100%, $\boldsymbol{D_h} = \mathbf{8}, W_h = 6$ (c) Vol 100%, $\boldsymbol{D_h} = \mathbf{10}, W_h = 6$

Figure 9: Cross-section views of z-pins with varying hole depth

3.4 Z-pin quality with various hole widths, hole depths, and pin volumes

In the previous sections (Sec. 3.1 – 3.3), the effects of individual z-pin parameters (pin volume, hole width, and hole depth) were shown while the other two parameters were fixed. However, all three parameters can have a combined influence the z-pin quality. Therefore, the z-pin quality was investigated across a variety of combinations of the three z-pin parameters. Multiple samples (> 60 samples) were prepared, each having different combination of pin volume, hole width and hole depth. Table 1-3 summarizes the z-pin quality for various hole widths, hole depths and pin volumes. "G" indicates good quality of z-pins with the upward convex shape or flat shape of z-pin top surface. Examples of good quality z-pins include Figure 5(b), Figure 6(a), Figure 8(a), and Figure 9(b, c). "G_{ov}" indicates good quality of z-pins with the downward convex shape of z-pin top surface. Examples of good quality z-pins include Figure 6(b,c), Figure 8(b,c), and Figure 9(a). "OK" indicates somewhat good quality of z-pins with some defects. An example of "OK" quality z-pins was shown in Figure 5(a).

Table 1: Z-pin quality for the pin volume of 80%

80% Vol		Hole width ratio, W_h			
		3	4	5	6
Hole depth ratio, D_h	4	OK	G	G	G
	6	OK	G	G	G
	8	B	OK	G	G
	10		B	B	G

G = good quality, B = bad quality, OK = quality between B and G

Table 2: Z-pin quality for the pin volume of 100%

100% Vol		Hole width ratio, W_h			
		3	4	5	6
Hole depth ratio, D_h	4	G	G	G_{ov}	G_{ov}
	6	OK	G	G	G
	8	B	G	G	G
	10		B	B	G

G = good quality, B = bad quality, OK = quality between B and G,
G_{ov} = good but over the nozzle tip level

Table 3: Z-pin quality for the pin volume of 120%

120% Vol		Hole width ratio, W_h			
		3	4	5	6
Hole depth ratio, D_h	4	B	G	G_{ov}	G_{ov}
	6	N/A	OK	G_{ov}	G_{ov}
	8	B	B	G	G_{ov}
	10		B	B	B

G = good quality, B = bad quality, OK = quality between B and G,
G_{ov} = good but over the nozzle tip level

As shown in Table 1-3, the z-pin quality tended to degrade when the hole depth was too deep for a given hole width. Likewise, the z-pin quality tended to degrade if the hole width was too narrow for a given hole depth. As the pin volume increased, some of the z-pin samples with marginally good quality degraded due to overflow of extra material from the z-pins.

4. SUMMARY

Z-pinning parameters (hole width, hole depth, and pin volume) have been investigated for carbon fiber-reinforced PLA. The base structure of the samples had a rectilinear pattern (0°/90° cross-ply) with the hole width varying from 1.5 mm to 3 mm (W_h = 3 – 6) and the hole depth varying from 8 layers to 20 layers (D_h = 4 – 10). The z-pin volume varied from 80% to 120% of the hole volume. The cross-section of all samples was analyzed and the quality of the z-pin structure was categorized as bad ("B"), good ("G"), good with the z-pin top surface higher than nozzle tip ("G_{ov}"), and moderate quality between good and bad ("OK"). Overall, the z-pin quality tended to degrade when the hole depth was too deep for a given hole width. Likewise, the z-pin quality tended to degrade when the hole width was too narrow for a given hole depth. As the pin volume increased, some of the z-pin samples with marginally good quality become bad. There is not yet sufficient evidence in this study to determine which category of z-pin shows the highest mechanical performance, so this remains an ongoing effort.

Acknowledgement

Research was sponsored by the U.S. Department of Energy, Office of Energy Efficiency and Renewable Energy, Advanced Manufacturing Office and used resources at the Manufacturing Demonstration Facility, a DOE-EERE User Facility at Oak Ridge National Laboratory, under contract DE-AC05-00OR22725 with UT-Battelle, LLC. Research was supported by an appointment to the Oak Ridge National Laboratory Advanced Short-Term Research Opportunity (ASTRO) Program and Higher Education Research Experiences (HERE) program, sponsored by the U.S. Department of Energy and administered by the Oak Ridge Institute for Science and Education. Research was also sponsored by the U.S. Army Combat Capabilities Development Command Aviation & Missile Center.

References

1. T.T. Wohlers, W. Associates, I. Campbell, T. Caffrey, O. Diegel, and J. Kowen, *Wohlers Report 2018: 3D Printing and Additive Manufacturing State of the Industry : Annual Worldwide Progress Report*. 2018: Wohlers Associates.
2. C. Duty, J. Failla, S. Kim, J. Lindahl, B. Post, L. Love, and V. Kunc. *Reducing Mechanical Anisotropy In Extrusion-Based Printed Parts*. in *Solid Freeform Fabrication Symposium*. 2017. Austin, TX.
3. C. Duty, J. Failla, S. Kim, T. Smith, J. Lindahl, and V. Kunc, *Z-Pinning approach for 3D printing mechanically isotropic materials*. Additive Manufacturing, 2019. **27**: p. 175-184.
4. S. Kim, T. Smith, J.A. Failla, J.M. Lindahl, V. Kunc, and C.E. Duty. *Parametric Analysis on Vertical Pins for Strengthening Extrusion-Based Printed Parts*. in *SAMPE 2018*. 2018. Long Beach, CA.

5. C. Duty, J. Failla, S. Kim, T. Smith, J. Lindahl, A. Roschli, . . . , and V. Kunc. *Z-Pinning Approach for Reducing Mechanical Anisotropy of 3D Printed Parts*. in *The 29th Annual International Solid Freeform Fabrication (SFF) Symposium*. 2018. Austin, TX.
6. C. Duty, S. Kim, V. Kunc, L. Love, B. Post, J. Failla, and J. Lindahl. *Z-axis improvement in additive manufacturing*. in *U.S. Patent Application 15/965,106*. Nov. 1, 2018.
7. *Slic3r Infill Patterns*. Available from:" https://manual.slic3r.org/expert-mode/print-settings", Cited.

PROCESS SENSITIVITY STUDY OF METAL FUSED FILAMENT FABRICATION (MF^3) USING FINITE ELEMENT-BASED PROCESS SIMULATION APPROACH

Mohammad Qasim Shaikh[1]*, Paramjot Singh[1], Pierre-Yves Lavertu[2], Kunal H Kate[1], Sundar V. Atre[1]

[1]Materials Innovation Guild, University of Louisville, KY 40208, USA.
[2] e-Xstream Engieering, MSC Software Corp (Hexagon), Newport Beach, CA 92660, USA.

Corresponding author: mohammadqasim.shaikh@louisville.edu

ABSTRACT

Metal fused filament fabrication (MF^3) combines fused filament fabrication and sintering processes to fabricate complex metal components. In design for MF^3 ($DfMF^3$), it is important to understand how part design, printing parameters and material properties of filament affect MF^3 printing process in terms of processability, part quality, and ensuing properties. Understanding the interrelationships between material-process-design variations is essential to leverage the potential of MF^3 and deploy it to industrial applications. The goal of the study is to conduct a sensitivity analysis to identify the significance of each input parameter in terms of influence on the output parameters. FEA-based simulations were used to estimate the process outcome in response to variable inputs. However, simulating and optimizing the MF^3 printing process is a complex problem having a number of linked geometry, process and material variables to be considered that influence the process output parameters like deflection, warpage, residual stress, thermal history and print time. Having multiple objective functions for process optimization makes it very difficult to identify critical parameters for designing part geometry, processing conditions and powder-binder feedstock for MF^3. Towards the goal of identifying the level of significance of various input variables during MF^3 process, a systematic procedure for sensitivity analysis has been successfully developed for the printing phase of the MF^3 process. In this sensitivity analysis, all geometry, process and material input variables were varied within a meaningful window one at a time, and influence on process outcome was noted in terms of output parameters. The present sensitivity analysis procedure is expected to be an invaluable tool not only for the materials engineer who has to develop an optimal formulation of MF^3 material, but also for the design engineer who has to determine the critical input parameters for given design targets, as well as for the production engineer who has to optimize and monitor the production stage.

Keywords: metal fused filament fabrication; FEA simulation; process sensitivity; estimations, additive manufacturing design for AM,

1. INTRODUCTION

Additive manufacturing (AM) is a technology for fabricating 3D objects by adding material layer upon layer, as opposed to subtractive and formative manufacturing technologies. AM is being increasingly used for fabricating physical objects using polymers, metals, or ceramics, which may be used in various applications [1]. Metal fused filament fabrication (MF^3) is a hybrid AM process used to fabricate custom 3D metal components [2]. The established metal additive manufacturing processes such as laser-powder bed fusion, direct metal laser sintering, and direct energy deposition are energy-intensive and not suited for certain specific requirements like in-space manufacturing. MF^3 provides the best alternative to address these challenges. MF^3 is a multi-step process that involves as shown in Figure 1: (a) mixing and extrusion of a powder-polymer mixture into filaments, (b) 3D printing of a green part, (c) polymer removal from the 3D printed green part by debinding to get a brown part, and (d) densifying the brown part to achieve a fully dense metal part by sintering. MF3 has been found capable of metal additive manufacturing of materials like Ti-6Al-4V, bronze, copper, 17–4 PH stainless steel, 316L stainless steel [3-6].

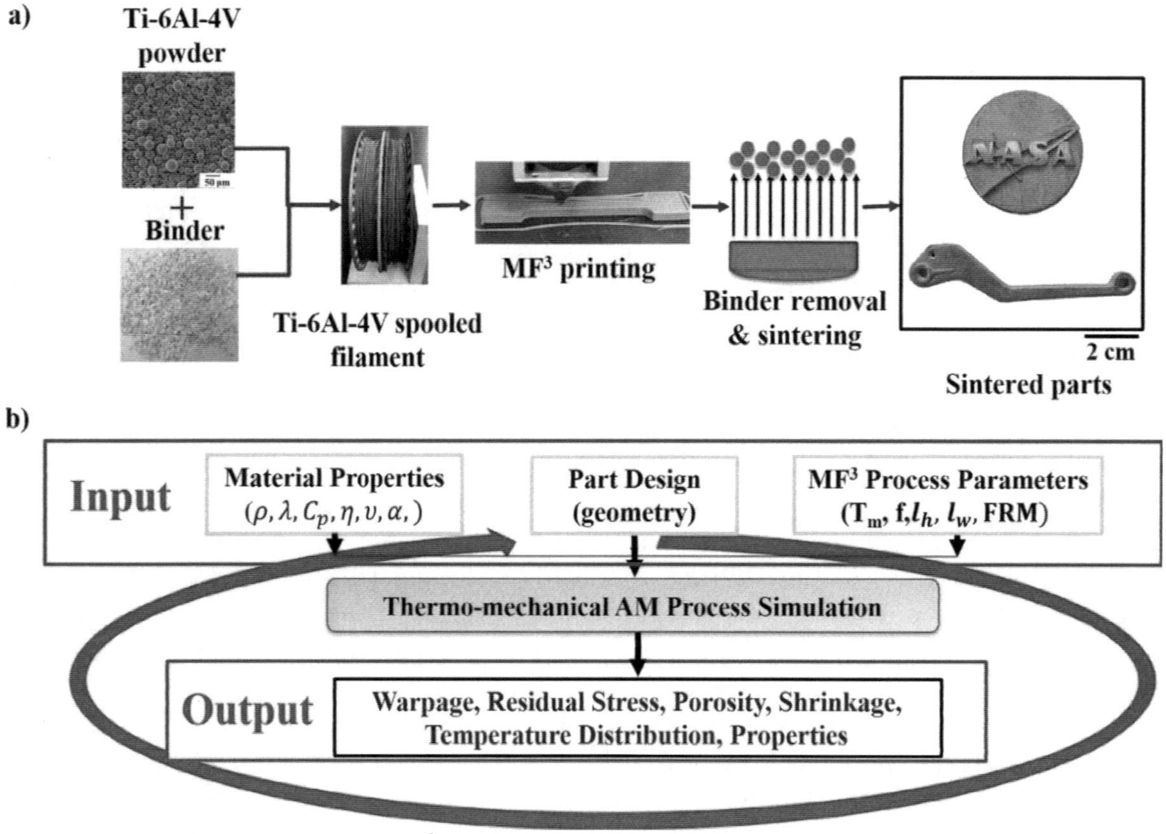

Fig. 1. **a)** Process overview for MF^3 with examples of fabricated Ti-6Al-4V parts by our group and **b) t**he present work for determining the input material parameters for conducting $DfMF^3$.

The material system used in MF^3 comprises an optimum formulation of metal powder mixed with custom polymer binder, like in powder injection molding (PIM). The solids loading and binder composition are varied for different material systems to enable efficient debinding and sintering, or to meet specific application needs [7]. Consequently, the properties of new material may vary considerably. Now, question arise as how these changes in material properties in MF^3 will influence the process outcome and performance of the end product. The powder-polymer composites are processed by fused filament fabrication (FFF), also known as fused deposition modeling (FDM), which is typically used to get a 3D shape with polymers [1]. This extrusion-based printing process has several variable input parameters like extrusion temperature, build plate temperature, printing speed, layer thickness, extrusion width, toolpath and slicing strategy. These process parameters have significant effects on the printing process as well as printed part quality and performance. The layer-by-layer material deposition develops temperature gradient leading to warpage and residual stresses in printed which are influenced by the process parameters [8,9]. Hence, there is a need for establishing optimized parameters settings for specific materials. The common trial-and-error experimental approach is often costly and time consuming, and sometimes not even feasible. The empirical relationships between input and output parameters are often not straightforward.

Numerical simulation techniques provide effective alternative means to predict the printing process outcome by estimating the thermo-mechanical behavior of 3D parts being printed. [10-14]. The availability of FEA-based simulation tools has shown some promise for aiding the engineer to resolve the inter-related problems involving material-process-geometry in FFF/ FDM process [15,16]. However, the use of such tools to perform material and process simulations of MF^3 is yet to be explored, that will favor the widespread use of the MF^3 process to manufacture parts with different materials for a variety of applications [17].

Understanding the constraints imposed by extrusion based additive manufacturing process and highly filled material used in MF^3 is crucial for successful design and manufacturing. Hence, there is a need for design for MF^3 ($DfMF^3$) approach supported by relevant design tools. Also, development of a product calls for several factors considerations to improve the functional and aesthetic attributes of the part. It may lead to multiple design changes. Here, the questions raise as to how small changes in product and tool dimensions influence process outcome variations. Moreover, any variations in powder-polymer composition can consequently affect filament properties, filament processing, 3D printing, debinding, and sintering. Accordingly, material compositional variations can affect the component geometrical attributes. However, little information is available on the sensitivity of various predictions of CAE simulation tools to variations in the material property values. Additionally, as several process settings need to be changed and controlled during the MF^3 printing process, it is important to understand how a CAE simulation tool captures the influence of such process variations in its predictions.

The above crucial questions are addressed in the current study by analyzing the MF^3 process for the sensitivity of the key output variables to the selected material, geometry and process variables. FEA-based simulations are used to estimate the process outcome in response to variable inputs. CAE simulation tool, Digimat (e-Xstream/ MSC Inc.), was used for integrated thermo-mechanical process simulation [4]. Process simulation helps in predicting and minimizing warpage and residual stresses, achieving intended dimensions through reverse warpage, and estimating the structural performance of the 3D printed parts. The quantitative relationships on the dependence of many MF^3 process output parameters on changes in component geometry, process parameters and material properties are presented. The dimensionless sensitivity factor allows the parameters with different units to be compared quantitatively with a single yardstick. It also facilitated the identification of dominant input parameters and relative contribution to each output parameter. The results provide important insights into the geometry-process-material interrelationships in the MF^3 process.

The study provides a powerful approach that can be applied to identify the parameters that need to be optimized in the design stage and carefully monitored and controlled in the production stage. We anticipate that the sensitivity analysis tool presented in this paper will be a valuable tool for the design and production engineer in the additive manufacturing industry. Also, during material formulation development the tool can provide insights about sensitivity evolution with different solids loading and binder compositions to optimize the formulations. The simulations in this study only focused on the printing stage of the MF^3 process, as simulation solutions for debinding and sintering process are under development.

1.1 FEA simulation of MF^3 process

In this study, the commercially available software, Digimat was used to carry out the finite element-based (FE) process simulations of MF^3 printing process. The tool provides the capability of using indirect sequentially coupled thermomechanical analysis for both stress and heat analysis [26]. It enables numerical analyses of the complex thermomechanical loadings that occur during the layer-by-layer deposition of the material, successive cooling of the part and thermal history of the material deposition that generates differential shrinkage between adjacent beads or layers that shapes the part. Through the element activation/deactivation function in the Digimat, a small part of the geometry was activated sequentially that represents filament-wise printing that progresses layer-by-layer [29]. Once all the elements were activated, the final results are then used for a thermomechanical coupled analysis in order to simulate the solidification process and the cooling phase. Extrusion based printing of higher filled powder polymer mixture in MF^3 exhibits strong dependency of the filament material thermal, mechanical and physical properties. In this study, the properties used for Ti-6Al-4V in custom polymer binder were obtained from estimation models [2]. The FE simulations were used to estimate the process time, deflection/ warpage, residual stresses and thermal history that take place in the printing process. These estimations can enable design and process engineers to not only come-up with optimal material, design and processing parameters but also identify potential issues and troubleshoot them at the early stage of design.

2. METHODOLOGY

The simulations in this study only focused on the printing stage of the MF^3 process. Simulation solutions for debinding and sintering stages of the MF^3 process are being developed and will be reported separately. Five different component geometries were selected and simulated: thin walled components (Figure 2A-D), and solid component (Figure 2E). Starting with a thin-walled hollow structure, designs were evolved to evaluate sensitivity with changing part size, volume, thickness, height, layer surface area, and structural stiffness. In the present study, only Design-I was studied in detail, while other four designs and evolution sensitivity with changing geometries will be presented in future.

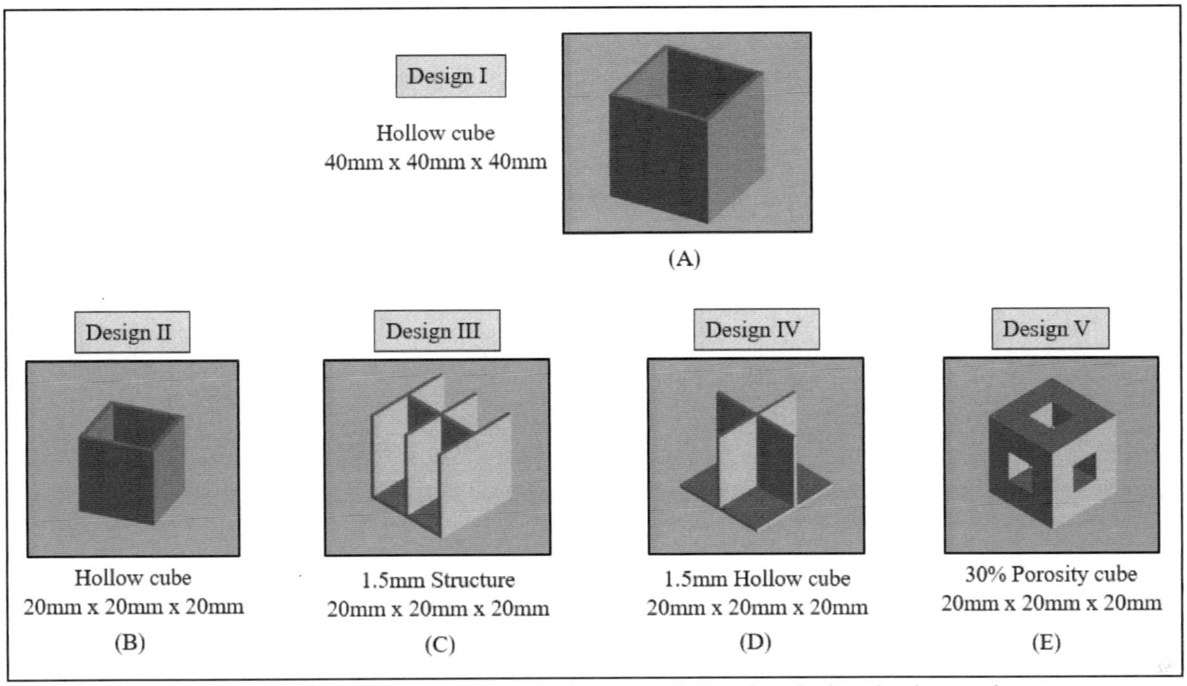

Figure 2. Five different part designs used for simulation in the study.
(A-D) Thin walled components (E) Solid component

An optimal formulation of Ti-6Al-4V in custom polymer binder was developed by our research group for MF^3 [2]. The same material was evaluated in this sensitivity study. The material definition for simulation includes mechanical and thermal properties over a range of temperature. A CAD model in STL format defines the part geometry. It was discretized into voxel mesh. Voxel mesh size is decided according to layer thickness and part geometry. A GCode file from a slicing tool, Simplify3D, defines toolpath, layer thickness and printing speed. Other printing parameters like layer width, extrusion temperature, build plate temperature, chamber temperature, convection coefficient are defined through simulation tool graphical user interface. Having material, geometry and process input defined, thermo-mechanical simulations were performed for each of the five geometries.

For each of the designs, first a base case MF^3 printing process was studied with given geometry, typical values of material properties, and process parameters as input parameters. Then, each of the geometry, process and material input variables was varied within a meaningful window one at a time, and the influence on process outcome was noted. For each input variation a simulation job was performed, and simulation results are reported as output parameters.

The input parameters for this study have been divided into three subgroups:

- component geometry parameters
- process parameters
- material properties

The output parameters evaluated in the study include:

- print time
- deflection
- Z-warpage
- residual stresses
- temperature distribution

The goal of this sensitivity analysis is to evaluate the sensitivity of output parameters to each of input parameters, and subsequently identify input parameters that have the greatest influence on each of the output parameters in MF^3 printing process.

Sensitivity analysis

For the sensitivity analysis, each of the input parameters was varied over a range of interest and feasibility. The response of each output parameter was recorded to evaluate its sensitivity to variation in input parameters. The sensitivity was calculated as the gradient or slope of the dimensionless dependent variable with respect to the dimensionless independent variable according to the following equation [25]:

$$\text{Sensitivity} = \frac{\text{percentage change in output}}{\text{percetage change in input}} = \frac{\text{increment in output/ initial value of output}}{\text{increment in input/ initial value of input}} \quad \ldots\ldots\ (1)$$

This definition of sensitivity was used to investigate input and output parameters having different units. The sensitivity factor in this study represents change in the output parameter for 10% change in given input parameter. For example, a sensitivity factor of -5 means the percentage decrement of the output is 5% if the percentage increment of the given input is 10%, and is independent of the units of the input and output parameters.

2.1 Input parameters

Component geometry parameters

Figure 2 shows geometries of the five designs used in the present study. Starting with a thin-walled hollow structure, designs were evolved to evaluate sensitivity with changing part volume, thickness, height, layer surface area, and structural stiffness. Part wall thickness and Z-height are varied for the thin-walled designs (Design-I to IV) to estimate how process sensitivity will evolve with changing geometry. For the solid part design (V), percentage of porosity was varied instead of wall thickness to account for changes in volume. The values of the geometry parameters selected for sensitivity analysis are presented in Table 1.

Table 1: Component geometry parameters variations

Component geometry parameters	Design I	Design II	Design III	Design IV	Design V
Part wall thickness (mm)	2 1.5, 2, 2.5	1.5 1, 1.5, 2	1.5 1, 1.5, 2	1.5 1, 1.5, 2	-
Part geometry porosity (%)	-	-	-	-	30 0, 30, 60
Part Z-height (mm)	40 20, 40, 60	20 10, 20, 30	20 10, 20, 30	20 10, 20, 30	20 10, 20, 30

Process parameters

The printing process in MF^3 involves several input parameters. In the present study, a set of key parameters are studied. Table 2 presents typical values of selected process parameters based on experience. Layer thickness defines the height of each layer of printing. While a smaller layer thickness discretizes the build into higher resolution leading to better accuracy and surface quality, it leads to higher print time. Typically, a layer thickness of 0.2mm was used in MF^3. Variations to 0.1mm and 0.3mm was investigated for sensitivity in this study. Layer width define discretization of each layer in XY plane. Extrusion temperature defines the extruder temperature for softening the material just enough to overcome its viscosity and extrude through the nozzle smoothly. An extrusion temperature of 240°C with variations to 220°C and 260°C was evaluated in this study. Build plate was heated to keep the substrate temperature high enough to enable efficient diffusion between two layers. It also provides slower cooling rate that helps minimizing part warpage. Influence of build plate temperature of 65°C with variations to 45°C and 85°C is evaluated in this study. Printing speed defines the travel speed of nozzle while it extrudes material to print. As it directly defines print time, it is always desirable to get higher print speed. But it is limited by several parameters like material viscosity, filament stiffness to push through the nozzle, and print quality. Typically, a print speed of 10mm/s is used in MF^3 of Ti-6Al-4V. Variations to 5mm/s and 15mm/s was investigated for sensitivity in this study. Finally, two toolpath 0 – 90 and 45 – 135 were investigated in this study.

Table 2: Printing process parameters variations

Process parameters	**Variations**
Layer thickness (mm)	0.2 0.1, 0.2, 0.3
Layer width (mm)	0.48 0.36, 0.48, 0.60
Extrusion temperature (°C)	240 220, 240, 260
Build plate temperature (°C)	65 45, 65, 85
Printing speed (mm/s)	10 5, 10, 15
Toolpath (°)	0 – 90 0 – 90, 45 – 135

Powder-binder material properties

In this study, the properties used for Ti-6Al-4V in custom polymer binder were obtained from estimation models. The material definition for simulation includes mechanical and thermal properties over a range of temperature for the input material as shown in Table 3. The given material properties were varied to +/- 20% to investigate printing process sensitivity. These variations can be looked at as outcome of change in material formulation in terms of solids loading and custom polymer binder.

Table 3: Material properties variations

Material properties (at 23°C)	**Variations**
Thermal conductivity (mW/mm·°C)	1.466 1.1728, 1.466, 1.7592
Coefficient of thermal expansion (1/°C)	2.83E-05 2.26E-05, 2.83E-05, 3.40E-05
Specific heat capacity (mJ/t·°C)	8.95E+08 7.16E+08, 8.95E+08, 1.07E+09
Young's modulus (MPa)	205 164, 205, 246
Specific volume (mm^3/t)	3.38E+08 2.70E+08, 3.38E+08, 4.06E+08

2.2 Simulation setup: Base case for Design-I

In this study the finite element analysis (FEA) based simulations are conducted using Digitmat to estimate the outcome from MF^3 printing process. Table 4 lists the geometry, process parameters and material properties used in the simulation of base case for Design-I.

Table 4: Input for process simulation for the base case of Design-I

Input Variables		
Component geometry parameters	Part wall thickness	2mm
	Part Z-height	40mm
Process parameters	Layer thickness	0.1mm
	Layer width	0.48mm
	Extrusion temperature	$240^{\circ}C$
	Build plate temperature	$65^{\circ}C$
	Printing speed	5mm/s
	Toolpath	$0^{\circ} - 90^{\circ}$
Material properties	Thermal conductivity	1.466 mW/mm·C
	Coefficient of thermal expansion	2.83E-05 $1/^{\circ}C$
	Specific heat capacity	8.95E+08 $mJ/t\cdot^{\circ}C$
	Young's modulus	205 MPa
	Specific volume	3.38E+08 mm^3/t

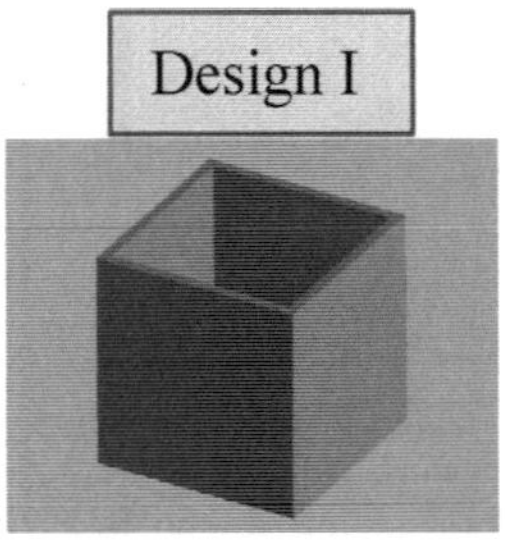

2mm hollow cube 40mm x 40mm x 40mm

The CAD model of a 2mm thick hollow cube of size 40mm x 40mm x 40mm was imported in STL format. It was discretized into voxel mesh. The mesh size will have an influence on the prediction of the residual stresses, but a rather minor influence on the displacement results (warpage) [29]. However, the mesh size cannot be smaller than the layer thickness. Considering the computational time, a mesh size of 0.3mm was used. A GCode file from Simplify3D defines toolpath, layer thickness and printing speed. Other printing parameters like layer width, extrusion temperature, build plate temperature, chamber temperature, convection coefficient are defined through simulation tool graphical user interface as shown in Figure 3. The material definition includes the mechanical and thermal properties over a range of temperatures as shown in Figure 3.

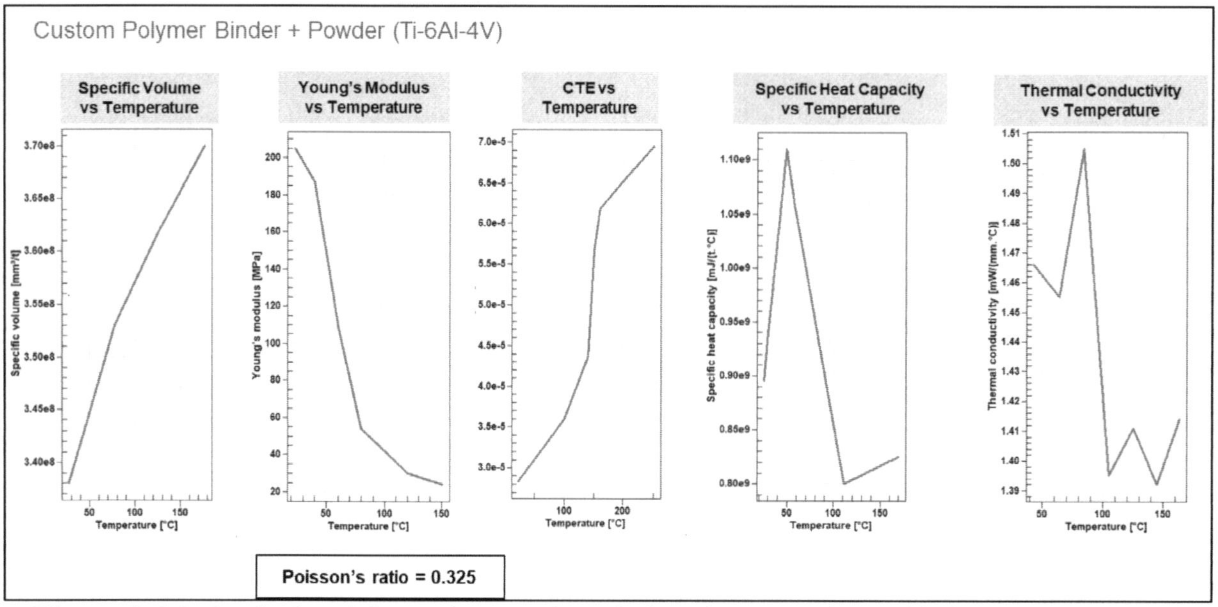

Figure 3. Mechanical and thermal properties of Ti-6Al-4V + polymer binder over a range of temperatures

Considering the computational time and number of simulation jobs, a layer-by-layer method activation method was used to simulate the printing process. Here, finite element layers are activated one at a time, whereas in filament discretization a chunk of filaments gets activated. Having defined the material, geometry and process inputs, the job was submitted for thermo-mechanical simulation.

3. RESULTS AND DISCUSSION

First, the results of Design-I base case are explained in detail. The simulation results are presented below for the individual output parameters, starting with print time, then deflection, Z-warpage, residual stress and substrate temperature. Then, the sensitivity analysis results of Design-I are discussed. Each of the input parameters was varied as discussed in previous section. For each individual variation, a simulation job was performed. The sensitivity factor (SF) was calculated using equation (1). Accordingly, the significance was termed 'high' for SF at levels >10, 'medium' in the range between 5 and 10, 'low' in the range between 1 and 5, and 'no' at levels <1.

Print time

Print time is an important aspect of any AM process. While lower print time indicates higher commercial viability, it is generally accompanied by quality compromise. Simulation provides an estimate of time required to print the given part under set process parameters. It is dependent upon component geometry parameters (part wall thickness, layer cross-sectional area, part maximum height) and process parameters (printing speed, part orientation/ build height, layer width, layer thickness). The estimated print time for the base case of Design-I was 60858s.

Deflection, all

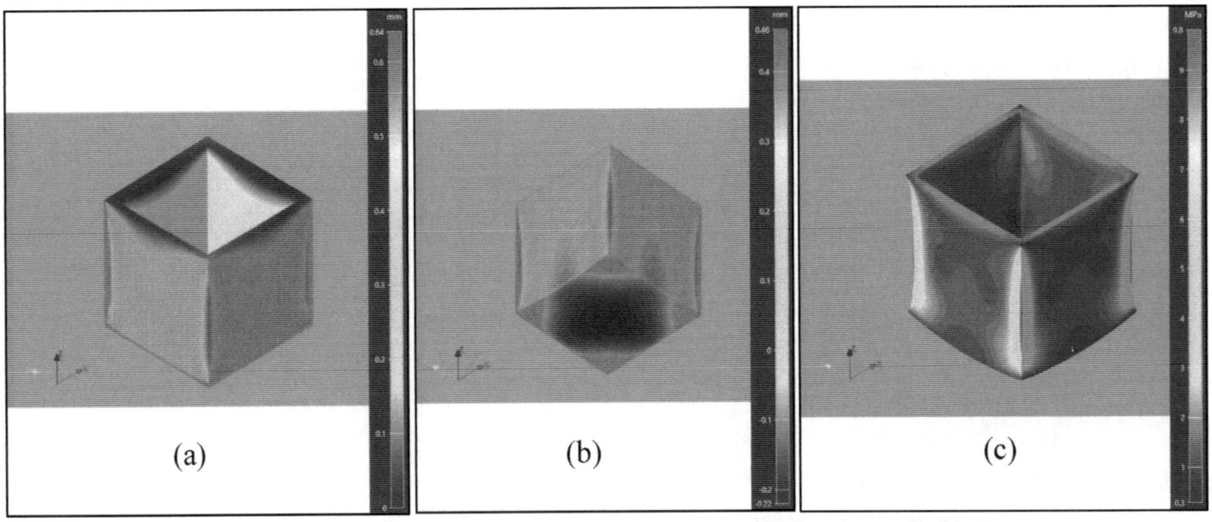

Figure 4. Design-I base case simulation results: (a) Deflection, all (b) Z-warpage (c) Residual stress (von Mises)

In MF^3 printing process, the material is subjected to softening/ partial melting while pushed through the extruder-nozzle, deposited on substrate layer-by-layer and then allowed to cool down and solidify. Afterwards, it is subjected to repeated cycles of heating and cooling with each new layer getting deposited. This leads to development of thermal gradient and inherent thermal strains in the printed part [29]. The extent of the strain varies according to coefficient of thermal expansion of the material. The thermal strain causes deflection and warpage, finally leading to deviations printed part dimensions opposed to defined CAD geometry. Hence, to achieve desired part quality and dimensional tolerances it is important to understand and control the deflection/ warpage phenomenon during MF^3 printing. Simulation provides an estimate of deflection in X, Y, Z direction as well as overall deflection (Deflection, all) results for given material, geometry and process parameters. It is primarily dependent upon temperature-related process parameters (extrusion temperature, build plate temperature) and material properties (coefficient of thermal expansion). The estimated maximum deflection (Deflection, all) for the base case of Design-I was 0.64mm as shown in Figure 4(a).

Z-warpage

As thermal strains lead to deflection in printed part, it is combined with nonuniform shrinkage resulting from nonuniform cooling due to layer-by-layer printing. This leads to warpage which is more significant across the printing plane along Z-axis. The estimated Z-warpage for the base case of Design-I was 0.69mm as shown in Figure 4(b).

Residual stress

The process simulation provides estimation of residual stress induced during and by end of the printing process due to non-uniform cooling. Figure 4(c) indicates von Mises stress as residual stress developed by the end that can be attributed to thermal gradient as a result of layer-by-layer stacking as well as differential heat loss from central zone compared to outer periphery. Lower thermal gradient, uniform and slower cooling helps in reducing residual stresses. The estimated von Mises stress for the base case of Design-I was 9.79MPa.

Substrate temperature

The simulation predicts the temperature history of the printing process by means of a transient thermal analysis. It is helpful to study the local temperature evolution during the process. Fig 5(a) shows the temperature variation indicating thermal gradient at the end of printing. The graph (b) shows the history of temperature evolution oi the layer at 5mm height. It can be seen that the layer is extruded and deposited at set extrusion temperature of 240°C. However, by the time next layer is deposited, the previous layer cools down to 55°C.

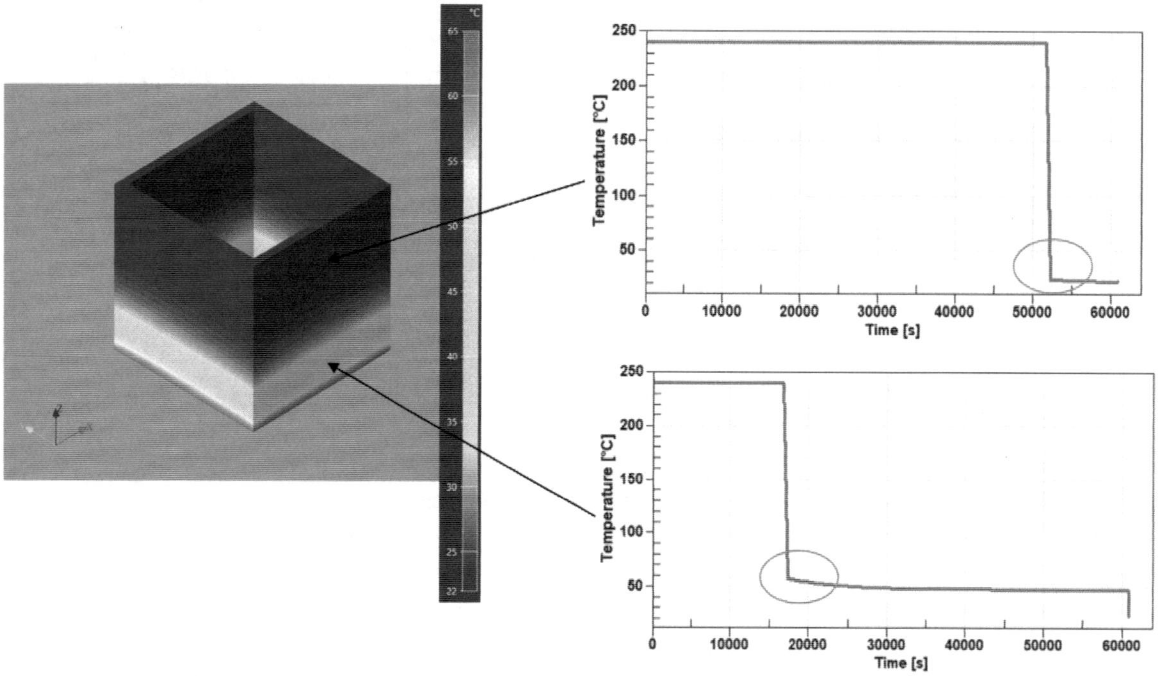

Figure 5. Design-I base case simulation results: (a) Temperature distribution (b) Substrate temperature at 35mm height (c) Substrate temperature at 5mm height

This is due to heat loss to the environment as an open chamber printer was considered in the study. The chamber temperature was set to ambient temperature, 20°C, in the simulation. However, the build plate was kept at 65°C. So, the printed layers were gaining heat from the heated bed. The heat flux depends upon the temperature gradient and thermal conductivity of the material. As a result of simultaneous heat loss and heat gain, the layer 5mm height is found to be at 55°C when the next layer comes in. We can say that the substrate temperature of this next layer is 55°C. Similarly, the graph (c) shows the history of temperature evolution in the layer at 35mm height. Here, the substrate temperature is found to be 22°C, which is close to the set chamber temperature 20°C. It this case the amount of heat gain from build plate was much lower than heat loss to environment. This is because of limited thermal conductivity of the material, while convection heat loss was more significant. The phenomenon of heat gain through conduction and heat loss through convection also depends upon surface area of exposure, Z-height, time taken to print each layer which ultimately depends on part geometry as well as printing parameters.

As build progresses, substrate temperature drops from 65°C (build plate temperature) to 20°C (ambient temperature). The importance of substrate temperature is that a lower substrate temperature leads to a lower amount of diffusion between two layers, hence, lower interlayer bond strength. Finally, a lower mechanical strength [27,28]. Hence, it is an important consideration not only from build setup or process parameters, but also from part and material design perspectives.

3.1 Sensitivity Analysis

As described, equation (1) provides the sensitivity factor (SF) calculation for every individual output parameter response to each variation of input parameters. Table 5 summarizes the SF results for Design-I. The sensitivity analysis provides means to identify significant input parameters for each of the output parameters. Sensitivity factor facilitates a quantitative yardstick to identify the significant parameters. A single SF is considered for each input-output combination for further analysis.

Table 5: Sensitivity factors (maximum) for Design-I

		Print Time	Max. Deflection	Max. Z-Warpage	Max. Residual Stress	Min. Substrate Temperature
Component geometry parameters	Part wall thickness	18.77	1.80	1.85	2.40	0.96
	Part Z-height	8.05	1.82	8.20	5.27	-6.67
Process parameters	Layer thickness	-19.65	-0.08	-0.19	-0.26	0.70
	Layer width	-18.56	-0.07	-0.11	-0.11	-0.13
	Extrusion temperature	0.00	17.06	17.25	16.78	0.32
	Build plate temperature	0.00	1.32	1.86	0.40	0.85
	Printing speed	-20.00	-0.07	-0.16	-0.16	0.70
	Toolpath	0.02	0.00	0.00	0.00	0.00
Material properties	Thermal conductivity	0.00	-0.12	-0.26	-0.16	1.08
	CTE	0.00	10.50	10.26	10.36	0.02
	Specific heat capacity	0.00	-0.05	-0.07	-0.08	0.45
	Young's modulus	0.00	0.00	0.00	10.00	0.00
	Specific volume	0.00	0.06	0.09	0.10	-0.52

Print time

According to the simulation results, print time was found to be sensitive to component geometry and process parameters. It was not influenced by any of the material properties. As part wall thickness increases, print time increases. A strong sensitivity was observed in print time towards part wall thickness. It was obvious that higher print area of each layer of the vertical walls and larger height of the bottom horizontal portion of the geometry led to larger print time. However, it was observed that print time increased even when wall thickness was reduced. This opposite response was difficult to understand logically. However, by investigating the toolpath it was learnt that in spite of same slicer setting when thickness was decreased to 1.5mm, a different infill with zig-zag path was generated which consumed more time, whereas in 2mm thickness, two straight line tracks formed the infill consuming lesser print time. Print time also showed high sensitivity to part Z-height or build height as number of layers vary proportionately. Print time is greatly influenced by layer thickness. Negative sign indicates that print increases when layer thickness decreases and vice versa. Part geometry having non-uniform cross-section along Z-axis, print time doesn't vary proportionately with layer thickness variation. Generally, layer width has significant influence on print time. With wider tracks, time to print infill of each layer decreases. However, in this design the thin walls are printed with minimal infill, hence no significant impact observed. Printing speed has a direct and significant influence on print time. Toolpath has negligible effect on print time in this design. However, based on infill required, in some cases toolpath plays considerable role in print time. None of the variations in material properties had any influence on print time.

Deflection, all

The deflection results showed sensitivity to the three sets of input parameters to different extent. Low sensitivity was noted towards component geometry parameters like part wall thickness and Z-height. No variation was observed at increasing Z-height from 40mm to 60mm. In terms of printing process, layer thickness and layer width showed no influence on deflection. Extrusion temperature influences deflection strongly as it defines the thermal gradient. Higher extrusion temperature leads to higher deflection, and vice versa. Sensitivity to build plate temperature was low. Printing speed and toolpath showed no influence on deflection. From material properties, while coefficient of thermal expansion (CTE) showed strong effect, all other properties had no influence at all. This is because CTE defines volumetric changes with temperature change, which is used for deflection calculation in simulation.

Z-warpage

Z-warpage showed sensitivity pattern similar to deflection as the physical phenomenon and computational approaches are similar for both. While part wall thickness showed low sensitivity, part Z-height had considerable influence on Z-warpage. This is due to varying deflection in build direction with variation build height. On processing front, layer thickness and layer width showed no influence. Extrusion temperature influenced Z-warpage strongly, same as that for deflection. Higher extrusion temperature led to higher Z-warpage. Sensitivity to build plate temperature was low. Printing speed and toolpath showed no influence. From material properties, coefficient of thermal expansion (CTE) again showed strong effect, while all other properties had no influence at all.

Residual stresses

The residual stress results showed sensitivity to most of variables from the three sets of input parameters. On component geometry, while sensitivity was low to wall thickness, it was considerable to Z-height. Variation in Z-warpage leads to residual stress variations with varying height. Layer thickness and layer width had negligible impact. A strong sensitivity was observed towards extrusion temperature same as it was for deflection and warpage. Lower extrusion temperature will help reduce residual stresses. Build plate temperature, printing speed and toolpath showed negligible influence on residual stresses. For the first time, all the material properties showed some influence, though it was very small for thermal conductivity, specific heat and specific volume. Coefficient of thermal expansion (CTE) showed same amount of influence as it had on deflection and Z-warpage. Young's was found influential only towards residual stress results. As the residual stress is the result of resistance to deflection from material, it is proportionate with material stiffness. Hence, higher E and higher CTE value leads to higher residual stresses, and vice versa.

Substrate temperature

It was observed that substrate temperature showed sensitivity to most of input parameters from all three sets, though it was considerable only for a few input variables. From component geometry, wall thickness had negligible influence in this case. However, if part thickness will increase significantly, print time will increase. This allows more time to dissipate heat before next layer comes in, leading to lower substrate temperature. Highest sensitivity was observed towards part Z-height. As part height increases, substrate temperature decreases. As mentioned, greater amount of heat is lost to environment than heat gained from build plate. This is because of limited thermal conductivity of the material, while convection heat loss was more significant, leading to lower temperature. While all the printing parameters showed some influence on substrate temperature, none of them was considerable enough in this case. Generally, build plate temperature has considerable influence up to certain build height. On material front, sensitivity was considerable for thermal conductivity, negligible for specific heat capacity and specific volume, and zero for CTE and Young's modulus.

4. CONCLUSIONS

The MF^3 of Ti-6Al-4V was analyzed for the first time using a systematic procedure based on sensitivity analysis principles. The procedure developed in this study was found to have the following key attributes:

1. The sensitivity analyses facilitated the identification of the dominant input parameters in MF^3 for the systems under consideration.
2. The quantitative relationships on the dependence of many MF^3 process output parameters on changes in component geometry, process parameters and material properties are presented.
3. The procedure clearly identified the relative importance of specific attributes of parts geometry, processing conditions, slicing strategies, powder-binder material properties based on the correlations between input parameters in each category and output parameters as process outcome.
4. The tool provides a powerful approach to identify the parameters that need to be optimized in the design stage and carefully monitored and controlled in the manufacturing stage of MF^3.
5. The insights gained on material-process-geometry interrelationships through the sensitivity study will enhance DfMF^3.
6. The present sensitivity analysis procedure is expected to be an invaluable tool not only for process optimization but also for going beyond from part design to materials development for MF^3.

5. ACKNOWLEDGEMENT

Authors acknowledge the financial support from Minority Business Development Agency, USA (MBDA) and National Aeronautics and Space Administration (NASA). The authors also recognize MSC Software (A Hexagon company) for providing the Digimat simulation tool used in this study.

6. REFERENCES

1. Joamin Gonzalez-Gutierrez, Santiago Cano, Stephan Schuschnigg, Christian Kukla, Janak Sapkota, Clemens Holzer, *Additive manufacturing of metallic and ceramic components by the material extrusion of highly-filled polymers: a review and future perspectives.* Materials 2018. 11(5): p. 840
2. Paramjot Singh, Qasim Shaikh, Vamsi K. Balla, Sundar V. Atre, Kunal H. Kate, *Estimating powder-polymer material properties used in design for metal fused filament fabrication (DfMF^3)*, JOM (2020) 72: 485.
3. Yvonne Thompson, Joamin Gonzalez-Gutierrez, Christian Kukla, Peter Felfer, *Fused filament fabrication, debinding and sintering as a low cost additive manufacturing method of 316L stainless steel,* Additive Manufacturing 30 (2019) 100861
4. Guohua Wu, Noshir A. Langrana, Rajendra Sadanji, Stephen Danforth, *Solid freeform fabrication of metal components using fused deposition of metals.* Materials & Design, 2002. 23(1): p. 97-105.

5. Walter Lengauer, Ivica Duretek, Markus Fürst, Viktoria Schwarz, et al., *Fabrication and properties of extrusion-based 3D-printed hardmetal and cermet components.* International Journal of Refractory Metals and Hard Materials, 2019. 82: p. 141-149.
6. Luquan Ren, Xueli Zhou, Zhengyi Song, Che Zhao, Qingping Liu, et al., *Process parameter optimization of extrusion-based 3D metal printing utilizing PW–LDPE–SA binder system.* Materials, 2017. 10(3): p. 305.
7. Seokyoung Ahn, Seong Jin Park, Shiwoo Lee, Sundar V. Atre, Randall M. German, *Effect of powders and binders on material properties and molding parameters in iron and stainless steel powder injection molding process.* Powder Technology 193 (2009) 162–169
8. Antonio Armillotta, Mattia Bellotti, Marco Cavallaro, *Warpage of FDM parts: experimental tests and analytic model.* Robotics and Computer–Integrated Manufacturing 50 (2018) 140–152
9. Pedram Parandoush, Dong Lin, *A review on additive manufacturing of polymer-fiber composites.* Composite Structures 182 (2017) 36–53
10. Rafael Quelho de Macedo, Rafael Thiago Luiz Ferreira, *Residual thermal stress in fused deposition modelling*, 24th edn. (2017), ABCM International Congress of Mechanical Engineering, Curitiba
11. Miquel Domingo-Espin, Josep M. Puigoriol-Forcada, Andres-Amador Garcia-Granada, Jordi Llumà, et al, *Mechanical property characterization and simulation of fused deposition modeling Polycarbonate parts.* Materials & Design 83 (2015) 670–677
12. Huanxiong Xia, Jiacai Lu, Sadegh Dabiri, Gretar Tryggvason, *Fully resolved numerical simulations of fused deposition modeling. Part I: fluid flow.* Rapid Prototyping Journal 24/2 (2018) 463–476
13. Yong Zhou Gang Xiong, Timo Nyberg, Dan Liu, *Temperature analysis in the fused deposition modeling process.* IEEE Computer Society ICISCE.2016.150
14. S.F. Costa, F.M. Duarte, J.A. Covas, *Thermal conditions affecting heat transfer in FDM/FFE: a contribution towards the numerical modelling of the process.* Virtual and Physical Prototyping, 2014. https://doi.org/10.1080/17452759.2014.984042.
15. Yizhuo Zhang, Y. Kevin Chou, *3D FEA simulations of fused deposition modeling process.* ASME International Conference on Manufacturing Science and Engineering, 2006. Ypsilanti, MI.
16. Alberto Cattenone, Simone Morganti, Gianluca Alaimo, Ferdinando Auricchio, *Finite element analysis of additive manufacturing based on fused deposition modeling (FDM): distortions prediction and comparison with experimental data.* Journal of Manufacturing Science and Engineering, 2018. https://doi.org/10.1115/1.4041626.
17. B. Barmore, *Fused Filament Fabrication of Filled Polymers for Metal Additive Manufacturing.* Mechanical Engineering, Oregon State University, 2016, pp. 55–60.
18. Tomas Webbe Kerekes, Hyoungjun Lim, Woong Yeol Joe, Gun Jin Yun, *Characterization of process–deformation/damage property relationship of fused deposition modeling (FDM) 3D-printed specimens.* Additive Manufacturing 25 (2019) 532–544
19. Filip Górski, Wiesław Kuczko, Radosław Wichniarek, *Influence of process parameters on dimensional accuracy of parts manufactured using fused deposition modelling technology.* Advances in Science and Technology Research Journal. DOI: 10.5604/20804075.1062340
20. J. Giannatsis, K. Sofos, V. Canellidis, D. Karalekas, V. Dedoussis, *Investigating the influence of build parameters on the mechanical properties of FDM parts.* Innovative Developments in Virtual and Physical Prototyping, Dec 2011. Publisher: CRC Press.
21. Gokulakrishnan Jothibabua, Saravana Kumar Gurunathanb, *Surrogate Based Sensitivity Analysis of Part Strength due to Process Parameters in Fused Deposition Modelling.* Procedia Computer Science 133 (2018) 772–778
22. B.H. Lee, J. Abdullah, Z.A. Khan, *Optimization of rapid prototyping parameters for production of flexible ABS object*, Journal of Materials Processing Technology 169 (2005) 54–61
23. Emily R. Fitzharris, Narumi Watanabe, David W. Rosen, Meisha L. Shofner, *Effects of material properties on warpage in fused deposition modeling parts*, The International Journal of Advanced Manufacturing Technology (2018) 95:2059–2070. https://doi.org/10.1007/s00170-017-1340-8

24. Bharath Vasudevarao, Dharma Prakash Natarajan, Mark Henderson, Anshuman Razdan, *Sensitivity of RP surface finish to process parameter variation*, https://www.researchgate.net/publication/2467464
25. Sundar.V. Atre, S.J. Park, R. Zauner, Randall M. German, *Process simulation of powder injection moulding: Identification of significant parameters during mould filling phase*, Powder Metallurgy (2007), 50:1, 76-85.
26. A. El Moumen, M. Tarfaoui, K. Lafdi, *Modelling of the temperature and residual stress fields during 3D printing of polymer composites*, The International Journal of Advanced Manufacturing Technology (2019) 104:1661–1676.
27. Q. Sun, G.M. Rizvi, C.T. Bellehumeur, P. Gu, *Effect of processing conditions on the bonding quality of FDM polymer filaments*, Rapid Prototyping Journal 14/2 (2008) 72–80.
28. Céline Bellehumeur, Longmei Li, Qian Sun, Peihua Gu, *Modeling of Bond Formation Between Polymer Filaments in the Fused Deposition Modeling Process,* SME Journal of Manufacturing Processes Vol. 6/No. 22004
29. e-Xstream, Digimat-AM simulation solution for Additive Manufacturing. https://www.e-xstream.com/product/digimat-am. Accessed 30 Dec 2019.
30. Alphastar, Genoa Additive Manufacturing design tool and software suite for polymers, metals and ceramics. http://www.alphastarcorp.com/products/genoa-3dp-simulation/. Accessed 30 Dec 2019.
31. Nikzad, M., S. Masood, and I. Sbarski, *Thermo-mechanical properties of a highly filled polymeric composites for fused deposition modeling.* Materials & Design, 2011. **32**(6): p. 3448-3456.

PROCESS INDUCED VARIATIONS IN POLYJET PRINTING UNDER TENSION-TENSION FATIGUE LOADING – INVESTIGATIONS VIA HOMOGENOUS 3-DIMENSIONAL TEST COUPONS

Ravi Pratap Singh. Tomar, Furkan. Ulu, Ram V. Mohan, Ajit D. Kelkar
North Carolina A&T State University
Greensboro, NC/USA 27411

ABSTRACT

The capability of additive manufacturing (AM) for making monolithic, multi-material structures allows the fabrication of complex parts with varying mechanical properties, material combinations, and phases. Due to the flexibility of orientating the digital design in AM, a part can be created in any orientation; however, processing-induced variations affect the mechanical properties of final parts as the way AM forms a part is different from traditional manufacturing and could lead to many artifacts. This paper investigates the effects of process variations on the tension – tension fatigue behavior in Polyjet by designing a homogeneous 3-D test coupon configuration. In traditional 2D coupons, one direction is thinner than the other two directions, which may not account for non-uniformity in a material deposition during printing. Further, the effect of normal strain is not effectively captured. To understand the process-induced variation and to capture the normal strain effect, this work proposed and evaluated a homogenized 3D configuration test coupon and evaluated material behavior under tension-tension fatigue loading.

Study findings provide insight on the limitation of traditional fatigue test coupons and allow for careful evaluation of process variation effects in PolyJet printing. The effect of print orientations on the tension-tension fatigue characteristics is studied and presented. Further test results help to develop a safe material design envelope that can be used as a design guideline for a specific material without worrying about build orientation.

Corresponding author: Ram V. Mohan

1. INTRODUCTION

Characterization of additive manufactured parts is essential for reasons such as developing new materials and processes, establishing design allowable, qualification of materials for a specific application, and quality control of production processes. During the service life of 3D printed parts, it is subjected to a different type of loading that can be broadly classified as static and dynamic loading. Static loading includes tensile, compressive, shear, flexural twisting, etc. whereas dynamic loading includes impact or fatigue (cyclic) loading. As of today, standard test methods are available for characterization of tensile, compressive, flexural, impact, and tension - tension fatigue behavior for metal and composite materials and are well established, which is not the case for the polymer-based additive manufacturing process. Among these characterization procedures, quasi-static methods such as tensile, flexural, etc. are comparatively easier and less time-consuming. However, the characterization of dynamic behavior particularly fatigue behavior is challenging and time-consuming. Additionally, test data for AM printed parts under cyclic loading are still slacking and require further understanding.

Variations in build characteristics with part orientation during layer-by-layer deposition is an important issue in additive manufacturing. Parts printed in different orientation appears to be similar physically but could behave very differently under the same loading conditions. This can be explained by variation in processing conditions in AM, specifically, build orientation. Further, these processing induced variations affect the performance of parts in AM, as the way AM parts are formed are different in comparison to traditional manufacturing that could lead to many artifacts. In AM, both material and structure are built at the same time during the process. Material and structure formation can lead to differences in properties based on how it is built. Any variation during processing can change material configuration and lead to process-induced artifacts. The literature cites studies that were performed to understand the process-induced variation in AM, in particular, tensile elongation. However, these studies are based on traditional ASTM D628 2D test coupons [1][2][3]. These traditional 2D test coupons are thinner in one direction, or the cross-section aspect ratio is not the same. This limits the ability to correctly bring out and highlight manufacturing-related defects in AM. Firstly, these test coupons may not represent the real-world material behavior by not capturing effectively build orientation variations, and secondly, the effect of normal strain, which is a compressive strain in thinner directions, is not accurately captured. To eliminate these influencing geometrical aspects during the building of AM parts in different orientations, the present work proposed a 3-dimensional tensile test configuration with the same cross section's aspect ratio providing a uniform material deposition and build. This 3D tensile test configuration eliminates the limitation of the traditional 2D test coupon and helps to understand the process-induced variation carefully in AM.

The homogenized configuration of 3D test coupons is studied further in the present work to investigate the fatigue characterization, and are printed with simulated polypropylene (PP). The manufacturing of fatigue test coupons is discussed first, followed by studies on the post-curing effect and characterization of tension-tension fatigue behavior of homogenized 3D test coupons in two print orientations, namely XY and YX orientation.

2. EXPERIMENTATION

2.1 Material System

The unique capabilities of Jetting multiple materials with different mechanical properties, a part can be manufactured with customized digital material properties, such as combining an elastomeric material TangoBlackPlus and a rigid polymer VeroPureWhite in a specific concentration. Therefore, AM with its capability of combining two or more photopolymers such as flexible and rigid material has a wide scope for mimicking natural and biological structures [4][5][6][7]. Digital material in PolyJet printing is a result of combining two or more PolyJet photopolymers in specific concentrations to create a composite material with hybrid characteristics in a PolyJet based AM process. Numerous studies investigated [8][9][10][11] the influence of process parameters on mechanical properties, but very little work has been focused on the behavior of materials formed by mixing of two photopolymers [12] as well as final parts formed with mixing two or more multiple UV curable base resins. This understanding is critically essential in AM, as a process, in terms of digital materials formed, and process parameters, etc., that can significantly affect the material properties, which may not be seen in other traditional fabrication.

The test specimens were printed using the Stratasys J750 printer, based on PolyJet material technology. For this study, VeroPureWhite (VW) and TangoBlackPlus (TP) are selected as base materials because both have significantly different mechanical properties at room temperature. VW is rigid at room temperature, and TP is rubbery at room temperature. The Stratasys J750 can mix two base materials to form different possible combinations of digital materials as shown in Figure 1. The exact composition of these photopolymers is proprietary information; therefore, the characteristics of PolyJet materials are not fully understood. For this work, RGD8530 is identified as a material of interest. RGD8530 is a simulated polypropylene (PP). From the material point of view, simulated PP is formed by mixing two photopolymers locally, not creating the homogenous material mixture, but is expected to have homogenized properties that are similar to standard polypropylene polymer.

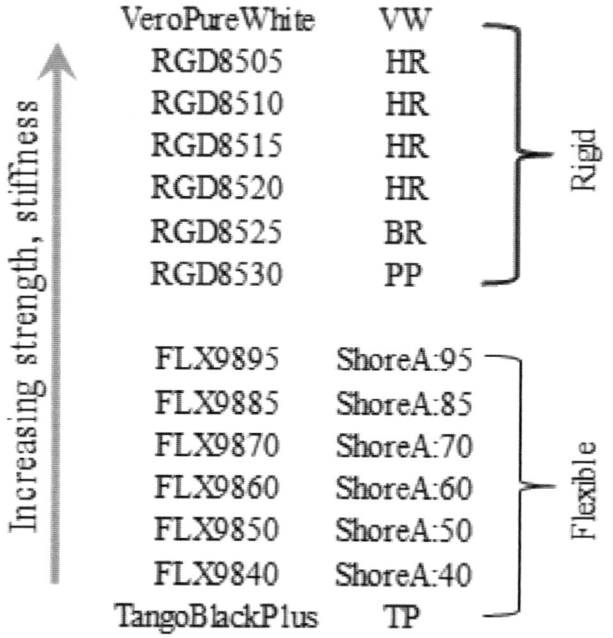

Figure 1. Different digital materials printable from two base materials (TangoBlackPlus and VeroWhitePlus) with Stratasys J750. Simulated Polypropylene is RDG8530.

2.2 Homogenized 3D Tensile Test Coupon

Various studies have been conducted based on 2D test coupons. In a traditional 2D test coupon configuration, one cross-sectional direction is thinner than the other two directions; therefore, the effect of normal strain, which is a compressive strain in thinner directions, is not effectively captured. To understand process-induced variations and to capture the normal strain effect, we recently considered and evaluated a homogenized 3D configuration test coupon [13]. A homogeneous 3-dimensional configuration is designed to keep the same cross section's aspect ratio, as shown in Figure 2, which allows for uniform material deposition during printing, and reduces process-induced artifacts. The objective of the 3D test configuration was to capture processing effects without artifacts potentially resulting from different aspect ratios of the cross-sectional dimension.

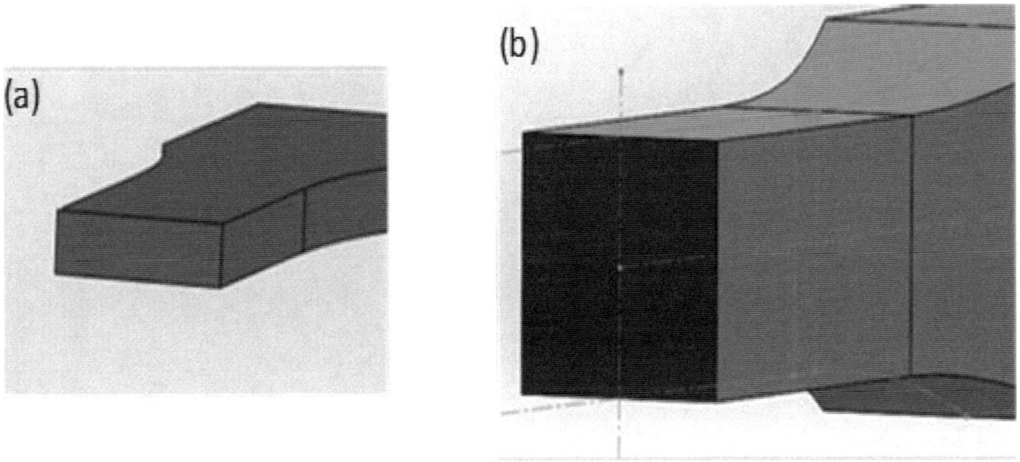

Figure 2. Cross-section (a) Traditional 2D test coupon (b) Homogenized 3D test coupon.

2.3 Fabrication of Panel and 3D Test Coupon

Fatigue test is very time-consuming, and fatigue becomes very critical in the case of materials that has property variations. Generally, in AM, individual coupons are printed, but to keep minimum property variation within test coupons, all coupons were obtained from a single large printed panel. Required fatigue test coupons were obtained by water jet cutting from the larger printed panel.

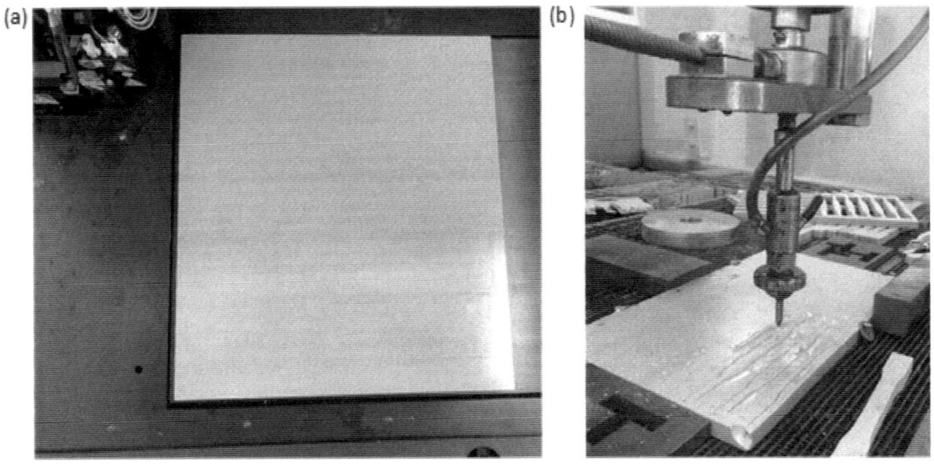

Figure 3. (a) Printing of large panel (b) Fabrication of coupon with the water jet.

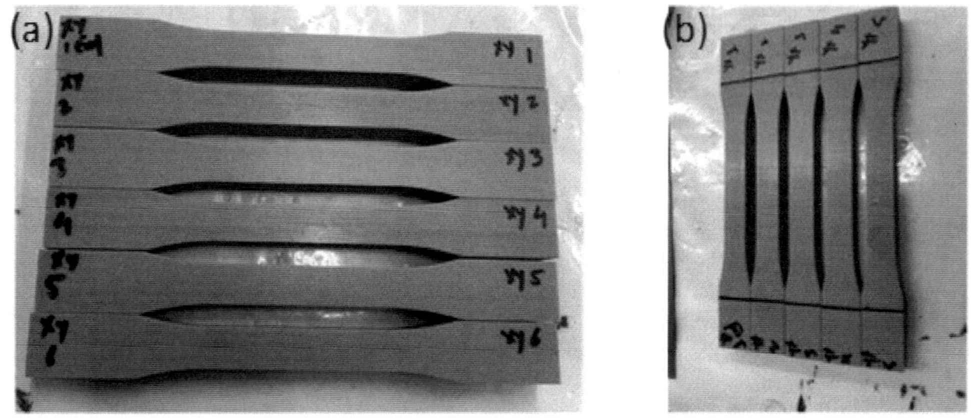

Figure 4. Water jet cut test coupon for fatigue testing (a) XY orientation (b) YX orientation

2.4 Thermal Characterization

Fatigue experiments generally take a longer time, and during the test, these test coupons need to be stored. One of the concerns with storing was that there might be little additional photopolymer cure continuing while storing them over time, as these polymers were light-sensitive, which can result in the modification of properties. Therefore, material samples from fatigue coupons were subjected to thermal characterization to determine the glass transition temperature (T_g) of digital polypropylene.

In order to check the full curing of test coupons printed with simulated polypropylene (PP) material, thermal characterization of simulated polypropylene is performed. Simulated polypropylene prior to 3D printing was characterized by Differential Scanning Calorimeter (DSC). The T_g for simulated calculated from the DSC analysis was found to be 42 °C to 43 °C.

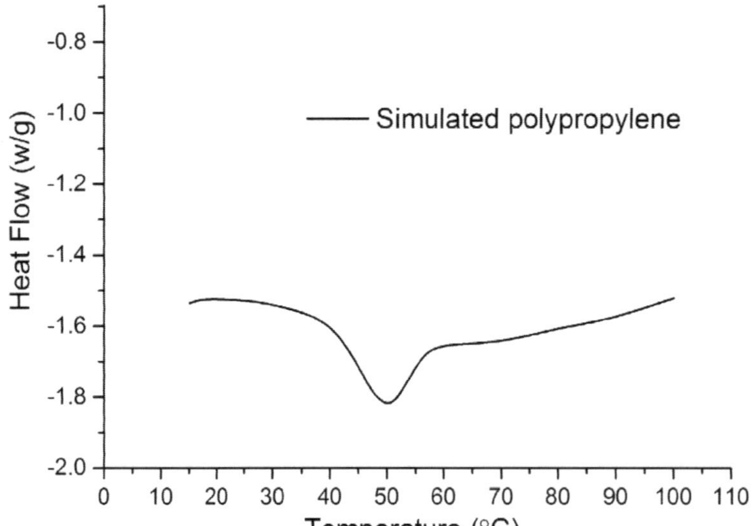

Figure 5. DSC graph for simulated polypropylene (PP).

2.4.1 Post curing

The calculated T_g of simulated polypropylene is used to perform the post-curing of the test coupons. The post-curing performed below the glass transition temperature (T_g). The glass transition temperature (T_g) of the material of interest, simulated polypropylene (PP), is very low, which is barely above the room temperature, so post-curing needs to be carefully performed.

2.4.1.1 Post oven-curing

The test coupon was post cured for two temperatures of 35 °C and 40 °C in the oven. The curing cycle selected was ramping to a predefined temperature and holding the predefined temperature for an hour, and subsequently cooling down for an hour. Tensile characterization was performed on post-cured specimens to understand the effect of post-curing.

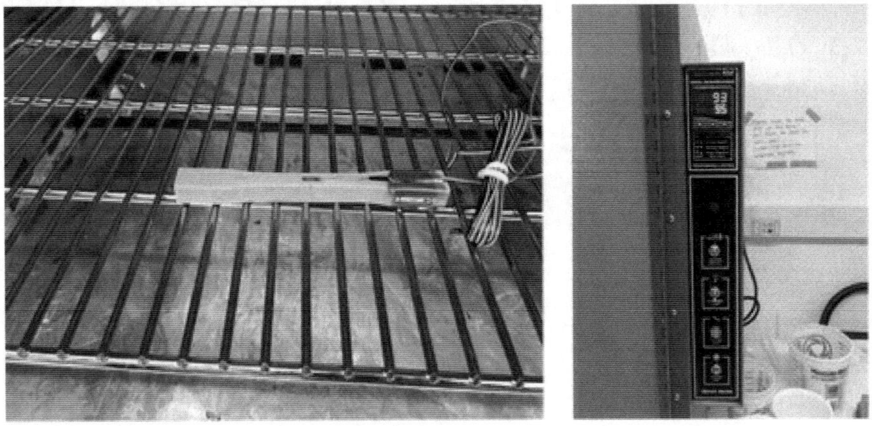

Figure 6. Oven cure setup.

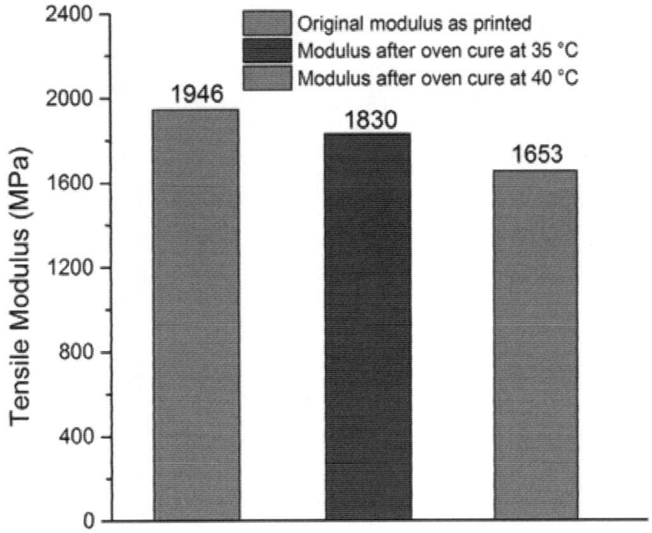

Figure 7. Stiffness degradation comparison before and after the oven cure.

There is a considerable reduction in the modulus observed after the oven cure. Oven Curing at 35 °C degraded the original modulus by almost 6%, whereas additional cure at 40 °C, the modulus degraded almost 15%. This implies that the test coupon as printed was fully cured, and no external oven curing is required.

2.4.1.2 Post UV curing

Both commercially available base photopolymer VeroPureWhite and TangoBlackPlus have strong photopolymerization from 350 nm to 410 nm wavelength. The light power intensity affects the penetration depth of curing light and the thickness of cured resin [14][15]. Therefore, the UV light source with a wavelength of 405 nm and power of 60 watts is used for the post-curing of test coupons.

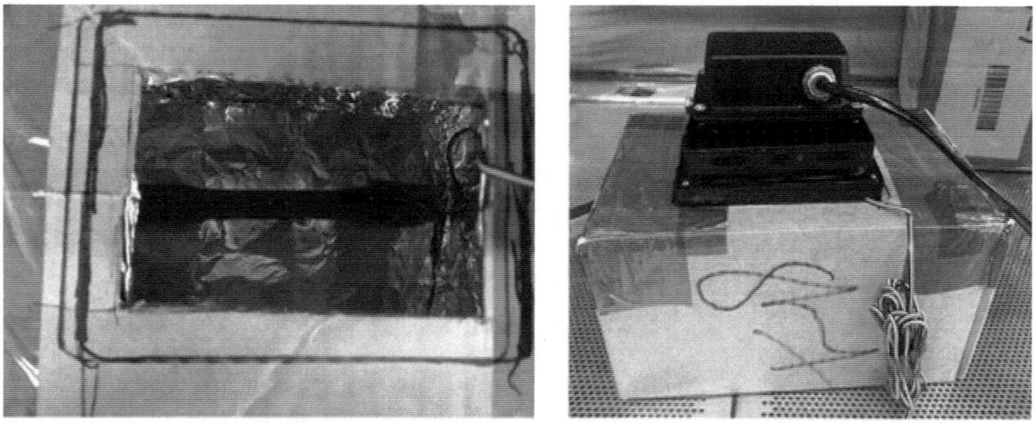

Figure 8. Setup for an external UV cure process.

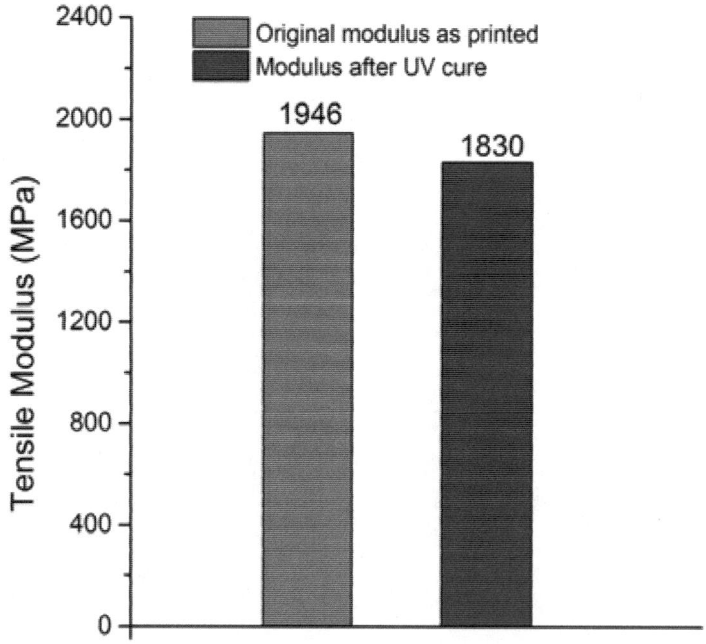

Figure 9. Stiffness degradation comparison before and after UV cure.

The modulus of the test coupon has degraded almost 5% from the initial modulus after exposing it with external UV light. This indicates that external UV curing is also not required, and the test coupon is already fully cured as printed.

3. RESULTS

A considerable amount of work has reported in the literature on fatigue characterization for FDM printed parts [16]. Since there is a lack of work related to fatigue life analysis of PolyJet printed parts and material systems, the goal of this work is to study the fatigue characteristics of PolyJet printed materials. Prior to performing the fatigue tests, the tensile strength of the 3D test coupon under static loading needs to be obtained.

3.1 Static testing

Tension tests were completed first as per ASTM D638 to estimate the mean ultimate tensile strength (UTS) for 3D coupons printed in two orientation XY and YX. Three specimens for each orientation were tested using an Instron 3384 system following the ASTM test standard. The pull rate was 5 mm/min. The extension was measured using an extensometer. The ultimate tensile strength (UTS) of both orientations are given in Table 1, and a representative stress-strain curve for tensile strength is shown in Figure 10.

It was observed that the UTS of the XY orientation specimens were slightly higher as compared with YX orientation specimens, and elongation of XY orientation specimens was longer than the YX orientation specimens. This corresponds to the printing of XY orientation specimens along the loading direction, whereas the loading direction is perpendicular to the printing direction of YX orientation specimens.

Table 1. Ultimate Tensile Strength of Orientation XY and YX Test Coupons.

Orientation	Property (MPa)	Specimen 1	Specimen 2	Specimen 3	Average	STDEV
XY	Tensile strength (UTS)	35.83	35.25	35.49	35.52	0.29
YX	Tensile strength (UTS)	33.80	33.45	33.32	33.51	0.25

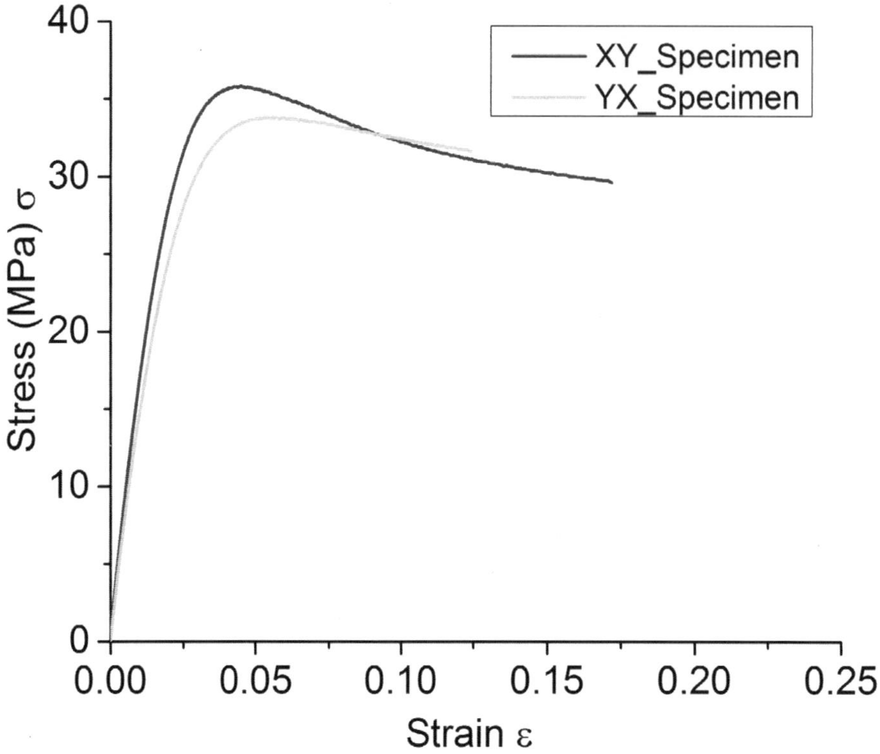

Figure 10. Stress-strain characteristic of XY and YX orientation test coupon.

3.2 Fatigue characterization

Tensile strength computed from the static tests is used to compute the load for each test coupon. The applied load to the specimen was calculated depending on the stress level. The minimum cycle load in each case was 1/10 of the maximum load (stress ratio R = 0.1) using a low frequency of 1 Hz. The tests were conducted following ASTM D7791 standards [87], with at least three specimens at each stress level. The results of the fatigue characterization for XY orientation and YX orientation are shown in Table 2.

The fatigue tests were stopped at 40,000 cycles for a 40% stress ratio. It can be noticed from the results of the fatigue characterization that average life for XY orientation is higher than that of YX orientation. The higher fatigue life can be attributed to continuous polymerization of XY orientation along the axis of the specimen as compared to discontinuous polymerization (more than one pass) for the YX orientation. Also, the soft inclusion in the rigid matrix of simulated polypropylene emulated a flexible fiber; these are along the loading direction in XY orientation. Flexible inclusions respond to cyclic loading differences elastically and provide a positive effect on fatigue characteristics. Therefore, the emulated soft fiber orientation relative to the loading axis substantially affects tension-tension fatigue properties.

Table 2. Results of Tension-Tension Fatigue Characterization.

Stress Ratio (s/s_u)	Number of cycles at failure			
	Specimen 1	Specimen 2	Specimen 3	Average
XY orientation				
0.4	40000	40000	40000	40000
0.5	13102	13015	10015	12044
0.6	4911	4971	4035	4639
0.7	2308	2126	2448	2294
0.8	786	745	748	759
YX orientation				
0.4	40000	40000	40000	40000
0.5	9052	12049	10638	10579
0.6	3823	4025	3565	3804
0.7	2046	2038	1653	1912
0.8	742	767	763	757

The standard way of showing fatigue behavior is in the form of the S-N diagram, as shown in Figure 11. It is observed that XY orientation specimens survived longer cycles than YX orientation.

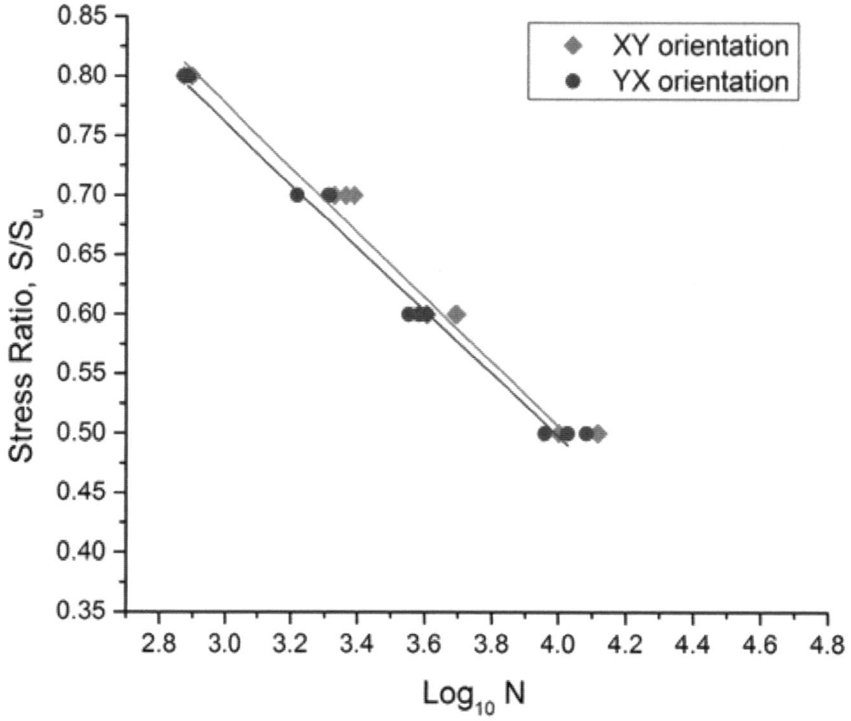

Figure 11. Comparison S-N diagram of fatigue for XY and YX orientation.

4. CONCLUSIONS

In this paper, the 3D homogenized test coupon printed with simulated Polypropylene (PP) in XY and YX orientation were studied for tension-tension fatigue behavior. The homogenized 3D test coupons eliminate the geometrical aspects during the building of AM parts in different orientations and allow for careful evaluation of process-induced variation in AM.

Tension-Tension fatigue characterization of digital material printed specimens were performed to evaluate its strength and fatigue life in XY and YX orientation. It was observed that the average fatigue life in the case of XY orientation was higher than YX orientation. This higher fatigue life corresponds to the printing of XY coupons along the loading direction, whereas, the loading direction is perpendicular to the printing direction in YX coupons. These variations are also attributed to the material configuration of simulated polypropylene, which shows characteristics of short flexible fiber-like inclusions of a soft photopolymer in a second hard photopolymer. These tension-tension fatigue results help to understand the material characteristic of simulated Polypropylene (Digital material) and help to design to withstand required loading cycles under fatigue loading without considering any build orientation.

5. REFERENCES

[1] J. T. Cantrell *et al.*, "Experimental characterization of the mechanical properties of 3D-printed ABS and polycarbonate parts," *Rapid Prototyp. J.*, vol. 23, no. 4, pp. 811–824, 2017.

[2] A. Cazón, P. Morer, and L. Matey, "PolyJet technology for product prototyping: Tensile strength and surface roughness properties," *Proc. Inst. Mech. Eng. Part B J. Eng. Manuf.*, vol. 228, no. 12, pp. 1664–1675, 2014.

[3] D. C. T. R. H. Crawford, "Optimizing Part Quality with Orientation," in *Solid Freeform Fabrication Symposium Proceedings*, 1995, p. 362.

[4] R. pratap singh Tomar, M. Ghazizadeh, E. H. Martin, and A. D. Kelkar, "Low Velocity Impact Response of Bio-inspired Fiberglass Woven Composites," in *American Society of Composites-30th Technical Conference*, 2015.

[5] R. pratap singh Tomar and F. Ulu, "Investigation of Digital Computer Aided Design Assembly and Tessellation Effects On Digital ABS Part Quality in Polyjet-3D Additive Printing," in *In International SAMPE Technical Conference*, 2018.

[6] R. pratap singh Tomar, F. Ulu, R. Mohan, and A. Kelkar, "Investigation of process induced variation in polyjet printing with digital polyproylene via homogeneous 3D tensile test coupon," in *ASME 2019 International Mechanical Engineering Congess and Exposition*, 2019.

[7] F. Ulu, R. pratap singh Tomar, and R. Mohan, "Development of Thermally Conductive Polymer/CNF -CNT Nanocomposite Materials via PolyJet Additive Manufacturing by Improvement of Digital Material Design," in *ASME 2019 International Mechanical Engineering Congess and Exposition*, 2019.

[8] L. Bass, N. A. Meisel, and C. B. Williams, "Exploring variability of orientation and aging effects in material properties of multi-material jetting parts," *Rapid Prototyp. J.*, vol. 22, no. 5, pp. 826–834, 2016.

[9] A. Kęsy and J. Kotliński, "Mechanical properties of parts produced by using polymer jetting technology," *Arch. Civ. Mech. Eng.*, vol. 10, no. 3, pp. 37–50, 2010.

[10] M. W. Barclift and C. B. Williams, "Examining Variability in the Mechanical Properties of Parts Manufactured Via Polyjet Direct 3D Printing," *Int. Solid Free. Fabr. Symp.*, pp. 876–890, 2012.

[11] A. I. Shiave, R. P. S. Tomar, I. P. Espinosa, and R. Mohan, "Deformation Mechanisms and Dislocations in Nickel–Cobalt Core–Shell Nanowires Under Uniaxial Tensile Loading—A Molecular Dynamics Modeling Analysis," *Adv. Sci. Eng. Med.*, vol. 11, no. 12, pp. 1187–1201, 2019.

[12] J. Mueller, K. Shea, and C. Daraio, "Mechanical properties of parts fabricated with inkjet 3D printing through efficient experimental design," *Mater. Des.*, vol. 86, pp. 902–912, 2015.

[13] R. pratap singh Tomar, F. Ulu, D. Ram Mohan, and A. Kelkar, "Investigation of Process Variation Effects Via a Homogeneous 3-Dimensional Tensile Test Coupon in Polyjet 3D Additive Printing," in *In International SAMPE Technical Conference*, 2019.

[14] J. Bennett, "Measuring UV curing parameters of commercial photopolymers used in additive manufacturing," *Addit. Manuf.*, vol. 18, pp. 203–212, 2017.

[15] S. Y. Hong *et al.*, "Experimental investigation of mechanical properties of UV-Curable 3D printing materials," *Polymer (Guildf).*, vol. 145, pp. 88–94, 2018.

[16] C. W. Ziemian, R. D. Ziemian, and K. V. Haile, "Characterization of stiffness degradation caused by fatigue damage of additive manufactured parts," *Mater. Des.*, vol. 109, pp. 209–218, 2016.

DIGITAL DESIGN FOR VOXEL BASED ADDITIVE MANUFACTURING AND PART CHARACTERISTICS-SOFTWARE TO PRINTER HARDWARE TRANSLATION CORRELATION AND EFFECTIVE SOFTWARE SOLUTIONS IN POLYJET PRINTING

Furkan I. Ulu, Ravi P. Tomar, Ram V. Mohan*
North Carolina A&T State University
Greensboro, North Carolina/USA

ABSTRACT

Additive Manufacturing (AM) as a processing technique is now enabling tailored material and structure optimization due to its capability for forming complex materials and geometries, material phases that can be at resolution levels of hundreds of microns. Fundamental to this is digital design that defines the associated material, complex geometries, and their translation in printer hardware during build. Voxel digital designs (similar to a pixel representing colors in a digital image) in PolyJet printing define and build composite parts with a wide range of mechanical properties, material phases and geometries. Voxel digital designs enable to define material at each voxel level. However, compatibility and correct translation between multi-material voxel digital design and 3D printing hardware is required. This paper investigates print quality of designed part configurations (complex shapes, associated materials for geometrical regions) defined via voxel digital designs to understand the correlation between voxel digital designs and AM hardware. Demonstrative AM part configurations with complex material and geometries (using representative geometrical shapes such as circular, gear, elliptical, diamond, hexagonal, etc., as test cases), were digitally designed using voxel printing. Success of additive manufacturing relies on such digital designs being translated correctly without any artifacts by printer hardware. To carefully understand the correlations, Vero White (Hard) material and Tango Black (Soft) material were used to define different materials regions of complex geometries, and built with a PolyJet AM system. Final printed part of each voxel digital design was analyzed. Results indicate that success of printing from our voxel digital designs depended on the hardware factors such as print resolution in different directions of 3D builds, which affected the final geometry and material configurations. Printing hardware factors can be overcome by redefining voxel digital design for printing purpose only. Careful understanding of digital design translation to printer hardware factors provide means to develop effective digital design software solutions. Details of our software solutions with voxel digital designs to obtain desired part quality and correct geometries during AM, without translational digital technology (DT) to hardware technology (HT) translational factors will be discussed and presented.

Keywords: Additive Manufacturing, Voxel-based printing, Digital Design
Corresponding author: rvmohan@ncat.edu

1. INTRODUCTION

Additive manufacturing (AM) is forming objects by addition of material layer by layer. AM allows to create high detailed complex shapes, part consolidation, and custom geometry that are difficult with conventional part production such as machining and injection molding[8-doubrovski]. AM has been used by designers for prototyping however the demand has increased rapidly since prices of AM devices are decreased, more effective-engineered materials can be created, and final printed part quality has been improved. These developments lead to increasing number of applications of AM to process end-used products [1]. Powder based binder jetting was the first AM method presented for fabrication of objects and scientific visualization [2]. This method enables the digital fabrication with desired colored and texture with high accuracy and resolution. In this approach, the color or material description was used as a parameter for encoding of information on an object's surface, and the digital design must be given as closed triangle mesh with associated texture or vertex attributes [3]. Therefore, general geometry description must be converted to boundary representations through a tessellation process, which may lead to partial loss of the design or alteration of data set. Later, selective laser sintering (SLS) was presented. SLS uses a laser as the power source to sinter powdered material layer by layer was introduced to form complex geometries as a solid object. This method has capability of printing complex geometries without support scaffolds, which enables forming intricate object. However, the processing approach of this method does not enable printing more than one material. PolyJet 3-D printers has capability of printing multiple materials with different mechanical properties. This method, also, enables printing complex hierarchical structures with high resolution (14 micron layer thickness). PolyJet 3D printing technology provides fabrication of 3D objects with heterogeneous compositions with capability of printing materials with different mechanical properties simultaneously [4]. The capability of creating structure with material property variations allows engineering the part composition that can be formed with specific functionality such as combination of hard and soft material can form complex objects with various mechanical properties [5–7].

Moreover, AM generally uses Stereolithography (STL) files that conduct tessellated data to define the digital design of geometry [8]. Tessellation of designed geometry may lead loss of spatial information since it is triangular discretization of each geometric surface. During AM process, each surface is considered a solid object that represents a single material since STL file has only surface information. Therefore, sub-models are required to be created and then assembled to print multi-material parts [8]. These limitations show the need of improvement 3D printing workflow.

In contrast to the methods described above another approach was presented to print multi-material parts through voxel printing [9]. Since PolyJet printing technology has capability of multiple materials simultaneously, voxel printing can be employed to print multi-material parts. Voxel is an extruded model of 2D pixel. Each voxel can represent specific material information as each pixel can represent different color code in a digital image. High level of spatial and material control in fabrication can be obtained by slicing and creating a set of layers with specific thickness in a bitmap raster file format and stacked them into a voxel matrix to build 3D object. Each pixel of raster file format defines the individual material information of a droplet and its spatial placement. However, compatibility and correct translation from multi-material voxel digital design to 3D printing hardware required. This paper focuses the print quality of designed part configurations defined via voxel designs to understand the correlation between voxel digital design. To investigate this correlation, printed part quality of designed part configurations such as complex shapes and defined materials for geometrical regions are inspected. Multi-material parts with circular shape inclusions, gear shape inclusions, elliptical shape inclusions, diamond shape inclusions, and hexagonal shape inclusions were digitally designed using voxel printing. Vero White (Hard) material and Tango Black (Soft) material are used as inclusion and outer region materials. Final printed part quality of each voxel digital design analyzed, and incompatibility is observed since hardware resolution was not meet with the digital design. This defect was eliminated by redefining the voxel digital design for voxel printing purpose but not for representing final printed part design.

2. EXPERIMENTATION

2.1 Experimental Set-up

Voxel is an extruded model of a pixel. Therefore, each voxel can represent specific color and material information with different mechanical properties as each pixel defines different color code in a 2D digital image. Combining and composing of voxels lead to various range of material properties to create complex materials. A workflow is presented and applied to print complex shaped multi-material parts in this research. Figure 1 represents the workflow of voxel printing used.

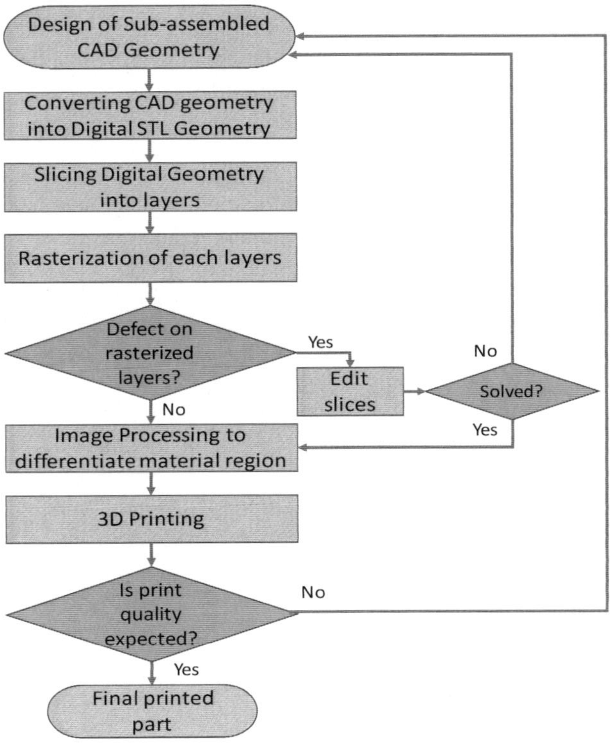

Figure 1. Flowchart of Voxel Digital Printing for Multi-Material Design

Using the flowchart above, the following design of experiments was prepared to inspect the correlation between voxel digital design and 3D printer with respect to print quality. Design of sub-assembled CAD geometry was prepared with representative geometrical shapes such as diamond, gear, horizontal and vertical elliptical, pentagonal, star shape to demonstrate AM part configurations with complex material and geometries. Vero White (VW) (Hard) material used in surrounding region and Tango Black Plus (TB+) (Soft) material used in inclusion region in representative sub-assemblies to define individual material regions. Inclusion shape assignment of this test part employed are designed as follows:

Case 1) Dimond inclusion, Case 2) Gear Inclusion, Case 3) Horizontal Elliptical inclusion, Case 4) Vertical Elliptical inclusion, Case 5) Pentagonal inclusion, Case 6) Star Shape inclusion.

All cases were designed as sub-assemblies and printed with voxel digital design printing. CAD model of the representative part configurations were designed using SOLIDWORKS® 2018. Desired test part dimensions are 17x17x3.5 mm in all x-, y-, and z- directions. All inclusions were designed to fit in a circle with 6 mm diameter as sub-modals and represented second material region. These designed test parts with multi-material regions for all cases are shown in figure 2.

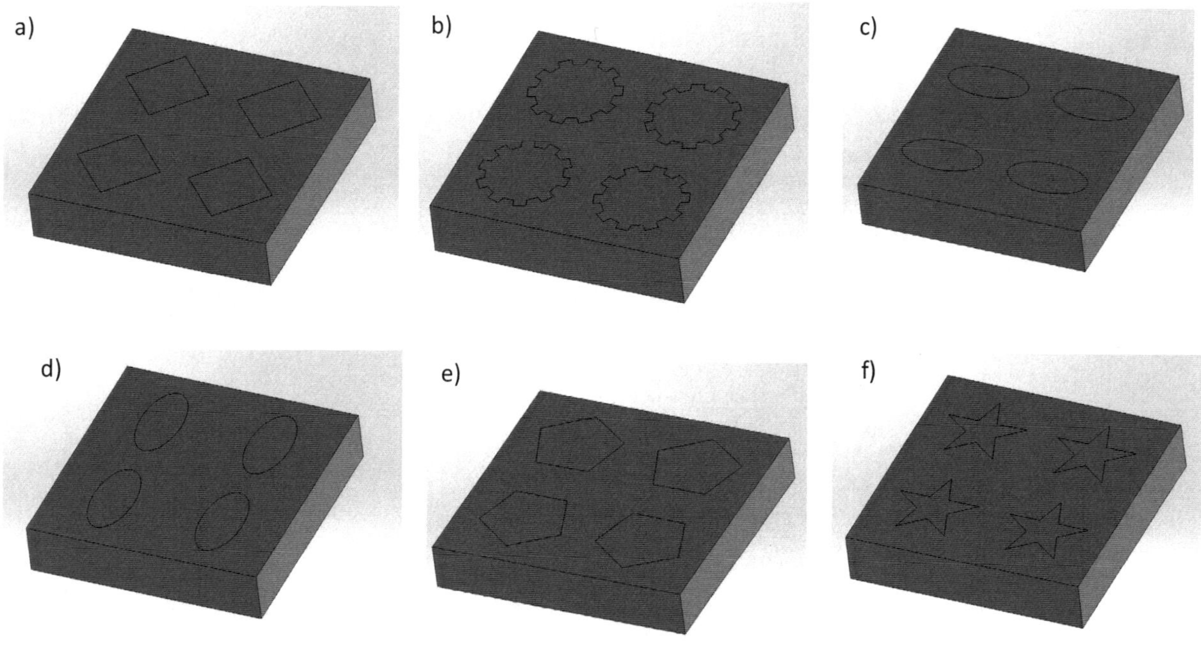

Figure 2. Sub-assemblies of multi-material test parts

CAD geometry of test part configurations are shown in figure 2. as diamond inclusion, gear inclusion, horizontal elliptical inclusion, vertical elliptical inclusion, pentagonal inclusion, and star shape inclusion, respectively. These assembled CAD geometries sliced and rasterized in Autodesk NETFABB software in z-direction. Each slice were exported as PNG file extension with 600 dpi resolution that is higher resolution of PolyJet system in x-direction. The resolution of created images for each slice was 402x402 pixels. Single layer thickness was 27 microns to print the parts with high mix mode of PolyJet 3D printer. Therefore, 130 slices were exported which equals 3.5 mm thickness of digital assembled CAD geometry. Then, image processing is applied to each PNG files that represents each layer of digital file to define different color information and separate material information in digital voxel printing file. Image processing is conducted by a MATLAB code that developed for material separation of voxel printing. Figure 3. represents the workflow of image processing of each slices.

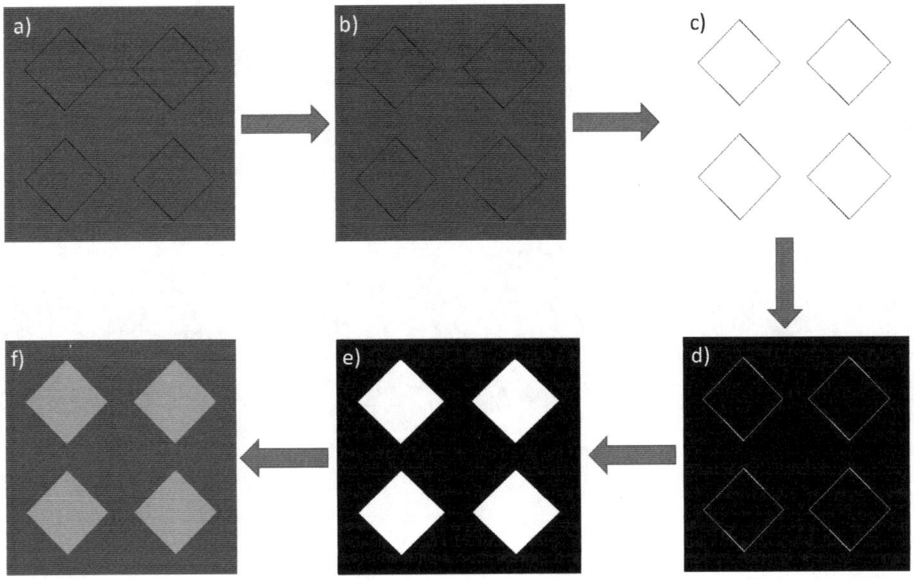

Figure 3. Image process workflow

There was only interface region information is captured during rasterization as shown in figure 3-a. Then, different RGB color code is defined at the regions that different material is associated. The effect of this new image processing for voxel printing is discussed with observation on printed parts in results section.

3. RESULTS

The final printed part from the digital designs was inspected. The printed part quality was not as expected. Dimensions of the printed part in x- and y- directions were not same although they were designed as same length. Moreover, sample with gear inclusion (case 2) showed different concentration of color for each inclusion geometry in same sample. Printed part configurations are shown in figure 4.

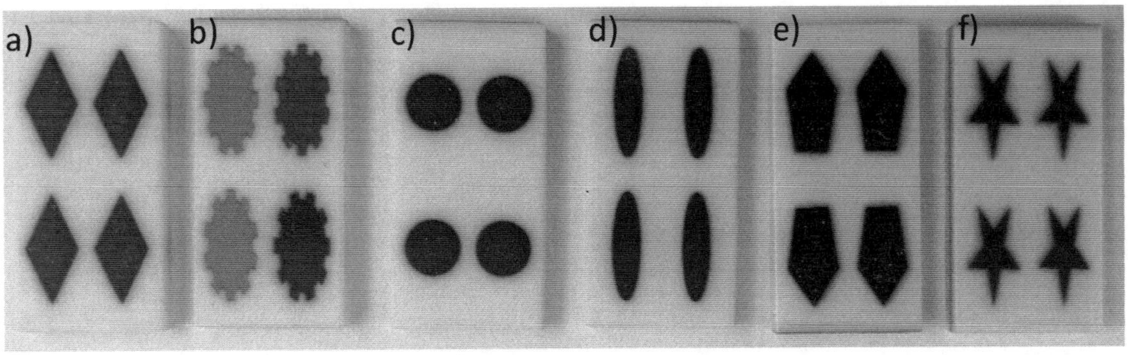

Figure 4. Printed parts from voxel printed parts

The reason of this variation in x and y direction is because of the difference of hardware 3D printer resolution in x and y direction. Since, PolyJet 3D printer have 300 dpi in y direction and 600 dpi in x direction for voxel printing and the 3D digital CAD file was rasterized in 600 dpi considering the STL file printing resolution of PolyJet 3D printer, these two times large variation in y direction were resulted. To eliminate this resolution error, CAD geometry of all test part were modified with reducing the dimension of every parameter to half in y direction. Moreover, some part designs were changed drastically to receive the desired final printed part. Modified CAD geometries of each test part configurations for AM voxel printing is shown in figure 5.

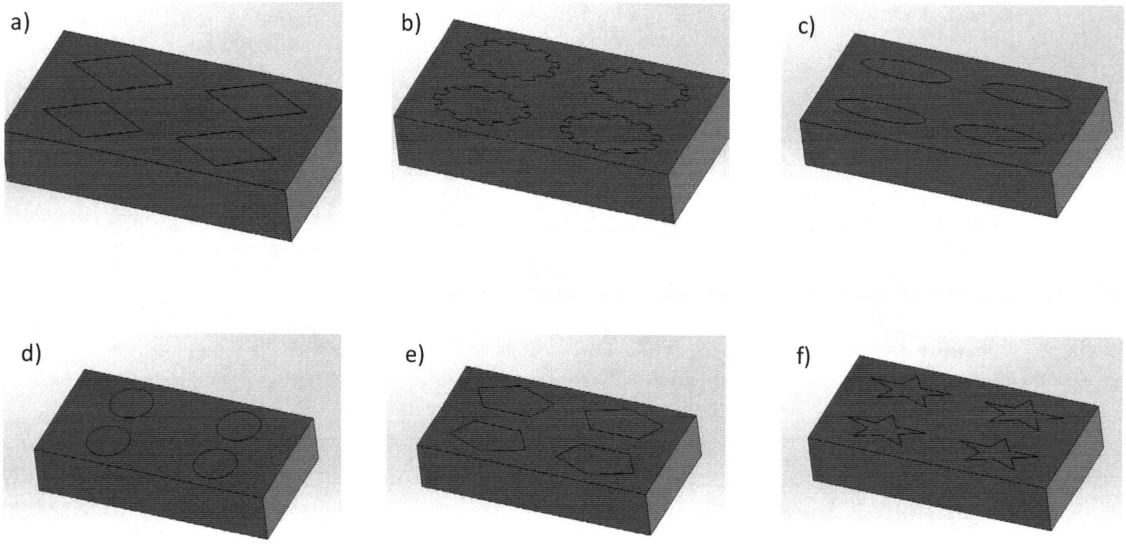

Figure 5. Modified CAD geometries of each test part configurations for AM Voxel printing

As shown in figure 4-b gear inclusion same defects and print quality was not as expected since the concentration of TB+ material was not identical for each inclusion in the final printed part. After analysis of each raster image slice, it is found that rasterization error is occurred in some inner slices. Since rasterization errors in inner slices it is hard to capture in digital imaging. Therefore, the developed MATLAB code was modified to eliminate these rasterization errors. Images of a slices with error and after modification with new MATLAB code were given in figure 6.

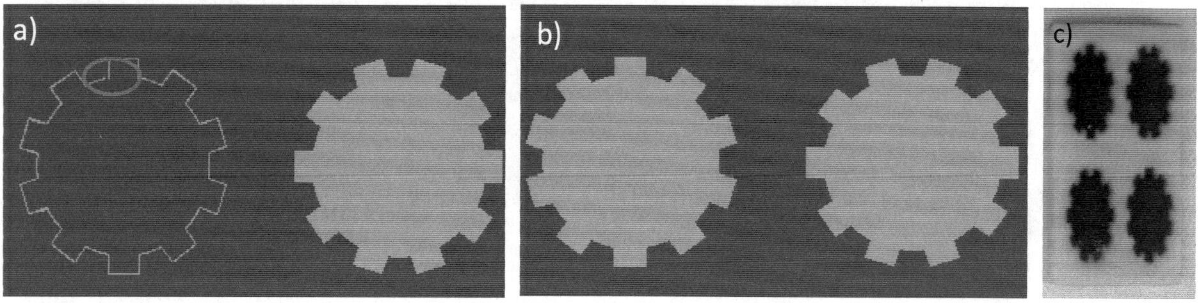

Figure 6-a. Slice with rasterization error, b. Error is eliminated, c. Printed part after error elimination

After CAD modification and MATLAB code modification of image processing for each slice, the printer parts were successful and print quality was as it was expected. Therefore, it is concluded that a digital software solution such as modification of CAD geometry design for only the purpose of voxel printing, but not for the part geometry is solution for voxel digital printing. Final printed part configurations are shown in figure 7.

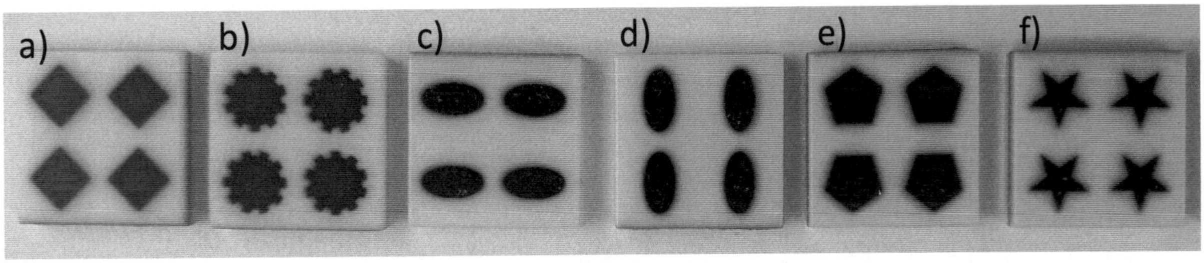

Figure 7. Final multi-material printed part configurations with expected part quality.

4. CONCLUSIONS

In this research the focus was to understand the correlation between the voxel digital design and the 3D printing hardware for multi-material part printing. To carefully understand the correlations, Vero White (Hard) material and Tango Black (Soft) material were used to define different materials regions of complex geometries, and built with a PolyJet AM system. Quality of each printed multi-material parts was analyzed. Initial printed part comes with the defect of resolution difference of hardware in x and y direction. A practical solution was proposed to solve this issue by reducing the dimensions in y direction to half. The digital rasterization error was captured and solved by modifying image processing code. Results show that successful printing with the voxel designs are highly depend on 3D printer hardware that effects final print quality of multi-material printed part. Printing hardware factors can be overcome by redefining voxel digital design for printing purpose only. Understanding of digital design translation to hardware is very crucial to improve the print quality by redefining the design and development of software.

5. REFERENCES

[1] T. Wohlers, Wohlers report 2012, Wohlers Associates, Inc, 2012.

[2] M.J. Bailey, Tele-manufacturing: rapid prototyping on the Internet, IEEE Computer Graphics and Applications. 15 (1995) 20–26.

[3] C. Bader, D. Kolb, J.C. Weaver, S. Sharma, A. Hosny, J. Costa, N. Oxman, Making data matter : Voxel printing for the digital fabrication of data across scales and domains, Science Advances. 4 (2018) eaas8652. doi:10.1126/sciadv.aas8652.

[4] F.I. Ulu, R. Mohan, R.P.S. Tomar, Development of Thermally Conductive Polymer/CNF Nanocomposite Materials via PolyJet Additive Manufacturing by Improvement of Digital Material Design, in: ASME 2019 International Mechanical Engineering Congress and Exposition International Mechanical Engineering Congress and Exposition, 2019. doi:10.1115/IMECE2019-11556.

[5] R. pratap singh Tomar, F.I. Ulu, R. V Mohan, A. Kelkar, Investigation of Process Variation Effects Via a Homogeneous 3-Dimensional Tensile Test Coupon in Polyjet 3D Additive Printing, in: SAMPE Technical Conference, 2019: p. p 11. doi:10.33599/nasampe/s.19.1546.

[6] R.P.S. Tomar, F.I. Ulu, A. Kelkar, R. V Mohan, Investigation of Process Induced Variations in PolyJet Printing With Digital Polypropylene via Homogeneous 3D Tensile Test Coupon, in: ASME 2019 International Mechanical Engineering Congress and Exposition International Mechanical Engineering Congress and Exposition, 2019. doi:10.1115/IMECE2019-11639.

[7] A.I. Shiave, R.P.S. Tomar, I.P. Espinosa, R. Mohan, Deformation Mechanisms and Dislocations in Nickel–Cobalt Core–Shell Nanowires Under Uniaxial Tensile Loading—A Molecular Dynamics Modeling Analysis, Advanced Science, Engineering and Medicine. 11 (2019) 1187–1201. doi:10.1166/asem.2019.2478.

[8] F.I. Ulu, S.P.R. Tomar, R. V Mohan, Investigation of Digital Computer Aided Design Assembly and Tessellation Effects on Digital ABS Part Quality in PolyJet-3D Additive Printing, in: SAMPE Technical Conference, 2018.

[9] E.L. Doubrovski, E.Y. Tsai, D. Dikovsky, J.M.P. Geraedts, H. Herr, N. Oxman, Voxel-based fabrication through material property mapping: A design method for bitmap printing, Computer-Aided Design. 60 (2015) 3–13. doi:10.1016/J.CAD.2014.05.010.

THE INFLUENCE OF PROCESSING PARAMETERS ON THE TRANSITION ZONE FOR BLENDED MATERIAL 3D PRINTING

James C. Brackett[1], Dakota Cauthen[1], Tyler C. Smith[1], Vlastimil Kunc[2], Chad Duty[1,2]

[1]University of Tennessee – Knoxville
1512 Middle Drive
Knoxville TN 37916

[2]Manufacturing Demonstration Facility
Oak Ridge National Laboratory
Oak Ridge, TN

ABSTRACT

The use of Multiple Materials (MM) in Additive Manufacturing (AM) is increasingly important for expanding the range of applications in the manufacturing industry, particularly for large-format processes. Typically, polymer-based AM incorporates MM transitions through discrete interfaces between layers. This arrangement significantly increases the occurrence of layer delamination failures due to decreased bonding between dissimilar polymers. Elimination of discrete material interfaces by continuously transitioning from Material A to B provides a possible solution. Such continuous gradients could be used to create functionally graded structures that take full advantage of AM's capability to deliberately impart site-specific properties. Cincinnati's Big Area Additive Manufacturing (BAAM) system at Oak Ridge National Lab has been equipped with a dual-hopper system that enables in-situ material switching specifically intended for functionally graded and MM printing. The resulting material transition exhibits varied behavior based on printing conditions, which can have an impact on part design and resulting mechanical properties. In this work, the transition zone is characterized as a function of the printing screw speed (related to volumetric flow) and the screw geometry.

Corresponding author: James Brackett, jbracke4@vols.utk.edu

1. INTRODUCTION

The recent rise of numerous Additive Manufacturing (AM) techniques has enabled precise production of complex geometries with minimal waste. Specialization of AM systems has made processing of polymers, metals, and ceramics possible. Depending on the system in use, material feedstock can appear in a variety of forms, from solid wires to fine powders.

1.1 Multiple Material Additive Manufacturing

A recent area of focus has been the implementation of Multiple Materials (MM) in a single AM system. Multiple Material Additive Manufacturing (MMAM) systems use more than one material, not including pre-mixing or pre-compositing, and does not rely on non-AM post-processing treatments to complete the product [1]. Exploring the full potential of MMAM systems is crucial to continuing the expansion of AM in general. For example, appropriate design considerations can be applied to a MMAM system to create Functionally Graded Materials (FGM). As defined by Loh et. al., an FGM seeks to fulfil a functional purpose through a defined variation of density (single material) or composition (MM) in one or more directions [2]. An FGM appears in either a stepwise or continuous form. The stepwise sub-category is characterized by multiple regions of differing composition or microstructure that are separated by discrete boundaries and arranged to form an overarching gradient. Alternatively, a continuous FGM forgoes discrete interfaces for a controlled, smooth gradient in microstructure or composition that fulfills a functional requirement [3].

1.2 Multi-Material Additive Manufacturing

1.2.1 Multi-Material Structures

MMAM systems have been used to create a variety of structures, but a common theme is the struggle to bond dissimilar materials in-situ. Even a small difference, such as the presence of matrix reinforcement or an additive to induce color change, between two otherwise identical polymer matrices can cause issues. Roger and Krawczak investigated the bonding of a carbon black-filled Acrylonitrile Butadiene Styrene (ABS) with virgin ABS in three different printing orientations using a multi-nozzle Fused Deposition Modelling (FDM) printer. Although utilizing an interpenetrating pattern to prevent the interface from occupying a single plane improved tensile strength compared to a typical wall-like interface, it still failed to achieve a tensile strength comparable to the weakest constituent [4]. Kim et. al. conducted a similar study on material interfaces using ABS and Poly-Lactic Acid (PLA) with a dual-nozzle FDM system. It focused on the number of interfaces, their location, and the volume percent of each material. Increasing PLA content by increasing the size of its designated region provided no significant improvement, indicating that the interface between the two dissimilar materials were hampering the print's tensile strength. Fractures along interfaces and voids observed prior to testing supported this conclusion. Fixing the ABS:PLA ratio at 1:1, using multiple interfaces in the build and z-directions highlighted the importance of structural arrangement and showed that boundaries in the z-direction significantly improved tensile strength while the number of interfaces had a minimal effect. Additionally, the interfaces parallel to build direction consistently resulted in voids vulnerable to delamination [5]. A similar MM study that also used a dual-nozzle FDM printer took a different approach. With a specialized nozzle that allowed deposition of material into voids left in previous layers, the "z-pinning" method improved interfacial strength of a PLA-Nylon MM print in tensile testing. Furthermore, the failure mode shifted from delamination to failure of the mechanical interlocking created by extruding the PLA "pins" into the voids left in the previously deposited layers of Nylon [6].

Another study using a multi-nozzle Fused Filament Fabrication (FFF) printer examined an ABS/PLA face sheet/core sandwich structure. Comparing the three-point bend performance of single-material PLA sandwiches to the dual material prints initially exhibited a significantly lower modulus due to delamination. However, printing a PLA sandwich with the two separate nozzles needed for dual-material FFF printing resulted in comparable mechanical properties to the ABS/PLA sandwich. For MMAM, this development implicates excessive cooling of previous layers may also play a role in interface vulnerability in addition to bonding of dissimilar materials [7]. Vu et. al. also investigated interface bonding in dual-material sandwich panels using an inkjet printing system. T-Peel testing showed fractures favored the material interfaces, but fracture resistance improved when printing the material interfaces perpendicular to build direction, similar to Kim et al. [8]. Additionally, Vu et. al. also observed greater fracture toughness in specimens printed with a stepwise FGM core compared to its single-material counterpart [9]. While there is an overlap between MM and FGM structures, a distinction and separate discussion of the effects of FGM design is worthwhile.

1.2.2 Stepwise Functionally Graded Materials

Other studies have been focused entirely on using stepwise FGM design in conjunction with MMAM to improve performance in MM structures. In one study, a stepwise FGM structure was created using nine different materials with elastic moduli varying from 1 MPa to 1 GPa to achieve a mix of flexibility and rigidity. Applying this technique to a custom robot shell resulted in the increased survivability observed in a flexible version while still maintaining its shape and function like the rigid version [10]. In other MM systems, software issues in placing more than one material in the same layer can limit their functionality. This issue is commonly found in boundary-representation approaches, so an investigation into printing FGM structures sought to develop a volumetric method. The resulting software successfully demonstrated MM placement suitable for stepwise FGM's by creating regions within layers of both controlled porosity and different materials [11]. However, FGM production has not been limited to small-scale systems. For instance, Cincinnati's Big Area Additive Manufacturing (BAAM) system can use its screw-based extrusion to blend pelletized feedstock, allowing for customizable control of stepwise gradients across a part. Sudbury et al. explored this potential through printing of modified sandwich structures. The study utilized a carbon-fiber filled ABS and unfilled ABS to explore the weight and cost savings of a stepwise FGM compared to a single-material print. The gradients successfully reduced both cost and weight of the final structure, highlighting some of the possible benefits of incorporating FGM into AM [12]. However, this approach limited the possible gradients to the z-direction and requires a single composition for each layer.

1.3 Exploration of Continuous Functionally Graded Materials

Addressing BAAM's initial limitations is necessary to fully capitalize on its MMAM potential. The extrusion-based process benefits from both its feedstock type (pellets) and extrusion mechanism (screw). The pelletized feedstock can be easily mixed and does not suffer from any continuity issues that would require a stoppage of the printing process. Similarly, screw-based extrusion ensures that the feedstock mixes inside the barrel to form a singular bead, again allowing for continuation of a print regardless of feedstock changes. A new configuration was developed to harness these unique advantages and applied to the BAAM-B system at Oak Ridge National Lab (ORNL).

1.3.1 A Dual-Hopper System for Multi-Material Additive Manufacturing

An innovative dual-hopper system was constructed and attached to BAAM-B's print-head to enable MM feeding and in-situ material changes. Shown in Figure 1, each hopper connects to a different feedstock source. Only one hopper feeds into the barrel at a time, but a simple manual or coded command will switch them, changing material on-the-fly without interrupting the print in any way. The switch is completed by a simple rocker mechanism that forces the currently feeding hopper closed while pushing the secondary hopper into position over the barrel. The BAAM-B's redesigned material feeding system allows for printing of MM, stepwise FGM, and continuous FGM structures.

Figure 1. Image of the dual-hopper attachment and feeding tubes for BAAM-B.

1.3.2 Continuous Transitions

With the dual-hopper system, the BAAM inherently creates a blended region, termed the transition zone. shortly after the switch caused by mixing of the pellets both above the barrel and in the melt. Thus, it can incorporate a continuous material transition in its prints by strategically placing hopper switches at pre-specified points. However, this requires knowing the characteristics of the material transition induced by the dual-hopper and extrusion screw. Describing the transition zone mathematically with a "shape function" would aid in defining design constraints and predicting mechanical properties around the zone. For this, the initial step is constructing a "shape curve" of the transition process. In general, this requires tracking the material composition, i.e. % Material A, over an arbitrary distance until the composition is 100 % Material B. Figure 2 shows a few possibilities. An "ideal" scenario could be represented by a weighted mean or rule of mixtures as is commonly done when predicting the material properties of simple mixtures [13]. However, this approach expects an even, consistent mixture. The abrupt change introduced by a hopper switch, in conjunction with screw-driven mixing, will likely result in a non-linear transition zone., which is represented by the dotted-lines in Figure 2. Considering the variety of factors present in AM and specifically on the BAAM, multiple factors could influence the final shape curve's characteristics. Each observed shape curve is expected to be unique to the material pair in question just as each composite material exhibits a different range of properties even if only one constituent is altered. For example, a material's rheological properties and reinforcement type and quantity have a large influence on how it will mix, making the three variables a likely source of change in the shape curve. From a system contributions perspective, the BAAM is compatible with several different screw geometries, all of which are operated at varying rotational speeds. Additionally, the total mass output (as determined by nozzle size and coded bead dimensions) will influence the observed distance or time to complete the transition zone. Each of these material and instrument parameters could lead to a change in the transition zone's length or shape.

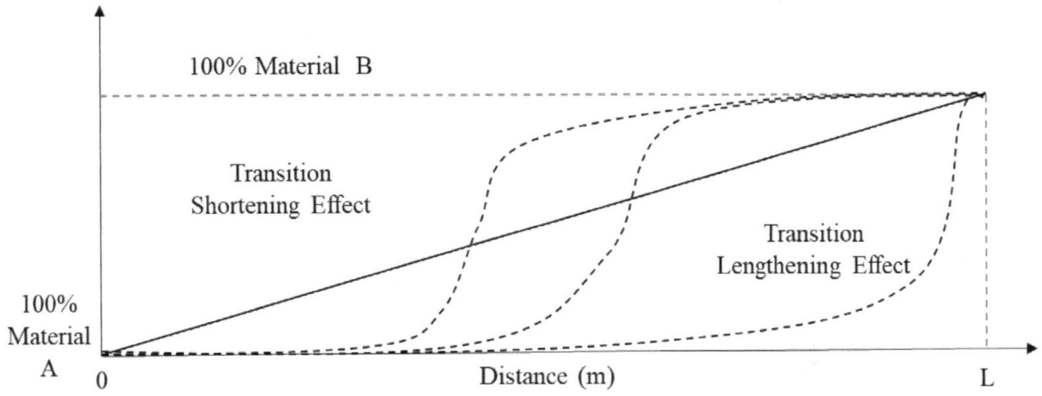

Figure 2. An illustration of a few possible transition behaviors. The solid line indicates a linear transition from Material A to Material B, while the dotted lines demonstrate some of the possibilities for the more likely non-linear behavior.

1.3.3 Continuous Transition Examples

A few previous studies have demonstrated the BAAM's capability for continuous transitions. The first examined a glass fiber-filled ABS to carbon fiber-filled ABS using manual feeding at 100 Rotations Per Minute (RPM). Obtained using Thermogravimetric Analysis, the fiber content at selected locations was plotted to develop an initial shape function, which indicated the system required a travel distance of 1.5 m (15 ft) to achieve steady-state conditions [14]. In a follow-up study, Sudbury et. al. examined the effect of screw speed on the transition time from a neat ABS to carbon fiber-filled ABS. With transition times of 52 s at 100 RPM and 5.85 s at 400 RPM, increasing the screw's rotational speed by 400% disproportionately reduced the transition time by almost 90% [15]. This indicates that the screw speed may influence the travel distance required for a material transition to occur. Another study examined the in-situ capability and process repeatability of the dual-hopper modification. It found a consistent and repeatable transition behavior when printing the same object multiple times and provided evidence that the transition zone maintains a consistent trend toward the secondary material with no deviation from surrounding areas. Furthermore, closer inspection revealed that the direction of transition also functions as an influencing parameter. In this case, transitioning from ABS to a carbon fiber-filled ABS completed earlier than its reverse in addition to having a difference in initiation and transition zone lengths [16]. Clearly, the BAAM is more than capable of producing continuous FGM structures, but the finer aspects to transition behavior still need to be characterized.

1.4 Importance of BAAM System Parameters

The previous work has focused on understanding the general process of the material transitions produced when using the modified BAAM's dual-hopper system. To further understand the contributing factors, an analysis of the system's possible affects is necessary. RPM is commonly changed from one print to another, so understanding its contribution will inform any future prints and guide print design. Similarly, the extrusion screw's geometry could alter the transition behavior and prove to be a valuable tool in controlling the shape function. As such, the effect of each is critical for accurately implementing site-specific properties via MM and FMG structures in large-scale MMAM with the BAAM. While a change in transition time was observed in [15] at different RPM's, a complete picture obtained by tracking both distance and time allows for a direct comparison to the previous work and a discussion on the print space required for each. The influence of the screw's geometry is examined by comparing transitions produced by a "non-mixing" and "mixing" screw. Upon completion, this work will be able to predict the effects of altering RPM and screw geometry.

2. EXPERIMENTATION

2.1 Sample Production

The pelletized feedstocks supplied by Techmer Engineered Solutions were a neat ABS (HIFILL ABS 1512 3DP) and a 20 wt % carbon fiber-filled ABS (ELECTRAFIL ABS 1501 3DP). Prior to printing with the dual-hopper BAAM, each material was dried at 80 °C for at least four hours. A single, continuous bead 5.97 m (235 in) long was printed as shown in Figure 3. Each print used a nozzle of 10.16 mm (0.4 in), a bead height of 5.08 mm (0.2 in), and a bead width of 13.97 mm (0.55 in). The rotational speed of the screw was set to either 100 or 300 RPM, and a print was completed at each speed, once with a "mixing" and once with a "non-mixing" screw geometry. The travel speeds were 4.3 cm/s (1.7 in/s) and 10.7 cm/s (4.2 in/s) for the 100 RPM and 300 RPM prints, respectively. Thermal conditions were maintained at a 100 °C bed temperature and 250 °C at the melt. With these conditions, the impact of both rotational speed and differing screw geometries on the total length and general shape of the transition curve. After printing, the sample beads were sectioned with a bandsaw and labeled for easy tracking of distance and time from the point the hopper switch occurred. As shown by [16], an 8.4 mm sample was consistently representative of the surrounding 25.4 mm. Maintaining this approach, an Isomet 1000 Precision Saw extracted single 8.4 mm samples from each location of interest for characterization.

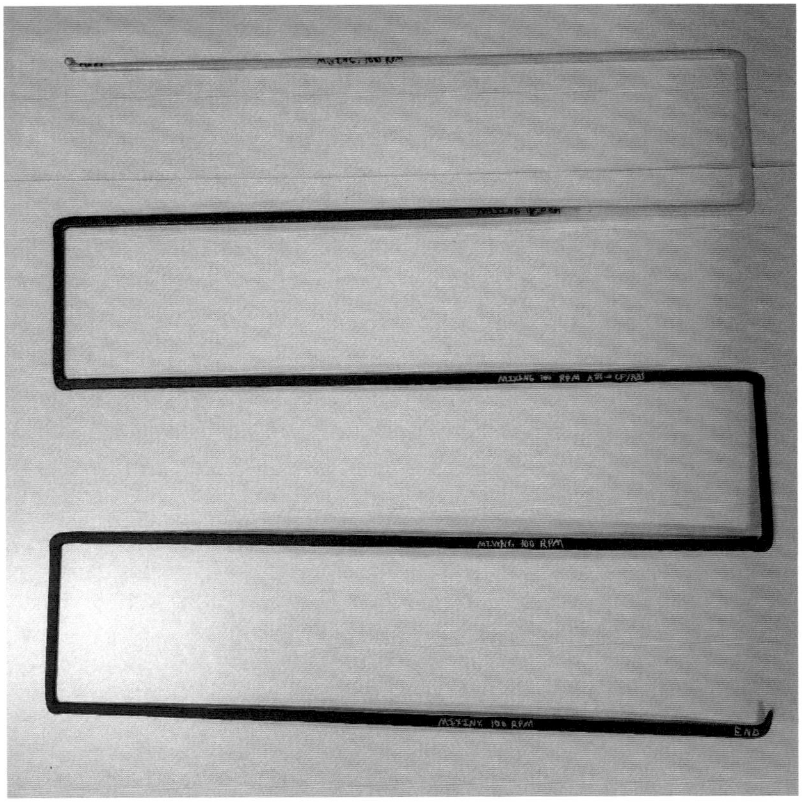

Figure 3. A 100 RPM sample printed with BAAM-B's dual-hopper.

2.2 Characterization Method

Charting the transition from neat to carbon fiber-filled ABS requires a means of tracking material composition at any given point in the bead. For this material pair, carbon fiber content by weight can be used as a reliable indicator. To obtain a standard residual value for each material, three batches of both ABS and CF/ABS as-provided pellets were submitted to characterization. The average resulting fiber content of the CF/ABS pellets will be used as a baseline. All mass measurements were conducted with a RADWAG AS 220.R2 analytical balance with a 0.1 mg accuracy.

ASTM D3171-15 provides several procedures for constituent content analysis, including some that utilize an acid to attack the matrix and hydrogen peroxide to accelerate the reaction [17]. Using this approach as inspiration, a custom method for digesting the polymer matrix was developed. Ultrasonic Assisted Acid Digestion (UAAD) replaces hydrogen peroxide with a sonicator bath that enables digestion of the ABS matrix with simple acetone. Prior to UAAD, samples are dried at 80 °C and their dry mass recorded. A Lab Safety Supply glass microfiber filter with 47 mm diameter and 1.5 μm pores, also dried and weighed, is assigned to each sample. For testing, specimens are placed in a borosilicate glass tube or vial and submerged in HPLC Acetone of $\geq$ 99.5 % purity. The vials are placed in the sonicator bath for thirty minutes. Upon complete dissolution of the matrix, any remaining fibers are separated from the solution using the previously mentioned filters in a vacuum-assisted filtration set-up. The filter is extracted from the apparatus and then dried at 80 °C overnight before final mass measurements.

The dry mass of the fibers is compared to the original dry mass of the sample to calculate the fiber content by weight, but it also borrows from ASTM D3171-15 by including a ratio that compensates for any polymer left behind. Originally intended as a "carbonization ratio" for procedures using combustion, this study will use an average Dissolution Ratio, *DR*, obtained from the neat ABS control samples to adjust for the typical amount of polymer captured in the filtering process. In Equation [1], m_{cr} is the measured mass of the pan and filter after UAAD, m_c is the measured mass of pan and filter prior to UAAD, and m_i is the dry mass of the specimen. Equation [2] shows how the *DR* is used in calculating the fiber content. Again, M_i is the mass of the specimen, M_{cr} is the mass of the pan and filter after UAAD, and M_c is the mass of the pan and filter prior to UAAD. Since this process is originally intended for analysis of polymer composites, measurements of neat or near-neat samples have the potential to experience greater degradation. Due to this slight variability, non-reinforced samples could return a negative fiber content up to -1.0 wt %.

$$DR = \frac{m_{cr} - m_c}{m_i} \quad [1]$$

$$\text{Fiber }\% = \left(\frac{M_i - \left(\frac{M_i - (M_{cr} - M_c)}{1 - DR} \right)}{M_i} \right) \times 100\ \% \quad [2]$$

2.3 Shape-curve construction

To examine the differences in transition behavior, the fiber content measurements for each sample set are plotted as a function of distance traveled from when the hopper switch was initiated. Each transition curve will begin with a Purge Zone, proceed to a Transition Zone, and eventually reach a Steady-State Zone. Assessing any variance between the curves in each zone's initiation, length, and termination will inform future work and provide design tools for controlling material transitions in full-scale FGM and MM printing.

3. RESULTS

3.1 Discussion of the Control Samples

As mentioned, unprocessed pellets were tested using UAAD to provide the necessary DR and a baseline for fiber content in the provided CF/ABS. The average *DR* from three batches of ABS pellets was 0.0201 and was used to calculate fiber content for all other samples. The carbon fiber-filled ABS pellets returned an average fiber content of 18.8 wt %, which is lower than the expected 20 wt %. Therefore, saturation will be considered as being at or above 19.0 wt % carbon fiber.

3.2 Parameter Influences

Utilizing UAAD, the fiber content of samples from four different parameter sets was determined. Plotting the fiber content of samples as a function of the distance from where the hopper switch was initiated provides an illustration of the transition behavior. Figure 4 demonstrates this process with the specimens from the 100 RPM printed with the mixing screw geometry. This one was chosen due to its limited variance and enables a clear designation of three different zones, as seen in Figure 5. As shown, a print running at 100 RPM with the mixing screw has a purge zone of about 1.51 m, a transition zone of 1.72 m, and thus takes 3.23 m to reach a steady-state CF/ABS composition.

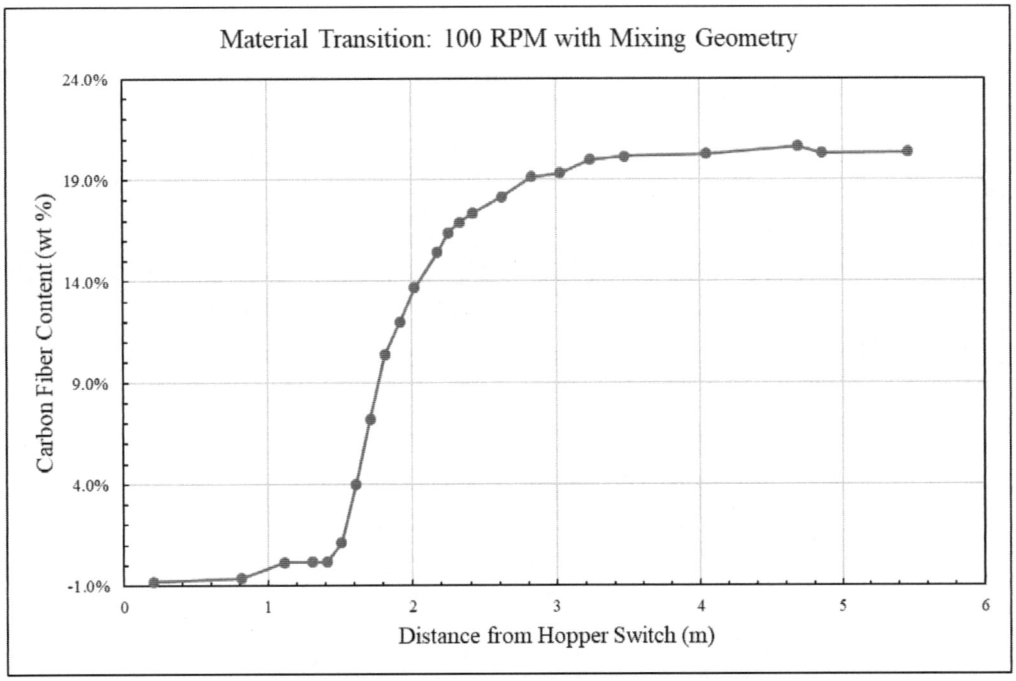

Figure 4. The shape curve for an ABS to 20 wt % CF/ABS conducted at a screw speed of 100 RPM with the mixing screw geometry.

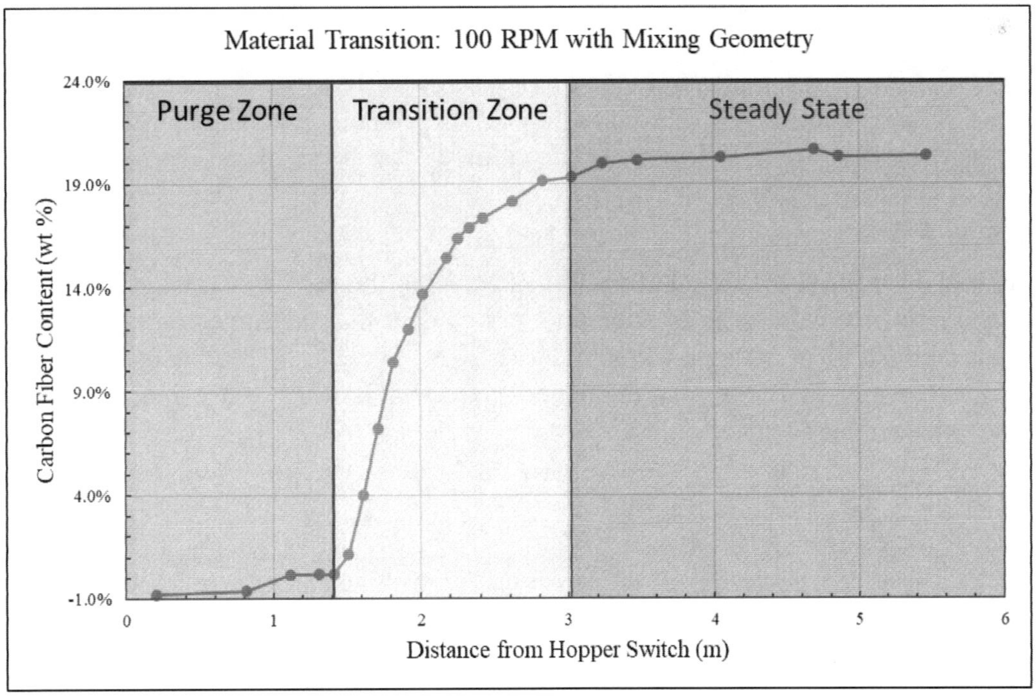

Figure 5. Definition of the three regions observed during a dual-hopper material transition.

While an in-depth microscopic analysis has not yet been completed, some preliminary images are of interest. As seen in Figure 6A and 6B, the neat ABS and CF/ABS have easily distinguishable structures. These two appearances are maintained throughout the Purge and Steady-State Zones, respectively. The Transition Zone example shown in 6C has a similar structure to the CF/ABS sample, despite only having a fiber weighting of 11.3 wt %.

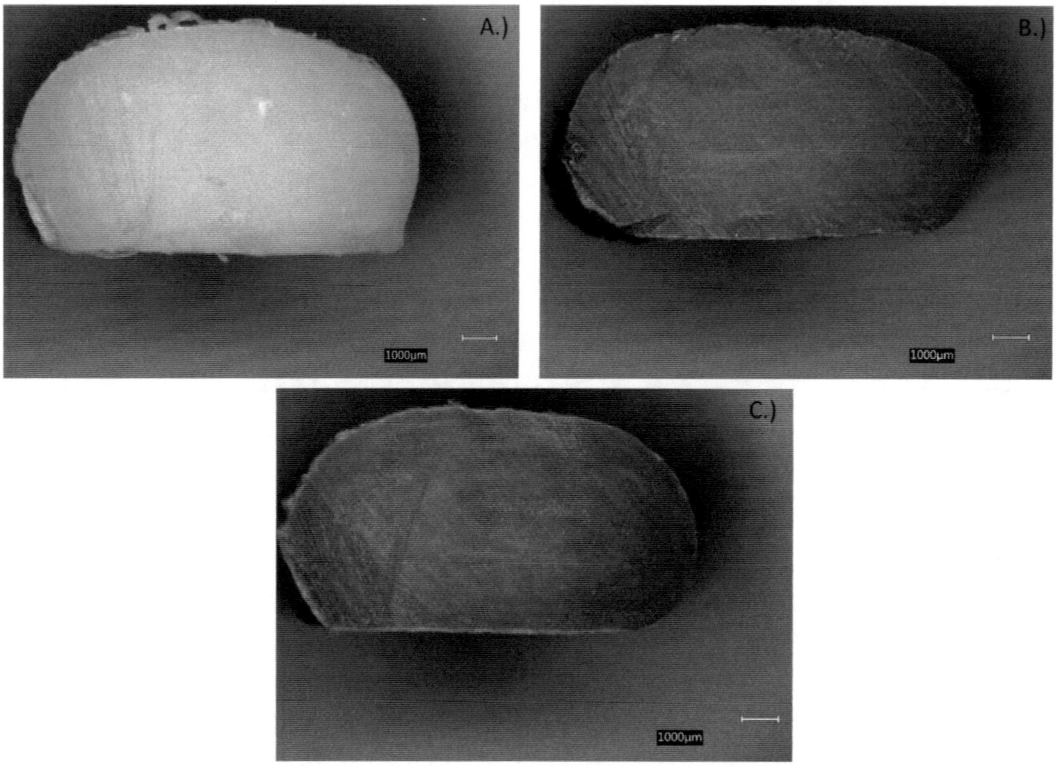

Figure 6. Microscopy of samples from the mixing screw at 300 RPM, as seen in Figure 7. A.) Purge Zone, 8th data point, b.) Steady-State Zone, 21st data point, and C.) Transition Zone, 11th Data point.

3.2.1 Effect of RPM

Figure 6 compares a speed of 100 RPM to 300 RPM using the mixing screw geometry. Treating the initial spike in fiber content as an anomaly, it is evident that the transition lags behind at the higher RPM, taking 2.02 m to begin compared to 1.51 m. The 300 RPM transition does exhibit a shorter transition region at 1.46 m, but the overall transition is still slightly longer at 3.48 m as a result of the extended purge zone.

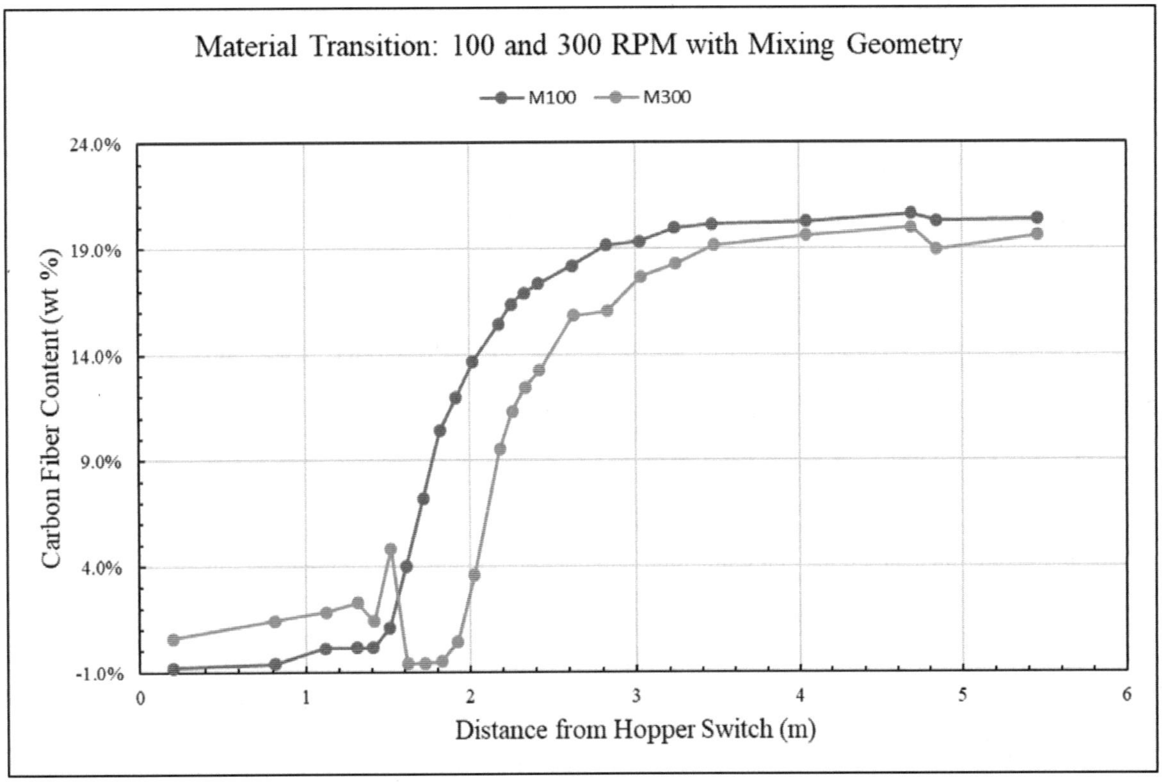

Figure 6. A comparison of the shape curves at 100 and 300 RPM. Both transitions used the mixing screw geometry.

Similarly, Figure 7 shows the same lag in transition behavior at a higher RPM, despite the change in screw geometry. Interestingly, a blip in fiber content is seen again early on in the 300 RPM transition, possibly indicating that the higher rotational speed has premature evidence of the material transition. The 100 RPM transition has a purge zone of 1.61 m compared to 1.81 m at 300 RPM, not as significant a difference as seen with the mixing screw geometry. Transition zone lengths were 1.4 m and 1.67 m. Interestingly, the distance required to reach steady state for 300 RPM was once again 3.48 m while 100 RPM was shorter at 3.01 m. Though the transitions for the non-mixing screw geometry closely resembled and tracked each other, they exhibited more variance, which caused the deviation in zone lengths. This is most evident just after 3 m where 300 RPM has a drop in fiber content. In general, the screw RPM could be valuable tool in tweaking when the transition occurs in a print and in controlling its length.

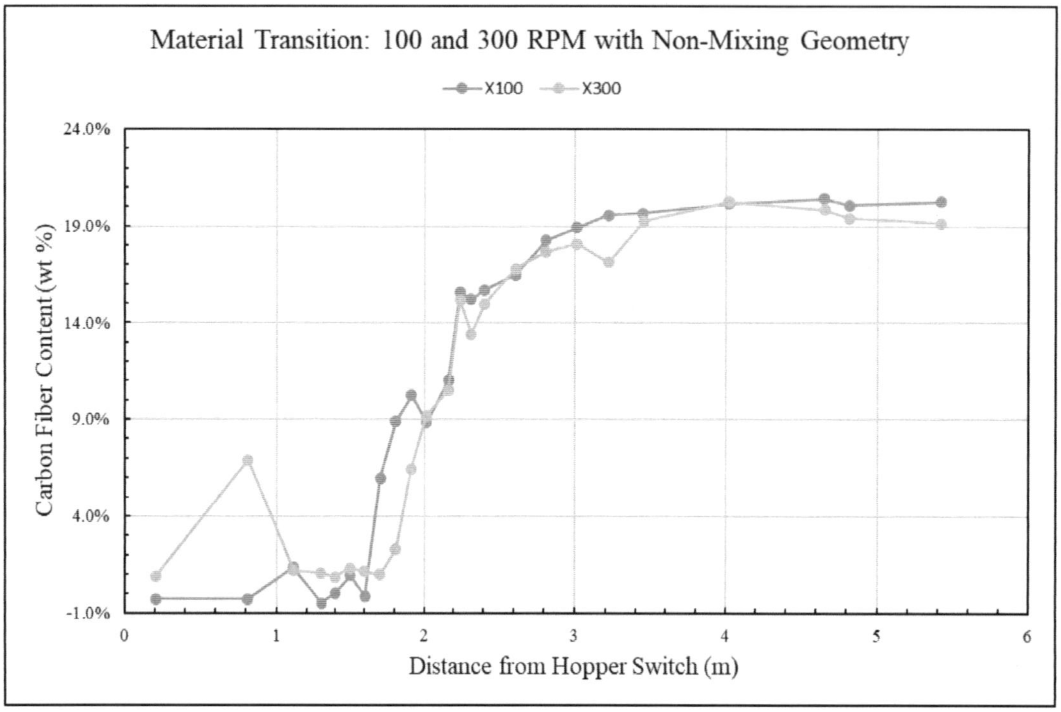

Figure 7. The shape curves obtained from specimens printed using the non-mixing screw at 100 and 300 RPM.

3.2.2 Effect of Screw Geometry

Combining Figures 6 and 7 provides a better opportunity to analyze screw geometry's influence on the transition curves. As Figure 8 demonstrates, the effect is largely negligible. The 100 RPM transitions are nearly identical, with the transition zone beginning just 0.1 m later for the non-mixing screw. A possible candidate for further investigation is screw geometry at higher rotational speed. Although a difference in initiation point is observed at 300 RPM, it is again small at 0.2 m the transition completes at the same distance. The greatest effect of screw geometry appears to be in the consistency of the transition, with a larger variance appearing in the non-mixing geometry curves compared to the mixing geometry curves. Regardless, choice of screw geometry appears to have little to no impact on material transitions, allowing design considerations to focus on which screw geometry best suits the materials being processed.

As mentioned, there are a few notable outliers in in the Purge Zone and noticeable variation in the non-mixing curves. Initial UAAD testing exhibited a standard deviation of only 0.08 wt %, and the CF/ABS pellets had a deviation of 0.29 wt %. Considering similar processes have reported testing error to be $\leq \pm$ 1 wt % [18], applying the same rule is prudent for UAAD. However, the anomalies observed are greater than the accepted error. Thus, their source could be attributed to sample variation or operator error, or it may indicate that the measurement process is still in need of some refinement. Either way, further analysis will be needed to accurately address the issue. Regardless, the data still supplies easily observable trends useful for characterizing the transition behavior.

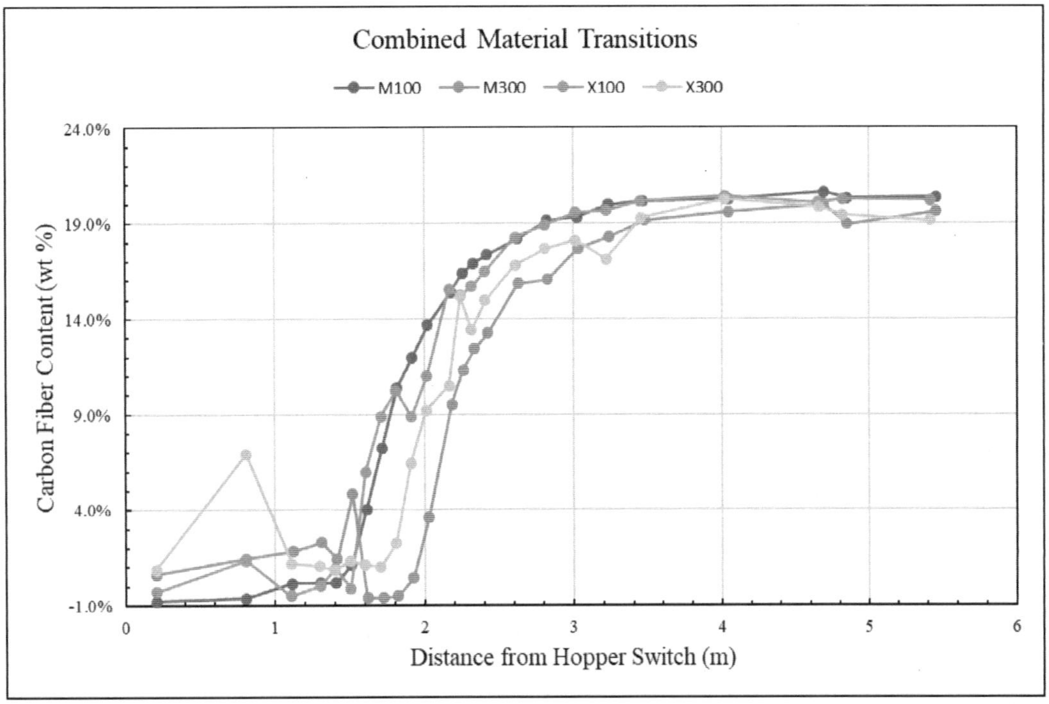

Figure 8. A combined plot of all four characterized transitions. All four transitions exhibit a similar pattern, with the slight alterations caused by parameters visible when side-by-side.

4. CONCLUSIONS

UAAD successfully characterized the fiber content by wt % of a multitude of samples. From this data, four distinct shape curves for an ABS to CF/ABS transition were constructed and compared. The rotational speed of the screw in a BAAM system slightly increases the distance required to reach steady state CF/ABS extrusion. This was true regardless of the screw geometry. Furthermore, the presence of a static mixer in a screw's geometrical profile had a negligible effect, as evidenced by the resulting shape curves of different screw geometries at the same RPM generally overlapping.

These effects further inform design for MMAM suing BAAM-B's dual-hopper system, especially when printing FGM structures. Given that the screw geometry had no observable influence on the shape curve, a screw's geometry can be chosen solely based on the characteristics needed for the materials involved. The screw RPM, however, had a noticeable effect, with a higher RPM delaying completion of the transition. Thus, screw RPM can be used as a tool to tweak the transition profile for a material pair to better suit the design's requirements.

5. ACKNOWLEDGEMENTS

Research sponsored by the U.S. Department of Energy, Office of Energy Efficiency and Renewable Energy, Industrial Technologies Program, under contract DE-AC05-00OR22725 with UT-Battelle, LLC. Thanks to Techmer Engineered Solutions for providing material for this study. Additional thanks to Cincinnati Incorporated for providing the BAAM printing system.

6. REFERENCES

[1] M. Vaezi, S. Chianrabutra, B. Mellor, and S. Yang, "Multiple material additive manufacturing – Part 1: a review," *Virtual & Physical Prototyping,* Article vol. 8, no. 1, pp. 19-50, 2013, DOI: 10.1080/17452759.2013.778175.

[2] G. H. Loh, E. Pei, D. Harrison, and M. D. Monzón, "An overview of functionally graded additive manufacturing," *Additive Manufacturing,* vol. 23, pp. 34-44, 2018/10/01/ 2018, DOI: https://DOI.org/10.1016/j.addma.2018.06.023.

[3] G. Udupa, S. S. Rao, and K. V. Gangadharan, "Functionally graded Composite materials: An overview," in *International Conference on Advances in Manufacturing and Materials Engineering*, vol. 5, S. Narendranath, M. R. Ramesh, D. Chakradhar, M. Doddamani, and S. Bontha Eds., (Procedia Materials Science. Amsterdam: Elsevier Science Bv, 2014, pp. 1291-1299.

[4] F. Roger and P. Krawczak, *3D-printing of thermoplastic structures by FDM using heterogeneous infill and multi-materials: An integrated design-advanced manufacturing approach for factories of the future*. 2015.

[5] H. Kim, E. Park, S. Kim, B. Park, N. Kim, and S. Lee, "Experimental Study on Mechanical Properties of Single- and Dual-material 3D Printed Products," *Procedia Manufacturing,* vol. 10, pp. 887-897, 2017/01/01/ 2017, DOI: https://DOI.org/10.1016/j.promfg.2017.07.076.

[6] C. Duty, J. Condon, T. Smith, A. Lambert, S. Kim, and V. Kunc, "Improving the 3D Printed Bond Strength at a Discrete Interface Between Dissimilar Materials," in *Society for the Advancement of Material and Process Engineering 2020*, Seattle, WA, May 4-7 2020, 2020.

[7] S. Brischetto, C. Ferro, R. Torre, and P. Maggiore, "3D FDM production and mechanical behavior of polymeric sandwich specimens embedding classical and honeycomb cores," *Curved and Layered Structures,* vol. 5, pp. 80-94, 04/01 2018, DOI: 10.1515/cls-2018-0007.

[8] I. Vu, L. Bass, N. Meisel, B. Orler, C. B. Williams, and D. A. Dillard, "Characterization of Mutli-Material Interfaces in PolyJet Additive Manufacturing," *Solid Freeform Fabrication Symposium Proceedings,* Conference Proceeding pp. 959-982, 2015.

[9] I. Q. Vu, L. B. Bass, C. B. Williams, and D. A. Dillard, "Characterizing the effect of print orientation on interface integrity of multi-material jetting additive manufacturing," *Additive Manufacturing,* vol. 22, pp. 447-461, 2018/08/01/ 2018, DOI: https://DOI.org/10.1016/j.addma.2018.05.036.

[10] N. W. Bartlett *et al.*, "A 3D-printed, functionally graded soft robot powered by combustion," (in English), *Science,* Article vol. 349, no. 6244, pp. 161-165, Jul 2015, DOI: 10.1126/science.aab0129.

[11] B. Ezair and G. Elber, "Fabricating Functionally Graded Material Objects Using Trimmed Trivariate Volumetric Representations," *Fabrication and Sculpting Event,* Conference Proceedings 2017.

[12] Z. Sudbury, C. Duty, and K. Vlastimil, "Expanding Material Property Space Maps with Functionally Graded Materials for Large Scale Additive Manufacturing," *Solid Freeform Fabrication Symposium Proceedings,* pp. 459-484, 2017.

[13] M. F. Ashby *, "Hybrids to fill holes in material property space," *Philosophical Magazine,* vol. 85, no. 26-27, pp. 3235-3257, 2005/09/11 2005, DOI: 10.1080/14786430500079892.

[14] Z. Sudbury, C. Duty, V. Kunc, V. Kishore, C. Ajinjeru, J. Failla, and J. Lindahl, "Characterizing Material Transition for Functionally Graded Material Using Big Area Additive Manufacturing," *Solid Freeform Fabrication Symposium Proceedings,* pp. 738-747, 2016.

[15] Z. Sudbury, C. Ajinjeru, V. Kishore, C. Duty, P. Liu, and V. Kunc, "Blending of Fiber Reinforced Materials Using Big Area Additive Manufacturing," in *Society for the Advancement of Material and Process Engineering 2017*, Seattle, WA, May 22-25 2017, 2017.

[16] J. Brackett, Y. Yan, D. Cauthen, V. Kishore, J. Lindahl, T. Smith, H. Ning, V. Kunc, and C. Duty, "Development of Functionally Graded Material Capabilities in Large-scale Extrusion Deposition Additive Manufacturing," *Solid Freeform Fabrication Symposium Proceedings,* pp. 1793-1803, 2019.

[17] ASTM D3171-15 "Standard Test Methods for Constiuent Content of Composite Materials," ASTM International, West Conshohocken, PA, 2015, DOI: 10.1520/D3171-15, www.astm.org.

[18] Q. Wang, H. Ning, U. Vaidya, S. Pillay, and L.-A. Nolen, "Fiber Content Measuremnt for Carbon Fiber-Reinforced Thermoplastic Composites Using Carbonization-In-Nitrogen Method," *Journal of Thermoplastic Composite Materials,* vol. 31, no. 1, pp. 79-90, 2018, doi: 10.1177/0892705716679481.

MECHANICAL AND ELECTRICAL PROPERTIES OF 3D PRINTED PA6 NANOCOMPOSITES

Jitendra S. Tate[1,2], Oluwasola K. Arigbabowo[2], Micheal LePere[2],
Henry Okeke[2], Jesse Garcia[2], and Aziza Nahar[2]

Texas State University
San Marcos, TX, 78666
[1]Materials Science, Engineering and Commercialization Program, Texas State University
[2]Ingram School of Engineering, Texas State University
Phone: 512-245-4872; Fax: 512-245-7771
Email: JT31@txstate.edu

ABSTRACT

Polyamide 6 (PA 6) nanocomposites are viable engineered nanocomposite materials with potential application in electrostatic discharge dissipation applications. Creating an electrically conductive path to dissipate electrostatic charges in such materials can be a viable solution to Electrostatic Discharge (ESD) concerns. The addition of nanofillers can also enhance mechanical properties of the parent polyamide 6, a structural thermoplastic ideal for 3D printing via fused deposition modelling (FDM). While improving the ESD capability, it is imperative to sustain the structural integrity of the nanocomposites. Hence, this study evaluated the mechanical, thermal and electrical properties of 3D printed PA6 Nanocomposites for electrostatic discharge applications. 3 and 5 wt.% of Carbon Nanofiber (CNF) was compounded with PA6 using co-rotating twin screw extruder) to produce 1.75mm diameter monofilaments for fused deposition modelling (FDM). The test samples were printed using commercial-off-the-shelf (COTS) 3D printer, Lulzbot TAZ 6 FDM printer. Mechanical, electrical and thermal characterization were carried out according to their respective ASTM standard. The tensile and flexural properties were enhanced by 3wt% addition of CNF, but no significant improvement was observed at 5wt%. The CNF nanocomposites exhibited good thermal stability and crystallization phenomenon at both loading levels. The volume resistivity of the PA6 matrix was reduced to order of 10^{11} and 10^{12} by 3wt% and 5wt% CNF addition respectively, which seems promising for manufacturing static discharge products.

Key Words: Polyamide 6, Carbon Nanofiber-CNF, Nanocomposites, Fused Deposition Modelling- FDM, Additive manufacturing

1. INTRODUCTION

Polyamides (PA) are known engineering thermoplastics with remarkable combination of mechanical, thermal and manufacturability properties for diverse applications. Depending on the composition of their main chain, polyamides have been widely categorized as either aromatic or aliphatic polyamides [1]. Aromatic polyamides, also referred to aramids, are high performance materials that possess beneficial properties such as low moisture sensitivity, oxidation resistant, excellent thermal and mechanical properties. However, they are very expensive, and their applications are limited to very demanding applications e.g. Nomex® and Kevlar® utilized in bullet proof vest and military body armor[2]. Aliphatic polyamides are the widely used polyamides for engineering and industrial applications. They belong to a family with the generic name of "Nylons" and are manufactured via condensation of copolymers obtained from the reaction between equal molecules of diamines and dicarboxylic acid. Nylon are semi crystalline in nature with presence of both crystalline and amorphous phases. They possess exceptional mechanical, wear and crystallinity properties that is widely exploited in a wide range of applications in commercial fibers, transportation, sports, electronics and automotive industries[3]. The common types of Nylons (aliphatic polyamides) are PA 12, PA11, PA 6,6 and PA 6.

Polyamide 6 (PA 6) is one of the most widely utilized thermoplastics in manufacturing. It is a cost effective and suitable material choice in producing end use functional parts via Additive Manufacturing, also known as 3D printing- creating parts layer by layer from a 3D geometrical model. It affords advantages of overall cost efficiency, flexibility and manufacturability of custom polymers without jeopardizing their intrinsic properties. Aside from being an ideal 3D printing material, easy modification of PA 6 has made them practical for vast uses and promising applications[4]. Despite their great potential in offering excellent thermal and mechanical properties, the lack of mobile electron carriers, typical of polymeric materials, has been a major drawback in expanding its use in applications demanding for good electrical properties. This can be addressed by creating a three-dimensional low resistance network for electron flow and modifying the Polyamide 6 matrix with conductive nanoparticles could be considered a plausible solution for electromagnetic shielding (EMI) and electrostatic discharge (ESD) concerns. Carbon fillers such as carbon black, carbon nanotubes (CNT), nanographene platelets (NGP) and carbon nanofiber (CNF) have been identified to be viable in this regard and incorporations of these fillers into PA 6 matrix could results in a novel type Polyamide 6 nanocomposites.

Bauhofer (2009) reported that the PA6 percolation threshold could be achieved at loading levels as low as 2.5wt.% or higher CNT loading [5]. Percolation threshold, as illustrated in Figure 1, is the minimum quantity of filler required to form a conductive path within the polymer. Furthermore, Zonder *et al.* (2011) affirmed that PA 12 can better disperse CNT due to its polar nature and electrical and rheological percolation was achieved around 1.4wt% CNT. This ability has been attributed to the large aspect ratio (length-diameter) typical of fillers such as carbon nanotubes and carbon nanofibers fillers[6]. In the work of Gaikwad *et al.* (2013), where NGP was successfully dispersed in PA11 using co-rotating twin-screw extruder, the percolation threshold was achieved at 7wt% NGP. They emphasized that manufacturing method of composites plays important role in percolation threshold as it affects orientation, dispersion, aspect ratio of filler and interparticle spacing within the polymer matrix [7].

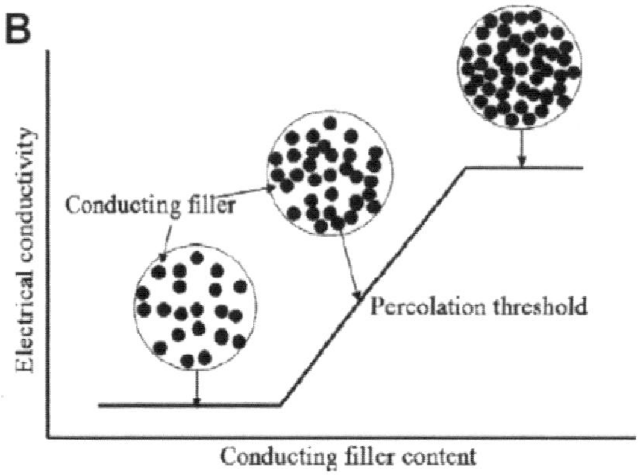

Figure 1: Percolation Phenomenom in Conducting Composites [8]

Carbon nanofibers has been identified to enhance the stiffening of thermoplastics matrix. It has the ability to improve the thermal and thermo-oxidative stabilities of the polymer matrix, and dispersion of CNF was reported to have acted as nucleation sites, thereby leading to formation of new crystalline domains [9, 10]. According to Goodridge (2011), the improvement in tensile properties obtained via laser sintering of PA12/Carbon Nanofiber (CNF) composites was attributed to the high aspect ratio of the nanofibers. Aspect ratio is the proportional ratio of length to diameter of the fibers. The diameter of CNF ranges between 20- 200nm which are quite larger than that of carbon nanotubes (2-100 nm) and their length can be more than a micron, resulting into high aspect ratio of CNF and CNT > 100 [11-13]. In addition to the low density and high aspect ratio of CNF, it has been considered more economical than CNT in terms of manufacturability and monetary cost [13], thereby proving more viable in developing advance nanocomposites. Hence, this paper was aimed to develop a carbon nanofiber polyamide 6 composite which could be viable for 3D printing and to evaluate the mechanical, thermal and electrical performance of CNF/PA6 nanocomposites.

2. MATERIALS SYSTEM AND MANUFACTURING

2.1 Materials System

Nylon 645, a thermoplastic monofilament that is ideal for fused deposition modelling 3D printing was chosen as the polymer matrix for the nanocomposites. Nylon 645 is a copolymer that consists of the purest form of a delta transition of Nylon 6/9, Nylon 6 and Nylon 6T, which is manufactured by Taulman 3D. Also, it possesses a high chemical resistance and is highly compatible in application where Nylon 6 and Nylon 6,6 are required.

The nanoparticles used for polymerization in this study were pyrograph III, PR-24-XT-LHT grade. The LHT grade is a low heat treated (LHT) carbon nanofiber grade which has been carbonized chemically by heating to temperatures of 1500°C to deposit the carbon present on the surface of Pyrograf®[14]. This heat treatment produces nanofibers which generally provides the highest electrical conductivity in nanocomposites. Pyrograph III-carbon nanofibers has a nanofiber density of 1.55 to 1.70 g/cc. Carbon nanofibers possess high length to diameter ratios which is essential for achieving an electrical resistivity near the percolation threshold and can be viable to develop nanocomposites for static dissipation applications. In addition, its lower cost per pound as compared to other conductive nanofillers such as carbon nanotubes makes it more economically viable. Carbon nanofibers are also considered for the enhancement of the mechanical properties of the polymer matrix as carbon nanofibers also contain excellent intrinsic mechanical properties.

2.2. Melt Compounding

The polymer nanocomposite was compounded via twin screw extruder. This was achieved with the help of Process 11 twin-screw extruder manufactured by Thermo Fisher Scientific. The compact co-rotating 11mm twin screw extruder is designed with a unique monocoque housing that is suitable for compounding thermoplastic polymers. In addition to its water-cooled primary feed port it contains 3 multifunction barrel ports for additional feeding and venting. Its barrel is separated into 8 sections to facilitate temperature profiles. The L/D ratio of this extruder is 40.

The extruder possesses an intermeshed co-rotating twin screws capable of both distributive and dispersive mixing into a homogeneous mass in a continuous process. There are 8 heating zones along the extruder barrel and the temperature profile employed from the primary hopper to the die are: 245°C, 250, 250 °C, 255 °C, 255 °C, 255 °C, 255 °C, 255 °C. The twin screws rotation rate was set to an optimized value of 150 rpm. The Nylon 645 was fed through the primary hopper while the Pyrograf III was side fed via the secondary hopper along the extruder barrel. The setup is shown in Figure 2 below.

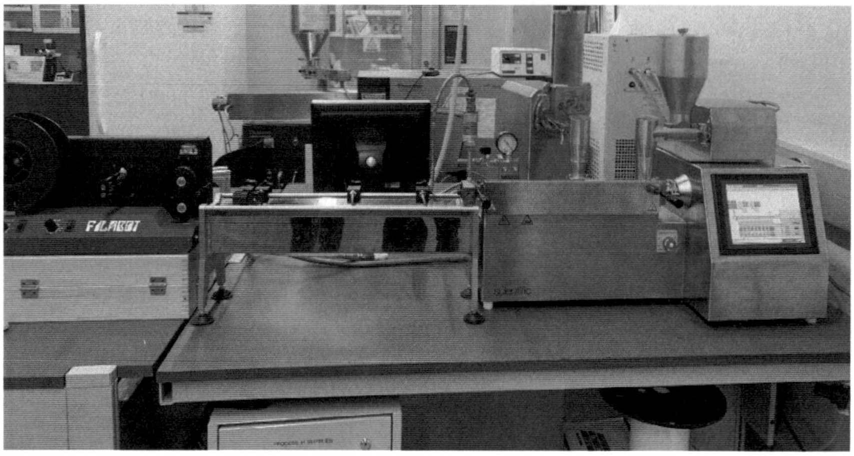

Figure 2. Twin Screw Extruder (11mm diameter)

2.3 Fused Deposition Modelling

The 3D printer used to print nanomodified filament samples was a LulzBot TAZ^{TM}. The LulzBot TAZ^{TM} was used to print tension, flexural, dynamic mechanical analyses, and resistivity test samples of carbon nanofiber modified Nylon at weight percentages of 0%, 3%, and 5%. Because Nylon 6 has a melting temperature of 207°C, a printing temperature between 255 °C and 260 °C were selected to ensure proper deposit viscosity for the filament to adhere to the bed and to itself. To ensure the foundation print was successful, we selected the bed temperature of 100 °C to get the nylon to adhere to the bed properly. To ensure the filament prints around the perimeter of the test specimens well, we have selected an outer edge fill speed to be half of that of the infill speed since less nylon filament adhere to on the edges of the specimen than that of the infill. We have also selected a nozzle with diameters ranging from 0.4mm to 0.8mm due to the high length to width aspect ratios of the carbon nanofibers extruding from the nozzle which can lead to clogging.

3. EXPERIMENTATION

3.1 Mechanical Characterization

Total five tension samples were printed according to ASTM D638-Standard Test Method for Tensile Properties of Plastics [15]. The tensile dog bone specimen "Type 1" has overall length, overall width and thickness as 165mm, 19mm, and 3.2mm, respectively. The tensile testing was performed as shown in Figure 3 using United Testing Machine (UTM) with a 50kN loading capacity (11 kips) and the strain values were collected using 2 inches gauge length extensometer with a measuring accuracy of 0.0002. The force-extension data was generated using the Datum 5i software of the UTM and tensile properties were evaluated.

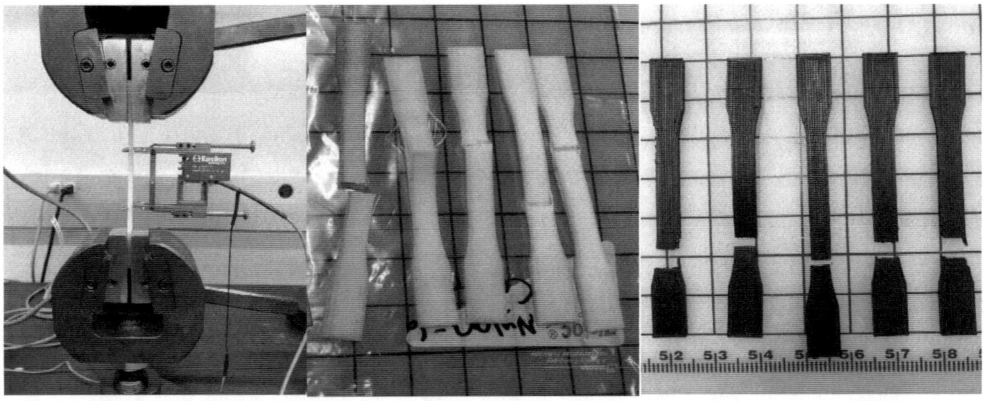

Figure 3: Tension test of 3D printed PA6/CNF nanocomposites

Total five flexural samples were printed and subjected to a 3-point bed test according to "ASTM D790- Standard Test Methods for Flexural Properties of Unreinforced and Reinforced Plastics and Electrical Insulating Materials" [16]. The flexural samples with a dimension of 12.7 mm wide, 3.2 mm thick, and 127 mm long were tested on the support span (span-to-depth ratio) of 16:1.These tests were performed according to Type I, which utilized the cross head position as the deflection measurements and Procedure B was followed, which is designed principally for those materials that do not break or yield in the outer surface of the test specimen within the 5.0 % strain limit when Procedure A conditions are used.

3.2 Electrical Testing

The electrical resistivity of control and nanomodified samples was measured according to ASTM D257 – "Standard Test Methods for DC Resistance or Conductance of Insulating Materials [17]. This is to evaluate the volume resistivity of the CNF conductive fillers. The test samples printed were circular disks with a dimension of 1mm thickness and 88.9 mm diameter. The volume resistivity was determined by Keithely Megohmeter setup comprising the Keithely 2450 source meter, Keithely 6514 Electrometer and the Linear four-point probe (Jandel Universal Probe) for housing the test sample. Voltage varying from 10V to 120 V was applied using the source meter for an electrification time of 60 seconds and the amount of current flowing was sensed by the electrometer. The volume resistivity values were evaluated according to Equation 1 below;

$$\rho_v = k_v * \frac{R}{t} \tag{1}$$

Where $\rho_v =$ volume resistivity ($\Omega.\text{cm}$),

$k_v =$ effective area of the circular electrode $= 22.881\ \text{cm}^2$,

R = measured resistance (Ω), and t = sample thickness (1mm).

3.3 Thermal Analysis

The material thermal stability was evaluated via ASTM - Standard Test Method for Thermal Stability by Thermogravimetry [18]. This was to determine the decomposition temperature of the materials and the degree of mass change via thermogravimetry. The mass of the sample was about 7.5mg for nanomodified composites. The samples were placed inside an alumina pan at ambient temperature after zeroing an empty and clean specimen pan. The test material was heated from ambient temperature to 600 °C at a rate of 10 °C /min. The testing material was heated in an alumina pan and maintained under inert atmosphere of 99.9% purity, purging the inert gas (Nitrogen) at a rate of 50 ml/min.

Thermal transitions of the nanomodified polyamide 6 were investigated via ASTM D3418-15- Standard Test Method for Transition Temperatures and Enthalpies of Fusion and Crystallization of Polymers by Differential Scanning Calorimetry (DSC)[19]. This test method was employed to evaluate the glass transition temperature, peak crystallization temperature and melting point of the material. The testing procedure was a single testing as discussed in the thermogravimetry procedure and was carried out with the aid of simultaneous differential thermogravimetry (SDT) analyzer capable of both TGA and DSC. The DSC sensing device of the SDT was used to monitor the difference in heat input between a reference material and a test material due to energy changes in the material, thereby resulting into different exothermic valleys and endothermic peaks, indicating different thermal transitions.

Dynamic mechanical data of the carbon nanofiber modified polyamide 6 was investigated according to ASTM D4065- Standard Practice for Plastics: Dynamic Mechanical Properties: Determination and Report of Procedures [20]. This test was aimed to determine the viscoelastic characteristics of the material by monitoring the elastic and loss modulus as a function of temperature. A rectangular cube with dimension 35mm X 12mm X 3 mm, maintained at a temperature of 23° C was subjected to a mechanical oscillation of a frequency of 1 HZ in a single cantilever test mode. The material was taken through a temperature ramp from 23° C to 200° C at a rate of 5 ° C/min via a Dynamic Mechanical Analyzer equipment TA- DMA 2980 and the obtained data was analyzed using Universal Analysis Software.

4. RESULTS AND DISCUSSION

4.1 Mechanical Properties

The mechanical response of the investigated material is of prime importance in quantitatively evaluating the structural integrity of the developed 3D printing nanocomposite filaments. This could prove viable as a fabricating 3D printing filament for end use industrial products, thereby replacing injection molded parts. The mechanical properties evaluated are the tensile and flexure properties as presented in Table 1

Table 1: Mechanical Properties Data

	0 wt.% CNF	3 wt.% CNF	5 wt.% CNF
Ultimate Tensile Strength (MPa)	37.21 (2.108)	41.44 (1.836)	30.14 (6.512)
Tensile Modulus (GPa)	1.574 (0.074)	2.943 (0.905)	1.559 (0.171)
Percentage of Elongation (%)	78.90 (0.055)	46.30 (0.107)	34.45 (0.229)
Equilibrium Toughness (MJ/m^3)	1.46 (0.158)	2.05 (0.649)	1.20 (0.083)
Flexural Strength (MPa)	33.93 (5.331)	39.54 (3.340)	40.39 (6.530)
Flexural Modulus (GPa)	0.768 (0.133)	1.130 (0.115)	1.173 (0.243)

***Standard Deviation in Parenthesis**

The trend in the observed tensile and flexural strength are presented graphically in Figure 4 below. In Figure 4a, tensile strength was observed to have appreciably increase at 3wt% CNF inclusion, which decreases as the CNF concentration was increased to 5wt%. It is well understood that high aspect ratio fillers like CNF and CNT prove difficult to disperse homogeneously because of their high surface area and a strong inclination to agglomerate [21]. Due to this reason, high shear dispersion techniques like twin screw extrusion has been recommended. However, the high shearing action associated with this technique could truncate the long fibers, thereby reducing its aspect ratio below a critical value. This could explain the deterioration of tensile strength observed at 5wt%. Although the tensile strength depreciates at 5wt%, the continuous improvement in the flexural strength of the nanocomposites, as shown in Figure 4b, is an indication that inclusion of CNF could prove viable in enhancing mechanical properties. These mechanical strengths indicators are closely related as they both measures the resistance to deformation depending on the type of applied load.

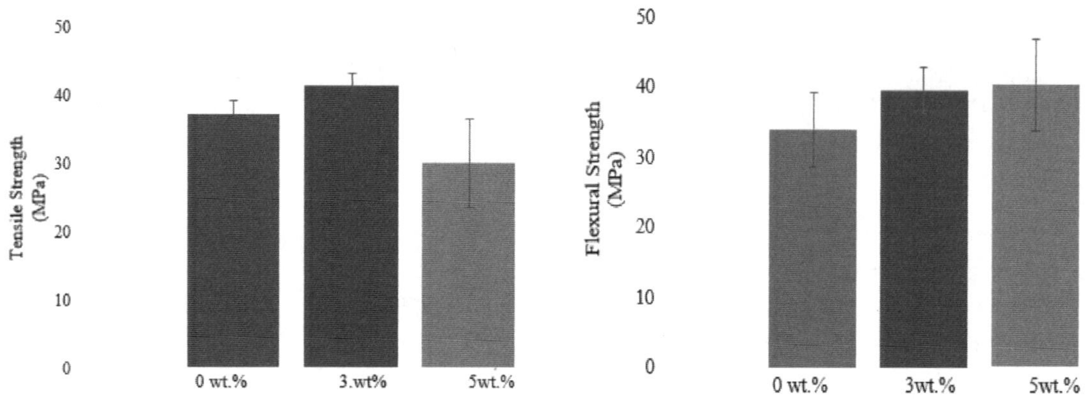

(a) *(b)*

Figure 4: Tensile and Flexural Strength of 3D printed PA6/CNF nanocomposites

Barrick (2011) reported that improvement in tensile properties of nanocomposites is dependent on the aspect ratio and the nanofiber-matrix interfacial strength. As shown in Figure 5a, the addition of 3wt.% CNF exhibits a strong reinforcing and stiffening effect of the Nylon and the enhancement observed in tensile modulus could be attributed to a strong interfacial strength which effects an efficient load transfer from the matrix to the reinforcement [22]. By increasing to 5wt% CNF inclusion, no appreciable effect was observed in tensile modulus, however the nylon matrix was increasingly stiffened by increasing CNF contents under bending load as shown in the flexural modulus plot in Figure 5b. The spatial orientation of CNF significantly impedes the movement of the amorphous phase of the Nylon matrix, thereby resulting into the stiffening of the material. It is noteworthy that higher loading levels is prone to entanglements of the fibers, promote agglomeration and consequently deteriorates the stiffness of the nanocomposites. Therefore, it is imperative to optimize the processing parameters of the extrusion process to maximize the reinforcing ability of CNF at high loading levels, like 5wt.% and above.

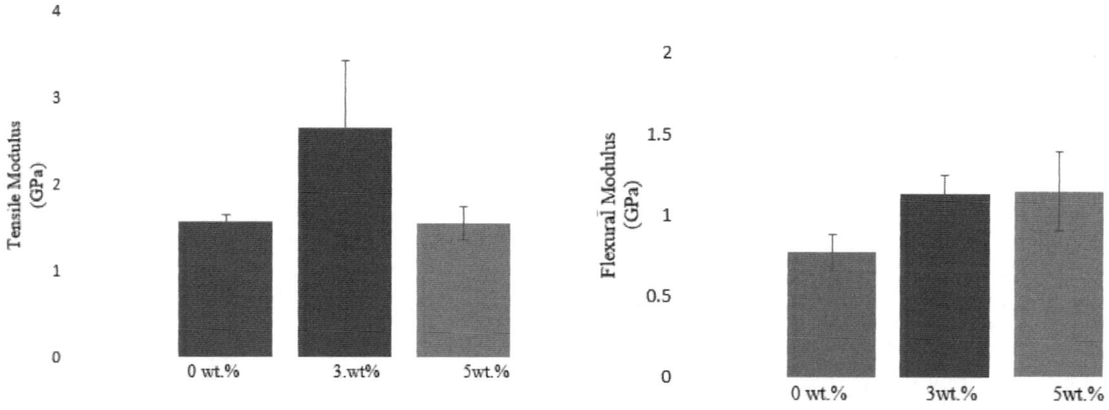

(a) *(b)*

Figure 5: Tensile and Flexural Modulus of 3D printed PA6/CNF nanocomposites

Figure 6 presents the trend in percentage (%) elongation and toughness of the investigated material. The percentage elongation reduces drastically with increasing CNF loading as presented in Figure 6a. % elongation is an indication of the ductility of the material and the continuous decline could be attributed to the restriction in the polymer chain movement. This restriction has been suggested to be due to the reduction in the free volume capacity of the Nylon 6 [22].

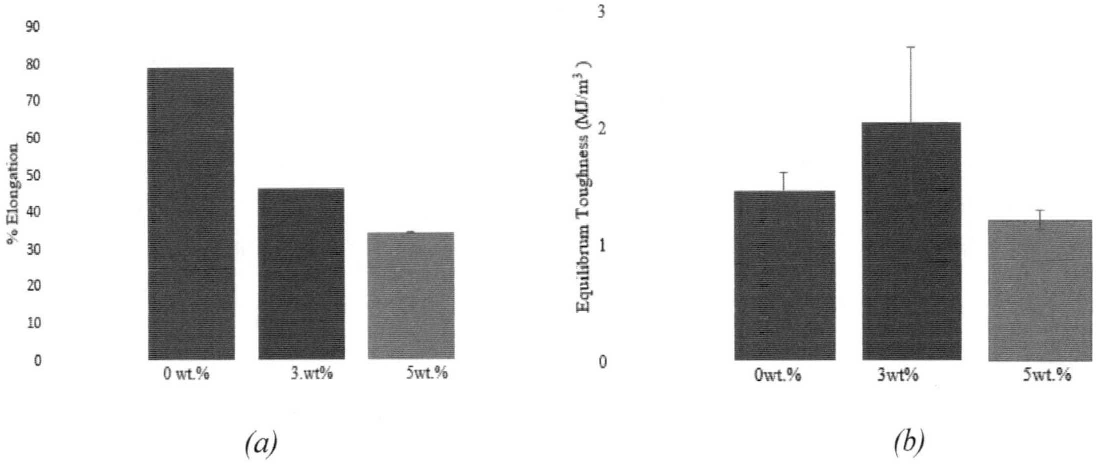

Figure 6: % Elongation and Toughness of 3D printed PA6/CNF nanocomposites

Figure 7 shows the tensile stress versus tensile strain plots which was obtained by integrating from zero to elongation at break to illustrates the area under the tensile curve. The area under the stress-strain curve was evaluated as the equilibrium toughness as shown in Figure 6b, which represents the total strain energy per unit volume that was induced by the applied stress. The similarity in the trend showed in the equilibrium toughness and tensile strength is an indication of the close relationship between the two tensile properties, which explains the fact that maximizing the strength is parallel to improving the toughness of the nanocomposites.

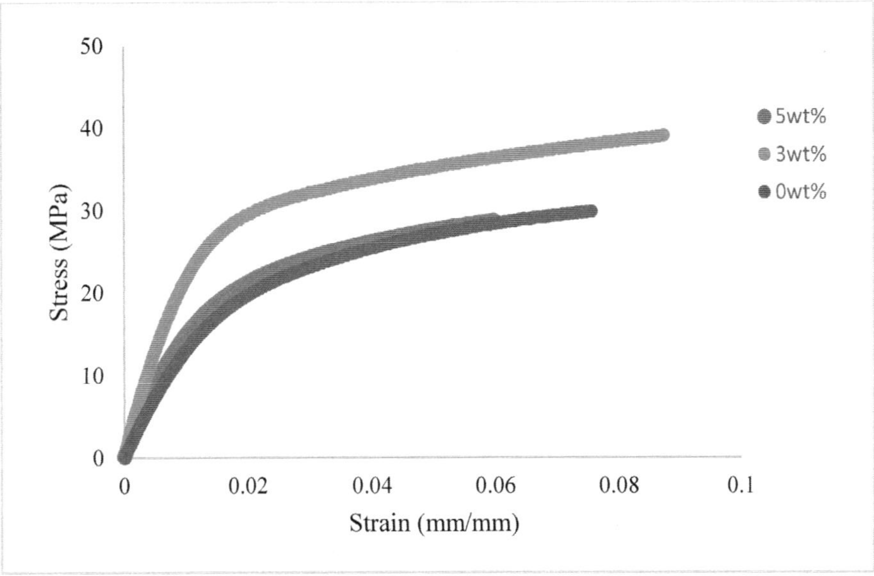

Figure 7: Tensile stress vs Axial strain of 3D printed PA6/CNF nanocomposites

4.3 Volume Resistivity

Table 2 presents the volume resistivity data collected in an electrical resistivity testing by varying the voltage source from 10 -120 v. The primary importance of the volume resistivity measurement was to evaluate how the carbon nanofiber impart the conductivity of the Nylon matrix which could prove viable in low resistivity applications such as static dissipation. The reduction in the resistivity of the nylon matrix was achieved via the attainment of a percolation nexus of CNF, which creates a conductive pathway throughout the matrix, thereby facilitating electron transfer [23].

Table 2: Volume Resistivity Values

	Resistivity (Ohm.cm)		
Voltage (V)	**3wt%**	**5wt%**	**0wt%**
10	6.21E+11	1.75E+12	5.40E+13
20	6.27E+11	1.52E+12	5.40E+13
30	7.79E+11	1.43E+12	5.40E+13
40	7.04E+11	1.38E+12	5.40E+13
50	7.33E+11	1.36E+12	4.35E+13
60	7.30E+11	1.33E+12	4.46E+13
70	5.46E+11	1.32E+12	4.48E+13
80	5.16E+11	1.30E+12	4.55E+13
90	5.08E+11	1.29E+12	4.40E+13
100	5.23E+11	1.31E+12	4.45E+13
120	5.06E+11	1.30E+12	4.30E+13

It could be observed in Figure 8 that the volume resistivity of the Nylon 6 matrix in the order of 10^{13} was lowered to the order of 10^{11} and 10^{12} for 3 and 5 wt.% respectively. In CNF nanocomposites, the tunneling effect facilitated by the high aspect ratio of the filler has been identified as the main mechanism of electrical conduction[24]. The observable reduction in the resistivity could be as a result of a conductive network being formed by the CNF, where electrons can tunnel along the length of one filler to another, thereby overcoming the high resistance imposed by the nylon matrix. In order to maximize the conductive ability of CNF, it is imperative to maintain its high aspect ratio. This could give an insight into the slight decrease of resistivity value observed at 3 wt. and 5wt.%. It is suspected that the CNF could have suffered from high shear forces of the twin-screw compounding process, resulting into shortening of the long fibers and consequently delaying the onset of electrical percolation. Although, the results obtained does not reflect a significant influence of the conductive CNF, the resistivity of 3wt.% still falls within the allowable resistivity values for ESD compliance rated between 10^{10} -10^{12} ohms.cm [25].

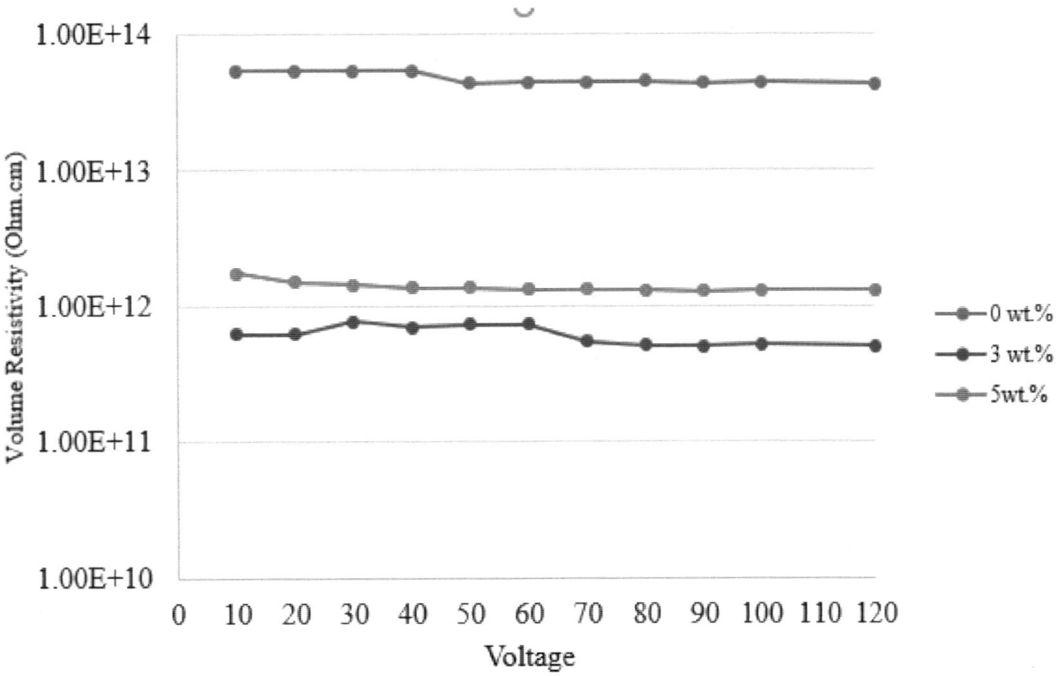

Figure 8: Volume resistivity of 3D printed PA6/CNF nanocomposite

4.4 Thermogravimetric Analysis

The thermogravimetric analyses (TGA) illustrates the thermal stability by measuring the weight loss as a function of temperature. TGA is of prime importance in evaluating the thermal stability of the nanocomposites. There is a close relationship between the thermal stability and the structural integrity of materials which require understanding the decomposition kinetics of our material in a temperature-controlled environment.

Table 3: TGA data

Thermal properties	0 wt.% CNF	3 wt.% CNF	5 wt.% CNF
Decomposition Temperature	400.95	402.53	402.68
Weight % loss	97.738%	97.357%	96.832%

The TGA curve in Figure 9 shows the onset decomposition temperature and the weight percent loss at the end of the heating profile. The extrapolated onset temperature signifies the temperature at which the weight loss begins which is an indication of the thermal stability of the material. The observed slight increase in thermal stability with increasing CNF content could be attributed to the restriction of the Nylon macromolecules by the network structure of the CNF. The intrinsic high thermal conductivity of CNF prevented any sort of heat accumulation and conducts heat uniformly, thereby increasing the thermal stability of the Nylon matrix. In addition, strong interfacial adhesion has been identified as a reason for the thermal stability of CNF based nanocomposites[21]. Strong interfacial bonds prevent the occurrence of interstices that could occur under thermal stresses where the diffusion of volatile gases from the decomposing material is prominent. However, the interface is prone to interstices buildup at higher CNF loading. Furthermore, no major degradation was observed with inclusion of CNF on the TGA curve and the thermal stability of the nanocomposites was sustained throughout the scanning temperature. The weight loss percentages at the final temperature reduces with increasing CNF loading and the pristine nylon being the highest with 97.738%. The reduction in weight loss observed in the nanocomposites could be ascribed to the restriction imposed by the CNF network structure, preventing the mobilization of the gaseous nylon macromolecules to encounter outer atmosphere during degradation [26].

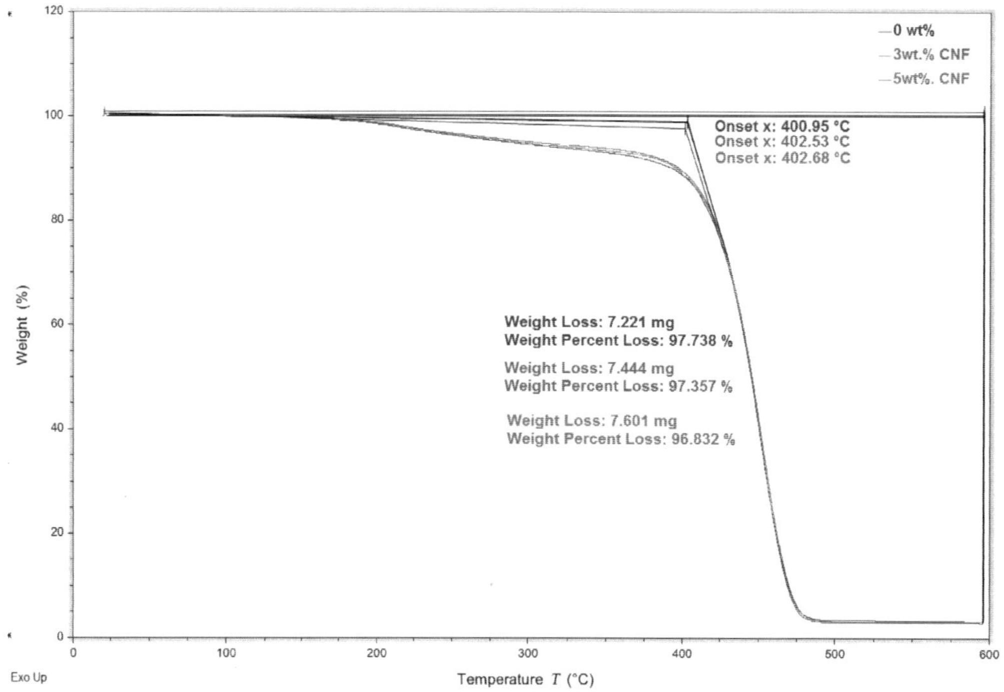

Figure 9: TGA curve of PA6/CNF nanocomposites

4.5 Differential Calorimetry

Table 4 highlights the thermal transition obtained in the differential calorimetry (DSC) first heat cycle of the investigated material. DSC is an important technique that can be used to qualitatively and quantitatively characterize the thermal transitions unique to a material.

Table 4: DSC Thermal transition

Transition parameters	0 wt.% CNF	3 wt.% CNF	5 wt.% CNF
Glass Transition temperature	59.39	60.35	59.93
Peak crystallization temperature	171.61	167.42	166.81
Melting temperature	214.35	213.31	213.32
Heat of crystallization	52.53	77.34	77.43
Degree of Crystallinity	22.8	34.6	35.4

Figure 10 presents the DSC scan obtained in the calorimetry study. The first observable signal on the DSC scan is the glasssy region which signifies the temperature region where the armouphous portion of the material is moving from a glassy to a rubbery state. The glass transition temperature (T_g) of the neat matrix occurs at 59.93°C and a minimal increase to 60.35 and 59.93 °C was observed at 3 and 5 wt.% reinforcement respectively. Dispersion of CNF within the Nylon matrix led to reduction in free volume movement of the polymer macromolecules due to the restriction imposed by the CNF, thereby delaying the onset of glass transition. The slight increase in T_g of the amorphous region of the Nylon/CNF composites is an indication of the polymer molecules interlocked into a rigid amorphous structure which could possibly translate to the enhancement of the mechanical properties of semicrystalline nylon matrix. In a semicrystalline polymer like Nylon, the armouphous region is reported to be responsible for the elasticity phenomenom [27]. During applied tensile stress, the amorphous molecular chain deforms elastically and the crystalline domains remains undisturbed until yielding begins. The presence of this armouphous region delays the onset of the plastic deformation of the crystalline phase, thereby resulting in the improvement of the strength and modulus of the semicrystalline polymer. Furthermore, the T_g could be used to predict the service temperature of polymeric materials depending on application and desired physical properties, especially for fully armouphous polymers. However, it is noteworthy that the Nylon 6, a semicrystalline polymer, being investigated in this study does not softening even above the T_g . This could be attributed to the presence of relatively strong intermolecular forces which is more sensitive to higher temperature, like melting temperature[28]

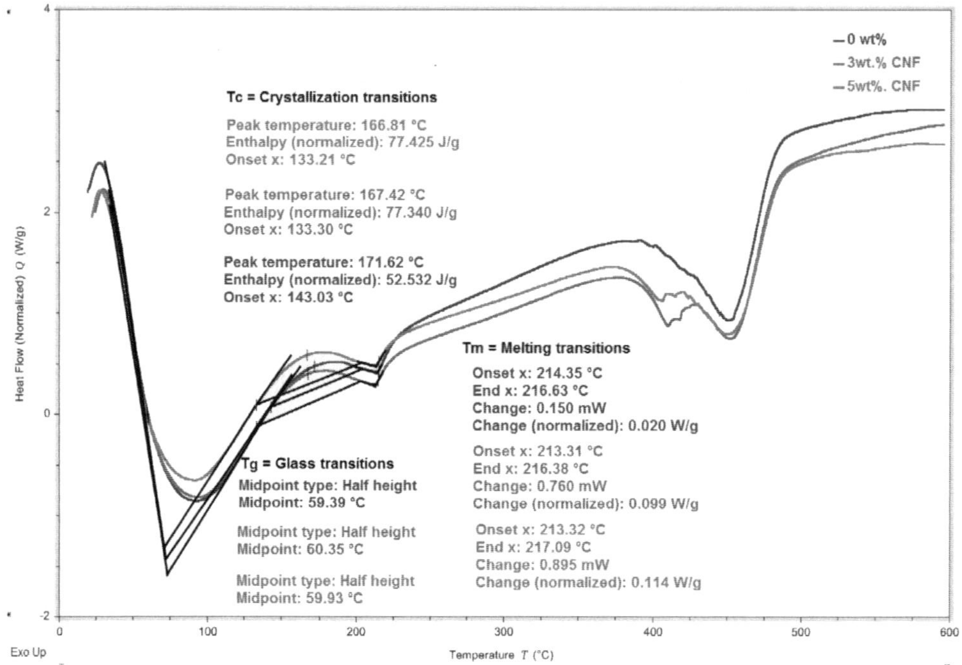

Figure 10: DSC Scan of PA6/CNF nanocomposites

The transition peak after the T_g illustrates the crystallization behavior of the investigated material system. The peak crystallization temperature was observed to be highest for the neat Nylon matrix, however, there was an early onset of crystallization at 3 and 5 wt.% respectively. This is expected and it is consistent with the Hoffman's nucleation theory which describes polymer crystallization in terms of kinetics and thermodynamics of surface nucleation[29]. The addition of CNF creates new surfaces in the nylon matrix thereby reducing the total surface energy of nuclei formation and consequently promoting heterogeneous nucleation which possesses a lower Gibbs free energy for crystallization. The heat of crystallization is the enthalpy associated with the crystallization process and could be used to determine the degree of crystallinity in the investigated material according to the following equation [26].

$$X_c (\%) = (\Delta Hm/(1\text{-}\varphi)\ \Delta H_o)\ X\ 100 \qquad (2)$$

Where ΔH_m is the enthalpy of crystallization melting shown in Figure 10, φ is the weight fraction of CNF, ΔH_o = 230 J/g is the consensual value reported in literature for the enthalpy of fusion of 100 % crystalline PA6 [30]. Using equation (1) the X_c (%) values have been computed as 22.8%, 34.6% and 35.4% of 0, 3 and 5 wt.% respectively of CNF loading. The observed increasing trend affirms the heterogeneous nucleation promoted by the carbon nanofibers. It is suspected that the higher degree of crystallinity experienced at 3 and 5wt% loading levels translated into the improvement of thermal stability and mechanical properties. However, the unusual deterioration of tensile properties of 5wt% could be due to possible weak interfacial strength vis-à-vis agglomeration effects.

The melting transitions in Figure 10 illustrates the onset of melting of the investigated material system. It could be observed that the onset of melting temperature slightly decreases with increasing CNF loadings. The presence of second phase CNF could promote increase in dislocation movements in the Nylon Matrix in form of interstices. These interstices could be perceived as weak impurities which lowers the melting temperature of the Nylon/CNF composites

4.6 Dynamic Mechanical Analysis

The DMA was used to evaluate the ability of the investigated material to return or store energy (storage modulus G'), dissipate energy (loss modulus G'') and the ratio of these effects (tan delta) signifies the transition from the glassy to rubbery state of the material. The storage modulus response observed in Figure 11 shows a trend like that of the tensile modulus plot. At 3wt.% loading of CNF, an improvement in the G' throughout the scanned temperature is an indication of the stiffening ability of CNF on the nylon matrix. The enhancements of storage modulus at 3wt% and 5wt% was more prominent at higher temperature, indicating the ability of CNF to enable the Nylon matrix to sustain its stiffness. This has been suggested to be due to interfacial interlocking between CNF and the polymer matrix which contributes to an additional reinforcement of restricted matrix molecular chain movement in the interface region of the nanocomposite[31]. The rapid decline in storage modulus in the transition region observed with increase in temperature could be attributed to a collective segmental motion of the chains in the polymer backbone, resulting into an energy dissipation phenomenon. The loss modulus illustrated in Figure 12 is an out-of-phase response which measures energy absorbed due to relaxation. The peak position of the loss modulus did not shift significantly at 3wt%, signifying that the inclusion of CNF has an insignificant effect on the dynamic relaxation behavior of the Nylon matrix [32]

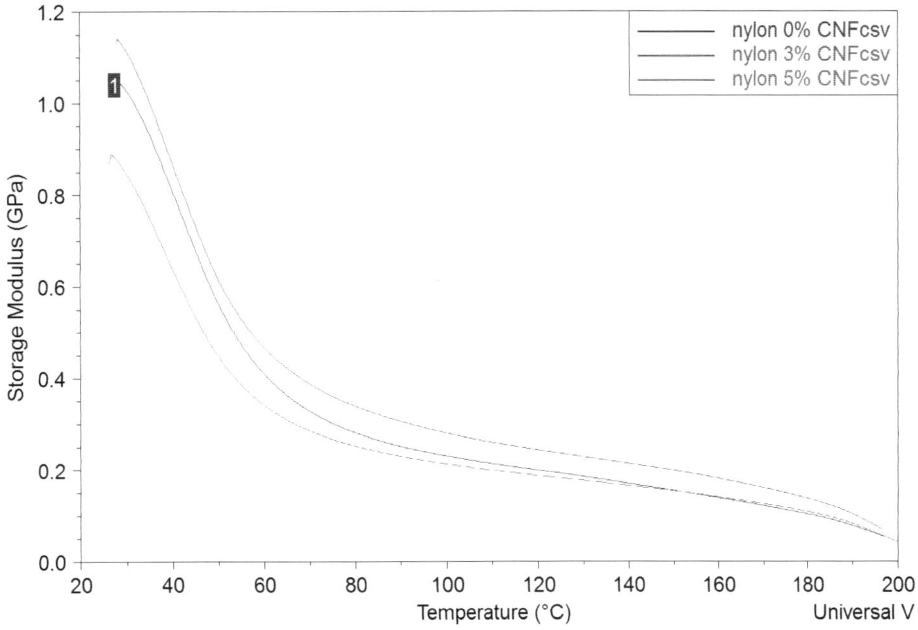

Figure 11: DMA thermograms of Storage modulus – Temperature Profile

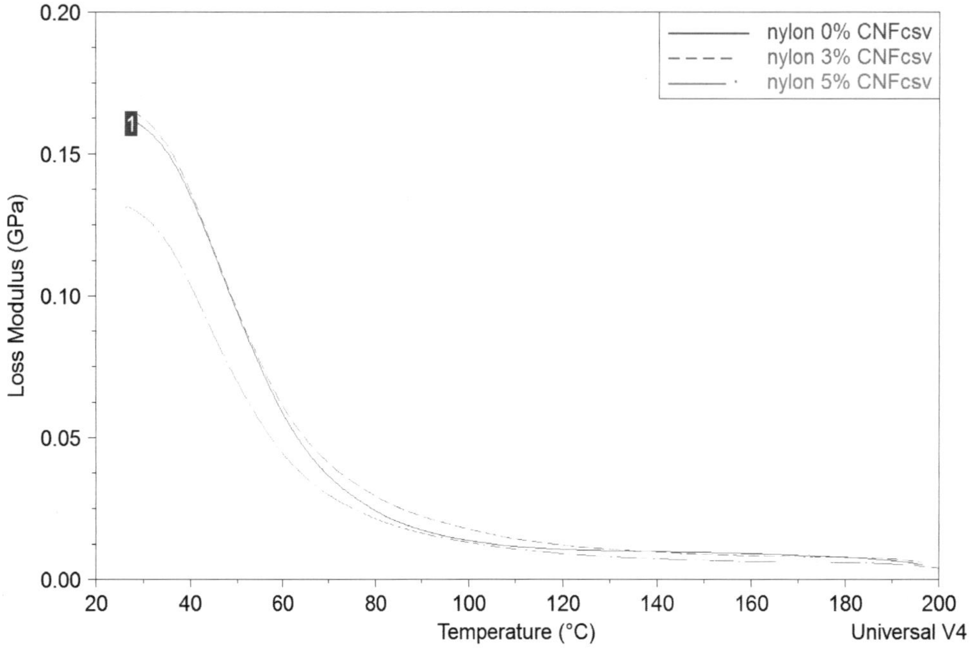

Figure 12: DMA thermograms of Loss modulus – Temperature Profile

The damping behavior of the investigated material were evaluated by the tangent of the phase angle (Tan Delta) which measures the ability of the material to absorbs energy. According to Figure 13, it could be observed that the area under the tan delta curve reduces with increasing CNF contents. This behavior is an indication of a homogeneous thermal energy distribution within the nylon/CNF nanocomposites that results from the high differential thermal conductivity between the CNF and Nylon matrix [26]

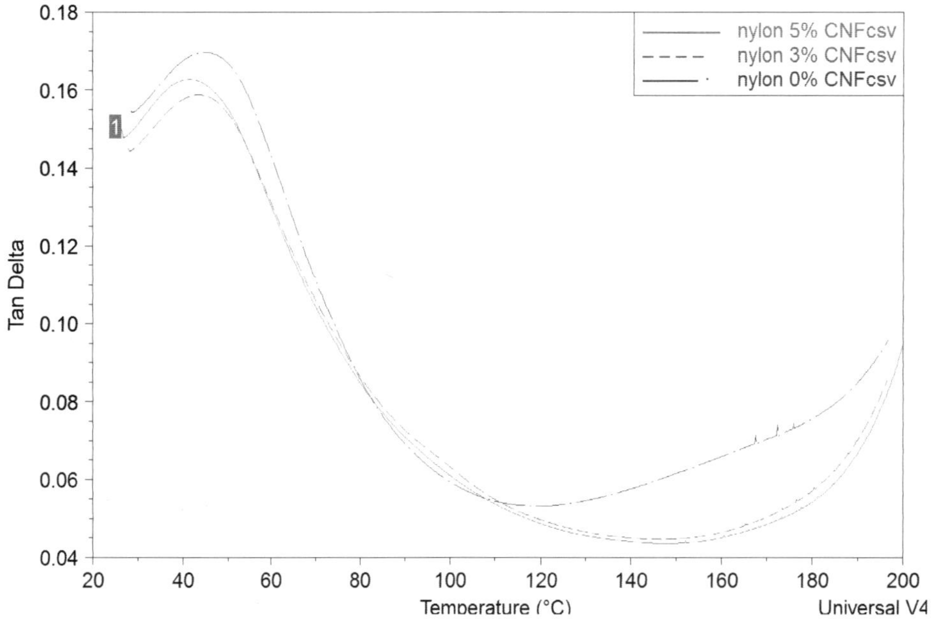

Figure 13: DMA thermograms of Tan Delta – Temperature Profile

5. CONCLUSION

PA6/CNF nanocomposite was successfully developed for 3D printing using co-rotating twin screw extruder for the compounding process. Fused deposition Modelling was used to manufacture test samples for mechanical, electrical and thermal properties investigation of the PA6/CNF nanocomposites. The following highlights are made from the results of this study.

i) There was an improvement of 11.4% tensile strength at 3wt% loading levels of CNF, however, 19% decrease of the tensile strength was observed when increasing the concentration of CNF to 5wt%. The agglomeration tendency of CNF is a well understood phenomenon and maximizing CNF potential at higher loading levels require optimizing the parameters of the manufacturing process.

ii) Significant improvement of 86.9% increment in the tensile modulus was observed at the 3wt% CNF loading. However, no significant effect was observed in the tensile modulus of 5wt%. This could be due to a porous interfacial strength for effective load transfer. There was an improvement of flexural properties with increasing CNF inclusion

iii) Carbon Nanofiber has proved to improve the electrical properties by lowering the volume resistivity of the insulating Nylon matrix. The volume resistivity observed at 3wt%. loading is quite significant and could be viable for electrostatic discharge application.

iv) The thermal properties of the nanocomposites were sustained with incorporation of CNF and no major deterioration was observed. No significant improvement in the thermal stability, weight percent loss, glass transition temperature of the nanocomposite. However, the crystallization transitions confirm higher degree of crystallinity with CNF inclusion.

v) It can be concluded from this study that the inclusion of CNF has shown high potential of developing a multifunctional PA6 nanocomposite for 3D printing, which could possess an excellent combination of mechanical, electrical and thermal properties. It is imperative to consider carrying out an extensive microstructural study via TEM to evaluate interfacial properties between the CNF and the Nylon matrix.

6. ACKNOWLEDGEMENTS

The authors would like to thank every members of Advanced Composites Lab, Texas State University for their accommodation while running the experiments of this research. We would also like to appreciate the help of Mr. Tom Taulman of 3D Taulman for providing us with adequate material.

7. REFERENCES

[1] L. W. McKeen, "7 - Polyamides (Nylons)," in *The Effect of UV Light and Weather on Plastics and Elastomers (Fourth Edition)*, L. W. McKeen, Ed.: William Andrew Publishing, 2019, pp. 185-222.

[2] (1998). *Aramids* [Online]. Available: http://www.pleo.com/dupont/nomex/.

[3] (2015). *Polyamides (Polylactams, Polyphthalamides and Polyaramids)* [Online]. Available: https://polymerdatabase.com/polymer%20classes/Polyamide%20type.html.

[4] X. Liu, R. Zhang, H. Zhu, and J. Wang, "Modeling method for battery distribution path optimization of EV charging and swapping service network," *Dianli Zidonghua Shebei/Electric Power Automation Equipment,* vol. 35, no. 6, pp. 10-16, 2015.

[5] W. Bauhofer and J. Z. Kovacs, "A review and analysis of electrical percolation in carbon nanotube polymer composites," *Composites Science and Technology,* vol. 69, no. 10, pp. 1486-1498, 2009/08/01/ 2009.

[6] L. Zonder, A. Ophir, S. Kenig, and S. McCarthy, "The effect of carbon nanotubes on the rheology and electrical resistivity of polyamide 12/high density polyethylene blends," *Polymer,* vol. 52, no. 22, pp. 5085-5091, 2011/10/13/ 2011.

[7] S. Gaikwad, J. Tate, N. Theodoropoulou, and J. Koo, "Electrical and mechanical properties of PA11 blended with nanographene platelets using industrial twin-screw extruder for selective laser sintering," *Journal of Composite Materials,* vol. 47, no. 23, pp. 2973-2986, 2013.

[8] P.-C. Ma, N. A. Siddiqui, G. Marom, and J.-K. Kim, "Dispersion and functionalization of carbon nanotubes for polymer-based nanocomposites: A review," *Composites Part A: Applied Science and Manufacturing,* vol. 41, no. 10, pp. 1345-1367, 2010/10/01/ 2010.

[9] A.-C. Brosse, S. Tencé-Girault, P. M. Piccione, and L. Leibler, "Effect of multi-walled carbon nanotubes on the lamellae morphology of polyamide-6," *Polymer,* vol. 49, no. 21, pp. 4680-4686, 2008/10/06/ 2008.

[10] R. Chávez-Medellín, L. A. S. d. A. Prado, and K. Schulte, "Polyamide-12/Functionalized Carbon Nanofiber Composites: Evaluation of Thermal and Mechanical Properties," *Macromolecular Materials and Engineering,* vol. 295, no. 4, pp. 397-405, 2010.

[11] W. Chanklin, "Electrical Properties Study of Carbon Fillers in Polymer Nanocomposites " Doctor of Philosophy Chemical Engineering, The University of New Brunswick, New Brunswick, Canada, 2016.

[12] A. Allaoui, S. V. Hoa, and M. D. Pugh, "The electronic transport properties and microstructure of carbon nanofiber/epoxy composites," *Composites Science and Technology,* vol. 68, no. 2, pp. 410-416, 2008/02/01/ 2008.

[13] F. K. Ko and Y. Wan, *Introduction to Nanofiber Materials*. Cambridge: Cambridge University Press, 2014.

[14] (2011). *Pyrograf III* [Online]. Available: http://pyrografproducts.com/Merchant5/merchant.mvc?Screen=cp_nanofiber.

[15] ASTM. (2014). *Standard Test Method for Tensile Properties of Plastics* [Online]. Available: https://www.astm.org/Standards/D638.htm.

[16] A. S. D790, " " Standard Test Methods for Flexural Properties of Unreinforced and Reinforced Plastics and Electrical Insulating Materials," ASTM International, West Conshohocken, PA," 2010.

[17] ASTM. (2014). *Standard Test Methods for DC Resistance or Conductance of Insulating Materials* [Online]. Available: https://www.astm.org/Standards/D257.htm.

[18] ASTM. (2017). *Standard Test Method for Thermal Stability by Thermogravimetry* [Online]. Available: https://www.astm.org/Standards/E2550.htm.

[19] ASTM. (2015). *Standard Test Method for Transition Temperatures and Enthalpies of Fusion and Crystallization of Polymers by Differential Scanning Calorimetry* [Online]. Available: https://www.astm.org/Standards/D3418.htm.

[20] ASTM, "Standard Practice for Plastics: Dynamic Mechanical Properties: Determination and Report of Procedures " 2012.

[21] G. Sui, W.-H. Zhong, M. A. Fuqua, and C. A. Ulven, "Crystalline Structure and Properties of Carbon Nanofiber Composites Prepared by Melt Extrusion," *Macromolecular Chemistry and Physics,* vol. 208, no. 17, pp. 1928-1936, 2007.

[22] N. Mahmood, M. Islam, A. Hameed, S. Saeed, and A. N. Khan, "Polyamide-6-based composites reinforced with pristine or functionalized multi-walled carbon nanotubes produced using melt extrusion technique," *Journal of Composite Materials,* vol. 48, no. 10, pp. 1197-1207, 2014.

[23] S. Y. Veena Choudhary and Anju Gupta , IntechOpen, . , "Polymer/Carbon Nanotube Nanocomposites, Carbon Nanotubes - Polymer Nanocomposites," 2011.

[24] L. Feng, N. Xie, and J. Zhong, "Carbon Nanofibers and Their Composites: A Review of Synthesizing, Properties and Applications," (in eng), *Materials (Basel),* vol. 7, no. 5, pp. 3919-3945, 2014.

[25] G. G. Tibbetts, I. C. Finegan, and C. Kwag, "Mechanical and electrical propertiesof vapor-grown carbon fiber thermoplastic composites," *Molecular Crystals and Liquid Crystals,* vol. 387, no. 1, pp. 129-133, 2002/01/01 2002.

[26] A. K. Barick and D. K. Tripathy, "Preparation and characterization of carbon nanofiber reinforced thermoplastic polyurethane nanocomposites," *Journal of Applied Polymer Science,* vol. 124, no. 1, pp. 765-780, 2012.

[27] T. H. Courtney, *Mechanical Behavior of Materials*, Second Edition ed. Waveland press, Inc, 2000.

[28] G. W. Ehrenstein and R. P. Theriault, *Polymeric materials : structure, properties, applications*. Munich; Cincinnati, OH: Hanser ; Hanser Gardner Publications, 2001.

[29] S. Z. D. Cheng and B. Lotz, "Enthalpic and entropic origins of nucleation barriers during polymer crystallization: the Hoffman–Lauritzen theory and beyond," *Polymer,* vol. 46, no. 20, pp. 8662-8681, 2005/09/23/ 2005.

[30] C. Millot, L.-A. Fillot, O. Lame, and P. Sotta, "Assessment of polyamide-6 crystallinity by DSC: Temperature dependence of the melting enthalpy," *Journal of Thermal Analysis and Calorimetry,* vol. 122, 04/09 2015.

[31] A. Eitan, F. T. Fisher, R. Andrews, L. C. Brinson, and L. S. Schadler, "Reinforcement mechanisms in MWCNT-filled polycarbonate," *Composites Science and Technology,* vol. 66, no. 9, pp. 1162-1173, 2006/07/01/ 2006.

[32] S. Yang, J. Taha-Tijerina, V. Serrato-Diaz, K. Hernandez, and K. Lozano, "Dynamic mechanical and thermal analysis of aligned vapor grown carbon nanofiber reinforced polyethylene," *Composites Part B: Engineering,* vol. 38, no. 2, pp. 228-235, 2007/03/01/ 2007.

CARBON NANOFIBER Z-THREADED CARBON FIBER REINFORCED POLYMER COMPOSITE (ZT-CFRP) LAMINATE PARTS PRODUCED USING A MAGNETIC COMPACTION FORCE ASSISTED ADDITIVE MANUFACTURING (MCFA-AM) TECHNIQUE

Bikash Ranabhat[1,2], Sebastian Kirmse[1,2], Michael Johnson [1], Kuang-Ting Hsiao[1,*]

University of South Alabama

Department of Mechanical Engineering, 150 Jaguar Drive, Shelby Hall, Suite 3130

Mobile, AL, 36608

1.Department of Mechanical Engineering

2.Systems Engineering Program

ABSTRACT

The main challenge in AM processes of continuous fiber-reinforced polymer (FRP) composites is the lack of matured technology, advanced polymers and nanocomposites that match the performance and fabrication requirements. This research addressed the fabrication technology and materials concerns. The research presented herein developed a fast-curing porous carbon nanofiber z-threaded CFRP (P-ZT-CFRP) prepreg tape using a radial flow alignment technique (RFA) and fabricated the P-ZT-CFRP laminate using MCFA-AM. For comparison, two porous CFRP (P-CFRP) laminates cured by the Out-Of-Autoclave Vacuum-Bag-Only (OOA-VBO) method were also manufactured. The interlaminar shear strengths (ILSS) for the two P-CFRPs were 38.5 MPa (with 50 % acetone in resin to create voids) and 43.8 MPa (with 20 % acetone) respectively. The P-ZT-CFRP (with 20 % acetone) manufactured with the MCFA-AM (without vacuum) produced the ILSS of 54.5 MPa. The 0.5 wt% CNF z-threads in resin provided an effective structural integrity enhancement in the porous laminate. The specific ILSS per weight of the P-ZT-CFRP was 83% and 87% of that from the control CFRP manufactured by OOA-VBO (75.5 MPa) and the regular CFRP produced by MCFA-AM (70.5MPa), respectively; which indicated the interesting potentials to tailor lightweight and high-strength CFRPs using AM and nanomaterials. Future improvement directions were discussed.

Keywords: Additive manufacturing, Interlaminar shear strength, Z-threaded CFRP, Porous.
Corresponding author: Email: kthsiao@southalabama.edu, Phone: +1 (251) 460-7889

SAMPE Virtual Conference Proceedings, 2020. Society for the Advancement of Material and Process Engineering – North America.

1. INTRODUCTION

Additive manufacturing (AM), also known as 3D printing, rapid prototyping (RP), or solid-freeform (SFF), is a "process of joining materials to make objects from the three-dimensional (3D) model data, usually layer upon layer, as opposed to subtractive manufacturing methodologies" [1], which was first described in 1986 by Charles Hull [2]. Numerous novel AM processes have been developed over more than 20 years of AM development with its applications in aerospace (engine components), automotive (cooling systems), architectural design (exact scale models), consumer goods (eyewear), etc. [3]. Recently, AM of polymer composites has been the center of attraction for academia, industries, and national research labs due to its potential to create the lightweight and high strength structures. AM of fiber-reinforced polymer composite has been found useful to enhance the mechanical properties of the composite compared to the unreinforced polymer composite [4]. Long fiber reinforcement has offered better strength and stiffness compared to unreinforced and short fiber reinforced composites [5,6,7,8,9]. However, the parts produced by the current AM processes are still inferior to conventional composite manufacturing by a single order of magnitude [10]. The common problems observed in AM of polymer composites are voids, physical gaps at the layer/layer interface, lack proper adhesion between fibers and matrix, low fiber volume fraction, and so on [9]. Thus, the main challenge in the field of AM of polymer composite is the lack of matured technology, suitable advanced polymer materials, and compatible nanocomposites that match the performance and fabrication requirements.

From the literature, fused deposition modeling (FDM) is the most widely used AM technique. It also holds 41.5 % of the market share [21]. Some of the recent and promising results of additively manufactured fiber-reinforced polymer composite are presented in Table 1. To the authors' best knowledge, the ILSS test has been rarely performed on the additively manufactured parts.

This study uses the novel magnetic compaction force assisted additive manufacturing to print the P-ZT-CFRP laminate. The MCFA-AM feasibility[19] was demonstrated and proved useful to fast-print continuous CFRP parts with a comparable ILSS to traditional composite manufacturing method like the time-consuming Out Of Autoclave-Vacuum Bag Only (OOA-VBO). The estimated magnetic compaction pressure of 0.21 MPa was used for consolidation during printing and the compaction pressure significantly reduces the size and content of voids despite no vacuum being applied during MCFA-AM process [19].

Table 1. Summary of mechanical properties of some additively manufactured fiber-reinforced polymer composites.

Manufacturing Technique	Fiber	Matrix	Mechanical properties reported
Fused Deposition modeling [22]	U-CF	PLA	• Flexural modulus 129.4 GPa
Fused filament fabrication [23]	U-CF	Nylon	• Tensile strength 600 MPa • Flexural strength 430 MPa • Flexural modulus 38.1 MPa
Fused Deposition modeling [24]	U-CF	Onyx	• Flexural strength 270.63 MPa
Laser assisted AM [8]	U-GF	Polypropylene	• Tensile strength 217.6 ± 28.6 MPa • Tensile modulus 5.9 ± 0.8 GPa • Flexural strength 169.7 ± 48.6 MPa • Flexural modulus 21.4 ±5.1 GPa

U-CF: Unidirectional Carbon Fiber
U-GF: Unidirectional Glass Fiber

Previous studies have shown that by using carbon nanofibers (CNFs) z-threads in CFRP laminates, mechanical properties (e.g., Mode-I interlaminar fracture toughness [12], interlaminar shear strength (ILSS) [13], and longitudinal compressive strength [14]), thermal properties (e.g., through-thickness thermal conductivity [15]) and electrical properties (e.g., through-thickness DC electrical conductivity [16-18]) can be significantly improved even when using low CNF concentrations (0.3 wt% to 2.0 wt%). This paper employs a radial flow alignment technique (RFA) [20] to make CNF z-threaded CFRP prepreg tape, thus produced prepreg tape is used by MCFA-AM to print parts. However, additional new challenge is rising due to the fast curing thermoset resin in the MCFA-AM process. Conducting the RFA with the fast curing resin was challenging because that the resin's high viscosity and fast curing nature conflict with the CNF dispersion and the RFA process time requirement. Hence, in this study, the fast curing epoxy was slightly modified using acetone, which solves the two problems, i.e., dispersion of CNFs in the resin and slows the curing process allowing the extra time to work during the RFA technique. However, it was hypothesized that the acetone's complete removal from the fast curing prepreg could be difficult and the residual acetone, if any, in the matrix would reduce the mechanical strength (due to voids/porosity caused by the residual acetone); in such cases, the Porous ZT-CFRP (i.e., P-ZT-CFRP) are expected to be produced and tested in the paper. However, the P-ZT-CFRP could have lower density than traditional CFRP and could have some advantages in the weight specific properties including the specific ILSS per weight. Therefore, the goals of this paper are to:

- Produce P-ZT-CFRP prepreg using RFA technique.
- Additively manufacture P-ZT-CFRP laminate specimens employing MCFA-AM.
- Compare the specific ILSS of additively manufactured P-ZT-CFRP composite laminate with the P-CFRP composite laminate manufactured by Out-Of-Autoclave Vacuum-Bag-

Only (OOA-VBO), traditional CFRP laminate manufactured by OOA-VBO [19] and CFRP laminate manufactured by MCFA-AM [19].

- Analyze the ILSS results associated with CNF reinforcement, the fiber volume fraction, the voids content and the curing methods (either OOA-VBO or MCFA-AM).

2. EXPERIMENTAL METHODOLOGY

2.1 Materials

Three samples were prepared, i.e., sample 1: porous CFRP laminate with 50 % acetone in resin to create porosity, sample 2: porous CFRP laminate with 20 % acetone, and sample 3: 0.5 wt% CNF z-threaded porous CFRP laminate with 20 % acetone (P-ZT-CFRP). Sample 1 and sample 2 were fabricated using OOA-VBO, while sample 3 was fabricated employing MCFA-AM (without vacuum). Each laminate sample was trimmed into five specimens to conduct the ILSS test. The materials used to prepare the samples are presented in Table 2.

Table 2. Materials used and the suppliers

Materials	Suppliers
Carbon fiber: Toray T-700 unidirectional (680 g/m^2 areal weight, 1.8 g/cm^3 fiber density, 12k tow size).	Toray
Polymer matrix: EPIKOTETM Epoxy Resin TRAC 06170 and EPIKURETM Curing Agent TRAC 06170 (100:16)	Hexion Inc.
Surfactants:BYK-9076 & BYK-9077	BYK USA, Inc
CNFs:PR-24-XT-HHT	Applied Science, Inc
Acetone	Fisher Scientific International, Inc

2.2 Methodology

2.2.1 Fabrication of porous CFRP laminate

The manufacturing steps for the preparation of sample 1 (50 % acetone) and sample 2 (20 % acetone) are shown in Figure 1. The resin, curing agent, and acetone x wt% (x = 50 % or 20 %) were mixed thoroughly by high shear mixing for 5-10 min until no clouding is visible. The mixture was poured on top of the fabric and spread out with a roller for uniform resin impregnation into the unidirectional carbon fabric. The prepreg was heated on top of the hot plate at 45 °C for 10-15 minutes until prepreg tackiness was reached. The prepreg was then transferred to polymer bagging and stored in the freezer at -18 °C to be later used for curing via the OOA-VBO process. The layup scheme used in the OOA-VBO process is shown in Figure 2. the curing cycle used to cure sample 1, and sample 2 is presented in Table 3.

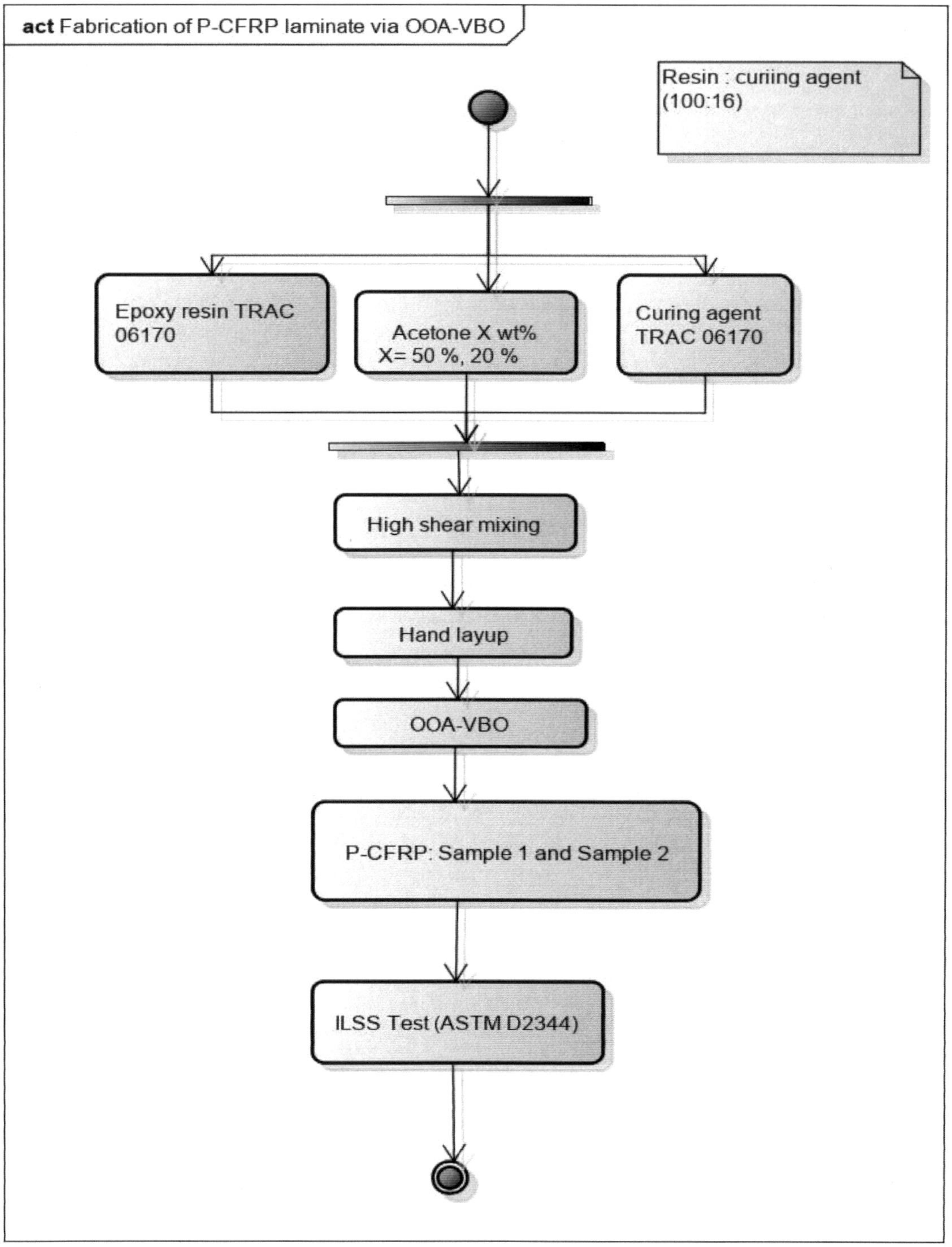

Figure 1. Fabrication of porous CFRP (samples 1 & 2)

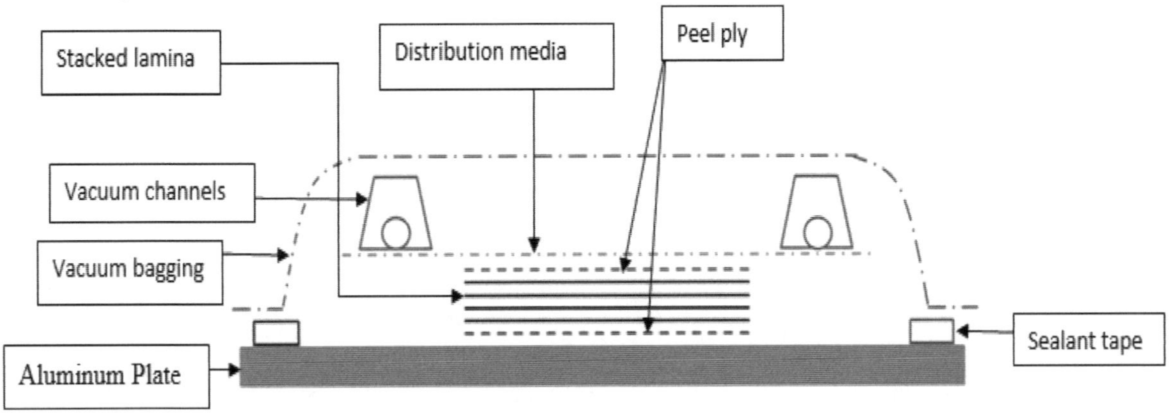

Figure 2. OOA-VBO Layup scheme used for the preparation of P-CFRP laminate [17]

Table 3. OOA-VBO cure cycle.

Time (min)	Temperature (°C)	Vacuum Pump
30	Room temperature (23)	On
30	45	On
20	55	On
60	120	On
60	140	Off

2.2.2 Fabrication of Porous CNF z-threaded CFRP Laminate

The fabrication of sample 3 (P-ZT-CFRP) laminate was achieved utilizing two novel techniques; (1) radial flow alignment (RFA) technique to z-thread CNFs into the unidirectional carbon fabric to form the P-ZT-CFRP prepreg and (2) magnetic compaction force assisted AM to fabricate P-ZT-CFRP laminate. To optimize the process, EPIKOTETM Epoxy Resin TRAC 06170 was heated at 60 °C then 1 wt% of each surfactant BYK-9076 and BYK-9077, 0.5 wt% of CNFs (PR-24-XT-HHT) was added in the resin, and the mixture was mechanically mixed by high shear mixing for an hour followed by 30 minutes of sonication. To prevent the curing reaction during RFA, 20 wt% of acetone was added in the mixture before adding the EPIKURETM Curing Agent TRAC 06170. The ratio used to mix resin to curing agent was 100:16. The mixture was then used by the radial flow alignment technique [20] to z-thread the CNFs into the carbon fiber fabric and prepared z-threaded prepreg tape. Thus, the porous CNFs z-threaded prepreg (P-ZT-CFRP) with acetone was stored in the freezer first and then transferred into the oven at 35 °C, -0.08 MPa (vacuum) for 20-25 minutes to remove some acetone without bubbling caused by acetone evaporation. Note that some residual acetone could still be in the prepreg. Then after the P-ZT-CFRP prepreg was heated at 40 °C to attain the desired B-staging, it was stored in the freezer for the future use in MCFA-AM. The MCFA-AM manufactured P-ZT-CFRP laminate sample (i.e., sample 3) was post-cured as described in [19] with the post curing cycle, as presented in Table 4. The Radial Flow Alignment (RFA) process and the Magnetic Compaction Force Assisted AM (MCFA-AM) are further explained in the next section.

Table 4. Cure cycle for P-ZT-CFRP laminate

Time (min)	Temperature (°C)
30	45
15	55
60	120
60	140

2.2.2.1 Radial flow alignment technique and Magnetic compaction force assisted AM

Radial flow alignment [20] is a proven rheological methodology to z-thread CNFs in carbon fabric [17,18]. In this paper, RFA was used to directionally stretching and radially align long carbon nanofibers (CNFs) suspended in liquid epoxy/acetone mixture through a unidirectional carbon fabric to form the P-ZT-CFRP prepreg tape. The schematic diagram of RFA is shown in Figure 3. Then the porous CNFs z-threaded CFRP prepreg tapes were used as the feedstock for the magnetic compaction force assisted additive manufacturing (MCFA-AM) [11] methodology to fabricate P-ZT-CFRP laminate by adding the prepreg tapes layer by layer on top of a moving backing-article, which was attracted by the magnetic force emitted from the printing head) and utilizing the magnetic compaction force to compact the prepreg during the prepreg consolidation/curing; therefore a CFRP laminate can be produced under desired compaction pressure and with desired 3D-shape and 3D-carbon fiber direction control without a mold nor a fixed support substrate/plate. A simplified MCFA-AM 3D printing head was developed using the basic principle described in patent-pending publications [11] and used in this study, as described in [19]. The simplified 3D printing head is shown in Figure 4. The overall P-ZT-CFRP laminate manufacturing processing steps are shown in Figure 5. The edge (i.e., side-view) of cured P-ZT-CFRP laminate observed in microscope is shown in the Figure 6.

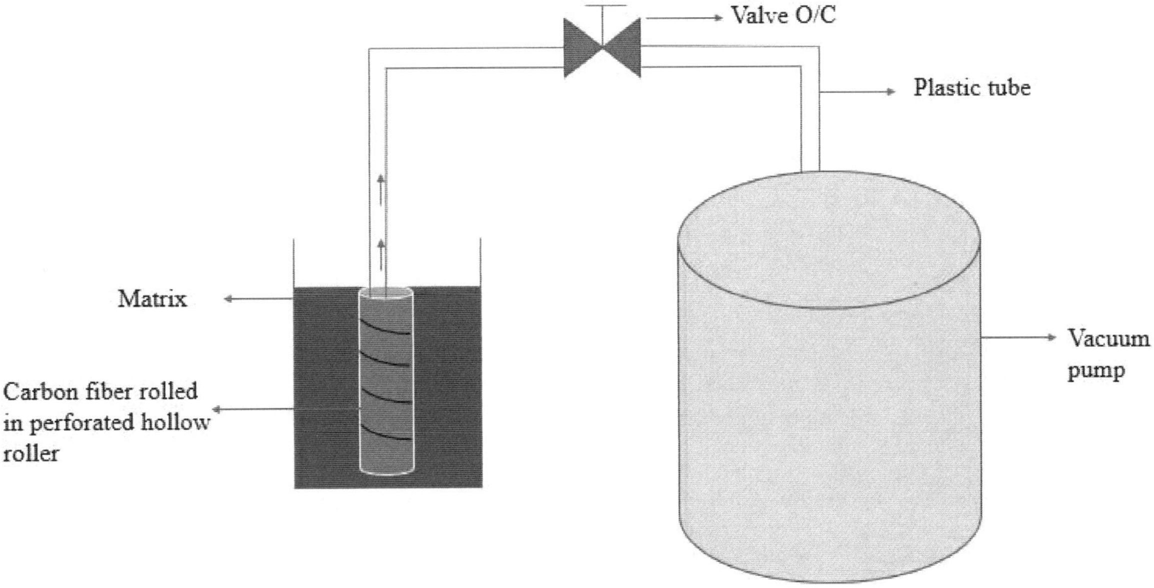

Figure 3. Schematic diagram for radial flow alignment technique

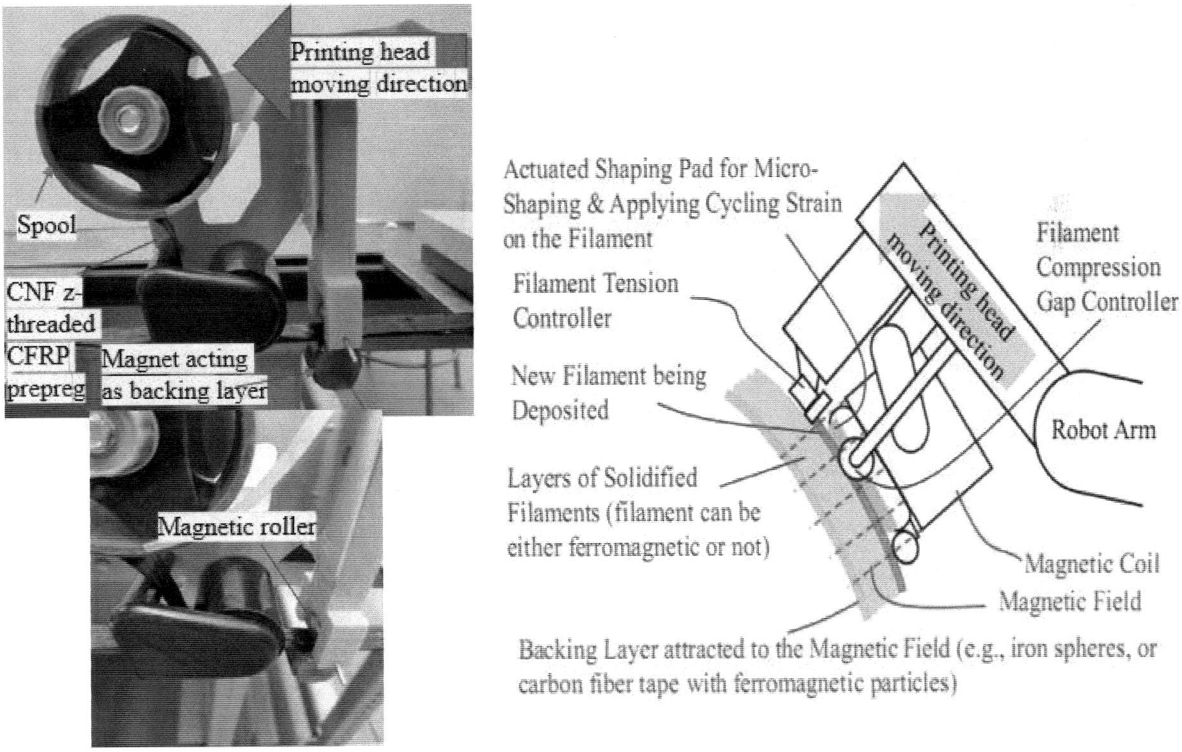

Figure 4. A simplified hand-hold printing head designed of MCFA-AM (left) and the MCFA-AM printing head schematic described in [11] (right). (modified from [19])

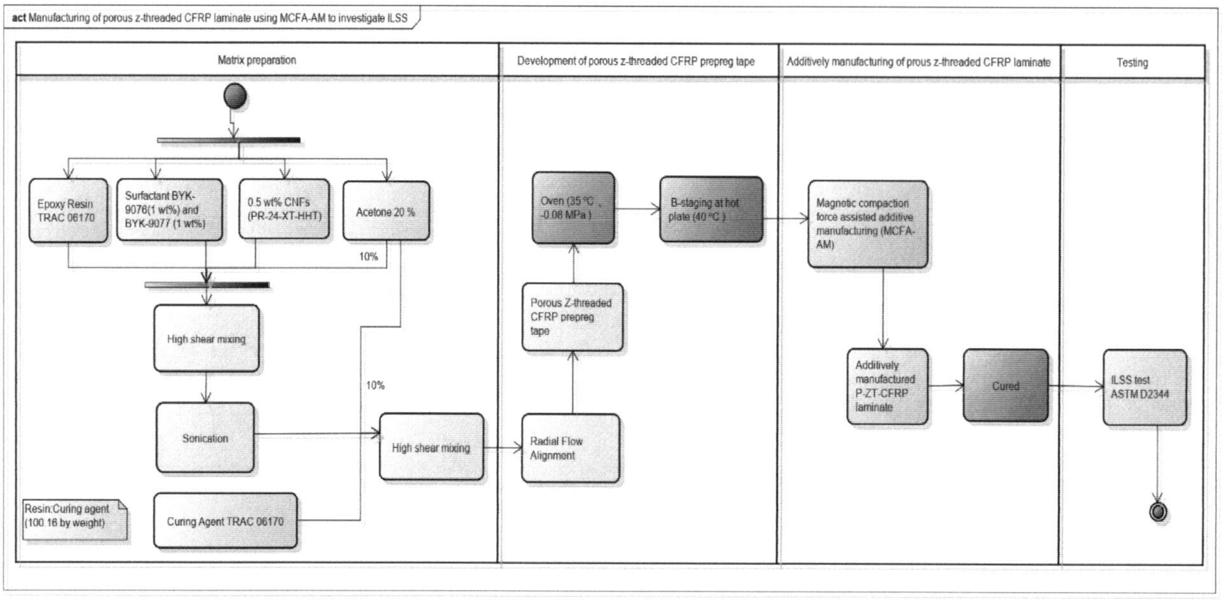

Figure 5. Overall P-ZT-CFRP laminate manufacturing.

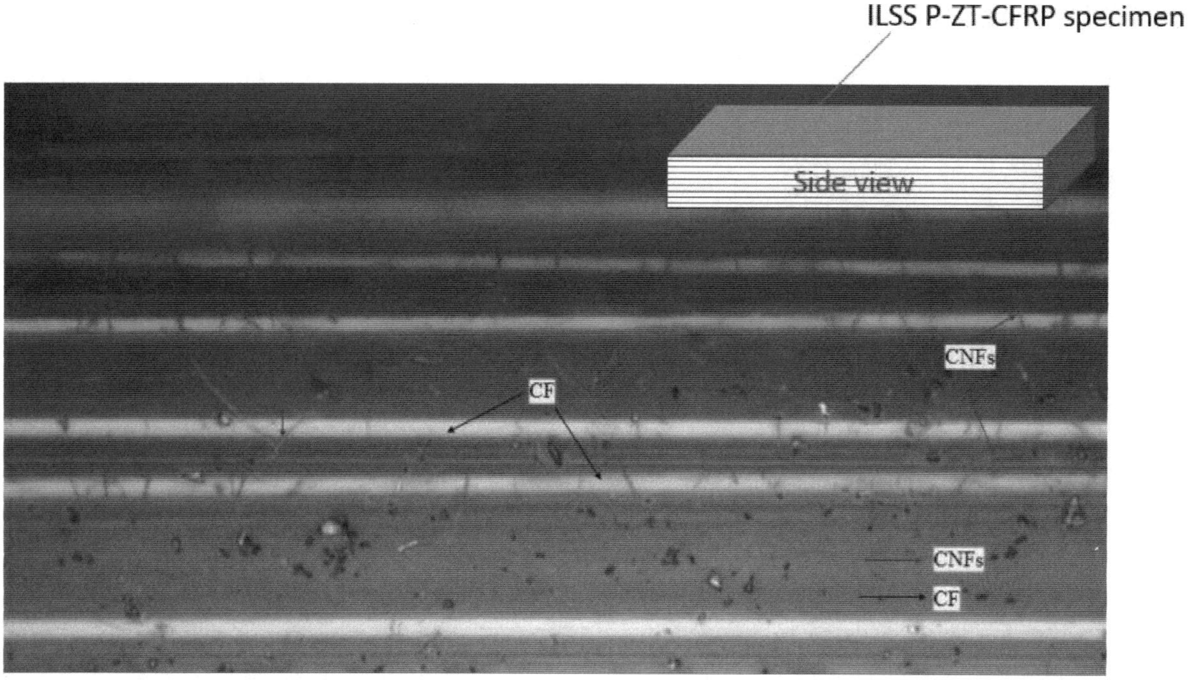

Figure 6. Microscopic image (extended-focused) taken at 1000X for the edge (i.e., side-view) of the CNC cut and hand-polished ILSS specimen from the P-ZT-CFRP laminate (sample 3).

2.3 Characterization techniques

Once samples 1, 2, and 3 were produced, a computer numerical control (CNC) router, CNC Shark Pro HD from the Next Wave Automation was used to cut five ILSS specimens (5 mm wide and 15 mm long) from each laminate sample. The specimens were sanded to the thickness of 2.5 mm.

The ILSS specimens obtained was characterized in terms of fiber content, void content, and interlaminar shear strength (ILSS). The fiber content and void content was estimated from the acetone fraction and the density measurement of the specimens as detailed in the Appendix. The ILSS was measured using the ASTM D2344 [25]. The test was conducted on TINIUS OLSEN Super "L" Universal Testing Machine, as shown in Figure 6. Short-beam strength was calculated using Equation 1.

$$F^{sbs} = 0.75 * \frac{P_m}{b * h} \tag{1}$$

Where F^{sbs} = short-beam (shear) strength (which represents ILSS), MPa
P_m = maximum load observed during the test, N
b = measured specimen width, mm
h= measured specimen thickness, mm

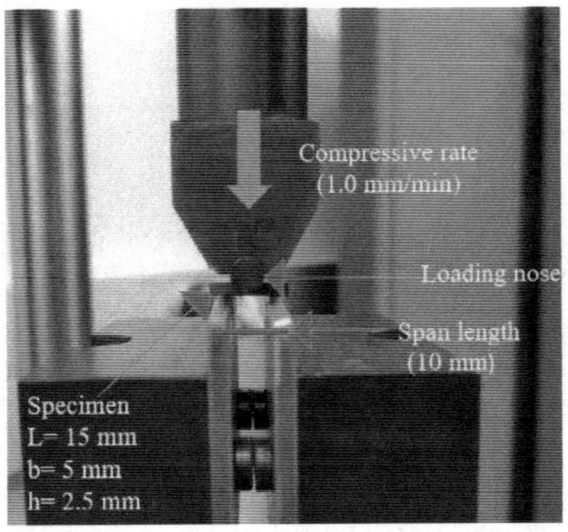

Figure 6. Short beam test setup [19]

3. RESULTS AND DISCUSSION

The results of the short beam shear tests for the OOA-VBO cured P-CFRP laminate and the MCFA-AM fabricated P-ZT-CFRP laminate are shown in Table 5, together with the OOA-VBO cured traditional CFRP and the MCFA-AM produced traditional CFRP laminate from a previous study [19].

Table 5. Mean ILSS tested by short beam strength test.

	Regular CFRP [19]		P-CFRP		P-ZT-CFRP
	Control CFRP	AM-CFRP	Sample 1	Sample 2	Sample 3
Manufacturing method	OOA-VBO	MCFA-AM	OOA-VBO	OOA-VBO	MCFA-AM
CNF (wt%)	0%	0%	0%	0%	0.50%
Acetone (wt%)	0%	0%	50%	20%	20%
Averaged Composite Density (g/cm^3)	1.49	1.46	1.52	1.21	1.3
Averaged void content (vol%)	NA	NA	6%	12%	10%
Averaged Carbon Fiber Volume Fraction (vol%)	50%	45%	70%	32%	44%
ILSS (MPa) tested by SBS test (average)	75.4	70.5	38.5	43.8	54.5
ILSS Normalized against Control Traditional CFRP		94%	51%	58%	72%
Specific ILSS per weight (10^3 m)	5.16	4.92	2.58	3.69	4.27
Specific ILSS per weight Normalized against Control Traditional CFRP		95%	50%	72%	83%
Estimated Specific Carbon Fiber Directional Tensile Strength per weight Normalized against Control Traditional CFRP (consider only CF volume fraction and density)		92%	137%	79%	101%

Fabrication of P-ZT-CFRP utilizing the RFA and MCFA was challenging due to the fast curing nature of the resin and dispersion of CNFs in the fast curing resin. This challenge was overcome by adding acetone to the mixture. The acetone helped to disperse the CNFs in the resin as well as slow the curing process allowing additional time to work on the P-ZT-CFRP prepreg tape fabrication during the RFA. The addition of acetone solved the main challenge but created another issue i.e., difficulty in the removal of acetone during the curing process. The amount of acetone used was a critical factor in this study. Therefore, this study was started with two different concentrations of acetone in the mixture (Sample 1 (50% acetone in resin) and Sample 2 (20%

acetone in resin) both cured by OOA-VBO). The results showed the ILSS for sample 1, and sample 2 was 38.5 MPa and 43.8 MPa, respectively; compared to the traditional control CFRP sample by OOA-VBO (ILSS= 75.4 MPa), the ILSS decreased by 49 % and 42 %, respectively. The decrease in the ILSS was mainly due to the voids created by the trapped acetone during the curing process. By reducing the amount of acetone in sample 2 to 20 %, the ILSS was increased by around 14 % compared to sample 1. However, it was also interesting to note that Sample 1 had higher density , higher carbon fiber volume fraction, and lower void content compared with Sample 2; this could be due to the lower viscosity and the acetone removal by vacuum caused the carbon fibers being packed more tightly. Due to the complexity in the RFA process when utilizing fast curing resin, sample 3 (P-ZT-CFRP) was fabricated by using 20 % acetone and 0.5 wt% CNF in the matrix mixture and cured with MCFA-AM technique. The ILSS value for sample 3 was improved by 42% and 24 % compared to sample 1 and sample 2, respectively. The increase in the ILSS value of sample 3 was due to the CNF z-threads showed in Figure 6. The sample 3 ILSS value is equivalent to 72 %, and 77 % of the control CFRP laminate (ILSS = 75.4 MPa) by OOA-VBO curing and the MCFA-AM cured CFRP laminate (ILSS = 70.5 MPa) reported in [19]. The sample 3 was lighter than the control CFRP and AM-CFRP. The specific ILSS per weight of sample 3 was recorded 4.27 while control CFRP and MCFA-AM CFRP have 5.16 and 4.92, respectively. In other words. The specific ILSS per weight of sample 3 is equivalent to 83 % and 87 % of control CFRP manufactured by OOA-VBO and the regular MCFA-AM CFRP, respectively. The fiber volume fraction for sample 1, sample 2, and sample 3 were reported 70%, 32 %, and 44 %, respectively while void content reported was 6 %, 12 % and 10 % respectively. Although the carbon fiber direction tensile strength was not tested in this paper, the tensile strength for unidirectional CFRP are typically predicted by rule of mixture and can be mainly proportional to carbon fiber volume fraction if one neglect the matrix's small contribution; therefore, one could roughly estimate the specific tensile strength per weight against the control CFRP cured by OOA-VBO. It is not a surprise some porous CFRP could have higher specific tensile strength per weight thanks to the lower density. However, it is intriguing to note that the P-ZT-CFRP laminate with 83% and 101% of the specific ILSS and specific tensile strength per weight, normalized against the control CFRP cured by OOA-VBO, indicated a lighter and stronger CFRP could be tailored by tailor a good balance of porosity and CNF contents in the future.

4. CONCLUSIONS

The P-ZT-CFRP laminate was fabricated for the first time utilizing the radial flow alignment, and magnetic compaction force assisted AM. The P-ZT-CFRP laminate (sample 3) fabricated by MCFA-AM showed an improved ILSS value compared to the P-CFRP samples cured by OOA-VBO (samples 1 and 2). The fiber volume fraction for sample 1, sample 2, and sample 3 were found to be 70%, 32 %, and 44 %, respectively. The voids contents in samples 1, sample 2, and sample 3 were 6%, 12% and 10% respectively. The sample 1 with higher fiber volume fraction was heavier than the sample 2 with low fiber volume fraction . This may be due to the high void content and lower carbon fiber volume fraction in sample 2. The improvement in the ILSS was achieved due to the use of CNFs z-threads in the CFRP prepreg. The CNFs z-threaded in the thickness direction helps to improve the ILSS of the P-ZT-CFRP laminate. The improvement was found to be 42 % and 24 % compared to sample 1 and sample 2. The porous nanocomposite (P-ZT-CFRP) showed great potential to overcome the effects of void on the mechanical properties of CFRPs. The voids, which are a common problem in AM, could be overcome through the use of CNF z-threaded CFRP prepreg. The CNF z-threads provides structural integrity in the composite laminate. The specific ILSS per weight of the P-ZT-CFRP sample manufactured by MCFA-AM technique was 0.83 times that of the control CFRP manufactured by the OOA-VBO method and 0.87 times that of the CFRP produced by MCFA-AM. The ILSS value could be improved by improving the techniques to prepare P-ZT-CFRP prepreg tape (e.g., full removal of acetone during the curing process) and application of high magnetic compaction pressure in the MCFA-AM process.

5. FUTURE WORKS

The future work of this study could be (i) finding the alternative chemical to replace acetone that does the same curing inhibiting and viscosity reduction functions as acetone in the matrix as discussed above, (ii) if acetone is still used, one could find the effective way to remove the acetone without creating much voids, (iii) finding the optimal concentration of CNFs and acetone/chemical to be used in the fast curing resin during RFA (iv) use considerable magnetic compaction pressure during fabrication and study the effects of magnetic compaction pressure on ILSS and tension test, (v) apply vacuum during the MCFA-AM process.

6. ACKNOWLWDGMENTS

The authors would like to acknowledge the financial support from the Alabama Department of Commerce through the Alabama Innovation Fund (Award number: 150436), and the Graduate Assistantship from the Systems Engineering Program at the University of South Alabama. The authors are grateful for the carbon fiber materials provided by Troy, Resins supplied from Hexion Inc.

7. REFERENCES

1. ASTM Standard ISO / ASTM52900-15, Standard Terminology for Additive Manufacturing – General Principles – Terminology, ASTM International, West Conshohocken, PA, 2015, www.astm.org

2. Hull, C.W., Apparatus for production of three-dimensional objects by stereolithography. 1986, Google Patents.

3. Attaran, Mohsen. "The rise of 3-D printing: The advantages of additive manufacturing over traditional manufacturing." Business Horizons 60, no. 5 (2017): 677-688. https://doi.org/10.1016/j.bushor.2017.05.011

4. Blok, L. G., Longana, M. L., Yu, H., & Woods, B. K. S. (2018). An investigation into 3D printing of fibre reinforced thermoplastic composites. Additive Manufacturing, 22, 176-186. https://doi.org/10.1016/j.addma.2018.04.039

5. Parandoush, P., & Lin, D. (2017). A review on additive manufacturing of polymer-fiber composites. Composite Structures, 182(2017), 36-53. https://doi.org/10.1016/j.compstruct.2017.08.088

6. Klosterman, D., Chartoff, R., Graves, G., Osborne, N., & Priore, B. (1998). Interfacial characteristics of composites fabricated by laminated object manufacturing. Composites Part A: Applied Science and Manufacturing, 29(9-10), 1165-1174. https://doi.org/10.1016/S1359-835X(98)00088-8

7. Yang, C., Tian, X., Liu, T., Cao, Y., & Li, D. (2017). 3D printing for continuous fiber reinforced thermoplastic composites: mechanism and performance. Rapid Prototyping Journal, 23(1), 209-215. https://doi.org/10.1108/RPJ-08-2015-0098

8. Parandoush, P., Tucker, L., Zhou, C., & Lin, D. (2017). Laser assisted additive manufacturing of continuous fiber reinforced thermoplastic composites. Materials & Design, 131, 186-195. https://doi.org/10.1016/j.matdes.2017.06.013

9. Ning, F., Cong, W., Qiu, J., Wei, J., & Wang, S. (2015). Additive manufacturing of carbon fiber reinforced thermoplastic composites using fused deposition modeling. Composites Part B: Engineering, 80, 369-378. https://doi.org/10.1016/j.compositesb.2015.06.013

10. Comer, A. J., D. Ray, W. O. Obande, D. Jones, J. Lyons, I. Rosca, R. M. O'higgins, and M. A. McCarthy. "Mechanical characterisation of carbon fibre–PEEK manufactured by laser-assisted automated-tape-placement and autoclave." Composites Part A: Applied Science and Manufacturing 69 (2015): 10-20. https://doi.org/10.1016/j.compositesa.2014.10.003

11. Hsiao, K-T., Method and apparatus for 3d printing, patent-pending WO/2018/057784. (https://patentscope.wipo.int/search/en/detail.jsf?docId=WO2018057784)

12. Hsiao, Kuang-Ting, Alexander M. Scruggs, John S. Brewer Jr, Gregory JS Hickman, Erin E. McDonald, and Kendrick Henderson. "Effect of carbon nanofiber z-threads on mode-I delamination toughness of carbon fiber reinforced plastic laminates." Composites Part A: Applied Science and Manufacturing 91 (2016): 324-335.doi:10.1016/j.compositesa.2016.10.022.

13. Kirmse, Sebastian, and Kuang-Ting Hsiao. "Enhancing the interlaminar shear strength of unidirectional carbon fiber reinforced plastic (CFRP) laminate using a nanofiber z-threading strategy." Proceedings of CAMX 2018 (The Composites and Advanced Materials Expo), Dallas, TX, Oct. 15 18 (2018): 2018.

14. Kirmse S, Kim K, Ranabhat B, Hsiao K-T. Effects of Carbon Nanofiber Z-Threads on the Longitudinal Compressive Strength of Unidirectional CFRP Laminates. SAMPE 2019 - Charlotte, NC, Charlotte, NC: SAMPE (Society for the Advancement of Material and Process Engineering); 2019, p. TP19--1531. doi:10.33599/nasampe/s.19.1531.

15. Scruggs, Alexander M., Sebastian Kirmse, and Kuang-Ting Hsiao. "Enhancement of Through-Thickness Thermal Transport in Unidirectional Carbon Fiber Reinforced Plastic Laminates due to the Synergetic Role of Carbon Nanofiber Z-Threads." Journal of Nanomaterials 2019 (2019). doi:10.1155/2019/8928917.

16. Scruggs, Alexander M., Kendrick Henderson, and Kuang-Ting Hsiao. "Characterization of electrical conductivity of a carbon fiber reinforced plastic laminate reinforced with z-aligned carbon nanofibers." Proceedings of CAMX (2016): 26-29.

17. Ranabhat, B., and Hsiao, K. "Improve the Through-Thickness Electrical Conductivity of CFRP Improve the Through-Thickness Electrical Conductivity of CFRP Laminate Using Flow-Aligned Carbon Nanofiber Z-Threads," in Proceedings of SAMPE 2018 (Society for the Advancement of Material and Process Engineering), Long Beach, CA, May, 21-24, 2018, SE18--1100.

18. Ranabhat, Bikash, Sebastian Kirmse, and Kuang-Ting Hsiao."Roles of carbon nanofiber alignment and concentration for the improvement of z-directional DC electrical conductivity

of CNF z-threaded CFRP laminates manufactured using radial flow alignment technique," in Proceedings of CAMX 2019, Anaheim, CA, 2019.

19. Ranabhat, B., Kirmse, S., and Hsiao, K. "Feasibility Study of Novel Magnetic Compaction Force Assisted Additive Manufacturing (MCFA-AM) Methodology for Continuous Carbon Fiber Reinforced Polymer (C-CFRP) Composites," in SAMPE 2019 - Charlotte, NC, 2019.

20. Hsiao, Kuang-Ting." Apparatus and method for directional alignment of nanofibers in a porous medium" US Provisional Serial NO. 62/003,982; PCT/US2015/033000; Pub.No.: WO/2015/184151 A1 (2015).

21. Wohlers, Terry. "3D printing and additive manufacturing state of the industry." Annual Worldwide Progress Report. Wohlers Associates(2014).

22. Damodaran, V., A. G. Castellanos, M. Milostan, and P. Prabhakar. "Improving the Mode-II interlaminar fracture toughness of polymeric matrix composites through additive manufacturing." Materials & Design 157 (2018): 60-73. https://doi.org/10.1016/j.matdes.2018.07.006

23. Zindani, Divya, and Kaushik Kumar. "An Insight into Additive Manufacturing of Fiber Reinforced Polymer Composite." International Journal of Lightweight Materials and Manufacture (2019). https://doi.org/10.1016/j.ijlmm.2019.08.004

24. Yu, Tianyu, Ziyang Zhang, Shutao Song, Yuanli Bai, and Dazhong Wu. "Tensile and flexural behaviors of additively manufactured continuous carbon fiber-reinforced polymer composites." Composite Structures 225 (2019): 111147. https://doi.org/10.1016/j.compstruct.2019.111147

25. ASTM D2344 / D2344M-16, Standard Test Method for Short-Beam Strength of Polymer Matrix Composite Materials and Their Laminates, ASTM International, West Conshohocken, PA, 2016, www.astm.org

8. APPENDIX

8.1 Fiber Volume fraction (FVF)

$$V_f = \frac{\rho_c - \rho_m}{\rho_f - \rho_m} \quad \text{(A)}$$

Where, ρ_c, ρ_f, ρ_m are the density of composite, fiber and matrix respectively.

$$\rho_m = v_{fr} * \rho_r + v_{act} * \rho_{act}$$

Where v_{fr}, ρ_r are volume fraction and density of resin respectively and v_{act}, ρ_{act} are volume fraction and density of acetone respectively.

8.2 Void content

$$V_{void} = \frac{\rho_t - \rho_c}{\rho_t} \tag{B}$$

Where, ρ_t, ρ_c are the theoretical and measured density.

$$\rho_c = \frac{mass}{volume}, \ \rho_t = V_f * \rho_f + (1 - V_f) * \rho_{matrix}$$

Effect of Infill Patterns on Heating for Additively Manufactured Compression Molds

Seokpum Kim[1], Pritesh Yeole[1,2,4], Ahmed Arabi Hassen[3], Vipin Kumar[3], Vlastimil Kunc[1], Uday Vaidya[1,2,4,*]

1. Energy and Transportation Science Division, Oak Ridge National Laboratory, Oak Ridge, TN 37830, USA
2. Department of Mechanical, Aerospace, and Biomedical Engineering, University of Tennessee, Estabrook Rd, Knoxville, TN 37996, USA
3. Materials Science and Technology Division, Oak Ridge National Laboratory, Oak Ridge, TN 37830, USA
4. Institute for Advanced Composites Manufacturing Innovation (IACMI), 2370 Cherahala Blvd, Knoxville TN 37932, USA

* Corresponding Author

E-mail: uvaidya@utk.edu, Tel: +1-865-974-7620

ABSTRACT

Carbon Fiber Reinforced Polymer (CFRP) composites have been used in additive manufacturing (AM) to increase the stiffness and strength of the parts. CFRP produced by AM is being considered for tooling applications. In AM, short carbon fibers are aligned along the deposition direction. However, it results in anisotropic thermal properties that affect the heat transfer and warpage of the final part. In this study, three male molds with different infill patterns were considered based on the slicing software for the extrusion deposition fabrication-additive manufacturing (EDF-AM) process. These include (a) 0°: infill pattern along the printing direction; (b) 90°: infill pattern perpendicular to the printing direction; (c) 0°/90°: alternate layers along and perpendicular directions. Finite Element Analysis (FEA) was conducted to understand the effect of infill pattern on a heat propagation. The results show that the directionality of the infill affects the heating performance.

This manuscript has been authored by UT-Battelle, LLC under Contract No. DE-AC05-00OR22725 with the U.S. Department of Energy. The United States Government retains and the publisher, by accepting the article for publication, acknowledges that the United States Government retains a non-exclusive, paid-up, irrevocable, worldwide license to publish or reproduce the published form of this manuscript, or allow others to do so, for United States Government purposes. The Department of Energy will provide public access to these results of federally sponsored research in accordance with the DOE Public Access Plan (http://energy.gov/downloads/doe-public-access-plan).

1. INTRODUCTION

The extrusion deposition fabrication – additive manufacturing (EDF-AM) system at Oak Ridge National Laboratory includes a single screw extruder in which the screw rotates inside a heated barrel and a molten material is extruded out from the deposition nozzle under high pressure [1]. Short carbon fibers in the extruded material are highly aligned along the deposition direction, and therefore, the thermal conductivity of a carbon fiber printed component is anisotropic [2]. Specifically, the extruded bead has higher conductivity in printing direction (deposition direction, or X direction) than the conductivity in the other two perpendicular directions (Y- and Z-directions). The anisotropy can be utilized to improve the performance of an AM produced part by combining the design and deposition direction control. The flexibility of the EDF-AM process allows precise control of the deposition direction (toolpath). Smith et al. [3] successfully achieve a 20% improvement in the performance of the heat sink by controlling the printing direction. Jia et al. [4] studied the specific orientation of graphite flakes to impart high through-plane thermal conductivity for maximum heat dissipation in the AM produced part. The orientation of discontinuous CFs also affects the thermal expansion of the part; expansion is more constrained along the deposition direction or fiber alignment direction due to very high thermal conductivity values of aligned CF to dissipate heat more efficiently. Deformation in the molds can lead to a significant distortion in the final parts made from these molds. Hence, maintaining a tight tolerance even at high working temperature is one of the most critical challenges with AM molds. Several studies were conducted on AM tools for high-temperature applications and promising results were obtained, where tool deformations stayed within acceptable limits [5, 6].

This work investigates the relationship between print orientation (deposition path), and part thermal performance made of 50% by weight carbon fiber reinforced polyphenylene sulfide composite pellets (50CF-PPS). A computational prediction of heat transfer is demonstrated in this study, based on the toolpaths generated from three different infill patterns. A multi-step approach is followed in the present work as (1) anisotropic thermal conductivity of the extruded bead (2) toolpath based on the infill pattern of the molds (determined by slicing software). This research concerns the different infill patterns and their effect on the heat transfer in composite molds manufactured by the EDF-AM process.

2. MATERIALS AND SAMPLE PREPARATION

In this study, 50CF-PPS was selected because PPS is a semi-crystalline thermoplastic polymer with a high heat deflection temperature [7]. The melting point of PPS is 285 °C. In this research, a mold, as shown in Figure 1, was manufactured. The infill pattern is the interior structure/geometry of the printed cross-section. Infill density and geometric patterns have a significant effect on the mechanical performance [8]. In BAAM printing, fibers are aligned along the deposition direction. Therefore, the structure of the infill pattern has a significant influence on the fiber orientation, which has a simultaneous effect on the thermal conductivity and the warpage of the final part. To understand the effect of infill pattern, three molds of the same geometry containing an infill pattern of 0° (along the deposition direction), 90° (transverse to the deposition direction) and 0°/90° (one layer along the deposition direction while the second transverse) were prepared. The outside perimeter with 3 layers was kept constant. The schematic of various infill patterns is shown in Figure 2.

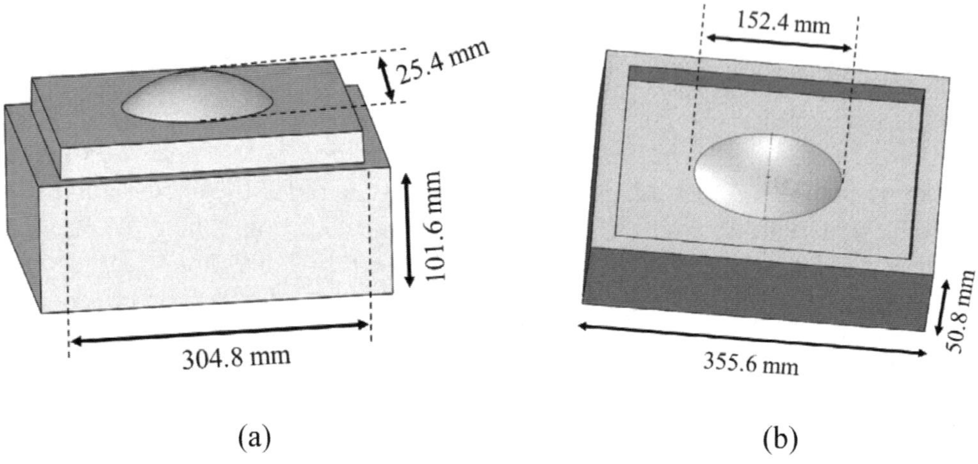

Figure 1: Dome-shaped AM printed Mold (a) Male Mold (b) Female Mold

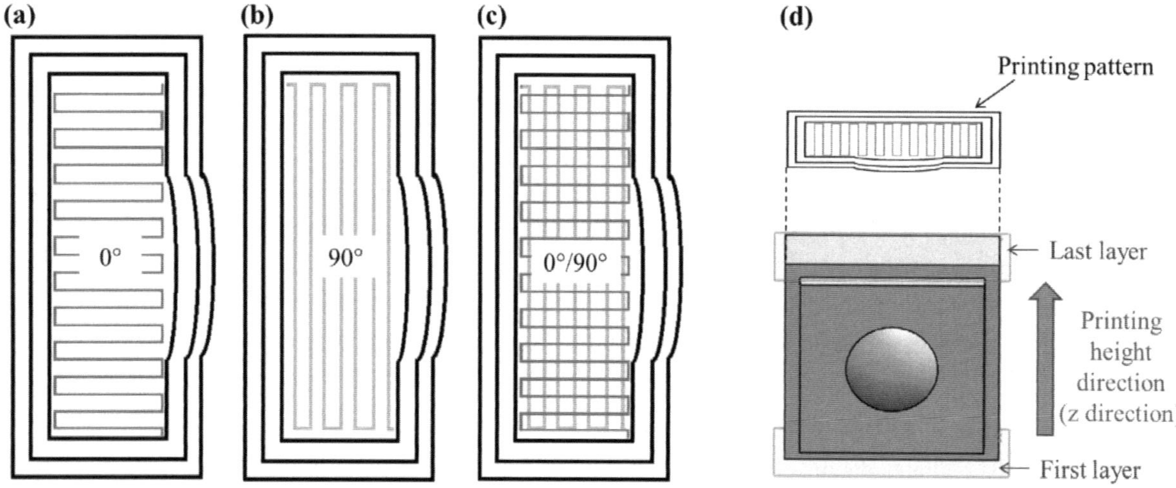

Figure 2: Infill Pattern (a) 0° (b) 90° (c) 0°/90°. The bead width: 13.97 mm (0.55"), bead height: 5.08 mm (0.2"), and infill overlap: 10% - 15% (d) Printing direction

In AM FRC samples, it is expected to have different thermal conductivity along with the three principal directions due to the fiber orientation: X-direction – parallel to the motion of extruder, Z-direction – interlayer direction, Y-direction – perpendicular to both extruder and interlayer direction. In this study, the transient plane source (TPS) technique [9] was used to measure the thermal conductivity of the samples. The BAAM screw extruder has five heating zones to melt the polymer; zone 1: 305 °C, zone 2: 321 °C, zone 3: 326 °C, zone 4: 337 °C, zone (tip): 337 °C. A block of 304.8 mm × 304.8 mm ×304.8 mm was printed with a unidirectional orientation of the deposited beads, and a smaller cube of 50.8 mm^3 was extracted from it for thermal conductivity measurement. The experimental procedure, results, and discussion of thermal conductivity measurement of 50CF-PPS are explained in Ref. [10].

3. THERMAL CONDUCTIVITY MEASUREMENT

Table 1: Thermal conductivity data at various temperatures

Sr. No.	Temperature (°C)	Thermal Conductivity (W/m K)		
		X-Direction	Z-Direction	Y-Direction
1	23	1.403 ± 0.040	0.397 ± 0.005	1.290 ± 0.017
2	58	1.479 ± 0.017	0.474 ± 0.004	1.239 ± 0.022
3	99	1.488 ±0.026	0.463 ± 0.009	1.349 ± 0.009
4	123	1.485 ± 0.040	0.445 ± 0.004	1.289 ± 0.061
5	159	1.498 ± 0.007	0.466 ± 0.006	1.267 ± 0.010

Thermal conductivity of 50CF-PPS samples was measured in the temperature range from room temperature to 170 °C, and it was observed that the thermal conductivity at any given temperature in X-direction was higher when compared to Y-direction and Z-direction. As explained in the previous section, CF tends to align in the deposition direction of BAAM extruder representing higher thermal conductivity. A tampering system was used in order to increase the bonding between deposited layers (Z- direction). During tampering, material partially squeezed out in Y-direction and forced some of the chopped fibers to align in Y-direction. Thus, it can be estimated that the thermal conductivity of BAAM printed samples in Y-direction was in between X-direction and Z-direction. Thermal conductivity values are given in Table 1.

4. FINITE ELEMENT ANALYSIS

Finite Element Analysis (FEA) was performed on the male molds with three different infill geometries, as shown in Fig. 2. A commercial FEA tool, ABAQUS 2018, was used for computational thermal analysis. The geometric model is discretized into 28.6k elements, including 13.8k tetrahedron elements and 14.8k hexahedron elements. The element size in the z-direction is 5.08 mm, which is the same as the layer height (= slicing thickness) of the 3D-printed mold. The average element size in x- and y-direction is 14 mm. A linear interpolation function is used for both types of elements. The analysis is for thermal simulation without deformation simulation, and therefore, shear locking is not a concern in this study.

The analysis was performed with two steps: AM process simulation and service loading simulation. The purpose of the first step (AM process) is to assign the directionality of the anisotropic material properties in every element. The fibers in a printed bead are highly aligned along the deposition direction, and therefore, the anisotropic thermal properties and mechanical properties of the printed bead have a directionality along the deposition direction. The toolpath direction varies location by location in the entire printed part, as shown in Figure 2. Therefore, in the FEA model, the directionality of the anisotropic material properties of each element should be assigned based on the local toolpath direction. During the AM process simulation in Abaqus, the local toolpath directions are assigned in the FEA elements. The directionality information is transferred to the next step simulation.

Table 2: Coefficients used in the thermal simulation

Constitutive Relation	Coefficient	Value	Unit
Conduction	κ_1	1.472	W/(m·K)
	κ_2	1.287	W/(m·K)
	κ_3	0.4528	W/(m·K)
Convection	h	5.0	W/(m²·K)
Radiation	ϵ	0.9	dimensionless

The second step is a service loading simulation. The printed mold will be placed on a heated platen of a hydraulic compression press at ORNL. The temperature of the mold will gradually increase and reach a steady state. The goal of the service loading simulation is to predict the surface temperature of the mold at a steady state. The prediction was performed via a thermal FEA simulation with the following three constitutive relations: thermal conduction, thermal convection, thermal radiation. For thermal conduction, the three-dimensional Fourier's law was used with the orthotropic conductivity coefficients ($\kappa_1, \kappa_2, \kappa_3$). For thermal convection, Newton's law of cooling was used with the convection coefficient (h). For thermal radiation, the Stefan-Boltzmann law was used with the emissivity coefficient (ϵ). The coefficients are listed in Table 2.

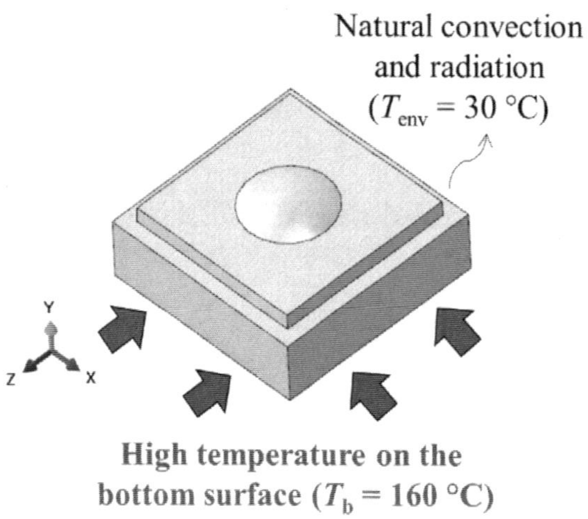

Figure 3: Boundary conditions for service load simulation (thermal simulation)

The boundary conditions are shown in Figure 3. The environment temperature was set to 30 °C, and the mold initially had a uniform temperature of 30 °C. The bottom surface temperature was set to 160 °C throughout the simulation. The convection and radiation were applied to the surface of the mold except for the bottom surface. The convection and radiation from the bottom surface were not considered. As the simulation proceeded, the temperature increased from the bottom of the mold, and the heat gradually propagated to the top surface of the mold.

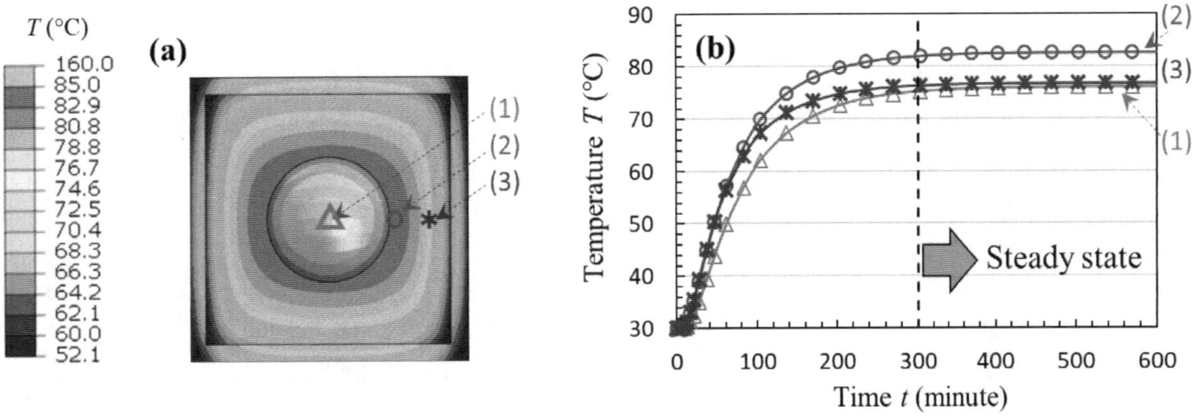

Figure 4: (a) Temperature field of the top surface at t = 600 minutes for the mold with 0°/90° infill lines and 3 perimeter lines, (b) Temperature history of the three locations indicated in Fig. (a). The temperature reached a steady state at t = 300 minutes.

Because of the heat loss via convection and radiation from the surface of the mold, the increase of the temperature slowed down and reached a steady state. Figure 4(a) shows the temperature field of the top surface at t = 600 minutes for the mold with 0°/90° infill lines and 3 perimeter lines. Three locations along the center vertical line were picked: (1) Center of the dome indicated in red '△', (2) A concave point at the interface line between the dome surface and the flat surface which is indicated in purple '○', (3) A point further away from the center which is indicated in blue '✱'. Figure 4(b) shows the temperature history of these three locations. The temperature increased rapidly up to t = 150 ~ 200 minutes and reached a steady-state at t = 300 minutes.

The high temperature near the circular interface line between the dome surface and top flat surface may be attributed to the following reasons: (1) The center area has high temperature because of the heat loss on the surfaces at the four sides of the mold. The dome surface is further from the bottom surface (heat source) when compared to the top flat surface. Hence the conduction to the dome surface is less effective when compared to the conduction to the top flat surface. Therefore, a large area near the center has a high temperature except for the dome surface. (2) The concave surface has high temperature as compared to the convex surface because the ratio of surface area to the volume on a concave area is smaller than the ratio on a convex area.

We have performed the thermal simulations for the three molds, as shown in Figure 2, and the temperature fields of the top surfaces from the three molds are shown in Figure 5.

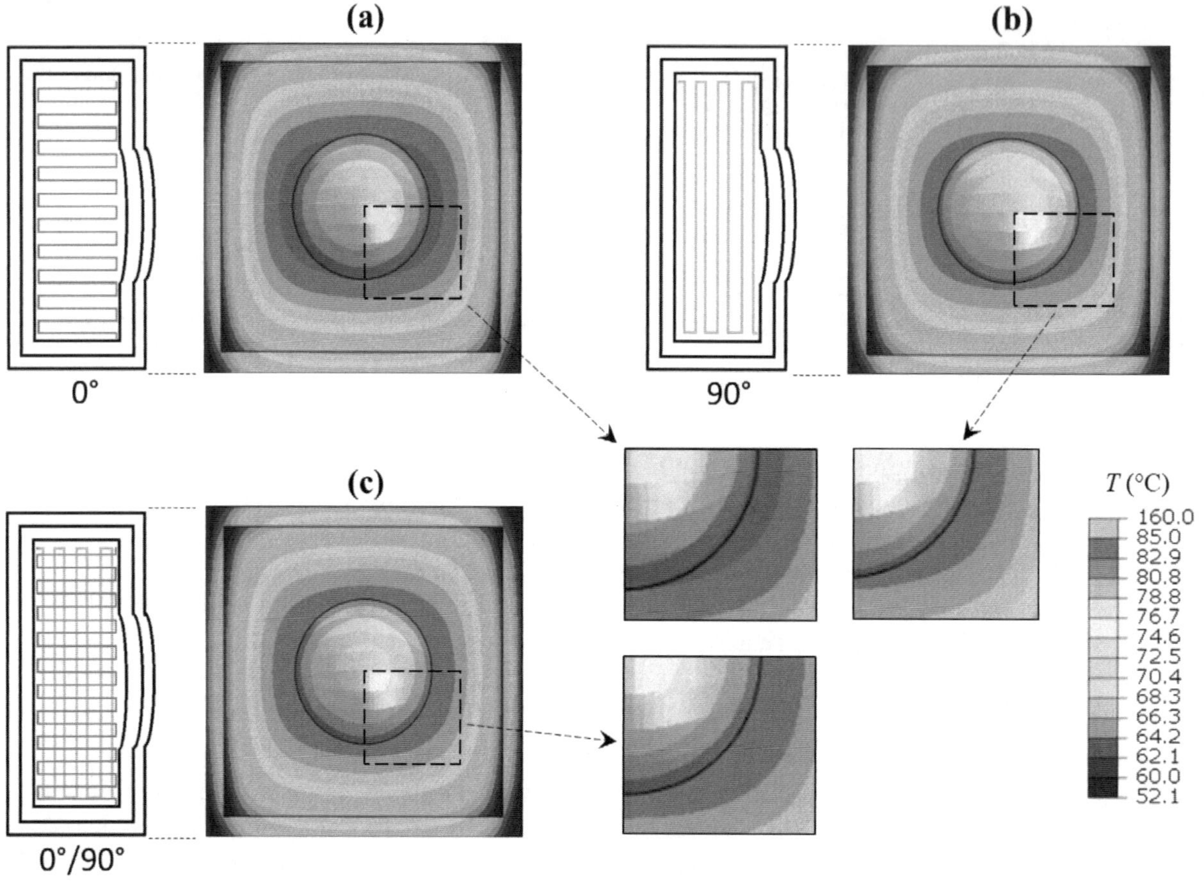

Figure 5: Temperature field of the top surface of the three molds. (a) Mold with 0° infill, (b) Mold with 90° infill, (c) Mold with 0°/90° infill.

At the interface line between the dome surface and the top flat surface, the mold with 0° infill has the highest temperature (83.9 °C), and the mold with 90° infill has the lowest temperature (82.2°C). The temperature of the mold with 0°/90° infill is in between the temperatures (82.8 °C) of 0° infill mold and 90° infill mold. Overall, the temperature of the three molds has similar profiles. The average temperatures on the surface of the interest are 79.3 °C, 77.7 °C, and 78.4 °C for 0° infill, 90° infill, and 0°/90° infill molds, respectively. The location of the peak temperature and the surface of interest is shown in Figure 6(a). Figure 6(b) shows the average temperatures of the surface of interest and the peak temperatures for the three molds with 0° infill, 90° infill, and 0/90° infill. The temperature difference between the three molds is attributed to the infill directions. An extruded material has highly aligned fibers along the deposition direction, and therefore, the conductivity along the deposition direction is higher than the conductivity in other directions. Heat transfer in the mold is mainly in the direction from the bottom surface of the mold to the top surface of the mold. If the infill direction is parallel to the heat transfer direction (0° infill), then due to the high conductivity along the heat transfer direction, the top surface holds high temperature. If the infill direction is perpendicular to the heat transfer (i.e 90° infill), then due to the low conductivity along the heat transfer direction, the top surface has low temperature.

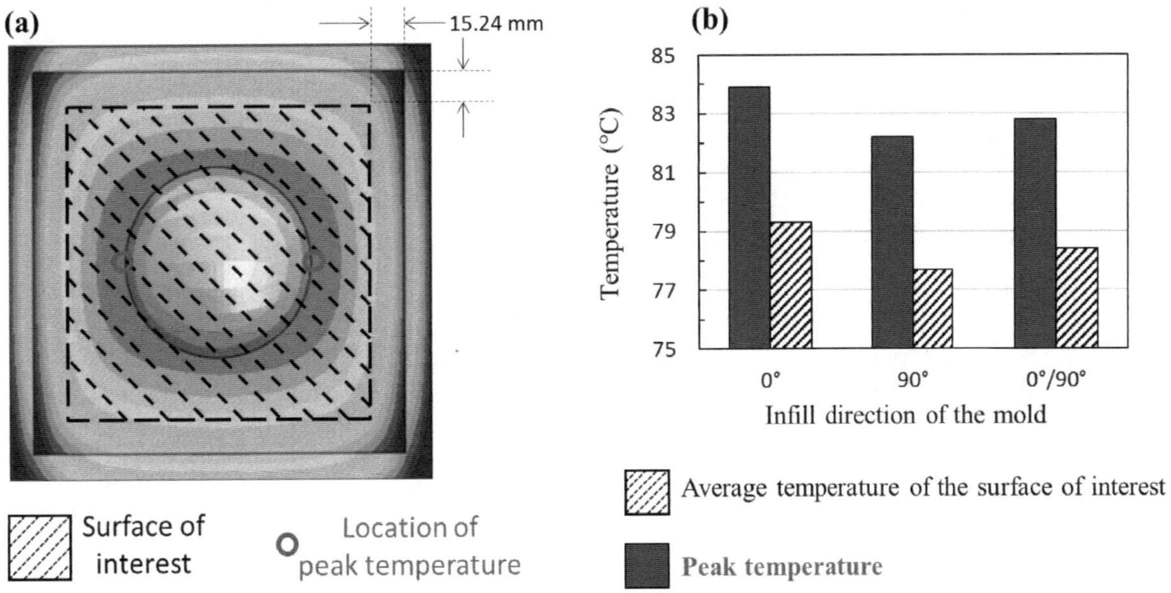

Figure 6: (a) Surface of interest (shaded area) and the location of peak temperature (purple circle) (b) Average temperatures of the surface of interest and the peak temperatures of the three molds with 0° infill, 90° infill, and 0/90° infill.

It should be noted that the current simulations have the following limitations. First, the simulation considers only the heat transfer behavior in the mold and ignores the mechanical deformation. Because the bottom surface of the mold has a high temperature (T = 160 °C) than the top surface (T = 65 - 85 °C), the thermal expansion at the bottom surface will be more significant than the thermal expansion at the top surface, which will result in warpage of the mold. The warpage of the mold will create a gap between the heated platen and the bottom surface of the mold, and the amount of the gap will be different among the three molds. Due to the gap, the bottom surface of the mold will have a non-uniform temperature distribution. Second, the simulation assumes an ideal case of infill printing in which the deposited beads have a perfect bonding. In actual large-scale printing, the deposited beads with high temperatures (extrusion temperature of 337°C), which were initially bonded to each other, might get separated as they shrink during cooling. The shrinkage of the printed beads can create a gap between infill lines, which can decrease the conductivity across the infill lines. The limitations addressed above should be considered in the simulation when we print actual molds in BAAM.

5. SUMMARY

The effect of infill pattern in AM produced molds is studied by considering the effect of CF orientation and its thermal expansion at higher temperature. The thermal conduction of the 50CF-PPS EDF-AM sample was higher in X-direction when compared with other directions (Y – and Z – direction). It was observed that the conductivity has a linear direct co-relation with temperature. FEA model was developed to understand the effect of infill pattern (0°, 90°, and 0°/90°) on the heat conduction when EDF-AM fabricated mold was placed on the heated platen. The average surface temperature of the mold was highest in 0° (79.3 °C) > 0°/90° (78.4 °C) > 90° (77.7 °C) because CFs were highly aligned along the deposition direction. The model was developed with some limitations/assumptions; therefore, the experimental validation of the model is under development.

ACKNOWLEDGEMENT

Research sponsored by the U.S. Department of Energy, Office of Energy Efficiency and Renewable Energy, Industrial Technologies Program, under contract DE-AC05-00OR22725 with UT-Battelle, LLC.

REFERENCE

1. P. Chesser, B. Post, A. Roschli, C. Carnal, R. Lind, M. Borish, and L.J.A.M. Love, *Extrusion control for high quality printing on Big Area Additive Manufacturing (BAAM) systems*. 2019. **28**: p. 445-455.
2. B.S. Richardson, L.J. Love, P.C. Chesser, J. Miller, and A. McCalip, *Enabling Technologies for Medium Additive Manufacturing (MAAM)*. Oak Ridge National Lab.(ORNL), Oak Ridge, TN (United States). 2018.
3. M. Smith, S. Kim, A. Lambert, M. Walde, J. Lindahl, K. Mungale, . . . , and V. Kunc. *Maximizing the Performance of a 3D Printed Heat Sink by Accounting for Anisotropic Thermal Conductivity During Filament Deposition*. in *2019 18th IEEE Intersociety Conference on Thermal and Thermomechanical Phenomena in Electronic Systems (ITherm)*. 2019. IEEE.
4. Y. Jia, H. He, Y. Geng, B. Huang, X.J.C.S. Peng, and Technology, *High through-plane thermal conductivity of polymer based product with vertical alignment of graphite flakes achieved via 3D printing*. 2017. **145**: p. 55-61.
5. E. Barocio, B. Brenken, A. Favaloro, and R. Pipes. *Extrusion deposition additive manufacturing of composite molds for high-temperature applications*. in *Proceedings of the Int. SAMPE Tech. Conf., Seattle, WA, USA*. 2017.
6. V. Kunc, J. Lindahl, R. Dinwiddie, B. Post, L. Love, M. Matlack, . . . , and A. Hassen. *Investigation of in-autoclave additive manufacturing composite tooling*. in *CAMX Conference, Anaheim, CA*. 2016.
7. A.A. Hassen, J. Lindahl, X. Chen, B. Post, L. Love, and V. Kunc. *Additive manufacturing of composite tooling using high temperature thermoplastic materials*. in *SAMPE Conference Proceedings, Long Beach, CA, May*. 2016.
8. S. Kim, *An integrated design approach for infill patterning of fused deposition modeling and its application to an airfoil*. Oak Ridge National Lab.(ORNL), Oak Ridge, TN (United States). 2017.
9. S.E. Gustafsson, E. Karawacki, and M.N.J.J.o.A.P. Khan, *Determination of the thermal‐conductivity tensor and the heat capacity of insulating solids with the transient hot‐strip method*. 1981. **52**(4): p. 2596-2600.
10. V.K. Ralph B. Dinwiddie, John M. Lindahl, Seokpum Kim, Pritesh Yeole, and L.J.L. Chad Duty, Ahmed Arabi Hassen, *Specific heat capacity and anisotropic thermal conductivity of additively manufactured carbon-fiber-reinforced thermoplastics; abs + 20% cf, pps + 50% cf and ppsu + 25% cf*. . 2020: Journal of Composite Science and Technology - Under Review.

DESIGN OF COMPOSITE COMPRESSION MOLDING TOOLS USING LARGE SCALE ADDITIVE MANUFACTURING

Michael Bogdanor[a], Himal Agrawal[a], Eduardo Barocio[a], Anthony Favaloro[a], Brian Smiddy[b], Ken Susnjara[b], R. Byron Pipes[a]
[a]Composites Manufacturing and Simulation Center, Purdue University
West Lafayette, IN, USA
[b]Thermwood Corporation
Dale, IN, USA

ABSTRACT

In this work, both pieces of a two-sided, compression molding tool are additively manufactured and post-processed using Thermwood's Large Scale Additive Manufacturing system. The printed geometries are machined to achieve net shape and for the insertion of cartridge heaters. The matched tools, made from 25% carbon fiber reinforced PESU, are used in the successful compression molding of a demonstration blocker door part made from a thermosetting prepreg platelet molding compound (PPMC). Design of the molded part and the additive manufactured tool was performed using the ADDITIVE3D and PPMC integrated workflow applications, respectively, to determine the manufacturing informed performance of the part and tool. Additionally, heat transfer simulations were performed to inform the molding process and requisite environmental conditions for successful molding including cartridge heater and platen temperature.

Corresponding author: Michael Bogdanor bogdanor@purdue.edu

1. INTRODUCTION

Large scale tooling is emerging as one of the most promising applications of the products of Extrusion Deposition Additive Manufacturing (EDAM). Single sided tools and fixtures for in- and out-of-autoclave composite manufacturing have seen significant development in recent years. Two-sided compression molding introduces several additional challenges beyond those seen in single-sided applications, but as will be discussed herein, modeling and simulation are excellent tools for addressing these challenges. Typical implementation of the EDAM process involves a gantry mounted extruder that deposits a fiber-filled thermoplastic bead onto a build plate. The combined motion of the gantry system and the build plate results in at least three axes of movement yielding parts with dimensions on the order of meters. Current commercial equipment is capable of depositing material at rates more than 500 lbs/hr or 225 kg/hr, typically using carbon fiber-filled, high performance polymers. Unsuccessful prints with large build volumes and high-grade materials result in significant waste in material, time, and cost. As such, there is a tremendous need to be able to predict the outcome of EDAM prints before any material is deposited to reduce the number of physical iterations and replace them with virtual design and optimization iterations. In this case study, the integrated workflow application, ADDITIVE3D, developed at Purdue

University in the Composites Manufacturing and Simulation Center (CMSC), will be exercised to demonstrate the design, modeling, and simulation of a two-sided EDAM tool, printed by Thermwood using the Large Scale Additive Manufacturing (LSAM) machine, for the compression molding of an aerospace blocker door manufactured with a prepreg platelet molding compound (PPMC) and designed using the PPMC3D application.

The EDAM process with fiber-reinforced polymers has enabled the application of additively manufactured (AM) tooling for traditional composite manufacturing processes that include hand layup [1], Vacuum Assisted Resin Transfer Molding (VARTM) [2,3], autoclave curing [4,5], and compression molding [5]. The relative small deformation reported for tools used in VARTM [2] and autoclave [4,5] processes indicates that the thermal stability of the printed composite materials is adequate for prototype tooling. The concept of AM tooling have been adopted in other fields, for instance, patterns for forming sand cavities used in foundries have been produced using the fused filament fabrication process [6], and inserts for injection molding have been produced with PolyJet® process [7].

EDAM tooling offers several benefits in composites manufacturing and prototyping efforts. First, EDAM tooling is generally lighter and more rapid to produce than the metal counterpart. Furthermore, AM can produce geometries for tooling that are not possible in subtractive manufacturing and can be utilized to place internal cavities and cutouts in large structures. The technology also offers the possibility of tailoring the printing path to match the coefficient of thermal expansion (CTE) of the part to be molded with the tool, this is particularly beneficial in molding of composite parts which themselves exhibit high degrees of anisotropy to minimize part deformation off the tool.

Modeling and simulation are key drivers in understanding the printing process and the performance of EDAM products. The previously developed ADDITIVE3D workflow [8–11] simulates the entire printing process using a time-dependent, coupled thermo-mechanical analysis that directly models the polymer kinetics, temperature dependent heat transfer behavior, and resulting viscoelastic mechanical behavior. The ADDITIVE3D toolkit includes model initialization based upon the actual machine path. The results can also be mapped to final, as-machined meshes for performance analyses as is the use case herein. The simulations are utilized to study the effects of the print direction on the deformations and interlayer stresses in the resulting part and to assist the designer in optimizing the heating and processing strategy for the part manufacturing process using the printed tools.

In the following sections, a brief overview of the ADDITIVE3D simulation approach is provided first. Next, the details of the case study of the compression molding of a PPMC blocker door using EDAM tooling will be discussed. The following section will describe the modeling and simulation efforts used to design the printed part and the manufacturing process. Then the results of the experimental study and molding trial are presented, demonstrating the successful molding of the part with the EDAM tools. The final section provides the conclusions from the study.

2. TOOLING DESIGN CASE STUDY

The blocker door part (Figure 1a) used in this case study presents a design case of an aerospace part that while not being a flight-critical component, has a design governed by its structural performance. The blocker door is 300 mm by 200 mm (11.8 in by 7.9 in) with a nominal surface thickness of 4 mm (0.16 in) and ribs up to 30 mm (1.18 in) deep. The ribs were designed with a draft angle of 5 degrees to facilitate demolding. In this representative design, 16 blocker doors are mounted on the outer circumference of the fan duct for a turbofan engine as shown in Figure 1b, shown in the deployed configuration. During typical operation, the blocker doors are stowed along the surface of the fan duct and have a smooth curved surface allowing the air to flow through the duct. When the thrust reversing system is deployed, the blocker doors rotate to restrict the airflow through the bypass and redirect the air through openings on the sides of the fan duct to reduce thrust and slow the aircraft. The blocker door is designed to withstand the pressure load from redirecting the air flow with a stiffness criterion such that the deformation of the geometry does not allow air flow through gaps between adjacent doors. This part is well suited to application of carbon fiber PPMC to meet the strength, stiffness, and weight requirements while maintaining high production rates. Compression molding of the carbon fiber and epoxy PPMC requires heated, two-sided tooling capable of achieving mold surface temperatures of 265 °F (130 °C) and pressure of 1500 psi (10.3 MPa) while maintaining dimensional stability for a 12 to 15-minute curing cycle.

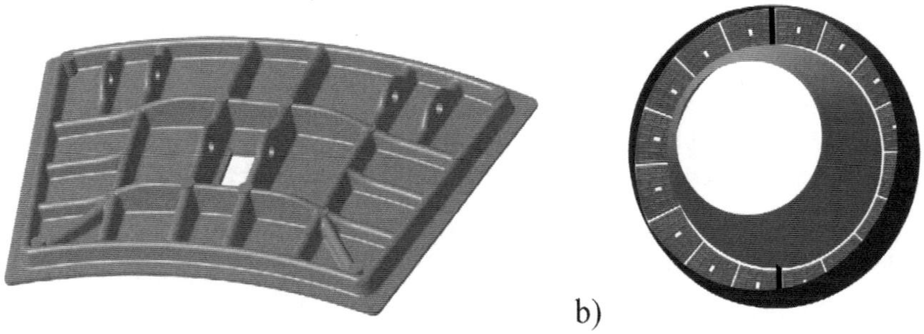

a) b)

Figure 1. Conceptual design for a) composite blocker door and b) blocker door assembly with fan duct.

The two sides of compression molding tool were additively manufactured using Thermwood's LSAM system using a 25wt% CF-PESU from Techmer PM. The efficacy of the compression molding system relies on manufacturing the parts within the required tolerances and the ability of the tool to withstand multiple compression molding cycles for prototype production. Anisotropic thermal expansion of the discontinuous fiber system in conjunction with in-plane printing lead to high anisotropic thermal expansion of the part. These anisotropic thermal expansions need to be accounted for while manufacturing a part to design for geometry change due to local thermal stresses in addition to the local residual stresses from the printing process. Tool shape compensation provides a way to design the shape of the tool to account for anisotropic thermal expansions and hence avoid shape distortion on cooling to maintain final part tolerances.

A common failure observed with additively manufactured tools is the delamination between the printed layers from interlaminar shear and/or tension due to low fracture toughness of the resin system. Hence the tools are typically printed in a way that tensile stress in the through thickness direction is limited to prevent the debonding between layers. In designing for the compression molding process, a tight shear edge is required between the positive and negative tools to ensure that the molding charge stays within the molding cavity. When the two halves of the tool are separated in demolding, significant tensile stress can develop in the tools in the molding direction. Another source of tensile stress that develops is from Poisson effects in the molding process where high compressive stress in the mold closing direction is accompanied by tensile stresses in the perpendicular directions. Designing for these stresses is critical for successful tooling approaches.

Furthermore, the low thermal conductivity of the printed part perpendicular to local fiber direction leads to high thermal gradients and reduced temperature distribution away from the heat source, be it platens or internal elements. Successful molding requires that the tooling surfaces maintain the resin processing temperature at steady state due to low thermal conductivities of the printed part. Increasing the temperature at the platen surfaces will increase the temperature at the tooling surfaces, but is limited by the glass transition temperature, T_g, of the resin. Two potential solutions to increase the temperature at the molding surface are to place cartridge heaters inside the tool near the molding surface and heat the tool using conduction or creating a heated chamber around the tool and heating the tool using convection. In the design utilized in this demonstration, cartridge heaters are placed inside the tool and the temperature at the platen and at the cartridge heaters is kept below the T_g of the resin to successfully mold PPMC thermoset charge using compression molding. Further, the tool is heated in a closed configuration to simulate some of the effect of having a fully enclosed convection chamber around the tooling surface.

The geometry of the two halves of the tool used for the experiments and simulations is shown in Figure 2, based on the geometry requirements of the blocker door part. The negative tool contains the molding cavity with integrated ribs to provide geometrical stiffness to the resulting part. The molding cavity has a short region of zero draft to create the shear edge between the positive and negative tools and to ensure that the press-closing force is applied to the molding material and not the tool itself. The depth of the zero draft region is based on the size of the charge and part thickness for the blocker door so that the molding material does not escape the cavity during processing. The overall dimensions of the negative tool are 690 mm by 340 mm by 125 mm (27.2 in by 13.4 in by 4.9 in) and the positive tool are 690 mm by 240 mm by 125 mm (27.2 in by 9.4 in by 4.9 in). The tools were designed and manufactured with flanges on the sides to facilitate connection to the platens of the 250-ton compression molding press used at the Indiana Manufacturing Institute (IMI) for the prototype molding trials. Printing and machining of the tools, including installation of the cartridge heater and manufacturing of the heater control system, was performed at Thermwood Corporation in Dale, IN.

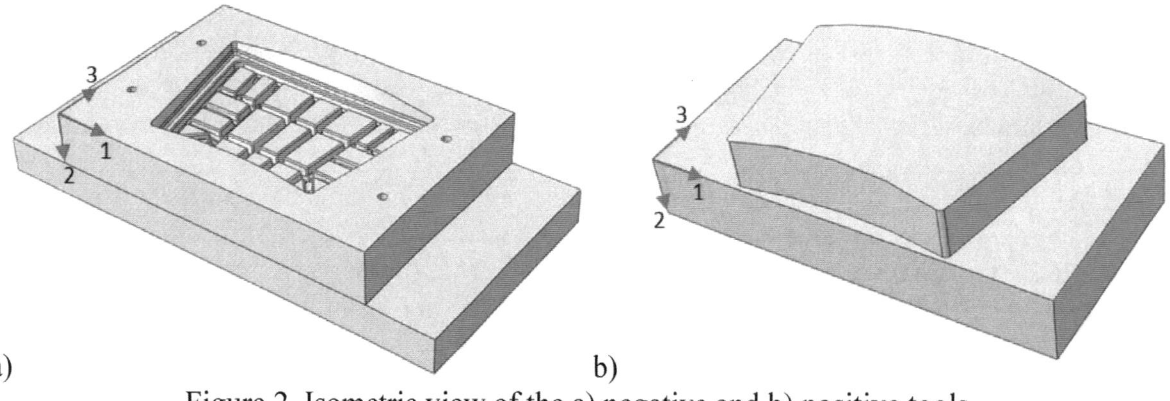

a) b)
Figure 2. Isometric view of the a) negative and b) positive tools.

3. SIMULATION RESULTS

Simulation was used to study the effects of different compression molding conditions and guide the design for AM of the compression molding tool. Characterizing a material for simulating the AM process is intensive and is still ongoing for the 25% CF-PESU. However, a material card for Celanese 50% CF-PPS (polyphenylene sulfide) [12] has been previously validated for the ADDITIVE3D simulation and was used in the study presented herein. There will be significant differences between the material properties of the two materials at the bead level from the difference in fiber loading and polymer kinetics. Thus, the presented simulation results should be considered as providing design guidance (both printing strategy and molding strategy), not quantitative results. Furthermore, as the polymers differ significantly, the effects of the printing process, particularly deformation, will not be well predicted with the current material card. However, the resulting local mechanical properties, thermal expansion, and heat transfer coefficients will be similar. Thus, the process simulation component of ADDITIVE3D was used primarily to map the local material properties of the printed material onto the analysis mesh of the following performance simulations to account for the anisotropic material properties at each integration point in the element as defined by the printing path. Nevertheless, the performed simulations were critical for achieving a successful molding using AM tooling.

Two concerns when designing an AM compression molding tool are the large CTE and the relatively low tensile strength of the printed material in the layer stacking direction (z-direction). While the large CTE is concerning in mold designs that can entrap the molded part upon cooling to room temperature, the low tensile strength is concerning due to stresses introduced in the tool during the process of ejecting the part from the tool. A stress analysis of the ejection process was carried out to study the effect of printing direction in the negative tool and evaluate the performance of the tool against the strength of the interlayer bonds. Figure 3 shows the two printing orientations considered. In either case, the printed geometries are fully-filled and near net shape for the final geometry. A machining step will occur after printing to achieve the final molding surfaces and machine in the molding cavity.

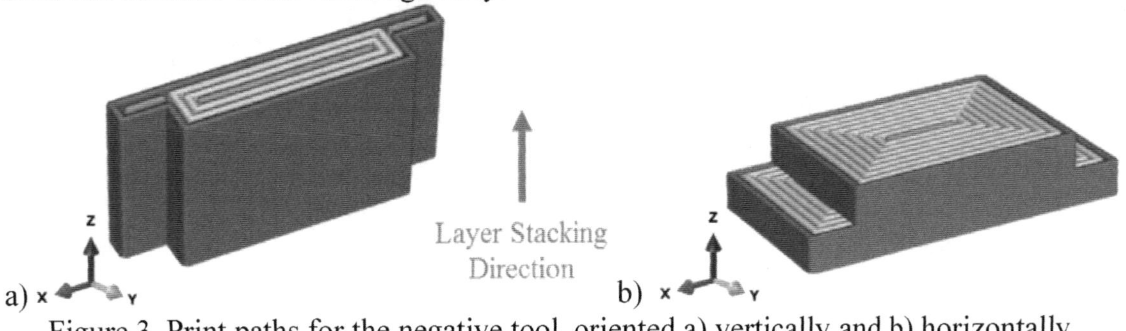

Figure 3. Print paths for the negative tool, oriented a) vertically and b) horizontally.

To simulate the ejection process of the part, the negative tool is clamped in four regions located at the flange of the tool, and a surface traction load was applied in the direction of part ejection. This simulation included effects of manufacturing such as local material orientation and residual stresses after the machining process. Figure 4 shows the stresses in the z-direction resulting from the part ejection process for the two print orientations. Considering a maximum allowable stress in the z-direction of around 30 MPa (4.35 ksi), this study showed that it was less likely to damage the tool during the ejection process if the tool is oriented vertically as shown in Figure 4a.

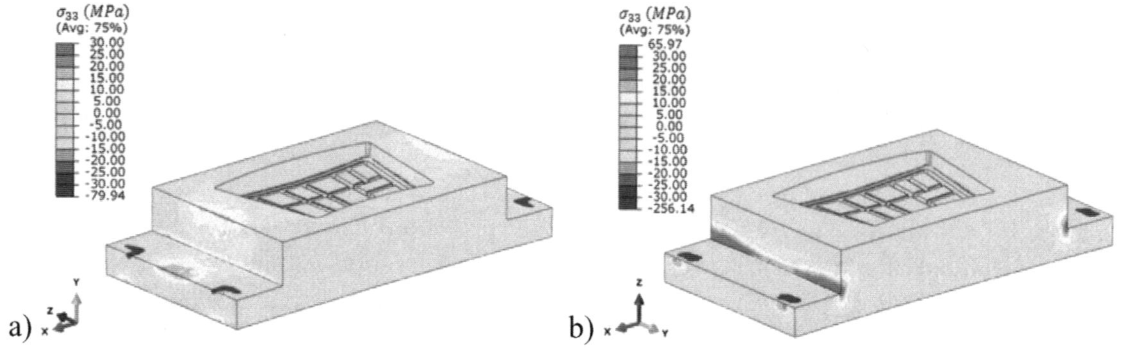

Figure 4. Z-direction normal stress resulting from the part ejection process for tool, oriented a) vertically and b) horizontally.

Prediction of the part behavior under the given molding conditions requires the understanding of the underlying physics behind the heat transfer of the anisotropic printed material. Heat transfer simulations under different thermal conditions were performed to understand and predict the performance of the tools under varying heating strategies. The most important factor for successful molding, beyond maintaining tool surface integrity, is the temperature of the tool at the molding surface which needs to be higher than the processing temperature of the resin for successful compression molding of the part.

Steady state heat transfer simulations were performed with different thermal conditions to find the temperature distribution of the part at the molding surface. Figure 5 shows the side view of the two halves of the tool and the temperature boundary condition at the bottom of the tool surface (marked in red) which is in contact with the platen. All the other surfaces allow natural convection with heat transfer coefficient of 10 W/m-K at an ambient temperature of 27 °C (81 °F).

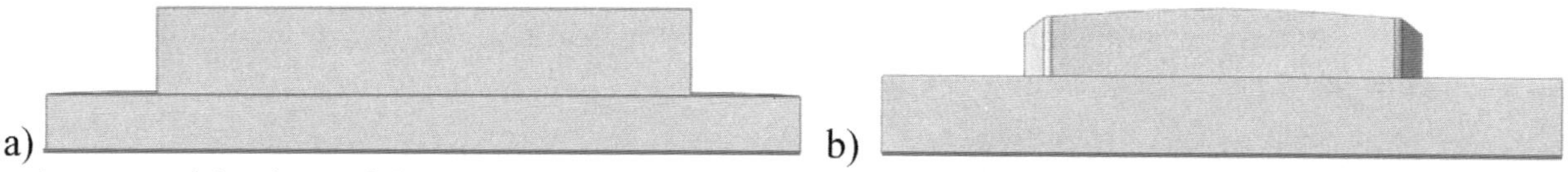

Figure 5. Side view of the a) negative and b) positive tool respectively with red line marking the temperature boundary condition surface.

The temperature applied at the bottom of the tool surface was 212 °C (414 °F), the T_g of the resin, whereas all the other surfaces convect naturally at the ambient temperature of 27 °C (81 °F). The steady state temperature distribution at the molding surface of the negative tool is shown in Figure 6 and that of positive tool in Figure 7.

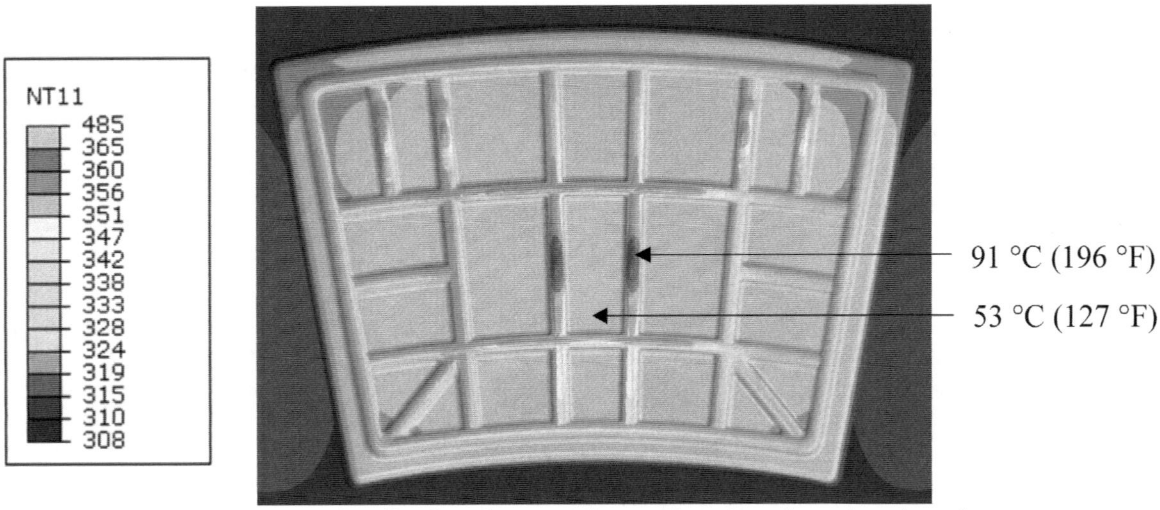

Figure 6. Steady state temperature distribution for negative tool under platen heating alone.

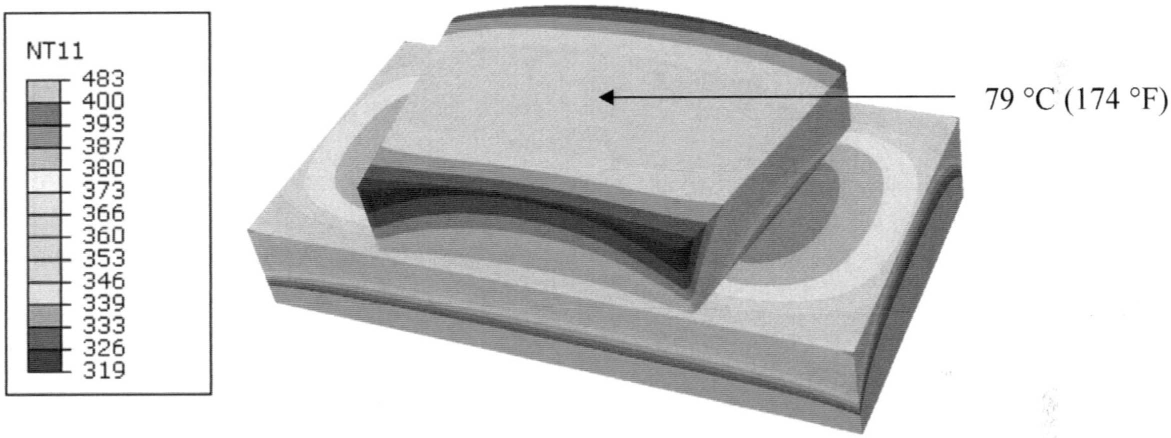

Figure 7. Steady state temperature distribution for positive tool under platen heating alone.

Figure 6 shows the steady state temperature distribution when heating the tool from the platens while the other surfaces convect naturally. The contour plot of the output value, NT11, from the Abaqus heat transfer analysis for surface temperature is reported in Kelvin. The highest temperature recorded in ribs is 91 °C (196 °F) which is less than the 130 °C processing temperature of the resin. A similar observation is made for the positive tool where the average temperature on the molding surface at the center is 77 °C (171 °F) and towards the edges 47 °C (117 °F). Accordingly, it is concluded that platen heating along will be insufficient for heating the tool to the temperatures required for the compression molding process.

Two options for increasing the temperature of the tooling surface are building a heating chamber around the tool and using forced convection or placing a heat source inside the tool and heating the tool using conduction. Both the possibilities have been analyzed with the help of heat transfer simulations using Abaqus. Firstly, the effect of forced convection from a heating chamber on the temperature distribution of the tool has been analyzed. The top surface of the negative and the positive tool have been assigned forced convection boundary conditions with convective heat transfer coefficient h of 20 W/m-K at elevated temperature of 127 °C (261 °F). The boundary conditions shown in Figure 8 correspond to the practical scenario of building a heating chamber on top of the negative tool and lowering the positive tool into the heating chamber so that the red surface of the positive tool shown is effectively heated through forced convection. The platen is still assigned a temperature boundary condition of 212 °C (414 °F) and the remaining surfaces convect naturally at ambient temperature of 27 °C (81 °F). The steady state temperature distribution is then analyzed for both the halves of the tool.

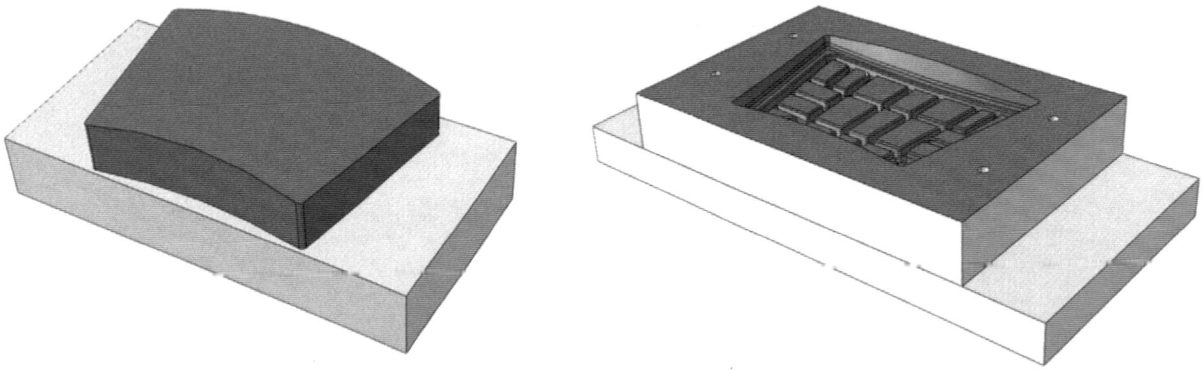

Figure 8. Surfaces assigned forced convection boundary conditions to model heating chamber.

Figure 9 and Figure 10 show the temperature distribution in the negative and the positive tool, respectively. This analysis indicates forced convection at elevated temperatures combined with heating the tool from the platens is expected to provide nearly adequate temperature to support the compression molding process. The average temperature on the molding surface of the negative tool is 127 °C (261 °F) which is close to the processing temperature of the resin with a high of 143 °C (289 °F) at the center ribs of the tool. The temperature distribution in the positive tool ranges from 138 °C (280 °F) at the center to 131 °C (268 °F) at the corners. While the average temperature across the tools is likely to provide enough heat to sufficiently cure the part, the distribution of temperature in the negative tool indicates that there is a potential that the corners of the part will not see sufficient temperature required to locally cure the material.

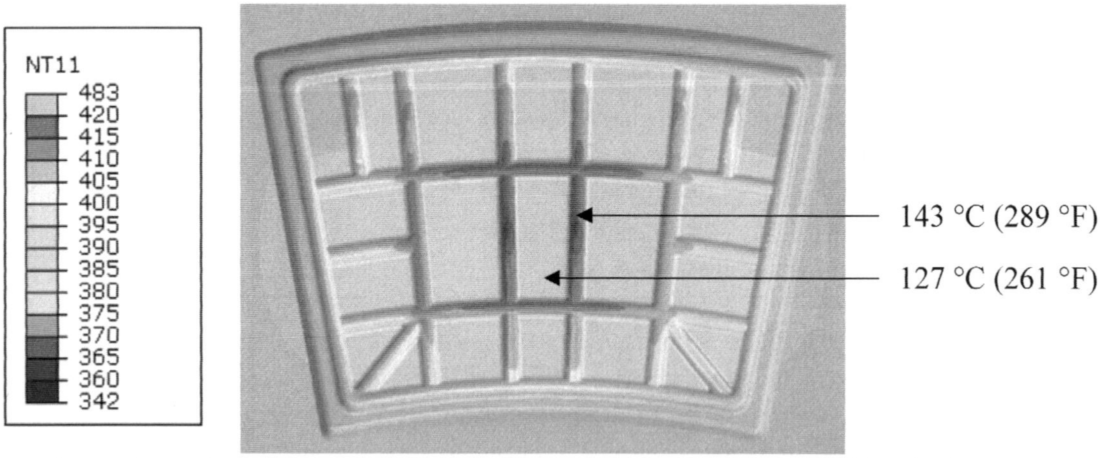

Figure 9. Negative tool steady state temperature distribution for convection combined with platen heating.

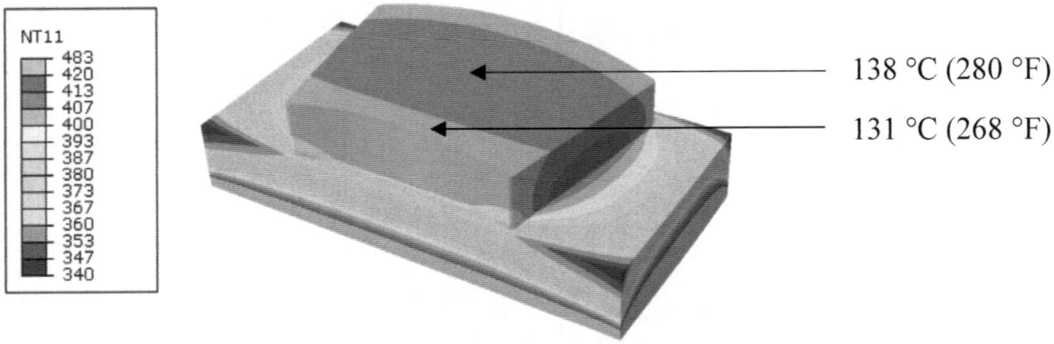

Figure 10. Positive tool steady state temperature distribution for convection combined with platen heating.

One of the challenges with platen heating is the distance that the heat travels from the heat source to the molding surface. Integrated cartridge heaters can be inserted into the tool to reduce this distance that the heat travels. In this study, 1 in (25.4 mm) diameter cartridge heaters were utilized with active controls to regulate the temperature in the heaters to remain below the glass T_g of the PESU. Figure 11 shows the location of cartridge heaters inside the negative and positive tools. Four heaters were inserted through bored holes into the negative tool and three were used in the positive tool. The locations in the negative tool were selected to avoid interference with the ribs in the tooling surface, and the locations in the positive tool were designed to allow the tool to fully close with the heaters engaged.

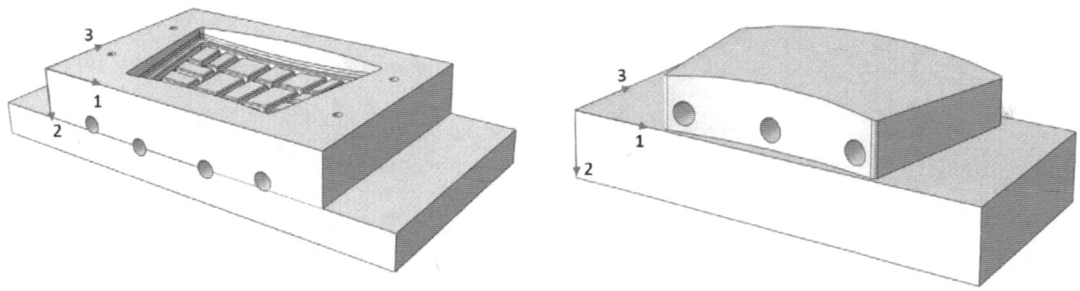

Figure 11. Location of cartridge heaters in the additive manufactured tools.

The temperature of the cartridge heaters was set to 193 °C (379 °F) in the simulations, which is below the T_g of the resin. The steady state temperature distribution is then obtained with the platen at 212 °C (414 °F) and the other surfaces convecting naturally at the ambient temperature of 27 °C (81 °F). Figure 12 and Figure 13 show the temperature distributions at the molding surface of the tool with combined heating from the cartridges and platens. The average temperature at the main surface of the negative tool is 131 °C (268 °F). Some localized regions with temperatures up to 161 °C (322 °F) are present in the ribs where the geometry feature is located close to the cartridge. The temperature distribution over the positive tool is significantly more uniform with the use of the cartridges, with an average temperature of 137 °C (279 °F). The results of the heat transfer analysis indicate that the use of the cartridge heaters is preferred for the compression molding process.

Figure 12. Negative tool steady state temperature distribution for cartridge heating combined with platen heating.

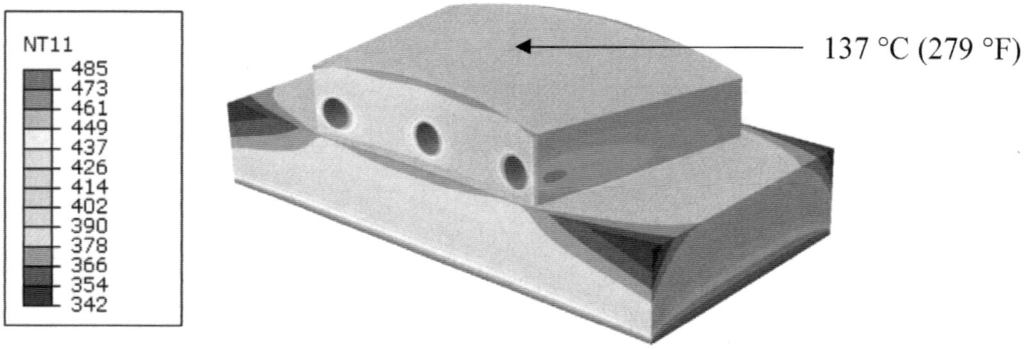

Figure 13. Positive tool steady state temperature distribution for cartridge heating combined with platen heating.

4. EXPERIMENTAL RESULTS

The experimental setup used for compression molding the blocker door part is shown in Figure 14. The seven cartridge heaters were independently controlled from a central unit powered by a 240 V supply. The negative tool was mounted on the lower platen and the positive tool to the upper platen. The printed tools are held in place with metal clamps attached to the side flanges. Two metal plates were affixed to the negative tool on either side and attached to each other with bolts running through the printed tool to provide a pre-compressive stress to the geometry and resist interlaminar tensile stress to maintain tool integrity. Both sides of the tool were pretreated with a mold-release coating. The upper and lower platens of the press were set to temperatures of 205 °C (400 °F) and 210 °C (410 °F), respectively, and the cartridge heaters were set to 193 °C (379 °F).

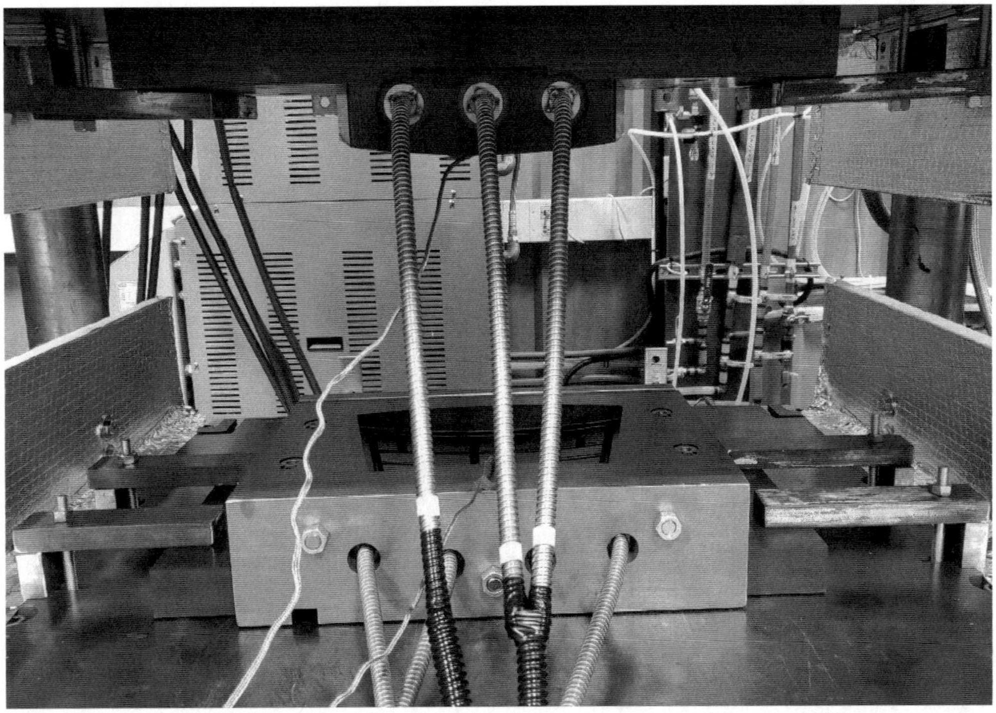

Figure 14. LSAM printed tools with cartridge heaters mounted on the 250-ton compression molding press at the IMI.

The temperature of the molding surface was monitored using an infrared camera and the temperature of the cartridge heaters was adjusted to maintain consistent temperature on the molding surfaces of the positive and negative tools to avoid residual stresses formation due to temperature difference. After the tools were mounted and aligned at room temperature, the system was heated with the two tools in near-contact. The upper and lower platens of the press were set to temperatures of 205 °C (400 °F) and 210 °C (410 °F), respectively, and the cartridge heaters were set to 193 °C (379 °F). Once the set temperatures were achieved, the tool was cycled to ensure the fit between the matched tools.

Prior to placing the molding charge into the negative tool, the temperature distribution was obtained for the positive and negative tool from the infrared camera as shown in Figure 15 and Figure 16 on a Celsius scale. The temperature distribution obtained from the infrared camera for the molding trial closely matches the temperature distribution obtained from the simulation. The surface temperature of the positive tool from infrared camera is in the range of 125 to 130 °C (257 to 266 °F) compared to the predicted average surface temperature from the simulations of 127 °C (261 °F). Similarly for the negative tool, the thermal imaging reveals local temperature increases in the ribs in the range of 160 °C (320 °F) and a temperature on the main surface closer to 130 °C (266 °F) following the trend of the simulation result.

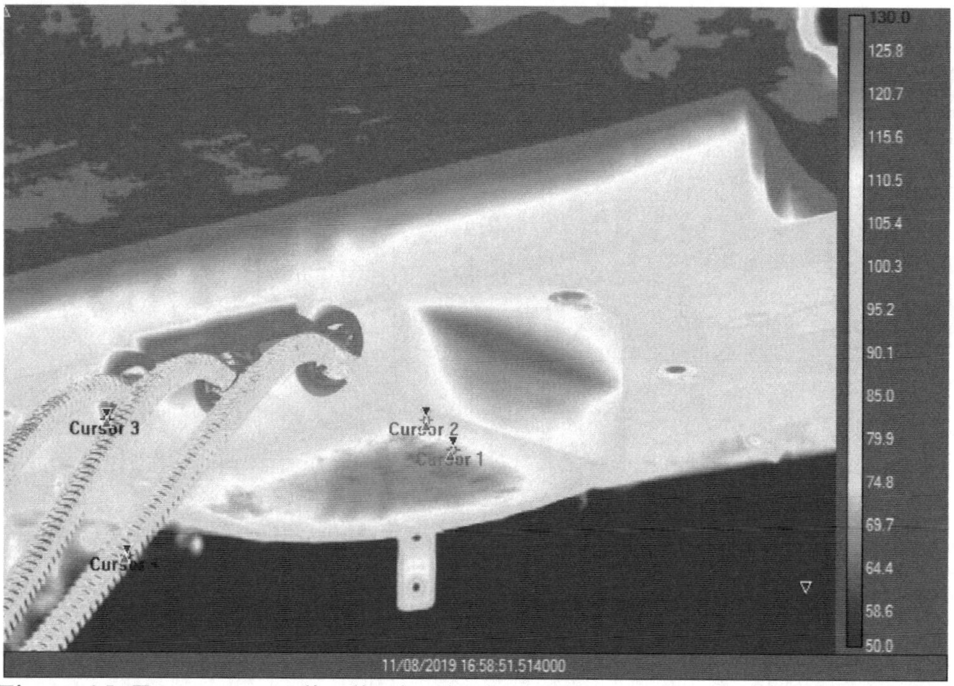

Figure 15. Temperature distribution in the positive tool before molding in °C.

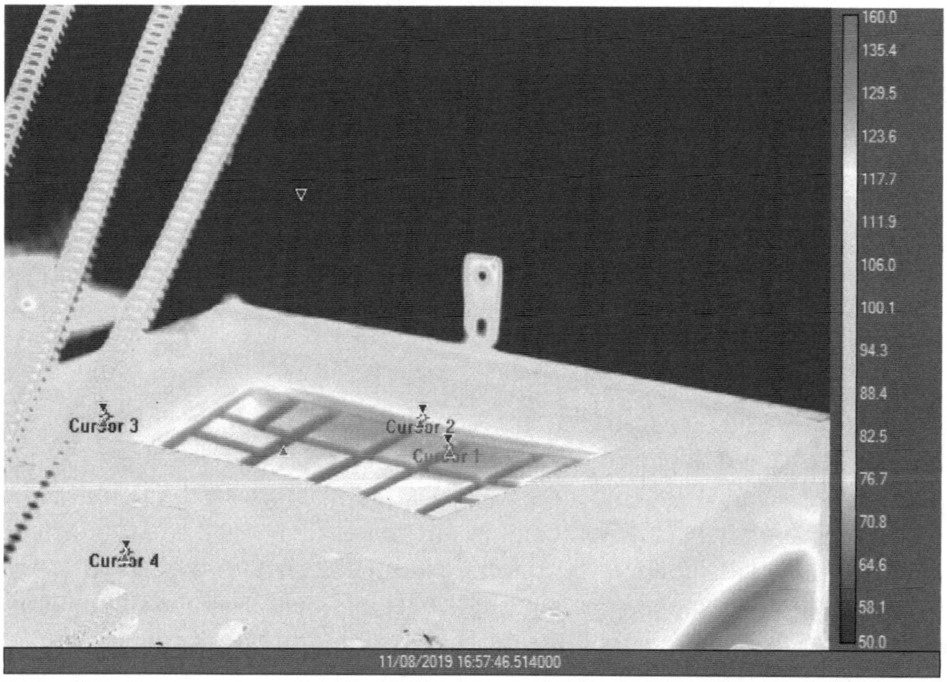

Figure 16. Temperature distribution in the negative tool before molding in °C.

A rectangular molding charge of the PPMC material with nominally 70% coverage was placed into the heated negative tool. The tool was closed for 10 sec at 10 tons (89 kN) and then the force was increased to 78 tons (694 kN) for 15 minutes. At the completion of the curing cycle, the top platen was elevated with the positive tool. When the tools separated, the part remained attached to the positive tool and was removed using hammers and chisels with manual effort. No ejector pins were utilized in the demolding process. Figure 16 shows the consolidated part after the compression molding.

Figure 17. Successful molded part top and bottom view.

The resulting part was highly successful from visual inspection. The charge was able to fully fill the mold with limited resin push out in localized regions of the deepest ribs. This effect is a common behavior in initial molding trials, regardless of the tooling material, and can be addressed through further process optimization. The surface finish of the part shows good consolidation. Furthermore, the AM tools themselves maintained integrity after the molding. Further investigation is planned for monitoring surface deformation through multiple molding cycles.

5. CONCLUSIONS

This study has demonstrated that EDAM is a viable method to produce rapid prototype tooling for thermoset composites compression molding of moderate complexity and that simulation plays a critical role in understanding the printing process and the ultimate performance of printed parts. The LSAM process using the 25% by weight carbon fiber PESU material produced a tool that was able to maintain the molding pressures of 1500 psi (10.3 MPa) for a representative prototype production process as required to consolidate the material. While the low thermal conductivity of the polymer and composite present challenges for heating the tool, this study has demonstrated the effective utilization of internal heating strategies to achieve the required processing temperatures without negatively impacting the state of the polymer.

Modeling and simulation, utilizing the ADDITIVE3D integrated simulation workflow, drove the design of the molded part, printed tool, and the manufacturing process. The coupled thermo-mechanical simulation of the printing process provides designers with detailed understanding of how the printing strategy impacts the residual stress and deformation in the part. Furthermore, this simulation approach provides a rigorous foundation for performance models of the printed part incorporating the local, orientation dependent material properties of the composite bead in predicting behavior of the tool in the molding process. Modeling and simulation enable the part, tool, and process designers to iterate virtually, reducing the time, money, and risk associated with optimization driven by physical experimentation alone.

6. REFERENCES

[1] Sudbury TZ, Springfield R, Kunc V, Duty C. An assessment of additive manufactured molds for hand-laid fiber reinforced composites. Int J Adv Manuf Technol 2017;90:1659–64. doi:10.1007/s00170-016-9464-9.

[2] Hassen AA, Springfield R, Lindahl J, Post B, Love L, Duty C, et al. The durability of large-scale additive manufacturing composite molds. CAMX 2016 - Compos. Adv. Mater. Expo, 2016, p. 26–9.

[3] Li H, Taylor G, Bheemreddy V, Iyibilgin O, Leu M, Chandrashekhara K. Modeling and characterization of fused deposition modeling tooling for vacuum assisted resin transfer molding process. Addit Manuf 2015;7:64–72. doi:http://dx.doi.org/10.1016/j.addma.2015.02.003.

[4] Kunc V, Lindahl J, Dinwiddie R, Post B, Love L, Duty C, et al. Investigation of in-autoclave additive manufacturing composite tooling. CAMX 2016 - Compos. Adv. Mater. Expo, 2016.

[5] Barocio E, Brenken B, Favaloro A, Pipes RB. Extrusion deposition additive manufacturing of composite molds for high-temperature applications. Int. SAMPE Tech. Conf., 2017, p. 1512–23.

[6] Kuczko W, Wichniarek R, Górski F, Buń P, Zawadzki P. Application of Additively Manufactured Polymer Composite Prototypes in Foundry. Adv Sci Technol Res J 2015;9:20–7. doi:10.12913/22998624/2360.

[7] Tábi T, Kovács NK, Sajó IE, Czigány T, Hajba S, Kovács JG. Comparison of thermal, mechanical and thermomechanical properties of poly(lactic acid) injection-molded into epoxy-based Rapid Prototyped (PolyJet) and conventional steel mold. J Therm Anal Calorim 2016;123:349–61. doi:10.1007/s10973-015-4997-y.

[8] Brenken B, Barocio E, Favaloro A, Kunc V, Pipes RB. Development and validation of extrusion deposition additive manufacturing process simulations. Addit Manuf 2019;25:218–26. doi:10.1016/j.addma.2018.10.041.

[9] Brenken B. Extrusion Deposition Additive Manufacturing of Fiber Reinforced Semi-crystalline Polymers. School of Aeronautics and Astronautics, Purdue University, 2017.

[10] Barocio E, Favaloro A, Brenken B, Barocio E, Pipes RB, Favaloro A, et al. Fusion Bonding of Fiber Reinforced Semi-Crystalline Polymers in Extrusion Deposition Additive Manufacturing. School of Materials Engineering, Purdue University, 2017. doi:https://doi.org/10.1016/j.addma.2018.10.041.

[11] Favaloro AJ, Brenken B, Barocio E, Pipes RB. Simulation of Polymeric Composites Additive Manufacturing using Abaqus. Sci Age Exp 2017:103–14.

[12] Brenken B, Barocio E, Favaloro A, Kunc V, Pipes RB. Development and validation of extrusion deposition additive manufacturing process simulations. Addit Manuf 2019;25:218–26. doi:10.1016/j.addma.2018.10.041.

DUAL MATERIAL SYSTEM FOR POLYMER LARGE SCALE ADDITIVE MANUFACTURING

Tyler Smith[1,3,4], Ahmed Arabi Hassen[1,3], Randall Lind[2,3], John Lindahl[1,3], Phillip Chesser[2,3], Alex Roschli[2,3], Vipin Kumar[1,3], Vidya Kishore[2,3], Brian Post[2,3], Jordan Failla, Chad Duty[3,4], Lonnie Love[2,3], Vlastimil Kunc[2,3]

[1]Materials Science and Technology Division (MSTD), Oak Ridge National Laboratory (ORNL), Oak Ridge, TN 37830, USA

[2]Energy & Transportation Science Division (ETSD), Oak Ridge National Laboratory (ORNL), Oak Ridge, TN 37830

[3] Manufacturing Demonstration Facility (MDF), Oak Ridge National Laboratory (ORNL), Knoxville, TN, USA

[4]Department of Mechanical, Aerospace and Biomedical Engineering, University of Tennessee, Knoxville, TN, USA

ABSTRACT

Big Area Additive Manufacturing (BAAM) technology allows for manufacturing of large-scale objects with a potential to reduce energy embedded in products, reduce or eliminate energy necessary for transportation of goods along with reducing the lead time and cost in some cases. Over the last few years, Oak Ridge National Laboratory (ORNL) has been focusing on large-scale printing of single material systems, typically un-reinforced or short fiber reinforced polymers, in order to address needs in stiffness-limited applications. This paper describes the development of a multi-material large-scale AM system through a collaboration with Cincinnati Inc. and Performance Feed Screw Inc. Modifications to the Big Area Additive Manufacturing (BAAM) system includes a new extruder design to accommodate a dual feed system, an expanded two-dryer system with a capacity of 273 kg/dryer, and a system that is capable of mixing pelletized materials up to 60Kg/hr. This article highlights the advantages and limitations of the multi-material system as well as potential applications.

Corresponding author: smithtc@ornl.gov

SAMPE Virtual Conference Proceedings, 2020. Society for the Advancement of Material and Process Engineering – North America.

1. INTRODUCTION

Additive Manufacturing (AM) or Rapid Prototyping (RP), utilizes metals, ceramics, and composites to produce 3D structures at a low cost and lead time. RP can be achieved through methods such as Fused Filament Fabrication (FFF), Digital Light Processing (DLP), Stereolithography (SLA), etc. to produce small-, mid- and large-scale structures using polymer materials and their reinforced composites [1,2]. Small-scale FFF printing is a standard system commonly used for small prototypes, flexible sensors, and metal printing [2-4]. The process involves heating a filament through a nozzle and depositing layers that range between 0.1 mm - 0.8 mm (0.004 inch - 0.028 inch) in height with nozzle diameters ranging from 0.2 mm - 1.2 mm (0.008 inch - 0.0472inch) with a build volume up to a 28317 cm^3 (1ft^3). Structures fabricated using FFF technique have low volumetric flowrates around 16387mm^3/hour (1$inch^3$/hour) resulting in long build times [5]. Several small-scale systems have recently used multiple extruders in order to print dual materials. Multi-material printing allows users to tune the color, cost, weight, and the mechanical properties throughout the structure [6,7]. Studies have shown that by using multi-materials within a structure, new mechanical responses can be achieved such as increased mechanical properties, soft and rigid robotic structures, impact resistant structures, etc. [6,7]. Other applications include embedded sensors, custom mechanical responses, and changing the color of the part [3,6].

In order to create large articles at a low cost and short time, large scale systems were introduced to print with nozzles ranging from 2.5 mm - 12.7 mm (0.1 inch - 0.5 inch) and a layer height range of 1.27 mm - 8.89 mm (0.05 inch - 0.35 inch). The Big Area Additive Manufacturing (BAAM) system designed by the collaborative efforts of Oak Ridge National Laboratory (ORNL) and Cincinnati Inc., is a large thermoplastic extrusion deposition system with a build volume of 1.5 m × 3.6 m × 2.4 m (5 ft × 12 ft × 8 ft) and a deposition rate up to 45 kg/hr (100 lb/hr) [8]. The BAAM system has been used to manufacture parts such as large-scale prototypes, molds and dies and large lightweight optimized core structures [9-11]. Large-scale AM is currently limited to printing structures with a single material which limits the ability to optimize structure weight, functionality, cost, and layer time.

High performance feedstock materials, such as Polyethersulfone (PESU) and Polyphenylsulfone (PPSU), have been used for 3D printing of high temperature molds and dies [11]. AM compression molds are currently printed completely solid, as the pressures in the compression molding process can exceed 10.3 MPa- 13.8 MPa (1500 psi-2000 psi). The infill material is typically less critical for the integrity of the structure during the printing process, but necessary to prevent deformation during the compression process. Infill in compression molding prevents distortion during the compression process by filling the structure with a packing material to prevent the load bearing perimeters from deforming. Traditionally, a single material is utilized in these infill patterns, where it drives the cost and weight of the manufactured parts up. Cost and weight of structures can be minimized by utilizing recycled materials or foams for the infill. The mechanical and thermal properties of the AM structure are highly dependent on the material, processing conditions, and infill pattern density and direction [5,12-14]. Incorporating multi material printing to the process will allow for tailoring of these mechanical and thermal properties [6,14].

2. DUAL MATERIAL SYSTEM DESIGN

2.1 Design Considerations

There are several approaches for designing a dual material system for large scale AM systems. However, there are several constraints and considerations that must be accounted for in a successful design. The primary considerations can be summarized as:

- Gantry system weight capacity: The BAAM gantry weight capacity was set to a limit of weight of 135 kg (300 lb). This limitation prevents excess inertia during gantry travel. Inertia concerns stem from rapid directional changes during sharp turns which could cause the extruder to shift or move under its own inertia. The standard deposition speed during printing is 0.28 m/second (11inches/second). The weight limitation prevents the addition of a second extruder to the gantry since the current extruder weights approximately 90 kg (200 lb).

- Space limitation: Additional capabilities typically occupy additional space. Since the maximum build volume of the printer needs to remain the same, the optimal design should occupy the minimal volume. Therefore, the new device should either be compact or added above the extrusion deposition line where build volume is not affected. Adding a second extruder would require a large space either on the same gantry or by adding a second gantry system to printer.

- Material switching and purging: Material that dwells in the extruder for prolonged periods of time can lead to problems such as polymer degradation and oxidation. As an example, it has been observed that foam materials tend to degrade when held at temperature for a prolonged time, which could lead to decreased mechanical performance and poor print quality (color changes and expansion decreases). In the case of adding a second extruder, material would need to reside in the extruder for the duration of the printing period of the first extruder. In this case, it would likely be necessary to purge the entire volume of the extruder in order to maintain consistent material properties. This would result in a significant amount of wasted material and excessive layer times. An ideal system would be capable of switching materials quickly with minimal dwell time in the extruder.

- Material type and processing conditions: The compatibility of the feedstock materials being used will be dictated by their chemical composition and processing conditions. A dual extruder system would allow for a wide variety of materials to be printed since they would not be directly blended or required to flow through the same extruder. If the materials were combined near the nozzle exit, a co-axial extrusion system could be used to create a core-shell structure where material A is extruded as an inner core while Material B coats the circumferential of Material A [15]. Although this system could deposit two materials at the same time and switch quickly, the material combinations would be limited to those with compatible chemistries that would enable bonding.

- Deposition rate: BAAM currently extrudes at 45 kg/hr (100 lb/hr) when depositing carbon fiber reinforced acrylonitrile butadiene styrene (CF/ABS). In order to maintain high deposition rates, the dryers and blenders must be able to process materials at a rate greater than the maximum deposition rate.

2.2 Design and Fabrication

In order to successfully develop a dual material system for the BAAM system, several modifications have been made. First, in order to comply with the weight constrains, a dual hopper with a sliding mechanism was designed, see Figure 1. In this system, the feedstock materials can be switched from Material A to Material B at the entrance of the extruder. This allowed for multi-material extrusion through a standard single screw extruder and avoided the need for an additional extruder (i.e. extra 90 kg). To reduce weight, all tubes and pivot brackets were fabricated using aluminum, and high strength steel was used in high stress locations. A sliding mechanism was attached at the entrance of the extruder to allow for material valves to open and close when switching material. The sliding mechanism consists of three primary components: the pivot mechanism, the piston, and maintenance/alignment features (Figure 1).

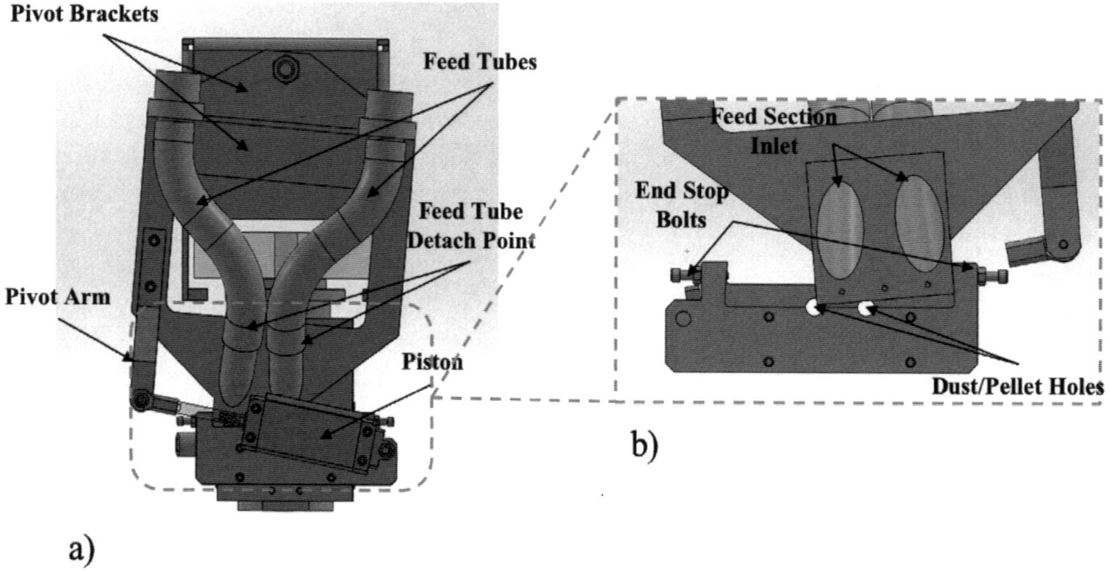

Figure 1. Design for dual material switching mechanism integrated to the BAAM system; a) CAD drawing showing the dual hopper/feeding tubes, sliding mechanism, and controlling piton, and b) Cross section at the extruder entrance showing alignment and dust removal features

Switching material during a build involves pushing or sliding a metal plate to open or close a feed line as seen in Figure 1. The sliding mechanism consists of a pneumatic driven piston that controls a pivot arm that is attached to the feed tubes. The pivot arm transfers the force from the piston to slide the feed line between the two hoppers (Figure 1). A thin profile flexible mount piston that requires 0.689 MPa (100 psi) driving pressure to deliver a force of 547 N (123lb) was used. The piston force must be sufficient to either move aside or shear through pellets that may be blocking the flow path to prevent jamming of the feed tubes. Gravity fed pellets are desired to be as close to a vertical path as possible to prevent bridging in the feed tubes. Small changes in the hopper angle are desired to minimize bridging in the feed tubes because the feed angle changes while switching. To reduce the angular change of the hoppers, the radius from pivot point to the feed entrance was increased. As sliding occurs, pellets blocking the motion must either move into the extruder or shear at the interface of the sliding metal. These locations were fabricated using hardened steel to prevent wear. The piston motion could be adjusted by tuning the location of the end stop bolts (Figure 1b) to align the interface between the feed tube and extruder entrance.

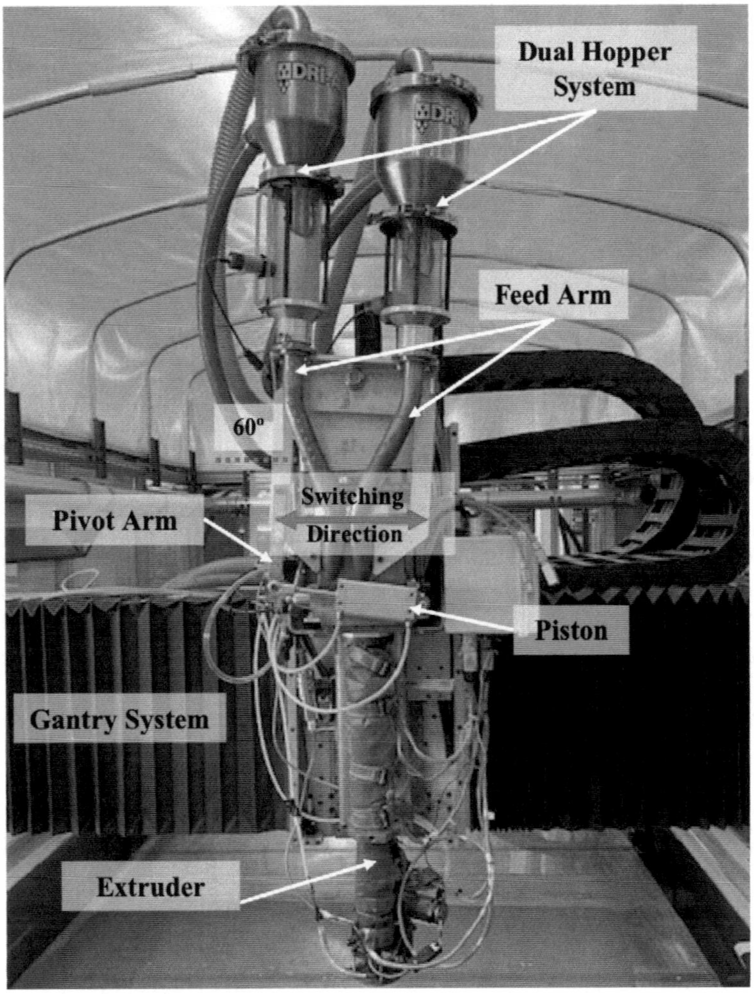

Figure 2. BAAM extruder equipped with dual hopper system

One of the common issues in working with a pelletized feedstock is the containment of the materials in the feed lines. CF dust and pellets are often found around the system throughout the build time, which can result in defects in the part. Also, dust and pellets can block the motion of the switching mechanism and jam the feed lines. To mitigate these issues, two holes were added to the sliding locations to allow for the excess dust and pellets (i.e. occurs during the switching process) to exit the system without damaging any of the system components, see Figure 1b. Applying the dual feed system design to the current BAAM extruder results in system capable of feeding from two separate dyers or blenders, enabling the deposition of functionally graded material or multi-material extrusion (Figure 2)

The BAAM system was fitted with two 272kg (600lb) dryers (DRI-AIR model RH600), used for drying the feedstock materials before the deposition process, as seen in Figure 3. A blender (Maguire model WSB-100 Series) with a precision mixer that is capable of mixing four different materials up to 59 kg/ hr (130lb/hr) was added to the system, see Figure 3. The blender is used for precision mixing of additives, expandable foaming spheres, chemical foaming agents, and various other functional materials. The mixer is equipped with four hoppers such that any ratio between various materials can be accomplished during a single build. After mixing, the material is shuttled to the dual hopper system integrated to the BAMM extruder.

Figure 3. BAAM equipped with dual dryer and blender with zoom out of blender showing hoppers and mixing chamber

3. DEMONSTRATION AND APPLICATION

Dual material deposition and blender capabilities with BAAM has allowed for the printing of structures that demonstrate properties which were unattainable with a single material system. Traditionally, structures that need rigid exteriors have been designed as thin-walled hollow structures, or alternatively a sparse infill has been used with the same material. Dual material printing enables solid or sparsely filled structures to achieve lower cost and weight requirements by strategically placing lightweight structural materials (e.g. foam) in areas where loading conditions are low (Figure 4b). For example, in molds, a high cost and heavy polymer such as PESU can be used on the exterior molding surface while a high temperature foam is used on the interior to reduce the weight and cost. Applications such as wings or turbine blades can be custom designed to use stiffer materials where loads are high and lighter weight materials where loads are low [9]. Dual material printing coupled with custom core structures allows for stiff and lightweight structures (Figure 4d).

.

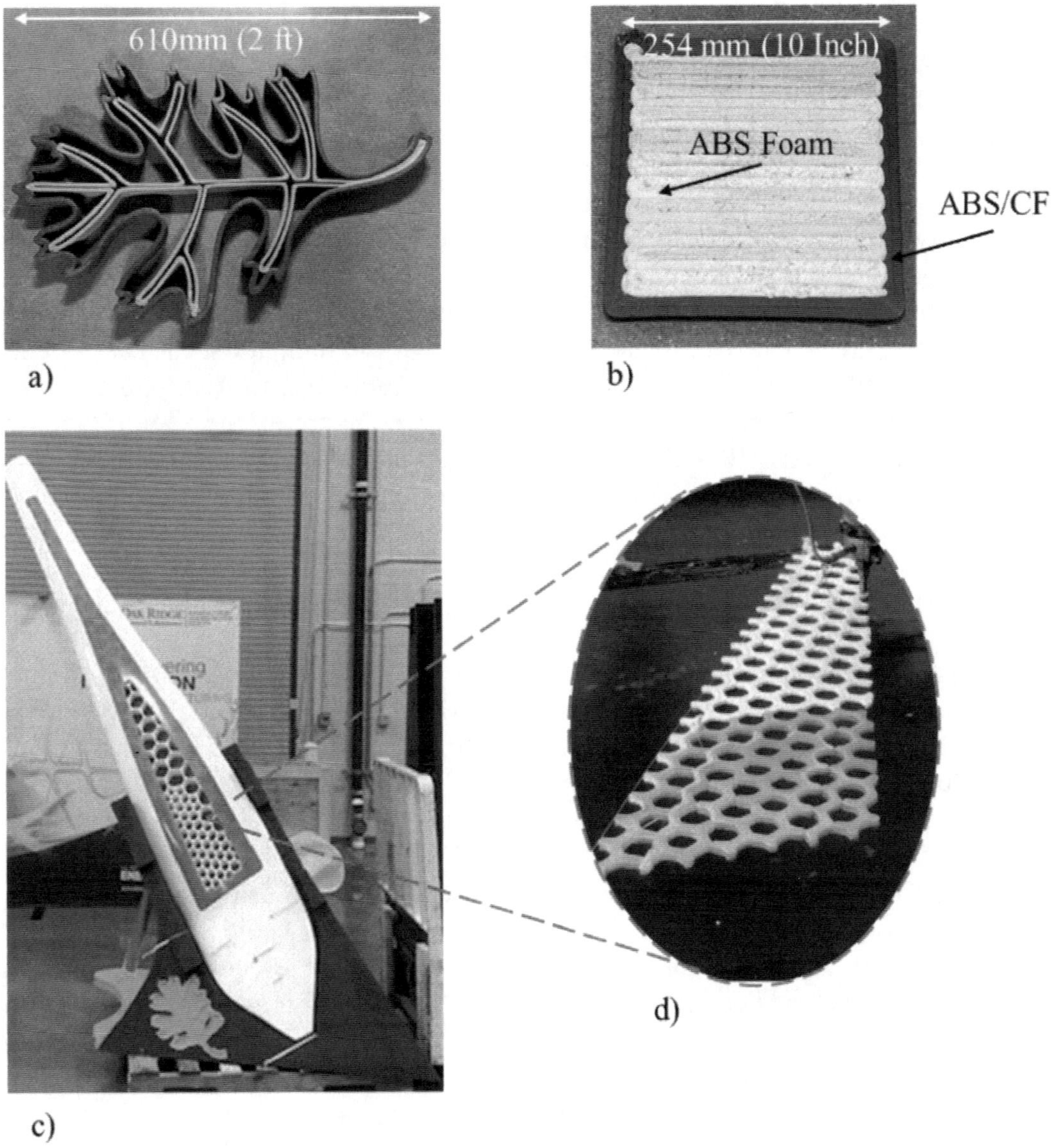

Figure 4. Applications for multi-Material structures showing variations in weight, cost, stiffness, and custom core designs; a) Multi material bio-derived leaf (Green: Bamboo/Polylactic Acid (PLA) and Brown: Wood flower/ PLA), b) Flat panel showing ABS foam used for lightweight structure by strategic placing of foam in infill pattern with a ridged CF/ABS exterior, c) Optimized PLA foam core structure for a turbine blade, and d) Zoom out for the functionally graded core during printing process

Reinforcing fibers such as CF, Bamboo Fiber (BF), and Glass Fiber (GF) are often added to base resins to provide higher stiffness and reduce warping of the printed parts. Fillers in polymers typically increase the cost and weight of the material and the impact on mechanical property performance depends on the type and amount of filler used. Dual Material printing enables placing materials with various fillers or neat resins throughout a part to tailor mechanical response in certain locations. Locations with high stiffness requirements can be printed with a CF filled material, and areas with low thermal properties and lower stiffness can be printed with a GF filled material. Similarly, Figure 4a shows how a fiber filled bio-derived material exterior and a low-cost bio filler interior are printed utilizing the dual hopper system. Utilizing the printing process and dual hopper system, structures with local reinforcement can be achieved as well [16].

Blended material compositions can also be altered during the build to define site-specific material properties. Modifying the material composition during printing creates some challenges as well. For example, printing with a combination of foaming and non-foaming materials can affect the geometry of the build (layer height, bead width) depending upon the local foaming conditions. Changes in bead and layer height must be accounted for in the toolpath such that material has enough space to deposit and does not overbuild and cause collisions. Currently, high expansion foams are extruded only on every other layer because the layer height for the foam is twice that of normal non-foamed materials. Blending four different materials can create structures where elasticity, strength, and weight are altered rapidly. For instance, building with a blend of thermoplastic polyurethane (TPU) and CF/ABS can result in highly stiff elastomers for high toughness applications. Recycled materials can also be blended with the virgin material to reduce waste in the AM and industry process to create a more eco-friendly printing environment.

As materials transition from one to another, a transition zone must be either purged (the process of clearing the printer's barrel for next transitioned material) or deposited into the structure before a virgin material can be deposited. If this material is purged, the layer time increases and allows the structure to cool. For many structures, an increase in layer time can lead to defects as well as poor layer-to-layer adhesion, causing delamination and failure. Similarly, in the case of foams, as material cools, the size of the layer height and bead width shrink, changing the proper spacing and placing defects throughout the structure. Rather than purging the transition material, it can be printed into a non-critical portion of the structure. Although this would remove defects from purging, it would modify the overall mechanical performance of the structure. Therefore, studies are currently being conducted to determine the impact of the material transition on the mechanical properties throughout the transition zone. For instance, it has been found that in the transition zone from CF ABS to neat ABS, a 96.68 cm^3 (5.9 in^3) volume of material is deposited, and mechanical properties may vary depending on filler content and base resin dominating in this zone [17]. Understanding how the transition occurs and the mechanical properties of the structure changes enables engineers to switch materials ahead of time such that the unstable mechanical properties appear in a non-critical section of the structure

4. CONCLUSION

BAAM extrusion-based printing can often manufacture large structures at a lower cost and faster rate than traditional manufacturing. However, the structures have been limited a single material, which limits the ability to have different local mechanical properties across the structure. Equipping the BAAM with a dual material system enables articles to be designed and printed with rigid and stiff exteriors with foam interiors, as well as different stiffness materials for custom impact and compression responses. Currently the dual hopper design is limited by temperature because both materials need to have similar processing conditions. The system also needs to account for a purge and/or transition zone to fully transition between materials. In future work, values for calculating the volume required to fully transition between materials will be determined and integrated into the slicing software such to minimize the waste during multi-material prints.

5. ACKNOWLEDGEMENTS

Research sponsored by the U.S. Department of Energy, Office of Energy Efficiency and Renewable Energy, Advanced Manufacturing Office, under contract DE-AC05-00OR22725 with UT-Battelle, LLC.

This research was supported in part by an appointment to the Oak Ridge National Laboratory Advanced Short-Term Research Opportunity (ASTRO) Program, sponsored by the U.S. Department of Energy and administered by the Oak Ridge Institute for Science and Education.

This research was supported in part by an appointment to the Oak Ridge National Laboratory Higher Education Research Experience (HERE) Program, sponsored by the U.S. Department of Energy and administered by the Oak Ridge Institute for Science and Education.

Large scale AM machine used in this research was sponsored by Cincinnati Inc., OH, USA. Feedstock materials used in this work were provided by Techmer PM., TN, USA.

6. REFERENCES

[1] Martínez-Pellitero, Susana, et al. "Analysis of influence factors on part quality in micro-SLA technology." Procedia manufacturing 13 (2017): 856-863

[2] Pham, Duc Truong, and Stefan Simeonov Dimov. "Rapid prototyping and rapid tooling—the key enablers for rapid manufacturing." Proceedings of the Institution of Mechanical Engineers, Part C: Journal of Mechanical Engineering Science 217.1 (2003): 1-23.

[3] Xu, Yuanyuan, et al. "The boom in 3D-printed sensor technology." Sensors 17.5 (2017): 1166.

[4] Rehman, Sadiq Ur, Muhammad Hasnain Raza, and Ahmad Raza Khan. "Delta 3D Printer: Metal Printing." Journal of Electrical Engineering, Electronics, Control and Computer Science 5.3 (2019): 19-24.

[5] Brenken, Bastian, et al. "Fused filament fabrication of fiber-reinforced polymers: A review." Additive Manufacturing 21 (2018): 1-16.

[6] Singh, Rupinder, et al. "Multi-material additive manufacturing of sustainable innovative materials and structures." Polymers 11.1 (2019): 62.

[7] Skylar-Scott, Mark A., et al. "Voxelated soft matter via multimaterial multinozzle 3D printing." Nature 575.7782 (2019): 330-335.

[8] Cincinnati. Additive Solutions 3D Print Your Way Everyday, http://wwwassets.e-ci.com/PDF/Products/Additive-Fact-Sheet.pdf [accessed January 24, 2020]

[9] Kim, Seokpum, et al. Graded Infill Structure of Wind Turbine Blade Accounting for Internal Stress In Big Area Additive Manufacturing. Oak Ridge National Lab.(ORNL), Oak Ridge, TN (United States), 2018.

[10] Kunc, Vlastimil, et al. "Large Scale Additively Manufactured Tooling For Composites." Proceedings of 15th Japan International SAMPE Symposium and Exhibition. 2017.

[11] Kunc, Vlastimil, et al. "Investigation of in-autoclave additive manufacturing composite tooling." CAMX Conference, Anaheim, CA. 2016.

[12] Hill, Charles, et al. Big Area Additive Manufacturing (BAAM) Materials Development and Reinforcement with Advanced Composites. No. IACMI/-0015-2017/3.6. Inst. for Advanced Composites Manufacturing Innovation (IACMI), Knoxville, TN (United States), 2018.

[13] Love, Lonnie & Kunc, Vlastamil & Rios, Orlando & Duty, Chad & Elliott, Amelia & Post, Brian & Smith, Rachel & Blue, Craig. (2014). The importance of carbon fiber to polymer additive manufacturing. Journal of Materials Research. 29. 1893-1898. 10.1557/jmr.2014.212.

[14] Kim, Heechang, et al. "Experimental study on mechanical properties of single-and dual-material 3D printed products." Procedia Manufacturing 10 (2017): 887-897.

[15] Cornock, Rhys, et al. "Coaxial additive manufacture of biomaterial composite scaffolds for tissue engineering." Biofabrication 6.2 (2014): 025002.

[16] Kumar V., Kim, S., Kishore, V., Love, V., Blue, C., Kunc, V., Hassen A.A. (2020) “Hybrid Manufacturing Technique Using Large-Scale Additive Manufacturing and Compression Molding for High Performance Composites,” The Annual Technical Conference for Thermoplastics Professionals. San Antonio, TX, USA (submitted).

[17] Brackett, James, et al. "Development of Functionally Graded Material Capabilities in Large-scale Extrusion Deposition Additive Manufacturing."

MECHANICAL METAMATERIAL CONTINUUM MATERIALS REALIZED THROUGH 3D PRINTING USING COMPLIANT MECHANISMS TECHNIQUES

Molly Carton, Duane Storti, Mark Ganter, Jeffrey Ian Lipton
University of Washington
Seattle, Washington

ABSTRACT

Recently developed classes of mechanical metamaterials that couple twisting with linear and volumetric expansion show great promise for the fields of soft robotics, deployable structures, and medical devices. These twisting auxetics can convert torque from conventional motors directly into expansion. Previously they have been realized only as assembled units, but in order to be usable in metamaterial structures they must be fabricable as a single continuum of material. However, the complex joints and relations between layers of material are not compatible with conventional fabrication processes. We use mono-material 3D printing of compliant materials and compliant mechanism design techniques to produce 3D printable continuum material versions of these twisting metamaterials. We show that we can produce continuum analogs of rotating bilayer joints, angle preserving joints, and multi-component joints and replicate the bilayer polyhedral auxetics. We show how these techniques can be used to fabricate planar structures that can be folded into 3D objects. These printing and design techniques allow for the easy and scalable fabrication of twisting auxetic metamaterials which can be actuated with motors.

Corresponding author: Jeffrey Lipton

1. INTRODUCTION

Volumetric expansion is used in many biological and soft robotic systems for movement. Soft robots often use fluid flow or thermal phase changes to drive expansion [14, 15, 16, 17]. While they have been useful for some limited grasping and movement tasks, thermal actuation is slow and pneumatic actuation is susceptible to leaks and punctures and is inefficient. Auxetic metamaterials are characterized by negative Poisson's ratio, causing them to expand under strain. These materials can be planar arrays of two-dimensional [18, 11, 13] or three-dimensional cells [8, 2, 7]. The Poisson's ratio and directional anisotropy of these materials are geometrically determined by the metamaterial structure and its underlying symmetries. The structure is scale and material agnostic, making these metamaterials widely applicable. The nature of the joints determines the scale of the meta-material cell. Three-dimensional auxetics that can be twist driven can be the basis for volumetric expanding robots powered by motors.

It was shown in [6] that bilayer cellular auxetic structures with polyhedral symmetries are ideally suited for motor-driven volumetric expansion. However, the work relied on discrete structures that have a high assembly cost. Bilayer auxetics consist of placing a mirror copy of a layer on top of another and aligning common points of rotation. Producing bilayer auxetics out of a continuum of material requires alignment preserving joints not previously deployed in continuum auxetic models.

Here we show how we can use compliant mechanism techniques and additive manufacturing to create compliant bilayer auxetic cells. These cells expand and contract volumetrically when antipodal points are twisted. We employ living hinge-based designs which approximate rigid links connected by pin joints. We use compliant mechanism design techniques [1, 5] to produce continuous material auxetic linkages a library of joint designs needed for these structures. We then use these techniques to produce a three-dimensional platonic auxetic unit cell.

In order to reduce the cost and time of production, the parts are designed to be printable with desktop fused filament fabrication (FFF) and require no support material, reducing post-processing time. To accomplish this, we fabricate 3D polyhedral auxetic cells from planar prints based on the 2D nets of the 3D polyhedral auxetic cell. Making net models that can be auxetic structures required us to develop a modified living hinge geometry

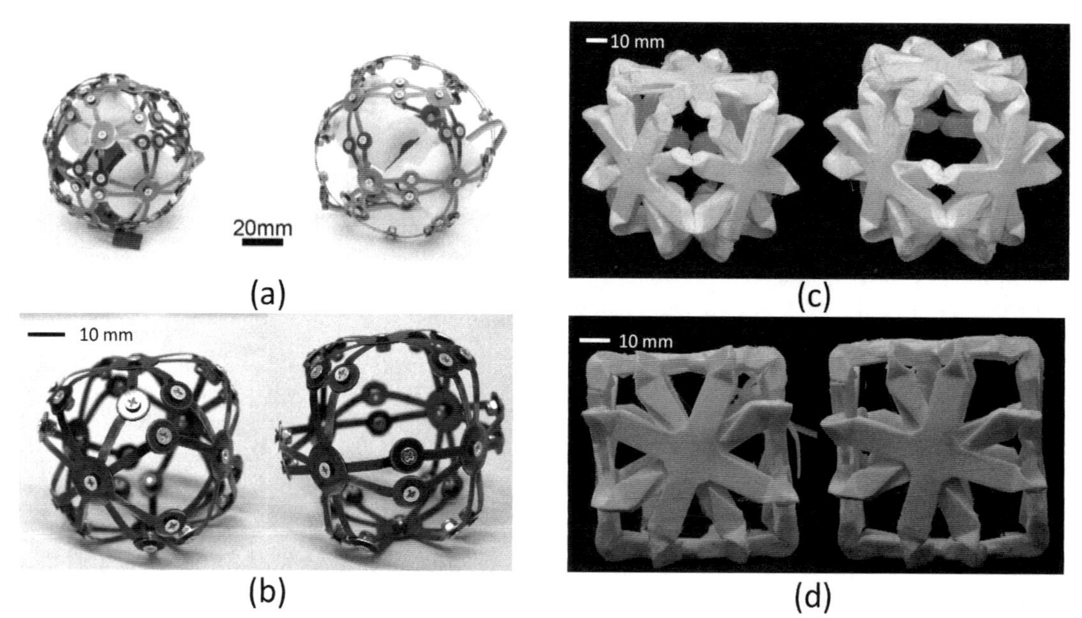

Figure 1: Bilayer auxetics can be used as motor driven actuators(a). Polyhedral bilayers can be fabricated from discrete components (b) or 3D printed as a continuous piece of material and assembled into an auxetic structure (c, d) .

that both enforces the design rules for support free FFF and allowed for significant defomation of the joint along one axis, while acting as a single degree of freedom hinge in another.

In this paper we:

- Develop the planar joints needed to replicate bilayer planar auxetic structures

- Show how to extend this method to produce 3D polyhedral auxetics using net based designs
- Demonstrate that these designs can be fabricated on low cost FFF printers without support material.

2. BACKGROUND

2.1 Compliance in Auxetics

Most auxetic cellular materials can be modeled as repeating arrays of linkages. These linkages form a single degree of freendom network between cells causes the material to expand and contract as a whole. Planar auxetic patterns include re-entrant honeycomb and arrows, chiral missing-rib cells, and corner-connected rotating polygons, all of which are modeled as rigid links or polygons connected by pin joints. With the right choice of internal pattern these lattices can obtain up to Poisson's ratio of -1 [23].

Compliant versions of these use either living hinges or bending of the links as the main mode of deformation [12]. The living hinge technique has been applied to several single layer 3D auxetics. Buckliballs, for instance, use cutout circles in a spherical surface to create a pattern of polygonal faces connected by simple joints, which rotate into their collapsed state when their interconnecting members buckle [9]. Through special geometry, all of the connections undergo the same buckling mode, keeping the collapse of the sphere radially uniform. These compliant spheres were possible to cast from a single material due to their geometric simplicity.

2.2 Compliant Mechanism approaches to complex joints

It is possible to convert pin joint models to continuous materials because of compliant mechanisms. A lumped-compliant system – one in which compliance is mostly localized at small-length joints rather than distributed over a larger flexing area – corresponds to a linkage mechanism with rigid links joined by hinges with specific degrees of freedom [5, 1, 19]. These consist of a combination of simple living hinges between thicker approximately rigid beams.

More complex joints can be constructed by joining simple elements into serial configurations. For example, when creating a compliant version of a known mechanism such as the Peaucellier–Lipkin linkage which requires more than two links to come together at a single point, these points are decomposed into two slightly offset simple living hinges [21]. This allows the compliant version to be made from simple joints. By taking advantage of the flexibility of additive manufacturing, we do not have this constraint, freeing us to making joints such as those in the following section that are not internally planar.

2.3 Actuation of Auxetics

There have been several approaches to actuate auxetic structures. Buckliballs use pneumatic evacuation of hollow spheres to drive twisting volumetric collapse [9]. This rely on having a closed evacuable chamber in the center of the sphere, creating a uniform inward pressure. As with most soft robotics actuation techniques, this mean combining the cells expands the number of failure points in the system, as a single puncture could prevent all connected cells from actuating. Other approaches have relied on thermally actuated shape memory alloys and polymers to expand or contract the structure. Like other thermal actuation methods, this makes for a slow actuation that is power intensive [26].

Handed shearing auxetic cylinders introduced twist-based expansion and contraction without pneumatic actuation [3]. These cylinders extend and shrink with motor-driven rotation. This motion is coupled with some radial expansion of the cylinder. While primarily employed as cylinders, the same principles were compatible with spherical structures. In [6] the concept was extended to show that twist based actuation could generate volumetric expansion by generating torques between antipodal points on a structure.

3. JOINTS FOR COMPLIANT AUXETICS

(a) (b) (c) (d)

Figure 2: Bilayer planar auxetic cellular patterns require joints connecting different numbers of elements. From left: (a) re-entrant honeycomb, single and bilayer (b) rotating triangles, which requires three segments per layer connected at the rotating center (c) and (d) chiral and missing rib, which require four segments connected at the rotating center.

There are a wide variety of known auxetic patterns, both planar and three-dimensional. All planar auxetic patterns have a bilayer counterpart connected between the layers at the points of rotation [8, 27]. Figure 2 shows some of the canonical planar auxetic patterns and their bilayer counterparts. Figure 2 c. shows the chiral bilayer auxetic described in [8] and figure 2 d. shows the auxetic described in [27]. The other bilayer auxetics are designed by analogy. For each bilayer there is an inter-layer joint at the point where the structures remain fixed relative to one another.

Bilayer planar structures are mirrored between horizontal layers, leading to mirrored rotational movement. The inter-layer joints thus need to have a single rotational degree of freedom. The bilayer chiral cells in f. are designed in [8] with a rotating joint consisting of angled beams connecting the layers. This joint adds an auxetic element in the third axis, but it constrains the direction of movement and in a volumetric cell would add unnecessary internal stress. We sought a design which could eliminate this extra movement and keep the layers a fixed distance apart while still maintaining auxetic behavior in plane.

These joints are determined by the number of segments that meet at the central point on each layer, forming a rigid body. In the re-entrant honeycomb pattern, these cross-layer joints consist of two simple crossed beams. For the rotating triangle pattern, each layer has three segments meeting at the central rotating point. The chiral pattern and the missing rib pattern each have four segments per layer meeting at the inter-layer joint.

3.1 Bilayer movement

The bilayer structures have a complex joint at their cross-layer bonds. For the simplest of these joints – two beams crossing at a point – a flexure joint consisting of a square of four simple flexures at the corners of four beams (a structure named the Q-joint in [1]) provides a planar alternative with similar behavior. As the hinge flexes the links stay approximately parallel but they are only colinear at the center of the range of movement; at the maximal angles the beams are offset up to the link length of the internal tetrahedron. This behavior does not mimic the pin joint closely enough to be used in these auxetic parts, and does not work at all for the joints with more crossing beams.

Our solution is a bilayer joint inspired by flex bearings (figure 3 d.). Flexure joints between a connecting central core and a concentric surrounding free layer allow the two layer elements to rotate with respect to one another. Joints are characterized by the number of beams attached to the central point for each of the two layers.

Because beams move in the same plane rather than in two planes, the width of the beams presents a limit to the angle range of these joints. The two layers cannot cross over one another and are constrained by hitting their neighboring beam. Cutting an angle in the beams to create a narrower region near the center increases the angle range of movement but weakens the rigidity of the layer. This problem is more significant the larger the number of segments in the layer, as crowding of the center increases.

	2-segment	3-segment	4-segment
Ideal	180°	120°	90°
Assembled	122°	52°	39°
Printed	70°	54°	28°

Table 1: The angle ranges for the printed bilayer joints are somewhat lower or comparable to the assembly structures [6], although both are short of the ideal range.

As seen in table 1, the two-segment joint pictured in figure 3 a. has approximately 70 degrees of range, a loss over the assembled structure in [6]. The range of the three-segment joint in b. is approximately 54 degrees of range and the range of the four-segment joint in c. is 28 degrees. The angle range decreases with the crowding of neighboring segments around the central core. These angle ranges are sufficient to produce some auxetic behavior but do not match that of the spring steel assembled cells.

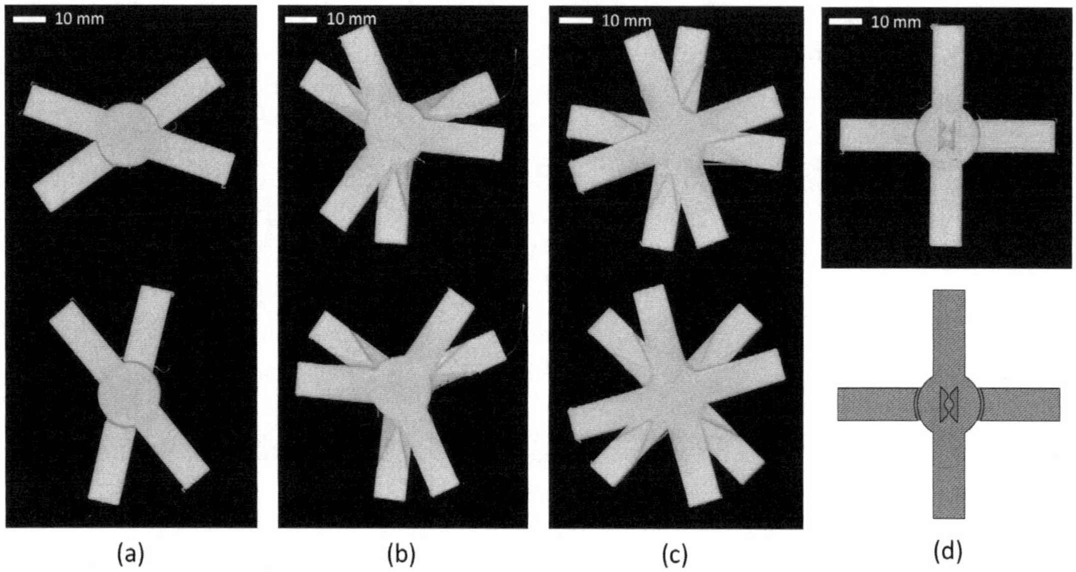

Figure 3: 3D printed TPU test parts of bilayer joints demonstrating range of motion. From left: (a) two segments per layer, (b) three segments, (c) four segments, (d) internal structure with living hinge attached to central core: 3D printed test part and CAD model.

4. EXTENDING TO THREE DIMENSIONS

4.1 Using Nets to Print Polyhedral Auxetic Cells

Producing bilayer polyhedra from the compliant bilayers developed in section 3 can be done in several ways. Firstly, complaint faces could be fabricated as individual sections and then bonded together. Secondly, the faces could be rotated and aligned in CAD, joined with a simple joint, and printed as a single piece. Finally, the system could be printed as a planar net, folded, and assembled.

All polyhedra can be segmented into a planar structure by selective cutting and flattening of edges to form a net. For each polyhedron, there are families of nets in which the polygonal sides are unfolded into non-overlapping planar connected figures. Figure 4 row 2 shows example nets for several polyhedra along with models of bilayer auxetics. Compared to assembling the cells or printing the individual faces separately these nets significantly reduce both the number of seams and the number of prints required to manufacture these polyhedral cells.

Compared with printing as a single unified part, the net method also significantly reduces the need for support material, and with it the need for significant post processing. If the structures from section 3 were applied to the patterns in figure 5 and printed, there would be a large internal void in each model. This void would need to be filled with a support structure to facilitate printing the top faces. While some processes such as selective laser sintering use the unused powder as support and thus do not require support structures, most processes require an additional printed support structure that needs post processing to remove and takes extra time to print.

We chose to use FFF fabrication as described in section 5 because it is widely accessible and cheap to use. FFF in particular requires the use of support material to fill the voids that direct fabrication would produce. On a mono-material FFF system, the support material is the build material, making the process of support removal a time consuming manual process. By eliminating the need for support material, we can use the cheapest and most widelyt available 3D printing method. In exchange we retain the post processing requirement of manually folding the edges and bonding the faces together. To use printed nets we need a joint which can both connect printed faces and be folded, and a joint that can be bonded after fabrication.

4.2 Net joints and bonds.

A joint in our net exists at the vertex of our polygons. It must constrain the length of the joint while allowing for bending along the edge of the net, and must be able to change angle as it moves along the edge of the net. This constrains us to using a version of a universal joint with two degrees of freedom. Previous polyhedral structures were based on circular notch living hinges or bending beams. Circular notches are effective at creating a single degree of freedom joint but are not suitable for being folded along the net. A bending beam structure would be a three degree of freedom joint, while the joints along the edges of the net need to have two degrees of freedom. As a result we are forced to use a different style of joint from other 2D and 3D auxetic structures.

While there are several compliant joints equivalent to universal joints, we chose to modify the circular notch style to allow for the bend. This modification allowed us to enforce the 45 degree angle restriction on overhangs commonly used in monomaterial 3D printing. As seen in figure 5B, we also extended the hinge upwards out of plane in an arc to better resist parasitic twisting. This constrains most of the movement to the bending mode that keeps the beams in plane with the faces of the polyhedron.

Nets place cuts along the edges of the polyhedral structure. Applying this to an auxetic structure would put the cut directly through a hinge, which is not an ideal place for bonding. As a result we moved the location of the cuts. The part was designed with seams within the beams and away from the hinges, reducing the stress on the seam. The seams could then be joined with thermoplastic adhesive and had significantly more surface area for bonding. The printing and bonding methods are discussed in the next section. As seen in Figure 1, the combination of cut structure, bilayer joint, edge joint geometry, and bonding methods allowed us to take a net and form a bilayer polyhedral auxetic.

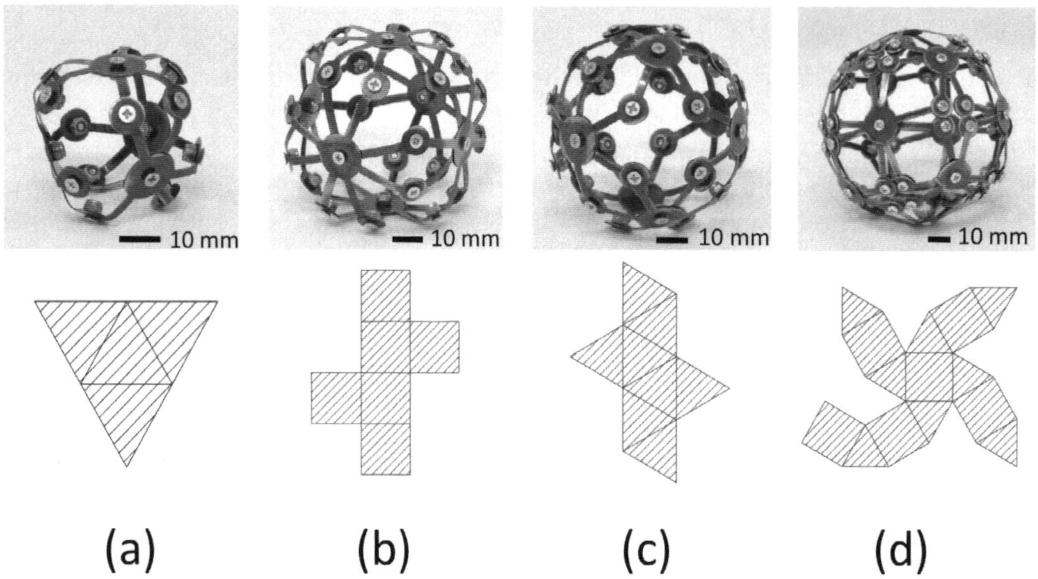

Figure 4: The polyhedral auxetic cells from [6], created by tiling a jointed pattern on the faces of polyhedra. From left, the cells are based on the rotating tetrahedron (a), cube (b), octahedron (c), and cuboctahedron (d). These cells can be made as a single layer (top) or bilayer (middle) and each polyhedron can be unfolded into a non-unique net (bottom).

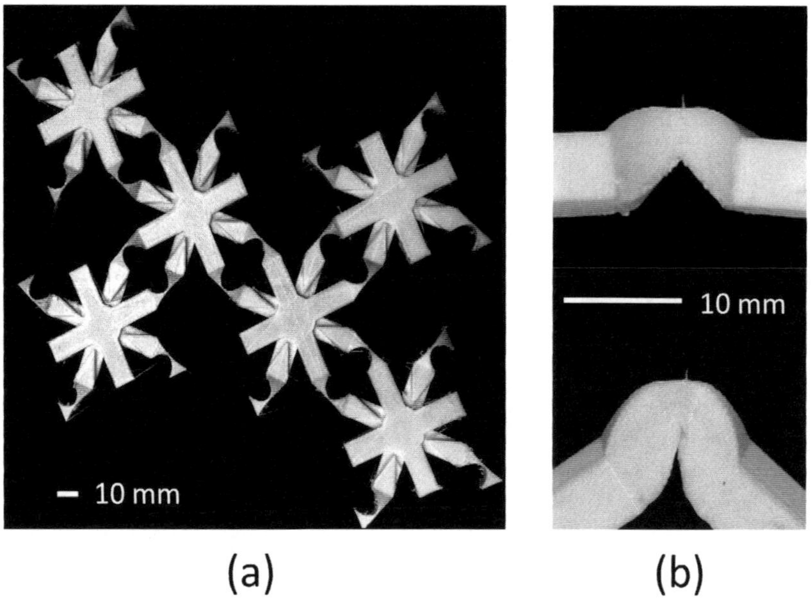

Figure 4: The print on a cubic net (a, background digitally removed) has joints placed on the folding edges between the square faces of the assembled cube. The joint at the edge (b) is notched in order to fold over the edge when the net is assembled.

5. PRINTING AND BONDING METHODS

We fabricated the final part in TPU (NinjaTek Cheetah in Snow) on a desktop FFF printer (Creality Ender 3 Pro). Extrusion was raised to 105% to maintain solidity and infill 100% using a concentric pattern, keeping the print lines aligned along the joints. Planar printing avoids issues of print anisotropy in the hinges. Layer thickness was .2 mm. For the regions requiring layers to remain separate, internal to the bilayer joints, a gap of two layers (.4 mm) was built into the part. This allows the lower printed layer to cool, making the inter-layer adhesion very weak. These layers can then be easily separated in the finished print. We printed the three-dimensional cell shown in figure 1 as a planar part, unfolded using the net in figure 4 j. We used thermoplastic adhesive to bond the net together.

6. CONCLUSIONS AND FUTURE WORK

We introduced a series of compliant joints used to replicate bilayer auxetic structures as continuous material complaint mechanisms. We then used these joints to replicate a bilayer polyhedral auxetic cell. These polyhedral auxetic cells can be fabricated in planar form using a commercial desktop mono-material FFF printer. We conclude that using net-based construction, these three-dimensional metamaterial cells can be manufactured with limited manual post processing and in a cheap and accessible manner. This process considerably reduces the assembly costs from the spring steel assemblies.

In making these compliant mechanisms we have sacrificed the material-agnostic nature of the auxetic metamaterial design. The geometric parameters of these hinges are designed for FFF printed TPU, which limits the range of compliance and the detail present in the parts. With other manufacturing processes it will be possible to create better detail and smaller or larger parts, but these designs will have to be reworked to suit other materials and processes. This will require using more expensive and less accessible processes such as CLIP, SLS, and HP Jet fusion.

We believe this work represents a significant step towards producing large arrays of auxetic cells whose expansion can be driven by motors. This will provide a more robust and controllable form of volumetric actuation for soft robots when compared to pneumatics and a faster and more controllable method than thermal driven processes.

7. REFERENCES

[1] L. L. Howell, et al., eds. *Handbook of compliant mechanisms.* John Wiley & Sons, Incorporated, 2013.

[2] Lakes, Roderic. "Foam structures with a negative Poisson's ratio." *Science* 235 (1987): 1038-1041. DOI: 10.1126/science.235.4792.1038

[3] Lipton, Jeffrey Ian, et al. "Handedness in shearing auxetics creates rigid and compliant structures." *Science* 360.6389 (2018): 632-635. DOI: 10.1126/science.aar4586

[4] Kovács, F., et al. "A class of expandable polyhedral structures." *International journal of solids and structures* 41.3-4 (2004): 1119-1137. DOI: 10.1016/j.ijsolstr.2003.09.046

[5] Gallego, Juan A., and Just Herder. "Synthesis methods in compliant mechanisms: An overview." *ASME 2009 International Design Engineering Technical Conferences and Computers and Information in Engineering Conference.* American Society of Mechanical Engineers Digital Collection, 2009. DOI: 10.1115/DETC2009-86845

[6] Lipton, J., et al. "Modular Volumetric Actuators Using Motorized Auxetics". *2019 IEEE/RSJ International Conference on Intelligent Robots and Systems (IROS)*. IEEE, 2019.

[7] Alderson, A., Rasburn, J., and Evans, K. E. "Mass transport properties of auxetic (negative Poisson's ratio) foams." *Phys. Status Solidi B,* 2007, 244(3), 817–827. DOI: 10.1002/pssb.200572701

[8] Fu, Minghui, Fengming Liu, and Lingling Hu. "A novel category of 3D chiral material with negative Poisson's ratio." *Composites Science and Technology* 160 (2018): 111-118. DOI: 10.1016/j.compscitech.2018.03.017

[9] Shim, Jongmin, et al. "Buckling-induced encapsulation of structured elastic shells under pressure." *Proceedings of the National Academy of Sciences* 109.16 (2012): 5978-5983. DOI: 10.1073/pnas.1115674109

[10] Mark, Andrew G., et al. "Auxetic metamaterial simplifies soft robot design." *2016 IEEE International Conference on Robotics and Automation (ICRA)*. IEEE, 2016. DOI: 10.1109/ICRA.2016.7487701

[11] Evans, Kenneth E., and Andrew Alderson. "Auxetic materials: functional materials and structures from lateral thinking!." *Advanced materials* 12.9 (2000): 617-628. DOI: 10.1002/(SICI)1521-4095(200005)12:9<617::AID-ADMA617>3.0.CO;2-3

[12] Farrugia, Pierre-Sandre, et al. "Different Deformation Mechanisms Leading to Auxetic Behavior Exhibited by Missing Rib Square Grid Structures." *physica status solidi (b)* 256.1 (2019): 1800186. DOI: 10.1002/pssb.201800186

[13] Alderson, A, and K. L. Alderson. "Auxetic materials." *Proceedings of the Institution of Mechanical Engineers, Part G: Journal of Aerospace Engineering* 221.4 (2007): 565-575. DOI: 10.1243/09544100JAERO185

[14] Mosadegh, Bobak, et al. "Pneumatic networks for soft robotics that actuate rapidly." *Advanced functional materials* 24.15 (2014): 2163-2170. DOI: 10.1002/adfm.201303288

[15] Robertson, Matthew Aaron, and Jamie Paik. "New soft robots really suck: Vacuum-powered systems empower diverse capabilities." *Science Robotics* 2 (2017): eaan6357. DOI: 10.1126/scirobotics.aan6357

[16] Li, Shuguang, et al. "Fluid-driven origami-inspired artificial muscles." *Proceedings of the National academy of Sciences* 114.50 (2017): 13132-13137. DOI: 10.1073/pnas.1713450114

[17] Terryn, Seppe, et al. "Self-healing soft pneumatic robots." *Science Robotics* 2.9 (2017): eaan4268. DOI: 10.1126/scirobotics.aan4268

[18] Lakes, Roderic. "Advances in negative Poisson's ratio materials." *Advanced Materials* 5.4 (1993): 293-296. DOI:10.1002/adma.19930050416

[19] N. Lobontiu: Compliant Mechanisms: Design of Flexure Hinges, CRC Press, 2003. ISBN 0849313678

[20] Howell, Larry L., and A. Midha. "A method for the design of compliant mechanisms with small-length flexural pivots." *ASME Journal of Mechanical Design.* (1994): 280-290. DOI: 10.1115/1.2919359

[21] Hricko, J. "Design of compliant micro-stage based on Peaucellier-Lipkin straight-line mechanism." *2014 23rd International Conference on Robotics in Alpe-Adria-Danube Region (RAAD)*. IEEE, 2014. DOI: 10.1109/RAAD.2014.7002262

[22] Awtar, Shorya, Alexander H. Slocum, and Edip Sevincer. "Characteristics of beam-based flexure modules." *ASME Journal of Mechanical Design* 129.6 (2007): 625-639. DOI:10.1115/1.2717231

[23] Milton, Graeme W., and Andrej V. Cherkaev. "Which elasticity tensors are realizable?." *ASME Journal of Engineering Materials and Technology* (1995): 483-493. DOI: 10.1115/1.2804743

[24] Novak, Nejc, Matej Vesenjak, and Zoran Ren. "Auxetic cellular materials-a review." *Strojniški vestnik-Journal of Mechanical Engineering* 62.9 (2016): 485-493. DOI: 10.5545/sv-jme.2016.3656

[25] Weisstein, E. W. 'Cuboctahedron.' 1999. Michigan State University, East Lansing, MI. Accessed January 2020 <https://archive.lib.msu.edu/crcmath/math/math/c/c832.htm>

[26] Jacobs, S., et al. "Deployable auxetic shape memory alloy cellular antenna demonstrator: design, manufacturing and modal testing." *Smart Materials and Structures* 21.7 (2012): 075013. DOI: 10.1088/0964-1726/21/7/075013

[27] Cabras, Luigi, and Michele Brun. "Auxetic two-dimensional lattices with Poisson's ratio arbitrarily close to− 1." *Proceedings of the Royal Society A: Mathematical, Physical and Engineering Sciences* 470.2172 (2014): 20140538. DOI: 0.1098/rspa.2014.0538

Additive Manufactured ULTEM 9085 Part Qualification and Allowable Generation

Reza Hajiha[1], Harsh Baid[1], Steven Floyd[2], Nicole Curtis-Brown[2], Nima Moazami[1], Frank Abdi[1,] and Elizabeth Clarkson[3]

[1]*AlphaSTAR Corporation, Long Beach, CA, USA*
[2]*Northrop Grumman Space Systems-Space Park, CA, USA*
[3]*National Institute for Aviation Research, Wichita State University, Wichita, KS, USA*

ABSTRACT

Part qualification of additive manufactured polymers has become more important due to the increasing tendency in industrial application of Fused Deposition Modeling (**FDM**) which yields to quality parts with low cost of manufacturing and high freedom of customization. To reduce printing errors and evaluate fabrication process designs before actual manufacturing, reliable virtual methods are required to model the material, predict part performance, determine the design allowables regarding the variations in material, printing errors and manufacturing facilities.

This case study demonstrates thermo-mechanical analysis, print process simulation, allowable generation and validation of FDM based 3D printed **ULTEM 9085** material system. A nano assisted micromechanics approach is implemented to predict room temperature mechanical properties of 3D printed material system considering micro voids as inclusion calculated using thermal transport phenomenon. The case study also includes prediction of temperature dependent mechanical properties of FDM printed unfilled ULTEM 9085 material using multi-scale material modeling, cure kinetics and multi-factor technique. From print process modeling point of view, geometrical and pathing error macro size defects, called "bald spot", were calculated and resulted in degradation of both modulus and strength properties in the element scale of the Finite Element Model (FEM). Next, thermal and coupled thermal structural analysis using Multi-scale Progressive Failure Analysis (MS-PFA) were performed to calculate temperature distribution, deformation and residual stress during and after 3D printing process. Finite Element Analysis of service loading integrated with MS-PFA was also performed on virtually 3D printed parts considering residual stress and deformed shape as well as geometrical and pathing errors. The simulation showed excellent agreement with available test data. Also, damage initiation and evolution were predicted as a result of MS-PFA. By performing virtual sampling, the tensile strength allowables and constituents' sensitivities for ULTEM9085 printed in multiple orientations were predicted and validated with NIAR's qualification and Stratasys's equivalency test data. The generic acceptance region for parts printed by RP+M and Stratasys printers were plotted to establish FDM ULTEM9085 generic acceptance region between those two distinguished manufacturing facilities.

Keywords: **1**) Fused Deposition Modeling; **2**) ULTEM 9085; **3**) Micro Void; **4**) Nano assisted Micro Mechanics; **5**) Temperature Dependent Mechanical Properties; **6**) Geometrical and Pathing Errors; **7**) Coupled Thermal-Structural Analysis; **8**) A-B and Generic Basis Allowables

NOMENCLATURE

Notation	Description
CDF:	Cumulative Distribution Function
***CTE*:**	Coefficient of Thermal Expansion
***D&DT*:**	Durability and Damage Tolerance
ET:	Elevated Temperature
***FDM*:**	Fused deposition Model
***FFF*:**	Fused Filament Fabrication
***ICME*:**	Integrated Computational Material Engineering
***MCQ*:**	Multi-scale material characterization and Qualification
***MS-PFA*:**	Multi Scale Progressive Failure Analysis
***RT*:**	Room Temperature
***TG*:**	Transient Glass Temperature
***TMg*:**	GENOA 3DP Thermal Management
***VVR*:**	Void Volume Ration

1. INTRODUCTION

Additive Manufacturing (AM) continues to show great promise. AM technology has demonstrated an ability to decrease costs and reduce manufacturing lead-time, while providing capability to fabricate complex geometries that are otherwise impossible using conventional manufacturing processes. *Yet challenges* remain with AM that makes part qualification difficult. Builds are reliant upon experimentally driven trial and error, and NDE (nondestructive evaluation) data driven analysis is not properly exploited to detect and possibly correct build anomalies. Common problems associated with **Polymers**, with and without inclusions, build problems include *porosity (gas bubbles, inter-bead voids, fiber pull out), interlayer delamination, wrinkling, warpage, anisotropic behavior, residual stress, fiber waviness/agglomeration and saturation of mechanical (strength/stiffness) properties* **(Figure 1)**. The manufacturing anomalies, in terms of effects of defects, are the source of scatter and uncertainty in mechanical and fatigue properties for both metal and polymer AM parts. In general AM process often produces part with unique microstructural irregularities in metals, plastics, and fibrous composite material. **Figure 7** shows typical inter-bead void formation for two different 3D printed meso-structures. **Table 1** describes the **FDM** machine parameters

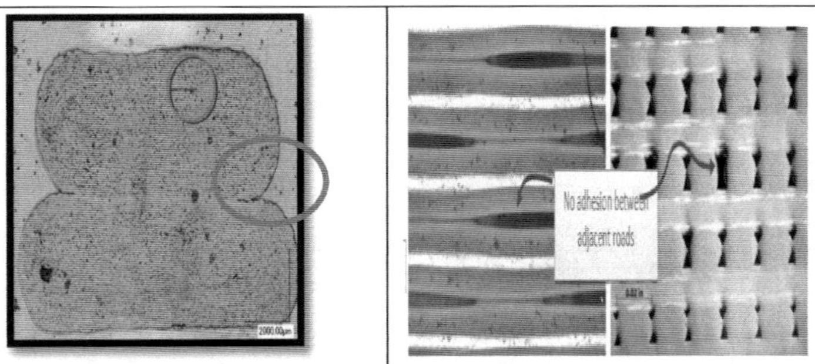

a) Cracks within/between beads　　b) Delamination & Porosity

Figure 1. Micrographs of polymer meso-structures, showing typical defect formation

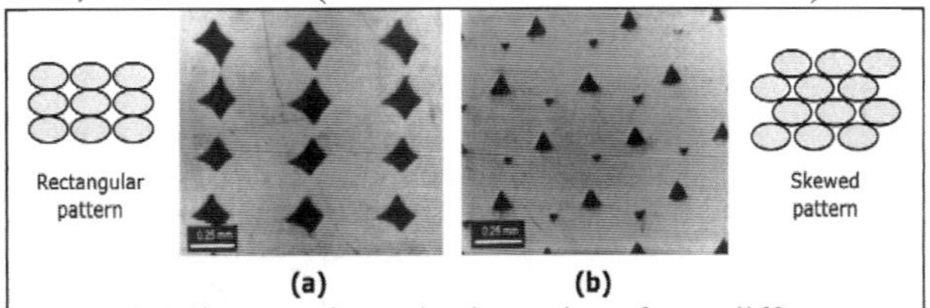

Figure 2. Micrographs and schematics of two different meso-structures, a) rectangular and b) skewed, showing typical triangular void formation [11]

PROBLEM: The AM process that is being considered in this paper is FDM. Typical FDM exhibits high porosity, sub-optimal bonding between roads and layers, poor interfacial bonding between polymer and fillers. Proper material modeling is needed to assist in process optimization and printed structure performance modeling. It is also observed that during the print process temperature gradient between two layers leads to: **1)** contraction of hot layer and compression of cold layer until equilibrium is achieved; and **2)** thermal stresses generated during solidification process induces thermal warping/distortion in 3D printed part.

Table 1. Thermoplastic AM machines primary printing parameters.

Items	Parameters	Description	Controlled or Predefined
1	Printing Speed	Velocity of nozzle across surface	Controlled
2	Print Pattern	Island, Ox, Concentric, rectangular	Controlled
3	Hatch Spacing	Distance between neighboring passes	Controlled
4	Layered Thickness	Hight of single layer	Controlled
5	Nozzle Temperature	Polymer Extrusion Temperature	Controlled
6	Nozzle Diameter	Extrusion dimension	Controlled
7	Extrusion Speed	Polymer feeding Velocity	Controlled
8	Baseplate temperature	Build plate temperature	Controlled
9	Baseplate Type	Material of Base plate	Controlled
10	Chamber Temperature	Enclosed Space Temperature	Controlled
11	Inclusion Volume Ratio	Material, and type	Controlled
12	Beed Width dimensions	Based on Hatch Spacing, Print speed, Nozzle diameter	Uncontrolled

The mechanical behavior of **FDM** specimen including the effect of build parameters is being extensively studied to understand the variation in mechanical and thermal properties [2-**4**]. Researchers have studied the effect of layer thickness, deposition angle as well as infill on the flexural properties [2]. Studies have been also conducted to see the compressive properties variation due to raster angle, air gap and wall thickness **[3].** Finite Element Analysis has also been studied in great extent by many researchers to perform additive manufacturing process [4-**5**]. Finite element method has been implemented by researchers [4] to evaluate the temperature distribution during the **FDM** simulation process. Others have developed FE methods [5] to calculate mechanical properties of **FDM** parts validated with test.

One of the primarily focus of this paper is simulation of 3D printing process using ULTEM 9085 material. For appropriate simulation of 3D printing process, it is very important to assign temperature dependent material properties (both thermal and mechanical). The temperature dependent mechanical properties are either obtained from test or assumed based on some degradation factor. The 3D printing simulation can lead to totally different directions depending on the assumed thermal dependent mechanical properties. The thermal stress calculated from 3D printing simulation process could be totally different depending on the assigned thermal dependent mechanical properties. The prediction of the structural performance due to service loading could be totally different depending of the residual stresses generated during the 3D printing simulation process. If the temperature dependent material property is not known, an appropriate methodology/technique is needed to be implemented to predict accurate properties. This paper also focuses on cure kinetics-based method coupled with multi-factor technique to predict temperature dependent material properties (stiffness and strength). The predicted temperature dependent material properties are implemented in FE model generated using 3D printing process to perform coupled structural and thermal analysis. A three-point bending simulation is carried out and the flexural stress-strain curve is validated with test available in the literature for **ULTEM 9085** material [6].

This paper describes the effort of **FDM printed *Neat (Unfilled) Ultem 9085 thermoplastic material, ASTM coupons, and sub-component utilizing ICME tool set for multi-scale material modeling, and process modeling.*** **Material modeling** included prediction of: 1) Thermal state and Material State using Micro thermal Management (**GENOA 3DP TMg**) thermal transport software, **2)** Multi-scale material characterization and Qualification (**MCQ**) modeling including prediction of: (i) mechanical properties (modulus, strength) Vs. Temperature, and **b)** Material qualification, and generation of probabilistic A_ B basis allowables and perform generic equivalency using different machines. **Process Modeling** included: **a)** detection of macro voids and defects due to print pattern error, prediction of loss of stiffness and strength and **b)** formation of residual stress, of dog bone, and cross beam box distortion during AM printing process considering effect of defects.

Machine Equivalency: Generic Basis Equivalency are determined in NIAR report [8-9] using Rapid prototyping and Manufacturing (RP+M) facilities as a part of material qualification. Stratasys AM machine then, performed some tests using its own manufacturing facilities as a part of equivalency process. This effort shows the validation of A & B Basis allowables for NIAR [9] Qualification report. Then, by performing virtual sampling, the allowables for parts printed by Stratasys facilities are predicted .Consequently, the generated virtual samples are used to define the Generic basis values and plot the Generic acceptance region for ULTEM 9085 FDM printed material system between two distinguished Additive manufacturing machines.

2. METHODOLOGY

The entire 3D printing process for XY direction (**Figure 2**) was carried out in three key steps: (**1**) Micro-Thermal Modeling (**2**) material modeling, and (**3**) 3D printing process modeling. The flow process of the material model, and printing process is shown in **Figure 3**. First, the air bubbles content appearing during the FDM printing was predicted using GENOA 3DP TMg (1 % air bubbles). Material modeling is required to assist in process optimization and printed structure performance modeling. Material modeling is conducted using nano-micro mechanics algorithm to predict the desired room temperature and elevated temperature (RT/ET) mechanical material properties (strength, stiffness) of ULTEM 9085 resin. Cure Kinetics was coupled with multi-factor technique to predict temperature dependent stiffness and strength of unfilled ULTEM 9085 material. From process modeling point of view, 3D printing simulation using the developed process parameter for FDM fabrication such as print pattern is performed to calculate the residual stress, and distortion of heat affected zone.

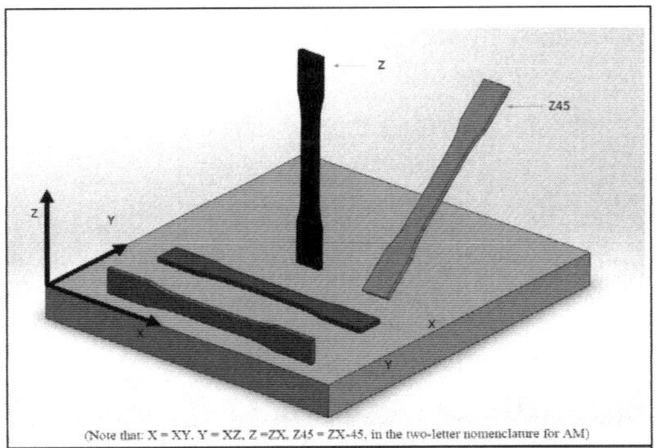

Figure 2. Demonstration of printing orientations including XY

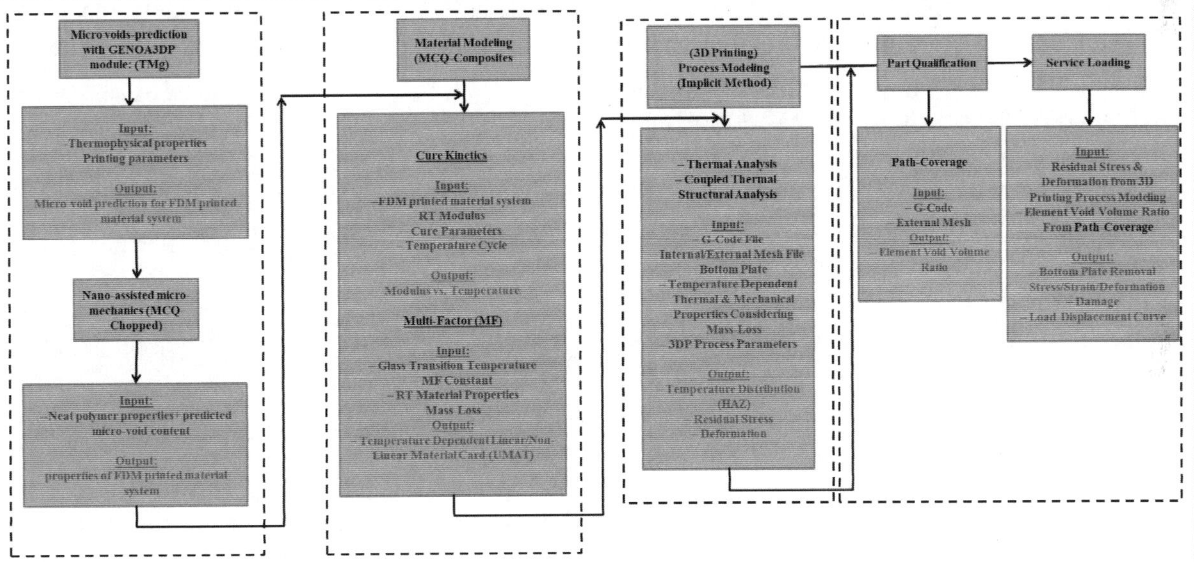

Figure 3. General schematic of the Process Path Flow (PPF) of 3D printing process for Dog bone specimen under tension loading using 3D printed ULTEM9085.

2.1 GENOA 3DP PathCoverage -Print Error Detection: *Macro Void Detection Software*

Before significant resources are committed to print a part, it is a good idea to assess the coverage of the printer paths on the printed part. This tool, GENOA 3DP PathCoverage, supports **a**) Quick visual assessment of printer GENOA 3DP PathCoverage, and **b**) Void ratio computation for all elements. This 2D graphics code displays one printed layer at a time. It supports quick switch (single key press) to the next or previous layer. The width and semi-circular end caps of each path are shown accurately on the part, given a user- specified bead width for the paths. Hence, a quick look can often reveal problems in GENOA 3DP PathCoverage. Further, a user can zoom in up close to verify minor coverage "leak". There is a mode to facilitate the identification of problematic paths. For better visualization of void detection, a part fabricated using Fused

Deposition Modeling (**FDM**) is compared with GENOA 3DP PathCoverage results in **Figure 4.** If the part is already printed by a defective set of paths, a user can choose to have the void ratio for each element of a FEM computed by this code, so that further quality assessment can be performed by a simulation code that accounts for such coverage voids.

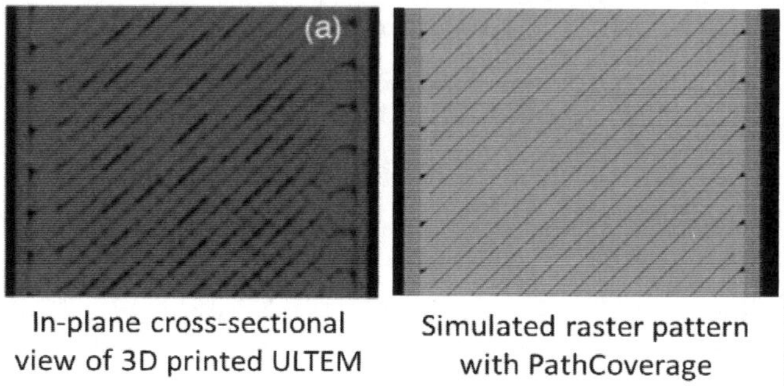

Figure 4. GENOA 3DP PathCoverage Bald Spot detection [1]

2.2 Material Modeling Thermo-Mechanical)

2.2.1 Micro Thermal Management (GENOA 3DP TMg)

A novel algorithm called Zeroth Order-Model (ZOM) and advanced Thermal code was developed based on systematic semi-analytical and Finite-Difference Explicit Method (FDEM)) approach. This multi-physics multi-phase (powder, solid, and liquid) model predicts real-time calculation of through-the-thickness temperature versus time and location in a more efficient way compare to other FEM. It has an advantage of extremely fast calculation and high accuracy (both macro- and microscale level). The algorithm is applicable to different class of materials such as metal-powder, Neat and reinforced polymer, and ceramic which proposes. Approach that addresses all thermal states/regions including i) heating, **ii**) melting, **iii**) melt-superheated **iv**) superheat-cooling v) solidification/sintering as well as **vi**) cooling depending on AM process condition and definition. The transient temperature gradients are calculated to predict the heat conduction, convection, radiation, axial and radial laser irradiation and absorption based on Gaussian distribution in laser-based concepts or a heat flux to the liquefier during **FFF** and **FDM** AM processes. This multi-physics multi-phase (solid, powder and liquid) module aims to compensate for the systematic physical property variabilities of the powder bed layer (e.g. density and volume changes, Volume of Solid (**VOS**), void formation, shrinkage, liquid formation and movement due to melting) and to determine process parameters for the subsequent layers to correct the defects in previous layers. This tool would be able to control the depth/width of the melt pool and Heat Affected Zone (**HAZ**) to deliver more energy to partially melted regions or less energy to overheated regions. **Figure 5** shows an example of FDM 3D Print Temp. Map. Equation 1 represents the main heat transfer equation that has been used in temperature calculation. In this equation, , ρ stands for density, $\boldsymbol{l}$ stands for material-phase, $\boldsymbol{C_p}$ is specific heat capacity, $\boldsymbol{V}$ is the volume of the element, $\boldsymbol{Q_s}$ is the heat source (laser or nozzle liquefier), $\boldsymbol{Q_k}$ is thermal conductivity, $\boldsymbol{Q_r}$ is thermal radiation, $\boldsymbol{Q_c}$ is thermal convection, and ***i, j, k, t*** stand for 3 different special directions and time step.

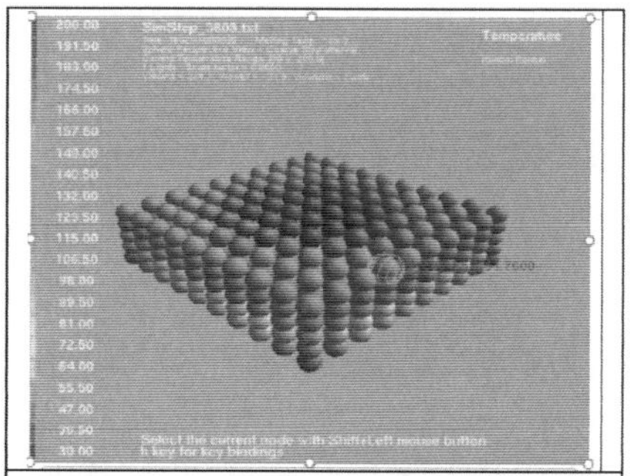
Figure 5. FDM 3D Print Temp. Map

$$\rho_l C_{pl} V \frac{dT(i,j,k)}{dt} = Q_s(i,j,k,t) - Q_k(i,j,k,t) - Q_r(i,j,k,t) - Q_c(i,j,k,t) \qquad [1]$$

2.2.2 Material Characterization and Qualification (MCQ):

The micromechanical composite models account for damages and defects that are typical of composites for as many extreme conditions as the parts will experience. MCQ's micromechanical material modules can compute responses of lamina and laminate levels and produce mathematical representations of the composite during any state of load progression (stress, strain, damage, no damage, etc.) and give those representations to a finite elements solver to run the necessary the analysis. It generates thermal, structural, electrical, and hygral properties and degradation under service environments for progressive failure analysis. Responses at each scale are automatically used as the loads to smaller scales (layer stresses are used to determine fiber/inclusion/matrix/interphase stresses – some micromechanics simulations or some layers will not use fibers). The composite damaged state stiffness and strength recomputed and are fed back into the FE model for the next load step (**Figure 6**). Material Characterization and Qualification (MCQNANO) includes **a**) Calibrate material properties using MCQNANO; and **b**) Predict the strength for each bead using the MCQNANO.

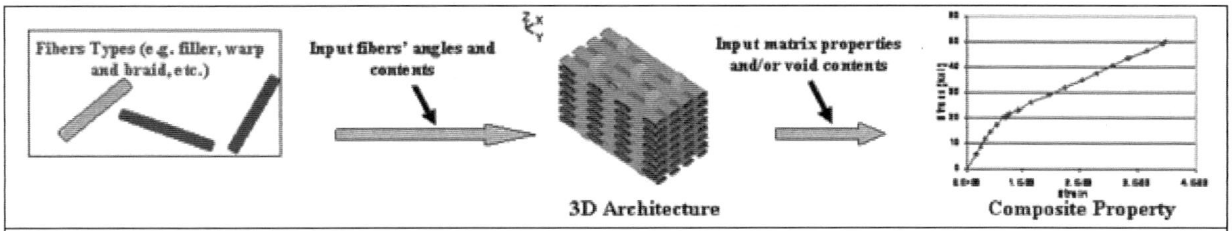

Figure 6. Micro-mechanics unit cell model considers manufacturing defects / fabrication parameters

2.2.3 Nano-Micro Mechanics based Material Property Prediction:

Many micromechanics approaches have been developed for the evaluation of the overall properties of heterogeneous materials and their dependence on the properties of constituents and the microstructure of materials. They are mainly based on Eshelby equivalent inclusion theory. A nano-micro mechanics-based material modeling approach is implemented to predict short fiber reinforced ABS material system Room Temperature (RT) mechanical properties including air particles as void and weak fiber matrix interface. The proposed multi-scale material modeling methodology combines Mori-Tanaka effective far-field theory [7] and micro-mechanics [10] to predict the desired material properties of 3D printed filled ABS material system.

In the suggested approach, the effective nanocomposite (or enhanced matrix) material properties, where carbon nanotubes of known weight fraction are aligned in a neat matrix, are determined first analytically. The analysis approach uses well-known Mori-Tanaka formulation [7] for calculating the anisotropic nanocomposite properties from isotropic matrix and nano-particles properties (stiffness, aspect ratio and volume fraction). In an independent review, Tucker and Liang (1999) found that Mori-Tanaka type models give the best results for large-aspect-ratio fillers (nanoparticles). For a composite material reinforced with aligned fiber-like particles, the Tandon and Weng (1984) prediction of the moduli E_{11} (aligned particle direction), E_{22} (transverse to the aligned particle direction), the in-plane shear modulus G_{12}, and the out-of-plane shear modulus G_{23} of the composite are:

$$E_{11} = E_m \Big/ \left(1 + f_p\left(A1 + 2\upsilon_m A2\right)/ A\right) \tag{2}$$

$$E_{22} = E_m \Big/ \left(1 + f_p\left(-2\upsilon_m A3 + (1-\upsilon_m)A4 + (1+\upsilon_m)A5 * A\right)/(2A)\right) \tag{3}$$

$$G_{12} = G_m \Big/ \left(1 + f_p\left(\mu_m / (\mu_p - \mu_m)\right) + 2(1-f_p)H_{1212}\right) \tag{4}$$

$$G_{12} = G_m \Big/ \left(1 + f_p\left(\mu_m / (\mu_p - \mu_m)\right) + 2(1-f_p)H_{2323}\right) \tag{5}$$

where A and A_i are constants depending on the components of the Eshelby tensor and the matrix/nano-particles properties, and H_{ijkl} are the Cartesian components of the Eshelby tensor.

2.2.4 Multi-Factor Technique Based Temperature Dependent Material Strength Properties Prediction

Mechanical properties (Tensile and Compressive Strength) are calculated as a function of temperature using multi-factor technique along with progressive failure analysis. The strength properties are calculated using Equation-6, a multi-factor technique as a function of glass transition temperature. In Equation-6 $\boldsymbol{\sigma_o}$ is room temperature strength property, $\mathbf{T_g}$ is the glass transition temperature, $\mathbf{T_o}$ is the temperature of interest and α is the multi-factor coefficient.

$$\sigma = \sigma_o \left[(T_g - T)/(T_g - T_o)\right]^{\alpha} \quad (6)$$

2.2.5 Cure Kinetics Based Temperature Dependent Material Properties Prediction

Mechanical property (Modulus) is calculated as function of temperature. The approach utilizes a micro-mechanics fiber/matrix-based progressive consolidation process. The constituent stiffness is updated based on degree of cure predicted by cure kinetics. The proposed simulation process will account for the transient heat transfer and temperature distribution through parts by time-history incremental simulation of the curing process. Viscosity of the resin is determined as a function of the degree of cure simulated by the Bogetti/Gillespie [12] model.

2.3 3D Printing Process Modeling; Thermal and Coupled Thermal-Structural Analysis

3D printing process modeling is categorized into two steps i.e. **thermal** and **coupled thermal-structural** analysis. 3D numerical model was developed to perform thermal and coupled thermal-structural analysis for XY build direction dog bone specimen with 0°/90° orientation. G-Code file was generated using Slic3r software for XY build direction. The G-Code file was used to specify the actual paths performed during the printing process. Finally, a detailed re-construction of the model was generated for FE analysis.

The coupling between the thermal and mechanical field can be obtained by a sequential solution procedure. The sequentially coupled thermal-structural analysis is used when the structural responses depend on the temperature field, but the temperature field can be found without the effect of stress/deformation field. This means that the heat transfer analysis is first performed to determine the temperature field. The temperature field is calculated considering conduction, boundary convection, and boundary radiation. Heat transfer numerical analysis is implemented to model solid body heat conduction with temperature dependent conductivity and specific heat, general convection, and radiation boundary conditions. The heat transfer analysis is nonlinear due to: **a**) temperature dependent material properties, **b**) nonlinear boundary conditions (film coefficient can be function of surface temperature). Then, the calculated temperature field is used as an input to the coupled analysis. Since the material is temperature dependent, the material properties change as a function of the temperature field during the fabrication simulation. For heat transfer analysis, FDM process modeling was used by assigning constant thermal boundary values to the added elements.

2.4 Test Reduction by Composite Strength Allowables Determination

A- and B-basis strength values determination for materials are vital to achieve the structural design with acceptable safety and confidence level in order to reduce the risk in aerospace industry. The possibility of failure is lowered for significant aircraft structures by considering A and B-basis design values throughout the whole design process. While the 3D printed parts properties show scatter due to constituents, environmental conditions and architectural variations some criteria are required to determine the Printed material tolerance with certain levels of confidence. For A-basis,

at least 99% of the material strength values population are anticipated to be equal or exceed this tolerance magnitude with a confidence level of 95%. B-basis values are applied to structures and components considered as redundant and less critical where failure results in safe load redistribution. For B-basis, 90% of the material strength population values are expected to be equal or exceed B-basis strength value with 95% of confidence level. The method to define strength allowables with test reduction is based on existing accepted standards and practices [8, 9]. It uses statistics from lamina level testing to reverse engineer uncertainties in composite constituents' properties variations and deviations derived from fabrication process (**Figure** 7).
Afterwards, the uncertainties are implemented in CLM-RM to generate random virtual test samples for un-notched specimens used instead of actual test samples. The methodology combines probabilistic methods with progressive failure analysis (PFA) and Bayesian statistics to reduce the number of tests needed to define of A- and B-basis strength values. [14]. There are three discrete proficiencies (options) to generate the A and B basis Allowables:

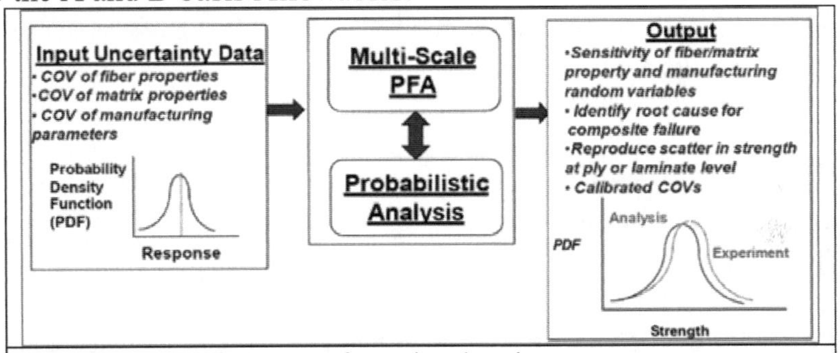

Figure 7. Recreating scatter from virtual testing process

1. **Test data procured by the user** – Method is a combination of deterministic and statistical approach elaborated in CMH-17(Composite materials handbook) and FAA CFR 14 (Aeronautics and space).

2. **Reduced test data** (1 coupon from each batch) –A combination of progressive failure analysis with probabilistic methods. Generic basis allowable is used to improve upon CVM allowable generation.

3. **Lack of test data** – This method is a combination of probabilistic approaches, based on standard practices, and progressive failure analysis. The calculation is based on the scatter recreation in lamina strength caused by variations in fiber/matrix properties and fabrication defects within an appropriate margin error (~±10%). The approach has been validated by comparing its A/B-basis predictions with values obtained from standard methods using test data from robust or reduced sampling. **Figure 8** shows technical sequences to define A- & B- basis allowables. The results of PFA analysis on unnotched virtual samples were the input to CMH-17 methodology for A-B Basis determination. The virtual random sampling generates a full Cumulative Distribution Function of Response (CDF) for the composite laminate strength. The A/B-basis values are then easily extracted from the CDF curve at 0.01 and 0.1 probabilities, respectively. [14].

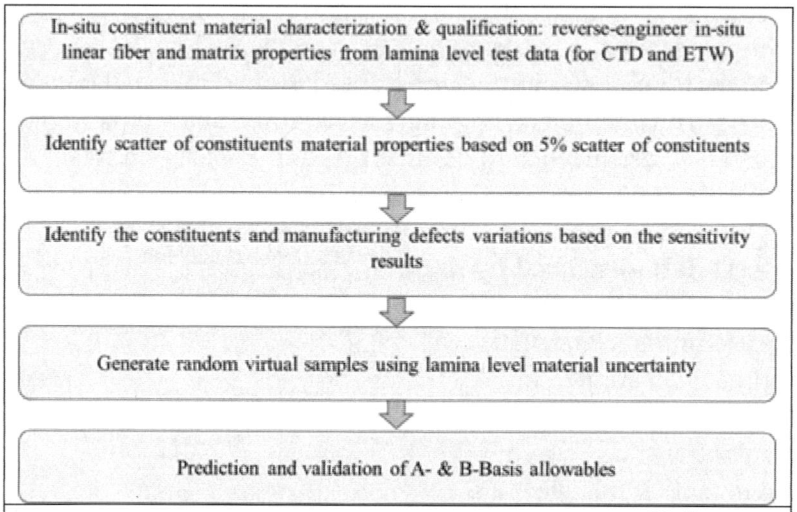

Figure 8. Technical sequences to determine A- & B-basis [14]

2.5 Computing the Generic Acceptance Region

The multivariate 95% confidence ellipsoid represents the generic acceptance region based on the NCAMP dataset. The boundary of the generic acceptance region for strength equivalence is defined to be lower edge of the multivariate 90% certainty ellipsoid for a single unit mean vector established on the NCAMP dataset. These regions are determined through the computation of the population covariance matrix, Σ, the population mean vector. The equation for the generic acceptance region **[14]**:

$$N(\nu - M_1)'\Sigma^{-1}(\nu - M_1) \leq c \quad (7)$$

Where,

- N = 1 because this is the region for a single unit
- v is any p-dimensional vector such that the equation is true (p is the number of different tests in a complete unit).
- M_1 is the p-dimensional mean vector computed from all units in the NCAMP database.
- Σ is the p×p covariance matrix computed from all complete units in the NCAMP database
- c is the appropriate percentile of a distribution

The acceptance and equivalency regions are ellipses when only two properties are in a group. The concept is extended to a multi-dimensional ellipsoid for more than two properties. This multidimensional ellipsoid is projected to a 2-dimensional ellipse for the graphical representation of those properties **[14]**.

3. RESULTS AND DISCUSSION

The entire 3D printing process was carried out in three key steps: **Material Modeling** including the prediction and implementing the micro void content to determine printed ULTEM 9085 properties as well as prediction and validation of elevated temperature properties, **Process Modeling** of Fused Deposition and implementation of print errors known as Macro voids, and **Service Loading** performance.

3.1. Micro Thermal Modeling

Thermal History (**Figure 9a-b**) predicted including heating, melting, melt-superheating, superheat-cooling, solidification/sintering, and cooling periods. It also predicted **Material State** (i.e. transient density, volume of solid (VOS), and void volume ratios), **Figure 9c-d**. The model can be extended to account for multi-part interaction due to thermal gradient, and residual stress and the effect of print build orientation.

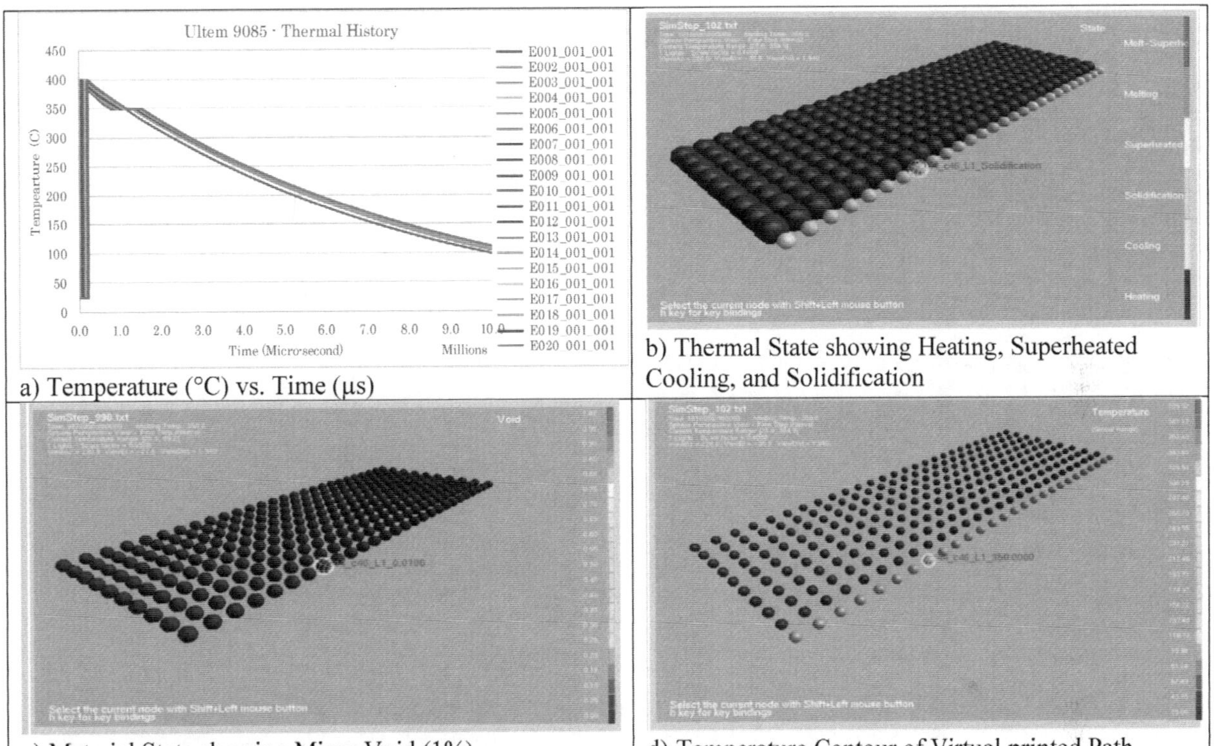

a) Temperature (°C) vs. Time (μs)

b) Thermal State showing Heating, Superheated Cooling, and Solidification

c) Material State showing Micro Void (1%)

d) Temperature Contour of Virtual printed Path

Figure 9. GENOA 3DP TMg: FDM Process, Thermal, Material State of ULTM9085

3.2 Micro-Mechanical Material Property Prediction Modeling at RT and ET):

The material modeling process includes calibration and validation of Room temperature printed ULTEM 9085 including the effect of micro voids predicted by **Micro Thermal Model** (1%). Further, it includes using neat material elevated temperature modulus to validate ULTEM 9085 Cure kinetics and predict the elevated temperature mechanical properties for the FDM extruded material which eventually contributed to generation of temperature dependent material card for the FEM solver.

3.2.1 Material Property Prediction at RT

A nano-micro mechanics-based material modeling approach is used to predict FDM extruded ULTEM 9085 Room Temperature (RT) mechanical properties including spherical air particles implementation as micro voids. The proposed multi-scale material modeling methodology combines Mori-Tanaka effective far-field theory [7] and micro-mechanics [10] to predict the desired material properties of extruded ULTEM 9085 resin. **Input:** The neat resin properties are obtained from San Diego Plastics Inc spec sheet [15-16]. **Assumption:** 1% of spherical air particles with 20 μm diameter is assumed and is uniformly distributed in the resin; Stress-based failure criteria is implemented.

Output: The code outputs aligned 2D random, and 3D random room temperature material properties. The inputs, assumptions and outputs for material modeling (RT) is shown in **Figure 10a-e**. For the simplicity of this material modeling process only 3D random properties are considered. **Figure 10f** shows the Predicted Mechanical Properties (modulus, and strength of ULTEM 9085 at RT)

a) ***Input:*** *Air bubble particles (1%) from GENOA 3DP TMg*

Property	Magnitude
Density (tonne/mm³)	1.205 E-12
Elastic Modulus (MPa)	0.1
Tensile Strength (MPa)	6.89 E-3
Compressive Strength (MPa)	6.89 E-3
Interface Shear Strength (MPa)	1.13 E2
Adhesion Factor	0.05

b) ***Input:*** *Neat Unfilled ULTEM 9085 Properties*

Property	Magnitude
Density (tonne/mm³)	1.34
Elastic Modulus (GPa)	2.7
Poisson's ratio	0.4
Tensile Strength (MPa)	118.80
Compressive Strength (MPa)	151.68
Shear Strength (MPa)	113.42

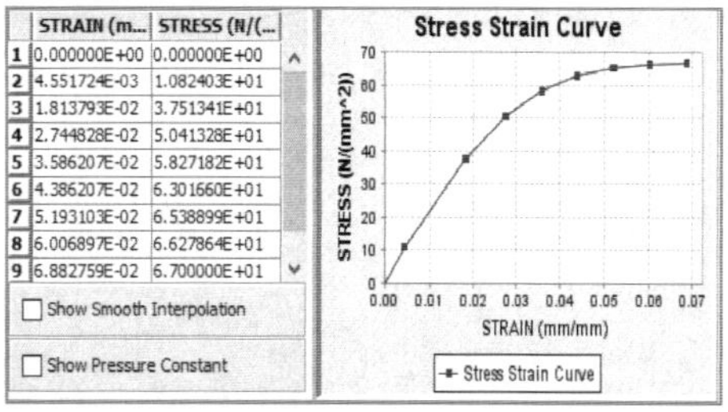

c) ***Input:*** *Nonlinearity of ULTEM 9085 with pressure constant of 2 [15, 16]*

d) ***Assumption:*** *Air bubbles geometry and content (1% micro void content predicted by GENOA 3DP TMg)*

Property	Value
Shape	Spherical
Weight Ratio	9.08 E-6
Volume Ratio	1% (Predicted by GENOA 3DP TMg)
D (μm)	20

e) ***Assumption:*** *Damage and Fracture Criteria used for Printed ULTEM 9085 characterization*

Failure Criteria	Damage	Failure
Longitudinal Tensile	Yes	Yes
Longitudinal Compressive	Yes	Yes
Transverse tensile	Yes	Yes
Transvers Compressive	Yes	Yes
Normal Compressive	Yes	Yes
In-Plane Shear	Yes	Yes
Normal Tensile	Yes	Yes
Transverse Normal Shear	Yes	Yes
Longitudinal Normal Shear	Yes	Yes

f) ***Output:*** *Printed ULTEM 9085 material mechanical properties prediction by multi-scale material/structural Nano-mechanics*

Property	Predicted Material Properties by MCQ (MPa)
S11T: Longitudinal Tensile Strength	118.46
S11C: Longitudinal Compressive Strength	152.77
S22T: Transverse Tensile Strength	118.46
S22C: Transverse Compressive Strength	152.77
S33T: Out-of-Plane Tensile Strength	118.46
S33C: Out-of-Plane Compressive Strength	151.25
S12S: In-Plane-Shear	104.24
S13S, S23S	103.21
E11, E22, E33: Longitudinal, Transverse & Out-of-Plane Modulus	2647
G12, G13, G23	947

Figure 10. Material Modeling: input to Nano-Assisted Micro-Mechanics analysis

3.2.2 Modulus Property Prediction at Elevated Temperature

Next, temperature dependent material properties are calculated using cure kinetics approach and compared with test provided in the literature [11]. The data was for a test on a printed FDM ULTEM 9085 cylinder. **Assumption:** The cure kinetics are dependable from the geometry and void content and the same parameters could be used for different geometries with the same material. Considering this assumption, the cure kinetics first was predicted and validated for an arbitrary geometry (FDM printed cylinder with micro and macro voids) and then applied for the baseline material properties (ULTEM 9085 including 1 % micro void content) to predict the modulus for elevated temperatures. **Assumption**: The cure parameters are calibrated to output temperature dependent modulus to show similar behavior as test. **Output:** Temperature dependent resin Modulus property even beyond glass transition temperature of ULTEM 9085 (~186 ^{o}C).

***Inputs:** FDM Cylinder Printed ULTEM 9085 validated cure kinetics*

Parameter	Value
Uncured Elastic Modulus	1.244 GPa
Complete Point of Resin Cure	0.8
Resin Chemical Shrinkage in Specific Volume of The Resin occurring Until Diffusion	1
A1-Pre exponential factor	3.5016 E7 Sec^{-1}
A2-Pre exponential factor	-2.5 E7 Sec^{-1}
A3-Pre exponential factor	3.6 E7 Sec^{-1}
DE1-Activation Energy	67 mJ/MOL
DE2-Activation Energy	61.5 mJ/MOL
DE3-Activation Energy	92 mJ/MOL
Time Determined with Degree of Cure	0.35
Total Volume Shrinkage in Curing 1-3 for Epoxy (percent)	2 %
Curve Fitting Constant	7.9 E-20 MPa* Sec^{-1}
Activation Energy for Viscosity	90.8 mJ/MOL
Viscosity Constant (independent of temperature)	141

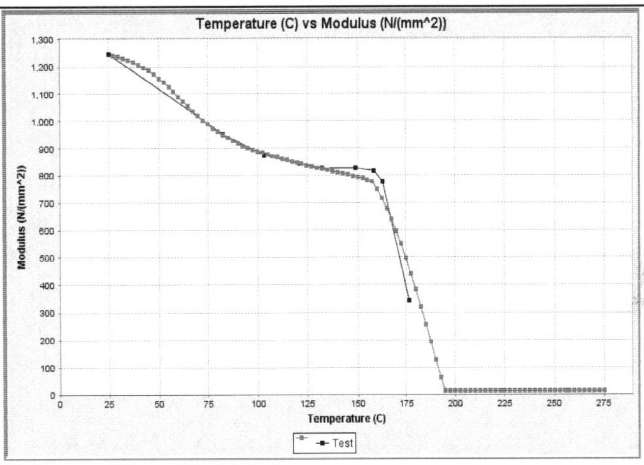

Prediction/Test [11]: Modulus Vs. Temp No Void

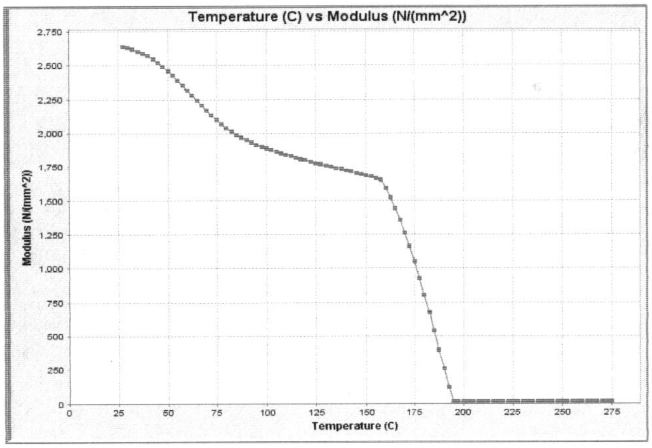

Prediction/Test: Modulus Vs. Temp with 1% Void (Porous)

Figure 11. Predicted Modulus of ULTEM 9085 at ET

3.2.3 Multi-Factor Plot of Strength

The multi-factor coefficients were varied for each temperature case ($\alpha_{49.9C} = 0.4$, $\alpha_{59.91C} = 0.48$, $\alpha_{77.41C} = 0.59$, $\alpha_{99.92C} = 0.52$ and $\alpha_{157.4C} = 0.269$, $\alpha_{180C} = 0.3605$). For each temperature case progressive failure was performed for both longitudinal and transverse tensile and compressive as well as shear condition and final strengths were estimated. The predicted strength properties are shown in **Figure 12.**

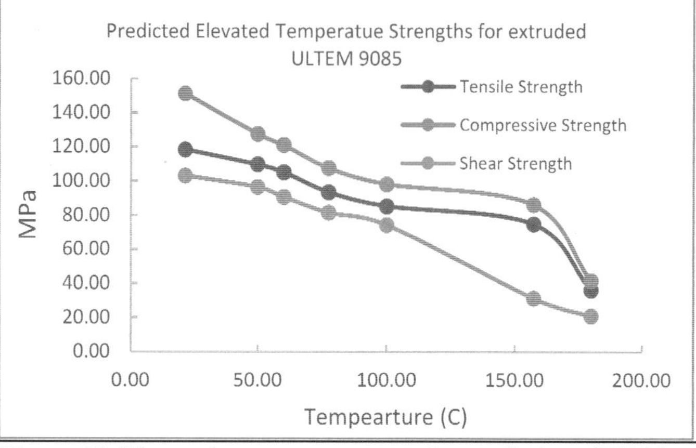

Figure 12. Elevated temperature strengths prediction of FDM extruded ULTEM 90845

The properties cannot be calculated beyond glass transition temperature based on multi-factor technique. The tensile and compressive strength properties are assumed for higher temperatures equal to ones in Tg. A UMAT card was generated based on temperature dependent ULTEM 9085 mechanical properties. The UMAT card was implemented in the FE model to perform coupled thermal and structural thermal analysis.

3.3 GENOA 3DP PATHCOVERAGE The result of accounting for the reduction of mechanical properties due to these macro voids is shown in the same figure. **Figure 13a**, GENOA 3DP PathCoverage predictions are corroborated by the presences of macro-voids in the printed part. **Figure 13b** provides a close-up of the predicted gaps and the size of each gap as a percentage area of that element. The remaining images provide validation for the accuracy of the software.
This tool was also implemented to predict bald spots and degrade both the modulus and strength accordingly in element level. Finally, the process was followed by service loading to qualify the part and to predict the part performance.

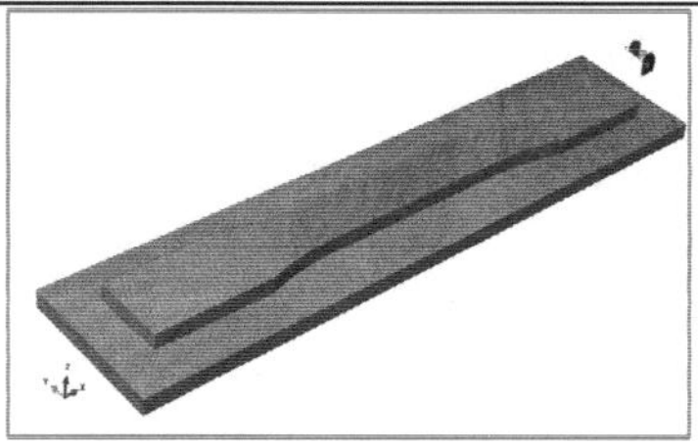

a) Dog-bone specimen and Base Plate FE Mesh

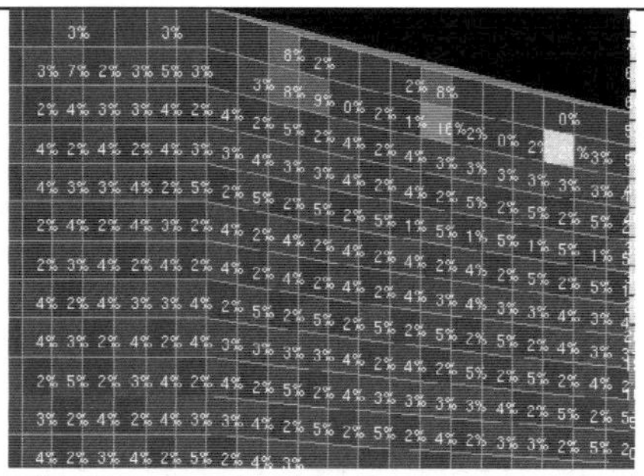

(b) Computed void volume ration due to anomalies and bald spots for all 6 layers

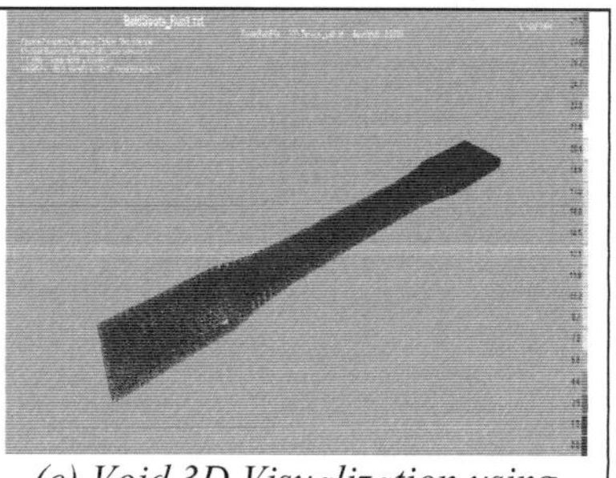

(c) Void 3D Visualization using GENOA 3DP TMg

Figure 13. FE Model generation, Macro Void Calculation and Visualization

3.4 FDM Printing Process Modeling

The process modeling includes generation of thermal analysis model and sequentially coupled structural thermal analysis model. GENOA 3DP PathCoverage tool was also used to implement the predicted bald spots and degrade both the modulus and strength according to the G-code. Both the analyses were performed to calculate the temperature distribution as well as deformation and residual stress during the printing process. Summary of the inputs for process modeling is shown in table.

Input: Summary of Slic3r settings	
Property	**Magnitude**
Slic3r Setting	**Adjustment**
Layer height & First layer height	0.536mm
Fill density	100%
Infill pattern	Rectilinear
Fill angle	-/+ 45°
Nozzle diameter	0.65mm
G-code Flavor	Rep-rap
Infill Print Speed	60 mm/s
Perimeter Print Speed	60 mm/s
Nozzle Diameter	0.65 mm

Assumptions: Print Parameters	
Property	**Magnitude**
Bead Width	0.6 mm
Nozzle Temperature	350 °C
Reference Temperature	186 °C
Chamber Temperature	100 °C
Printed Part Convection Coefficient	30 W/(m^{2}K)
Emissivity Coefficient	0.9
Bottom Plate Temperature	160 °C
Bottom Plate Thickness	3.251 mm
Bottom Plate Convection Coefficient	12.5 W/(m^{2}K)
Bottom Plate Emissivity Coefficient	0.45

Figure 14. Inputs to generate G-code and Finite Element Model for FDM print process modeling

3.4.1 Thermal Analysis

A commercially available tools was used to generate an input file for thermal and structural-thermal analyses composed of meshed model based on the imported G-code and user defined Temperature dependent material card. An FE solver was then used to execute thermal analysis and determine the temperature distribution on the printed part during FDM printing process. The output from thermal analysis was used as an input to the coupled thermal-structural analysis. The temperature contour fields for build direction (XY) at the end of last printed layer (layer #6)_ is shown in **Figure 15a**. Also, the thermal history of the locations on 1st, 4th and last layers are shown in **Figure 15b**.

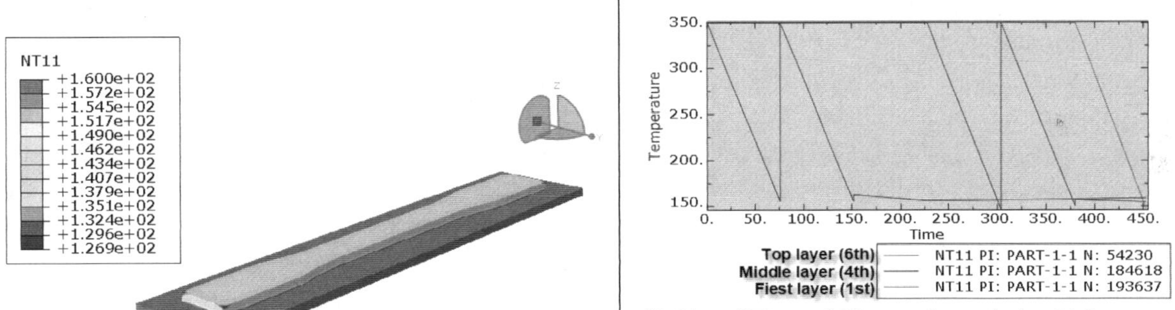

a) Temperature contour at end of print process (XY dir)

b) Plot of Thermal History through the thickness at center of specimen

Figure 15. Thermal Profile (FEM) and History for build direction XY

3.4.2 Coupled Thermal-Structural Analysis – Residual Stress & Deformation

The uncoupled thermal analysis results were used in a sequentially coupled thermal-structural analysis to predict possible residual stress, deformation, and damage/delamination. The deformation results for different build directions are illustrated in **Figure 16a**. The residual stress of the specimen is shown in **Figure 16b**. Damages on the printed part derived from delamination is also predicted and shown in **Figure 16c.** The predicted residual stresses and deformations from coupled structural-thermal analysis are carried over to service load analysis to provide useful information about the performance of the structure under service loading.

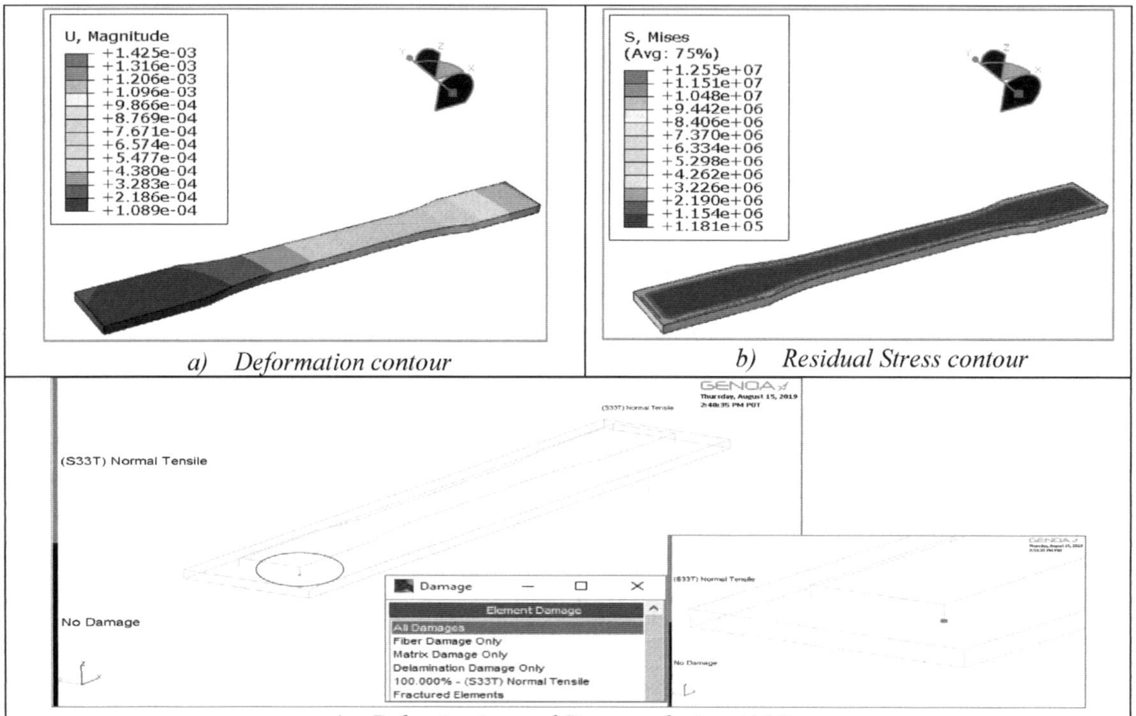

a) Deformation contour *b) Residual Stress contour*

c) Delamination and Damage during AM Process

Figure 16. Process Model: Coupled Thermal-Structural Analysis – Residual Stress, Deformation, and Damage

3.5 Service Load – Damage/Delamination

3.5.1 Dog Bone Specimen

In service loading step, tensile load was implemented to 3D printed part to calculate the load-displacement curve and deformation and was validated with available test results. Additionally, damage initiation and evolution were also calculated. **Figure 17** Shows the gage length, distance between grips (ASTM D638), fixed and load nodes selections representing the edges held by the grips. Tensile test simulation was performed at room temperature for flat (XY) build direction. The FE model description, assumptions are also shown in **Figure 17**. **Figure 18** shows Dog bone deformation and stress contours, load displacement curve, validation of stress-strain curves with the tests at Room Temperature Dry (RTD) from **[8]** and **[9]**. Also, the fracture location and the damage index and plot are shown once the fracture occurs in the simulation (**Figure 18**).

Assumption: Damage and Fracture Criteria used for Printed ULTEM 9085 characterization		
Failure Criteria	**Damage**	**Failure**
Longitudinal Tensile	Yes	Yes
Longitudinal Compressive	Yes	Yes
Transverse tensile	Yes	Yes
Transvers Compressive	Yes	Yes
Normal Compressive	Yes	Yes
In-Plane Shear	Yes	Yes
Normal Tensile	Yes	Yes
Transverse Normal Shear	Yes	Yes
Longitudinal Normal Shear	Yes	Yes

Input: FE model for ASTM D638	
Number of Nodes	233742
Number of Elements	55800
Element type	C3D8R

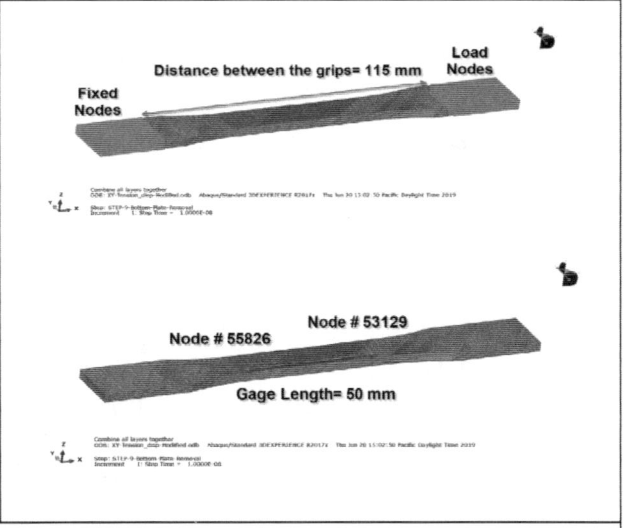

Figure 17. Illustration of FE models specifications, gage length, distance between grips (ASTM 6038). Fixed and load nodes selections and assumptions: Failure criteria

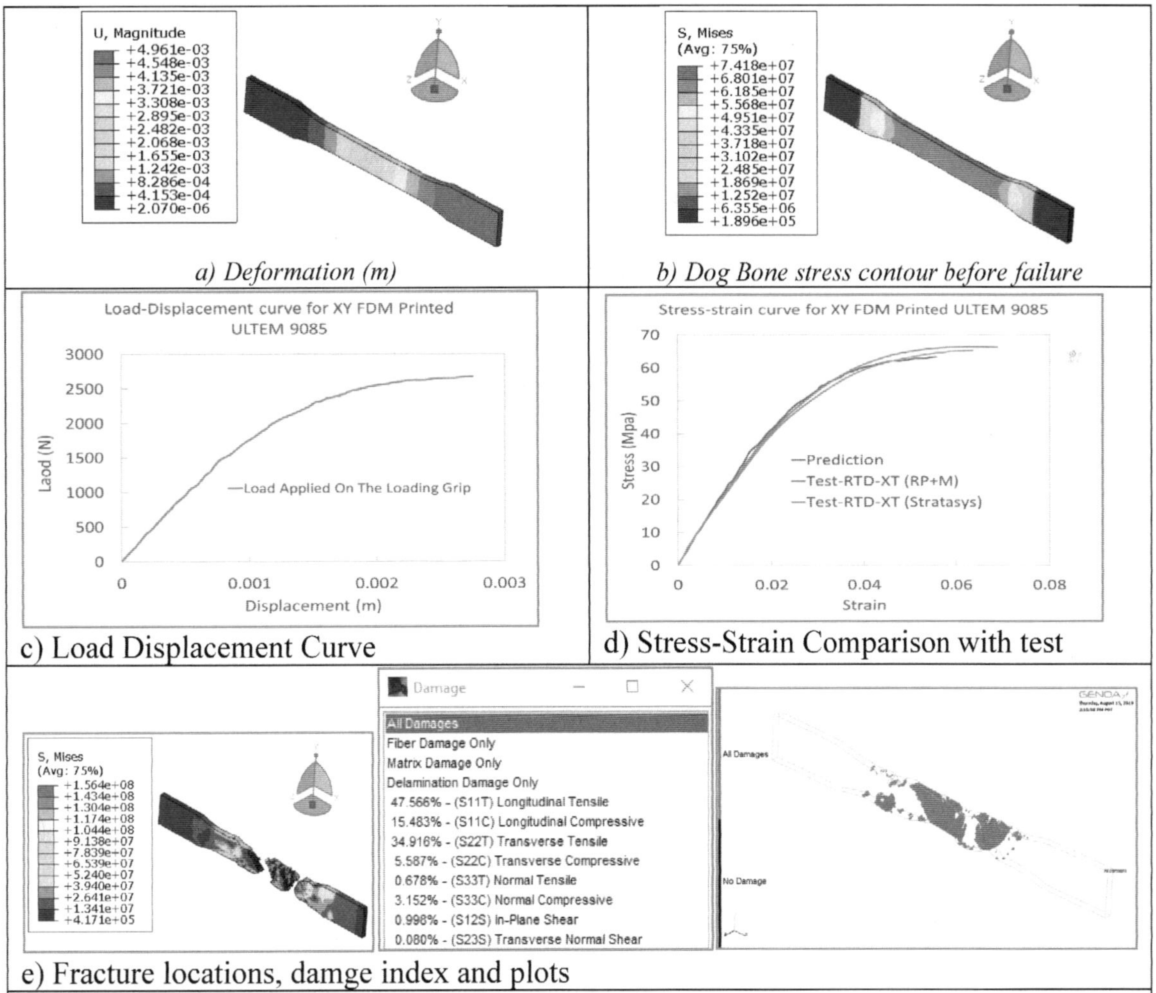

Figure 18. Outputs: Service Loading results

3.5.2 Cross Box Beam Subcomponent

A cross box beam (design courtesy of Made In Space, Inc.) was simulated using the same material system. The inputs and assumptions for the 3D printing simulation as well as the key results (residual stress and part deformation) are shown in **Figure 19.**

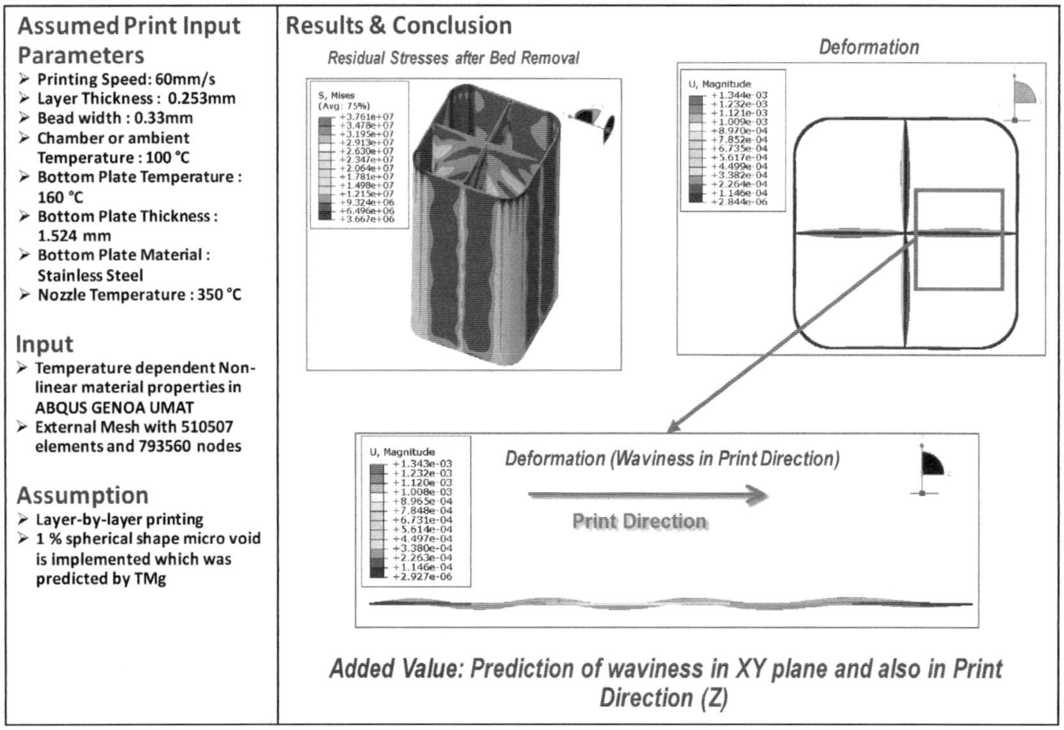

Figure 19. Cross box beam simulation input, assumptions and key outputs

This deformation profile was exhibited in one of the test articles printed by Northrop Grumman Corporation, see **Figure 20**. The deformed wavy pattern shape shown in **Figure 19** shows a very good comparison to actual printed part (**Figure 20**) from quantitative point of view. Further investigation into this behavior and predictive capability is necessary to make any conclusions of accuracy.

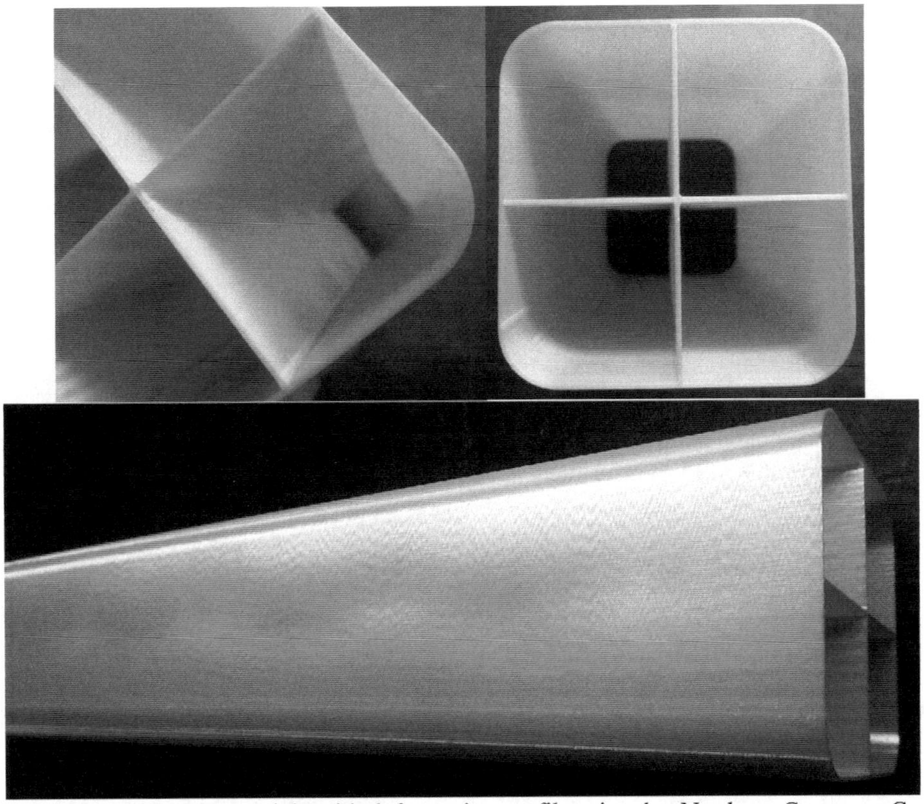

Figure 20. Cross box beam test article with deformation profile printed at Northrop Grumman Corporation

3.6 Material Allowable Generation, and Generic Basis Equivalency

Figure 21 compare the predicted stress-strain of AM printed specimen at XY, XZ, ZX, and ZX45 at RT using Multiscale Progressive Failure Analysis with test curves (Black=Test, Red=Simulation). The inputs are the extruded ULTEM 9085 properties (**Figure 10f**) and the assumptions which is showed in **Table 2** are applied.

Table 2. Assumptions: Calibration of material and void properties to predict the effect of print orientation on the Stress-Strain curve of Printed ULTEM 9085

Print Orientation	Pressure constant	EPS11T (Longitudinal tensile strain limit)	Average void volume ratio
XY	0	0.0623	0.056
XZ	-0.45	0.055	0.02
ZX	0.25	0.04	0.03
ZX45	0.15	0.0303	0.03

Table 3 and Table 4 show the coefficient of variation and distribution type for the mechanical properties of extruded ULTEM 9085 as well as the overall void in the structure of printed specimens by **RP+M** and **Stratasys** Machines respectively.

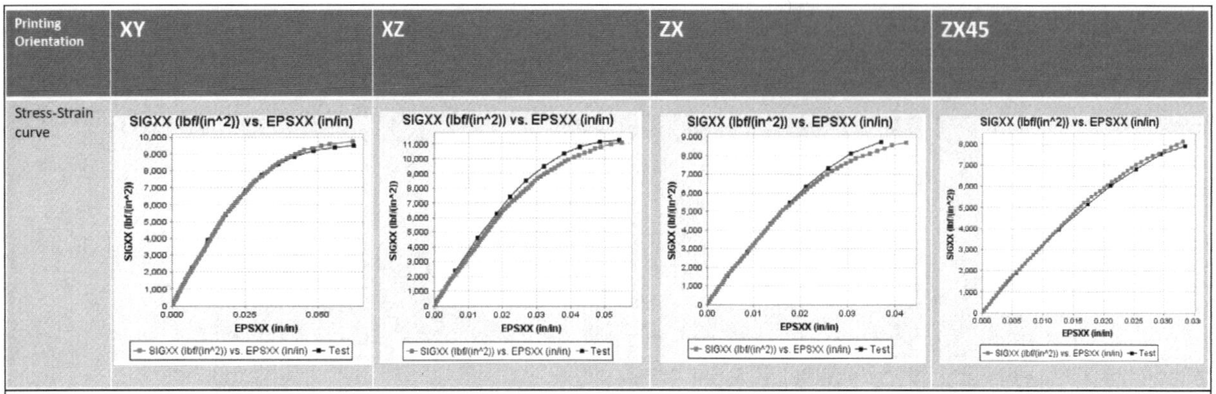

Figure 21. Calibration and validation of Stress-Strain curves of Multiple FDM Print orientation of ULTEM 9085

Table 3. XY Printed Ultem9085 RP+M machine: Coefficient of variables of constituents

Property	Mean Value	Coefficient of Variation (Factor)	Standard Deviation	Distribution type
Tensile Strength (lbf /in^2)	1.718204E+04	0.05	8.591020E+02	Normal
Compressive Strength (lbf /in^2)	2.193826E+04	0.05	1.096913E+03	Normal
Shear Strength (lbf /in^2)	1.496978E+04	0.05	7.484890E+02	Normal
Poisson's ratio	3.975040E-01	0.05	1.987520E-02	Normal
Elastic Modulus ((lbf /in^2)	3.839236E+05	0.05	1.919618E+04	Normal
EPS11T (Longitudinal tensile strain limit)	6.239009E-02	0.28	1.746923E-02	Normal
Void Volume Ratio	0.056000	0.05	0.002800	Log Normal

Table 4. XY Printed Ultem9085 Stratasys machine: Coefficient of variables of constituents

Property	Mean Value	Coefficient of Variation (Factor)	Standard Deviation	Distribution type
Tensile Strength (lbf /in^2)	1.718204E+04	0.05	8.591020E+02	Normal
Compressive Strength (lbf /in^2)	2.193826E+04	0.05	1.096913E+03	Normal
Shear Strength (lbf /in^2)	1.496978E+04	0.05	7.484890E+02	Normal
Poisson's ratio	3.975040E-01	0.05	1.987520E-02	Normal
Elastic Modulus ((lbf /in^2)	3.839236E+05	0.05	1.919618E+04	Normal
EPS11T (Longitudinal tensile strain limit)	6.239009E-02	0.35	1.746923E-02	Normal
Void Volume Ratio	0.056000	0.15	0.008400	Log Normal

Table 5 shows the Cumulative Distribution Function (CDF) for XY printing direction using *RP+M* and *Stratasys* FDM machines. The sensitivity of mechanical properties of Extruded ULTEM 9085 and the overall void content role to determine the ultimate strength of the specimens for each machine is also showed in Table 5.

Table 5. ***Results of CDF and Sensitivities in all printing orientations***

Machine	Cumulative Distributution Function (CFD)	Sensistivity Results
RP+M	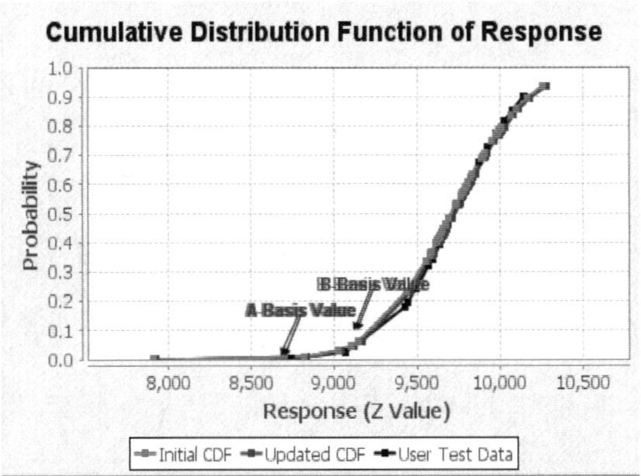	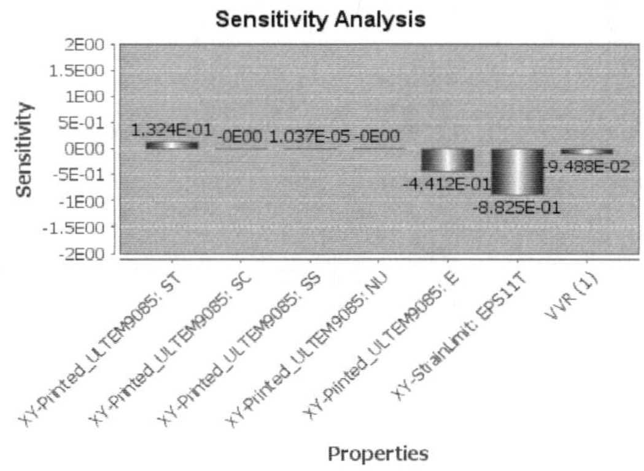
Stratasys	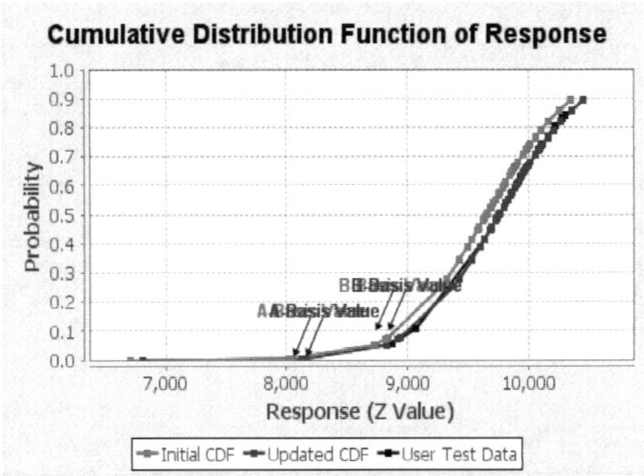	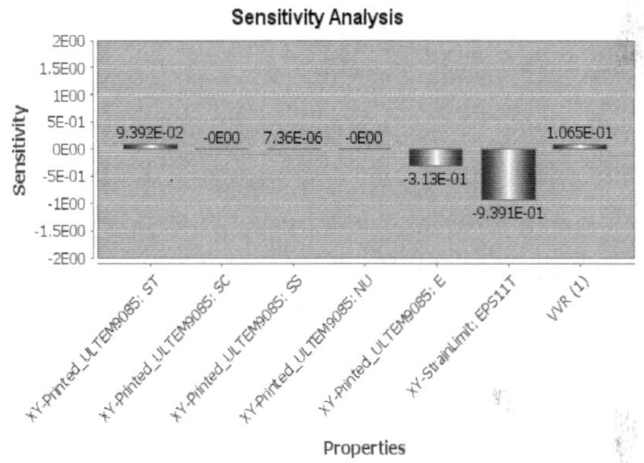

Table 6 shows the comparison of virtual samples generation method and using test samples to determine the Printed ULTEM 9085 allowables in different printing orientation. **Table 6** shows that there is a good agreement between the allowables predicted and those measured in NIAR test report. The CFD of the generated samples was also updated using Bayesian method to comply with the test samples trend, however, did not change the magnitudes drastically and the initial predictions were in good range. Any outlier sample must be investigated regarding cause and effect on application before deciding what to do or rejected as not being equivalent.

The generated virtual sample parts strengths printed by *Stratasys* and *RP+M* were used to plot the generic acceptance region. This region indicated inside the ellipses defined by blue dots is the generic acceptance region (**Figure 22**).

Table 6. test prediction and validation of A- and B-basis allowables in tension for two different Machines (RP+M, and Stratasys) * Different methods to define allowables in test and simulation

Printing Orientation case	RP+M Fabrication Facility (Psi)										Stratasys Fabrication Facility (Psi)			
	A-Allowable					B-Allowable					A-Allowable		B-Allowable	
	Prediction by MCQ (Normal)		Estimation in NIAR report	test results Analytical method	Error %	Prediction by MCQ (Normal)		Measured in NIAR report	test results Analytical method	Error %	Prediction by MCQ (Normal)		Prediction by MCQ (Normal)	
	Initial	Updated by Bayesian method	Basis values and estimates			Initial	Updated by Bayesian method	Basis values and estimates			Initial	Updated by Bayesian method	Initial	Updated by Bayesian method
XY	8917.181	8928.883	8205	ANOVA	8.82*	9257.459	9269.162	8827	ANOVA	5*	8059.2673	8164.115	8737.368	8842.217
XZ	9952.455	10011.574	9145	ANOVA	9.47*	10452.119	10511.238	9977	ANOVA	5.35*	9950.087	10215.312	10450.927	10716.152
ZX	8070.224	8052.497	7972	Normal	1	8283.923	8266.196	8214	Normal	0.63	6849.338	6883.071	7533.272	7567.006
ZX45	6987.190	6769.837	6009	ANOVA	12.66*	7492.988	7275.635	6802	ANOVA	6.96*	6850.399	7187.431	7388.452	7725.484

4. CONCLUSIONS

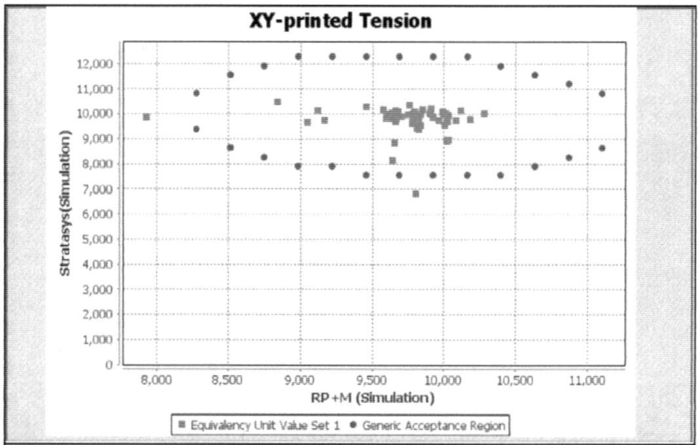

Figure 22. Generic basis Allowables and acceptance region for FDM printed ULTEM 9085 resulted from MCQ (with virtual sampling) for RP+M and Stratasys manufacturing facilities

An entire 3D printing process of ULTEM 9085 material was presented in this paper. The elevated temperature modulus of Printed ULTEM 9085 was predicted and validated and temperature dependent material properties were calculated using predicted coupled cure kinetics and multi factor technique. The calculated properties were implemented in 3D printing process modeling (coupled structural-thermal analysis) to calculate residual stresses and deformation. This method would let the manufacturers to predict the possible deformation in the printed parts and adjust the machine parameters to compensate or minimize it. Further, tensile service loading was applied on the simulated printed dog-bone in XY orientation and the results of stress curve was validated. The whole method could be used to simulate printing of parts with different geometries and apply service loading to evaluate the design virtually which contributes to cost reduction. The strength allowables of printed ULTEM 9085 using RP+M and Stratasys were predicted and the results were validated with tests. Also, the generic acceptance region lets the designers know the probable strength region of two (or multiple) printed machines or manufacturing facilities so they can design the parts and components accordingly considering the effect of machines on the quality of printed parts.

5. REFERENCES

1. Zaldivar, R. J., D. B. Witkin, T. McLouth, D. N. Patel, K. Schmitt, and J. P. Nokes. "Influence of processing and orientation print effects on the mechanical and thermal behavior of 3D-Printed ULTEM® 9085 Material." Additive Manufacturing 13 (2017): 71-80.
2. Lužanin, Ognjan, Dejan Movrin, and Miroslav Plančak. "Effect of layer thickness, deposition angle, and infill on maximum flexural force in FDM-built specimens." Journal for Technology of Plasticity 39, no. 1 (2014): 49-58.
3. Taylor, G., Wang, X., Mason, L., Leu, M.C., and Chandrashekhara, K., "Investigation of Ultem 1010 FDM sparse-build parts using design of experiments and numerical simulation," in Proceedings of The Composites and Advanced Materials Expo, Anaheim, CA, September 26-29, 2016.
4. Ji, L.B., and Zhou, T.R., "Finite element simulation of temperature field in fused deposition modeling." Advanced Materials Research, Vol. 97, pp. 2585-2588, 2010.
5. Górski, F., Kuczko, W., Wichniarek, R., and Hamrol, A., "Computation of mechanical properties of parts manufactured by Fused Deposition Modeling using finite element method." In 10th International Conference on Soft Computing Models in Industrial and Environmental Applications, Springer International Publishing, pp. 403-413, 2015.
6. Taylor, G., Wang, X., Mason, L., Leu, M.C., and Chandrashekhara, K., "Optimization of ULTEM 1010 fused deposition modeling specimens for flexural behavior," in Proceedings of The Composites and Advanced Materials Expo, Orlando, FL, December 11-14, 2017.
7. Mori T, Tanaka K. Average Stress in the Matrix and Average Elastic Energy of Materials with Misfitting Inclusions. Acta Metallurgica 1973; 21, p. 571-574.
8. https://www.wichita.edu/research/NIAR/Documents/NCP-RP-2018-007-Rev-NC-ULTEM-9085-Stat-Analysis-Report-April-16-2019.pdf
9. https://www.wichita.edu/research/NIAR/Documents/CAM-RP-2018-027-NC-Stratasys-Certified-ULTEM-Equivalency-FINAL-MPDR-4.17.2019.pdf
10. Chamis, Christos C., Frank Abdi, Mohit Garg, Levon Minnetyan, Harsh Baid, Dade Huang, Jerry Housner, and Farid Talagani. "Micromechanics-based progressive failure analysis prediction for WWFE-III composite coupon test cases." Journal of Composite Materials 47, no. 20-21 (2013): 2695-2712.
11. Li, H., G. Taylor, V. Bheemreddy, O. Iyibilgin, M. Leu, and K. Chandrashekhara. "Modeling and characterization of fused deposition modeling tooling for vacuum assisted resin transfer molding process." Additive Manufacturing 7 (2015): 64-72.
12. T. A. Bogetti and J. W. Gillespie, "Process-induced stress and deformation in thick-section thermoset composite laminates," Journal of composite materials, vol. 26, pp. 626-660, 1992.
13. Blok, L. G., M. L. Longana, H. Yu, and B. K. S. Woods. "An investigation into 3D printing of fibre reinforced thermoplastic composites." *Additive Manufacturing* 22 (2018): 176-186.
14. Hajiha, R, Godines, C, De Fenza, A, Baid, H, Clarkson, E, Moazami, N, Abdi, F. "Allowable Generation and Generic Basis Value for Multiple Loading and Environmental Conditions". CAMX 2019, Anaheim, CA.
15. http://www.sdplastics.com/ensinger/ultem.pdf
16. El-Gizawy, A. Sherif, Shan Corl, and Brian Graybill. "Process-induced properties of fdm products." In Proceedings of the ICMET, International Conference on Mechanical Engineerings and Technology Congress & Exposition. 2011.

MACROSCOPIC MODELING OF THE LINEAR-VISCOELASTIC VIBRATION BEHAVIOR OF SHORT-FIBER REINFORCED PLASTICS

Fabian Urban[a]*, Peter Middendorf[b]
[a]NVH Powertrain, Mercedes-Benz AG, Stuttgart, Germany
[b]IFB – Institute of Aircraft Design, University of Stuttgart, Stuttgart, Germany

ABSTRACT

At present, the simulation cannot satisfactorily reproduce the actual vibration behavior of short-fiber reinforced plastics. This is because the required viscoelastic material data, in particular damping, is often not available. For this purpose, a new experimental method was developed which can characterize the exact stiffness and damping of the material for a wide frequency range, taking into account environmental conditions such as temperature and humidity as well as fiber orientation. Additionally an orthotropic material model was developed, which can describe the linear-viscoelastic vibration behavior on a macroscopic level with the measured material data and a 2nd order orientation tensor from the injection molding simulation. A validation was performed on cuboid test specimens and engine mounting brackets. Results showed that the approach achieves an accuracy in calculating the resonance frequencies of <4 %. It works with thermoplastics and thermosets without an iterative reverse engineering approach but with a significant effort needed to characterize the materials.

*Corresponding author: fabian.urban@daimler.com

1. INTRODUCTION

More and more components made of short-fiber reinforced plastics (SFRP) are used in modern powertrains. They are not only of interest for the NVH (noise, vibration, harshness) improvement of conventional combustion engines, but also for high-frequency electric drives in the future. The main reason for this is the good acoustic properties due to the lower stiffness and higher damping compared to classic metallic materials. Engine mounting brackets made of SFRP are perfectly suited for this application and established in series production for several years. They transmit less structure-borne noise from the engine block to the carbody and are therefore a crucial element in the reduction of vehicle noise. In order to meet the increased customer demands regarding the acoustic value of a vehicle, a precise prediction of the vibration behavior of such components is necessary at an early stage of the development process. [1–5]

However, the simulation cannot satisfactorily predict the actual vibration behavior of fiber-reinforced plastics at present. One reason for this is that the required material data, in particular the damping, often originate from purely static tests or are completely unknown. However, dynamic, orientation- and frequency-dependent material properties are required in order to be able to reproduce the special viscoelastic properties of the SFRP. [4,6–8] For this purpose, however, there is a lack of suitable test methods to determine the exact stiffness and damping of the material for a large frequency range under consideration of environmental conditions such as temperature, humidity and fiber orientation as well as a proper material model, which can process the material data in the simulation. [9]

The present study focuses on the description of a macroscopic material model, which can process material data characterized with a new experimental method. Therefore, short-fiber reinforced thermoplastic and thermoset materials are investigated as test specimens and as an engine mounting bracket. At first, the state of the art regarding characterization and currently used methods for the description of fiber orientation and viscoelasticity is presented. The proposed approach is described beginning with the injection molding simulation followed by the orthotropic material model processing fiber orientation as well as material data. Thereafter, the material characterization is explained concluded by a method to create the material card by means of an averaged fiber orientation. Finally, the boundary conditions of the experiments and simulations carried out are described and their results are compared for validation.

2. STATE OF THE ART

2.1 Experimental Characterization

Various methods exist for the characterization of dynamic material properties. The standard ISO 6721-1 [10] generally describes methods which allow the determination of the viscoelastic parameters storage and loss modulus of plastics at different temperatures or frequencies.

The most common approach is the dynamic mechanical analysis (DMA) described in standard ISO 6721-5 [11]. In this method, a sample is excited with an oscillating force and its response behavior is measured. Force and displacement response yield stress and strain, which, together with their phase shift, allow the calculation of the storage and loss modulus. In addition to the temperature, the excitation frequency can be varied to characterize a frequency-dependent behavior. [12] The method is limited to a frequency range up to approx. 200 Hz. By using calculation approaches such as those of William-Landel-Ferry or Arrhenius, the time-temperature shift allows to conclude from a temperature-dependent to a time- and therefore frequency-dependent material behavior, thus greatly extending the usable frequency range of the material properties. [13–15] This assumption only partially applies to thermorheologically simple materials. Since plastics show phase transitions like e.g. the glass transition, they generally behave thermorheologically complex. This means that there is no clear relation between time and temperature at all and therefore no steady curve for the complex modulus by the mathematical shift approaches in the frequency domain. [1,2]

An alternative approach is the flexural resonance method. The standard ISO 6721-3 [16] describes how material properties can be derived from resonance curves of specimens. The frequency position of the resonances and the geometry of the test specimen are used to determine the storage modulus and the width of the resonance peak is used to determine the loss modulus using the 3dB-method. However, it is shown that the mounting method and contactless excitation proposed in the standard only delivers good results up to frequencies of approx. 1000 Hz. [17–19] More recent approaches to mounting, excitation and measurement for a much larger frequency range, by using e.g. a shaker and a laser vibrometer, can be found in [20–23].

2.2 Modelling of the fiber-orientation dependent material properties

There are many methods in the literature to describe the influence of short-fibers on the mechanical properties of plastics. Usually, these mathematical and physical models consist of a multi-stage homogenization method. [1]

In recent years, models based on the mean-field theory have been established. According to Böhm [24], these are approaches that describe the material in so-called Representative Volume Elements (RVE). It establishes a relationship between the outer macroscopic and the inner microscopic stresses and strains of the RVE. Using a mixing rule, the stiffness of the composite material is calculated from the material properties of the fiber and matrix material, the fiber volume fraction and a concentration tensor.

To determine the concentration tensor there are several procedures in the literature. Often the Mori-Tanaka theory [25] is applied. The idea is that the fibers are sufficiently far apart and thus decoupled from each other, so that an embedded fiber only sees the properties of the surrounding matrix. Using the fundamental Eshelby solution [26] for the mechanical relationship between fiber and matrix, the composite stiffness can be determined. One limitation, however, is the necessarily low fiber volume fraction due to the decoupling of the fibers to each other, Christensen [27] showed the applicability for a fraction of up to 30 %, Kaiser [28] also for values greater than 30 %. Possibilities for direct calculation of the engineering constants are the method according to Halpin-Tsai [29], which does not require an Eshelby solution, and the more precise method according to Tandon-Weng [30], which is based on Mori-Tanaka. Besides these 2-phase models, which only consider fiber and matrix, there are more precise approaches which allow a 3-phase consideration of fiber, matrix and interphase in between. The Reconsidered-Double-Inclusion-Model (RDI-Model) by Schöneich [31] is worth mentioning here.

However, all approaches only provide information on microstructures which fibers are all unidirectional oriented in the RVE under consideration. However, this is not the case for injection molded components. Therefore, the different fiber orientations from the injection molding simulation are divided into unidirectional, equal sections by means of 2-step homogenization, and are then weighted and superposed to a direction-dependent total stiffness. A prerequisite, however, is sufficient knowledge of the existing injection molded fiber orientation distribution in the examined object. Nowadays, this can be extracted from the injection molding simulation using process-structure coupling software that allows a transfer to the finite element (FE) mesh of the structure simulation. [1]

2.3 Modelling of the frequency-dependent viscoelastic material properties

In the literature, the description of the frequency-dependent stiffness and damping of plastics is done by mathematical formulations or approximate solutions, which however distinguish to the realistic physical behavior of the composite material. This is due to the fact that a direct characterization of the material properties is not yet possible for a large frequency range without problems. Therefore, the viscoelastic material behavior is approximated by means of substitute models. A good overview is provided by Treviso in [9]. In the following, the possibilities that are available in the common FE-programs are discussed.

If the calculation of the vibration simulation takes place in modal space, linearity must be assumed. This linearization leads to the fact that viscoelastic material models are not useable at all. Therefore, substitute models must be applied. This can for example be modal damping. It is defined as a relative part of the critical damping, is frequency-dependent but difficult to determine and can only be used isotropically. [1] Alternatively Rayleigh damping can be used. It is a linear combination of mass and stiffness matrix, i.e. a purely mathematical approach without physical considerations. It can only be defined globally for the entire model and only at exactly two frequencies. [32] However, these approaches are very simply suitable for adapting the unknown material parameters to an experiment by means of reverse engineering until the results show good agreement. This was shown for example in [2,20].

If, on the other hand, the calculation can be done by directly solving the linear system of equations, there are other possibilities. Using Prony series, the frequency-dependent material behavior can be approximated by a sum function with any number of Prony elements. For a high accuracy, however, a relatively large number is required, which in return requires a large experimental and simulative effort. [33,34] Alternatively, the viscoelastic material behavior can be transferred directly to the FE-program as storage and loss modulus in a Jacobian matrix. [32]

3. PROPOSED APPROACH

3.1 Overview

The existing methods for the vibration simulation of SFRP have been successfully using microscopic approaches. A major problem is the lack of a characterization method to measure viscoelastic material properties for a wide frequency range. The data is still mathematically extrapolated from low frequencies or determined by means of a reverse engineering approach and not directly measured.

The proposed approach therefore follows a different path. By means of a new more advanced experimental method, SFRP can be characterized in a wide frequency range as a function of temperature, humidity and fiber orientation. The viscoelastic material data can be used directly in the FE-simulation by using a macroscopic material model. Figure 1 gives an overview of the approach.

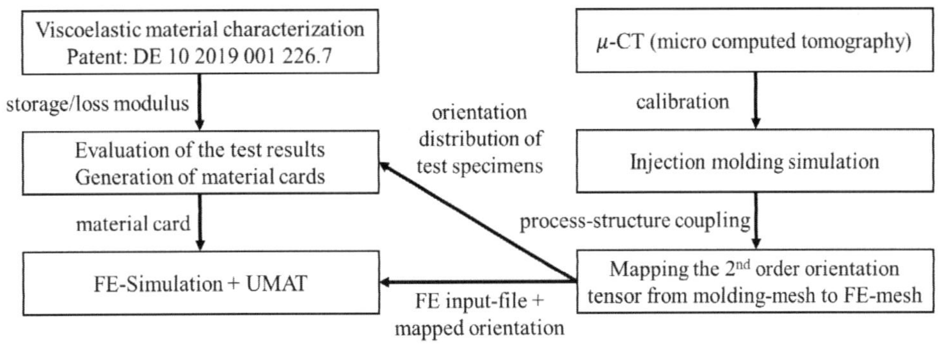

Figure 1. Overview of the proposed approach

The injection molding simulation is calibrated with the micro computer tomography (μ-CT) and the fiber orientation is then mapped onto the FE-mesh of the structure simulation. The flexural vibration of the test specimens is measured and analyzed. Storage and loss modulus are determined for every resonance frequency. Together with the averaged fiber orientation for every measured specimen the frequency and orientation dependent material card is calculated. It is then read and processed by the FE user-material (UMAT). For every calculated frequency step the implemented material model fills an own complex orthotropic stiffness matrix for every element of the structure simulation FE-mesh.

3.2 Fiber orientation

During the injection molding process a characteristic cross-sectional profile of the injection molded component is observed. It can be approximated using a 3-layer model. In the outer layers, the shear effects at the borders of the injection mold cause the fibers to align in the direction of flow. In the middle of the flow path, an alignment of the fibers transverse to the flow direction can be observed. The thickness of the middle layer depends essentially on the velocity profile of the flow and thus on the viscous properties of the plastic used and its process parameters. [1]

The common tools of injection molding simulation use variations of the Folgar-Tucker model and its extension the Reduced-Strain-Closure model. Advani and Tucker [35] developed the possibility to describe the orientation distribution in the form of a local orientation tensor of 2nd or 4th order. For the elastic consideration, the 2nd order orientation tensor is sufficient. [1,36] The calculation of especially thermosets is still problematic. The curing reaction must be considered and characteristic values are hardly available. Therefore, the fiber orientation distribution of the injection molding simulation should be validated with a μ-CT analysis and calibrated if necessary.

Since the FE-mesh of process and structure simulation are completely different, the calculated fiber orientation distributions, i.e. the orientation tensor, must be transferred from the finer injection molding mesh to the coarser structure simulation mesh using a process-structure coupling software. For this purpose, the "Nearest-Neighbor method" is used to enable a corresponding assignment. The orientation tensor then provides information about the orientation in the local element in the form of a local eigenvector pointing in the main direction with the most fibers. The eigenvalues describe the percentage of fibers in the main and the other directions of the Cartesian coordinate system, which differs from element to element. [1]

3.3 Material model

The aim is to use the viscoelastic material properties from the experiment directly and macroscopically. To assign an orthotropic stiffness matrix to each FE-element, the material data is taken directly from a large, universally valid material card using an angle-dependent description. The 2nd order orientation tensor from the injection molding simulation provides eigenvectors and eigenvalues that locally describe the main direction of the fibers and their percentage in the respective directions. Therefore, a material model linking eigenvalues and angle-dependence is required.

First, the different coordinate systems shown in Figure 2 must be explained. In the global FE-coordinate system (x,y,z) each element receives a local orientation coordinate system consisting of three eigenvectors (1,2,3) defined in the x,y,z-system. The 1-vector points in the direction in which the largest percentage of fibers in the element is oriented. The eigenvalues (EV_1,EV_2,EV_3) indicate the percentage of fiber in the respective directions of the 1,2,3-system and have the sum of 1. Now a homogenization takes place by forming the resulting vector $\vec{p}_{hom}$ in the 1,2,3-system with the three eigenvalues in order to consider fiber influences of all directions in the local element. The relative deviation of the homogenized vector to the local 1,2,3-system thus contains the information about the stiffness of the element.

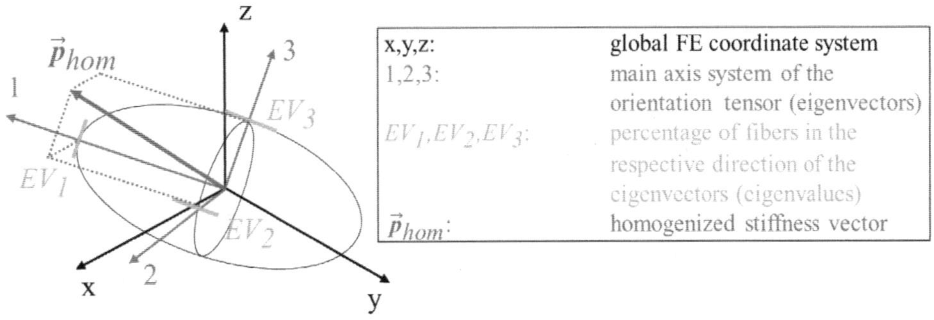

Figure 2. Coordinate systems of a local macroscopic FE-element

Figure 3 shows how to conclude an angle-dependent local element description from the deviation of $\vec{p}_{hom}$ to the 1,2,3-system. Assuming that the eigenvalue 1 is equal to 1, the other eigenvalues have to be equal to 0. In this case, the homogenized vector points in the direction of the 1-axis and all fibers in the element are oriented in this direction. Thus, the element would have an ideal relative angle of 0°-stiffness with all fibers in this direction, which corresponds to the ideal high stiffness of the 1-axis. If, on the other hand, the eigenvalues 2 or 3 are equal to 1, the homogenized vector would point in the corresponding direction. The stiffness of the local element relative to the 1-direction would thus have an ideal relative angle 90°-stiffness with all fibers in those directions, which therefore corresponds to the ideal low stiffness of the 2- and 3-axis. The real homogenized vector moves in between. By means of projection in the 12- and 13-plane, a relative deviation angle ϕ and θ from the 1-axis is obtained, which varies between 0° and 90°.

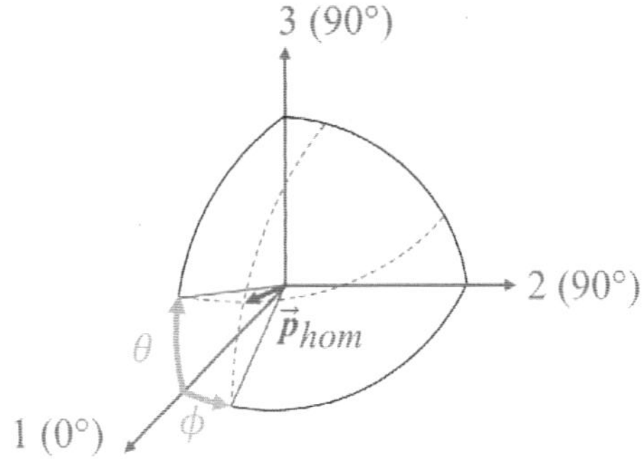

Figure 3. Local orientation projection into the 12- and 13-plane

Therefore, the local stiffness in the 1,2,3-coordinate system in the 1-direction results from the relative deviation of ϕ and θ from the 1-axis with the storage modulus $E'(\phi)$ and $E'(\theta)$ taken from the angle-dependent material card to

$$E_1' = \frac{E'(\phi) \cdot EV_2 + E'(\theta) \cdot EV_3}{EV_2 + EV_3}. \quad (1)$$

The eigenvalues are used to allow a relative weighting of the fiber quantity for the respective direction. To calculate the stiffness in the 2- and 3-directions, the quadrants have to be switched and the homogenized vector has to be completed to a Cartesian coordinate system. Its projected angular deviation from the 2-axis and to the 3-axis results in the respective stiffness

$$E_2' = \frac{E'(90° - \phi) \cdot EV_1 + E'(90°) \cdot EV_3}{EV_1 + EV_3}, \qquad (2)$$

$$E_3' = \frac{E'(90°) \cdot EV_2 + E'(90° - \theta) \cdot EV_1}{EV_2 + EV_1}. \qquad (3)$$

Equations (1-3) can be used in the same way for the loss modulus. Thus, the complex orthotropic stiffness matrix in Figure 4 can now be filled locally for the considered element. Finally, it must be transformed back into the global x,y,z-coordinate system with the coordinates of the eigenvectors in order to perform the FE-calculation.

$$\begin{bmatrix}\sigma_{11}\\ \sigma_{22}\\ \sigma_{33}\\ \sigma_{12}\\ \sigma_{13}\\ \sigma_{23}\end{bmatrix} = \begin{pmatrix} \frac{1-\nu_{23}\nu_{32}}{D}E_1^* & \frac{\nu_{13}\nu_{32}+\nu_{12}}{D}E_2^* & \frac{\nu_{12}\nu_{23}+\nu_{13}}{D}E_3^* & 0 & 0 & 0 \\ & \frac{1-\nu_{13}\nu_{31}}{D}E_2^* & \frac{\nu_{21}\nu_{13}+\nu_{23}}{D}E_3^* & 0 & 0 & 0 \\ & & \frac{1-\nu_{12}\nu_{21}}{D}E_3^* & 0 & 0 & 0 \\ & & & G_{12}^* & 0 & 0 \\ & sym & & 0 & G_{13}^* & 0 \\ & & & 0 & 0 & G_{23}^* \end{pmatrix} \begin{bmatrix}\epsilon_{11}\\ \epsilon_{22}\\ \epsilon_{33}\\ 2\epsilon_{12}\\ 2\epsilon_{13}\\ 2\epsilon_{23}\end{bmatrix}$$

$$D = 1 - \nu_{12}\nu_{21} - \nu_{13}\nu_{31} - \nu_{23}\nu_{32} - 2\nu_{21}\nu_{32}\nu_{13}$$

$$\nu_{21} = \frac{E_2^*}{E_1^*}\nu_{12}, \qquad \nu_{31} = \frac{E_3^*}{E_1^*}\nu_{13}, \qquad \nu_{32} = \frac{E_3^*}{E_2^*}\nu_{23}$$

$$\text{Approximation: } G_{jk}^* = \frac{E_j^* E_k^*}{E_j^* + 2\nu_{jk}E_j^* + E_k^*}; \text{j, k} := \{1,2,3\}$$

Figure 4. Complex orthotropic stiffness matrix [32,37]

3.4 Experimental Characterization

At first, cuboid test specimens are cut out of a flat injection molded plate in different orientations relative to the flow direction, see Figure 5.

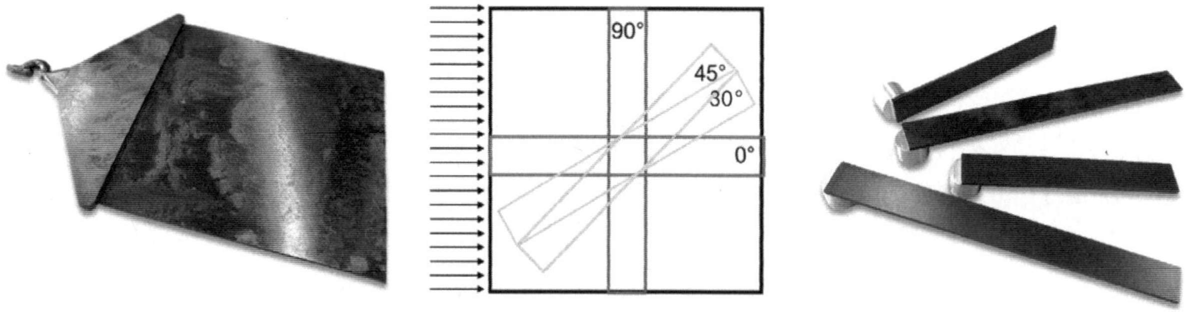

Figure 5. Preparation of test specimens (left to right: injection molded plate, cutting orientations, test specimens with specimen-adapter)

The experimental characterization method of the viscoelastic material properties originates from the self-developed patent [38] which is based on the flexural resonance method in Chapter 2.1. The functional principle and the real mounting structure is shown in Figure 6.

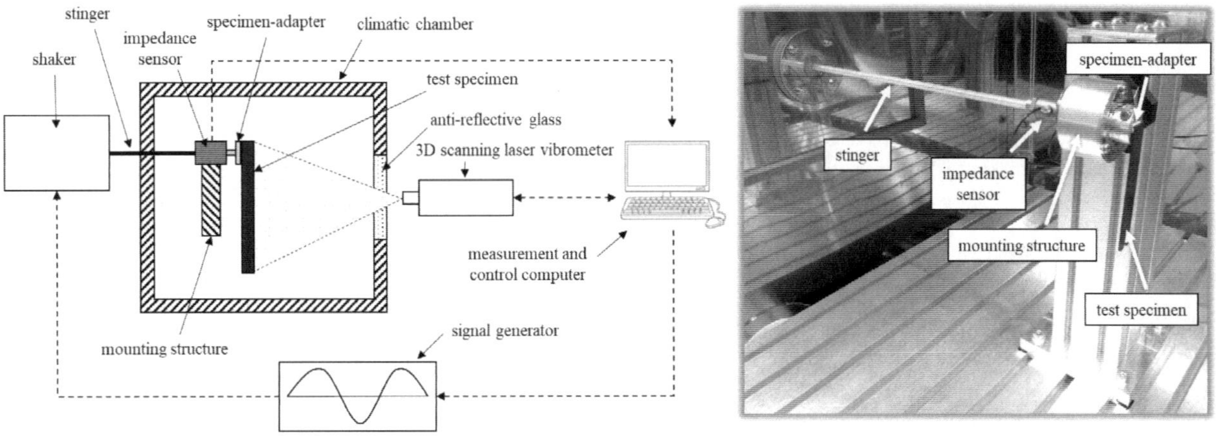

Figure 6. left: principal test setup, right: mounted specimen in climatic chamber

The properties of plastics are very sensitive to even small changes in temperature and humidity. A climatic chamber allows constant ambient conditions during the tests. A possible operating range of -30 °C to 220 °C requires that the sensitive equipment is located outside the climatic chamber. The sensor measuring the excitation, as well as the mounting structure must be located inside the climatic chamber and therefore be temperature and humidity resistant.

The specimens are mounted in a fixed-free suspension inside the climatic chamber and excited with a sinusoidal sweep by a shaker. The stinger transmits the excitation into the chamber and through the impedance sensor measuring force and acceleration into the mounted specimen. Using a laser vibrometer, the flexural vibration response is measured at the free end of the specimen. Dividing the resulting acceleration a with the excitation force F yields the accelerance frequency response function (FRF) a/F $[\frac{\mathrm{m}}{\mathrm{s}^2}/\,\mathrm{N}]$. Since linear-viscoelastic material behavior is assumed, there is a linear relationship between the excitation force as energy introduced into the system and the resulting acceleration of the test specimen. By using the force normalization, the measured and evaluated FRF does not change despite a fluctuation of the excitation force in the test. The FRF is therefore comparable and reproducible.

By varying the excitation frequency, any number of resonance orders up to a frequency of 10 kHz can be evaluated from the FRF. From the i-th resonance frequency f_i of the flexural vibration, together with the length l and thickness h of the test specimen, the density ρ of the material and the fixed-free suspension condition k_i, the storage modulus E' in each resonance frequency can be determined by means of a transformed Bernoulli bar theory according to [16] with

$$E_f' = \frac{48\pi^2\rho l^4}{h^2}\frac{f_i^2}{k_i{}^4};\; k_i = (i - 0{,}5)^2\pi^2. \quad (4)$$

The loss factor $tan(\delta)$ is determined with the 3dB-method from the i-th resonance frequency width Δf_i at which the amplitude of the FRF resonance peak has fallen on both sides by 3 dB and the resonance frequency f_i with

$$tan(\delta)_f = \frac{\Delta f_i}{f_i}. \tag{5}$$

The loss modulus E'' can now be calculated using (4) and (5) with

$$E_f'' = E_f' \cdot tan(\delta)_f. \tag{6}$$

Storage and loss modulus can only be determined in the resonance frequencies of the test specimens. Depending on stiffness, geometry and density, this can include more than a dozen resonances up to a frequency of 10 kHz. Viscoelastic material data can be calculated for each of the resonance frequencies. By interpolating between these discrete data points, a frequency-dependent curve can be determined. However, the sample has to show a pure flexural vibration without a superimposed, e.g. torsional mode. In addition, the test setup should not show any resonance behavior itself, the laser has to measure a good quality signal on the sample and the energy introduced by the shaker must be sufficiently high even at high frequencies.

Although the test setup can be used universally for non-reinforced plastics or other classes of materials, the focus is on SFRP. For an orientation dependence, test specimens are cut out in an orientation relative to the flow direction from an injection molded plate. However, experience shows only 75 % of the fibers are oriented in the flow direction, the remaining fibers are oriented in the transverse and thickness direction. Therefore, fully oriented test specimens are not available for injection molding but rather a mixed composition varying with the cutting direction of the specimens.

3.5 Generating a material card

In order to be able to fill equations (1-3) with angle-dependent storage and loss modulus, a comprehensive material card is required. The characterization is carried out using specimens cut-out from injection molded plates. However, the cutting direction has nothing to do with the fiber orientation angles required in the material model. Therefore, a method has to be found to conclude from a cutting orientation to a fiber orientation, which can be used to fill the material card with the measured data.

First, the evaluation procedure and the generation of material cards must be explained. Since viscoelastic material data has to be calculated directly in the FE-simulation, an update of the stiffness matrix can take place for every frequency step. Thus, for each a material card with angle-dependent storage and loss modulus is needed. The procedure is shown in Figure 7. The measured FRF is evaluated in its resonance frequencies. The material data determined on several test specimens of the same cutting orientation are averaged and interpolated in the frequency domain. Afterwards a swap of the order of values takes place and for each interpolated frequency, an angle-dependent function is generated by interpolation between the measured orientations. These data are stored as discrete points in a material card for a given temperature and humidity, so that the FE-simulation can search for the required material properties.

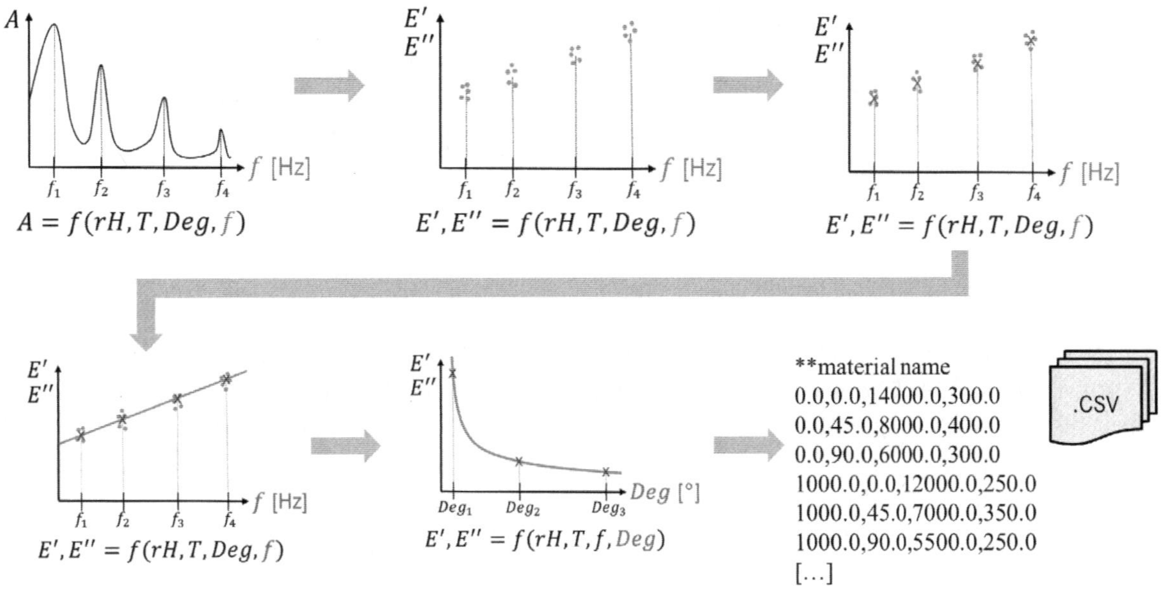

Figure 7. Generation of a material card from a FRF

The measured specimens have an averaged stiffness of all fiber orientation distributions that occur in them. It is assumed that the averaged deviation of all macroscopic elements of a test specimen corresponds to the stiffness value measured with it. In order to determine the correct averaged orientation angle, the material model must now be calculated backwards. To do this, the averaged absolute orientation direction in the global coordinate system is determined for each cut-out specimen using the injection molding simulation of the plate from which the specimens are prepared. These directions are shown as arrows in Figure 8 and differ from each other. The angular deviation from their respective cutting orientation yields the averaged fiber orientation angle. It corresponds to the measured specimen stiffness. This is shown in Figure 8 for example for the 45°-specimen. The determined averaged angles can now be used for inter- and extrapolation of the material card in the angle domain and thus in the FE-calculation.

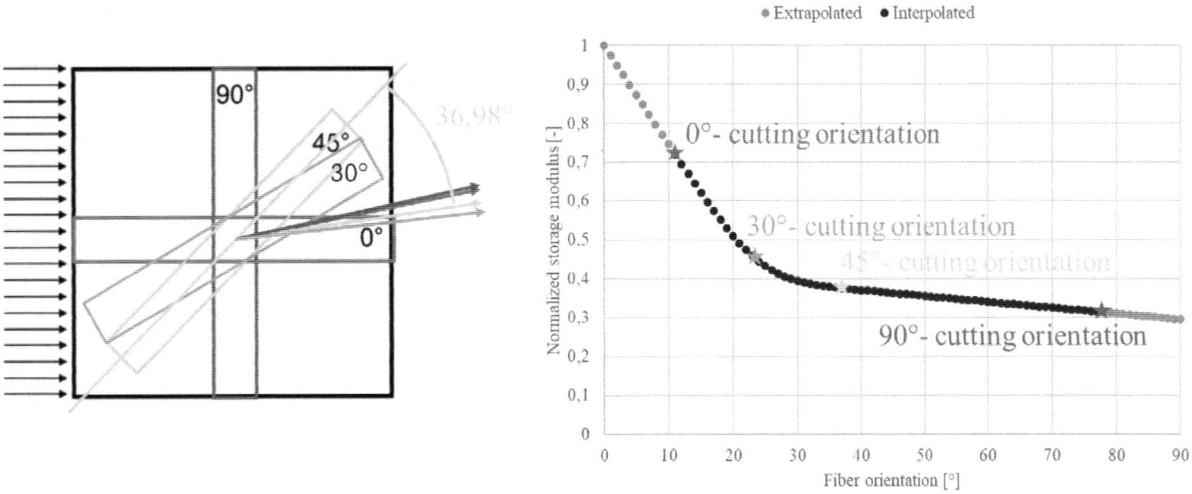

Figure 8. Conclusion from a cutting orientation to an averaged fiber orientation

The injection molding simulation is deliberately used to determine the average fiber orientation of the test specimens, rather than the more precise information from a CT analysis. This is because only a few small sections of the specimens can be examined in the µ-CT and not their entirety, which can lead to a significant error. In addition, all further calculations are carried out with the same injection molding model that has already been calibrated with a CT analysis.

4. RESULTS

4.1 Experimentation

The tests are carried out with two different injection moldable materials. On the one hand, a polyamide 6.6 with 50 weight-% short glass fibers (PA66-GF50) is used as a representative of the thermoplastics, on the other hand, a phenolic resin with 55 weight-% short glass fibers (PF-GF55) is used as a representative of the thermoset plastics in order to show a universal applicability of the approach. Cuboid test specimens are prepared from injection molded plates. The thermoplastic specimens are prepared in the directions 0°, 30°, 45°and 90° with the lengths 180 mm and 140 mm as well as a width of 15 mm and a thickness of 2 mm from a 180x180x2 mm plate. The thermoset samples are prepared in the directions 0°, 45° and 90° with the lengths of 170 mm and 130 mm as well as a width of 15 mm and a thickness of 2 mm from a 171,5x171,5x2 mm plate. For each orientation and length, three specimens are prepared, each of which is cut out at the identical positions from the injection-molded plate. They are then at an end on an area of 5x15 mm attached to a specimen-adapter with a two-component epoxy resin adhesive and cured. Afterwards, the real geometry, i.e. the freely oscillating length as well as the thickness and the density, are measured for each sample in order to determine the storage modulus free of geometric or material fluctuations.

A Weiss-Technik ClimeEvent C/600/40/3 climatic test chamber and a Polytec PSV-500-3D-Xtra scanning laser vibrometer are used for the experiments. The samples are now dried or wet conditioned for the respective test according to the material manufacturer's specifications and left in a constant climatic environment for some time before the tests. Temperature and humidity investigated range between 0 °C and 130 °C for 0 %RH (relative humidity) and between 0 °C and 50 °C for 50 %RH. Engine mounting brackets freely suspended in the climatic chamber are also examined, which are conditioned and stored according to the same specifications. The Polytec PSV Software is used for the analysis of the laser measurements, the calculation of the material card is performed by a VBA-Script (Visual Basic for Applications).

For calibration and validation of the injection molding simulation, µ-CT investigations are performed on samples with the size 5x5x2 mm. They are taken from the injection molded plate so that each specimen is covered by three different positions along its length.

4.2 Simulation

The injection molding simulation is performed with CADMOULD, the mapping with CONVERSE. Hereby up to ten different fiber orientation layers over the thickness of the examined object can be transferred to a structure simulation mesh. The meshing is carried out for the test specimens with ABAQUS and for the engine mounting brackets with SIMLAB. The simulation is performed as a direct-solution steady state dynamic analysis with a UMAT programmed in FORTRAN. The UMAT reads in the material card and the 2nd order orientation tensor, sorts them and stores them for the simulation. In the material card, the data set for storage and loss modulus closest to the calculated frequency step is then searched for and the required data is interpolated for each element from the angular distribution to fill the complex stiffness matrix. Then the eigenvectors for each element are passed to ABAQUS to rotate the matrix into the global coordinate system.

The specimen is meshed with the easily stretchable C3D20 elements, see Figure 9. This is because it is very important to use all ten fiber orientation layers, in order to correctly reproduce the stiffness of the overall composite without causing too much computing time. The engine mounting bracket is meshed with Tet elements, because the geometry is much more complex and thus a sufficient number of layers can be transferred from the injection molding simulation.

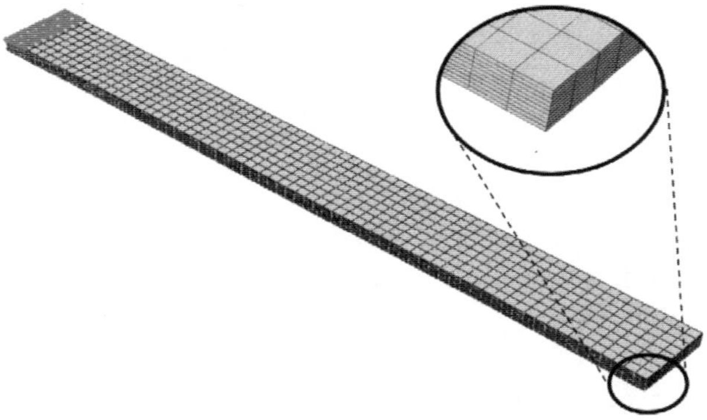

Figure 9. FE-mesh of the test specimen

Excitation and mounting conditions of the test specimen are modelled according to the test setup. Its movement is blocked at one end on an area of 5x15 mm in all spatial directions except in the thickness direction, where it is excited for all frequencies with constant acceleration of 1 mm/s². The FRF is calculated in a comparable way by evaluating the acceleration at the same point on the specimen as in the experiment and normalizing it with the excitation force. Furthermore, linear-viscoelastic material behavior, a constant fiber orientation in each element, constant material properties across a FE-element for all integration points and equal treatment of thermoplastic and thermoset are assumed. Additionally, the orthotropic shear modulus in Figure 4 can be calculated as an approximate function of Young's modulus and Poisson's ratio. Due to the limited availability of data and the little influence on the results, the Poisson's ratio is assumed directional but frequency independent and taken from a material database.

4.3 Results

First, the comparison of μ-CT and injection molding simulation takes place. Figure 10 shows the orientation tensor over the normalized thickness of the specimen at an exemplary location near the sprue. The solid line is taken from the CT measurement. Green represents the inflow, blue the crossflow and yellow the thickness direction. The crosses show the positions across the thickness at which data from the injection molding simulation is extracted, in total at a maximum of 10 possible locations.

The thermoplastic shows a narrow middle layer with a rapidly shifting orientation towards this area. Over the remaining thickness, the orientation is somewhat constant. The line from the CT examination is approximated very well, but the exact mid-layer cannot be analyzed due to the algorithm of the injection molding simulation. The thermoset, on the other hand, drops continuously towards the middle and therefore does not have a clearly shaped middle layer but rather a continuous transition of the orientation distribution. The CT measurement is approximated well by the injection molding simulation inflow, crossflow is slightly overestimated and the thickness direction is slightly underestimated.

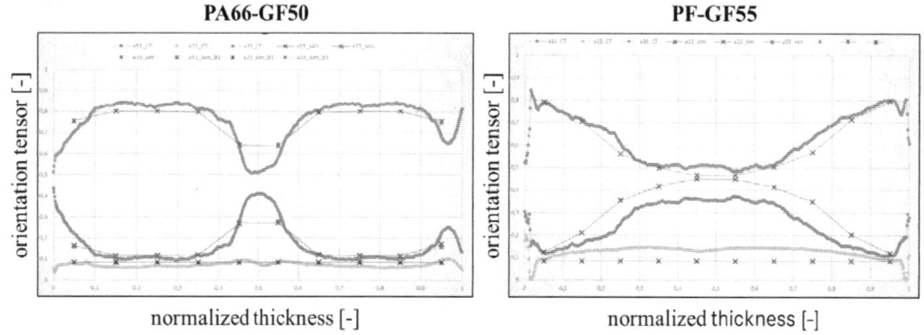

Figure 10. Results of the measured and simulated orientation tensor (green/blue/yellow line: inflow/crossflow/thickness direction from the μ-CT analysis, crosses with lines: corresponding data from the injection molding simulation)

In the following, the comparison between experiment and simulation of the test specimens used to create the material card is carried out. Figure 11 shows the thermoplastic PA66-GF50. The comparison illustrates the double logarithmic FRF exemplary for 23 °C and 0 %RH for the orientations 0°, 45° and 90° and a specimen length of 140 mm, in red the simulation and in blue the experiment. Both the resonance frequency positions and the amplitude peaks match very good at almost all resonances. Deviations in the amplitude can result from superposition of other modes of vibration in the experiment. The comparison shows that the maximum deviation of the resonance frequencies is <3 % and the amplitude peak <10 % for the 0°-specimen. The other orientations show an even better agreement.

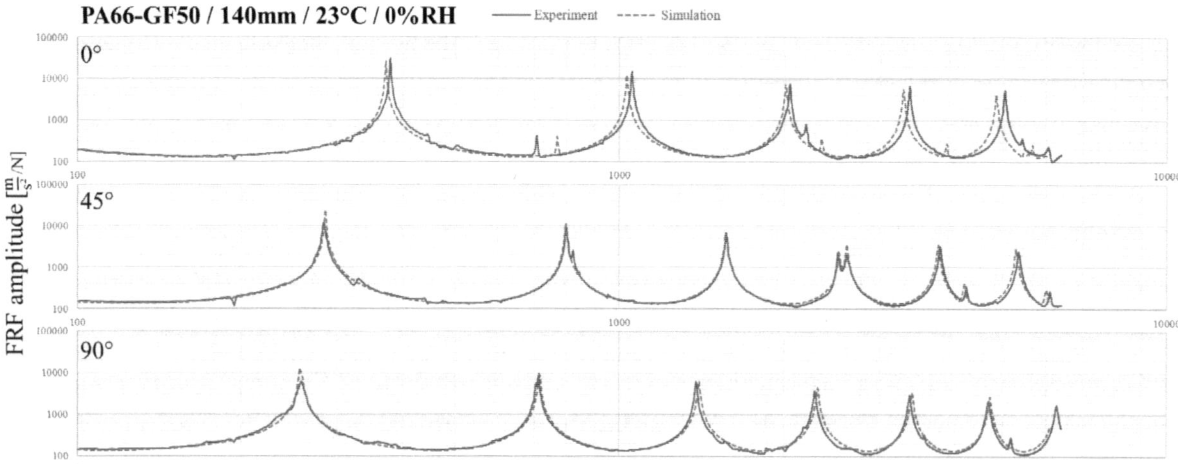

Figure 11. Results experiment/simulation of thermoplastic PA66-GF50

The comparison of the thermoset material is analogous to the thermoplastic but with a 170 mm specimen and shown in Figure 12. Resonance frequency positions and amplitude peaks fit very well. Like with the thermoplastic the 45° sample has a good accuracy. The largest deviation of the resonance frequencies is <4 % and the amplitude peak is <15 % for the 90°-specimen.

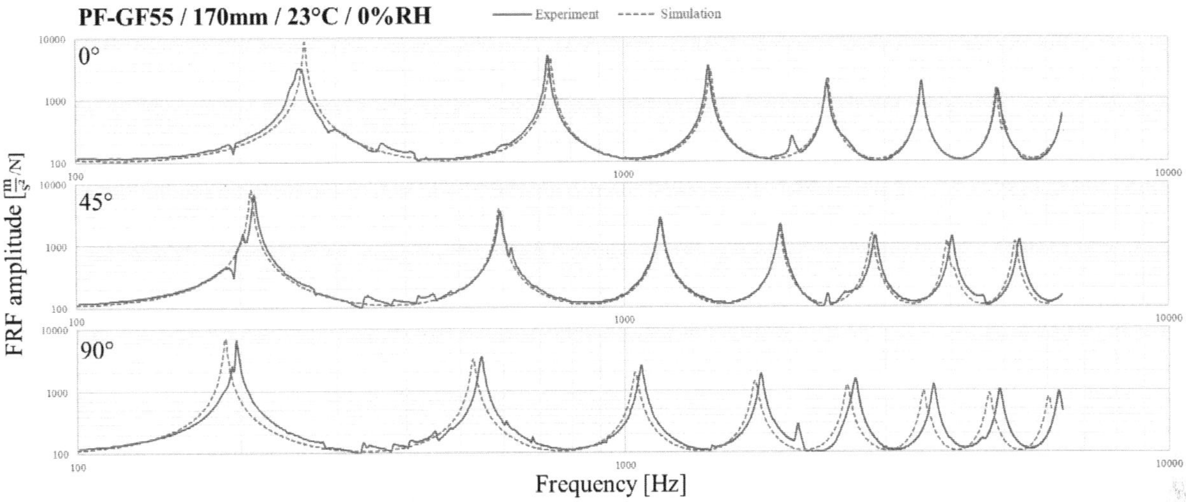

Figure 12. Results experiment/simulation of thermoset PF-GF55

Finally, an engine mounting bracket made out of PA66-GF50 is simulated. This component is quite complex and comprehensive in terms of geometry and fiber distribution. Particular difficulty is due to the correct mounting in the experiment and the modelling in the simulation. The results in Figure 13 are very significant. The deviation of the resonance frequencies from simulation to experiment is <3 %.

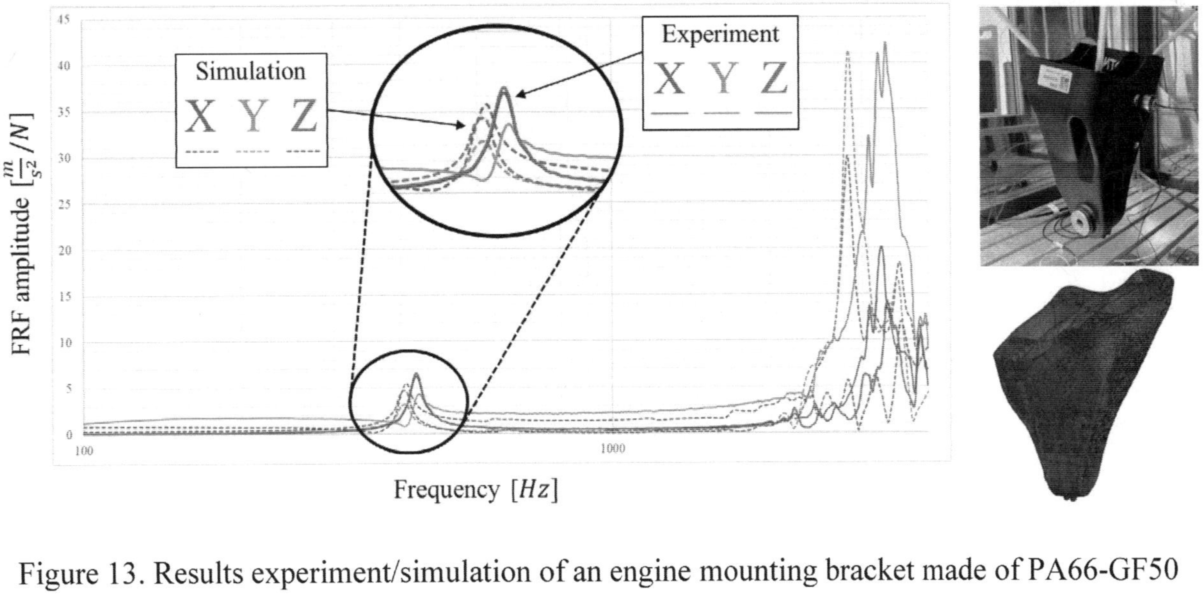

Figure 13. Results experiment/simulation of an engine mounting bracket made of PA66-GF50

5. CONCLUSIONS

A novel approach for vibration calculation of SFRP was presented. It was shown that macroscopic composite data can be used to model the linear-viscoelastic material behavior. The results show a good agreement with the experiment without mathematical approximation solutions or iterative reverse engineering approaches.

It should be noted that the results of the test specimens do not necessarily have to agree, as they serve as a basis for the material card. Rather, it was shown that the approach succeeds in creating the angle-dependent function of the storage and loss modulus and reassembling it again to create an overall composite of different orientation distributions and thus direction-dependent stiffness. Especially the orientation of 45° is to be emphasized, since the shear modulus and its approximate solution is particularly influential here and the results have shown a good agreement. Furthermore, a 2nd order orientation tensor was sufficient for this approach, as well as the maximum of 10 layers from the injection molding simulation to correctly reproduce the overall composite of the material. Finally, it was shown that the approach works well not only for thermoplastics but also for thermosets. The latter had been difficult to model so far, because the material had to be understood on a microscopic level. By means of the new approach, this is no longer necessary.

The disadvantage, however, was that the effort required to create the material card as well as the resource consumption of the simulation is relatively high. Since the material card can be used universally, the method is suitable for areas where the same material is used for different components.

6. REFERENCES

[1] Stommel, M., Stojek, M., Korte, W. *FEM zur Berechnung von Kunststoff- und Elastomerbauteilen*, München: Hanser, 2nd ed., 2018. DOI: 10.3139/9783446452831.

[2] Kremer, H. *Materialdatenermittlung thermoplastischer Kunststoffe für Körperschallsimulationen auf Basis von Reverse Engineering*. PhD thesis, Aachen: Mainz, 2014.

[3] Raschke, K., Korte, W. "Faserverstärkte Motorbauteile besser berechnen." *Kunststoffe* 109 (2019): 184–189.

[4] Laak, M. op de, Hauth, M. "Noch schneller zur Zylinderkopfhaube." *Kunststoffe* 94 (2004): 126–130.

[5] Michaeli, W., Hopmann, C., Kremer, H. "Materialdatenermittlung für Akustiksimulationen mittels Reverse Engineering: Akustisches Verhalten von Kunststoffen." *Kunststoffe* 102 (2012): 5–7.

[6] Giersbeck, M., Hornberger, K., Kech, A. "Virtuelle Bauteilentwicklung." *Kunststoffe* 101 (2011): 50–53.

[7] Hopmann, C., Michaeli, W., Kremer, H. "Frequenzabhängiges Verhalten von Kunststoffen." *Kunststoffe* 102 (2012): 64–66.

[8] Pischinger, S., Michaeli, W., Joerres, M., Steffens, C., Atzler, M., Arping, T. "Verfahren zur akustischen Simulation von Kunststoffbauteilen." *MTZ - Motortechnische Zeitschrift* 70 (2009): 692–701. DOI: 10.1007/BF03225522.

[9] Treviso, A., van Genechten, B., Mundo, D., Tournour, M. "Damping in composite materials: Properties and models." *Composites Part B: Engineering* 78 (2015): 144–152. DOI: 10.1016/j.compositesb.2015.03.081.

[10] ISO 6721-1, 2019. "Plastics — Determination of dynamic mechanical properties — Part 1: General principles" International Organization for Standardization, Geneva, 2019, https://www.iso.org/obp/ui/#iso:std:iso:6721:-1:ed-4:v1:en.
[11] ISO 6721-5, 2019. "Plastics — Determination of dynamic mechanical properties — Part 5: Flexural vibration — Non-resonance method" International Organization for Standardization, Geneva, 2019, https://www.iso.org/obp/ui/#iso:std:iso:6721:-5:ed-2:v1:en.
[12] Menard, K. P. *Dynamic Mechanical Analysis: A Practical Introduction*, Hoboken: Taylor & Francis, 2nd ed., 2008. DOI: 10.1201/9781420053135.
[13] Menges, G., Haberstroh, E., Michaeli, W., Schmachtenberg, E. *Menges Werkstoffkunde Kunststoffe*, München: Hanser, 6th ed., 2011. DOI: 10.3139/9783446443532.
[14] Keuerleber, M. *Bestimmung des Elastizitätsmoduls von Kunststoffen bei hohen Dehnraten am Beispiel von PP*. PhD thesis, Stuttgart, 2006.
[15] Arping, T. W. *Werkstoffgerechte Charakterisierung und Modellierung des akustischen Verhaltens thermoplastischer Kunststoffe für Körperschallsimulationen*. PhD thesis, Aachen: Mainz, 2010.
[16] ISO 6721-3, 1994. "Plastics — Determination of dynamic mechanical properties — Part 3: Flexural vibration — Resonance-curve method" International Organization for Standardization, Geneva, 1994, https://www.iso.org/obp/ui/#iso:std:iso:6721:-3:ed-1:v1:en.
[17] Berthelot, J.-M., Assarar, M., Sefrani, Y., Mahi, A. E. "Damping analysis of composite materials and structures." *Composite Structures* 85 (2008): 189–204. DOI: 10.1016/j.compstruct.2007.10.024.
[18] Bonfiglio, P., Pompoli, F. "Determination of the dynamic complex modulus of viscoelastic materials using a time domain approach." *Polymer Testing* 48 (2015): 89–96. DOI: 10.1016/j.polymertesting.2015.09.016.
[19] Crane, R. M., Gillespie, J. W. "Characterization of the vibration damping loss factor of glass and graphite fiber composites." *Composites Science and Technology* 40 (1991): 355–375. DOI: 10.1016/0266-3538(91)90030-S.
[20] Ilg, J. *Bestimmung, Verifikation und Anwendung frequenzabhängiger mechanischer Materialkennwerte*. PhD thesis, München: Dr. Hut, 2015.
[21] Belder, K. de, Pintelon, R., Demol, C., Roose, P. "Estimation of the equivalent complex modulus of laminated glass beams and its application to sound transmission loss prediction." *Mechanical Systems and Signal Processing* 24 (2010): 809–822. DOI: 10.1016/j.ymssp.2009.11.001.
[22] Cortés, F., Elejabarrieta, M. J. "Viscoelastic materials characterisation using the seismic response." *Materials & Design* 28 (2007): 2054–2062. DOI: 10.1016/j.matdes.2006.05.032.
[23] El-Hafidi, A., Gning, P. B., Piezel, B., Belaïd, M., Fontaine, S. "Determination of dynamic properties of flax fibres reinforced laminate using vibration measurements." *Polymer Testing* 57 (2017): 219–225. DOI: 10.1016/j.polymertesting.2016.11.035.
[24] Böhm, H. J. "A Short Introduction to Continuum Micromechanics." *Mechanics of Microstructured Materials*. Ed. Helmut J. Böhm. Vienna: Springer Vienna, 2004. pp. 1–40. DOI: 10.1007/978-3-7091-2776-6_1.
[25] Mori, T., Tanaka, K. "Average stress in matrix and average elastic energy of materials with misfitting inclusions." *Acta Metallurgica* 21 (1973): 571–574. DOI: 10.1016/0001-6160(73)90064-3.

[26] Eshelby, J. D. "The determination of the elastic field of an ellipsoidal inclusion, and related problems." *Proceedings of the Royal Society of London. Series A. Mathematical and Physical Sciences* 241 (1957): 376–396. DOI: 10.1098/rspa.1957.0133.

[27] Christensen, R. M. "A critical evaluation for a class of micro-mechanics models." *Journal of the Mechanics and Physics of Solids* 38 (1990): 379–404. DOI: 10.1016/0022-5096(90)90005-O.

[28] Kaiser, J.-M. *Beitrag zur mikromechanischen Berechnung kurzfaserverstärkter Kunststoffe - Deformation und Versagen*. PhD thesis, Saarbrücken, 2013.

[29] J. C. Halpin, Kardos, J. L. "The Halpin-Tsai equations: A review." *Polymer Engineering and Science* 16 (1976): 344–352. DOI: 10.1002/pen.760160512.

[30] Tandon, G. P., Weng, G. J. "The effect of aspect ratio of inclusions on the elastic properties of unidirectionally aligned composites." *Polymer Composites* 5 (1984): 327–333. DOI: 10.1002/pc.750050413.

[31] Schöneich, M. M. *Charakterisierung und Modellierung viskoelastischer Eigenschaften von kurzglasfaserverstärkten Thermoplasten mit Faser-Matrix Interphase*. PhD thesis, Saarbrücken, 2016. DOI: 10.22028/D291-23208.

[32] Dassault Systèmes. *Abaqus 2018 Theory Guide*, 2017.

[33] Gieß, M. *Untersuchungen zur akustischen Formteilauslegung*. PhD thesis, Aachen: Shaker, 2019.

[34] Barbero, E. J. *Finite element analysis of composite materials using ANSYS*, Boca Raton Fla.: CRC Press, 2nd ed., 2014.

[35] Advani, S. G., Tucker, C. L. "The Use of Tensors to Describe and Predict Fiber Orientation in Short Fiber Composites." *Journal of Rheology* 31 (1987): 751–784. DOI: 10.1122/1.549945.

[36] Abdin, Y. *Micro-mechanics based Fatigue Modelling of Composites Reinforced with Straight and Wavy Short Fibers*. PhD thesis, Leuven, 2015.

[37] Kohnke, P. C. *ANSYS - engineering analysis system theoretical manual for Rev. 4.4*, Houston: Swanson Analysis Systems Inc., 5th ed., 1989.

[38] Urban (né Pfeifer), F., Armbruster, B. "Vorrichtung zur Schwingungsanregung eines Probekörpers." Patent DE 10 2019 001 226.7, 2019, https://register.dpma.de/DPMAregister/pat/register?AKZ=1020190012267.

MODELING THE TEMPERATURE-DEPENDENT STRUCTURAL PERFORMANCE OF RIGID POLYURETHANE AND POLYISOCYANURATE FOAMS USING MEAN-FIELD HOMOGENIZATION AND FINITE ELEMENT ANALYSIS

Hieu T. McElroy[(1)] and Luca Guj[(2)]
The Dow Chemical Company
[(1)]Lake Jackson, TX, United States
[(2)]Horgen, Switzerland

ABSTRACT

In recent years, rigid Polyurethane (PUR) and Polyisocyanurate (PIR) foams are increasingly used as insulation materials for wall and roof applications in high energy-efficient homes and commercial buildings around the world due to their superior thermal performance. Metal-faced PUR/PIR sandwich panels are widely used as structural insulated panels in commercial building applications to reduce construction time as well as materials and costs associated with structural framing. Thus, material manufacturers must understand the structural performance of PUR/PIR foams to ensure the foam panels meet the quality requirements and industry standards. Because manufacturing and characterization are expensive, time-consuming, and labor-intensive, Dow Research and Development (R&D) has developed a combined experimental-numerical approach to model the anisotropic temperature-dependent properties of rigid PUR/PIR foams. This approach allows predictions of thermal-mechanical properties, i.e., tensile and compression behaviors, to end-use performance, e.g., metal-faced sandwich solutions for commercial buildings. The numerical model is based on the "double-inclusion" mean-field homogenization technique and finite element analysis. Finally, the model was experimentally validated using tensile and four-point bending tests as well as digital image correlation analysis.

Corresponding author: Hieu T. McElroy (htruong@dow.com)

1. INTRODUCTION

Rigid Polyurethane (PUR) foam panels have been used for insulation in the construction industry for a few decades. With recent technical advancements, the next-generation rigid foam products, so-called Polyisocyanurate (PIR) foams, were developed. PIR foams have superior performance regarding processability, weight reduction, structural strength, adhesion to metal facings, thermal insulation and, fire performance. Therefore, PIR foams have gained shares in different market segments, e.g., metal-faced sandwich solutions for commercial and institutional buildings and pipe-in-pipe for oil and gas applications. In recent years, metal-faced PIR sandwich panels have been widely used as structural insulated panels in energy-efficient homes and commercial building applications around the world to reduce construction time as well as materials and costs associated with structural framing.

SAMPE Virtual Conference Proceedings, 2020. Society for the Advancement of Material and Process Engineering – North America.

It is essential to understand the structural performance of PUR/PIR foams to ensure the foam panels meet the quality requirements and industry standards [1]. Manufacturing and characterization are often expensive, time-consuming, and labor-intensive. Thus, Dow Research and Development (R&D) has developed a combined experimental-numerical approach to model the anisotropic temperature-dependent properties of rigid PUR/PIR foams. This approach allows predictions of thermal-mechanical properties, i.e., tensile and compression behaviors, to end-use performance, e.g., metal-faced sandwich solutions for commercial buildings. The numerical model is based on the "double-inclusion" mean-field homogenization (MFH) technique and finite element analysis (FEA). Finally, the model was experimentally validated using tensile and four-point bending tests as well as digital image correlation (DIC) analysis.

2. BACKGROUND: FUNDAMENTALS OF RIGID FOAM PANELS' PROCESSING AND CHARACTERISTICS

Rigid PUR/PIR foams are made by mixing and chemical reactions of a number of components: polyol, isocyanate, blowing agent, catalysts, surfactants, and other additives. The result is a closed-cell foam structure containing PUR/PIR cell walls enclosing a combination of gas (the majority is CO_2). Figure 1 illustrates the components and products of the reaction.

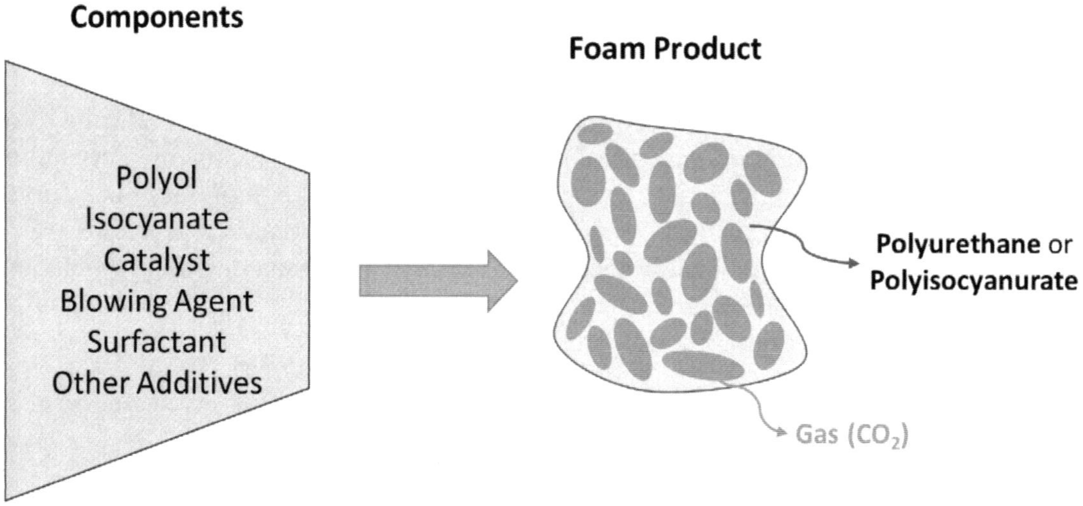

Figure 1. Components and products of PUR/PIR foams.

Metal-faced PUR/PIR sandwich panels are made using the Rigid-Faced Double Belt Lamination (RFDBL) process, demonstrated in Figure 2(a). The process consists of mixing the components described previously to obtain a reacting mixture. This mixture is then uniformly distributed on the bottom metal sheet across the panel width using a distributor, often referred to as poker, while both the top and bottom metal sheets are continuously pulled along the advance (length) direction. The physicochemical blowing process occurs, leading to the formation of foam, and as the foam rises up, the gap between the two moving metal sheets is filled. Finally, panels of the desired length are cut out. An example of the final product is shown in Figure 2(b).

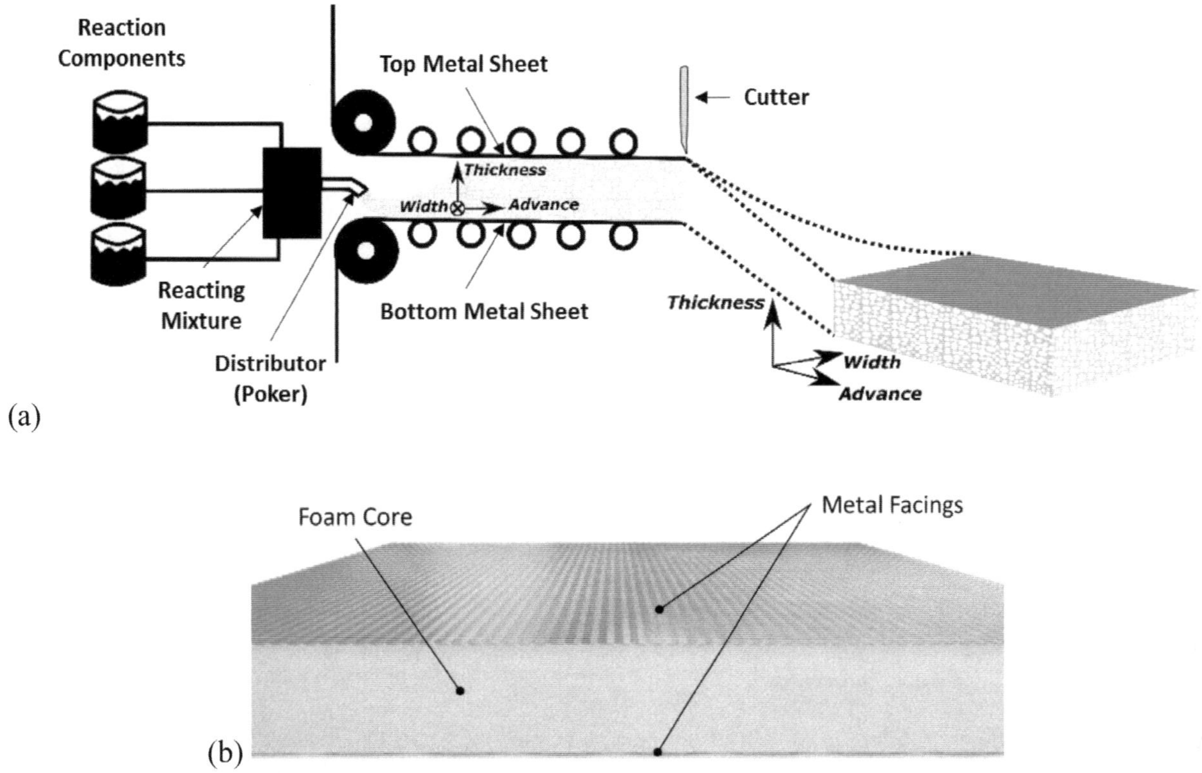

Figure 2. (a) PUR/PIR foams' directions related to the RFDBL production process (b) Typical insulated metal panel for constructions.

As a result of the manufacturing process, the foam cells are elongated and oriented along the advance direction. In addition, the foam microstructure (cell size and foam density) varies through the panel's thickness, i.e., the density decreases, and cell size increases from the bottom to the top of the panel. Nevertheless, the foam microstructure is relatively uniform across the panel's width. The advance, width, thickness directions in the metal foam panels are illustrated in Figure 2(a). The variations in the foam microstructure lead to anisotropic thermal-mechanical behaviors. Generally, PUR/PIR rigid foams are brittle and exhibit relatively linear-elastic behavior in tensile loadings. An example of the anisotropic stress-strain response of PIR foam in tension at room temperature is shown in Figure 3(a). Because the foam cells are oriented in the advance direction, the tensile modulus of samples tested in the advance direction is significantly higher than the tensile modulus of samples tested in the width and thickness directions. In the width and thickness directions, the foam has similar stiffness and strains at failure. For simplicity of modeling and analysis, in this work, it is assumed that the rigid foams exhibit transversely isotropic behavior in the width and thickness (W&T) directions, exemplified by the average tensile behavior among the two directions, i.e., W&T behavior in Figure 3(a). The schematic of rigid foams' transversely isotropic microstructure is shown in Figure 3(b).

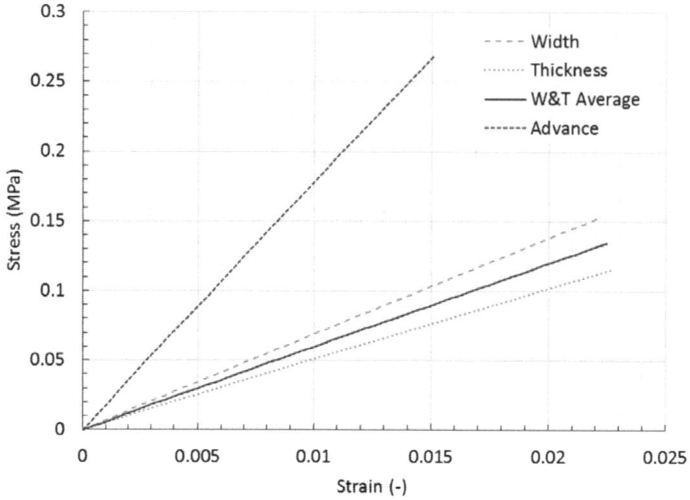

(a)

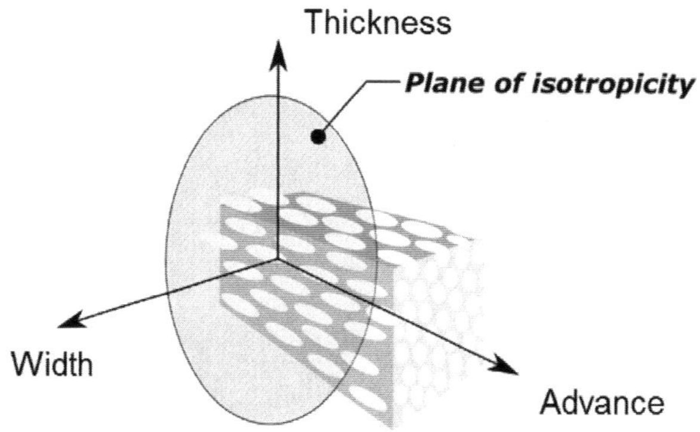

(b)

Figure 3. (a) Anisotropic behavior of a PIR foam in tension at room temperature (b) Schematic of rigid foams' transversely isotropic microstructure.

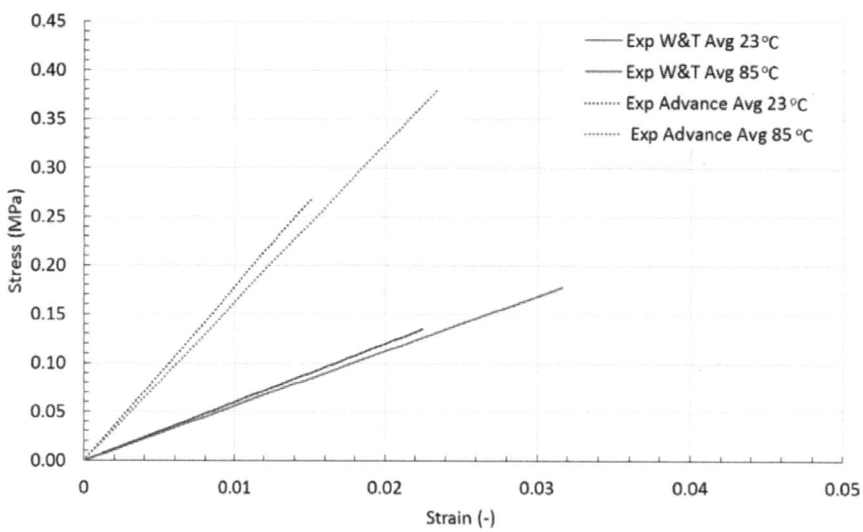

Figure 4: Tensile behavior: Advance vs. W&T direction at 23°C and 85°C.

At high temperatures, the foams still exhibit relatively linear-elastic and transversely isotropic behavior. The moduli are, however, decreased while the values of strain at break increase. Figure 4 shows the comparison of tensile stress-strain responses obtained at 23 °C and 85 °C of a PIR foam.

3. MODELING

3.1 Mean-Field Homogenization Approach

The constitutive model proposed by Dow R&D was designed to represent the structural behavior of PUR/PIR rigid foams within different end-applications. The structural performance of PUR/PIR foams is measured not only in terms of stiffness but also in terms of strength. During service, metal-faced rigid foam panels experience failures induced by thermal cycles. These thermal cycles are caused by either weather conditions (i.e., solar radiation and rain wash) or the after-production cooling cycles, especially when the operating temperature is well below freezing. The proposed model can depict all of the defined features driving the rigid foams' structural performance, including stress-strain and failure behaviors, temperature dependency, and orientations due to the manufacturing process. These are the minimum requirements to represent the real influence of the production process as well as the foam morphology on the performance in the end-use applications.

A further criterion for the model, used in the selection of the technical approach to foam modeling, is the ability to represent the material failure in an implicit finite element analysis. In reality, failures of PUR/PIR foams in the aforementioned applications occur over several hours to several days. Moreover, the foam failure is due to differential thermal dilation of the materials considered in end-use applications such as PIR foam core and steel faces. Therefore, only implicit numerical techniques can be used to represent this phenomenon. On the contrary, numerical techniques for modeling crack initiation and propagation pertain to explicit numerical solvers tailored for simulating fast and explosive events like crash tests and air-bag opening. In short, the constitutive model requirements are:

- Ability to depict orientation in stress-strain and failure behavior as well as their dependence upon temperature, and
- Adaptability to implicit analysis.

Based on the criteria above, the Double-Inclusion mean-field homogenization technique developed by Nemat-Nasser and Hori [2] was selected. The Double-Inclusion homogenization technique is implemented in the commercial software Digimat [3], and it is seamlessly integrated with the Abaqus [4] finite element software used for the thermo-mechanical simulation of the PUR/PIR foam-based products. In addition, the Double-Inclusion method was selected because it can consider high inclusions' volume fraction. This is due to the fact that rigid PUR/PIR foams exhibit closed-cell microstructure with a volume fraction of voids or porosity up to 95 %.

The Double-Inclusion homogenization technique was developed for heterogeneous materials consisting of two or more constituents. Therefore, the model requires describing the constituents of the material under study in terms of structural and thermal behavior: Young's modulus as a function of temperature $E(T)$, Poisson's ratio ν, density ρ, and coefficient of linear thermal expansion CLTE(T). In this work, the proposed method, inspired by Gibson and Ashby [5], considers the rigid foam as a composite with two constituents: gas in the cells (i.e., voids) and the enclosing cell walls structure (i.e., matrix). Figure 5 illustrates the two-constituent composite concept of the PUR/PIR rigid foams in this study. First, the structural behavior of the void phase (Young's modulus E and Poisson's ratio ν) as well as the general orientation of the foam are defined. Then, the matrix morphology and structural behavior can be identified. In particular, the PUR/PIR matrix material elastic behavior, i.e., Young's modulus E and Poisson's ratio ν, as well as morphology, i.e., cell aspect ratio AR, are identified by reverse engineering capability provided by the Digimat-MX module. Digimat-MX provides a "Reverse engineering and parameter identification" tool to optimize and obtain the set of material parameters for the matrix, which minimizes the error between the experimental and numerical tensile stress-strain curves.

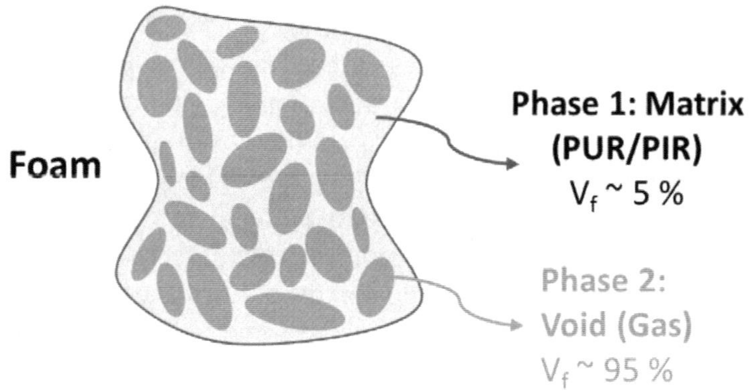

Figure 5. Double-Inclusion homogenization model's schematic for PUR/PIR foams.

3.2 Failure criterion

The failure behavior of the PUR/PIR foams is defined using the Tsai-Wu 3D transversally isotropic failure criterion [6]:

$$f_i = \sqrt{\mathcal{F}_i(\boldsymbol{\varepsilon})} \; with: \tag{1}$$
$$\mathcal{F}_i(\boldsymbol{\varepsilon}) = \frac{\varepsilon_{11}^2}{X_t X_c} + \frac{\varepsilon_{22}^2 + \varepsilon_{33}^2}{Y_t Y_c} + \frac{(2\varepsilon_{12})^2 + (2\varepsilon_{13})^2}{S^2} + \frac{(2\varepsilon_{23})^2}{Y_t Y_c} - \frac{\varepsilon_{11}\varepsilon_{22} + \varepsilon_{11}\varepsilon_{33}}{2X_t X_c} - \frac{2\varepsilon_{22}\varepsilon_{33}}{Y_t Y_c} + \left(\frac{1}{X_t} - \frac{1}{X_c}\right)\varepsilon_{11} + \left(\frac{1}{Y_t} - \frac{1}{Y_c}\right)(\varepsilon_{22} + \varepsilon_{33})$$

where failure initiation occurs when $f_i = 1$. This failure criterion exhibits five different allowable values, as demonstrated in Figure 6:

- longitudinal tension X_t and compression X_c (direction 1 in Figure 6 or Advance in Figure 3(b))
- transverse tension Y_t and compression Y_c (directions 2 and 3 in Figure 6 or Thickness and Width in Figure 3(b))
- transverse shear S.

Therefore, the model allows for capturing different failure modes, i.e., tension vs. compression vs. shear, as well as failure behaviors in different directions, e.g., along vs. normal to the axis of cell orientation. The Tsai-Wu 3D transversally isotropic failure criteria (equation [1] and Figure 6) for the PUR/PIR foams modeled in this work are calibrated using data from tension, compression and four-point bending tests performed at 23 °C and 85 °C in the advance (longitudinal) as well as width and thickness (transverse) directions.

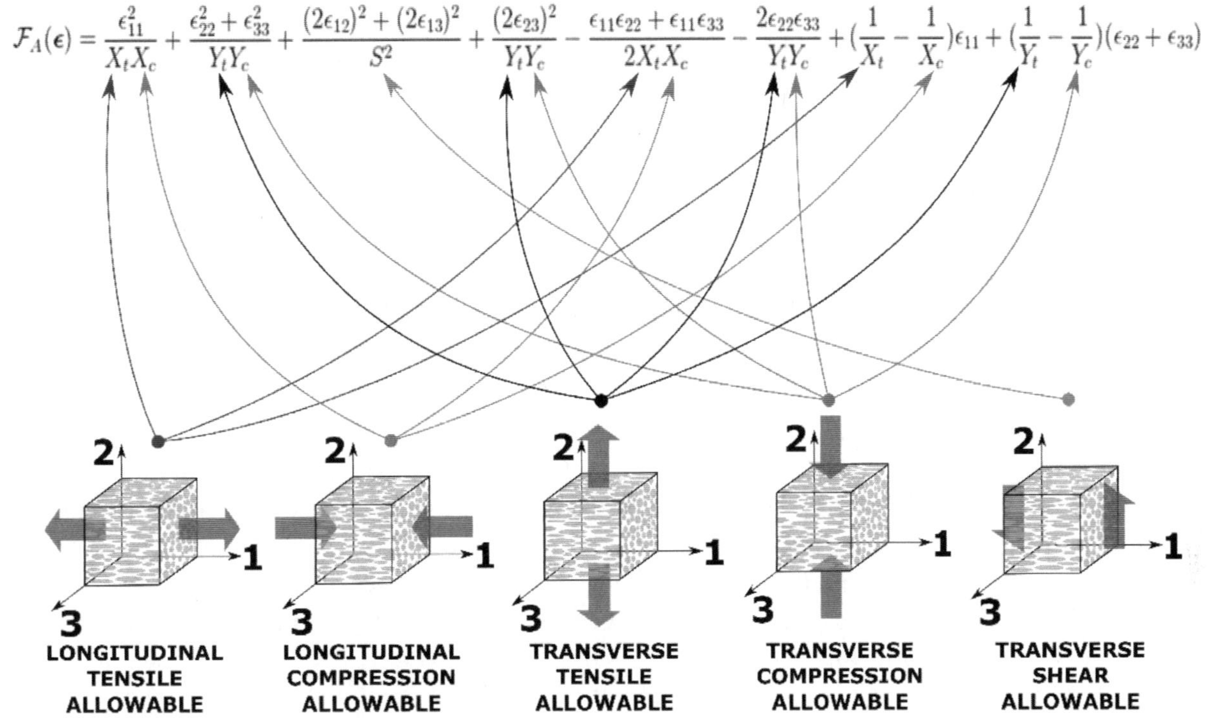

Figure 6. Tsai-Wu 3D Transversally Isotropic failure criterion allowable values.

Once the failure initiation is reached, a linear damage evolution law is used.

$$\acute{\sigma}_0 = (1 - D)\sigma_0$$
$$D = \frac{D_{max}}{\delta\varepsilon_{lim}}\delta\varepsilon \text{ with } \delta\varepsilon_{lim} = \varepsilon - \varepsilon_{ini} \quad [2]$$

Where $\acute{\sigma}_0$ is the stiffness of the damaged material, D is the linear stiffness degradation law, D_{max} is the maximum damage at failure, ε_{ini} is the strain at damage initiation, $\delta\varepsilon_{lim}$ is the change in the strain at which the failure is completely developed.

3.3 Modeling Workflow

The workflow of the proposed modeling approach for rigid foams is demonstrated in detail in Figure 7. Multiple steps are undertaken beginning with the tension, compression, and shear characterization of the foams in the advance, width, thickness directions, and performed at room (23 °C) and elevated (85 °C) temperatures. Then, different modules within the Digimat framework are used to calibrate the thermal-mechanical material models based on the experimental data input. Here, the microstructure and orientation dependence of the foam is taken into account via mean-field homogenization. The temperature-dependent thermal-mechanical properties of the foams are also accounted for. The material models are then used in the commercial FEA software Abaqus to evaluate the thermal-structural performance of the end-use applications.

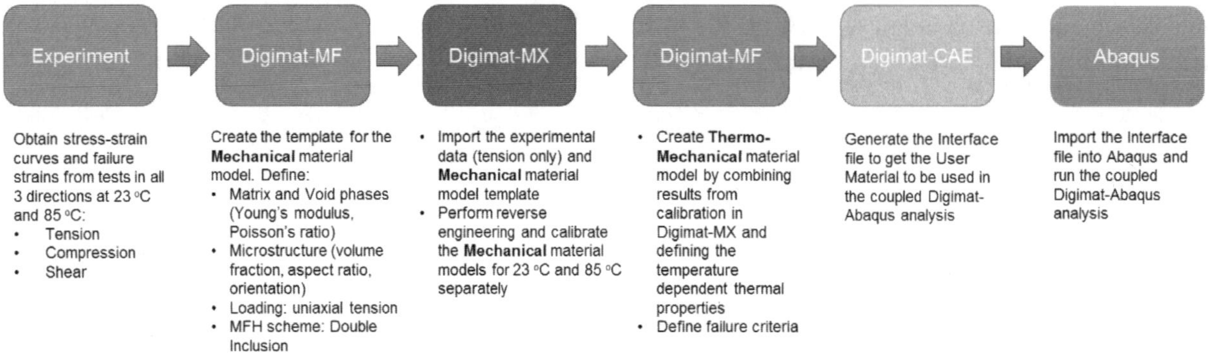

Figure 7. Workflow of the proposed modeling approach.

3.4 Applications

Using the approach and workflow discussed above, the material models for the foams were calibrated in Digimat, then used in macroscale FE models in Abaqus to simulate the tensile and four-point bending tests. The model geometries were created so that they match the validation experiments, which will be discussed in detail in the next section. The model setups are shown in Figure 8. It should be noted that the tensile test model was performed for both 23 °C and 85 °C, and the loading directions are parallel to the advance as well as the width and thickness directions of the foams. For the four-point bending tests, the length, width, thickness directions of the specimens correspond to the advance, width, thickness directions of the metal-faced panels.

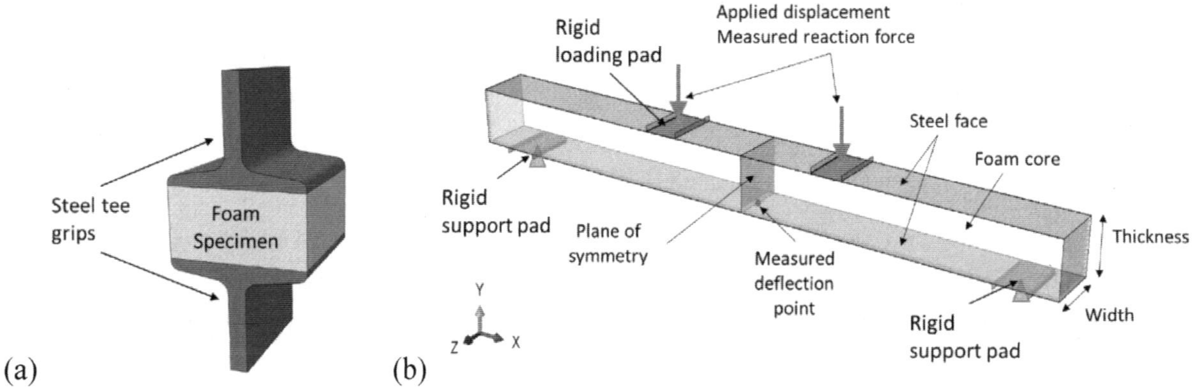

Figure 8. (a) Tensile test model setup (b) Four-point bending test model setup.

4. VALIDATION EXPERIMENTS

4.1 Tensile Tests with Digital Image Correlation

Tensile tests were conducted at 23 °C and 85 °C. Foam specimens of dimensions 50 mm x 50 mm x 25 mm were cut from the panel so that the 25-mm dimension is parallel to the advance, width, thickness direction of the panel. Each specimen was adhesively bonded to a pair of steel tee grips. The steel tee grips have dimensions of 50 mm (length) x 50 mm (width) x 50 mm (height) x 6.35 mm (flange thickness). The grips underwent a surface treatment process prior to bonding to ensure proper adhesion between the grips and foam. An Instron frame model 3365 with a 5 kN load cell and pneumatic loading grips were used. All specimens were loaded in tension at a rate of 1.3 mm/min. Figure 9(a) shows the tensile test setup conducted in this work. The tensile stress-strain responses, as well as modes of failure, were compared to the model's results. In-situ digital image correlation was conducted at room temperature to measure the displacement and strain fields on the foam during loading. Two PointGrey cameras were used to obtain the 3D-DIC measurements. Prior to testing with DIC, the front surfaces of the bonded foam specimen assemblies were spray-coated with black paint to create speckle patterns for the DIC analysis. Poisson's ratios on the foam samples were calculated using the strain fields from the DIC analysis and compared to the model predictions.

4.2 Four-point Bending Tests with Digital Image Correlation

Four-point bending tests were conducted at room temperature, according to European Norm EN 14509 [1] recommendations. The specimens have nominal thickness, width, and length dimensions of 100 mm, 100 mm, and 1050 mm, respectively. The loading and support spans are respectively 300 mm and 900 mm. The width and thickness of the steel loading plates are 100 mm and 10 mm. The tests were performed using an Instron frame model 3365 with a 5 kN load cell and at a loading rate of 12 mm/min. In-situ DIC measurements were conducted to obtain the full-field strain distribution on the foam during loading and at failure. The four-point bending test setup is shown in Figure 9(b).

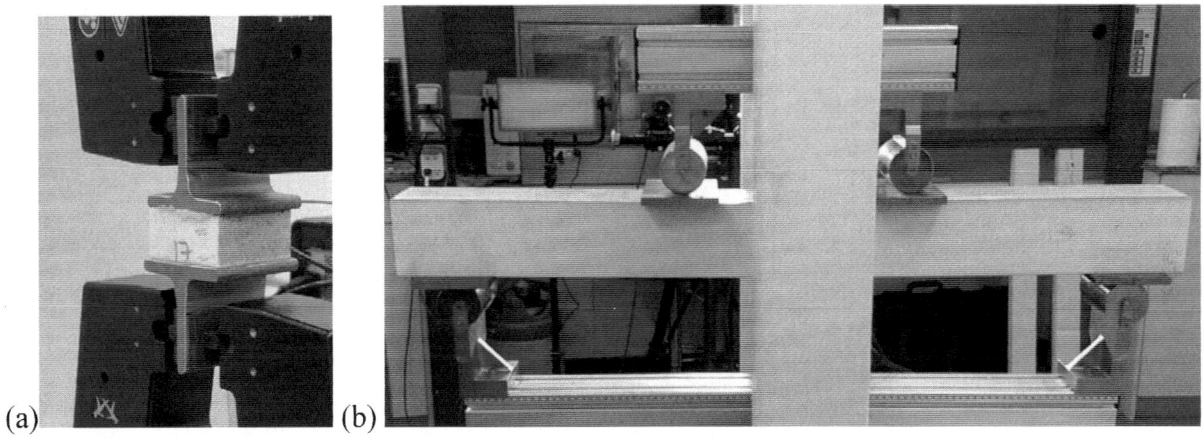

Figure 9. (a) Tensile test setup (b) Four-point bending test setup.

5. RESULTS

It should be noted that the proposed modeling of approach works for rigid foams in general. For this paper, we selected a PIR foam to demonstrate the results of the models and validation experiments.

5.1 Tensile Stress-Strain Responses

Figure 10 shows the comparison between experimental and numerical tensile stress-strain responses in the advance and average of thickness and width directions at 23°C and 85°C of PIR foam. It can be seen that the proposed model represents the tensile behavior of the PIR foam with a high degree of accuracy in terms of stiffness response. Particularly, it captures the foam's higher stiffness in the advance direction compared to that of the width and thickness directions. In addition, the model is able to represent the relevant variation with temperature of the foam failure behavior at different temperatures. In summary, the proposed Double-Inclusion constitutive model for rigid PUR/PIR foams can accurately represent the thermo-structural behavior of the PIR foam, encompassing orientation, failure, and dependence upon temperature.

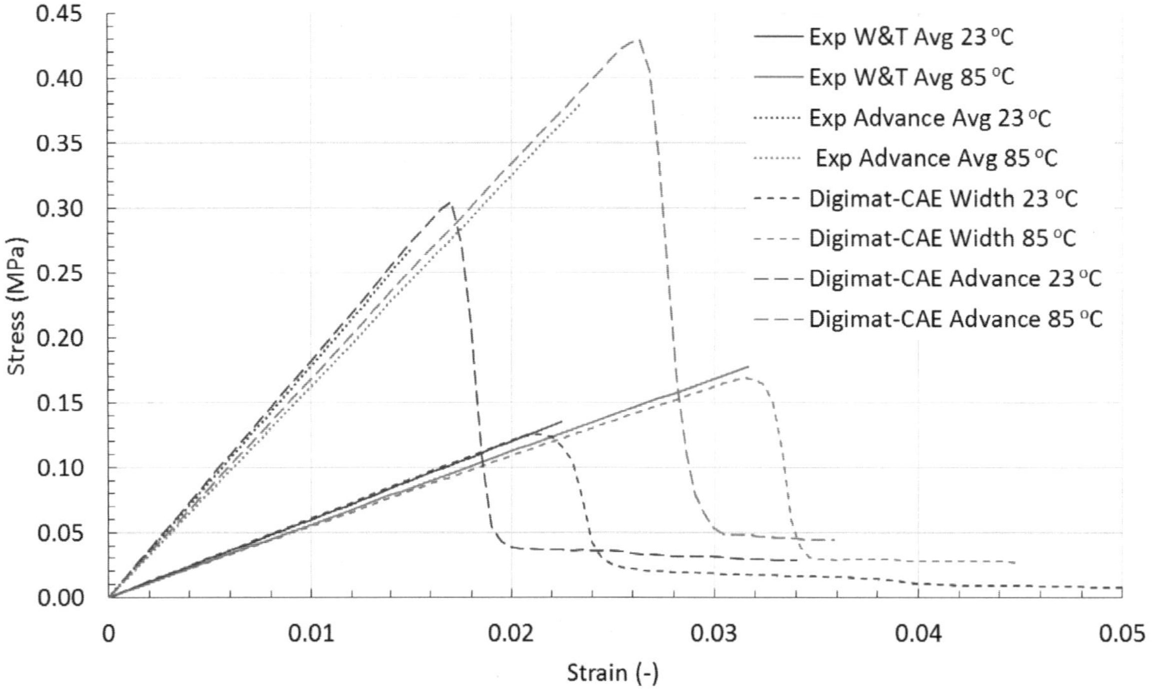

Figure 10. Experimental vs. Numerical: Tensile stress-strain responses in the advance and average of thickness and width directions at 23°C and 85°C of PIR foam.

5.2 Failure Modes and Failure Envelopes

The failure modes of the tensile specimens observed in the experiment and obtained from modeling are shown in Figure 11. The model predicted that failures occur inside the foam near both the top and bottom surfaces of the foam specimen, where it is bonded to the steel tee grips. This corresponds to the high strain levels in the specimen near the grips, as seen in the maximum principal strain contour plots in Figure 11. In reality, each specimen has only one major crack that spans the entire cross-section and locates close to either the top or bottom metal grip. This is because of the random distribution of defects and their sizes inside the foam specimen, and the first defect that causes failure initiation in the locations experiencing high strain will propagate leading to final failure.

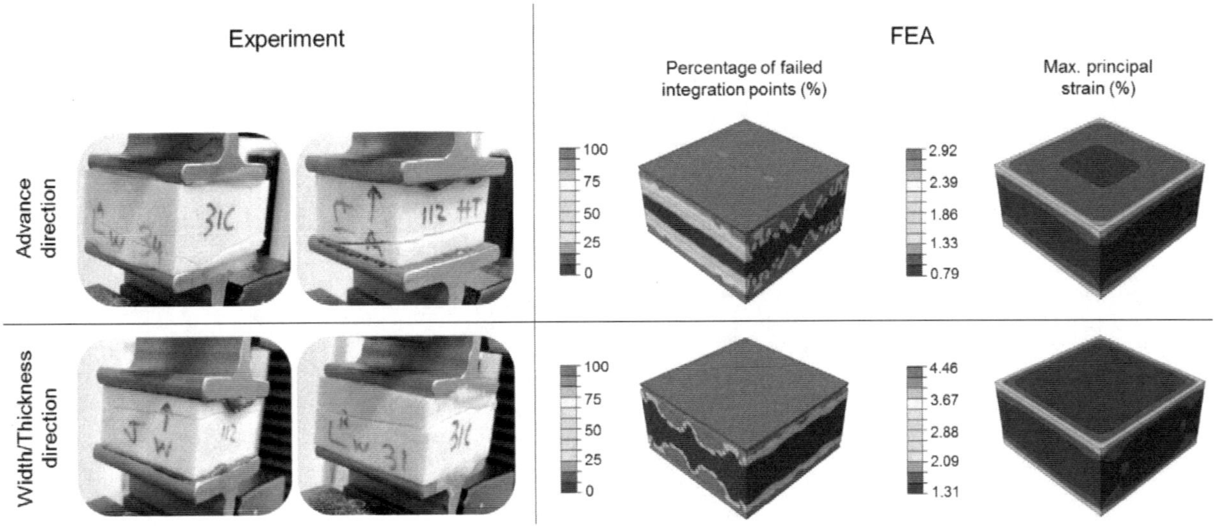

Figure 11. Experimental vs. Numerical prediction: Failure modes.

Table 1 summarizes the allowable parameters obtained from calibration and validation for the Tsai-Wu failure criteria of PIR foam. The identified allowable values in Table 1 reflect the behavior of the foam in different failure modes. PIR foam tends to fail more prematurely in tension than in compression or shear, which is reflected by the X_c, Y_c, S being noticeably higher than X_t and Y_t. This failure behavior is exemplified by the failure envelopes at different temperatures shown in Figure 12.

Table 1. Allowables for the Tsai-Wu failure criterion of PIR foam.

Mode	**Symbol**	**Temperature**	
		23 °C	**85 °C**
Tension	Xt	0.016	0.023
	Yt	0.022	0.032
Compression	Xc	0.065	0.090
	Yc	0.095	0.120
Shear	S	0.075	0.100

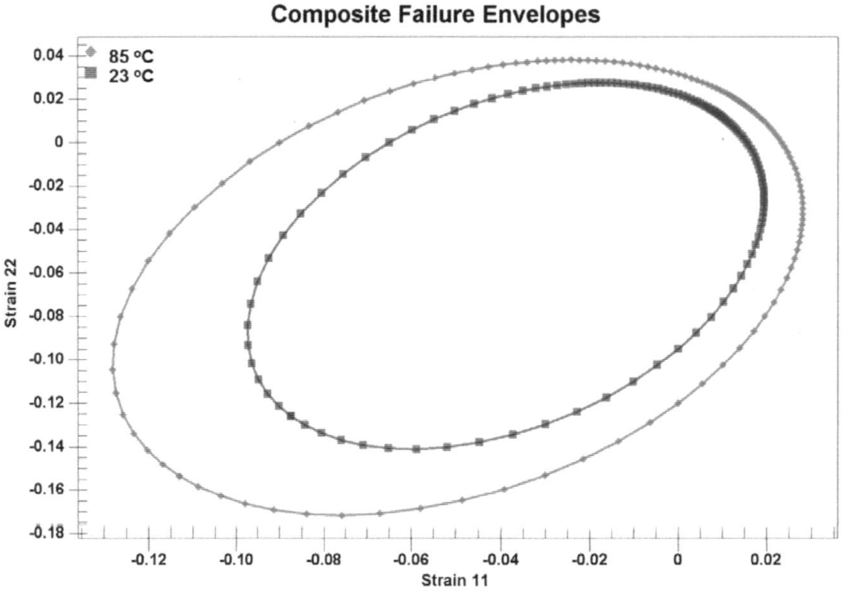

Figure 12. Failure envelopes projected on the $\{\varepsilon_{11} \equiv \varepsilon_{Advance}, \varepsilon_{22} \equiv \varepsilon_{W\&T}\}$ plane.

5.3 Poisson's Ratio

DIC strain measurements were used to calculate and construct full-field distributions of Poisson's ratios. Six different Poisson's ratios were measured and compared to the model predictions. Examples of Poisson's ratios ν_{21} and ν_{23} maps are presented in Figure 13. Figure 14 shows comparisons of the average ν_{21} and ν_{23} measured via digital image correlation and numerical values computed using the Double-Inclusion homogenization method, highlighting the excellent agreement between the predicted and measured Poisson's ratios.

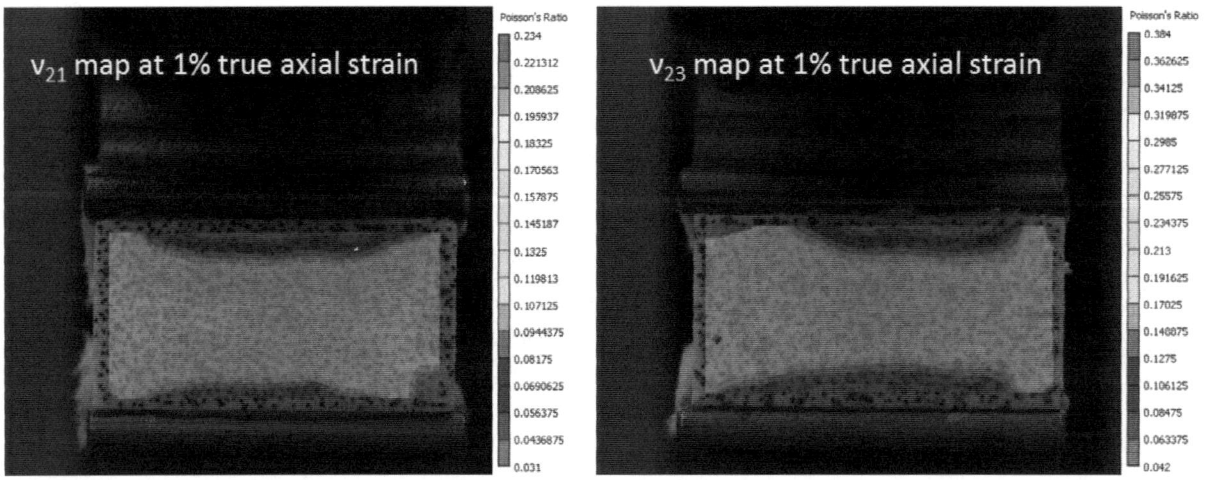

Figure 13. Examples of Poisson's ratio distributions on PIR foam during tensile loading.

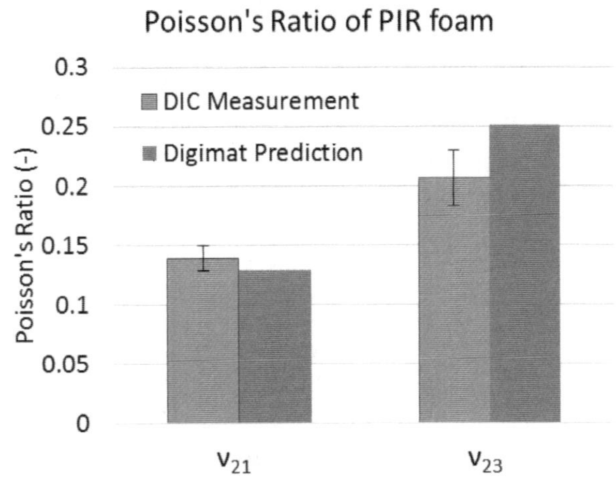

Figure 14. Comparison between model predictions and DIC measurements of PIR foam's Poisson's ratios.

5.4 Four-point Bending Measurements

The load-displacement curves obtained from the four-point bending experiment and numerical model are shown in Figure 15(a). The experimental load-displacement curve is non-linear. The model response is, however, linear until failure. This is because the PIR material model calibrated using the MFH approach in this work assumes the linear-elastic behavior of the foam. The stiffness and strain at failure values from the experiment and numerical model are in good agreement. Figure 15(b) shows the agreement between failure initiations predicted by the FE model and captured in the experiment. The predicted location and initial shape of the shear failure, highlighted by the red dashed boxes in Figure 15(b), are accurate considering the implicit solver used in this work.

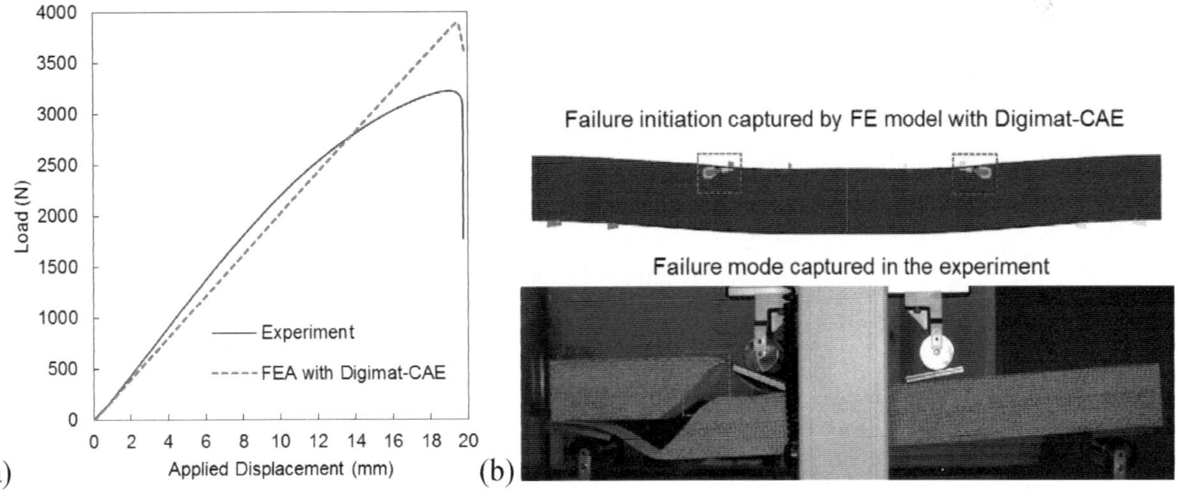

(a) (b)

Figure 15. (a) Experimental and numerical comparison of load-displacement curves (b) Failure initiation predicted by the FE model and failure mode captured in the four-point bending experiment.

Full-field distributions of shear and compression strains on the foam obtained from FEA and DIC were compared, as shown in Figure 16. It can be seen that the model predictions are in good agreement with the experimental results. In the shear strain maps in Figure 16(a), the location of maximum shear strain corresponds to the location of failure initiation seen previously in Figure **15**(b). During loading, the foam experience some level of compression. This is captured in both the model and the experiment (Figure 16(b)).

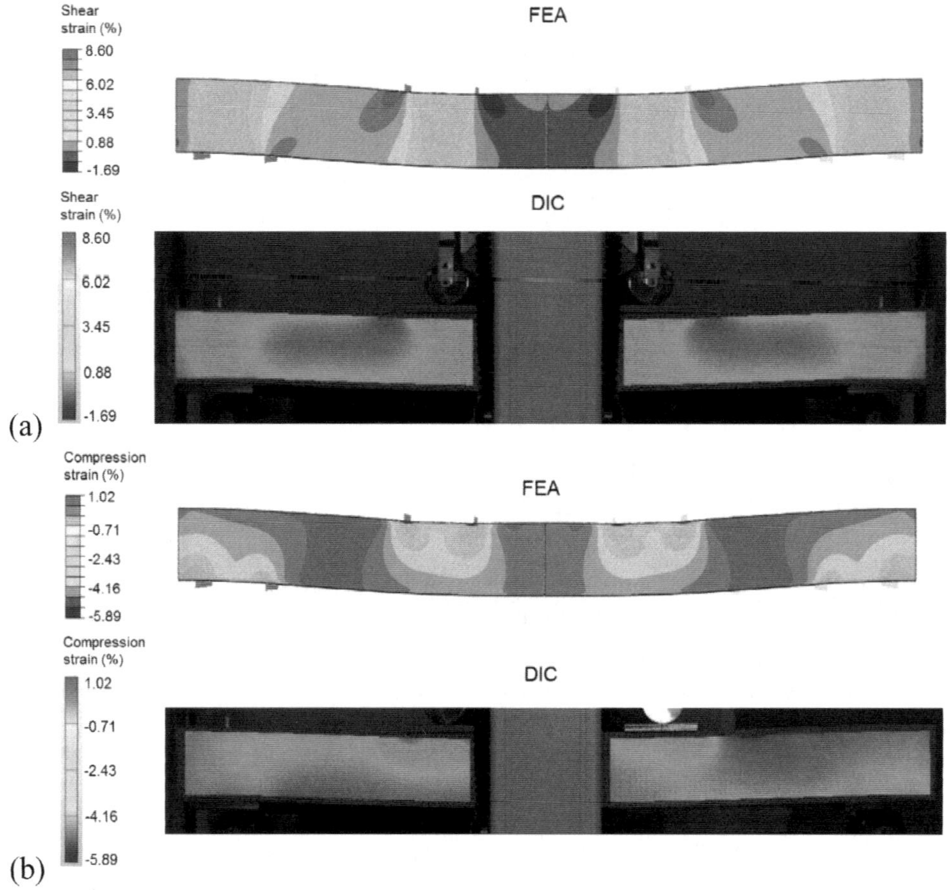

Figure 16. FEA vs. DIC: (a) full-field shear strain before failure (b) full-field compression strain before failure.

6. CONCLUSIONS

In conclusion, this study illustrates a modeling approach, based on the "Double-Inclusion" mean-field homogenization method, proposed by Nemat-Nasser and Hori [2], for modeling rigid PUR/PIR foams encompassing:

- Material orientation due to production;
- Failure;
- Temperature dependence of stiffness and failure.

The proposed approach is suitable for both explicit and implicit solver within Abaqus, allowing one to address fast and explosive events (i.e., crash tests, air-bag opening) as well as slow developing phenomena (i.e., post-production crack creation and propagation due to thermal differential deformation). This approach can be used to evaluate the thermo-structural behavior of

PUR/PIR rigid foams in end-use products such as metal-faced sandwich panels for construction applications.

The proposed model has been successfully validated by means of digital image correlation, direct comparison with tensile tests as well as four-point testing of metal-faced sandwich panels. The model was able to predict with high accuracy the Poisson's ratios of a PIR foam that were measured using digital image correlation. The model for the PIR foam was further validated by comparing numerical and experimental tensile stress-strain data. In addition, the model is able to capture the PIR foam orientation in terms of stiffness as well as failure behavior. Moreover, the effect of temperature variation was represented correctly.

7. REFERENCES

[1] European Norm BS EN 14509:2013, "Self-supporting double skin metal faced insulating panels - Factory made products – Specifications." European Committee for Standardization, 2013.

[2] Nemat-Nasser, S., and Hori, M. *Micromechanics: Overall Properties of Heterogeneous Materials*. Amsterdam: North-Holland, 1993.

[3] Digimat 2018.1 User's Manual.

[4] Abaqus 2018 User's Manual.

[5] Gibson, L.J, and Ashby, M.F. *Cellular Solids: Structure and Properties*. Cambridge: Cambridge University Press, 1997.

[6] Tsai, S.W. and Wu, E.M. "A General Theory of Strength for Anisotropic Materials," *Journal of Composite Materials* 5(1) (1971): 58-80.

MAGNETIC AND ELASTOMERIC DAMPING EFFECTS ON THE VIBRATION AMPLITUDE OF A VIBRATION ISOLATION SYSTEM

Toan Le, Dr. Gregory Scott, Dr. Ashraf Elbarbary, Dr. Eltahry Elghandour
California Polytechnic State University
San Luis Obispo, California 93410

ABSTRACT

This technical paper describes the design and experimental process for a single-stage multi-degree-of-freedom vibration isolation system for a scanning tunneling microscope. The mathematical model used for the mass-spring system analysis will be presented, along with various design iterations for the mass-spring system and the frame. Based on the mathematical model, a MATLAB script is developed to study hundreds of springs available online so that a spring that satisfies important specifications can be chosen. Finite Element software is used to investigate the natural frequencies of the frames, which will assist the frame's design selection. Throughout the design process, magnetic and elastomeric damping methods are actively considered to simplify the design due to tight constrain of the vacuum chamber.

1. INTRODUCTION

In the early 1980s, Binnig, Rohrer, Gerber, and Wiebel successfully built a scanning tunneling microscope (STM) in an IBM Research Laboratory in Zurich, Switzerland[1]. The STM uses the quantum tunneling effect to study the surface of materials on atomic scale[2]. When the two conducting electrodes are in very close proximity but not actually in physical contact, the tunneling effect of electrons occurs due to the wave-particle duality of electrons. Since the quantum tunneling effect decays exponentially with the increasing distance between the scanning tip and the conducting material's surface, the distance between two electrodes must be kept within a few ångstroms. Therefore, the device is prone to noises due to vibration from the surroundings, so a vibration isolation system is necessary to allow the microscope to function properly and produce images of high resolution. Passive vibration isolation is a commonly used technique due to its low cost and simpler design compared to active vibration isolation, especially when the entire vibration isolation system (VIS) stays inside an Ultra High Vacuum (UHV) environment. Multi-stage VIS have been used previously, but this design is usually too complex and not cost-effective. This research aims to successfully build a multiple-degree-of-freedom single-stage passive VIS for a STM, which centers on a circular copper stage suspended by a combination of six springs. Section 2 will discuss support excitation model used to select springs and weight of the system's main stage as well as the experiment process to validate the design of mass-spring system. This section will also discuss the design of the VIS frame where the system will be mounted on before putting into the vacuum chamber. Section 3 will present the experimental data for testing of the mass-spring system as well as the simulation results using Finite Element Method. Section 4 will present the conclusions for the current design and discuss future steps in including magnetic and elastomeric damping.

2. DESIGN & EXPERIMENTATION

The design process starts with studying previous designs of VIS. Figure 1 below shows a two-stage VIS. This design uses vertical springs to suspend the main stage where the microscope will be attached. An issue with this design is that it does not provide torsional stiffness to the system, which will make the vibration isolation less effective on the main stage.

Figure 1. A two-stage VIS

There are two methods to add torsional stiffness to the system: adding a torsional spring or changing the configuration of the springs. While adding a torsional spring seems feasible, it adds more complexity into the design in terms of where and how the mounting structure for the torsional spring will be. In addition, since the system will be put into an UHV environment, most torsional springs on the market are incompatible because they are made from music wire steel, which is a restricted material to be used in vacuum environment[3]. A more effective method is to change the configuration from vertical springs to angled springs. Figure 2 indicates a section view of the main stage and two springs connected to the stage with an angle. With this design, a part of the spring stiffness will be used for the vertical mode while the rest of the stiffness will be used when the stage is twisted by an external torsional source.

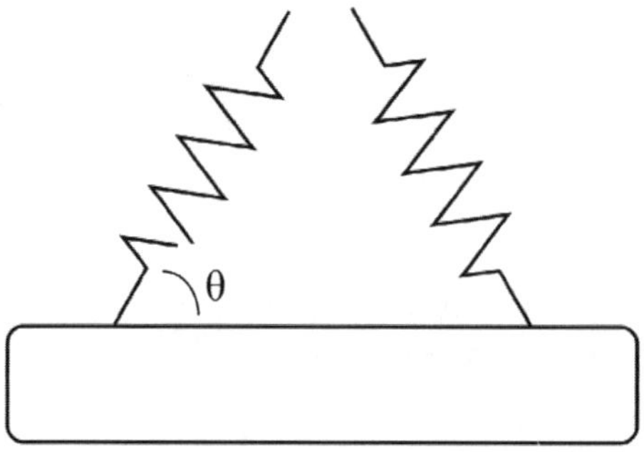

Figure 2. Configuration for 2 adjacent angled springs forming angle of θ above the main stage

2.1 VIBRATION DESIGN

Since the springs in the passive VIS are the softest compared to rigid stainless-steel vacuum chamber or aluminum 6061 frame, the mass-spring system has the lowest natural frequencies in the whole system. Hence, the analysis should start from the mass-spring system using support excitation model. Figure 3 depicts the mathematical support excitation model used to derive the equation of motion for the mass-spring system where $y, \dot{y}, \ddot{y}$ are input displacement, velocity, and acceleration respectively, and $x, \dot{x}, \ddot{x}$ are displacement, velocity, and acceleration of the main stage respectively.

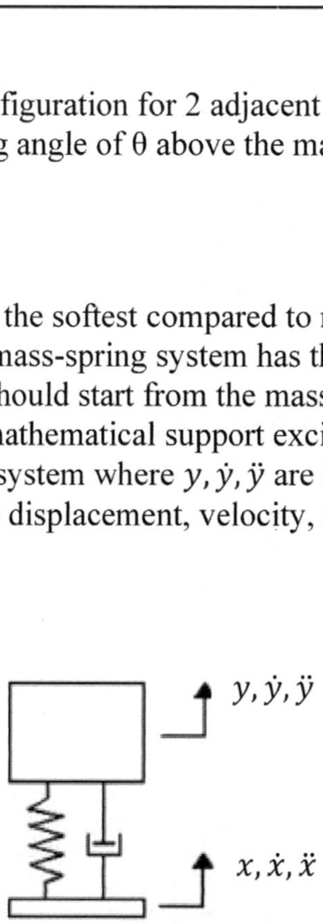

Figure 3. Support excitation model used for mass-spring system

Once the model is developed, the equation of motion can be derived as follows

$$M_{eq}\ddot{x} + B_{eq}\dot{x} + K_{eq}x = f(y, \dot{y}, \ddot{y}) \quad [1]$$

To achieve vibration isolation, the ratio of input frequency and natural frequency of the mass-spring system must be greater than $\sqrt{2}$. For a system with six angled springs, the natural frequencies can be derived as below, where k is the stiffness of one spring, and m_{stage} is the mass of the main stage.

$$\omega_n = \sqrt{\frac{K_{eq}}{M_{eq}}} = \sqrt{\frac{6 \times k \times sin\theta}{m_{stage}}} \quad [2]$$

2.1.1 *SPRING SELECTION*

From equation [2] and the criteria for vibration isolation, a MATLAB script is used to plot the theoretical lines of weight per spring versus spring stiffness needed to achieve vibration isolation for input frequencies of 1 Hz, 1.5 Hz, 2 Hz, 2.5 Hz, and 3 Hz. Data from two spring manufacturers are also plotted to help selecting springs. For these plots, if the spring data point appears above a theoretical solid line, vibration isolation can be achieved for this lowest frequency. Figure 4 shows the data from Century Spring website while Figure 5 shows the data from The Spring Store website. Since the manufacture uses English parameters for their springs, Figure 4 and 5 keep these units to show consistency. An important criterion for spring selection is that the total deflection must be within 25.8-28 cm (10-11 inches) due to the constraint in height of vacuum chamber, as well as the desire to be able to see the microscope working through a small window flange of 15 cm (6 inches) that is located 26.5 cm (10.5 inches) below the top flange.

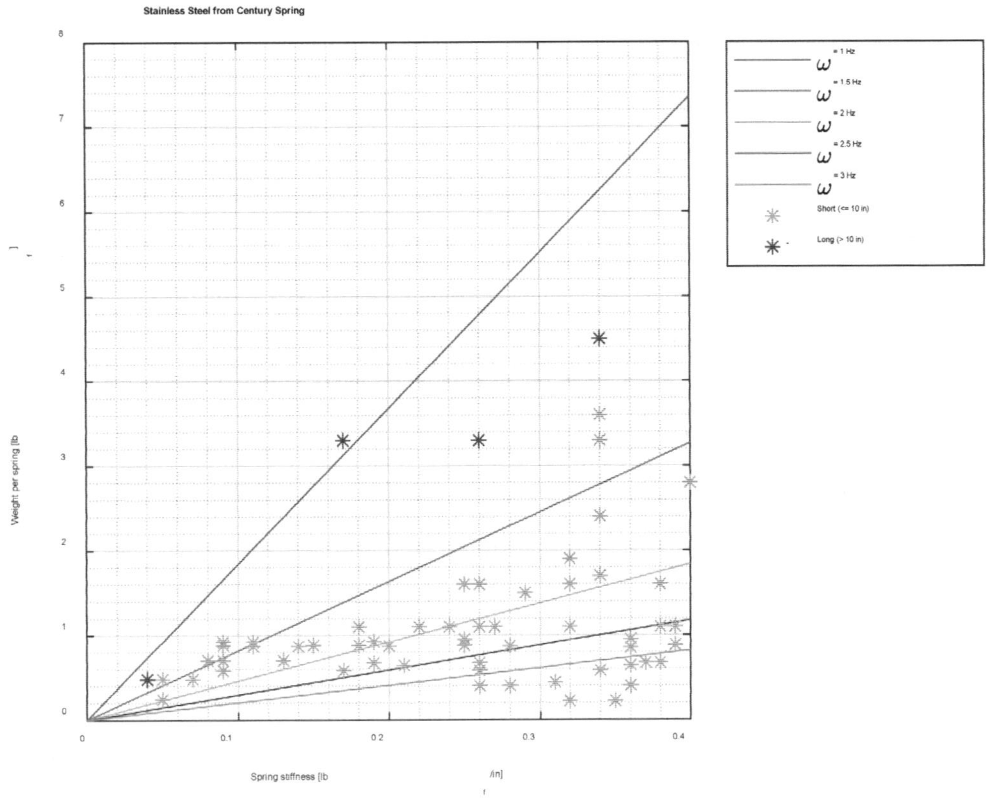

Figure 4. Long and short stainless-steel spring data from Century Spring plotted against the theoretical lines of frequencies indicating which isolation frequency could be achieved

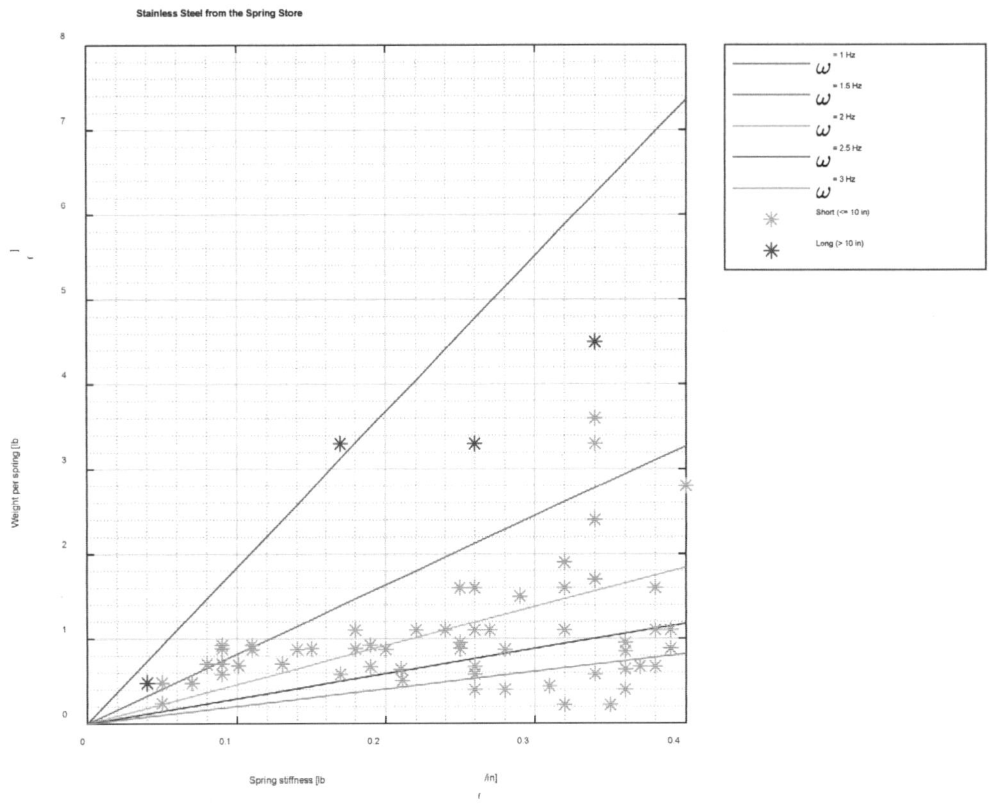

Figure 5. Long and short stainless-steel spring data from The Spring Store plotted against the theoretical lines of frequencies indicating which isolation frequency could be achieved

It is important to note that if the springs are loaded with higher loads than their max allowable load, we risk bending the hook or subject the springs to the non-linear region that Hooke's Law for springs might not work. Figure 4 and 5 show that by selecting long springs and underloading them with the desired weight, we can keep the springs operating in linear region and keep the total deflection of the main stage to be between 25.8 and 28 cm (10-11 inches) as discussed above. The selected spring is from Century Spring with the specifications shown in Table 1.

Table 1: Parameters of the selected spring

Material	Stainless-steel
Part Number	80657S
Initial Tension (N)	1.33
Length (mm)	127
Wire Diameter (mm)	0.94
Outer Diameter (mm)	12.70

2.1.2 Main Stage Selection

With the selected spring in mind, the calculated total weight of the main stage needs to be 4 kg to achieve vibration isolation for the lowest input frequency of 1.5 Hz. Due to material restriction to be used in vacuum chamber, there are two materials that the main stage can be made of: copper or aluminum. Using aluminum 6061 like previous designs seems like a good option initially, but the volume is too large to fit into the current small vacuum chamber while leaving room for magnetic damping accessories. Moreover, using aluminum main stage still requires copper componenets to facilitate magnetic damping, which is one of a few effective methods of damping used in UHV. Figure 6 shows a typical design for magnetic damping where a copper plate is mounted vertically on the aluminum main stage with two magnets are positioned on both sides.

Figure 6. A previous design of magnetic damping using copper plate hanging down from main stage

This design also poses a challenge for damping the rotational mode of vibration because the net change in magnetic flux through copper is minimal. Eventually, to simplify the design, to save space, and to increase magnetic damping effectiveness for all modes of vibration, a copper stage is selected. One of the challenges in using copper material is that it is much harder to machine copper because copper is much softer than aluminum. In addition, copper costs more than aluminum.

2.2 FRAME DESIGN

Since the natural frequencies of the mass-spring system are low and the input frequencies from human activities and the environment are less than 100 Hz[4], the frame for the system needs to be rigid so that its natural frequencies are outside of the working range of the mass-spring system. Figure 7 shows different designs for the frame. The natural frequencies for theses frame will be studied through Finite Element Software to determine the best option.

Figure 7. Frame 1 with vertical circular rods (left), Frame 2 with aluminum L brackets (middle), and Frame 3 with hollow tube and cutout windows (right)

2.3 TESTING SETUP

Before purchasing materials and machining the stage and the frame, a testing structure is set up using plywood material. Using available springs with stiffness of approximately 0.019 N/mm, we can verify the mathematical model by using Random Noise vibration test on a shaker and recording data for different weights of main stage. Figure 8 shows the plywood testing structure with added weights to simulate the target weight to achieve vibration isolation for 1.5 Hz lowest input frequency.

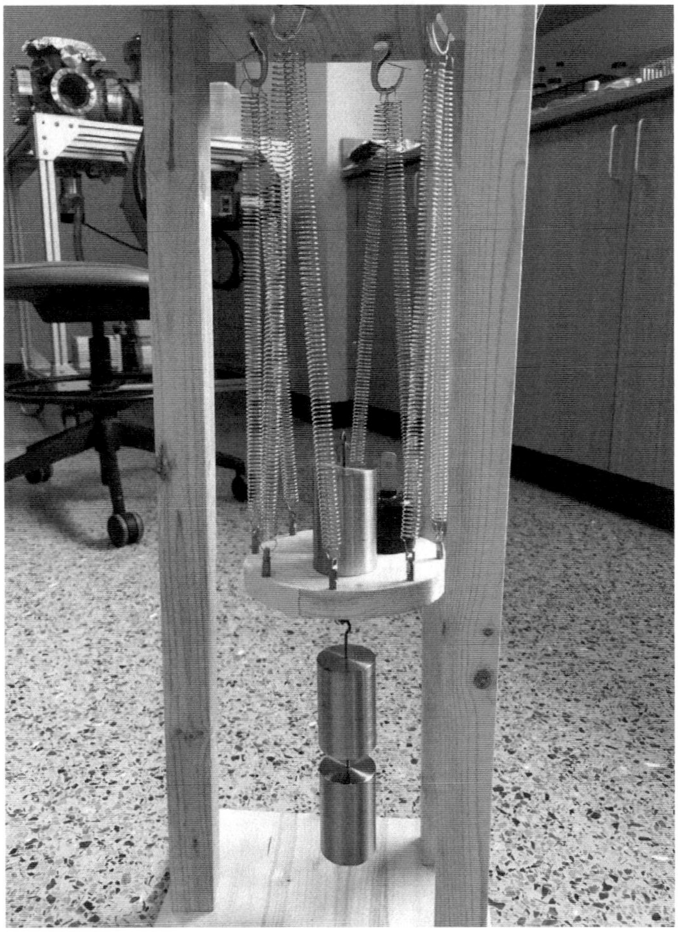

Figure 8. A testing structure made of wood and springs
with stiffness of k = 0.019 N/mm

3. RESULTS

The results are separated in two sections: the experimental data to validate the mass-spring system design and the Finite Element Analysis simulation results of three frame designs.

3.1 FREQUENCY RESPONSE OF TESTING STAGE

To accurately demonstrate the effectiveness of the mass-spring system in attenuating input frequencies from the surroundings, the Random Noise test is truncated to intervals of 0.5 Hz, stretching from 0.5 Hz to 10 Hz. The data are stored in a floppy disk in a Spectrum Analyzer, converted to MATLAB data files, and presented in Figure 9 below.

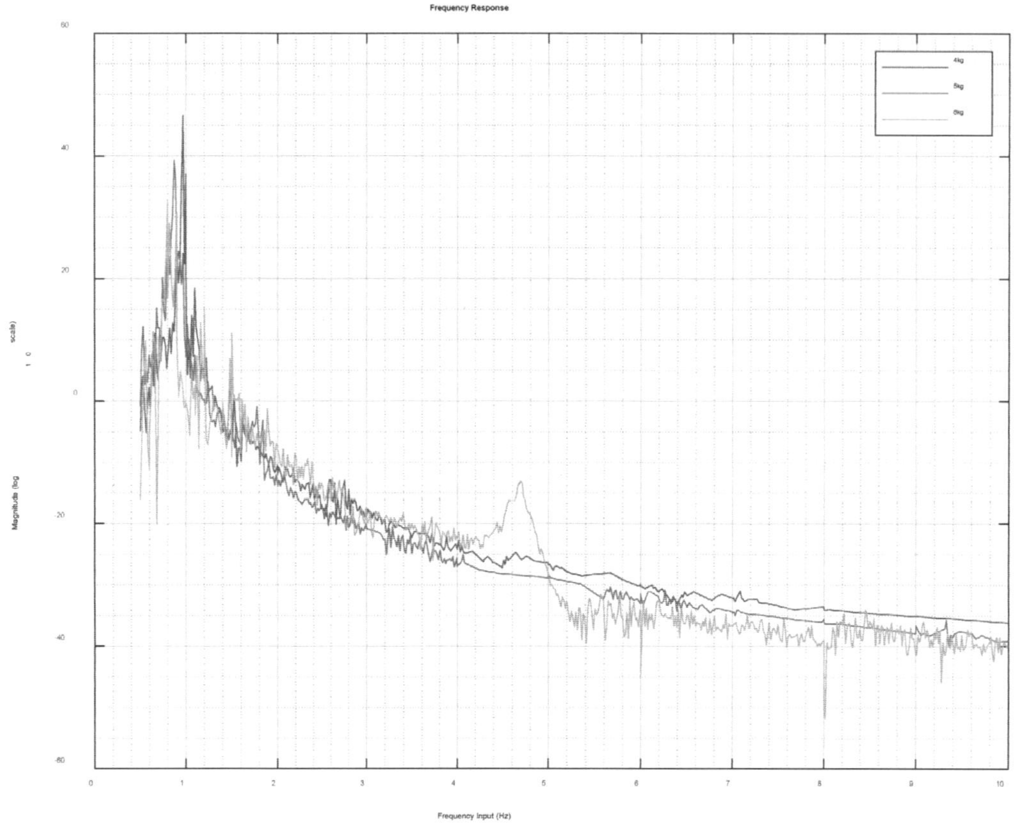

Figure 9. Frequency response of the testing structure with 4 kg, 5 kg, and 6 kg stage

Figure 9 shows that for input frequencies of 1.5 Hz and above, the system transmissibility ratio is less than 0 on a log scale, which indicates vibration isolation is being achieved. The irregularities in testing with 6 kg stage is because the plywood frame is too soft and flimsy to support such heavy stage.

3.2 Natural frequencies of frame iterations

Finite Element software shows the first four mode shape frequencies for each frame design, which are shown in Table 2. Overall, frame 3 with hollow tube design and cutout windows is the most promising design because all the theoretical natural frequencies of the frame are significantly outside of the working range of frequencies of the mass-spring system.

Table 2. Frequency of the first four mode shapes for three frame designs

	Mode 1 (Hz)	Mode 2 (Hz)	Mode 3 (Hz)	Mode 4 (Hz)
Frame 1	74.9	74.9	122.1	345.3
Frame 2	49.6	49.6	98.0	313.1
Frame 3	478.9	650.6	1,135.4	1,244.8

4. CONCLUSIONS

The mathematical model for the mass-spring system is derived and validated. In addition, FEA models for the frames are also studied. With the main stage is made of copper, the magnetic damping accessories are reduced so that the overall design becomes simpler and less bulky. Moreover, the copper stage will facilitate magnetic damping in all modes of vibrations (vertical, rotational, and wobbly) more effectively because the change in magnetic flux in all modes of vibration is maximized. Elastomeric damping will be added as cushions between the vacuum chamber and the hollow tube frame to absorb the vibrations before they enter the mass-spring system. Finally, soft elastomeric materials will be added in joints and connectors within the system to further reduce the amplitude of input vibrations. Final tests will be executed once the final assembly is finished to further study the magnetic damping effect. Additionally, these tests will also serve to confirm whether more elastomeric damping should be used or which locations in the structure the elastomeric damping should be added.

5. REFERENCES

1. Binnig, G., Rohrer, H., Gerber, C. & Weibel, E. Tunneling through a controllable vacuum gap. *Appl. Phys. Lett.* **40**, 178–180 (1982).
2. Jannick, J. Digital Signal Processor Based Control of a Scanning Tunneling Microscope. (1994).
3. Coyne, D. LIGO Vacuum Compatible Materials. *LIGO Sci. Collab.* **Advanced L**, (2011).
4. Chikkamaranahalli, S. B., Vallance, R. R., Damazo, B. N. & Silver, R. M. Damping mechanisms for precision applications in UHV environment. *Proc. - ASPE Spring Top. Meet. Challenges Intersect. Precis. Eng. Vac. Technol. ASPE 2006* **38**, 25–32 (2006).
5. Elghandour, Eltahry and Kolkailah, Faysal" Location effects of passive damping material in cross-ply laminates on natural frequency and mode shape" International Conference on Computational Methods and Experimental Measurements: Naples, IT, 1999.
6. Akers, T, Bach, J, Ward, C, Wait, D, Biezad, D, Elghandour, E, Kolkailah, F"Active control techniques of structural modes using piezoelectric actuators" WIT Transactions on Modelling and Simulation, Volume 7, 2001/5/23.

VOXELIZED SKELETAL MODELING TECHNIQUES VIA COMPLEMENTAL SKELETONS

Tyler J. Williams, Prof. Mark A. Ganter, Prof. Duane W. Storti
University of Washington Mechanical Engineering Department
Seattle, WA/USA

ABSTRACT

Skeletal modeling is a modeling methodology based on finding and manipulating the medial surfaces of a 3D model. One of the key strengths of skeletal modeling is the simplicity of manipulating the thickness of objects, a task that can be difficult with standard 3D modeling techniques. However, due to the nature of medial surfaces, changes made to one section of a skeletal model can alter multiple regions on the surface of the resulting object, making it difficult to perform small edits to the model without adding more branches to the skeleton. This paper develops a novel twist on skeletal modeling: the 'Complemental Skeleton.' Through the usage of a complemental skeleton, localized changes to object thickness can be made to the model by modifying the radial data, allowing the modeler to leave the geometry of the skeleton unaltered. This feature greatly lowers the amount of work required to create asymmetrical edits.

Corresponding author: Tyler Williams (tjwill@uw.edu)

1. INTRODUCTION

There are two main types of skeletons that are used for 3D objects. The curve skeleton, a collection of 1D curves that follow the shape of the object, lends itself well to surface reconstruction, shape classification, and 3D animation. The curve skeleton lacks a well-defined methodology of creation, and, more importantly, is often a one-way transformation; the original object cannot be reconstructed from the curve skeleton alone [1-3].

The alternative to curve skeletons, often referred to as surface skeletons, are collections of 2D sheets, 1D curves, and 0D points that follow the shape of the object. They lend themselves well to surface modification, geometry editing, and thickness modification [4-8]. In contrast to the curve skeleton, the surface skeleton is strictly defined as the collection of the object's medial surfaces. In addition, the skeletal data can be used to completely recreate the original object, allowing for an object to be turned into its surface skeleton, modified in the skeletal state, and then transformed back into a fully 3D model [7-9]. An example of this process can be seen in Figure 1. The radial data associated with the skeleton can be scaled numerically in order to increase or decrease the overall thickness of the object.

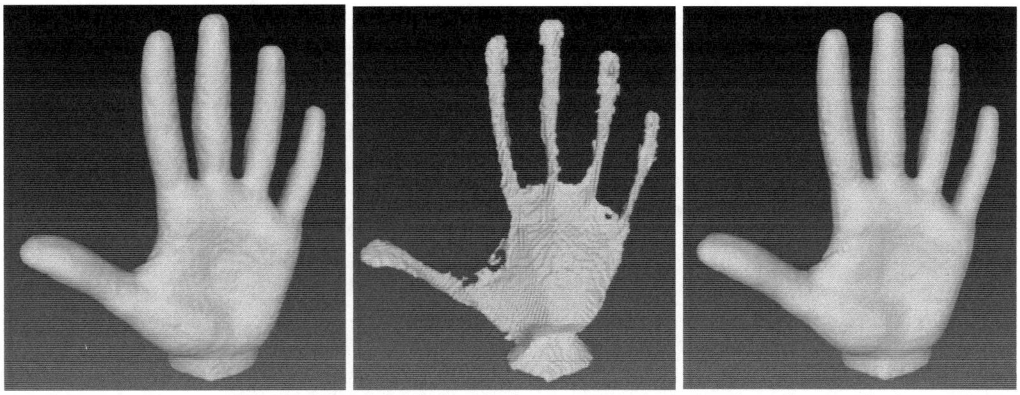

Figure 1. Left to Right: Input Model, Skeleton, Reconstruction from Skeleton [10].

1.1 Definitions

In this section, we define the foundational concepts and keywords used in this paper.

1.1.1 Medial Surface

The medial surface can be defined as the collection of points inside the object that have multiple 'equally closest' points on the surface of the object. For any point on the medial surface, the largest sphere centered on that point that is fully contained within the object will touch at least two points on the surface of the object. We will term these spheres to be 'Maximal Spheres'. For the purposes of this paper, the medial surfaces of an object can be seen as the 'skeleton' of the 3D model. The medial surface itself is a four-dimensional object, as each point on the surface has the three spatial coordinates (x,y,z) and a radial component (r). Recreating the object, termed 'refleshing' in this paper, consists of unioning together skeletal spheres centered on each point along the skeleton with their radii set to match the 'r' coordinate at that point in space [7].

1.1.2 Voxel Model

There are many ways that 3D geometry can be stored and manipulated. Voxel geometry stores 3D geometry based on a discrete grid, where each voxel in the grid is assigned a value that states whether the voxel is inside the object, outside the object, or on the surface of the object. Conventionally, internal voxels have negative values, external voxels have positive values, and voxels that land on the surface of the object are set to zero. The magnitude of the value is irrelevant for the purposes of storing the object itself, allowing for the values to store some additional data about the object, such as color, material properties, or distance information [9, 11]. An example of a relatively low resolution voxel model is shown in Figure 2. As only the sign of the voxel value determines the geometry of the object, a given matrix storing a voxel model is not unique.

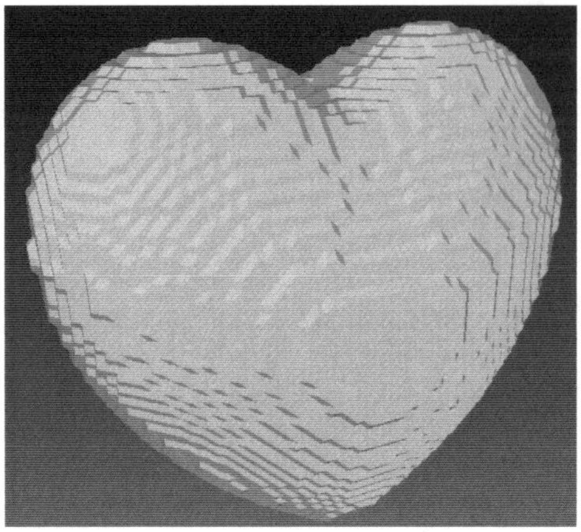

Figure 2. Taubin's Heart Surface in 75^3 Voxelspace, converted to an STL for rendering via the Marching Cubes algorithm. [13]

1.1.3 Signed Distance Field

For our purposes, a signed distance field (SDF) can be used to describe a voxel model. It follows the same sign conventions as standard voxel models, but now the magnitude of each voxel stores the distance from its point in space to the nearest point on the surface of the object [12]. This added restriction enforces uniqueness. Any given voxel model has only one SDF.

1.1.4 Parallel Computing

Normally, a computer functions in a serial manner, doing one task then moving to the next and repeating until it is finished. Parallel computing allows for tasks to be divvied up into independent computations that can be run simultaneously. This practice works particularly well when a single algorithm must be solved for each cell of a large matrix, allowing for all the solutions to be computed in lock-step and cutting down on computation time dramatically [9, 12]. For this paper, PyCUDA, a Python binding for the CUDA parallel computing platform developed by Nvidia, was used as our primary language for implementation.

1.1.5 Skeletal Data

For simplicity and accuracy in naming, the skeleton of a 3D model is going to be divided into two distinct parts. The 'Skeleton Geometry' will refer to the geometric location data. This represents the shape of the skeleton itself. The 'Radial Field' is going to refer to the radii of the maximal spheres as defined by the spatial coordinates. The 'Skeletal Data' refers to the combination of these parts, as both are required to recreate the object.

1.2 Features of Using Skeletal Modeling

Some editing tasks are most easily performed by transforming the model into its skeletal data, making modifications to the radial data, then refleshing the skeletal data to retrieve the desired object. Changing object thickness is often one of the easiest modifications that can be performed via skeletal modeling, as it consists of changing the magnitude of the Radial Field by adding some constant or multiplying any number of values by some constant. An example is visible in Figure 3.

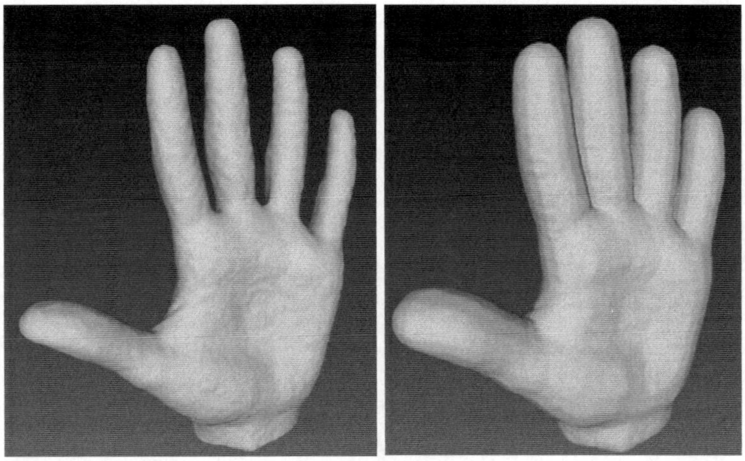

Figure 3. Left: Original input model [10], Right: Recreated after increasing each radial value by a constant.

It is important to note that changes made in this manner can affect multiple regions on the surface of the resulting model, as the Skeleton Geometry remains centered within the object. Given this limitation, using skeletal modeling techniques to make localized edits can be cumbersome. New surfaces, curves, or points would have to be added to the Skeleton Geometry if the desired modifications are intended to only affect specific regions. This is not as simple to perform as modifying the Radial Field.

1.3 Proposed Alternative

This paper introduces the concept of the Complemental Skeleton, a new type of skeletal entity that allows for localized modifications to the surface of the resulting object. The complemental skeleton lies between the original medial surface and the original surface and has its skeletal dataset. When combined with the original skeletal data, it allows for the complete object to be reconstructed. Images from the process are visible in Figure 4.

The key benefit of the complemental skeleton is that modifications applied to its Radial Field will generally only affect the regions of the surface on the corresponding side of the object, leaving the opposite side untouched. This property allows for asymmetric edits to be made without requiring modifications to the Skeletal Geometry.

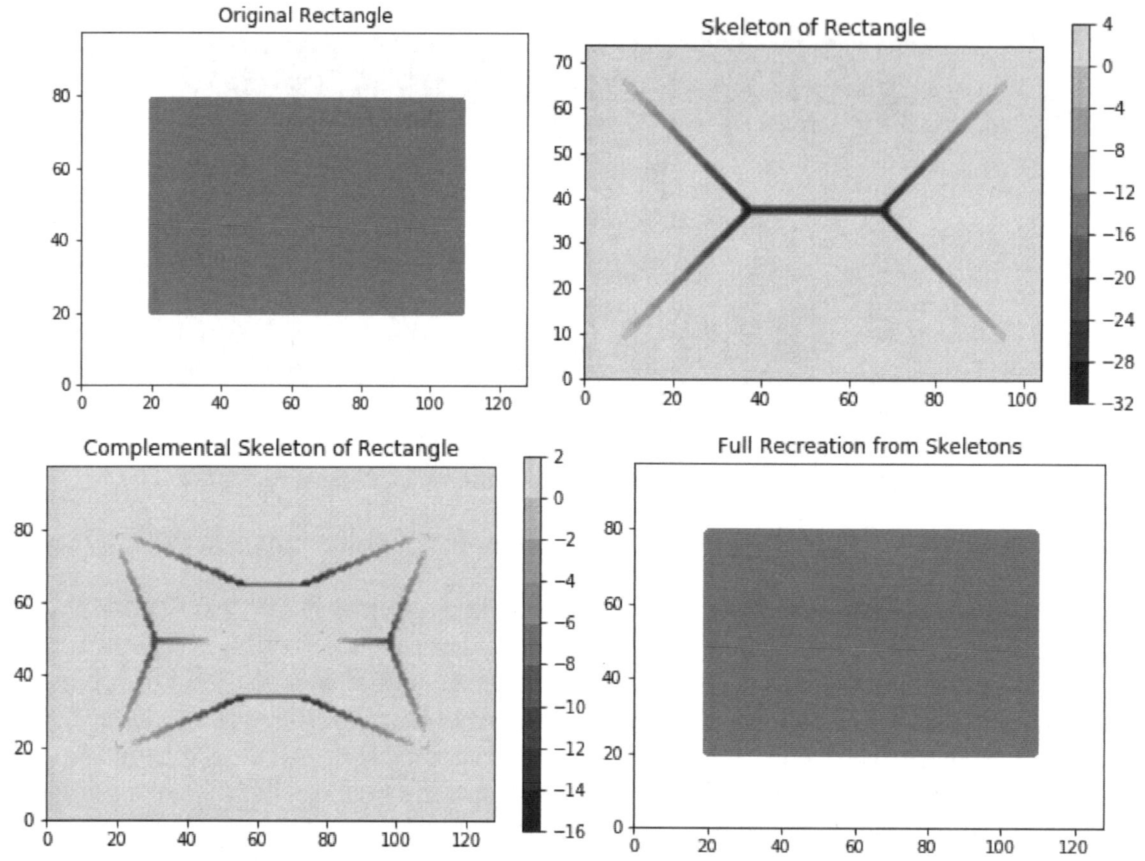

Figure 4. Left to Right, Top to Bottom: Original Object, Original Skeleton, Complemental Skeleton, Object recreated from Complemental Skeleton.

1.4 Motivations

As mentioned earlier, skeletal modeling is particularly useful in modifying the thickness of objects [4-8]. One application where model thickness is incredibly important is 3D printing, as 3D printers often have a minimum model thickness. Using the complemental skeleton in tandem with standard skeletal modeling techniques increases the utility of the methodology and allows for a larger variety of changes to be made to the model without switching from skeletal modeling to a different modeling technique. In short, the complemental skeleton has the potential to smooth the workflow of editing 3D models for 3D printing.

2. METHODS OF IMPLEMENTATION

For brevity in the main body of the paper, pseudo-code for each **bolded** algorithm is available in the appendix. All code used is freely available for download and inspection at https://github.com/tjwill95/complemental_skeletonizer. Additionally, this process also functions for finding the curve skeleton (and complemental curve skeleton) of 2D geometry. For clarity, the 2D geometry is used in the diagrams, but the process does operate on 3D data, as shown in the pseudocode.

2.1 Finding the Skeletal Data

Let's begin with a voxel model. The definition of the medial surface gives us a clue as to how to find it: If we can find the SDF of the object, the medial surface will pass through all points along the 'peaks' of the SDF. The rectangle and the SDF of the rectangle is shown in Figure 5.

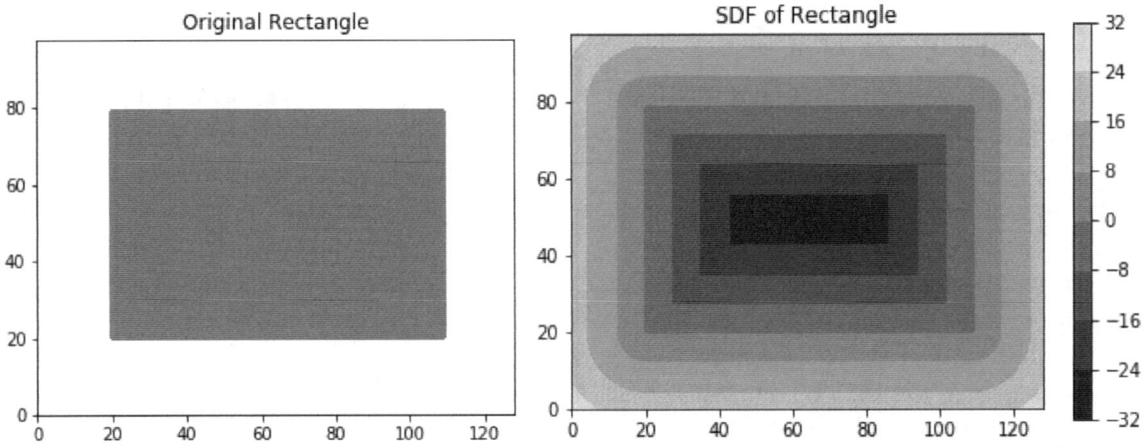

Figure 5. Left: Original Rectangle. Right: Contour Plot of the SDF of the Rectangle.

These 'peaks' can be found mathematically by discovering where the gradient of the SDF does not exist. The gradient of the rectangle's SDF is shown on the left in Figure 6, and it represents the Skeletal Geometry.

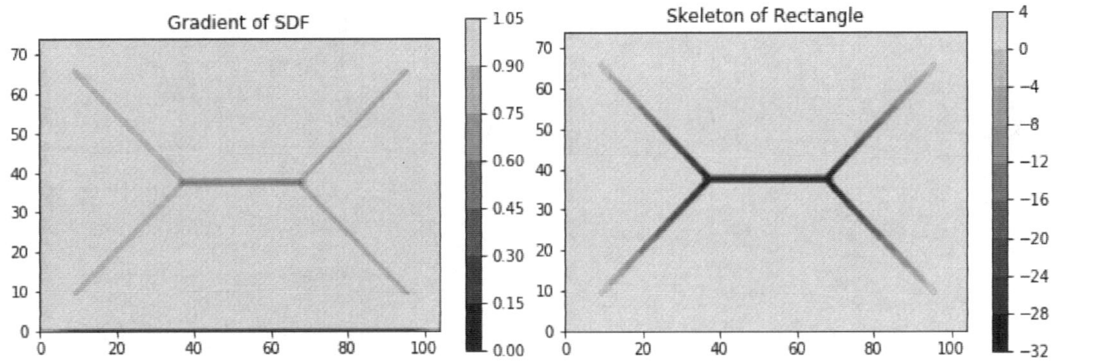

Figure 6. Left: Contour Plot of the magnitude of the gradient of the SDF of a rectangle.
Right: Skeleton of a rectangle.

However, as voxel models are defined in discrete space rather than in continuous space, the gradient function will always return some value, even if inaccurate. To solve this problem, we will instead define the Skeletal Geometry as all points where the magnitude of the gradient of the distance field at that point is less than a threshold value. As the distance field usually has a gradient magnitude of one (moving one voxel away from the surface should increase the distance by one unit), we should set the threshold to be a value less than one. Empirically, the value of 0.85 leads to good results, as this includes the relevant points while removing many of the numerical errors that can arise in more complex models.

If we stop here, we are left with a loss of information, making it impossible to reconstruct the object without the Radial Data. As the magnitudes of the voxels that make up the Skeletal Geometry have no effect on the skeleton's shape, we can store the negative of the Radial Field in the voxels of the Skeletal Geometry, allowing for the original object to be recreated. A plot of the skeletal data is shown on the right of Figure 6.

2.2 Modifying the Skeleton

We now have both the Radial Field and the Skeletal Geometry. The simplest modification we can make is to the Radial Field. Changing the magnitude of the Radial Field at any point is going to result in changes to the thickness of the object based on that portion of the skeleton. Depending on the object, many voxels in the Radial Field may need to be modified in order to achieve the desired effect, especially if the magnitude is being reduced. The internal geometry of a model may be within the maximal spheres of several points on the Skeletal Geometry, so reducing the Radial Field at just one point on the skeleton may have nearly imperceivable effects on the resulting geometry. As can be seen in Figure 7, reducing the magnitude of the entire central section of the skeletal data to zero had a much lower effect on the resulting shape than increasing the magnitude of one point in the Radial Field.

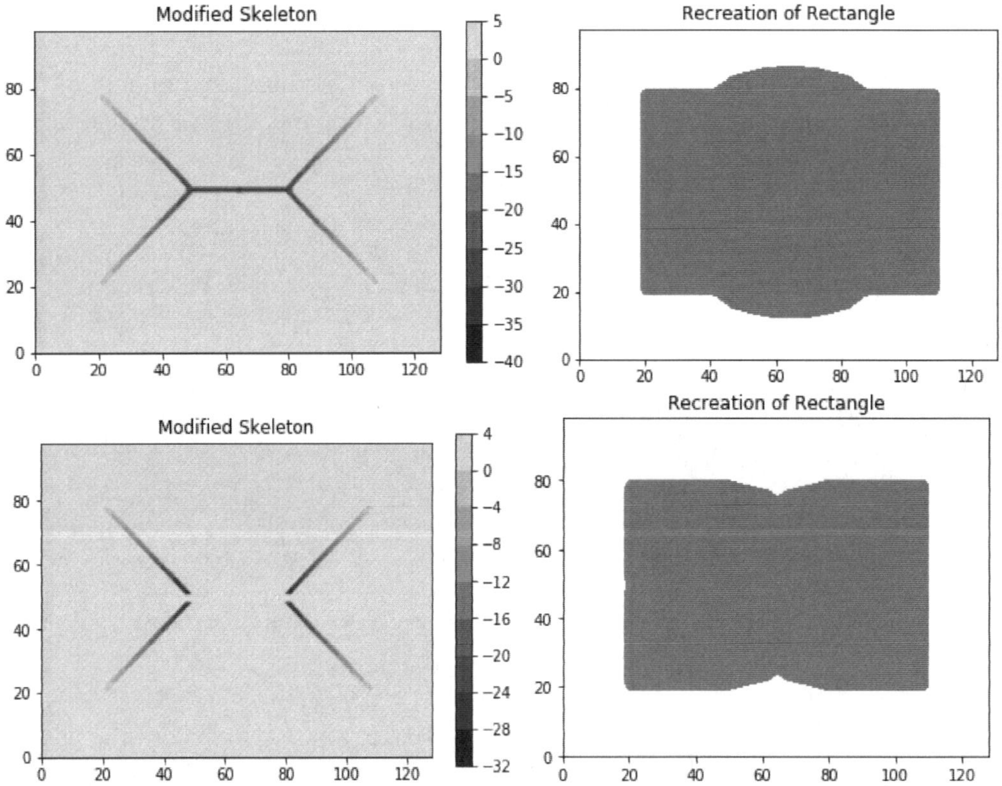

Figure 7. Left: Modified Skeletons. Right: Resulting Models. Top: The central voxels have their magnitudes increased. Bottom: The central row of voxels has been removed.

It should be noted that making any modifications to the Skeletal Geometry or the Radial Field may result in an object that no longer counts as a 'legal' skeleton, i.e. maximal spheres could be drawn that no longer center on the Skeletal Geometry. The true skeletal data of the refleshed object is generally going to be different from the modified skeletal data used to generate it. This is especially apparent when adding voxels to the Skeletal Geometry, as the true medial surface will generally have branches to help define any protrusions, or may bend to accommodate for changes. An example where new voxels have been added to the Skeletal Geometry can be seen in Figure 8.

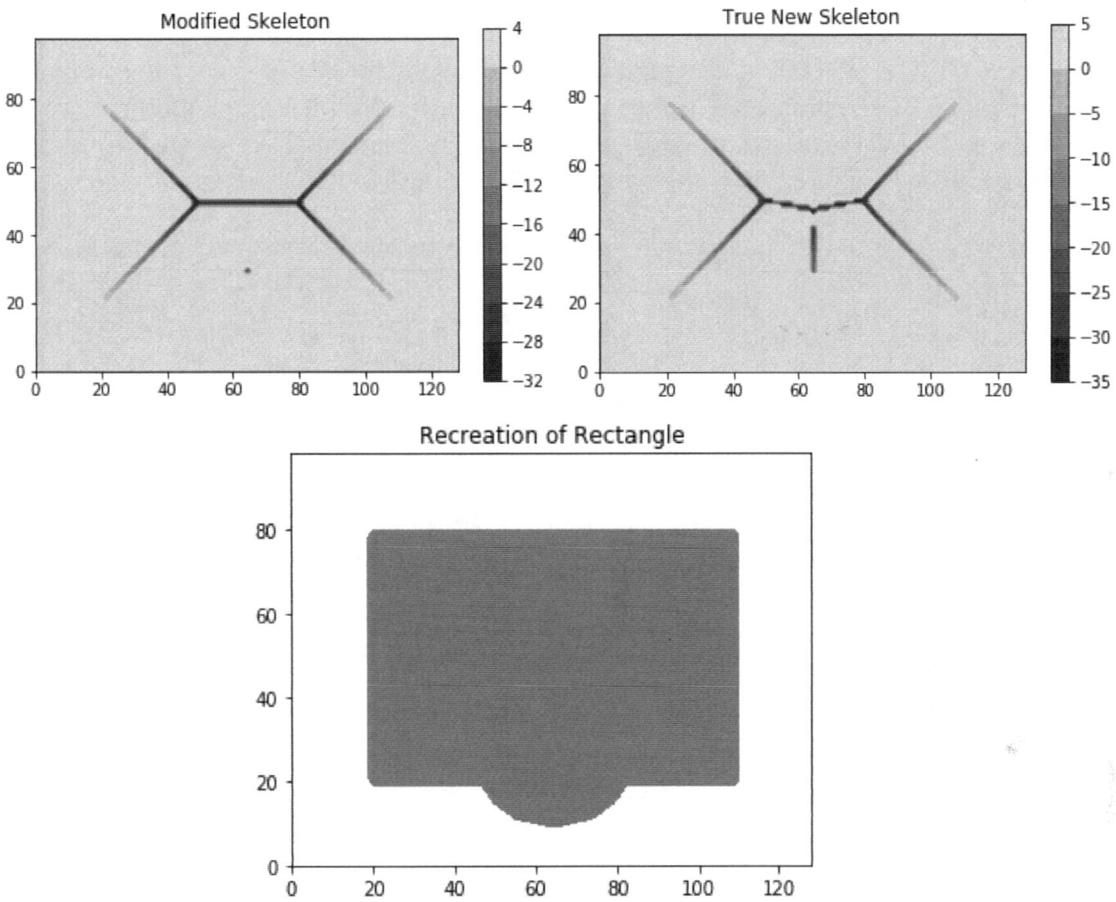

Figure 8. A dot has been manually added to the original skeletal data of the rectangle (top left), resulting in a circular bulge to the bottom of the shape (bottom). The 'true' skeletal data of the updated shape is shown to have more complex changes (top right).

As can be seen in Figure 8, new voxels had to be added to the skeletal data in order to produce the asymmetric modification to the rectangle.

2.3 Recreating the Object

Now that we have our adjusted skeleton, we would like to reconstruct our original object with the changes that would result from the modifications that we've made to the skeletal data. As mentioned earlier, we could generate a sphere centered on each voxel of the medial surface, using the values stored in the voxels to determine their radii, and union them all together. However, this solution is inelegant and computationally expensive. A more elegant solution is to propagate outwards from the medial surface. This process lends itself well to being parallelized and iterated, allowing for each voxel to compute its own intermediary value independently by checking the input states of its neighbors. The resulting matrix is then sent back through the algorithm several more times as the data gets propagated outwards.

2.4 Finding the Complemental Skeleton

Changes to the Radial Field are often going to affect the resulting object in multiple regions. This enforced symmetry is not always the desired behavior, as it makes it difficult to implement asymmetrical changes without modifying the Skeletal Geometry itself, which can be cumbersome for the modeler. The complemental skeleton, as mentioned earlier, acts as an intermediary between the object's skeletal data and the complete object. The complemental skeleton of the afore shown rectangle is visible in Figure 9 in combination with the original skeletal data.

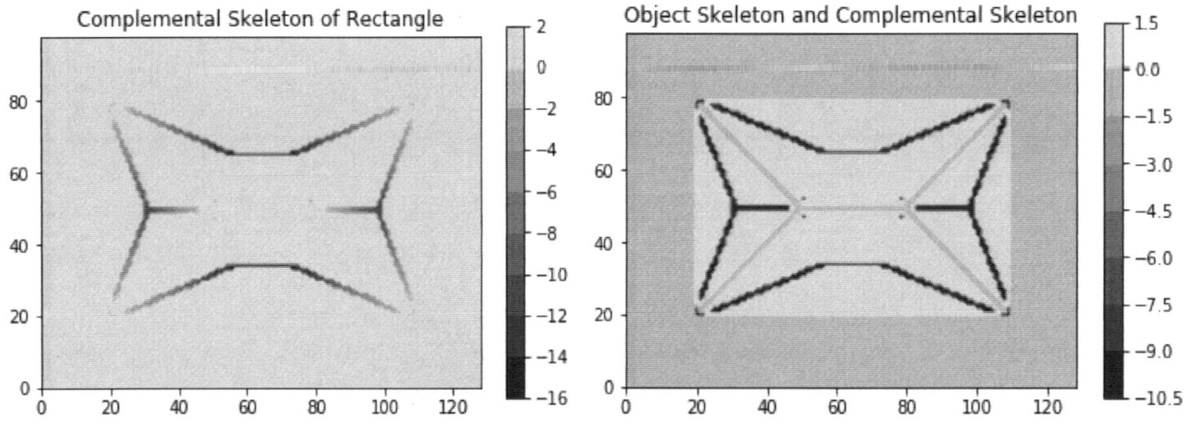

Figure 9. Left: Complemental Skeleton of Rectangle. Right: Complemental Skeletal Geometry of Rectangle (Blue) superimposed over Skeletal Geometry of Rectangle and Inverse of Rectangle (Green)

To find the complemental skeleton, we must take the initial voxelized model, remove every voxel that it shares with its Skeletal Geometry (performing Boolean **subtraction**), then use the resulting geometry as the input for our skeleton-finding algorithm. The output of this process is our complemental skeleton; the skeletal data of the complement of the original skeletal data.

2.5 Modifying the Complemental Skeleton

Modifying the Complemental Skeleton

The complemental skeletal data can be modified in all the same ways as the standard skeletal data. The Radial Field can be modified, and voxels can be added to or removed from the Skeletal Geometry of the complemental skeleton. The key concept to keep in mind is that the Skeletal Geometry of the complemental skeleton now represents opposing regions of the object, so small changes to one section of the complemental skeleton's Radial Field are unlikely to affect the opposite region of the model's surface when recreated. However, this new freedom does have limits. Reducing the magnitude of the Radial Field on the complemental skeleton may now produce voids within the object. Some examples of modifications are shown in Figure 10.

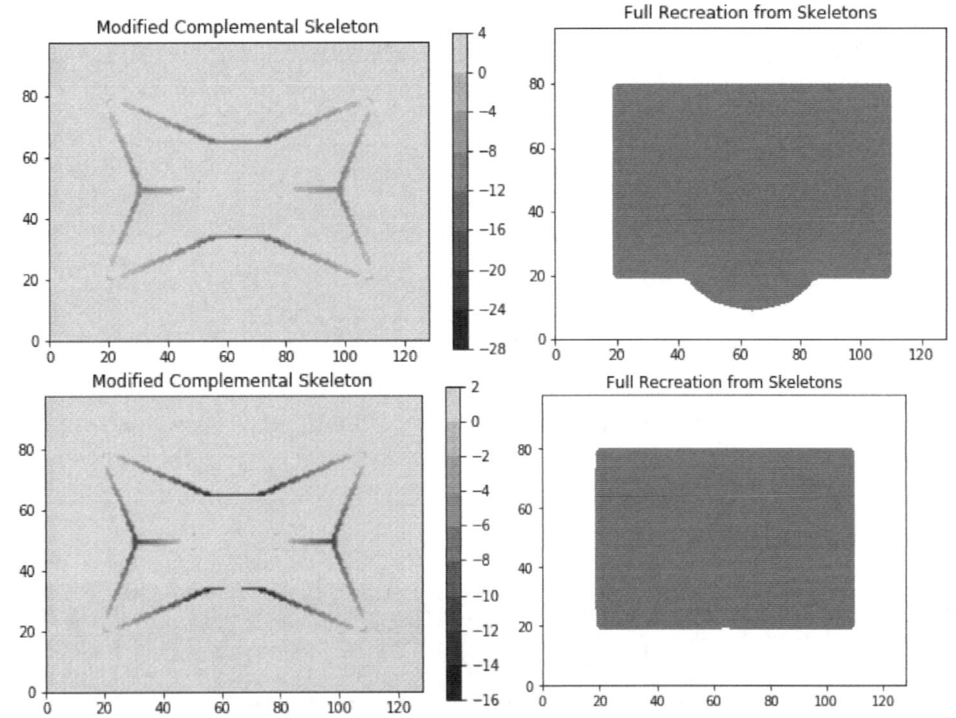

Figure 10. Left: Modified Radial Fields of the complemental skeletons. Right: Reconstructed models. Top: Several voxels on the lower portion have had their magnitudes increased. Bottom: Those voxels have been removed, to little effect.

2.6 Recreating the Object

After modifications have been made to the complemental Radial Field, the complemental skeletal data can then be used as the input for the same refleshing algorithm used for the original skeletal data. In continuous space, this process would result in creating the original However, in our discretized domain, repeating the same process as before leaves a void in the output object of the same shape as the initial Skeletal Geometry. This is seen on the left in Figure 11.

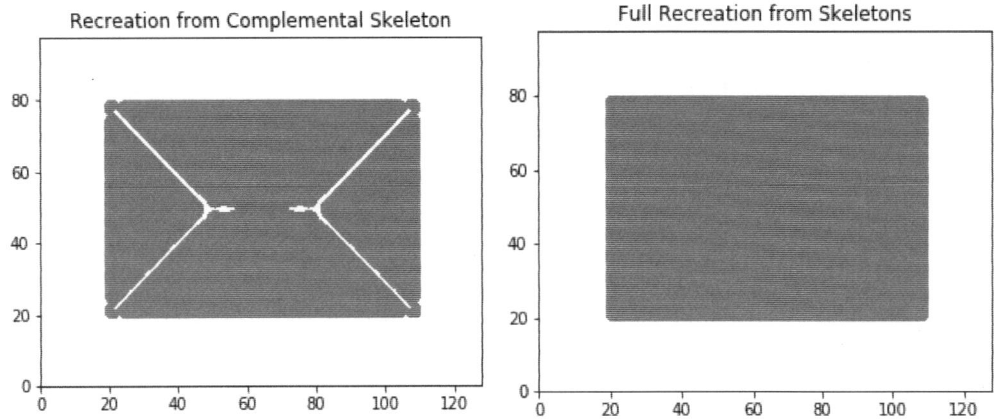

Figure 11. Left: Direct Reflesh of Complemental Skeletal Data. Right: Complete Model Reflesh.

Fortunately, this void can be repaired by simply performing a **union** operation between the original Skeletal Geometry with the geometry that results from using the **refleshing** algorithm on the complemental skeletal Data. The resulting object is seen on the right in Figure 11.

3. DISCUSSION

3.1 Strengths

The process of using the complemental skeleton provides skeletal modelers with the ability to make more changes to the resulting object without modifying the Skeletal Geometry. In particular, it is difficult to make changes to the Radial Field that won't result in changes to multiple regions of the object. Such changes traditionally require modifications to be made to the Skeletal Geometry in order to affect only one side of the object. However, simple changes made to the Radial Field of the complemental skeleton will only affect their corresponding regions, allowing for these asymmetric changes to be made without modifying the Skeletal Geometry. An example is shown in Figure 12.

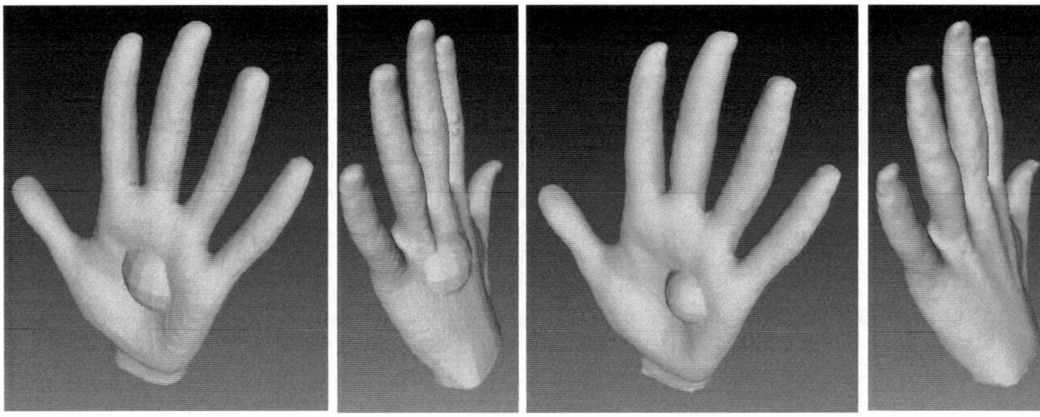

Figure 12. In both sets of images, a modification is made by altering the magnitude of a single voxel in the Radial Field. For the left two, the altered voxel is on the skeleton. For the right two, the altered voxel is on the complemental skeleton.

As seen in Figure 12, the resulting object has an area of higher thickness protruding from the palm, caused by the change to the magnitude of the Radial Field at that point. When the sphere is centered on the Skeletal Geometry, the extents of the sphere protrude from both sides of the object. When the sphere is centered on the complemental Skeletal Geometry, the extents of the sphere only protrude from one side of the object, leaving the other side untouched.

3.2 Current Limitations

The fundamental issue with this methodology is that it is discretized, which means that the Skeletal Geometry is made of voxels, each with a finite size. The true Skeletal Geometry of an object is infinitely thin, which does lead to more accurate results and recreations. The basic concept of the complemental skeleton could be created for continuous space, but that goes beyond the scope of this report.

As mentioned previously, the complemental skeleton works well for increasing local thicknesses, but has difficulties when reducing object thickness. As there is now an internal set of faces that must be accounted for, reducing the thickness can lead to voids within the object. These voids still result in a valid object, but may not be intended. Additionally, the same conditions that apply to standard skeletal model manipulation apply to the complemental skeleton. Modifications made to complemental Skeleton Geometry are unlikely to result in an accurate Complemental Skeleton for that object.

On an algorithmic level, it is important to know of the limitations of the gradient function, as it is key to this method of locating the Skeletal Geometry. It is very sensitive to surface texture, meaning noise or fine details on the surface of the model can lead to extraneous voxels being classified as part of the Skeletal Geometry. A partial solution to this is to smooth the model before applying the gradient function, which reduces the noise and significantly lowers the number of false positives. However, this methodology does not discriminate between noise and intended detail, and can lead to thicker skeletal geometry than is necessary.

Finally, skeletal modeling in discrete space has limitations. In particular, models recreated from voxelized skeletal data tend to have corners rounded off. This is because the true medial surfaces of the model go completely up to any edges or corners, but the extents of these skeletal branches can be difficult to accurately capture in voxel space.

3.3 Future Improvement

The most impactful next step is to improve the algorithm used to generate the Skeletal Geometry. The gradient function works reasonably well, but its limitations require accommodations that are sometimes less than ideal. To minimize noise, the surface should be smoothed, which decreases the accuracy of the SDF. Increasing the grid size will produce thinner skeletons, allowing for a closer approximation of the true medial surfaces. The thinner the Skeletal Geometry, the more accurate the reflesh of the complemental skeletal data can be.

4. CONCLUSIONS

The inclusion of the complemental skeleton allows skeletal modeling to be used for asymmetric edits without necessitating cumbersome modifications to the Skeletal Geometry. To achieve similar results, new voxels would have to be added to the Skeletal Geometry, where the complemental skeleton allows the modeler to simply modify the Radial Field for similar results. This is not a replacement for standard skeletal modeling practices, but instead a new way to use some of the same basic techniques of skeletal modeling to make localized edits to 2D and 3D models. As the complemental skeleton is still a new construct, its full set of uses has yet to be explored.

5. REFERENCES

[1] Cornea, N.D., Silver, D., & Min, P., "Curve-Skeleton Properties, Applications and Algorithms." *IEEE Trans Vis Comput. Graph., 13*(3) (2007): pp.530-48

[2] Huang, H., Wu, S., Cohen-Or, D., Gong, M., Zhang, H., Li, G., & Chen, B., "L_1-Medial Skeleton of Point Cloud." *ACM Trans. Graph., 32*(4) (2013): pp.65-1

[3] Attali, D. & Montanvert, A., "Computing and Simplifying 2D and 3D Continuous Skeletons." *Computer Vision and Image Understanding 67*(3) (1997): pp.261-273

[4] Lee, Y. & Lee, K., "Computing the medial surface of a 3-D boundary representation model." *Advances in Engineering Software 28* (1997): pp.593-605

[5] Dey, T.K. & Zhao, W., "Approximate Medial Axis as a Voronoi Subcomplex." *Proceedings of the seventh ACM symposium on Solid Modeling and Applications* (2002): pp. 356-366
[6] Puig Puig, Anna, "Discrete Medial Axis Transform for Discrete Objects." (1997)
[7] Delamé, T., Roudet, C., & Faudot, D., "From A Medial Surface To A Mesh" *Eurographics Symposium on Geometry Processing 31*(5) (2012): pp.1637-46
[8] Blanding, R., Brooking, C., Ganter, M., & Storti, D., "A Skeletal-Based Solid Editor." *Proceedings of the fifth ACM symposium on Solid modeling and applications* (1999): pp. 141-50
[9] Zhang, Di, *A GPU Accelerated Signed Distance Voxel Modeling System*. 2016. University of Washington Seattle, PhD dissertation. <http://hdl.handle.net/1773/38175>
[10] Artec Space Spider, "Hand" *Artec3D* Santa Clara, CA, Accessed 1/7/2020 <www.artec3d.com/3d-models/hand>.
[11] Ensz, M., Storti, D., & Ganter, M., "Implicit Methods for Geometry Creation." *International Journal of Computational Geometry & Applications 8*(3) (1998): pp.509-36
[12] Rong, G., & Tan, T., "Jump Flooding in GPU with Applications to Voronoi Diagram and Distance Transform." *Symposium on Interactive 3D graphics and Games* (2006): pp.109-16
[13] van der Walt, S., Schönberger, J.L., Nunez-Iglesias, J., Boulogne, F., Warner, J. D., Yager, N., Gouillart, E., Yu, T., & The scikit-image contributors, "scikit-image: Image processing in Python." PeerJ 2:e453 (2014)
[14] Williams, T., Langehennig, S., Ganter, M., & Storti, D., "Using Parallel Computing Techniques to Algorithmically Generate Voronoi Support and Infill Structures for 3D Printed Objects." *Solid Freeform Fabrication Symposium Proceedings 30* (2019): pp.1830-52

6. APPENDICIES

Several algorithms are mentioned in the Methods of Implementation that were not fully described in order to preserve the flow of the section. Here is a description of the process for each of the referenced algorithms in the following format:

Main Function (Inputs: A, B, C; Outputs: X, Y, Z):
PSEUDOCODE

Sub Function (Inputs: A, B, C; Outputs: X, Y, Z):
PSEUDOCODE

We are using Python as a baseline for pseudocode, meaning that the index of an array is dictated by square brackets (u[i,j,k] = element stored at index i,j,k), a colon selects all values within that index, (u[i,:,:] = 2D array), and the # symbol is used to dictate comments.

SDF (Inputs: Voxel Model of Object; Outputs: SDF of Object):
The Signed Distance Field function used for this paper uses the methodology described in [12], as it uses the Jump Flooding technique on a parallel computing platform. Detailed pseudocode can be found in [14], as the same code was used between the projects.

Boolean Operations (Inputs: Voxel Model 1, Voxel Model 2; Outputs: Voxel Model 3):

The Boolean operations (Union and subtraction) used in this paper follow the same methodology as described in [9] and [11]. Detailed pseudocode can be found in [14], as the same code was used between the projects.

Gradient (Inputs: SDF of Object; Outputs: Gradient Magnitude Field):

```
Gradient Magnitude Field = Copy of SDF of Object
Gradient Kernel (SDF of Object, Gradient Magnitude Field)
#Done in parallel, exact setup depends on the language/platform
Return Gradient Magnitude Field
```

Central Finite Difference (P, N, H):

```
#P = value one step forward from our coordinate
#N = value one step backward from our coordinate
#H = step size
return ((P-N)/(2*H))^2 #Standard central difference equation for first derivative
```

Gradient Kernel (Output, Input):

```
i, j, k = thread indices
dx = Central Finite Difference(Input[i-1,j,k], Input[i+1,j,k], 1)
dy = Central Finite Difference(Input[i,j-1,k], Input[i,j+1,k], 1)
dz = Central Finite Difference(Input[i,j,k-1], Input[i,j,k+1], 1)
#1D derivatives
dPxPy = Central Finite Difference(Input[i-1,j-1,k], Input[i+1,j+1,k], 2^(0.5))
dNxPy = Central Finite Difference(Input[i-1,j+1,k], Input[i+1,j-1,k], 2^(0.5))
dPyPz = Central Finite Difference(Input[i,j-1,k-1], Input[i,j+1,k+1], 2^(0.5))
dNyPz = Central Finite Difference(Input[i,j-1,k+1], Input[i,j+1,k-1], 2^(0.5))
dPxPz = Central Finite Difference(Input[i-1,j,k-1], Input[i+1,j,k+1], 2^(0.5))
dNxPz = Central Finite Difference(Input[i-1,j,k+1], Input[i+1,j,k-1], 2^(0.5))
#Diagonal (2D) derivatives
d1 = (dx+dy+dz)
d2 = (dPxPy+dNxPy+dz)
d3 = (dx+dPyPz+dNyPz)
d4 = (dPxPz+dy+dNxPz)
Output[i,j,k] = min(d1,d2,d3,d4)^(0.5)
#Finds the squares of the magnitudes, then compares them to find the minimum and
#takes the square root, storing the output in the corresponding cell of the output matrix.
```

Skeleton (Input: SDF of Model, Grad Magnitude Field, Grad Threshold, SDF Threshold; Output: Skeleton):

```
#Gradient threshold is usually 0.85, as a magnitude below this indicates that the gradient
#function would return an 'invalid' value.
    #SDF Threshold helps get rid of false positives, if the SDF value is greater than the
#Threshold (ie, positive or low magnitude negative), the voxel can be ignored with
#minimal impact on the resulting skeleton. Usually set to around -1 voxel lengths.
Skeleton = Copy of SDF of Model
```

Skeleton Kernel(Skeleton, Grad Magnitude Field, Grad Threshold, SDF Threshold)
#Done in parallel, exact setup depends on the language/platform
Return Skeleton

Skeleton Kernel(Skeleton, Grad Field, Grad Threshold, SDF Threshold):
i, j, k = thread indices
if (skeleton[i,j,k] > SDF Threshold) or (Grad[i,j,k]>Grad Threshold):
Skeleton[i,j,k] = 1
#If the gradient is above the threshold, we are necessarily outside of the skeleton,
#so any positive value would remove us from the object.
#If the skeleton value at that point (currently set to match the SDF value) is above
#the threshold, we are either outside the object, so any positive value maintains
#that, or we are inside with a low magnitude and can be safely ignored.

Reflesh (Input: Skeleton, Iterations; Output: Recreated Object):
Read = Skeleton
Write = Skeleton
For Iterations number of loops:
Reflesh Kernel (Read, Write)
#Done in parallel, exact setup depends on the language/platform
Read, Write = Write, Read
#Ensures that the input value isn't being overwritten while it's still being used,
#then swaps the matrices around to make the loop easy to keep track of.
Recreated Object = Write
Return Recreated Object

Reflesh Kernel (Read, Write):
i, j, k = thread indices
For index from 0 to 26:
checkPos = (i+Mod3(index//9)-1, j+Mod3(index//3)-1, k+Mod3(index)-1)
#Double backslash is integer division, divide then round down to nearest integer.
#That line cycles through all coordinates in a 3x3 grid centered on i,j,k
Updated = Read[checkPos]+
((i-checkPos[0])^2+(j-checkPos[1])^2+(k-checkPos[2])^2)^(1/2)
#Calculates the distance from the current cell to the checked cell, then adds the
#result to the depth value at the checked point in Read.
Write[i,j,k]=min(Write[i,j,k],Updated)

Transverse Impact Simulation of CFRP Panels Using a Continuum Damage Mechanics Model in LS-DYNA

Scott J. Nesbitt, Sahar Abouali, Johannes Reiner*, Anoush Poursartip, Reza Vaziri
Composites Research Network; The University of British Columbia
*School of Engineering; Faculty of Science, Engineering and Built Environment; Deakin University
Vancouver, Canada; *Geelong, Australia

ABSTRACT

Impact damage prediction of non-crimp fabric CFRP panels is an area of interest for both automotive and aeronautical manufacturers. The University of British Columbia has been developing models to predict damage in composite materials since the mid 1990's [1]–[3] and has had the CODAM2 model implemented as MAT219 in the commercial FE program LS-DYNA since 2011.

In this paper, an end-to-end view is given of how to characterize composite materials to provide the inputs necessary for modelling with CODAM2 and the cohesive zone method. This includes an overview of the necessary test methods, post-test calibration procedures of the model, and subsequent application of the model to transverse impact simulations of a non-crimp fabric CFRP plate.

The results from the simulations are compared with experimental data [4], [5] in order to demonstrate the effectiveness of the model and provide insight into where further improvements can be made in modelling technology.

Corresponding author: Reza Vaziri – reza@composites.ubc.ca

1. INTRODUCTION

The second implementation UBC Composite Damage Model CODAM2 is a transversely isotropic continuum damage mechanics model based on a strain softening approach to predict the intra-laminar response of composite materials in finite element (FE) simulations. It is considered an elastic damage model and includes the capability for non-local damage implementation in order to eliminate mesh dependencies (element size and orientation) on the predicted size and direction of damage propagation. When combined with the cohesive zone method to model delamination, it forms a capable modelling tool for transverse impact simulation.

As with all damage models for orthotropic composite materials, several material parameters must be entered to fully describe the material. These include elastic constants, damage initiation values

SAMPE Virtual Conference Proceedings, 2020. Society for the Advancement of Material and Process Engineering – North America.

(strain for CODAM2 and peak stress/traction for cohesive zones), and fracture energy measures (provided as damage saturation strain values in CODAM2 and fracture energy for cohesive zones). By utilizing standard and non-standard test methods and applying the obtained values to an FE model of a transversely impacted composite plate, a method of determining the structural response of the plate and final damage state is developed. This paper provides an end-to-end discussion of how the inputs for this model are determined and implemented for transverse impact simulation. The outcome is then compared to experimental results.

In this paper, the conventions for principal material directions within the CODAM2 model in LS-DYNA are used. This consists of A (in-plane primary axis), B (in-plane transverse axis) and C (out-of-plane transverse axis). These are equivalent to the customary x, y, z or, – for unidirectional lamina – 1 (fiber axis), 2, 3 directions.

2. MATERIAL CHARACTERIZATION & IMPACT TEST DESCRIPTION

The material used in this work is a warp-knitted multiaxial non-crimp fabric (NCF) with a $[0_2/60/-60]_{2S}$ layup and therefore the properties of the laminate are not quasi-isotropic. The stitching was done at the sub-laminate level $[0_2/60/-60]$; therefore, the stitches do not provide any continuous through-thickness reinforcement other than at the sub-laminate level. The samples have an average tow width of 5 mm (as measured from stitch to stitch) and an average thickness of 2.56 mm. The fiber volume fraction is estimated to be between 0.45 and 0.50 based on optical micrographs, and the density was measured as 1523 kg/m^3 with ASTM D792 [6]. Impact samples were cut to be 100 mm x 100 mm in planar size and were impacted with 12.7 mm diameter spherical impactors having varying masses and energies. Details of the experimental work are not provided here, but can be found in [4] and [5].

3. CHARACTERIZATION OF INTRA-LAMINAR PROPERTIES

Standard Tests

3.1 Tensile Testing

Tensile testing is the most common material characterization test. Therefore, it will not be covered in great deal here except to say that obtaining actual in-situ strengths and fracture strains is difficult at best. Challenges of this test have been highlighted by numerous testing experts [7], [8] and often relate to challenges with gripping the sample to induce sufficiently high loads, or issues with elastic constraint in the transverse direction leading to matrix damage and splitting. The test is covered in ASTM D 3039 [9]. For the CODAM2 model the most important parameters are the tensile moduli and tensile damage initiation strains in both in-plane directions, as well as the in-plane Poisson's ratio.

3.2 Compression Testing

While the CODAM2 model assumes that the elastic modulus in tension and compression modes of loading are identical, it does include mode specific damage initiation strains. These values are commonly produced in methods such as the IITRI method [10], and the more convenient combined loading compression method [11]. When tested, it is common that the elastic compressive modulus is lower than that of the elastic tensile modulus.

3.3 Flexural Testing

For models that are constructed using a sub-laminate or laminate method (i.e. the individual lamina are not resolved such that their orthotropic material properties and position in the thickness of the laminate are taken into account) a judgement must be made on using either the in-plane tensile/compressive modulus as measured above, or replacing these with the flexural modulus instead. For monolithic (non-layered) materials, the two properties are identical. Bending stiffness for both in-plane directions of the composite can be measured using 4-point bend tests and is covered in ASTM D7264 [12].

3.4 Intra-laminar Shear Testing

The number of shear property test methods is considerable and has been summarized succinctly in a column in CompositesWorld Magazine [13]; however, for intra-laminar shear properties the most common methods utilized include the Iosipescu (V-Notched Beam) method [14] and the V-Notched Rail Shear method [15], both of which have been compared and contrasted in another article [16].

The shear modulus is straightforward to attain from the strain gauges applied to the specimens and is the only shear property necessary for CODAM2. Enhancements to the CODAM2 model implemented as user subroutines – but not yet implemented directly in LS-DYNA – include the ability to model non-linear shear.

Non-Standard Tests

3.5 Intra-laminar Fracture Toughness

When composites fail in-plane, the total dissipated energy is higher than that represented by the area under the stress-strain curve of the linear loading region. Even for materials that appear to fail suddenly in a brittle fashion, there is energy that is dissipated post-peak in the creation of fracture surfaces, friction as material slides past each other, and to some extent plastic deformation in the polymer matrix. All these interacting failure modes serve to delay catastrophic failure by changing the notch-tip stress concentration within a fracture process zone (analogous to the plastic zone in metals) and redistributing stress to other regions. As elements in a simulation fail, it is important that a physically realistic amount of energy is absorbed. This is typically accomplished with a strain softening approach, where the stress-strain response after damage initiation results in a reduced elastic modulus for increasing strain as shown in Figure 1. The calibration of this strain-softening model is done through the characterization of intra-laminar fracture toughness which involves the use of tests like those found in fracture mechanics of metals. In the case of CODAM2 the actual model input is known as a damage saturation strain and is the strain value at which the material is fully damaged (secant modulus is zero). For CODAM2 in its local damage form, this can be readily calculated according to the formula:

$$\varepsilon_s = \frac{2g_c}{E\varepsilon_i} = \frac{2G_c}{EL_c\varepsilon_i} \qquad [1]$$

where the variables are as shown in Figure 1 and described below:
G_c = fracture toughness of the material (energy per unit area)

g_c = fracture energy density
L_c = characteristic length of the element
E = initial (undamaged) elastic modulus
ε_i = damage initiation strain

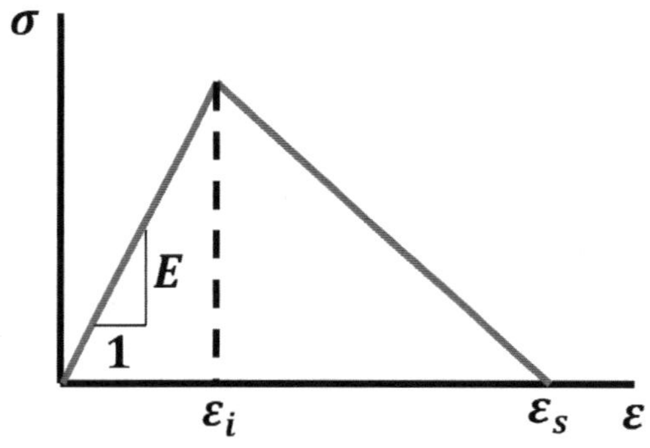

Figure 1. Bilinear stress-strain curve used in CODAM2

The characteristic length of the element is taken as the in-plane side length of a hex element and is used to convert the fracture toughness of the material (energy per area) into a fracture energy density (energy per volume) associated with an element of a given size. This is consistent with the crack band theory of Bazant [17] and ensures that the appropriate amount of energy is absorbed during failure of a given element size. Therefore, the fracture energy properties are a function of element size for the local method and make this method more suitable for implementation in models with elements of a constant size. The non-local method available in CODAM2 is not affected by this size dependency.

3.5.1 Intra-laminar Tensile Fracture Toughness

In the study of the fracture mechanics of metals, the use of compact tension specimens is common. Scaled up compact tension samples were made for plastic in ASTM D5014 [18]. Unfortunately, when scaled versions of this sample are applied to highly orthotropic composite samples, the response of the sample may not be as desired. Due to the orthotropic nature of the samples, damage can be initiated elsewhere outside of the intended fracture process zone. For this reason the over height compact tension specimen shown in Figure 2 and 3 was developed by Kongshavn and Poursartip [19]. The extended height increases the maximum load that the arms of the sample can sustain without damage. The oversized sample also ensures that there is sufficient linearly elastic material surrounding the propagating damage to ensure that linear elastic fracture mechanics is valid.

These samples are loaded quasi-statically under displacement control and may be stopped intermittently throughout the damage process to allow for ultrasonic c-scans to be made in order to characterize the size of the damage process zone during damage development. The height of the process zone is an important parameter for the calibration of non-local damage methods. It ensures that an accurate quantity of material around the simulated crack tip undergoes damage and provides a damage pattern that is not influenced by mesh orientation.

To determine the fracture toughness in both primary planar axes, tests must be conducted in two perpendicular directions to fully characterize the orthotropic specimen. In the case of the NCF material with a $[0_2/60/-60]_{2S}$ layup this corresponds to loading in the 0° direction (Case A) and loading in the 90° direction (Case B). The experimental results are shown in Figure 4.

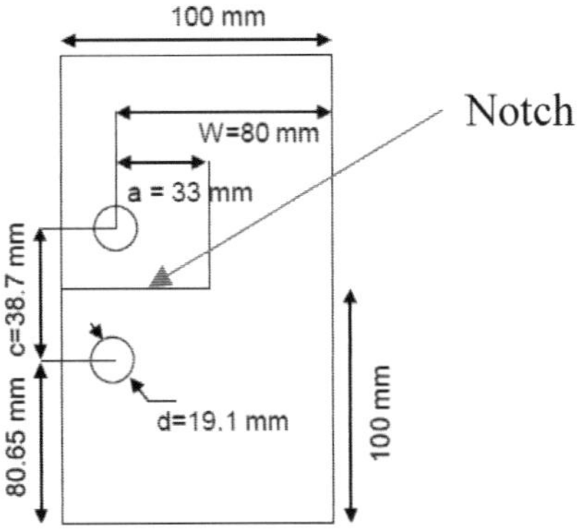

Figure 2. OCT specimen dimensions; the height is twice that of the width to ensure that damage initiates at the notch tip prior to the "arm" region

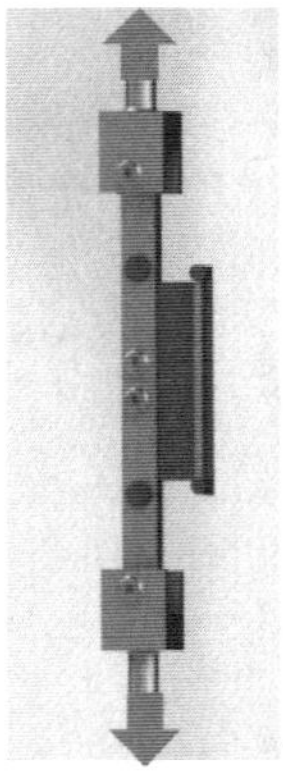

Figure 3. OCT specimen fixture configuration and loading direction; the metal bars clamped to the right side of the specimen limits out of plane bending

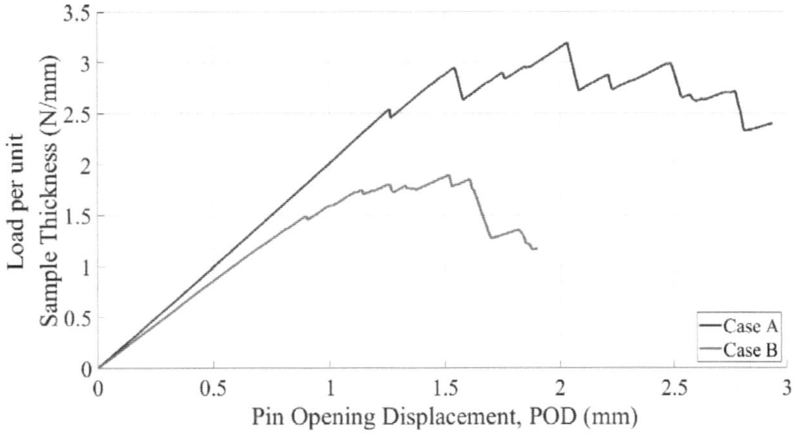

Figure 4. Load versus pin opening displacement of two OCT specimens showing the initial linear elastic response of the coupon followed by the progression of intra-laminar damage; Intra-laminar damage models should be calibrated in order to re-create similar load vs POD responses in a virtual test

To make this information useful for numerical analysis, an identical virtual test is often performed in order to calibrate the model to fit the observed experimental load-POD curve (inverse method); however, there are other methods involving non-contact strain field measurement (digital image correlation) in order to directly calculate the fracture energy values [20]. Methods such as these are only valid when the fracture process zone is visible from the exterior and therefore samples exhibiting substantial internal splitting and delamination must utilize the model calibration (inverse method). When the inverse method is used, the damage saturation strain value is found directly.

When modelling these virtual tests, care must be taken to ensure that the failure modes seen in the experiments are reproducible in the simulation. For example, Case B in Figure 4 exhibits damage due to self-similar crack growth largely as a result of matrix cracking and fiber breakage in the ±60° plies. This is seen in the ultrasonic inspection image of Figure 6 where the fracture process zone extends steadily towards the right (along the plane of the notch). In this case, the intra-laminar fracture of matrix and fiber is the dominant source of energy dissipation. With Case A the damage mode is not the same. In Figure 5 it is seen that there exists a large delamination zone extending perpendicular to the notch. This delamination is initiated by splitting that occurs in the matrix of the 0° lamina. Therefore, it is important that the model used to represent this material be capable of reproducing the interaction between matrix cracks and delamination initiation as an uncoupled model will not predict this delamination.

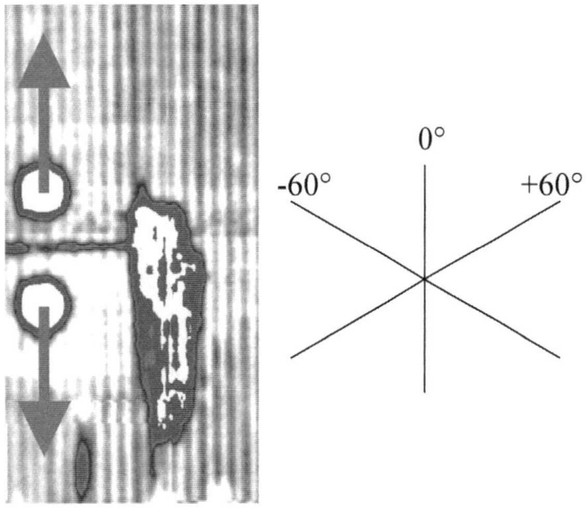

Figure 5. Ultrasonic inspection of Case A specimen; large region of splitting and non-self-similar matrix crack-initiated delamination is observed

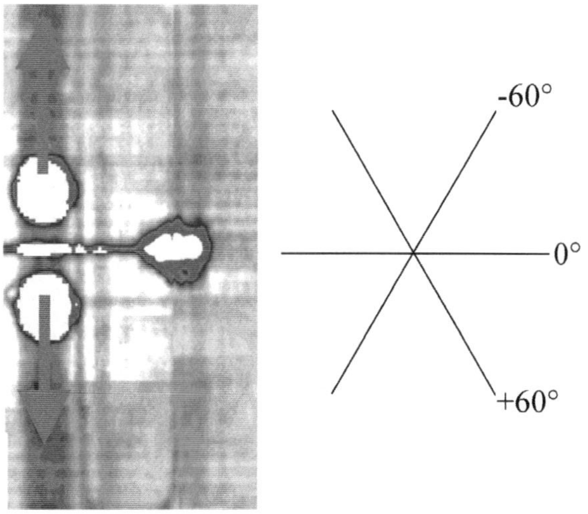

Figure 6. Ultrasonic Inspection of Case B specimen; self-similar crack growth consisting of matrix cracking and fiber breakage of the ±60° fibers is observed with minimal delamination

An example comparison of a virtual test with the experimental data is shown in Figure 7 below. The parameters of the virtual test (saturation strain in the fiber direction – and by extension, fracture toughness) are adjusted to achieve the best fit of data. The over-estimation of peak load in Figure 7 can be attributed to the lack of non-linear shear capability and the lack of interaction between matrix cracking and delamination initiation in this version of the model.

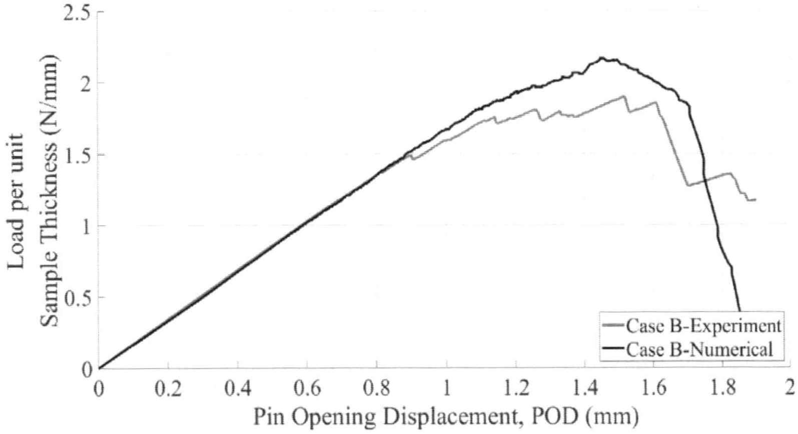

Figure 7. Load versus pin opening displacement of the Case B OCT specimen comparing experimental results with the numerical virtual test. The virtual test has been adjusted to best match the experimental response, thus calibrating the intra-laminar damage model

3.5.2 Intra-laminar Compression Fracture Toughness

Owing to the wide and varied damage modes present in composite materials, the intra-laminar fracture toughness of a laminate in compression is not the same as it is in tension. To characterize the intra-laminar fracture toughness in compression, a compact compression test is utilized. Under compression, it is known that kink band formation and widening is the primary damage mechanism for unidirectional composites, and in the case of a multidirectional laminate the effects of delamination and matrix cracking are also present. It is these different damage modes that result in the difference of fracture energies between tension and compression.

Due to kink band widening in the latter stages of compressive damage, the stress-strain response is known to contain a plateau region [21] as shown in Figure 8. CODAM2 does have the ability to account for this using the RESIDS parameter, However, in the current modelling of transverse impact we assume the simple bi-linear stress-strain response of Figure 1 in both tension and compression. A more independent material response has been implemented through user subroutines [22] and any future direct implementations of CODAM2 in LS-DYNA may feature this ability as well. If a plateau region is used in the stress-strain response, the non-contact method of fracture energy determination [20] provides a convenient method to optimize the shape of this response, but with additional effort the inverse method of model calibration can also work well.

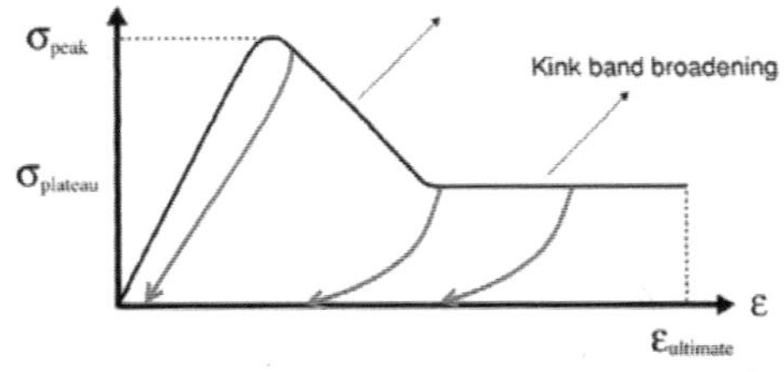

Figure 8. Compression-strain curve; kink band widening results in the plateau of stress at higher strains, giving a different strain softening response than that of tension shown in Figure 1

4. CHARACTERIZATION OF INTER-LAMINAR PROPERTIES

The inclusion of inter-laminar damage (commonly called delamination) is of critical importance for determining the structural response of a laminated composite, particularly for those that are transversely loaded and undergo substantial flexure. The orthotropic nature of composite lamina results in a mismatch of bending stiffness coefficients giving rise to substantial inter-laminar shear stress at the interfaces between layers. Delaminations in laminated composites can often extend much further beyond the typical intra-laminar damage region, particularly under high-velocity impact conditions [23]–[25].

4.1 Inter-laminar Tensile Testing (Mode I)

While CODAM2 only requires the transverse Poisson's ratio ν_{CB} – which may often reasonably be estimated from handbook values – the cohesive zone method does require a good estimate of the inter-laminar tensile strength to determine when delamination initiates in Mode I crack opening. In the tiebreak contact of LS-DYNA, this parameter is known as NFLS (normal failure stress) and a method to retrieve this value is covered in ASTM D7291 [26]. Through-thickness tensile modulus is not necessary for CODAM2 as it assumes transverse isotropy (which is applicable to unidirectionally reinforced composites) and this modulus is the same as E_B.

It should be noted that due to the finite size of an element being generally larger than the length scales at which fracture initiates, the interlaminar tensile strength from ASTM D7291 provides an upper limit of what should be used in the model. As element size increases, strength should be reduced to compensate for the inability to resolve localized high stress that causes initiation. Therefore, as an alternative to tests per ASTM D7291, a model with a certain mesh size can be calibrated using a virtual test of the end-notch flexure test covered in section 4.3.2 below. The inter-laminar strength is then adjusted until crack initiation occurs at the same load as the experiment.

4.2 Inter-laminar Shear Testing (Mode II)

While the V-notched rail shear or short beam shear test methods are applicable to determining inter-laminar shear properties (and preferable for modulus determination), the manufacture of their samples requires the production of at least 19 mm thick laminates for the V-Notched Beam method and 56 mm for the V-Notched Rail Shear method. At best this consumes a substantial amount of material, and at worst the extra thickness during processing results in a material that is not representative of the thinner laminate that may be used in the final product. For these reasons the short beam shear method described in ASTM D2344 [27] is commonly used. The method can characterize the inter-laminar shear strength required by the tiebreak contact of LS-DYNA (SFLS – shear failure strength); and the inter-laminar shear modulus is not required. The same comment on strength and element size as in the paragraph above also applies here, and calibration via a virtual test of the double-cantilever beam test may be used instead.

4.3 Inter-laminar Fracture Toughness

Just as an accurate value for intra-laminar fracture toughness was critical for determining the quantity of energy absorbed by the failing material under damage, so too is the inter-laminar fracture toughness for the interface. The two tests used for determining these material properties are described further below.

4.3.1 Mode I Inter-laminar Fracture Toughness

Mode I fracture is the smaller of the two most common modes (both in terms of fracture toughness value and in the amount of damage energy absorbed by this mode during an impact). This is generally due to the fact that impact produces through-thickness compressive stress waves which only become tensile once reflected from a free boundary. If these tensile stresses do cause delamination it is usually localized near the point of impact in thin samples because out of plane tensile stress cannot be sustained further away from the impact site owing to stress equilibrium. The exception being if substantial splitting occurs, thus making a layer more prone to delamination. Mode I fracture toughness for composite materials is found using the double cantilever beam test as outlined in ASTM D5528 [28]. A fully automated measurement of the Mode I fracture toughness has been recently developed in a modified double cantilever beam test using digital image correlation [29].

4.3.2 Mode II Inter-laminar Fracture Toughness

Mode II fracture is the dominant mode of delamination in composite panels under impact load. This is largely due to the orthotropic bending stiffness mismatch between adjacent layers, which under flexure gives rise to inter-laminar shear stresses [30]. Mode II fracture toughness for composite materials is found using the double cantilever beam test as outlined in ASTM D7905 [31].

5. NUMERICAL MODEL

Owing to the complexity of modelling composite materials (orthotropic nature, numerous interconnected failure modes, etc.) several methods have been proposed and developed to account for the nature of composite materials. Summaries of many of these methods can be found in various sources such as Comprehensive Composite Materials II [32], Numerical Modelling of Failure in Advanced Composites [33] and of course the World Wide Failure Exercises [34]. As discussed above, this paper focuses on the use of CODAM2 for the simulation of intra-laminar damage and the cohesive zone method for inter-laminar damage.

The model geometry is shown in Figure 9 and 10. It consists of a 100 mm x 100 mm x 2.56 mm composite part with several layers. This part is modelled with 1 mm square solid, constant stress (ELFORM 1) elements. Solid elements were chosen over shell elements to include through-thickness elastic displacements locally in the region of the impactor. This part is sitting on a support frame modelled as a rigid body with a 75 mm x 75 mm square cutout. The impactor is a 12.7 mm diameter hemisphere that is also modelled as a rigid body. Nodal constraints in the out-of-plane direction are applied in four locations where the sample is clamped during the experiments.

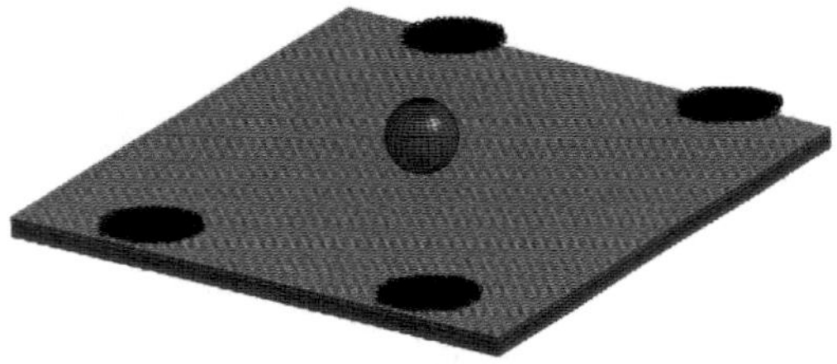

Figure 9. Overview of entire LS-DYNA Model domain; Impactor is the brown sphere; the black circular regions are nodal constraints to mimic the effect of the specimen clamps

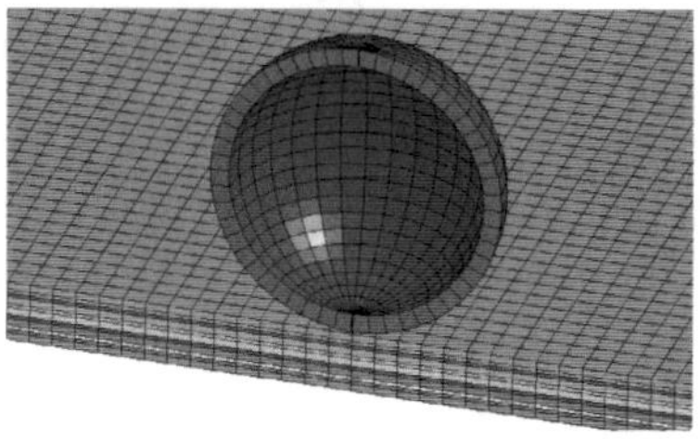

Figure 10. Closeup cross-section view near the impact site; the 11 layers of composite can be clearly seen; the spherical indenter is modelled as a rigid body and therefore the hollow geometry does not affect the results

5.1 Intra-laminar Model

The UBC CODAM2 model is available as a native material model (MAT219) in LS-DYNA. The CODAM model was originally developed by Williams, et. al. [1] and CODAM2 was developed later by Forghani, et. al. [35] to enable non-local methods of damage development.

For this work, the local version of MAT219 is used to predict the structural response and subsequent impact damage in transversely impacted NCF plates. It was found that for these transverse impacts on NCF material it was important to model the structure at the lamina level in order to account for the absorption of energy by inter-laminar delamination. This delamination is due to the high inter-laminar shear stress generated as a result of the flexure of the panels and their mismatch of bending stiffness between adjacent layers as discussed by Liu [30].

After carrying out material characterization tests on the NCF laminate, the elastic parameters for the individual lamina were back-calculated using classical laminate plate theory such that the re-produced laminate as a whole had the elastic properties measured in the material characterization tests. In-plane damage initiation strains are taken directly from laminate tensile test failure strains and in-plane saturation strains (fracture energies) are based on the inverse method described in section 3.5.1. The parameters used for the numerical model are shown in Table 1 below.

Table 1. Intra-laminar material parameters for LS-DYNA MAT219 impact model

ρ (g/mm^3)	E_A (MPa)	E_B (MPa)	ν_{BA} (-)	ν_{CB} (-)	G_{AB} (MPa)	Initiation Strain (-)				Fracture Energy (kJ/m^2)	
						Matrix		Fiber		Matrix	Fiber
						Tens.	Comp.	Tens.	Comp.		
0.001523	116000	14000	0.03	0.5	3000	0.0046	0.0150	0.0157	0.0160	3.5	95

5.2 Inter-laminar Model

Automatic One-Way Surface-to-Surface Tiebreak contacts are used to model the interface between adjacent layers of the laminate. Option 9 is selected, which corresponds to a model that is based on the MAT138 cohesive element and accounts for mixed mode delamination. This model requires the specification of the peak normal traction (inter-laminar tensile strength) and peak shear traction (inter-laminar shear strength) to determine the onset of damage coupled with the inter-laminar fracture toughness for Mode I and Mode II (G_{IC} & G_{IIC}). Like the strain-softening approach used in the intra-laminar model of CODAM2, the stiffness between nodes on either side of the interface is reduced as their separation is increased in a way that ensures the energy dissipation per unit area of interface is consistent with the fracture energy of the mode (mostly mode II) causing the delamination. The parameters used for the numerical model are shown in Table 2 below.

Table 2. Inter-laminar material parameters for LS-DYNA MAT219 impact model

Inter-laminar Tensile Strength (MPa)	**Inter-laminar Shear Strength (MPa)**	**Mode I Fracture Energy (kJ/m^2)**	**Mode II Fracture Energy (kJ/m^2)**
60.0	63.6	0.49	1.80

6. RESULTS

The model was tested with three different impactor masses at various impact velocities, resulting in numerous impact energies. The following results show an example for a drop eight impact test with a mass of 2818 g and an impact velocity of 2.70 m/s, giving an impact energy of 10.3 J or 4.0 J per millimeter of sample thickness. The predicted force versus displacement response is shown in Figure 11 in comparison with the experimental impact and quasistatic loading results. It can be seen that this impact is representative of a low-velocity impact by virtue of the response closely following that of the quasistatic test. Furthermore, the model is capable of accurately predicting the loading response of the sample, but is seen to give a slightly higher peak load (4750 N vs. 4600 N) and slightly lower peak displacement (4.43 mm vs. 4.69 mm) indicating some lack of energy absorbance.

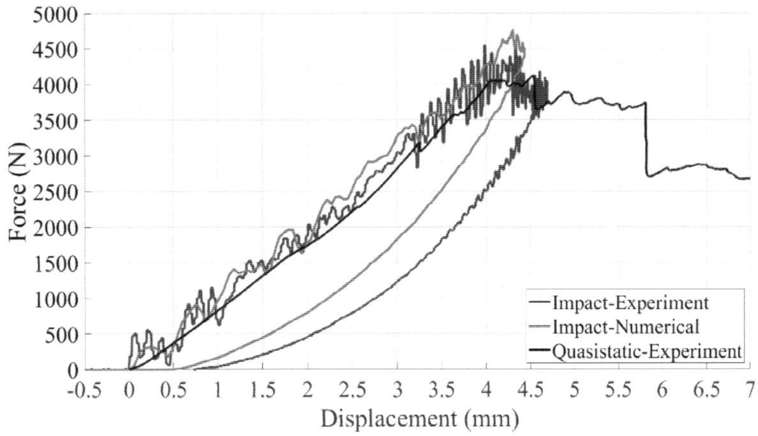

Figure 11. Comparison of force versus displacement curves between the numerical impact simulation and experimental drop tower impact of 10.3 J with an impact velocity of 2.70 m/s; results agree closely and almost overlap the curve corresponding to the quasistatic experiment

This discrepancy can be attributed to the inability of the model to accurately capture the splitting and tow delamination phenomenon observed in the experiment as shown in Figure 12 below. In this figure it can be seen that two tows have nearly completely delaminated from the coupon. It is believed that the effect of the warp direction stitches on the surface of the laminate contribute towards this failure mode as they interrupt the load path between adjacent tows, making splitting more likely to occur. It can be seen in Figure 12c that the model does not predict this region of damage, and therefore the peak load is higher than expected and the maximum deformation of the plate slightly lower.

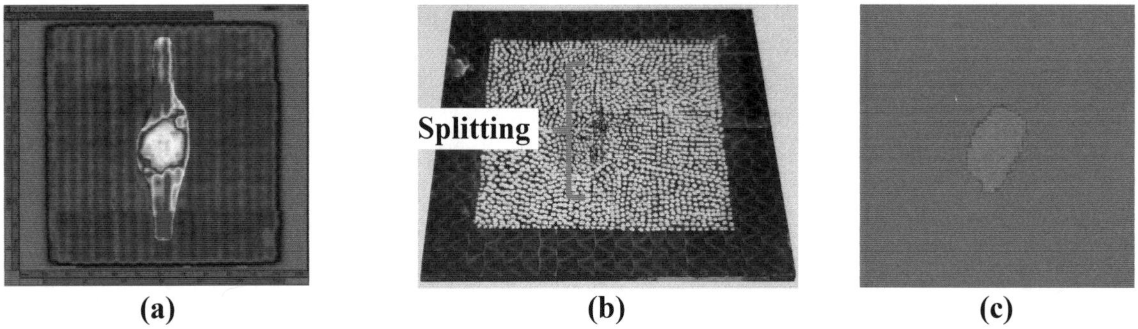

Figure 12. Delamination damage observed in experiments (a) and (b), and in the numerical model (c); the numerical model correctly predicts delamination below the impact site, but does not account for mesoscopic delamination associated with substantial tow delamination delineated by the NCF stitching

7. CONCLUSIONS

An end-to-end view of the material characterization tests and calibration procedures for orthotropic composite damage models has been described with specific emphasis on characterization for use in the UBC CODAM2 damage model. This included volumetric tests (density), intra-laminar tests and inter-laminar tests to fully characterize the material. These material values were then applied to a numerical model utilizing the LS-DYNA built-in MAT219 representation of the CODAM2 intra-laminar damage model capable of modelling a transversely isotropic elastic material. Inter-laminar damage was modelled with the use of the tiebreak contact in a similar fashion to that of the MAT138 mixed mode cohesive zone element. The structural response obtained from impact simulations were found to be very close to that measured from impact experiments, but slightly too stiff, possibly owing to the inability of the model to accurately capture the matrix splitting effect seen in the experiment. The model shows good promise for predicting damage in transverse impact cases, but additional refinement to include the effect of splitting in surface layers (aided by NCF stitching) would provide additional improvements to the model.

8. REFERENCES

[1] K. V. Williams, R. Vaziri, and A. Poursartip, "A physically based continuum damage mechanics model for thin laminated composite structures," *Int. J. Solids Struct.*, vol. 40, no. 9, pp. 2267–2300, May 2003, doi: 10.1016/S0020-7683(03)00016-7.

[2] A. Forghani, A. Poursartip, and R. Vaziri, "An orthotropic non-local approach to modeling intra-laminar damage progression in laminated composites," *Int. J. Solids Struct.*, vol. 180–181, pp. 160–175, Dec. 2019, doi: 10.1016/J.IJSOLSTR.2019.07.015.

[3] M. Shahbazi, "An efficient virtual testing framework to simulate the progression of damage in notched composite laminates," The University of British Columbia, 2017, doi: 10.14288/1.0354263.

[4] S. Nesbitt, R. Vaziri, and A. Poursartip, "Characterization of the Structural Response and Energy Absorption Characteristics of Braided and Non-Crimp Fabric CFRP Panels Subjected to Transverse Impact Loadings at Variable Strain Rates," in *4th International Aerospace Structural Impact Dynamics Conference*, 2019.

[5] S. Nesbitt, M. Waimer, N. Toso-Pentecôte, R. Vaziri, and A. Poursartip, "Structural Response and Damage Characterisation of Non-Crimp Fabric CFRP Panels under Impact Loading," in *11th Canadian-International Conference on Composites*, 2019.

[6] "D792-13 Standard Test Methods for Density and Specific Gravity (Relative Density) of Plastics by Displacement." ASTM International, West Conshohocken, PA, 2013, doi: https://doi.org/10.1520/D0792-13.

[7] D. Adams, "Unidirectional composite axial tensile specimens," *CompositesWorld*, 2006. [Online]. Available: https://www.compositesworld.com/articles/unidirectional-composite-axial-tensile-specimens. [Accessed: 15-Jan-2020].

[8] D. Adams, "Composite material testing: How do I know if my measured composite properties are correct, or even reasonable?," *CompositesWorld*, 2018. [Online]. Available: https://www.compositesworld.com/blog/post/how-do-i-know-if-my-measured-composite-properties-are-correct-or-even-reasonable. [Accessed: 15-Jan-2020].

[9] "D3039-17 Standard Test Method for Tensile Properties of Polymer Matrix Composite Materials." ASTM International, West Conshohocken, PA, 2017, doi: https://doi.org/10.1520/D3039_D3039M-17.

[10] "D3410-16 Standard Test Method for Compressive Properties of Polymer Matrix Composite Materials with Unsupported Gage Section by Shear Loading." ASTM International, West Conshohocken, PA, 2016, doi: https://doi.org/10.1520/D3410_D3410M-16.

[11] "D6641-16 Standard Test Method for Compressive Properties of Polymer Matrix Composite Materials Using a Combined Loading Compression (CLC) Test Fixture." ASTM International, West Conshohocken, PA, 2016, doi: https://doi.org/10.1520/D6641_D6641M-16E01.

[12] "D7264-15 Standard Test Method for Flexural Properties of Polymer Matrix Composite Materials." 2015, doi: https://doi.org/10.1520/D7264_D7264M-15.

[13] D. Adams, "A comparison of shear test methods," *CompositesWorld*, 2009. [Online]. Available: https://www.compositesworld.com/articles/a-comparison-of-shear-test-methods. [Accessed: 14-Jan-2020].

[14] "D5379-12 Standard Test Method for Shear Properties of Composite Materials by the V-

Notched Beam Method." ASTM International, West Conshohocken, PA, 2012, doi: https://doi.org/10.1520/D5379_D5379M-19.

[15] "D7078-19 Standard Test Method for Shear Properties of Composite Materials by V-Notched Rail Shear Method." ASTM International, West Conshohocken, PA, 2019, doi: https://doi.org/10.1520/D7078_D7078M-19.

[16] D. Adams, "Shear test methods: Iosipescu vs. V-Notched Rail," *CompositesWorld*, 2009. [Online]. Available: https://www.compositesworld.com/articles/shear-test-methods-iosipescu-vs-v-notched-rail. [Accessed: 14-Jan-2020].

[17] Z. P. Bazant and J. Planas, *Fracture and Size Effect in Concrete and Other Quasibrittle Materials*. Boca Raton: CRC Press LLC, 1998.

[18] "D5045-14 Standard Test Methods for Plane-Strain Fracture Toughness and Strain Energy Release Rate of Plastic Materials." ASTM International, West Conshohocken, PA, 2014, doi: https://doi.org/10.1520/D5045-14.

[19] I. Kongshavn and A. Poursartip, "Experimental investigation of a strain-softening approach to predicting failure in notched fibre-reinforced composite laminates," *Compos. Sci. Technol.*, vol. 59, no. 1, pp. 29–40, Jan. 1999, doi: 10.1016/S0266-3538(98)00034-7.

[20] N. Zobeiry, R. Vaziri, and A. Poursartip, "Characterization of strain-softening behavior and failure mechanisms of composites under tension and compression," *Compos. Part A Appl. Sci. Manuf.*, vol. 68, pp. 29–41, Jan. 2015, doi: 10.1016/J.COMPOSITESA.2014.09.009.

[21] Y. Yuan, K. Niu, and Z. Zhang, "Compressive damage mode manipulation of fiber-reinforced polymer composites," *Eng. Fract. Mech.*, vol. 223, p. 106799, Jan. 2020, doi: 10.1016/J.ENGFRACMECH.2019.106799.

[22] J. Reiner and R. Vaziri, "Local and Nonlocal Continuum Damage Simulation of Impact and Compression After Impact Tests on CFRP Laminates," in *22nd International Conference on Composite Materials (ICCM22)*, 2019.

[23] R. Olsson, "Mass criterion for wave controlled impact response of composite plates," *Compos. Part A Appl. Sci. Manuf.*, vol. 31, no. 8, pp. 879–887, Aug. 2000, doi: 10.1016/S1359-835X(00)00020-8.

[24] W. J. Cantwell and J. Morton, "Comparison of the low and high velocity impact response of CFRP," *Composites*, vol. 20, no. 6, pp. 545–551, Nov. 1989, doi: 10.1016/0010-4361(89)90913-0.

[25] H. Morita, T. Adachi, Y. Tateishi, and H. Matsumot, "Characterization of Impact Damage Resistance of CF/PEEK and CF/Toughened Epoxy Laminates under Low and High Velocity Impact Tests," *J. Reinf. Plast. Compos.*, vol. 16, no. 2, pp. 131–143, Jan. 1997, doi: 10.1177/073168449701600203.

[26] "D7291-15 Standard Test Method for Through-Thickness 'Flatwise' Tensile Strength and

Elastic Modulus of a Fiber-Reinforced Polymer Matrix Composite Material." ASTM, West Conshohocken, PA, 2015, doi: https://doi.org/10.1520/D7291_D7291M-15.

[27] "D2344-16 Standard Test Method for Short-Beam Strength of Polymer Matrix Composite Materials and Their Laminates." ASTM International, West Conshohocken, PA, 2016, doi: https://doi.org/10.1520/D2344_D2344M-16.

[28] "D5528-13 Standard Test Method for Mode I Interlaminar Fracture Toughness of Unidirectional Fiber-Reinforced Polymer Matrix Composites BT - Standard Test Method for Mode I Interlaminar Fracture Toughness of Unidirectional Fiber-Reinforced Polymer Matrix." ASTM International, West Conshohocken, PA, 2013, doi: https://doi.org/10.1520/D5528-13.

[29] J. Reiner, J. P. Torres, and M. Veidt, "A novel Top Surface Analysis method for Mode I interface characterisation using Digital Image Correlation," *Eng. Fract. Mech.*, vol. 173, pp. 107–117, Mar. 2017, doi: 10.1016/J.ENGFRACMECH.2016.12.022.

[30] D. Liu, "Impact-Induced Delamination—A View of Bending Stiffness Mismatching," *J. Compos. Mater.*, vol. 22, no. 7, pp. 674–692, Jul. 1988, doi: 10.1177/002199838802200706.

[31] "D7905-14 Standard Test Method for Determination of the Mode II Interlaminar Fracture Toughness of Unidirectional Fiber-Reinforced Polymer Matrix Composites BT - Standard Test Method for Determination of the Mode II Interlaminar Fracture Toughness of Uni." ASTM International, West Conshohocken, PA, 2014, doi: https://doi.org/10.1520/D7905_D7905M-19E01.

[32] J. Reiner and R. Vaziri, "Structural Analysis of Composites With Finite Element Codes: An Overview of Commonly Used Computational Methods," in *Comprehensive Composite Materials II*, vol. 8, P. W. R. Beaumont and C. H. Zweben, Eds. Elsevier Ltd., 2018, pp. 61–84.

[33] A. Forghani, M. Shahbazi, N. Zobeiry, A. Poursartip, and R. Vaziri, "An overview of continuum damage models used to simulate intralaminar failure mechanisms in advanced composite materials," in *Numerical Modelling of Failure in Advanced Composite Materials*, P. P. Camanho and S. R. Hallett, Eds. Woodhead Publishing, 2015, pp. 151–173.

[34] A. S. Kaddour, M. J. Hinton, P. A. Smith, and S. Li, "The background to the third world-wide failure exercise," *J. Compos. Mater.*, vol. 47, no. 20–21, pp. 2417–2426, Sep. 2013, doi: 10.1177/0021998313499475.

[35] A. Forghani, N. Zobeiry, A. Poursartip, and R. Vaziri, "A structural modelling framework for prediction of damage development and failure of composite laminates," *J. Compos. Mater.*, vol. 47, no. 20–21, pp. 2553–2573, Sep. 2013, doi: 10.1177/0021998312474044.

OPTIMIZATION OF CURE CYCLES USING AN ESR (ENCAPSULATED SAMPLE RHEOMETER)

Richard Hanzlik (Applications Engineer) & Henry Pawlowski (Senior Scientist)
Alpha Technologies
6279 Hudson Crossing Pkwy
Hudson, OH 44236

Louis C. Dorworth (Direct Services Manager)
Abaris Training Resources, Inc.
5401 Longley Lane, Suite 49
Reno, NV 89511

ABSTRACT

One of the most critical steps in the production of composite material product is the cure cycle. The recommended cure cycles are usually provided by the prepreg suppliers. These cycles are often adjusted by trial and error until optimum final product properties are achieved. Operating an autoclave or oven while simultaneously measuring material properties can prove difficult due to size and cost restraints. The introduction of the ESR (encapsulated sample rheometer), which meets the ASTM 7750, has provided a cost-effective method to cure composite materials under any cure cycle necessary. In addition, the ESR is capable of measuring gel, cure, and final physical properties from those conditions [1].

This paper evaluates simulated cure cycles with varying temperature ramps and hold temperatures, time at temperature, and the application of pressure while measuring the inherent cure properties of several resin and fiber packages. These simulations lead to the conditions for the optimum full cure. Next, the physical properties were measured and compared at the same temperature to determine the impact of cure conditions on the final product. Comparison of physical properties and Tg leads to optimization of the process, thus balancing efficiency and ensuring reliable material properties.

Corresponding Author: Richard Hanzlik, rhanzlik@gmail.com

INTRODUCTION

The measurement of cure in a thermoset polymer matrix composite which produces a very high modulus laminate is difficult. Many different techniques have been developed to measure cure in such systems. Methods used for prepreg include measurement of thermal properties during cure with a Differential Scanning Calorimeter (DSC) that measures heat of reaction during cure of the matrix. DSC can be used to determine activation energies via Kissinger's method.[2] however, it can be difficult to model the kinetic impact of multiple layers using this method. Thermal Mechanical Analysis (TMA) of cured laminates is another method commonly used to determine state of cure and glass transition temperature of prepregs.[3] Newer alternative methods include measurement of the change in electrical properties of the matrix during cure such as dielectric properties, and change in sound transmission such as ultrasonic techniques.[4]

The original ASTM method for measuring the gel point on prepregs is described in D3532 (Standard Test Method for Gel Time of Carbon Fiber-Epoxy Prepreg).[5] It has been the legacy standard for the past fifty years. Recently, the introduction of ASTM D7750 [6] has expanded the capabilities available for measuring gel and cure. This approach has been developed to use viscoelastic properties measured after the application of a small mechanical strain. There are a variety of mathematical models used to describe such reactions using either DSC or mechanical data. A popular simple model [6] is shown in Equation 1:

$$\frac{d\alpha}{dt} = k(1-\alpha)^n \qquad \text{Equation 1}$$

Where α is the degree of cure, t is time, k is the rate constant and n is the order of the reaction. The solution to this equation when n=1 is given by Equation 2:

$$\ln(1-\alpha) = \ln A - k(t-t_i) \qquad \text{Equation 2}$$

the value of (1- α) represents the unreacted material, A is a constant, k is the rate constant, t is time and t_i is the time at which the reaction is initiated. The slope in the plot of ln(1- α) versus t is the rate constant at the test temperature and the x-intercept is the t_i. The rate constant k often has a dependence on temperature. The rate constant k should increase as the test temperature increases. The rate constant often follows the Arrhenius Model as a function of temperature:

$$k = Ae^{-Ea/RT} \qquad \text{Equation 3}$$

Where k is the Rate Constant, A is the pre-exponential factor, E_a is the Energy of Activation, R is the Universal gas constant and T is the Temperature (1/K).

Taking the natural log of each side of Equation 3 gives Equation 4:

$$\ln k = \ln A - (E_a/RT) \qquad \text{Equation 4}$$

The slope in the plot of ln k and 1/T is equal to E_a/R.

This study looked at the application of these models to the actual measurement of the cure reaction by mechanical changes in the prepreg during cure. There have been several resin studies using similar models based on viscosity.[6] The expectation in this study was that the addition of fibers in a prepreg should have minimal effect on the cure kinetics compared to neat resins.

The studies in this paper look at the cure reaction in the prepreg without removing the resin from the prepreg and using multiple layers of prepreg that result in a coupon that simulates the manufacture of a part.

Recent improvements in this system have also allowed a thorough characterization of the state of cure by the measurement of Tg.

DESCRIPTION OF THE APPARATUS

The instrument is a commercially available DMA called the ESR (Figure 1).

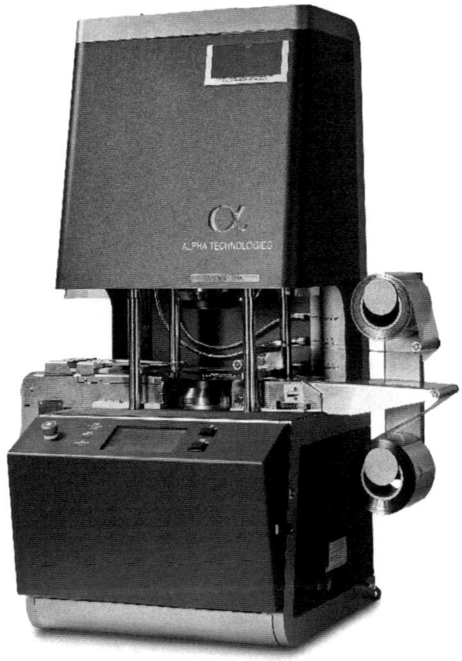

Figure 1. The Encapsulated Specimen Rheometer (ESR).

The system has two parallel plate dies (Figure 2). The die surface is grooved to help hold high modulus samples and prevent slippage.

Figure 2. Closeup of instrument dies.

The outer edge of the sample cavity contains two seal plates and two seals, which completely seal the sample cavity when the dies are closed. The sample is placed onto the lower die and then the upper die is brought down. An o-ring in the sample chamber prevents the escape of resin to maintain a consistent fiber / resin ratio and helps build the pressure in the cavity to limit voids, prevent slippage and thus improve the repeatability of the test. The upper die is attached to a torque transducer. This placement of the transducer eliminates the measurement of lower die seal and bearing frictions. A cure test is typically run continuously with a strain of 0.7% and an oscillation frequency of 1 Hz or 1.667 Hz. The die dimensions, sample thickness, frequency of oscillation and strain are all used with appropriate rheological equations to convert torque to shear modulus. A Fourier Transform of the data separates the torque signal into an elastic and a viscous or loss component. This torque signal is then converted to viscoelastic properties of the material under test: elastic shear modulus G', loss shear modulus G", complex shear modulus G*, and Tan(Delta) [=G"/G'].

The digitally controlled thermal system allows the measurement of material properties from room temperature to 350 °C. The temperature can be ramped upwards from very low rates of 0.2 °C /minute to a very rapid rate of 50 °C/minute. A forced air cooling system allows the temperature to be reduced at rates of up to 30 °C/minute. The rapid response to temperature changes also enables the system to follow the temperature changes in most production systems and enables a more accurate estimation of the true state of the material during any cure profile.

Figure 3 shows a sample after processing with a cure profile. The o-ring is visible at the outer diameter and the sample is contained within two layers of film for ease of removal.

Figure 3. Prepreg sample after processing with a temperature profile

VISCOELASTIC PROPERTIES OF POLYMERS

In practical terms, the process of manufacturing composites is to form (layup) the prepreg to the surface of the tool while in a predominately viscous state and capture the resin by various bleeder, breather and barrier films. It is important create a test specimen that contains multiple layers and resembles a part created by a layup. The changes in the viscoelastic properties during processing are illustrated in Figure 4. While in the low viscosity (viscous) state, gases are removed, and voids are compressed. This is followed by an increase in modulus (cure) until the material is solid (elastic) and retains the shape that is required in a structure. As noted above, layup, degassing and consolidation occur while the material is in the predominately liquid state as measured by a low value of the storage modulus G', a relatively large value for the loss modulus (*G*") and a relatively large value of Tan(Delta) [=G"/G']. The final cure stage is measured by achieving the appropriate final storage modulus (*G*') at the required temperature. Conceptually G' can be considered a measure of how solid (elastic) the material is while G" is a measure of its viscosity. While this is not an entirely accurate description of G' and G", it is empirically useful. *G*', *G*" and Tan(Delta) provide tools for monitoring the transition of the prepreg from a 'liquid' prepreg to a solid state laminate. These parameters are, in fact, the technical basis for tests that support current design limits.[7] Existing specifications for flow, tack, drape, gel time, cure time and glass transition temperatures are all measures of viscoelastic state; however, they are difficult to quantify in real time by legacy-specified methods and cannot be incorporated as triggers for automated processing.

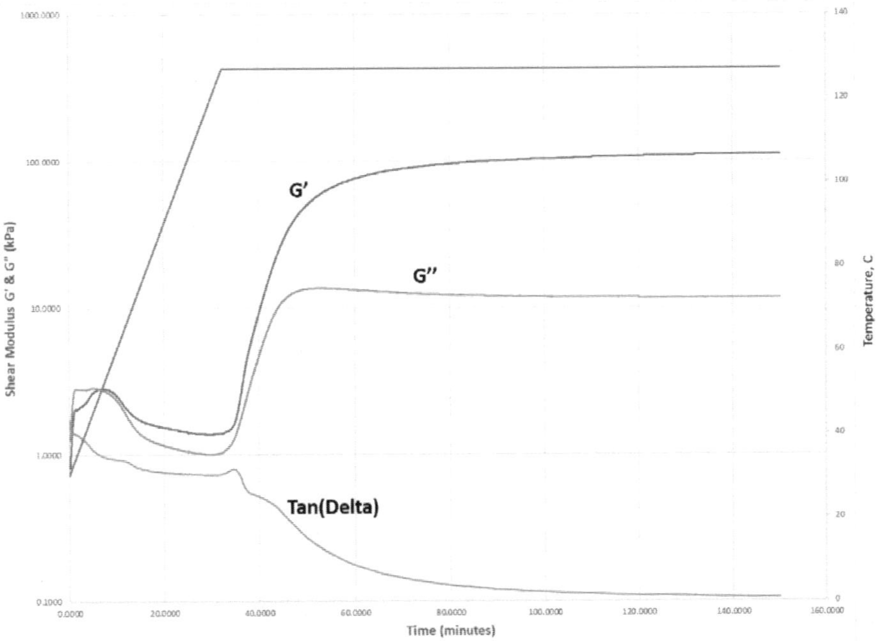

Figure 4. An example of the changes in viscoelastic properties that occur during the processing of a prepreg. The applied temperature profile is also shown.

Understanding the viscoelastic properties allows for a better assessment of the current state of the material for informed decisions regarding quality, processability and any possible changes to processing conditions.

For those not familiar with the complex shear modulus (*G**), it is sufficient to think of it as the force required to change the shape of a prepreg. This can be further divided into the force required for a temporary change in shape (elastic modulus *G'*) where the material will 'snap back' after the force is removed and the force that is converted to heat within the material (loss modulus *G"*). A term used to evaluate the liquid versus solid character of the material including the ability to recover its shape is the Tan(Delta) which is equal to the flow characteristic divided by the elastic characteristic. The Tan(Delta) value can indicate the ability of the resin to flow. When the value is greater than 0.500, then the material can flow and allow the prepreg layup to change its shape during processing. When the value is less than 0.300, then the material will resist changes in its shape. After cure, the Tan(Delta) value often is significantly lower than 0.100 and the prepreg is now very stiff and will not change its shape unless the temperature is raised above its Tg. The Tan(Delta) peak observed near the gel point is also a useful point that is repeatable and occurs in the region that has the empirical definition of gel. The important point is that for a given fiber/resin combination the data is quantitative, repeatable and informative with respect to the material state.

In addition to measuring the state of cure using mechanical properties, the final state of cure using Tg of a sample after cure. The Tg value is determined by either the peak in G", the peak in Tan(Delta) or the knee in the G' / G* curves.

EXPERIMENTAL

Materials

1. Prepreg No. 1
2. Prepreg No. 2
3. Unidirectional Tape

Test Conditions

The tests used in these studies are described below:

1. Processing Conditions: Most prepregs are processed by ramping the temperature from room temperature up to the hold temperature. After leaving the prepreg at the hold temperature for a sufficient time to finish the thermoset reaction, the temperature is reduced back to room temperature. This portion of the study determined the effect of the hold temperature on the final cure state. It also determined the effect of ramp rate on the final cure state.
    a. Ramp at 3 C/minute and hold at 117 °C for 2 hours
    b. Ramp at 3 °C/minute and hold at 107 °C for 2 hours
    c. Ramp at 3 °C/minute and hold at 127 °C for 2 hours
    d. Ramp at 1 °C/minute and hold at 127 °C for 2 hours
2. Isothermal Processing: The introduction of the ESR has allowed the testing of prepregs under isothermal processing. The purpose for the isothermal testing is to apply the results at 2 or more temperatures to reaction models in order to predict the processing of prepregs under non-isothermal conditions or other isothermal temperatures. The prepregs in this study were tested at 107 °C, 117 °C, 127 °C, 137 °C and 147 °C. The unidirectional tape was tested at 167 °C, 177 °C and 187 °C.

RESULTS

Non-isothermal Studies

Figure 5 shows the cure of Prepreg No. 1 under three different processing conditions. All the temperature profiles used a ramp of 3 °C/minute, but each had a different hold temperature. As expected, the higher temperature generally cured faster. However, the higher hold temperature also showed a reduction in the G' value at the hold temperature. The Tg values from these tests are given in Table 1. The Tg value from all three hold temperatures varied only slightly or up to 4 C over a range of 20 °C in the hold temperature.

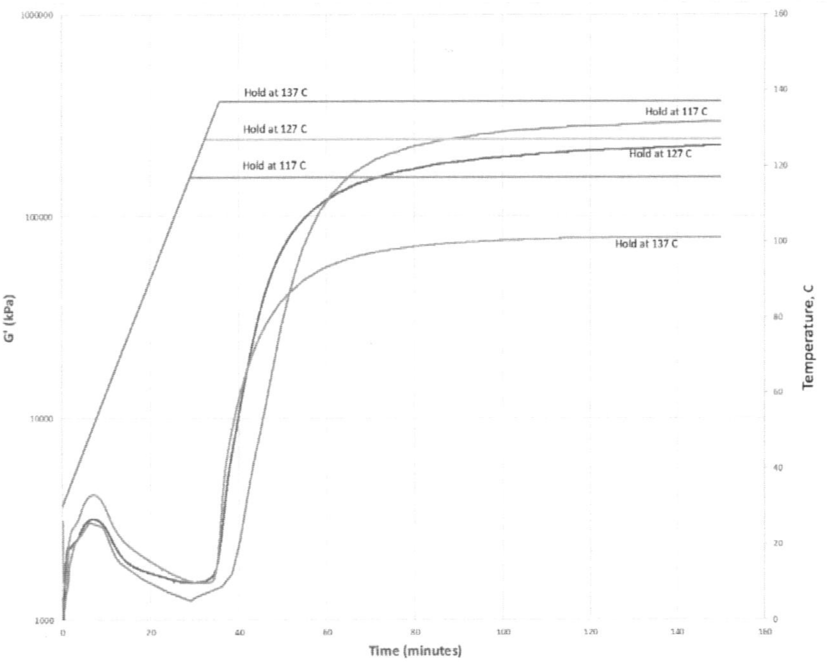

Figure 5. Changes in the elastic shear modulus G' during the processing of a prepreg with three different hold temperatures. Higher temperatures speed up the processing and reduce the final G' value at the hold temperature.

Figure 6 shows the G’ results from one hold temperature but with two different ramp rates using Prepreg No. 1. The cure curve gel point with the slower ramp is shifted to a longer time as expected. The Tg differences in Table 1 are around 2 °C or very small.

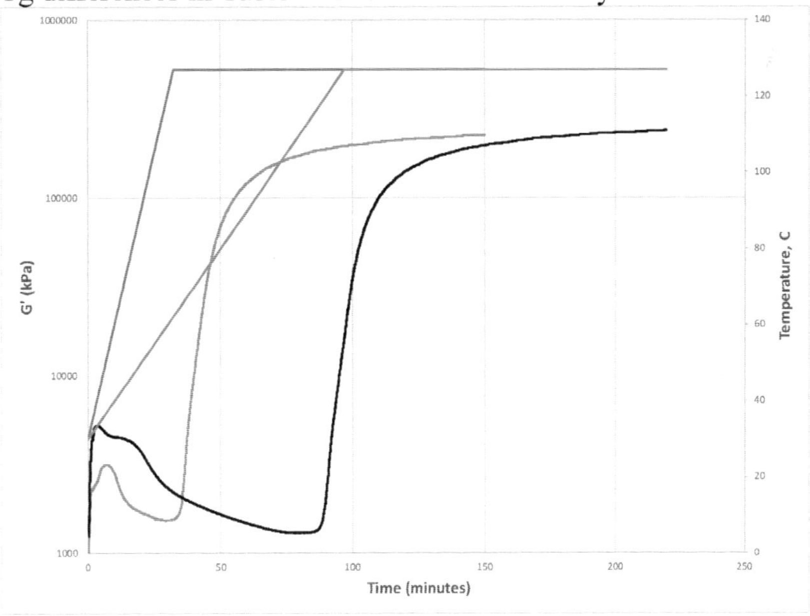

Figure 6. Cure of a prepreg at the same hold temperature but with two different ramp rates. Results show a similar cure curve shifted to longer cure times.

Figure 7 shows the cure of Prepreg No. 1 and No. 2 using the same ramp and hold temperature condition. The results clearly show that Prepreg No. 2 cures slower and produces a lower G’ at the end of cure at the processing temperature. Each prepreg can produce a distinct result to help characterize different prepregs and determine any effect from differences in the fiber, resin or aging.

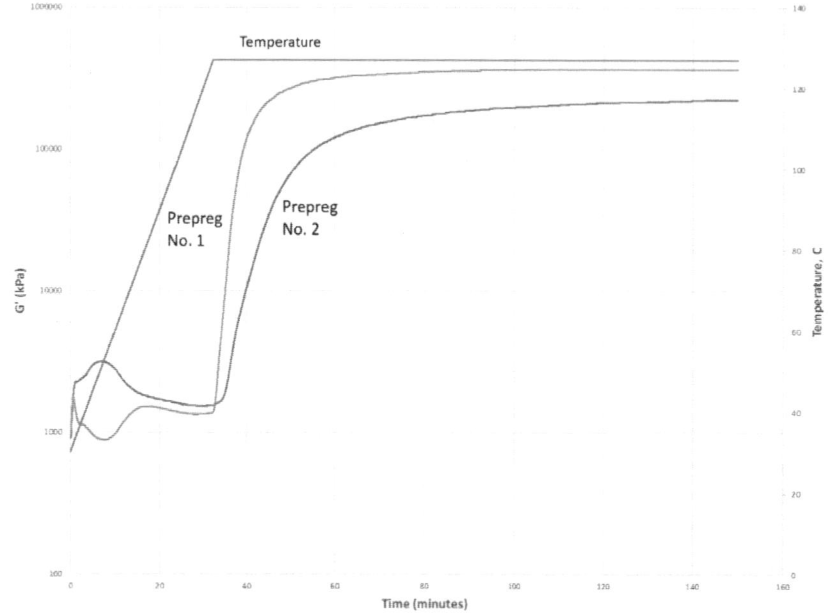

Figure 7. G' results during a ramp and hold for two different prepregs. Results show significant differences in the viscosity, cure rate and final modulus.

Isothermal Studies

Figure 8 shows Prepreg No. 2 tested under isothermal test conditions at four different test temperatures.

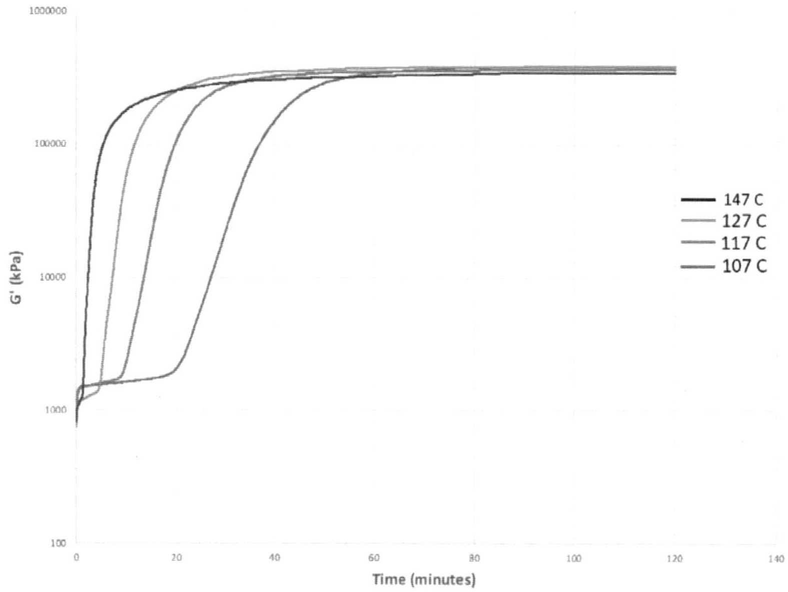

Figure 8. Cure of prepreg No. 2 at four different temperatures

As expected, the higher temperature produces the faster cure while the higher temperature also reduces the maximum G’ value at the end of the cure at the hold temperature. Table 1 shows that the Tg changes significantly with increasing test temperatures. The Tg goes from 142 °C with cure at 107 °C to 184.1 C with cure at 147 °C. The result is a 42 °C increase in Tg with a 40 °C increase in the test temperature.

The advantage of isothermal processing is that the data can be applied to the Arrhenius Model of reactivity. The first step in analyzing the changes in mechanical properties shown in Figure 8 and calculating rate constants for the Arrehnius Model is to normalize the reaction curves [8]. The lowest G’ during a test becomes 0.0 and the highest becomes 1.00 as shown in Figure 9.

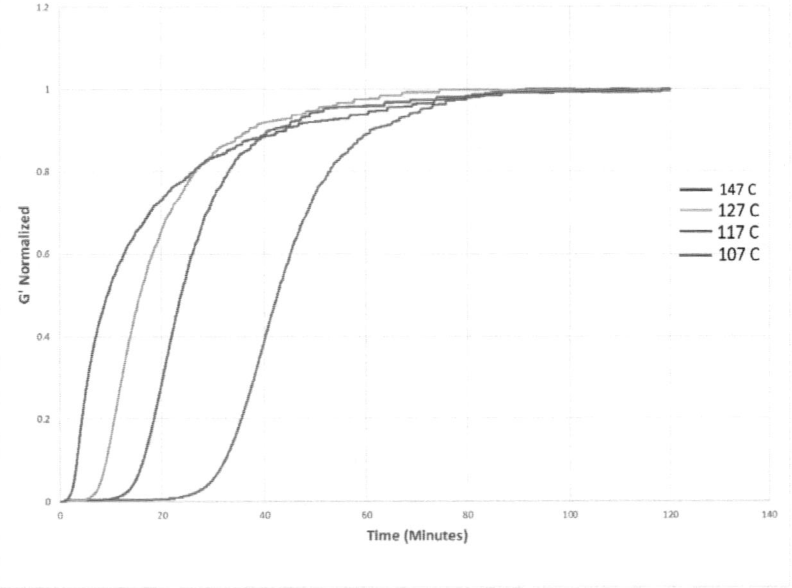

Figure 9. Normalized cure curves from Prepreg No. 2 determined from the results in Figure 8.

The fraction reacted is referred to as α so that the fraction not reacted becomes (1-α). For example, if 40% were reacted then α=0.40 and (1- α) becomes 0.60. This is needed because many reaction rates are directly affected by the unreacted material and not the final product.

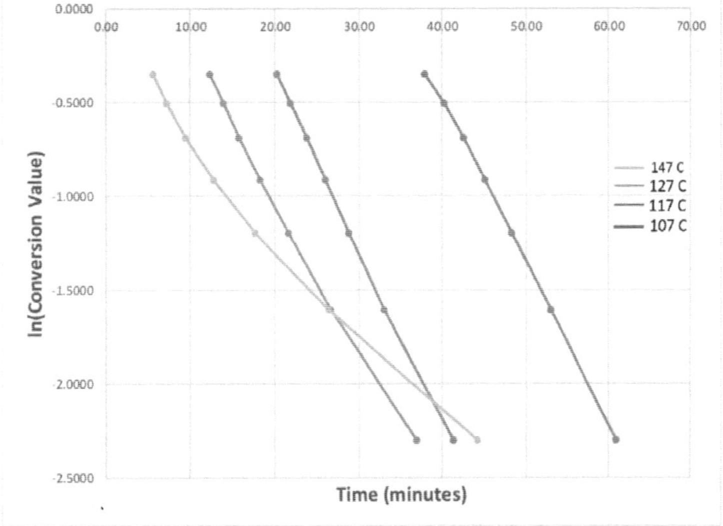

Figure 10. Ln(1-X) versus Time for Prepreg No.2

The value of ln(1- α) is plotted versus time with the expectation that it will be linear. The slope of the plot is the reaction rate constant k as shown in Figure 10. Table 2 contains the rate constants and the initiation times. Surprisingly, the rate constant k does not show a clear increase with every increase in temperature. The slopes are very similar. What changes more dramatically with temperature is the x intercept or t_i or initiation time. Figure 11 shows a plot of $\ln(t_i)$ versus the inverse Kelvin Temperature. The results for the three lowest test temperatures are very linear. The highest temperature had a distorted plot in Figure 10 and did not produce an initiation time that aligned with the other temperatures.

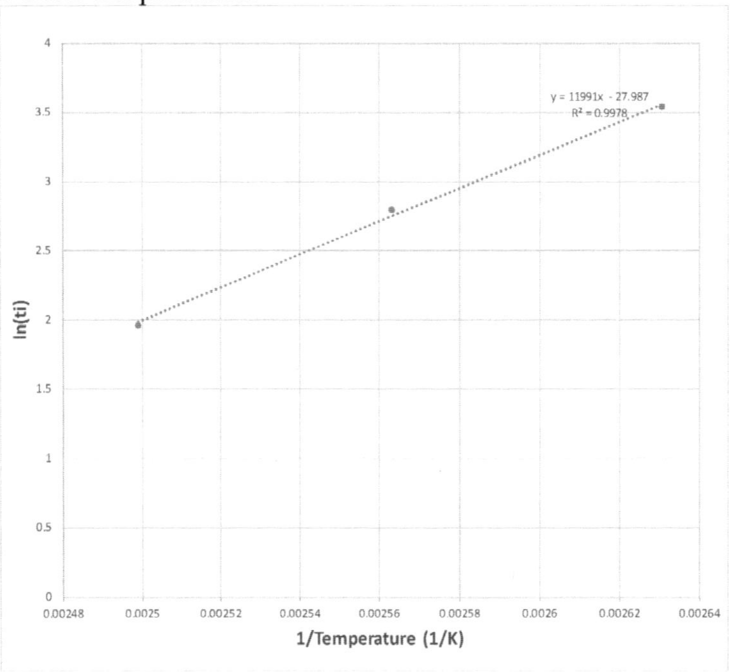

Figure 11. The initiation time for the reaction of prepreg No. 2 follows the Arrhenius Model even though the rate constants did not.

Figure 12 shows the isothermal cure curves from the Unidirectional Tape at three temperatures.

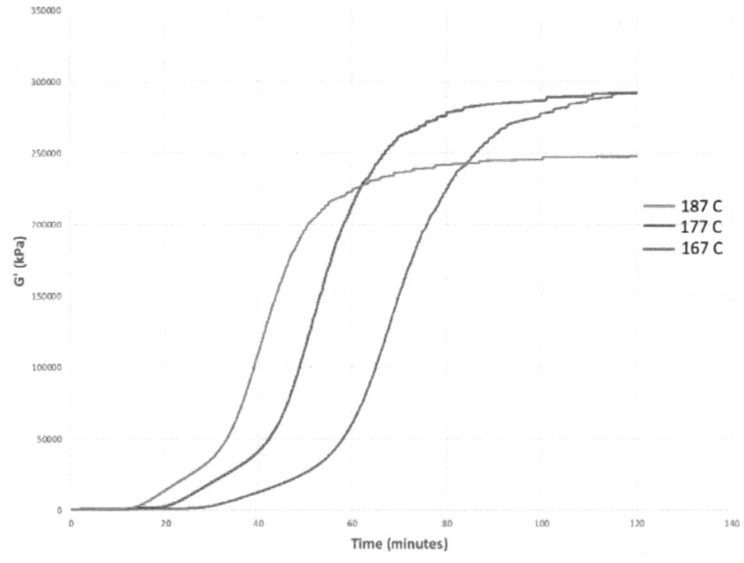

Figure 12. Isothermal cure curves from the Unidirectional Tape.

The Tg values in Table 1 show a significant increase with an increase in the isothermal test temperature.

The Tg value increased by 17 C for a 20 °C increase in the test temperature or slightly less than seen in Prepreg No. 2. Figure 13 shows the normalized cure curves and Figure 14 shows the plot of ln(1- α) versus Time for the Unidirectional Tape.

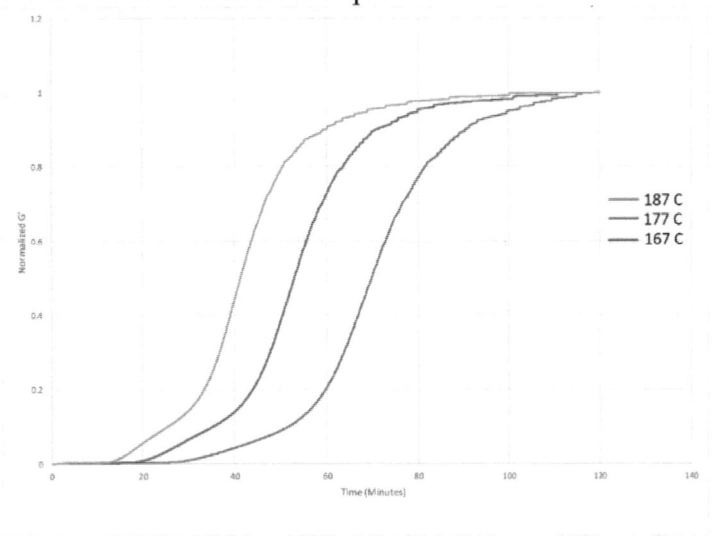

Figure 13. Normalized cure curves from the Unidirecitonal Tape results in Figure 12.

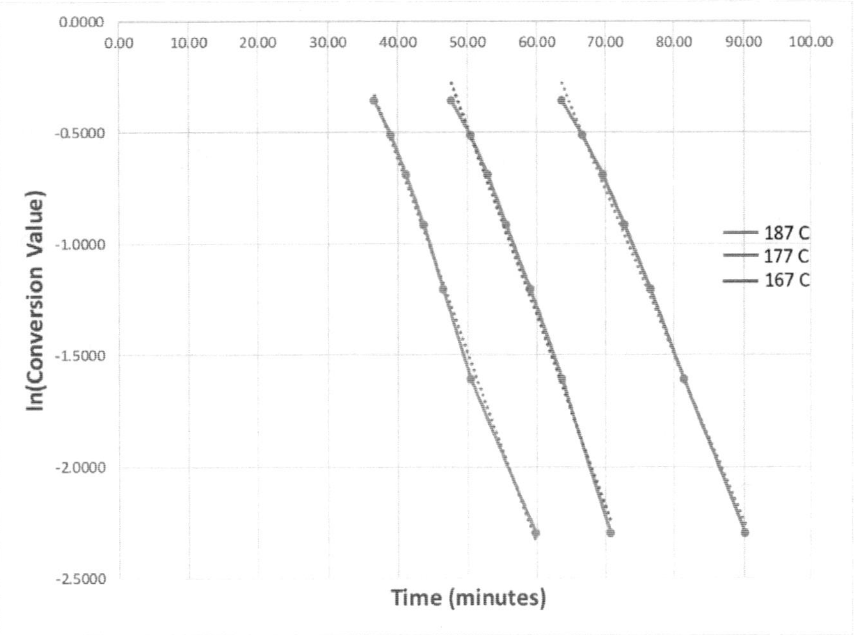

Figure 14. These plots were very linear, but the slopes did not change much with changes in temperature.

The rate constants taken from the very linear slopes did not show the expected change with increasing temperature. The rate constants are also in Table 2. Figure 15 shows the plot of the $\ln(t_i)$

versus the inverse Kelvin Temperature for the Unidirectional Tape sample. Again, the Initiation Time follows the Arrhenius Model.

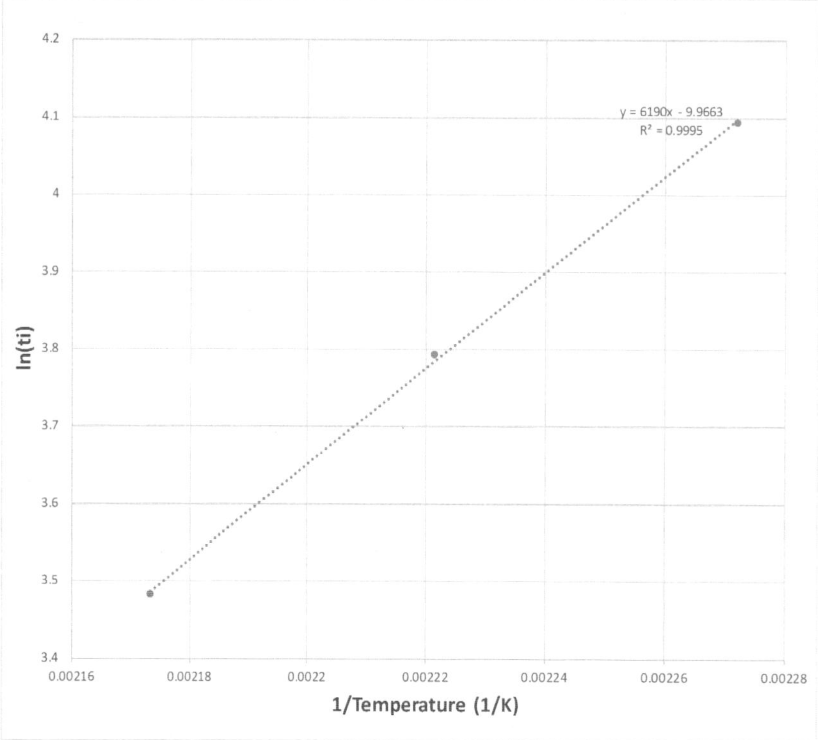

Figure 15. Again, the Initiation Time followed the Arrhenius Model very closely.

CONCLUSION

The ESR was used to analyze the cure of prepregs and a Unidirectional Tape. The results showed the viscoelastic properties of prepregs throughout the cure cycle while building a coupon that resembles a composite part. It can also characterize different prepregs based on the speed of the reaction and the mechanical properties at the processing temperature. An attempt was made to fit the reaction curves to a simple Arrhenius Model. The results showed a lot of evidence of a 1st order reaction but the model failed on a prepreg and a Unidirectional Tape sample because the rate constants did not increase as the temperature increased. However, the calculated Initiation Times did follow the Arrhenius Model. This result suggests that mechanical testing done by an ESR or similar instrument may be necessary to verify the actual kinetics of the prepreg cure, especially when dealing with multiple layers of a prepreg.

One prepreg showed very little change in Tg with change in the hold temperature. A second prepreg showed a significant change in Tg with a change in the isothermal cure temperature. The Unidirectional Tape sample Tg was also very sensitive to the isothermal test temperature and showed an increase with an increase in the processing temperature. The ESR appears to be capable of determining the presence of any unreacted materials in a prepreg cured under standard conditions using the final measured Tg. Further study can be conducted to evaluate the influence unreacted components may have on final product properties.

REFERENCES

1. H. Pawlowski, T. Rose; "Improving the Measurement of prepreg Viscosity, Gel, and Cure Using an Encapsulated specimen rheometer"
2. Sang-Beom Shim, James C. Seferis, Yong Sung Eom, Yi Taek Shim, "Thermal characterization and comparison of structural prepregs with different cure temperatures", Thermochimica Acta, Volume 291, Issues 1–2, 1997, Pages 73-79
3. Kim, J., Moon, T. J., & Howell, J. R. (2002). Cure Kinetic Model, Heat of Reaction, and Glass Transition Temperature of AS4/3501-6 Graphite–Epoxy Prepregs. Journal of Composite Materials, 36(21), 2479–2498.
4. W. Stark, W. Bohmeyer,7 – "Non-destructive evaluation (NDE) of composites: using ultrasound to monitor the curing of composites," Woodhead Publishing Series in Composites Science and Engineering, Non-Destructive Evaluation (NDE) of Polymer Matrix Composites, Woodhead Publishing,2013,Pages 136-181,
5. DIN53529 Measurement of Vulcanization Characteristics
6. ASTM International. (2019). ASTM D 7750-12: Standard Test Method for Cure Behavior of Thermosetting Resins by Dynamic Mechanical Procedures using an Encapsulated Specimen Rheometer. In Annual book of ASTM standards 2019 Volume 15. West Conshohocken, PA: American Society for Testing and Materials.
7. Borchardt, H.J.; Daniels, F.J, The Application of Differential Thermal Analysis to the Study of Reaction Kinetics., Am. Chem. Soc. 1956, 79, 41
8. Soltani, A Seyed, "Properties of a CFRPC Cured at Staged Cure Cycles" 21-24

Appendix

Table 1. Tg Values from All Samples and Temperature Profiles

	Prepreg Sample	Temperature Ramp C/minute	Hold Temperature C	Hold Time minute	Tg Max Tand C	Tg Max G" C
Non-Isothermal	1	3	127	150.0	137.3	134.5
	1	1	127	220.0	139.3	136.1
	1	3	117	150.0	138.8	136.8
	1	3	137	150.0	135.6	132.8
	2	3	127	150.0	177.3	164.4
Isothermal	2	0	107	120.0	144.4	142.0
	2	0	117	120.0	157.5	153.8
	2	0	127	120.0	179.9	164.9
	2	0	147	120.0	187.6	184.1
	Unitape	0	167	160.0	205.1	201.3
	Unitape	0	177	120.0	215.7	211.2
	Unitape	0	187	120.0	222.1	218.4

Table 2, Constants from 1st Order Arrhenius Model

Sample	Temperature C	Slope (k)	Intercept	Inititiation Time minutes	R
Prepreg No. 1	107	-0.0860	2.9501	34.30	0.9996
	117	-0.0929	1.5126	16.27	0.9992
	127	-0.0788	0.5583	7.08	0.9972
	147	-0.0494	-0.2140	------	0.9897
Prepreg No. 2	167	-0.0744	4.4559	59.92	0.9989
	177	-0.0845	3.7504	44.38	0.9989
	187	-0.0838	2.7238	32.51	0.9984

STOCHASTIC COMPUTATIONAL MODELING OF THE FRACTURING BEHAVIOR IN DISCONTINUOUS FIBER COMPOSITE STRUCTURES

Seunghyun Ko[1], Jinkyu Yang[1], Mark E. Tuttle[2], Marco Salviato[1]

[1]Department of Aeronautics and Astronautics, University of Washington, Seattle, WA

[2]Department of Mechanical Engineering, University of Washington, Seattle, WA

ABSTRACT

In this study, we numerically investigate the mode I intra-laminar fracture and size effect of Discontinuous Fiber Composites (DFCs) as a function of the platelet sizes and the structure thicknesses. First, the complex 3D mesostructures of DFCs are generated using a random platelet generation algorithm. The algorithm explicitly generates stacks of platelets into pre-defined partitions. Each partition stores unique platelet orientation and associated thickness. The partitioned morphology is transferred to a Finite Element (FE) model. The in-plane damage mechanisms are captured by a quasibrittle failure criterion whereas the out-of-plane inter-platelet damage is captured by cohesive elements. The stochastic FE model is calibrated using size effect experiments of a single platelet size and thickness. After the calibration, the model captures the effects of different platelet sizes as well as the change in thickness. Additionally, the model provides unique insights on the change in damage mechanisms of DFCs with respect to the platelet morphology. By simulating different mesostructures, we can further extend the effects of platelet sizes and the thicknesses beyond the experimental results.

Keywords: Discontinuous fiber composites, Quasibrittle, Stochastic finite element model
Corresponding author: Marco Salviato, salviato@aa.washington.edu

1. INTRODUCTION

Discontinuous Fiber Composites (DFCs) provide unique opportunities to the composite industry to fulfill the need for large volume manufacturing capability with complex geometries [1]. This is possible because DFCs are made of chopped prepreg tapes. Short, chopped prepreg tapes can follow the contour of the mold more easily than continuous fibers [2]. Therefore, the compression molding technique becomes an ideal manufacturing method for DFC, unlike continuous fiber composites. Consequently, DFCs are suitable to fit into the complex molds within a short manufacturing period. Because DFCs are made of prepreg tapes, the loss of fiber volume fraction is minimized (50 ~ 55% [3]) compared to the short fiber composites. Therefore, DFCs possess comparable strength (~60%) and stiffness (~80%) with typical quasi-isotropic laminate composites [4]. Additionally, by controlling the size of the platelets, the material properties of DFCs can be engineered. In the early stage of studying DFCs, researchers have focused on finding the relationship between the material responses and the corresponding mesostructure often defined by the platelet size, platelet aspect ratio, tow thickness, and structure thickness [4-10]. The most studied aspect is the platelet size. The mechanical response of the various platelet sizes, tow thicknesses, and aspect ratios have been investigated. The experimental results show that the strength of DFCs increased with the larger platelet size, narrow aspect ratio, and thinner tow thickness. However, the optimum platelet size for the certain structural applications must be chosen based on the mechanical performance and processability [25]. As the platelet size increases, the performance of DFCs also improves such as strength [5, 7] and fracture energy [10]. However, shorter and smaller platelet sizes are much easier to follow the contour of the molds. Those platelets have advantages towards the manufacturing capability. Therefore, to find the optimum size of the platelets, we should compromise between the performance and manufacturing capability. Structure thickness and platelet tow thickness are also widely studied. A strong anisotropic material like DFCs has a non-negligible size effect in a thickness direction. As the structure thickness increases, we observe the strength and fracture energy increases as well while the variation of material properties saturates [5,6,7,9]. The tow thickness has an opposite trend. The structure strength increases as the tow thickness decreases [7,16]. However, the associated fracture behaviors with changing tow thicknesses have not fully characterized. This is because executing experiments in such that distinguishing the material size effect from the structural size effect is a big challenge. Besides the platelet and structure morphology, the manufacturing and curing process of DFCs have also been widely studied. Preetam and Lloyd [27] expressed the importance of the adhesion between the plies and found that the fracture energy can be doubled depending on the surface bonding process. Wan and Takahashi [28], and Nilakantan and Nutt [28] investigated the molding pressure and temperature during the curing process. They found the strength of DFCs increased as the temperature and pressure increased.

Based on the experimental findings, researchers now focus on constructing the computational models. One of the key aspects of modeling DFCs is to capture the mesostructure composed of randomly distributed platelets of finite length. There are multiple ways to characterize the mesostructure. First, direct measurements of the DFC structures are possible using X-ray Micro Computed Tomography (μCT) [11-13]. This technique provides an excellent measurement of platelets' orientations although it may not be the most accessible option considering the significant costs. Denos *et al.* [11] successfully integrated direct measurement from X-ray scan with the finite element model to create a digital copy of the scanned DFC structure. The simulation results match precisely with the experiment results including the location of the crack propagation. Utilizing the measurements from CT scan such as the average platelet orientation tensors, Kravchenko *et al.* [6,7] created the stochastic finite element model adapting a commercial software Digimat [30]. The FE model captures the effects of tow thicknesses, structure thicknesses, and platelet orientation distributions. Another possibility is to obtain an indirect measurement via optical microscopy of a statistically significant number of materials cross sections [14]. However, this approach can be extremely time consuming since its accuracy strongly correlated with the number of samples investigated and their locations. Finally, numerical generations of the platelets using the pre-defined probability of the fiber orientations are presented [9,10,14,15,16,17]. By calibrating the distributions of the fiber orientations obtained by the experimental observations using the μCT scanner and microscopes, the computational approach provides a robust method to study the mesostructures of DFCs.

This manuscript is an extension of previous studies by Ko *et al.* [9,10,18]. Therein, the authors found the fracture energy of mode I intralaminar fracture energy of DFCs with respect to the platelet size and structure thickness, elaborating the strong size effects possessed in DFCs. This paper provides a novel approach to build a computational model for DFCs capturing the size effects which is shown to be efficient and accurate. The model is calibrated leveraging the previous experimental results. Then, we study the failure mechanisms of DFCs and predict the strength of different mesostructure of DFCs.

2. COMPUTATIONAL MODELING

2.1 Mesostructure generation algorithm

To accurately create mesostructures of DFCs, it is critical to explicitly model the depositions of the platelets. The developed random platelet generation algorithm from the previous studies [9,10] explicitly generates the platelets into the given space with the assigned probability distributions in fiber orientations and the locations of the platelets' center point. Researchers have found that the in-plane orientations of the platelets in parts featuring relatively low flow can be well described by a uniform distribution, and therefore, the algorithm uses the uniform distribution from 0 to 180° [11, 14]. However, the algorithm can take any probability distributions depending on its manufacturing process. The platelet information is saved into the partitioned space. The partition-based method is an efficient method to save the random distributions of the platelets [9,10,11,15,16]. We use square, uniform, 1 mm × 1 mm size partitions for simplicity and robustness of the computational model. However, it is worth mentioning that other options are available. Li *et al.* [17], for instance, utilized Voronoi diagram shaped partitions to mimic the surface of DFCs. The algorithm generates the platelets until the number of average platelets through the thickness in the whole domain reaches the desired value. Each platelet thickness is adjusted in order to indirectly simulate the flow of the platelets. At this step, we also introduce layers of resin where there is a lack of platelets through the thickness. A detailed description of the generation process can be found in references [9,10].

2.2 Build a stochastic finite element model

The stored platelet information inside the partitions is transferred to a finite element model in Abaqus CAE [19] using a Python script. The model contains 2D layers of partitioned parts with the corresponding platelet information. The partitioned layers are modeled using S4R shell elements which is a 4-node, quadrilateral conventional shell elements with reduced integration. The in-plane damage behaviors such as the fiber and matrix failure are captured by the Hashin failure criteria [20]. We use a linear softening damage evolution to model the post-peak behaviors of the platelets. From previous studies [9,10], we have observed extensive damage developed in-between the platelets. Therefore, layers of cohesive elements are inserted between the platelets to capture the out-of-plane interlaminar failures. We choose COH3D8, 8-node 3D cohesive elements, with quadratic stress criteria as the failure criteria and a linear traction-separation law for the post-peak behaviors. The utilized cohesive laws are regularized such that they are mesh-objective and guarantee the correct dissipation of the fracture energy. A schematic of the model is in Fig. 1. By taking advantage of the shell elements along with the 3D cohesive elements, the model contains the necessary degrees of freedom to capture the failure mechanisms of DFCs with a relatively low computational cost. The size of the mesh is restricted by the size of the partition since an element can't overlap between the partitions. At the location of the notch, a size of mesh is reduced to 1/10 of the size of fracture process zone (FPZ) to capture the nonlinear behavior in front of the notch tip (see Fig. 2). The effective FPZ size for DFCs with the platelet size of 50×8 mm and thickness of 2.2 mm is found to be 3.63 mm [9]. A larger number of elements is desirable to fully capture the cohesive law [26]. The size of the mesh located outside of the notch region is set to be equal to the partition size. A mesh sensitive study is performed and showed minimum change with reduced size of the mesh.

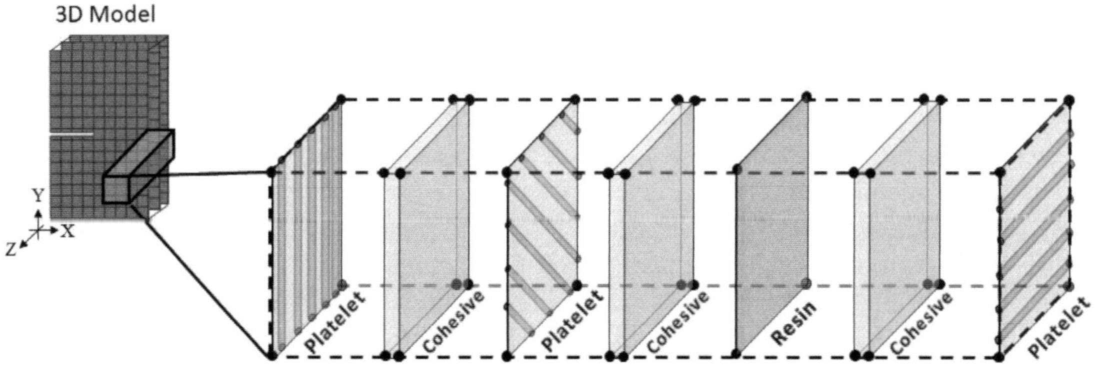

Figure 1. The schematic of a 3D stochastic finite element model. Shell elements are used to model the platelets and resin layers. Cohesive elements are used to model the cohesive layers. Tie constraints are applied in-between each different type of elements.

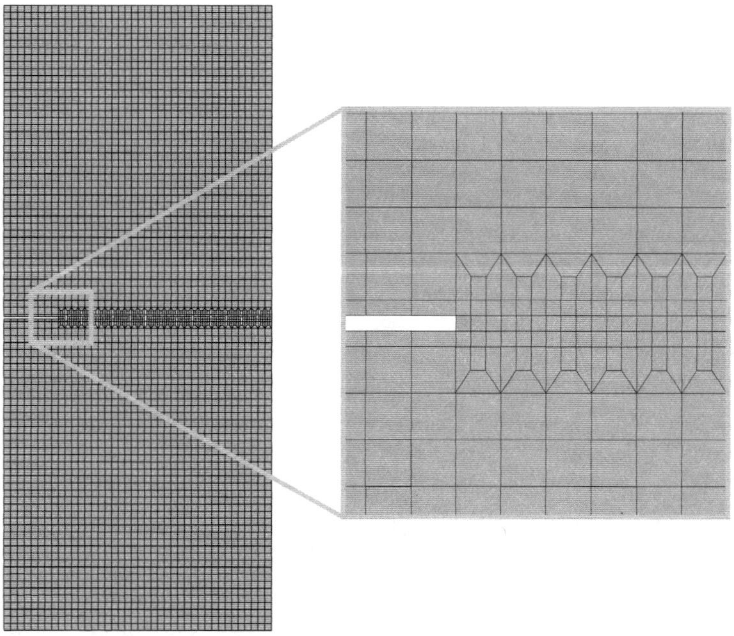

Figure 2. (a) Overview of the finite element model, (b) detailed mesh in front of the notch tip.

To calibrate the model, a set of the size effect experimental results from the previous study [9] is used. The size effect tests are carried on four different sizes of geometrically scaled single edge notch tension specimens. Size effect testing of notched structures is the only way to fully characterize the energetic size effect in quasibrittle composites, as abundantly shown in several recent publications [21-23]

We only need to calibrate once with a single type of DFC size effect tests. For the baseline, we use the platelet size of 50 × 8 mm and thickness of 2.2 mm. The calibrated material properties of the platelet and resin are in Table 1, the cohesive elements are in Table 2.

Table 1. Calibrated material properties for the platelets, and resin layers.

Properties	**Platelet**	**Resin**
Initial thickness, t [mm]	0.139	Varies
Longitudinal modulus, E_1 [GPa]	135	3
Transverse modulus, E_2 [GPa]	10	3
Shear modulus, G_{12} [GPa]	5	1.1
Poisson's ratio, ν_{12}	0.3	0.35
Longitudinal tensile strength [MPa]	1550	68
Transverse tensile strength [MPa]	85	68
Longitudinal shear strength [MPa]	110.6	68
Transverse shear strength [MPa]	61.49	68
Longitudinal tensile fracture energy [N/mm]	30	0.8
Transverse tensile fracture energy [N/mm]	0.6	0.8

Table 2. Calibrated material properties for the cohesive elements

Properties	**Value**
Mode I fracture energy [N/mm]	0.8
Mode II, III fracture energy [N/mm]	1.6

In Fig. 3, calibrated simulation results of the DFCs are plotted along with the experiments. The parameter D represents the width of the specimen and σ_{NC} corresponds to the nominal strength of the coupon. The nominal strength is defined as a $P/(Dt)$ where P is the maximum load and t is the coupon thickness. The size effect law extrapolated from the simulation results matches well with the experiments. The fracture energy obtained from the calibrated simulations is only 1.3% different from the value obtained from the experiments. For the future works, all specimen sizes will be calibrated to ensure the matching size effect curve.

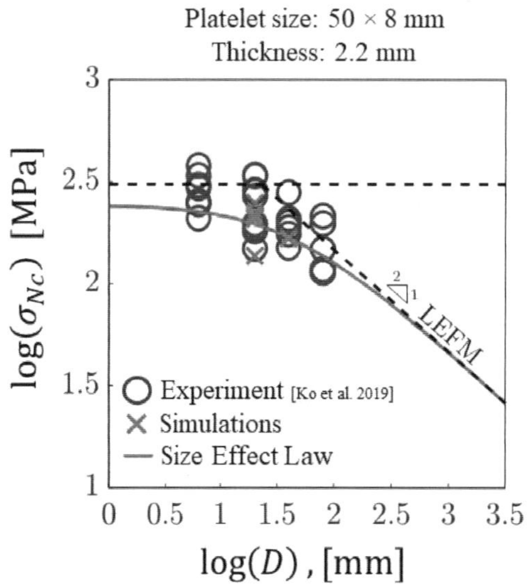

Figure 3. Calibrated FE model results show a good match with the experimental results.

3. RESULTS

3.1 Failure mechanisms of DFCs

After calibrating the model, we observe the failure mechanisms of DFCs layer by layer. First, the model captures the notch insensitivity of the smaller specimens frequently observed during the experiments [9, 10]. As shown in Fig. 4 (a) and (b), the model has a matching final failure location of the 20 mm width specimen. In Fig. 4 (b), the matrix damage is shown using the colorbar representing 0 as no damage and 1.0 as fully damaged element. At some layers, such as layer 16, the damaged elements are along with the final failure location. But at the layer 1 and 4, the damage is also accumulated near the notch tip. However, even if the damage is localized at the notch tip followed by the linear elastic fracture mechanics, DFCs can fracture away from the notch due to poorly oriented platelets or resin pockets. Using the computational model, we can easily capture this unique behavior of DFCs.

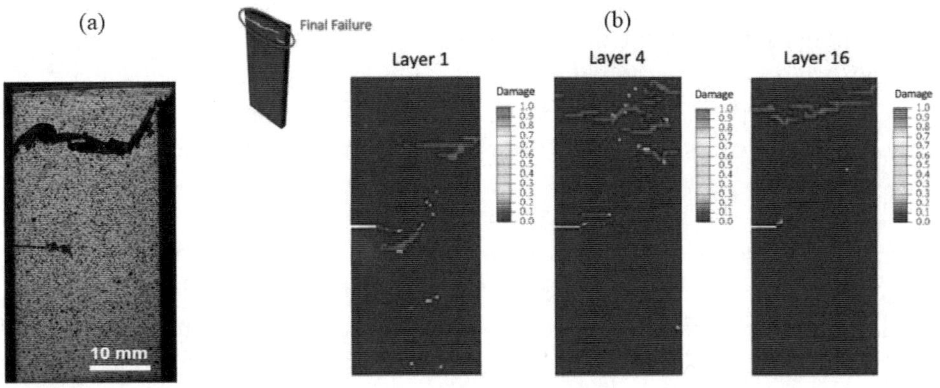

Figure 4. Comparison of the notch insensitive fracture behavior between (a) the experiment [9] and (b) the computational model.

To fully understand the damage mechanisms in a different scale, we use a JEOL JSM-7000F field emission Scanning Electron Microscope (SEM) to observe the fractured regions of DFCs. In Fig. 5, multiple damage mechanisms are present. As Fig. 5a shows, fiber breakage, platelet splitting, inter-platelet delamination are the most frequently observed failures. In Fig. 5b, we can clearly observe the matrix cracking contributing to the fracture. Such complex damage mechanisms lead to a high fracture energy in DFCs [9,10] compare to the continuous fiber composites.

(a) (b)

Figure 5. Images of the fracture surfaces captured by SEM. (a) Fiber breakage, platelet splitting, and delamination between platelets. (b) matrix cracking.

A key advantage of the computational model presented in this work is that, since we model the interactions between platelets explicitly, we can study the individual damage mechanisms and understand the evolution of each damage accumulation in-situ with the loading conditions. In this study, we consider three main failure mechanisms of DFCs. They are fiber failure in tension, matrix failure in tension, and out-of-plane delamination. Damage due to shear and compression also play important roles [24] but they are not considered here. Each damage can lead to final failure depending on the mesostructure. To study the evolution of these failures, we synchronize the load-displacement curve with the damage indices over the structure. The index starts from 0 to 1 meaning no damage to fully damaged elements. We obtain the damage in every element from every layer, then combine into a 2D image without averaging them. We distinguish the individual damages using different colors, black for the delamination, blue for the matrix damage, and red for the fiber damage.

In Fig. 6, a small specimen size of 44.5 × 20 mm is plotted. Figure 6b, c, and d represents a sequence of the damage evolution with respect to the displacement. In a linear region shown in Fig. 6b, a small amount of in-plane damages is accumulated near the notch tip (showing in red and blue). There is widespread delamination damage across the structure, but it does not lead to a critical fracture. When the structure reaches the peak as it shown in Fig. 6c, a large matrix damage is formed starting from the notch tip. Eventually, the structure fails with the fiber failure starting at the notch tip.

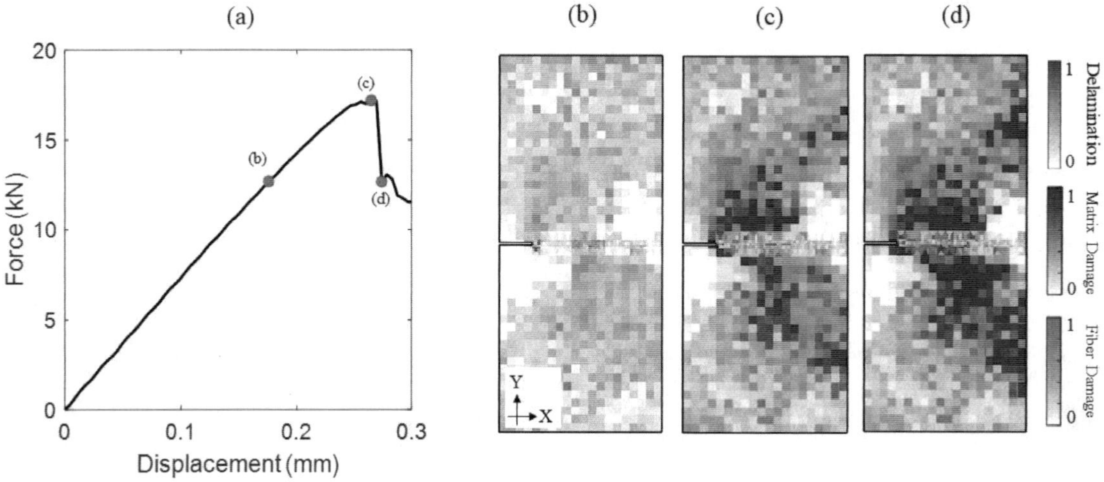

Figure 6. Load-displacement curve and the damage evolution sequence of a small size DFC coupon. A widespread matrix damage is shown over the structure.

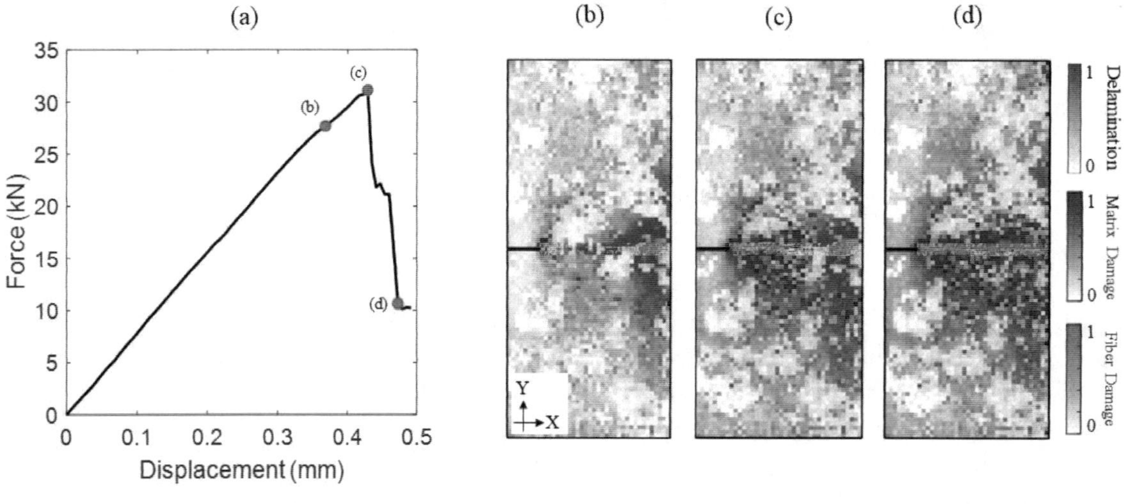

Figure 7. Load-displacement curve and the damage evolution sequence of a medium size DFC coupon. A localized matrix and fiber damage are shown near the notch tip.

In Fig. 7, there is a distinct trend of damage localizations near the notch tip. Similar to the small specimen, a widespread delamination is developed first. At the peak load, fiber damage is accumulated heavily near the notch tip and propagate perpendicular to the loading direction (direction in Y). This is a different fracture phenomenon compared to the small size coupon, where it shows more of widespread damage by the matrix direction. This observation is important because it provides qualitative data, indicating that as the structure size increases, the structure fails in more brittle manner. Therefore, we can apply the linear elastic fracture mechanism where the pre-existing notch is the strongest source of the fracture. This change in fracture behavior matches well with the size effect curve in Fig. 3 where the larger size specimen follows the linear slope of the LEFM curve. This is an important aspect to capture when studying the strong quasibrittle material such as DFCs. As shown in the previous study [9], DFCs contain a significantly large area of fracture process zone (FPZ) compared to the continuous fiber composites. The tested results show that for the DFCs with the platelet size of 75 × 12 mm, the size of effective FPZ can be larger than the quasi-isotropic layup in an order of magnitude. Therefore, the correct computational model should capture the effect of FPZ size by matching the size effect.

3.2 Prediction of the size effect law

The proposed stochastic finite element model is calibrated using the platelet size of 50 × 8 mm and thickness of 2.2 mm. If the model correctly represents the necessary details of the mesostructures, it should capture the size effect of the different platelet sizes with thickness. Also, it should be able to do it without recalibration of the material properties. To confirm, we create mesostructures with the platelet size of 25 × 4 mm and thickness of 3.3 mm. The calibrated material properties remain unchanged. The mesostructures are transferred to the FE models. We numerically test two different sizes of the coupons (small and medium). The results are plotted along with the experimental data from [9,10] in Fig. 8. The size effect curve from the numerical simulations matches well with the trend of the experimental data. Even though there is a high scatter in the experimental results, the FE model captures the size effect satisfactorily. This shows that the model captures the fracturing behaviors of the different platelet sizes and thickness without adjusting the input material properties. With the stochastic FE model, we can find the associated strength and moreover, estimate the fracture energy of a given platelet size and thickness without running the size effect experiments. For the future work, we expect to run additional simulations in different sizes to confirm the matching size effect curve. Additionally, we will observe the effect of the other parameters such as platelet tow thickness, platelet aspect ratio, and resin pocket volume percentage.

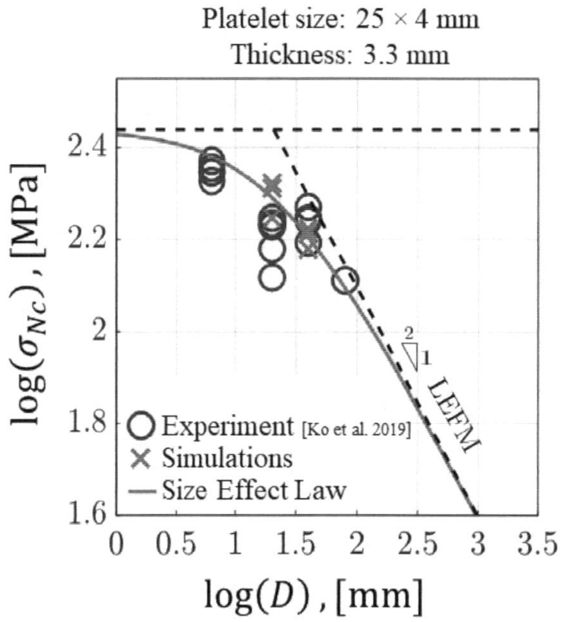

Figure 8. Experimental results are plotted along with the numerically predicted size effect curve for the DFC with platelet size of 25×4 mm and thickness of 3.3 mm.

4. CONCLUSIONS

We introduce the computational stochastic finite element model that utilizes a random platelet generation algorithm to simulate the complex material mesostructure. Using the model, we study the fracturing behaviors of the discontinuous fiber composites with three main failure mechanisms: fiber tension, matrix tension, and out-of-plane delamination. The model presents in-situ, unique damage evolutions of the DFC structures. The small-size structure shows a widespread matrix failure at the peak load whereas the larger-size structure has more localized damage accumulated near the notch end. These observations support the strong size effect of DFCs where the effective fracture process zone plays a significant role in the failure mechanisms. The implemented model provides the predictions of strength for the different mesostructures without recalibrating the material properties. This will bring a significant reduction of the experimental cost to validate the material properties and will help engineers to choose the optimized platelet morphology for their DFC structures.

5. ACKNOWLEDGEMENTS

We thank FAA-funded Center of Excellence for Advanced Materials in Transport Aircraft Structures (AMTAS) and the Boeing Company for the financial supports. Also, we acknowledge the partial support from the Joint Center for Aerospace Technology Innovation (JCATI). We thank Ahmet Oztekin, Cindy Ashforth, and Larry Ilcewicz from the FAA, and William Avery from the Boeing Company for their advice and support. We also thank Bruno Boursier from the Hexcel Corporation for the technical support.

6. REFERENCES

1. J. Fudge, "Chopped prepregs – compelling performance and cost effective material." SAMPE conference 41st ITSC, Wichita KS Oct 19-22nd, 2009.
2. GP Picher-Martel, A. Levy, P. Huber. "Compression moulding of Carbon/PEEK randomly-oriented strands composites: A 2D finite element model to predict the squeeze flow behavior." Comp Part A 2016;81:67-77
3. B. Boursier, A. Lopez. "Failure initiation and effect of defects in structural discontinuous fiber composites", Hexcel Research and Technology, 2010.
4. P. Feraboli, E. Peitso, F. Deleo, T. Cleveland. "Characterization of prepreg-based discontinuous carbon fiber/epoxy systems." J Reinf Plast Compos 2009;28(10):1191–214.
5. M. Selezneva, L. Lessard. "Characterization of mechanical properties of randomly oriented strand thermoplastic composites." J Compos mater 2016;50(20):2833-51.
6. S. Kravchenko, D. Sommer, B. Denos, W. Avery, R.B. Pipes. "Structure-property relationship for a prepreg platelet molded composite with engineered meso-morphology." Compos Struct 2019;210:430-45.
7. S. Kravchenko, D. Sommer, B. Denos, A. Favaloro, C. Tow, W. Avery, R.B. Pipes. "Tensile properties of a stochastic prepreg platelet molded composite." Comp Part A 2019;124:105507.
8. S. Kravchenko, D. Sommer, R.B. Pipes. "Uniaxial strength of a composite array of overlaid and aligned prepreg platelets." Comp Part A 2018;109:31-47.
9. S. Ko, J. Davey, S. Douglass, J. Yang, M.E. Tuttle, M. Salviato. "Effect of the thickness on the fracturing behavior of discontinuous fiber composite structures." Comp Part A 2019;125:105520.
10. S. Ko, J. Yang, M.E. Tuttle, M. Salviato. "Effect of the platelet size on the fracturing behavior and size effect of discontinuous fiber composite structures." Comp Struct 2019;227:111245.
11. B. Denos, D. Sommer, A. Favaloro, R.B. Pipes, W. Avery. "Simulation of prepreg platelet compression molding: method and orientation validation." J Rheol 2018;62(2):1443-55.
12. L. Martulli, L. Muyshondt, M. Kerschbaum, S. Pimenta, S.V. Lomov, Y. Swolfs. "Carbon fibre sheet moulding compounds with high in-mould flow: Linking morphology to tensile and compressive properties." Comp Part A 2019;126:105600
13. Y. Wan, I. Straumit, J. Takahashi, S.V. Lomov. "Micro-CT analysis of the orientation unevenness in randomly chopped strand composites in relation to the strand length." Comp Struct 2018;206:865-875.
14. Z. Chen, T. Huang, Y. Shao, Y. Li, H. Xu, K. Avery, D. Zeng, W. Chen, X. Su. "Multiscale finite element modeling of sheet molding compound (SMC) composite structure based on stochastic mesostructure reconstruction." Comp Struct 2018;188:25-38
15. M. Selezneva, S. Roy, S. Meldrum, L. Lessard, A. Yousefpour. "Modelling of mechanical properties of randomly oriented strand thermoplastic composites." J Compos Mater 2017;51:831-45.

16. Y. Li, S. Pimenta, J. Singgih, S. Nothdurfter, K. Schuffenhauer. "Experimental investigation of randomly-oriented tow-based discontinuous composites and their equivalent laminates." Comp Part A 2017;102:64-75

17. Y. Li, Z. Chen, L. Su, W. Chen, X. Jin, H. Xu. "Stochastic reconstruction and microstructure modeling of SMC chopped fiber composites." Comp Struct 2018;200:153-164.

18. S. Ko, K. Chan, R. Hawkins, R. Jayaram, C. Lynch, R. El Mamoune, *et al.* "Characterization and computational modeling of the fracture behavior in discontinuous fiber composite structure." SAMPE Conference, Long Beach, CA. 21-24 May 2018.

19. Dassault Systemes ABAQUS. ABAQUS Documentation. Providence, RI; 2018.

20. Z. Hashin. "Failure criteria for unidirectional fiber composites." J Applied Mechanics 1980;47:329-34.

21. Z.P. Bažant, J. Planas. "Fracture and size effect in concrete and other quasibrittle materials." CRC press; 1998.

22. M. Salviato, K. Kirane, Z.P. Bažant, G. Cusatis. "Experimental and numerical investigation of intra-laminar size effect in textile composites." Comps Sci Technology 2016;135:67-75.

23. M. Salviato, K. Kirane, Z.P. Bažant, G. Cusatis. "Mode I and II interlaminar fracture in laminated composites: A size effect study." Journal of Applied Mechanics 2019;86:091008.

24. Y. Li, S. Pimenta. "Development and assessment of modelling strategies to predict failure in tow-based discontinuous composites." Comp Struct 2019;209:1005-1021.

25. M. Selezneva. "Experimental and theoretical investigations of mechanical properties of randomly-oriented strand (ROS) composites." PhD dissertation, McGill University, Montreal, 2015.

26. Y. Qiao, M. Salviato. "Strength and cohesive behavior of thermoset polymers at the microscale: A size-effect study." Eng Fract Mech 2019;213:100-117.

27. M. C. Preetam, S.V. Lloyd. "Effect of bond quality on crack growth resistance of adhesively bonded composite joints under static and cyclic loading." SAMPE Conference, Seattle, WA. 22-25 May 2017.

28. Y. Wan, J. Takahashi "Tensile and compressive properties of chopped carbon fiber tapes reinforced thermoplastics with different fiber lengths and molding pressures." Comp Part A. 2016;87:271-281.

29. G. Milakantan, S. Nutt. "Reuse and upcycling of thermoset prepreg scrap: Case study with out-of-autoclave carbon fiber/epoxy prepreg." J Compos Mater 2018;52:341-360.

30. Digimat User's Manual. Version 2017.0, e-Xstream Engineering. MSC.

HOLISTIC DESIGN ANALYSIS APPROACH FOR ADDITIVE TAPE PLACEMENT PROCESS

Anthony Cheruet, Pierre-Yves Lavertu
e-Xstream Engineering
ZAE Robert Steichen – 5 rue Bommel, Hautcharage, L-4940, Luxembourg

Yvan Blanchard
Coriolis Composites
Am Technologiezentrum 5, D-86159 Augsburg, Germany

ABSTRACT

This paper is focused on the design optimization of complex 3D composites structures made by additive manufacturing processes. There are commercial CAD-CAM software solutions for detailed offline path programming, but there is a growing need for innovative tools and methodologies for doing trade off studies very early at design stage. Most of the technology available on the market are not really suited to easily check and compare various design proposals, with an efficient compromise between accuracy and performances, especially for thick and large laminates. A new innovative solution has been developed, allowing both designers and stress engineers to quickly analyze complex double-curved geometries, including variable stiffness approach and structural analysis of manufacturing defects. Automatic Fiber Placement (AFP) is a fast and efficient deposition process of carbon prepreg for large component applications. To accommodate the composite strips onto a double curved surface, the tows can be cut, restarted and slightly mis-oriented resulting in the apparition of gaps between tows. These defects, the gaps and the misalignment of the tows, affect the mechanical performance of the final part.

Keywords: AFP, Automatic Fiber Placement, Automated Fiber Placement, Additive Manufacturing, AM, Multi-Scale Analysis, Simulation
Corresponding author: Pierre-Yves Lavertu – Pierre-yves.lavertu@hexagon.com

1. INTRODUCTION

Today, more and more CFRP structures are manufactured by automated processes such as fiber placement robotic systems (Figure 1).

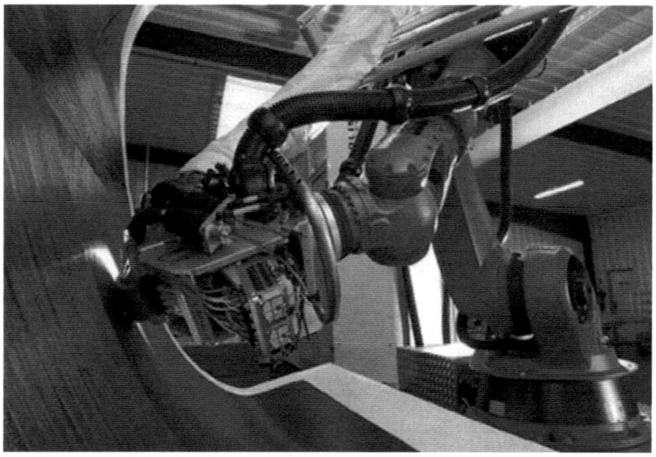

Figure 1. Automated Fiber Placement robotic system (Coriolis Composites).

Those structures are designed in standard CAD environments (such as CATIA CPD or NX/FiberSim), and then offline path programming is done on separate CAM software systems. The problem is that iterations between composites design definition and manufacturing path programming model, especially for doing conceptual trade-off studies, can't be efficiently done. Another issue is that design optimization tools for composites manufacturing are often complex and difficult to implement in the designer environment. The expected system should be very easy to use and to customize (i.e. Excel spreadsheet, with VB or Python scripting and automation). Moreover, such analysis systems are not able to reach a good compromise between performances, level of details and results accuracy (close to detailed offline NC programming tools). The expected computation time is about seconds or minutes, and several analysis runs should be done in parallel in a few hours only. Designers and stress engineers need robust tools and methodologies to help them testing large number of configurations (material width, maximum number of tows by course, maximum fiber deviation angle, etc) of representative laminates, i.e. large aerospace panels with double curvature and hundreds of plies. The target system should also be able to address various layup strategies, such as steered-fiber design. The design choices must also be validated by accurate FE structural analysis, taking local layup defects into account.

2. USER INTERFACE AND CORE ARCHITECTURE

The implemented solution is based on a Microsoft EXCEL spreadsheet, for defining all the user input parameters (material, process, strategy) and to launch the background batch processes in silent mode.

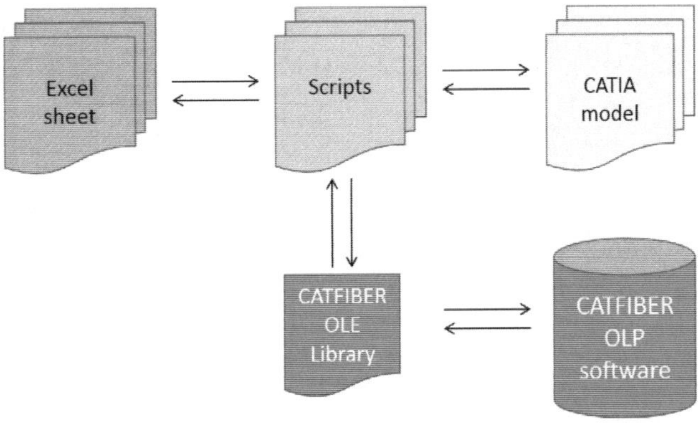

Figure 2. Core architecture of the implemented software solution.

This spreadsheet is also used to collect the output results data, and ease comparative studies between several runs. All the standard Excel graphical tools help to better visualize results (histograms, 2D and 3D plots, etc). The core system is based on CATIA V5 software and the CATFiber© standard solution for composites offline path programming. All the geometrical outputs (surfaces, wireframe geometry) are then contained in a standard CATPART document. It also allows to have a unique environment for CAD, CAE and CAM analysis. A specific OLE-based CATIA library was built, to drive the different additive manufacturing algorithms of CATFiber© and collect the results data. This generic library is directly used through a VBA or Python scripting, in order to easily customize the analysis, depending on target application (Figure 2).

2.1 Additive Manufacturability analysis

The manufacturability (or producibility) analysis of plies is based on the existing Automatic Ply Sectorization algorithm already implemented in the CATFIBER OLP software. The technique is based on an iterative algorithm and consists at doing path parallels on the layup surface, and then split the ply boundary in several pieces if the engineering or manufacturing requirements are no longer respected (fiber deviation, steering radius). A new course centerline path is then computed, optimized to minimize triangle of gaps and maximize the number of parallel strips. This step is repeated to fill the entire ply boundary (Figure 3). Here are the main input parameters for the algorithm:

- Geometrical layup surface
- Engineering rosette axis
- Ply contour
- Initial seed-point

- Material width
- Maximum number of tows by course
- Maximal angular deviation
- Minimum steering radius

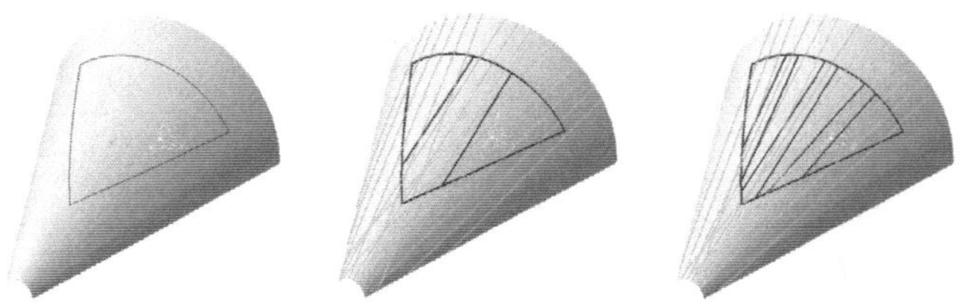

Figure 3. Left: initial conical surface with boundary, rosette and initial seed-point, Middle: splicing with 10 deg of deviation tolerance leading to 3 sectors, Right: splicing with 5 deg of deviation tolerance leading to 7 sectors.

The output result is a list of ply sectors, each made of a closed contour and one or several optimized guide curves. The system was also improved to integrate the following additional engineering and manufacturing requirements:

- Take results inside the engineering edge of ply only
- Avoid ply cuts and tow drop off located in a specified zone (example: stringer bonding area).
- Adjust the tape width (number of tows) if roller conformance is not compliant (excessive crush and/or fiber bridging defects)
- Minimize the lateral gap and overlap between adjacent tape courses, with automatic tow length adjustment (Figure 4)
- Automatic management of sequence of several plies in a same layer ("patch" areas)

It is also possible to define a (father) cutting pattern for plies with same orientation, and apply it to the children plies, for a better control of splicing curves location through the laminate thickness.

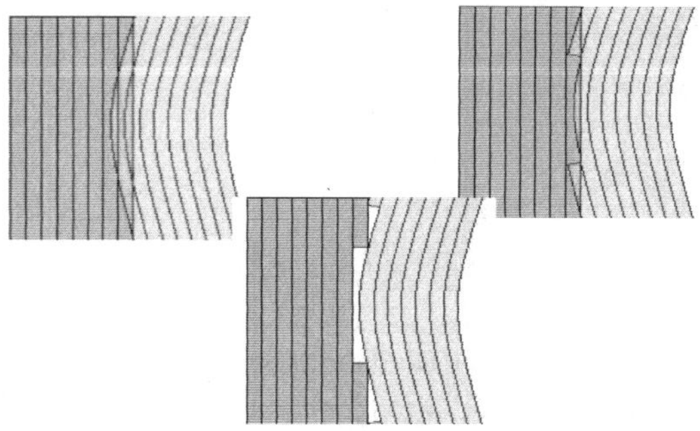

Figure 4. Gap & Lap management between tape courses (first image is initial design, and others are possible tow adjustments depending on engineering requirement.

2.2 Material covering analysis

In order to accurately compute and analyze the engineering requirements, the same covering algorithm as the one available in the full CATFIBER NC programming (OLP) software is used. Starting from the optimized course centerlines and ply cuts (sectors) computed using the automatic splicing algorithm, the ply boundary is filled with (tow) strip surfaces. This allows to capture all the process and material specificities, such as triangle of gaps, tow overlaps, minimum course length (MCL), minimum distance between tow cuts (Figure 5). It is important to know that parallel course strips can be time consuming, especially for very large layup areas. But the coverage algorithm is very fast thanks to a surface tessellation definition and highly parallelized computations.

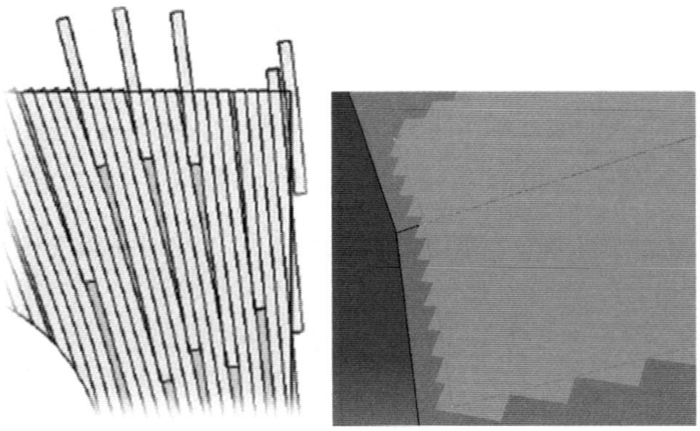

Figure 5. Tape coverage examples, with tow drops and minimum course length (left) and tessellation surface rendering (right).

2.3 Variable-stiffness design

The variable-stiffness design approach consists at defining curvilinear course path within layer, allowing optimized non-conventional (tow-steered) composites stacking (Figure 6). This is possible using fiber placement process.

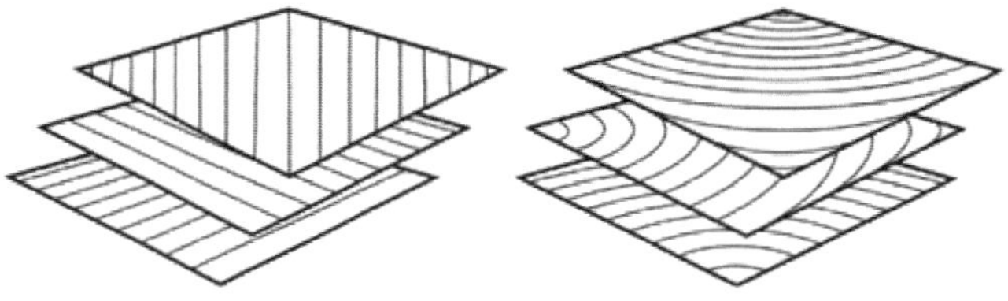

Figure 6. Conventional (left) and tow-steered (right) laminates.

The path programming and manufacturability analysis tools are based on advanced rosette transfer features. Those features enable user to easily customize the engineering rosette to better reflect the internal load path of the structures. The rosette transfer can integrate one or several guiding curves, extracted from the geometry (stringer or spar location curves for example). The transfer algorithm will then compute by interpolation the new updated fiber direction (Figure 7).

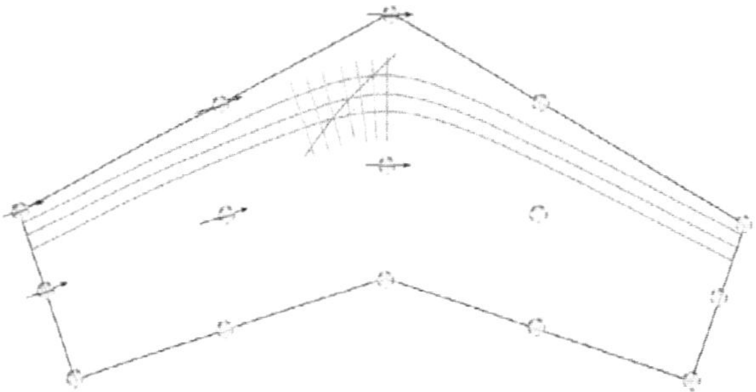

Figure 7. Example of rosette curves (in red), for horizontal stabilizer application [4].

This generic rosette operation feature is based on a grid structure made of individual and independent local rosettes [5]. Thus, it is possible to import input discrete fiber angle distribution, either from FEA principal stress vectors, or from an external genetic optimization algorithm. With such a rosette grid, the course coverage algorithm can generate smoothed interpolated tow-steering paths (Figures 8a & 8b).

We could also notice that ply boundaries can be computed by topological optimization solvers (Simulia Tosca, Altair Optistruct, or others), in order to define the optimized plies shape and stacking thickness (Figure 8c).

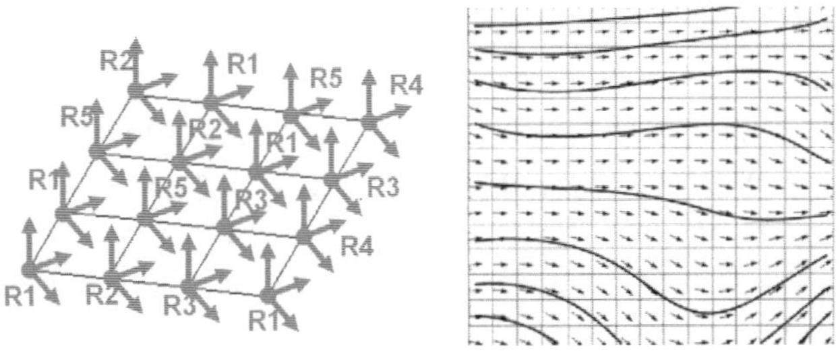

Figure 8a,b. Grid of rosettes (left), and example of smoothed courses built from load paths (right).

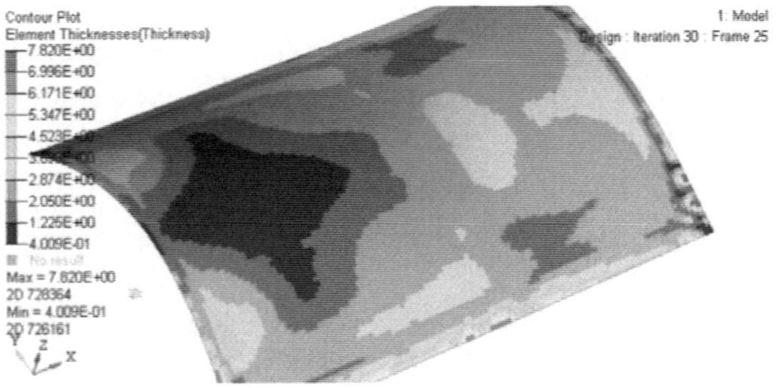

Figure 8c. Topological optimization of Aerospace panel for Automated layup (courtesy of Altair Engineering).

2.4 Manufacturing cost estimation

The design analysis system is also able to compute and output several cost indicators, to help designer to better sort the manufactural design proposals. Since the final manufacturing cost is too complex to determine (layup cycle time, non-productive time, tooling preparation, inspection, etc), and because the layup time strongly depends on target machine and material, we decided to define the following intermediate cost estimators:

- Number of tape courses with average centerline length (gives a rough overview of productivity)
- Number of ply cuts (indicates the number of tow drop-off within the layer)
- Number of tow drops (helpful for identifying the less productive courses / sequences, and the machine cutting knives maintenance cost)
- Buy-to-fly ratio (split of ply material area with engineering edge of ply)

There is also the possibility (as an option) to compute an estimation of the toolpath, using generic AFP & ATL machine layer settings. This allows to get an additional cost indicator for layup time, taking variable feed-rate into account (feed and cut speed limitation) and part curvature as well.

2.5 Structural Analysis validation

It is important to check the structural strength of the manufacturable and cost-optimized design proposals. That's why the software solution integrates an add-on to transfer all the manufacturing data (tape courses) for accurate Finite Element Analysis. This allows stress engineers to validate the layup strategies, especially if using the variable-stiffness approach, by taking true "as-manufacturing" fiber directions and effect of the presence of gaps and resin-rich areas. This is done by using a data mapping between the tape triangle mesh and the structural shell mesh, through a core sampling technique. This mapping is done using the Digimat Platform and concerns the transfer of the as-manufactured fiber orientation, the exact location of the gap and resin-rich area. Such information can be transferred automatically to the Finite Element Model used for sizing activities by Stress Engineers. In this case, Digimat also generated the composite layup command cards with the local as-manufactured fiber orientation over each ply. In addition, the effect of the gaps on the local stiffness can be handled in two ways, depending on the reconsolidation process. The first one considers that the gaps affect the local thickness while the second one considers that the gaps are filled by resin and affects the local fiber volume fraction of the composite. Using a micro-mechanical model of the material, this local variation of the fiber content is computed at each Gauss point of the FE Model. With such approach, verification of the global stiffness and strength of the component under working conditions become extremely efficient. Local post-processing can be performed as well with the verification of the local stress field discontinuities due to the presence of gaps within the plies (Figure 16).
This current validation simulation approach operates under important assumptions that need to be highlighted:

- There is no process analysis being conducted and therefore, the bonding between each tape layers are considered ideal (perfectly bonded).
- In case of thermoset prepregs being used as tape, no curing analysis of any kind is being conducted which again will lead to the assumption that the layers were fully cured and bonded together.
- The validation simulation includes fiber steering but does not take tape wrinkle into considerations.

It is also known that when working with FEA, the element size impacts in some ways the results of the analysis. When mechanical performances are evaluated, geometrical features will impact the part mechanical response locally and elements may generate geometrical artifact. In the current approach, two new important geometrical features must be captured: the tape strips and the gaps on top of all other part geometrical features. Consequently, a maximum element size must be defined and should be smaller than the gaps to represent. In the current approach, while the element size isn't fixed, it must be equal or smaller than the gaps. No element size sensitivity study has been done in this case and but is highly recommended when working on a new application.

3. APPLICATION ON AIRCRAFT WING SKIN

The software solution for design analysis was applied on a typical aircraft wing extrados skin (Figure 9). The CATIA composites stacking is made of 100 sequences (about 24 m2 of layup area), using the following nominal process and material settings (defined through the Excel spreadsheet):

- 8 tows of 38.1mm (1.5in) width

- Full tapes only (8 tows min)
- Fixed angle path strategy (with parallels)
- Minimum course length of 100mm
- No tow overlaps
- Nominal gap between courses of 0.5 mm
- Maximum gap allowed of 1.5 mm

The part integrates a rosette transfer made of guiding taken from the sub-structure (spars), allowing to better catch the load paths of the structure, leading to variable fiber directions in the layer. The wing skin surface is slightly double-curved, but the selected rosette transfer induced tow drops in the middle and on the edges of the ply (Figure 10). An advanced capture of those triangle of gaps (and resin) was done, to accurately check the impact on global and local structural strength (Figures 14 & 15).The full part analysis was launched in automatic using the batch framework in about 250 minutes (including all the output results data collection).
At the end of the process, all the output geometrical results (course centerline curves, ply boundary splices, ply layup surfaces) are stored in the CATIA part model and can be easily visualized and inspected (Figures 11a & 11b). It was also possible to launch a study on the influence of the maximum fiber deviation allowed on the total gap area for a given tow material width, and directly plot the results in the Excel file (Figure 12). Another example of design study is the influence of maximum deviation angle and minimum steering (turning) radius targets on manufacturing indicators. Some results on first full plies are given in Tables 1 & 2, with comparative results for the 0 deg ply. It appears that the input targets (especially the steering radius) can't be respected here since we ask for fixed angle paths and full tapes only. Those data are automatically consumed to the CATFIBER NC Programming OLP system (in the same CAD-CAM environment) for tool-path simulation with machine kinematics (Figure 13).

Configuration	Nb tows by course	Tow width (mm)	Max deviation Target (deg)	Min Steering radius Target (mm)
A	6	63.5	5	20
B	6	63.5	3	1000
C	6	63.5	3	20

Table 1. Design configurations (main input parameters).

Ply Angle	Config	Nb tapes	Nb ply cuts	Max angle dev (deg)	Min steering (mm)	Gap max (in)	Coverage (%)
90	A	49	0	5.3	1227.33	0.51	100.34
45	A	38	2	12.05	3690.37	38.1	101.48
135	A	47	3	9.01	1157.48	34.54	101.2
0	A	21	3	4.41	3025.14	18.03	101.6
0	B	21	16	2.48	2890.52	19.05	101.6
0	C	21	7	2.87	5097.78	19.05	101.6

Table 2. Analysis results for 4 full plies

Figure 9. Wing skin CAD model with an example of ply boundary.

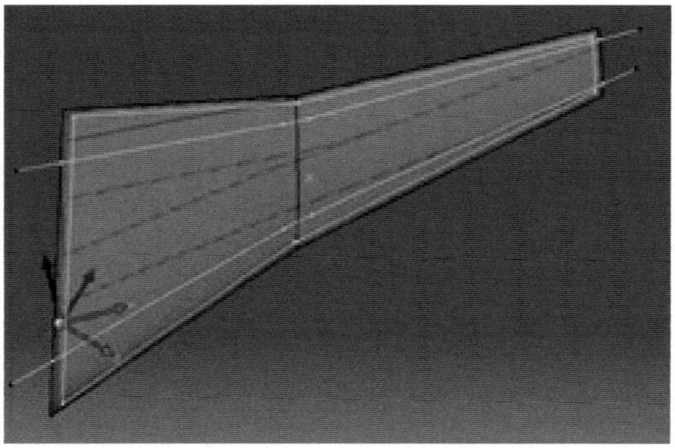

Figure 10. Wing skin with global Cartesian rosette (on the bottom left) and advanced rosette transfer with guiding curves (in white), and the resulting material coverage for a full 0 deg ply (with tow drops).

Figure 11a and 11b. Complete wing skin fiber-placed material build-up (one color by sequence).

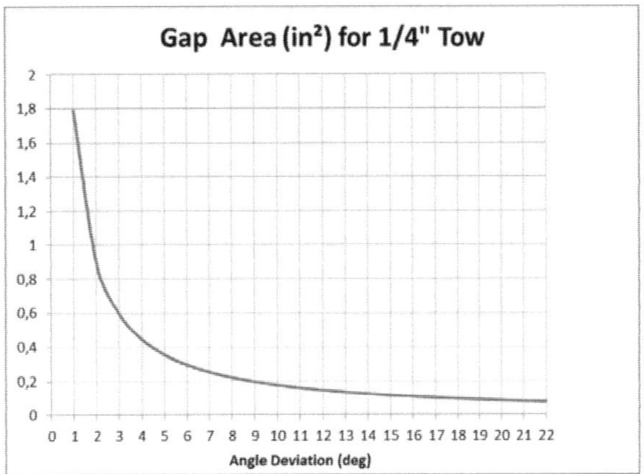

Figure 12. Example of analysis study on the influence of angle deviation on total gap area (Excel chart).

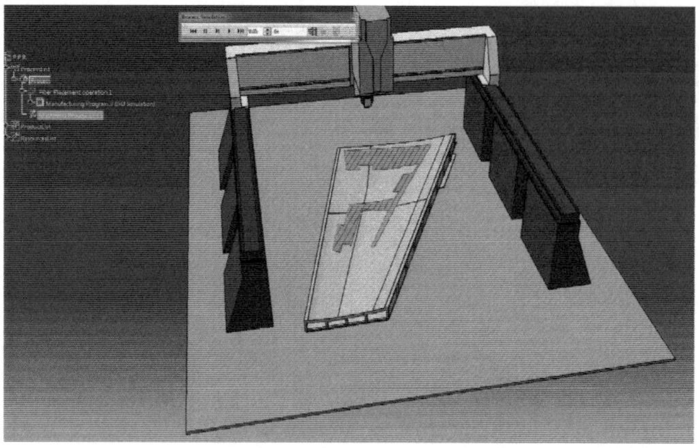

Figure 13. Layup simulation with a CNC Gantry AFP machine kinematics in CATFIBER OLP software.

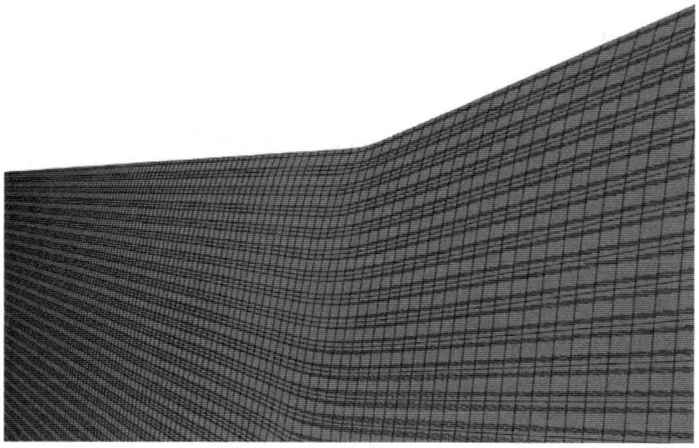

Figure 14. Fiber directions mapping on FE structural shell mesh (2D Mesh).

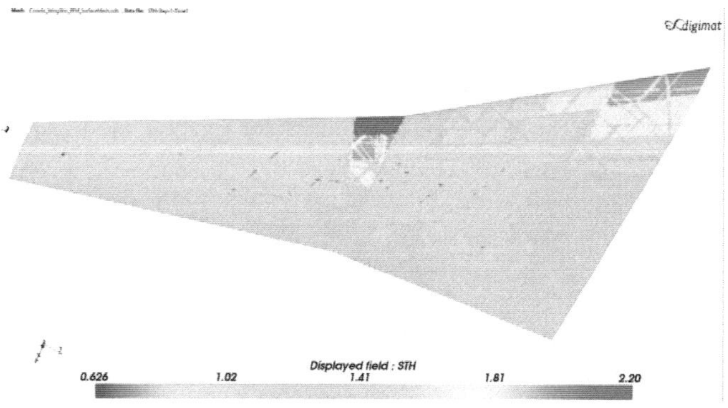

Figure 15. Thickness map analysis (quasi-isotropic laminate) with tow gaps.

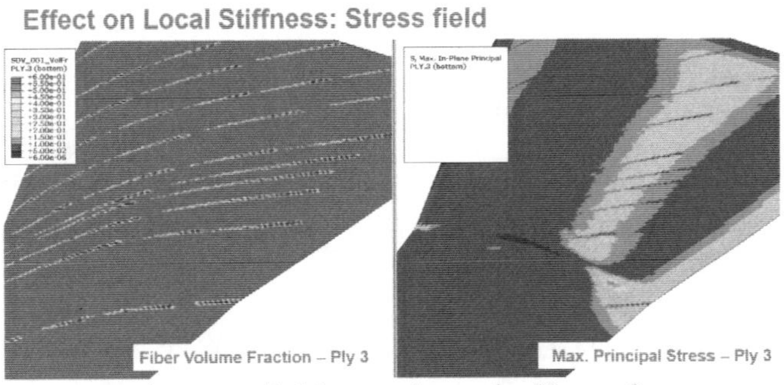

Figure 16. Local stress field per ply and effect of gaps computed.

4. CONCLUSION

Design optimization of composites structures made by additive manufacturing process is efficient and reliable by taking into account engineering requirements, material, machine limitations, and manufacturing cost. The offline programming system, traditionally used by NC programmers for the detailed layup program, should be packaged for designers and stress engineers, in order to eliminate useless iterations between CAD and CAM systems. Such a design system must be easy to use and flexible enough, with the possibility to check and compare various design proposals, including tow-steering capability and advanced structural strength validation. Such a solution, based on automation and scripting, was successfully implemented on a typical aero panel case with a large laminate size. Improvements will be done as a next step, and process simulations such as curing, layer bonding, crystallization can be added to help predict any defects during the manufacturing to better optimize the results and enhance the global performances. In addition to that, in-situ tape placement data could be added for digital twin generation. Finally, Other complex geometries will be also tested (wing spars, fuselage frames, engine fan blades).

5. REFERENCES

1. Wu, KC. "Design and analysis of tow-steered composite shells using fiber placement". *American Society for Composites 23rd Technical Conference*. Memphis, TN, USA. 2008, DOI: 10.4271/2012-01-0082

2. Croft, K., Lessard, L., Pasini, D., Hojjati, M., Chen, J., Yousefpour, A. "Experimental study of the effect of automated fiber placement induced defects on performance of composite laminates". *Journal of Composite Part A, Composites*: Part A, vol 42, pp 484–491, 2011, DOI: 10.4271/2012-01-0082

3. Fayazbakhsh, K., Arian Nik, M.,Pasini, D., Lessard, L., "Defect layer method to capture effect of gaps and overlaps in variable stiffness laminates made by Automated Fiber Placement", DOI: 10.4271/2012-01-0082

4. NLR., "Fiber steered skin design of composite thermoplastic horizontal stabilizer torsion box", Feb. 2017.

5. Olivier Munaux, 2016, "Method for Defining Fiber Trajectories from a Transfer Surface", PCT/FR2014/000108.

6. G. Gonzalez Lozano, A. Tiwari, C. Turner, "A design algorithm to model fiber paths for manufacturing of structurally optimized composite laminates". *Composite Structures* (2018), DOI:10.1016/2018.07.088

A STUDY OF CUTTING EDGE GEOMETRY CHANGE AND SURFACE ROUGHNESS IN DRILLING OF CARBON FIBER REINFORCED PLASTICS (CFRP) USING UNCOATED AND COATED TOOLS

Mohammad Sayem Bin Abdullah[1], Dinh Nguyen[2], Dave Kim[3]

[1]Department of Mechanical Engineering, University of Washington, Seattle, WA 98105.

[2]Department of Mechanical Engineering, Michigan State University, East Lansing, MI 48824.

[3]School of Engineering and Computer Science, Washington State University, Vancouver, WA 98686.

ABSTRACT

Cutting edge geometry is a great research concern in machining. This paper investigates the effects of tool wear on cutting edge geometry change, consequent effect of cutting edge geometry on inner hole surface roughness in drilling of CFRP. A drilling experiment has been conducted on CFRP using one uncoated carbide drill bit and two coated carbide tools, namely BAM (BAlMg4) and (AlCrSi/Ti)N. Sharp, honed, chamfered, chamfered induced honed cutting edge has been found in CFRP drilling in this investigation. The sharp tool edge transforms into honed and chamfered due to tool wear in subsequent drilling. The coatings delay the edge geometry changes. The surface roughness of the holes has been measured across the half hole at three consecutive holes in every 40 holes interval holes up to 120 holes. Upon clustering the cutting edges of the three drill bits, the paper investigates the tool edge geometry relation with surface roughness parameters (Ra, Rq, Rz, Rt, Rp, Rv) for each drill bits. The results show that cutting edge geometry has a strong influence on the surface roughness in the drilling of CFRP. Chamfered edge created the largest roughness, while the roughness decreased when the chamfered edges turn to chamfered and honed.

Keywords: Carbon Fiber Reinforced Plastics (CFRP), Tungsten carbide (WC), BAM, (AlCrSi/Ti)N, Drilling, Cutting edge geometry
Corresponding author: *Mohammad Sayem Bin Abdullah, Email: sayemab@uw.edu
* A previous graduate student at Washington State University, Vancouver, WA

1. INTRODUCTION

The application of Carbon fiber reinforced plastics (CFRP) is increasing in many industries, including aerospace and automobile due to its high specific strength, high specific stiffness, high fatigue resistance, low density, low thermal expansion, corrosion resistance, and excellent damping properties [[1–7]. As polymer matrix is used as the matrix material with carbon fiber as reinforcement, CFRP cannot be welded like metal and other counterparts [8]. Hence, mechanical fastening is the most widespread means of assembling CFRP to other structures due to its simplicity, lower-risk involved in joining, and ease of repair [8,9]. Conventional drilling is the most widely used process for making holes in CFRP plates and many industries, i.e., aerospace, require strict tolerance on CFRP holes [10,11]. One of the important tolerance is posed on surface roughness as the very rougher surface may affect strength, load-bearing capacity, and fatigue life of CFRP[12,13]. The highest acceptable Ra value of in drilling and edge trimming of CFRP-metal stack and CFRP are 3.2 μm and 4.8 μm [11,14]. Defective holes generated by drilling are reported because of the removal of 60% of the composite parts [9,13]. The surface roughness is believed to be the result of severe tool wear as the tool becomes dull and loses its sharpness. Such a dull tool eventually pushes the fiber instead of cutting or fracturing, thereby creates a rougher surface. The cutting edge of cutting tool interacts with the workpiece; thus, the geometry changes due to tool wear and continuous interactions. The tool edge geometry, also known as microgeometry of the cutting tool, may have a relationship with surface roughness. Zhao et al. and Denkena et al. reported that various types of micro-geometry in the machining of metal like steel and metal matrix composite[13–18]. They categorized the cutting edge in sharp, honed, chamfered, chamfered and honed. An illustration of their finding has been shown in a later section of this paper, in Figure 3. Zhao et al. studied the relation of microgeometry on surface roughness in machining metal [21,22]. They reported that the cutting edge strength becomes stronger and creates a rougher surface. Samuel Raj and L. Karunamoorthy the effects of microgeometry on surface roughness along with other parameters, where they reported the proportion relation of peak flattening and cutting edge flattening with surface roughness [23].

In this paper, the relationship of surface roughness of Carbon Fiber Reinforced Plastics (CFRP) hole and cutting edge geometry obtained due to tool wear is investigated in a drilling experiment. This drilling experiment was conducted to compare the effectiveness of the ceramic coated (BAM coating, (AlCrSi/Ti)N coating, and the uncoated tool and the results are published in our previous publication [4]. However, the goal of this paper is to investigate any potential relation of tool micro geometry (cutting edge geometry) and surface roughness. This paper provides a hint to the future direction of evaluating cutting edge geometry and roughness relationship in composite machining.

2. EXPERIMENTATION

2.1 Experimental Set-Up

The drilling experiment was carried out in a three-axis vertical CNC milling machine (Mini-Mill, HAAS, USA). A quasi-isotropic CFRP plate of [(0/45/90/-45)4/90/02/0]s was drilled using three different types of coated and uncoated tool. The CFRP panel provided by the Boeing Company. The thickness of the CFRP plate was 7.62 mm, with an average ply thickness of 0.1885 mm and an average fiber diameter of 5 microns. The tools used were; uncoated (Grain size = 0.7 μm and 10%, Manufactured by Precorp Co., Spanish Fork, UT); BAM coated WC; and (AlCrSi/Ti)N coated WC tool(by Fraunhofer USA, MI). The BAM coating was 3.5 μm thick deposited on the uncoated WC by the sputtering method. The (AlCrSi/Ti)N coated tool was WC 7.3 μm thick deposited on the uncoated WC by the cathodic arc deposition technology. The spindle speed was 6000 RPM, and the feed was 0.0762mm/rev in CFRP. A water-soluble coolant (Kool Mist. Formula #78) was used as spray mist during the experiment. Each drill bits was used to drill 140 holes.

The tool profile has been measured using a VHX-6000 confocal laser microscope (Itasca, Illinois, USA), which has been reported in previous publications [4,20]. Surface Roughness Parameters were measured using has been measured using Roughness Measuring Profilometer (Mituyoto, Kanagawa, Japan) according to ISO 1997 roughness standard. The evaluation length was 6 mm to obtain as much as roughness possible along with the hole thickness. Surface roughness has been measured across the half holes in three consecutive holes every 40-hole interval, at holes 1-3, 39-41, 79-81, 119-121 of the CFRP hole, as shown in the schematic in Figure 2. The SEM images of the machined surfaces have been taken using Quanta 3D 200i (Thermo Fisher Scientific, Waltham, MA, USA).

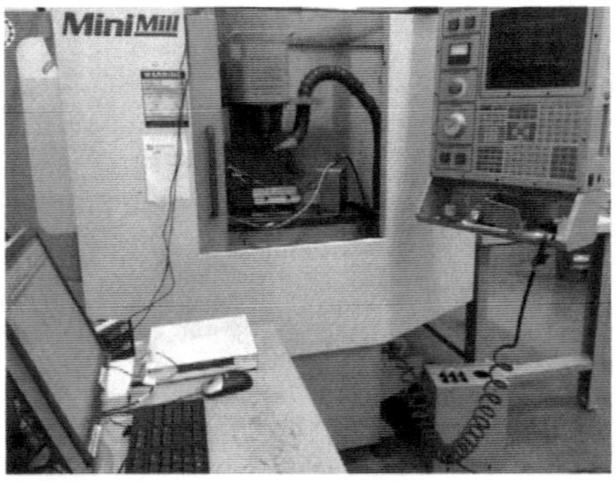

Figure 1 The experimental set up [4]

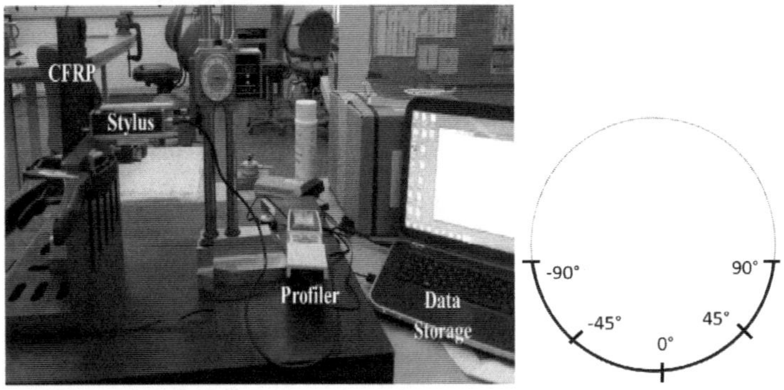

Figure 2 Measuring of surface roughness in half hole (left), location of the measurement (right)

3. RESULTS

3.1 Categorization of Tool Edge Geometry

This section aims to categorize the cutting edge of the tool. Continuous drilling with a tool causes changes in cutting-edge geometry. The tool edge geometry has been categorized into sharp, honed, and honed and chamfered in the machining of metal and metal matrix composite[13–17]. A schematics of tool edge geometry is shown in Figure 3. This approach is applied in this section to categorize the cutting-edge geometry as the cutting edges obtained from the tool profile show a similar geometric feature, as discussed in the literature. The cutting edge geometries obtained at hole 1, hole 40, hole 80 and hole 120 of this experiment are shown in Figure 4 according to categories described.

Figure 3 The microgeometry of cutting tool reported by Zhao et al. (2017) [21]

Initially, the sharp tools are sharp and become honed because of tool wear, which becomes chamfered due to further tool wear. In the beginning, the sharp edge receives the forces concentrated at a point, and the endure higher stress leading to higher materials removal. Hence, the sharp tool edge becomes honed, which eventually becomes chamfered and honed because of further tool wear. The edge strength of the tool edge varies depending on the edge geometry. Experimentally, the sharp edge creates a finer surface, while the honed edge creates rougher surfaces in metal machining [13–17]. Comparing Figure 3 and Figure 4, the existence of honed edge, chamfered edges and chamfered and honed edges are proven in CFRP machining.

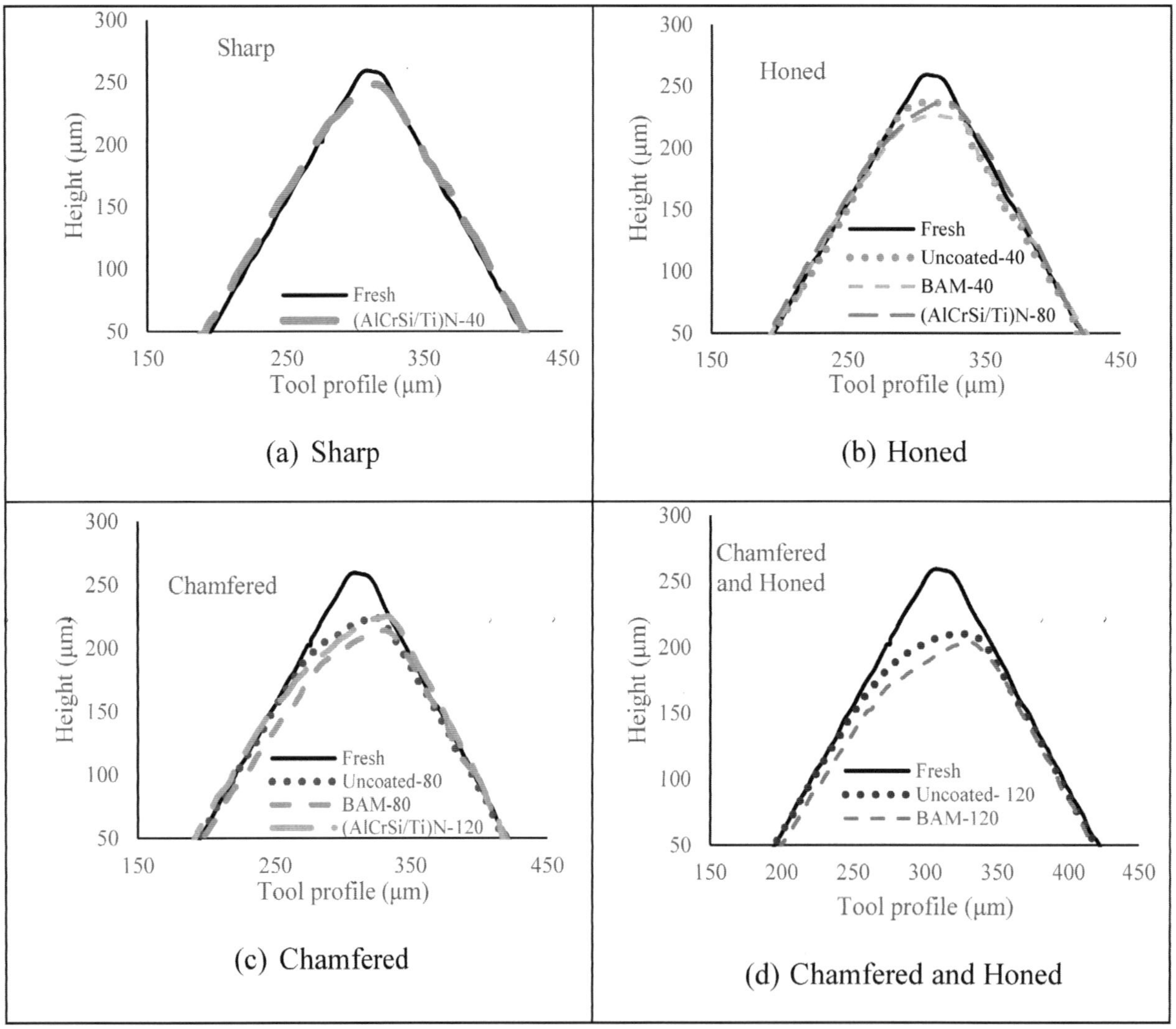

Figure 4 The Tool Edge Geometry according to the categorization for the uncoated, BAM coated and (AlCrSi/Ti)N coated tool from hole 1 to hole 120

Table 1 represents the flank wear area of the three drills obtained from the previous experiment along with the tool geometry type mentioned above [43]. All the tools had sharp edges before drilling any hole. The uncoated tool edge transformed into honed shape at hole-40 because of edge rounding and flank wear of 48 μm and started to become chamfered and honed followingly with further flank wear. The values of flank wear of uncoated tool at hole 80 and hole 120 are respectively 106 μm and 135 μm. Observing the tool contour of BAM coated tools in Figure 4, the BAM coated tool showed both the honed and the chamfered and honed edges at different stages of this experiment. The sharp BAM coated tool became honed due at hole 40 because of excess tool wear in BAM coated tool. From Table 1, it is seen that the BAM coated tool experienced 99 μm flank wear, which is large in both flank wear and wear area than the uncoated tool at hole 40. However, after hole-40, the BAM coated tool became more chamfered and honed due to larger flank wear of more than 200 μm, while the uncoated tool has flank wear of fewer than 140 μm. Therefore, larger chamfering in BAM coated tool is noticed compared to the uncoated tool at hole 80 and hole 120. As seen in Figure 4 (b); both the uncoated and the BAM coated tool edge are honed at hole 40, the (AlCrSi/Ti)N-coated tool at hole 80. The (AlCrSi/Ti)N-coated tool pertained to the sharp edge until hole 40 because of the least material removal due to better adhesion to the tool substrate. It is noted that the (AlCrSi/Ti)N coated tool at hole 40 is characterized by the least amount of wear area of 636 μm^2 according to Table 1. The better adhesion results from the existence of alternating layers of nano-coating and comparable substrate coating stiffness of the (AlCrSi/Ti)N coated tool resulted in the least amount of flank wear. Because of the aforementioned reasons, the (AlCrSi/Ti)N coated tool edge experienced honing and chamfering much slower compared to the BAM coated and the uncoated tool. That is why, this tool was still sharp until hole 40, as seen in Figure 4 (a); and became honed at hole 80 as seen in Figure 4 (b); the honed edge was induced by chamfering at hole 120 ensured by the tool profile in Figure 4 (c).

On the other hand, the cutting edge of other tools (BAM and Uncoated) became honed and chamfered much earlier compared to the (AlCrSi/Ti)N coated tool. Finally, the sharp tool edges are obtained for hole 1 for all drill and at hole 40 for (AlCrSi/Ti)N coated drill, shown in Figure 4 (a). After honing, the tool enters into chamfering, the third stage of geometry transformation. The honed cutting edges are obtained for uncoated hole 40, BAM hole 40 and (AlCrSi/Ti)N hole 80 in Figure 4 (b). Then chamfered cutting edges are identified at uncoated hole 80, BAM hole 80 and (AlCrSi/Ti)N 120, in Figure 4 (c). The chamfering occurs due to further material removal from the tool. Following to chamfering of the tool, further tool wear causes the tool to enter the fourth geometric transformation, when the tool becomes chamfered and honed, as seen in Figure 3. The uncoated tool and BAM coated tool edges at hole-120 are chamfered and honed. Meanwhile, the (AlCrSi/Ti)N coated tool even did not become hone and chamfered until hole 120. The tool geometries are clustered according to the categories stated in Table 1.

Table 1 The categorization of the cutting edge geometry at different holes of the experiment for the uncoated, BAM coated and (AlCrSi/Ti)N coated tool along with tool wear area

	Uncoated			BAM			(AlCrSi/Ti)N		
Hole	Edge Geometry	Flank wear (μm)	Wear Area (μm^2)	Edge Geometry	Flank wear (μm)	Wear Area (μm^2)	Edge Geometry	Flank wear (μm)	Wear Area (μm^2)
1	Sharp	-	-	Sharp	-	-	Sharp	-	-
40	Honed	48	1822	honed	99	2417	Sharp	47	636
80	Chamfered	106	2955	Chamfered	235	4453	Honed	73	1346
120	Hone and Chamfered	135	3804	Hone and Chamfered	281	5560	Chamfered	121	2426

3.2 Tool Edge Geometry and Surface Roughness

The surface roughness is believed to be related to cutting tool geometry. The surface roughness of the holes 1-3, 39-41, 79-81, and 119-121 drilled by each tool have been measured. Likewise, the roughness holes corresponding to the specific tool geometries have been analyzed in this section. Table 2 represents the changes of average surface roughness (Ra) and mean surface roughness (Rq) of the tool edge categories; 'sharp', 'honed', 'chamfered' and 'chamfered and honed'. The Average Ra and Average Rq in this table is the average of all the Ra and Rq values obtained from half-holes (five measurements in each hole) for all corresponding hole drilled. According to Table 2, the sharp tool edges create at average Ra of 2.48 μm, and the average Rq of 4.52 μm. As the tool becomes honed, the average Ra and average Rq increases to 2.68 μm and 4.70 μm. And, the chamfered tool edges generate machined surface of 3.08 μm (average Ra) and 5.44 μm (average Rq). It is noticeable that the roughness of the machined surfaces increases again as the tool edge becomes chamfered from honed. A similar phenomenon has been reported in metal machining when the roughness increased as the tool become chamfered from honed. The edge strength increases as the tool become chamfered from hone [13–17] and create a rougher surface. In addition to that as the tool becomes more chamfered, the increased tool edge radius causes further pushes to the fibers instead of cutting. This interplay causes roughness to increase. According to Table 2, the chamfered and honed edges generates surfaces of roughness values of 2.85 μm (average Ra) and 5.07 μm (average Rq). Both the average roughness value and mean roughness value decreased from their respective values (Average Ra= 3.08 μm, Average Rq=5.44 μm). Such decrease in the roughness may have been caused due to matrix smearing in CFRP machined surface due to softening of the matrix due to increased heat generation [9,24]

Table 2 The Average Surface Roughness (Ra) and Mean Surface Roughness (Rq) with the standard deviation for different types of cutting edge geometry

Tool Edge Geometry type	Average Ra (μm)	Average Rq (μm)
Sharp	2.48 ± 1.64	4.52 ± 3.55
Honed	2.68 ± 1.38	4.70 ± 2.67
Chamfered	3.08 ± 1.43	5.44 ± 2.67
Chamfered and Honed	2.85 ± 1.24	5.07 ± 2.66

Total height (Rt) and 10-point height (Rz) are also important parameters in CFRP machining. The bar chart in Figure 5 shows the change of average total height (Average Rt value) and average 10-point height (Average Rz value) for all types of tool edge geometry. The Rt value is 46.38 μm for sharp cutting edges, while the value is 58.38 μm and 63.13 μm respectively for honed edge and chamfered edge. This resembles that the Rt value larger for the honed edge than the sharp edge, and larger for the chamfered edge than the honed edge. The Rt value increases due to deeper and frequent fibers pulled out. However, the Rt value decreased down to 55.16 μm from 63.13 μm. Such a decrease can be attributed to the matrix smearing, as the smeared matrix fills the gap caused by fiber pullouts and other occasional found in CFRP machining. With respect to the tool edge geometry, the 10-point height (Rz) follows a similar trend like Rt of Rt as seen in Figure 5 due to the stated reasons above.

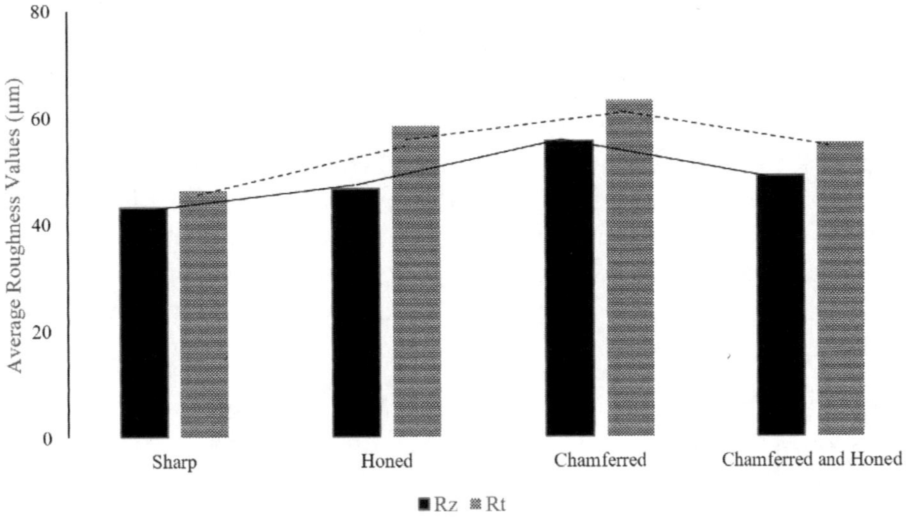

Figure 5 The Average Surface Roughness (Ra) and Mean Surface Roughness (Rq) with the standard deviation for different types of cutting edge geometry (The Average Ra and Average Rq in this table is the average of all the Ra and Rq values obtained from half-holes (five measurements in each hole) for all corresponding hole drilled)

Table 3 presents the average peak height (Rp) and average valley depth (Rv) measured for different edge geometry. The peak height, Rp, physically refers to the potential furthest position of the uncut fibers in CFRP, and the valley depth (Rv), refers to the deepest fiber pull-outs on CFRP machined surface[25,26]. For sharp tools, the average Rp value is 12.13 μm, while the Rp value is

15.57 µm and 18.78 µm respectively for honed edges, and chamfered edges. Such increment refers to the larger uncut fiber, which is true as the dull tool leaves more uncut fibers in CFRP. The Rp values again drop down for the chamfered and honed tools to 16.67 µm from 18.78 µm in chamfered edges. This again hints to sever matrix smearing that reduces the roughness. Additionally, the average Rv value increased steadily from sharp edge to honed edge and chamfered edge; again decreased down when the honing induced the chamfering.

Table 3 The average peak height (Rp) and Average valley depth (Rv) in the experiment

Tool Edge Geometry type	Average Rp (µm)	Average Rv (µm)
Sharp	12.13 ± 12.84	31.09 ± 29.65
Honed	15.57 ± 12.71	31.16 ± 20.32
Chamfered	18.78 ± 12.20	36.82 ± 21.08
Chamfered and Honed	16.74 ± 13.55	32.33 ± 17.83

3.3 Tool Edge Geometry and Surface Morphology from SEM Images Observation

The SEM images of the machined surface have been taken to observe the surface characteristics. Figure 6 represents the SEM images of the machined surface for every cutting edge. According to Figure 6 (a), the sharp cutting edge machined surface looks smoother with few locations of fiber pullout in a few plies in 90° direction, which is intrinsic in CFRP machining. Usually, in drilling of CFRP fiber pull out happens when the relative angle between fibers and the cutting tool is around 135° due to bending causing fiber fracture happening in the beneath the surface [6,27]. For machined surface obtained by honed edge [in Figure 6 (b)] and for chamfered edges [in Figure 6 (c)], the presence of fiber pullout is more frequent and fiber pull out depth looks deeper. These SEM images are the evident cause of the increased roughness for honed edge and the chamfered edge of the CFRP surface discussed in the previous section. However, Figure 6 (d) is the SEM image of the machined surface of the honed and chamfered edge. It is evident that the matrix is smeared on the CFRP surface, and deep pockets are less visible, especially in the direction of 90° and -90°. These pieces of evidence of matrix smearing portray the underlying reason for roughness decrease in the previous section.

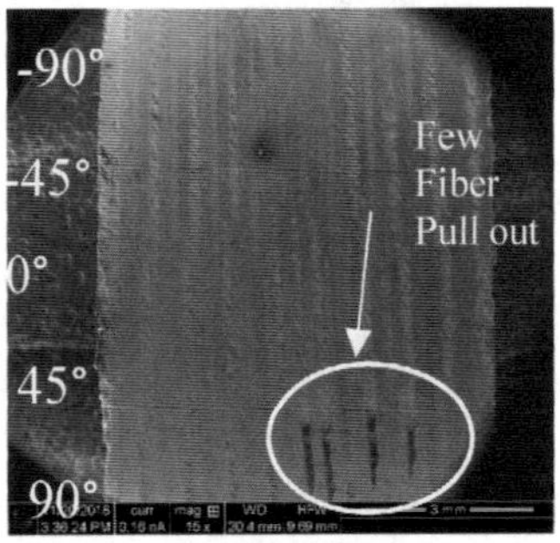

(a) Sharp

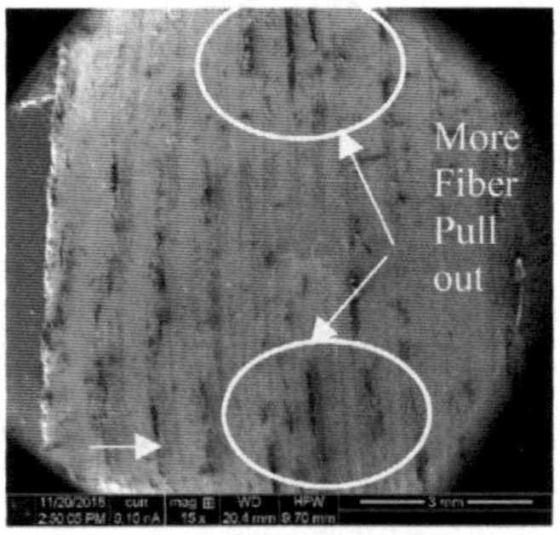

(b) Honed

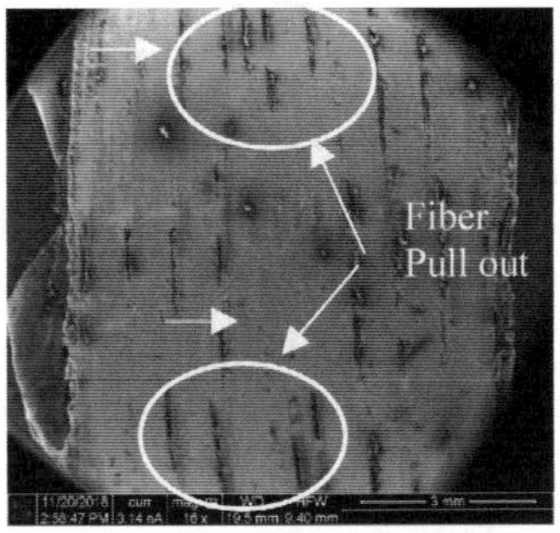

(c) Chamfered

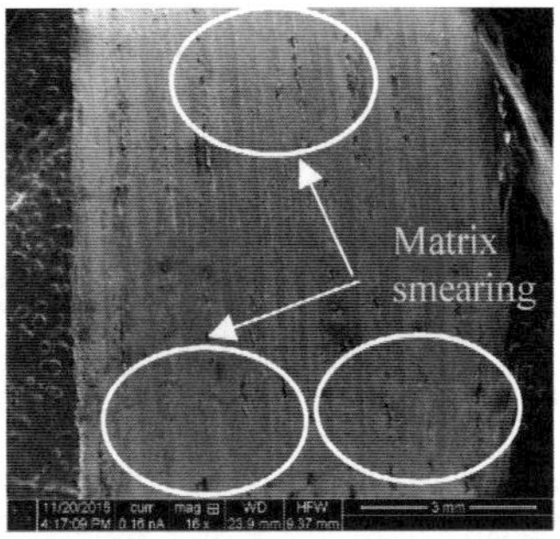

(d) Honed and Chamfered

Figure 6 SEM images of the drilled holes using (a) Sharp cutting edge, (b) honed cutting edge[Presence of Fiber Pull out in Yellow ellipse], (c) chamfered cutting edge [Presence of Fiber Pull out in Yellow ellipse and (d) honed and chamfered cutting edge [Presence of matrix smearing in white ellipse]

4. CONCLUSIONS

This paper presented a primary investigation to study the relation of cutting edge geometry in CFRP drilling with the uncoated WC tool, BAM Coated and (AlCrSi/Ti)N coated tool. The followings are the conclusions inferred from the experimental findings.

1. Like the metal and metal matrix composite machining, honed, chamfered, chamfered and honed cutting edges are observed in CFRP machining. The sharp tool edge first turned into a honed edge due to tool wear. And, the additional flank wear creates chamfer on the honed edges. The coating, especially the (AlCrSi/Ti)N, delayed the changes in cutting edge geometry. The (AlCrSi/Ti) coated drill remained almost sharp until hole 40, while the edge geometry changed into honed shape at hole 80 followed by chamfering at hole 120. On the other hand, the uncoated tool and the BAM coated tool, severely susceptible to tool wear, became honed at hole 40. Meanwhile, the uncoated tool became chamfered at hole 80 and BAM coated tool edge was chamfered and honed at hole 80. In the end, at 120, all the tools had chamfered and honed cutting edge except the (AlCrSi/Ti)N coated tool, which remained chamfered.
2. The roughness of the machined surface has a strong relation to cutting edge geometry. The surface roughness increase if the tool edge is transformed from sharp to honed, and increased further if the cutting edge is chamfered. The chamfered edge creates the largest roughness (Ra, Rq, Rt, Rz, Rp, Rv) on the machined CFRP surface. However, when the tool edge is 'chamfered and honed', the roughness decreases. with the onset of matrix smearing.
3. Further extensive study is required to understand the surface roughness and tool geometry relationship.

5. REFERENCES

[1] Holmes M. Global carbon fibre market remains on upward trend. Reinforced Plastics 2014;58:38–45. https://doi.org/10.1016/S0034-3617(14)70251-6.

[2] Hale J. Boeing 787 from the Ground Up. Aero 2006;4:17–23. https://www.boeing.com/commercial/aeromagazine/articles/qtr_4_06/article_04_1.html

[3] Schade DW. Analysis of the light weight fiber reinforced plastics value chain with regard to the German industry in its global context n.d.:54. https://www.m-five.de/pdf/M_Five_Light_Weight_Composites_Value_Chain_in_GER_FINAL_2017.pdf
[4] Swan S, Bin Abdullah MS, Kim D, Nguyen D, Kwon P. Tool Wear of Advanced Coated Tools in Drilling of CFRP. Journal of Manufacturing Science and Engineering 2018;140:111018-111018–10. https://doi.org/10.1115/1.4040916.
[5] Nguyen D, Voznyuk V, Bin Abdullah MS, Kim D, Kwon PY. Tool Wear of Superhard Ceramic Coated Tools in Drilling of CFRP/Ti Stacks. MSEC2019, Volume 2: Processes; Materials: 2019. https://doi.org/10.1115/MSEC2019-2843.
[6] Hou G, Qiu J, Zhang K, Cao S, Cheng H, Luo B, et al. Comparative tool wear and hole quality investigation in drilling of aerospace grade T800 CFRP using different external cooling lubricants. The International Journal of Advanced Manufacturing Technology 2020;106:937–51. https://doi.org/10.1007/s00170-019-04554-9.
[7] Ahsan MA, Hosur M, Tareq SH, Hasan SMK. Quasi-static compression characterization of binary nanoclay/graphene reinforced carbon/epoxy composites subjected to seawater conditioning. Materials Research Express 2020;7:015033. https://doi.org/10.1088/2053-1591/ab62fb.
[8] Soutis C. Fibre reinforced composites in aircraft construction. Progress in Aerospace Sciences 2005;41:143–51. https://doi.org/10.1016/j.paerosci.2005.02.004.
[9] Kim D, Beal A, Kwon P. Effect of Tool Wear on Hole Quality in Drilling of Carbon Fiber Reinforced Plastic–Titanium Alloy Stacks Using Tungsten Carbide and Polycrystalline Diamond Tools. Journal of Manufacturing Science and Engineering 2015;138:031006. https://doi.org/10.1115/1.4031052.
[10] Wang F, Qian B, Jia Z, Fu R, Cheng D. Secondary cutting edge wear of one-shot drill bit in drilling CFRP and its impact on hole quality. Composite Structures 2017;178:341–52. https://doi.org/10.1016/j.compstruct.2017.04.024.
[11] Montoya M, Calamaz M, Gehin D, Girot F. Evaluation of the performance of coated and uncoated carbide tools in drilling thick CFRP/aluminium alloy stacks. The International Journal of Advanced Manufacturing Technology 2013;68:2111–20. https://doi.org/10.1007/s00170-013-4817-0.
[12] Liu D, Tang Y, Cong WL. A review of mechanical drilling for composite laminates. Composite Structures 2012;94:1265–79. https://doi.org/10.1016/j.compstruct.2011.11.024.
[13] Rahim EA, Mohid Z, Hamzah MR, Yusuf AF, Rahman NA. Performance of Tools Design when Helical Milling on Carbon Fiber Reinforced Plastics (CFRP) Aluminum (Al) Stack. Applied Mechanics and Materials 2013;465–466:1075–9. https://doi.org/10.4028/www.scientific.net/AMM.465-466.1075.
[14] Sandvik Coromant. User's Guide- Machining carbon fbre materials n.d. https://www.sandvik.coromant.com/sitecollectiondocuments/downloads/global/technical%20guides/en-gb/c-2920-30.pdf
[15] Denkena B, Lucas A, Bassett E. Effects of the cutting edge microgeometry on tool wear and its thermo-mechanical load. CIRP Annals 2011;60:73–6. https://doi.org/10.1016/j.cirp.2011.03.098.
[16] Denkena B, Biermann D. Cutting edge geometries. CIRP Annals 2014;63:631–53. https://doi.org/10.1016/j.cirp.2014.05.009.

[17] Zhang L, Ren C, Ji C, Wang Z, Chen G. Effect of fiber orientations on surface grinding process of unidirectional C/SiC composites. Applied Surface Science 2016;366:424–31. https://doi.org/10.1016/j.apsusc.2016.01.142.
[18] Zhao T, Agmell M, Persson J, Bushlya V, Ståhl JE, Zhou JM. Correlation between edge radius of the cBN cutting tool and surface quality in hard turning. Journal of Superhard Materials 2017;39:251–8. https://doi.org/10.3103/S1063457617040050.
[19] Zhao B, Ding WF, Dai JB, Xi XX, Xu JH. A comparison between conventional speed grinding and super-high speed grinding of (TiCp + TiBw) / Ti–6Al–4V composites using vitrified CBN wheel. The International Journal of Advanced Manufacturing Technology 2014;72:69–75. https://doi.org/10.1007/s00170-014-5656-3.
[20] Abdullah MSB. EFFECTS OF FIBER ORIENTATIONS AND PROPERTIES ON TOOL WEAR AND MACHINED SURFACE QUALITY IN MACHINING OF CARBON FIBER REINFORCED PLASTICS LAMINATE. Washington State University, 2019. https://www.researchgate.net/publication/338701977_EFFECTS_OF_FIBER_ORIENTATIONS_AND_PROPERTIES_ON_TOOL_WEAR_AND_MACHINED_SURFACE_QUALITY_IN_MACHINING_OF_CARBON_FIBER_REINFORCED_PLASTICS_LAMINATE
[21] Zhao T, Zhou JM, Bushlya V, Ståhl JE. Effect of cutting edge radius on surface roughness and tool wear in hard turning of AISI 52100 steel. The International Journal of Advanced Manufacturing Technology 2017;91:3611–8. https://doi.org/10.1007/s00170-017-0065-z.
[22] �zel T, Hsu T-K, Zeren E. Effects of cutting edge geometry, workpiece hardness, feed rate and cutting speed on surface roughness and forces in finish turning of hardened AISI H13 steel. The International Journal of Advanced Manufacturing Technology 2005;25:262–9. https://doi.org/10.1007/s00170-003-1878-5.
[23] Raj DS, Karunamoorthy L. Study of the Effect of Tool Wear on Hole Quality in Drilling CFRP to Select a Suitable Drill for Multi-Criteria Hole Quality. Materials and Manufacturing Processes 2016;31:587–92. https://doi.org/10.1080/10426914.2015.1004713.
[24] Ramulu M, Wern CW, Garbini JL. Effect of fibre direction on surface roughness measurements of machined graphite/epoxy composite. Composites Manufacturing 1993;4:39–51. https://doi.org/10.1016/0956-7143(93)90015-Z.
[25] Landon Y, Cherif M. Characterization of the Surface Quality of Holes Drilled in CFRP Laminates. AMR 2013;698:107–16. https://doi.org/10.4028/www.scientific.net/AMR.698.107.
[26] Rimpault X, Chatelain J-F, Klemberg-Sapieha JE, Balazinski M. Surface profile texture characterization of trimmed laminated composite in the stacking sequence direction. Measurement 2016;91:84–92. https://doi.org/10.1016/j.measurement.2016.05.039.
[27] Ramulu M, Kim D, Choi G. Frequency analysis and characterization in orthogonal cutting of glass fiber reinforced composites. Composites Part A: Applied Science and Manufacturing 2003;34:949–62. https://doi.org/10.1016/S1359-835X(03)00203-3.

ON THE DEVELOPMENT OF A COMPUTATIONAL DESIGN METHODOLOGY FOR TAILORED FIBER PLACEMENT PREFORMS

Daniel Rapking, Eric Zhou, Bert Liu, Michael Braginsky, Gyaneshwar Tandon, Scott Huelskamp
University of Dayton Research Institute
300 College Park
Dayton Ohio 45469

ABSTRACT

The promise of the automated forming process known as tailored fiber placement (TFP) is that it can produce composite preforms by systematically placing fiber in such a way as to optimize the part's performance while greatly reducing weight with the additional benefit of improved out-of-plane performance due TO localized stitching. This promise is hugely dependent on the designers' ability to develop the fiber placement and stitching designs for optimal mechanical performance, and current methodologies for such designs rely mostly on trial-and-error. Our group aims to develop a computational design methodology that will optimize mechanical performance in order to reduce part development time and costs. This paper focuses on the methodology for modelling of the inter-ply interface as part of this effort. Two approaches – a total homogenization of the interface region or separate modeling of the effective regions of each stitch at the interface between plies – are compared with double cantilever beam (DCB) tests in order to develop manufacturing guidelines regarding desired stitch densities required to improve performance.

Corresponding author: Daniel Rapking

1. INTRODUCTION

One of the main design limiting aspects of composite materials is low through thickness strength. Aerospace manufacturers have explored different techniques to improve through thickness properties: through stitches, z-stitching, tailored interfaces, etc. These techniques have demonstrated improvements in out-of-plane performance, but generally at the cost of reductions in in-plane mechanical properties or an increase in associated fabrication costs. The use of conventional methods of improving out-of-plane performance has been extensively researched [1- 3] and has shown many potential benefits and drawbacks of various methods. The general agreement in literature shows that the inclusion of z-stitching will increase interlaminar properties (toughness/strength) at the cost of reducing in-plane properties (modulus/strength). This, coupled with the increase in cost compared with simply stitching traditional composite structures, has ultimately limited the widespread adoption of z-stitching.

Tailored fiber placement (TFP) is a process akin to embroidery that can be used to produce near net-shape composite preforms, which are stitched together as a result of the fabrication process [4]. The TFP process uses tow steering in order to produce a composite structure by layering tows upon each other, with all layers being attached to a carrier cloth. This composite can be constructed of multiple layers of preforms consisting of various curvilinear tow paths. Due to the ability to explicitly steer each individual tow path, it is possible to optimize [5] the tow paths to maximize part performance. Previous work has shown that the TFP forming process is compatible with a wide array of tow materials, and sizes [6].

Other works have explored computational methods for optimizing and modeling TFP preforms [7]. This work will focus on the methods utilized here to begin building a computational framework that will accurately capture the impact of the TFP preforming process on part performance. This will include mechanisms for capturing the impact of stitches on interfacial toughness, part modulus, and ultimate strength. The computational framework utilizes the finite element analysis tool BSAM [8], which is a progressive damage analysis tool that incorporates Discrete Damage Modeling (DDM) techniques.

2. EXPERIMENTAL FABRICATION

2.1 Mode I Fracture Toughness Part Fabrication

This work evaluates experimental test data from previous work by Huelskamp et al [9] in order to test the preliminary capabilities of this computational framework. A detailed accounting of the part fabrication process can be found in the aforementioned paper, but is summarized here. TFP preforms were fabricated for tensile testing according to ASTM D3039, mode I fracture toughness testing according to ASTM D5528. Baseline test panels that did not have any stitching from the TFP process were produced from unidirectional carbon fabric (PX35UD300-0611, Zoltek) spread to an areal weight of 300 g/sq. m. Three separate TFP preforms were fabricated with 50k tow carbon (PX350098T-13, Zoltek). The stitching material used was a 40 tex Kevlar thread (K-Tech 75, AMANN), with the backing clothe consisting of a 123 g/sq. m plain weave fiberglass fabric. Parts were infused with liquid resin via RTM using Epon 862/Epikure W.

The description of each of the three stitched preforms that were tested, can be seen below. The characteristic of interest is the Through Center stitch density parameter. This corresponds to the density of Kevlar stitches at the Teflon insert mid-plane. This is due to the method fabrication that requires the stacking of two separate halves together as seen in Figure 1.

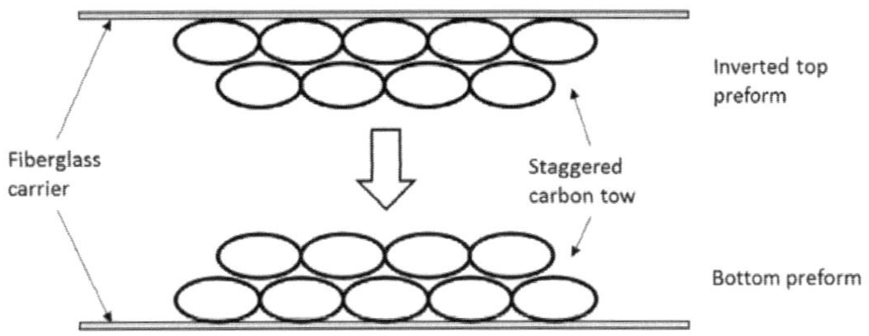

Figure 1. Cross section of TFP preform

Each of these preforms were constructed with a peak to peak spacing of 11 mm, allowing the tows to sufficiently spread during infusion to ensure a uniform fiber volume fraction distribution across the composite.

Table 1. Stitch Density of Various TFP Preforms

	Preform 1	**Preform 2**	**Preform 3**
Preform Stitch Spacing (mm)	4.4	4.0	3.6
Joining Stitch Spacing (mm)	3.7	3.5	3.3
Per Layer (stitch/sq.cm.)	3.5	3.8	4.3
Through Center (stitch/sq.cm.)	7.3	8.2	9.2
Through Layer 1 (stitch/sq.cm.)	14.3	15.9	17.7
Through Layer 2 (stitch/sq.cm.)	10.8	12.0	13.5

2.2 Components Constructed for Micro CT

In addition to simulating the mode I fracture data developed by Huelskamp et al, work was done to characterize the stitched structure in order to feed information back to the computational model. This required the construction of components that would be scanned under a microcomputed tomography system. These components were constructed utilizing the same fabrication methods described in the above section, but had several important variations. The first component was constructed to match Preform 2, or the nominal stitch density. Additional components were constructed utilizing a stitch density of 16 stitches/cm^2 as well as a stitch density of 30 stitches/cm^2. These components were constructed with two different peak to peak spacing's, the nominal or "loose" spacing of 11 mm and a "tight" spacing of 6 mm. The nominal stitch spacing of 8.2 stitches/cm^2 can be seen in the in situ μCT fixture in Figure 2 below along-side representative samples of the increased stitch densities.

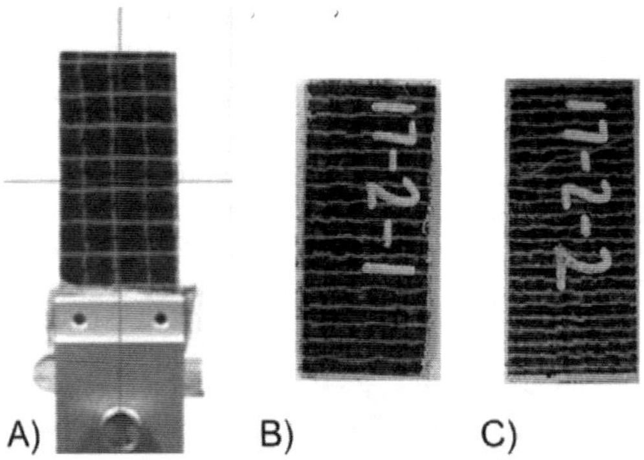

Figure 2. A) Preform 2 mounted for μCT characterization, B) 16 stitches/cm^2 specimen for μCT, C) 30 stitches/cm^2 specimen for μCT

3. RESULTS

3.1 Experimental Test Data

The results of the experimental tests on mode I fracture and tensile strength are summarized at length in the work by Huelskamp et al, and are only briefly summarized here. Testing of each of the stitched preforms showed significant increase in mode I fracture toughness across all three stitch densities. The baseline fracture toughness of the Epon 862/Epikure W system was found to be 0.718 kJ/m^2. Toughness was increased by over 100% at even the lowest level of stitch density. The paper discussed the difficulties in testing the preforms that made total data reduction difficult. Samples from preform 3 experienced difficulty during testing, where stitches failed non-uniformly leading to some specimens "snapping" or failing during testing, without evolving damage at the mid-plane of the test specimen. This premature failure can be seen in Figure 3.

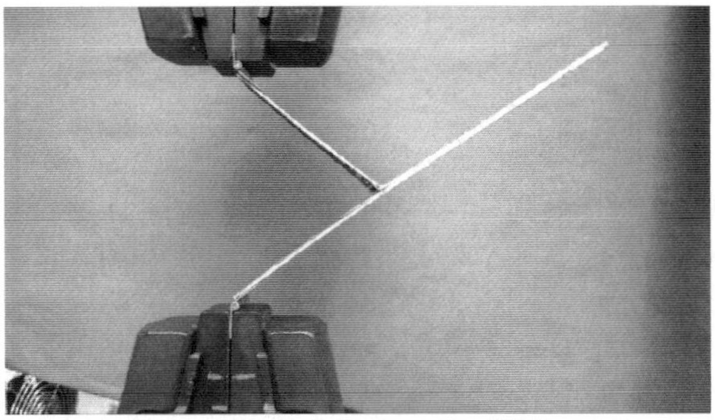

Figure 3. Premature failure of a TFP preform

Tensile results were normalized to a fiber volume fraction of 60% and results are summarized in Table 2 below alongside fracture toughness data. As expected the stitched preforms exhibited lower ultimate tensile strengths. This is expected to occur due to localization of resin rich pockets near stitches, in addition to the distortion of local fiber paths from the principal fiber direction. An interesting outcome of the test data was that no preform saw a statistically significant reduction in tensile modulus. Preform 3 actually exhibited a slightly higher tensile modulus after normalizing compared to the baseline. This is assumed to be the result of fibers not being perfectly oriented in any of the preforms.

Table 2. Summary of ASTM standard testing

ASTM D5528 Results		ASTM D3039	
Stitch Density (stitches/cm^2)	G1c (kJ/m^2)	Modulus (GPa)	UTS (GPa)
Baseline	0.718	107.28	1.36
Preform 1 (7.3)	1.686	105.35	1.04
Preform 2 (8.2) - Nominal	2.288	107.1	1.18
Preform 3 (9.2)	1.965	114.87	1.15

The primary take-away from the test data was the significant improvement in mode I fracture toughness (greater than 100% in all cases), without a significant reduction in tensile strength(less than 15%). These results and the computational methods used to approximate this behavior is discussed in the following sections.

3.2 Modeling Mode I Fracture Testing

When starting the modeling efforts on this program several different mechanisms were discussed as possibilities for capturing the impact of stitching on the mechanical performance of the test coupons. The first method discussed was the construction of models that discretely modeled the entirety of the model. This would require the use of the Virtual Textile Morphology Suite (VTMS) to generate a realistic approximation of the tows and stitches from each preform, and then generate a complex finite element mesh for analysis. This method was quickly discounted due to the difficulty in accurately representing the geometry, in addition to this method producing models too large to solve in a timely fashion. The second method discussed was the inclusion of 1D rod elements at locations of stitches, but this was excluded because it would require the modification of the current finite element code in BSAM.

Ultimately it was decided to explore the modeling of two first order approximation methods. Method one would utilize standard cohesive surfaces between plies that had the fracture toughness of the traction-separation law governing opening increased to the experimentally measured value. This would simplify the model complexity greatly, but is limited by the over simplified response of the model. An additional limitation is the requirement of measuring the toughness of every potential variation of stitch densities. The second methodology discussed was the modeling of stitch locations as separate cohesive zones compared to the "nominal" cohesive zone governing the standard ply interface. This method utilizes an automated search algorithm to generate node sets used to create cohesive surfaces based on an x/y/z coordinate list of the stitches produced by the ZSK machine utilized to fabricate the preforms. This method presented the potential benefit of simply needing the strength/toughness of the Kevlar stitching used to reinforce the composite in order to estimate the impact on toughness across varied regions of the part. In addition this method allows the user to create "effective" stitch radii that can capture the distortion in the local fiber direction caused by the stitching fiber. This allows the model to change local fiber direction as a function of fiber distortions.

Models of the mode I fracture testing were constructed and modeled with both methods described above. Figure 4 shows the construction of the homogenized model alongside a model with discrete cohesive zones at each stitch location.

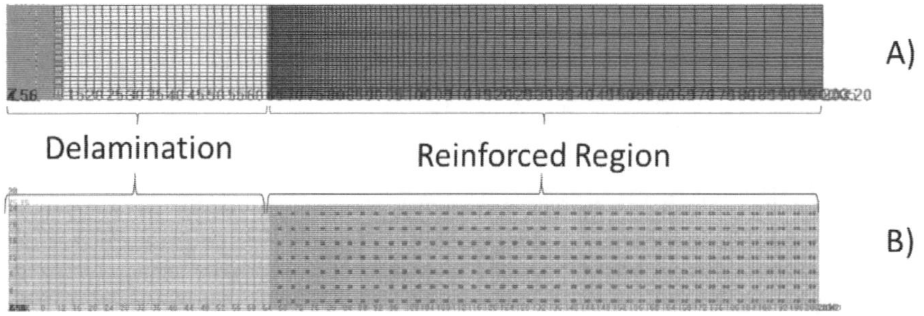

Figure 4. A) Homogenized reinforcement region, B) Discretely modeled cohesive zones at each stitch location

Models with cohesive zones at each stitch location were run with effective radii varying from 0 mm (i.e. only within the element the stitch falls within) up to 1.4 mm. This range was chosen based arbitrarily. The models consisting of a totally homogenized interface showed good agreement with the average peak experimental load for each of the preforms simulated. The models showed linear increase in load until the onset of delamination at the mid-plane and showed uniform growth of the delamination as the load was applied. This produced a very "smooth" response compared to the experimental data. Initial modeling of the discrete cohesive zones interlaminar strength was an estimate, as no data on the strength of the Kevlar stitching thread had been collected yet. This, unsurprisingly, resulted in under-predicting the peak load of the experimental data for mode I fracture testing. The results of these two model constructions can be seen in Figure 5. Figure 5a consists of the totally homogenized interface models for all three stitch preforms in addition to the baseline test data, the average experimental peak is highlighted in the red box. Figure 5b consists of the test data for stitched preform 2 in blue/grey/yellow, while the computational model is in shades of orange.

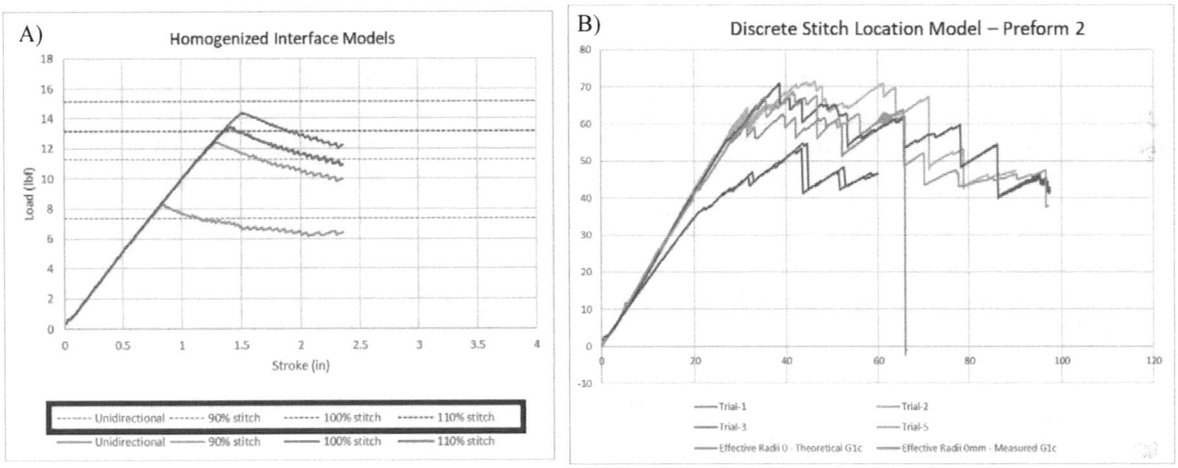

Figure 5. A) Homogenized interface model results, B) Discrete stitch location model results for preform 2

3.2.1 Kevlar Stitching Thread Test Data

In order to accurately modify the cohesive strength of the discretely modeled stitch locations Kevlar thread was tested according to ASTM D7269. Kevlar thread sections were tested at gage lengths of 50.8 mm (2 in). Specimens were monitored for maximum extension, peak load and an average break force recorded. The average fiber diameter was calculated based on the fiber denier to be 188 microns. After calculating the average fiber diameter the average failure stress was calculated to be 1909 MPa. A summary of failure data from the Kevlar thread is seen in Table 3.

Table 3. Kevlar stitching thread failure loads

Specimen #	Length (mm)	Extension (mm)	Failure Load (N)
1	50.8	6.80466	52.42018204
2	50.8	6.644132	51.94600178
3	50.8	7.347204	54.20218345
4	50.8	6.309614	51.71335988
5	50.8	6.195822	54.90945043

3.2.2 Analytical Estimate of Stitch Influence on Toughness

With the test data on thread strength collected a method of estimating the improvement in fracture toughness was devised. This method was devised in order to estimate the potential improvement in toughness for any variation on stitch density. The analytical model starts at the assumption that each Kevlar thread is uniformly strained in conjunction with the top layer of composite material. Crack growth in the matrix is assumed to be self-similar within the Kevlar. The Kevlar exhibits linear-brittle fracture, where the fracture of Kevlar leads to release of all strain energy. Equation 1 shows the analytical method used to estimate the fracture toughness associated with Kevlar thread stitches. This calculation considers the failure strength and strain of each fiber, in addition to the total cross sectional area of fibers in the laminate. Equation 2 shows that the total estimated fracture toughness for the stitched area is the summation of the toughness due to stitches coupled and baseline resin. Table 4 shows the comparison between the analytically predicted toughness and the measured toughness. A principle driver of the difference is the assumption in the analytical method that there are no defects present and strain is uniformly applied. This is done to estimate the theoretical upper bound possible with a variety of stitch densities.

$$G_{IC-Eff\ Fiber} = \sigma_{Stich} * \varepsilon_{Stitch} * \frac{\#\ Fibers}{Unit\ Area} * A_{Fiber} * t_{laminate} \quad [1]$$

$$G_{IC-Total} = G_{IC-Eff\ Fiber} + G_{IC-Resin} \quad [2]$$

Table 4. Comparison calculated fracture toughness and measured values

Comparison of Fracture Toughness Values (kJ/m^2)		
Preform	Measured	Calculated
Baseline (unstitched)	0.718	N/A
Preform 1	1.686	2.272
Preform 2	2.288	2.464
Preform 3	1.986	2.677

3.2.3 Determination of Cohesive Zone Strength Based on Effective Stitch Radii

Taking the thread strength and analytical toughness calculations, models were constructed to determine how to scale the failure strength of the discrete stitch cohesive zones as a function of effective radii. This "calibration" of cohesive strength was completed for each effective radii. For simplicity this process is shown for the 0 mm effective radii model in Figure 6. Each effective radii model had the cohesive strength scaled as a function of area captured by the cohesive zone compared to cross sectional area of Kevlar fibers. Models were run until peak load was within 1% of average peak load from the experimental test data. This process can be seen by moving from left to right in Figure 6, where the cohesive strength is varied until the model has honed in on the appropriate value. After determining the appropriate cohesive strength for the stitch region fracture toughness in that region is varied in order to qualitatively fit the experimental data. This process is completed across all the effective radii in order to produce a calibration curve for fitting cohesive strength to each effective radii.

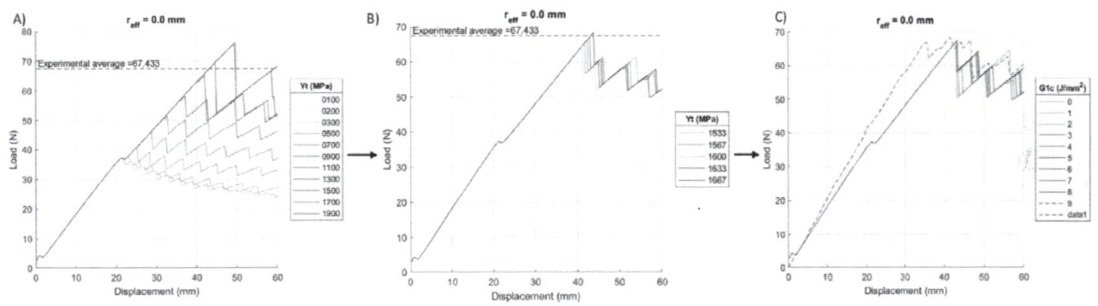

Figure 6. Calibration process for the cohesive strength at stitch locations A) Initial variation of cohesive strengths, B) Finely tuned cohesive strengths, C) Calibrated cohesive strength with varied fracture toughness parameters

After this was completed for each of the varied effective radii a rough "calibration curve was developed (Figure 7). This calibration curve shows consistent values, where the cohesive strength is nominally the same even as the effective radii is increased. This is due to the fact that the increase in effective area did not increase the total number of elements that approximated the stitch region. This would be taken into consideration if mesh density of the models was continuously increased until every subtle increase in effective radii changed the number of elements represented by the stitch region. This data was fit with a linear trend in order to produce a mechanism for determining what cohesive strength should be utilized as a function of effective stitch area.

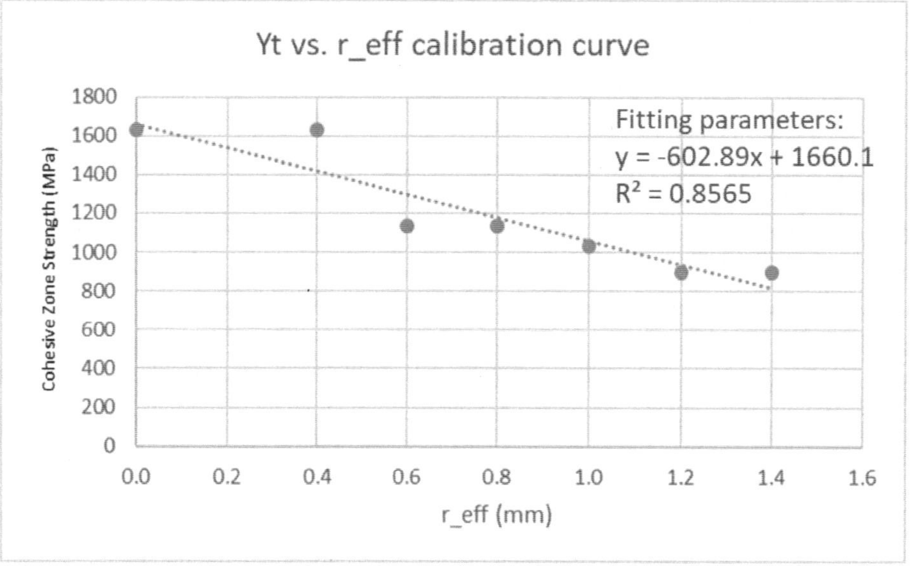

Figure 7. Stitch location cohesive zone calibration curve

3.3 Characterization of Fiber Distortion Due to Stitching Process

Each model looked at incorporating an effective radii in order to accommodate any potential distortion in fiber due to the fabrication process. Future modeling efforts will overlay tow paths on homogenized meshes and assign fiber directions on an element by element basis. This allows the FEA tool to capture local stiffness variations due to misaligned fibers. At this stage the models simply assumed perfect fiber orientation with various cohesive zone sizes only impacting damage growth at the interface. Looking at the cross-section photomicrographs in the previous work it is clear that distortions in the fiber direction occurs as a result of the stitching process.

In order to understand how significant these distortions are various laminates were constructed and µCT'd to characterize the fiber distortion. These laminates consisted of $[0/90]_s$ laminates. Stitch density in these laminates were 8.2 stitches/cm^2, 16 stitches/cm^2, and 30 stitches/cm^2 (30 stitches/cm^2 was analytically estimated to produce an increase in interlaminar tensile strength of 50 %). These preforms were cut into samples approximately 6 cm x 4 cm. They were then µCT'd in an Xradia Versa 520 with a voxel size of approximately 10 microns. CT data was then evaluated for three primary characteristics consisting of distortion width, length, and mis-alignment angle. An example of a defect can be seen in Figure 8. This location shows a Kevlar stitch and the resin rich region surrounding the fiber in the material system.

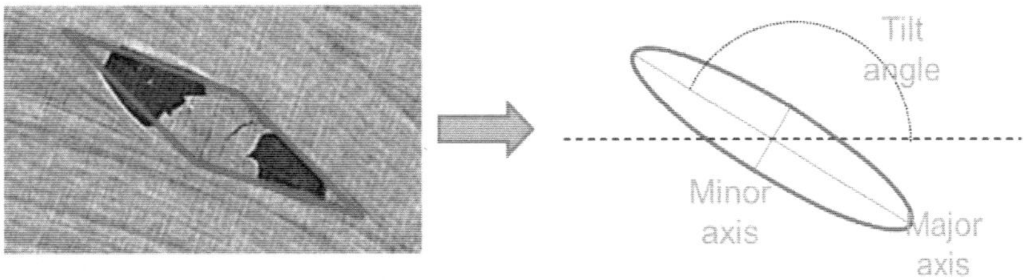

Figure 8. Characterization of fiber distortion due to Kevlar stitching

Samples of varied stitch density (16/30 stitches/cm^2) were analyzed at serval points of interest. This data was collected and summary results are seen in Table 5. Tilt angle corresponds to the deviation in angle from the principal fiber direction. Table 5 presents average observed distortions as well as the maximum distortions observed. Maximum distortions are always found at the interface between 0° and 90° fiber directions. This work was unable to determine the root cause of these distortion, but it is assumed to be a function of friction between the needle and fiber, in addition to the speed at which stitching occurred. These large scale distortions occurred with very low frequency throughout the μCT'd volume. In addition to collecting data on the fiber distortion, the assumption that increased stitch density would generate resin rich "roads" between fibers was verified (Figure 9).

Table 5. μCT fiber distortion data

	Sample 1	**Sample 2**	**Sample 3**	**Sample 4**
Major axis avg (mm)	2.33	2.28	1.97	2.13
Minor axis avg (mm)	0.22	0.22	0.28	0.26
Average Tilt angle (°)	1.438	2.567	2.545	1.698
Max distortion angle (°)	33.45	18.32	30.98	24.01
Stitch Density (s/cm^2)	16	30	16	30

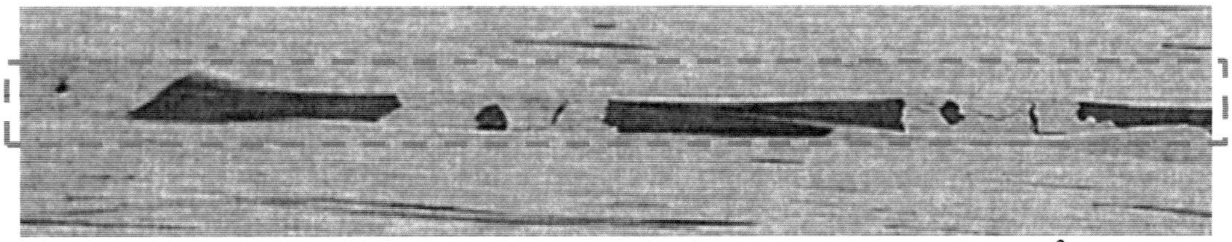

Figure 9. Resin rich "road" between fiber tows (30 stitches/cm^2)

4. CONCLUSIONS

The presented work is just one important building block in the overall framework for predicting the performance of complex parts produced with TFP. Models simulating the mode I fracture tests have shown the limitations of utilizing a total interface homogenization. Discretely modeling stitch locations with varied cohesive zones has shown the ability to capture varied load response during delamination propagation as is clearly seen from comparisons with experimental data. Additionally, the discrete modeling approach has the flexibility of scaling the "effective" region of the fiber distortion to modify in-plane mechanical properties. High fidelity CT data was collected to characterize the impact of varied levels of stitch density on fiber distortion. Ultimately, increasing the levels of stitch density did not impact the total degree of fiber alignment in the region of reinforcing stitches significantly. It is clear that increasing the stitch density does ultimately impact the total volume of resin rich regions.

While this initial simulation work has shown good agreement between the mode I fracture toughness results, there is a significant amount of work that still needs to be completed. This includes simulating the impact of fiber distortion on tensile modulus and strength. The distortion data collected by high fidelity μCT will be used to assign mechanical properties, not just stiffness variations, on the element-by-element basis producing numerical models capable of capturing stiffness variations due to fiber distortions, improved interlaminar tensile strength and toughness due to Kevlar reinforcing stitching (via discrete cohesive surfaces between plies). These simulations will be compared to mechanical test data to evaluate and increase the accuracy of the methodology.

Once the methodology has been verified to capture the impact of distortions on stiffness and strength, the plan is to validate this methodology using Double Bearing Shear (DBS) (ASTM D5961), in addition to 4 point bending of angled brackets (ASTM D6415) with the ultimate goal of using it to optimize composites' design for TFP.

5. REFERENCES

[1] A. Mouritz and B. Cox, "A mechanistic approach to the properties of stitched laminates," Composites: Part A, pp. 1-27, 2000 [10.1016/S1359-835X(99)00056-1]

[2] Mattheij, P., K. Gliesche, and D. Feltin. "3D reinforced stitched carbon/epoxy laminates made by tailored fibre placement." Composites Part A: Applied Science and Manufacturing 31.6 (2000): 571-581 [10.1016/S1359-835X(99)00096-2]

[3] Mouritz, A. P., K. H. Leong, and I. Herszberg. "A review of the effect of stitching on the in-plane mechanical properties of fibre-reinforced polymer composites." Composites Part A: applied science and manufacturing 28.12 (1997): 979-991 [10.1016/S1359-835X(97)00057-2]

[4] G. Gardiner, "Tailored Fiber Placement: Besting metal in volume production," CompositesWorld, 2 September 2013. [Online]. Available: http://www.compositesworld.com/articles/tailored-fiber-placement-besting-metal-in-volume-production

[5]Almeida Jr, José Humberto S., et al. "Cross-section optimization of topologically-optimized variable-axial anisotropic composite structures." Composite Structures 225 (2019): 111150.[10.1016/j.compstruct.2019.111150]

[6] E. Koricho, A. Khomenko, T. Fristedt and M. Haq, "Innovative tailored fiber placement technique for enhanced damage resistance in notched composite laminate," Composite Structures, vol. 120, pp. 378-385, 2015] [10.1016/j.compstruct.2014.10.016]

[7] Almeida Jr, José Humberto S., Lars Bittrich, and Axel Spickenheuer. "Improving the open-hole tension characteristics with variable-axial composite laminates: Optimization, progressive damage modeling and experimental observations." Composites Science and Technology 185 (2020): 107889.] [10.1016/j.compscitech.2019.107889]

[8] Hoos, Kevin, et al. "Static strength prediction in laminated composites by using discrete damage modeling." Journal of Composite Materials 51.10 (2017): 1473-1492.] [10.1177/0021998316651986]

[9] Huelskamp, S., C. Tanner, and J. Stonecash. "Effects of Z-Stitching in Heavy Tow Carbon Laminates via Tailored Fiber Placement (TFP)." SAMPE JOURNAL 54.1 (2018): 44-52

ON THE RELEVANCE OF THERMOMECHANICS AND CRYSTALLIZATION KINETICS FOR FINITE ELEMENT THERMOFORMING SIMULATION

Dominik Dörr[1,2], Luise Kärger[2], Frank Henning[2,3]

[1]SIMUTENCE GmbH, Karlsruhe, Germany
[2]Karlsruhe Institute of Technology, Karlsruhe, Germany
[3]Fraunhofer Institute for Chemical Technology, Pfinztal, Germany

ABSTRACT

Thermoforming of thermoplastic tapes is currently of great interest for the automotive industry, due to low cycle times, material efficiency and recyclability. Depending on material parameters and process conditions, however, manufacturing defects may occur. Finite Element (FE) forming simulation offers the possibility of a detailed a priori analysis of the deformation behavior of multilayered thermoplastic blanks during forming, considering material behavior and process conditions by means of constitutive equations and boundary conditions, respectively. Usually, thermoforming simulation is assumed to be iso-thermal, which is a reasonable assumption for temperatures above the onset of crystallization for semi-crystalline thermoplastics. Especially in a process design phase, however, the onset of crystallization cannot be excluded in general. This study presents a fully coupled thermomechanical approach for finite element forming simulation of thermoforming processes, predicting the evolution of temperature and crystallization of semi-crystalline thermoplastics. The approach is successfully validated for a generic geometry with a high agreement to experimental thermoforming tests. Finally, the relevance of including thermomechanics and crystallization kinetics is analyzed by means of a virtual sensitivity study. The study reveals that only by including those effects, the influence of all process parameters on formability can be predicted.

Keywords: Thermoplastic, Forming, Process simulation, Thermomechanical, FEA
Corresponding author: dominik.doerr@simutence.de

1. INTRODUCTION

Composite materials reveal a particularly high lightweight potential, due to their excellent weight-specific material properties and their capability to be tailored to specific applications [1]. While the automotive industry still mostly uses commodity materials such as steel and short glass fiber-reinforced composites [2], the aerospace industry is using high-performance materials since decades. This is attributable to automotive mass production, where the material dominates the overall costs. Hence, automotive industry requires highly automated and robust production technologies, which make a high investment affordable as long as material costs are sufficiently low.

In general, it is necessary that system-efficient composite solutions are developed, which consider besides technical and technological targets also economical boundary conditions. For this reason, the increasing demand for the reduction of green house gas emissions can be an enabler for composites in automotive industry. However, too high production costs and a limited understanding of automotive-specific processes still hinder high-performance composites to enter the automotive mass-production market.

Continuously fiber-reinforced plastics (CoFRP), however, reveal an exceptional lightweight potential and are particularly suited for structural components, due to their excellent weight-specific properties [3]. Nevertheless, manufacturing of CoFRP for both automotive and aerospace applications involves different challenges. First, materials are mostly expensive. Therefore, material consumption needs to be optimized. Second, forming of two-dimensional blanks into a three-dimensional geometries substantially influences the fiber structure, which in turn influences dimensional stability in terms of spring-in and warpage, and structural performance. For these reasons, a reliable virtual prediction of forming behavior by means of forming simulation is of utmost importance to enable time- and cost-effective process and product design. Beyond that, prediction accuracy of subsequent simulation steps is in conjunction with a virtual process chain significantly increased [4]. Within a virtual process chain (cf. Figure 1), relevant information is virtually predicted, retained and passed on successively to subsequent simulation steps.

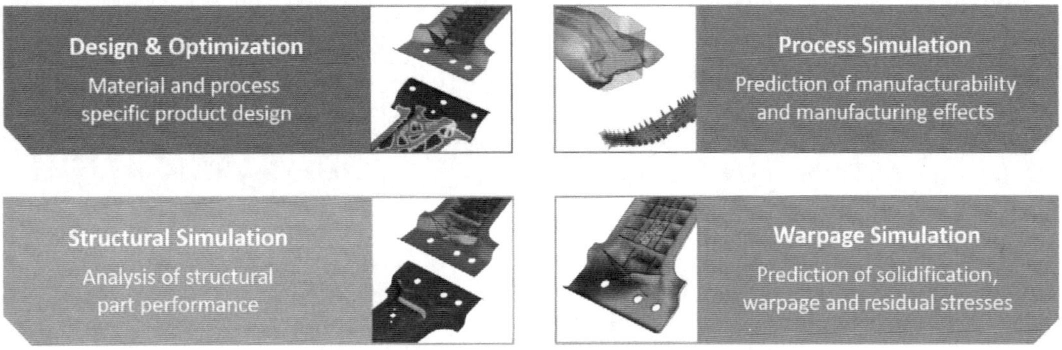

Figure 1: Schematic illustration of the sub-steps of a virtual process chain for the reliable virtual analysis of manufacturing and structural behavior of fiber-reinforced plastics.

Finally, achieving low cycle times is a challenge, which becomes increasingly important also for aerospace industry. Therefore, thermoforming of continuously fiber-reinforced thermoplastics (CoFRTP), also referred to as thermoplastic tapes, play an increasingly important role, due to low cycle times, material efficiency and recyclability [5,6].

Thermoforming processes are influenced by several parameters, like layup sequence, tool and laminate temperature, forming velocity or laminate gripping, as shown in experimental thermoforming studies by several authors [7,8,9]. Depending on these parameters and the formed geometry, manufacturing defects such as wrinkling are possible. These defects, as well as the modification of the fiber structure due to forming, are predictable by means of Finite Element (FE) thermoforming simulation. For this purpose, material behavior and process conditions are taken into account based on constitutive equations and boundary conditions, respectively.

In relation to the current interest in forming of CoFRP, commercial codes for macroscopic FE forming simulation have been developed. These are in particular the PAM-FORM [8] and the ANIFORM code [10]. Moreover, several codes for FE forming simulation have been presented in literature [11-14]. Thereby, recent approaches for thermoforming simulation of CoFRTP mostly assume thermoforming processes to be isothermal. Only few approaches consider yet the thermal behavior as well as the temperature-dependent forming behavior in terms of a thermomechanical coupling [15,16]. Both studies are restricted to temperatures well above the onset of crystallization of the investigated semi-crystalline thermoplastics. The onset of crystallization, however, induces a significant increase of mechanical properties [17,18]. Thus, it is to be expected that crystallization distinctively influences forming behavior. Since the onset of crystallization during forming cannot be excluded in general and especially prior to process optimization, a thermomechanical approach for thermoforming simulation considering also the crystallization kinetics of semi-crystalline thermoplastics is desirable for the virtual process design.

This paper presents the application of a coupled thermomechanical approach for thermoforming simulation considering besides temperature evolution also crystallization kinetics and developed by the authors [19,20]. Based on this, the phase transition from the molten to the solid material state is predicted and is taken into account for modeling of temperature- and rate-dependent thermoforming behavior.

The approach was originally developed at the Karlsruhe Institute of Technology (KIT) in Karlsruhe Germany [19,20] and is now being industrialized by the spin-off company SIMUTENCE. The software package called SIMUDRAPE comes as add-on for the commercially available Finite Element software ABAQUS/EXPLICIT and includes a fully automatized pre- and postprocessing as well as the above outlined methods for advanced contact and material modeling.

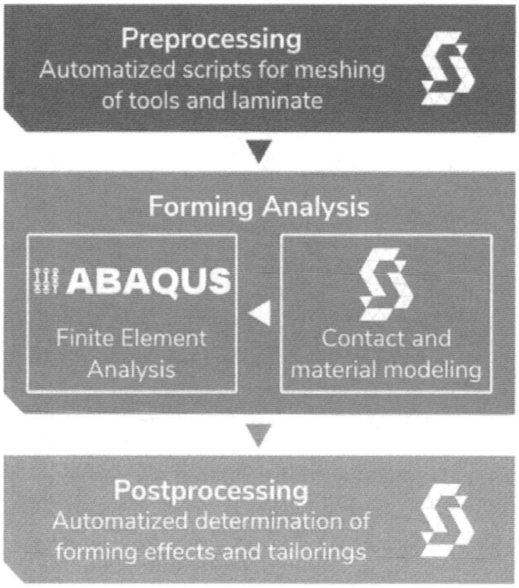

Figure 2: Schematic illustration of the working principle of SimuDrape add-ons for ABAQUS.

In the following, the coupled thermomechanical approach for thermoforming simulation is outlined in Section 2. Subsequently, Section 3 presents simulation results for a generic geometry for the isothermal and the thermomechanical case, including the comparison to experimental tests. Based on this, the necessity to account for thermal effects and crystallization kinetics is investigated.

2. SIMULATION APPROACH

FE forming simulation is based on constitutive modeling of so-called intra- and inter-ply deformation mechanisms (cf. Figure 3). Intra-ply mechanisms include membrane and bending behavior of the single ply and are usually modeled layer-by-layer by conventional shell elements, due to the high slenderness ratio of the single plies. Inter-ply mechanisms include the frictional behavior between the tool and the laminate, as well as frictional behavior and tacking between the single plies of the stacked laminate and are modeled by means of contact constitutive equations.

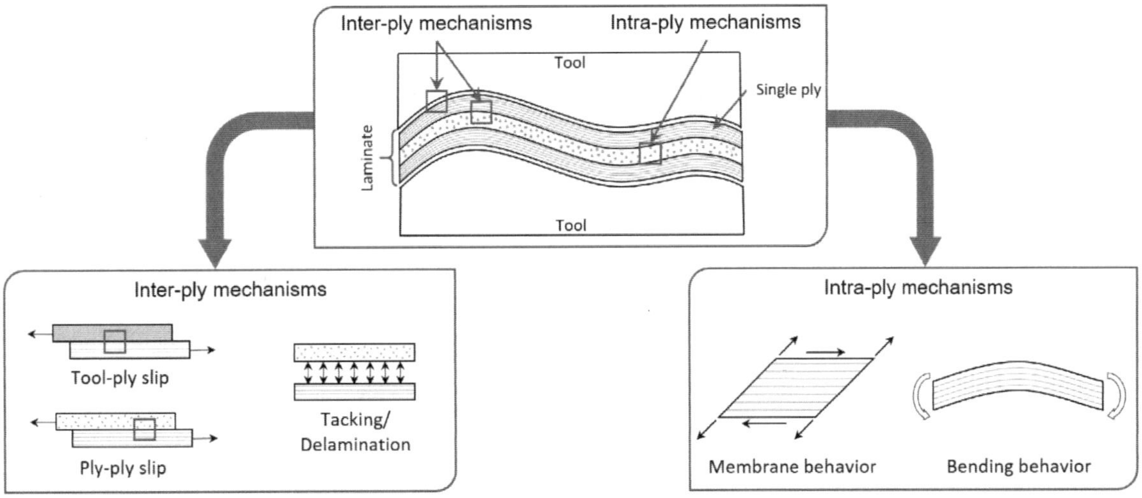

Figure 3: Schematic illustration of the deformation mechanisms during composite forming categorized into intra- and inter-ply mechanisms.

For the approach applied for thermoforming simulation in this study, the commercial FE solver ABAQUS/EXPLICIT in combination with several user-subroutines is applied. In the following, the governing equations for thermal and mechanical modeling, as well as for thermomechanical coupling are outlined.

2.1 Modeling of thermal behavior

The governing equation for modeling of thermal behavior is the heat balance equation, in combination with the generalized Fourier's law for heat conductivity:

$$\underbrace{\int_{\Omega}(\rho c_p \dot{T} \delta T)\mathrm{d}V}_{\delta W^{\mathrm{cap}}} = \underbrace{\int_{\Omega}(\lambda \cdot \mathrm{grad}(T)) \cdot (\mathrm{grad}(\delta T))\mathrm{d}V}_{\delta W^{\mathrm{cond}}} + \underbrace{\int_{\Gamma_{\Omega}}(s\delta T)\mathrm{d}A}_{\delta W^{\mathrm{surf}}} - \underbrace{\int_{\Omega}(r\delta T)\mathrm{d}V}_{\delta W^{\mathrm{source}}}, \quad (1)$$

where T is the temperature, ρ the material density, c_p the heat capacity, $\boldsymbol{\lambda}$ the orthotropic heat conductivity tensor, s a surface flux on the boundary and r a volumetric heat flux. To predict thermal behavior in thermoforming simulation, the capacity term δW^{cap}, the conductivity term δW^{cond}, and the source term $\delta W^{\mathrm{source}}$ are used to model intra-ply heat capacity, conductivity and latent crystallization heat, respectively. In contrast, the surface flux term δW^{surf} is adopted to account for the inter-ply heat conductivity as well as for convection and radiation at the free surfaces, due to the exposure of the hot laminate to a cooler environment. Thereby, inter-ply heat conductivity includes the heat exchange between the single plies of the laminate as well as between the tool and the laminate. The latter is also referred to as tool-ply gap conductance.

During cooling of a semi-crystalline thermoplastic from a temperature well above the melting temperature, the thermoplastic undergoes a partial alignment of the molecular chains during solidification, which is called crystallization, and which defines the transition from the molten material state to the solid material state. In order to predict the relative crystallinity α throughout thermoforming, a wide range of cooling rates has to be captured. A suitable approach under this requirement is the approach originally presented by Ziabicki [21] and the enhancement to several cooling rates by Sierra et al. [22], as shown in our previous work by Kugele et al. [23]. This approach combines Nakamura's equation [24]

$$\dot{\alpha} = nK(T)\big(1-\alpha(t)\big)\ln\Big(-\big(1-\alpha(t)\big)\Big)^{\frac{n-1}{n}}, \tag{2}$$

based on the Avrami index n and the crystallization rate constant K, with an empirical description of the crystallization rate constant given by [21]

$$K(T) = K_{\text{max}} \exp\left(\frac{-4\ln(2)\,(T - T_{\text{max}})}{D^2}\right). \tag{3}$$

Here, K_{max}, T_{max} and D are material parameters for one specific cooling rate. By means of the determination of these parameters for various cooling rates, crystallization is predictable over a wide range of cooling rates [23].

2.2 Modeling of mechanical behavior

The coupled thermomechanical approach includes besides modeling of thermal behavior also temperature- and crystallization-dependent mechanical behavior, as outlined in the following.

2.2.1 Intra-ply deformation behavior

Mechanical modeling of intra-ply deformation is based on a conventional shell approach, which is implemented in the commercial FE solver ABAQUS in combination with several user-subroutines. The related mechanical virtual internal energy is given by

$$\delta\Pi = \int_V \left[\underbrace{\boldsymbol{S}^{\text{mem}} \left(\frac{\partial \boldsymbol{E}}{\partial \boldsymbol{r}_i}\right) \delta \boldsymbol{r}_i}_{\delta\Pi^{\text{mem}}} + \underbrace{\boldsymbol{S}^{\text{ben}} \left(\frac{\partial \boldsymbol{E}}{\partial \boldsymbol{d}_i}\right) \delta \boldsymbol{d}_i}_{\delta\Pi^{\text{ben}}} \right] \mathrm{d}V. \tag{4}$$

where $\boldsymbol{E}$ is the Green–Lagrange strain, $\boldsymbol{r}_i$ the nodal displacements, $\boldsymbol{d}_i$ the nodal unit director base, and $\boldsymbol{S}^{\text{mem}}$ and $\boldsymbol{S}^{\text{ben}}$ the second Piola–Kirchhoff (PK2) stress for membrane and bending behavior, respectively. Different constitutive equations are evaluated for membrane and bending behavior, resulting in a physical decoupling of membrane and bending behavior. This decoupling is essential for intra-ply modeling in thermoforming above the recrystallization temperature (molten material state), due to the high stiffness in fiber direction and the comparably low bending stiffness, which is induced by the possible relative motion between the fibers at processing conditions, implying that conventional plate theories are not applicable. In contrast, conventional plate theory is applied below the crystallization temperature.

For material modeling, it is distinguished between the molten and the solid material state. In each case, the same constitutive equations are applied for both membrane and bending behavior but with different material parameters. For the molten material state, hyperviscoelastic constitutive equations are applied, to account for the rate- and temperature-dependent intra-ply deformation behavior. In contrast, hyperelastic constitutive equations are applied for the solid material state, since rate-dependency is expected to be negligible compared to the molten material state. Moreover, it is assumed that the free energy stored in the molten material state fully dissipates with the onset of crystallization. Based on this, the total deformation gradient $\boldsymbol{F}$ is multiplicatively decomposed into a solid part $\boldsymbol{F}^{\text{solid}}$ and a molten $\boldsymbol{F}^{\text{molten}}$ part by

$$\boldsymbol{F} = \boldsymbol{F}^{\text{solid}} \cdot \boldsymbol{F}^{\text{molten}}, \tag{5}$$

where the molten part is defined by the onset of crystallization. Based on this, the deformation state at the onset of crystallization $\boldsymbol{F}^{\text{molten}}$ is introduced as new reference configuration for the evaluation of the solid constitutive equations.

2.2.2 Inter-ply deformation behavior

Inter-ply deformation behavior, including the tacking in normal direction and friction in tangential direction, is modeled by means of contact constitutive equations implemented in user-subroutines. This is necessary, since based on ABAQUS built-in methods, the simultaneous tacking in normal direction and sliding in tangential direction at the tool-ply interfaces cannot be accounted for.

The related contact constitutive equations are regularized by means of the Penalty method in both normal and tangential direction. The tacking in normal direction is modeled by means of an additional stiffness, if the contact is under tension and as long as a maximum admissible distance between the interacting surfaces is not exceeded. In tangential direction, pressure- and rate-dependent friction is modeled for the molten material state by $\boldsymbol{\tau}^{\text{molten}}$, whereas purely elastic tacking is modeled for the solid material state by $\boldsymbol{\tau}^{\text{solid}}$:

$$\boldsymbol{\tau}^{\text{molten}} = \left(\mu p + \eta|\dot{\boldsymbol{\xi}}| + \tau_0\right)\frac{\dot{\boldsymbol{\xi}}}{|\dot{\boldsymbol{\xi}}|}, \qquad \boldsymbol{\tau}^{\text{solid}} = \epsilon\boldsymbol{\xi}, \tag{6}$$

where $\boldsymbol{\xi}$ is the tangential displacement vector, $\boldsymbol{\tau}$ the tangential stress, p the transversal pressure, μ, η and τ_0, material parameters, respectively, and ϵ the tangential solid stiffness.

2.3 Thermomechanical coupling

To account for temperature-dependent thermoforming behavior, related constitutive equations are evaluated as a function of temperature in the molten material state. To account additionally for a possible phase transition during forming, constitutive equations are evaluated for the solid and the molten material state simultaneously. For intra-ply modeling, the resulting stress tensors for the molten material state $\boldsymbol{S}^{\text{molten}}$ and the solid material state $\boldsymbol{S}^{\text{solid}}$ are superimposed based on the relative crystallinity α by

$$\boldsymbol{S} = (1-\alpha)\boldsymbol{S}^{\text{molten}} + \alpha\boldsymbol{S}^{\text{solid}}, \tag{7}$$

which guarantees a smooth transition between the molten and the solid material state. A similar approach is applied for inter-ply behavior for the constitutive equations given in Equation 6.

3. VALIDATION

The above outlined coupled thermomechanical approach is finally applied to thermoforming simulation of a generic geometry for an orthotropic and a quasi-isotropic layup. Each of the 8 single plies is modeled by means of separate element layers with 42,672 elements per layer. The tools are modeled as discrete rigid surfaces with a constant temperature, since their deformation and heating during forming is expected to be negligible. The tool kinematics is modeled by a displacement boundary condition in real-time, to accurately account for rate-dependency and thermal behavior of the laminate. Laminate gripping is modeled following the approach proposed by Poppe et al. [28] based on so-called connector elements, which is necessary to accurately account for the kinematics of the pneumatic needle grippers, since these can rotate and elongate during forming. For modeling of thermal and mechanical behavior, the approach outlined in Section 2 is adopted and parameterized for a PA6-CF tape. Details on the parameterization can be found in our previous studies [19,20]. For validation of the above outlined thermoforming simulation approach, the simulation results are compared to surface measurements of experimental tests. Thereby, the results of the thermomechanical approach as well as the results of an isothermal approach, which is achieved by assuming a constant laminate temperature of 270 °C, are considered. Based on this, the necessity to account for thermal effects and crystallization kinetics in thermoforming simulation is investigated.

3.1 Orthotropic layup

Figure 4 shows the comparison of thermoforming simulation results to surface measurements of the according experimental results under application of the thermomechanical and related isothermal approaches at several remaining tool travels Δz and for the orthotropic layup. A good agreement of the obtained overall shape and outer contour is observed for both layups. However, in a more detailed comparison it is observed that some local defects are predicted only by the thermomechanical approach, in particular for the large evolving fold above the bigger corner bending. This forming defect is clearly predicted by the thermomechanical approach, especially in Stage 2. Moreover, local wrinkling around the beads is predicted more pronounced by the thermomechanical approach. This is in good agreement with the experimental tests and the difference can be reduced to local cooling and crystallization of the laminate, as shown for several remaining tool travels Δz in Figure 5 for the lowest ply. Here, comparably low cooling due to free convection and radiation is observed for the free surfaces. In contrast, distinct local cooling due to tool-ply contact (gap conductance) is observed. Thereby, local crystallization is observed in these areas already during forming. Besides this, local wrinkling is directly reflected in the temperature distribution, due to the local tool-ply contact.

Stage 1: $\Delta z = 17.5$ mm

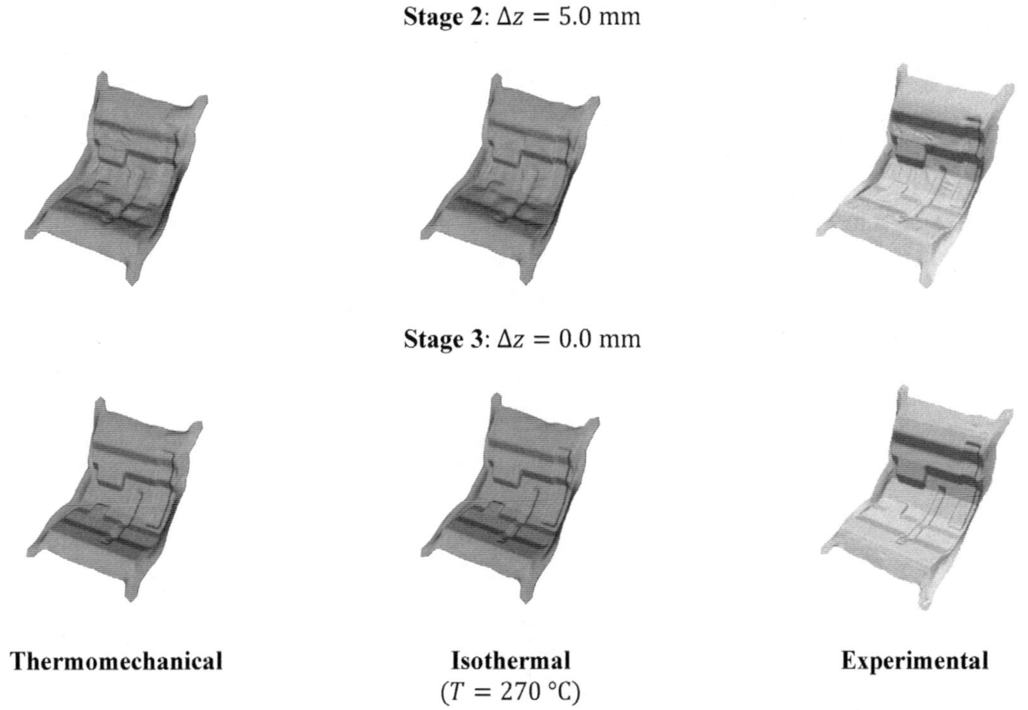

Figure 4: Comparison of thermoforming simulation, under application of the thermomechanical and the isothermal (T = 270 °C) approach to 3D measurements of experimental tests at several remaining tool travels Δz, for the orthotropic layup ($[0;90]_{2s}$), and for a tool temperature of 110 °C [20].

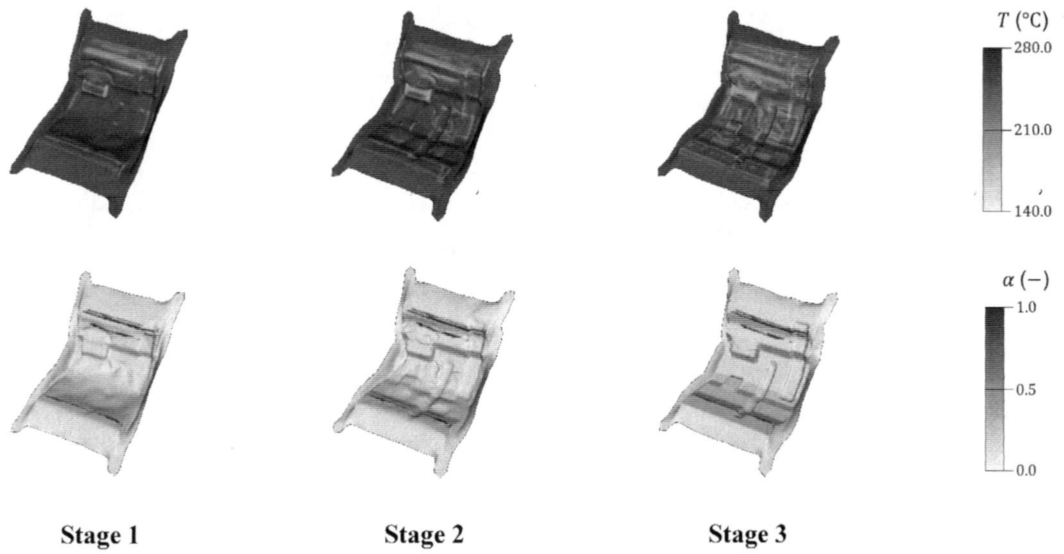

Figure 5: Temperature T (top row) and relative crystallinity α (bottom row) for the orthotropic layup ($[0;90]_{2s}$) at several remaining tool travels Δz, obtained from the thermomechanical approach [19].

3.2 Quasi-isotropic layup

Figure 6 shows the comparison of thermoforming simulation results under application of the thermomechanical and the isothermal approach to surface measurements of the according experimental results for the quasi-isotropic layup at several remaining tool travels Δz. Distinctive wrinkling behavior is observed throughout forming in the experimental test, which is captured clearly by both simulation approaches. Increased wrinkling behavior, however, is predicted by the thermomechanical approach. Here, a high degree of agreement to the experimental tests for the amount, direction and location of wrinkles is observed. In contrast, less pronounced wrinkling is predicted under application of the isothermal approach. Additionally, slight deviations are observed for the unfolding and sag of the laminate in the outer areas, especially for Stage 3.

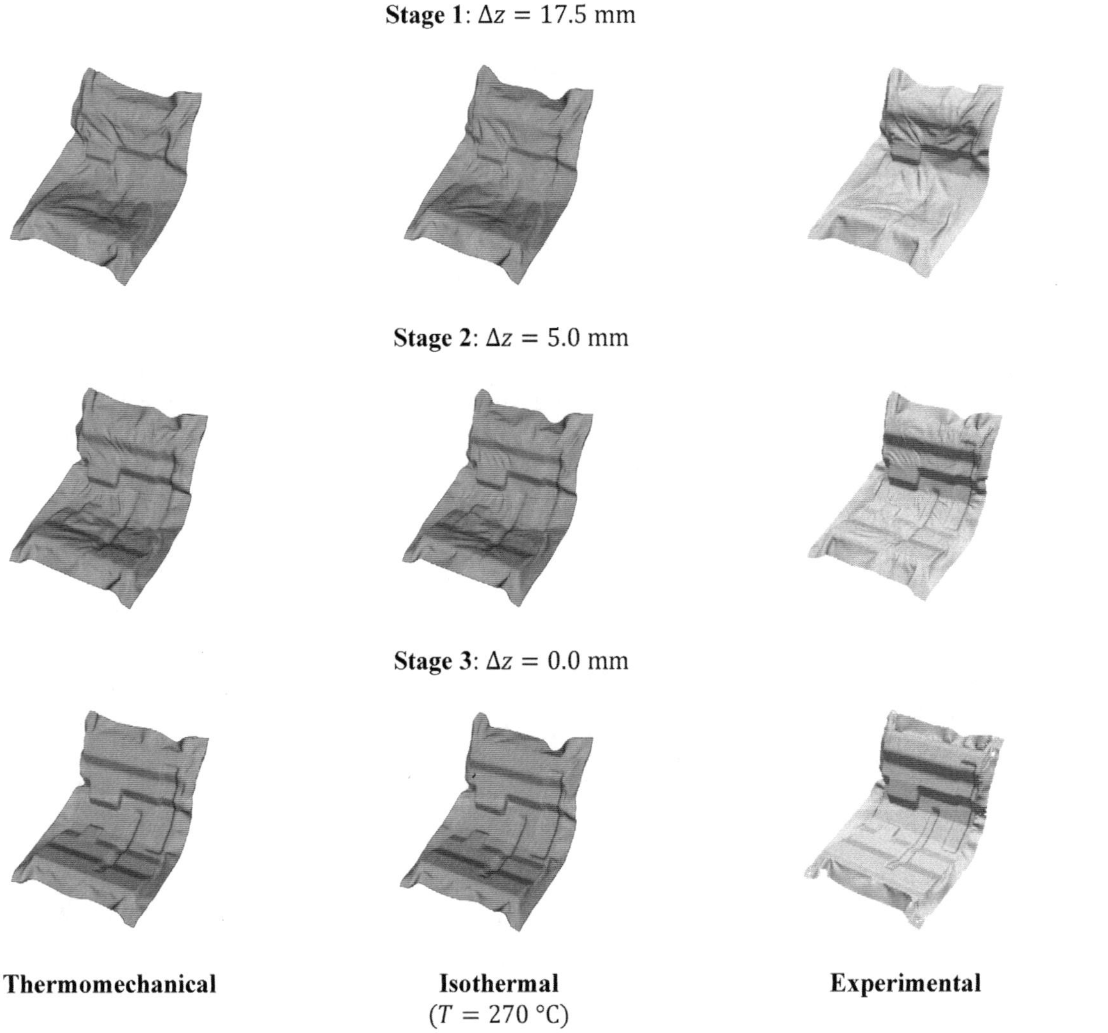

Figure 6: Comparison of thermoforming simulation, under application of the thermomechanical and the isothermal ($T = 270$ °C) approach, to 3D measurements of experimental tests at several remaining tool travels Δz, for the quasi-isotropic layup ([0; 45;90;45]s), and for a tool temperature of 110 °C [20].

4. CONCLUSIONS

A coupled thermomechanical approach for thermoforming simulation of continuously fibre-reinforced thermoplastics implemented in the commercial FE solver ABAQUS in combination with several user-subroutines is presented. This approach is now being industrialized under the product name SIMUDRAPE by the KIT spin-opff company SIMUTENCE. Following this approach, mechanical behavior is modeled by means of hyperelastic and hyperviscoelastic constitutive equations. Modeling of thermal behavior takes heat loss due to radiation, convection and heat conduction as well as the prediction of crystallization kinetics during forming into account. The phase transition is predicted by the relative crytallinity and is taken into account for mechanical modeling by means of temperature- and crystallinity-dependent material properties.

The application of the coupled thermomechanical approach to a generic geometry shows an improved agreement to the according experimental forming tests compared to the related isothermal approach. For the orthotropic layup, it is observed that some forming defects are captured by the thermomechanical approach only. In contrast, distinct wrinkling behavior of the quasi-isotropic layup observed in the experimental tests is captured basically by both approaches. Wrinkling behavior, however, is predicted too less pronounced for the isothermal approach. This results from distinct local cooling of the laminate, obviously not captured by the isothermal approach, but observed for the thermomechanical approach locally at the tool-ply contact areas. Besides this, local recrystallization is observed already during forming. In this study, recrystallized areas result mainly from the initial positioning of the laminate onto the lower (female) tool at the part flanges. This could possibly be prevented by an adapted process strategy. Nevertheless, the simulation results reveal that recrystallization already during forming cannot be excluded in general, especially prior to process optimization, which is expected to be an important finding for the virtual process design.

With an eye towards industry 4.0, the presented approach is considered as a next step in direction of a holistic thermoforming simulation approach, by taking into account temperature- and rate-dependency of the formed material, as well as the phase transition from the molten to the solid state. These material characteristics are usually not yet taken into account for thermoforming simulation. However, the validation of the presented approach has proven to improve the prediction accuracy. Therefore, the capability of thermoforming simulation to be applied to an a priori virtual process design is improved. Nonetheless, the proposed model has some limitations. First, only adapted heat loss in the vicinity of the tools, but no radiative heat transfer between the tool and the laminate is accounted for in thermal modeling. Taking into account this effect might improve thermal modeling, but would require further advanced modeling techniques and a distinctively higher computational effort. Second, no thermal expansion and shrinkage due to crystallization is taken into account so far. These effects are expected to not significantly influence forming behavior itself. However, consideration of these effects would enable the prediction of residual stresses and shape distortions, if recrystallization is occuring already during forming. Third, it is assumed in the presented approach that the free energy of the molten phase fully dissipates during solidification. This approach is expected to be reasonable for formability prediction, since local solidification represents the loss of formability. The free energy of the fibers, however, would most likely influence the prediction of residual stresses and shape distortion. Finally, the solid phase is assumed to behave purely. The consideration of plasticity and damage could be interesting to be investigated for locally solidified areas.

5. REFERENCES

1. F. Henning, L. Kärger, D. Dörr, F. J. Schirmaier, J. Seuffert, and A. Bernath. *Fast processing and continuous simulation of automotive structural composite components*. Composites Science and Technology, (171):261–279, 2019.

2. E. R. Fuchs, F. R. Field, R. Roth, and R. E. Kirchain. *Strategic materials selection in the automobile body: Economic opportunities for polymer composite design*. Composites Science and Technology, 68(9):1989–2002, 2008.

3. M. F. Ashby. *Materials selection in mechanical design*. Elsevier/Butterworth-Heinemann, Amsterdam, 4. edition, 2011.

4. L. Kärger, A. Bernath, F. Fritz, S. Galkin, D. Magagnato, A. Oeckerath, A. Schön, and F. Henning. *Development and validation of a CAE chain for unidirectional fibre reinforced composite components*. Composite Structures, (132):350–358, 2015.

5. K. Friedrich and A. A. Almajid. *Manufacturing Aspects of Advanced Polymer Composites for Automotive Applications*. Applied Composite Materials, 20(2): 107–128, 2013.

6. P. K. Mallick, editor. *Materials, design and manufacturing for lightweight vehicles*. Woodhead Publishing in materials. CRC Press, Boca Raton, 2010.

7. T. Joppich, D. Doerr, L. van der Meulen, T. Link, B. Hangs, F. Henning. *Layup and process dependent wrinkling behavior of PPS/CF UD tape-laminates during nonisothermal press forming into a complex component*. AIP Conference Proceedings 1769: 170011, 2016.

8. P. de Luca, P. Lefébure, A.K. Pickett. *Numerical and experimental investigation of some press forming parameters of two fibre reinforced thermoplastics: APC2-AS4 and PEI-CETEX*. Composites Part A: Applied Science and Manufacturing, 29(1–2): 101–10, 1998.

9. A. Benkaddour, G. Lebrun, L. Laberge-Lebel. *Thermostamping of [0/90] n carbon/peek laminates: influence of support configuration and demolding temperature on part consolidation*. Polymer Composites: 22-42, 1998.

10. S. P. Haanappel, R. ten Thije, U. Sachs, B. Rietman, and R. Akkerman. *Formability analyses of uni-directional and textile reinforced thermoplastics*. Composites Part A: Applied Science and Manufacturing (56): 80–92, 1998.

11. Y. Gong, X. Peng, Y. Yao, Z. Guo. *An anisotropic hyperelastic constitutive model for thermoplastic woven composite prepregs*. Composites Science and Technology (128): 17–24, 2016.

12. P. Harrison, R. Gomes, N. Curado-Correia N. *Press forming a 0/90 cross-ply advanced thermoplastic composite using the double-dome benchmark geometry*. Composites Part A: Applied Science and Manufacturing (54): 56–69, 2013.

13. P. Boisse, N. Hamila, A. Madeo. *The difficulties in modeling the mechanical behavior of textile composite reinforcements with standard continuum mechanics of Cauchy. Some possible remedies*. International Journal of Solids and Structures (154): 55–65. 2016.

14. L. M. Dangora, C. J. Mitchell, J. A. Sherwood. *Predictive model for the detection of out-of-plane defects formed during textile-composite manufacture*. Composites Part A: Applied Science and Manufacturing (78): 102–12, 2015.

15. M. Machado, L. Murenu, M. Fischlschweiger, Z. Major. *Analysis of the thermomechanical shear behaviour of woven-reinforced thermoplastic-matrix composites during forming.* Composites Part A: Applied Science and Manufacturing (86):39–48, 2016.

16. E. Guzman-Maldonado, N. Hamila, N. Naouar, G. Moulin, P. Boisse. *Simulation of thermoplastic prepreg thermoforming based on a visco-hyperelastic model and a thermal homogenization.* Materials & Design (93):431–42, 2016.

17. T.G. Donderwinkel, B. Rietman, S.P. Haanappel, R. Akkerman. *Stamp forming optimization for formability and crystallinity.* AIP Conf Proc 1769: 170029, 2016.

18. S. Ropers. *Bending behavior of thermoplastic composite sheets: viscoelasticity and temperature dependency in the draping process.* AutoUni - Schriftenreihe Vol. 99. Wiesbaden and s.l.: Springer Fachmedien Wiesbaden; 2017.

19. D. Dörr. *Simulation of the thermoforming process of UD fiber-reinforced thermoplastic tape laminate*s, Doctoral thesis, Karlsruhe Institute of Technology (KIT), 2019.

20. D. Dörr, T. Joppich, D. Kugele, F. Henning, L. Kärger: *A coupled thermomechanical approach for finite element forming simulation of continuously fiber-reinforced semi-crystalline thermoplastics*, Composites Part A: Applied Science and Manufacturing (125): 105508, 2019.

21. A. Ziabicki. *Fundamentals of fibre formation: the science of fibre spinning and drawing.* John Wiley & Sons Ltd; 1976.

22. J.D. Sierra, M.d.P. Noriega, J.F. Gómez, J.M. Pastor. *Isothermal and non-isothermal crystallization kinetics for blends of polyamide 6 and polypropylene.* International Journal of Plastics Technology 2(5):1–5, 2006.

23. D. Kugele, D. Dörr, F. Wittemann, B. Hangs, J. Rausch, L. Kärger, et al. *Modeling of the non-isothermal crystallization kinetics of polyamide 6 composites during thermoforming.* AIP Conf Proc. 1869: 030005, 2017.

24. K. Nakamura, T. Watanabe, K. Katayama, T. Amano. *Some aspects of nonisothermal crystallization of polymers. I. Relationship between crystallization temperature, crystallinity, and cooling conditions.* Journal of Applied Polymer Science 16(5):1077–91, 1972.

25. D.P.H. Hasselman, K.Y. Donaldson, J.R. Thomas. *Effective thermal conductivity of uniaxial composite with cylindrically orthotropic carbon fibers and interfacial thermal barrier.* Journal of Composite Materials 27(6):637–44, 1993.

26. D. Kugele, J. Rausch, J. Kriegeseis, K. Gündisch, L. Kärger, F. Henning. *On the thermal behavior of thermoplastic laminates during transfer - a novel wind-tunnel approach.* Proceedings ECCM17, München, Germany.

27. D. Kugele, J. Rausch, P. Müller, L. Kärger, F. Henning. *Temperature distribution in thickness direction of thermoplastic laminates during thermoforming.* Proceedings international conference of automotive composites (IJAUTOC 2016), Lisbon, Portugal.

28. C. Poppe, T. Joppich, D. Dörr, L. Kärger, F. Henning. *Modeling and validation of gripper induced membrane forces in finite element forming simulation of continuously reinforced composites.* AIP Conf Proc 1896: 030002, 2017.

METHODOLOGY FOR VALIDATION OF MATERIAL RESPONSE MODELS USING IN-SITU ABLATION SENSING METHODS

Colin Yee[1], Jon Langston[1], Hao Wu[2], and Joseph H. Koo[1,2*]

[1]The University of Texas at Austin, Dept. of Mechanical Engineering, Austin, TX 78712

[2]KAI, LLC, Austin, TX 78739

*jkoo@mail.utexas.edu

ABSTRACT

A complete and efficient methodology for characterizing the pyrolysis response of ablative Thermal Protection Systems (TPS) materials is developed. The end goal of acquiring the inputs necessary for material response (MR) modeling in extreme hyperthermal environments is to permit rapid down-selection of TPS materials. The example case in this study focuses on the characterization of the MXB-360 glass/phenolic ablative for use in the Insulation Thermal Response and Ablation Code (ITRAC) developed by Northrop Grumman Innovation Systems. Aerothermal ablation testing was conducted on an oxyacetylene test bed with the use of small test models. Emphasis is placed on determining pyrolysis kinetic modeling constants by use of TGA, pyrolysis gas composition and enthalpy tables by energy dispersive x-ray (EDX) spectroscopy, and in-depth temperature profile matching by in-situ ablation sensing methods. Sensor construction methods, oxidizing versus non-oxidizing environment considerations, data analysis of in-situ ablation sensing methods, and validation criteria for MR modeling are briefly discussed.

1. INTRODUCTION

1.1 Background

Increased commercialization of space and renewed interest in hypersonic in-atmosphere flight has spurred the need for light weight, high-performance composite Thermal Protection Systems (TPS). TPS are critical in protecting components from heating during hypersonic flight or reentry in order to ensure reusability of the flight chassis or safe delivery of the payload [1]. Flight testing and post flight analysis is traditionally utilized to test TPS as it is the most reliable way to subject material samples to the combination of heat, vibration, and shock associated with hypersonic flight [2, 3], but ablation and heating models must be made more reliable in order to minimize expensive flight testing and lengthy design cycle times [4].

Low cost testing of small material samples greatly streamlines the down selection process of materials for TPS use. This testing typically involves the use of either heated plasma [5, 6] or oxyacetylene torches [7, 8] to apply high heat flux to the material sample. Ablation testbed operation still requires a significant time investment however, and access to high-temperature hypersonic wind tunnels is gated by both scheduling and cost. Repeat testing under differing conditions or high numbers of samples may still be costly, so the use of reliable numerical modeling to simulate alternate test conditions streamlines the TPS development cycle.

The modeling of ablation phenomena is complicated due to the moving boundary conditions combined with mechanical erosion, thermochemical erosion, and changing density. Material Response (MR) modeling codes make use of thermal and mechanical material properties, as well as pyrolysis Kinetic Modeling Parameters (KMPs) in combination with the Arrhenius equation, in order to simulate the decay of materials in high temperature environments [9]. While it is common practice to mechanically and thermally characterize materials as they are produced, KMPs and other ablation specific characteristics, such as net gaseous outflow (B') tables, are generally not provided by manufacturers and must be determined by the end user. Common mechanical and thermal property characterization will not be covered. As such, this methodology is developed as a training introduction for those familiar with engineering materials but with limited experience in designing or testing pyrolyzing ablative materials.

The decay state of a material acts as a function by which material properties between the virgin material and charred material can be bridged. KMPs are typically determined via Thermogravimetric Analysis (TGA) in either an inert or oxidizing environment and then validated against experimental data. Depending on the complexity of the material, KMPs may be determined for each reaction step that occurs during decomposition, which is also dependent on the availability of reactants in the environment. Accurate KMPs and heats of formation are critical in modeling the ablation of TPS materials as they are the foundation of an accurate decay state function and determine the rate of energy consumption or release while the material is decomposing.

Figure 1 shows the dominant effects associated with the ablation of TPS materials. Heat input to the material comes primarily from frictional heating of air at hypersonic speeds as well as radiation and convection heat transfer from the resulting plasma sheath [4]. The material pyrolyzes as it is heated, releasing decomposition gases which mostly diffuse back through the lower density material near the ablation front. The gases may then flow through the char layer, providing further heat input to the material as they either ignite or radiate. The diffusion of pyrolysis gases additionally results in internal stresses throughout the TPS material.

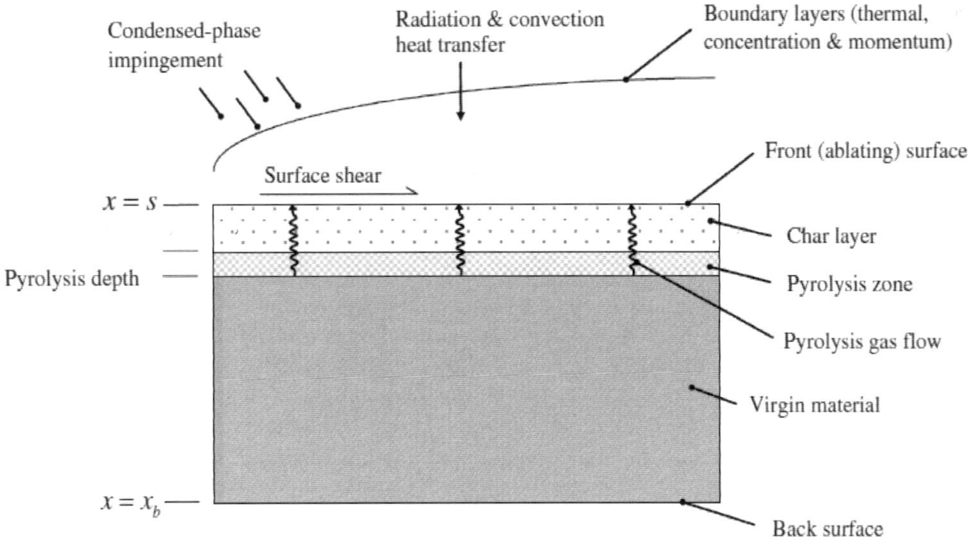

Figure 1. Ablation phenomena [10].

As the material continues to heat, the fully pyrolyzed char is removed by shear forces and the pyrolysis zone penetrates further into the virgin material. The virgin material and the back surface maintain a lower temperature than the ablating surface due to heat being removed by the heat of transformation, mass loss through the removal of heated char, and mass loss through the diffusion of pyrolysis gases. Thus, structural and electrical components are maintained at a safe temperature while the ablating front is subjected to the extreme conditions of hypersonic travel through air. Predicting the expected mass loss of the shield, internal temperature gradient, and depth of ablation are critical for design engineers in order to ensure that the shielded components are adequately protected. Accurate simulation of TPS ablation reduces the time for iterative design. Additionally, high confidence MR modeling may permit a reduced safety factor and thus considerably reduce flight weight.

1.2 Arrhenius Equation

While heat and mass conservation equations can be utilized for fully virgin or char state material ablation, the density of a given section of material and its state of thermal decay are determined by the decay form of the Arrhenius reaction equation. The Arrhenius decay equation for change in extent of reaction with respect to temperature is described as follows:

$$\frac{d\alpha}{dT} = Ae^{-\frac{E_A}{RT}} f(\alpha) \qquad [1]$$

Where α is the extent of reaction between virgin and char states, A is the Arrhenius pre-exponential, E_A is the activation energy of the material, R is the universal gas constant, and f(α) is the reaction model function [9]. The extent of reaction α can be described as a ratio of the density difference of the material between the virgin and char states compared to the current density:

$$\alpha = \frac{\rho_v - \rho_s}{\rho_v - \rho_c} \qquad [2]$$

With ρ_v being the virgin density, ρ_s being the current density, and ρ_c being the char density. At a virgin state, ρ_s is equal to ρ_v and α is equal to zero; at a fully charred state, α is one. The changes in all material properties from the virgin to the charred material state are approximated by α, which acts as a linear interpolation term. The reaction model function can be described differently depending on what model is being utilized. Commonly, either the Friedman or Kissinger model are utilized: [11]

$$f(\alpha) = (1 - \alpha)^n \alpha^m [-\ln(1 - \alpha)]^P \qquad [3]$$

$$f(\alpha) = (1 - \alpha)^m \qquad [4]$$

Where m, n, and p are experimentally determined curve fit parameters. Another method described by Ozawa [12] uses an arbitrary curve fitted to TGA data for the reaction model function f(α). KMPs as a whole refer to these reaction model function curve fit parameters (n, m, p, etc.) along with Arrhenius pre-exponential and the activation energy.

2. TEST METHODS

2.1 Thermogravimetric Analysis (TGA)

TGA is performed by weighing a sample of material inside an insulated furnace with a scale while its temperature is increased by a constant heating rate. The temperature is raised to a high temperature (typically in the range of 1,000 to 1,600 °C) and the sample is weighed at a fixed time interval. A purge gas, normally either air or nitrogen, is used to remove pyrolysis gases and ensure that only the solid material is weighted. This permits the user to observe how the material degrades in increasingly high temperature environments while also separating the degradation reactions. It is also more expeditious than multiple material degradation tests at constant temperature and compensates for degradation that occurs during the heat up to a constant temperature test [12]. Structure changes in polymers at constant high temperatures also makes constant heating rate TGA the preferred method of determining KMPs.

The number of degradation reactions in a material undergoing TGA is normally observed by the number of inflection points in the mass versus temperature curve, pictured below in Figure 2. While polymer degradation is very complex, bulk KMP determination focuses on analyzing the dominant reaction chains as a combined group.

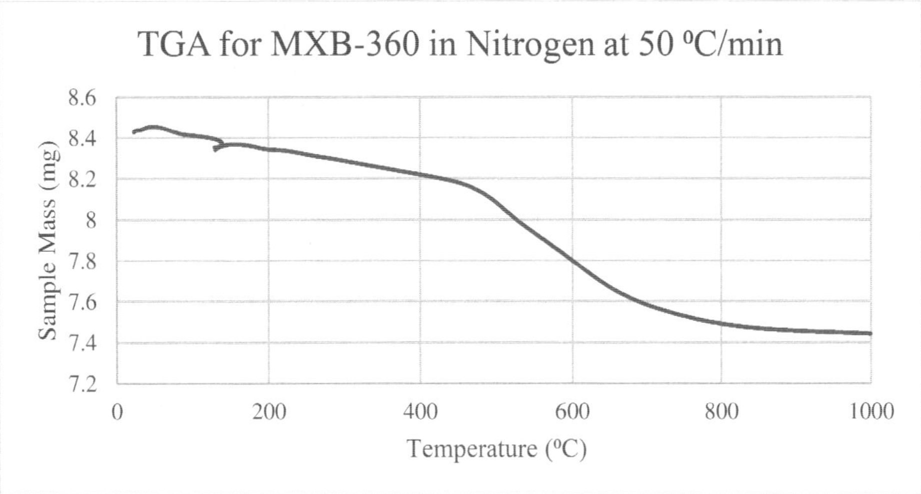

Figure 2. TGA data for MXB-360 glass/phenolic at a heating rate of 50 °C/min using nitrogen as a purge gas with a flow rate of 50 ml/min.

The TGA data in Figure 2 was generated using a TGA-50 from Shimadzu Scientific Instruments. The 10 mg MXB-360 samples were heated in either a nitrogen or air environment from room temperature to 130 °C and then maintained at that temperature for 30 minutes in order to remove any moisture. At the end of the drying period, the TGA-50 directly heats the sample at a constant heating rate up to 1,000 °C. The purge rate of the air or nitrogen is maintained at 50 ml/min in order to adequately remove pyrolysis gases while minimizing risk of disturbing the material sample in the furnace.

2.2 Aerothermal Heating Test Beds

For destructive testing of charring materials, aerothermal test beds provide low cost alternatives to heating during flight testing. Aerothermal test beds use heated gas flows to enable a combination of convective and radiative heat transfer to the test model. Common choices for aerothermal testing are oxyacetylene test beds (OTB) used to simulate oxidizing conditions (*i.e.,* solid rocket exhaust) [13] or inductively coupled plasma (ICP) torches shown in Figure 3 which can simulate non-oxidizing or less-oxidizing (*i.e.,* atmospheric reentry) environments with the use of inert gases.

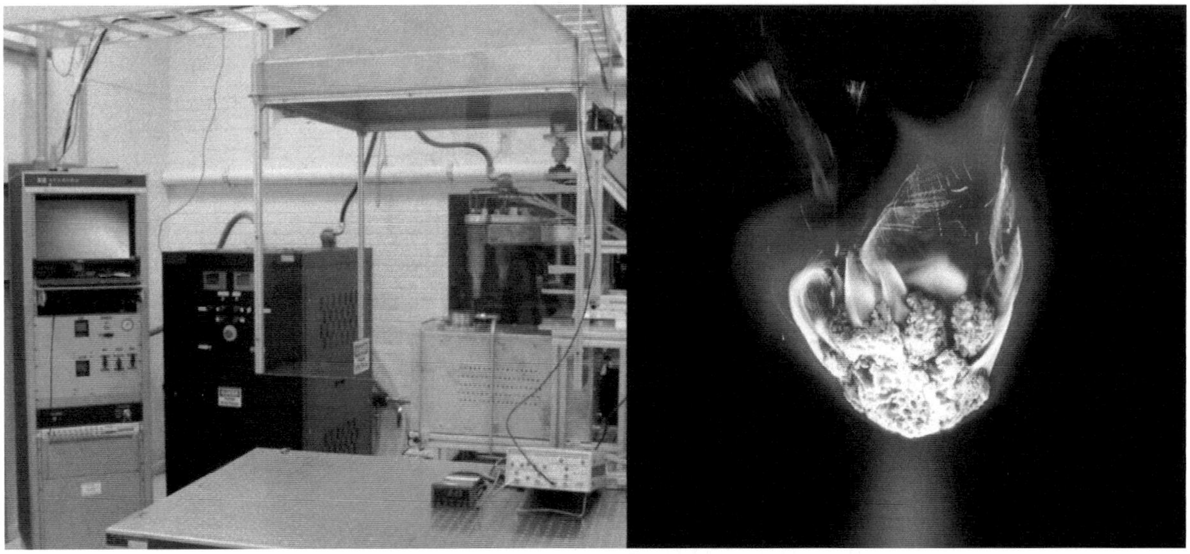

Figure 3. An ICP torch with robotic armature located at the University of Texas Pickle Research Campus (left). A thermoplastic test model undergoing testing in the ICP in an argon environment (right).

While the heat flux applied to the test model may be accurately measured by a Gardon gage, many MR codes require the separation of convective heat flux from radiative heat flux. Instead of extensive use of instrumentation, which may not be feasible, computational flow dynamics (CFD) analyses may be used to extract flow speed, film coefficients, recovery enthalpy, and any other necessary flow characteristics required by the MR code.

Aerothermal testing is advantageous in that it produces many measurables from a single test, such as surface temperature, back-side temperature, in-depth temperature, mass change, volume change, depth of ablation, ablation rate, and pyrolysis zone depth. Morphological and compositional changes in the charred and pyrolyzing material can be examined after testing.

It is important to select a test bed which closely represents the use conditions, as oxidizing flow can impact the temperature at which a material begins to decompose. For example, carbon fibers begin to decompose much sooner in an oxidizing flowfield than in a non-oxidizing flowfield; the result is that an oxidizing flow will cause a carbon fiber composite to char completely rather than delaminate, which will result in a significantly different internal and back-side temperature profile.

Figure 4. MXB-360 test model undergoing OTB aerothermal testing.

The in-situ ablation sensing data later presented for analysis was taken from a 30 mm diameter, 15 mm thick cylinder of MXB-360 glass/phenolic composite subjected to a 1,000 W/cm^2 heat flux for 60 seconds from the OTB pictured above in Figure 4. As OTB tests are inexpensive to construct and operate, and the test model used a minimal amount of material, as a result the cost of testing was minimized.

2.3 Energy Dispersive X-Ray (EDX) Spectroscopy

EDX utilizes the electron beam from a scanning electron microscope (SEM) to excite the electrons of the atoms of the observed material. As the electrons move back into a lower energy state shell, they emit X-rays, which are then counted by a sensor and sorted by energy. Different elements are then correlated to their signature x-ray energy and counts per electron volt, providing a quantitative elemental composition of the material.

While EDX is not able to measure lighter elements, such as hydrogen and nitrogen, it can be used to establish a base ratio and mass fraction for other elements. Ratios can then be used to determine light elements assuming there is not significant contamination by lighter elements, that the compounds in the material are relatively pure, and the general composition is known. Software can determine the accuracy of the counts, enabling EDX to give a rough elemental composition in a single low-cost step as opposed to using multiple mass spectrometry methods to obtain a complete elemental composition.

The EDX data used for surface thermochemistry simulation in this study was taken from a Hitachi S5500 STEM. Several samples of virgin MXB-360 were examined to establish an elemental composition baseline and identify contaminants present from the production process. Because the silica in MXB-360 separates from the phenolic during ablation, EDX spectroscopy was performed on the glass char and the carbonaceous char separately. Only the carbonaceous char was compared against the virgin material for the purposes of determining the pyrolysis gas composition, as it is assumed that the silica did not contribute significantly to the elemental composition of the gas, as discussed later.

3. ANALYSIS

3.1 In-Situ Temperature Sensing Data

Thermocouples embedded inside the material test model and on the backside are common methods of contact measurement during thermal ablation testing. Non-contact measurement is necessary for front-side temperature and can be achieved via IR pyrometer or IR camera. Based on front side, in-depth, back-side, and final heat-soaked temperature, the thermal state of the material can be well known throughout the duration of the test, providing a large amount of data for property determination and validation [13]. These temperature profiles can then be used to validate the MR model for the given material.

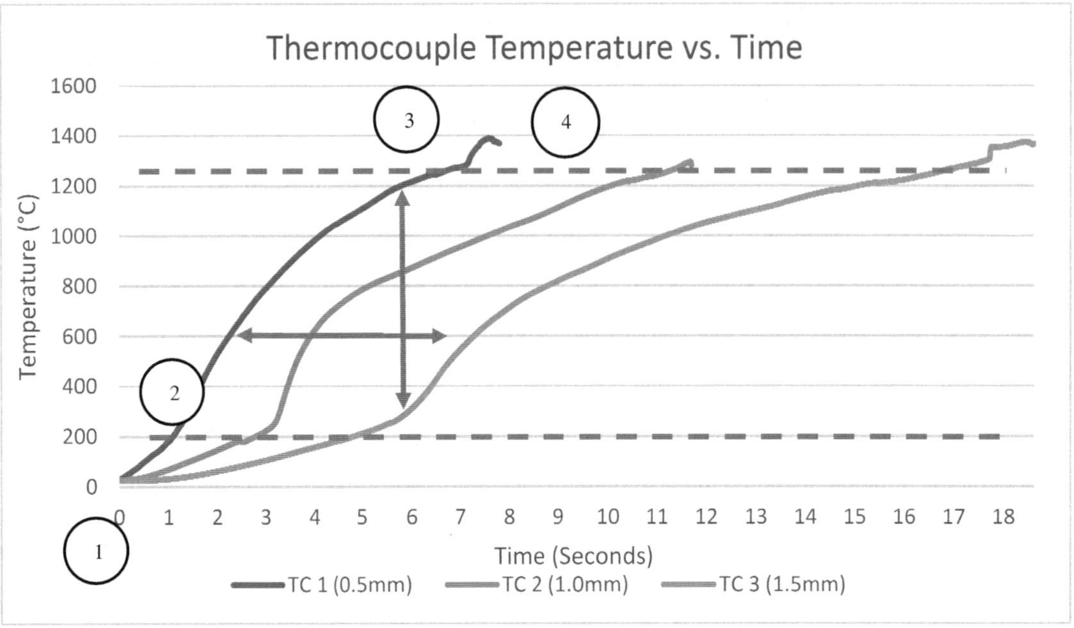

Figure 5. In-situ ablation sensing temperature data from three embedded thermocouples in an MXB-360 test model.

Figure 5 shows an example data set of three K-type thermocouples embedded at 0.5 mm, 1.0 mm, and 1.5 mm from the ablating surface of an MXB-360 glass/phenolic test model in an OTB calibrated to a heat flux of 1,000 W/cm^2. (1) Initial heating of the material occurs along with one minor reaction. (2) Between the dotted red lines, pyrolysis is ongoing for a single dominating reaction chain and a minor secondary reaction, resulting in a decrease in heating rate as the reactions consumes energy. At higher temperatures the reaction rate increases, consuming more energy and damping the heating rate further. Peak temperature difference (vertical red arrow) and ablation rate (horizontal red arrow) are simple to extrapolate from the curves. (3) The material has completely transformed to char and the last pyrolysis reaction no longer damps the heating rate. (4) The thermocouple extends out of the char layer as it erodes, and the thermocouple is destroyed by the front side heat flux.

Inflection points in heating can be used to help gage the number of reactions ongoing in the material during pyrolysis in the environment simulated by the test bed. Care should be taken not to disregard minor reactions, however, which will generally not appear as inflection points but will still influence the overall heating rate.

3.2 Mass Spectrometry for Pyrolysis Gas Enthalpy and Composition

Radiation and convection heating from the mixture of environmental gases and pyrolysis gases escaping the charring ablative material constitutes a significant amount of heat and mass transfer at the surface. In a subsonic flow field, convection will generally dominate as the primary mode of heat transfer, but radiation from gases at a bow shock boundary will dominate at higher Mach numbers. In order to adequately model the heat and mass transfer conditions at the boundary, the composition of the pyrolysis gas resulting from decomposition of the material must be well modeled. Net gaseous outflow (B') must be tabulated as a function of pyrolysis gas (B_g') and char consumption (B_c') for a given material through thermochemical B-prime tables and T-chem tables. These tables may be generated from a surface thermochemistry program, such as Aerothermal Chemical Equilibrium (ACE) program, but only after the elemental composition of the pyrolysis gas and char are roughly understood [10].

Pyrolysis gas evolution at lower temperatures can be captured and analyzed through TGA/gas chromatography instrumentation. However, determining properties of pyrolysis gas evolved at very high heat fluxes is problematic. Pyrolysis gases in extreme hyperthermal environments will have differing compositions due to the amount of heat available and the potentially oxidizing environment; additionally, they immediately undergo further decomposition reactions and mix with the test bed flow gases upon escaping out of the front side of the material and then travel into the hot flow. Direct measurement would be extremely difficult, so a simpler method is to determine the pyrolysis products by simulating the decomposition of the mass fraction undergoing reaction to char.

Simulating pyrolysis gas formation at high temperature requires mass conservation, quantitative mass spectrometry, and a validated surface thermochemistry simulation code. The elemental composition differences between the virgin and charred state of the material are taken and the compared against mass fraction data from TGA. The reacting materials are fed into a reaction analysis code which can output the mixed pyrolysis gas enthalpies along with the chemical species that compose the pyrolysis gas. These enthalpy tables and gas composition tables are then utilized along with the elemental composition to determine thermodynamic properties and the equilibrium conditions at the surface at each step of the MR model [14].

If a high level of compositional fidelity is required, combinations of mass spectrometry methods can be combined in order to develop a detailed molecular and elemental composition. For example, combining carbon, hydrogen, nitrogen (CHN) analysis with a CHN analyzer; inductively coupled plasma-optical emission spectroscopy (ICP/OES) to determine trace metals; and ion chromatography (IC) to determine chlorine and sulfur content would yield a complete quantitative atomic composition. If basic assumptions about hydrogen content and lighter elements are permissible, single step methods, such as energy-dispersive X-ray (EDX) spectroscopy shown in Figure 6 may be utilized to save time.

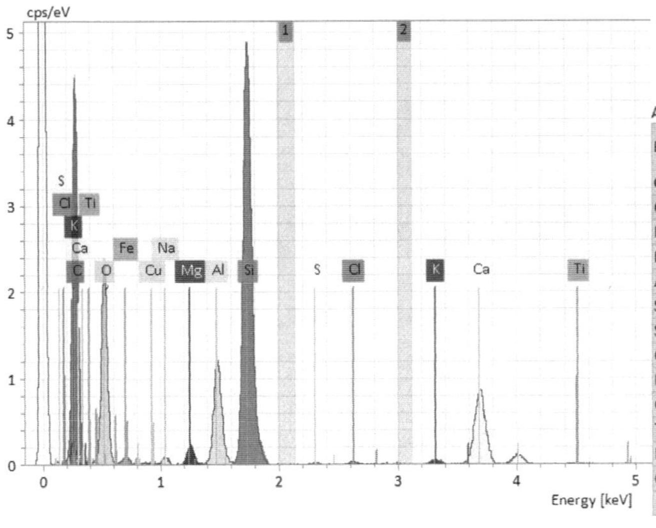

Acquisition

Element	At. No.	Mass [%]	Mass Norm. [%]	Atom [%]	abs. error [%] (3 sigma)	rel. error [%] (3 sigma)
C	6	64.48	48.87	61.01	22.43	34.78
O	8	41.24	31.26	29.29	14.86	36.03
Na	11	0.57	0.43	0.28	0.19	34.20
Mg	12	0.74	0.56	0.35	0.20	27.28
Al	13	3.47	2.63	1.46	0.57	16.28
Si	14	14.14	10.72	5.72	1.84	13.03
S	16	0.07	0.05	0.03	0.09	121.25
Cl	17	0.11	0.09	0.04	0.09	81.65
K	19	0.20	0.15	0.06	0.10	50.63
Ca	20	4.53	3.43	1.28	0.49	10.78
Ti	22	0.07	0.05	0.02	0.09	131.30
Fe	26	2.17	1.64	0.44	0.29	13.26
Cu	29	0.14	0.10	0.02	0.11	80.24
		131.92	100.00	100.00		

Figure 6. Two example read-outs from Energy-dispersive X-ray (EDX) spectroscopy of a glass/phenolic compound on a Hitachi S5500 STEM. Left: A graph of the X-ray peaks sorted by energy and tagged at the top with the most likely source element. Right: A chart showing the calculated elemental composition generated automatically by EDX software.

Once elemental composition differences between virgin and char materials are obtained, they may be combined with density properties and the mass fraction data from burn or TGA tests, then input into a chemical equilibrium code. The resultant B' and T-chem tables may then be formatted for use by the desired MR model and serves as the primary form of surface thermochemistry and mass transfer reference for a material.

3.3 Kinetic Modeling Parameter (KMP) Development

Mass versus temperature data from constant heating rate TGA testing can be normalized to provide both percent mass loss and extent of reaction versus time. By rearranging the Arrhenius decay equation and the reaction model function, the KMPs for each decay reaction can be backed out and utilized to predict decay behavior in differing heating conditions, though the KMPs and decay reactions will change for differing environmental oxidative conditions. As the environment is assumed to have a fixed composition for these calculations, the decay reaction is not a function of environmental conditions, and thus KMPs would need to be derived for each expected environment that the material would be exposed to.

$$\frac{d\alpha}{dT} = Ae^{-\frac{E_A}{RT}}f(\alpha) \qquad [1]$$

$$f(\alpha) = (1-\alpha)^m \qquad [4]$$

For the remaining analysis, the single reaction order parameter form is used as given by Kissinger in equation (4). Utilizing the Kissinger equation instead of the Friedman equation will result in singular activation energy (E_A) and reaction order (m) values while the pre-exponential (A) will change dependent on the heating rate. While the Kissinger decomposition equation may not provide adequate simulation data above α values of 0.7, it is significantly simpler to numerically optimize the KMPs and is less computationally intensive than the Friedman method [11].

As the pre-exponential changes with heating rate for the Kissinger method, we can change the form of (1) in order to get another form of the Arrhenius equation:

$$\frac{d\alpha}{dT} = \frac{A}{\beta}e^{-\frac{E_A}{RT}}f(\alpha) \qquad [5]$$

With β representing the heating rate dT/dt. We then further rearrange equation (5) such that:

$$\ln\left[\beta\frac{\frac{d\alpha}{dt}}{f(\alpha)}\right] = -\frac{E_A}{RT} + \ln A \qquad [6]$$

E_A and A can be extracted from a linear curve fit of the left side of equation (6) against 1/T based on TGA data. For this example, TGA data is taken from MXB-360 glass/phenolic compound with heating rates of 50, 20, and 10 °C/min using air as a purge gas, providing an oxidative environment. Commercially available TGA instruments typically are not capable of exceeding heating rates of 100 °C/min due to difficulty in construction and low demand for very high heating rates in the biomedical and industrial polymer fields. The first heating rate of 50 °C/min was chosen to be as high as the given TGA instrument can safely handle, while subsequent heating rates are approximately half as much as the previous in order to ensure significant separation on a logarithmic scale. While these heating rates are not representative of the approximately 200 °C/sec peak heating rate experienced during OTB testing, the use of three or more TGA data sets from differing heating rates can be used to predict realistic KMPs for all heating rates as demonstrated by Sihn et al. [11] The temperature versus normalized mass loss graph in an oxidizing environment is shown below in Figure 7.

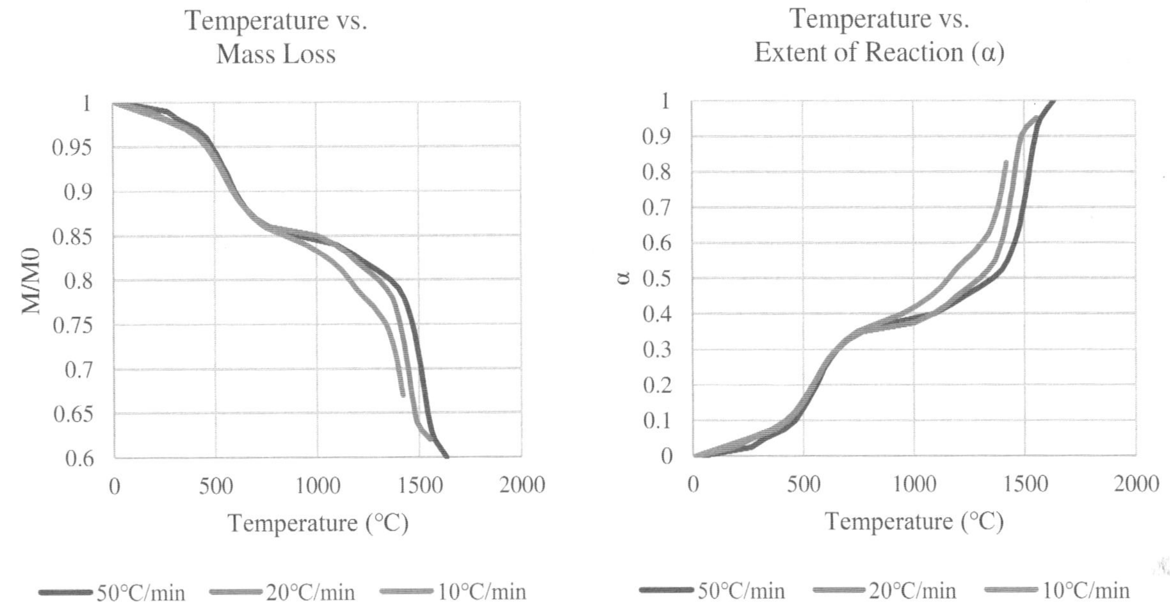

Figure 7. Left: TGA temperature versus normalized mass loss. Right: TGA temperature versus extent of reaction. Note that only two curves are immediately apparent in the data, one below 1,000 °C and one above.

By normalizing the data based on the initial and final mass fractions using equation (7), we can also develop the temperature versus overall extent of reaction (α) graph from Figure 7.

$$\alpha = \frac{M_0 - M}{M_0 - M_f} \quad [7]$$

Where M_0 is the initial mass fraction, M_f is the final mass fraction, and M is the mass fraction as a function of temperature. Equation 7 is also used to describe the extent of reaction for individual decomposition reactions ongoing in the material based on the mass fractions at which they initiate and terminate.

Based on these two graphs there are two distinct reaction regimes within the material, one below 1,000 °C and one above 1,000 °C. By determining the derivative of α with respect to temperature, the individual reactions can be observed directly. Three reactions take place, the first of which peaks at about 550 °C, the second at 1,200 °C, and the third at 1,450 °C. Figure 8 shows the combined and separated dα/dT data sets.

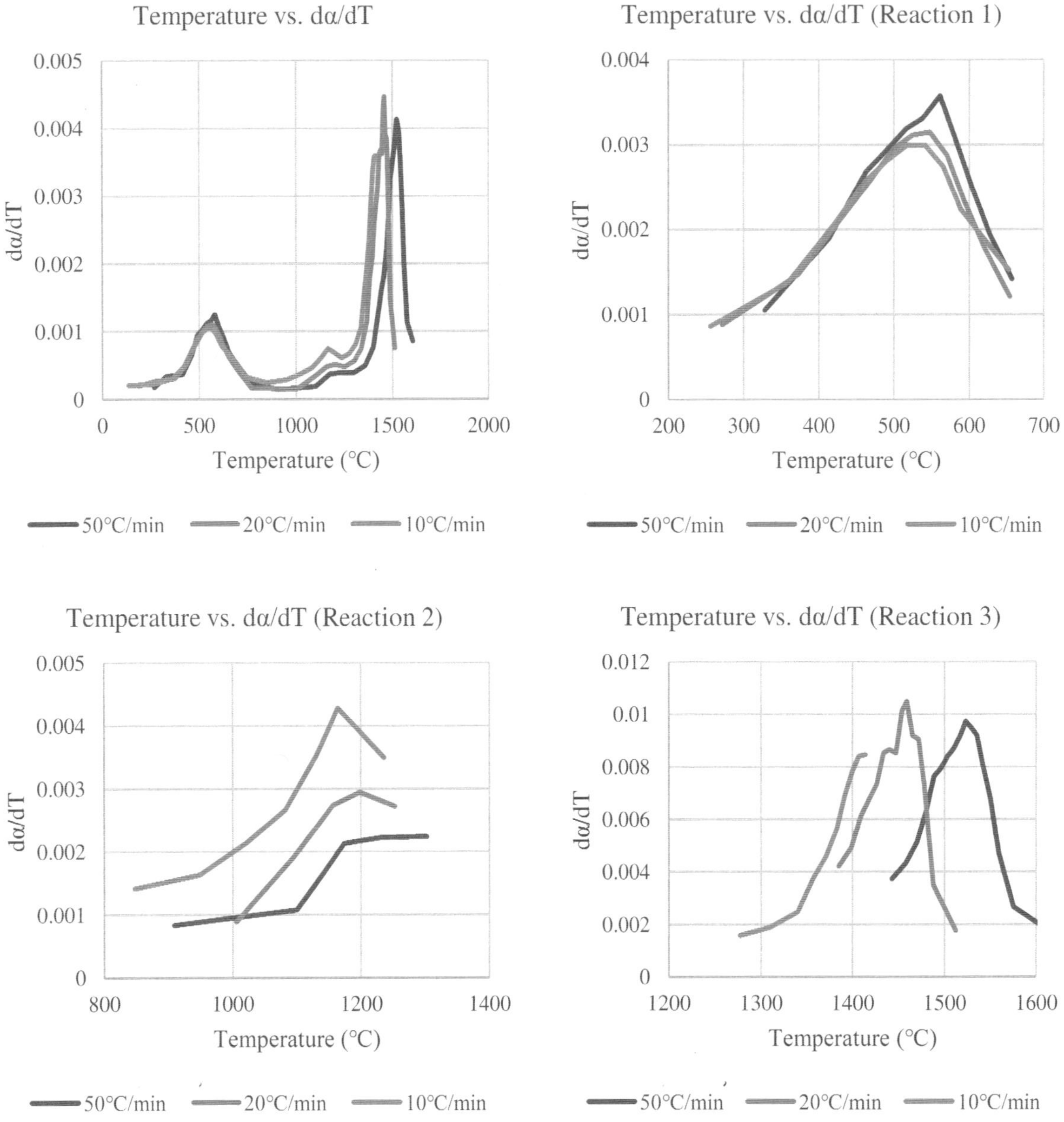

Figure 8. The combined temperature versus dα/dT curve for the MXB-360 TGA data and the individual temperature versus dα/dT curves for reactions one, two, and three.

With the β values known, only the value of f(α) is needed in order to determine values for the left side of (6). Using numerical methods, f(α) can be chosen to linearize the curves as much as possible as shown in Figure 9. Note that because of the decision to utilize the Kissinger equation (4) vice the Friedman equation (3) for f(α), the resultant curve will not reduce to a completely linear solution because the reactions overlap and the results fit better to a spline curve with more degrees of freedom. This will produce more error in activation energy E_A and pre-exponential factor A while trying to determine the reaction order parameter *m*.

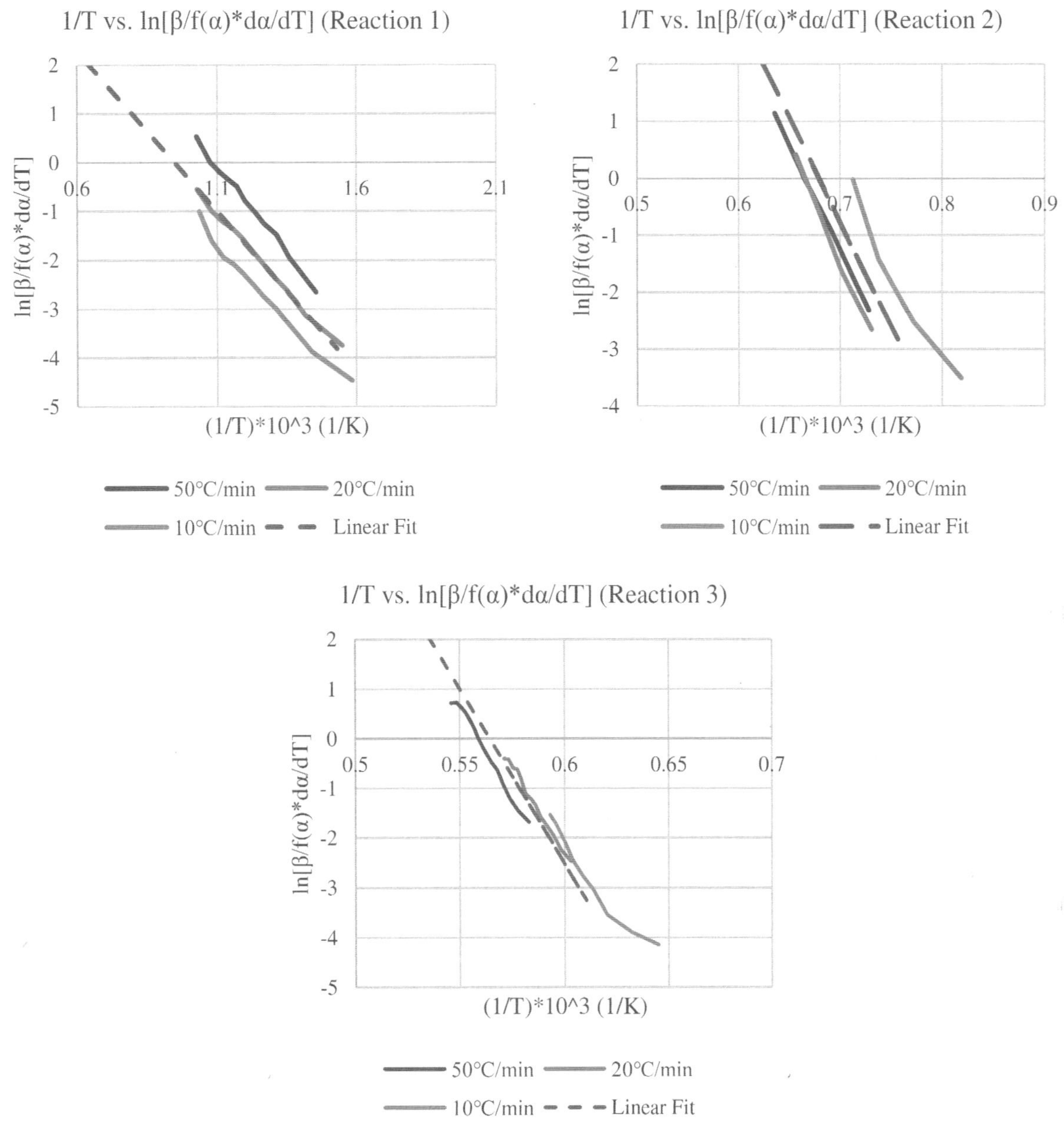

Figure 9. 1/T versus ln[β/f(α)*dα/dT] for reactions one, two, and three. A linear curve fit is performed on the average of the three curves for each reaction, generating the red dotted line, whose intercept is logarithmically proportional to the pre-exponential term and whose slope is proportional to the activation energy.

A least squares linear curve is separately fit to the data for each of the reactions. These fits are averaged together to generate the activation energy and pre-exponential term. In order to linearize the data as much as possible, the reaction order term m from equation (4) is varied in order to minimize the sum of the squares of Y-axis error from points on the fitted curves to points derived from data. This rudimentary method is fast and simple to implement using default tools provided with Microsoft Excel, though any number of numerical optimization methods may be employed to determine a reaction order term that best linearizes the data, and thus, is closest as possible to the best value for modeling the material. The individual reactions may then be assigned mass fractions according to the data shown in Figure 8 and then validated by using an MR model and comparing the output against ablation test temperature data.

Table 1. KMPs for the three decomposition reaction chains based on the TGA data in an oxidative environment and the Kissinger method

Reaction	1	2	3
m (unitless)	1.205	3.927	1.244
E_A (kJ/mol)	54.6	301.9	587.2
A (1/sec)	$5.04x10^2$	$5.02x10^{10}$	$2.02x10^{17}$

With the reaction order term for $f(\alpha)$ fixed, the slope of the curve fit (1/RT) and Y-intercept [ln(A)] give the E_A and A terms as shown in Table 1. While β is known and fixed for the OTB testing heating rate conditions and the Kissinger decay function, Sihn et al. [11] describe how additional TGA data sets at different heating rates can be used to project variations in these parameters for any given heating rate β and any reaction model function.

4. RESULTS

4.1 Material Response Modeling

With the analytically difficult material properties determined (B' tables, pyrolysis gas enthalpy, and KMPs), data can be fed into the MR code of choice, ITRAC [14] in this case, and compared to the original in-situ temperature data acquired during OTB testing. Figure 10 below shows the one-dimensional simulated results from ITRAC of the MXB-360 glass/phenolic material compared against the OTB test data.

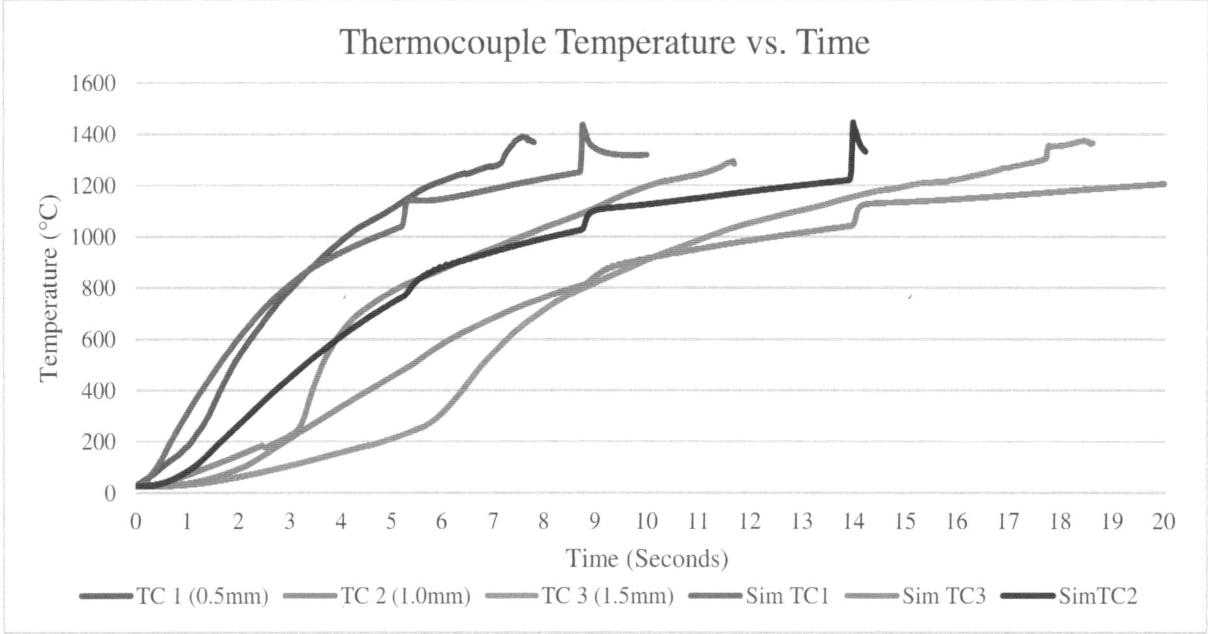

Figure 10. Temperature versus time graph for MXB-360 comparing OTB test data to simulated data in red from ITRAC. Note the temperature jumps on the simulated thermocouples: this is due to how ITRAC moves the ablation boundaries in steps.

While the simulation trends appropriately with respect to ablation behavior, the poor separation of pyrolysis reactions two and three during KMP determination results in differing heating rates, and slight insulation from the initial char layer results in an overstated initial heating rate. Gaussian separation programs may be utilized to predict curves that correspond to reactions two and three to separate them further and provide better estimations of the mass fraction they compose. Code limitations prevent matching the rapid heating of the TC junctions as they extend out of the char layer. In this instance, the trends appear generally similar and both sets of curves share two inflection points. Fine tuning of KMPs and material properties would help provide a closer fit, but the given values may be appropriate for some users.

Given reliable TGA data, poorly matched simulation data are often the result of operator error due to the number of steps involved in KMP development. There is additional opportunity for errors to occur when inputting results into a MR model, such as leaving placeholder tables in the input code or manual data entry mistakes. Temperature curves are good tools for diagnosing these issues; a mismatch in the number of inflection points in the curve points directly to issues with KMP values or the incorrect number of reactions occurring in the simulation, while large temperature spikes bring attention to heat capacitance, density, and other material property tables, especially if they occur at a material interface within the test model. Care should be taken to diagnose problems with all inputs prior to dismissing a model for design or testing purposes.

4.2 Determining Validation Criteria

Finding an endpoint for acceptably accurate KMPs and MR code output is important to marking the end of analysis. A perfect simulation of test or historical results is good to strive for but may not be feasible for a given working timeline, especially considering that design teams may require the MR code output to continue with their work or may only require a limited level of fidelity for a decision point.

Mass fraction remaining or mass loss tends to track experimental results more reliably than recession depth in general, as recession depth can be accelerated greatly due to factors, such as material fracturing, flaws in the production process, large pores in the heating zone, aerothermal channeling effects, and so on. Recession depth may not be useable for materials with significant swelling properties, such as engineering thermoplastics. Advection processes are not simple to control during test conditions, and average material thickness is simple to correlate to mass remaining, so mass fraction accuracy is an acceptable compromise unless working with a highly minimized safety factor.

In-depth temperature profile matching, as shown previously in Figure 10, can work well but is highly sensitive to sensor placement and to environmental effects. For example, a backside temperature sensor may show instantaneous cooling as the test model or torch are moved away from each other due to air flow, which may not be immediately obvious and may cause the analyst to dismiss the code output as improper when it is not. The simulated curves in Figure 10 demonstrate that the simulated ablation boundary is approximately 0.25 mm shallower than the OTB ablation boundary within the first 20 seconds, which may be within acceptable error for the purposes of considering the model to be validated for design or selection purposes.

5. SUMMARY

A rapid methodology for characterizing the ablation properties of materials was outlined in order to aid in reducing the cost and scheduling burden normally associated with flight testing materials or using hypersonic wind tunnels for TPS down selection. Data analysis was detailed for the use of TGA to determine reaction KMPs, EDX for elemental composition, and in-situ ablation sensing on an aerothermal test bed for KMP validation. As KMPs are dependent on the user's choice of Kissinger, Friedman, or Ozawa reaction order functions, KMP validation is based on temperature curve trends compared against real world test data in order to determine their predictive capability. Both environmental composition and oxidative level are assumed to be fixed over time, so any changes would introduce significant error to MR modeling, requiring KMPs to be recalculated. Validation of a given MR model must consider the ability to match mass loss, surface temperature, backside temperature, and recession data when subjected to aerothermal testing using a variety of heat fluxes. Additional validations of the MR model are in progress.

6. REFERENCES

1. Starkey, R.P. and M.J. Lewis, *Critical Design Issues for Airbreathing Hypersonic Waverider Missiles.* Journal of Spacecraft and Rockets, 2001. **38**(4): p. 510-519.
2. Szczepanowska, H. and T.G. Mathia, *Space Heritage: The Apollo Heat Shield; Atmospheric Reentry Imprint on Materials' Surface.* MRS Proceedings, 2011. **1319**.
3. Suzuki, T., et al., *Postflight Thermal Protection System Analysis of Hayabusa Reentry Capsule.* Journal of Spacecraft and Rockets, 2014. **51**(1): p. 96-105.
4. Knight, D., et al., *Assessment of predictive capabilities for aerodynamic heating in hypersonic flow.* Progress in Aerospace Sciences, 2017. **90**: p. 39-53.
5. Paglia, L., et al., *Carbon-phenolic ablative materials for re-entry space vehicles: plasma wind tunnel test and finite element modeling.* Materials & Design, 2016. **90**: p. 1170-1180.
6. MacDonald, M.E., C.M. Jacobs, and C.O. Laux, *Interaction of Air Plasma With Ablating Heat Shield Material.* IEEE Transactions on Plasma Science, 2014. **42**(10): p. 2658-2659.
7. Natali, M., et al., *An in-situ ablation recession sensor for carbon/carbon ablatives based on commercial ultra-miniature thermocouples.* Sensors and Actuators B: Chemical, 2014. **196**: p. 46-56.
8. Koo, J.H., et al., *In Situ Ablation Recession and Thermal Sensor for Thermal Protection Systems.* Journal of Spacecraft and Rockets, 2018. **55**(4): p. 783-796.
9. Ewing, M.E. and B. Pincock, *Heat Transfer Modeling of a Charring Material Using Isoconversional Kinetics.* Heat Transfer Engineering, 2016. **38**(13): p. 1189-1197.
10. Ewing, M.E., T.S. Laker, and D.T. Walker, *Numerical Modeling of Ablation Heat Transfer.* Journal of Thermophysics and Heat Transfer, 2013. **27**(4): p. 615-632.
11. Sihn, S., et al., *Identifying unified kinetic model parameters for thermal decomposition of polymer matrix composites.* Journal of Composite Materials, 2018. **53**(20): p. 2875-2890.
12. Ozawa, T., *A New Method of Analysing Thermogravimetric Data.* Bulletin of the Chemical Society of Japan, 1965. **38**.
13. Yee, C., et al., *In Situ Ablation Recession Sensor for Ablative Materials Based on Ultraminiature Thermocouples.* Journal of Spacecraft and Rockets, 2014. **51**(6): p. 1789-1796.
14. Ewing, M.E. and D.A. Isaac, *Thermodynamic Property Calculations for Equilibrium Mixtures.* Journal of Thermophysics and Heat Transfer, 2018. **32**(1): p. 118-128.

Damage Tolerant Composites Reinforced with Z-axis Milled Carbon Fiber

Anvesh Gurijala, Dr. Rasam Soheilian, Derek Mossop

Boston Materials, Inc.
Boston, MA

ABSTRACT

Despite having high specific strength and stiffness, carbon fiber composites are prone to brittle and catastrophic failure. This behavior poses challenges when re-designing metal components with light and stiff composites, especially components with thin walls that are common in electronic devices. In this work, thin (~0.8 mm thickness) panels made with 30% Z-axis milled carbon fiber and 70% unidirectional (0°) carbon fiber (tradename: Supercomp™ 1015 Axial) are shown to match the flexural modulus and strength (per ASTM D-790) of 100% unidirectional (0°) carbon fiber panels with similar thickness and weight. Unlike the pure unidirectional fiber laminates, which fail catastrophically, the panels reinforced with Z-axis milled fiber can be loaded again after initial failure and they retain 42% of their original flexural strength. This work demonstrates a new approach to produce damage tolerant composites using Z-axis milled fiber reinforcement and is relevant for light and stiff enclosure designs for electronic devices.

Keywords: Supercomp, Z-axis, Milled Fiber, Damage Tolerant, Electronics
Corresponding author: Anvesh Gurijala, anvesh@bomaterials.com

1. INTRODUCTION

The form and functionality of electronic devices, such as tablets, laptops, and smartphones, are rapidly evolving to meet consumer demands for higher processing capabilities, longer battery life, and lower cost. Many of these demands rely on innovations in processing, connectivity, display, and battery technologies. However, a shared requirement for all electronic devices, especially portable consumer electronics, are stiff, thin, and durable enclosures to package and protect the internal electronics. Today, aluminum is the incumbent material choice for electronic device enclosures, especially in high-end portable electronics.[1]

Thin and light enclosures can increase the volume and weight allowances of the internal electronic components. However, the continued use of aluminum in enclosures with thin walls can lead to design failures, such as the susceptibility of the enclosure to bending during manufacturing or end use by the consumer (**Figure 1**).[2] Materials that have a higher specific stiffness than aluminum are required to maintain the structural stability of thinner and lighter enclosures. The requirement for stiff and light materials is an opportunity for carbon fiber composites to displace metals like aluminum. Unlike most metals which undergo a plastic deformation prior to fracture, conventional carbon fiber composites in thin structures are brittle and catastrophically fail.

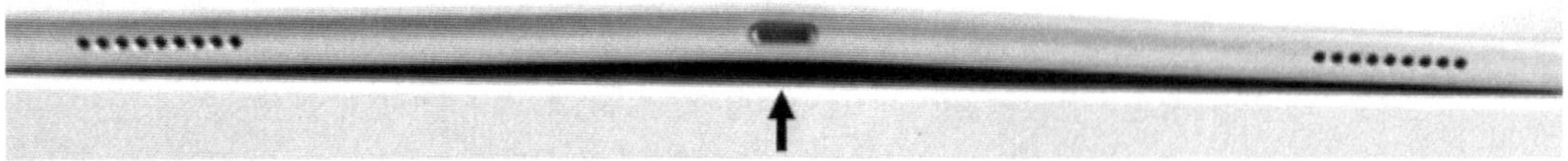

Figure 1. A new high-end tablet with an aluminum enclosure is delivered bent to the consumer[2]

In order to eliminate catastrophic failure in new composite enclosures, manufacturers have utilized a combination of fiberglass and carbon fiber.[3] Fiberglass composites have 40% higher density and 40% lower flexural stiffness than carbon fiber composites.[4] Thus, incorporating fiberglass decreases the risk of catastrophic failure but also dilutes the lightness and stiffness benefits of carbon fiber. Similar techniques are used in many recreational applications where highly oriented polypropylene fibers (tradename: Innegra™) are incorporated with carbon fiber to create a hybrid composite that does not catastrophically fail but it also reduces the flexural stiffness compared to a 100% carbon fiber composite. Boston Materials' Supercomp maintains the high specific flexural stiffness properties of a pure carbon fiber composite and eliminates the risk of catastrophic failure.[4] Therefore, Supercomp is an ideal material to use in thin, light, and damage tolerant composite enclosures for electronic devices.

2. EXPERIMENTATION

2.1 Materials

Braided unidirectional (UD) dry carbon fiber tapes were sourced from A&P Technology (ZERO® Uni-4.0SM). The dry UD tapes were made using Toray standard modulus, PAN-based carbon fiber (T-700). Both the baseline and Supercomp samples utilized the same dry UD tape. Supercomp 1015 Axial Dry was manufactured by Boston Materials by applying Z-axis oriented milled carbon fibers to the dry UD tape. The milled carbon fibers were sourced from Zoltek (PX30 MF150). Using a proprietary technique, the milled fibers are magnetically functionalized and then dispersed in a water-based slurry. The milled carbon fiber slurry is then applied to the dry UD tape and then the milled fibers are oriented in the Z-axis using vertical magnetic fields. After the fiber orientation step, the resultant material is dried, and the Supercomp 1015 Axial Dry material is spooled. This entire process is implemented in a 12"-wide roll-to-roll production process (**Figure 1**).

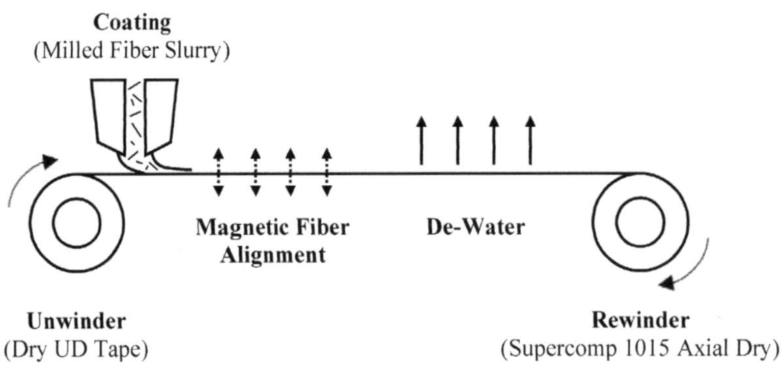

Figure 1. Schematic of the process used to produce Supercomp 1015 Axial Dry

Both the Supercomp 1015 Axial Dry and the baseline dry UD tape were processed into prepregs by infusing films of 250 °F cure epoxy sourced from Mitsubishi Chemical Carbon Fiber & Composites (Newport 301). The epoxy films are infused into the carbon fiber materials by using a series of heated calender rollers, forming Supercomp 1015 Axial prepreg (**Figure 2**) and the baseline UD prepreg. The film infusion was performed by Aldila Composite Materials.

Figure 2. Scanning electron microscope (SEM) cross-section of Supercomp 1015 Axial prepreg

2.2 Coupon Fabrication

The test panels were made by curing laminated prepregs between PTFE release films at 250 °F under 50 psi compaction for one hour using a press (Carver 25 Ton Laboratory Press Model M). The layup schedules and physical properties of the two different types of test panels are shown in **Table 1**. The cured panels were cooled to ambient temperature before removing from the press.

Table 1: Layup schedules and physical properties of the Supercomp and baseline test panels

Panel Type	Layer #	Orientation	Target Thickness (mm)	Target Areal Weight (g/m²)
Supercomp	4	0°	0.14	145
	3	0° w/ Z-axis fiber ↓	0.28	280
	2	0° w/ Z-axis fiber ↑	0.28	280
	1	0°	0.14	145
TOTAL			**0.84**	**850**
Baseline	6	0°	0.14	145
	5	0°	0.14	145
	4	0°	0.14	145
	3	0°	0.14	145
	2	0°	0.14	145
	1	0°	0.14	145
TOTAL			**0.84**	**870**

The resultant test panels were cut into coupons using an abrasive waterjet (OMAX Protomax). The dimensions of the coupons are shown in **Table 2**. The coupons maintain a span-to-thickness ratio of 32:1, in accordance with the ASTM D790 test standard.[5] The 0° fibers were oriented parallel to the span and length of the coupons.

Table 2: Coupon dimensions (width is out-of-plane in the schematic)

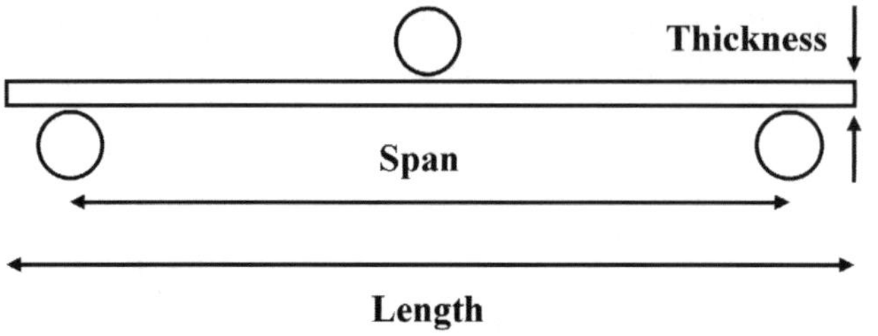

Coupon #	Width (mm)	Measured Thickness (mm)	Calculated Span (mm)	Length (mm)
Supercomp-1	11.8	0.86	28	33
Supercomp-2	11.8	0.84	27	33
Supercomp-3	11.8	0.81	26	33
Baseline-1	11.8	0.86	28	33
Baseline-2	11.8	0.86	28	33
Baseline-3	11.8	0.87	28	33

2.3 Coupon Testing

The prepared coupons were tested under three-point flexural loading (ASTM D790) to determine 0° flexural strength and modulus using a universal testing system (Instron 3369 with 5 kN load cell). The baseline coupons were subject to an initial loading, after which they failed catastrophically and could not be loaded a second time. Unlike the baseline coupons, the Supercomp coupons had residual flexural strength for a second loading. Strain measurements were taken from crosshead displacement to calculate flexural modulus.

3. RESULTS

The prepared coupons were tested under three-point flexural loading (ASTM D790) to determine 0° flexural strength and modulus. The baseline coupons were subject to an initial loading, after which they failed catastrophically and could not be loaded a second time. Unlike the baseline coupons, the Supercomp coupons had residual flexural strength for a second loading.

Three-point bending was selected for this work to demonstrate the end-use behavior of Supercomp in applications such as electronics enclosures, skis, and hockey sticks. In these applications, manufacturers utilize a three-point or cantilevered deflection test to gauge the stiffness, strength, and failure behavior of the composite part. Accordingly, the three-point bending performed in this work closely matches the testing that component manufacturers use to test their parts.

3.1 Initial Properties

The initial loading on the baseline and Supercomp coupons showed similar results (within 5%) for both flexural modulus and strength (**Figure 3**). The baseline coupons (n=3) registered 82.03 ± 3.60 GPa and 1680.13 ± 165.96 MPa for flexural modulus and strength, respectively. The Supercomp coupons (n=3) registered 81.54 ± 5.47 GPa and 1606.51 ± 48.98 MPa for flexural modulus and strength, respectively. The initial loading results demonstrate that the innermost plies of a coupon under three-point flexural loading can be replaced with Z-axis milled fiber without a reduction in flexural strength or modulus. This data was not normalized to a target fiber volume.

There was a discrepancy between the flexural strength and modulus reported by Toray for the T-700 carbon fiber cured with epoxy and the coupons that were tested in this work. Toray reports 120 GPa and 1670 MPa for flexural modulus and strength, respectively. However, the data reported by Toray is normalized to 60% fiber volume.[6]

The expected fiber volume for the Supercomp and baseline coupons tested in this work is 55%. Normalizing the flexural strength of the Supercomp and baseline coupon to 60% fiber volume yields similar values to the flexural strength reported by Toray. However, the flexural modulus values are nearly 18% lower than the value reported by Toray, even after normalizing to 60% fiber volume.

The knockdown in the flexural modulus is attributed to the braided unidirectional carbon fiber that is used to produce the Supercomp and baseline coupons. The braiding process does not adequately tension the carbon fibers and allows the fibers to relax. This relaxation is visually apparent as a subtle in-plane distortion of the carbon fiber off the 0° direction. Alternatively, the production process used by Toray, and other prepreggers, applies a high tension on the carbon fibers during the tow spreading process. The spread carbon fiber tows are then immediately captivated by hot-melt epoxy during the prepregging process, which preserves the applied tension on the carbon fibers and holds them uniformly in the 0° direction. Implementing highly oriented carbon fibers in the 0° direction can increase the flexural modulus of the baseline and Supercomp coupons.

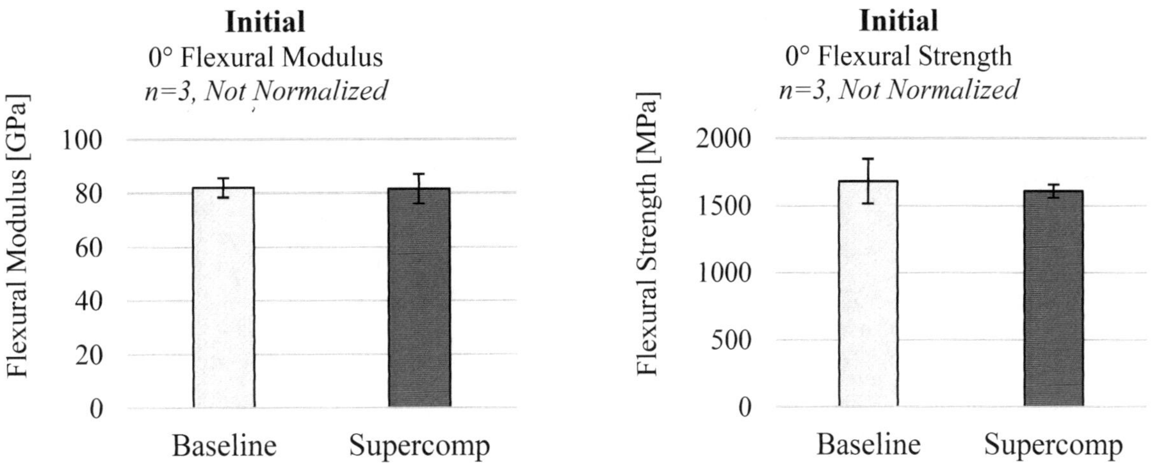

Figure 3. Flexural modulus and strength results after the initial three-point flexural loading

3.2 Residual Properties

After the initial loading, the baseline coupons catastrophically fractured into two pieces. The Supercomp coupons remained in one piece but had a full fracture of the UD 0° fibers at the top surface of the coupon, which in compression during the three-point flexural loading (**Figure 4**). This local fracture is from buckling of the skin plies during compression, which is like the failure mechanism of a composite sandwich panel under the three-point flexural loading (**Figure 5**).[7]

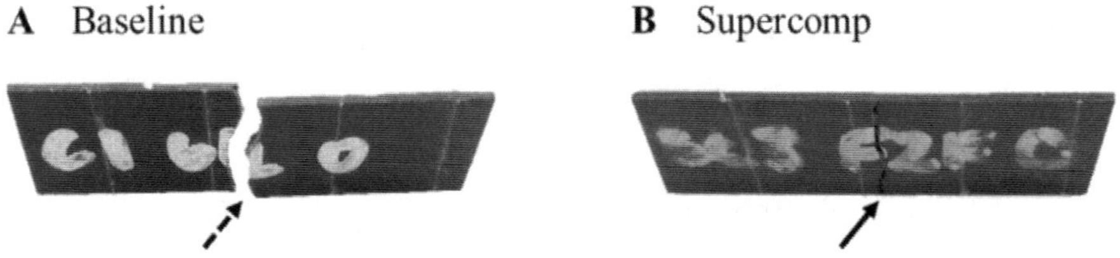

Figure 4. A) The baseline coupon catastrophically failed after the initial loading (dotted arrow) and B) the Supercomp coupon remained intact after the initial loading but had fractured UD 0° fibers at the top surface, which are under compression in three-point flexural loading (solid arrow).

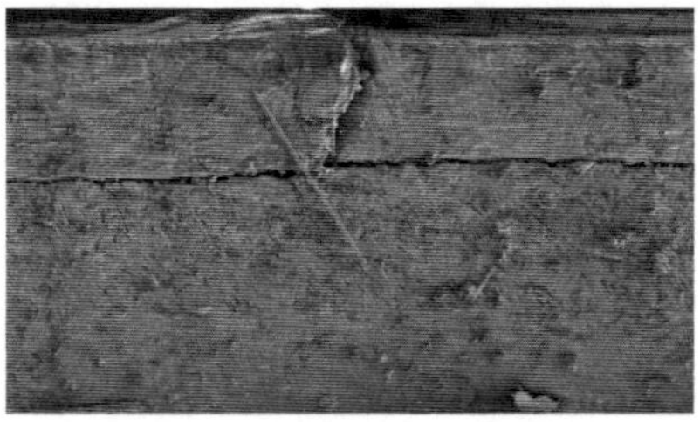

Figure 5. Post-test fracture morphology in the Supercomp coupons after the initial loading. The fracture is caused by buckling of the topmost plies, which are under compression during three-point flexural loading. The baseline coupons fully fractured after the initial loading.

The Supercomp coupons (n=3) demonstrated a residual flexural strength of 677.42 ± 51.35 MPa. The residual flexural strength is 42% of the initial value (**Figure 5**). Residual strength for the baseline coupons could not be collected because the they fully fractured after the initial loading.

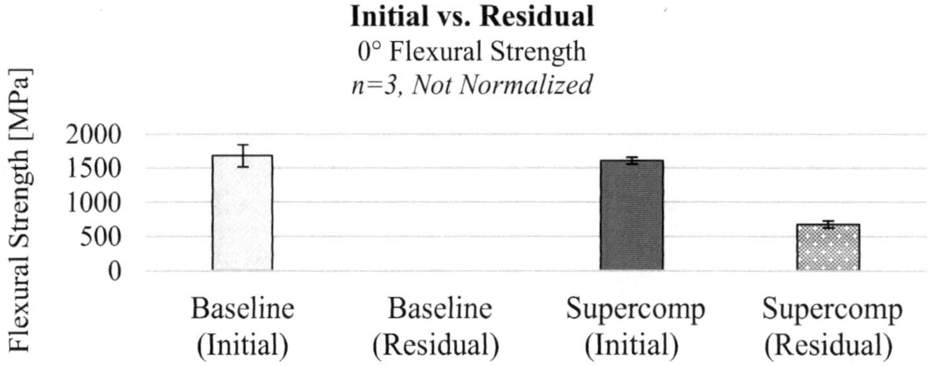

Figure 6. Flexural properties before/after the second loading (Note: residual properties could not be collected for the baseline coupons because they catastrophically failed after the initial loading)

4. CONCLUSIONS

In this work, it was demonstrated that the innermost plies of a thin (~0.8 mm thick) laminated composite panel with 30% (by weight) Z-axis milled fiber and 70% UD 0° fiber (Supercomp) has comparable flexural strength and modulus as a 100% UD 0° fiber (baseline) panel with similar physical dimensions and weight. This result alone is important for cost optimization since the milled carbon fiber is over 50% lower cost than the continuous fiber used in UD reinforcements.

After the initial three-point flexural loading, coupons cut from the baseline panel failed catastrophically, which is expected. Under similar loadings, the coupons cut from the Supercomp panels remained primarily intact after the initial loading. While there was fractured UD 0° fiber on the topmost skin, which is under compression in three-point flexural loading, the Supercomp coupons could be loaded for a second time and they retained 42% of the initial flexural strength.

In follow-up work, the Z-axis milled fiber reinforcement will be directly applied to off-the-shelf carbon fiber prepreg that has highly oriented fiber in the 0° direction. This process will be implemented by coating a release substrate with the Z-axis milled carbon fiber. The coated substrate will be infused with epoxy, forming a Z-axis milled carbon fiber prepreg. This material will then be transferred to an off-the-shelf carbon fiber prepreg. This process will avoid the flexural stiffness knockdown caused by the braided unidirectional tape.

Additionally, the initial and residual properties of Supercomp panels will be benchmarked against carbon fiber and fiberglass hybrid composite panels that are currently used in consumer electronics applications. The thermal and electrical properties will also be characterized to develop thermal management and electromagnetic shielding functionalities with Supercomp. Lastly, the milled fiber used as the Z-axis reinforcement will be sourced from post-consumer recycled carbon fiber to demonstrate complete circularity for the Supercomp material. Complete circularity is especially important for consumer electronic devices with relatively short lifecycles (which can be as low as 3 years for smartphones) in order to re-capture potential lost value from scrap and reduce reliance on expensive virgin feedstocks.[8]

5. REFERENCES

[1] *Contract consumer electronics, injection molding, enclosure opportunities.* Venture Outsource, 2013.

[2] *Apple says iPad Pros have no design flaw.* TechNewsToday, 2019.

[3] P. Malnati, *Consumer electronics: hybrid composite covers.* CompositesWorld, 2019.

[4] *301 Technical Data Sheet.* Mitsubishi Carbon Fiber and Composites, 2019.

[5] *Standard Test Methods for Flexural Properties of Unreinforced and Reinforced Plastics and Electrical Insulating Materials.* ASTM International, 2019.

[6] *Torayca T700S Data Sheet.* Toray, 2019.

[7] *Materials* 2019, 12(18), 2959; https://doi.org/10.3390/ma12182959

[8] A. Holst, *Replacement cycle length of smartphones in the United States 2014-2023, in years.* Statista, 2019.

TENSILE STRENGTH AND FAILURE MODES OF CONTINUOUS CARBON NANOTUBE YARNS FOR STRUCTURAL COMPOSITE APPLICATION

Yourri-Samuel Dessureault, Claire Jolowsky, Samantha Bell, Stefan Spiric, Josiah Molyneux, Dr. Jin Gyu Park, Dr. Ayou Hao, Dr. Zhiyong Liang
High-Performance Materials Institute, Department of Industrial & Manufacturing Engineering, Florida A&M University – Florida State University College of Engineering
Tallahassee, FL 32310, USA

ABSTRACT

Carbon nanotubes (CNT) have shown very attractive mechanical properties. Recently, CNT yarns (CNTy) have become commercially available and are becoming a promising choice for composite structures. To gain a fundamental understanding of CNTy mechanical performance, three variations of CNTy were tested at long gauge lengths (100 – 120 mm). The as-received CNTy demonstrated the highest specific tensile strength (0.911 ± 0.137 N/tex), while the CNTy modified to maximize CNT packing showed the lowest specific tensile strength (0.337 ± 0.217 N/tex). The results were the fit to 2-parameter Weibull distributions with near 0.99 goodness of fit. This revealed that long-range CNT alignment acted as the key contributor to the tensile strength, while defects influenced the failure mode. The distribution analysis at a 95 % confidence interval of the CNTy tensile properties were conducted for potential engineering applications.

Corresponding author: Yourri-Samuel Dessureault

1. INTRODUCTION

High-performance composites have been a boon to the development of advanced aerospace systems, which currently rely on high-strength fibers, such as carbon fibers. However, improvements to these fiber systems are becoming ever increasingly difficult to achieve. Meanwhile, carbon nanotubes (CNTs) and CNT fibers provide a potential route for the evolution of fiber reinforcements due to their mechanical and multifunctional properties [1,2]. As such, CNTs have been studied since Iijima's work demonstrated a material that could exhibit unrivaled strength-to-weight ratios [3–5]. CNT fibers have exhibited exceptional mechanical performance when tested at low length scales, but when testing lengths have been increased, the observable strength decrease dramatically [6,7]. Existing testing standards point out that in manmade fibers loading conditions must be varied to accurately examine the mechanical performance. This suggests that CNT fiber studies may need to expand the testing envelop to fit industry applications if CNT fiber systems are to ever replace carbon fibers.

Continuous CNT fibers are now commercially available for scaled-up application studies [8,9], which may open the door to evently realize high-mechnical performance and engineering applications of CNT material systems. However, as a newer reinforcement material, highly systematic studies are sparce, and as such a fundamental understanding of CNTy microstrucureproperty relationships at the engineering scale are underdeveloped. This study carried out scaled-up tests with large gauge lengths and morphological analysis paired with reliability statistics to reveal tensile properties, failure modes, and expectation of performance of the commercially available continuous CNT yarns in pristine and modified forms.

2. EXPERIMENTATION

2.1 Material preparation and characterization methodology

2.1.1 Raw material selection and preparation

Miralon yarns produced by Nanocomp Technologies, Inc were examined. The as received Miralon will hence forth be referred to as LCY standing for large cross-section yarn. The LCY was chemically purified at the High-Performance Materials Institute using a proprietary hot solvent process to open the LCY's dense pore structures and remove residual catalyst without causing capillary densification. Upon purification, the LCY was mechanically modified into the second yarn system used in this study – small cross-section yarns (SCY). The last yarn variant investigated was a highly densified SCY, referred to as a flattened SCY (fSCY), which was produced using a proprietary high-pressure roller device to collapse the SCY microstructure.

2.1.2 Sample structure characterization

The linear density, g/km or tex, of each yarn samples was determined by measuring the mass of a 305 mm yarn length on a Mettler Toledo XP105 analytical balance. An x-ray diffraction system was used to measure the degree of alignment of the CNTy via wide angle x-ray diffraction (WAXD). WAXD measurements were obtained on a Bruker NanoSTAR system equipped with an Incoatec IlS microfocus x-ray source. The WAXD intensity was captured by a Fuji Photo Film image plate. Each yarn type was cross-sectioned using a JEOL IB-19500CP ion beam polisher. The cross section of these samples was imaged on a Zeiss 1540EsB or JEOL 7401F field emission microscope, and yarn dimensions were measured by using area analysis in the ImageJ program.

2.1.3 Mechanical testing

To prevent out of gauge length failures, tensile samples were tested on a Shimadzu mechanical testing system equipped with a TH76-1+Ko capstan grip fixture from Grip-Engineering Thümler GmbH and 500 N load cell. The yarns were cut to 305 mm in length and tabbed to a gauge length of 100 – 120 mm with 1500 grit sandpaper tabs and Loctite 4851 adhesive. These length scales differ from most reported tests [10–12] by ~ 150 %. All samples were preloaded between 0.1 – 0.15 N, tested at a loading rate of 3 N/min, and were considered completely fractured once the measured stress dropped below 5 % of the peak stress. Load and displacement values were recorded during testing to produce the stress-strain curves. The specific stresses were calculated from the average linear density, and the strains were calculated from the elongation of the initial fiber length as measured by crosshead displacement of the mechanical testing system.

2.2 Statistical analysis procedure of mechanical results

The statistics software Minitab 19 was used for all statistical analysis. The distribution of best fit was identified from the ultimate tensile strengths for each yarn variation. Right censored parametric distribution analysis was conducted using the known failure points, and the shape, scale, and threshold parameters were determined and recorded. The distribution curves were imposed onto histograms for simplified viewing.

3. RESULTS

3.1 Microstructure characteristics of carbon nanotube yarn

3.1.1 Yarn morphology

Figure 1 shows the cross-section of the LCY, SCY, and fSCY. Notable features are the large voids present throughout the LCY and SCY samples, caused by the continuous yarn manufacturing process [13]. Aligning CNT network textiles is known to yield an increase in mechanical strength and stiffness. It is fair to suspect that the large pockets could act as defects in the LCY and SCY. Neat CNT networks rely on the van der Wahl forces and mechanical entanglement to provide homogenous load transfer [14]. Therefore, these voids would prevent CNT bundles across the gap from entangling or van der Wahl interactions. Smaller voids may exist in both LCY and SCY samples but were most likely filled due to re-deposition during cross-section preparation, which is known to happen during ion beam milling.

As the samples were modified, each yielded a different exterior surface area and morphology. From the image analysis via ImageJ, the perimeter length of each yarn variant was measured. Extrapolation using the known linear density values discussed in Section 3.1.2 of each yarn type resulted in SCY and fSCY having a 166.96 % and 363.05 % greater surface area per tex than LCY, respectively. The varying surface areas could significantly affect a polymer matrix's ability to bond to yarn and can be expected to play a critical role in interfacial strength if these yarns are to be used in future composite applications [15,16]. A circumstance of the SCY modification process was recognized in the Figure 1B – the lower portion of the SCY appears to have separated from the larger yarn mass, which indicates potential damage to the yarn surface or to the CNT bundle entanglement internally. This is could potentially exacerbate the preexisting concern of heterogeneous load transfer present from the internal voids.

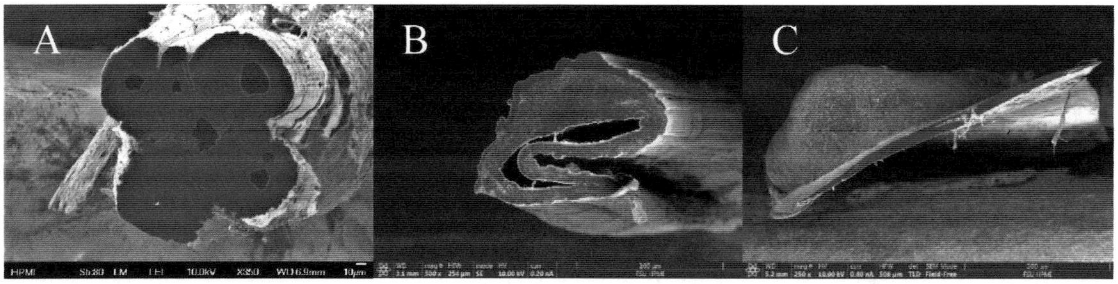

Figure 1: Scanning electron microscopy of ion beam polisher cross-sectioned LCY (A), SCY (B), and fSCY (C). The scale bars in A, B, and C are 10, 100, and 200 μm, respectively.

3.1.2 Changes in long-range order

Alignment degree plays a critical role in a CNT textile's tensile strength [17,18]. As the LCY was modified into SCY and fSCY, a concern was that the alignment degree would be negatively affected and could lead to reductions in mechanical performance during tensile tests. The associated alignment variations were characterized using the measured WAXD intensity of a crystallographic material against its crystal planes [19], and the WAXD intensity data for all three yarn types was fit to Lorentz functions. The fitted curves were procedurally applied as described in another publication [20] to Herman's orientation factor (HOF) for graphitic crystallites [21]:

$$\cos^2(\varphi) = \frac{\sum_0^{\pi/2} I(\phi) \cdot sin\,\varphi \cdot \cos^2(\phi)}{\sum_0^{\pi/2} I(\phi) \cdot sin\,\varphi} \tag{1}$$

$$f = \frac{3 \cdot \cos^2(\phi) - 1}{2} \tag{2}$$

where $I(\phi)$ is the normalized Lorentzian fitted x-ray intensity at the respective azimuthal scan angle (ϕ).

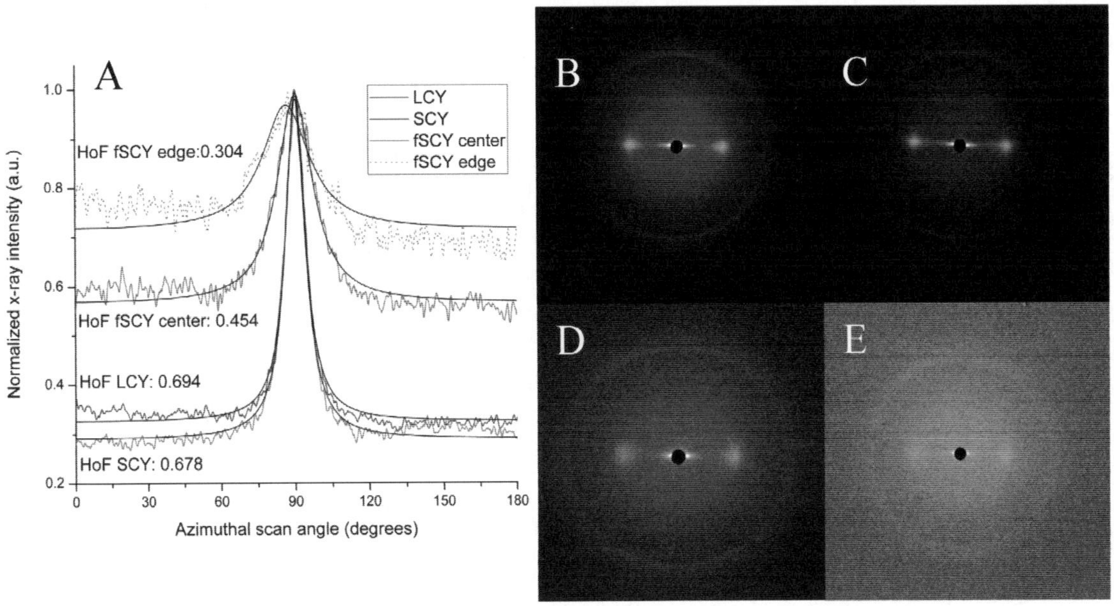

Figure 2: Combined normalized WAXD x-ray signal sets with respective fitted Lorentzian functions used in alignment analysis by Herman's orientation (A). WAXD patterns of LCY (B), SCY (C), fSCY center of yarn (D), and fSCY edge of yarn (E).

From the WAXD patterns and HOF results shown in Figure 2, the LCY and SCY exhibited similar alignment degrees at ~ 70 %, and any variations in strength or modulus were not a consequence of alignment loss from the modification process. However, fSCY was greatly affected by the densification process and exhibited a reduction of alignment across its width. From the HOF, the center of the fSCY width was calculated to be ~ 45 % aligned; however, the alignment was nonuniform across the fSCY width as the edge was ~ 30 % aligned. This noticeable drop in alignment from center to edge was determined to be a result of high compression load during the condensation process. The densification process clearly yielded increased CNT bundle packing or self-assemblage. However, it is unlikely that any benefits caused by the densification processes were unlikely to overcome the degradation of alignment.

3.2 Tensile behavior of carbon nanotube yarns

3.2.1 Capstan measurements of tensile strength

Representative specific tensile strength curves for the LCY, SCY, and fSCY are shown in Figure 5 and summarized in Table 2. The LCY exhibited the highest average specific tensile strength at 0.911 N/tex, while SCY had the highest specific stiffness at 49.634 N/tex. The SCY did not behave much differently than the LCY with the exception of a lower total strain to escape the s-tail phenomenon, discussed later in this section. The SCY specific strength on average was ~0.1 N/tex lower than LCY, while the specific stiffness increased in comparison. Interestingly, the SCY, which was expected to achieve higher specific strengths due to the improved CNT packing showed the lowest specific strength. The lack of changes in mechanical properties between LCY to SCY was minimal, so despite the reduced strength, the SCY may be better suited for composites as the modification process increased the yarn surface area, as seen in Figure 2. The fSCY performed significantly worse that the other yarns, most likely due to the previously mentioned edge defects.

The fSCY experienced the largest failure strain standard deviation, but this was expected as fSCY had a lower average degree of alignment. In cases where CNT networks were allowed to experience large deformation under loads, CNT bundles started to align, improving the mechanical properties of the CNT textile [17,22]. In fSCY samples that reached failure strains over 3 %, the stress-strain curves began to increase exponentially, agreeing with the previous CNT strain-hardening work [14,17]. As previously state, any benefits caused by densification processes were unlikely to overcome the degradation of alignment, and that CNT textiles favor network alignment compared to self-assemblage when determining mechanical performance.

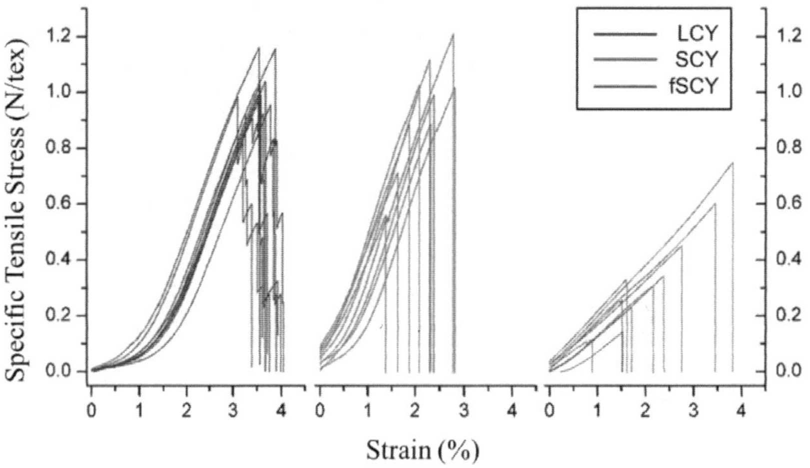

Figure 3: Specific stress-strain of random sample group of LCY, SCY, and fSCY.

Table 1: Average specific tensile properties of LCY, SCY, and fSCY.

Sample Type	Avg. Specific Tensile Strength		Avg. Specific Stiffness		Avg. Failure Strain	
	N/tex		*N/tex*		*%*	
LCY	0.911	± 0.137	46.845	± 9.778	3.186	± 0.632
SCY	0.838	± 0.211	49.634	± 7.624	2.181	± 0.542

fSCY	0.337	± 0.217	15.595	± 3.907	2.169	± 1.073

An interesting occurrence that was recorded in most samples was an s-tail at low strain percentages. This sort of behavior could be expected in tests that were started without a preload, but in extended trials, this tail was still measured in preloads as high as 1.6 % of the load at failure. This s-tail phenomenon is hypothesized to be a result of the constricting of the large pores that existed in LCY and SCY samples. From the stress-strain curves in Figure 3, this phenomenon was clearly isolated to the first ~ 1.5 % and ~ 0.5 % strains in LCY and SCY samples, respectively. This hypothesis is bolstered by the lack of the s-tail in the high density fSCY samples, which, as seen in Figure 1C, exhibited no visible pores. However, without density measurements or internal structure images of the LCY or SCY under strain, it may not be possible to confirm that the yarn pores were to blame for the stress-strain s-tail at low strains. If the yarns' loose morphology can be mitigated, the specific tensile strengths and stiffness could increase in the LCY and SCY samples.

Figure 4 graphs the specific mechanical performance of every sample, clearly showing that the densification process reduced the tensile properties while also reducing stiffness variations. Also, mechanical results showed that in their current states, the LCY, SCY, and fSCY were weaker and less stiff when compared to carbon fiber variants but similar to glass fibers available in the market.

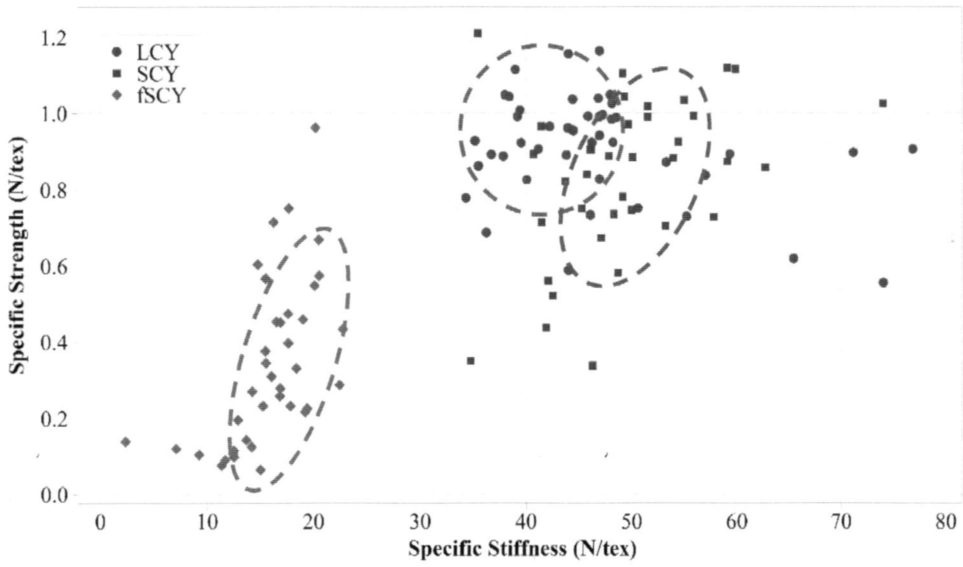

Figure 4: Specific mechanical performance distributions of tested yarn systems.

3.2.2 Fracture type categorization

Failure modes were characterized as either brittle edge or mixed mode edge. Brittle fracture was the dominant failure mode in all yarn types and can be recognized in Figure 5A,B by the distinct failure plane across the LCY and SCY. Some CNT pull out at the fracture edge was present in all brittle failures. Failure through only pullout is another unique form of failure in CNTs, higher strain or trailing force tails at failure are indicative pull out behavior. This pull-out failure is caused when the CNT network partially untangles and slides apart. Mixed mode edge failure is classified as when the failure edge shows no evident plane or exhibits multiple planes of failure and/or displayed a high degree of the pull out, as seen in Figure 5C. Figure 5D provides a simplified failure rendering to clarify how each failure typically takes shape in the yarn samples.

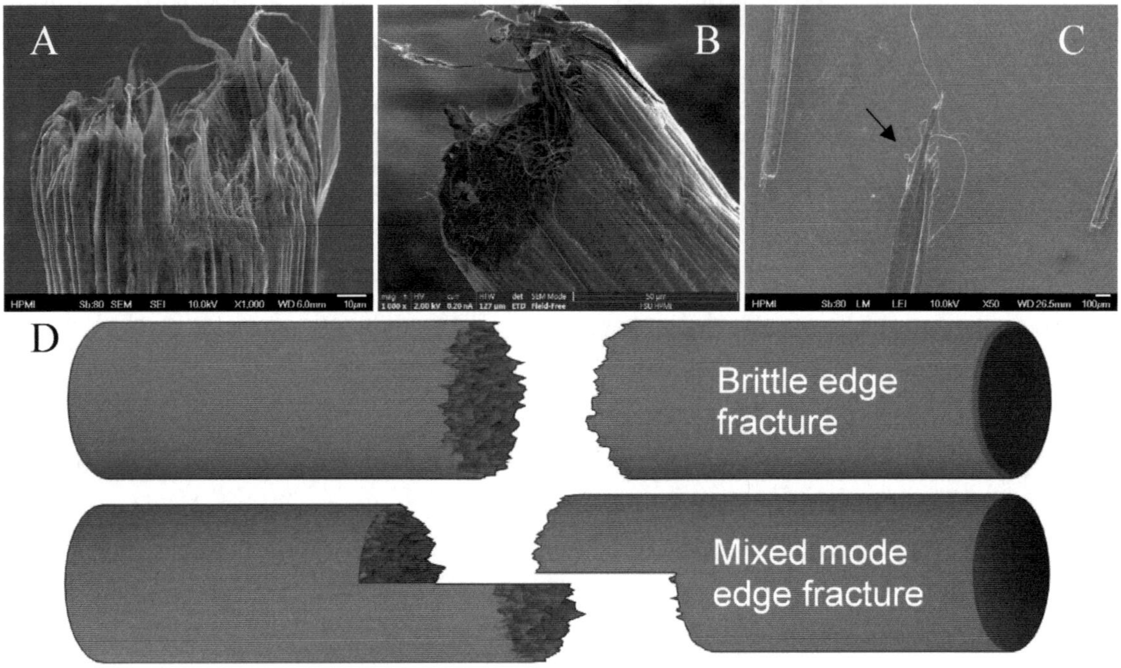

Figure 5: Brittle edge fracture of LCY quarter section (A) and SCY (B); Mixed mode edge fracture of fSCY (C); Rendering of failure modes (D).

An interesting note is that yarns which failed in brittle edge mode repetitively resulted in higher ultimate tensile strength when compared to yarns which failued by mixed mode edge. On average, the LCY, SCY, and fSCY samples that experienced brittle failures demonstrated 19.98 %, 7.89%, and 8.52 % higher ultimate tensile strengths, respectively. However, the ratio of the different failure types between yarn variants was not consistent. LCY and SCY failed in a brittle edge mode 160 % and 40 % more often than in a mixed mode edge failure, respectively; however, fCSY were 100 % more likely to fail in a mixed mode edge as opposed to a brittle edge mode. Figure 6 displays these distributions.

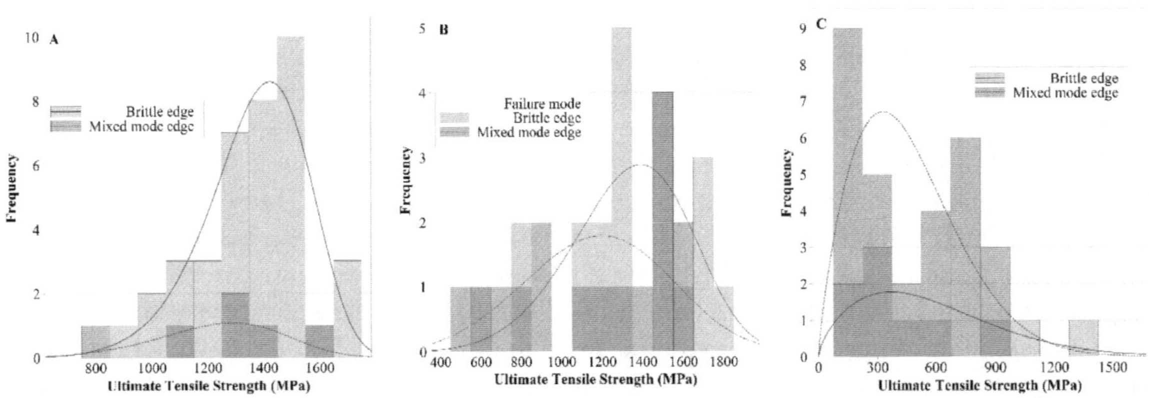

Figure 6: Histogram of ultimate tensile strength with fitted 2-parameter Weibull distribution of LCY, SCY, and fSCY grouped by brittle edge mode and mixed mode edge fractures.

Variations of CNTy surface and in-situ defects are likely the cause of the discrepancies of the different failure modes between samples. Structurally, LCY and SCY are nearly identical, apart from the yarn surfaces which differed from the mechanical modification process that SCY undergo. The high degree of edge defects that the fSCY process creates supports this claim as well. Additionally, internal impurities were recently determined to have a positive effect on mechanical properties of CNT networks in shorter length tests, while having an inverse effect as lengths increased [7]. The behavior is simplified as describing the impurities as a sort of internal sandpaper that increase friction between carbon nanotubes and their bundles. This provides support for the performance differences between the LCY or SCY and the fSCY as it is reasonable to assume that the densification process could amplify these friction effects. Moreover, this counter-effective friction behavior is likely present in the both LCY and SCY as the yarns escape the s-tail and the nonporous structure of the yarns begin to close.

3.3 Weibull analysis of tensile performance

Each specific strength data point was input into Minitab 19 to provide a probability distribution identification. Several distributions achieved excellent goodness-of-fit including smallest extreme value, 2-/3-parameter log-logistical, and 2-/3-parameter Weibull. Weibull distributions are built upon extreme value theory that serves as the basis for the weakest link model [23]. The principle driving the concept of the weakest link is that many defects compete to be the eventual site of failure. This can also provide an effective model for brittle fibers. Based on strain hardening behavior that CNT networks exhibit as their long-range alignment is increased and the abrupt failure behavior in Figure 3, the 2-parameter Weibull distribution was chosen over other distributions with similar goodness-of-fit. Furthermore, the Weibull moduli of several materials are documented, so the specific strength distributions could be easily compared.

The 2-parameter Weibull probability density function (pdf) was obtained from the pdf for a 3-parameter Weibull distribution that is displayed below:

$$f(t) = \frac{\beta}{\eta}\left(\frac{t-\gamma}{\eta}\right)^{\beta-1} e^{-\left(\frac{t-\gamma}{\eta}\right)^{\beta}} \tag{3}$$

where β is the shape parameter, η represents the scale parameter, and γ is the threshold parameter. In the 2-parameter, Weibull pdf, γ is set to zero, which in reference to this study indicates that without loading the sample cannot fail. Note that 3-parameter Weibull statistics achieved a slightly better fit than the standard 2-parameter; however, when the threshold parameter was included, it was calculated to be less than zero. This suggests that the minimum force could result in failure without any applied load; therefore, the more traditional 2-parameter Weibull was chosen to describe the failure distribution. Once the yarn distributions were chosen, the parameter probability plots were made, and the specific strengths were plotted. Figure 7 presents the statistical results. Note that β is referred to as the Weibull moduli that was used to characterize what sort of failure can be expected. For example, silicon carbide and aluminum oxide have low Weibull moduli of between 5 – 10, whereas metals like steel and aluminum range between 90 – 100 [24]. From the recorded shape parameters on the CNT yarns, the Weibull moduli closely resemble those of ceramics. The recognized Weibull shape of each yarn type agreed with previous hypothesis that the quantity and magnitude of defects are to blame for the failure in the yarns.

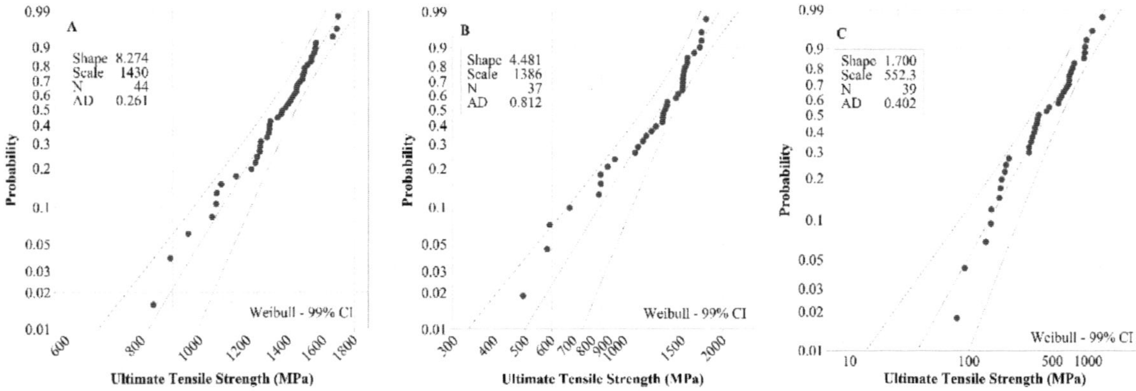

Figure 7: 99 % confidence interval 2-parameter Weibull distribution probability plot of LCY (A), SCY (B), and fSCY (C).

Despite visible micro-pull out of CNT bundles shown in Figure 1, which seem to indicate non-brittle characteristics, the Weibull distributions indicate that the failures may be driven by defects which initiate complete fractures of entangled CNTs. Consequently, at the time of failure, the van der Wahl forces holding CNTs aligned along the loading axis were unable to overcome the force suddenly applied and would slide out instantly without measurable resistance. In short, the failure of the CNT yarns as characterized under SEM appears to contain both brittle and ductile failure of the CNT network, however through Weibull distribution analysis of the ultimate tensile strength indicated that the samples failed nearly identically to materials which are traditionally viewed as brittle and showed no sign of ductile behavior.

4. CONCLUSIONS

Three variations of a commercially available CNTy were systematically studied to reveal their microstructure-mechanical property relationships and further analyzed via a combination of fracture imaging and reliability statistics. Regardless of the differing morphologies, mechanical results of the LCY and SCY show very similar performances. Self-assembly was promoted in the fSCY through mechanical densification; however, this change resulted in a reduction of alignment. Consequently, the tensile properties of fSCY were much lower than other CNTy variants. An s-tail phenomenon which was recorded in LCY and SCY samples was likely caused by the large pores, as seen in the cross-sectional SEM images. Alignment clearly is a major governing factor affecting mechanical properties, while surface defects influence how samples are expected to fail further affecting the ultimate tensile strength. Of the analyzed failure modes, the brittle edge fractures resulted in higher CNTy strength. Each data set of tensile tests was fit to an appropriate probability distribution, and the identified distribution was utilized to fit the probability density function grouped by fracture mode and probability plot for each yarn type. The experimental results for ultimate tensile strength showed a very good fit to 2-parameter Weibull statistics. The CNT yarns were concluded to exhibit brittle behavior since specific stress-strain curves showed that the CNT yarns experienced abrupt failures and the ultimate tensile strengths showed excellent fit to Weibull distributions, which is characteristic of known brittle fibers.

The experimental results show that as of now commercially available CNT yarns still exhibit lower mechanical strength and moduli compared to industrial scale production carbon fibers and other reported lab scale CNT textiles. However, a factor to consider is the lack of resin-CNT interactions that may improve load transfer throughout the CNT network. From the captured porosity images, LCY and SCY may be excellent candidates for resin infiltration; however, infiltration of fSCY may pose a challenge. These questions and observations are being addressed in on-going investigations of CNT micro-composite yarns.

5. REFERENCES

[1] Shokrieh MM, Rafiee R. A review of the mechanical properties of isolated carbon nanotubes and carbon nanotube composites. Mech Compos Mater 2010;46:155–72. https://doi.org/10.1007/s11029-010-9135-0.

[2] Al-Saleh MH, Sundararaj U. Review of the mechanical properties of carbon nanofiber/polymer composites. Compos Part A Appl Sci Manuf 2011;42:2126–42. https://doi.org/10.1016/j.compositesa.2011.08.005.

[3] Baughman RH. Carbon Nanotubes--the Route Toward Applications. Science (80-) 2002;297:787–92. https://doi.org/10.1126/science.1060928.

[4] Bai Y, Zhang R, Ye X, Zhu Z, Xie H, Shen B, et al. Carbon nanotube bundles with tensile strength over 80 GPa. Nat Nanotechnol 2018;13:589–95. https://doi.org/10.1038/s41565-018-0141-z.

[5] Ding H, Zhang Q, Wang F, Tian Y, Wang L, Shi Y, et al. Structure control of polyphenylene sulfide membrane prepared by thermally induced phase separation. J Appl Polym Sci 2007;105:3280–6. https://doi.org/10.1002/app.26595.

[6] Koziol K, Vilatela J, Moisala A, Motta M, Cunniff P, Sennett M, et al. High-Performance Carbon Nanotube Fiber. Science (80-) 2007;318:1892–5. https://doi.org/10.1126/science.1147635.

[7] Gspann TS, Montinaro N, Pantano A, Elliott JA, Windle AH. Mechanical properties of carbon nanotube fibres: St Venant's principle at the limit and the role of imperfections. Carbon N Y 2015;93:1021–33. https://doi.org/10.1016/j.carbon.2015.05.065.

[8] Siochi EJ, Harrison JS. Structural nanocomposites for aerospace applications. MRS Bull 2015;40:829–35. https://doi.org/10.1557/mrs.2015.228.

[9] Kim J-W, Sauti G, Wincheski RA, Cano RJ, Jensen BD, Smith Jr. JG, et al. Undirectional Carbon Nanotube Yarn/Polymer Composites. Hampton, Virginia: 2018.

[10] Evora MC, Lu X, Hiremath N, Kang N-G, Hong K, Uribe R, et al. Single-step process to improve the mechanical properties of carbon nanotube yarn. Beilstein J Nanotechnol 2018;9:545–54. https://doi.org/10.3762/bjnano.9.52.

[11] Humble J. Modeling of nonisothermal crystallization kinetics of semilinear polyphenylene sulfide for process simulation. Annu Tech Conf - ANTEC, Conf Proc 2016:632–8.

[12] Sugano K, Kurata M, Kawada H. Evaluation of mechanical properties of untwisted carbon nanotube yarn for application to composite materials. Carbon N Y 2014;78:356–65. https://doi.org/10.1016/j.carbon.2014.07.012.

[13] Lashmore DS, Braden R, Hart AJ, Welch J. Chemically-assisted alignment of nanotubes within extensible structures. US 8,246,886 B2, 2012.

[14] Downes RD, Hao A, Park JG, Su Y-F, Liang R, Jensen BD, et al. Geometrically constrained self-assembly and crystal packing of flattened and aligned carbon nanotubes. Carbon N Y 2015;93:953–66. https://doi.org/10.1016/j.carbon.2015.06.012.

[15] Peng Q, He X, Li Y, Wang C, Wang R, Hu P, et al. Chemically and uniformly grafting carbon nanotubes onto carbon fibers by poly(amidoamine) for enhancing interfacial strength in carbon fiber composites. J Mater Chem 2012;22:5928. https://doi.org/10.1039/c2jm16723a.

[16] Xiao T, Liu J, Xiong H. Effects of different functionalization schemes on the interfacial strength of carbon nanotube polyethylene composite. Acta Mech Solida Sin 2015;28:277–84. https://doi.org/10.1016/S0894-9166(15)30014-8.

[17] Jolowsky C, Sweat R, Park JG, Hao A, Liang R. Microstructure evolution and self-assembling of CNT networks during mechanical stretching and mechanical properties of highly aligned CNT composites. Compos Sci Technol 2018;166:125–30. https://doi.org/10.1016/j.compscitech.2018.04.003.

[18] Ma W, Liu L, Zhang Z, Yang R, Liu G, Zhang T, et al. High-Strength Composite Fibers: Realizing True Potential of Carbon Nanotubes in Polymer Matrix through Continuous Reticulate Architecture and Molecular Level Couplings. Nano Lett 2009;9:2855–61. https://doi.org/10.1021/nl901035v.

[19] White JL, Spruiell JE. The specification of orientation and its development in polymer processing. Polym Eng Sci 1983;23:247–56. https://doi.org/10.1002/pen.760230503.

[20] Moharir AV, Vijayraghavan KM, Panda BC, Gupta VB. Dependence of Stiffness and Strength of Cotton Fibers on Crystallite Orientation. Text Res J 1982;52:756–60.

https://doi.org/10.1177/004051758205201204.

[21] Hermans PH, Weidinger A. Quantitative X-Ray Investigations on the Crystallinity of Cellulose Fibers. A Background Analysis. J Appl Phys 1948;19:491–506. https://doi.org/10.1063/1.1698162.

[22] Downes R, Wang S, Haldane D, Moench A, Liang R. Strain-Induced Alignment Mechanisms of Carbon Nanotube Networks. Adv Eng Mater 2015;17:349–58. https://doi.org/10.1002/adem.201400045.

[23] Zok FW. On weakest link theory and Weibull statistics. J Am Ceram Soc 2017;100:1265–8. https://doi.org/10.1111/jace.14665.

[24] Meyers MA, Chawla KK. Statistical Analysis of Failure Strength. Mech. Behav. Mater. 2nd ed., New York City, NY: Cambridge University Press; 2009, p. 449–60.

LOW COST GLASS-CERAMIC MATRIX COMPOSITES FOR HARSH ENVIRONMENT HEAT EXCHANGER APPLICATIONS

John Gangloff[1], Justin Alms[1], John Holowczak[1], John Needham[1], Paul Sheedy[1], Thomas Yun[1], Daniel Mosher[1]
John Podhiny[2], Brian Sullivan[2]
James Haynes[3], Brian Jolly[3]

[1]United Technologies Research Center
East Hartford, CT, USA
[2]Materials Research & Design, Inc.
Wayne, PA, USA
[3]Materials Science and Technology Division, Oak Ridge National Laboratory
Oak Ridge, TN, USA

ABSTRACT

High-temperature, high-pressure heat exchange is a ubiquitous need in power generation, industrial components and aerospace systems, including supercritical CO2 cycles. In these applications, higher cycle temperatures and pressures provide a significant driver to overall thermal efficiency and fuel consumption as well as a challenge to metallic material capability. To enable operation in these harsh operating regimes, United Technologies Research Center (UTRC), Materials Research and Design (MR&D) and Oak Ridge National Laboratory (ORNL) are developing a high temperature, lightweight, low-cost heat exchanger technology made from a Glass-Ceramic Matrix Composite (GCMC) material system, denoted UT-16, which is capable of long operational life over a wide range of harsh environments. While this GCMC material system is broadly applicable to a variety of high temperature components, this effort focuses on the development and demonstration of a Counterflow Tube Heat Exchanger (CTHX). The CTHX geometry has been configured specifically for the GCMC material which can be fabricated orders of magnitude faster than most conventional CMCs. The as-fabricated material has extremely low porosity resulting in very low leakage rates and can also be coated with a chemically inert layer for additional protection.

1. INTRODUCTION AND MOTIVATION

1.1 Introduction

The utilization of high-temperature, high-pressure heat exchange is an abundant requirement in power generation and industrial applications, including steam cycles, supercritical CO2 (SCO2) cycles, industrial gas turbines, and thermal rectifiers. In these applications, higher cycle temperatures and pressures remain a significant driver to overall thermal efficiency and fuel consumption. To enable operation in these harsh operating regimes, United Technologies Research

This document does not contain any export controlled technical data.

SAMPE Virtual Conference Proceedings, 2020. Society for the Advancement of Material and Process Engineering – North America.

Center (UTRC), Materials Research and Development (MR&D) and Oak Ridge National Laboratory (ORNL) are developing a high temperature, high strength, and low-cost heat exchanger (HX) technology made from a Glass-Ceramic Matrix Composite (GCMC) material system, denoted UT-16, which is capable of long operational life over a wide range of harsh environments. While this GCMC material system is broadly applicable to power generation components, this development program utilizes a Counterflow Tube Heat Exchanger (CTHX) configuration as the specific demonstration application. Figure 1 overviews the GCMC heat exchanger concept and enabling technologies under development.

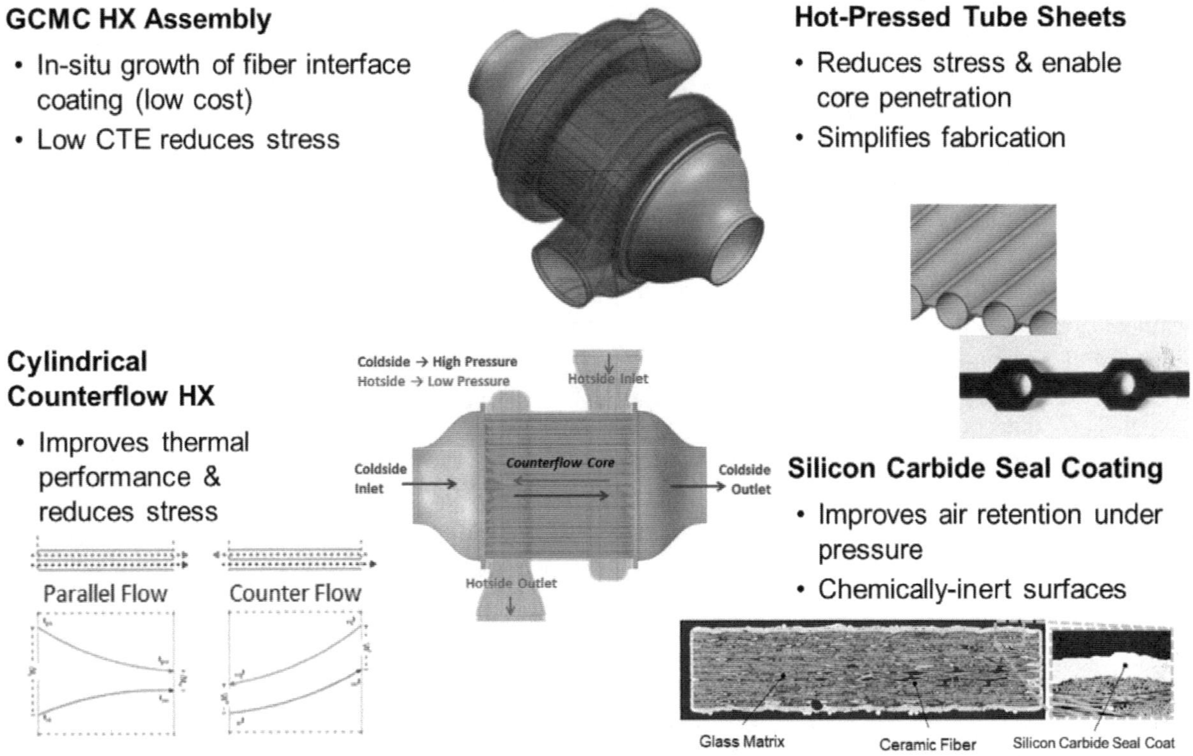

Figure 1. GCMC heat exchanger concept overview showing key technologies.

The CTHX geometry has been configured specifically for the GCMC material and its manufacturing approach. The CTHX features cylindrical components and arrays of simple webbed-tubes and baffles to simplify manufacturing and assembly and to reduce stresses from pressure loads and thermal growth. The CTHX eliminates nearly all secondary surfaces in the HX, increasing thermodynamic performance and substantially deemphasizing material thermal conductivity, thereby reducing the cost of the GCMC by eliminating the need for costly additives.

Chemical Vapor Infiltration (CVI) of Silicon Carbide (SiC) is also being explored as a multifunctional surface coating to provide a chemically-inert and air-tight seal on HX surfaces. The program is being supported by the U.S. Department of Energy-Advanced Manufacturing Office (DOE-AMO) and leverages prior U.S. Department of Defense and United Technologies Corporation (UTC) investments in low-cost GCMC materials, high-effectiveness HX architectures, and chemical-resistant air-barrier seal coatings.

1.2 Motivation

The project goal is to develop and demonstrate heat exchanger technology capable of long operational life in an approximately 1600°F (871°C) and approximately 1000 psi (6.89 MPa) differential pressure (between hot and cold HX sides) operating environment at a cost and reliability consistent with commercialization requirements.

Reaching the higher efficiencies of SCO2 cycles has been challenging due to the high temperature and pressure limitations of metallic superalloy HX materials, so if successful, it is expected that GCMC HX technology will help to increase the impact of SCO2-based power generation systems. The higher operating temperatures enabled by GCMC can also reduce HX volume by 2.5X compared to a lower-temp Inconel HX of equivalent capacity and increase cycle thermal efficiency by 3-5 points based on system model predictions. In addition, the use of GCMC may provide a system volume reduction ~ 60%, weight reduction ~ 85%, superior durability, and lower cost.

Baseline performance goals for the program also includes demonstration of low internal cold-to-hot side leakage <0.1% to ensure negligible impact on recuperator effectiveness and cycle efficiency. The program has performed a preliminary demonstration of HX durability over 20 high temperature thermal cycles with leakage rates remaining orders of magnitude below 0.1% target.

2. MATERIAL R&D BACKGROUND

UTRC is the central R&D organization for UTC, and has extensive expertise in the development of CMC material systems and their fabrication in support of UTC's aerospace businesses (Pratt & Whitney, Collins Aerospace). UT-16 and similar fiber-reinforced GCMC material systems were originally developed at UTRC in the 1980s for high performance aerospace components [1-3].

Figure 2 shows several UT-16 components successfully produced for these applications, replacing nickel-based superalloy components. More recently, HX tube-sheets were successfully prototyped. The CTHX architecture has been studied extensively at UTRC using CMCs as well as conventional materials/manufacturing and 3D printing. Thermal-fluid models have been developed and validated for low temperature applications, improving the confidence of performance predictions in the program.

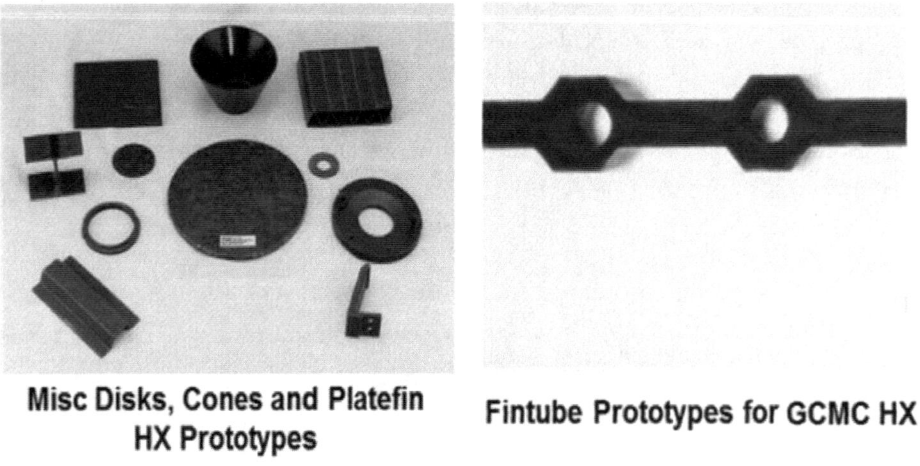

Figure 2. Various UT-16 GCMC components, including tube-sheet prototypes for GCMC HX.

ORNL has extensive experience in Chemical Vapor Deposition (CVD) and CVI of SiC and other high temperature ceramic coatings for a variety of applications, including advanced nuclear fuel kernels, SiC/SiC CMCs for gas turbine engine applications, and fuel elements for nuclear thermal propulsion for space power. ORNL also has extensive experience in CVI of SiC matrices for SiC/SiC CMCs, and participated heavily in developing the modern CMCs that are currently in-service on aircraft.

For thermo-mechanical composites design, MR&D has decades of experience in CMC design and manufacturing. MR&D has successfully designed CMCs and components for extreme environment applications including engine turbine vanes, combustion liners, thrust chambers, exit cones, nozzle extensions, and thermal protection and hot structure control surface components for numerous hypersonic vehicles.

One can compare potential material choices for these operating conditions between metallic superalloys, monolithic ceramics, and ceramic matrix composites (CMCs). Metallic superalloys represent the conventional approach to high temperature and pressure HXs, but are expensive, have temperature limitations, and have limited operational life due to large thermal expansion and creep. Monolithic ceramics offer higher temperature capability, low thermal expansion and excellent chemical resistance, but are flaw sensitive (brittle), difficult to shape into intricate features, and present scaling challenges. Conventional CMCs have favorable properties, but are extremely expensive and issues with air retention precluded their commercialization.

3. MANUFACTURING

3.1 Processing Overview

Figure 3 shows a flowchart of the GCMC manufacturing process. As illustrated in Figure 3, ceramic fibers are impregnated with an aqueous slurry – typically containing glass powder and an organic binder – and wound into tapes to create a pre-impregnated or "prepreg" tape material. After the prepreg tape is created, individual plies are cut from the prepreg tapes and stacked in a laminated structure. The binder is then removed and the tape layup is loaded into the tooling die.

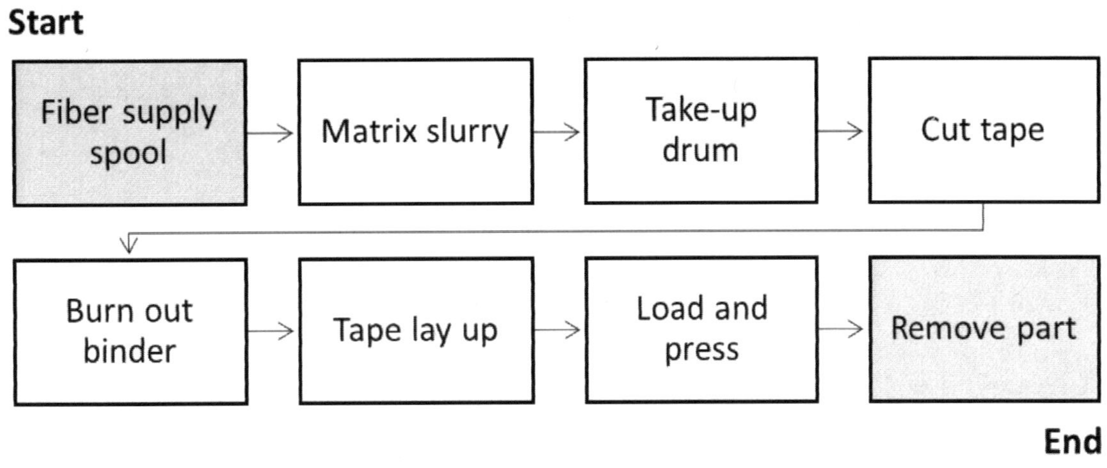

Figure 3. A flowchart of the GCMC manufacturing process.

The die is placed into a hot press and heated under high vacuum until the glass is molten. Pressure is applied and the GCMC is consolidated via molten glass flow. After consolidation, secondary operations may include heat treatments and final machining if necessary. Parts can additionally be subjected to a CVI SiC seal coat. Much of the tooling dies are also typically reusable.

3.2 Die Assembly and Tube-Sheet Forming

Figure 4 illustrates the design, build, and analysis workflow of HX tube sheets. Prototype tube-sheet samples have been fabricated and analyzed using optical microscopy and microCT imaging. A set of tube-sheets was manufactured using tooling to define passages and create external geometry. Multiple samples have also been produced with different fiber architectures. One of these panels is shown in the as-molded condition in Figure 4 (Center).

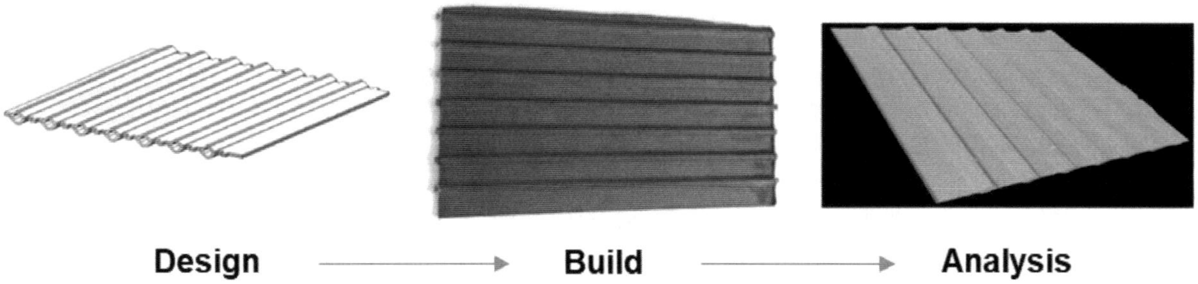

Figure 4. A schematic of tube-sheet design, build, and analysis workflow. Tooling is used to form GCMC HX geometry during hot pressing.

Figure 5 (Left) shows an image of a headered tube sheet. The header regions provide mechanical support for pressure testing. Figure 5 (Right) shows a schematic view of a headered tube sheet in a pressure testing assembly. Initial pressure testing was performed at ambient temperatures with water. Future testing aims to demonstrate high temperature and high pressure testing with gas.

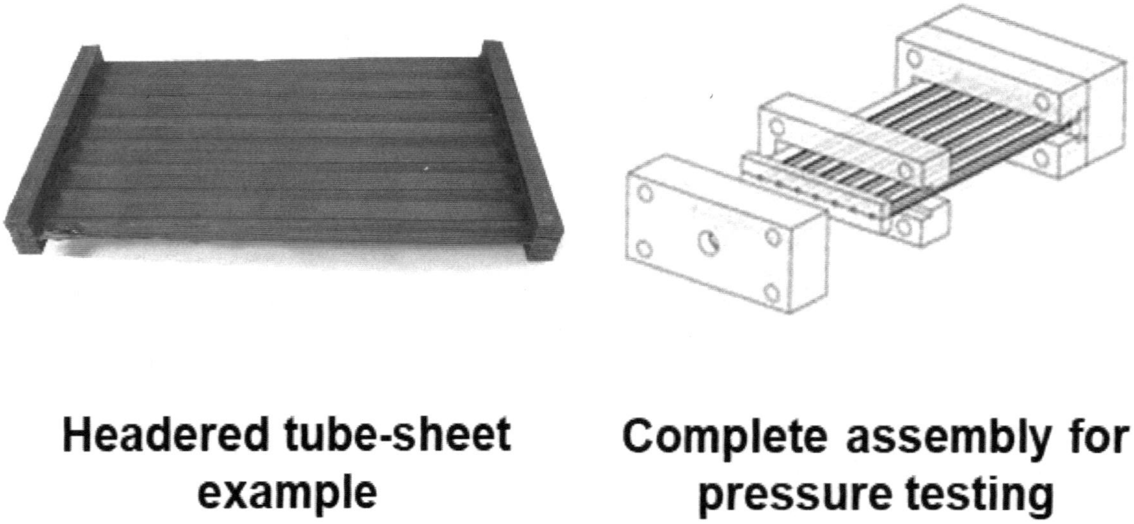

Figure 5. Headered tube-sheet joining and assembly for proof pressure testing.

4. COATING

A CVI study is ongoing to optimize coating GCMC composites with SiC. Figure 6 shows photographs of the CVI reactor setup at ORNL and an example of uncoated and coated GCMC samples. In addition, a CVD reactor is available for smaller test coupons.

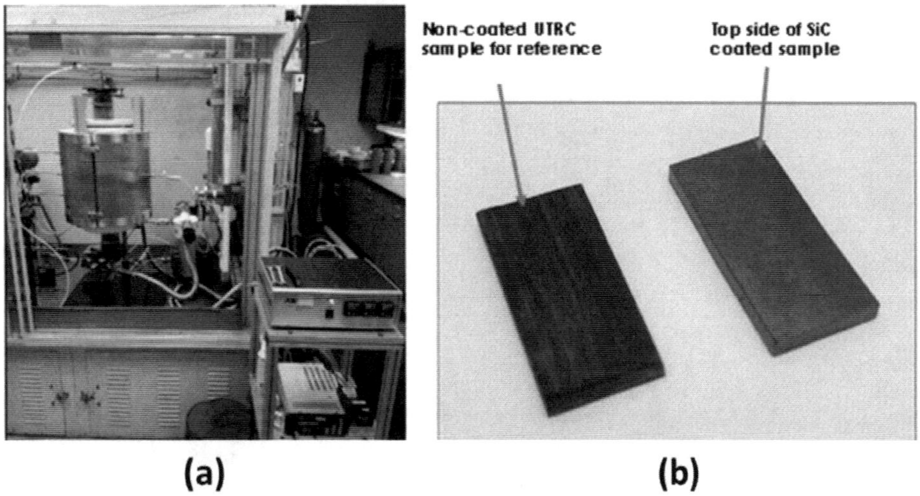

Figure 6. (a) A Chemical Vapor Infiltration (CVI) reactor setup at ORNL; (b) SiC coating of CMC Coupons.

Figure 7 shows optical images of a coated GCMC cross section that was performed to analyze the SiC seal coat. Shown is an original rectangular sample cut in half for metallography. The coating appears well adherent with no separation observed at the interface. Initial trials indicated some periodic cracking in the seal coat, possibly due to CTE mismatch.

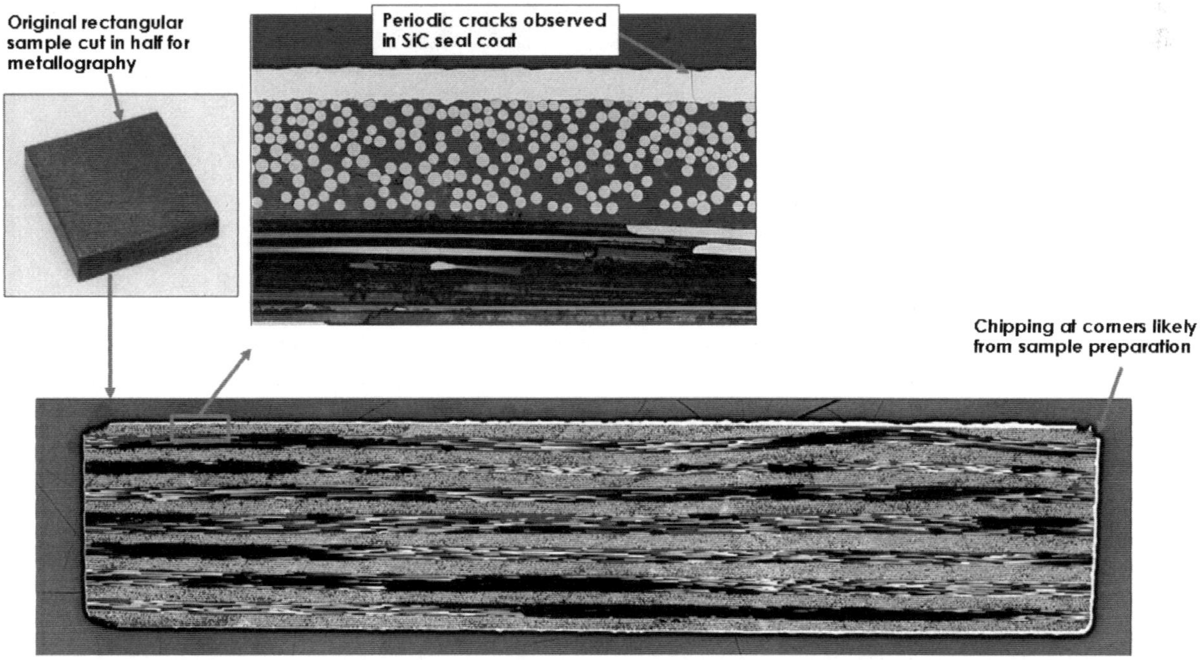

Figure 7. A SiC-coated GCMC example for optical image analysis of coating process quality.

5. CHARACTERIZATION

Permeability testing was performed on a series of GCMC coupons. Figure 8 shows elements of the permeability testing development, including coupon configuration and heat treatment. The aim was to create simulated GCMC joints subjected to manufacturing process conditions and perform testing of gas leakage before and after thermal cycling at the simulated GCMC joints.

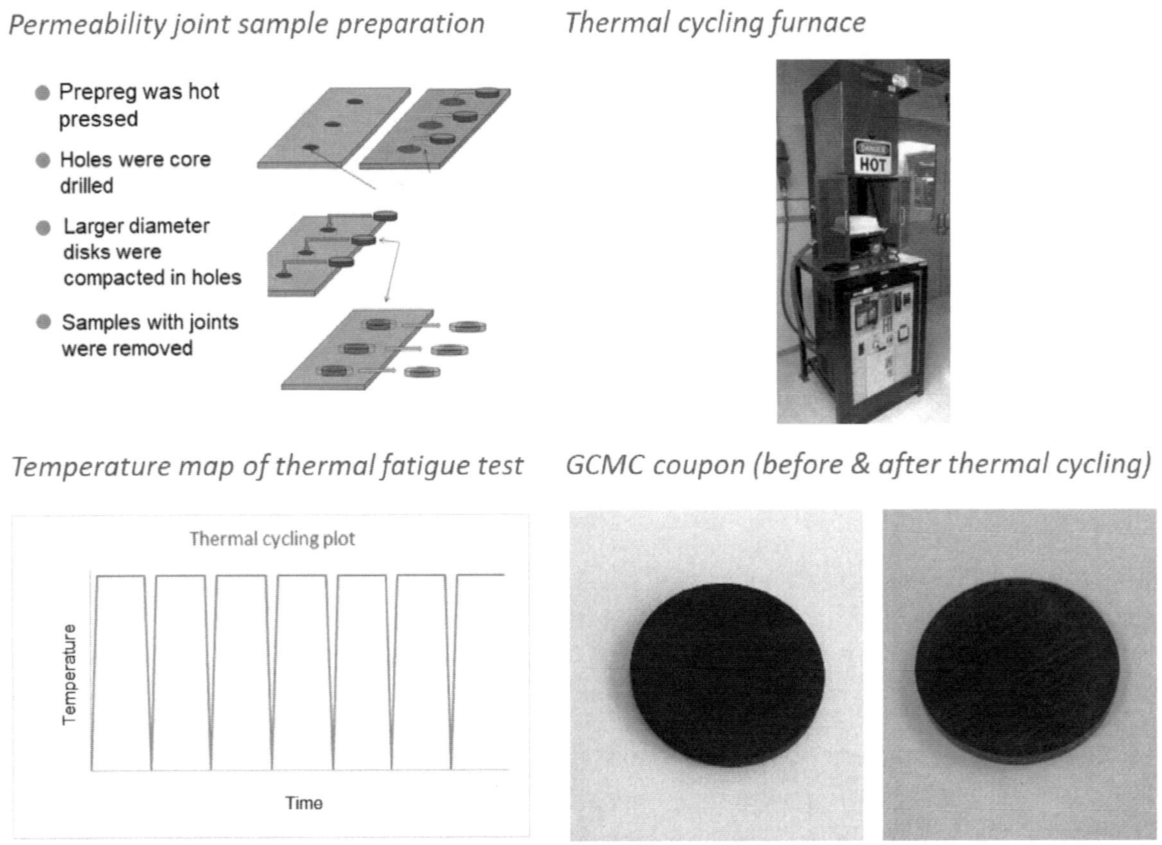

Figure 8. GCMC gas permeability test sample preparation and thermal cycling before testing.

Permeability joint samples were first prepared by hot pressing two flat panels. Circular holes of one diameter were drilled in one flat panel, while circular core samples of a larger diameter were core drilled and extracted from the other panel. The larger diameter core samples were then hot pressed into the panel with the smaller diameter holes to create a panel with circular joints. The circular joint samples were then core-drilled and extracted for gas permeability testing. Some circular joint samples were subjected to thermal cycling before permeability testing.

Figure 9 shows images documenting the permeability test setup and some results compared to the project baseline. The permeability test setup consists of a gas flow with pressure control into a permeability cell that holds the GCMC samples. The permeability is calculated based on rate of change of the gas pressure before and after the permeability cell.

Initial permeability tests show ultra-low permeability when compared to the program's target. Also, the thermally cycled gas permeability samples showed ultra-low permeability when compared to the program's target. The measured leakage rate can be considered negligible compared to the program target – it is believed that it may be likely system-level leakage. Another key observation was no reduction in bending strength after 100 cycles at the program criteria temperature. Samples that were coated did not show a permeability change.

Note that, the permeability joint sample represents how the tube-sheet will be joined to other tube-sheets and to gas flow headers. Establishing that the GCMC process could produce samples of acceptable gas loss limits was key for progressing to larger scale heat exchanger manufacturing process development per program targets.

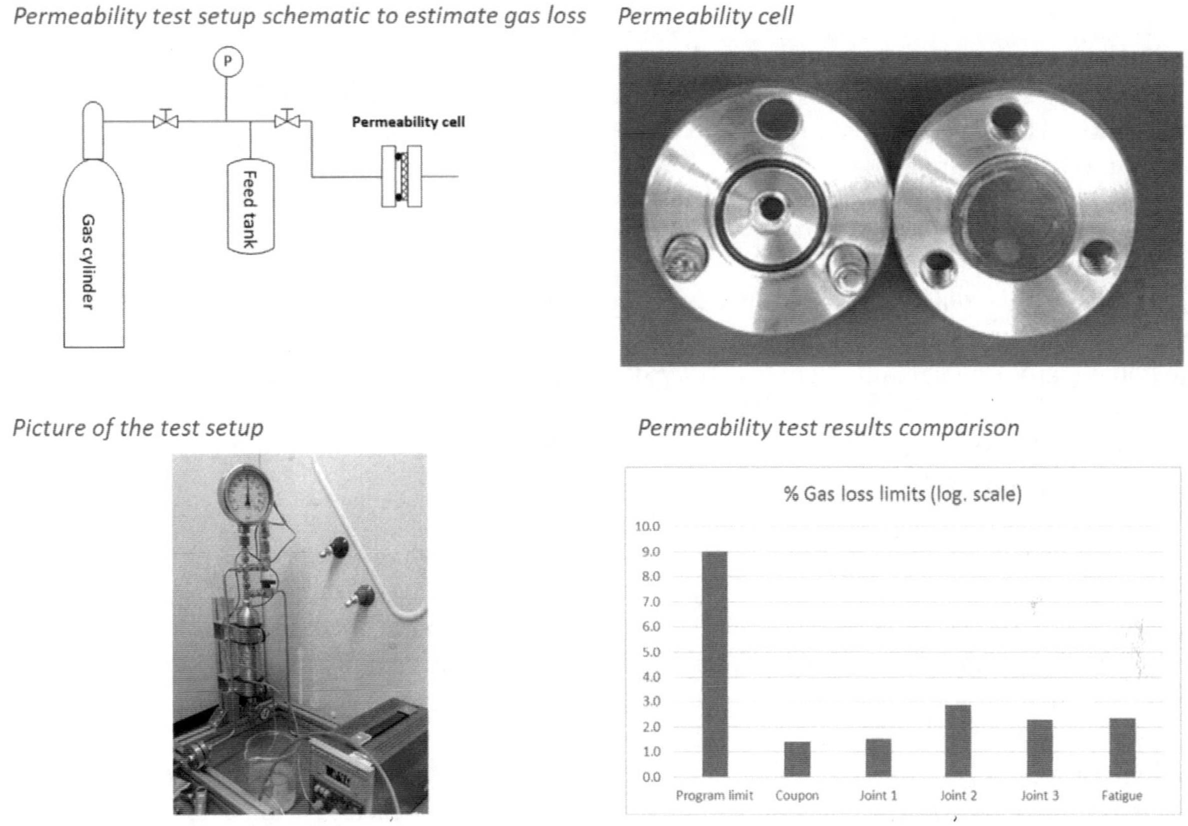

Figure 9. GCMC gas permeability test setup. The results show very low gas losses through joints.

6. SUMMARY

Low-cost glass-ceramic matrix composites are under development for harsh environment heat exchanger applications. Experiments have been completed for coupon fabrication, with the aim of demonstrating part reliability and reproducibility. Tube-sheet manufacturing trials are underway, with a focus on how to replicate complex flow features and part joining.

The benefits of the GCMC material system under investigation are potentially lower cost and lighter weight applications for harsh environments, compared to conventional nickel-super alloys or other similar materials. GCMCs are able to be manufactured in a near-net shape and with complex features, such as the heat exchanger components explored in this work.

Chemical Vapor Infiltration (CVI) trials have been demonstrated with initial SiC seal coating results analyzed. This effort is being led by Oak Ridge National Laboratory. CVD and CVI reactors have been setup to coat different length scales of GCMC samples and coupons. Evaluation is ongoing to study seal coating uniformity.

GCMC gas permeability samples have been pressed by UTRC and coated by ORNL. The gas permeability samples are designed to capture tube-sheet joint interfaces. The aim of the permeability testing is to establish that the formed heat exchanger joints between tube-sheets and header components will not leak beyond a design limit. The gas permeability test results have shown that joints are performing multiple orders-of-magnitude lower than the design target.

Key next steps for this program are further refinement of tube sheet manufacturing regarding repeatability and efficiency, CVI coating of prepared tube sheets with demonstration of coating uniformity. These next steps with aid in the evaluation of tube sheet HX performance under relevant SCO2 conditions.

7. ACKNOWLEDGEMENTS

The project team would like to acknowledge Ram Ranjan of United Technologies Research Center for his project leadership. The project team would also like to acknowledge and thank Brian St. Rock and Rajiv Ranjan, both formerly of United Technologies Research Center, for their project management and technical contributions. The project team would also like to acknowledge and thank Steven Sikirica, Gibson Asuquo, and Chad Sapp of U.S. Department of Energy – Advanced Manufacturing Office for their program management and support.

This material is based upon work supported by the U.S. Department of Energy's Office of Energy Efficiency and Renewable Energy (EERE) under the Advanced Manufacturing Office Award Number DE-EE0008318. This report was prepared as an account of work sponsored by an agency of the United States Government. Neither the United States Government nor any agency thereof, nor any of their employees, makes any warranty, express or implied, or assumes any legal liability or responsibility for the accuracy, completeness, or usefulness of any information, apparatus, product, or process disclosed, or represents that its use would not infringe privately owned rights. Reference herein to any specific commercial product, process, or service by trade name, trademark, manufacturer, or otherwise does not necessarily constitute or imply its endorsement, recommendation, or favoring by the United States Government or any agency thereof. The views and opinions of authors expressed herein do not necessarily state or reflect those of the United States Government or any agency thereof.

8. REFERENCES

1. Prewo, K.M., Brennan, J.J., and Layden, G.K. "Fiber Reinforced Glasses and Glass-Ceramics for High Performance Applications." *Ceramic Bulletin* 65(2) (1986): 305-313.
2. Brennan, J.J. "Interfacial Characteristics of Glass-Ceramic Matrix/SiC Fiber Composites." *Journal de Physique Colloques* 49(C5) (1988): 791-809.
3. Brennan, J.J. "Interfacial Studies of Refractory Glass-Ceramic Matrix/Advanced SiC Fiber Reinforced Composites." United States. Department of the Navy Office of Naval Research. Contract N00014-87-C-0699. *Final Report R93-918246-6*. 1993.

THIN PLY SHEET MOLDING COMPOUND ENABLES THIN LAMINATES WITH PROPERTIES AS OF CONTINUOUS FIBERS

M. Alves[1], S. Pimenta[1], D. Carlstedt[2], L. E. Asp[2], F. Ohlsson[3], F. Rinn[3]

[1]meComposites, Mechanical Engineering Department, Imperial College London, South Kensington Campus, London, SW7 2AZ, UK

[2]Department of Industrial and Materials Science, Chalmers University of Technology, SE-41296, Gothenburg, Sweden

[3]Oxeon AB, Företagsgatan 24, SE-504 64, Borås, Sweden

ABSTRACT

In a research collaboration with Chalmers University of Technology, KTH Royal Institute of Technology (SWE) and Imperial College (UK), Oxeon AB has developed ultrathin carbon fiber sheet molding compound (SMC) products for high performance, thin laminate and ease of manufacture composite applications. Said SMC products are based on ultrathin (0.02 mm or 0.0008 in) Spread Tow Tapes that are cut by length and randomly distributed in order to obtain a preform with an in-plane isotropy in either dry or pre-impregnated form displaying the previously never seen minimum cured plate thickness of about 0.5 mm. Recent work has demonstrated that these SMC materials can reach high stiffness and strength making them good alternatives to metal alloys or even continuous fiber composites. In the conduced work, SMC composites were manufactured using ultra-thin tapes of high modulus and ultrahigh modulus fibers having a large length-to-thickness ratio. The paper will describe how these SMC products lead to high stiffness and to significant increase in strength compared to commercially available SMC materials.

Corresponding author: Florence.rinn@textreme.com (F. Rinn)

1. INTRODUCTION

Tow Based Sheet Molding Compounds (TB-SMCs) are an up and coming class of high-performance composite materials that consist of long chopped carbon fibre tows (20-50 mm length) randomly distributed in a polymeric matrix. TB-SMCs are actually also called Tow Based Discontinuous Composites (TBDCs) [1-5]. The random and discontinuous nature of the material makes it more manufacturable than conventional continuous-fibre composites; in addition, high fibre volume fractions (up to 60 %) are achievable with this type of microstructure, which leads to high mechanical properties [1, 6-11]. Some examples of current commercial applications of TB-SMCs include window frames of the 787 Dreamliner [12] and helicopter tailboom frames [13]. Compression moulded TB-SMC specimens have shown stiffness values comparable to those of quasi-isotropic continuous-fibre laminates [7, 8], hence making them a good alternative to continuous fibre composites for stiffness-driven designs. Moreover, due to the material architecture, with stiff inclusions embedded in a soft matrix, high damage tolerance has been achieved [2, 14]. However, the discontinuous architecture of TB-SMCs leads to stress concentrations near the tow-ends which generate damage at the interface between tows, potentially leading to failure due to tow debonding and pull-out. For this reason, the strength values of commercially available TB-SMCs are significantly (between 30-60 %) lower than those of quasi-isotropic continuous-fibre laminates [7]. It has been shown that the thickness of the tows has a significant effect on the mechanical properties of TB-SMCs [1, 3, 9]. Wan et al. [9] found that manufacturing TB-SMCs with thin tows (of approximately 50 μm) resulted in higher ultimate strength values (500 MPa), closer to those of continuous fibre composites. Li et al. [1] manufactured TB-SMC specimens using two different tow thicknesses (but similar specimen thickness) and found that the specimens with thicker tows lead to lower mechanical properties, both in terms of modulus and strength. This effect of increasing TB-SMCs strength for decreasing tow thicknesses has been successfully modelled by Li and Pimenta [4].

So far, only TB-SMCs material systems with relatively high tow thicknesses (50-200 μm) and standard-or intermediate-modulus fibres (below 300 GPa) have been tested and reported in the literature. This leaves room in the design space of the material's microstructure to investigate whether it is possible to achieve higher mechanical properties with these types of architectures. In this work, TB-SMCs manufactured with ultra-thin (20 μm) and high modulus carbon fibres tows (425-750 GPa) are explored, with the objective to maximise the material's mechanical properties, especially its ultimate strength. Section 2 details the manufacturing and testing procedures of TB-SMC plates with different microstructural configurations selected. Moreover, experimental studies are performed to independently assess how the materials' mechanical properties are affected by the thickness and stiffness of the carbon-fibre tows. The experimental results are analysed and discussed in Sections 3 and 4, and the main conclusions are drawn in Section 5.

2. MATERIALS AND EXPERIMENTAL METHODS

It has been suggested [4, 5] that tensile failure of TB-SMCs occurs through tow fracture (governed by the tensile strength of the fibres) and tow pull-out (governed by the fibre stiffness, tow thickness, and mode-II fracture toughness of the interface between tows). For typical tow dimensions, tow pull-out is the dominant failure mode, meaning that TB-SMCs tend to fail at much lower strains than those required to break the fibres. Therefore, it should be possible to increase the resistance of TB-SMCs to tow pull-out by using thinner tows with stiffer fibres and with a tougher fibre-matrix interface; achieving this goal could suppress failure of TB-SMCs via tow pull-out, allow for the properties of the fibres in the composite to be fully exploited, and ultimately increase the tensile strength of TB-SMC materials.

2.1 Materials and manufacturing procedure

In order to demonstrate the potential improvements in the mechanical properties of TB-SMCs as identified numerically by Li and Pimenta [5], TB-SMC plates were manufactured using tows of high-modulus carbon-fibres spread into ultra-thin tapes and epoxy films, as described next. The spread tapes were produced by TeXtreme® spread tow technology (a state-of-the-art procedure for spreading tows into thin flat tapes with desired thickness and width) at Oxeon AB, using HS40 (High Modulus - HM) PAN-based and K13916 (Ultra High Modulus - UHM) pitch-based carbon-fibres. The nominal mechanical properties of the fibres (modulus E^f, strength X^f and strain to failure ϵ_{fail}^{f}) are presented in Table 1. The spread tapes had a width of 20 mm, and an areal weight of 21 gsm (corresponding to a thickness of approximately 20 μm) or 105 gsm (corresponding to a thickness of approximately 100 μm) for the HS40 fibres and of 105 gsm for the K13916 fibres; the spread tapes were cut in 50 mm long pieces, hereby referred to solely as tapes. The epoxy resin used was supplied by Oxeon AB in the form of thin resin films, with an areal weight of 30 gsm. The tapes were manually placed at random locations and orientation on resin films, forming one equivalent lamina (EL), as schematically represented in Figure 1a. After the tape placement, heated de-bulks were performed on each EL; this consists in placing each EL in a vacuum bag on a heated RIFT table (at 70 °C) for 10 min, in order to ensure a better resin infusion. The final TB-SMC plate was obtained by stacking several ELs (in order to produce plates with the number of tapes specified in Table 2), as shown in Figure 1a. During this stacking process, the reference direction of each EL was incrementally rotated by 90° in the stack, in order to further improve the randomness of tape orientations and the in-plane isotropy of the material. This stacking process was followed by a final heated de-bulk of the plate (ensuring a more uniform distribution of the resin along the through-the-thickness direction) before final curing; the plates were cured in an autoclave at 6 bars and 120 °C for 1 hour, after an initial 30 min stage of at 75 °C.

Table 1: Mechanical properties of the different fibre types used in the experiments

Fibre type	E^f (GPa)	X^f (MPa)	$\epsilon_{\text{fail}}^{\text{f}}$ (%)	Ref.
HS40 (HM)	425	4640	1.01	[15]
K13916 (UHM)	760	3000	0.40	[15]

Five different plate configurations (described in Table 2 under IDs coded as <fibre type> - <nominal tow thickness> - <number of tows in plate> - <testing direction>) were manufactured and tested in order to independently assess different effects:

Table 2: Configurations of the different manufactured TB-SMC plates

Plate ID	Fibre type	Nominal t^t(μm)	Nominal N_{tows}	Measured thickness (mm)	Measured V_f (%)	Testing direction* (°)
HM-20-800-0	HS40	20	800	0.187 ± 0.021	47 ± 3.7	0
HM-20-800-90	HS40	20	800	0.194 ± 0.029	46 ± 3.5	90
HM-20-1600	HS40	20	1600	0.480 ± 0.040	46 ± 1.7	0
HM-100-800	HS40	100	800	0.902 ± 0.064	54 ± 3.2	0
UHM-100-800	K13916	100	800	0.715 ± 0.075	55 ± 5.0	0

*Testing direction of the specimens relative to the reference working direction.

- In-plane isotropy (HM-20-800-0 and HM-20-800-90): both plates were manufactured with the same specifications, although the specimens of each plate were tested at different orientations (0° and 90°) to the testing direction.

- Effect of the areal tape coverage (HM-20-800 and HM-20-1600): both plates were manufactured with the same fibre type and tape thickness, but with different total number of tows in the plate (N_{tows} = 800 and N_{tows} = 1600, respectively).

- Effect of the tape thickness (HM-20-800 and HM-100-800): both plates were manufactured with the same fibre type and number of tows, but with different tape thicknesses (20 and 100 μm).

- Effect of the fibre stiffness (HM-100-800 and UHM-100-800): both plates were manufactured with the same tape thickness and number of tows, but with different fibre types (HM and UHM).

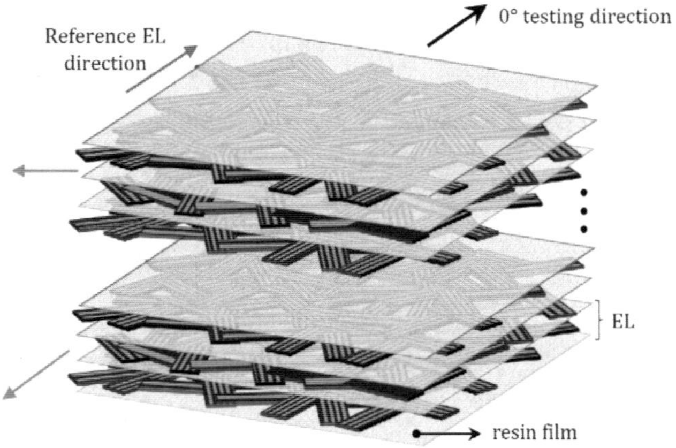

(a) Schematic representation of TB-SMC plate.

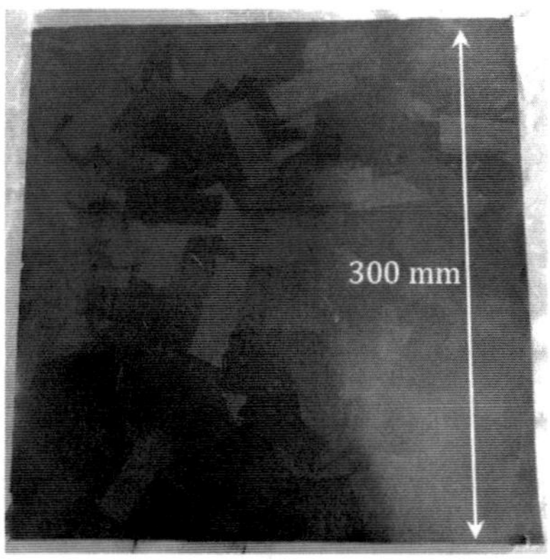

(b) Final manufactured TB-SMC plate

Figure 1: Manufacturing procedure of the TB-SMC plates.

2.2 Testing and analysis

Each plate was cut into 5 rectangular specimens, 250 x 50 mm in length and width respectively, which were then subjected to a standard uni-axial tensile test, with a gauge length of 150 mm (following the ASTM-D3039 standard [16], and with a head displacement of 1 mm/min). The strain in the specimens was measured with an optical strain gauge [17]; the total strain was taken as the average strain measured by three virtual extensometers, each with a strain gauge length of 75 mm, evenly spaced along the specimen's width. Furthermore, the modulus of each specimen was taken as the (best fit) slope of the measured stress-strain curve between 0.05 % and 0.2 % strain. Optical micrographs were used to analyse the resulting microstructure of each plate configuration at three distinct cross-sections; the volume fraction of each cross-section was obtained from processing these three micrographs ([0.15 x 0.40] mm) using simple threshold algorithms. Images of the fracture surface of each specimen were taken, in order to identify the predominant failure mechanism in each plate configuration.

3. RESULTS

3.1 Tensile test results

All modulus and strength results presented in this section have been linearly scaled to a V_f = 57 % (which is a typical value for commercially available TB-SMCs [18]), in order to allow for a direct comparison between different material systems. Figure 2a shows the measured stiffness of the specimens from each plate configuration. The modulus values of the plates made with the HM is relatively constant across the different plate configurations. The plate with the UHM fibres yielded the highest average stiffness value, as expected, although, with a large standard deviation.

Figure 2b presents both the average strength value measured for each plate configuration as well as the standard deviation:

- The highest tensile strength value of 600 MPa was achieved with the thin HS40 tapes, and in particular for the plate configuration with 16 ELs.
- The strength difference between the HM-20-1600 plate and the HM-20-800 is quite significant, with a p-value = 0.06, (since it only differs in the total number of tows).
- The HM-100-800 plate, with the thicker tape configuration, presents a significantly lower average strength value when compared to the thinner tape configurations (HM-20-800), with a p-value = 0.05.
- The UHM-100-800 plate was the one that presented the lowest average strength. This was also the configuration with the most significant variability in the measured results, likely due to the fact that only 3 (out of 5) of the specimens produced valid failures (and only those 3 are included in Figures 2b- 2d).

Figure 2c shows the failure strain of the TB-SMCs (ϵ_{fail}^{TB-SMC}) tested; once again the HM-20-1600 configuration was the one that presented the highest ϵ_{fail}^{TB-SMC} and the UHM-100-800 configuration presented the lowest value. However, by analysing the ratio between the TB-SMC failure strain and the fibre failure strain $\epsilon_{fail}^{TB-SMC}/\epsilon_{fail}^{f}$, as done in Figure 2d one can better quantify the exploitation of the fibre strength within the TB-SMC, for each plate configuration: this suggests that the configuration with the UHM fibres (UHM-100-800) achieved a better exploitation of the fibre strength than the corresponding configuration with HM fibres, due to a more efficient stress transfer mechanism promoted by the stiffer fibres in the former. However, the lower fibre strength of the UHM fibres (3000 MPa) compared to the HM fibres (4640 MPa), is limiting the TB-SMC strength value of the UHM material.

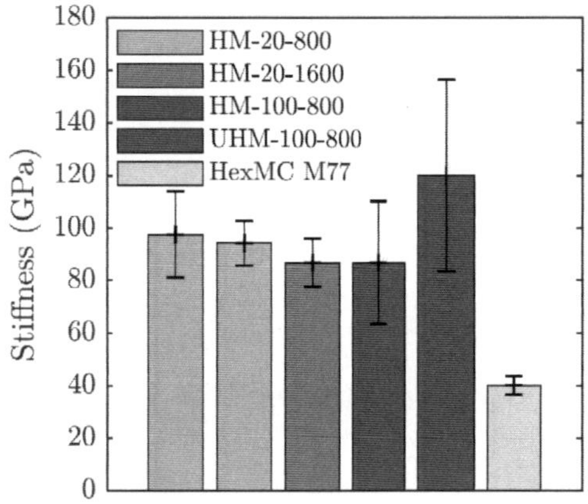

(a) Experimental stiffness values.

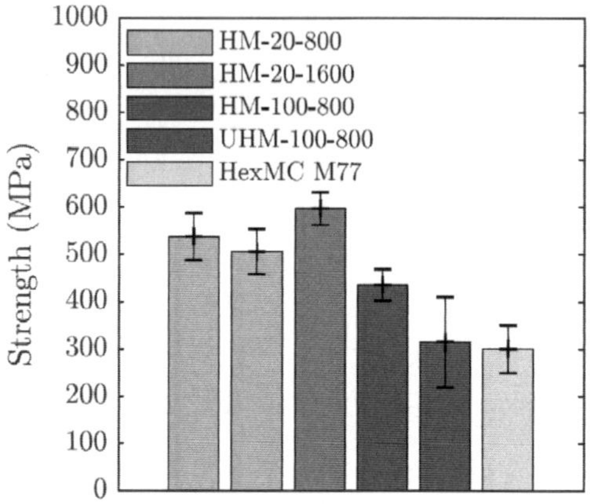

(b) Experimental strength values.

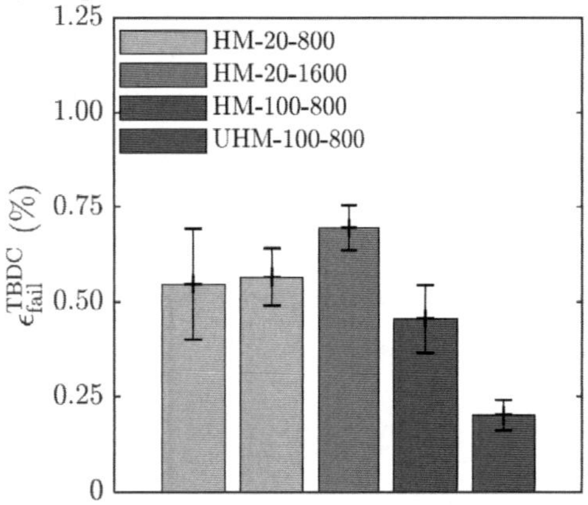

(c) Experimental failure strain of the different material configurations.

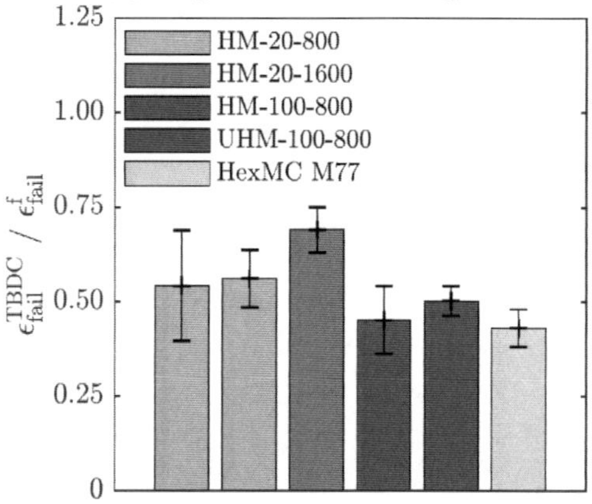

(d) Ratio between the experimental and respective fibre failure strain.

Figure 2: Experimental tensile test results of the different TB-SMC configurations tested (Vf = 57 %), and comparison against the commercially available HexMC-M77 [18]; the error bars represent one standard deviation

3.2 Fracture surfaces

The fracture surfaces of a representative specimen taken from each plate configuration are shown Figure 3. The fracture surfaces for the thin tape configurations (Figures 3a and 3b) show no evidence of tape-pull out, and a clear evidence of tape fracture. For the thicker tape configurations (Figures 3c and 3d), a mixture of tape-pull out and tape fracture was observed, reflecting the less efficient stress transfer that occurs for thicker tape configurations.

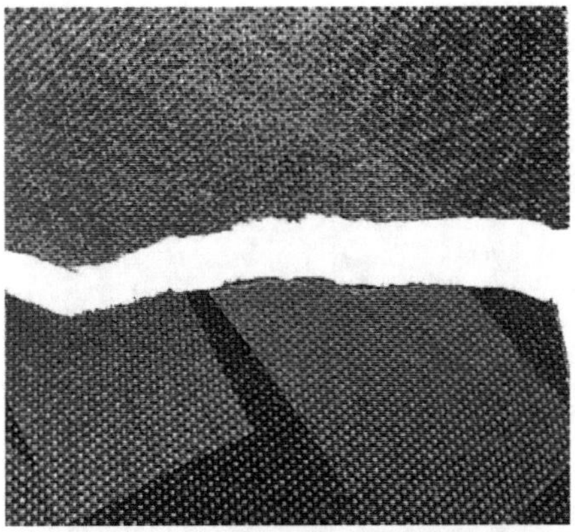

(a) Fracture surface of a HM-20-800 specimen.

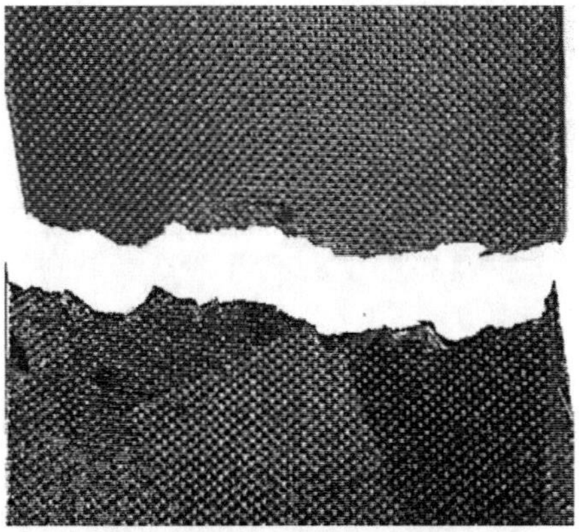

(b) Fracture surface of a HM-20-1600 specimen.

(c) Fracture surface of a HM-100-800 specimen.

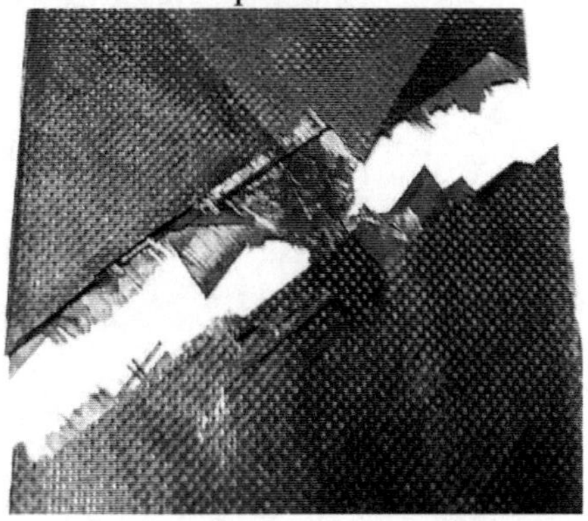

(d) Fracture surface of a UHM-100-800 specimen.

Figure 3: Fracture surfaces of specimens from the different plate configurations.

3.3 Comparison with other material systems

Figure 4 compares micrographs of four different materials systems: Figures 4a and 4b are cross-sections of two commercially available TB-SMC materials systems [18, 19]; the microstructures of these materials are very irregular, with the carbon fibre tows presenting significant through-the-thickness waviness and large resin rich regions. Figure 4c corresponds to a cross-section of a conventional continuous-fibre composite [20], with its conventional regular and well-structured microstructure. Finally, Figure 4d shows the cross-section of a HM-20-1600 specimen developed in this paper; the microstructure of this material system presents very little amount of tape waviness, and is visually more similar to the microstructure of the quasi-isotropic material system in Figure 4c, than to the commercially available TB-SMCs in Figures 4a and 4b.

The mechanical properties of the HM-20-1600 and UHM-100-800 configurations (which are the material systems that performed the best from the ones tested in this work) are compared to other carbon-fibre based material systems specified in Table 3; these include TB-SMC systems presented in the literature [9, 21, 22], the HexMC R-M77 [18] and the AMC R8593 [23] commercially available TB-SMC material systems, and a widely-used aerospace-grade CF/epoxy continuous QI laminate.

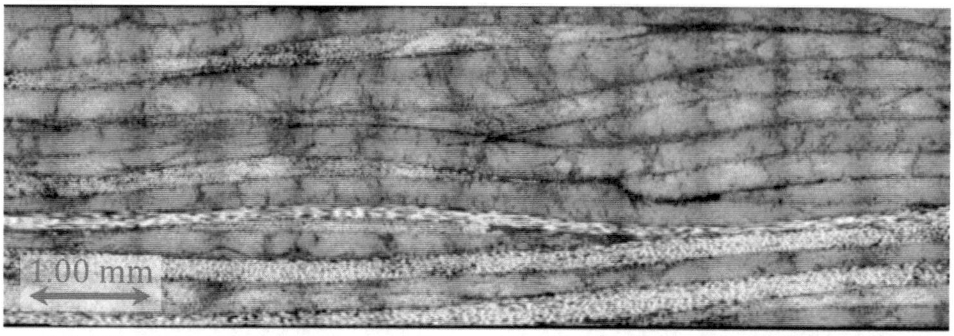

(a) Micrograph of the cross-section of a HexMC-M77 [18] specimen.

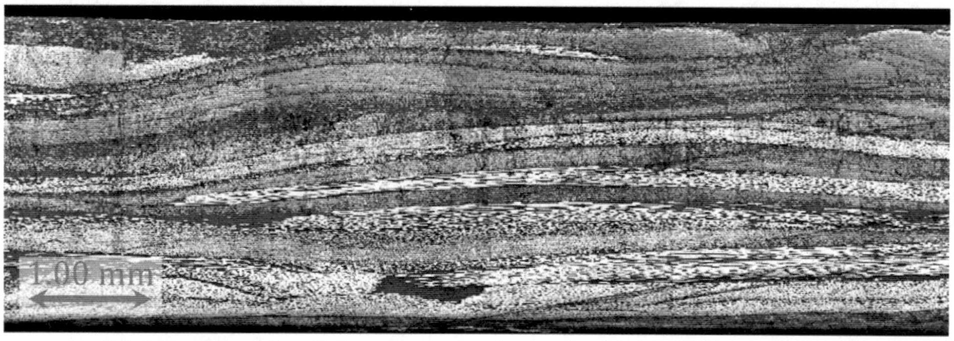

(b) Micrograph of the cross-section of a SMC [19] specimen.

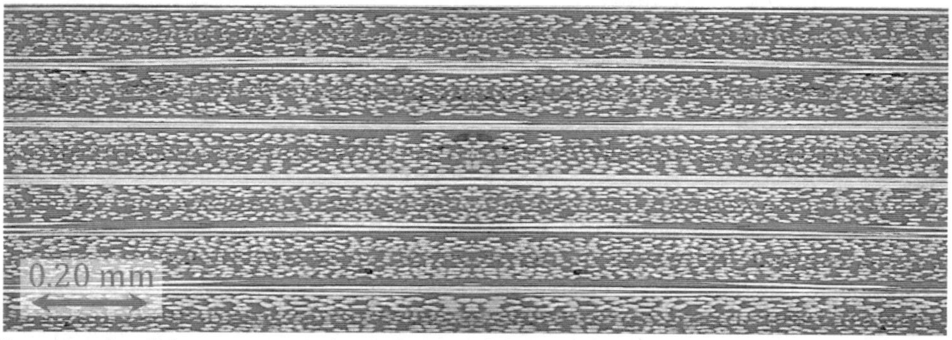

(c) Micrograph of the cross-section of a quasi-isotropic continuous-fibre specimen [20].

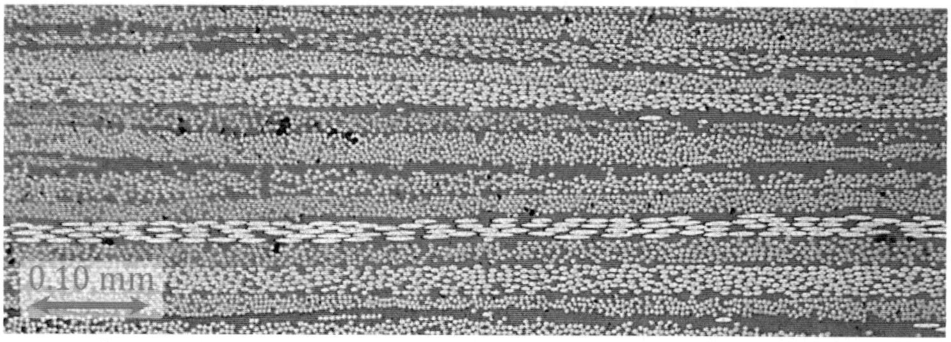

(d) Micrograph of the cross-section of a HM-20-1600 specimen.

Figure 4: Micrographs of cross-sections of different material systems.

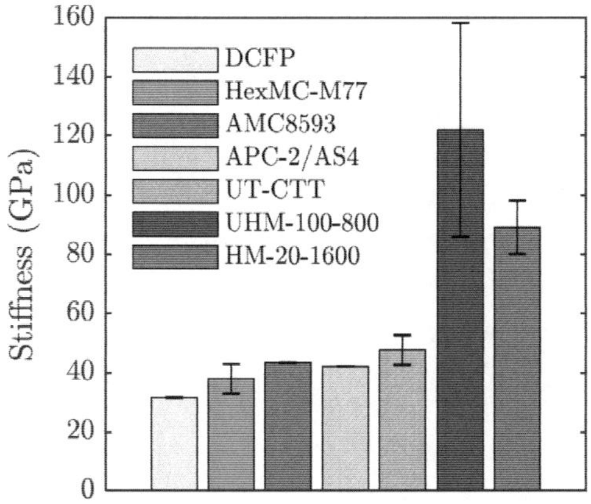

(a) Comparison between the stiffness of different TB-SMCs.

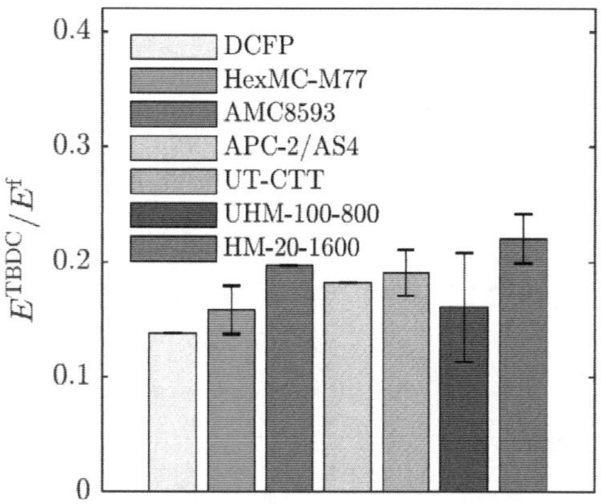

(b) Ratio between the measured stiffness of the TB-SMC and the respective fibre stiffness.

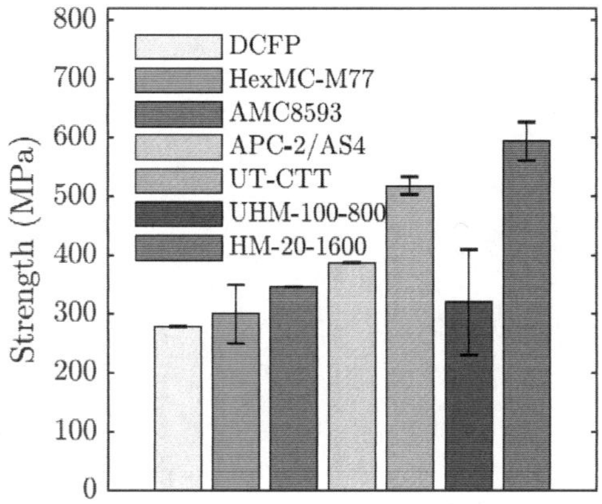

(c) Comparison between the strength of different TB-SMCs.

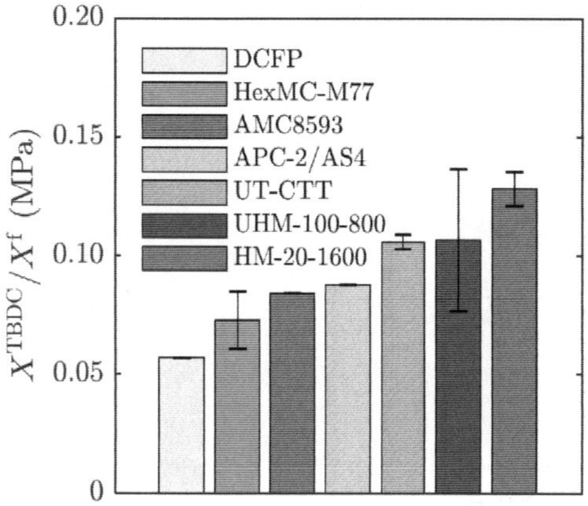

(d) Ratio between the measured strength of the TB-SMC and the respective fibre strength.

Figure 5: Comparison of the mechanical properties between the HM-20-1600 material proposed in this study, and discontinuous material systems reported in the literature (specified in Table 3). All mechanical properties have been linearly normalised to V_f = 57 % for comparison purposes.

Table 3: Configurations of the different CFRP systems compared in Figure 5.

Material ID	Material system description	V_f (%)	Ref.
DCFP	T700 carbon fibre/epoxy, manufactured with modified DCFP	45	[21]
HexMC-M77	Carbon epoxy material system for compression moulding	57	[18]
AMC8593	PAN-based carbon epoxy material system (moulding compound)	47	[23]
APC-2/AS4	Carbon/PEEK CYTEC APC-2/AS4 material system	61	[22]
UT-CTT	Ultra-thin (50 μm) carbon fibre/thermoplastic system	55	[9]
AS4 8552	Continuous fibre quasi-isotropic laminate	60	[24]
UHM-100-800	Thin (100 μm) and UHM carbon fibre/epoxy system	54	
HM-20-1600	Ultra-thin (20 μm) and HM carbon fibre/epoxy system	46	

Figure 5a shows that the HM-20-1600 material system has a significantly higher average stiffness when compared to all other TB-SMC material systems. In Figure 5c the strength comparison of the selected TB-SMC material systems shows that, once again, the HM-20-1600 material system turned out to be the one that presents the highest average strength of all TB-SMC-type materials.

Figures 5b and 5d show how the ratio between the stiffness/strength of the TB-SMC (E^{TB-SMC}/E^{f}) and the fibre stiffness/strength (X^{TB-SMC}/X^{f}) differs for each material system. These results clearly show that the proposed HM-20-1600 configuration is the one that better exploits the fibre properties, due to enhanced stress transfer mechanism enabled by the ultra-thin and stiff tapes.

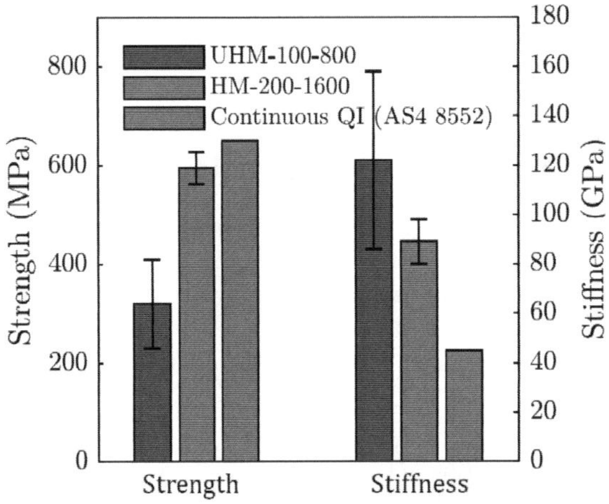

Figure 6: Comparison between the mechanical properties of the HM-20-1600 and UHM-100-800 configurations and a continuous fibre QI laminate (specified in Table 3).

The mechanical properties of the UHM-100-800 and HM-20-1600 configurations are compared to those of a quasi-isotropic laminate of continuous fibres (AS4 8552 [24]) in Figure 6. Remarkably, the strength of HM-20-1600 discontinuous material is practically the same as the one of the continuous fibre laminate. Moreover, due to the high modulus fibres used in both discontinuous configurations, the measured stiffness values are much higher than that of the continuous fibre laminate.

4. DISCUSSION

4.1 Effect of microstructure on mechanical properties

This work shows that, using ultra-thin tapes, TB-SMCs with a more regular microstructure are achieved (compared to conventional TB-SMCs with conventional tows), as shown in Figure 4d. This leads to improvements in stiffness (as previously suggested in the literature [1, 8-9], and corroborated by the slightly higher Young's modulus of the HM-20-800 material compared to the HM-100-800 in Figure 2a) as the tapes present less out-of-plane orientations, and hence lower reductions of in-plane stiffness due to tow/tape waviness [3]. Furthermore, this more regular microstructure is also less prone to weak spots within the material (such as resin pockets) which generally trigger premature failure, and hence also contributes to the significant increase in the ultimate strength of the material, observed in Figure 2b (compare HM-20-800 with HM-100-800).

In addition to the effect on the material's microstructure, the use of thin and stiff tapes also leads a more efficient stress transfer between the tows, and hence to a better exploitation of the mechanical properties of the fibres, as shown by the HM-20-1600 and UHM-100-800 configurations in Figure 2d.

4.2 Improvement over existing materials

4.2.1 Stiffness

The stiffness improvement achieved with the ultra-thin and stiff tape based TB-SMCs is essentially due to the higher stiffness of the HM and UHM fibres. However, there are small differences between the stiffness of the HM-20-800 and HM-100-800 plate configurations, with the later presenting lower stiffness values (see Figure 2a). This may be due to the wavier disposition of the thicker tows in comparison to the thinner tows, leading to higher stiffness knockdowns, which is in agreement with the numerical predictions done by Li and Pimenta [5].

4.2.2 Tensile strength

The average strength value of 600 MPa, obtained with the HM-20-1600 plate configuration (see Figure 2b), represents an increase of 100 % when compared to the strength of the commercially available materials and of ≈ 20 % when compared to the UT-CTT material [9], which was the TB-SMC system with the highest strength previously reported in the literature. The enhancement of the stress-transfer mechanisms enabled by the use of ultra-thin and stiff tapes delays the on-set of tow pull-out (governed by toughness criterion) and promotes failure by tape fracture, as evidenced by the fracture surfaces of Figures 4a and 4b.

The comparison between the mechanical properties of the best proposed configurations (HM-20-1600 and UHM-100-800) against those of a continuous QI laminate in Figure 6, shows that is possible to match the strength of a continuous fibre laminate and surpass its stiffness using a composite with a discontinuous microstructure. This highlights the great potential that this new class of high-performance discontinuous composites offer, as they are able to bridge the gap in mechanical properties that typically exists between continuous and discontinuous fibre composites, while still retaining the advantages of design flexibility and manufacturability of discontinuous composites, as discussed in the next section.

4.2.3 Design flexibility and manufacturability

It has been suggested in the literature [1, 8] that the strength of TB-SMCs increases with plate thickness (for a constant tow thickness); this is corroborated by the slightly higher strength and failure strain of the HM-20-1600 configuration when compared to the HM-20-800 configuration (see Figures 2b, 2c and 2d). This effect can be explained by two factors: firstly, thicker specimens have a smaller fraction of tows/tapes at the surface of the specimen, and these surface tows/tapes cannot transfer stresses as effectively as those well-embedded in the through-the-thickness direction; secondly, thicker specimens have a larger average number of tows in the through-the-thickness direction, which increases the homogeneity of mechanical properties in comparison with thinner specimens [3]. Consequently, there is a limitation for the minimum thickness that can be accepted in components manufactured with TB-SMC systems with conventional tow thicknesses. This limitation can be relaxed significantly through the use of ultra-thin tapes, as shown by the results in Figures 2 and 5:

- Firstly, the mechanical properties of the plates manufactured with ultra-thin tapes were considerably higher than those of conventional TB-SMCs (see Figures 2 and 5), even though the plate thickness of the former was below 0.5 mm.

- Secondly, the similar properties observed with the two HM-20-800 plates tested at different directions (see Figure 2) suggest that using ultra-thin tapes to make TB-SMCs enables the manufacture of shell-type structures, with wall thicknesses lower than 0.2 mm, with in-plane isotropy.

Consequently, TB-SMCs manufactured with ultra-thin tapes offer much more design flexibility than conventional TB-SMCs, and allow for production of shell-type structures with very thin walls (which is useful e.g. for body-in-white parts of cars). Figure 6 shows that TB-SMCs made of ultra-thin tapes of HM fibres combine (i) a similar strength and even higher stiffness than those of aerospace-graded composites with (ii) the improved manufacturability and design flexibility of a discontinuous and random microstructure. The latter can be realised by creating SMC-type intermediate materials (similar to HexMC [18] and carbon-SMC [19]) but using ultra-thin (spread) tapes of (ultra-)high modulus fibres and fast-curing matrices; these intermediate materials could then be directly used in compression moulding, with several advantages over conventional continuous-fibre composites:

- Allowing for the production of parts within a matter of minutes and out-of-the-autoclave, and with minimal manual work;

- Allowing for the production of complex 3D geometries (rather than 2D shells) with local embedded reinforcements where they are needed most (rather than relying on ply-drops and bonded stiffeners, as often done with continuous-fibre composites).

- Possibility of using UHM fibres to manufacture parts/components with complex geometries; this is enabled by the discontinuous architecture of TB-SMCs, and is not easily achievable with continuous-fibre architectures.

- Reducing significantly the amount of waste generated by material cut-offs.

5. CONCLUSIONS

This work has successfully proposed and demonstrated a pathway to increase the stiffness and ultimate tensile strength of tow-based discontinuous composites. The main conclusions are summarised below:

- Quasi-isotropic TB-SMCs manufactured with ultra-thin (20 μm) and high-modulus (450 GPa) carbon-fibre tapes reached a tensile strength of 600 MPa and stiffness of 90 GPa for a V_f of 57 %. This represents a strength increase of 100 % when compared to the strength values of commercially available TB-SMC materials, and of 20 % when compared to the TB-SMC material system with the highest strength previously reported in the literature.

- This work demonstrates that TB-SMCs manufactured using ultra-thin and high-modulus carbon-fibre tapes can reach the tensile strength of conventional aerospace-grade QI laminated CFRP, and overcome their stiffness.

- The use of thin and stiff tapes in the manufacture of TB-SMCs leads to this significant improvement in mechanical properties by enhancing the stress transfer mechanisms within the material and suppressing matrix/interface dominated failure mechanisms. This allows for a better exploitation of the mechanical properties of the fibres, which has been demonstrated by the features of the fracture surfaces of these TB-SMCs, which were dominated by fibre-failure rather than by tape pull-out.

- Test results suggest that there is a reduction in tensile strength for specimens with a small ratio between the specimen thickness and tape thickness. This indicates that TB-SMCs with ultra-thin tapes can be used to manufacture components with very thin walls, removing several design barriers and offering more design flexibility over conventional TB-SMCs (based on tows with conventional thicknesses).

This work shows the potential of ultra-thin and stiff TB-SMCs as a high-performance and highly-manufacturable material, able to comply with the stiffness and strength requirements of primary structural applications and suitable for a wider range of part geometries and more automated manufacturing processes when compared to continuous-fibre composites. These materials can potentially change the way in which composite materials are used in several industries (automotive, aerospace, sport and others), and at the same time open the use of high-performance composite materials to new industries with different manufacturability constraints.

Acknowledgements

The research leading to these results has been done within the framework of the FiBreMoD project and has received funding from the European Union's Horizon 2020 research and innovation programme under the Marie Sklodowska-Curie grant agreement no. 722626. S. Pimenta also acknowledges the support from the Royal Academy of Engineering in the scope of her Research Fellowship on `Multiscale discontinuous composites for large scale and sustainable structural applications' (2015-2019). L. E. Asp acknowledges the support from VINNOVA via the LIGHTer Academy.

6. REFERENCES

[1] Li, Y., Pimenta, S., Singgih, J., Nothdurfter, S., and Schuffenhauer, K., "Experimental investigation of randomly-oriented tow-based discontinuous composites and their equivalent laminates", *Composites Part A: Applied Science and Manufacturing*, vol. 102 (2017): 64-75.

[2] Pimenta, S., Ahuja, A. and Yong, A., "Damage tolerant tow-based discontinuous composites", *ICCM20*, Copenhagen, Denmark, July 2015.

[3] Alves, M., and Pimenta, S., "Analysing the effect of fibre waviness on the stiffness of tow-based discontinuous composites", *ECCM18*, Athens, Greece, June 2018.

[4] Li, Y., and Pimenta, S., "Development and assessment of modelling strategies to predict failure in tow-based discontinuous composites", *Composite Structures*, vol. 209 (2019): 1005-1021.

[5] Alves, M., Carlstedt, D., Ohlsson, F., Asp, L. E., and Pimenta, S., "Ultra-strong and stiff randomly-oriented discontinuous composites: closing the gap to quasi-isotropic continuous-fibre laminates", *Composite Part A: Applied Science and Manufacturing [under review]*

[6] Harper, L. T., "Discontinuous carbon fibre composites for automotive applications". University of Nottingham, Ph.D. thesis, 2006.

[7] Selezneva, M. and Lessard L., "Characterization of mechanical properties of randomly oriented strand thermoplastic composites", *Journal of Composite Materials*, vol. 50, no. 20 (2016): 2833-2851.

[8] Feraboli, P., Peitso, E., Deleo, F., Cleveland, T. and Stickler, P. B., "Characterization of Prepreg-Based Discontinuous Carbon Fiber/Epoxy Systems", *Journal of Reinforced Plastics and Composites*, vol. 28, no. 10 (2009): 1191-1214.

[9] Wan, Y. and Takahashi, J., "Tensile properties and aspect ratio simulation of transversely isotropic discontinuous carbon fiber reinforced thermoplastics", *Composites Science and Technology*, vol. 137 (2016): 167-176.

[10] Darvell, M., "Randomized CFRTP tape piece laminates in structural automotive applications". KTH Royal Institute of Technology, MSc Thesis, 2015.

[11] Visweswaraiah, S. B., Selezneva, M., Lessard, L. and Hubert, P., "Mechanical characterization and modelling of randomly oriented strand architecture and their hybrids - A general review", *Journal of Reinforced Plastics and Composites*, vol. 37, no. 8 (2018): 548-580.

[12] *ASDNews*. "Nordam Pioneers Hexmc Composite Window Frames for Boeing 787 Dreamliner", 2007.

[13] Roy, S., Yousefpour, A., Bednar, F., and Beaulieu, P. "Compression Molding of Composite Tailboom Frames", *70th American Helicopter Society International Annual Forum*, Vol. 3, Montreal, May 2014.

[14] Pimenta S., and Robinson, P., "An analytical shear-lag model for composites with 'brick-and-mortar' architecture considering non-linear matrix response and failure", *Composites Science and Technology*, vol. 104 (2014): 111-124.

[15] Mitsubishi Chemical Carbon Fiber and Composites. 2019. Accessed 03 Oct. 2019 <http://mccfc.com>.

[16] ASTM D3039, "Standard test method for tensile properties of polymer matrix composite materials" ASTM International, West Conshohocken, PA, 2002.

[17] "IMetrum, iMetrum LTD, Non Contact Precision Measurement. Simulia, 2018".

[18] Hexcel, "HexMC User Guide." 2014. Accessed 04 Apr. 2019 <https://www.hexcel.com/user_area/content_media/raw/HexMC_UserGuide.pdf>

[19] Mitsubishi Chemical Corporation, "Pyrofil STR120 datasheet."

[20] Mencattelli, L., and Pinho, S. T., "Ultra-thin-ply cfrp bouligand bio-inspired structures with enhanced load-bearing capacity, delayed catastrophic failure and high energy dissipation capability", *Composites Part A: Applied Science and Manufacturing* (2019): 105655.

[21] Harper, L., Burn, D., Johnson, M., and Warrior, N., "Long discontinuous carbon fibre/polypropylene composites for high volume structural applications", *Journal of Composite Materials*, vol. 52, no. 9 (2018): 1155-1170.

[22] Selezneva, M., Kouwonou, K., Lessard, L. and Hubert, P., "Mechanical properties of randomly oriented strands thermoplastic composites", *ICCM19*, Montreal, Canada, July 2018.

[23] Q. Composites, "AMC8590 Technical Data Sheet." 2014. Accessed 08 Apr. 2019 <https://www.quantumcomposites.com/pdf/datasheets/amc/Quantum-AMC-8590-20126-76-8.pdf>

[24] Feraboli, P., Peitso, E., Deleo, F., Cleveland, T. and Stickler, P. B., "Characterization of prepreg-based discontinuous carbon fiber/epoxy systems", *Journal of Reinforced Plastics and Composites*, vol. 28, no. 10 (2009): 1191-1214.

ADVANCED PHTHALONITRILE RESIN SYSTEMS FOR VACUUM PROCESSING

Boris A. Bulgakov, Alexandr V. Babkin, Alexey V. Kepman, Viktor V. Avdeev

M.V. Lomonosov Moscow State University
Leninskie gory, 1-11
Moscow, 119991

JSC Institute of New Carbon Materials and Technologies

Leninskie gory, 1-11
Moscow, 119991

ABSTRACT

Advanced phthalonitrile resin systems based on recently introduced low-melting phosphorus-containing monomers were developed. A new class of bis-benzonitirle reactive plastisizers allowed to obtain resin systems possessing improved processability (melt viscosity < 100 mPa·s at temperatures down to 120 °C) suitable for manufacturing composites by vacuum infusion molding process or RTM at the lowest reported processing temperature of 120 °C on carbon fabric (3K HTA40, twill). The curing cycle was adjusted to achieve the best mechanical properties of the composites reaching 852 MPa in compressive strength. The influence of post-curing conditions on the resulting composites properties was established. After post-curing at 375 °C the obtained composites demonstrated up to 90% mechanical properties retention at 400 °C (τ_{12} = 60-80 MPa). Solvent-free phthalonitrile prepregs were developed for the first time. Prepreg consolidation yielded high-quality composites. Flame retardant properties of investigated composites demonstrated extremely high LOI > 80%. Thus easy-processable phthalonitriles for out-of-autoclave manufacturing techniques with convenient curing conditions and excellent heat resistant and flame-retardant properties were developed.

1. INTRODUCTION

Fiber reinforced plastics (FRP) gradually replace common materials such as metals and alloys in aerospace industries. Modern aircrafts consist of polymer composites for more than 50% and more new materials and techniques are developing to take their place in the industry. Contemporary trends in composites manufacturing are aimed to use cost-effective resin injection methods instead of autoclave processing. Thus, in emergent Russian aircraft MS-21 wingbox is constructed from 25 meters long single CFRP parts produced by vacuum infusion. The most demanded technology for automotive industry nowadays is high-pressure RTM, which in combination with microwave heating can reduce production cycle down to 20 mins per part [1]. At the same time new materials are required to extend application range of the composites. One of the limitations of the polymer composites is quite low operation temperatures caused by the nature of matrices. Even typical heat-resistant matrices based on bis-maleimides [2,3], polyimides [4,5], benzoxazines [6,7], cyanate ether [8,9] can be use at temperatures up to 250-300 °C.

Polymer matrices based on phthalonitriles are known for their resistance to high temperatures (T_g > 400 °C, $T_{5\%}$ > 500 °C) at the same time for poor processability due to which the only possible way to manufacture composites was to use prepregs obtained from the resin solution [10–15]. The development of low-melting monomers and oligomers was a long-term trend in a field of phthalonitrile resins [16–18]. In the case of oligomeric bis-phthalonitriles a good processability suitable for cost-effective resin injection methods of composite manufacturing was reached but at the same time a significant decrease in Young's modulus of the thermosets was observed [16,19]. With the development of siloxane- and phosphate-bridged bis-phthalonitriles [20,21] a manufacturing of carbon fiber reinforced plastics (CFRP) by cost-effective technologies such as VIMP and RTM (impregnation at 150 °C) was reported [22,23]. However, the reported composites demonstrated moderate mechanical properties which retained at 300 °C for 70-80%. Thus, an improvement of these materials was still demanded.

In the current report we introduce new bis-benzonitrile reactive diluents for phthalonitrile resins and describe resin formulations based on the phosphorus-containing monomers. Using VIMP at 120 °C and hot-pressing prepreg consolidation CFRP samples were manufactured and characterized in detail.

These results can widely extend the application range for composite materials in aerospace (for production of complex-shaped parts, e.g. jet engine blades or skin of ultrasonic vehicles) and other applications such as high-temperature composite tooling for thermoplastics processing, flame-retardant materials for interior parts of submarines, engine cowls, exhaust pipes etc.

2. EXPERIMENTATION

2.1 Materials and methods

All the monomers were obtained according to the procedures reported earlier. 1,3-bis (3,4-dicyanophenoxy)benzene (**5**) was synthesized according to the know procedure [24] with quantitative yield (98 %). Bis(3-(3,4-dicyanophenoxy)phenyl) phenyl phosphate was synthesized following the procedure described in [21]. Bis(4-cyanophenyl) phenyl phosphate was obtained as described in [25]. 3,3'-(1,3-phenylenbis(oxy))dianiline (APB) was purchased from SynAsia and used as received. Quartz flour Silverbond 30 and was purchased from Sibelco as used as it is. Carbon fabric 22502 from JSC INCMaT 2x2 twill wave from 3k HTA40 TohoTenax from Teijin fiber was used for CFRP preparation. Auxiliary materials for vacuum bagging were purchased from Airtech.

Differential scanning calorimetry (DSC) was performed on Netzsch DSC 204 Phoenix, at a heating rate of 10°C/min and an Ar purge rate of 50 mL/min and was applied for Tg of the uncured resins determination and curing study. Thermal stability was evaluated by TGA on Netzsch TG 209 P3 Tarsus, at a heating rate of 10°C/min in range 40–980 °C and an argon or air purge rate of 50 mL/min. Heat deflection temperature (HDT) was measured by the three-point bending method according to ASTM-E2092-03 on Netzsch TMA 402 with 1.82 MPa loading. DMA was performed using TA Instruments DMA Q800 in three-point bending regime with 1 Hz frequency and amplitude 40 μm with heating rate of 5 K/min. Mechanical tests were carried out using Hounsfield H100KS, H5KS and Instron 5985 testing machine. The reported mechanical property values were based on an average of at least five tests. Determination of flexural mechanical properties was carried out using the three-point bending method according to the ASTM D790 at 1.28 mm/min crosshead rate

2.2 Curing

Curing of phthalonitrile resins for primary properties estimation was performed in a glass vials in a glovebox under atmosphere of Ar. The mixture was placed into a vial, melted and degassed at 120–180 °C by stirring under vacuum. Then the temperature was raised to 180 °C at the rate of 2 °C/min and kept for 6 hours. Then the temperature was raised up to 330 °C or 375 °C (10 K/h) and the samples were post-cured for 8h.

Samples for mechanicals tests were manufactured as follows: resin **PNT** (500 g) was placed into a 2 L glass reactor, then melted and degassed by mechanical stirring (200 rpm) under vacuum at 150 °C for 15 min. The melt was then poured into a metal mold, which was placed into an air circulated heated oven and cured for 6 hours at 180 °C. After that, the mold was disassembled, and the cured laminate was cut in two equal parts. One part was studied in cured state, the other was post-cured at up to 300-375 °C as described above.

2.3 Vacuum infusion processing

Stainless steel plate with dimensions 500×700×4 mm^3 coated with release agent (except for the edges) was used to produce CFRP by vacuum infusion process. Sealing tape was secured along the perimeter. Silicone rubber tubes for resin feed and evacuating channels were installed on the opposite edges of the plate. At the side of the feed channel 10 layers of 200 g/m^2 carbon fabric twill 2x2 (toho tenax HTA40) were placed and covered with peel ply and distribution medium. Further, vacuum film was attached to the sealing tape, the vacuum bag was placed in the oven, connected to the vacuum pump and heated to 120 °C. Upon reaching 10 mm. Hg inside the vacuum bag, degassed resin preheated to 120 °C was infused via feed channel. After 10–15 minutes the resin exited the vacuum bag into the vacuum channel, which indicated the end of impregnation. After the completion of the infusion process, the temperature was raised to 180 °C at the rate of 1 °C/min and held for 6 hours. Then the package was disassembled, and the cured sample was post-cured at 330-375 °C as described above.

2.4 PNT-P preparation

To prepare prepreg resin **PNT-P**, 25 kg of **PNT** was placed into a 100 l metal reactor, melted at 120 °C under vacuum (<1 mmHg) and stirring. Then 10 kg of quartz flour were mixed with the resin by adding a 2500 g parts of the flour followed by vigorous stirring under vacuum for 15 minutes. After mixing of each portion stirrer was stopped and vacuum was released to add the next portion of the flour. This procedure was repeated four times with final stirring for 30 mins under vacuum. After chilling down, 35 kg of light-green paste was obtained.

2.5 Prepreg fabrication and consolidation

To fabricate prepreg **PNT-P1** was placed on a carbon or glass fabric piece 30 × 30 cm and spread uniformly on its surface by hot ironing. Then 10 layers of carbon prepreg(8 layers of glass prepreg) were plied on a flat metal mold and sealed in a vacuum bag. Pre-form was placed into a heating press and plates temperature was increased up to 140 °C with heating rate 2 °C per minute. Than pressure of 8.75 bar was applied to a sample and temperature was raised up to 180 °C with heating rate 2 °C per minute. When the temperature reached 180 °C the applied pressure was increased to 20 bar and the sample was treated this way for 6 hours following with cooling down the apparatus down to 50 °C with rate 5 °C/min. After that the mold was disassembled and the sample post-cured in a free state according to procedure mentioned above with top temperature up to 330 °C for 8 hours.

3. RESULTS

In our recent works we reported synthesis of low-melting phosphate-bridged pthhalonitrile monomers featured by increased thermo-oxidative stability [21]. Monomer **1** (Figure 1) was chosen among the phosphorus-containing as the most prospective for further applications due to the low T_g = 22 °C and optimal synthesis procedure. Monomer **1** (Figure 1) cured in the presence of 4 mol. % of 4,4′-(1,3-Phenylenedioxy)dianiline (APB) with the final temperature step of 375 °C for 8 hours possessed thermal properties at the same level as the common phthalonitrile monomer **2** (Figure 1) described earlier [24] and cured in the same conditions as **1**.

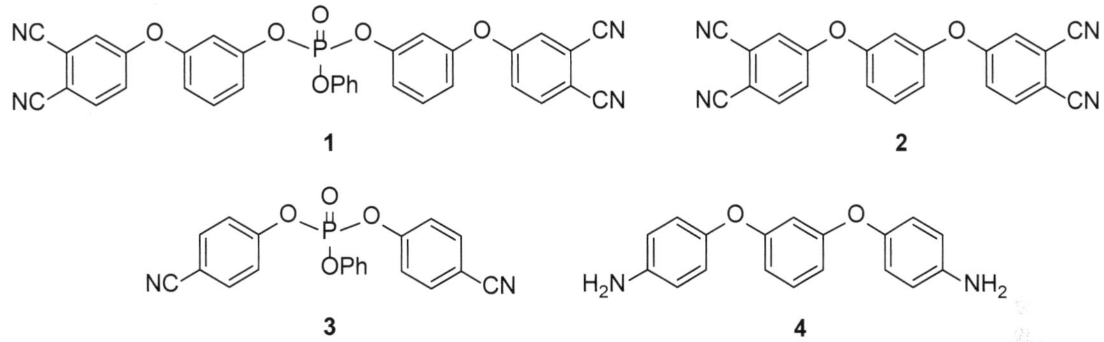

Figure 1. The monomers comprising PNT formulation

It is notable that the cured matrix obtained from **1** monomer demonstrated increased thermo-oxidative stability in comparison to cured monomer **2**. This effect could be explained by the presence of a phosphate group in the molecular structure of **1**. In the experimental conditions TGA the sample of cured **1** retained almost 30 % of the initial mass at 1000 °C in contrast to the sample of cured **2** which totally combusted at 800 °C (Figure 1).

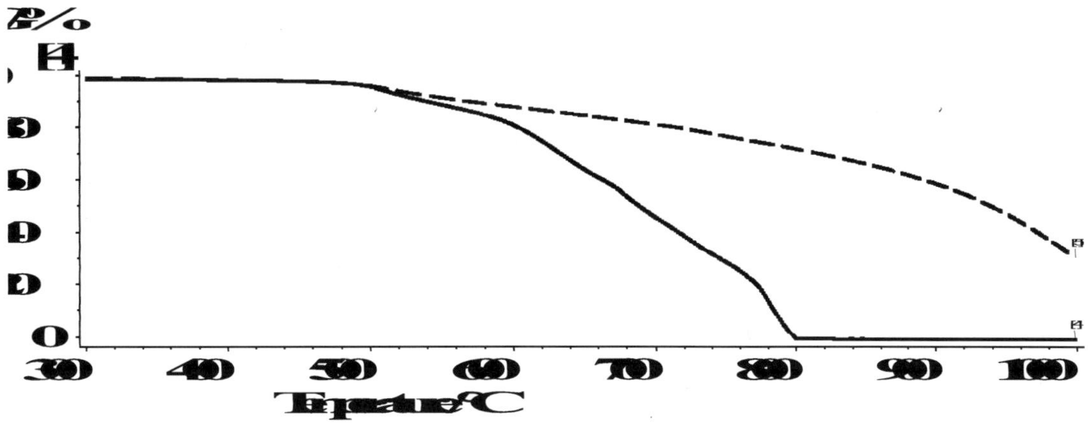

Figure 2. TGA curves of the cured monomers **1** (dashes) and **2** (solid) in air.

Using monomers **1** and **2** in combination with reactive diluent **3** (Figure 1) recently synthesized in our lab [25] and diamine **4** (APB) as curing agent, a formulation named **PNT** was developed.

Blended resin **PNT** demonstrated melt viscosity 1000 mPa·s 100 °C with plateau at temperatures higher than 150 °C with viscosity value 10 mPa·s (Figure 3). Isothermal experiments showed good processing time for at 120 °C enough for CFRP manufacturing (>24 hours) and quick gelation at 180–200 °C (30-60 min). Such properties allow to consider resin **PNT** for CFRP manufacturing by cost-effective injection techniques such as VIMP or RTM.

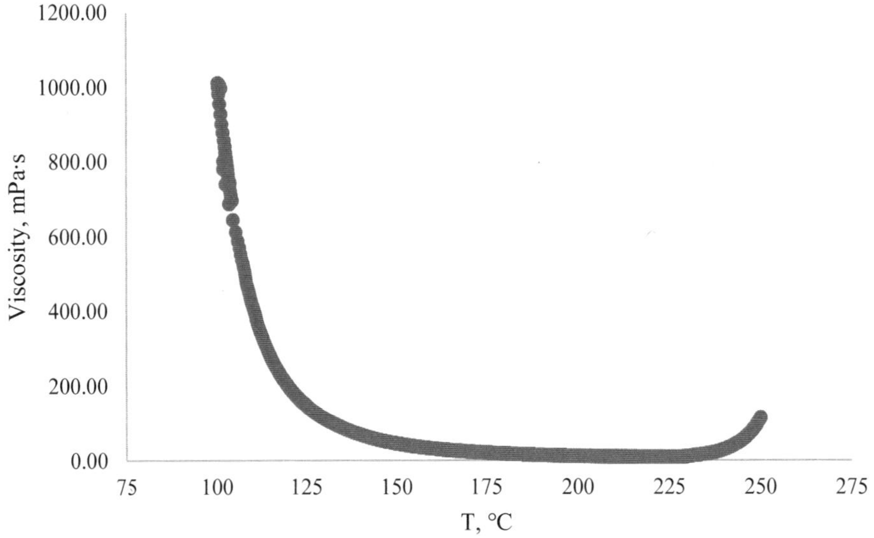

Figure 3. Viscosity profile of **PNT**

Curing of **PNT** was performed with a two-step heating program. Based on the isothermal viscosity curves the first step was curing at 180 °C with HDT determination by TMA in penetration mode after certain time periods. Temperature of 180 °C was chosen due to respectively quick gelation of the resin at this temperature and possibility of using standard auxiliary materials such as polymeric films and sealing tape for vacuum infusion processing. It was revealed that after 3 hours at 180 °C the resin demonstrated HDT = 167 °C (determined by DMA) and 198 °C after 6 hours. Thus, the latter conditions were picked for the first step of curing because HDT higher than curing temperature was reached. The second step was heating up to 330-375 °C with heating rate 10 °C/h and holding the temperature at final step for 8 hours.

Cured resin plates were obtained to estimate mechanical properties of the matrix. The results are presented in Table 2. It is seen that mechanical properties of the matrix post-cured at 375 °C are significantly lower than for post-cured at 330 °C which can be explained by the shrinkage causing strong inner stress in the thermoset. At the same time post-curing increases Young's modulus of the thermosets and it is important to note that the key point in this case is that the full-cycle cured matrix retained high Young's modulus at increased temperatures. No phase transition was detected on the DMA curve for the sample post-cured at 375 °C. Storage modulus of this thermoset E' = 1 GPa retained even at 450 °C indicating a possibility of CFRP maintenance at this temperature. One of the important properties for high-temperature materials is a low coefficient of thermal expansion (CTE): post-cured samples demonstrated CTE = 39-63 10^{-6} K^{-1}.

Table 1. Thermal properties of thermosets derived from **PNT**

Parameter	**Curing conditions**		
	6 h, 180 °C	**8 h, 330 °C**	**8 h, 375 °C**
T_g, °C	198	347	>450
Tensile strength, MPa	59	43	–
Tensile modulus, GPa	4.2	4.8	–
Flexural strength, MPa	104	95	55
Flexural modulus, GPa	4.6	5.1	4.9
CTE, 10^{-6} K^{-1}	63	50	39
Density, g/cm^3	1.316	1.324	1.386

3.1 CFRP manufacturing and characterization

Impregnation was performed in a programmable oven equipped with outlets for a vacuum pump at 120 °C and took about 30 minutes. After the resin appeared in outlet tube outside the oven it was sealed and heated for 6 hours at 180 °C. The sample was disassembled and post-cured at 330-375 °C resulting in heat-resistant CFRP formation with the matrix content of 30–32 %.

Mechanical characteristics of CFRP are shown in Table 3. It is seen, that curing at 330°C results in composite possessing higher mechanical properties than the one cured at 375 °C. The reason of this phenomenon is high shrinkage after curing at 375 °C which causes cracks formation in the composite (observed by SEM).

Table 2. Mechanical properties of the composites obtained by vacuum infusion

Parameter		Test method	Post-curing conditions	
			8 h, 330 °C	**8 h, 375 °C**
Tensile strength	σ_{11}^{+}, MPa	ASTM D3039	834	685
	E, GPa	ASTM D3039	60	61
Compressive strength	σ_{11}^{-}, MPa	ASTM D6641	852	538
	E, GPa	ASTM D3039	67	60
Interlaminar shear strength	τ_{13}, MPa	ASTM D2344	53	44
In plane shear strength	τ_{12}, MPa	D3579	106	75
	E, GPa	D3579	6.0	5.7

Both obtained CFRP samples were tested for interlaminar shear strength at elevated temperatures up to 450 °C (Table 4). At 450 °C an observed shear strength was τ_{13} = 38 MPa, demonstrating retention of mechanical properties 86% of the value at room temperature.

Table 3. Mechanical properties of the composites

T, °C	Post-curing conditions			
	8 h, 330 °C		8 h, 375 °C	
	τ_{13}, MPa	τ_{12}, MPa	τ_{13}, MPa	τ_{12}, MPa
300	50	95	43	75
350	46	88	41	69
400	–	–	38	68

3.2 Inflammability tests

Samples of the cured matrix and CFRP were tested for LOI according to the Russian standard GOST- 21793-76. Both materials showed high LOI values in comparison to the known materials. For matrix LOI was 63% and for CFRP more than 80% (the experiment was stopped due to the safety reasons) which are the highest values among the reported for any plastics.

Table 4. Flammability of CFRP with matrices of different types

Sample	**LOI, %**
PNT	63
CFRP (PNT based, VI)	>80

3.3 Prepreg fabrication and consolidation.

Resin **PNT** possessed T_g = 21°C and thus was suggested suitable for prepreg fabrication since at room temperature it was ductile and allowed to roll the prepreg. In order to reduce material cost and decrease the shrinkage effects and keep thermal performance, 40% of quartz flour was added to **PNT** to form resin assigned **PNT-P** as white green paste. The paste was spread on carbon fabric sheets by hot ironing and then 10-layer thick CFRP samples containing 42-45% of the matrix and 8-layer GFRP containing 37-40% of the matrix were obtained by hot-pressing processing. ILSS data depending on the curing conditions is given in Table 5. It is seen that reduction of the curing temperature results in increase of mechanical properties of the composites.

Table 5. ILSS values collected for prepreg originated CFRPs

T, °C	Vacuum infusion		CF, prepreg consolidation, vacuum	CF, prepreg consolidation, hot-pressing		GF, prepreg consolidation, hot-pressing
	8 h, 330 °C	8 h, 375 °C	8 h, 330 °C	8 h, 300 °C	8 h, 330 °C	8 h, 330 °C
25	53	44	41	50	41	50.7
300	50	43	35	44	41	38.0
350	–	41	31	–	33	27.7
400	–	38	21	–	26	–

4. CONCLUSIONS

Application of new class of reactive plasticizers allowed to obtain low-melting resin formulation **PNT** perfectly suitable for cost-effective injection methods of FRP manufacturing and for the first time for solvent-free prepreg obtaining. Cured resin PNT demonstrated attractive mechanical and thermal properties that allowed considering it as matrix for high-temperature CFRP. Due to epoxy-like processing, it was possible to develop convenient method of CFRP formation by vacuum infusion molding process. Composites demonstrated high mechanical properties retained for up to 90 % at 450 °C and exceptional fire retardant properties due to the use of phosphorus-containing monomers. Therefore, the reported CFRP could substitute aluminum alloys or titanium in the parts of aircrafts subjected to high temperatures, space ships and other fields of industry for production of large-scaled and complex-shaped parts operating at elevated temperatures

First experiments of solvent-free prepreg fabrication and consolidation showed the possibility to apply this wide-spread technique for phthalonitrile resins. At the same time, more data should be collected for approval of the technology. Additional information on the topic will be presented in the conference.

ACKNOWLEDGMENTS

The work was supported by Government of the Russian Federation Ministry of Science and Higher Education. Contract No. 14.607.21.0204 (RFMEFI60718X0204)

5. REFERENCES

[1] J. McHugh, F. Grasse, CHARACTERIZATION OF RELEASE AGENTS USED IN RESIN TRANSFER MOULD-ING (RTM) AND LIQUID COMPRESSION MOULDING (LCM) PROCESSES, in: Proc. THERMOSETTING RESINS 2018 Conf. 25 – 27 Sept. 2018, BERLIN, Ger., 2018: p. 117.

[2] S.E. Evsyukov, T. Pohlmann, H.D. Stenzenberger, m-Xylylene bismaleimide: a versatile building block for high-performance thermosets, Polym. Adv. Technol. 26 (2015) 574–580. doi:10.1002/pat.3488.

[3] A. V. Babkin, E.M. Erdni-Goryaev, A. V. Solopchenko, A. V. Kepman, V. V. Avdeev, Mechanical and thermal properties of modified bismaleimide matrices toughened by polyetherimides and polyimide, Polym. Adv. Technol. 27 (2016) 774–780. doi:10.1002/pat.3711.

[4] H.R.. Lubowitz, C.H.. Sheppard, POLYIMIDE OLIGOMERS AND BLENDS AND METHOD OF CURING, US5116935, 1992. https://patentimages.storage.googleapis.com/pdfs/882329da991b91855fcc/US5116935.pdf (accessed June 16, 2017).

[5] H.D. Stenzenberger, Addition polyimides, in: High Perform. Polym., Springer-Verlag, Berlin/Heidelberg, 1994: pp. 165–220. doi:10.1007/BFb0021199.

[6] M. Xu, J. Hu, X. Zou, M. Liu, S. Dong, Y. zou, X. Liu, Mechanical and thermal enhancements of benzoxazine-based GF composite laminated by *in situ* reaction with carboxyl functionalized CNTs, J. Appl. Polym. Sci. 129 (2013) 2629–2637. doi:10.1002/app.38988.

[7] A. Chernykh, J. Liu, H. Ishida, Synthesis and properties of a new crosslinkable polymer containing benzoxazine moiety in the main chain, Polymer (Guildf). 47 (2006) 7664–7669. doi:10.1016/j.polymer.2006.08.041.

[8] M. Laskoski, D.D. Dominguez, T.M. Keller, Development of an oligomeric cyanate ester resin with enhanced processability, J. Mater. Chem. 15 (2005) 1611–1613. doi:10.1039/b500126a.

[9] I. Hamerton, Chemistry and technology of cyanate ester resins, Springer Science & Business Media, 2012.

[10] S.B. Sastri, J.P. Armistead, T.M. Keller, Phthalonitrile-carbon fiber composites, Polym. Compos. 17 (1996) 816–822. doi:10.1002/pc.10674.

[11] S.B. Sastri, J.P. Armistead, T.M. Keller, U. Sorathia, Phthalonitrile-glass fabric composites, Polym. Compos. 18 (1997) 48–54. doi:10.1002/pc.10260.

[12] D. Augustine, D. Mathew, C. Reghunadhan Nair, End-functionalized thermoplastic-toughened phthalonitrile composites: influence on cure reaction and mechanical and thermal properties, Polym. Int. 64 (2015) 146–153. doi:10.1002/pi.4774.

[13] L. Zong, C. Liu, S. Zhang, J. Wang, X. Jian, Enhanced thermal properties of phthalonitrile networks by cooperating phenyl-s-triazine moieties in backbones, Polymer (Guildf). 77 (2015) 177–188. doi:10.1016/j.polymer.2015.09.035.

[14] Y. Luo, M. Xu, H. Pan, K. Jia, X. Liu, Effect of ortho-diallyl bisphenol A on the processability of phthalonitrile-based resin and their fiber-reinforced laminates, Polym. Eng. Sci. 56 (2016) 150–157. doi:10.1002/pen.24237.

[15] D.D. Dominguez, H.N. Jones, T.M. Keller, The effect of curing additive on the mechanical properties of phthalonitrile-carbon fiber composites, Polym. Compos. 25 (2004) 554–561. doi:10.1002/pc.20049.

[16] D.D. Dominguez, T.M. Keller, Low-melting Phthalonitrile Oligomers: Preparation, Polymerization and Polymer Properties, High Perform. Polym. 18 (2006) 283–304. doi:10.1177/0954008306060143.

[17] M. Laskoski, J.S. Clarke, A. Neal, H.L. Ricks-Laskoski, W.J. Hervey, T.M. Keller, Synthesis of bisphenol A-free oligomeric phthalonitrile resins with sulfone and sulfone-ketone containing backbones, J. Polym. Sci. Part A Polym. Chem. 54 (2016) 1639–1646. doi:10.1002/pola.28020.

[18] Z. Zhang, Z. Li, H. Zhou, X. Lin, T. Zhao, M. Zhang, C. Xu, Self-catalyzed silicon-

containing phthalonitrile resins with low melting point, excellent solubility and thermal stability, J. Appl. Polym. Sci. 131 (2014) n/a-n/a. doi:10.1002/app.40919.

[19] M. Laskoski, A. Neal, T.M. Keller, D. Dominguez, C.A. Klug, A.P. Saab, Improved synthesis of oligomeric phthalonitriles and studies designed for low temperature cure, J. Polym. Sci. Part A Polym. Chem. 52 (2014) 1662–1668. doi:10.1002/pola.27161.

[20] A.V. Babkin, E.B. Zodbinov, B.A. Bulgakov, A.V. Kepman, V.V. Avdeev, Low-melting siloxane-bridged phthalonitriles for heat-resistant matrices, Eur. Polym. J. 66 (2015) 452–457. doi:10.1016/j.eurpolymj.2015.03.015.

[21] B.A. Bulgakov, A.V. Babkin, P.B. Dzhevakov, A.A. Bogolyubov, A.V. Sulimov, A.V. Kepman, Y.G.Y.G. Kolyagin, D.V. Guseva, V.Y. Rudyak, A.V. Chertovich, Low-melting phthalonitrile thermosetting monomers with siloxane- and phosphate bridges, Eur. Polym. J. 84 (2016) 205–217. doi:10.1016/j.eurpolymj.2016.09.013.

[22] B. Bulgakov, A. Sulimov, A. Babkin, I. Timoshkin, A. Solopchenko, A. Kepman, V. Avdeev, Phthalonitrile-carbon fiber composites produced by vacuum infusion process, J. Compos. Mater. 51 (2017) 4157–4164. doi:10.1177/0021998317699452.

[23] B.A. Bulgakov, K.S. Belsky, S.S. Nechausov, E.S. Afanaseva, A. V. Babkin, A. V. Kepman, V. V. Avdeev, Carbon fabric reinforced propargyl ether/phthalonitrile composites produced by vacuum infusion, Mendeleev Commun. 28 (2018) 44–46. doi:10.1016/j.mencom.2018.01.014.

[24] T.M. Keller, D.D. Dominguez, High temperature resorcinol-based phthalonitrile polymer, Polymer (Guildf). 46 (2005) 4614–4618. doi:10.1016/j.polymer.2005.03.068.

[25] V.E. Terekhov, V.V. Aleshkevich, E.S. Afanaseva, S.S. Nechausov, A.V. Babkin, B.A. Bulgakov, A.V. Kepman, V.V. Avdeev, Bis(4-cyanophenyl) phenyl phosphate as viscosity reducing comonomer for phthalonitrile resins, React. Funct. Polym. 139 (2019) 34–41. doi:10.1016/J.REACTFUNCTPOLYM.2019.03.010.

[26] A. V. Babkin, E.B. Zodbinov, B.A. Bulgakov, A. V. Kepman, V. V. Avdeev, Low-melting siloxane-bridged phthalonitriles for heat-resistant matrices, Eur. Polym. J. 66 (2015) 452–457. doi:10.1016/j.eurpolymj.2015.03.015.

LIGHTWEIGHT AND FLEXIBLE THERMAL PROTECTION SYSTEMS FOR HIGH TEMPERATURE COMPOSITE APPLICATIONS

Zhe Liu, Yourri-Samuel Dessureault, Matthew Lundblad, Ayou Hao and Zhiyong Liang
High-Performance Materials Institute, Department of Industrial & Manufacturing Engineering, Florida State University, Tallahassee, FL 32310, USA

Youssef Aider and Yeqing Wang
Department of Aerospace Engineering, Mississippi State University, Advanced Composites Institute, Mississippi State University, Starkville, MS 39759, USA

ABSTRACT

Carbon fiber reinforced polymers (CFRP) are increasingly used in aerospace applications which demand lightweight and stability at elevated temperatures. This paper discusses a method of fabricating a lightweight and flexible skin layer that made of carbon nanotube (CNT)/phenolic nanocomposite and using it as a thermal protection layer (TPL) to improve the heat resistance of CFRP. The presence of TPLs in the hybrid composites resulted in a 17% decrease of through-thickness thermal conductivity. Residual flexural strength and modulus after a flame torch test were 39% and 70% respectively of the initial value, a substantial improvement over 11% and 21% of a control sample. We also discuss the effects of TPLs on the coefficient of thermal expansion (CTE) and protection mechanisms. A heat transfer model is developed using finite element analysis to understand the thermal protection mechanism of the developed TPL layer. In the model, the heat conduction equation is solved with temperature-dependent material properties of the different composite material layers. The heat flux flowing from the flame torch to the material surface is estimated through a calibration procedure that compares the predicted back surface temperature against the experimental data. The model was used to conduct a parametric study to investigate the effects of TPL layer thicknesses and volume fractions on the thermal response of the hybrid composite material. This research could lead to the scalable manufacturing of CFRPs with enhanced performance characteristic at elevated temperatures for aerospace applications.

Keywords: thermal protection system, carbon/carbon, high-temperature
Corresponding author: Ayou Hao

1. INTRODUCTION

Advanced carbon fiber reinforced polymers (CFRP) are being increasingly used in current aircraft, satellites, launch vehicles, and missiles, which require to display thermal stability at high temperatures. However, the most commonly used matrices in structural composites are epoxy and bismaleimide (BMI) resins with a short-term service temperature only up to 177 and 232°C, respectively [1, 2]. At sufficiently high temperatures, resin matrices will decompose leading to performance degradations or even damage. Solving this problem has required the development of novel CFRP composite systems with lightweight, affordable, compatible and higher thermal durability. This has brought a surge of using thermal protection (TP) materials in CFRPs to reduce or eliminate heat damage during thermal exposure.

Thermal protection systems (TPS) are a critical material to protect composite for astronautical applications because of its effective protect materials from severe heating environments [3, 4]. There are three types of TP materials: metal and ceramic coating TP system [5-7], carbon/carbon (C/C) composites, and polymeric TP materials [8]. The advantages of ceramic coating and C/C composites are ultra-high temperature resistance, but it was challenged by the thermal mismatching, cost, and complicated fabrication process, which limit its wide applications. Due to these limitations, most ablative TP systems are using the reinforced filler with organic resins, such as phenolic resins. Phenolics play an important role in the aerospace industry because of its high thermal stability, ablative properties, dimension stability, and chemical resistance [9], such as Hexion SC-1008.

The residual char can be mechanically disintegrated by aerodynamic friction. Thus, reinforcements must be used to prevent aerodynamic friction. The reinforcements usually include fibers, fillers, carbon nanotubes (CNTs), or honeycomb materials, which promoted the retention of the charred material in TPS. Consequently, heat and friction can be carried out effectively since the TP materials can continue to absorb heat endothermically. CNTs have been extensively used as the reinforcement in composite materials because of their high thermal and electrical properties [10]. some researchers have confirmed that CNTs can significantly improve the mechanical, electrical, and thermal properties of nanocomposites. All these advantages make CNTs promising fillers in composites, but the dispersion of nanoscale CNTs in polymer matrices still is a critical challenge. Buckypaper (BP) is a relatively uniform CNT network made by freestanding entangled CNT bundles. BP has been widely used as a reinforcement in polymeric composites[11]. Some researchers studied BP to enhance the fire retardancy of polymer matrix composites [12]. BP has been reported as a fire shield for the epoxy carbon fiber composites to improve the flame-retardant properties of composites [13, 14]. The main mechanism of BP as fire-retardant is the CNTs network act as a protective layer, which reduces the heat transfer to the polymer matrix substrate. Also, the combustible gases can be physically separated from the oxygen preventing the combustion process in the underlayer composites [12].

There is always an urge to improve the thermal resistance of TP materials to produce thinner and lighter composite structures to reduce the overall dimension and weight. This study attempts to add a thin layer of buckypaper reinforced phenolic resin (BP/phenolic) composite to the surface of a CFRP to effectively reduce thermal degradation when the composites under high heat flux situation. This method avoids using complicated manufacturing processes of TPS materials using nanoparticle dispersion. Compared to conventional TP systems, this new TP material has excellent mechanical properties with less thermal shield thickness, which provides additional weight savings due to attacking advantages of thermal protection properties of both CNTs and phenolic resin for high char yielding and good char properties as discussed before. In this study, the purpose of a thin BP/phenolic TPL is to endure for a short period of time exposed in a low heating environment (<20 W/cm^2 heat flux) and evaluate resultant TP performance.

In this work, we established a simple two-step curing procedure to fabricate BMI composites with a BP/phenolic thermal protection layers (TPLs). An ultra-thin unidirectional (UD) carbon fiber were designed as a bonding layer between the TPLs and BMI composites to provide better adhesion. Through this method, the BP/phenolic TPLs were successfully bonded onto CFRP composites to ensure strong interfacing with the two polymer matrices of the hybrid composites. The new hybrid composites exhibited excellent mechanical properties and higher thermal resistance.

2. EXPERIMENTATION

2.1 Materials

Random BP were provided by Nanocomp Technologies (Merrimack, USA). The prepreg of plain-weave carbon fiber with BMI resin matrix was supplied by Alidila Composite Materials. The unidirectional (UD) carbon fiber tapes (CFs) with two different aerial densities (40 and 80 gsm) were used. The resol-type phenolic resin (SC-1008) manufactured by Hexion with isopropyl alcohol (IPA) as the solvent (solid content between 60-64 %).

2.2 Fabrication

The buckypaper/phenolic TPLs were produced using an infiltration and hot-plate curing process. Fig.1 shows the manufacturing process. The buckypapers (BP) were immersed in the phenolic resin solutions ((1, 3, 5, 10, and 20 wt.% resin concentrations) for 24 hours, then dried for 24 hours. Stacks of three or six layers of saturated nanotube sheets were made to form the TPLs of varying thicknesses. One-layer of UD CF prepreg or one-layer dry buckypaper was selected as the bonding layer to be applied on the top of the stacked TPLs. The phenolic resin solution was added into the UD CF bonding layer at a 50% resin volume fraction, which was designed to having a partially impregnated the UD CF preform. The dry nanotube sheet bonding layers were used as-received without further treatment. The samples were cured with vacuum bag using hydraulic hot-plates under 100 psi of pressure [15].The curing temperature was shown in Fig. 1 step 5.

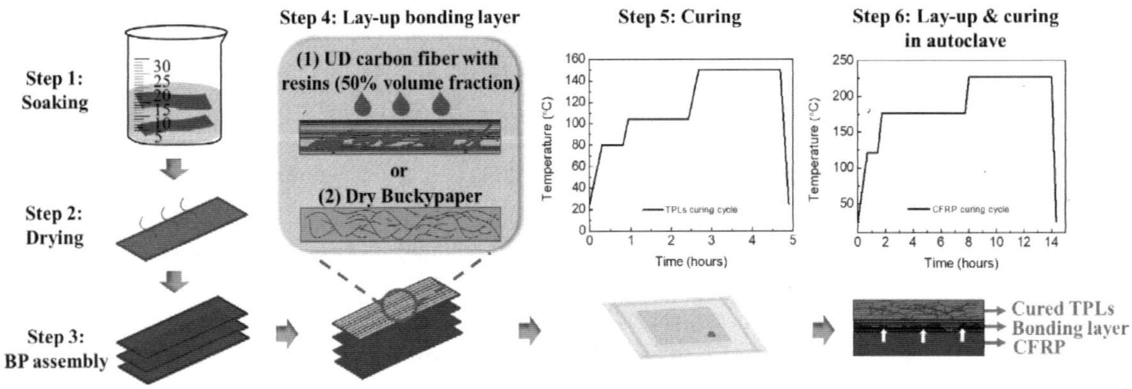

Figure 1 Fabrication of the BP/phenolic TPL nanocomposites and hybrid composites

Table 1 shows the designs and compositions of the sample lay-ups. The bonding layer was designed to improve the bonding strength between TPLs and CFRP substrates and to reduce the discrepancies between their CTEs. Designs without the bonding layer were also fabricated as the control. To prepare the hybrid composites, cured buckypaper/phenolic TPLs and CFRP prepreg were laminated on a steel mold. The assembly was vacuum-bagged and cured in an autoclave. The autoclave curing temperature and pressure were shown in Fig. 1 step 6. By heating and vacuum in the autoclave, the BMI resin in the prepreg was forced into the bonding layer. Table 1. Design and composition of the TPLs and hybrid composites.

Table 1 Design and composition of the TPLs and hybrid composites (all with PW IM7/BMI CFRP)

Materials with volume fractions	BP	Bonding layer	
CFRP-BP(6.7%)	Yes	/	/
CFRP-BP(9.0%)	Yes	Dry BP	/
CFRP-UD(3.5%)-BP(6.3%)	Yes	/	UD CF (40gsm)
CFRP-UD(3.5%)-BP(11.6%)	Yes	/	UD CF (40gsm)
CFRP-UD(8.1%)-BP(6.9%)	Yes	/	UD CF (80gsm)

2.3 Characterization

2.3.1 Morphology

The morphology of the composites cross-section was observed using a scanning electron microscopy (SEM) (JEOL. JSM-7401F USA, Inc.) The samples were coated with the gold particle using sputter coating.

2.3.2 Thermogravimetric analysis (TGA)

Thermogravimetric analysis (TGA) was carried out using Q50 TA Instruments. The test was heated from 25 °C to 800 °C in a nitrogen atmosphere at a heating rate of 10 ℃ min^{-1}.

2.3.3 Thermal Conductivity

The thermal conductivities were measured by a laser flash method (microflash LFA457, NETZSCH). The specimens were prepared as 10 mm × 10 mm × 1 mm using a diamond saw and coated with graphite. Testing was performed over a temperature range of 40-400 °C.

2.3.4 Mechanical test

Three-point bending test was studied by according to the ASTM D7264 standard using a SHIMADZU AGS-J system at a crosshead speed of 1 mm min^{-1}. The specimens were prepared with the dimensions of 50.8 × 12.7 × 1 mm^3. The flexural stress σ, strain ε, and modulus E were calculated using:

$$\sigma = \frac{3PL}{2wt^2} \quad (1)$$

$$\varepsilon = \frac{6ty}{L^2} \quad (2)$$

$$E = \frac{PL^3}{4wt^3y} \quad (3)$$

Where P is the applied load, L is the lower span length, w is the specimen width, t is the thickness, and y is the deflection at the center of the beam.

2.3.4 Flame test setup and heat flux calculation

Flame test was used to simulate thermal exposure circumstances, as shown in Fig. 3(a). A propane torch was used as the heat source for the flame test (Worthington Cylinder Corp, 14.1 oz. propane hand torch cylinder). Propane torches provide very high temperatures burning in air with a flame temperature of about 1982 °C [16]. As Fig. 2 shows, the flame is perpendicular to the hybrid CFRP composite faced with BP/phenolic TPLs. A K-type thermocouple was applied to the back of composites and connected to a Keithley digital multimeter to test the back-surface temperature. The edges of the composite samples were protected by a stainless-steel mold.

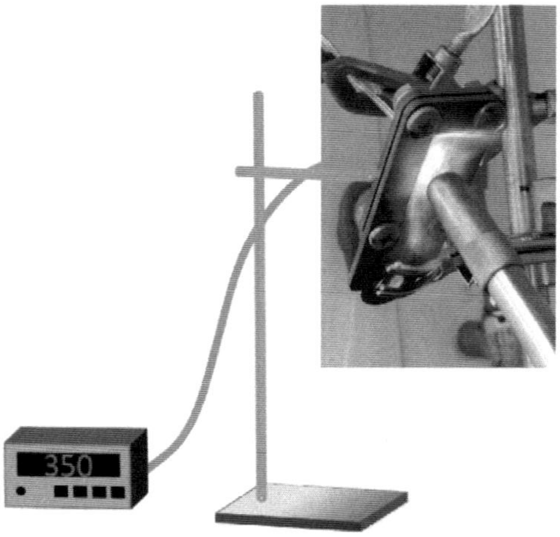

Figure 2 Flame test set-up

3. RESULTS

3.1 Properties of BP/phenolic nanocomposites

The mechanical properties of the BP/phenolic nanocomposites with different resin solution concentrations were studied. The strain-stress curves of cured BP/phenolic nanocomposites are shown in Fig. 3(a) and the related tensile properties and density are summarized in Fig. 3(b). The tensile strength of the random buckypaper cured with 1 wt.% resin solutions was approximately 250 MPa, and Young's modulus was 10 GPa. With the increase the resin solution concentration, the tensile strength and Young's modulus increased as the rise of resin solution concentration to 5 wt.% and then decreased. The highest average tensile strength and modulus of cured nanocomposites were 340 MPa and 24 GPa, respectively. The improved failure stress implied an enhanced load carrying capacity of the system. With continuously increasing the resin concentration to 20 wt.%, more resin was infiltrated into the BP, resulting in an increased density and thickness of composites as seen in Fig.3(b). This possibly was the reason for decreased tensile properties due to more defects compared to the thinner sample. Therefore, the 5 wt.% resin solution was used to fabricate BP/phenolic TPLs. The morphology of BP/phenolic nanocomposites with various resin solutions is shown in Fig. 4. As the resin concentration increased, more pores and voids within the BP were filled by the cured resin. Thus, the BP/phenolic nanocomposites involved fewer voids which is efficient in load transfer.

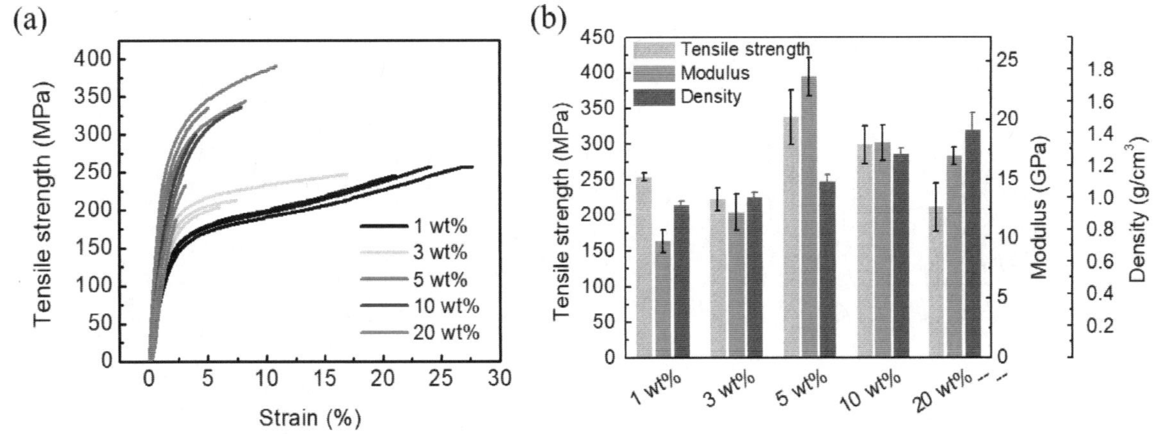

Figure 3 Tensile strength and modulus of BP/phenolic nanocomposites with different resin solution concentrations (a) stress-strain curve of tensile properties (b) summarized tensile properties and densities

Figure 4 SEM images of cured BP/phenolic nanocomposites (a) 1 wt.% (2) 3 wt.% (c) 5 wt.% phenolic resin solution

3.2 Morphology of hybrid composites

Fig. 5 shows the cross-sectional of hybrid composites with BP/phenolic TPLs by different volume fractions. No obvious voids were observed for hybrid composites in SEM images. Fig 5(a) clearly showed two components: the CFRP and the BP/phenolic TPLs. The measured thickness of BP/phenolic TPLs in Figs. 5(a) was 88 μm. Fig. 5(b), (c), and (d) displayed the cross-sections of hybrid composites with the different bonding layer between the CFRP and the TPLs. Fig. 5(b) showed the hybrid composites with the UD CF of 80 μm bonding layer. However, for the case of, Fig. 5(b), shows a significant number of voids in the thicker UD CF indicating poor interfacial bonding strength between the CFRP and the BP/phenolic TPLs. These voids would likely induce poor interaction between the CFRP and BP/phenolic TPLs further reducing the mechanical properties. Therefore, the hybrid composite design with an 80 μm UD CF bonding layer was not selected for further study. Fig. 5(c) and (d) show the 40 μm UDCF bonding layer and the dry buckypaper bonding layer, respectively, which is no void observed. The thickness of TPLs and the bonding layer was calculated based on the SEM images, as shown in Fig. 6. The volume fraction of each component was summarized in Table 2.

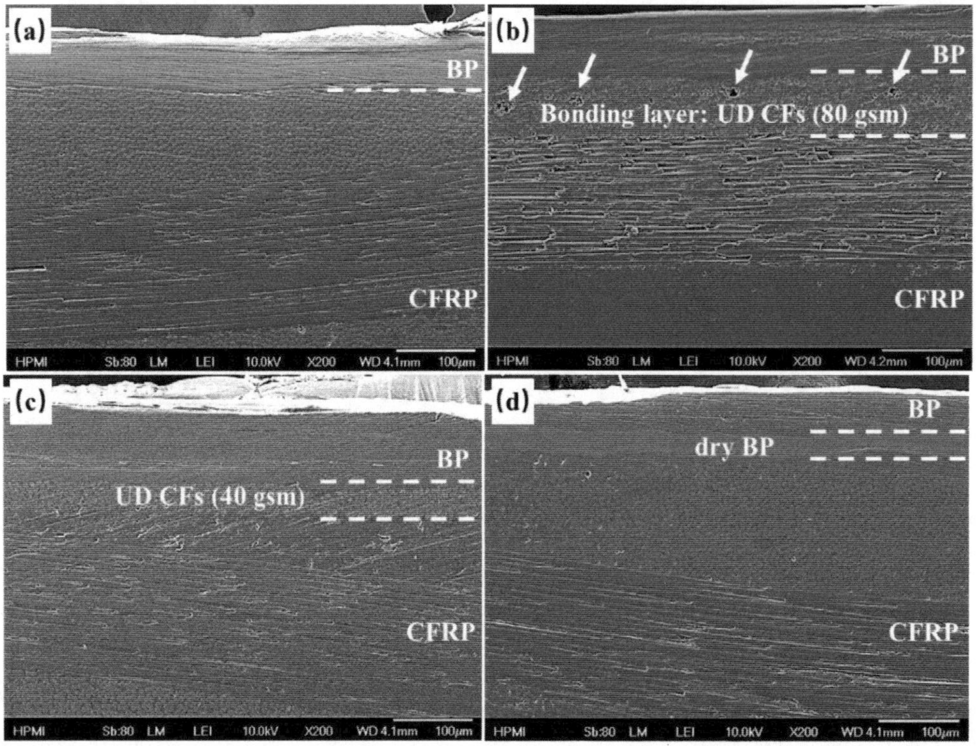

Figure 5 SEM images of cross-section of hybrid composites (a) CFRP-BP(6.7%) (b) CFRP-UD(8.1%)-BP(6.9%) (c) CFRP-UD(3.5%)-BP(6.3%) (d) CFRP-BP(9.0%)

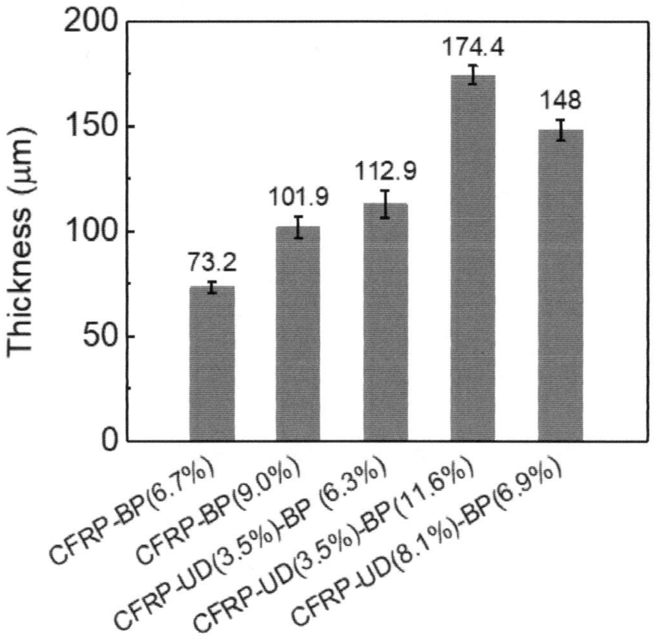

Figure 6 Thickness of BP/phenolic thermal protection layer

Table 2 The composition of hybrid composites

Materials	Vol % of TPL*	Vol% of UD CF*
CFRP control	0	0
CFRP-BP(6.7%)	6.7	0
CFRP-BP(9.0%)	9.0	0
CFRP-UD(3.5%)-BP (6.3%)	6.3	3.5
CFRP-UD(3.5%)-BP(11.6%)	11.6	3.5
CFRP-UD(8.1%)-BP(6.9%)	6.9	8.1

*The volume fraction is defined as the thickness of UD CF or TPL divided by the entire hybrid composite thickness

3.3 Thermal properties

Fig. 7 shows the thermal conductivity of CFRP and the hybrid composite from RT to 400 °C in the through-thickness direction. The thermal conductivity of CFRP control was approximately 1.106 W $m^{-1}K^{-1}$ at RT. The BP/phenolic TPL lowered the through-thickness thermal conductivity about 17% due to the nanoscale pores introduced into the TPLs. The heat transfer in the through-thickness direction was affected by the pore volume fraction, as well as the size, shape, orientation, and emissivity. Besides the thermal conductivity of CNT sheets in the in-plane direction is excellent (20-30 W $m^{-1}K^{-1}$ at RT [17], which is significantly higher than the phenolic resin (0.28 W $m^{-1}K^{-1}$ at RT). Thus, the buckypaper/phenolic exhibits good in-plane thermal conductivity with the CNT network, which allows the heat to dissipate along the in-plane direction instead of the through-thickness direction.

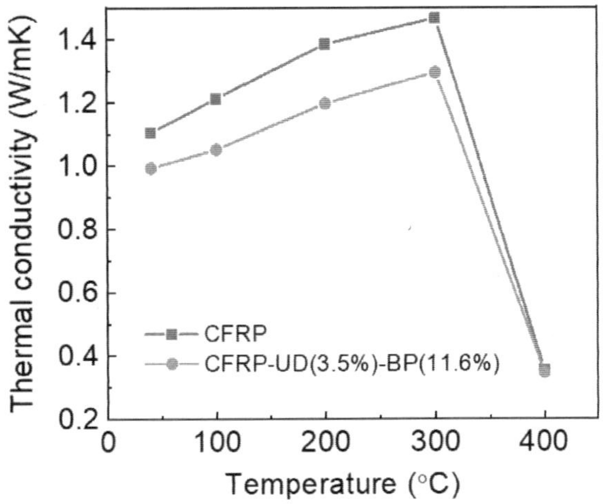

Figure 7 Through-thickness thermal conductivity

The CTE is a critical thermophysical property of composites in structural applications. Fig. 8 shows the CTE of the CFRP and the hybrid composites. The CTE values of hybrid composites were found to be lower than the CTE of the CFRP control both in in-plane and through-thickness directions. The CNTs can be beneficial in reducing the CTE of polymer composites, which may be due to the CNTs interaction with the polymer causing restriction of polymer chain movement. The lower CTE indicates lower thermomechanical stresses when the material was subjected to the thermal environment. Also, a lower CTE is often desirable to achieve good dimensional stability. Therefore, the CNT sheets incorporated into CFRP composites could potentially reduce the CTE of composites.

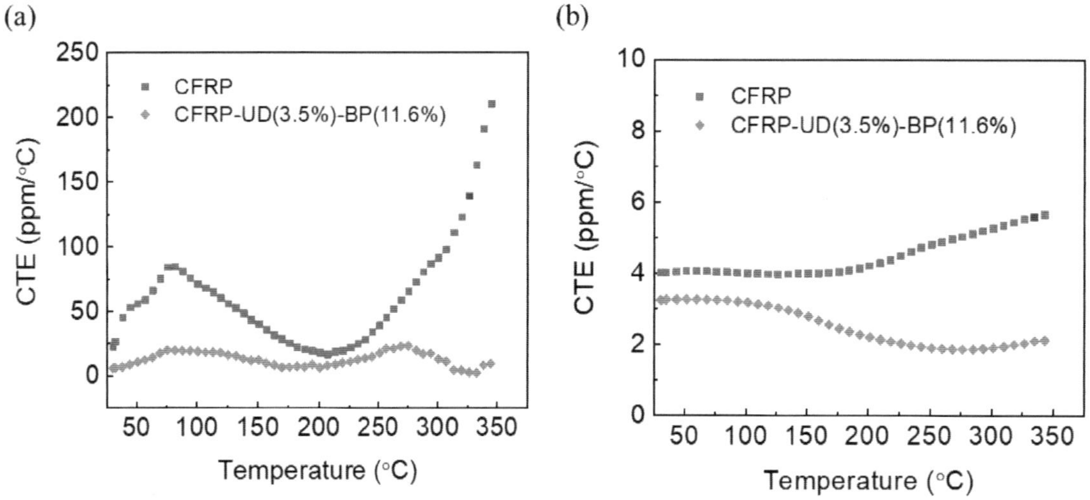

Figure 8 CTE of hybrid composites from RT to 350 °C (a) through-thickness (b) in-plane

3.4 Flame test

Fig. 9 shows the back-surface temperatures of the CFRP and hybrid composites. The temperature curves provide insight into how well the hybrid composites performed with the TPLs. For a 30 sec burning time, the back-surface temperature of CFRP elevated to 325 °C whereas the hybrid composites peaked at a back-surface temperature of about 200-220 °C. With increasing BP volume fractions to 11.6 %, the maximum back-surface temperature decreased to 200 °C. The results show that the TPLs effectively reduce heat penetration into the CFRP substrate.

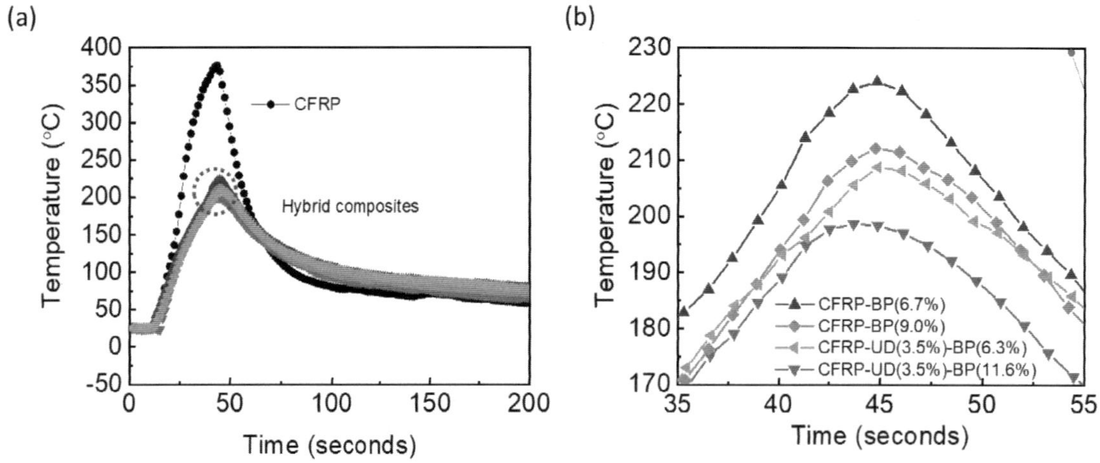

Figure 9 Back-surface temperatures for flame test (a) CFRP (b) Hybrid composites

3.5 Mechanical properties and residual strength after burning

To determine the mechanical properties of the hybrid CFRP and the thermal protection capability, 3PB was conducted on the pristine and burned hybrid CFRP composites. The mechanical properties of hybrid composites prepared with different volume fraction of TPLs were presented in Fig. 10(a). It is observed that the CFRP control had a flexural strength and modulus of 1034 MPa and 43 GPa, respectively. The TPLs benefit from a UD CF bonding layer by minimizing strength and modulus loss. This phenomenon is witness in the hybrid composite with UD CF volume fraction 3.5 %, which sustained 94% of the strength and modulus compared to the CFRP value. In short, this is an indication that the UD CF bonding layer has positive influence on mechanical properties. However, if the volume fraction of UD CF bonding layer increased to 8.1%, the flexural strength decreased to 67% of the CFRP control. It may be because the thicker UD CF bonding layer induced a high quantity of voids (as shown in Fig. 5b), which decrease the mechanical properties.

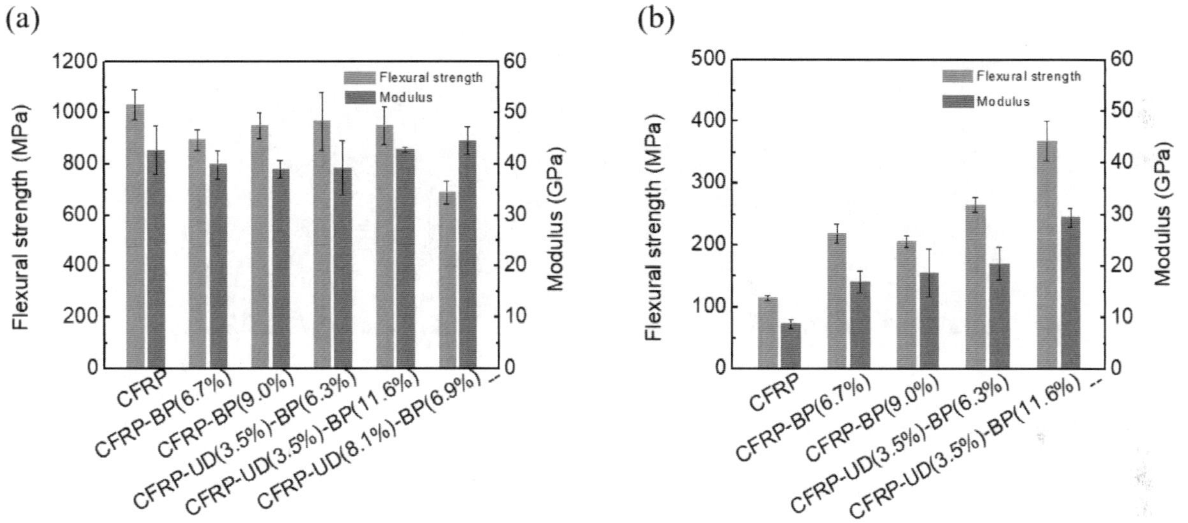

Figure 10 Flexural strength and modulus of 3PB specimens (a) before flame test (b) after flame test for 30 sec

After flame test, all samples were tested under 3PB test to study the residual strength and modulus, as shown in Fig. 10(b). After a 30 sec burn duration, the CFRP control had a strength and modulus of 115 MPa and 9 GPa, respectively. The measured residual strength and modulus properties were 11% and 21%, respectively, compared to the properties of the CFRP control prior to burning. With an increase of TPLs volume fraction, the residual mechanical properties improved significantly. The CFRP-UD CF (3.5%)-BP(11.6%) retained the highest residual mechanical properties. The result showed that the TPLs play a vital role to improve thermal protection capability.

3.6 SEM analysis after flame tests

Fig. 11 shows the morphology of the post-test samples. During the exposure to the high-temperature environment, the pristine CFRP was significantly damaged (Fig. 11(a)). The carbon fibers experienced a high degree of fracture, and most of the resin content was decomposed. The morphology of the zones underneath the TPL was also investigated, as shown in Fig. 11(b). By comparison of the CFRP control, the insulative behavior of TPLs may influence the decomposition behavior of the CFRP substrate. Therefore, the hybrid composites with TPLs was no greatly damaged, indicating that the TPLs reduce the heat penetration into the inside layers. The SEM images at high magnification, Fig. 11(c) and (d), show the residual CNTs still maintain 3D network preventing nano-sized char particles to blow away.

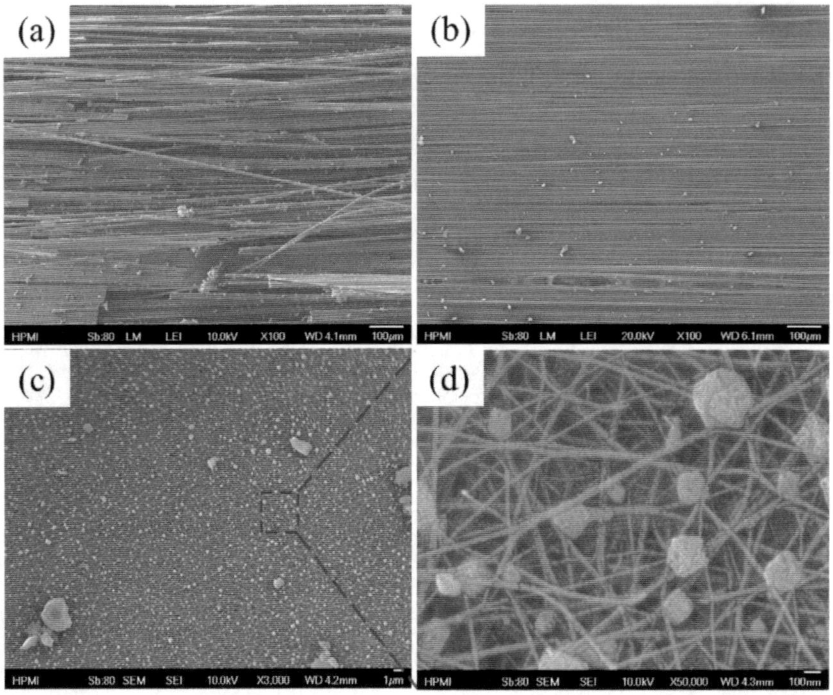

Figure 11 SEM images of composites after burning 30 sec, (a) CFRP control burning side, (b) hybrid BMI composites under the BP/phenolic TPLs burning side (c) and (d) BP/phenolic TPLs surface

3.7 Numerical Simulation

Finite Element Method (FEM) was used to model the hybrid composite material. The 1D transient heat equation was solved along the through-the-thickness direction of the composite. Temperature-dependent material properties (i.e. density, thermal conductivity, and specific heat capacity) were used in the equation as inputs. When solved, the heat equation provides the time-dependent temperature distribution in the hybrid composite, and the calculated back-surface temperature can be used to assess the overall performance of the thermal protection of the composite. When validated, the model can be an efficient and easy way to test different heat protection material designs and configurations such as the thickness, volume fraction, and location of each composite layer [17, 18].

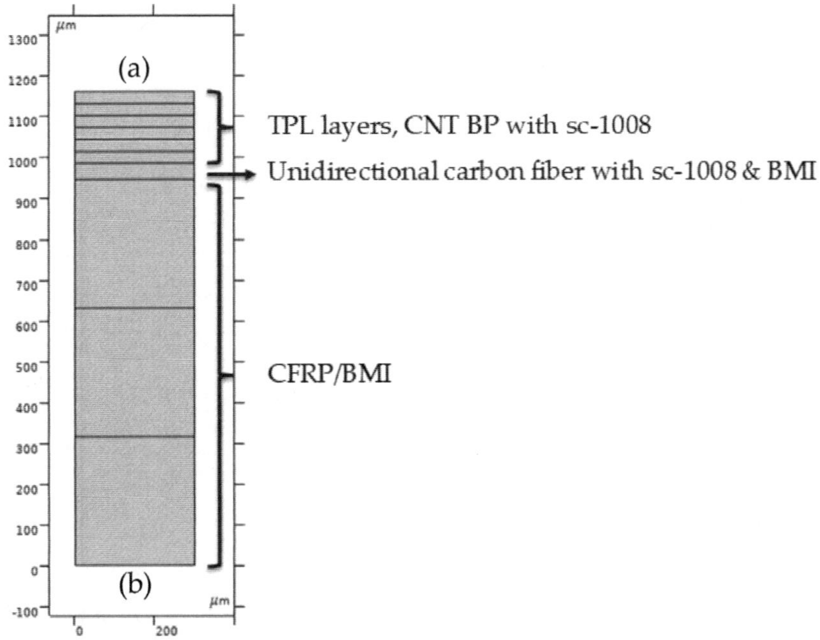

Figure 12 Model geometry dimensions and material composition.

The model is made of a total of 10 layers as shown in Fig.12. The top 6 layers represent the TPL layer with Phenolic SC-1008 resin and the carbon nanotube buckypaper material properties. Following the TPL layers, there is a single layer of unidirectional fiber-reinforced composite with material properties of carbon fiber, phenolic sc-1008, and BMI resin. Finally the 3 layers of CFRP/BMI composite are modeled at the end. The model boundary conditions are the input heat flux $-\mathbf{n} \cdot \mathbf{q} = q_0$ at the boundary (a) and the cooling convection and radiation losses heat flux $-\mathbf{n} \cdot \mathbf{q} = \varepsilon(T_{amb}^4 - T^4) + h(T_{amb} - T)$ at boundary (b). Since the heat flux was not measured during the experiment, a calibration study was made by solving the same 1D heat equation using the overall thermal conductivity of the hybrid composite [15]and treating the whole composite as a single homogeneous material. The input (flame) heat flux can then be calibrated to match the temperature versus time curve measured in the experimental results in Fig.9. The result of the calibration results gives a heat flux of 1.2 kW/m^2.

Using the same heat flux, the simulation is done with material properties of the different composite layer taken from literature [19-23]. The back-surface temperature result is plotted with respect to time.

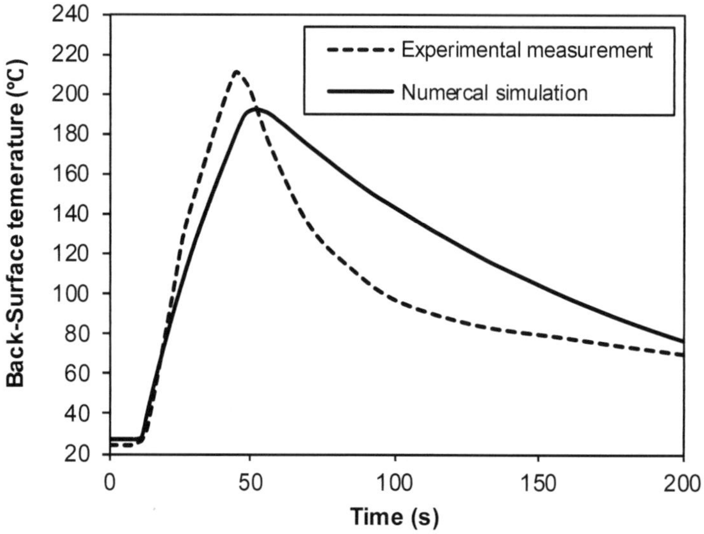

Figure 13 Back-surface temperature result versus time comparison between the experimental measurement and the numerical model.

We can see in the graph Fig. 13, the peak temperature of 197 °C at time t = 50 s is in good agreement with the experiment. Similarly, the temperature of 85 °C at t = 200 s also agrees well with the experiment. However, the curve of the graph is not the same, since the experiment shows a rapid cooling for the period between 50 s and 100 s and after that, the cooling becomes slower. Unlike the experiment, the simulation cooling is a linear line from time 50s to 200s and the radiation and convection boundary conditions can only affect the slope of the line but cannot make a big difference in the cooling curvature. The reason for such difference can be partially related to the accuracy of material properties found in the literature.

Even though the model is very fast and easy to configure for different geometries and materials, finding the material properties especially temperature-dependent in the three composite directions (i.e. longitudinal, transverse, and the-through-the thickness directions) can be challenging and hard to validate. Therefore, since the model is very sensitive to the material properties given, there is still a lack of confidence in the accuracy of the results if the material properties are not well validated by experiments. In addition to that, the heat loss due to charring is not taken into account, which has an important impact on the overall thermal performance of the composite. All these limitations can be improved in future work by taking into account the material phase change to account for charring [24].

4. CONCLUSIONS

The hybrid composites with BP/phenolic TPLs were successfully manufactured. The thermal stability, burning behavior, and mechanical properties have been evaluated. From the 3PB test, the CFRP protected with TPLs had similar strength and modulus compared to the CFRP control which implied the fabrication process did not decrease the mechanical properties of composites. The flame test results showed that the TPLs reduced heat penetration into the underneath of the CFRP.

Also, the post-burning SEM images of the hybrid composites indicated that the sacrificial BP/phenolic TPLs significantly decreased the amount of resin decomposition in the CFRP. As the results, the strength and modulus of hybrid composites retained 40% of initial mechanical performance whereas the control sample only sustained 8% after burning. The results show that the TPLs have provided effective thermal protection capability resulting in decreased thermal damage of interior CFRP materials. A preliminary finite element model was built and found good agreement with flame test results on back surface temperature. The developed approach is only using a simple manufacturing process and thinner TPLs, which can be attractive for the future integrated TP system for polymeric composites.

5. ACKNOWLEDGMENT

This work was partially supported by AFOSR FA9550-17-1-0005.

6. REFERENCES

[1] *CYCOM® 977-3 Epoxy Resin System.* [cited 2019 Jan 28]; Available from: www.cytec.com.

[2] *CYCOM®5250-4 Prepreg System.* 2011 [cited 2019 Jan 15]; Available from: www.cytec.com.

[3] A. R. Bahramian, M. Kokabi, M. H. N. Famili, et al., Ablation and thermal degradation behaviour of a composite based on resol type phenolic resin: Process modeling and experimental. *Polymer*, 2006. **47**(10): pp. 3661-3673.

[4] H. Cheng, H. Xue, C. Hong, et al., Preparation, mechanical, thermal and ablative properties of lightweight needled carbon fibre felt/phenolic resin aerogel composite with a bird's nest structure. *Composites Science and Technology*, 2017. **140**: pp. 63-72.

[5] C. Xu and W. C. Nickerson, Hybrid multifunctional composite material and method of making the same. 2018, Google Patents.

[6] L. Ju, J. Yang, A. Hao, et al., A hybrid ceramic-polymer composite fabricated by co-curing lay-up process for a strong bonding and enhanced transient thermal protection. *Ceramics International*, 2018. **44**(10): pp. 11497-11504.

[7] Z. Guo, Z. Liu, L. Ye, et al., The production of lignin-phenol-formaldehyde resin derived carbon fibers stabilized by BN preceramic polymer. *Materials Letters*, 2015. **142**: pp. 49-51.

[8] G. Pulci, J. Tirillò, F. Marra, et al., Carbon–phenolic ablative materials for re-entry space vehicles: Manufacturing and properties. *Composites Part A: Applied Science and Manufacturing*, 2010. **41**(10): pp. 1483-1490.

[9] L. Pilato, Phenolic resins: 100 Years and still going strong. *Reactive and Functional Polymers*, 2013. **73**(2): pp. 270-277.

[10] J. N. Coleman, U. Khan, W. J. Blau, et al., Small but strong: A review of the mechanical properties of carbon nanotube–polymer composites. *Carbon*, 2006. **44**(9): pp. 1624-1652.

[11] S. Zhang, A. Hao, N. Nguyen, et al., Carbon nanotube/carbon composite fiber with improved strength and electrical conductivity via interface engineering. *Carbon*, 2019. **144**: pp. 628-638.

[12] Q. Wu, W. Zhu, C. Zhang, et al., Study of fire retardant behavior of carbon nanotube membranes and carbon nanofiber paper in carbon fiber reinforced epoxy composites. *Carbon*, 2010. **48**(6): pp. 1799-1806.

[13] Q. Wu, C. Zhang, R. Liang, et al., Fire retardancy of a buckypaper membrane. *Carbon*, 2008. **46**(8): pp. 1164-1165.

[14] Z. Zhao and J. Gou, Improved fire retardancy of thermoset composites modified with carbon nanofibers. *Science and technology of advanced materials*, 2009. **10**(1): pp. 015005.

[15] Z. Liu, A. Hao, S. Zhang, et al., Lightweight carbon nanotube surface thermal shielding for carbon fiber/bismaleimide composites. *Carbon*, 2019. **153**: pp. 320-329.

[16] *Propane Hand Torch Cylinder.* [cited 2019 Jan 28]; Available from: https://www.bernzomatic.com/Products/Fuel-Cylinders/Hand-Torch-Cylinders/TX9.

[17] S. Torquato, S. Hyun, and A. Donev, Multifunctional composites: optimizing microstructures for simultaneous transport of heat and electricity. *Physical review letters*, 2002. **89**(26): pp. 266601.

[18] K.-S. Na and J.-H. Kim, Volume fraction optimization of functionally graded composite panels for stress reduction and critical temperature. *Finite Elements in Analysis and Design*, 2009. **45**(11): pp. 845-851.

[19] J. Mottram and R. Taylor, Thermal conductivity of fibre-phenolic resin composites. Part II: numerical evaluation. *Composites Science and Technology*, 1987. **29**(3): pp. 211-232.

[20] J. Mottram and R. Taylor, Thermal conductivity of fibre-phenolic resin composites. Part I: Thermal diffusivity measurements. *Composites science and technology*, 1987. **29**(3): pp. 189-210.

[21] L. Wan, X. Zhang, G. Wu, et al., Thermal conductivity and dielectric properties of bismaleimide/cyanate ester copolymer. *High voltage*, 2017. **2**(3): pp. 167-171.

[22] N. Nguyen, S. Zhang, A. Oluwalowo, et al., High-Performance and Lightweight Thermal Management Devices by 3D Printing and Assembly of Continuous Carbon Nanotube Sheets. *ACS applied materials & interfaces*, 2018. **10**(32): pp. 27171-27177.

[23] B. Kumanek and D. Janas, Thermal conductivity of carbon nanotube networks: A review. *Journal of materials science*, 2019. **54**(10): pp. 7397-7427.

[24] Y. Wang, T. K. Risch, and J. H. Koo, Assessment of a one-dimensional finite element charring ablation material response model for phenolic-impregnated carbon ablator. *Aerospace Science and Technology*, 2019. **91**: pp. 301-309.

CLOSING THE LOOP ON AUTOMOTIVE CARBON FIBER PREPREG MANUFACTURING SCRAP - FENDER CASE STUDY

Jordan Harris[1], Sean Kline[1], Jeffrey L. Wheeler[2], Doug Bradley[3], Komal Kooduvalli[4,5], Alan Franc[6] and Dale Brosius[7]

[1]Vartega, Inc., 15000 W 44th Ave. Golden, CO. USA

[2]Colorado School of Mines, 1500 Illinois St. Golden, CO. USA

[3]Michigan State University, 220 Trowbridge Rd, East Lansing, MI. USA

[4]Oak Ridge National Laboratory, 1 Bethel Valley Rd, Oak Ridge, TN. USA

[5]The University of Tennessee, 1321 White Ave, Knoxville, TN. USA

[6]Techmer PM, 1 Quality Cir, Clinton, TN. USA

[7]IACMI The Composites Institute, 2360 Cherahala Blvd, Knoxville, TN. USA

ABSTRACT

This case study demonstrates the successful use of recycled automotive carbon fiber in a thermoplastic injection molding automotive application. Vartega and project partners have shown that a closed-loop supply chain for automotive composites is viable. The source of the feedstock for recycling was unidirectional prepreg that was produced by Michigan State University (MSU) at the IACMI Scale-up Research Facility (SuRF) in Detroit, Michigan. Prepreg trim scrap was sent by MSU to Vartega in Golden, CO for recycling, value-added formatting, and fiber optimization via sizing application. The recycled carbon fiber (rCF) was then sent to Techmer in Clinton, TN for compounding at 10% loading in PA66. The finished compound was returned to the SuRF for injection molding into an automotive fender by MSU. The finished part exhibited excellent surface finish. Additional development is ongoing via IACMI Project 6.20 to produce 40% recycled carbon fiber reinforced PA66 for structural applications.

1. INTRODUCTION

This report details an IACMI project to demonstrate the feasibility of reclaiming carbon fiber manufacturing scrap and manufacturing new automotive parts using materials derived from this reclaimed scrap. The reclaimed carbon fiber originated as unidirectional prepreg that was produced by Michigan State University (MSU) for a separate IACMI demonstration project undertaken by Ford Motor Company, Dow Chemical and DowAksa. The prepreg trim was subjected to a chemical recycling process by Vartega that extracted the fibers from the semi-cured resin. These fibers were post-processed, compounded, and molded to create fiber reinforced PA66 automotive parts. This document details the process for converting scrap to a part, a life cycle assessment (LCA) comparing recycled to typical virgin material, and mechanical testing results of the recycled compound to a virgin baseline. In so doing, the viability of recycled carbon fiber (rCF) use in the context of injection molded parts can be quantified and evaluated.

This document is primarily intended for consumption by professionals in the manufacturing and sustainability industries, as a background in plastics processing and an understanding of mechanics of materials is helpful for contextualizing results. However, the processes and results are approachable for a broad audience.

1.1 Background

Carbon fiber is a strong and stiff synthetic fiber typically used for structural reinforcements of thermoset (usually epoxy) resins. Major applications include aerospace, performance automotive, and sporting goods where high material stiffness must be coupled with reduced weight.

While carbon fiber composites can be used to manufacture light-weight, high-strength parts, there are many potential manufacturing applications where carbon fiber composites are not currently used due to the high cost of carbon fiber. The high cost of carbon fiber composites is driven partly by the complex and energy intensive process of manufacturing carbon fiber and partly by the complexity of manufacturing parts from the carbon fiber using classical techniques such as lay-ups [1]. These economic and environmental costs can be mitigated respectively by using a recycled instead of virgin carbon fiber and by using a carbon fiber filled thermoplastic material.

The total life cycle emissions of a carbon fiber part can be reduced by using a recycled carbon fiber rather than virgin carbon fiber [2]. The energy-intensive carbonization and oxidation steps required to convert the fossil fuel-derived precursor material (typically polyacrylonitrile [PAN]) to carbon fiber correspond to a significant initial greenhouse gas footprint. This environmental effect is mitigated by fuel savings due to light-weighting in transport applications but the initial energy investment required to manufacture virgin carbon fiber is so large that the emissions benefits of using carbon fiber instead of a metallic material are minimal or possibly even negative in automotive applications [3]. By reusing fiber that would otherwise be disposed of in landfills, it is anticipated that this emissions burden could be reduced. Using a recycled carbon fiber also avoids the costly manufacturing process for virgin carbon fiber and reduces the economic cost of the composite part, thereby expanding market availability in automotive and other cost or environment-sensitive applications.

The second barrier to widespread use of carbon fiber in high volume automotive manufacturing is the cost of the complex manufacturing processes used to make parts from continuous carbon fiber. Classical use of carbon fiber composites involves performing a hand lay-up of sheets of carbon fiber composite. This process allows optimal utilization of the directional properties of long carbon fiber but is much slower than processes such as injection molding or progressive stamping that are ubiquitous in high-throughput manufacturing industries such as the automotive industry [1]. Carbon fiber may be adapted for use in high throughput manufacturing applications by compounding short (0.5-8 mm) fibers with a thermoplastic resin. While the use of carbon fiber in a compounded thermoplastic for injection molding results in a largely-isotropic short-fibered material with reduced strength relative to classical techniques, the ability to apply the material in a high-throughput cost-effective process can outweigh these drawbacks for high volume applications. As the fibers need to be chopped prior to compounding, recovered or recycled fiber may provide an ideal feedstock for producing carbon fiber filled thermoplastics.

This project demonstrates the engineering, economic, and environmental feasibility of manufacturing automotive parts from recycled carbon fiber filled thermoplastic materials. The project documents a case study where carbon fiber was reclaimed from manufacturing scrap using a novel chemical process, compounded into a thermoplastic material, and injection molded into a finished part. The mechanical properties of the recycled material are examined and a LCA of the part's emissions impact is presented. The results demonstrate that the use of recycled carbon fiber significantly reduces the environmental impact of the manufactured part.

2. EXPERIMENTATION

2.1 Feedstock Material

Unidirectional carbon fiber-epoxy prepreg was created at the IACMI Scale-up Research Facility (SuRF) located in Detroit, Michigan, for the purpose of making compression molded parts of interest to the automotive industry. Scrap for recycling comes from off-cuts generated during the manufacturing process. The prepreg scrap for this project consists of a standard modulus carbon fiber (Dow Aksa) and epoxy resin (Dow) at a loading of 65wt% fiber and 35wt% resin. A paper backing is attached to the prepreg system to assist in handling and layup. The scrap prepreg, including paper backing, was provided to Vartega for recycling.

2.2 Recycling

The feedstock prepreg trim scrap from the SuRF facility was delivered to Vartega for recycling. In order to prepare the feedstock for recycling, the material was roughly size reduced and the backing paper was removed from the prepreg. Resin was removed from the carbon fiber via Vartega's proprietary chemistry-based process. The efficacy of resin removal was less than desired due to temporary limitations of the existing process equipment at the time of the project, resulting in somewhat higher residue on the fiber than would otherwise be anticipated based on historical results. Thermogravimetric Analysis (TGA) of the fibers is shown in Figure 1. Samples were heated in high purity air at a ramp rate of 20 °C/min from 50 to 750 °C. Mass loss attributed to residual resin left on fibers is measured between 120-500°C. Mass loss above 500°C is attributed to degradation of the carbon fiber.

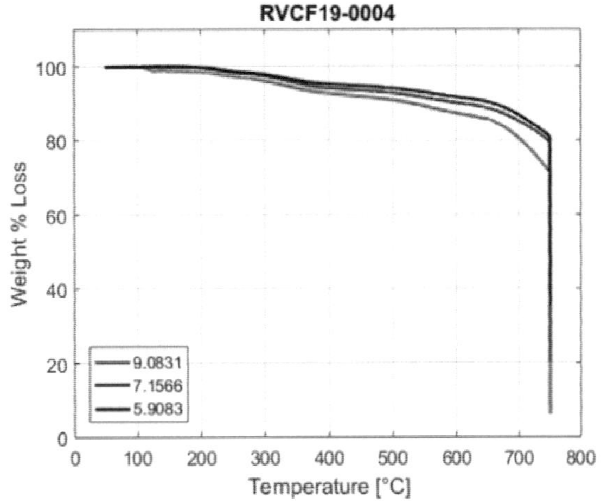

Figure 1: TGA testing showing an average residual resin account of 7.38%

Recycled fibers were chopped into nominal 6 mm lengths, although the actual length distribution is not quantified. It is understood from discussions with commercial compounders, that this semi-randomly oriented product format is difficult to work with due to both the tendency to bridge (i.e. interlock and jam) when feeding and the release of irritating dust during handling. To convert the carbon fiber into a more readily usable material, the fibers were overcoated with a water-soluble coating ("sizing") and formed into small bundles. The bundles are more uniform and denser than the raw cut fiber and are therefore less prone to bridging. The sizing serves to maintain bundle integrity, limit the dust produced during handling of the bundles, and act as an interface between the fiber and the resin matrix. The selected sizing had previously been tested and showed good interfacial properties with polyamides (PAs). Fiber bundles are dried to remove any moisture to levels below 0.2 wt%. Moisture values were determined using a Mettler Toledo HB43-S halogen moisture analyzer. Once dried bundles are sent for compounding.

2.3 Compounding

The PA66-rCF was compounded on a Coperion ZSK 40 mm co-rotating twin screw extruder at Techmer PM in Clinton, TN. The polymer and additives were fed by a gravimetric K-Tron feeder to the feed throat and the rCF was fed by a gravimetric K-tron feeder downstream. The extruder had an atmospheric vent at the side feeder and a vacuum vent prior to the die. The material was extruded into strands, cooled in a water bath, and chopped into pellets. The output rate of the extruder was approximately 135 kg/hr. Due to the relatively small quantity of rCF available, little optimization of the processing conditions and output rate was possible. The rCF format provided satisfactory feeding through the feed system (single screw fiber feeder with a pigtail auger). Fiber bundle integrity and flow properties were satisfactory for thermoplastic compounding processes. There were a small number of larger fiber clumps which did create occasional feeding difficulties. Those clumps of fibers were largely able to be manually sorted from the rCF and did not create significant processing issues. The compounding run yielded 240 kg of finished compound for molding trials. Ash burn-off testing of the finished compound showed that the fiber fill content was approximately 8.7%.

2.4 Injection Molding

Injection molding for this effort was conducted at the IACMI / Michigan State University Scale-up Research facility located in Corktown, a historic district of Detroit. This facility is a 100,000 square foot facility co-located with Lightweight Innovations for Tomorrow (LIFT). A Milacron MGS 3000 injection molding machine with a low shear screw design and an integrated Yaskawa Motoman robot was utilized to demonstrate manufacturing representative processing of the Vartega rCF-PA66 material.

Figure 2: Injection molding machine used to make fenders

A donated passenger side fender tool (courtesy of SABIC) from a Saturn 'L' series sedan was utilized to demonstrate representative production molding for a class 'A' exterior body panel application.

Figure 3: Fender tool used in injection molding process

The fender tool was initially developed for the Noryl GTX material system with a hot runner manifold system. For the PA66-rCF demonstration, single sided fill was utilized in a single cavity mold with the following molding parameters:

- Injection pressure: ~1200 psi
- Barrel temps: 580-590°F
- Tool temps: 230°F
- Backpressure: 150 psi

These molding conditions were derived based on visual appearance optimization for a class 'A' surface finish with the amount of material available. An overall cycle time of 74 sec was attained based on an average over ~70 consecutive parts. No gate or tooling modifications were made to the tool to tune for the PA66-rCF material. Based on previous industrial molding experience, it is believed that further reductions in cycle time and minor improvements to surface finish could be attained pending further tooling and process improvements.

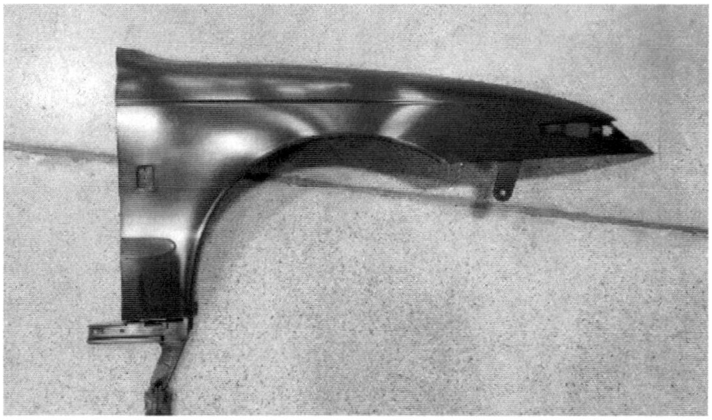

Figure 4: The finished PA66-rCF fender

2.5 Mechanical Testing

Finished compound pellets were dried in an air circulating oven for 2 hours in preparation for molding test specimens for mechanical testing. ASTM test specimens were molded on a family mold on a 110 ton Arburg injection molding machine with a hot oil mold heating system. Molding machine barrel temperatures were approximately 530-560 F, and the mold temperature was 200F. Mechanical properties were tested dry-as-molded (DAM). Tensile properties were measured on an Instron tensile specimen test unit, flexural properties were tested on an Instron 3 point bend test unit. Notched Izod Impact resistance was measured using a Tinius-Olsen pendulum impact tester, heat deflection temperature was tested with a Tinius-Olsen deflection temperature/Vicat temperature test unit, and surface resistivity was tested with an Electro-tech Systems resistivity unit with a concentric ring electrode. The applicable ASTM test standard methods were used for each test.

3. RESULTS

3.1 Mechanical Properties

Mechanical test data generated from the PA66-rCF product is compared to expected values of a 10% oversized and chopped standard modulus carbon fiber in PA66. Mechanical performance of compound made with recycled fiber tested lower compared to the non-recycled product. Previous testing has shown that on a per fiber basis, recycled fibers maintain 100% of the strength as a non-recycled equivalent [4]. Similar testing that was completed, but not a part of this report, showed essentially equal performance between the recycled and non-recycled products. As such, the reduction of performance attributed to the recycled compound was unexpected. As mentioned above, the amount of residual thermoset resin on the rCF was higher than expected due to limitations in the experimental process equipment. This excess thermoset resin was included in the mass of fiber that was metered into the compounding process, resulting in a lower than targeted total fiber content in the finished compound. Therefore, it is expected that the mechanical properties of the PA66-rCF compound will be reduced compared to the PA66-vCF material. The higher than desired residual resin may also result in less optimal adhesion between the PA66 and rCF, negatively impacting performance.

Property	Unit	Test Method	10% SURF rCF in PA66	10% commercial CF Typical properties
Tensile Strength @ Yield	Psi	ASTM D638	17,200	21,700
Tensile Elongation @ Break	%	ASTM D638	2.2	2.4
Notched Izod Impact	ft*lb/in	ASTM D256	0.46	0.50
Flexural Strength	Psi	ASTM D790	22,800	28,000
Flexural Modulus	Psi	ASTM D790	913,000	1,000,800
Surface Resistivity	Log ohm/sq	ASTM D257	12.0	6-10
Heat Deflection Temperature @264 PSI	F	ASTM D648	455	462
Density	g/cm3		1.17	1.19

Table 1: Mechanical Performance of a 10% rCF-PA66 compared to typical commercial values of a non-recycled comparable

3.2 Life Cycle Assessment

A cradle-to-grave LCA was performed to evaluate the embodied energy (EE) within the supply chain of a Saturn thermoplastic fender containing recycled carbon fiber content from Vartega's solvolysis process. The functional unit considered for analysis is 1 injection molded fender part weighing 1.63 kg. The model considers the pathway of carbon fiber from its previous generation to account for the embodied energy of the recycled fiber used in the second-generation part. The SimaPro v.9.0.0.33 LCA software [5] was used for analysis with input for the various materials, processes, and waste scenario details obtained through extensive industry survey, complimented by literature, data from published library databases (Ecoinvent 3 –allocation, cut-off by classification; US-EI 2.2; USLCI [5-8]), and the FRPC Energy Estimator Tool developed by the Oak Ridge National Laboratory [9]. Medium voltage US electricity grid dataset from the DATASMART package, provided by Long Trail Sustainability (2015) is considered for all process steps with the United States considered the geographical focus of this study [7]. The Cumulative Energy Demand (CED) [10, 11] methodology was applied to calculate embodied energy values.

3.2.1 Estimations and Assumptions

The following are the assumptions within the study:

1. Transportation is not considered in this model as the idea is that Vartega will potentially possess modular systems across the country to cater to scrap content at the location of various manufacturers that will generate such scrap material.
2. The fuel blending process for generating the fuel from the captured epoxy is not considered as specific process steps and pathways are unknown at this stage.
3. Generally, 80% of machine load is considered for calculation, unless otherwise specified (as in the case of the prepreg machine).

4. No losses occur in terms of CF or epoxy during the solvolysis process as both are assumed to be 100% reclaimed and reused in later steps.
5. Other environmental impacts associated with the solvolysis process are not included in this study due to the proprietary nature of the solvents in use. There is scope for an additional analysis in the future to understand the other environmental impacts (greenhouse gas emission, acidification potential, effect on climate change, respiratory effects, etc.) in relation to the solvolysis process.
6. Use phase of the final fender product is outside the scope of this analysis.
7. Although the prepreg from the pre-pregging machine step goes on to become a cured product, the steps associated with such processes (SMC, trimming, compression molding, etc.) is not considered within the boundary of this study as it has no direct influence on the prepreg that is sent to Vartega.
8. Values have been back calculated considering the weight of the fender in order to associate those impacts directly associated with producing the amount of (CF) material from the first generation to the second.

3.2.2 Results

Vartega collects the uncured prepreg scrap from various industries as feedstock for the solvolysis process. In the model, the prepreg machine produces a prepreg with 65% carbon fiber (CF) and 35% epoxy content whose scrap is sent to Vartega for the fiber recovery process. Vartega then recovers 100% of the CF and epoxy wherein the recycled CF (rCF) is sent to be compounded with virgin PA 66 to generate pellets for the fender containing 10% rCF (90% resin). The recovered epoxy has three pathways for consideration, which are explained below. The pellets are then fed into the injection molding machine where they are used to form the final fender part. Although there is potential for the fender to potentially be recycled, the second generation considers this part to be landfilled in this model, marking the end of its life cycle. Table 1 demonstrates the various process steps and exported embodied energy values for every material, process, and disposal option considered in both generations (Life 1 & Life 2).

While modeling recycling systems, various approaches may be applied. In this case, the cut-off by classification approach was applied to the product system [9]. With this approach, the recycler is rewarded (burden is not counted for that life), all burden relating to landfilling and virgin material are considered in every generation, and refurbished material take on the value of the recycling process depending on the amount of refurbished material used in the subsequent lives. Considering such an approach, it is seen that Life 1 has a total EE of 203 MJ while considering the amount of CF that is required from the prepreg scrap (0.163 kg as 65% from a total prepreg amount of 0.2507 kg) to accommodate the fender part (as 10% of the fender weight is 0.163 kg rCF) in the later downstream process. There is a small amount (25%) that is disposed to the landfill in Life 1 in the form of the prepreg backing which may contain material such as PE, PET, or wax paper. The EE of Life 2 is found to be 241 MJ, which is 19% more than Life 1. This is owing to the fact that only 10% of rCF is used in the fender with 90% coming from virgin epoxy resin and the part considered to be landfilled at the EOL stage. Energy decreases as more recycled content is used in Life 2. Ultimately, the reduction in the EE of carbon fiber from Life 1 to 2 as a result of undergoing the solvolysis process is overwhelming. Virgin CF exhibits EE at 1155 MJ/kg [8]. This is drastically reduced by 96% in Life 2 with EE of 43 MJ/kg, proving that recycled carbon fiber via the solvolysis process makes a compelling case for both energy and cost savings.

	Category	Process Steps	Embodied Energy (MJ)	Notes	Data source	Data
Life 1	Raw material	Carbon fiber (65%)	188	1 kg of CF has an EE of 1155 MJ => 0.163 kg (65% of 0.2507 kg = 0.163 kg) is 188.265 MJ ~188 MJ	FRPC	Commodity PAN, 50k tow
		Epoxy (35%)	12	1 kg of epoxy has an EE of 135 MJ => (35% of 0.2507 kg = 0.0877 kg) is 11.8395 MJ ~12 MJ	SimaPro	Epoxy resin, liquid, at plant/US- US-EI U Cradle-to-gate (until delivery at plant)
	Prepregging	Heating elements & compaction units	3	2 oil heaters heat up the compaction rollers, pre-impregnation bed and resin bath; 4 compaction rolls work simultaneously. This is run 3 times to create 2 epoxy films with fabric then being sandwiched = 10.77 MJ/kg = 2.7 MJ/0.2507 kg prepreg~3 MJ	SURF Data + SimaPro	Electricity, medium voltage, consumer mix, at grid/US* US-EI U
	EOL	Landfill	0.01	25% of the prepreg backing material that comes with the prepreg scrap is landfilled. This amounts to a minute amount of energy= 0.0097 MJ/0.2507 kg of prepreg~0.11 MJ	Vartega Data + SimaPro	Disposal, plastics, mixture, to US sanitary landfill/US US-EI U
	EOL	Solvolysis & Post Processing	11	Taken from the Solvolysis process LCA. This step will not be considered in the model as the final product is considered to be 100% recycled and cut-off by classification approach rewards the recycler for the amount of product that is recycled in every generation = 43.16 MJ/kg = 10.8218 MJ/0.2507 kg prepreg (43 MJ/kg)	Vartega Data + SimaPro	Electricity, medium voltage, consumer mix, at grid/US* US-EI U
Life 2	Refurbished material	Carbon fiber (10%)	7	Considering 10% refurbished material is used in Life 2, that amounts to 0.163 kg of rCF taken from the solvolysis step. [This value changes according to the percentage of virgin-to-refurbished raw materials considered] Therefore, 0.163 kg rCF has EE of 7.042 MJ	FRPC	Commodity PAN, 50k tow
	Virgin Raw material	PA 6/6 (90%)	201	1 kg of PA 6/6 has an EE of 137 MJ => EE of (90% of 1.63 kg = 1.467 kg) is 201.2245 MJ ~201 MJ	SimaPro	Nylon 66, at plant/US-US-EI U Cradle-to-gate (until delivery at plant)
		Compounding	22	1 kg of twin-screw compounding at Techmer has an EE of 13.4058 MJ/kg. Compounding 1.63 kg required for the fender equates to 21.8515 MJ/part ~22 MJ	Techmer Data + SimaPro	Electricity, medium voltage, consumer mix, at grid/US* US-EI U
	Manufacturing	Injection molding	11	1 kg of Injection Molding has an EE of 6.8332 MJ/kg. IM the fender part weighing 1.63 kg has an EE of 11.1382 MJ/part ~11 MJ	SURF Data + SimaPro	Electricity, medium voltage, consumer mix, at grid/US* US-EI U
	EOL	Landfill	0.3	Plastics mixture was taken for modeling consideration of Generation 2 part (1.63 kg) =>0.2522 MJ. All burden of landfilling is taken up in this generation as there is no mention of recycling at this stage.	SimaPro	Disposal, plastics, mixture, to US sanitary landfill/US US-EI U

Legend

Material	EOL Recycling
Manufacturing process	EOL Disposal

Table 2. The embodied energy of carbon fiber as a prepreg scrap in Life 1 and thermoplastic fender in Life 2

3.2.3 Reuse Options

As mentioned previously, Vartega has the capability to capture epoxy and CF entirely from the prepreg scrap. Some alternative options considered for potential reuse of the recovered epoxy are as follows:

1. Option 1: Resin is sent for fuel blending with the combustion energy captured and used.
2. Option 2: Resin is blended into a virgin epoxy system to offset a portion of the virgin amount.
3. Option 3: Resin is used directly as a product in composite or other applications.

Considering option 1, industries likely to use said fuel after blending are cement kilns. Portland cement data from Ecoinvent was taken to compare how Vartega's potential fuel would decrease the overall energy consumption within a cement kiln process. The default EE of Portland cement is found to be 6 MJ/kg. Data obtained from a partner of Vartega who performed testing on their resins mentioned a heat of combustion of 13,000 BTU/lb (30.49 MJ/kg taking 1 lb = 0.45 kg) tested by bomb calorimeter method. This value when taken as avoided energy (electricity offset) substantially decreases the overall energy of the cement kiln process resulting in excess energy of -96 MJ/kg being produced with a difference of 102 MJ (as 30.49 MJ translates to a CED of -95.65 MJ) and a 1700% decrease in energy due to its high calorific content. Further practical testing is required along with the understanding and implementation of the fuel blending process and transportation considered for analysis.

Similarly, option 2 considers replacement of part of the virgin epoxy with recycled epoxy recovered from the solvolysis process. Vartega has successfully made blends with up to 90% recycled resin thus far. This value was taken to project potential savings in energy in relation to 1 kg of virgin epoxy. Default EE of virgin epoxy is 135 MJ/kg. Incorporating 90% of the recycled epoxy, the new EE becomes 52 MJ/kg with a 61% decrease (83 MJ difference) in embodied energy.

Option 3 would be dependent on the specific raw materials used and the process considered. Research is currently ongoing to understand future applications and other capabilities to achieving this sort of circularity.

4. CONCLUSIONS

The feasibility of producing an automotive part using rCF sourced from automotive prepreg scrap via injection molding has been demonstrated. While the PA66-rCF material produced in this first experimental demonstration showed reduced mechanical properties relative to a standard PA66-vCF material, it is anticipated that the material properties of the PA66-rCF will be improved with further work that is currently ongoing. The mechanical properties of the present demonstration material were sufficient to injection mold a functional automotive body part. Furthermore, using rCF in the injection molding compound significantly reduced the EE of the injection molding compound. These results demonstrate the overall feasibility of the Vartega solvolysis prepreg recycling process as a means to return automotive prepreg scrap to further automotive manufacturing applications.

5. ACKNOWLEDGEMENTS

Institute for Advanced Composites Manufacturing Innovation, a Manufacturing USA Institute is supported by the U.S. Department of Energy's (DOE's) Advanced Manufacturing Office (AMO), within the Office of Energy Efficiency and Renewable Energy.

6. REFERENCES

[1] Hagnell, M. K., and Åkermo, M. (2015). *A composite cost model for the aeronautical industry: Methodology and case study.* Composites Part B: Engineering, 79, 254-261. DOI: 10.1016/j.compositesb.2015.04.043

[2] Hagnell, M. K., and Åkermo, M. (2019). *The economic and mechanical potential of closed loop material usage and recycling of fibre-reinforced composite materials.* Journal of cleaner production, 223, 957-968. DOI: 10.1016/j.jclepro.2019.03.156

[3] Witik, R. A., Payet, J., Michaud, V., Ludwig, C., and Månson, J. A. E. (2011). *Assessing the life cycle costs and environmental performance of lightweight materials in automobile applications.* Composites Part A: Applied Science and Manufacturing, 42(11), 1694-1709. DOI: 10.1016/j.compositesa.2011.07.024

[4] Maxey, A., Kline, S., Schrader, J., Iqbal, M., Fath, A, and Bassetti, S., *Fabrication of a recycled tow carbon fiber overwrapped pressure vessel.* SPE-ACCE Proceedings, 2017

[5] "SimaPro," 9.0.0.33 ed: PRé Consultants, 2018. (June 9). *Allocation cut-off by classification.* Available: https://www.ecoinvent.org/database/system-models-in-ecoinvent-3/cut-off-system-model/allocation-cut-off-by-classification.html

[6] "DATASMART LCI Package," ed: Long Trail Sustainability, 2015.

[7] G. Wernet, C. Bauer, B. Steubing, J. Reinhard, E. Moreno-Ruiz, and B. Weidema, *The ecoinvent database version 3 (part I): overview and methodology.* The International Journal of Life Cycle Assessment, vol. 21, no. 9, pp. 1218-1230, 2016. DOI: 10.1007/s11367-016-1087-8

[8] S. A. Das, Kristina. FRPC Energy Use Estimation Tool [Online]. Available: http://energytoolestimator.com/

[9] R. Hischier *et al.*, *Implementation of Life Cycle: Impact Assessment Methods, Final Report Ecoinvent v2. 2 No. 3* Swiss Centre for Life Cycle Inventories, Dubendorf, Switzerland: 2010.

[10] V. D. Ingenienure, *Cumulative energy demand-Terms, Definitions, Methods of Calculation.* VDI-Richtlinie, vol. 4600, p. 5, 1997.

[11] T. Ekvall and A.-M. Tillman, "Open-loop recycling: criteria for allocation procedures," *The international journal of life cycle assessment,* vol. 2, no. 3, p. 155, 1997.

A Mechanical Test Frame for Property Evaluations at Cryogenic Temperature

Daniel R. Merkel, Gayaneh Petrossian, Dustin Clelland, Kevin Simmons

Pacific Northwest National Laboratory
Richland, WA 99352, USA

ABSTRACT

On-board storage of hydrogen fuel has been designated as a limiting factor in the advancement of fuel cell technologies in the automotive industry. The use of cryo-compressed Type III pressure vessels presents one option to overcome this barrier. These multi-material vessels will be expected to perform at high pressure and extreme low temperatures, which creates a complex engineering design challenge. Many materials exhibit highly temperature-dependent properties, which must be considered for the efficient design of cryo-compressed pressure vessels. Complicating this issue, mechanical property data at cryogenic temperatures is sparse. This technical paper provides a description of a test apparatus commissioned specifically to help address this shortcoming. A mechanical test frame was retrofitted with a continuous flow cryostat capable to evaluate various mechanical properties throughout a broad temperature range of 25 °C to -269 °C with a load limit of 10 kN. An investigation of the thermomechanical properties of an epoxy resin was carried out as a demonstration.

Corresponding author: Daniel R. Merkel (daniel.merkel@pnnl.gov)

1. INTRODUCTION

The Department of Energy Fuel Cell Technology Office has designated the on-board storage of hydrogen fuel as a limiting factor in the advancement of fuel cell technologies in the automotive industry[1]. The key challenge in this regard is overcoming the low energy density of hydrogen at ambient temperature and pressure. One approach to improve this metric is cryo-compressed hydrogen storage vessels[2]–[4]. Target design specifications for such vessels are pressures up to 700 bar and temperatures as low as -253 °C[5]. Simultaneously, on-board cryo-compressed vessels must also meet weight constraints associated with automotive components to maintain vehicle efficiency[6]. Given these criteria, recent analyses deemed Type III pressure vessels to be of particular interest for this design challenge[7], [8].

Type III pressure vessels are commonly used in industrial applications for gas storage when low system weight is of importance[6], [7]. The vessels represent a multimaterial system consisting of a metal tank liner with a filament wound carbon fiber composite overwrap. This combination of dissimilar materials provides an efficient means to contain high pressure while conserving weight[7]. Pressure vessel designers rely on available material properties data to design efficient vessels; however, this data is generally limited at extreme low temperatures. As a result, designers must rely on best-estimates interpolated from just a few existing data points. This is especially problematic considering that the available data is reliable only for industrially relevant materials and has largely been collected at the boil-off temperatures of cryogenic liquids, *i.e.* -269 °C (helium) and -196 °C (nitrogen) [9]–[11].

This technical paper documents the recent development of a cryogenic test system commissioned specifically to address the current needs of cryo-compressed vessel designers. The system is designed to evaluate mechanical properties in a continuous temperature range of -269 °C to 25 °C at loads up to 10 kN. This paper provides an overview of the design and operation of the instrument. The tensile properties of a generic epoxy resin were investigated at -233 °C, -125 °C, and 25 °C. Techniques for geometric design of tensile specimens and fixturing methods to overcome complications associated with cryogenic testing are described.

2. Experimental Methods

2.1 Cryogenic Mechanical Test Apparatus

Mechanical test specimens were cooled to the desired temperature in a continuous flow cryostat purpose-built for mechanical testing (STVP-200, Janis Research Co., Woburn, MA). The cryostat was mounted on a 222 kN (50,000 lbf) limit mechanical test frame (MTS, Eden Prairie, MN) modified to accommodate the cryostat. The hydraulic actuator and main load train were removed from the frame to accommodate the cryostat. The bottom platen of the frame was lowered and fitted with an adapting plate to allow fixturing to the cryostat. The upper crosshead was fitted with a second adapting plate such that a servoelectric actuator (10K Servo Creep Controller, Interactive Instruments, Scotia, NY) could be mounted with the main load train extending from the actuator down through a center hole in the crosshead. A 44.5 kN (10,000 lbf) load cell was mounted directly to the actuator in-line with the main load train. The apparatus is shown in Figure 1.

The cryostat consisted of an external mechanical load frame (labeled in Figure 1a), an insulating vacuum jacket, and an internal mechanical load frame (shown in Figure 1b). The external frame remained permanently fixtured to the insulating jacket; the internal frame was removed from the cryostat between each test to remove and replace test specimens. A load rod extended from the test space of the internal frame through the top plate of the cryostat, supplying a direct mechanical linkage between the test specimen inside the cryostat to the main load train of the test frame. The load rod was linked to the load train by a pin-clevis type linkage, which allowed the cryostat to be mounted/removed from the test frame rapidly between tests. Zero-load conditions were maintained on the test specimen during mounting/removal of the cryostat using a collar limiting vertical motion of the load rod, along with button-head style specimen grips designed to prevent compression of the test specimen. The cryostat was rated for a maximum load of 10 kN (2,250 lbf.).

Tensile load was applied to test specimens via the load rod linked to the servoelectric actuator. The reaction force was carried through the bottom specimen grip, transferred to the internal load frame through the bottom plate, then transferred to the external load frame by the top plate. During a tensile test, the internal frame and external frame assumed compressive and tensile states, respectively.

Figure 1. (a) Mechanical testing cryostat assembled in MTS load frame with cryogen storage tank. (b) Internal mechanical frame of the cryostat.

2.2 Temperature control

Temperature was manipulated by coordination of manually controlled cryogen flow and digital PID controlled heating elements. Cooling was achieved by the continuous flow of liquid cryogen from a cryogen storage dewar to the cryostat via a purpose-built vacuum insulated transfer line (FHT-ST, Janis Research Co.). The insulating vacuum jacket of the transfer line was evacuated to 10^{-4} Torr using a turbomolecular vacuum pump. Either liquid helium or liquid nitrogen was used as the cryogen with minimum temperatures corresponding to the boiling points of each species, *i.e.* -269 °C and -196 °C, respectively. Cryogen flow into the cryostat was encouraged by creating a pressure differential between the cryogen storage dewar and the sample space of cryostat. In the case of liquid helium, slight vacuum was generated in the sample space of the cryostat using a mechanical vacuum pump. In the case of liquid nitrogen, pressurized nitrogen gas was used to maintain 27.6 kPa (4 psi) in the storage dewar while the sample space of the cryostat was maintained at ambient pressure. A manual valve in the transfer line was used to control the flow rate of cryogen. Liquid cryogen flowed through the transfer line to a small compartment at the bottom of the cryostat where the cryogen was vaporized to gas phase. The cryogen gas then flowed into the sample space to cool the test specimen and internal load frame.

Silicon diode heaters/temperatures sensors were mounted in the cryostat in two spaces. Two heaters/sensors mounted in the vaporizer space controlled the rate of vaporization of liquid cryogen as it flowed into the cryostat. Another two heaters/sensors mounted on the bottom plate of the internal frame of the cryostat near the test specimen controlled the temperature of the specimen. The temperatures of the vaporizer and sample spaces were controlled independently by a temperature controller (Model 336, Lake Shore Cryotronics, Westerville, OH). Typically, the temperature of the vaporizer space was set to 2 °C less than that of the sample space.

2.3 Tensile Geometry

Two specimen geometries were explored for tensile testing. First, the ASTM D638 Type II specimen, shown Figure 2a, was explored. Several issues were discovered during preliminary testing. First, the large material volume and surface area proved problematic at lower temperatures due to defect sensitivity of the epoxy. Rarely could such large specimens be made without significant defects such as air bubbles occurring in the narrow section of the specimen. When defects were influential, tensile fracture always initiated from a corner of the square cross-section. Additionally, gripping the flat-type geometry using self-tightening wedge grips at sub-ambient temperatures was often complicated by ice formation on the grip surface or stress risers developing outside the narrow section of the specimen due to gripping. Complications also arose due to thermal contraction during cooling, which imposed significant and undesirable tension on the specimen during test setup.

To mitigate these issues, a sub-size ASTM E8 cylindrical geometry was adopted, shown in Figure 2. The small volume and circular cross-section reduced the number of defects in each specimen. This geometry also allowed a button-head style gripping method that was not complicated by icing or stress risers outside the narrow section. This gripping technique also eliminated issues related to thermal contraction during cooling. Test specimens could be setup in the grips and instrumented with an extensometer without imposing a tensile state. Thus, the test assembly could be cooled to extreme low temperatures without inducing an undesired tensile state in the specimen. Fixturing grips were designed accordingly with special consideration of mechanical properties throughout the temperature of 0 °C to 25 °C. Nitronic 60 steel was chosen as an alternative to 300-series stainless steel to minimize mass of the grip while maintaining ductility at extreme low temperatures.

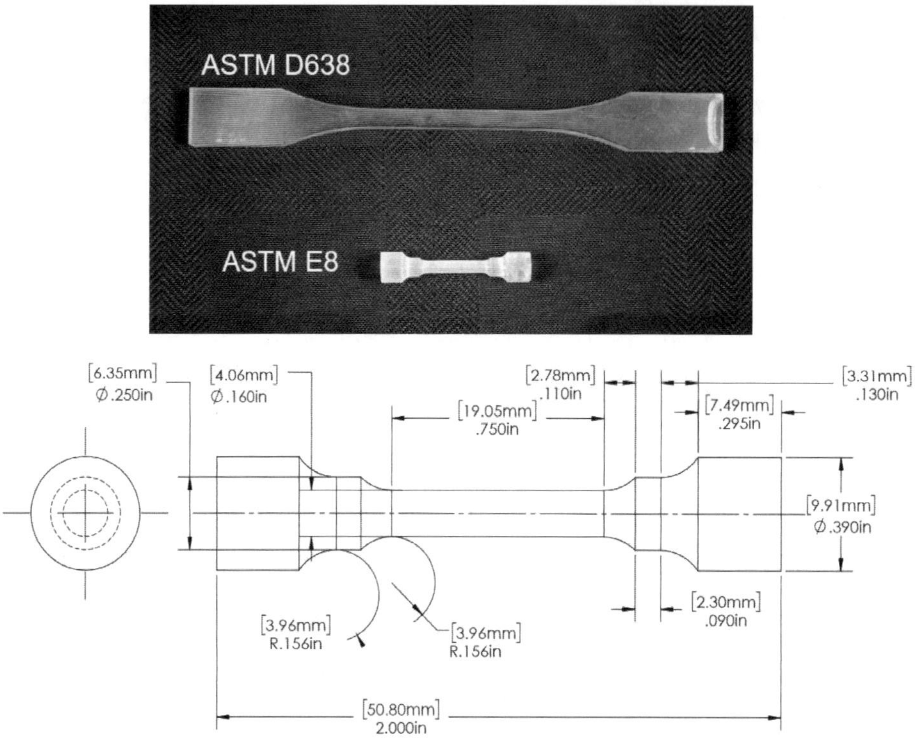

Figure 2. Tensile geometry for ASTM D638 Type II and sub-size E8.

2.4 Tensile Property Evaluation

Tensile properties were evaluated at a constant machine rate of 1.91 mm/min (0.075 in/min). A mechanical extensometer (3442-0064-025T-LHT, Epsilon Technology Corp, Jackson, WY) rated for minimum temperatures of -270 °C was used to evaluated strain during each tensile test. Tensile modulus was determined by linear fit of stress-strain curves in the initial portion (typically less than 1 % strain) of each test.

2.5 Materials

Commercially available bisphenol-A epichlorohydrin epoxy resin (EPON resin 828, Hexion Specialty Chemicals, Columbus, OH) having a density of 1.16 g/ml and a viscosity of 110-150 P at 25 °C was used as the base resin. Technical grade triethylenetetraamine (TETA) curing agent was purchased from Sigma-Aldrich (St. Louis, MO) with 60 % purity and a molecular weight of 146.23 g/mol. A stoichiometric amount of TETA was added to the epoxy resin at 18.1 phr and the mixture was degassed under 30 in-Hg vacuum for approximately 5 minutes. Prior to mixing, the resin was heated at 60 °C for 30 minutes. The resin mixture was then poured into silicone rubber molds and cured overnight at room temperature. The partially cured samples were taken out of the molds and post-cured in the oven at 100 °C for 3h. The samples were then heated to 125 °C for 1h and quench cooled to room temperature to relieve internal stress due to curing.

Differential scanning calorimetry (DSC Q2000, TA instruments, New Castle, DE) was used to establish the cure schedule and confirm full cure. A small amount (approximately 5 mg) of cured epoxy was cut and packaged in a closed lid aluminum Tzero DSC pan. Samples were heated to 200 °C at 10 °C/min, kept for 5 minutes to reach equilibrium, then cooled down to 25 °C at 10 °C/min, kept for another 5 minutes, and finally heated to 200 °C again to confirm the full curing of the sample.

Dynamic mechanical analysis (DMA Q800, TA instruments) was performed to better understand the thermomechanical behavior of the resin mixture. All measurements were performed using a heating rate of 2 °C/min starting from -140 °C. Dual cantilever grips were used at a fixed strain, with an amplitude of 5 μm and an oscillation frequency of 1 Hz. TA Universal Analysis software was used for analyzing the DSC and DMA data.

3. Results

3.1 Thermal and Dynamic Properties

A comprehensive differential scanning calorimetry study was carried out, both isothermal and dynamic, in order to establish a cure schedule for the mixture and eliminate any existing thermal stress. For the first step of the analysis, a mixture of the resin and curing agent was heated to a maximum temperature resulting in full cure without reaching the degradation temperature. The resin and the curing agent were mixed at room temperature and tested immediately afterwards. An exothermic curing peak was observed around 100 °C, as shown in Figure 3a. During cooling, a glass transition of 114 °C was recorded. Figure 3b demonstrates the DSC curve of the mixture after curing. No exothermic peaks were observed, which indicates the full curing of the sample. The glass transition was determined as the middle of the sloped region, which varies slightly during heating and cooling. The glass transition temperature noted during heating and cooling for were at 127 °C and 113 °C, respectively.

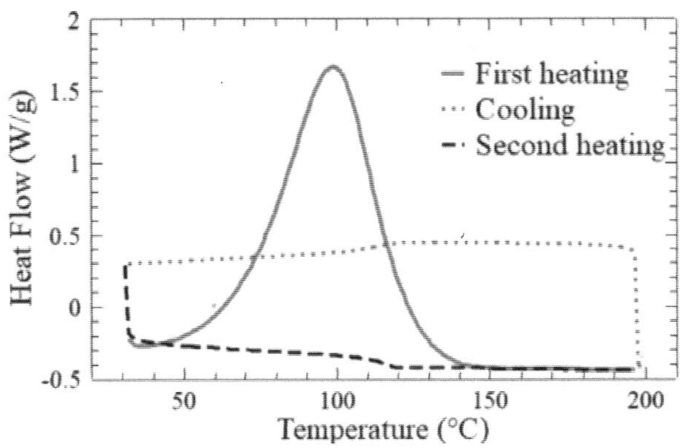

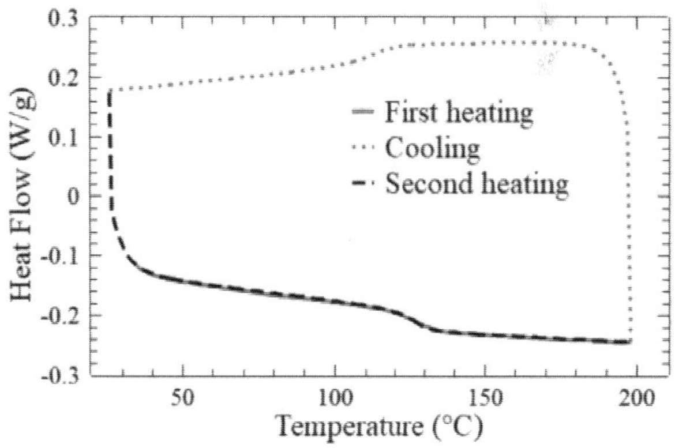

Figure 3. Differential scanning calorimetry (DSC) thermographs of (a) resin and curing agent immediately after mixing and (b) fully cured sample. Exothermic heat flow is oriented upward on the vertical axis.

Figure 4 shows the DMA curve for the fully cured epoxy. Three thermal transitions were observed, as indicated by the loss modulus peaks at 120 °C, 46 °C, and -42 °C. The high temperature peak at 120 °C corresponds to the glass transition or α-transition. Above this point, the samples were rubbery due to increased mobility of the molecular network chain. This transition appeared at a slightly higher temperature when evaluated by DSC owing to the significantly faster heating rate of 10 °C/min used in the DSC method. A subtle intermediate relaxation called the ω-transition was observed at 46 °C attributed to mobility of larger segment of the network structure[12]. Finally, the β-transition was observed at -42 °C representing the local and short segment motion of the chains. Below this temperature, the sample was significantly more brittle.

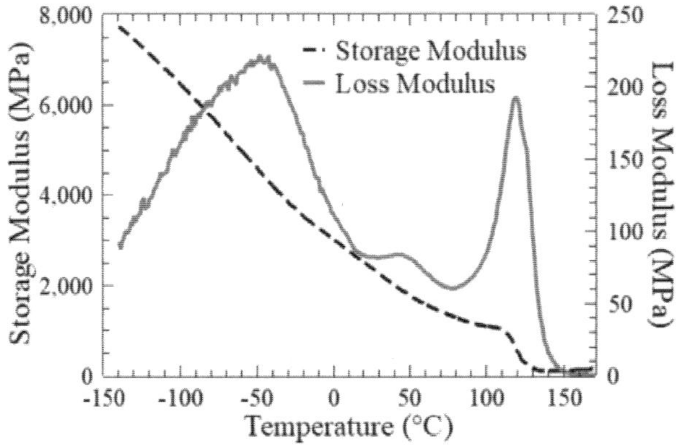

Figure 4. Dynamic mechanical analysis (DMA) of fully cured epoxy.

3.2 Tensile Properties

The cryostat mechanical test frame was used to evaluate tensile properties of epoxy specimens at a fixed crosshead rate. Tests were conducted at 25, -125, and -233 °C to examine the influence of temperature of tensile modulus. Stress-strain curves are shown in Figure 5 and tensile modulus is shown tabulated in Table 1. At 25 °C, the tensile response was approximately linear up to about 2 % strain. Continued deformation caused yielding as ductile deformation took place until failure. The tensile response at -125 °C and -233 °C were strictly brittle and linear to the point of failure. This transition from ductile to brittle behavior was consistent with DMA results, which showed significant thermal transition at -42°C. Likewise, tensile modulus increased as temperature decreased.

Tensile strength observed at low temperatures was significantly lower compared to the strength observed at 25 °C. Considering the brittle nature of failure, defect sensitivity appears to be a limiting factor causing premature failure. This is consistent with previous observations of epoxies at extreme low temperatures [13]. Additional stresses induced by the mechanical extensometer may also have played a role in limiting observable tensile strengths. These stresses would have been induced at the points where the extensometer arms attach to the specimen and result from a combination of the attachment method and extensometer weight. Typically, these influences are negligible compared to the strength of the material and are not expected to influence material performance. However, materials such as epoxy become increasingly brittle and sensitive to these influences at lower temperatures, making this a particularly difficult issue to resolve.

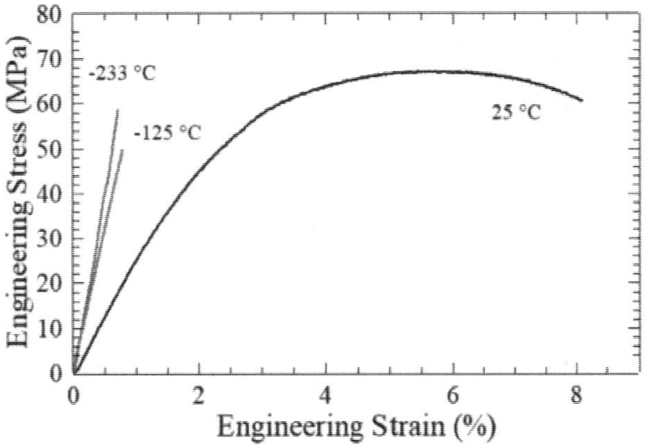

Figure 5. Representative tensile curves of epoxy at three temperatures.

Table 1. Tensile modulus of epoxy.

Temperature (°C)	Tensile Modulus (GPa) ± SD	Tensile Modulus (ksi) ± SD
25	2.76 ± 0.14 (*n*=6)	400 ± 21 (*n*=6)
-125	6.14 ± 0.26 (*n*=2)	890 ± 37 (*n*=2)
-233	8.29 (*n*=1)	1203 (*n*=1)

4. Conclusions

A mechanical test frame was retrofitted to accommodate a continuous flow liquid helium cryostat. The system was designed to evaluate mechanical properties at temperatures of -269 °C to 25 °C and loads up to 10 kN. A cylindrical geometry based on ASTM E8 was selected to minimize material defects, facilitate specimen gripping at cryogenic temperatures, and mitigate issues with thermal contraction of the specimen and/or instrument during cool-down.

Thermomechanical properties of an epoxy consisting of a bisphenol-A epichlorohydrin resin cured with TETA were studied. A curing schedule was established by DSC analysis of the epoxy mixture during and after cure. Thermal transitions at 120 °C, 46 °C, and -42 °C were observed by DMA, indicated by peaks of the loss modulus curve. Tensile properties were evaluated using the cryogenic mechanical test frame at three temperatures: 25 °C, -125 °C, and -233 °C. At 25 °C, epoxy deformed non-linearly with yielding beyond approximately 2% strain. The tensile response at -125 °C and -233 °C was linear with brittle failure. The transition from ductile to brittle behavior with temperature decrease agreed with the transition observed at -42 °C by DMA. Tensile modulus increased with decreasing temperature.

5. Acknowledgements

This work was fully supported by the U.S. Department of Energy (US DOE), Office of Energy Efficiency and Renewable Energy (EERE), Fuel Cell Technologies Office (FCTO) under Contract No. DE-AC05-76RL01830.

6. Disclaimers

any agency thereof, nor any of their employees, makes any warranty, expressed or implied, or assumes any legal liability or responsibility for the accuracy, completeness, or usefulness of any information, apparatus, product, or process disclosed, or represents that its use would not infringe privately owned rights.

7. References

[1] U.S. DRIVE Partnership, "Hydrogen Storage Tech Team Roadmap," 2017.

[2] J. Moreno-Blanco, G. Petitpas, F. Espinosa-Loza, F. Elizalde-Blancas, J. Martinez-Frias, and S. M. Aceves, "The storage performance of automotive cryo-compressed hydrogen vessels," *Int. J. Hydrogen Energy*, vol. 44, no. 31, pp. 16841–16851, 2019.

[3] D. J. Durbin and C. Malardier-Jugroot, "Review of hydrogen storage techniques for on board vehicle applications," *Int. J. Hydrogen Energy*, vol. 38, no. 34, pp. 14595–14617, 2013.

[4] H. Barthelemy, M. Weber, and F. Barbier, "Hydrogen storage: Recent improvements and industrial perspectives," *Int. J. Hydrogen Energy*, vol. 42, no. 11, pp. 7254–7262, 2017.

[5] R. K. Ahluwalia, J. K. Peng, H. S. Roh, T. Q. Hua, C. Houchins, and B. D. James, "Supercritical cryo-compressed hydrogen storage for fuel cell electric buses," *Int. J. Hydrogen Energy*, vol. 43, no. 22, pp. 10215–10231, 2018.

[6] S. Aceves, "Analytical and experimental evaluation of insulated pressure vessels for cryogenic hydrogen storage," *Int. J. Hydrogen Energy*, vol. 25, no. 11, pp. 1075–1085, Nov. 2000.

[7] J. H. S. Almeida, H. Faria, A. T. Marques, and S. C. Amico, "Load sharing ability of the liner in type III composite pressure vessels under internal pressure," *J. Reinf. Plast. Compos.*, vol. 33, no. 24, pp. 2274–2286, Dec. 2014.

[8] B. N. Nguyen, D. R. Merkel, K. I. Johnson, D. W. Gotthold, K. L. Simmons, and H. S. Roh, "Modeling the effects of loading scenario and thermal expansion coefficient on potential failure of cryo-compressed hydrogen vessels," *Int. J. Hydrogen Energy*, no. xxxx, Oct. 2019.

[9] B. Atli-Veltin, "Cryogenic composite fuel tanks: The mechanical performance of advanced composites at low temperatures," *2018 AIAA/ASCE/AHS/ASC Struct. Struct. Dyn. Mater. Conf.*, no. January, 2018.

[10] J. Schneider, M. Dyess, C. Hastings, J. Patterson, J. Noorda, and T. DeLay, "Lightweight Cryogenic Composit Over-wrapped Pressure Vessels (COPVS) for Launch Vehicle Applications," in *48th AIAA/ASME/ASCE/AHS/ASC Structures, Structural Dynamics, and Materials Conference*, 2007, no. April, pp. 1–19.

[11] T. K. DeLay, J. Patterson, J. Noorda, and J. Schneider, "Development of Cryogenic Composite Over-Wrapped Pressure Vessels (COPV's)," *SAMPE J.*, vol. 44, no. 3, 2008.

[12] R. Ramsdale-Capper and J. P. Foreman, "Internal antiplasticisation in highly crosslinked amine cured multifunctional epoxy resins," *Polymer (Guildf).*, vol. 146, pp. 321–330, 2018.

[13] R. P. Reed and J. D. McColskey, "NISTIR5024: Properties of Neat-Resin Systems," in *Composite Struts for SMES Plants*, NIST, 1994, pp. 17–200.

HIGH VELOCITY IMPACT RESPONSE OF HYBRID PSEUDO-WOVEN FIBER-REINFORCED COMPOSITE LAMINATES

Cyrus Vakili Rad [a,b], Julie Roark [a,b], Duane Revilock [c], Ramy Harik [a,b], Subramani Sockalingam[a,b] , Zafer Gurdal [a,b],

[a] McNair Center for Aerospace Innovation and Research, University of South Carolina, SC, USA
[b] Department of Mechanical Engineering, University of South Carolina, SC, USA
[c] Impact Dynamics Group, NASA Glenn Research Center, OH, USA

ABSTRACT

An intermediary architecture of woven laminates and unidirectional laminates, referred to as pseudo-woven laminates, is proposed as an alternative hybridized structure to improve impact properties of composite laminates. Pseudo-woven laminates make use of an automated fiber placement (AFP) manufacturing process to produce a unique laminate architecture. These laminates are heterogeneous with spatially defined features where elasticity tensor is spatially variable unlike traditional composites. The heterogeneity is associated with interply locking and topological variations in the fiber orientations within individual layers resulting in multiple interfaces and an expanded design space.

In this study, carbon/epoxy pseudo-woven subassemblies are hybridized with traditional unidirectional layups to enhance impact and damage tolerance. Three different laminate configurations are assessed under high velocity impact according to ASTM D8101. During the experiments digital image correlation (DIC) and high-speed video were employed to evaluate the laminate's response. Preliminary results on 24 ply laminates indicate pseudo-woven architecture exhibits increased penetration resistance and reduced back face surface damage compared to traditional laminates.

Corresponding author: Subramani Sockalingam

1. INTRODUCTION

Traditional carbon fiber reinforced polymer matrix composite material systems play a critical role in various engineering applications. These composites exhibit brittle failure, limited toughness, and poor damage tolerance (stiffness-strength-toughness dilemma) [1]. These systems are often overbuilt, thus any increase in damage tolerance can yield significant cost and weight savings. Creating novel composite architectures aided by advanced manufacturing is an attractive approach to enhance toughness and damage tolerance in laminated composites. These composite systems used in aerospace structures need to be designed to withstand low and high velocity impact events [2].

3D woven laminates provide through thickness reinforcement and have been shown to prevent delamination, resist crack propagation, and increase toughness. 3D woven laminates are shown to have a stepwise fracture in tensile tests and a higher failure strain than their 2D counterparts. However, 3D woven laminate architecture reduces in-plane tensile strength and stiffness due to the reorientation of fibers [3]. Pseudo-woven (PW) laminates are proposed as an alternative laminate configuration to overcome the shortcomings of 2D and 3D laminates. PW laminates have been shown to have similar tensile properties while exhibiting enhanced damage tolerance compared to traditional composite configurations [3], [4], [5].

In this preliminary study, a woven-like composite laminate referred to as pseudo-woven was manufactured using unidirectional slit tapes. This laminate architecture uses an Automated Fiber Placement (AFP) technique to selectively alter the placement of tows in a layer resulting in a woven like architecture [3], [4], [5].

2. MANUFACTURING

2.1 Materials and curing

In this study three laminate configurations were manufactured using IM7 G/8552-1 slit tapes and were cured at 176 °C and 6.2 Bar. The first with PW laminates on the outer surfaces, referred to as Weave – Traditional – Weave or P1-WTW. The second with PW laminates on the inside called Traditional – Weave – Traditional or P2-TWT, and the third quasi-isotropic control traditional laminate called P3 - CONT. Each panel was manufactured using automated fiber placement with care taken to ensure proper tow alignment during layup, a critical aspect of PW manufacturing. These layup configurations will be discussed in section 2.4.

2.2 Pseudo-woven laminates

2.2.1 AFP Notation

PW laminates are manufactured through a manipulation of the layup process typically used in AFP manufacturing. Traditionally AFP manufacturing utilizes full tow coverage when laying tows placing full bands resulting in the conventional laminate construction. When manufacturing a PW laminate, tows are intentionally skipped to produce gaps in the bands. The gaps produced during skipping are filled in the subsequent passes compounding to produce the woven-like architecture. To understand the structure/architecture of the PW laminates, its manufacturing parameters and naming scheme are first discussed. The structure of a PW laminate is described using its AFP notation shown in Figure 1 [4], [5].

$$[\theta^1,\theta^2,\theta^3,\theta^4,..\theta^m][x000x000]_n[w][ds]$$

a b c d

Figure 1. A generic example of AFP notation (a) directional set (b) tow mask (c) physical tow width (d) directional shift [5]

AFP notation is broken down into four parts to describe the architecture of a PW laminate. Figure 1 (a) denotes fiber orientations used in the laminate, called the directional set. The directional set provides the orientations used in the laminate and the order in which orientations are laid. Using Figure 1 as an example, 'θ^{1}' is laid, then 'θ^{2}' to 'θ^{m}', where 'm' is the number of fiber orientations in the laminate. The conventional definition of a ply is a layer of full coverage occupying a discrete Z position. Because of the apparent departure from conventional manufacturing for the PW laminates the authors have defined a ply as a completed directional set. The second grouping in Figure 1, subsection (b) represents the tow mask. This serves as a map of active, *x*, and inactive channels, *0*, of the AFP's head. The subscript *n* on the tow mask denotes the number of times the roller must pass over the domain to produce a panel of a specified width. For example, if a panel is sixteen tows wide then for this eight tow roller, n=2. The tow mask then is used to convey what channels are active and feed tows when the AFP's head makes its passes. Manipulation of this tow mask produces the woven-like architecture of the PW laminates. Figure 1 (c) represents the physical tow width, in this case 6.35 mm. It is possible for layups to have two adjacent active channels, where tow width is then defined as 12.7 mm.

Figure 1 (d) denotes directional shift, which takes effect after each directional set has been laid. Once a directional set has been laid, for a directional shift of *1*, the first fiber orientation in the previous directional set will shift to the end of the new set thus changing the directional set for the next ply. An example of directional shift is shown in Table 1, where column 'A' shows the effect of a directional shift of *1*, while column 'B' shows a shift of *0*. Note that the tow mask shifts irrespective of directional shift.

Table 1. An example of directional shift effect [5]

	A. Directional shift [1]		B. Directional shift [0]
A) Pass 1	$[\theta^1,\theta^2,\theta^3,\theta^4]$ [x000x000]n [6.35mm][1]	B) Pass 1	$[\theta^1,\theta^2,\theta^3,\theta^4]$ [x000x000]n [6.35mm][0]
A) Pass 2	$[\theta^2,\theta^3,\theta^4,\theta^1]$ [0x000x00]n [6.35mm][1]	B) Pass 2	$[\theta^1,\theta^2,\theta^3,\theta^4]$ [0x000x00]n [6.35mm][0]
A) Pass 3	$[\theta^3,\theta^4,\theta^1,\theta^2]$ [00x000x0]n [6.35mm][1]	B) Pass 3	$[\theta^1,\theta^2,\theta^3,\theta^4]$ [00x000x0]n [6.35mm][0]
A) Pass 4	$[\theta^4,\theta^3,\theta^2,\theta^1]$ [000x000x]n [6.35mm][1]	B) Pass 4	$[\theta^1,\theta^2,\theta^3,\theta^4]$ [000x000x]n [6.35mm][0]

Using the AFP notation, [45, 90,-45,0][x000x000]$_n$[6.35mm][0], the manufacturing of a PW panel used in the experiments is shown in Figure 2. In Figure 2 (a) the first directional set is laid onto the tool. The next pass, shown in Figure 2 (b) is laid in the same order as the previous set except the active channel shifts over one space. This process repeats in Figure 2 (c) and Figure 2 (d) completing the PW laminate sub-assembly.

a) b) c) d)

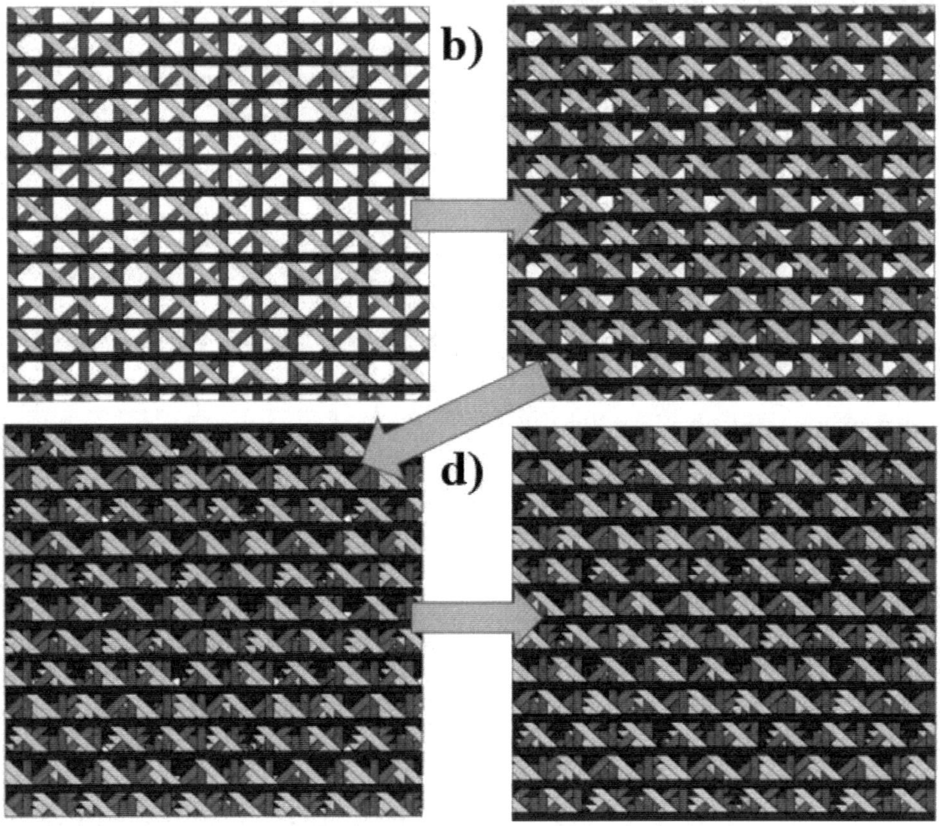

Figure 2. A depiction of the manufacturing process a) Pass 1 b) Pass 2 c) Pass 3 d) Pass 4

2.2.2 PW Naming Convention

The naming scheme for PW laminates directly represents their manufacturing parameters described in the AFP notation, for example: 4D-DS0-CS3 is a 4 ply laminate with 4 fiber orientations with adjacent active channels producing a double width tow represented here as *D*. The alternative to this would be a single width tow represented with an *S*. Directional shift is represented as *DS*, the number directly following *DS* is the quantity of the directional shift. The tow mask is represented by *CS* and is the number of inactive channels, or channels skipped, between active channels. The PW laminates manufactured for these experiments are 4S-DS0-CS3 and INV 4S-DS0-CS3. Their AFP notations are [45, 90, -45, 0][x000x000]n[6.35mm][0] and [0, -45, 90, 45][x000x000]n[6.35mm][0] respectively.

2.3 Hybridized panels

The manufacture of PW laminates is a time intensive process, PW laminates with 3 inactive channels between active channels take roughly four times longer to lay than a traditional laminate. As such, manufacturing an entirely PW laminate is an extremely time-consuming process. However, hybridizing a traditional laminate structure with PW laminates can be used to reduce manufacturing time while gaining attributes of PW laminates.

2.4 Laminate Manufacturing

For these experiments two hybrid structures are manufactured, the first configuration known as P1 - TWT is a panel with PW laminates sandwiched between traditional. Its stacking sequence is $[(45/90/-45/0)_2/4S\text{-}DS0\text{-}CS3/INV4S\text{-}DS0\text{-}CS3/(0/-45/90/45)_2]$. The other hybridized PW panel is a traditional laminate sandwiched between PW laminates, called P2 – WTW whose stacking sequence is $[4S\text{-}DS0\text{-}CS3_2/45/90/-45/0/0/-45/90/45/INV4S\text{-}DS0\text{-}CS3_2]$. The control sample's stacking sequence is a $[45/90/-45/0]_{3s}$ and is referred to as P3 - CONT. All panels manufactured for these tests are 24 plies thick. Panels made for this study were constructed using two different AFP machines, multiple panels of P1 - TWT, P2 - WTW, and P3 - CONT were made at the McNAIR Center University of South Carolina while one P2 - WTW panel was made at the NASA Langley Research Center. The University of South Carolina's AFP machine is a gantry machine with a refrigerated creel made by Ingersoll Machine Tools. The machine at the NASA Langley Research Center is made by Electroimpact, the machine is a robotic arm with an AFP head fixed to the end.

2.5 Specimen Preparation

Laminates were made into oversized sheets and trimmed to the 305mm (12 inch) by 305mm (12 inch) specimen specifications in ASTM Standard D8101 using a waterjet. After waterjet trimming edges were hand sanded to reduce the effect of boundary damage caused by cutting. A circular hole pattern was milled using a 4 flute carbide endmill on a CNC machine to ensure a proper fit on the test fixture. After specimens were trimmed and brought into geometrical specifications a coarse speckle pattern, Figure 3 was added onto the backside of the laminate. The speckle pattern was applied with spray paint using a stencil for each specimen.

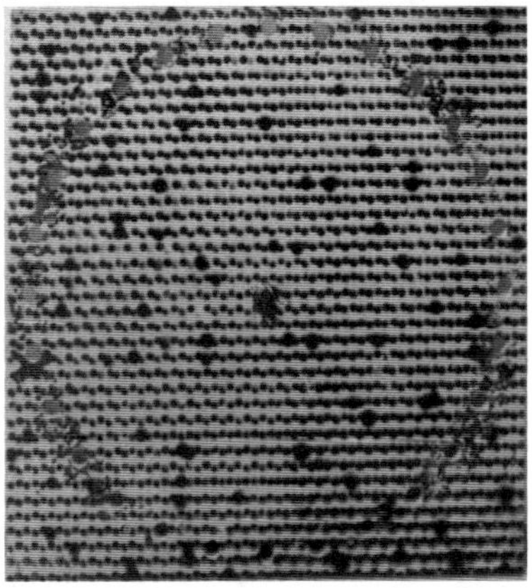

Figure 3. A typical example of the speckle pattern used in experimentation.

3. EXPERIMENTATION

3.1.1 Experimental Setup

Impact tests were performed in accordance to the ASTM Standard D8101 at the NASA Glenn Research Center Ballistic Impact Lab. In this experiment, composite flat panels were impacted in the normal direction with hemispherical aluminum projectiles, Figure 4, weighing approximately 51 g. Specimens were clamped to the rear frame of the test fixture, Figure 5, by a circular frame with an inner diameter of 254.064 mm to ensure a rigid connection with the rear frame [6].

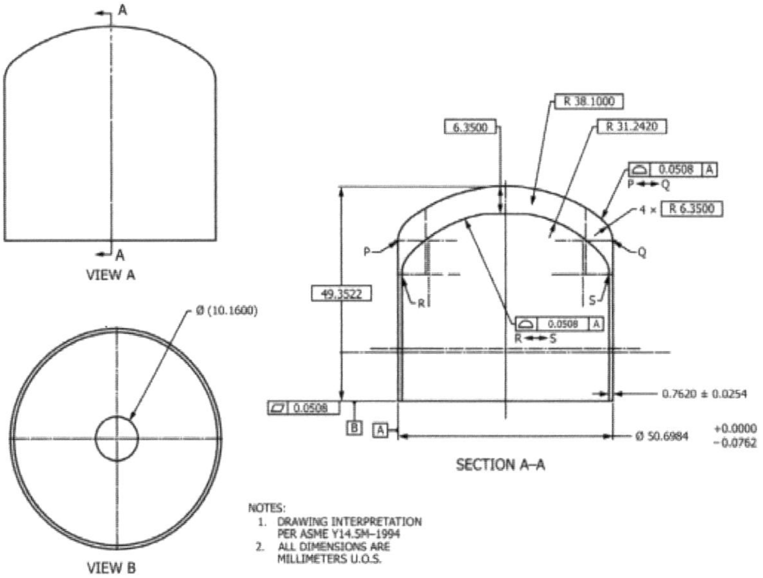

Figure 4. The technical drawing of the impactor used [6].

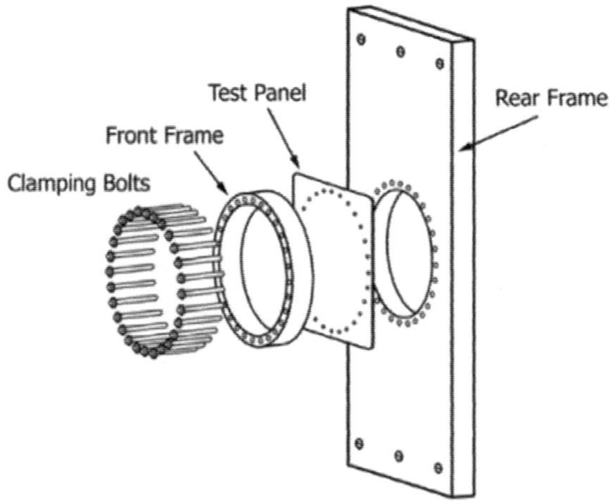

Figure 5. Exploded view of the test fixture [6].

Impactor velocities were determined using the TEMA motion analysis software by tracking a speckle pattern on the impactor. Multiple high-speed cameras were used to extract experimental data with the assistance of ARAMIS and TEMA. The camera setup for experimentation is shown in Figure 6 multiple cameras are used to collect data from multiple viewpoints. A single camera on the side of the test housing was used to capture projectile's impact and rebound velocity. Another camera at the front of the impact housing, above the projectile barrel, was used to track the strike face. Camera pairs on the backside of the test housing were used to capture backside displacement using DIC. The DIC data was processed using ARAMIS.

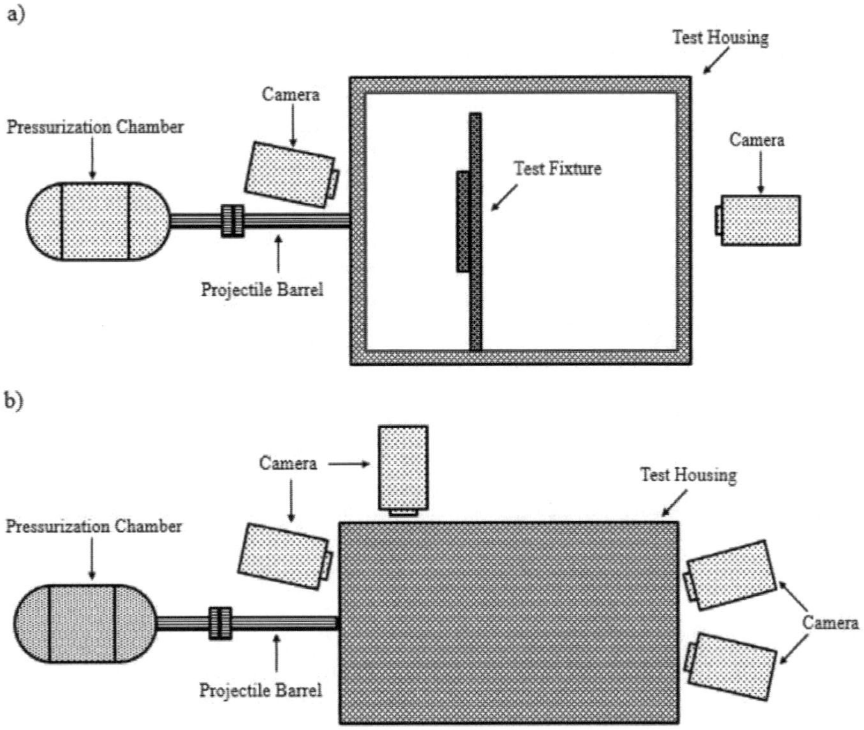

Figure 6. Multiple angles of the experimental camera setup. a) side view b) top view.

4. RESULTS AND DISCUSSION

4.1 Experimental Results

In this section, kinetic energy loss, V_{50}, and displacement are used to assess the response of the laminate configurations. Due to the cost of the panels the sample size for the impacted specimens is relatively low. Five velocities were targeted for this series of tests, 76.20 m/sec (250 ft/sec), 91.44 m/sec (300 ft/sec), 106.68 m/sec (350 ft/sec), 114.30 m/sec (375 ft/sec), and 121.92 m/sec (400 ft/sec). Table 2 shows a summary of raw test results this data is used to tabulate results in the sections below.

Table 2. A table including the target impact velocities, test IDs, panel descriptions, projectile masses, impact velocities, rebound velocities, exit velocities, and results. 'Con' means that the projectile was contained while 'pen' means that the projectile penetrated the specimen. Where applicable U.S. customary units are shown in parenthesis.

Targeted Impact Velocity m/s (ft/s)	Test ID	Panel Description	Projectile Mass (g)	Impact Vel m/s (ft/s)	Rebound Vel m/s (ft/s)	Exit Vel m/s (ft/s)	Result
76.20 (250)	LVG1305	P1 - TWT	52.16	71.29 (233.9)	-30.67 (-100.65)	-	con
	LVG1301	P2 - WTW	51.89	78.05 (256.1)	-31.30 (-102.7)	-	con
	LVG1326	P3 - CONT	49.84	78.42 (257.3)	-33.23 (-109)	-	con
91.44 (300)	LVG1302	P1 - TWT	51.87	93.81 (307.8)	-40.78 (-133.8)	-	con
	LVG1297	P2 - WTW	52.03	89.15 (292.5)	-28.65 (-94.0)	-	con
	LVG1306	P3 - CONT	52.02	90.98 (298.5)	-35.14 (-115.3)	-	con
106.68 (350)	LVG1303	P1 - TWT	52.15	107.50 (352.7)	-34.77 (-114.1)	-	con
	LVG1298	P2 - WTW	51.99	109.48 (359.2)	-29.38 (-96.4)	-	con
	LVG1307	P3 - CONT	52.22	107.16 (351.6)	-33.30 (-109.27)	-	con
114.30 (375)	LVG1304	P1 - TWT	52.29	112.74 (369.9)	-37.49 (-123)	-	con
	LVG1300	P2 - WTW	51.94	112.07 (367.7)	-31.27 (-102.6)	-	con
	LVG1308	P3 - CONT	52.05	113.47 (372.3)	-	39.56 (129.81)	pen
121.92 (400)	LVG1327	P1 - TWT	50.33	120.67 (392.2)	-46.02 (-151)	-	con
	LVG1299	P2 - WTW	51.8	119.54 (395.9)	-	23.87 (78.34)	pen

4.1.1 V_{50}

The V_{50} for the different composite configurations is calculated with Equation 1. Where V_{RM} is the maximum impact velocity at which the laminate still rebounded the projectile. The variable V_P is the lowest velocity at which penetration occurred.

$$V_{50} = \frac{V_{RM} + V_P}{2} \qquad [1]$$

Table 3 shows the computed V_{50} values, with P1 - TWT being greater than 120.67 m/s (392.2 ft/s) as the laminate did not penetrate at that impact velocity. V_{50} of P1 - TWT is around 8.5% higher than P3 - CONT and the V_{50} of P2 - WTW is 5.5% higher than P3 - CONT. This suggests that the incorporation of PW laminates increases the penetration resistance.

Table 3. Calculated V_{50} data

Laminate	V_{50} m/s (ft/s)
P1 – TWT	> 120.67 (392.20)
P2 - WTW	116.14 (381.05)
P3 - CONT	110.32 (361.95)

4.1.2 *Kinetic energy loss*

Residual kinetic energy was calculated as following ASTM standard D8101 with Equation 2. E_r is the residual kinetic energy and is computed from the projectile mass, *M*, and the residual velocity, V_r.

$$E_r = \frac{MV_r^2}{2} \qquad [2]$$

Impact Kinetic energy was calculated using Equation 3 from ASTM standard D8101. Impact kinetic energy, E_i, is computed from the projectile mass, *M*, and the projectile impact velocity, V_i.

$$E_i = \frac{MV_i^2}{2} \qquad [3]$$

Kinetic energy loss was calculated as demonstrated in ASTM standard D8101 with Equation 4 where loss in kinetic energy of the projectile as a result of impact is computed.

$$E_a = E_i - E_r \qquad [4]$$

Specimens were impacted at 5 different velocities resulting in clusters around the 150 J, 220 J, 300 J, 330 J, and 360 J impact kinetic energy levels, test data is plotted in Figure 7. The first cluster at 150 J shows that P2 - WTW had the most kinetic energy loss while P1 - TWT had the least. The 220 J cluster has all configurations showing similar results with the P3 - CONT being marginally lower than the rest. The 300 J cluster shows almost indistinguishable differences between the P1 - TWT and P3 - CONT configurations while the P2 - WTW has the highest loss. At the 330 J cluster all configurations seem to be nearly equal with the P2 - WTW configuration being marginally higher than the rest. It is important to note that the P3 - CONT configuration exhibits complete penetration at this impact energy. In the 375 J group P1 - TWT has the lowest loss in impactor kinetic energy while the P2 - WTW configuration has a higher loss and exhibits penetration. The PW hybrid laminates see higher energy absorption than their conventional counterpart in both configurations at every impact energy.

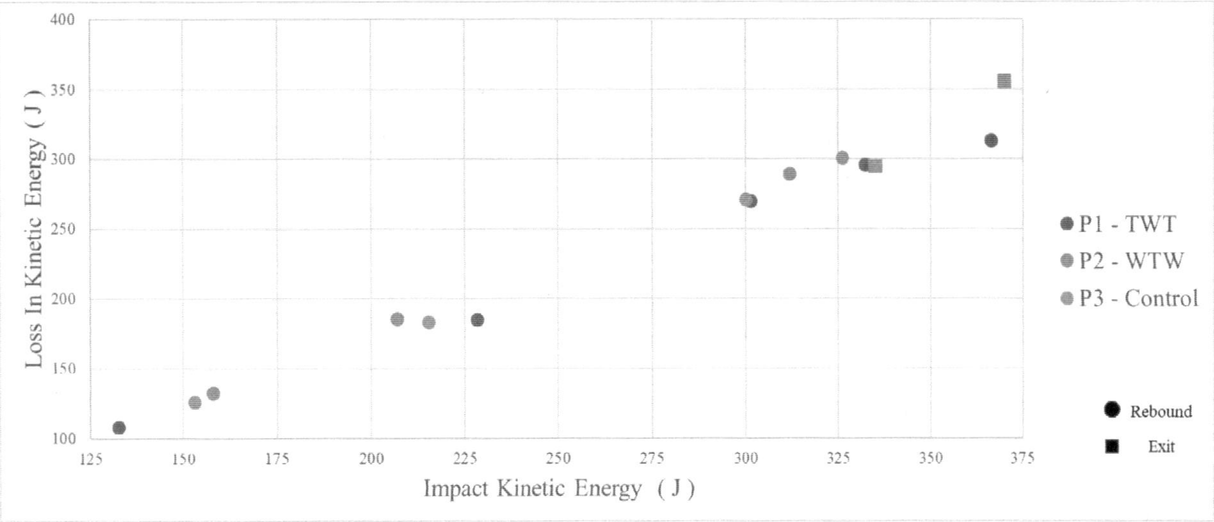

Figure 7. Residual Kinetic Energy vs Impact Kinetic Energy. Circular data points represent non-penetrating or rebounded impactors. Square data points represent penetrating or exiting impactors.

4.1.3 Displacement

Back face displacement was determined using the DIC data taken from the ARAMIS software. The DIC software mapped displacement data into the contours seen in Figure 8. The contours selected for this figure are at 3 different impact velocities, 106.68 m/s (350 ft/sec), 114.30 m/s (375 ft/sec), and 121.92 m/s (400 ft/sec). The images shown in Figure 8 represent peak displacement in the impact event. Peak displacement was determined using a plot output in ARAMIS seen in Figure 9 where Points are placed along a line drawn through the center of the specimen seen in Figure 8. When all points show peak displacement, as shown by the red x's in Figure 9 the corresponding video frame is extracted and is used as peak. Discontinuities in the displacement data are due to the paint breaking from the backside of the specimen during the impact event. Because of the nature of impact some of the contours have an artificially high displacement due to specimen and paint breakage. This value does not represent the behavior of the laminate but rather a small failed portion.

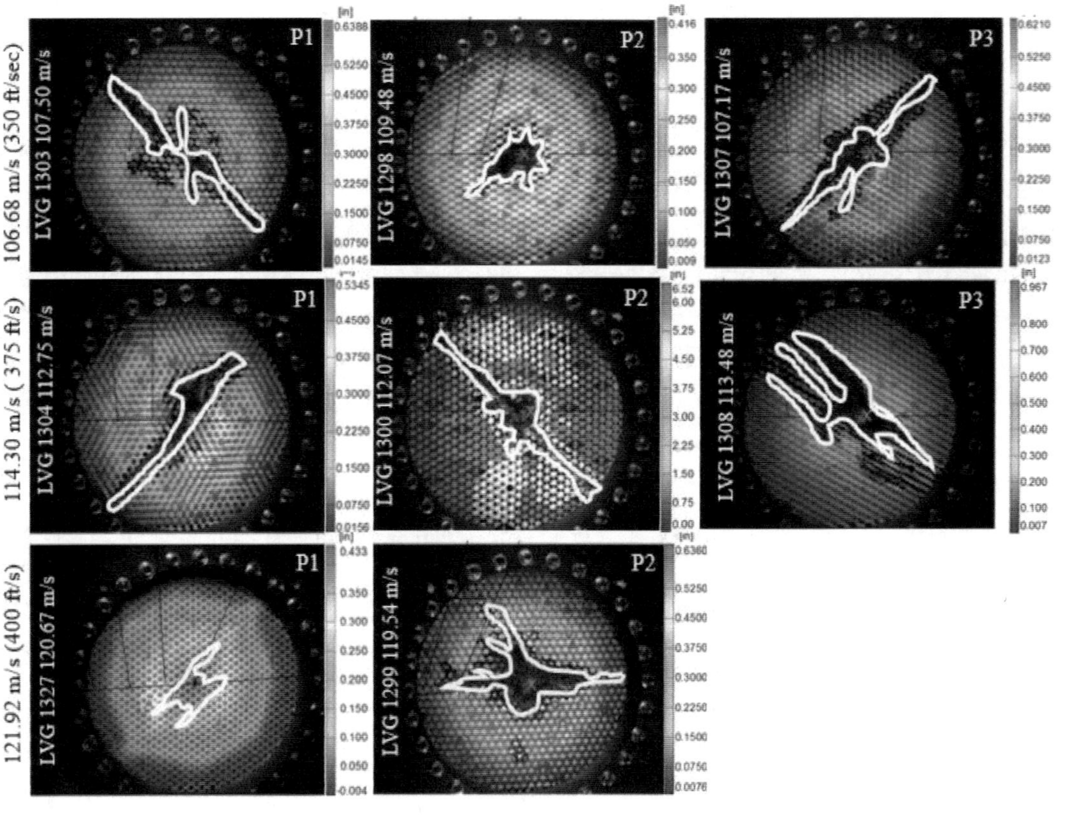

Figure 8. DIC contours and outlined backside surface damage for three different target impact velocities. Displacement contour units are in inches.

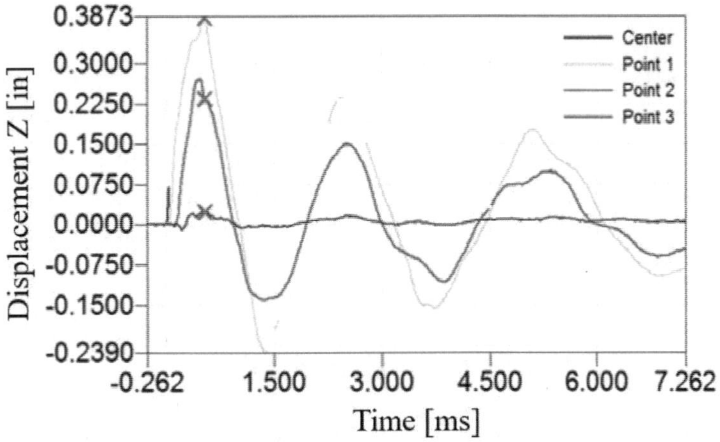

Figure 9. The points used to map displacement and detect peak displacement, shown by red x's, for test LVG 1297

The displacement along the cross-sectional area of specimens at peak displacement at the 91.44 m/s (300 ft/sec) impact velocity are given in Figure 10. Displacement data was taken from the black line bisecting the specimens seen in Figure 8 this is referred to as section 0. This impact velocity serves as a representative example of the entire sample set which all follow the same trend. Displacement along the section 0 at peak displacement shows P3 - CONT with the highest displacement while P1 - TWT has the lowest. All specimens show a gap where displacement data was lost due to the impactor breaking the speckle pattern from the backside of the specimen. The experimental data suggests that the back face deflection of PW hybridized configurations are lower than their conventional counterpart.

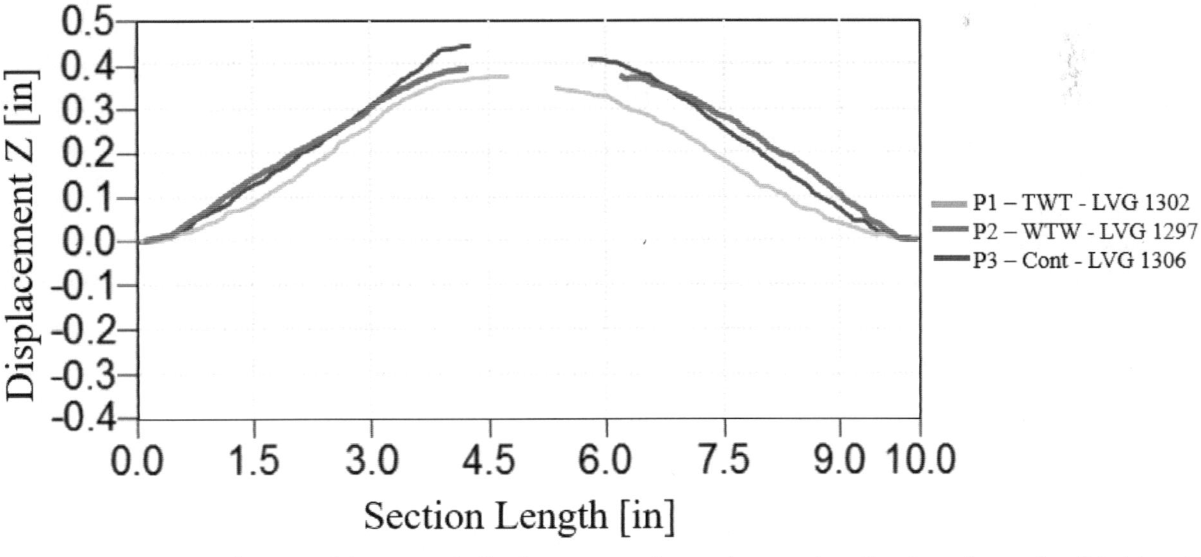

Figure 10. A superimposed image of displacement along the section 0 taken from the 91.44 m/s (300 ft/sec) impact velocity. The orange, red, and blue lines represent P1 – TWT, P2 – WTW, and P3 – CONT respectively.

4.1.4 Damage footprint

Superimposed onto the displacement contours given in Figure 8 is a white outline indicating the surface damage on the back face of the specimen. The hybridized PW laminates sustain a smaller damaged surface area at each impact velocity. There is not an P3 - CONT impact test for the 121.91 m/s (400 ft/s) due to penetration occurring at the 114.30 m/s (375 ft/s) impact velocity. It is interesting to note that both the P1 - TWT and P2 - WTW configurations demonstrate smaller surface area damage than the traditional configuration.

5. CONCLUSION

Three 24 ply carbon/epoxy hybrid pseudo-woven laminate configurations are manufactured in this study using AFP. One with semi-woven laminate on the inside (named as P1-TWT), second with semi-woven laminate on the outer surfaces (named as P2-WTW) and third quasi-isotropic control laminate $[45/90/-45/0]_{3s}$ named as P3-CONT. High velocity impact tests are conducted according to the ASTM D8101 standard. The hybridized configuration P1-TWT where the PW laminates are on the inside showed the highest penetration resistance V50, 8.5% greater than the control. The experimental results also indicate reduced back face damage and back face deflection for the hybridized laminates compared to the control layups. Further investigation using nondestructive evaluation is required to better understand the damage within the laminates.

6. ACKNOWLEDGEMENTS

The authors would like to thank Dr. Michael Pereira of NASA Glenn Research Center for the high velocity impact testing support, Dr. Dawn Jegley and the ISAAC team at NASA Langley Research Center for their contributions towards manufacturing one of the pseudo-woven panels. The authors CVR and SS gratefully acknowledge financial support from the SC Space Grant Consortium. The authors would also like to acknowledge the University of South Carolina's McNAIR Center for their support in AFP manufacturing.

7. REFERENCES

[1]. Czél, G., Jalalvand, M., Wisnom, MR. "Demonstration of pseudo-ductility in unidirectional hybrid composites made of discontinuous carbon/epoxy and continuous glass/epoxy plies." *Composites Part A* (2015). 75- 84.10.1016/j.compositesa.2015.01.019

[2]. Cantwell, W. J., Morton, J. "The impact resistance of composite materials---a review." *Composites* 1991: 347- 362. doi:10.1016/0010-4361(91)90549-V

[3]. Nagelsmit, M., Kassapoglou, C., Gurdal, Z. "A New Fibre Placement Architecture for Improved Damage Tolerance" (2010).

[4]. Vakili Rad, C., Thomas, F., Sockalingam, S., Gurdal, Z. "Low Velocity impact response of hybrid pseudo-woven fiber-reinforced composite laminates" SAMPE Conference (2019)

[5]. Vakili Rad, C., Thomas, F., Seay, B., Van Tooren, M., Sockalingam, S. "Manufacturing and characterization of novel clutch non-conventional fiber-reinforced composite laminates." *Composite Structures* (2019). 10.1016/j.compstruct.2019.02.018.

[6]. ASTM Standard D8101/D8101M-17, 2017. "Stand Test Method for Measuring the Resistance of Composite Materials to Impact by a Blunt Projectile" ASTM International, West Conshohocken, PA, 2017 DOI: 10.1520/D8101_D8101M-17, www.astm.org

High-Char Phthalonitrile Resin Blends

Tristan Butler[1], Boris Dyatkin[2], Brian Chaloux[3], Albert Epshteyn[4] and Matthew Laskoski[5]

[1]NRC Post-Doctoral Researcher Code 6127
[2]Federal Employee Code 6127
[3]Federal Employee Code 6123
[4]Acting Section Head Code 6123
[5]Acting Section Head Code 6127

U.S. Naval Research Laboratory
4555 Overlook Ave., S. W.
Washington, DC 20375
Phone: 202-767-3095

ABSTRACT

Oligomeric phthalonitrile resins exhibit excellent char retention when heated to 1000 °C under nitrogen (> 75%). However, very little is known about the properties and performance of these materials when heated to higher temperatures (> 1500 °C). In addition, blended systems that contain phthalonitrile resins and secondary phases specifically included to enhance char formation at temperatures in excess of 1000 °C have not been previously synthesized. In order to better understand the char forming properties of phthalonitrile resins and identify conditions that maximize char yield, thermogravimetric analysis scans to temperatures exceeding 2000 °C were conducted on a series of different phthalonitrile resins and blended systems. The resulting char from these studies was then characterized *via* powder X-ray diffraction, Raman spectroscopy, and scanning electron microscopy. Ddifferential scanning calorimetry and rheological characterizations were used to determine the proper thermal treatments for curing blended systems and to probe the interactions between different phases of the blend.

1. INTRODUCTION

Because of its use in the aerospace, automobile, alternative energy industries, and many others, the demand for highly graphitic composites is constantlygrowing.[1-2] Given this large scale, new processes which seek to economically produce synthetic graphite is becoming increasingly important.[3] The most common strategy for manufacturing synthetic graphite is to induce graphitization of amorphous carbon, or non-graphitizable carbon (NGC). For example, methods which utilize different metal catalysts (e.g. Fe, Ni, Co, etc.) to precipitate graphitic carbon (~1200 °C) at low temperature have been developed. Because these processes require direct contact between catalyst particles and carbon atoms, only regions near the catalyst form graphitic domains, which limits the practicality of this method.[4]

Graphitizable carbon (GC) phases such as mesophase pitch exhibit a high degree of graphitization when heated above 2800 °C. As opposed to NGCs, this class of carbon forms highly structured microdomains as they are heated, eventually forming graphitic carbon at sufficiently high temperatures. Graphitic phases, once blended with NGC materials, can be formed given the right conditions. For instance, electrospun nanofiber mats derived from polyimide-mesophase pitch blends form of graphite fibers when heated above 1600 °C.[5] Alternatively, highly graphitized carbon has been formed *via* the mixing of NaOH-treated spherical phenolic resins and various graphitizable cokes.[4]

While these examples demonstrate the ability to form graphitic carbon from NGC, the scope of NGC phases that can be utilized for these processes remains relatively unexplored. Since the char forming properties of NGCs are highly dependent on their chemical structure, it is likely that careful selection of NGC materials could optimize production of synthetic graphite. Along those lines, oligomeric phthalonitrile resins (PN) are highly aromatic and exhibit high char yields as compared to the phenolic resins that are often used as carbon precursor materials.[6] By using higher mass retention carbon precursors, which can undergo successful graphitization through the incorporation of GC phases, the overall yield of graphitic carbon could be greatly improved. This study seeks to evaluate the potential for producing graphitic carbon using a bisphenol A-based PN resins (BisA) blended with mesophase pitch. This resin was selected for study because of its high char yield (> 75%). Before charring at 1000 °C, the thermal properties of the blends were studied. After heating, the resulting carbon materials were characterized using a combination of powder X-ray diffraction (XRD), X-ray photoelectron spectroscopy (XPS), and Raman spectroscopy and scanning electron microscopy (SEM). The thermal properties of the each char was also investigated using thermogravimetric analysis in an oxidizing environment.

2. EXPERIMENTATION

2.1 Materials and Methods

All starting materials were reagent grade and used without further purification. The BisA resins used to fabricate blended samples were synthesized from a previously described method.[7] Mesophase pitch was obtained from Zoltek. Char yield experiments of the polymer or carbonized blends were conducted using thermogravimetric analysis (TGA) on a TA Instruments TGA Q50. Samples were heated to 1000 °C at a constant rate (either 5 or 10 °C min^{-1}) under a nitrogen purge (100 cm^3 min^{-1}). All DSC and TGA thermograms were analyzed using Thermal Analysis software from TA instruments. Raman spectroscopic analysis was completed using a Renishaw inVia spectrometer. The measurement used a 514 nm excitation wavelength in the 900 – 3200 cm^{-1} range and accumulated 12 collections with 10 s exposure time per each collection. WIRE software fit and reduced the resulting data. X-ray diffraction (XRD) was performed with a Rigaku diffractometer in the 20° – 80° 2θ region. PDXL software fit and reduced the resulting data. X-ray photoelectron spectroscopy (XPS) was collected using a K-Alpha (Thermo Scientific) XPS instrument. Scanning electron microscopy (SEM) images were collected using a JEOL SEM. Material samples were mounted on copper tape without sputtering and analyzed using 15 kV accelerating voltage. Skeleton density was measured with a pycnometer (Quantachrome Instruments) and used helium gas to measure the cell volume. Volumetric density was derived by measuring the apparent weight of samples in air (M_{dry}) and while submerged in water (Ns_{ubm}) and relied on the following calculation (based on Archimedes' principle):

$$Density = 0.0012 + \frac{M_{dry}}{0.9983} \cdot \frac{D_{water} - 0.0012}{M_{dry} - M_{submerged}}$$

Porosity was calculated using the following equation:

$$Porosity = 1 - \frac{Volumetric\ density}{Skeleton\ density}$$

2.2 Preparation Blended Samples

Prior to preparation of blended samples, a BisA pre-polymer was prepared by combining the neat BisA resin with 2.7 wt. % of bis[4-(3-aminophenoxy)phenyl]sulfone (m-BAPS) and heating at 200 °C for 60 min. The staged BisA resin was then blended with mesophase pitch in the proper ratios by grinding using a mortar and pestle for 5 min. Blends were prepared in the following BisA/Pitch ratios: 1:1, 3:1, and 5:1. The ground powder was then placed in a vacuum oven heated to 200 °C and evacuated for 30 min to degas each sample. The blends were then placed in an oven heated to 225 °C in are for an additional 2 hours in order to gel each sample. For comparison, a pure BisA sample was prepared using the same method. Each sample containing BisA was postcured by slowly heating to 350 °C under flowing nitrogen in a tube furnace. The temperature was then held at 350 °C for 2 hours before cooling to room temperature.

2.3 Charring of BisA/Pitch Blends

All samples were charred by heating at 0.8 °C min^{-1} in a tube furnace under flowing nitrogen up to 1000 °C. The temperature was held at 1000 °C for 1 hour before slowly cooling to room temperature over the course of 8 hours. All samples remained intact after charring however, the pure pitch sample exhibited significant foaming after being heated to 1000 °C.

3. RESULTS AND DISCUSSION

3.1 Thermal Properties of BisA/Pitch Blends Heated to 350 °C

Prior to carbonization, the BisA/pitch blends were heated to 700 °C in a nitrogen atmosphere to determine if the concentration of pitch in each sample affected the char yield. According to the TGA results (Figure 1), the char yield ranged between 75% and 78% regardless of the pitch concentration. While the sample containing only BisA begins to lose mass at a lower temperature (425 °C) than the other samples (435-450 °C), the differences are marginal. Given that these TGA scans are similar, this data suggests that that the addition of pitch has very little effect on the mass retention of samples when heated to 700 °C under an inert atmosphere, however, heating to higher temperatures may affect the mass retention of BisA/pitch blends.

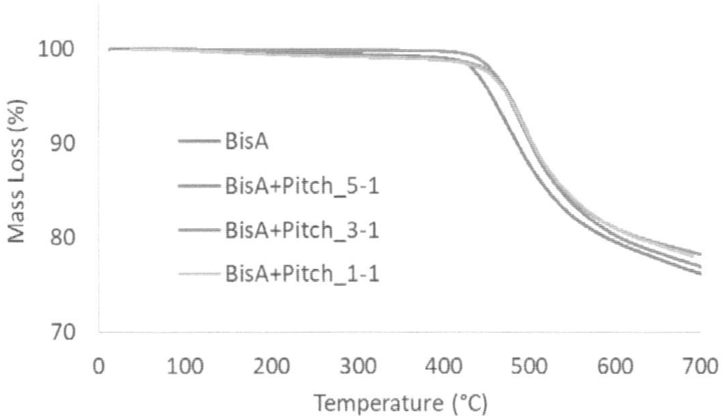

Figure 1. TGA thermograms of BisA/pitch blends heated to 700 °C under nitrogen. (heating rate = 10°C min^{-1}).

3.2 Thermal Characterization of BisA/Pitch Blends Heated to 1000 °C

In order to probe the effect of pitch has on the NGC produced from BisA char, the samples were heated to 1000 °C in a tube furnace. While heating to this temperature will not result in full graphitization, characterization of samples heated to 1000 °C will provide valuable insight into the beginning stages of the graphitization process for BisA/pitch blends. Inspection of the samples by eye after heating showed that porosity developed. In particular, the pure pitch sample formed a carbon foam and therefore could not be characterized using certain techniques. Because heating pitch to 350 °C results in formation of a foam, it is possible that volatile compounds are contained within the pitch which are liberated when heated.

Table 1. Comparison of Volumetric and Skeletal Densities of BisA/Pitch Blends Heated to 350 °C and 1000 °C.

	350 °C			1000 °C		
Material	Volumetric Density	Skeleton Density	Porosity	Volumetric Density	Skeleton Density	Porosity
1:1 BisA:Pitch	0.89 g/cm³	1.06 g/cm³	15.4%	1.13 g/cm³	1.41 g/cm³	19.9%
3:1 BisA:Pitch	1.02 g/cm³	1.14 g/cm³	10.4%	1.29 g/cm³	1.43 g/cm³	9.6%
5:1 BisA:Pitch	1.14 g/cm³	1.16 g/cm³	2.1%	1.37 g/cm³	1.49 g/cm³	8.0%
BisA	1.21 g/cm³	1.21 g/cm³	< 1.0%	1.55 g/cm³	1.56 g/cm³	1.2%

The density of each sample (with the exception of pure pitch) was determined using pychnometry and Archimedes' method before and after carbonization (Table 1). Examination of the data for the polymer samples heated to 350 °C shows that both the skeleton and volumetric densities decrease with the addition of pitch. The decrease in skeleton density can likely be explained by the rule of mixing, however, the density of pitch heat treated to 350 °C could not be measured due to foaming of the sample during heating. The addition of pitch also results in the formation of porosity. The pure BisA sample shows less than 1.0 % porosity, which indicates that there is very little off gassing when it as heated at 350 °C. Conversely, the 1:1 BisA/Pitch sample exhibits 15.4% porosity when heated to this temperature. Given this trend, it is likely that the evolution of volatiles produced by heating induces porosity.

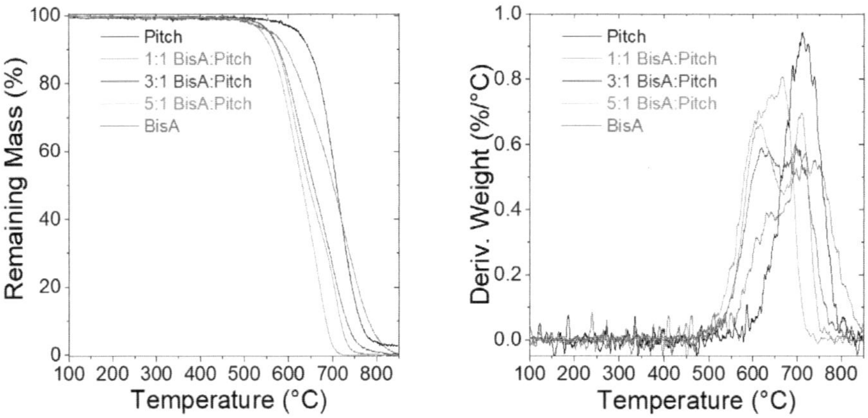

Figure 2. Thermogravimetric analysis scans of BisA/Pitch bledns (left) and their corresponding derivative (right).

Heating to 1000 °C increased the sample density relative to the 350 °C, which corresponded with the samples becoming more graphitized and transitioning closer to the 2.2 g/cm^3 density of graphite. Of note, higher treatment temperatures increased porosity for most compositions. Evolution of unstable oxygen-, hydrogen- and nitrogen- containing organic compounds likely increased the free volume in each material. Furthermore, higher mass fractions of pitch in the blend lowered the densities and increased the porosities of the resulting composites at both 350 °C and 1000 °C curing conditions.

Table 2. Relevant Oxidation Properties of BisA/Pitch Blends Heated to 1000 °C

Material	Oxidation Onset	Peak Oxidation	Oxidation Completion	Ash Content
Pitch	586 °C	711 °C	832 °C	2.81 wt.%
1:1 BisA:Pitch	529 °C	710 °C	812 °C	0.03 wt.%
3:1 BisA:Pitch	533 °C	696 °C	827 °C	0.26 wt.%
5:1 BisA:Pitch	514 °C	667 °C	819 °C	< 0.01 wt.%
BisA	510 °C	718 °C	847 °C	0.35 wt.%

In order to probe the oxidative stability of BisA/pitch blends, TGA scans were collected of each sample (Figure 2, Table 2). Increased pitch content in the composite blend increased the oxidation onset temperature and improved the stability of the material. On the other hand, higher wt.% of BisA extended the temperature range (from onset to completion) during which the composites burned in air. This behavior, in turn, suggests a slower oxidation kinetic process for the cured resin. The derivative TGA plot (Figure 2) shows that the 3:1 and 1:1 BisA/Pitch blends exhibited bimodal oxidation peaks, which suggested that some phase separation was present in those composites.

3.3 Structural and Morphology of BisA/Pitch Blends Heated to 1000 °C

Initial structural characterization was done *via* powder XRD (Figure 3, Table 3). For these measurements, each sample was ground into a fine powder before exposing them to X-rays. The BisA pattern shows two broad peaks centered near 26° and 43° which correspond to [002] and [101]/[100] planes, respectively. By comparison, the pitch pattern clearly contains the [002] peak, however, the peak near 43° is only weakly visible. While the blended samples show evidence of both peaks, the addition of pitch decreases the intensity of the [101]/[100] peak.

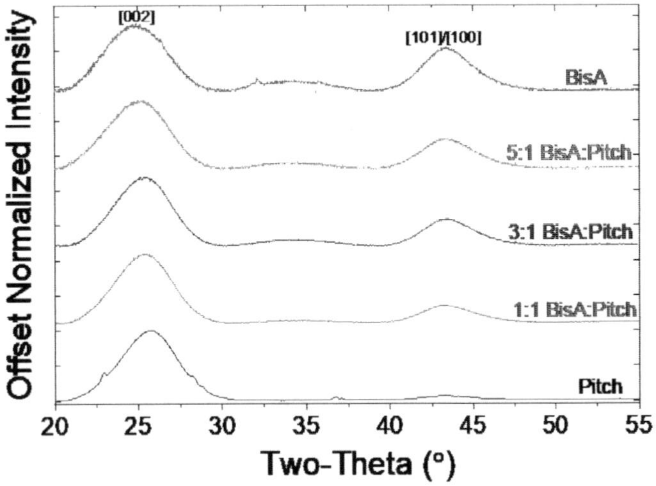

Figure 3. Powder X-ray diffraction (XRD) patterns for BisA/Pitch Samples

The grain sizes were also determined using the powder patterns for BisA/Pitch samples heated to 1000 °C. According to these results, the pitch sample contained the largest grains (1.90 nm), whereas the grains observed for BisA (0.81 nm) were the smallest. Although the background contribution precluded a determination of the grain size for the 3:1 and 5:1 BisA/pitch samples, the grain size for the 1:1 sample (1.63 nm) fell between those extremes.

Table 3. Grain Sizes of BisA/Pitch Samples Determined by Powder XRD

Material	Grain Size
Pitch	1.90 nm
1:1 BisA:Pitch	1.63 nm
3:1 BisA:Pitch	N/A
5:1 BisA:Pitch	N/A
BisA	0.81 nm

The degree of graphitization for the composites treated at 1000 °C was analyzed using Raman spectroscopy (Figure 4). According to these measurements, all materials, including the pure BisA and Pitch, exhibit a strongly graphitized structure after undergoing 1000 °C treatment. However, none of the spectra indicated any resonant 2D or D+G modes (in the 2700 cm^{-1} range).

Subsequently, graphitization of the composited was not complete at 1000 °C and no graphite or graphene-like structure evolved from the blends as a result.

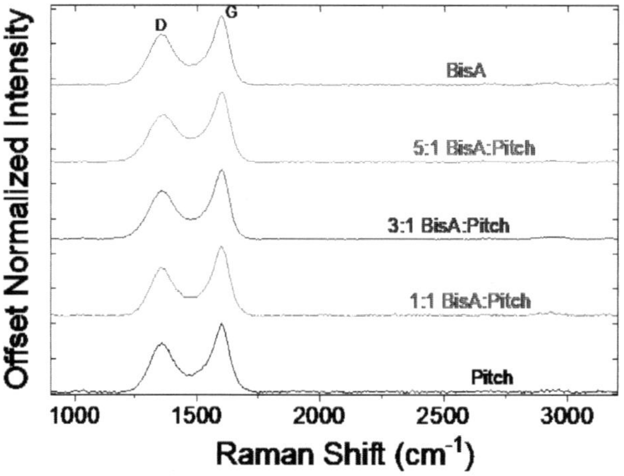

Figure 4. Raman Characterization of BisA/Pitch Samples

The key D and G band properties are summarized in Table 4. The I_D/I_G ratios for the blends of BisA and pitch are slightly lower than those of the respective standalone components, which suggests a possible synergistic influence of the blends on the overall graphitization of the resulting composites. However, neither a positive nor a negative correlation between the weight ratios of Pitch to BisA and the I_D/I_G ratio exists. The full widths at half maxima of the G band do not indicate any influence of pitch composition on the ordering of the graphitic domains in the resulting composites.

Table 4. Raman Data Summary of BisA/Pitch Blends.

Material	I_D/I_G	FWHM of G Band
Pitch	0.75	81.2 cm^{-1}
1:1 BisA:Pitch	0.71	80.7 cm^{-1}
3:1 BisA:Pitch	0.72	77.3 cm^{-1}
5:1 BisA:Pitch	0.71	83.4 cm^{-1}
BisA	0.76	78.1 cm^{-1}

The XPS-derived elemental composition of BisA, Pitch, and several blends after 350 °C and 1000 °C curing cycles is shown in Table 5. As expected, BisA contained a significant amount of oxygen and nitrogen (in the form of functional groups) following the 350 °C cure and, subsequently, lost approximately 40% of those groups during heating up to 1000 °C. Although inclusions of pitch (which has a relatively higher initial carbon content) into the resin blend reduced the overall oxygen and nitrogen compositions, the elemental ratio of carbon to non-carbon did not directly correlate with the mass fraction of pitch in the 350 °C or the 1000 °C blends. While the 1000 °C blends that included BisA had demonstratively higher oxygen than those with higher pitch amounts, the nitrogen content of blends, including 100% BisA and 100% Pitch baselines, was nearly identical for all compositions. Therefore, both the resin and pitch systems contained strong carbon-nitrogen bonds that withstood 1000 °C thermal treatments. Table 5. Elemental Analysis of Select BisA/Pitch Samples Determined by XPS.

	350 °C						1000 °C					
Material	Carbon		Oxygen		Nitrogen		Carbon		Oxygen		Nitrogen	
	at.%	wt.%	at.%	wt.%	at.%	wt.%	at.%	wt.%	at.%	wt.%	at.%	wt.%
Pitch							94.9	92.9	3.2	4.2	1.3	1.5
1:1 BisA:Pitch	82.5	75.2	12.3	14.9	0.9	1.0	89.2	85.3	7.7	9.8	1.7	1.9
3:1 BisA:Pitch	89.0	85.5	6.0	7.7	3.9	4.3	93.6	90.9	4.7	6.1	1.1	1.3
BisA	81.2	73.8	13.7	16.6	1.9	2.1	88.7	84.3	8.6	10.9	1.7	1.9

The C1s bonding peak for blends cured at 1000 °C is shown in Figure 5. The bonding peak is centered at 284.4 eV for all analyzed blends and suggests a similar carbon-carbon bonding profile for blends treated at that temperature. However, the 100% BisA C1s peak also exhibits a broad shoulder at ~285 eV; this feature is absent in all other peaks, including both 100% pitch and various Pitch/resin blends. This suggests that pure BisA maintained a wide range of chemical functionalities following the 1000 °C cure process. The absence of this bonding behavior in the 3:1 and 1:1 BisA/Pitch blends suggests that the presence of pitch had altered the bonding behavior, and, possibly, growth of graphitic domains in the composite during the 1000 °C thermal treatment.

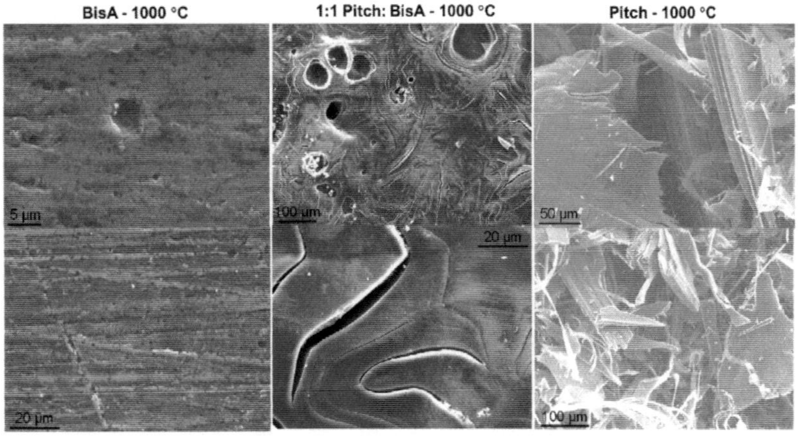

Figure 6. Scanning electron microscope (SEM) images of select BisA/Pitch blends.

The morphology of BisA, pitch, and 1:1 BisA/Pitch was characterized using SEM (Figure 6). The 1:1 BisA/Pitch sample was selected because potential phase separation would be more likely given the larger concentration of pitch compared to the other blends. Examination of the pure BisA sample shows that it is uniform and only a small degree of microporosity was observed. Conversely the pitch sample, which formed a foam when heated to 1000 °C, shows a large degree of microporosity which is consistent with the evolution of gas during heating. In addition, high aspect ratio features were observed, which were not evident in the BisA sample. The SEM images for the 1:1 BisA/pitch sample display characteristics present in both BisA and pitch samples. While the sample appears to be homogenous like BisA, there is significantly more porosity which is presumably due to volatilization of the pitch. Because there is no clear phase separation observed in the SEM, it suggests that the pitch phase is well dispersed in the blended sample, but there is no co-curing between pitch and BisA.

4. CONCLUSIONS

Blended polymer samples were produced based on combinations of a bisphenol A-based phthalonitrile resin (BisA), a non-graphitizable resin, and graphitizable mesophase pitch. These samples were then heated to 1000 °C in order to convert them to carbon and initiate the graphitization process. Structural characterization of these samples using powder XRD indicated that BisA has smaller grain sizes as compared to pitch after charring. Raman spectroscopy showed that some graphitization had occurred in every sample after heating to 1000 °C. However, it does not indicate that the presence of pitch has any effect on the ordering of graphitic domains in these samples. In addition, XPS characterization showed the presence of strong carbon and nitrogen bonds in BisA even after heating to 1000 °C. Because these bonds were not observed in the blended samples, it suggests that pitch does have some effect on the curing of BisA resin. Although incorporated pitch only marginally enhanced BisA graphitization, initial findings suggest that, with the inclusion of pitch, samples must be heated to high temperatures in order to fully describe the properties of these systems.

5. REFERENCES

1. Dang, A.; Zhao, Z.; Tang, C.; Fang, C.; Kong, S.; Khan, M.; Li, T.; Zhao, T.; Li, H., Effect of Phenolic Resin on Micropores Development in Carbon Foam with High Performance. *Materials (Basel)* **2019,** *12* (8).
2. Abrahamson, J. P.; Jain, A.; van Duin, A. C. T.; Vander Wal, R. L., Carbon structure and the resulting graphitizability upon oxygen evolution. *Carbon* **2018,** *135*, 171-179.
3. Kim, T.; Lee, J.; Lee, K.-H., Full graphitization of amorphous carbon by microwave heating. *RSC Advances* **2016,** *6* (29), 24667-24674.
4. Kim, D.-W.; Kil, H.-S.; Kim, J.; Mochida, I.; Nakabayashi, K.; Rhee, C. K.; Miyawaki, J.; Yoon, S.-H., Highly graphitized carbon from non-graphitizable raw material and its formation mechanism based on domain theory. *Carbon* **2017,** *121*, 301-308.
5. Yan, H.; Mahanta, N. K.; Wang, B.; Wang, S.; Abramson, A. R.; Cakmak, M., Structural evolution in graphitization of nanofibers and mats from electrospun polyimide–mesophase pitch blends. *Carbon* **2014,** *71*, 303-318.
6. Butler, T.; Alden, S. E.; Taylor, M.; Deese, S.; Rider, D. A.; Laskoski, M., Oligomeric phthalonitriles and tetrakis(phenylethynyl)benzene blends with improved processing and thermal properties. *Journal of Polymer Science Part A: Polymer Chemistry* **2018,** *56* (23), 2630-2640.
7. Keller, T. M.; Dominguez, D. D.; Laskoski, M., Oligomeric Bisphenol A-Based PEEK-like Phthalonitrile-Cure and Polymer Properties. *Journal of Polymer Science Part a-Polymer Chemistry* **2016,** *54* (23), 3769-3777.

LIGHT WEIGHT REINFORCED THERMOPLASTIC COMPOSITES WITH ADVANCED ACOUSTIC, FLAMMABILITY, TOXICITY, AND SMOKE PERFORMANCE

Ruomiao Wang
Hanwha Azdel, Inc.
Forest, VA

ABSTRACT

Light weight reinforced thermoplastic (LWRT) composites are broadly used by different industries, including automotive, building and construction, office furniture, heavy truck, recreational vehicle and, potentially, aerospace, among others. LWRT composites can have the classic advantages of low coefficient of thermal expansion, high strength-to-weight ratio, high impact with ductile failure, good acoustic performance, intrinsic mold and rot resistance, relatively low cost part forming, and easy installation. Recent developments in LWRT composites add more functions to this material, such as improved flame retardancy, improved sound absorption, and low toxicity and smoke release. In this paper, LWRT composites with two families of new formulations will be reported. In addition to traditional physical properties and mechanical properties of the materials, acoustic performance tested by ASTM E1050 and ASTM C423-17, flame retardant (FR) performance tested by ASTM E84, FMVSS 302 and SAE J369, smoke density performance tested by ASTM E662, and gas toxicity performance tested by BSS 7239 will be discussed. The comparison between standard LWRT composite and these newly development LWRT composite materials indicates another step forward in LWRT composite development.

1. INTRODUCTION

Composite materials have broad applications in different industries. Examples could be automotive, heavy truck, building and construction, office furniture, and recreational vehicles, and potentially aerospace and others. LWRT composite represents one type of composite materials, which has the advantages of low coefficient of thermal expansion, high strength-to-weight ratio, high impact performance with ductile failure, good acoustic performance, intrinsic mold and rot resistance, relatively low cost and easy part forming, easy installation and potential recyclability, and others. For more than two decades, LWRT composite material has had broad applications in the automotive industry, including different soft trims for both interior and exterior applications. Another broad emerging category of applications includes interior and exterior panels for recreational vehicles and commercial truck trailers. Applications for office furniture and building and construction industries became more realistic after the improvement in flame retardancy performances of LWRT composites was achieved [1, 2]. A further stretch for LWRT composite usage would be considered for commercial transportation and aerospace applications, where higher standards and more complicated tests are required beyond flame retardancy tests.

In this paper, LWRT composite sheets, which are produced by a wet-lay method [3], will be described. The LWRT composite sheets are polyolefin based and reinforced by chopped glass fibers. Resin and glass fiber were mixed in water and agitated into a uniform suspension. The suspension was deposited on a forming wire and the water was removed from the material to form a web. Further heating removed the moisture in the composite and melted the resin. A solid composite sheet is formed by compressing the web and holding the material under pressure till the temperature of the material is below the melting point of the resin. Figure 1 shows a schematic drawing of the wet-lay process. Surface laminates with desired functions can be applied to one side or both sides of the web during the stage when the web is under pressure. LWRT composite sheets can be used as in the status of flat panels as they are produced, or they can be molded to shapes with more complicated geometries after the material goes through a separate thermo-forming process.

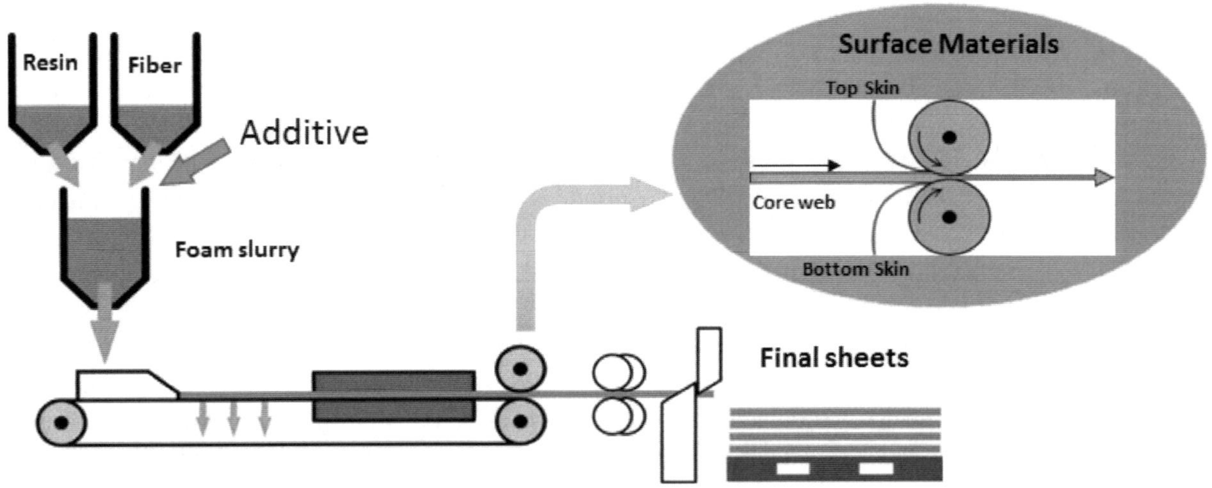

Figure 1. Schematic drawing of a wet-lay process

The standard LWRT composite was composed of resin, glass fiber and sometimes a functional additive. Two new grades of LWRT composites are reported in this paper, using two different flame retardancy functional additives, namely expandable graphite (EG) [1] and magnesium hydroxide (MDH) [2]. When tested by ASTM E84 method, both grades meet the Class A requirements. The grade using EG as the FR additive meets SE-0 requirement tested by SAE J369 method. Beyond FR performances, the smoke density performances of both grades were also tested by ASTM E662 standard and the combustion toxicity performances were tested by BSS 7239 standard. In traditional automotive LWRT composite applications, a more desired acoustic performance of LWRT composite would be achieved after thermo-forming process. In this study, further evaluation has been conducted on the LWRT composite in the status they were produced. Improved sound absorption, particularly tested by ASTM E1050 and ASTM C423-17 method, has been achieved and will be reported.

2. EXPERIMENTATION

2.1 Material and Material Preparation

Materials including standard LWRT composites and two new developmental families, were made by the wet-lay process. All the targeted material conditions are listed in Table 1. The differences among the materials are in the areal density, FR additive, and also the surface laminates. The materials are grouped in three different groups based on the substrate areal densities. In all three groups, both developmental grades are included, and Control A and Control C are two samples of standard LWRT composite. Materials in Group A are generally based on LWRT composite substrates with areal densities around 2800 gram per square meter (gsm); the potential application market for these materials is the building and construction industry, based on their thicknesses. A second group of materials, Group B, has about 1500 gsm in the substrate, and this is the group that has been tested for smoke and toxicity performances. The materials in Group C are the even lighter ones, with the substrate areal densities around 1200 gsm or 1000 gsm. These materials are more appropriate for automotive applications and office furniture. Group C is the only group of materials that has been through the thermoforming process, with the consideration for potential automotive parts made out of this grade. The choices of skin laminates are determined by the additional functions required for the skins for the applications. For example, an adhesive film layer would be required, if further surface layers would be laminated onto the substrate.

Table 1. Formulations and target conditions of tested materials

	Material	**FR Additive**	**Substrate Areal Density (gsm)**	**Surface Laminates**
Group A	Control A	N/A	2800	92gsm film/scrim laminate; 20gsm scrim
	Sample A1	MDH	2800	20gsm scrim; 20gsm scrim
	Sample A2	MDH	2800	88gsm perforated film; 20gsm scrim
	Sample A3	MDH	2800	92gsm film/scrim laminate; 20gsm scrim
	Sample A4	EG	2800	20gsm scrim; 20gsm scrim
Group B	Sample B1	MDH	1500	bare/bare
	Sample B2	EG	1500	bare/bare
Group C	Control C	N/A	1200	90gsm scrim; 20gsm scrim
	Sample C1	MDH	1200	20gsm scrim; 20gsm scrim
	Sample C2	EG	1200	20gsm scrim; 20gsm scrim
	Sample C3	EG	1200	90gsm scrim; 20gsm scrim
	Sample C4	EG	1000	90gsm scrim; 20gsm scrim

In some of the LWRT composite applications, the materials are applied in the way that they were manufactured. However, in some other applications, a thermoforming step is necessary to allow parts generated with more complicated geometries. Loft is the unique characteristic to allow the thermoforming process. When heated above the melting temperature of the resin, the thickness of an LWRT composite sheet will increase, and this phenomenon is called loft. The reason for the thickness increase is that the glass fibers, which were held together by the resin during the consolidation stage, spring back to their normal status. In this paper, some of the materials were evaluated as the status that the composite sheets were made, and some of the materials were first thermoformed into flat plaques with the desired thicknesses before any evaluations. In every evaluation, the specimens are described as either "as-produced" or "molded to a certain thickness". Some demonstration parts will also be shown in the latter part of this paper.

2.2 Physical Property Evaluation

Physical properties, including areal density, thickness, and loft thickness, were measured by internal test methods. Ash contents were measured following ASTM D5630 standard. Disks, with 99 mm in diameter, were punched from LWRT composite sheets and used as the specimens for all the above mentioned tests. The minimal sample size was three and five specimens were more commonly used for all the tests.

The oven was set up at 204 °C for loft testing. Depending on the substrate areal density, the specimens were kept in the oven for 4 – 6 minutes. A dial indicator with a 1" diameter round foot was used in all thickness measurements.

2.3 Flammability, Smoke and Toxicity Evaluation

ASTM E84 and SAE J369 were the two standards followed in the flammability tests. ASTM E84 is commonly used by the building and construction industry and SAE J369 is specific for automotive industry. According to ASTM E84 standard, in order to meet Class A performance the sample can have a flame spread index (FSI) no more than 25 and a smoke developed index (SDI) no more than 450. The composite sheets were tested in the as-produced status when tested by ASTM E84 method. Meanwhile, all the tests following SAE J369 method were conducted on thermoformed specimens. Part of the specimens went through the oil soaking process before the flammability tests.

The smoke density tests and toxicity tests were only conducted on Group B samples following ASTM E662 and BSS 7239 procedures. The materials were produced at 4.0 mm thickness, and the tests were conducted on substrate materials only.

2.4 Acoustic Performance Evaluation

Similar to flammability tests, two test standards were followed to test the acoustic performances. Materials in Group A were developed for potential building and construction applications, and therefore ASTM C423-17 standard was followed due to the acceptance for the standard by the industry. All materials from Group A were tested in their as-produced status for ASTM C423-17 testing. Broader testing was conducted following ASTM E1050 standard. Materials from Group A and Group C were tested by ASTM E1050 standard, and evaluation was conducted on some as-produced LWRT composite sheets and some other thermoformed LWRT composite sheets.

2.5 Mechanical Property Evaluation

Flexural property testing was conducted following ASTM D790 or ISO 178, respectively. Because Group A materials were intended for use in the way they were manufactured, Group A materials were tested without a thermoforming step. In contrast, materials from Group C were tested after thermoforming, and the materials were molded to 4 mm or 6 mm.

3. RESULTS

3.1 Physical Properties

Physical properties of all three groups of materials are summarized in Table 2. The test data confirm the materials were manufactured in the way they were designed and also ensure a fair comparison is conducted among materials with similar areal density.

Table 2. Physical properties of tested materials

Material	Substrate Areal Density (gsm)	Ash Contents (%)	Thickness (mm)	Loft Thickness (mm)
Control A	2842	57.5	13.1	14.5
Sample A1	2900	60.0	10.9	13.5
Sample A2	3006	58.4	11.4	14.3
Sample A3	2752	58.9	11.9	13.4
Sample A4	2765	51	13.2	16.1
Sample B1	1506	61.7	4.0	6.7
Sample B2	1522	47.2	4.0	6.7
Control C	1200	35.6	3.0	13.5
Sample C1	1273	59.9	3.3	6.2
Sample C2	1189	44.8	3.5	7.2
Sample C3	1187	43.2	3.6	7.3
Sample C4	1020	42.4	3.2	6.0

3.2 Flammability, Smoke and Toxicity

Flammability, smoke and toxicity tests focus on the developmental grades. The standard LWRT composite is not considered a grade with flame retardancy capability. Therefore, the evaluation effort is around the new grades.

Table 3 summarizes the flammability performances tested on different formulations by ASTM E84 standard, and Figure 2 shows examples of the testing curves obtained on Sample A1 and A4 during the ASTM E84 flammability test. The standard LWRT composite materials can meet ASTM E84 Class B requirements. Both developmental grades are capable to meet Class A requirements. EG is considered a more efficient FR additive because ASTM E84 Class A performance could be achieved at a relatively low additive loading level. Moreover, the materials using EG as the FR additive shows a significantly lower smoke development index than the materials using MDH as the FR additive. Materials using MDH as the additive possesses the advantage of broader color option. The mineral based FR additive has a color very similar to the natural color of polyolefin resins. Therefore, any color selections available for polyolefin resins will not be eliminated for this grade. On the other hand, the black color of EG limits the grade using EG as the FR additive to only grey or black colors.

Table 3. Flammability performances tested by ASTM E84 standard

Material	FSI	SDI
Sample A1	25	110

Sample A2	20	60
Sample A3	25	105
Sample A4	10	30
Sample C1	20	90
Sample C2	15	35
Sample C4	25	40

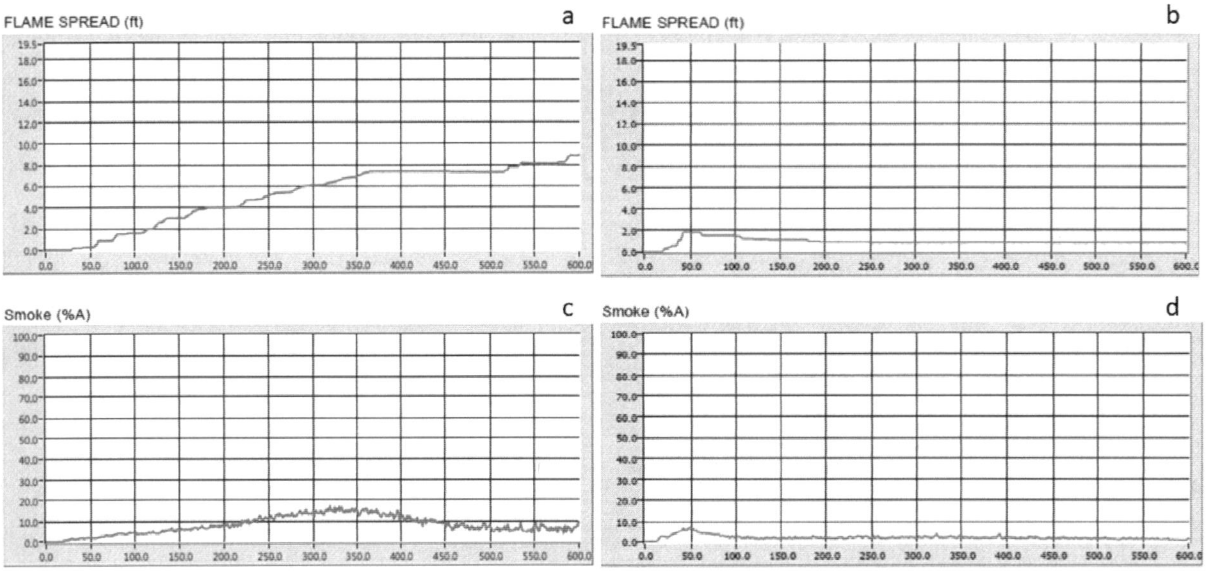

Figure 2. Burning curves used to calculate FSI and SDI obtained on Sample A1 and Sample A4. a) the curve to calculate FSI for Sample A1; b) the curve to calculate SDI for Sample A1; c) curve to calculate FSI for Sample A4; and d) curve to calculate SDI for Sample A4.

Figure 3a shows a molded automotive part out of Sample C3, which demonstrates the thermoformability for more complicated surface and thickness profiles of the developmental grade of LWRT composite. Figure 3b and 3c are the examples of specimens to compare the flammability performances of molded plaques generated out of Control C (Figure 3b) and Sample C3 (Figure 3c). Both sets of specimens were tested against SAE J369 standard after being through the oil soaking process. Specimens were able to self-extinguish within very limited amount of burning distance. The burning distances of the specimens being through the oil soaking process are typically similar to the ones obtained on the specimens without oil soaking process. The materials made out of EG formulation are capable to meet SE/0 requirements by SAE J369 standard.

Figure 3. An underbody panel molded out of Sample C3 (a), and examples of the tested specimen by SAE J369 after oil soaking process of Control C (b) and Sample C3 (c).

The smoke density performances of materials listed in Group B are compared in Table 4. Sample B1 shows advantage over Sample B2 in both flaming mode and non-flaming mode. The study on further improving the smoke density of the new LWRT composite is at an early stage. It is believed that the product thickness, material density, surface laminates and any secondary operation, such as consolidation or thermoforming, will affect the smoke density performance of the product. Sample B1 and Sample B2 pass the toxicity requirements listed in BSS 7239 standard.

Table 4. The smoke density performances of Sample B1 and Sample B2 tested by ASTM E662 standard

Property	Sample B1		Sample B2	
	Flaming Mode	Non Flaming Mode	Flaming Mode	Non Flaming Mode
Maximum Specific Optical Density (90 seconds)	0	19	4	26
Maximum Specific Optical Density (4 minutes)	9	54	13	71

3.3 Acoustic

The sound absorption coefficients of the samples were tested following ASTM E1050 standard. Figure 4a compares the acoustic performances among Group A samples, and Figure 4b compares the performances among Group C samples. Different from Group A samples, Group C samples were compared both in the status the composite sheets were manufactured and after being thermoformed to 4 mm and 6 mm. Due to the high areal density, Group A materials all show outstanding sound absorption performance. Materials show sound absorption coefficients above 0.6 at a frequency of 2000 Hz or higher.

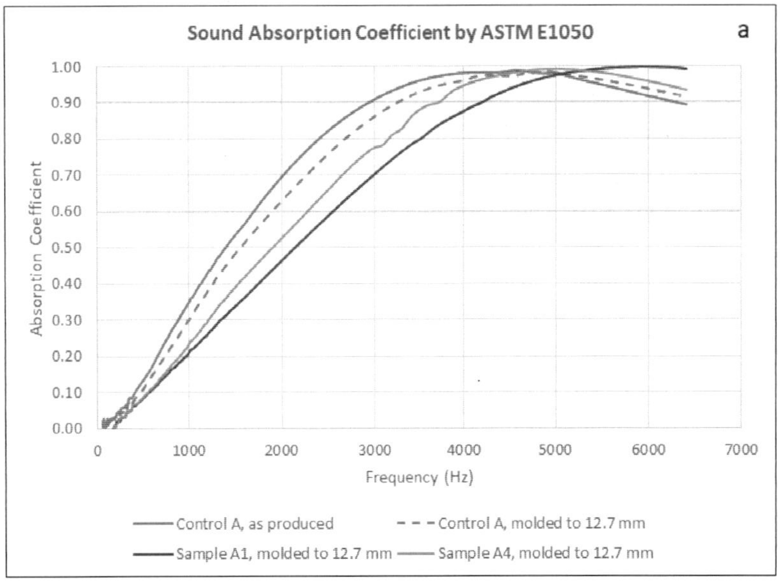

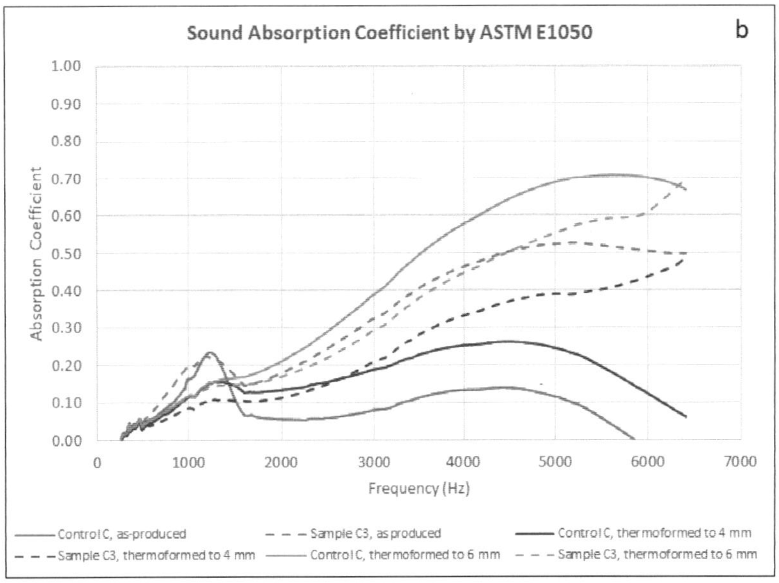

Figure 4. Sound absorption performances tested by ASTM E1050 standard. a) Group A materials and b) Group C materials.

Different relations between sound absorption coefficient and frequency were noticed on as-produced Group C samples and thermoformed Group C samples. Control C material shows higher sound absorption coefficients on the thermoformed plaques, which are typical for LWRT composite materials. However, materials made with EG as the additive show a special acoustic performance different from a standard LWRT composite. The as-produced Sample C3 shows sound absorption closer to a thermoformed standard LWRT material. This is believed to be the benefit from EG, which also has a function for sound absorption in addition to FR.

Some samples in Group A were chosen to be tested following ASTM C423-17 standard with the consideration for potential applications for building and construction industry in mind. Table 5 summarizes the test results. With the similarity in substrates among Sample A1, A2 and A3, the differences among surface laminates contributes to the resultant acoustic performance difference among the specimens. Sample A1 has a fully open structure on both sides of the composite. The perforated film on Sample A2 has an open area about 10%. However, the 92 gsm film/scrim laminate structure used on Sample A3 has a fully close structure. Even though LWRT composite material is considered as a material with outstanding sound absorption, a surface laminate with some level of open area is necessary to realize the sound absorption function of the composite panel.

Table 5. Acoustic performances of Sample A1, A2 and A3 tested by ASTM C423-17 standard

Material	Noise reduction coefficient (NRC)	Sound absorption average (SAA)
Sample A1	0.85	0.88
Sample A2	0.85	0.88
Sample A3	0.25	0.27

3.4 Mechanical Properties

Figure 5 compares the flexural strength of Control A and Sample A1 materials in the way they were produced. It is noticed that both flexural strength and flexural modulus in both machine direction (MD) and cross-machine direction (CD) were negatively affected by adding MDH to the formulation. The mechanical strength drop of the MDH based developmental grade is expected, since the MDH is not able to contribute mechanical strength to the overall structure and the high loading of MDH caused an effect beyond the negligible range.

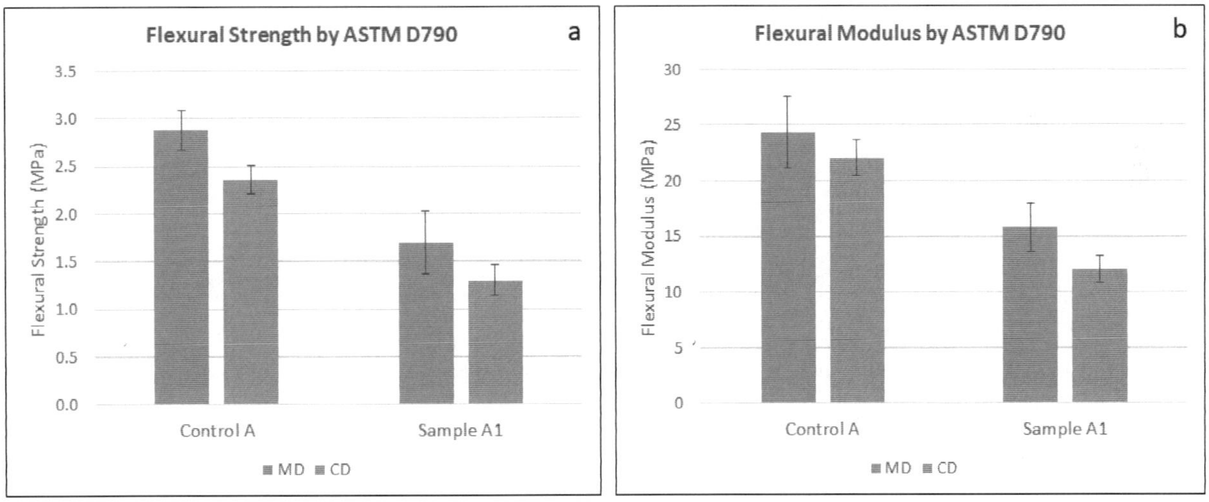

Figure 5. Flexural properties of Control A and Sample A1 tested as-produced.

Figure 6 compares the flexural properties of Control C, Sample C1 and Sample C3. The tested materials were molded to 4 mm and 6 mm. The formulation of Sample C3 is considered less affected, and this could be due to the relatively low loading level of additive.

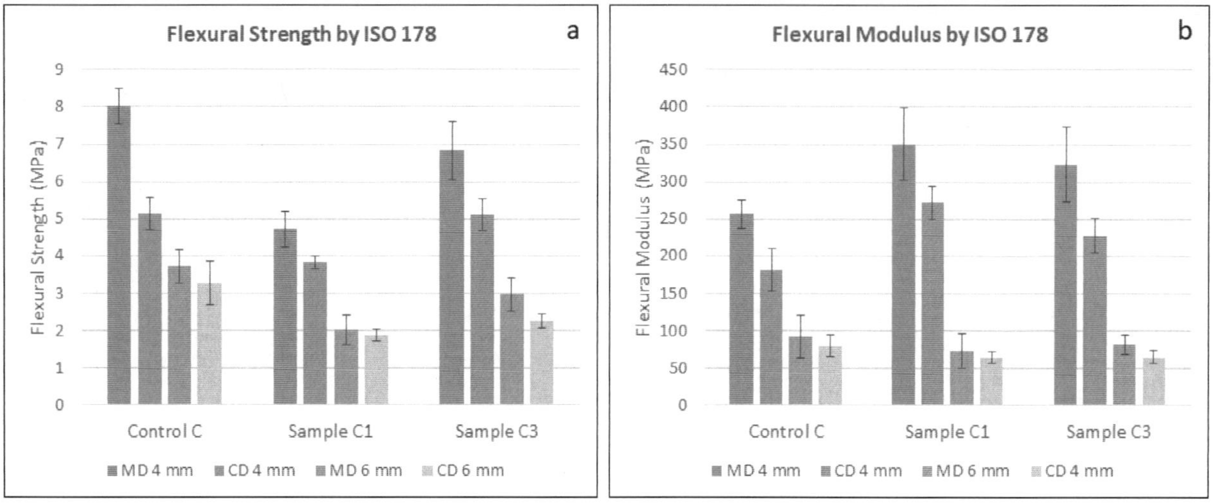

Figure 6. Flexural properties of Control C, Sample C1 and Sample C3 tested after thermoforming.

4. CONCLUSIONS

Two new grades of LWRT composites are under development. Compared with standard LWRT, the new grades focus on the improvements in flammability, smoke, toxicity and acoustic performances. Significant performance improvements have been realized in all considered aspects. The grade with EG as the functional additive shows more promising performances in flammability and acoustic performances over the grade with MDH. Meanwhile, the grade with MDH shows more potential for smoke performance. With the establishments achieved on ASTM E84 and SAE J369 flammability and ASTM C423-17 and ASTM E1050 acoustic performances, the focus of future investigation will shift towards further improvements in even higher standard flammability performances, such as aerospace requirements, and expanding the understanding in smoke and toxicity performances.

5. REFERENCES

1. Y. Tseng, R. Wang, M. O. Mason, T. D. Robertz, E. J. Vorenkamp, US Patent Application US20160168350 A1 (2016)
2. Z. Yu, R. Wang, Y. Yang, M. O Mason, US Patent Application US20170225429 A1 (2017)
3. P. L. Wallace, A. J. Willis, U. S. Patent US4944843 A (1990)

CHARACTERIZING THE EFFECT OF TRANSVERSE STRAIN ON CARBON NANOTUBE BASED SENSING SKINS FOR STRUCTURAL HEALTH MONITORING

Sagar M. Doshi[1], Michael Coatney[2], Madison Weiss[1], Amit Chaudhari[1], Dae Han Sung[1], Erik T. Thostenson[1], Asha Hall[2]

1- Center for Composite Materials and Department of Mechanical Engineering, University of Delaware, Newark, Delaware
2 – Vehicle Technology Directorate, Army Research Laboratory, Aberdeen, Maryland

ABSTRACT

Due to the advancement in the science of processing and characterization of nanostructured materials, along with the industrial-scale production and decreasing cost of carbon nanotubes, their use in technologies as sensors and sensing skins is gaining popularity. With increasing Technology Readiness Levels (TRL) and field trials implemented by researchers, it is critical to characterize the sensing response of the carbon nanotube sensing skins under complex loading scenarios in order to simulate the real-life conditions. Typically, researchers describe the sensitivity (gage factor, GF) and sensing response of a carbon nanotube sensing skin based on the electrical resistance change and the longitudinal strain (ε_1), not taking into consideration the effect of transverse strain (ε_2) due to the Poisson's ratio of the material. Since the sensitivity of the sensor is dependent on the transverse strain, the sensitivity is affected by the properties of the substrate material. As a result, the calibration of the sensing skin is challenging. The principal objective of our research is to characterize the effect of transverse strain and develop a methodology to quantify the influence of transverse strain on the sensing response. The carbon nanotube-based sensing skins are manufactured using a scalable manufacturing technique and the effect of different loading scenarios on the sensing response are investigated. Carbon nanotubes are deposited on non-woven aramid fabrics with randomly oriented fibers and the resulting carbon nanotube-based sensing skin is bonded to steel and composite substrates which are subjected to flexural and axial loads to characterize the sensing response. A cruciform shaped specimen with a carbon nanotube sensor is tested using a biaxial testing machine and change in resistance under varying transverse loads is characterized. Key results indicate that the sensitivity of the carbon nanotube sensing skin is significantly affected by the transverse strain due to the Poisson effect which should be taken into consideration when calculating the gage factor or calibrating the sensors.

Keywords: Carbon Nanotubes, Strain Sensors, Structural Health Monitoring, Biaxial Testing, Poisson's Effect, Piezoresistive Sensors.

Corresponding author: Sagar M. Doshi – smdoshi@udel.edu

SAMPE Virtual Conference Proceedings, 2020. Society for the Advancement of Material and Process Engineering – North America.

1. INTRODUCTION

Traditional techniques for monitoring the condition of civil infrastructure are often time consuming, inaccurate, expensive and ineffective. For example, a lot of bridges are still inspected at specific time intervals with manual markings made for fatigue cracks and other visible defects. These may often be unnoticed if they are in hard to reach areas or fatigue cracks initiate between the scheduled inspections. Multiple organizations have pointed out the deteriorating condition of the infrastructure in USA, for example, the American Society of Civil Engineers, in its report card (www.infrastructurereportcard.org) gave a grade of D+ for infrastructure. As a result, there exists a critical need for developing innovative sensors and sensing systems for effective structural health monitoring.

Typically strain gages can be used for monitoring the integrity of structures and the loads they are subjected to. A key challenge with the use of strain gages is that they are quasi point sensors may not be able to identify the damage if they are not in close vicinity or directly above the location of the damage. Field application of strain gages poses additional challenges with potential performance degradation over time due to environmental effects and wiring complexities. Researchers are working towards using acoustic emission sensors but their data analysis is involved and the interpretation of results is often not quantitative. Fiber optic sensors have an advantage over the point sensors and have been successfully used for distributed sensing [1,2], but there remain challenges with respect to compensation with temperature sensitivity, sensitivity for local strain and the ability to adequately cover the whole structure.

The rapid growth in the fundamental understanding, characterization techniques and the processing science in the field of nanotechnology in the last two decades has enabled specific tailoring and manipulation of nanomaterials. This has facilitated tremendous advancement in developing nanostructured hierarchical composites and textiles. Specifically, researchers have widely used carbon nanotubes due to their extraordinary properties for applications such as improving the mechanical properties of composites [3-5], self-sensing composites [6-8], energy harvesting composites [9,10], wearable strain sensors and flexible pressure sensors [11-15], sensing skin for structural health monitoring [16-18] and fatigue crack monitoring [19,20] and damage monitoring of joints [21].

With increasing focus on the research and development of carbon nanotube-based sensing skins, improving Technology Readiness Levels (TRL) and the rising number of field trials conducted, it is more important than ever to characterize the sensing response under different loading scenarios and to simulate real-life conditions addressing challenges such as calibration. For piezoresistive carbon nanotube-based sensors, researchers typically describe the sensitivity (gage factor GF) as the ratio of normalized change in electrical resistance ($\Delta R/R$) to longitudinal strain (ε_1), not taking into consideration the effect of transverse strain (ε_2).

$$Gage\ Factor\ (\boldsymbol{GF}) = \frac{\Delta R}{R} \times \frac{1}{\varepsilon_1}$$

While this works for sensors like metallic strain gages which are highly directional, but it is a significant source of inaccuracy when the transverse strain is not accounted for a carbon nanotube sensing skin which is sensitive to multi-axial strain. As a result, the sensor calibration is dependent on the substrate and its specific Poisson's ratio. For example, the sensor response when bonded to aluminum will be different than when bonded to steel or composite due to different Poisson's ratio and the transverse strain due to it. Therefore, it is necessary to characterize the effect of transverse strain on the sensing response and investigate the sensor under different loading conditions.

In this research, a facile, scalable manufacturing technique of dip coating is used to create carbon nanotube-based sensing skin. The carbon nanotubes are deposited on a non-woven aramid fabric with randomly oriented fibers. The sensing response is characterized under different loading scenarios such as flexural and axial loading. The influence of Poisson effect (and the transverse strain due to it) on the sensing response is investigated by bonding the sensing skin to materials having different Poisson's ratio and on a cruciform shaped specimen loaded using a biaxial testing machine.

2. EXPERIMENTAL

2.1 Materials

A non-woven aramid fabric (20601 50 g/m^2, Technical Fiber Products) with fibers oriented in a random direction is used as a carrier of the carbon nanotubes for the sensing skin. A commercially available dispersion of carbon nanotubes (Aquacyl AQ0302, Nanocyl) is used for the dip coating process. A 2-part epoxy adhesive (Hysol 9309, Loctite) is applied for creating an insulating layer and bonding the sensing skin on substrates. The composite coupons and the cruciform specimen are manufactured using M40J unidirectional carbon fiber prepreg (Toray Carbon Fibers America). A flash dry silver paint (SPI supplies) and a 2-part conductive epoxy (Epoxies etc.) is used to connect the wires to reduce the contact resistance while measuring the electrical resistance.

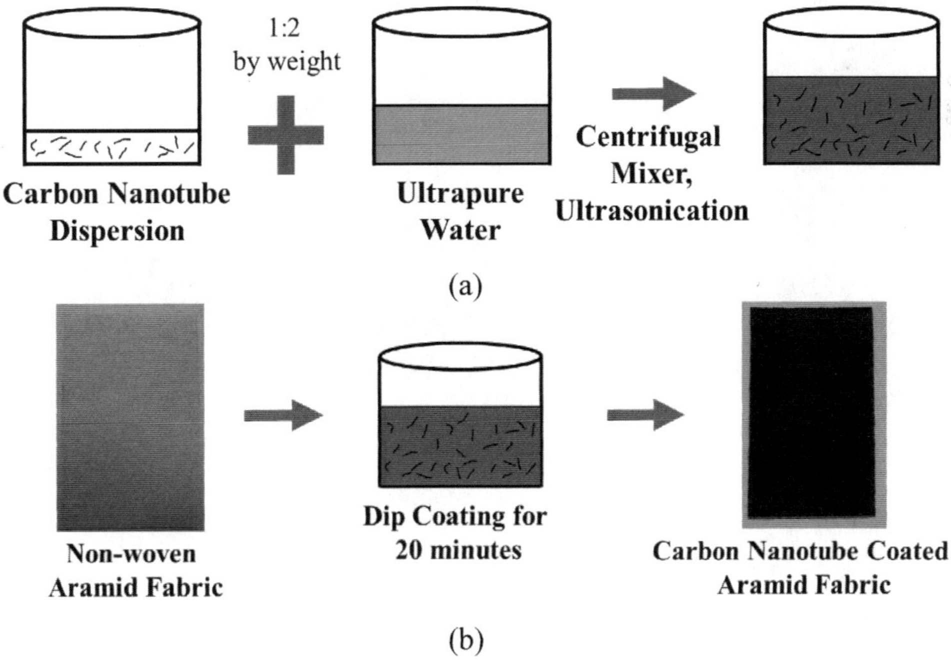

Figure 1. A schematic diagram of the manufacturing process for the carbon nanotube sensors. (a) shows the steps to create a uniform dispersion of carbon nanotubes and (b) shows the procedure followed in the dip coating process

2.2 Sensor Manufacturing using Dip Coating

The carbon nanotube-based sensing skins are created using a simple and easy to scale dip coating technique. The carbon nanotube dispersion, Aquacyl is diluted with ultrapure water in a ratio of 1:2 by weight. The diluted dispersion is then mixed using a planetary centrifugal mixer (ARM-310, Thinky) for 150 seconds at 2000 rpm. An ultrasonic bath sonicator is used for 30 minutes and the aramid fabric is dip coated for 20 minutes following drying in a convection oven at 150 °C for 30 minutes. A schematic of the sensor manufacturing process is shown in Figure 1. The length of the sensing skin is 62.5 mm (2.5 inches) and the width 18.75 mm (3/4 inches).

2.3 Specimen Manufacturing

The carbon nanotube sensing skin is bonded to different substrates such as steel coupons, composite coupons and a composite cruciform for different types of tests. A similar procedure is followed for manufacturing all kinds of specimens. The surface area where the carbon nanotube sensor will be bonded is sandblasted and cleaned with acetone. A thin layer of the Hysol adhesive is applied to create an insulating layer to prevent direct contact of the sensing layer to steel or composite and avoid the electrical shorting of the sensor. Once the adhesive layer is partially cured, the sensing layer is bonded using the same adhesive and cured in an oven at 85 °C for 1 hour and pressure is applied using clamps. A schematic of the steel specimen is shown in Figure 2. The part of the sensing skin where electrodes will be applied is masked using tape to prevent the adhesive from seeping in. The electrodes are then painted using a flash dry silver paint and the lead wires are attached using a 2-part conductive silver epoxy.

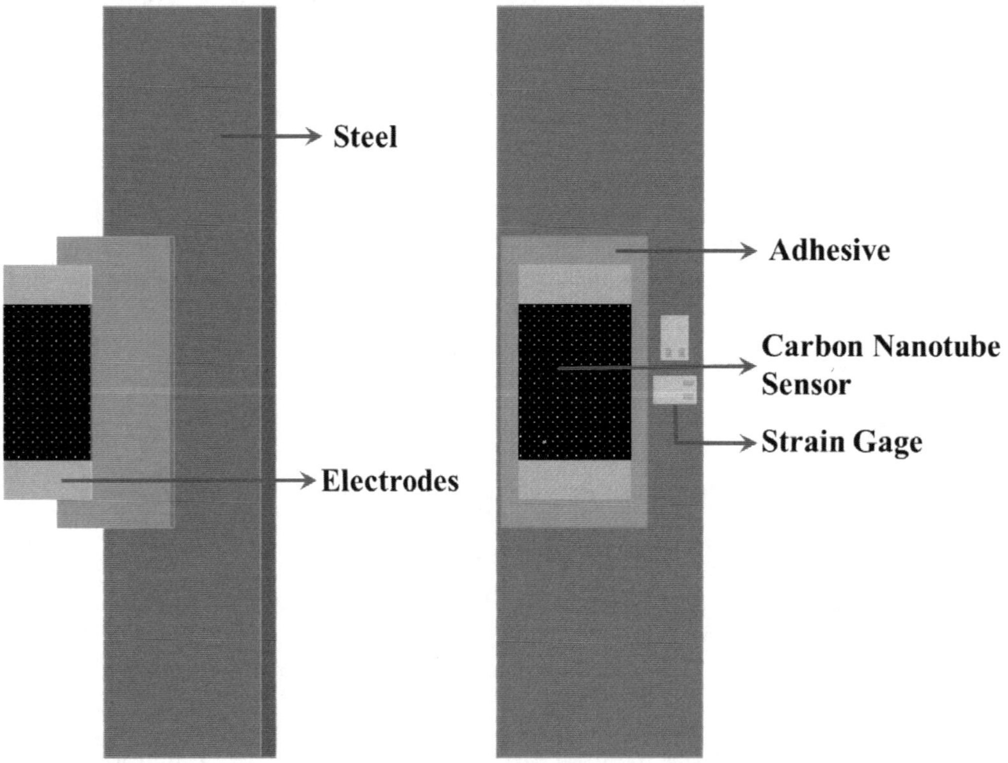

Figure 2. A schematic of the steel specimen with carbon nanotube sensing skin bonded.

2.4 Unidirectional Axial and Flexural Testing

The specimen is tested using a screw-driven mechanical load frame Instron 5985. For the flexural tests for steel specimens, a 4-point bending test is performed with the specimen supported at a distance of 300 mm (12 inches) and the load is applied at a distance of 50 mm (2 inches) from the center. A 4-point bending test is chosen so that the bending moment is constant within the loading area. The specimens are tested only in the elastic range to a strain of 0.1%. The maximum load in the axial test is 60 kN and the flexural test is 1.5 kN so that the specimens are subjected to similar strain levels of ~0.1%.

2.5 Biaxial Testing

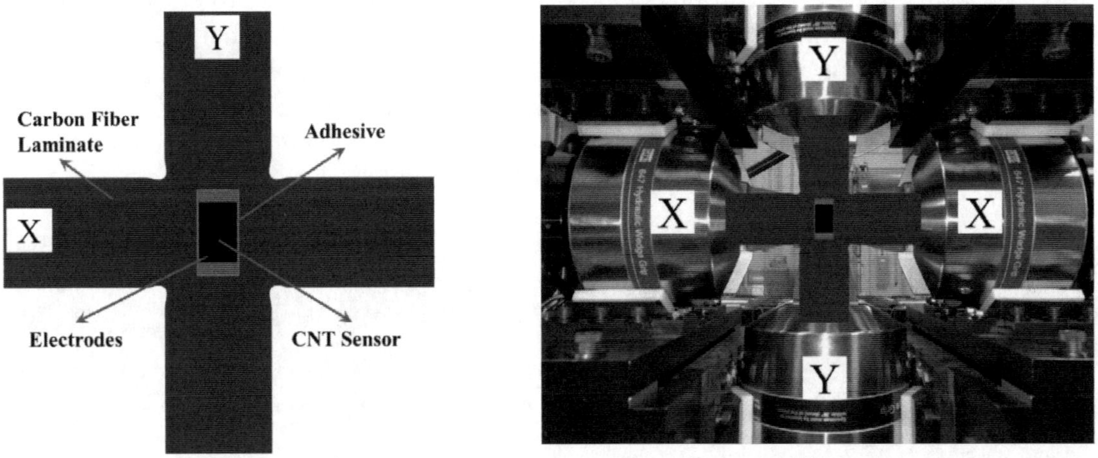

(a) (b)

Figure 3 (a) A schematic illustration of the cruciform specimen with the carbon nanotube sensor and (b) a photograph of the biaxial testing machine with the schematic of the specimen overlaid showing the orientation of the sensor with respect to the machine Y and X axes.

The biaxial composite specimens are manufactured with a ply layup of $[0_4/90_4]_s$ with a total of 16 plies. A temperature of 135 ºC and pressure of 296 kPa is used for curing the composite in an autoclave for 120 minutes. The cruciform specimen is waterjet cut from a square laminate to dimensions as described by Skinner *et al.* [22] to have a uniform strain field in the center of the specimen. A schematic of the cruciform specimen is shown in Figure 3(a). The specimen is tested using an MTS planar biaxial servohydraulic testing machine (Figure 3(b)). This load frame is instrumented with 6 independent actuators. The maximum capacity of the load cell is 100 kN in the Y and X directions and 1100 N-m for a torsional load. Our specimen with the carbon nanotube sensor is was subjected to an in-plane load only. Maximum loads in the Y and X directions are varied to artificially create a different ratio of transverse strain/longitudinal strain. The maximum loads were limited to 9 kN to test the specimen in the elastic range with a maximum strain of just under 0.1%.

2.6 Electron Microscopy

Scanning electron microscopy is conducted using a Zeiss Auriga 60 High Resolution Focused Ion Beam and Scanning Electron Microscope. An accelerating voltage of 3 kV and an SESI (Secondary Electron Secondary Ion) detector is used. The specimens are sputter coated with a thin conductive layer of Pd/Au to minimize the charging using a sputter coating machine (Denton Desk IV).

3. RESULTS

3.1 Carbon Nanotube Film Characterization using Electron Microscopy

Figure 4 shows the scanning electron micrographs of the carbon nanotube-based sensing skin. The non-woven aramid fibers act as a carrier for the carbon nanotubes and create the backbone structure on which the nanotube forms a percolating conductive network. Figure 4(a) is a low magnification image showing the microstructure of the non-woven aramid fabric and the randomly oriented fibers in it. The binder used by the manufacturer to create the fabric forms web like structures between some fibers and is also coated with carbon nanotubes. Figure 4(b) shows a high magnification image of a fiber within the non-woven fabric with a uniform coating of carbon nanotubes. The carbon nanotubes are deposited on the entire fiber and on all the fibers within the fabric. It is due to this homogenous coating; the nanotubes are able to form a conductive network that enables piezoresistive sensing.

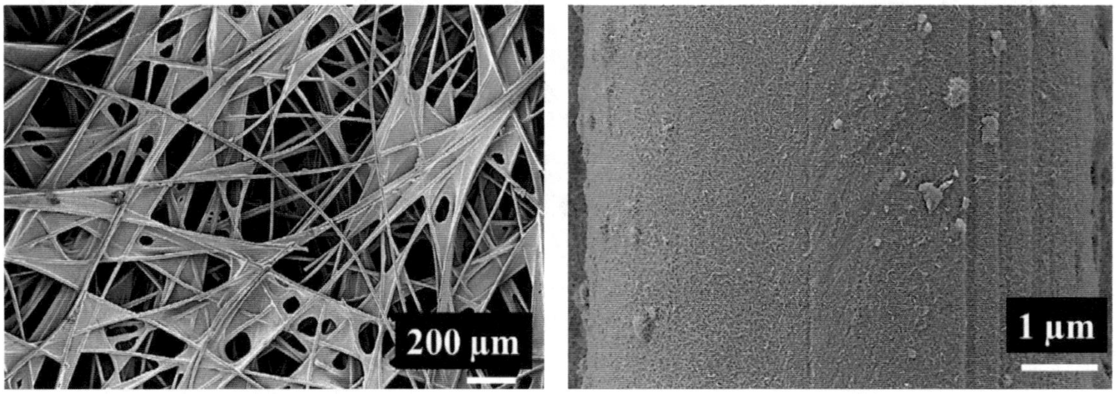

(a) (b)

Figure 4. Scanning electron micrographs of the carbon nanotube coated sensing skin. (a) Low magnification image showing the randomly oriented fibers in the non-woven fabric with binder forming web like structures between fibers and (b) high magnification image showing one of the fibers in the fabric which is uniformly coated with carbon nanotubes.

3.2 Electromechanical Characterization under Flexural and Axial Loads

To characterize the sensing response of the carbon nanotube-based sensing skin under different types of loads, the sensing skin was bonded to a steel coupon that was subjected to axial and flexural loads. The maximum loads for each test were chosen so that the applied longitudinal strain on the sensors would be $\sim 0.1\%$, significantly below the yield strain. Figure 5 shows the strain, resistance change and stress versus time response for the loading and unloading cycle for the flexural and axial bending. The load applied for the 4-point bending was 1.5 kN and the stress was calculated using the flexure formula. The applied load for the axial test was 60 kN and the stress was calculated by dividing the load by area. The effect of the thin sensing layer is neglected for these calculations.

Figure 5 shows that the sensing response is independent of the type of loading. The applied stress for both types of loading is ~200 MPa and the maximum strain measured in the longitudinal direction is ~0.1% for both. The maximum percentage change in resistance is about 0.9 % for each case. Since the applied load is in the elastic range, upon unloading there is no permanent change in electrical resistance.

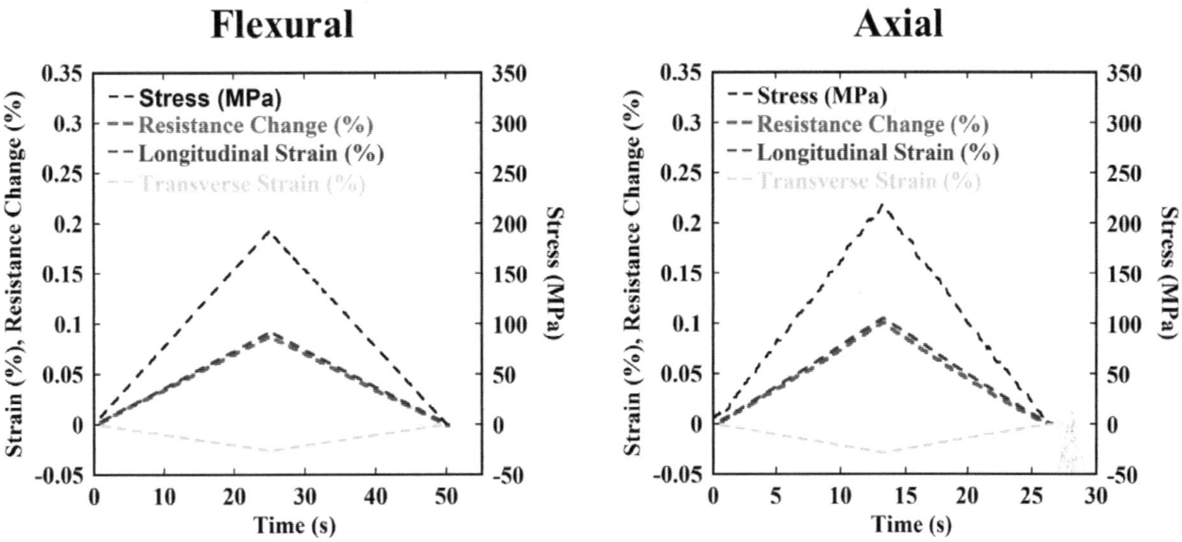

(a) (b)

Figure 5. Strain, resistance change and stress response vs time for the steel specimen when subjected to (a) flexural loading or 4 point bending and (b) axial loading. The resistance response is very similar and not dependent on the type loading.

3.3 Influence of Poisson's Ratio on the Sensing Response using Different Composite Laminates

To understand the influence of the substrate's Poisson's ratio on the sensing response, the carbon nanotube-based sensing skin is glued to 0° and 90° composite coupons and loaded in axial tension. Figures 6(a) and 6(b) show the load, resistance change (%) and strain (%) for carbon nanotube sensing skins when glued to a 0° and 90° composite coupon respectively. The maximum strain for both specimens is about 0.11%. The applied load for 0° specimen is 3000 N and for 90° specimen is 250 N for the same longitudinal strain because the Young's modulus in the direction of the fibers (0°) is significantly higher than the transverse direction (90°).

Since the sensing skin is sensitive to strains in all directions, the transverse contraction due to the Poisson effect also contributes to the bulk electrical resistance change. The contraction in the direction perpendicular to the loading direction will cause a decrease in electrical resistance. Therefore, the total percentage of electrical resistance change measured is the sum of resistance change due to longitudinal strain (positive and roughly equal for both specimens) and transverse strain (negative for 0° specimen and negligible for 90° specimen as shown in Figure 6(c) and 6(d)). As a result, the total percentage change in resistance is significantly higher for the 90° specimen than the 0° specimen even though both the specimens are subjected to similar longitudinal strain.

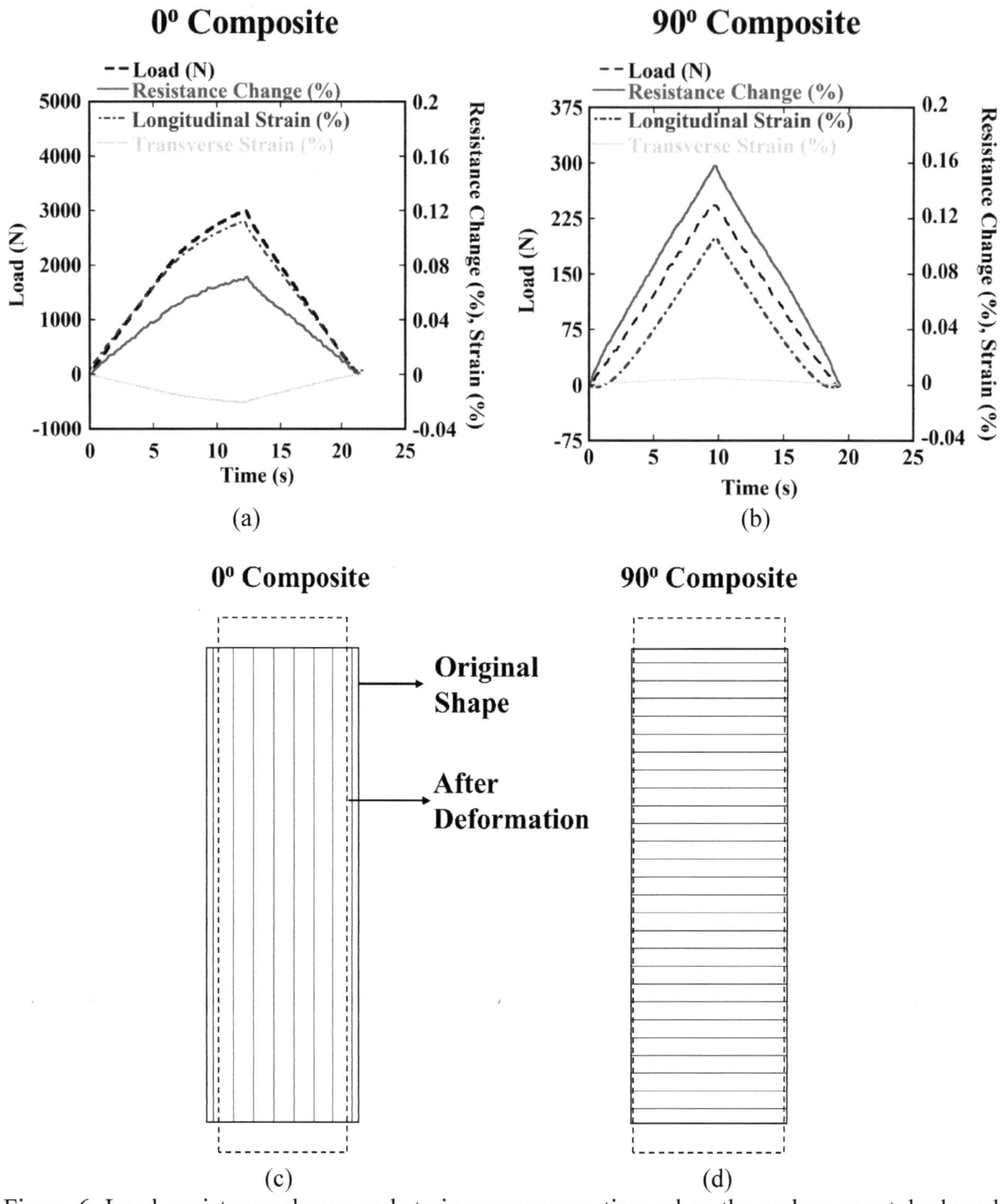

Figure 6. Load, resistance change and strain response vs. time when the carbon nanotube-based sensing skin is attached to (a) 0° composite coupon and (b) 90° composite coupon. For about the same maximum longitudinal strain, the percentage change in resistance is significantly greater for the 90° composite. (c) and (d) is a schematic showing the deformation of a typical 0° and 90° composite. The compression in 0° composite in transverse degree is due to a larger Poisson's ratio.

3.4 Sensing Response for a Cruciform Specimen using a Biaxial Testing Machine

To further investigate the effect of transverse strain on the sensing response, the carbon nanotube-based sensing skin is attached to a cruciform shaped specimen and tested using a biaxial testing machine. Three different types of loading conditions are applied. The load on the Y axis is +9 kN in all three cases whereas the load on the X axis (perpendicular to the direction of the sensor) is varied from -1.25 kN, +3 kN and to +9 kN. The loading curves for all three cases are shown in Figures 7(a), 7(b) and 7(c). The effective Poisson's ratio for each loading scenario is 0.14, -0.33 and -1, respectively. The strains in the Y and X direction for each loading case are shown in Figures 7(d), 7(e) and 7(f).

As observed in the previous experiments where the transverse strain plays a significant role in the total percentage resistance change and the sensitivity, in this case too, the load in the X direction (perpendicular to the sensor) plays a critical role. The percentage resistance change for three loading cases with decreasing Poisson's ratio is shown in figure 7(g), 7(h) and 7(i). In the first case, the load applied in the X direction was -1.25 kN compressive load. Therefore, the effective Poisson's ratio is positive 0.14. The total change in resistance is the sum of resistance change due to strain in the Y direction (positive) and X direction (negative). For the next two loading cases, the load in the X direction is 3 kN and 9 kN in tension. For these cases, the strain in both Y and X direction is positive and as a result the total resistance change is higher than the previous example where the load in X direction is compressive. A positive and higher effective Poisson's ratio indicates more compression and less total resistance change. A negative Poisson's ratio indicates tension in the transverse direction and results in a higher sensitivity and change in electrical resistance. For the three cases where Poisson's ratios are 0.14, -0.33 and -1 the total percentage resistance change is 0.09%, 0.13% and 0.2%, respectively.

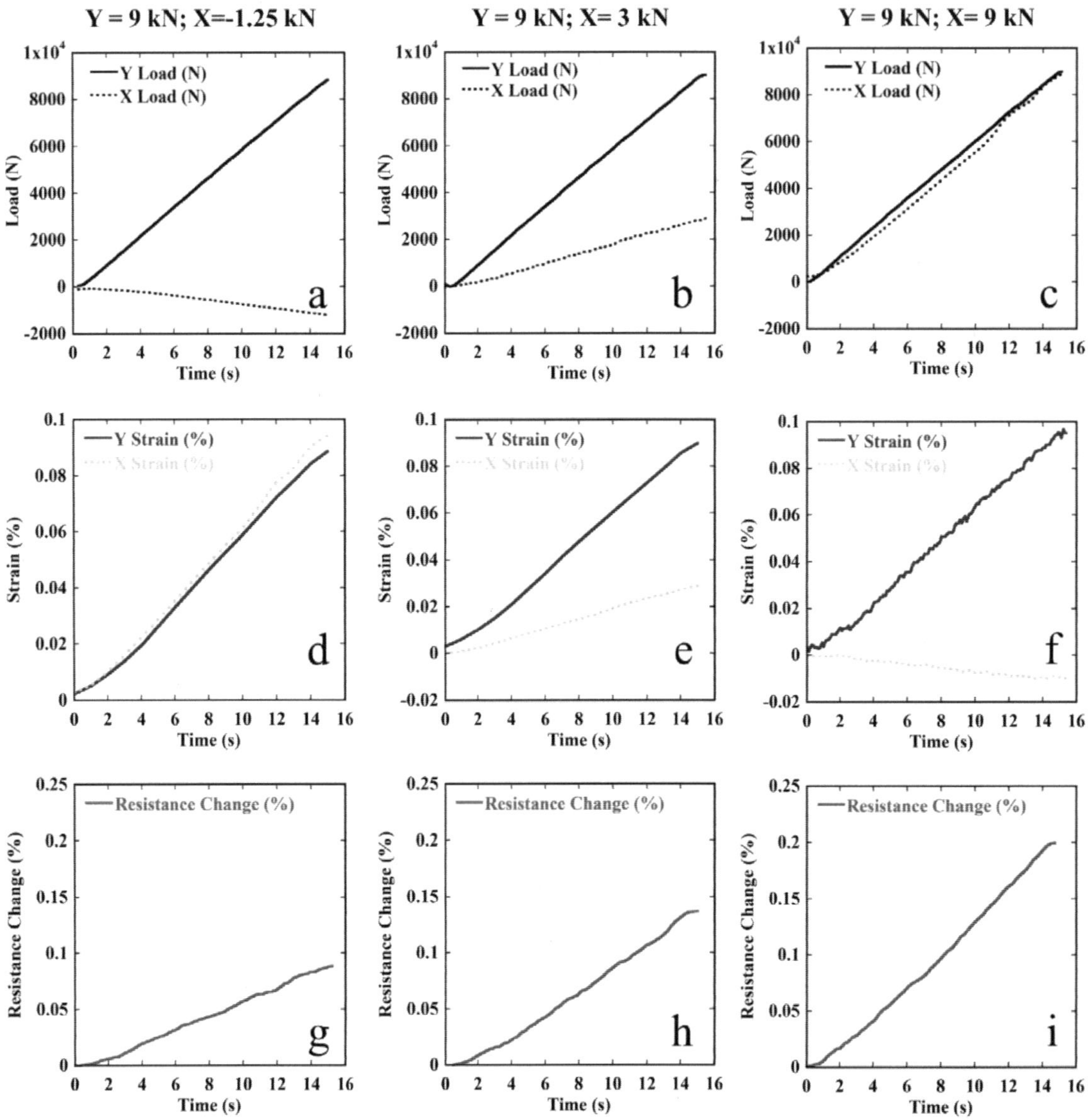

Figure 7. Load in Y and X direction vs. time for the three loading scenarios where Y = 9kN for all and (a) X= -1.25 kN, (b) X= +3 kN, (c) X= +9kN. Strain in Y and X direction vs. time for the three loading scenarios Y = 9kN for all and (d) X= -1.25 kN, (e) X= +3 kN, (f) X= +9kN. Percentage resistance change vs. time for the three loading scenarios Y = 9kN for all and (g) X= -1.25 kN, (h) X= +3 kN, (i) X= +9kN. The strain in transverse direction significantly affects the resistance response of the carbon nanotube sensing skin.

4. SUMMARY AND CONCLUSIONS

In summary, a scalable and facile manufacturing technique for carbon nanotube-based sensing skin is demonstrated. The sensing skin when attached to a steel coupon is subjected to axial and flexural loads and the resistance response is very similar when subjected to the same amount of longitudinal strain. This shows that the sensing response of the carbon nanotube sensor is independent of the type of load applied. Unlike a metallic strain gage, the carbon nanotube sensor is not directional and sensitive to deformation in all directions. As a result, the sensitivity of the sensor is significantly affected by Poisson's ratio of the substrate. When bonded to a 90^{0}

composite (low Poisson's ratio) the resistance change is greater than when glued to 0^0 composite (high Poisson's ratio). Experiments were also conducted by attaching the sensor to a cruciform shaped specimen and tested using a biaxial testing machine where the load in the transverse direction was varied and the resistance response was significantly different for each case. The sensitivity of the carbon nanotube-based sensing skin is significantly affected by the transverse strain due to the Poisson's ratio of the substrate material and should be taken into consideration when calibrating sensors for field applications.

5. ACKNOWLEDGMENTS

This study is part of a research project funded under the Federal Highway Administration's Exploratory Advanced Research Program, Award No. DTFH61-13-H-00010. Frank Jalinoos is the project manager and his support is much appreciated. Any opinions, findings, and conclusions or recommendations expressed in this publication are those of the authors and do not necessarily reflect the view of the Federal Highway Administration. Funding from the UNIDEL Foundation and Army Research Laboratory is also acknowledged. The authors would like to thank Dr. Siddhant Datta, Mechanical Engineering Department of Arizona State University for guidance in performing the cruciform tests. Key technical inputs from Dr. Jaret Riddick and Dr. Mark Bundy are acknowledged. A special thanks to Ms. Lavern Coleman for facilitating tests at the Army Research Laboratory.

6. REFERENCES

1. Bassil, Antoine, et al. "Distributed fiber optics sensing and coda wave interferometry techniques for damage monitoring in concrete structures." Sensors 19.2 (2019): 356. doi.org/10.3390/s19020356
2. Enckell, Merit, et al. "Evaluation of a large-scale bridge strain, temperature and crack monitoring with distributed fibre optic sensors." Journal of Civil Structural Health Monitoring 1.1-2 (2011): 37-46. doi.org/10.1007/s13349-011-0004-x
3. Coleman, Jonathan N., et al. "Small but strong: a review of the mechanical properties of carbon nanotube–polymer composites." Carbon 44.9 (2006): 1624-1652. https://doi.org/10.1016/j.carbon.2006.02.038
4. Kausar, Ayesha, Irum Rafique, and Bakhtiar Muhammad. "Review of applications of polymer/carbon nanotubes and epoxy/CNT composites." Polymer-Plastics Technology and Engineering 55.11 (2016): 1167-1191. https://doi.org/10.1080/03602559.2016.1163588
5. An, Qi, et al. "Tailored glass fiber interphases via electrophoretic deposition of carbon nanotubes: fiber and interphase characterization." Composites Science and Technology 166 (2018): 131-139. https://doi.org/10.1016/j.compscitech.2018.01.003
6. Doshi SM, Thostenson ET. Self-sensing carbon nanotube composites: processing and characterization. Multifunctionality of Polymer Composites: Challenges and New Solutions; 2015; Chapter 25, 752–84 https://doi.org/10.1016/C2013-0-13006-1
7. Nataraj, Latha, et al. "Early-Stage Damage Detection in Advanced Multifunctional Aerospace Composites Using Embedded Carbon Nanotubes and Flocked Carbon Fibers." Multidisciplinary Digital Publishing Institute Proceedings. Vol. 2. No. 8. 2018.

https://doi.org/10.3390/ICEM18-05386

8. Gao, Shang-lin, et al. "Glass fibers with carbon nanotube networks as multifunctional sensors." Advanced Functional Materials 20.12 (2010): 1885-1893. https://doi.org/10.1002/adfm.201000283
9. Sung, Dae Han, et al. "Characterization of thermoelectric properties of multifunctional multiscale composites and fiber-reinforced composites for thermal energy harvesting." Composites Part B: Engineering 92 (2016): 202-209. https://doi.org/10.1016/j.compositesb.2016.02.050
10. Hu, Renchong, et al. "Harvesting waste thermal energy using a carbon-nanotube-based thermo-electrochemical cell." Nano letters 10.3 (2010): 838-846. https://doi.org/10.1021/nl903267n
11. Doshi, Sagar M., and Erik T. Thostenson. "Thin and Flexible Carbon Nanotube-Based Pressure Sensors with Ultrawide Sensing Range." ACS sensors 3.7 (2018): 1276-1282. https://doi.org/10.1021/acssensors.8b00378
12. Jiang, Shuwen, et al. "An Ultra-wide Sensing Range and Highly Sensitive Flexible Pressure Sensor Based on Percolative Thin Film with Knoll-like Micro-structured Surface." ACS applied materials & interfaces (2019). https://doi.org/10.1021/acsami.9b02659
13. Zhang, Hao, et al. "Flexible and Pressure-Responsive Sensors from Cellulose Fibers Coated with Multiwalled Carbon Nanotubes." ACS Applied Electronic Materials 1.7 (2019): 1179-1188. https://doi.org/10.1021/acsaelm.9b00182
14. Doshi, Sagar M., et al. "Carbon Nanotube Coated Textile Sensors with Ultrahigh Sensitivity for Human Motion Detection" Proceedings of the IEEE Sensors Conference, Montreal, Oct 27th-30th, 2019.
15. Doshi, Sagar M., et al. "Carbon Nanotube Based Flexible Sensors for Human Motion Analysis." Proceedings of the American Society for Composites—Thirty-fourth Technical Conference. Atlanta, Sep 23rd-25th, 2019.
16. Ahmed, Shafique, et al. "Development of a novel integrated strengthening and sensing methodology for steel structures using CNT-based composites." Journal of Structural Engineering 143.4 (2016): 04016202. https://doi.org/10.1061/(ASCE)ST.1943-541X.0001697
17. Schumacher, Thomas, and Erik T. Thostenson. "Development of structural carbon nanotube–based sensing composites for concrete structures." Journal of Intelligent Material Systems and Structures 25.11 (2014): 1331-1339. https://doi.org/10.1177/1045389X13505252
18. Dai, Hongbo, et al. "Processing and characterization of a novel distributed strain sensor using carbon nanotube-based nonwoven composites." Sensors 15.7 (2015): 17728-17747. https://doi.org/10.3390/s150717728
19. Ahmed, Shafique, et al. "Integration of carbon nanotube sensing skins and carbon fiber composites for monitoring and structural repair of fatigue cracked metal structures." Composite Structures 203 (2018): 182-192.

https://doi.org/10.1016/j.compstruct.2018.07.005

20. Kwon, Dong-Jun, et al. "Damage sensing and fracture detection of CNT paste using electrical resistance measurements." Composites Part B: Engineering 90 (2016): 386-391. https://doi.org/10.1016/j.compositesb.2016.01.020

21. Ahmed, Shafique, et al. "Experimental and numerical investigation on the bond strength of self-sensing composite joints." International Journal of Adhesion and Adhesives 84 (2018): 227-237. https://doi.org/10.1016/j.ijadhadh.2018.03.014

22. Skinner, Travis, et al. "Fatigue damage behavior in carbon fiber polymer composites under biaxial loading." Composites Part B: Engineering (2019): 106942. https://doi.org/10.1016/j.compositesb.2019.106942

MULTIFUNCTIONAL FILMS FOR FUSION BONDING AND STRUCTURAL HEALTH MONITORING OF THERMOPLASTIC COMPOSITE JOINTS

H. Frederick[a], W. Li[a], W. Sands[b], E. Tsai[c] and G. Palardy[a*]

[a] Department of Mechanical and Industrial Engineering, Louisiana State University
3261 Patrick F. Taylor Hall
Baton Rouge, LA 70803

[b] Materials Science and Engineering, Clemson University
161 Sirrine Hall
Clemson, SC 29634

[c] Department of Chemistry, Louisiana State University
232 Choppin Hall
Baton Rouge, LA 70803

ABSTRACT

Ultrasonic welding is a promising method to quickly bond thermoplastic composite parts. In this process, a thermoplastic film, called an "energy director", is placed between two pieces of composite materials, and melted during the welding process to facilitate bonding. This study seeks to develop a multifunctional energy director suitable for three purposes: 1) ultrasonic welding, 2) structural health monitoring (SHM) via electrical resistance measurements, and 3) heating element for disassembly and repair of the damaged bond. Various materials and manufacturing methods were investigated to develop multifunctional films: 1) magnetic or conductive nanoparticles deposited onto polymer films, and 2) multi-walled carbon nanotubes (MWNTs) dispersed into polymers. The latter was selected as a promising candidate because of its ease of manufacturability through compression molding and its compatibility with the welding process. The thermo-electrical and electro-mechanical behavior of compression molded films containing up to 20 wt% MWNT in a polypropylene (PP) matrix was assessed via a sourcemeter and dynamic mechanical analyzer. To demonstrate multifunctionality, glass fiber/PP specimens were ultrasonically welded with the MWNT/PP films as energy directors. While the multifunctional films reduced lap shear strength by up to 10%, they did not significantly affect the welding process. Moreover, welded specimens with the MWNT/PP films were successfully tested under cyclic bending for SHM through resistance measurements.

Corresponding author: Genevieve Palardy (gpalardy@lsu.edu)

SAMPE Virtual Conference Proceedings, 2020. Society for the Advancement of Material and Process Engineering – North America.

1. INTRODUCTION

Lightweight composite materials are of great interest in several fields of applications, especially aerospace and ground transportation. The most popular types of conventional fiber-reinforced composites are thermoset composites (TSCs) and thermoplastic composites (TPCs). The main difference between thermoset and thermoplastic polymers is that the latter soften and may be formed again when heated up above a certain temperature. This makes thermoplastic composites more sustainable in the long-term as they can be reused and recycled. An additional benefit is that they can be joined through fusion bonding, otherwise known as "welding", eliminating the use of mechanical fasteners [1]. This feature makes them competitive for high-end, high-performance sectors where the weight reduction translates into important cost savings.

An upcoming joining technique for fiber-reinforced thermoplastic composites, ultrasonic welding (USW), is generating considerable interest, most particularly in the aerospace and automotive industries due to its fast cycle times and flexibility [2]. While this technique has been successfully demonstrated on lab-scale and medium-sized composite structures, most research is directed toward process development and optimization through static strength data [3-7]. There is still a lack of confidence, notably, in the fundamental prediction of failure behavior at the joint. In order for ultrasonically welded composite joints to be considered in industrial applications, it is imperative to find innovative ways to enable structural health monitoring (SHM), as well as predict loading and failure modes.

SHM of thermoset composites and adhesive joints is a popular field of research, but welded TPC joints are not widely studied [8, 9]. This work aims to integrate multifunctional films at the interface of ultrasonically welded thermoplastic composites, which would serve three purposes, as illustrated in Figure 1: i) energy director to enable ultrasonic welding, ii) sensor to monitor strains, and iii) heating element for disassembly and repair. Promising multifunctional films for strain sensing and self-heating are nanocomposite-based, where electrically conductive nanoparticles/nanofibers (e.g. carbon nanotubes (CNTs), CNT buckypaper) are combined with a polymer matrix [10-13]. However, an additional challenge unique to the USW process, over other common fusion bonding methods, is the application of vibrations disrupting the interface to be joined. As the interface melts, polymer is squeezed out in less than 1 second, meaning fragile sensors (e.g. optical fibers) or geometrically sensitive sensors (e.g. 3D printed patterns) are not suitable.

This paper focuses on multifunctional films characterization, ultrasonic welding and structural health monitoring of welded joints under bending (Figure 1). It will first summarize the preparation methods that have been investigated to create multifunctional films suitable for USW. Second, it will present thermo-electrical and electro-mechanical characterization of the most suitable films, through resistance heating and dynamic mechanical analysis (DMA). Third, the effect of multifunctional films on welded TPC samples will be assessed through welding curves, outputs and lap shear strength. Finally, potential for SHM will be demonstrated during periodic bending of welded joints in a single lap configuration.

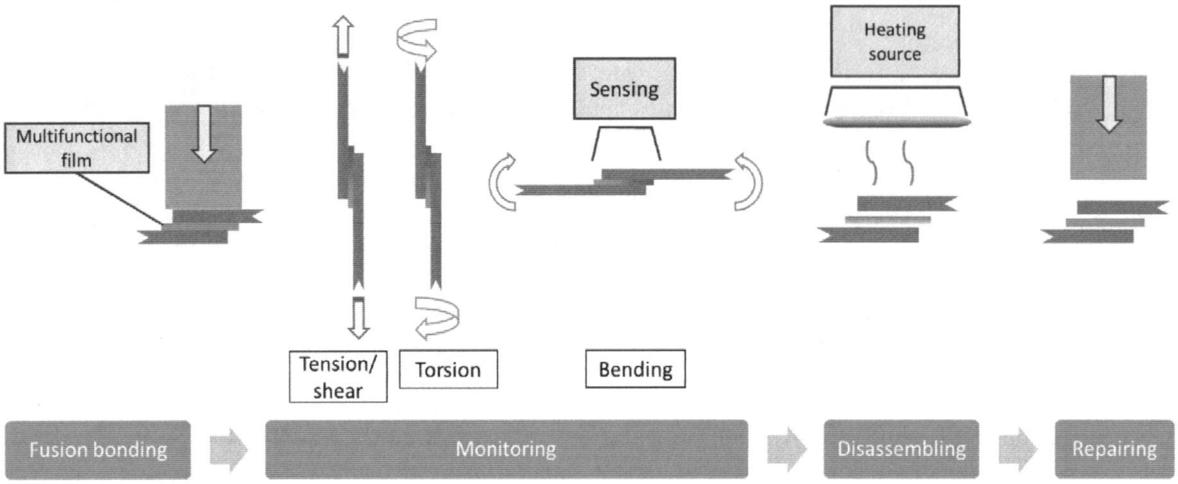

Figure 1. Global, long-term vision for the proposed research. The current paper focuses on multifunctional films characterization, fusion bonding and structural health monitoring of welded joints under bending.

2. EXPERIMENTATION

2.1 Materials and manufacturing for multifunctional films

Several materials and manufacturing methods were considered in this study, as listed in Table 1. Five types of polymers were investigated: PP, PA6, PEI, PPS and PEEK (see below Table 1 for definitions). Electrically conductive and magnetic nanoparticles were used: silver and iron oxide nanoparticles in aqueous or hexane solutions, and multi-walled carbon nanotubes (MWNT) dispersed in polymer pellets (masterbatches). Those particles were considered because strain monitoring can be carried out through resistance measurements and self-heating through resistance or induction heating.

Two manufacturing methods were explored: 1) drop casting, and 2) compression molding. Drop casting is depicted in Figure 2 and was studied first for its ease of implementation, versatility, low cost and demonstrated use for induction welding of TPCs [14]. It consists of depositing drops of a solution containing dispersed nanoparticles on a polymer film, then placing the film in an oven or on a hot plate to evaporate the solvent (at temperatures listed in Table 1). Nanoparticles were diluted to concentrations between 2,000 ppm and 30,000 ppm. One to three layers were deposited on the polymer films. The quality of the films was visually assessed and observed under scanning electron microscope (SEM) for defects.

The second film manufacturing method, compression molding, consists of placing thermoplastic pellets, granules or powder containing nanoparticles (masterbatches) on a heated hydraulic press and increasing the pressure to form films. To test out film manufacturing and potential for multifunctionality, commercial masterbatches were purchased as listed in Table 1. MWNT were selected because of their electrical properties and availability as masterbatches with a wide range of thermoplastics. PP masterbatches containing 15 wt% and 20 wt% MWNT were chosen based on their low melting temperature and cost to demonstrate strain sensing/self-heating capabilities with welded PP composites. Stainless steel shims were used to manufacture films with three thickness values: 0.06 mm, 0.25 mm and 0.50 mm. Those values were selected based on previous research on ultrasonic welding of thermoplastic composites [15]. The films were then cut into rectangular strips for characterization and ultrasonic welding.

Table 1. Materials and manufacturing methods considered for multifunctional films.

Polymers	Nanoparticles	Manufacturing methods
PA6[1], PEI[2], PPS[3], PEEK[4] (0.07 mm to 0.25 mm-thick films)	Silver nanoparticles in aqueous solution[6]	Dilution to concentrations from 2,000 to 30,000 ppm Drop casting + oven heating between 50 °C and 90 °C
	Magnetic nanoparticles (iron oxide) in aqueous solution[7]	
	Magnetic nanoparticles (iron oxide) in hexane[8]	
PP[5] (pellets)	MWNT[9] (15 wt% and 20 wt%)	Compression molding at 170-200 °C for 15 min

[1]Polyamide 6 (Goodfellow), [2]Poly ether imide (ULTEM®, McMaster-Carr), [3]Poly phenylene sulphide (Fortron®, CS Hyde Company), [4]Poly ether ether ketone (McMaster-Carr), [5]Polypropylene (pure pellets purchased from Goodfellow), [6]Purchased from US Research Nanomaterials, Inc., [7]Iron(II,III) oxide nanopowder with 50-100 nm particle size (Sigma-Aldrich), [8]Nanoparticles synthesized and provided by Dr. James Dorman and Natalia Da Silva Moura (Department of Chemical Engineering, LSU), [9]Multi-walled carbon nanotubes masterbatches (Cheap Tubes Inc.).

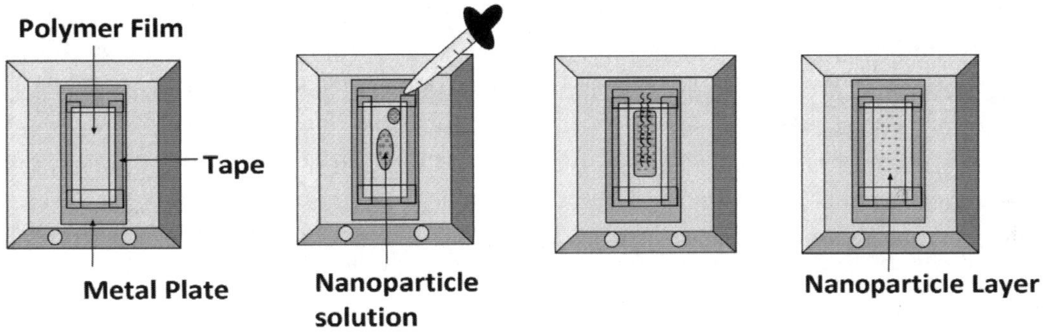

Figure 2. Film preparation using the drop casting method.

2.2 Materials and manufacturing for thermoplastic composites

In this study, glass fiber/polypropylene (GF/PP) 6030 unitape Polystrand™ (PolyOne) prepregs were used. A laminate with 1.86 mm final thickness, manufactured by placing eight GF/PP prepreg layers (0.33 mm thickness for each layer) in 0° orientation, was consolidated by hot pressing at 180 °C and 1 MPa for 15 min. After cooling, 101.6 mm × 25.4 mm rectangular strips were cut with a PICO 155 Precision Saw (Pace Technologies), with the longest dimension along the direction of the fibers.

2.3 Thermo-electrical and electro-mechanical characterization of multifunctional films

For electrical resistance and resistance heating tests, a voltage was applied through copper electrodes placed at both ends of the films with a Keithley Sourcemeter 2604B. It has a maximum voltage and current of 40 V and 1 A, respectively. Temperature during resistance heating was measured with a FLIR A325sc camera placed above the film. The DC voltages used to gather electrical and thermo-electrical data were 1 V, 2 V, 4 V, 6 V, 8 V, and 10 V. Each test was carried out on five film samples over a period of three minutes while the Keithley KickStart software acquired resistance, power and current data.

To determine if the films can be used as strain sensors, electro-mechanical characterization was carried out in a dynamic mechanical analyzer (DMA Q800, TA Instruments). The films were clamped with the tension fixture while electrodes from the Keithley Sourcemeter were attached to the steel grips of the DMA. For these tests, a constant 4 V was applied to the films. The loading cycles applied 1 N to 15 N tensile force, in 1 N increments, at a rate of 3 N/min. For each cycle, once the load was fully applied or removed, it was held constant for 30 seconds. Strain and stress curves were monitored at the same time as resistance change.

2.4 Ultrasonic welding of thermoplastic composite joints

A Dynamic 3000 ultrasonic welder (RINCO ULTRASONICS AG) with a 40 mm diameter sonotrode was used to weld samples in a single lap configuration with an overlap of 25.4 mm × 12.7 mm. This microprocessor-controlled ultrasonic welder has a maximum power of 3000 W and is operated at a frequency of 20 kHz. Two custom-designed clamps were used to prevent bending of the top adherend, while allowing its vertical displacement. As Figure 3 shows, samples consisting of a top adherend, bottom adherend and 0.5 mm-thick multifunctional film (also referred to as flat energy director (ED)) were welded. PP EDs were used (pure PP, 15 wt%, and 20 wt% MWNT), as listed in Table 1. The vibration amplitude was set to 38.1 μm and a welding force of 1000 N was used for all welds. Solidification force and holding time were set to 1000 N and 4000 ms, respectively. Three different vertical displacement values (also known as "travel") were chosen: 40%, 60%, and 80% of the initial film thickness. After welding, the ultrasonic welder output four main parameters: vibration time (ms), energy (J), maximum power (%), and weld distance (mm). In addition, power and travel curves were extracted from the welding data. For each travel and ED type, five samples were welded.

2.5 Mechanical testing of welded joints

After welding, tensile testing was carried out by using a 25 kN test machine (TestResources 313) for lap shear strength (LSS), according to the ASTM D1002 standard. Single-lap joint samples were clamped by hydraulic wedge grips at a distance of 60 mm and a loading rate of 1.3 mm/min. LSS was obtained by dividing the maximum load (F_{max}) by the overlap area (25.4 mm × 12.7 mm). Fracture surfaces and failure modes were observed through visual inspection. Three to five samples were tested for each ED type and travel value.

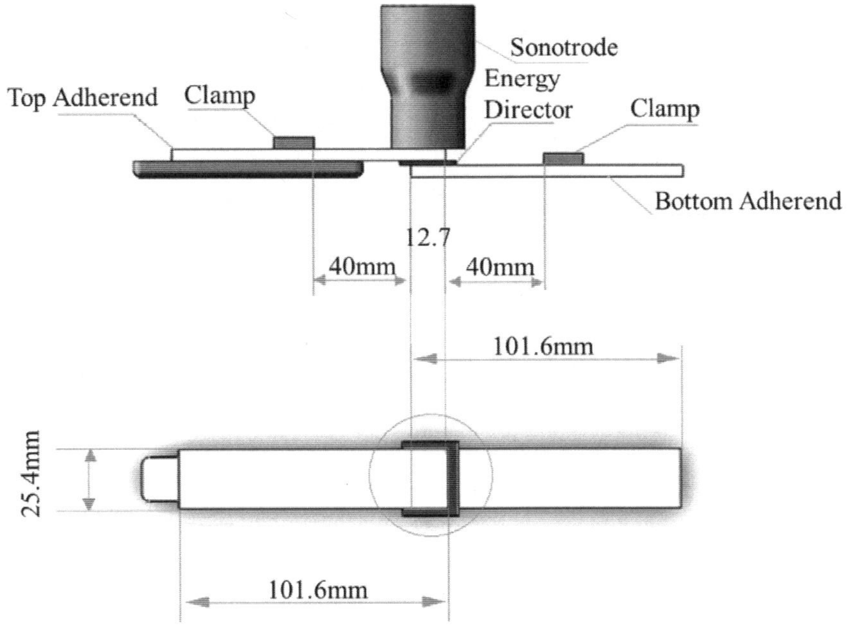

Figure 3. Schematic of the welded samples used in this study (all dimensions are in mm).

2.6 Structural health monitoring of welded joints

Samples welded with 15 wt% and 20 wt% MWNT/PP films at 60% and 80% travel were monitored in bending tests. Before testing, silver conductive paint (SPI #05002-AB) was applied on both sides of the overlap parallel to the direction of the fibers. Two 30 AWG copper wires were placed on the silver coating before drying and fixed with adhesive flash tape. Samples were dried overnight at room temperature. A vise grip was used to hold the welded samples in place and a downward load was periodically applied at 35 s intervals at the end of the cantilever beam with a maximum vertical displacement of 50 mm. Samples were held for 3 s at the maximum displacement before releasing the load. A Keithley Sourcemeter 2604B was used to apply a voltage of 3 V and measure resistance during bending. Resistance curves were recorded using the Keithley KickStart software.

3. RESULTS

3.1 Multifunctional film fabrication and selection

Figure 4 shows representative examples of the films created by drop casting. Polymers with the lowest glass transition temperature, such as PA6, were not well suited for this method: the films warped during the evaporation process as the temperature was close or above their glass transition. This resulted in uneven nanoparticle distribution and thickness, as seen in Figure 4a. More consistent films were produced with PEEK (Figure 4b), PEI (Figure 4c) and PPS. Magnetic particles dispersed in an aqueous solution also produced generally uniform layers on the polymer films (Figure 4d), but appeared brittle with low adhesion to the substrate, as indicated by the white arrow. Iron oxide nanoparticles in hexane (Figure 4e), while allowing for the highest dispersed concentration, left an oily residue after evaporation. SEM images showed micro-cracks in the films containing silver nanoparticles (Figure 5a) and confirmed brittleness of the iron oxide nanoparticles films with poor adhesion to the substrate (Figure 5b).

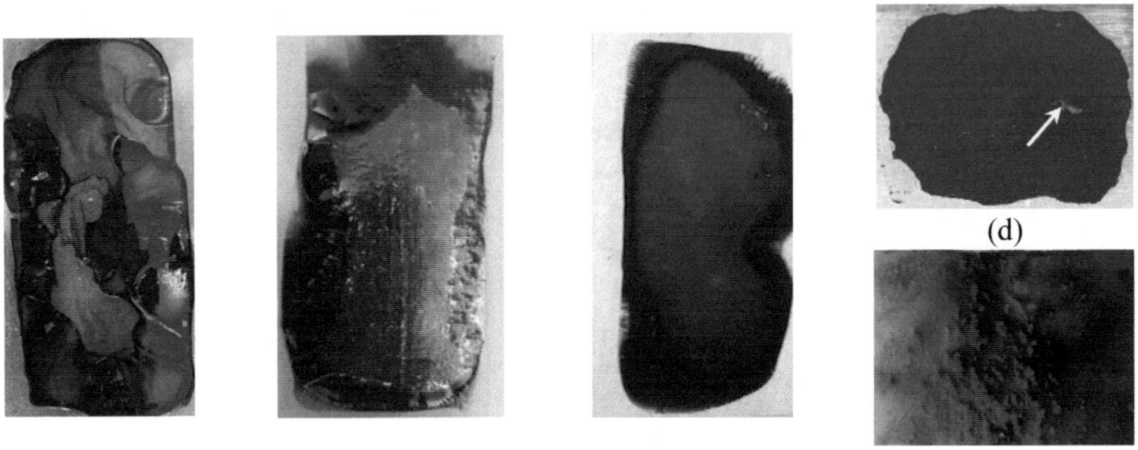

(a) (b) (c) (d) (e)

Figure 4. Examples of films coated with nanoparticles with the drop casting method: a) silver nanoparticles on PA6 film, b) silver nanoparticles on PEEK film, c) silver nanoparticles on PEI, d) magnetic nanoparticles on PEI film, and e) synthesized iron oxide nanoparticles dispersed in hexane on PEI. The white arrow on d) indicated a missing piece of film due to brittleness and poor adhesion to the substrate.

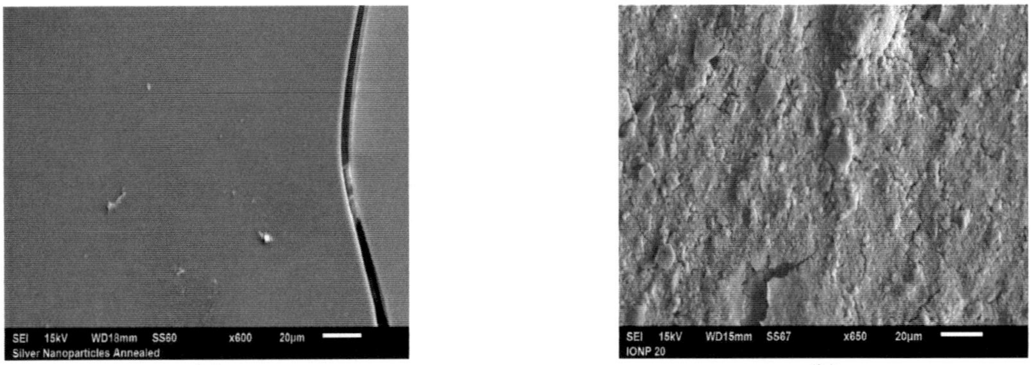

(a) (b)

Figure 5. SEM images of drop cast films with: a) silver nanoparticles (Figure 3b), and b) synthesized iron oxide nanoparticles (Figure 3e). Scale bar: 20 μm.

While the drop casting method was a versatile and low cost technique, it possessed several disadvantages:

1) If the glass transition temperature of the polymer films was below 100 °C, film distortion occurred during the evaporation phase;
2) In many cases, nanoparticle layers were uneven in both distribution and thickness (see Figure 4a - c);
3) Nanoparticle layers were brittle with low adhesion with the polymer films:
   a) Silver nanoparticles produced a solid layer, but the latter cracked easily, separating itself from the substrate;
   b) Magnetic nanoparticles dispersed in an aqueous solution produced even layers, but poor adhesion made it difficult to maintain their integrity;
4) Magnetic nanoparticles dispersed in hexane produced even layers, but hexane, when evaporated, left an oily residue, which would not be suitable to bond interfaces.

In addition, due to the brittleness and low adhesion of the nanoparticle layers, this manufacturing method would not be suitable for the ultrasonic welding process. The layers' integrity might be compromised under ultrasonic vibrations, rendering the strain sensing and disassembly capabilities ineffective.

On the other hand, compression molding produced films with consistent thickness with nanoparticles directly embedded and dispersed into the polymer matrix. One potential downside was the films' brittleness, depending on the polymer (crystallinity) and thermal history. When re-pressed more than once, the polymer films became more brittle. Nonetheless, this manufacturing method was considered suitable to proceed further in this project.

3.2 Multifunctional PP/MWNT films characterization

3.2.1 Thermo-electrical behavior

Figure 6 shows representative results for conductivity measurements of PP films containing 15 wt% and 20 wt% MWNT. As the voltage increased, average conductivity and standard deviation both decreased. Generally, electrical conductivity converged below 6 S/m for both films, with a higher value for 20 wt% MWNT, as expected from the denser conductive path network. It is in the same order of magnitude as previously observed in the literature for MWNT polymer films [12].

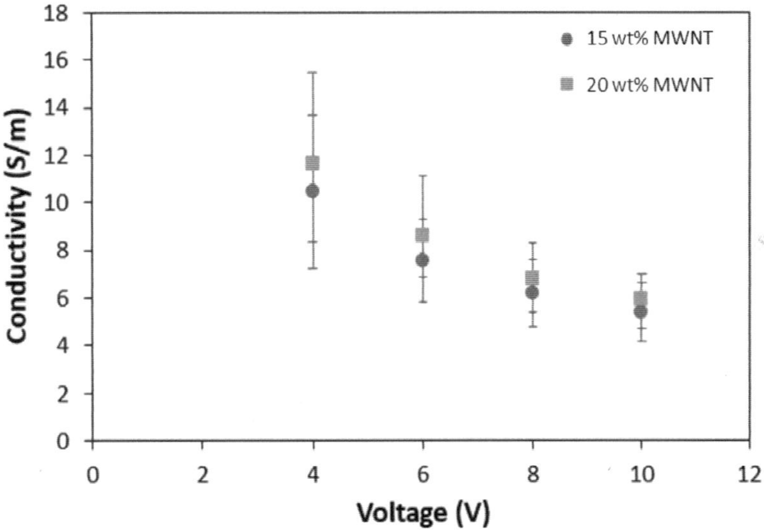

Figure 6. Representative average electrical conductivity of 15 wt% MWNT/PP and 20 wt% MWNT/PP films at different voltage values.

Figure 7 was obtained by plotting the average maximum temperature of the three film thicknesses containing 20 wt% MWNT. This data shows that little to no heating occurred with an input voltage of 4 V while, in some experiments, an input of 10 V caused the film temperature to reach values above 120 °C, approaching the melting temperature of polypropylene (160 °C). While there is some deviation from sample to sample, most likely due to the temperatures not being normalized for geometric differences, it can be seen that the thicker films tended to reach a higher temperature. A similar behavior was observed for 15 wt% MWNT films, with a maximum average temperature of 114°C. Further tests will be carried out at higher voltages to assess potential for disassembly and repair when the films are used as energy directors in ultrasonic welding. For electro-mechanical tests, a maximum voltage equal to 4 was selected to eliminate potential resistance change caused by the temperature [12].

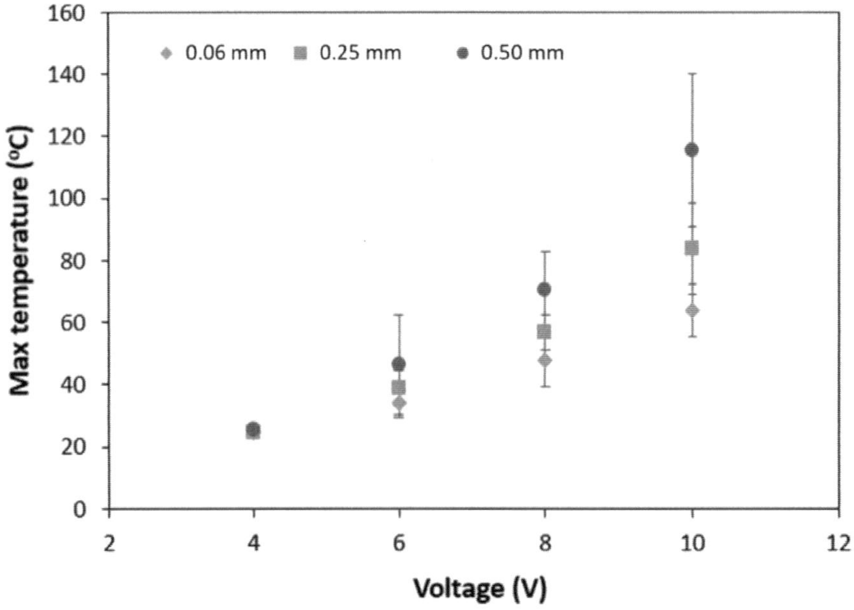

Figure 7. Average maximum temperature of 20 wt% MWNT/PP films with three thicknesses: 0.06 mm, 0.25 mm and 0.50 mm.

3.2.2 Electro-mechanical behavior with DMA

Films containing 15 wt% and 20 wt% MWNT were tested under tension in a DMA. Figure 8 shows representative results for one complete test run. As the films were strained, the resistance followed a similar cyclic behavior for both specimens. This behavior was however not noticeable below 0.1% strain. Comparing the two MWNT content values, the signal for the 20 wt% film appeared more sensitive to low strains while also maintaining a more consistent resistance reading.

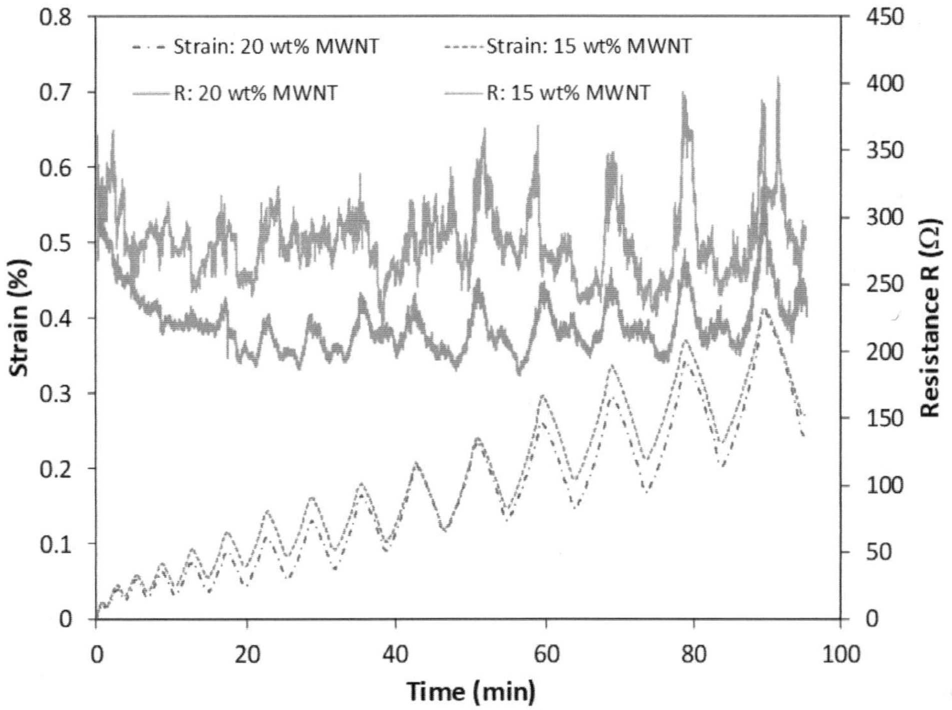

Figure 8. Relationship between strain and resistance of 15 wt% and 20 wt% MWNT/PP films.

3.3 Ultrasonically welded thermoplastic composite joints

3.3.1 Effect of multifunctional films on welding process

Typical power and displacement curves for welded samples at 60% travel with all three types of EDs (pure PP, 15 wt% and 20 wt% MWNT) are shown in Figure 9. A positive displacement value represents downward movement of the sonotrode. The shape of the displacement curves follows a similar behavior as seen in the literature, as the energy director is squeezed out at the interface [3, 15]. Table 2 compares the average values of vibration time (US-time), energy, and maximum power (out of 3000 W) for samples welded under different conditions. Generally, as the travel increased, time, energy and power increased as well. An increase in MWNT content resulted into nominal changes in the output values, suggesting that the multifunctional energy directors did not significantly affect the welding process.

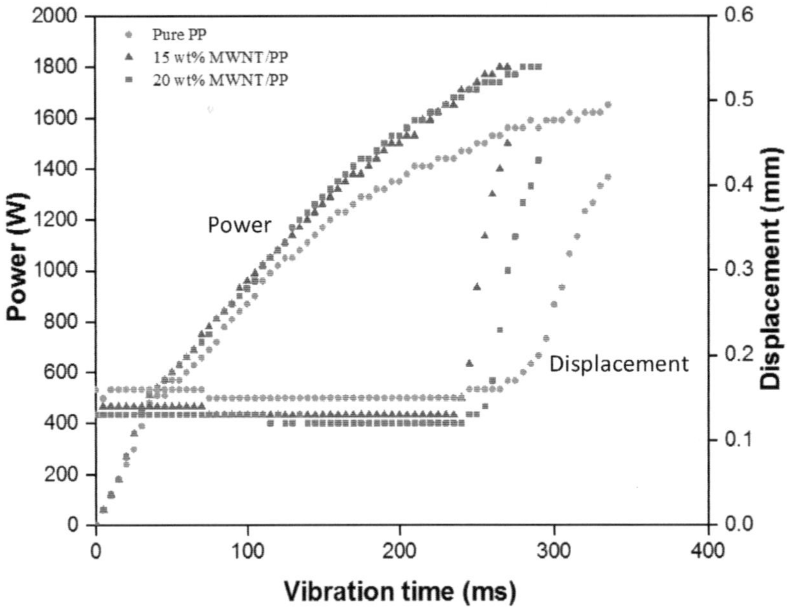

Figure 9. Comparison of power and displacement curves for GF/PP samples welded with pure PP films (circular), 15% wt MWNT/PP films (triangle), and 20 wt% MWNT/PP films (square) at 60% travel.

Table 2. Average values for vibration time (US-time), energy, and maximum power.

Welding Output / Welding Travel	US-Time (ms)	Energy (J)	Maximum Power (%)
40% (pure PP)	330 (±28)	316 (±22)	56 (±2)
60% (pure PP)	357 (±15)	366 (±11)	59 (±3)
60% (15 wt% MWNT)	311 (±11)	309 (±12)	58 (±2)
60% (20 wt% MWNT)	321 (±22)	321 (±44)	59 (±2)
80% (pure PP)	383 (±22)	406 (±38)	60 (±3)
80% (15 wt% MWNT)	384 (±10)	425 (±28)	59 (±2)
80% (20 wt% MWNT)	404 (±35)	473 (±46)	63 (±3)

3.3.2 Effect of multifunctional films on lap shear strength and failure modes

Average lap shear strength (LSS) values versus travel for the three different EDs are shown in Figure 10. A maximum LSS of 14.9 ± 1.59 MPa was obtained for GF/PP samples with pure PP films at 80% travel. Overall, pure PP EDs led to slightly higher LSS compared to multifunctional films. The lowest LSS values were obtained with 15 wt% MWNT films. Their average LSS was 9.8% lower at 60% travel and 9.4% lower at 80% travel, respectively, compared with pure PP films. A comparison of fracture surfaces between pure PP, 15 wt% MWNT, and 20 wt% MWNT films at 80% travel is shown in Figure 11. Generally, failure within the top ply of the substrates was observed for all samples, but fibers were deformed and squeezed out to the sides for both 15 wt% and 20 wt% MWNT films. This may have led to a decrease in LSS, compared to pure PP films, as observed in Figure 10. Nonetheless, multifunctional films did not significantly influence the weld quality or output parameters, with a maximum LSS decrease below 10%.

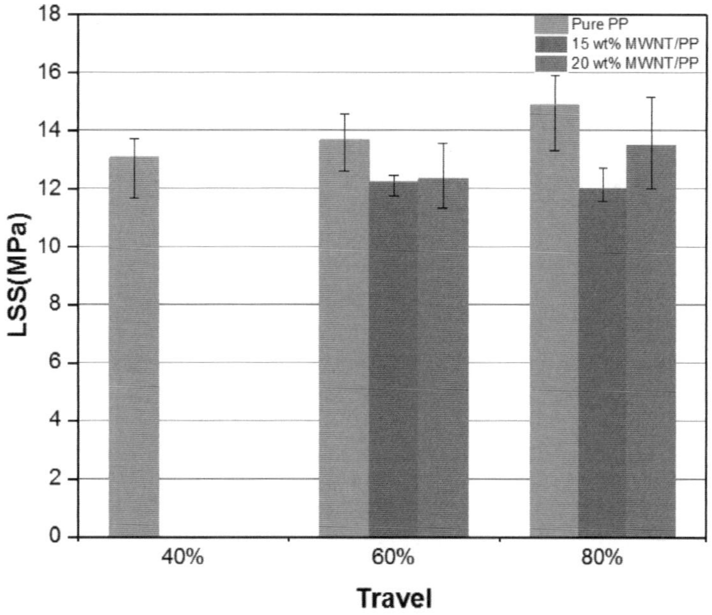

Figure 10. Comparison between the LSS of pure PP (orange), 15 wt% MWNT/PP (blue), and 20 wt% MWNT/PP (gray) films at 40%, 60%, and 80% travel.

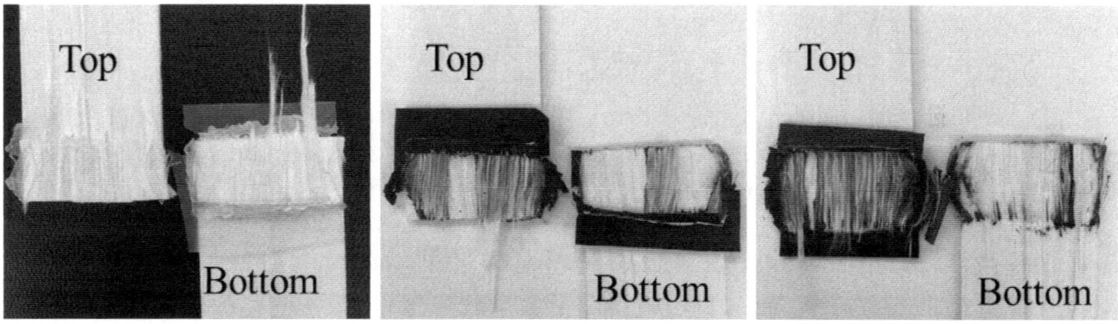

Figure 11. Representative fracture surfaces for GF/PP samples welded at 80% travel: pure PP films (left), 15 wt% MWNT films (middle), and 20 wt% MWNT films (right).

3.3.3 Potential for structural health monitoring

GF/PP samples welded according to 60% and 80% travel were tested in bending for structural health monitoring. Figure 12 shows representative curves for the measured resistance for welded joints at 60% travel, under the influence of periodic pressure, as explained in Section 2.5. It is observed, for both 15 wt% and 20 wt% MWNT films, that resistance changed with the application of external displacement, in the order of 0.2 Ω. This behavior was also observed in the literature for buckypaper-based sensors embedded into carbon fiber/epoxy rectangular samples subjected to bending [10]. Similarly to the samples tested in the DMA (Figure 8), multifunctional films with a higher MWNT content displayed more consistent readings. Overall, the films proposed in this study show promise as strain sensors for structural health monitoring of ultrasonically welded thermoplastic composite joints.

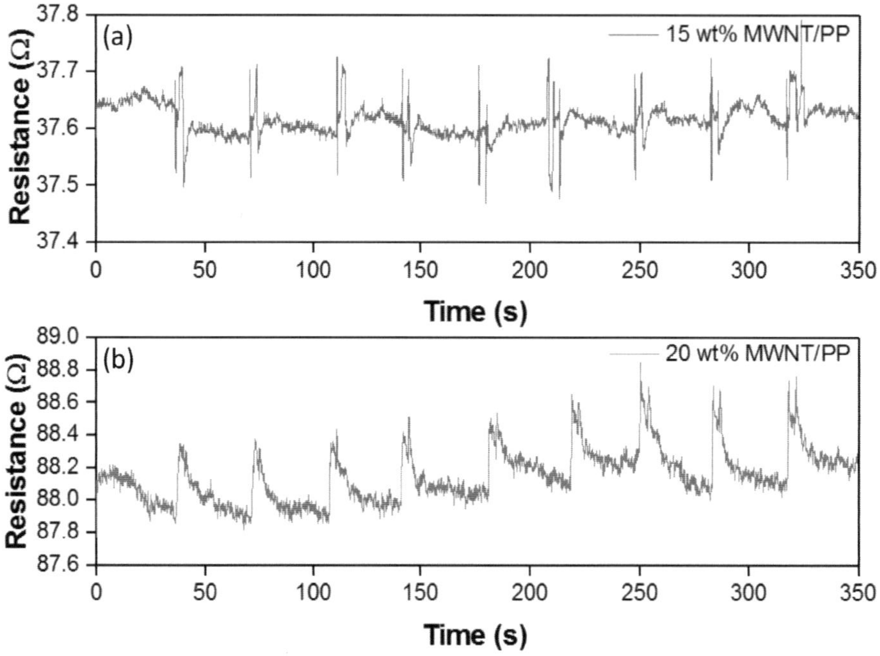

Figure 12. Representative curves for resistance versus time with periodic bending: (a) 20 wt% MWNT/PP films at 60% travel; (b) 15 wt% MWNT/PP films at 60% travel.

4. CONCLUSIONS

In this paper, the potential of various multifunctional polymer films for structural health monitoring of ultrasonically welded thermoplastic composite joints was investigated. The following observations and conclusions were noted:

- Compression molding is a simple method leading to high quality films with even thickness. Drop casting, on the other hand, produced films with uneven nanoparticles distribution, with poor adhesion to the polymer substrates, not suitable for ultrasonic welding.
- Thermo-electrical characterization of MWNT/PP films showed that a maximum temperature above 120 °C was reached with 20 wt% MWNT at 10 V. It would be possible to reach higher temperatures for disassembly and repair of welded joints at higher voltage.
- Electro-mechanical behavior of the films was assessed under cyclic tensile loading in a DMA. Films containing 20 wt% MWNT were more sensitive to low strains and displayed a more consistent resistance reading.
- Multifunctional films did not significantly influence the welding outputs (time, energy and power). Lap shear strength varied between 12.0 MPa and 14.9 MPa. It decreased by a maximum of 10% for EDs containing 15 wt% MWNT, compared to samples welded with pure PP EDs. Fracture surface analysis revealed failure in the top ply of the adherends, but deformed fibers, squeezed out to the sides, for samples welded with 15 wt% and 20 wt% MWNT EDs may explain lower LSS values.

- The potential for SHM of welded joints using MWNT films was demonstrated through cyclic bending tests. A change of resistance at the welded interface was measured according to the cyclic application of an external displacement in bending.

Future directions include several topics: i) DMA cantilever beam testing of welded interfaces with resistance monitoring, ii) better understanding of failure modes for welded samples with multifunctional energy directors, iii) evaluation of the effect of bond thickness (travel parameter) on SHM capabilities, iv) structural health monitoring of welded GF/PP samples until failure for bending, tension and torsion tests, and v) assessment of ED's sensitivity as strain sensors and comparison with CNT buckypaper-based films.

5. REFERENCES

1. Yousefpour A, Hojjati M, Immarigeon J-P. "Fusion Bonding/Welding of Thermoplastic Composites". *Journal of Thermoplastic Composite Materials.* 2004;17(4):303-41.DOI:10.1177/0892705704045187
2. Palardy G, Shi H, Levy A, Le Corre S, Fernandez Villegas I. "A study on amplitude transmission in ultrasonic welding of thermoplastic composites". *Composites Part A: Applied Science and Manufacturing*. 2018;113:339-49.DOI:10.1016/j.compositesa.2018.07.033
3. Villegas IF. "In situ monitoring of ultrasonic welding of thermoplastic composites through power and displacement data". *Journal of Thermoplastic Composite Materials*. 2015;28(1):66-85. DOI:10.1177/0892705712475015
4. Goto K, Imai K, Arai M, Ishikawa T. "Shear and tensile joint strengths of carbon fiber-reinforced thermoplastics using ultrasonic welding". *Composites Part A: Applied Science and Manufacturing*. 2019;116:126-37.DOI:10.1016/j.compositesa.2018.10.032
5. Li Y, Arinez J, Liu Z, Hwa Lee T, Fan H-T, Xiao G, et al. "Ultrasonic Welding of Carbon Fiber Reinforced Composite With Variable Blank Holding Force". *Journal of Manufacturing Science and Engineering*. 2018;140(9).DOI:10.1115/1.4040427
6. Zhao T, Palardy G, Villegas IF, Rans C, Martinez M, Benedictus R. "Mechanical behaviour of thermoplastic composites spot-welded and mechanically fastened joints: A preliminary comparison". *Composites Part B-Engineering.* 2017;112:224-34.DOI:10.1016/j.compositesb.2016.12.028
7. Zhao T, Broek C, Palardy G, Villegas IF, Benedictus R. "Towards robust sequential ultrasonic spot welding of thermoplastic composites: Welding process control strategy for consistent weld quality". *Composites Part A: Applied Science and Manufacturing*. 2018;109:355-67.DOI:10.1016/j.compositesa.2018.03.024
8. Shimada S, Tanaka H, Hasebe K, Hayashi N, Ochi Y, Matsui T, et al. "Ultrasonic welding of polymer optical fibres onto composite materials". *Electronics Letters* [Internet]. 2016; 52(17):[1472-4 pp.].DOI:10.1049/el.2016.0905
9. Ochôa P, Villegas IF, Groves RM, Benedictus R. "Diagnostic of manufacturing defects in ultrasonically welded thermoplastic composite joints using ultrasonic guided waves". *NDT & E International*. 2019;107:102126.DOI:10.1016/j.ndteint.2019.102126
10. Siddique S, Park JG, Andrei P, Liang R. "M3D aerosol jet printed buckypaper multifunctional sensors for composite structural health monitoring". *Results in Physics*. 2019;13. DOI:10.1016/j.rinp.2019.02.030

11. DeGraff J, Liang R, Le MQ, Capsal J-F, Ganet F, Cottinet P-J. "Printable low-cost and flexible carbon nanotube buckypaper motion sensors". *Materials & Design*. 2017;133:47-53.DOI:10.1016/j.matdes.2017.07.048
12. Jang S-H, Park Y-L. "Carbon nanotube-reinforced smart composites for sensing freezing temperature and deicing by self-heating". *Nanomaterials and Nanotechnology*. 2018;8.DOI:10.1177/1847980418776473
13. Costa P, Carvalho MF, Correia V, Viana JC, Lanceros-Mendez S. "Polymer Nanocomposite-Based Strain Sensors with Tailored Processability and Improved Device Integration". *ACS Applied Nano Materials*. 2018;1(6):3015-25.DOI:10.1021/acsanm.8b00647
14. Farahani RD, Janier M, Dubé M. "Conductive films of silver nanoparticles as novel susceptors for induction welding of thermoplastic composites". *Nanotechnology*. 2018;29(12):125701.DOI:10.1088/1361-6528/aaa93c
15. Palardy G, Villegas IF. "On the effect of flat energy directors thickness on heat generation during ultrasonic welding of thermoplastic composites". *Composite Interfaces*. 2016;24(2):203-14. DOI: 10.1080/09276440.2016.1199149

AUTHOR INDEX

International SAMPE Conference and Exhibition 2020

Online
1 June 2020

Volume 2 of 2

ISBN: 978-1-7138-2963-8

Printed from e-media with permission by:

Curran Associates, Inc.
57 Morehouse Lane
Red Hook, NY 12571

CURRAN ASSOCIATES INC. proceedings .com

Some format issues inherent in the e-media version may also appear in this print version.

SAMPE
21680 Gateway Center Drive, Suite 300
Diamond Bar, California, USA
91765-2454

Phone: +1-626-521-9460

www.sampe.org

Additional copies of this publication are available from:

Curran Associates, Inc.
57 Morehouse Lane
Red Hook, NY 12571 USA
Phone: 845-758-0400
Fax: 845-758-2633
Email: curran@proceedings.com
Web: www.proceedings.com

TABLE OF CONTENTS

VOLUME 1

ADDITIVE MANUFACTURING

ADVANCED DESIGN AND ANALYSIS

AEROSPACE APPLICATIONS

EMERGING MATERIALS AND PROCESSES

INSPECTION & REPAIR

VOLUME 2

MANUFACTURING TECHNOLOGY

MARKET APPLICATIONS

MULTIFUNCTIONAL MATERIALS AND STRUCTURES

WORKFORCE DEVELOPMENT

A QUANTITATIVE EVALUATION OF AFP STEERED COURSES THROUGH INSPECTION

Christopher Sacco, Roudy Wehbe, Anis Baz Radwan, Mazen Albazzan, Andrew Anderson, Ramy Harik
University of South Carolina
Columbia, South Carolina, United States

ABSTRACT

This article will discuss the use of a comprehensive methodology to inspect and track defects of steered Automated Fiber Placement (AFP) tows on a cylindrical surface. The high degree of automation in the AFP process makes the manufacturing method an excellent platform to produce variable stiffness composite structures. A key method in their production is the use of tow steering to create desired stiffness properties. However, with tow steering, there is an increased likelihood for the production of defects such as wrinkles and folds. A profilometry-based inspection method is utilized with a hand-crafted data processing technique to create accurate measures of tow displacement and tow deformation. This information is then used to create a quality metric which can be matched with processing parameters at the time of layup.

Keywords: Inspection, Automated Fiber Placement, Quality
Corresponding author: Ramy Harik (harik@cec.sc.edu)

1. INTRODUCTION

Automated Fiber Placement is a manufacturing process for the creation of large composite structures. Defect tracking in AFP production has largely been an exercise in qualitative observation. Particularly with respect to process parameter studies, the current state of the art does not incorporate any attempts at instantaneous defect measurement on a given course. To the authors' knowledge there exists no concrete quantitative measurement of defect quality derived directly from defect production. Attempts at automated inspection systems in the past, particularly those involving some machine learning approach, while reasonably effective and accurate, fail to produce the fidelity required for fine measurements. Therefore, a system capable of capturing instantaneous defect measurements for inclusion in process parameter studies is highly relevant.

In effect, the problem is three-fold. First, a data collection process capable of capturing relevant features in an AFP manufactured structure must be selected or developed. Relevant features are defined as any set of input vectors that are capable of encoding useful information about the quality of a layup, with certain sets of input vectors having a higher relevance than others[1]. Secondly, some method of processing the relevant feature set must be derived to extract defects. Lastly, it

[1] It is important to note that traditional imaging approaches have mixed success due to the low contrast nature of graphite composites. Therefore standard RGB images may not have a feature set relevant to defect identification.

must be ensured that this method for defect identification can map the input to both defect type and defect characteristics. Simply being left with a region in which a defect exists is not pertinent to this exercise. What is required is a solution with finer detail that can precisely characterize a given defect.

The following document outlines the author's approach to this problem. A comprehensive approach to gathering the quality conditions of a layup will be demonstrated with respect to defect detection on steered tows across a cylinder. Analysis and extraction of course geometry features will also be demonstrated.

1.1 Purpose

Steering of AFP tows on ruled surfaces presents tremendous opportunities for structural performance and stiffness tailoring at fractions of the weights currently achieved with standard design paradigms [1]–[3]. Unfortunately, steering in the context of AFP will trend towards inducing a number of major defects, notably wrinkles and folds [4]–[7]. While defect production is a phenomenon that is generally known in fiber steering, exacting correlation between processing parameters, steering radius, surface geometry, and defect production are not well developed.

To address these questions, a series of experiments were run at the University of South Carolina at the McNair Center for Aerospace Innovation. Steered paths were created across a 1219.2mm diameter cylinder in a design of experiment (DOE) intended to explore a complete solution space for processing parameters. Finding the effect of these design and manufacturing features on part quality requires that defects are tracked precisely and then encoded into useable knowledge to make statements on the layup quality and parameter relationships to layup quality. It is the author's intention to show that the inspection and data processing solution contained in this document are suitable to answer this question.

1.2 Literature Review

AFP inspection is an open research topic that has been explored extensively in several previous works by the authors [8]–[10]. Traditional manual inspection used currently in industry is a considerable drain on time, with some estimates of inspection accounting for more than 30% of machine operation [11]–[13]. Manual inspection is also principally a qualitative exercise, with somewhat arbitrary judgements about layup quality being rendered without precise measurement. It is in many ways a "gut" determination if a given part is acceptable.

Novel hardware approaches to the imaging or data collection of a composite part has been the subject of significant research. Thermography has become a leading technique for the inspection of composite parts. Thermographic inspection tracks the flow of heat through a part after thermal excitation. When defects are present, apparent material properties will differ from the reference background, thus affecting heat transmission through the part at the affected area [14]–[17]. Thermography has also been directly explored in AFP inspection, yielding some novel solutions to online AFP inspection [18], [19].

Once data has been collected, it must have relevant features extracted and then processed to determine damaged areas automatically. This is quickly becoming the realm of machine learning (ML); and with this development semantic meaning can be assigned to given feature sets in data.

This is apparent in the world of computer vision. Remarkable advances have been made following the introduction of the deep convolutional neural network in the AlexNet algorithm [20]. ML and particularly deep neural networks appear to be the future stalwarts of data analysis and processing. It should be noted that the strength of deep networks lies in the many layers of processing units, allowing for features to be automatically extracted and transformed from a raw input vector.

1.3 Proposed Solution

The creation of our system relies on a bedrock of several previous explorations into the area. Previously, an ML-based inspection method was developed for the automated identification of defects on an AFP manufactured structure [9]. The principle of precise defect characterization was first forwarded in this paper and resulted in an algorithm capable of determining both the class and the exact shape of a defect. Later additions to the system included a comprehensive user interface for creating user-defined defects and correcting defects misplaced by the automated system [8]. For data collection, the Ingersoll Machine Tools (IMT) Advanced Composite Structures Inspection System (ACSIS) [Figure 1] was utilized to scan a part and capture precise data about the height profile of the part. ACSIS data collection is a profilometry system utilizing 4 Keyence LJ-7080 blue-light profilometers.

Figure 1: ACSIS AFP Inspection System

Profilometry has the capability to rapidly produce height profiles of a part, which ACSIS then maps to pixel values to create a grey scale image to be processed. Coupled with the nature of the defects relevant to the steering study, this offers a unique opportunity for analysis. ML is becoming the preferred tool on the principle that features can automatically be pulled from raw data. However, if features are readily available, then it may often be more convenient to use a hand-crafted approach to data processing. In the case of profilometry scans, height profile is a potential mechanism to highlight defects.

Coupling this with the UI developed for previous inspection systems, the mechanism for potential error in a simple hand crafted solution become apparent. Simple marking of the course boundary allows for recreation of the course geometry, and from it a centerline description of the course can be determined.

2. CONCEPTUAL OUTLINE

The analysis algorithm is subdivided into several parts: (1) the extraction of defects from the height profile, (2) determining the centerline of each course, and (3) the iteration of a measurement line over the length of the course to determine the percentage of course width at a given instance is occupied by defects. Figure 2 shows the dataflow to from inspection to quality measurement.

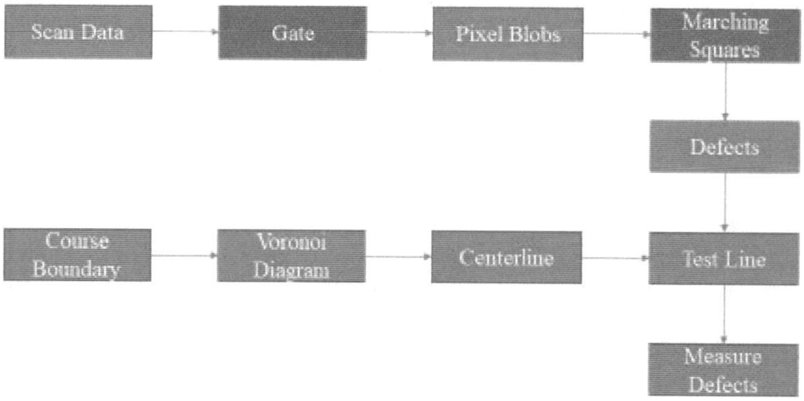

Figure 2: Quality Measurement Algorithm

2.1 Height to Defect

As mentioned, the ready abundance of height data draws us towards a relatively simple method for defect identification. By applying simple gates to the height profiles and shading in pixels, out of plane deformations and tow displacement can be determined. The lower gate represents the tow displacement and the upper gate the out-of-plane tow deformation. The resulting pixel blobs can be extracted using the marching squares algorithm [21], resulting in a bounding polygon around each collection of pixels. From this, small polygons generated from artifacts in the image were filtered out, resulting in a precise bounding polygon being placed around defects of interest. In the case that a significant deformation may cast a shadow that could trigger a false positive for a displacement. This is easily accounted for with a quick adjustment through the UI. The resulting defects map for the courses steered using different process parameters is shown in Figure 3 where tow displacements are represented in green and out-of-plane deformations in orange.

2.2 Reconstruction of Course Geometry

Before quality measures can be gathered, the defect data must first be placed in context to the course that the defects are on. In order to better represent the course geometry, a centerline is determined such that a general "direction" to the course can be used for further evaluation. First the Voronoi Diagram [22] is generated with the course boundaries as seed points [Figure 4]. From this, a set of Voronoi edges is constructed. The set is then filtered such that only edges within the course boundary remain. What is left is a series of unsorted edges lying within the course. Figure 5 gives an example of these unsorted edges. Note that while a general centerline description can easily be seen, there are several edge artifacts that require more processing to remove.

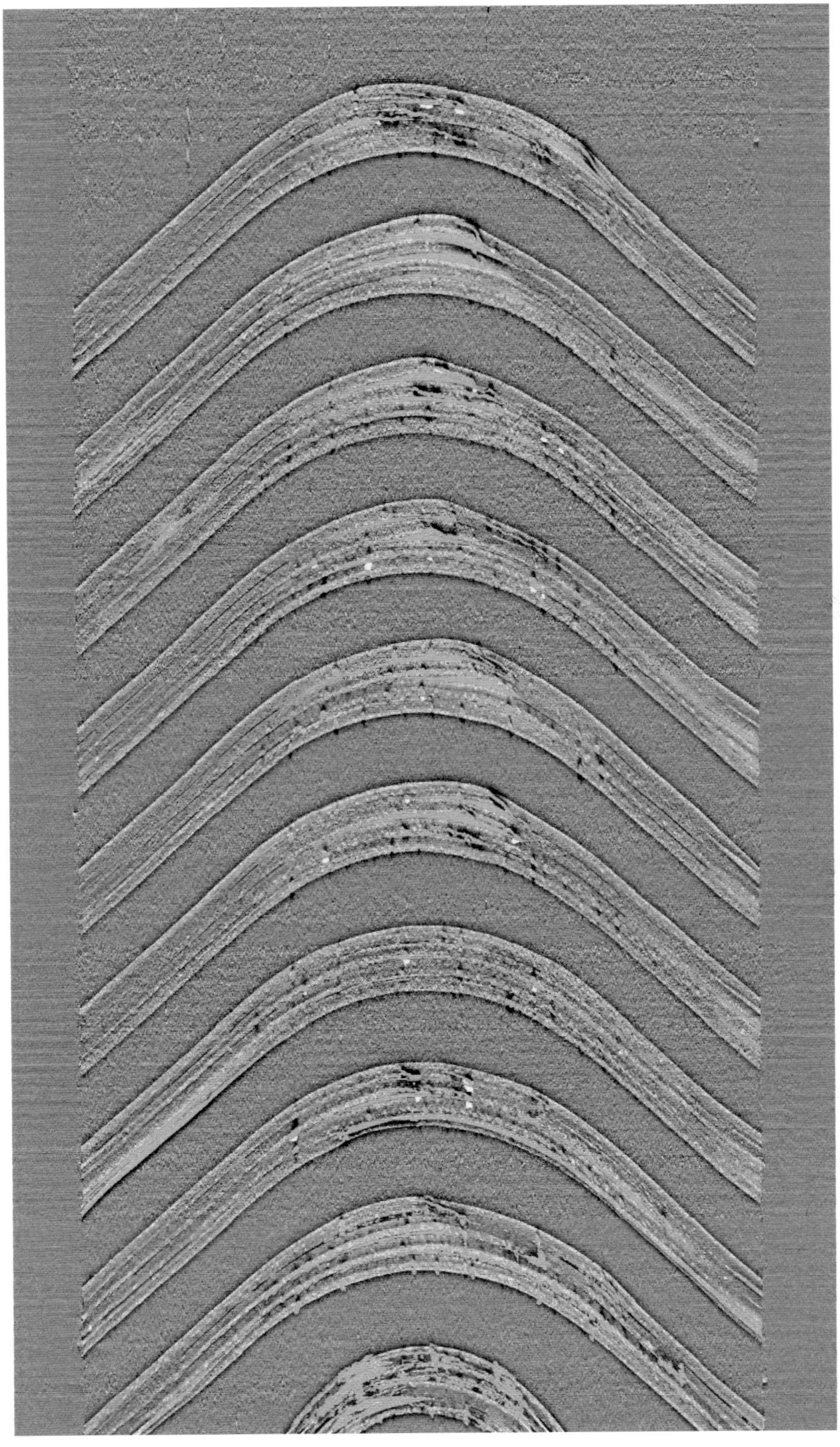

Figure 3: Defects Identified on Steered Courses

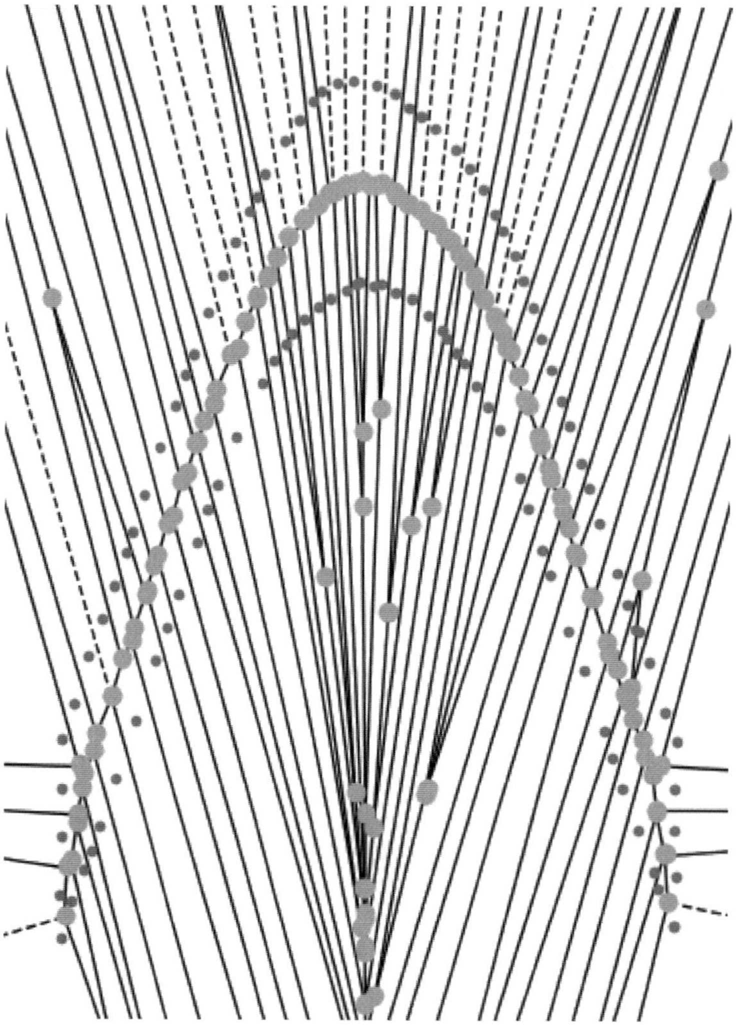

Figure 4: Voronoi Diagram Generated from Course Boundary

These edge artifacts can be removed through a number of potential methods. The authors can speculate about a number of transformational or clustering approaches using the vertices of the edges could result in good results. However, if we create a graph reflecting the connectivity of each of the edges, then a number of fast graph theoretic approaches can be deployed.

Principally, if the assumption is made that these edge artifacts do not bridge across the interior of our pseudo-centerline, then by extension the centerline approximation is the shortest path from the leftmost to the rightmost edge vertex. This is easily and quickly accomplished using Dijkstra's Algorithm [23] and assuming that all of the edges in the graph have equal weight. Therefore, so long as the number of edges is low enough such that the process of mapping back and forth between graph space and coordinate space is computationally reasonable, then the general centerline [Figure 5b] can be extracted.

If one wishes to improve this description of the centerline, then a simple smoothing procedure by approximating the extracted centerline by a piecewise linear function [Figure 6a] achieves a nice result and is an important step for preprocessing when iterating a line normal to the center for final defect measurement.

Figure 5: a. Set of Voronoi Edges in Course Boundary. b. Extracted Centerline

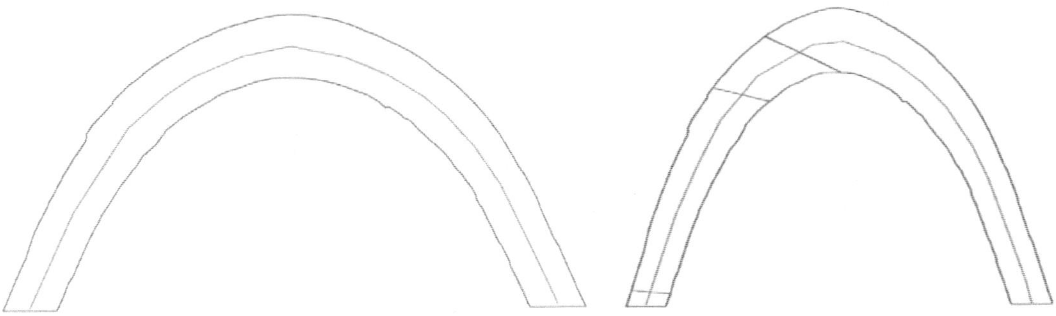

Figure 6: a. Piecewise Linear Function Fit to Centerline. b. Intersection Lines Normal to Course Center

2.3 Measuring Layup Quality

Thus far, the discussion of the analysis present in this document has centered on the identification of defects and the creation of corresponding geometry markers that enable locating the defects in context of the course. To link these two portions of our analysis together for the creation of a quality metric, a definition of the instantaneous effect of defects on the course must be created. To do this, the centerline extracted in the previous sections is used as a reference through which a line normal to the course can be created [Figure 6b]. This normal line is then iterated over the length of the course, where the length of that line intersecting with defects is determined. The length of the intersection is normalized against the width of the course at that section, resulting in the percentage of the course at a point that contains defects. Building this description over the course, and entire profile of manufacturing defect production across the entire course can be constructed. For each of the classes, a profile can respectively be built. In the case of the steered tow defects in this experiment, out of plane deformation, tow displacement, and total defects can be measured. However, this system is applicable to any set of defects that one might wish to identify.

3. RESULTS

In the case of a course generated in the steering experiments, a defect profile can be produced according to the methods outlined in the previous section. This defect profile [Figure 9] shows defect production as a function of the length of the course. One can note that as expected, the number of defects increases as the course moved from the linear regime to the steered section.

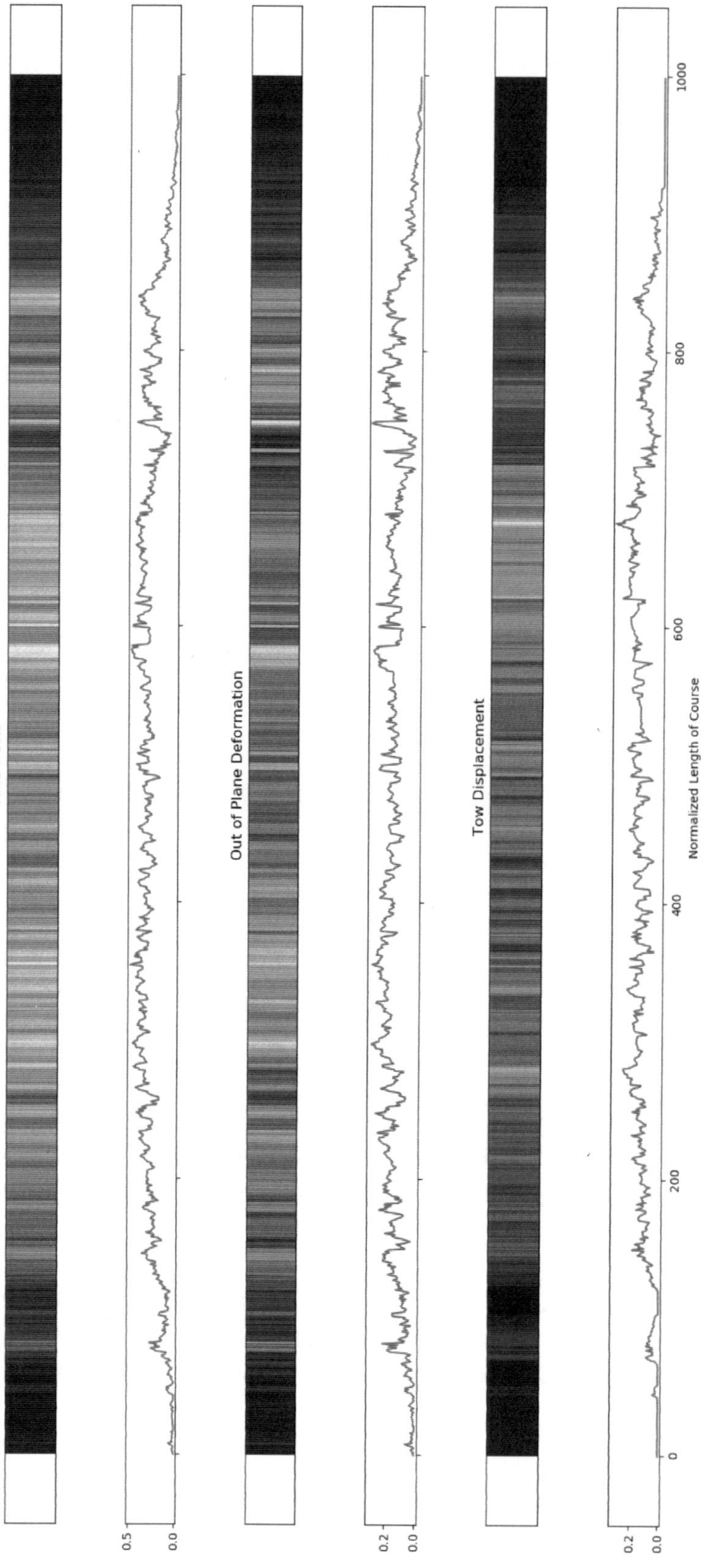

Figure 9: Defect Production Across a Steered Course

4. CONCLUSION & FUTURE WORKS

The method for quality assessment outlined in this document represents a novel method for understanding the production of defects instantaneously across a course under steering. Profilometry scans provided a manner by which defects could be identified such that the precise boundary of the defects could be characterized through the gating of height profiles for the identification of tow displacement and out-of-plane deformation. A process based on principles of computational geometry was developed to extract the centerline of a course from inspection data using the Voronoi Diagram and graph theoretic relationships between Voronoi elements. Using this information a measurement of the amount of a course width that was occupied by defects could be made. An instantaneous measurement of defect production was then created.

The quality measurements of the steered courses was done while keeping detailed process parameter information. Processing parameters for AFP manufacturing such as heating temperature, compaction pressure, and feed rate are poorly understood in the context of layup quality. It is the author's intent to use this new quality assessment tool to correlate process parameters along with design variables such as steering radius; therefore enabling a qualitative statement about the rank, optimality, and effect of processing parameters on layup quality.

5. REFERENCES

[1] T. R. Brooks, J. R. R. A. Martins, and G. J. Kennedy, "High-fidelity aerostructural optimization of tow-steered composite wings," *J. Fluids Struct.*, vol. 88, pp. 122–147, 2019. 10.1016/j.jfluidstructs.2019.04.005.

[2] M. Albazzan *et al.*, "Design of Variable Stiffness Cylinder with Holes Under Bending for Maximum Buckling Load Using Lamination Parameters," *SAMPE 2019 Conf. Exhib. Charlotte, North Carolina, US, 20 – 23 May 2019*, pp. 1–18, 2019. 10.33599/nasampe/s.19.1378.

[3] A. Khani, M. M. Abdalla, Z. Gürdal, J. Sinke, A. Buitenhuis, and M. J. L. Van Tooren, "Design, manufacturing and testing of a fibre steered panel with a large cut-out," *Compos. Struct.*, vol. 180, pp. 821–830, 2017. 10.1016/j.compstruct.2017.07.086.

[4] R. Wehbe, R. Harik, and Z. Gürdal, "In-Plane Tow Deformations Due to Steering in Automated Fiber Placement," in *AIAA Scitech 2019 Forum*, 2019, pp. 1–13. 10.2514/6.2019-1271.

[5] R. Wehbe, B. Tatting, R. Harik, Z. Gurdal, A. Halbritter, and A. Wanthal, "Tow-Path Based Modeling of Wrinkling During the Automated Fiber Placement Process," *Compos. Adv. Mater. Expo 2017 (CAMX 2017)*, no. September 11-14, Florida, United States, 2017.

[6] R. Wehbe, B. Tatting, Z. Gürdal, and R. Harik, "Fiber Tow Deformations During Layup of Steered Paths Using Automated Fiber Placement Process," *SAMPE 2019 Conf. Exhib. Charlotte, North Carolina, US, 20 – 23 May 2019*, 2019. 10.33599/nasampe/s.19.1591.

[7] S. Rajan *et al.*, "Experimental investigation of prepreg slit tape wrinkling during automated fiber placement process using StereoDIC," *Compos. Part B Eng.*, vol. 160, no. July 2018, pp. 546–557, 2019. 10.1016/j.compositesb.2018.12.017.

[8] C. Sacco, A. B. Radwan, T. Beatty, and R. Harik, "Machine Learning Based AFP Inspection : A Tool for Characterization and Integration," *SAMPE Conf. Proc.*, 2019. 10.33599/nasampe/s.19.1594.

[9] C. Sacco, A. B. Radwan, R. Harik, and M. Van Tooren, "Automated fiber placement defects: Automated inspection and characterization," in *International SAMPE Technical Conference*, 2018, vol. 2018–May.

[10] C. Sacco, "Machine Learning Methods for Rapid Inspection of Automated Fiber Placement Manufactured Composite Structures," (Master's thesis). University of South Carolina, 2019.

[11] J. Cemenska, T. Rudberg, and M. Henscheid, "Automated In-Process Inspection System for AFP Machines," *SAE Int. J. Aerosp.*, 2015. 10.4271/2015-01-2608.

[12] T. Rudberg, J. Nielson, M. Henscheid, and J. Cemenska, "Improving AFP Cell Performance," *SAE Int. J. Aerosp.*, 2014. 10.4271/2014-01-2272.

[13] D. Maass, "Progress in automated ply inspection of AFP layups," *Reinf. Plast.*, vol. 59, no. 5, pp. 242–245, 2015. 10.1016/j.repl.2015.05.002.

[14] J. Y. Wu, S. Sfarra, and Y. Yao, "Sparse Principal Component Thermography for Subsurface Defect Detection in Composite Products," *IEEE Trans. Ind. Informatics*, vol. 51, no. 24, pp. 855–860, 2018. 10.1109/TII.2018.2817520.

[15] A. P. Chrysafi, N. Athanasopoulos, and N. J. Siakavellas, "Damage detection on composite materials with active thermography and digital image processing," *Int. J. Therm. Sci.*, vol. 116, pp. 242–253, 2017. 10.1016/j.ijthermalsci.2017.02.017.

[16] X. Meng, Y. Wang, J. Liu, and W. He, "Nondestructive inspection of curved clad composites with subsurface defects by combination active thermography and three-dimensional (3D) structural optical imaging," *Infrared Phys. Technol.*, vol. 97, no. February, pp. 424–431, 2019. 10.1016/j.infrared.2019.01.026.

[17] V. Kalyanavalli, T. K. A. Ramadhas, and D. Sastikumar, "Long pulse thermography investigations of basalt fiber reinforced composite," *NDT E Int.*, vol. 100, no. December 2017, pp. 84–91, 2018. 10.1016/j.ndteint.2018.08.007.

[18] E. D. Gregory and P. D. Juarez, "In-situ thermography of automated fiber placement parts," in *AIP Conference Proceedings*, 2018. 10.1063/1.5031551.

[19] J. Brüning, B. Denkena, M. A. Dittrich, and T. Hocke, "Machine Learning Approach for Optimization of Automated Fiber Placement Processes," *Procedia CIRP*, vol. 66, pp. 74–78, 2017. 10.1016/j.procir.2017.03.295.

[20] A. Krizhevsky, I. Sutskever, and G. E. Hinton, "ImageNet Classification with Deep Convolutional Neural Networks," *Adv. Neural Inf. Process. Syst.*, pp. 1–9, 2012. 10.1016/j.protcy.2014.09.007.

[21] C. Maple, "Geometric design and space planning using the marching squares and marching cube algorithms," in *Proceedings - 2003 International Conference on Geometric Modeling and Graphics, GMAG 2003*, 2003. 10.1109/GMAG.2003.1219671.

[22] M. Erwig, "The graph Voronoi diagram with applications," *Networks*, 2000. 10.1002/1097-0037(200010)36:3<156::AID-NET2>3.0.CO;2-L.

[23] K. Rohila, P. Gouthami, and P. M, "Dijkstra ' s Shortest Path Algorithm," *Int. J. Innov. Res. Comput. Commun. Eng.*, 2014.

NDE INSPECTION OF AFP MANUFACTURED CYLINDERS USING AN INTELLIGENT SEGMENTATION ALGORITHM

Christopher Sacco, Anis Baz Radwan, Andrew Anderson, Ramy Harik
University of South Carolina
Columbia, South Carolina

ABSTRACT

This article will discuss the approach and results for the identification of defects on an Automated Fiber Placement (AFP) manufactured cylinder. While the increase in productivity in AFP manufactured structures has allowed for the large-scale production of composite parts, imprecision in process can lead to the production of defects. A comprehensive platform for the inspection of AFP manufactured cylinders for utilizing a profilometry-based data collection approach was used. This data is then processed by a novel machine learning method. The machine learning method is based on the creation of fully convolutional neural networks and is used to fully characterize defects developed on cylindrical parts. Defect information was used to inform the repair of the cylinders and capture defects on cylinders that were intended to have hand placed defects. Cylinders were then used for validation of structural analysis tools.

Keywords: Inspection, AFP, Machine Learning
Corresponding author: Ramy Harik

1. INTRODUCTION

1.1 Purpose

The gaining popularity of AFP and its many emerging subsets has led to the need to generate expert knowledge in the field. A particular concern for those attempting to create composite structures using the AFP process is the potential for the inclusion of manufacturing induced defects in a part. The time considerations produced by manual inspection are notorious [1]. Cemenska et al. [2] suggests that as much as 20% of part manufacturing time is occupied in inspection of parts for defects. A study by Electroimpact has demonstrated that 32% of cycle time during manufacturing is from inspection and rework [3].

In service of improving this process, the authors have developed an automated inspection system to be used to aide AFP operators in the process of identifying defects for rework. This system has been outlined in detail in three previous works and demonstrated on a number of small-scale structures [4]–[6]. The inspection process developed differs from other approaches in a distinct way: the defect data produced includes precise boundary data necessary for a comprehensive evaluation for the significance of the defects. In this document, we will demonstrate the use of our system on the manufacturing of several large cylinders and demonstrate the capability for our system to inform the repair of a large structure using the precise data obtained through the inspection process. Through this, the overall repair time was improved, and targeted repair strategies could be implemented.

1.2 Cylinder Design and Testing

The development of an analysis tool at the McNair Center for the rapid prediction of the structural performance impact of manufacturing defect necessitated the development of test structures to test the model. Cylinders were manufactured on-site at the McNair Center utilizing the Ingersoll Machine Tools Lynx AFP Machine in conjunction with our inspection system. In service to the task, two cylinder dimensions, a 1219.2mm diameter cylinder and a 609.6mm diameter cylinder, were produced in both a defective and pristine configuration. For the defective configuration, defects were hand-placed such that a significant degradation in structural performance could be produced. In both the areas of the defective cylinders that were not intended to be affected by the hand-placed defects and the entirety of the pristine cylinders required large scale inspection and repair for the development of an appropriate test article. Figure 1 shows how the repair process of the cylinders is influenced by the inspection of each ply.

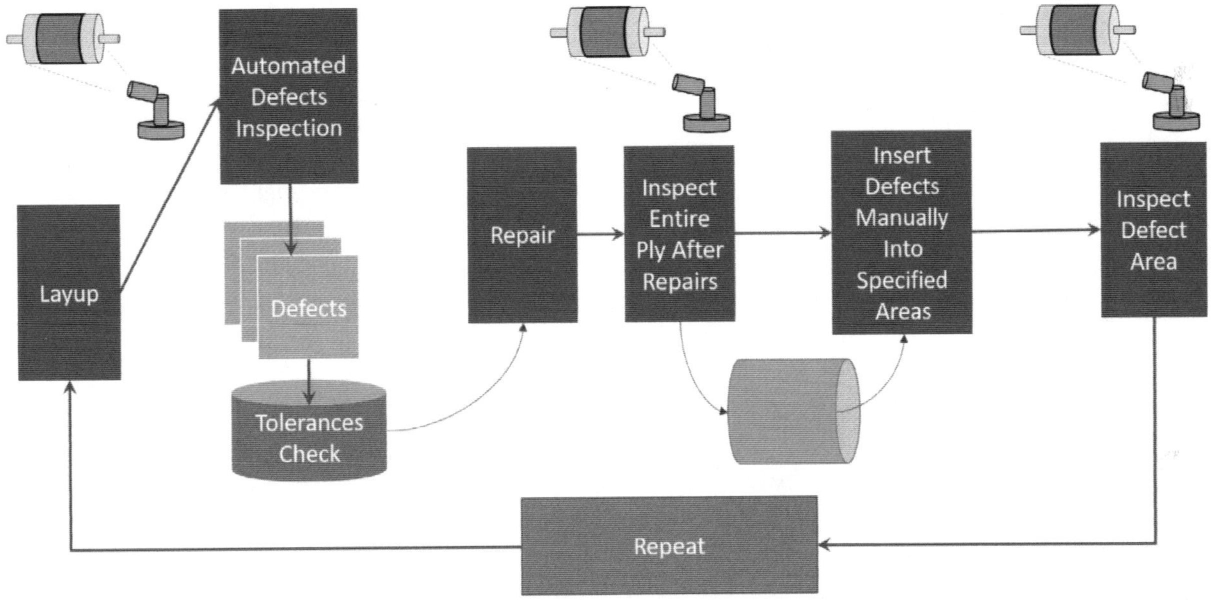

Figure 1: Workflow of Inspection in Manufacturing Process

2. LITERATURE REVIEW

In service of facilitating a discussion on automated inspection of the AFP process, it is useful to discuss composite inspection techniques both during and post-manufacturing. In addition, an introduction to a few key works in the machine learning and computer vision space will also be introduced. In general, the process for automated inspection of composite structures is clustered into two principle domains of research: (1) the physical data collection hardware and (2) the analysis algorithms for processing said data. When combining these two systems together, an automated inspection system for composite parts can be fully realized. To date, there have been a number of key approaches across each domain, with an assorted selection of systems attempting to make the leap towards full automation.

The low visual contrast nature of composite materials implies that traditional imaging techniques are, under most circumstances, ineffective. Therefore, novel alternatives to traditional optical imaging have emerged for the collection of data both across the surface and through a composite part. Thermography uses a source of thermal excitation to monitor the propagation of heat through a part [7]–[11]. Traditionally, halogen bulbs have been used to accomplish the excitation process, though LED sources have shown some positive benefits in the application [12]. By noting the reference propagation of heat through a structure, deviations from this reference can imply a change in material properties or structure. These deviations can denote common composite laminate defects such as voids or delaminations. In the context of AFP, thermography has been used as an in-situ method for the identification of common manufacturing errors, including gaps, overlaps, and twists [13].

Ultrasonic techniques are popular for post-manufacturing inspection, especially with respect to the identification of major sub-surface defects. Ultrasonic inspection uses concentrated sound waves propagated through a material to create an internal mapping of the object. Similar to thermography, where sound propagates differently against a reference, changes in structure are indicated [14]. Common variants of ultrasonic imaging include through transmission and pulsed-echo techniques [15].

While the acquisition of inspection data is a key parameter, the unsorted feature lists produced only push the manual sorting of data through a proxy rather than with human eyes. This improves some of the issues with visual inspection of low contrast composites. However, to move towards an automated system, most or all of this feature extraction and defect identification must happen automatically. One potential solution for this process is the increasingly prevalent use of machine learning (ML) in the analysis of data. A subset of ML, denoted as supervised learning, uses hand-labeled training data to create models that can infer the relationships between data without explicitly modeling the underlying principles connecting said data. This makes ML a strong tool for problems that lack clear definitions for analysis.

ML has become an incredibly useful tool for applications ranging from object detection [16] to natural language processing [17]. The introduction of AlexNet [18] and the massive improvements in image classification accuracy ushered in with it mean that industry ready computer vision tools can be generated readily. Thus, in the creation of our inspection system, ML became a natural tool for the identification of AFP defects.

3. SOLUTION

3.1 Data Acquisition

The hardware for data acquisition was provided by Ingersoll Machine Tools (IMT) in the form of the Advance Composite Structures Inspection System (ACSIS). ACSIS is an AFP inspection system that utilizes profilometry, or rapid 3-dimension height profiling [19], to scan an AFP manufactured part ply-by-ply. The profilometers are actuated on a Kuka KR120 robotic arm that can reach across the entire width of the AFP tooling [Figure 1].

Figure 2: IMT ACSIS Inspection System

The height profiles generated by the profilometers are gated, and then scaled such that each point on the height profile can be embedded into a greyscale image, with the pixel intensity encoding the height information [Figure 2]. The result is a rapidly profiled and generated image that is capable of conveying high resolution surface features along the surface of a ply.

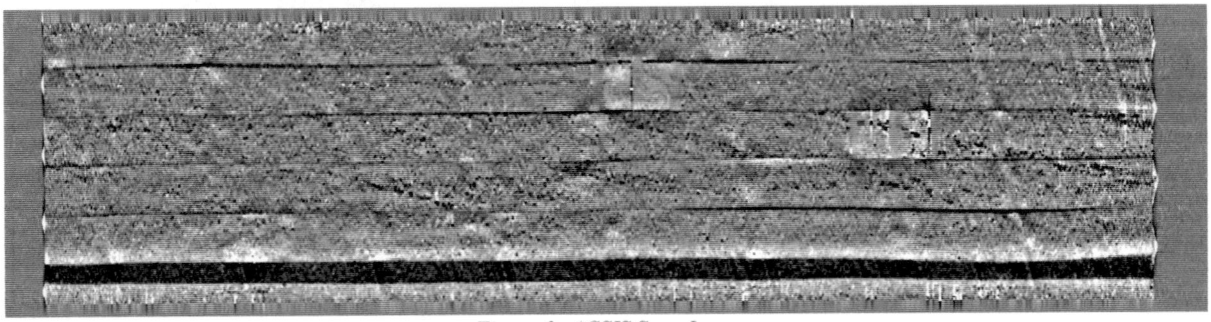

Figure 3: ACSIS Scan Image

3.2 Image Analysis

The essential nature of the analysis algorithms presented in the inspection system have been covered extensively in previous works. For the convenience of the reader, a detailed discussion of the machine learning engine and the user interface developed for operator integration will be left to further reading of the authors previous publications. Rather, a brief discussion of key concepts will be presented with an overview of how the inspection system outlined integrates with the larger AFP manufacturing environment.

3.2.1 Fully Convolutional Neural Network

One key element in the development of an AFP inspection system is the fidelity of defect details provided. While the sliding window approach utilized in a number of previous attempts at machine learning-based inspection can provide high classification accuracy and general location information, more precise information is elusive. For integration into finite element solvers or other third-party applications, precise boundary and location information may be necessary.

To accomplish this, the concept of semantic image segmentation was used as an ML task to derive a blob of pixels locating a defect rather than a patch of image. A Fully Convolutional Network (FCN) has been demonstrated to be an ideal tool for this application [20]. FCNs are entirely consisting of convolution layers that down-sample and then up-sample input vectors to have output vector size match input vector size. This creates a network capable of one-shot pixel-by-pixel classification [Figure 3]. The result embeds semantic meaning into entire pixel regions of an image and generate output with precise boundaries for each object class in an image.

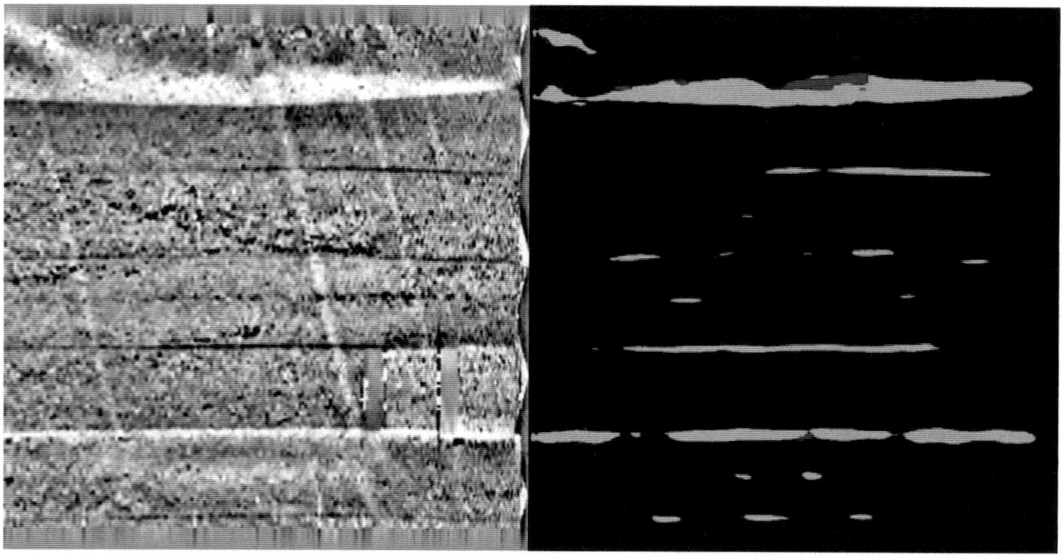

Figure 4: Pixel Output of Network

3.2.2 Operator User Interface

The increased detail afforded by the FCN structure of our ML algorithm also requires an equally elegant strategy for the display of this data to an operator. The large amount of data provided by the inspection system implies that observing defects from raw output is not feasible. Therefore, a user interface was created to improve the interaction between the operator and the system.

Defect boundaries were translated into bounding polygons through the Marching Squares Algorithm [21]. The location of the vertices of the bounding polygons were recorded and then displayed over the original scan image [Figure 4]. The operator has the capability to delete, change the class of, or define their own defects. When the operator is satisfied with the inspection data, they can automatically embed the data in a JavaScript Object Notation (JSON) file and export it to a remote server to be processed by third-party applications.

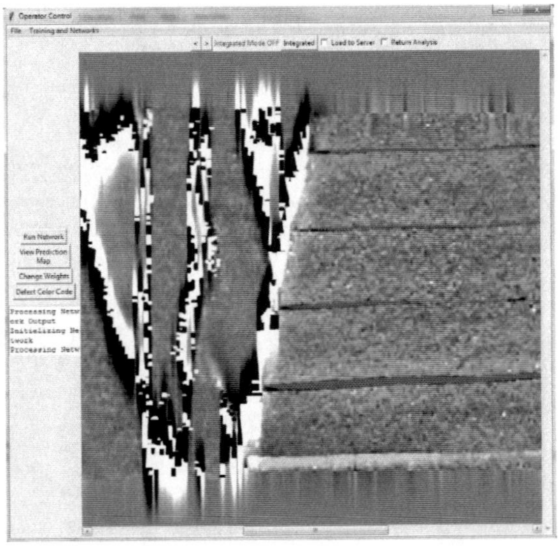

Figure 5: UI Display of Gaps and Overlaps

4. RESULTS

4.1 Inspection Informed Rework

For each ply produced on a given cylinder, inspection was performed, and defects were identified. Defect information was used to identify locations requiring rework, thus creating a system by which inspection directly influenced the rework process. Cylinder repair and rework is demonstrated in

Figure 6: Rework on Cylinder from Inspection Data

4.2 Inspection Generated Data

A selection of scans from the cylinders is displayed in

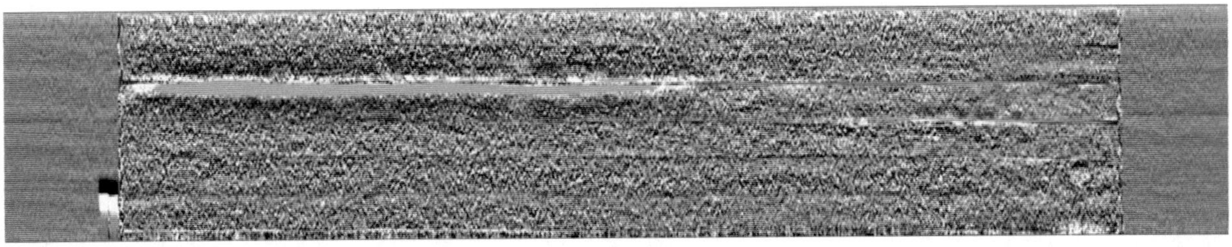

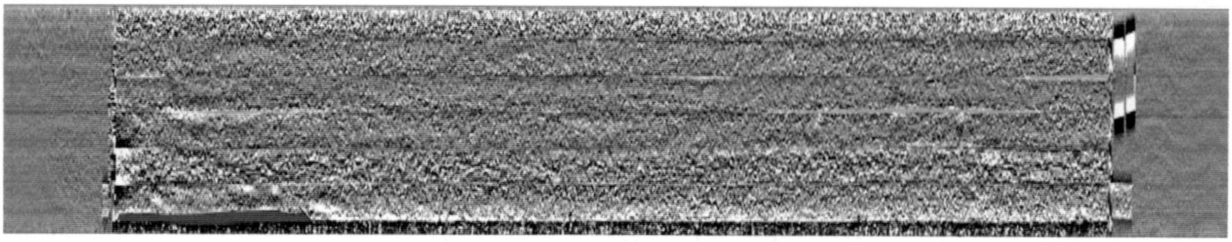

Figure 7: Defect Identified on Cylinder

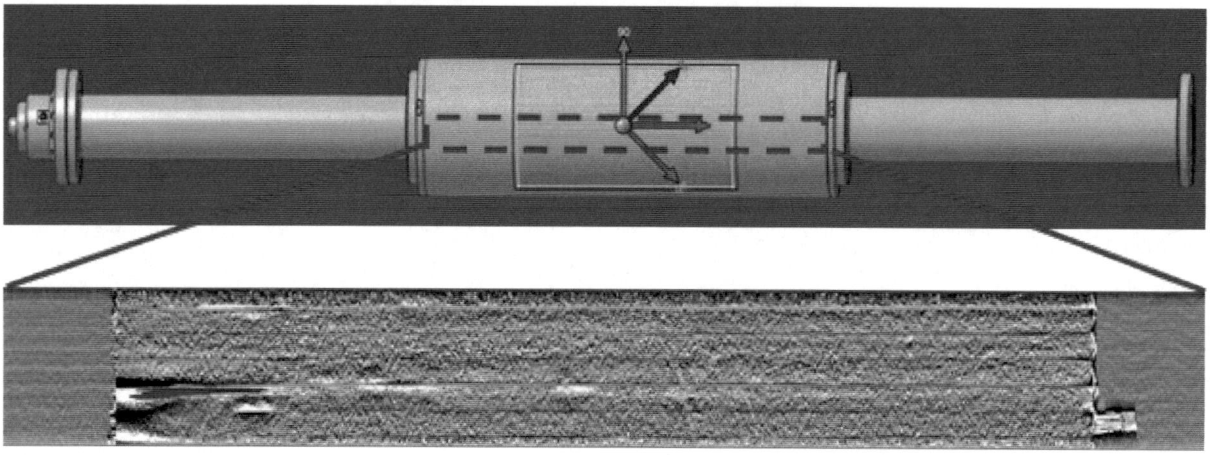

Figure 8: Defects Located on Cylinder

Custom code was written to translate the pixel positions into coordinates from the output of the CAM tool used to program the robot. With paths determined on the cylinders, it is relatively simple to determine spatial coordinates for each defect. From this point, defect data was transferred onto the remote server for use with third-party applications.

5. CONCLUSIONS

In this document a comprehensive inspection system developed in previous works was outlined and demonstrated to inform the repair of AFP manufactured cylinders. The data was logged and exported to a remote server for use with FE tools for the determination of the significance of each defect. In addition, cylinders were processed and tested at NIAR to validate the rapid analysis FE tools.

5.1 Future Work

In exercising the inspection system, it has become clear that a significant advantage in rework and repair could be derived by eliminating the need to interface directly with the inspection computer. A great deal of time is used in the process of translating what is seen on the screen of the inspection software UI to the physical part, even with coordinate positions to rely upon. Thus, it might behoove future work to eliminate this time expensive step.

One potential solution involves the use of an augmented reality system to overlay the defects on the physical part, making location and identification significantly faster. This interface could theoretically be accomplished using data already presented in the inspection solutions coupled with off-the-shelf hardware and software. The authors have taken several steps in the process of accomplishing this next phase.

Further, the ready identification of defects on a layup has developed to a point that the construction of quality metrics is becoming possible. Soon the days of qualitative analysis of layups in a research setting will likely be gone, and with a new paradigm of quantifiable defect data reliable metrics of layup quality will need to be developed. Integration with AFP systems will require it, as the increased knowledge associated with a quality metric will precede the identification of processing windows or machine parameters that increase part quality.

5.2 Acknowledgements

The material is based upon work supported by NASA under Award Nos.515 NNL09AA00A and 80LARC17C0004.

6. REFERENCES

[1] D. Maass, "Progress in automated ply inspection of AFP layups," *Reinf. Plast.*, vol. 59, no. 5, pp. 242–245, 2015. 10.1016/j.repl.2015.05.002.

[2] J. Cemenska, T. Rudberg, and M. Henscheid, "Automated In-Process Inspection System for AFP Machines," *SAE Int. J. Aerosp.*, 2015.

[3] T. Rudberg, J. Nielson, M. Henscheid, and J. Cemenska, "Improving AFP Cell Performance," *SAE Int. J. Aerosp.*, 2014. 10.4271/2014-01-2272.

[4] C. Sacco, A. B. Radwan, T. Beatty, and R. Harik, "Machine Learning Based AFP Inspection : A Tool for Characterization and Integration," *SAMPE Conf. Proc.*, 2019.

[5] C. Sacco, A. B. Radwan, R. Harik, and M. Van Tooren, "Automated fiber placement defects: Automated inspection and characterization," in *International SAMPE Technical Conference*, 2018, vol. 2018–May.

[6] C. Sacco, "Machine Learning Methods for Rapid Inspection of Automated Fiber Placement Manufactured Composite Structures," University of South Carolina, 2019.

[7] J. Y. Wu, S. Sfarra, and Y. Yao, "Sparse Principal Component Thermography for Subsurface Defect Detection in Composite Products," *IEEE Trans. Ind. Informatics*, vol. 51, no. 24, pp. 855–860, 2018.

[8] X. Meng, Y. Wang, J. Liu, and W. He, "Nondestructive inspection of curved clad composites with subsurface defects by combination active thermography and three-dimensional (3D) structural optical imaging," *Infrared Phys. Technol.*, vol. 97, no. February, pp. 424–431, 2019.

[9] A. P. Chrysafi, N. Athanasopoulos, and N. J. Siakavellas, "Damage detection on composite materials with active thermography and digital image processing," *Int. J. Therm. Sci.*, vol. 116, pp. 242–253, 2017.

[10] V. Kalyanavalli, T. K. A. Ramadhas, and D. Sastikumar, "Long pulse thermography investigations of basalt fiber reinforced composite," *NDT E Int.*, vol. 100, no. December 2017, pp. 84–91, 2018.

[11] M. Jolly *et al.*, "Review of Non-destructive Testing (NDT) Techniques and their Applicability to Thick Walled Composites," *Procedia CIRP*, vol. 38, pp. 129–136, 2015.

[12] S. G. Pickering, K. Chatterjee, D. P. Almond, and S. Tuli, "LED optical excitation for the long pulse and lock-in thermographic techniques," *NDT E Int.*, vol. 58, pp. 72–77, 2013.

[13] E. D. Gregory and P. D. Juarez, "In-situ thermography of automated fiber placement parts," in *AIP Conference Proceedings*, 2018.

[14] M. Meng, Y. J. Chua, E. Wouterson, and C. P. K. Ong, "Ultrasonic signal classification and imaging system for composite materials via deep convolutional neural networks," *Neurocomputing*, vol. 257, pp. 128–135, 2017.

[15] A. Wronkowicz, K. Dragan, and K. Lis, "Assessment of uncertainty in damage evaluation by ultrasonic testing of composite structures," *Compos. Struct.*, vol. 203, no. July, pp. 71–84, 2018.

[16] A. R. Jac Fredo, R. S. Abilash, R. Femi, A. Mythili, and C. S. Kumar, "Classification of damages in composite images using Zernike moments and support vector machines," *Compos. Part B Eng.*, vol. 168, no. December 2018, pp. 77–86, 2019.

[17] M. Sharp, R. Ak, and T. Hedberg, "A survey of the advancing use and development of machine learning in smart manufacturing," *J. Manuf. Syst.*, 2018.

[18] A. Krizhevsky, I. Sutskever, and G. E. Hinton, "ImageNet Classification with Deep Convolutional Neural Networks," *Adv. Neural Inf. Process. Syst.*, pp. 1–9, 2012.

[19] S. Van der Jeught and J. J. J. Dirckx, "Real-time structured light profilometry: A review," *Opt. Lasers Eng.*, vol. 87, no. 2016, pp. 18–31, 2015.

[20] J. Long, E. Shelhamer, and T. Darrell, "Fully convolutional networks for semantic segmentation," *Proc. IEEE Comput. Soc. Conf. Comput. Vis. Pattern Recognit.*, vol. 07–12–June, pp. 3431–3440, 2015.

[21] C. Maple, "Geometric design and space planning using the marching squares and marching cube algorithms," in *Proceedings - 2003 International Conference on Geometric Modeling and Graphics, GMAG 2003*, 2003.

Evolution of Composite Defects during Manufacturing: Wrinkles & Delamination

Sandeep Chava[1], Sirish Namilae[2]

[1]PhD Student, [2]Associate Professor, Department of Aerospace Engineering
Embry-Riddle Aeronautical University
Daytona Beach, Florida/USA

ABSTRACT

The ubiquitous usage of polymer matrix composites in many applications demands a comprehensive understanding of composite interfaces, which critically affect both the manufacturing processes and the deformation mechanisms. Processing-induced defects in composite structures such as wrinkles and delaminations are primarily a result of inter-ply interfacial movement during manufacturing. In this paper, a new in-situ experimental approach and an ex-situ X-Ray characterization are proposed for developing a fundamental understanding of ply interfaces during composite manufacturing. A carbon fiber laminate is cured in a specially designed autoclave with viewports with plies laid-up on a mold with cylindrical tooling setup to simulate the maximum movement of plies, resulting in the formation of wrinkles and delamination. Three cylindrical tools of radius 9.5mm, 12.7mm and 15.9mm are used in preparing three different molds for the layup. Ply-movement is measured in-situ using Digital Image Correlation (DIC) during the cure cycle through the viewports of the autoclave. In addition, the resulting defects are characterized post-cure using X-Ray Micro-CT. Results show that at wrinkle the maximum out-of-plane movement of 1.32 mm is happening for a 4-ply unidirectional laminate laid up on a mold with 15.9mm tool diameter.

1. INTRODUCTION

The pervasive usage of polymer matrix composites in aerospace [1], automobile [2], and other applications [3] necessitates a comprehensive understanding of composite interfaces that critically affect both the manufacturing processes and the deformation mechanisms in composites. Interfaces are two-dimensional transition regions in materials and are typically associated with higher entropy [4]. Because of this higher energy state, interfaces not only exhibit increased chemical activity [4, 5] but also play a key role in limiting deformation processes (e.g., fracture) [6-8]. This is especially true in composites where two dissimilar materials interact, for example, load transfer to fibers and hence the mechanical behavior of composites is essentially defined by the well-studied fiber-matrix interface [6-9]. A relatively less studied aspect of the composite interfacial behavior is that of the evolution of the inter-ply interfaces during the manufacturing of aerospace composites. Effective chemical bonding at these interfaces during fabrication would result in a defect-free structure, while the movement of these ply interfaces leads to processing induced defects, including porosity, wrinkles, and delaminations.

As shown in the schematic in figure 1, defects such as porosity, debonding, voids, resin affluent areas, foreign objects, broken fibers, wrinkling, matrix cracking, delamination, and blisters are observed during composite processing [10]. Defects like delaminations and wrinkling which are associated with the interfacial ply movement are common defects in composite manufacturing. Manufacturing companies like Boeing [11] and Airbus[12] have suffered substantial economic impacts due to these processing-induced defects in the manufacturing of airplane parts in the past. Inter-ply friction and slippage are identified as responsible for both delamination and wrinkles. This can either be observed in the manufacturing of parts with curvature or while trying to achieve a specific shape by creating a mold with a foreign object underneath plies [13].

In this context, it is essential to understand the evolution of the interfaces between plies and related processing defects like wrinkles and delaminations. Consider wrinkles for example; they are studied in various fields such as skin wrinkles [14], thin films [15], sheet metal forming [16], and fabrics [17], consequently theoretical developments such as scaling behavior [18], kinematic and geometric modeling [19], shear locking behavior [20], etc. can be applied to composite wrinkles. Processing-induced delamination and wrinkling are closely related. Inter-ply shear deformation and restricted motion are responsible for wrinkles, while inter-ply interface displacement in the normal direction causes delaminations. The formation of both of these defects, as well as other processing-induced defects, are primarily defined by the inter-ply adhesion and slip behavior in response to the processing conditions

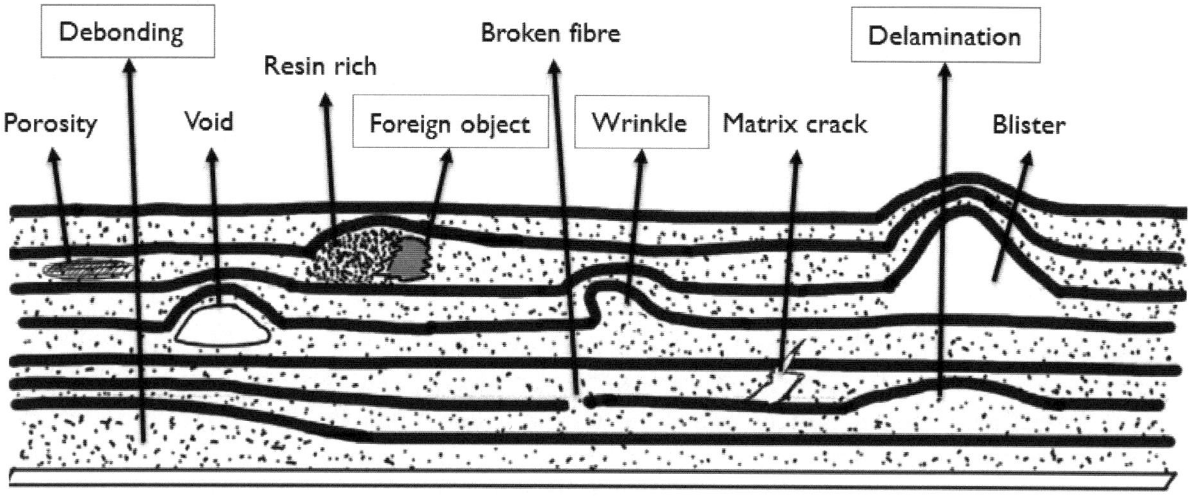

Figure 1. Schematic of ply layup showing all the defects in composites.

Published studies from literature focused on the effect of these manufacturing defects in reducing both the strength and the stiffness of the composite structures [21]. Some of the studies predicted and analyzed the formation of some of the processing-induced defects [22]. Despite these studies, there is a further need to understand the fundamental behavior of ply interfaces to engineer methods that reduce the occurrence of processing-induced defects like wrinkles and delamination. This inter-ply behavior depends on both the evolution of chemical crosslinking between partially cured polymer layers at the interface and the interfacial response to the thermomechanical loads during processing. In many cases, a design requirement like double curvature of the molds [23] can lead to these defects. Many of these studies are applied to the composite structure post-manufacturing. Current research focusses on identifying the conditions that cause ply-movement during manufacturing using 3D Digital Image Correlation (DIC). We quantify the interfacial ply-movement during the composite processing to cure cycle, thereby observe the formation of defects. Here we focus on the lateral movement of plies that leads to the formation of wrinkles.

2. EXPERIMENTAL SETUP

The new experimental approach devised in this paper uses a high temperature/pressure autoclave (from ASC systems), instrumented with 3D digital image correlation (DIC) to measure the surface ply deformations and strain during the composite processing. The autoclave is equipped with a borosilicate glass viewport and a light, which enables this measurement while the autoclave is operating on a cure cycle (typically 120 C and 100 psi pressure) for composite fabrication. The DIC system from Correlated Solutions has two 6-megapixel high-resolution cameras and is capable of measuring both in-plane and out of plane displacements and strains. Figure 2 shows the autoclave with the camera setup. A speckle pattern is spray-painted on to the specimen. The DIC is then used to capture timed pictures of the specimen during a cure. The relative movement of the speckle pattern is analyzed in VIC-3D software to determine surface displacements and strains. Results indicate that in the case of prepreg composites these surface displacements are directly correlated to the kinematics of ply interfaces.

While there have been some methods for in-situ monitoring during processing that use embedded piezoceramics [24], fiber Bragg grating sensors [25] or fiber optic sensors [26], these are often limited to discrete points and alter the structure of the material, the in-situ monitoring approach used in this research provides spatially and temporally continuous data and does not modify the fibers or matrix, therefore does not alter the thermomechanical properties. Moreover, this approach is based on a pressurized autoclave, which is the preferred manufacturing environment for composites, therefore the understanding developed here can be translated to practice more readily.

(a) (b)

Figure 2. (a) Autoclave with 3D DIC. (b) Autoclave viewports and DIC cameras.

Composite specimens of dimensions 20.32x15.24x10^{-2} m (8"x6") are prepared for fabrication using four $[90/90]_s$ oriented plies of unidirectional carbon fiber prepreg from Fibre Glast

Corporation. A white speckle pattern is sprayed on the top layer of the layup as shown in figure 3(c) for creating high contrast, random speckle pattern for digital image correlation. The layup and the tool configuration using cylindrical tools of three diameters in this research are designed to maximize the inter ply-movement as shown in the schematic in figure 3(a). The plies are laid upon these three molds and prepared for curing as seen in figure 3(b). The cure cycle by Fibre Glast consists of a ramp up and ramp down temperatures of less than 5°C per min to a target temperature of 120 °C at 690 kpa pressure. The hold time of 1hr along with other parameters was programmed into the autoclave composite processing control (CPC).

(a) (b) (c)

Figure 3. (a): Schematic of layup; (b) Layup over the cylindrical tool; (c) Specimen after cure.

After transferring the layup to the autoclave, the DIC cameras are set up pointing at the composite layup inside the autoclave through the viewport as shown in figure 2 (b). Both the autoclave cure cycle and the camera system are started at the same time capturing pictures throughout the cure cycle. Due to the ply-movement, the relative movement in the speckles on the top layer of the layup is analyzed in VIC-3D software to determine surface displacements and strains as shown in figure 4 (a). The results of this experimental approach provide an insight into the interfacial motion between plies resulting in surface deformations like wrinkles Further, it will lead to the quantification of the process conditions (temperature, pressure, cure cycle) and the tool geometry that lead to the movement of plies and manufacturing defects.

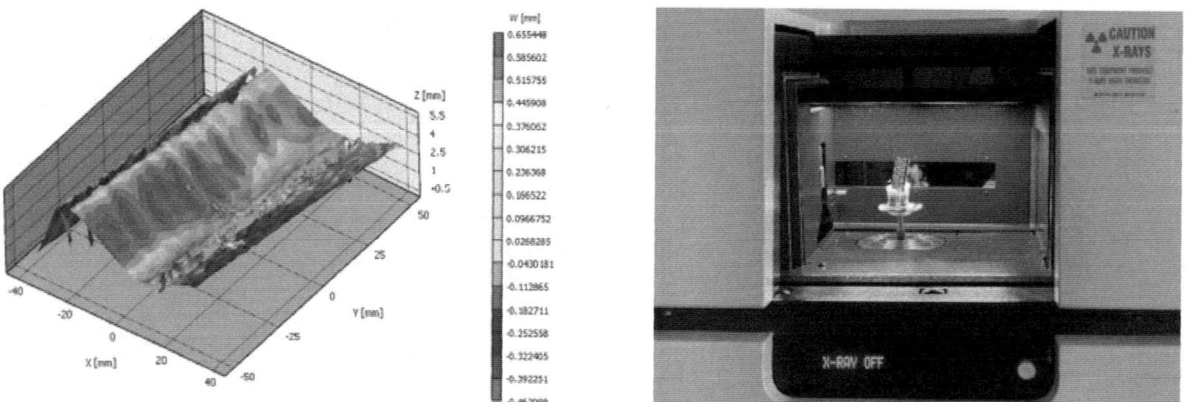

(a) (b)

Figure 4. (a): DIC VIC-3D analysis capturing wrinkles; (b) Specimen installed in X-ray

The interfacial kinematics and the related defect formation post-curing will be validated and characterized using SkyScan 1275 Desktop X-ray Microtomograph from Bruker. Due to the sample size limitations of this X-ray machine, a new small specimen of dimensions 5.08x3.81x10-2 m (2"x1.5") was taken from the original specimen. After installing the sample in the SkyScan as shown in figure 4 (b), X-ray energy of 100kV is used to scan the samples from different angles. The scan is then reconstructed using NRecon software to get the full structural detail of the specimen. A variety of problems in materials research [27] applied X-ray Micro-CT with a few recent studies reporting on polymer matrix composites [28,29]. X-ray Micro-CT in this research will provide a 3D structural detail post-curing to correlate with the in-situ interface movement measurements obtained during the processing.

3. RESULTS AND DISCUSSION

The pictures obtained from the DIC cameras for three different molds are analyzed in VIC-3D software for change in displacements in the out-of-plane Z direction. These displacements are the result of the interfacial ply movement happening due to the combination of the thermomechanical loads during the cure as well as residual stresses generated due to the crosslinking of the polymer bonds [24]. The average out of plane (Z) displacement at the end of the cure and the maximum displacement at a wrinkle in Z-direction are plotted in figure 5. Shrinkage can be observed in figure 5 (a) with an average displacement of $-0.058\text{x}10^{-3}$ m is observed for mold with a 9.5mm radius and $-0.046\text{x}10^{-3}$ m is observed for mold with 12.7mm diameter cylindrical tool. Even after shrinkage, the average out of plane displacement for mold with 15.9mm diameter is $0.005\text{x}10^{-3}$ m due to wrinkles affecting the average.

After the ramp-up stage of cure cycle and gelation, the interfacial ply slippage next to the cylindrical tool slowly increases eventually forming wrinkles towards the end of processing. Figure 5 (b) shows the height of the biggest wrinkle developed for all three molds. For the mold with a cylindrical tool of diameter 9.5mm, the biggest wrinkle is measured to be $0.54\text{x}10^{-3}$ m. For diameter 12.7mm, the wrinkle height is measured at $0.81\text{x}10^{-3}$ m and the biggest wrinkle of all is measured at $1.32\text{x}10^{-3}$ m for mold with a cylindrical tool of diameter 15.9mm. These results show a pattern of higher ply-movement for mold with higher curvature resulting in the biggest wrinkle observed for mold with a cylindrical diameter of 15.9mm.

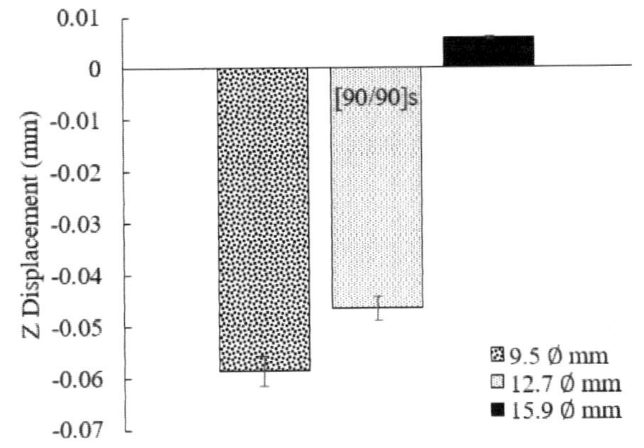

(a)

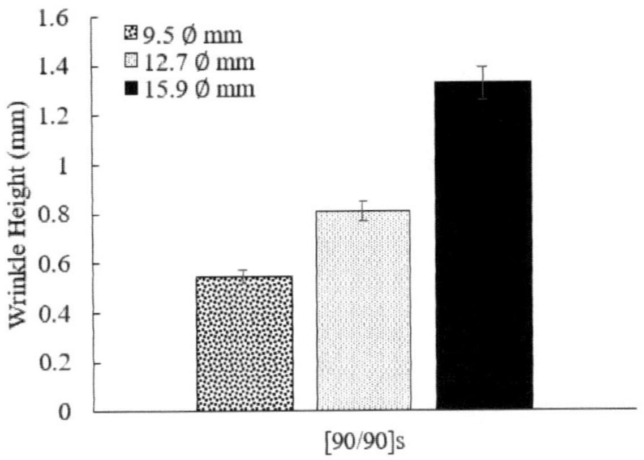

(b)

Figure 5. (a) Average out of plane displacement; (b) Biggest wrinkle height.

Along with mechanisms like buckling of fibers [30], internal residual stresses [31] and shear between plies [32], the effect of cure cycle [33], tool geometry [34], part geometry [35] and layup orientation of the plies [36] also play a key role in the formation of wrinkles and delamination. Fernlund et al. [37] conducted experiments to understand the effect of tool geometry and curing parameters on L and C shaped composite parts observing that high CTE tooling material gives C-shaped parts more spring-in than L-shaped parts. They also found that thin parts have more spring-in than thick parts. Tool-part geometry is highly considered in creating the mold used in this paper to elevate the effect of curvature on the ply movement. Pandey and Sun [38] analyzed the mechanisms of wrinkle formation during the processing of continuous-fiber polymeric composites and found out that wrinkling is triggered by the buckling of one layer experiencing maximum compression. They observed that the other layers in the layup provide support to the unwrinkled part of the wrinkling layer. This analysis also supports the experiments conducted in this paper creating three different compression forces on the same layup with three different curvatures of the mold.

X-ray Micro-CT has been utilized in the literature to characterize various features in materials research including fracture [39], fiber breakage [40], porosity [41], and crack growth [42] along with fiber waviness [43]. Almost all of these studies are for metal-matrix composites but in this paper, it is applied to polymer matrix composites. Figure 6 shows the 3D reconstruction of the sample from X-ray Micro-CT with figure 6(a) showing wrinkles on the top surface along with internal delaminations. The length of the wrinkle measured from X-ray as shown in figure 6(b), internal sectional view, is 1.32×10^{-3} m, which is the same as the out-of-plane displacement at biggest wrinkle measured from the DIC (figure 6) at the end of the cure. This independent ex-situ measurement validates the in-situ experimental approach.

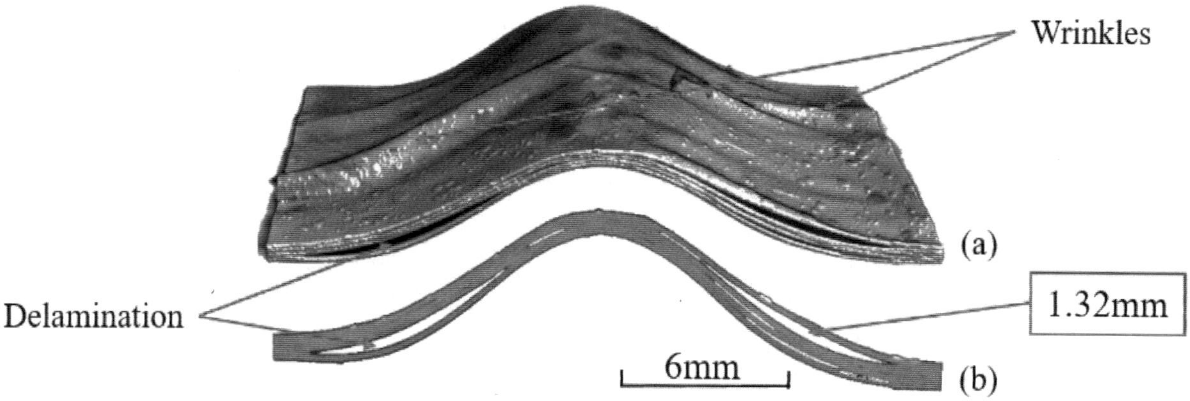

Figure 6. X-ray Micro-CT showing surface wrinkles & internal delamination

Various modeling techniques have been used in the past to model wrinkles and delamination. Zhu et al. [44] used theoretical modeling to model shear deformation and wrinkling of a plain-woven composite. They formulated mathematical expressions to predict the shear energy and critical shear angle of composite fabrics. Hasiao and Daniel [45] developed an analytical model for investigating the effects of fiber waviness on stiffness and strength of unidirectional composites under compressive loading. In their research, they modeled uniform, graded and localized waviness patterns. Slippage coefficient [46], Reynolds equation [47] and power-law approximations [48] are a few other methods used in the literature to model interply adhesion and ply slippage. While these models effectively identified the effects of fiber wrinkling and delamination, the in-situ capability presented in this paper can quantitatively identify the formation of wrinkles during the processing of composites. From figure 6 it is evident that there are wrinkles formed on the specimen after cure and from figure 5 it is also observed that curvature of the tool/part geometry plays a key role in the generation of bigger wrinkles and higher delamination. The experimental approach developed in this research can be utilized to understand how various processing parameters affect the manufacturing of composite parts and approaches can be devised to adjust these parameters to minimize processing induced defects like wrinkles and delaminations.

4. SUMMARY

In summary, a total of three mold configurations were developed using three different diameter cylindrical tools. Composite samples were prepared by curing a four-ply layup upon these mold configurations in an autoclave with viewports. Digital Image Correlation (DIC) cameras are set up through the viewports to capture and measure the out-of-plane ply movement. Resulting defects wrinkles and delamination are characterized post-cure through X-ray Micro-CT. The research approach used in this paper can be used to understand the evolution of composite manufacturing defects like wrinkles and delamination and device approaches to reduce these defects.

Acknowledgment

The financial support from ERAU internal grant is gratefully acknowledged.

5. REFERENCES

1. Marsh, G. (2005). Airframers exploit composites in battle for supremacy. Reinforced plastics, 49(3), 26-32. https://doi.org/10.1016/S0034-3617(05)00577-1
2. Witik, R. A., Payet, J., Michaud, V., Ludwig, C., & Månson, J. A. E. (2011). Assessing the life cycle costs and environmental performance of lightweight materials in automobile applications. Composites Part A: Applied Science and Manufacturing, 42(11), 1694-1709. https://doi.org/10.1016/j.compositesa.2011.07.024
3. Nanni, A. (1993). FRP reinforcement for prestressed and non-prestressed concrete structures. In Fiber-Reinforced-Plastic (FRP) Reinforcement for Concrete Structures (pp. 3-12). Elsevier https://doi.org/10.1016/B978-0-444-89689-6.50005-7
4. Sutton, A. P. (1995). Interfaces in crystalline materials. Monographs on the Physice and Chemistry of Materials, 414-423.
5. Sanchez, I. C. (Ed.). (2013). Physics of polymer surfaces and interfaces. Butterworth-Heinemann.
6. Plueddemann, E. P. (Ed.). (2016). Interfaces in Polymer Matrix Composites: Composite Materials (Vol. 6). Elsevier.
7. Greszczuk, L. (1969). Theoretical studies of the mechanics of the fiber-matrix interface in composites. In Interfaces in composites. ASTM International. https://doi.org/10.1520/STP44699S
8. Hsueh, C. H. (1990). Interfacial debonding and fiber pull-out stresses of fiber-reinforced composites. Materials Science and Engineering: A, 123(1), 1-11. https://doi.org/10.1016/0921-5093(90)90203-F

9. Chandra, N., Li, H., Shet, C., & Ghonem, H. (2002). Some issues in the application of cohesive zone models for metal–ceramic interfaces. International Journal of Solids and Structures, 39(10), 2827-2855. https://doi.org/10.1016/S0020-7683(02)00149-X
10. Smith, R. A. "Composite defects and their detection." Materials science and engineering 3 (2009): 103-143.
11. Petrescu, R.V., Aversa, R., Akash, B., Corchado, J., Berto, F., Apicella, A. and Petrescu, F.I., "When boeing is dreaming–a review", 2017. https://doi.org/10.3844/jastsp.2017.149.161
12. Heslehurst, Rikard Benton. "Defects and damage in composite materials and structures", CRC Press, 2014. https://doi.org/10.1201/b16765
13. Allaoui, S., Hivet, G., Soulat, D., Wendling, A., Ouagne, P. and Chatel, S., "Experimental preforming of highly double curved shapes with a case corner using an interlock reinforcement". International Journal of Material Forming 7(2), (2014), pp.155-165.. https://doi.org/10.1007/s12289-012-1116-5
14. Tallinen, T., Ojajärvi, J., Åström, J. A., & Timonen, J. (2010). Scaling behavior in non-hookean compression of thin-walled structures. Physical review letters, 105(6), 066102. https://doi.org/10.1103/PhysRevLett.105.066102
15. Prodromou, A. G., & Chen, J. (1997). On the relationship between shear angle and wrinkling of textile composite preforms. Composites Part A: Applied Science and Manufacturing, 28(5), 491-503. https://doi.org/10.1016/S1359-835X(96)00150-9
16. Zhu, B., Yu, T. X., & Tao, X. M. (2007). An experimental study of in-plane large shear deformation of woven fabric composite. Composites Science and Technology, 67(2), 252-261. https://doi.org/10.1016/j.compscitech.2006.08.011
17. Pandey, R. K., & Sun, C. T. (1999). Mechanisms of wrinkle formation during the processing of composite laminates. Composites science and technology, 59(3), 405-417. https://doi.org/10.1016/S0266-3538(98)00080-3
18. Fish, J., LeMonds, J., & Shek, K. L. (1999). Modeling of wrinkling in compression molding of composites. Journal of engineering mechanics, 125(8), 951-955. https://doi.org/10.1061/(ASCE)0733-9399(1999)125:8(951)
19. Dodwell, T. J., Butler, R., & Hunt, G. W. (2014). Out-of-plane ply wrinkling defects during consolidation over an external radius. Composites Science and Technology, 105, 151-159. https://doi.org/10.1016/j.compscitech.2014.10.007
20. Erland, S., Dodwell, T. J., & Butler, R. (2014). Inter and intra-ply shearing of uncured carbon fibre laminates. In ECCM16-16TH European Conference on Composite Materials, June(pp. 22-26).
21. Croft, K., Lessard, L., Pasini, D., Hojjati, M., Chen, J. and Yousefpour, A. "Experimental study of the effect of automated fiber placement induced defects on performance of composite laminates". Composites Part A: Applied Science and Manufacturing, 42(5), (2011), pp.484-491. https://doi.org/10.1016/j.compositesa.2011.01.007
22. Meola, C. and Carlomagno, G.M., "Infrared thermography to evaluate impact damage in glass/epoxy with manufacturing defects". International Journal of Impact Engineering 67, (2014): pp.1-11. https://doi.org/10.1016/j.ijimpeng.2013.12.010
23. Yin, H., Peng, X., Du, T. and Guo, Z., "Draping of plain woven carbon fabrics over a double-curvature mold". Composites Science and Technology, 92, (2014): pp.64-69. https://doi.org/10.1016/j.compscitech.2013.12.013

24. Lin, M., & Chang, F. K. (2002). The manufacture of composite structures with a built-in network of piezoceramics. Composites Science and Technology, 62(7-8), 919-939. https://doi.org/10.1016/S0266-3538(02)00007-6
25. Kuang, K., Kenny, R., Whelan, M. P., Cantwell, W. J., & Chalker, P. R. (2001). Embedded fibre Bragg grating sensors in advanced composite materials. Composites Science and Technology, 61(10), 1379-1387. https://doi.org/10.1016/S0266-3538(01)00037-9
26. Crosby, P. A., Powell, G. R., Fernando, G. F., France, C. M., Spooncer, R. C., & Waters, D. N. (1996). In situ cure monitoring of epoxy resins using optical fibre sensors. Smart materials and structures, 5(4), 415. https://doi.org/10.1088/0964-1726/5/4/005
27. Stock, S. R. "X-ray microtomography of materials." International Materials Reviews 44.4 (1999): 141-164. https://doi.org/10.1179/095066099101528261
28. Croom, B., Wang, W.M., Li, J. and Li, X., "Unveiling 3D deformations in polymer composites by coupled micro x-ray computed tomography and volumetric digital image correlation". Experimental Mechanics, 56(6), (2016): pp.999-1016. https://doi.org/10.1007/s11340-016-0140-7
29. Senck, S., Scheerer, M., Revol, V., Dobes, K., Plank, B. and Kastner, J., "Non-destructive evaluation of defects in polymer matrix composites for aerospace applications using x-ray Talbot-Lau interferometry and micro CT". 58th AIAA/ASCE/AHS/ASC Structures, Structural Dynamics, and Materials Conference, 2017, (p. 0355). https://doi.org/10.2514/6.2017-0355
30. Potter, K. D. "Understanding the origins of defects and variability in composites manufacture." International Conference on Composite Materials (ICCM)-17, Edinburgh, UK. 2009.
31. Breuer, U., Neitzel, M., Ketzer, V. and Reinicke, R., "Deep drawing of fabric-reinforced thermoplastics: Wrinkle formation and their reduction". Polymer composites, 17(4), (1996): pp.643-647. https://doi.org/10.1002/pc.10655
32. Hallander, P., Akermo, M., Mattei, C., Petersson, M. and Nyman, T., "An experimental study of mechanisms behind wrinkle development during forming of composite laminates". Composites Part A: Applied Science and Manufacturing, 50, (2013): pp.54-64. https://doi.org/10.1016/j.compositesa.2013.03.013
33. Liu, L., Zhang, B. M., Wang, D. F., & Wu, Z. J. Effects of cure cycles on void content and mechanical properties of composite laminates. Composite structures, 73(3), (2006) 303-309. https://doi.org/10.1016/j.compstruct.2005.02.001
34. Haddad, M., Zitoune, R., Eyma, F., & Castanie, B. Study of the surface defects and dust generated during trimming of CFRP: Influence of tool geometry, machining parameters and cutting speed range. Composites Part A: Applied Science and Manufacturing, 66, (2014) 142-154. https://doi.org/10.1016/j.compositesa.2014.07.005
35. De Moura, M. F. S. F., Daniaud, R., & Magalhaes, A. G. Simulation of mechanical behaviour of composite bonded joints containing strip defects. International journal of adhesion and adhesives, 26(6), (2006) 464-473. https://doi.org/10.1016/j.ijadhadh.2005.06.010
36. Lightfoot, J. S., Wisnom, M. R., & Potter, K. Defects in woven preforms: Formation mechanisms and the effects of laminate design and layup protocol. Composites Part A: Applied Science and Manufacturing, 51, (2013) 99-107. https://doi.org/10.1016/j.compositesa.2013.04.004

37. Fernlund, G., Rahman, N., Courdji, R., Bresslauer, M., Poursartip, A., Willden, K., & Nelson, K. Experimental and numerical study of the effect of cure cycle, tool surface, geometry, and lay-up on the dimensional fidelity of autoclave-processed composite parts. Composites part A: applied science and manufacturing, 33(3), (2002) 341-351. https://doi.org/10.1016/S1359-835X(01)00123-3
38. Pandey, R. K., & Sun, C. T. Mechanisms of wrinkle formation during the processing of composite laminates. Composites science and technology, 59(3), (1999) 405-417. https://doi.org/10.1016/S0266-3538(98)00080-3
39. Borbély, A., Biermann, H., Hartmann, O., & Buffière, J. Y. The influence of the free surface on the fracture of alumina particles in an Al–Al2O3 metal–matrix composite. Computational materials science, 26, (2003) 183-188. https://doi.org/10.1016/S0927-0256(02)00397-X
40. Maire, E., Babout, L., Buffiere, J. Y., & Fougeres, R. Recent results on 3D characterisation of microstructure and damage of metal matrix composites and a metallic foam using X-ray tomography. Materials Science and Engineering: A, 319, (2001) 216-219. https://doi.org/10.1016/S0921-5093(01)00924-8
41. Mummery, P. M., Derby, B., Anderson, P., Davis, G. R., & Elliott, J. C. X-ray microtomographic studies of metal matrix composites using laboratory X-ray sources. Journal of Microscopy, 177(3), (1995) 399-406. https://doi.org/10.1111/j.1365-2818.1995.tb03570.x
42. McDonald, S. A., Preuss, M., Maire, E., Buffiere, J. Y., Mummery, P. M., & Withers, P. J. X-ray tomographic imaging of Ti/SiC composites. Journal of microscopy, 209(2), (2003) 102-112. https://doi.org/10.1046/j.1365-2818.2003.01105.x
43. Baaklini, G. Y., Bhatt, R. T., Eckel, A. J., Engler, P., Rauser, R. W., & Castelli, M. G. X-ray microtomography of ceramic and metal matrix composites. Materials evaluation, 53(9) (1995).
44. Zhu, B., Yu, T. X., Teng, J., & Tao, X. M. Theoretical modeling of large shear deformation and wrinkling of plain woven composite. Journal of Composite Materials, 43(2), (2009) 125-138. https://doi.org/10.1177/0021998308098237
45. Hsiao, H. M., & Daniel, I. M. Effect of fiber waviness on stiffness and strength reduction of unidirectional composites under compressive loading. Composites science and technology, 56(5), (1996) 581-593. https://doi.org/10.1016/0266-3538(96)00045-0
46. Cho, M., Kim, M.H., Choi, H.S., Chung, C.H., Ahn, K.J. and Eom, Y.S., “A study on the room-temperature curvature shapes of unsymmetric laminates including slippage effects”. Journal of composite materials, 32(5), (1998): pp.460-482. https://doi.org/10.1177/002199839803200503
47. Hanna, E.G., Poitou, A. and Casari, P., “MODELING THE INTERPLY SLIP DURING FORMING OF THERMOPLASTIC LAMINATES”. Materials Physics and Mechanics, 40, (2018): pp.22-36.
48. Morris, S.R. and Sun, C.T., “An investigation of interply slip behaviour in AS4/PEEK at forming temperatures”. Composites Manufacturing, 5(4), (1994): pp.217-224. https://doi.org/10.1016/0956-7143(94)90136-8

A COMPARISON OF SURFACE PREPARATION TECHNIQUES FOR WIND TURBINE FIELD REPAIRS

Ariel F. Lusty, Douglas S. Cairns, David A. Miller, Daniel D. Samborsky
Montana State University
Bozeman, MT 59717

ABSTRACT

Wind turbine service lifetimes can exceed 20 years. This extreme operational demand necessitates reliable and consistent field repairs. However, the effects of current wind turbine field repair surface preparation techniques are not well-documented and are thus not well-understood. This leads to reliance on technician experience rather than scientific data for repair procedures, which causes variability in the quality of repairs. Solvent wiping is a common procedure for contaminant removal after damaged material is removed from a wind turbine blade, but it was unknown if solvent wiping reduces the surface energy and consequently the likelihood of a durable repair. The goal of this study was to quantify the effects various surface preparation techniques have on the overall strength and reliability of wind turbine repairs. Type of reagent-grade solvent, fiber direction, matrix, and adhesive were varied for contact angle and lap shear testing. It was found that the type of solvent used affects neither the surface energy nor the maximum lap shear stress, but the fiber direction and matrix-adhesive combinations significantly affected maximum lap shear stress values.

1. INTRODUCTION

Fixing structural damage is of utmost priority for owners and operators of wind turbines. Structural damage is typically from manufacturing defects, but may include in-service damage from mechanical fatigue, bird collisions, or lightning strikes. Damage that inhibits a wind turbine from efficiently harvesting energy should be repaired if the repair does not exceed the cost of blade replacement or if the procedure is economically viable for other reasons. The goal of any repair is to return the structure to its original strength, stiffness, shape, and surface finish [1]. The most common and preferred structural wind turbine blade repair method is the scarf repair due to the need to restore strength, avoid load concentration and eccentricities, and maintain aerodynamic efficiency.

Regardless of the repair design, contamination will be present before, during, and after damage removal. A commonly used technique for contaminant removal involves wiping a surface with reagent grade solvents to remove contamination, and wiping again with a dry cloth to remove the solvent before it evaporates. It is essential to ensure that the solvent being used is capable of dissolving the contaminant without further damaging the composite structure [2]. If the use of solvents damages the surface of the composite structure, bond strengths and repair durability could be compromised. Consequently, contact angle and lap shear testing were conducted to investigate the effects of solvents on composite bond strengths.

Contact angle testing can be used to calculate surface energy, which is the work spent in forming a unit area of a solid surface. Contact angle testing involves the placement of a water droplet on a

SAMPE Virtual Conference Proceedings, 2020. Society for the Advancement of Material and Process Engineering – North America.

surface and measuring the resultant angle the droplet makes with the surface. Low contact angle values correspond with hydrophilicity, high wettability, and higher surface energy. Alternatively, high contact angles indicate hydrophobicity, low wettability and lower surface energy [3]. Additionally, low surface energy has been shown to result in more adhesive failure, and high surface energy will typically result in more cohesive failure [4].

Vacuum assisted resin transfer molding (VARTM) was used to manufacture glass fiber reinforced polymer (GFRP) composite substrates. Profilometry was used to examine differences in substrate surfaces after manufacturing as well as after sanding. Sessile drop contact angle testing was conducted after composite samples were sanded and wiped with reagent grade solvents including acetone, methyl ethyl ketone (MEK), methyl isobutyl ketone (MIBK), or isopropyl alcohol (ISP), as these are commonly used solvents in composite repairs [1]. Contact angle testing was also conducted on plasma-treated specimens, as this is a common surface treatment method to increase surface energy [5]. The contact angle decreased significantly for plasma-treated specimens.

In addition to contact angle testing, lap shear testing using two different adhesives was used to compare the effects of wiping substrate surfaces with different solvents on the resulting bond strengths. It is important to note that lap shear testing does not produce true ultimate stress because the stress distribution in the adhesive is not uniform over the bond area [6]. Additionally, the stiffness of adherends influences peel stress concentrations at the end of the joint in lap shear testing. Increasing the modulus or thickness of adherends will lead to an increase in apparent strength of the bond. Nevertheless, lap shear data does have value from a comparative standpoint and can be used to check the effectiveness of different surface preparations [7]. In addition to differences in lap shear strengths, adhesive failure modes were visually examined and photographed. Dillingham demonstrated using double cantilever beam (DCB) testing that cohesive failure corresponded with higher mode I fracture toughness values, and adhesive failure corresponded with lower mode I fracture toughness values [4]. Similarly, there appears to be a correlation between failure modes and maximum lap shear stresses among lap shear results in this study.

If a particular solvent had deleterious effects on bond strength more than another, it would be reflected in differences in lap shear strengths and adhesive failure modes. Overall, the type of solvent used affected neither the contact angle nor the maximum lap shear stress, but the fiber direction and matrix-adhesive combinations significantly affected maximum lap shear stress values due to their differences in thicknesses and moduli.

2. MATERIALS

Materials were selected to replicate the surfaces that would be encountered in a wind turbine blade field repair. Wind turbine blades are typically made using sandwich composites, which consist of foam or balsa cores sandwiched between glass fiber reinforced polymer face sheets as shown in Figure 1. The core adds stiffness and increases buckling resistance while maintaining a high strength to weight ratio at a low cost.

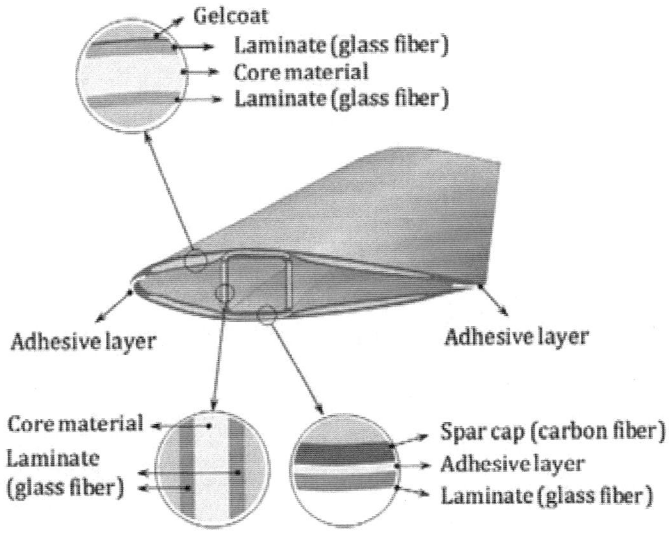

Figure 1. Typical cross-section of a wind turbine blade [8].

Face sheets are often manufactured using both 0° and ±45° glass fibers, with either an epoxy or a vinyl ester matrix. Accordingly, VARTM was used to make $[0]_4$ epoxy, $[45]_6$ epoxy, $[45]_6$ vinyl ester, and $[0]_4$ vinyl ester composites for contact angle and lap shear testing. Unidirectional Vectorply E-LT-2900 fabric was used for the $[0]_4$ layups, and ±45° PPG D810 fabric was used for $[45]_6$ layups. Hexion Epikote MGS RIMR 135/ Epicure MGS RIMH 1366 epoxy was used for the epoxy composites, as this is a commonly used matrix for wind turbine blade manufacturing. Fibreglast 1110 vinyl ester resin was used for vinyl ester composites and was selected for its low viscosity, which is appropriate for VARTM manufacturing.

Polyester peel ply was used between the layup and the vacuum bag during the VARTM manufacturing process. The top of the layup, or where the layup was in contact with the peel ply will be considered as the "rough side" in Table 1. The "rough side" of the laminate obtains its roughness from the textured indentation of the peel ply. The "smooth side" of the laminate is where the layup was in contact with the caul plate. Loctite Frekote 770-NC mold release was used to ensure the laminate would be able to be removed from the caul plate. Furthermore, sanding samples not only energized the surfaces but also removed mold release from samples. To ensure more surface regularity, the "smooth side" of laminates were used for all contact angle and lap shear testing.

Two adhesives were compared in lap shear testing: Huntsman Araldite 2051 and Loctite EA 9396 Aero. Huntsman Araldite 2051 is a two component, toughened methacrylate general purpose adhesive that cures after 15 minutes at room temperature. No post-cure was used for the Huntsman Araldite 2051. Loctite EA 9396 Aero is a two-component, low viscosity room temperature curing adhesive system. Loctite EA 9396 Aero adhesives were post-cured at 66°C for one hour.

Wypall X60 cloths were used for solvent wiping. These cloths were selected for their low lint content and lack of adhesive binders, both features that will decrease the likelihood of contamination and residues being left on a surface when used.

3. EXPERIMENTATION

3.1 Profilometry Testing

Profilometry was conducted to examine differences in the rough and smooth sides of samples as well as the consistency in sanding. The roughness average (R_a) was found using a Mitutoyo SJ-201 profilometer for the sanding methods that were used for contact angle and lap shear testing.

3.2 Solvent Wipe Contact Angle Tests

Sessile drop contact angle testing was conducted after samples were sanded and wiped with solvents. A diagram of a typical sessile-drop contact angle test depicting the three interfacial tension forces is illustrated in Figure 2.

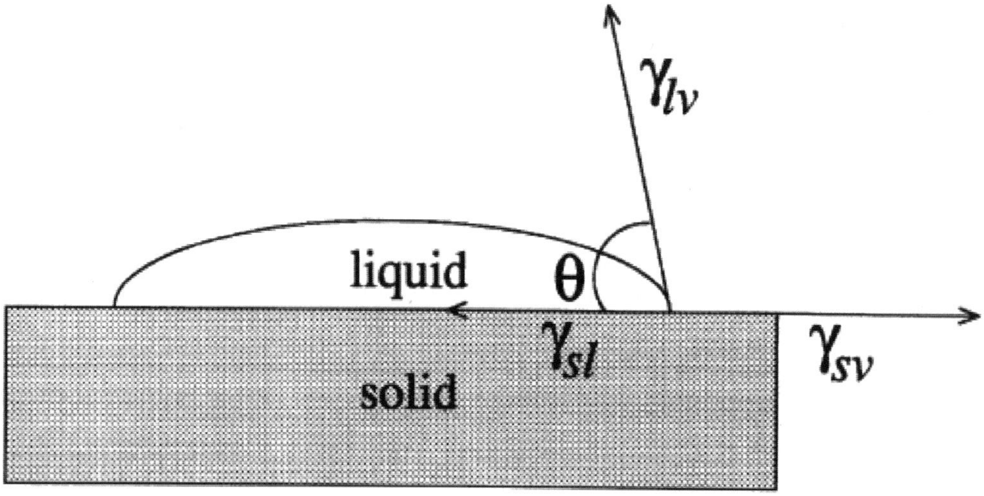

Figure 2. Schematic of a sessile-drop contact angle system at equilibrium [9]. The four variables θ, γ_{sl} , γ_{lv}, and γ_{sv} are defined as contact angle and interfacial tensions between solid and liquid, between liquid and vapor, and between solid and vapor, respectively [3].

For contact angle testing, samples were first wet sanded using Federation of European Producers of Abrasives (FEPA) 1000 grit sandpaper, then wiped with a dry cloth to remove water. The water break test was used to ensure mold release was removed from composite surfaces [10]. Next, the samples were wiped with another cloth that had been saturated with either distilled water, reagent-grade ISP, acetone, MIBK, or MEK. Then, samples were wiped with a dry cloth. Six contact angle tests were conducted for each sample in the locations illustrated in Figure 3.

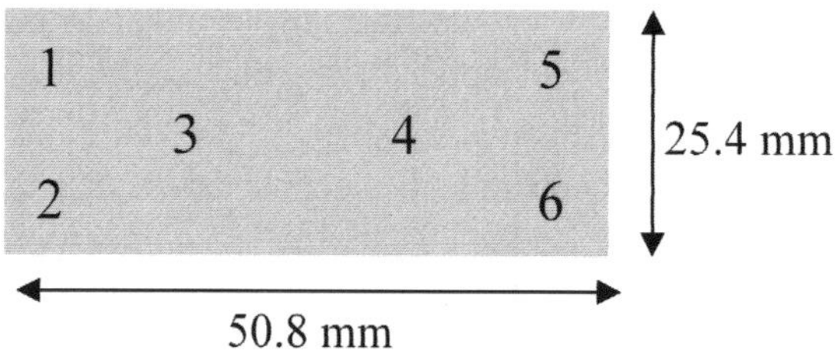

Figure 3. The six locations of contact angle test droplets on a 25.4 x 50.8 mm^2 composite surface.

Contact angle testing was conducted immediately after solvent wiping. A 1 µL droplet of MilliQ 0.2 µm water was formed at the tip of the needle. Next, the substrate was slowly brought up to the droplet. Once the droplet contacted the surface, the substrate was lowered. The contact angle decreases linearly with time after a water droplet is placed on a surface. Conclusions from preliminary testing demonstrated that water droplets would stabilize, but not completely evaporate, after approximately 10 seconds. Thus, after the droplet was on the composite surface for 10 seconds, an image was captured using VCA Optima software. Contact angles were also measured using VCA Optima.

3.3 Plasma Contact Angle Tests

Specimens cut from the same composite plates as the solvent wipe contact angle tests were shipped to Enercon Industries Corporation for plasma treatments, followed immediately by contact angle testing. Dimensions in Figure 3 were used. Specimens were sanded with FEPA 1000 grit sandpaper prior to shipment to remove mold release. Two areas on each sample were contact angle tested before and after being treated by two surface treatment types: a blown ion treatment and flame treatment. Line speed was run at 30.5 cm per minute for both blown ion and flame treatments. The output used for the blown ion treatment was at 0.5 kW, and the output for the flame burner was 200 liters per minute (LPM). The distance between the blown ion treatment and the sample was 6.35 mm, and the distance between the flame treatment and the sample was 50.8 mm.

3.4 Lap Shear Tests

Instead of the FEPA 1000 sand grit that was used for contact angle testing, FEPA 320 grit sandpaper was used to promote adhesion for lap shear testing. Solvent wipe techniques identical to those used in contact angle testing were used prior to adhesive bonding. Wire with a 0.33 mm diameter was used as spacers, and weights were used to reduce variation in adhesive thicknesses. The maximum lap shear stress is reported as the failure stress in the adhesive, which is found by dividing the failure load by the bond area. The $[45]_6$ laminates were tested at 0.5 in/min load rate to minimize creep, and $[0]_4$ laminates were tested at 0.05 in/min load rate [11]. No significant differences in maximum lap shear stress were found between the two load rates.

4. RESULTS

4.1 Profilometry Results

Results from profilometry tests are shown in Table 1 below.

Table 1. Average surface roughness (μm) and standard deviations of specimens subjected to several surface treatments. Two measurements were made for each tabulated value.

Substrate	Unsanded smooth side	Unsanded Rough Side	320 Grit Sanded Smooth Side	1000 Grit Sanded Smooth Side
$[0]_4$ Epoxy	0.460 ± 0.089	7.407 ± 0.483	0.818 ± 0.030	0.417 ± 0.168
$[45]_6$ Epoxy	0.353 ± 0.033	8.814 ± 1.173	0.815 ± 0.069	0.259 ± 0.069
$[0]_4$ Vinyl Ester	0.927 ± 0.292	9.058 ± 1.168	0.610 ± 0.229	0.335 ± 0.102
$[45]_6$ Vinyl Ester	0.475 ± 0.003	8.735 ± 2.690	0.815 ± 0.107	0.191 ± 0.020

As expected, samples that were sanded with 320 grit sandpaper had higher R_a values than samples sanded with 1000 grit sandpaper. The rough side had the highest R_a values from the rough indentation left after peel ply removal. Standard deviations remained about the same between sanded and unsanded smooth-sided samples.

4.2 Solvent Wipe Contact Angle Results

Results for contact angle testing after wiping with solvents are summarized in Table 2. The "unprepared" treated samples were neither sanded nor solvent wiped, and still had mold release on these surfaces. The "dry wipe" treated samples were wet sanded and wiped with dry cloths before contact angle testing.

Table 2. Average contact angle (degrees) for each surface treatment.

Treatment	$[0]_4$ Epoxy	$[45]_6$ Epoxy	$[0]_4$ Vinyl Ester	$[45]_6$ Vinyl Ester
Unprepared	87.7 ± 3.5	73.4 ± 5.0	87.8 ± 9.6	70.6 ± 12.8
Dry Wipe	64.3 ± 5.4	66.8 ± 9.7	67.1 ± 6.8	67.6 ± 6.3
Distilled Water	56.5 ± 5.9	79.0 ± 2.1	59.5 ± 9.9	82.1 ± 5.6
ISP	59.4 ± 5.4	70.2 ± 6.4	65.0 ± 9.1	72.0 ± 6.7
Acetone	54.5 ± 7.2	73.0 ± 2.6	67.3 ± 9.4	68.1 ± 3.7
MIBK	62.3 ± 6.1	58.7 ± 6.4	61.1 ± 5.2	71.8 ± 3.7
MEK	62.4 ± 7.3	65.5 ± 8.5	60.9 ± 8.9	68.3 ± 5.8

The contact angle was not significantly altered by wiping samples' surfaces with solvents. However, removing mold release through sanding is crucial to decreasing the contact angle, as shown in the drops in contact angle between "unprepared" and solvent wiped samples in the $[0]_4$ epoxy and vinyl ester columns. Variation in contact angle results among sanded samples may be in part due to variations in fiber exposure.

4.3 Plasma Treatment Contact Angle Results

Contact angles across all solvent-wiped specimens were fairly high, indicating poor wettability and poor adhesion. To decrease contact angle, samples were plasma treated followed by contact angle testing.

Table 3. Average contact angles and standard deviations on composite surfaces before and after blown ion treatment.

Treatment	$[0]_4$ Epoxy	$[45]_6$ Epoxy	$[0]_4$ Vinyl Ester	$[45]_6$ Vinyl Ester
Initial (before blown ion treatment)	73.5 ± 0.2	71.1 ± 0.7	75.0 ± 3.7	70.4 ± 3.7
After blown ion treatment	30.2 ± 2.2	30.6 ± 2.3	16.2 ± 0.0	22.9 ± 0.8

Across all composite surface types in Table 3, there was a 66% drop between the average contact angles before and after blown ion treatment.

Table 4. Average contact angles and standard deviations on composite surfaces before and after flame treatment.

Treatment	$[0]_4$ Epoxy	$[45]_6$ Epoxy	$[0]_4$ Vinyl Ester	$[45]_6$ Vinyl Ester
Initial (before flame treatment)	70.2 ± 0.1	78.5 ± 2.5	63.5 ± 0.2	77.7 ± 1.3
After flame treatment	25.8 ± 0.4	28.9 ± 0.9	15.9 ± 0.4	18.8 ± 2.5

Across all composite surface types in Table 4, there was a 69% drop between the average contact angles before and after flame treatment. Overall, contact angle dropped significantly between initial and plasma treated composite surfaces for both blown ion and flame treatments.

4.4 Lap Shear Results

The moduli of the $[0]_4$ and $[45]_6$ adherends were 39.4 and 13.3 GPa, respectively. Lap shear test results for the Huntsman Araldite 2051 adhesive are shown in Table 5 and 6 in addition to Figures 4 and 5. Lap shear test results for the Loctite EA 9396 Aero adhesive are shown in Tables 7 and 8 as well as Figure 6.

Table 5. Average maximum lap shear stress (MPa) and standard deviations for 658 ± 19 mm^2 average bond area of Huntsman Araldite 2051.

Surface Treatment	$[0]_4$ Epoxy	$[45]_6$ Epoxy	$[0]_4$ Vinyl Ester	$[45]_6$ Vinyl Ester
Sanded	14.89 ± 0.77	6.89 ± 0.29	22.61 ± 1.56	7.69 ± 0.63
ISP	14.52 ± 0.55	7.31 ± 0.58	22.17 ± 1.45	6.57 ± 0.93
Acetone	13.67 ± 2.33	6.88 ± 0.78	22.80 ± 0.46	8.48 ± 0.48
MIBK	14.13 ± 1.38	6.99 ± 0.94	22.75 ± 1.65	7.96 ± 0.13
MEK	13.49 ± 2.14	6.83 ± 1.21	23.63 ± 0.86	8.07 ± 0.38

The $[0]_4$ vinyl ester specimens had higher maximum lap shear stresses than the $[0]_4$ epoxy specimens. The epoxy and vinyl ester $[45]_6$ specimens all had around the same maximum lap shear stresses. There were no significant differences in maximum lap shear stresses from different solvents.

Table 6. Average adhesive and adherend thicknesses, adhesive and adherend standard deviations, and lap shear failure modes for Huntsman Araldite 2051 adhesive.

Substrate	Adhesive thickness (mm)	Adherend thickness (mm)	Primary Failure Mode
$[0]_4$ Epoxy	0.49 ± 0.14	3.21 ± 0.07	Adhesive failure
$[45]_6$ Epoxy	0.46 ± 0.18	2.09 ± 0.06	Fiber-tear failure
$[0]_4$ Vinyl Ester	0.33 ± 0.15	3.12 ± 0.07	Adhesive failure
$[45]_6$ Vinyl Ester	0.57 ± 0.20	2.24 ± 0.06	Fiber-tear failure

Adhesive thickness was generally higher than the diameter of the spacing wire used, likely from the high viscosity of the adhesive. The unidirectional specimens were about 1 mm thicker than the cross-ply specimens, which is the likely source of differences among lap shear results among different substrates along with differences in adherend moduli. The $[0]_4$ specimens resulted in mostly adhesive failure, and the $[45]_6$ specimens had mostly fiber-tear failure. Figure 4 shows the most common failure modes for each substrate. Besides specimens wiped with MEK, failure modes among solvent wipes did not vary significantly.

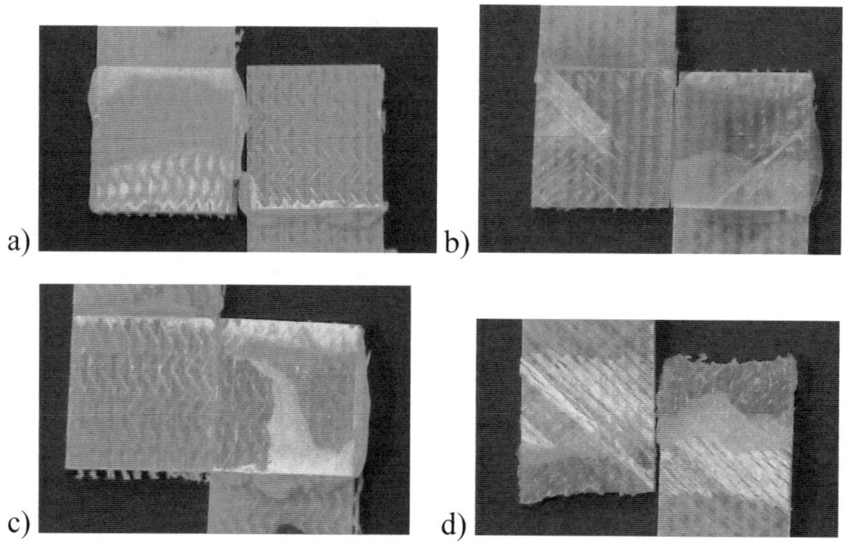

Figure 4. Photographs of the primary failure modes for each substrate: a) $[0]_4$ epoxy adhesive fracture surface (ISP wipe) with adhesive failure mode; b) $[45]_6$ epoxy adhesive fracture surface (MIBK wipe) with fiber-tear failure mode; c) $[0]_4$ vinyl ester fracture surface (ISP wipe) with adhesive failure mode; d) $[45]_6$ vinyl ester adhesive fracture surface (MEK wipe) with fiber-tear failure mode.

Compared to Figure 4 c, Figure 5 shows more cohesive failure.

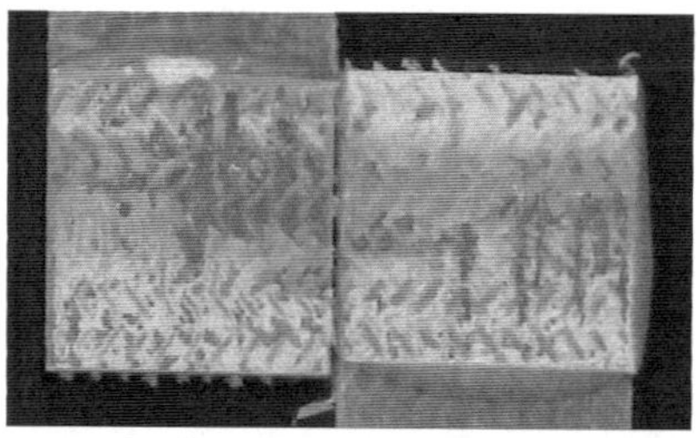

Figure 5. Primary failure mode for $[0]_4$ vinyl ester specimens wiped with MEK.

Although most $[0]_4$ vinyl ester specimens exhibited adhesive failure modes, those wiped with MEK exhibited the most instances of cohesive failure modes but did not have significantly higher maximum lap shear stresses. Overall, though, the layup-matrix combinations had the most differences in resulting maximum lap shear strengths. The $[0]_4$ specimens resulted in higher maximum lap shear strengths and $[45]_6$ layups resulted in the lowest maximum lap shear strengths. The $[45]_6$ specimens failed at lower ultimate tensile stresses than the $[0]_4$ specimens due to their higher susceptibility to bending in tension.

Table 7. Average maximum shear stress (MPa) for 658 ± 16 mm^2 average bond area of Loctite EA 9396 Aero Adhesive.

Surface Treatment	$[0]_4$ Epoxy	$[45]_6$ Epoxy	$[0]_4$ Vinyl Ester	$[45]_6$ Vinyl Ester
Sanded	13.84 ± 0.82	8.12 ± 0.47	5.81 ± 0.26	4.62 ± 0.50
ISP	13.05 ± 0.91	8.23 ± 0.30	6.56 ± 0.48	5.67 ± 0.32
Acetone	12.54 ± 0.27	8.16 ± 0.19	5.40 ± 0.28	4.96 ± 0.62
MIBK	13.39 ± 0.72	8.21 ± 0.14	5.63 ± 0.61	4.94 ± 0.48
MEK	12.62 ± 0.60	8.24 ± 0.11	5.81 ± 0.47	4.77 ± 0.34

When using the Loctite EA 9396 Aero adhesive, the $[0]_4$ epoxy specimens had the highest maximum lap shear stresses, and the vinyl ester specimens had the lowest maximum lap shear stresses.

Table 8. Average adhesive and adherend thicknesses and lap shear failure modes for epoxy specimens with Loctite EA 9396 Aero Adhesive.

Substrate	Adhesive thickness (mm)	Adherend thickness (mm)	Primary Failure Mode
$[0]_4$ Epoxy	0.15 ± 0.12	3.14 ± 0.03	Cohesive failure
$[45]_6$ Epoxy	0.11 ± 0.08	2.29 ± 0.04	Fiber-tear failure
$[0]_4$ Vinyl Ester	0.13 ± 0.13	3.16 ± 0.06	Adhesive failure
$[45]_6$ Vinyl Ester	0.20 ± 0.18	2.39 ± 0.04	Fiber-tear failure

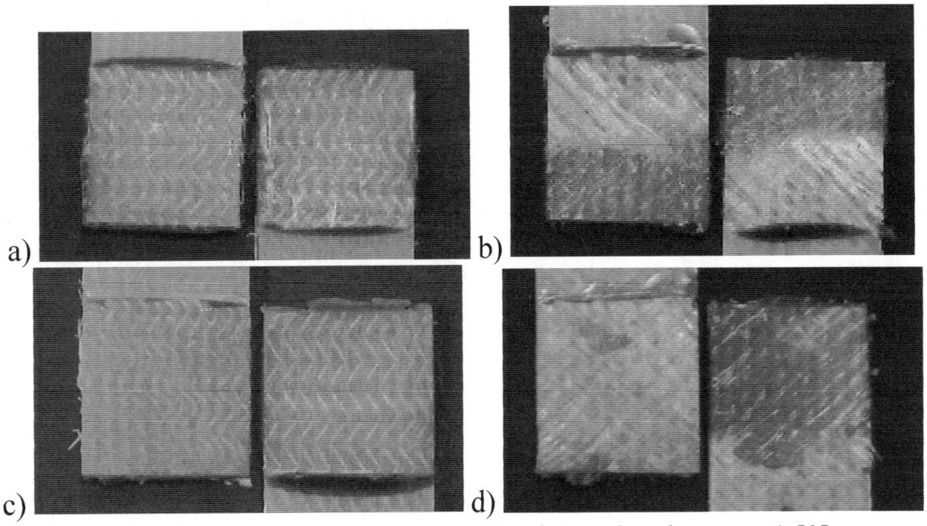

Figure 6. Photographs of the primary failure modes for each substrate: a) $[0]_4$ epoxy adhesive fracture surface (ISP wipe) with cohesive failure mode; b) $[45]_6$ epoxy adhesive fracture surface (MEK wipe) with fiber-tear failure mode; c) $[0]_4$ vinyl ester fracture surface (acetone wipe) with

adhesive failure mode; d) $[45]_6$ vinyl ester adhesive fracture surface (MIBK wipe) with fiber-tear failure mode.

The $[0]_4$ epoxy specimens resulted in mostly cohesive failure, while the $[0]_4$ vinyl ester specimens resulted in mostly adhesive failure. Typically, substrates that have similar molecular structures as the adhesives that are used with them result in higher bond strengths. Since the Loctite EA 9396 Aero adhesive is an epoxy, the $[0]_4$ epoxy specimens likely had the highest bond strengths and optimal failure modes because the epoxy substrates had a similar molecular structure as the epoxy adhesive.

5. CONCLUSIONS

Generally, wiping a composite substrate with either distilled water, ISP, Acetone, MIBK, or MEK neither significantly affects the resulting water contact angle, nor the lap shear strength. Unidirectional vinyl ester specimens resulted in more cohesive failure when wiped with MEK than for those wiped with other solvents, but maximum lap shear strengths were not significantly higher for these specimens. Even so, the causes for the increases in cohesive failure modes for the MEK-wiped specimens merit further investigation. Consequently, DCB and end-notch flexure (ENF) testing will be conducted to examine effects of solvent wiping on modes I and mode II fracture toughness. Since lap shear testing does not give the actual adhesive shear stress, DCB and ENF testing will also provide data that may be used in modeling software such as ABAQUS.

In addition to DCB and ENF testing, chemical testing such as Fourier transform infrared (FTIR) spectroscopy and energy-dispersive X-ray (EDX) spectroscopy will be used to further examine spectral differences in surfaces of composite specimens wiped with solvents. In addition to more mechanical and chemical testing and analysis of solvent effects on the composite surfaces, equally important testing of the effects of common contaminants on repair bond strengths and durability will be executed. Some common contaminants include hydraulic fluid and composite dust, and the effectiveness of each solvent on the removal of these contaminants will be studied.

Contact angle testing demonstrated not only the importance of removing mold release from composites prior to bonding, but also variation in contact angles among sanded samples. This may be due to variation from sanding techniques and also variation in fiber exposure. Plasma treatment delivered promising results for decreasing contact angle, but may be challenging to implicate into wind turbine field repairs due to already-present challenges with wind turbine blade accessibility. Furthermore, the effectiveness of applying sizing to the surface of a composite in increasing resultant repair bond strengths will be examined.

Differences observed when testing different lap shear substrates were mostly due to differences in adherend thicknesses and moduli rather than the type of solvent used. Adhesive-matrix combination also had a significant effect on the maximum lap shear stresses. The Huntsman Araldite 2051 adhesive resulted in 45% higher maximum lap shear stresses with vinyl ester specimens than epoxy specimens. The Loctite EA 9396 Aero adhesive resulted in 96% higher maximum lap shear stresses with epoxy specimens than with vinyl ester specimens, and the Huntsman Araldite 2051 adhesive resulted in 61% higher maximum lap shear strengths than the Loctite EA 9396 Aero adhesive particularly when used with the $[0]_4$ vinyl ester specimens.

Additional adhesives such as Fibreglast 1101 Epoxy Paste Bonding Adhesive and Plexus MA 310 methacrylate adhesive will be used in the future contamination and sizing effect studies.

After additional testing using different adhesives and studying the effects of common contaminants and surface treatments using mechanical and chemical testing methods, procedures will be developed for the evaluation of solvent effects on composite surfaces, the evaluation of the effectiveness of solvents on the removal of common contaminants, and the evaluation of surface treatment effectiveness on increasing repair bond strength and durability. The development of these procedures is crucial to the advancement of wind turbine blade repair technology and thus the extension of wind turbine blade lifetimes.

6. REFERENCES

[1] L. C. Dorworth, "Repairs for Advanced Composite Structures," in *Automotive Composites Conference and Exhibition*, 2013.

[2] L. C. Dorworth, G. L. Gardiner, and G. M. Mellema, *Essentials of Advanced Composite Fabrication and Repair*. Aviation Supplies & Academics, Inc., 2009.

[3] H. Y. Erbil, *Surface Chemistry of Solid and Liquid Interfaces*. Oxford, UK: Blackwell Publishing Ltd, 2006.

[4] R. G. Dillingham and B. R. Oakley, "Surface Energy and Adhesion in Composite–Composite Adhesive Bonds," *J. Adhes.*, vol. 82, no. 4, pp. 407–426, 2006.

[5] D. M. Mattox, "Chapter 2 - Substrate ('Real') Surfaces and Surface Modification," D. M. B. T.-H. of P. V. D. (PVD) P. (Second E. Mattox, Ed. Boston: William Andrew Publishing, 2010, pp. 25–72.

[6] F. C. Campbell, "Manfacturing Processes for Advanced Composites." New York : Elsevier, New York, 2004.

[7] B. Duncan and L. Crocker, "Review of Tests for Adhesion Strength," National Physics Laboratory Materials Centre, Teddington, Middlesex, UK, 2001.

[8] K. B. Katnam, A. J. Comer, D. Roy, L. F. M. da Silva, and T. M. Young, "Composite Repair in Wind Turbine Blades: An Overview," *J. Adhes.*, vol. 91, no. 1–2, pp. 113–139, 2015.

[9] D. Y. Kwok and A. W. Neumann, "Contact Angle Measurement and Contact Angle Interpretation," *Adv. Colloid Interface Sci.*, 1999.

[10] L. J. Hart-Smith, R. W. Ochnsner, and R. L. Radecky, "Surface Preparation of Composites for Adhesive-Bonded Repair," in *Engineered Materials Handbook: Adhesives and Sealants*, ASM International, 1990.

[11] ASTM Standard D5868-01, 2014, "Standard Test Method for Lap Shear Adhesion for Fiber Reinforced Plastic (FRP) Bonding." ASTM International, West Conshohocken, PA, 2014, DOI: 10.1520/D5868-01R14, www.astm.org.

REMEDIATION OF CONTAMINATED BOND SURFACES: CONTAMINANT MIXTURES

R. Giles Dillingham, Brooke Campbell, Matthew Nichols
BTG Labs
Cincinnati, OH/USA

ABSTRACT

Aerospace manufacturing and repair environments include the very real possibility that bond surfaces presented to the adhesive will not always have the correct composition to ensure acceptable bond performance. There are myriad avenues for contamination of bond surfaces. Contact contamination occurs from unsanitary material handling. Common sources for these contaminants are through substandard wipers, contaminated solvents, and abrasives which contain release agents such as stearates. Aerosol contamination events from airborne substances such as cutting fluids, lubricants, and mold releases are well documented. Another less frequent but real source of contamination is from peel ply that was unintentionally coated with release agent from the manufacturer. Control of adhesive bonding operations to guarantee reliable bond performance requires careful, quantitative bond surface control along with methods for successful remediation of contaminated surfaces. This work describes the development of quantifiable, reliable, and easily deployable techniques for controlling bond surface properties in manufacturing and repair by i. ensuring that the composite surface has been prepared properly for bonding, and ii. if the surface has been identified as having been improperly prepared (e.g. contaminated), establishing quantifiable methods for remediation of the surface.

1. INTRODUCTION

There currently does not exist the same level of confidence in the reliability of composite-composite adhesive bonds that exists in metal-metal adhesive bonds. The low reactivity of the composite surface compared with that of a properly prepared metal surface means that a good interface can be difficult to establish without pretreatments that increase the chemical activity of the surface. These pretreatments are either mechanical (peel ply removal, hand abrasion, grit blasting) or chemical (plasma treatment). However, unlike metal-metal adhesive bonds, composite-composite bonds are relatively insensitive to environmental degradation. Once a good interface has been established between a composite material and an organic adhesive, it is stable to the environment without needing to resort to passivating pretreatments.

Surface treatment processes for composite-composite bonding in aerospace manufacturing and repair are still largely manual, and probably will continue to be for the near future. Partly because these processes are manual, most manufacturing operations and repair procedures do not utilize quantitative control of these processes. Bond failures in composite aircraft are well known, and the lack of clear explanations for the causes of most of these failures results directly from the lack of quantified control over the bonding processes. This unexplained bond performance variability has resulted in a lack of confidence in the ability to manufacture bonded composite aircraft with a reliability that would permit certification. The economic and performance advantages of adhesively bonded primary aircraft structures are well-recognized and have given rise to significant programs such as the DARPA Open Manufacturing TRUST Program [1,6] as well as large internally funded programs at several manufacturers and research organizations [2-4] aimed at developing manufacturing processes and procedures that can provide quantifiable reliability of bonded primary structures.

These programs as well as other recent work in surface and adhesion science outside of the aerospace industry have established the firm scientific foundation necessary for developing truly reliable adhesive bonding processes for composite materials that can be practically deployed in both manufacturing and repair. 'Practically deployed' means that these processes need to be able to accommodate the realities of aerospace manufacturing and repair: bonding environments that are not always well-controlled, and limited training and capabilities of the technicians who exercise these processes.

Aerospace manufacturing and repair environments include the very real possibility that bond surfaces presented to the adhesive will not always have suitable chemical composition and physical properties to ensure acceptable bond performance. There are myriad of avenues for contamination of bond surfaces. Contact contamination occurs from unsanitary material handling. A common source for these contaminants is through the use of substandard wipers, contaminated solvents, and abrasives which contain release agents such as stearates [5]. Aerosol contamination events from airborne substances such as cutting fluids, lubricants, and mold releases have been well documented [6]. Another less frequent but real source of contamination is peel ply that has been unintentionally coated with release agent from the manufacturer [7].

It is currently difficult to protect against unintended contamination events because most surface preparation and bonding procedures consist of following prescribed process steps with no verification of the quality (i.e. cleanliness, surface composition) of the resulting surfaces.

Control of adhesive bonding operations to guarantee reliable bond performance requires careful, quantitative bond surface control. This includes four key components: i. knowledge of the detrimental contaminants present in the bonding environment, ii. Understanding of the threshold amounts of these contaminants that cause degradation of bond performance, iii. reliable methods for detecting these contaminants on the bond surface that can be practically deployed in manufacturing, and iv. methods for successful remediation of contaminated surfaces.

This paper describes some of the results of a recently completed Phase I SBIR that has laid the groundwork for development of quantifiable, reliable, and easily deployable techniques for controlling bond surface properties by i. ensuring the composite surface has been prepared properly for bonding, and ii. if the surface has been identified as having been improperly prepared (e.g. contaminated), establishing methods for remediation of the surface. Successful completion of this work will result in the removal significant barriers to a more widespread adoption both of primary bonded aircraft structures and bonded structural repairs. While previous work in the TRUST program focused on model contaminants consisting of pure, well characterized substances. The current work addresses in part the possibility of contamination by mixtures of pure components, for example mold release + lubricating oils.

Optimum surface treatment and bonding procedures are materials system dependent; i.e. the best treatment and bonding process for a polyimide composite and adhesive will likely be different from that for an epoxy composite and adhesive system, and certainly different from an optimized process for a bonded thermoplastic composite like polyether ether ketone (PEEK). The work described here has focused on a single materials system, surface preparation process, and bonding procedure. The results will be particular to this materials system. However, the concepts, experimental methodologies, and analysis tools are generally applicable to any materials system.

1.1 Remediation of contaminated surfaces

In general, remediation of a compromised surface requires removal of contamination. Contamination can be removed by three approaches: physical, chemical, or mechanical.

An example of a physical approach is removal by dissolving the contaminant in a solvent followed by removal of the resulting solution. In general, this approach cannot remove 100% of a contaminant, although if it reduces the amount of contaminant on the surface below the critical level that affects adhesion, it can likely regenerate a bondable surface.

An example of a chemical approach to contaminant removal is conversion of the contaminant to gaseous species by oxidation using a plasma. This can be effective for small amounts of hydrocarbons, which are eventually oxidized to CO_2 and H_2O, but this approach is frequently ineffective for silicone contaminants if present on the surface in more than miniscule quantities. Silicones oxidize to non-volatile silicas, and when thicker layers of silicone contaminants are exposed to plasmas, the surface of the contaminant film oxidizes to form a silica crust above an unoxidized layer of silicone. Surface energy of the silica crust can be high, but bonding to these surfaces is poor.

Mechanical removal methods include media blasting, abrasion, and even laser treatments, and these approaches depend on ablation of contaminated layers to carry contaminant away from the surface. Media blasting can be very effective at this removal. Abrasion, using abrasive coated paper or an abrasive coated nonwoven fabric (e.g. Scotch-Brite) is not as effective at transporting contaminated particles away from the surface as media blasting, and the process may require more careful pre-cleaning of the contaminated surface prior to abrasion to ensure that contaminant is not simply re-distributed and worked into the surface. Furthermore, mechanical methods have the potential to created physical damage to the laminate surface which can reduce the interlaminar fracture toughness of the composite and reduce the overall fracture toughness.

This work has focused on a combination of physical and mechanical remediation techniques: solvent cleaning of the surface followed by abrasion, with an emphasis on developing techniques with quantifiable endpoints. Successful remediation is defined as creating a surface that results in cohesive failure of DCB specimens in the adhesive or the laminate, with a toughness equivalent to that obtained from uncontaminated specimens. Identifying quantifiable endpoints for the remediation process will ensure that the process is reproducible when deployed in manufacturing and repair scenarios.

Removal of a contaminant by a solvent is a two-step process: i. the solvent is applied to the surface and the contaminant is allowed to dissolve into the solvent and ii. the solution (solvent + contaminant) is removed from the surface.

The dissolution of the contaminant in the solvent is determined by two factors: i. the solubility of the contaminant in the chosen solvent and ii. the ability of the solvent application process (e.g. spray, immersion, agitation, wiping) to facilitate the solution process by agitation. For a manual wiping process on a composite surface, dissolution will be inhibited by a rough surface that prevents contact of the wiper with the contaminated surface. A peel ply surface has deep hills and valleys in the peel ply fabric imprint in the resin cap which can prevent a wiper from contact with the contaminated surface. A peel ply surface that has been sanded has less roughness (as measured by the arithmetic average of the roughness profile, RA) but can still have very high aspect ratio features such as deep scratches and broken fibers that can prevent contact of contaminant with the wiper. The contaminant can "hide" in the nooks and crannies of the abraded surface.

In the case of a manual wiping process, the removal of the contaminant + solvent solution by a dry wipe is critical. This prevents re-deposition of the contaminant as the solvent flashes off. Again, surface roughness inhibits the ability of a dry wipe to contact and remove the solvent solution.

This discussion serves to highlight the fact that the ability of a solvent cleaning process to remediate a contaminated composite laminate surface will be affected by the following:

i. Surface topography (not simply RA, but other aspects of the surface topography)

ii. Solubility of contaminant in the chosen solvent

iii. Nature of the wiper used to apply and remove the solvent

iv. Method (and operator-to-operator reproducibility) of the solvent application and wiping process

2. EXPERIMENTATION

2.1 Application of controlled amounts of contaminants

The contaminant for this phase (Loctite® Frekote 700-NC™, Henkel Corp. blended with equal volumes of 30 wt non-detergent motor oil) was diluted to a fraction of a percent in xylene. Surfaces were contaminated by passing panels through the spray pattern of a high-volume low pressure (HVLP) spray gun at controlled speed. Directly measuring the amount of contaminant on a composite panel is difficult. For this reason, gravimetric measurements were used to determine the areal density (mass of deposited contaminants per unit area) as a function of concentration and spray gun parameters (air pressure and mixture adjustments). 1" x 3" aluminum foil coupons, were passed through the spray pattern at various speeds, and then weighed after allowing 20 minutes for solvent evaporation. Areal density was determined by comparing mass of aluminum foil samples of known area before and after spray deposition of contaminant solutions. For a given concentration and set of spray parameters, the areal density increased in a linear manner with the residence time of the surface in the spray. The residence time is the amount of time each unit area of surface is exposed to the spray; it is calculated as the width of spray pattern divided by the rate of sample traverse.

2.2 Selection of solvent for silicone remediation

Motor oil is readily removable by a variety of solvents, both polar and non-polar. Silicones are much more problematical due to their lack of solubility in most solvents. Solvents were prescreened for their ability to solubilize silicones by blending 3 ml each of silicone oil and the candidate solvent in a glass vial, capping and shaking until dispersed. The mixture was then allowed to sit for several minutes and then visually evaluated for phase separation (Figure 1). Solvent/silicone mixtures that showed evidence for phase separation were not evaluated further. Quantification of the ability of these solvents to remove silicone from a surface was evaluated by comparing infrared spectra from contaminated ferrotype plate (chrome-plated steel) before and after successive wet + dry wiping cycles. Absorbance is directly proportional to the thickness of the silicone film on the surface and is a quantitative indicator of the ability of the given solvent to remove silicone from the surface. American Fiber and Finishing Rymplecloth™ was the wiper used for both wet and dry wiping. The best performers (MEK, decane, hexane, and ethyl acetate) all gave very similar results; ethyl acetate was chosen for further evaluation because it was the best performer.

Figure 1. Initial evaluation of silicone solubility in candidate solvent. Left: silicone immiscible in solvent. Right: silicone miscible in solvent.

2.3 Selection of mechanical test geometry: lap joints vs double cantilever beams

The stress state in a lap joint under load is complex, including both Mode I and Mode II stresses, and is highest at the ends of the joint. The lack of a uniform stress state along with absence of a defined precrack makes the data very geometry dependent. Sensitivity to the interface is not great. While lap joints are easy to construct and test, they are poor for evaluating the effect of surface treatments and contaminants. DCB specimens, with a defined precrack and pure Mode I stress state, are known to correlate directly with in-service performance [7]. Figure 2 shows failure surfaces for both lap joints and DCB specimens created from DGEBA composites that were either solvent wiped, grit blasted, or plasma treated. All lap joints showed interlaminar failure regardless of surface treatment; DCB failure modes were strongly dependent on surface treatment. The increased sensitivity of the DCB's over lap joints to the specific surface treatment process shown in Figure 2 confirmed the choice of DCB specimens as the appropriate test method for this effort.

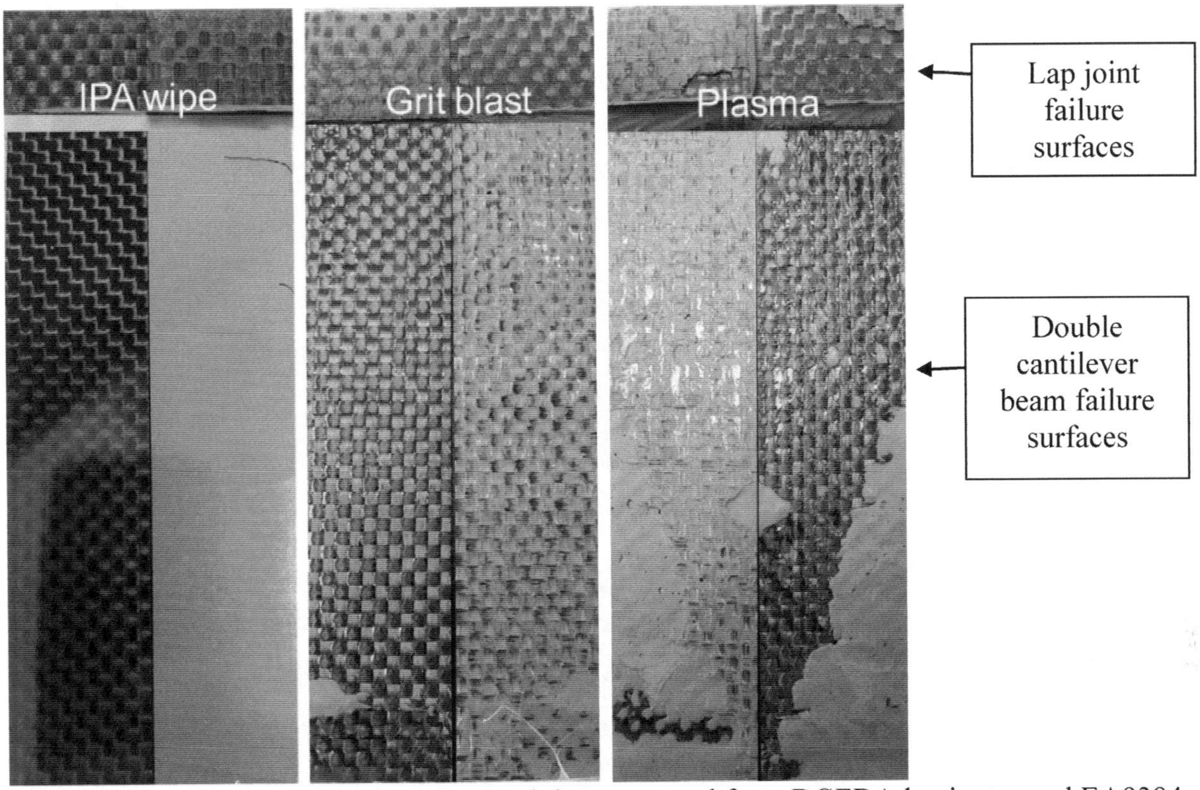

Figure 2. Failure surface for adhesive joints prepared from DGEBA laminates and EA9394 adhesive.

3. RESULTS

3.1 Response of material surface to solvent cleaning and abrasion

Figure 3 shows the effect of the surface preparation process on surface energy as measured by contact angle. The freshly exposed peel ply surface has a very low contact angle (~18°), showing that peel ply removal has created a clean, high-energy surface. Subjecting this surface to solvent cleaning results in an *increase* in the contact angle to about 41°. This is a typical response of a pristine high-energy surface to almost any kind of contact or handling: reactive sites created by the fracture during peel ply removal have been at least partially consumed by reaction and/or adsorption of a few molecular layers of contaminant. Although a well-intentioned solvent cleaning of a highly reactive surface frequently results in a degradation of its reactivity, experience shows that surfaces demonstrating contact angles in this range are still highly bondable. Abrasion + subsequent solvent wiping of this surface does not completely restore the original energy. but does not appear to degrade the surface. Contamination with silicone creates a low-energy surface with ~80° contact angle.

The exposed peel ply surface is fundamentally different from the abraded one: while the unabraded peel ply surface is mostly epoxy resin, the abraded surface consists largely of exposed and fractured graphite fibers.

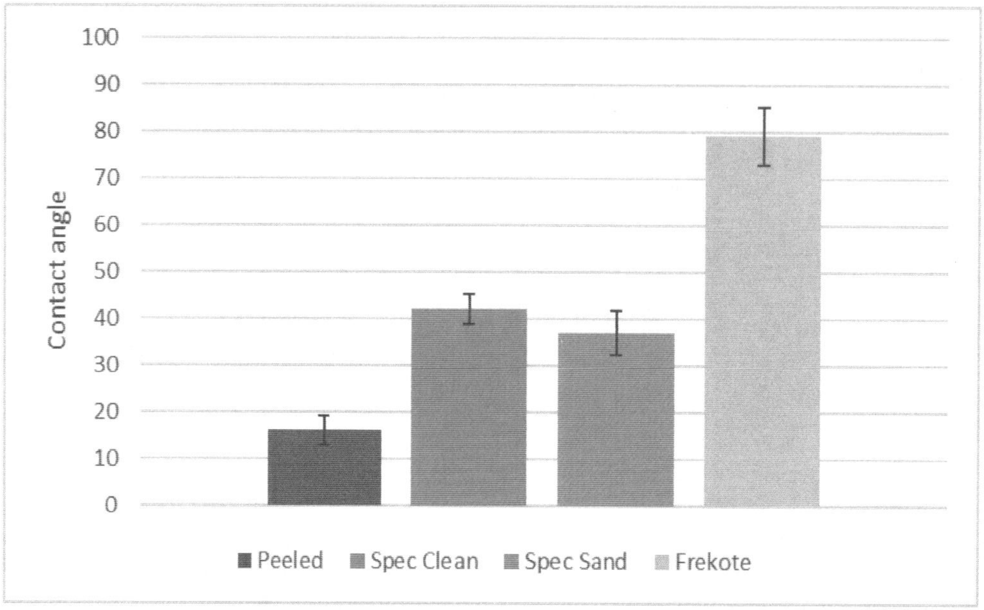

Figure 3. Evolution of contact angle on a freshly exposed peel ply surface with solvent cleaning, abrasion + solvent wiping, and subsequent contamination with silicone.

Profilometry is frequently used to characterize abraded surfaces. Figure 4 shows the effect of the manual abrasion process on the R_A value of these laminates. This surface profile metric (RMS height of surface asperities) is very high for the peel ply surface and is reduced proportional to the amount of abrasion as the asperities are removed. Figure 5 shows an electron micrograph of a manually abraded surface corresponding to fully abraded surface (R_A ~60μin). A large fraction of this surface now consists of abraded (i.e. physically damaged) graphite fiber; the contact angle measured on these surfaces is determined largely by that of the graphite fiber surfaces, not the epoxy matrix resin.

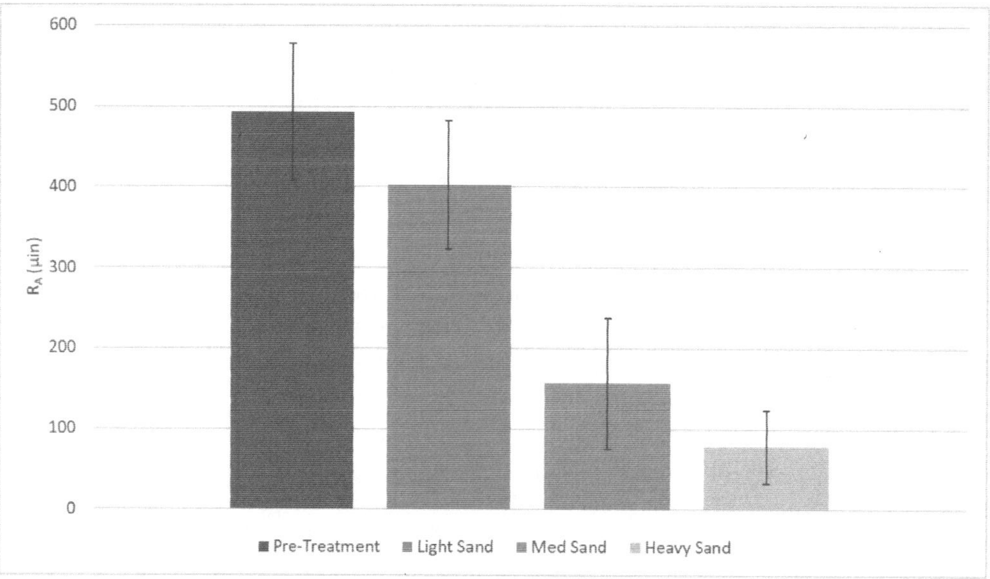

Figure 4. Evolution of R_A with successive sanding cycles.

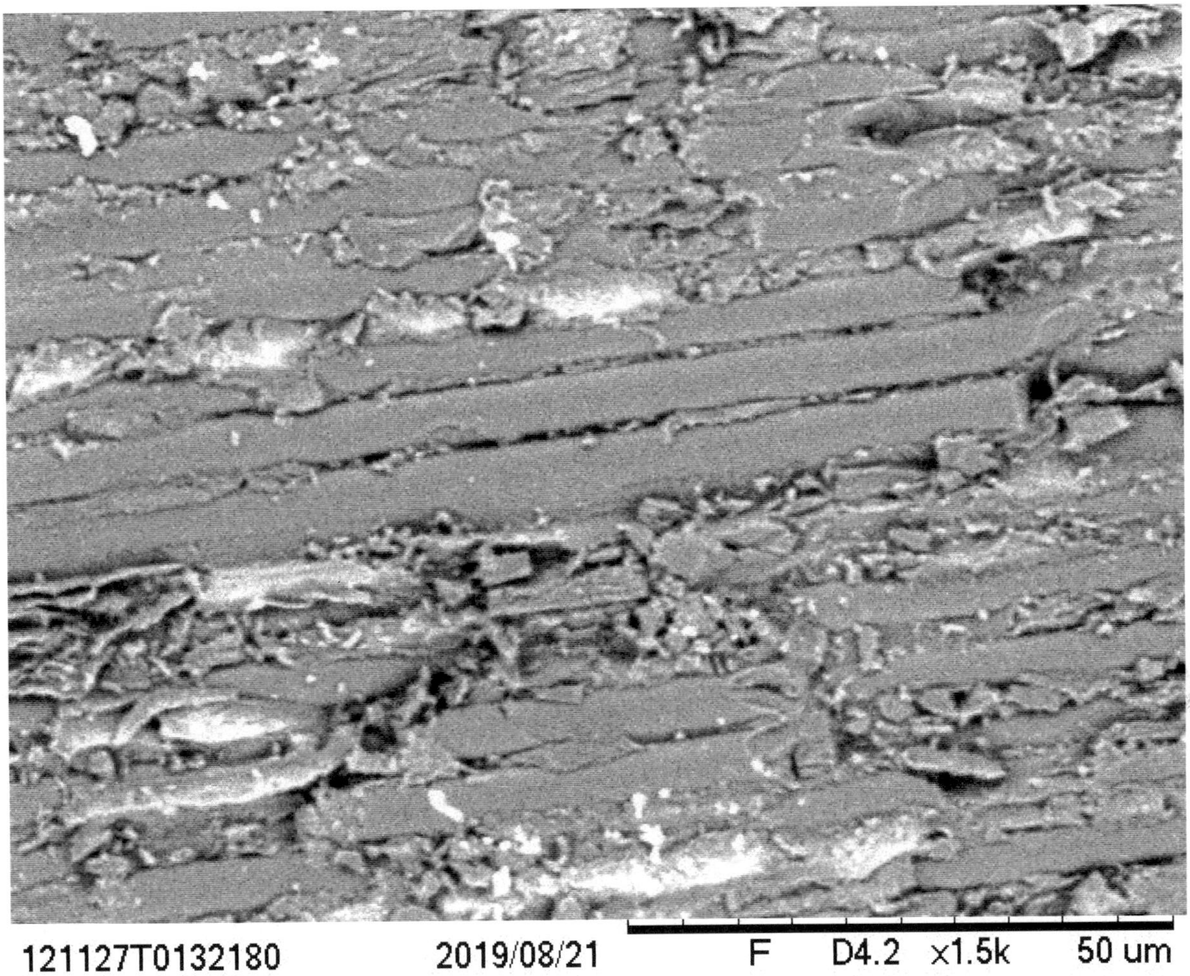

Figure 5. Electron micrograph of an abraded laminate surface corresponding to an R_A of about 60 µin

3.2 Controlled contamination

Figure 6 shows the linearity of the areal density with residence time for silicone. Similar curves were generated for each contaminant mixture and repeated whenever a change in contaminant or adjustment of the apparatus was performed.

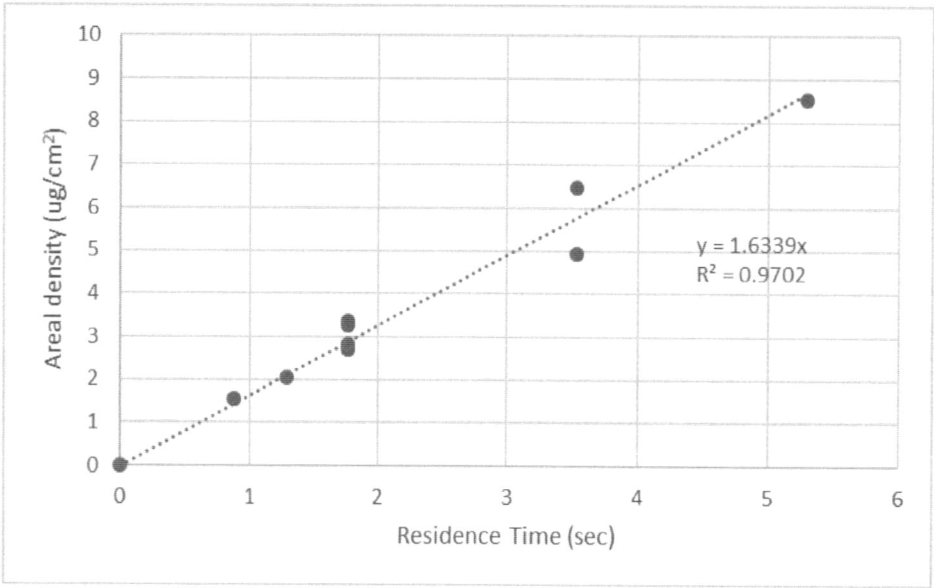

Figure 6. Areal density (mass/unit area) of Frekote 700NC dissolved in xylene.

Gravimetric measurements represent the average areal density over the entire surface; uniformity of deposition was determined by grazing angle Fourier Transform infrared spectroscopy (FTIR) on ferrotype plate (chrome-plated steel) witness coupons that were coated with contaminant at the same time as the foil samples and placed in various locations of the spray pattern.

In order to develop a fundamental understanding of the effects of contamination on adhesive bonding along with remediation, most of the work to date has focused on single component contaminants. However, mixtures of multiple contaminants, particularly immiscible contaminants, have the potential to segregate on a surface. This could have an effect on the ability to detect a contaminant. For example, if a highly detrimental contaminant (such as a silicone) is overlaid by a relatively benign contaminant (such as a mineral oil), the detectability of the detrimental contaminant could be compromised. (This would depend on the detection method used.) In this scenario, a seriously compromised bond surface could possibly be accepted as being suitable for bonding. To initiate investigation into the ability to both detect combined contaminants and to remediate surfaces contaminated in this way, a blend of 50/50 (v/v) Frekote 700/motor oil was used to contaminate composite surfaces. These contaminants are immiscible when neat, but form a single-phase solution in xylene at the dilution used. Figure 7 shows the effect of increasing contaminant level on the measured contact angles on 5320-1 composite surfaces. Although it is unknown whether the contaminant films phase separated to form island structures or discrete layers, the high contact angles for the contaminated surfaces (~90°) suggest that the silicone component has segregated to the top, presenting a pure silicone surface to the water droplet during contact angle measurement. This is expected from a purely energetic standpoint, as the system will tend to minimize surface free energy, which favors low surface tension silicone on the free surface.

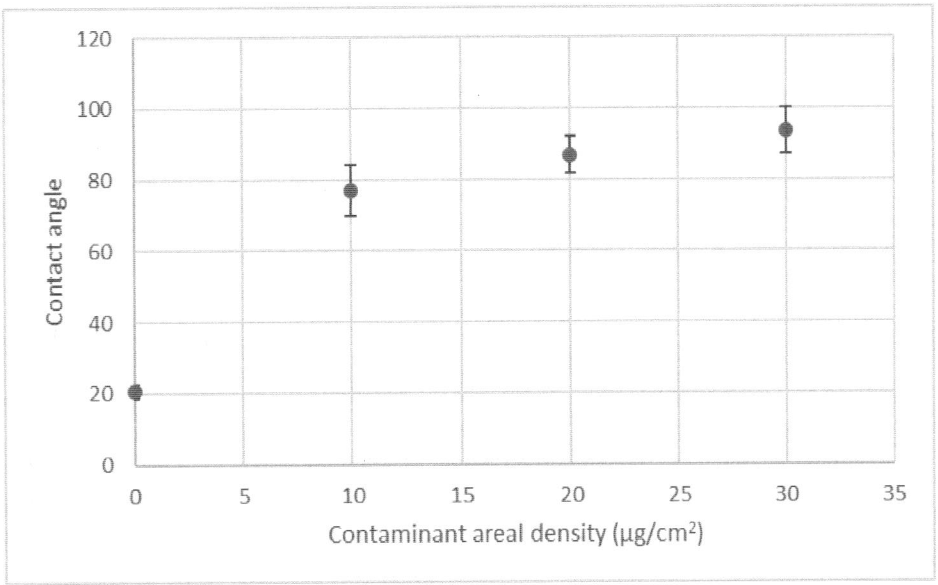

Figure 7. Contact angle on composite surface contaminated with a blend of 50/50 (v/v) Frekote 700NC/30 wt motor oil

3.3 Remediation of surfaces contaminated with mixed contaminants to a quantifiable endpoint

Further contact angle measurements indicated that these contaminated surfaces were readily remediated using solvent cleaning and abrasion process. As an example, Figure 8 shows contact angles obtained from a freshly exposed peel ply surface, after contamination with Frekote 700NC, and after multiple solvent cleaning cycles, and finally after subsequent abrasion.

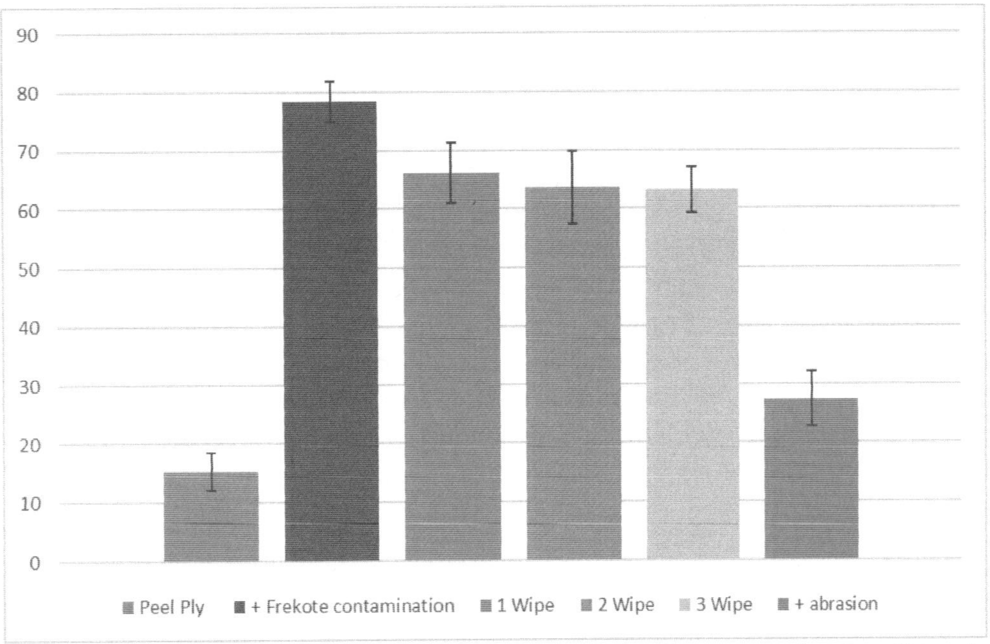

Figure 8. Contact angle on composite surface as-peeled, after contamination (11 µg/cm^2 of Frekote 700 NC), multiple solvent cleaning cycles, and abrasion.

Multiple solvent cleaning cycles were unable to restore the original surface properties. However, removal of the bulk of the contaminant via solvent wiping allowed restoration of the original abraded surface properties.

DCB testing of remediated surfaces corresponding to the samples in Figure 8 showed that subjecting the surfaces contaminated with oil + silicone to the standard solvent cleaning + abrasion preparation process successfully regenerated the original bondability. Figure 9 shows the fracture toughness results; Figure 10 shows the fracture surfaces. The contaminated samples failed 100% interfacially at a fracture toughness very close to zero. The fracture toughness values achieved with the remediated surfaces were equivalent to the samples that had been bonded in the freshly peel, uncontaminated condition.

Figure 8 shows that although the fracture toughness for the remediated samples was indistinguishable from the pristine sample, the failure mode appears to have shifted to be more cohesive in the laminate. This may indicate that laminate damage from abrasion has somewhat compromised the cohesive properties of the first ply. This effect has been documented with grit blasted composite laminates

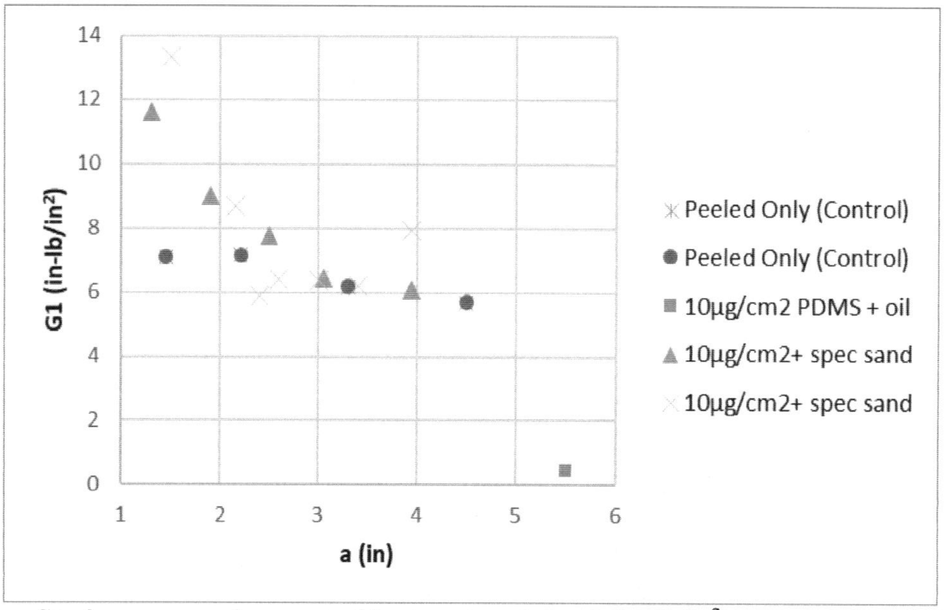

Figure 9. Strain energy release rate (fracture toughness) (in-lb/in^2) vs crack length (in) for composite surfaces as-peeled, after contamination (10µg/cm^2 of 50/50 (v/v) silicone/mineral oil), and solvent cleaning + abrasion. The spec solvent cleaning + abrasion process successfully regenerated a bondable surface.

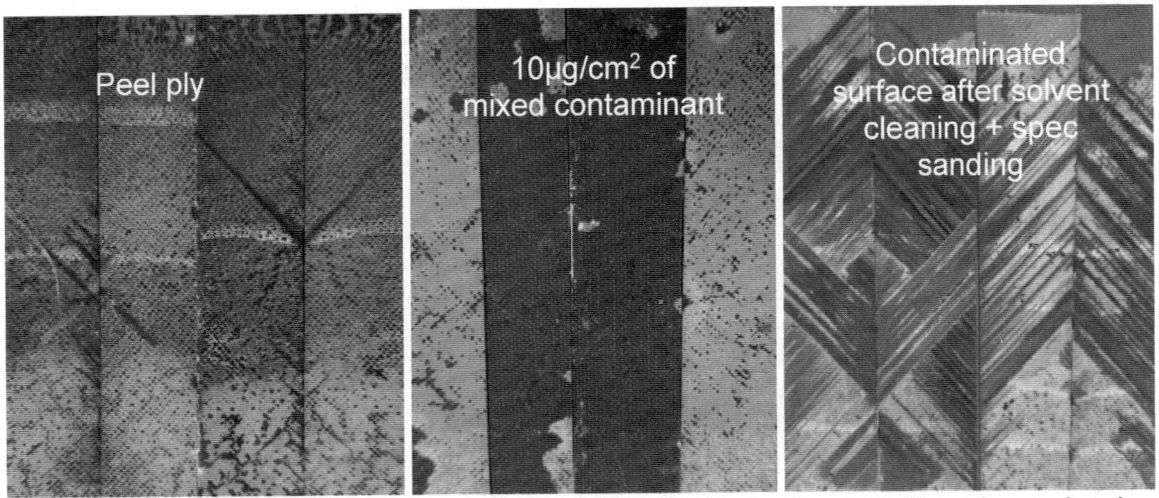

Figure 10. Representative fracture surfaces from samples in Figure 9. The solvent cleaning + abrasion process successfully regenerated a bondable surface.

4. CONCLUSIONS

- All uncontaminated specimens failed essentially 100% cohesively in the adhesive or within the composite
  - Solvent wiping and sanding fresh peel ply surfaces reduced surface energy
  - Sanded peel ply surfaces showed more tendency towards cohesive failure in the first ply
    - Fiber and resin damage from sanding may contribute to lower fracture toughness
- Contamination with ~10µg/cm^2 of either Frekote 700NC or mixtures of Frekote with motor oil resulted in 100% interfacial failures and negligible fracture toughness
  - Contaminant levels on composite surfaces were readily quantified via contact angle measurements
  - Contaminated surfaces failed interfacially with G1c values ~0
- Remediation of silicone/oil contaminated surfaces:
  - Multiple solvent wiping steps were insufficient to remediate a silicone contaminated, abraded surface
  - A combination of solvent cleaning and hand sanding successfully remediated highly contaminated surfaces to restore the original bondability

5. REFERENCES

1. www.compositesworld.com/articles/building-trust-in-bonded-primary-structures
2. Kay Blohowiak, The Boeing Co., personal communication.
3. Upul Palligarayu, NIAR, personal communication.
4. John Webster, Northrop Grumman, personal communication.
5. Steve Twaddle, Brett Bolan, James Mazza, Justin Massey, Ed Harris, personal communications.

6. Johnson Palmer, M., Hanson, B., Clarkson, S., Engelstad, S., Bichon, B., Rousseau, C., Norwood, S., McFarland, J., Popelar, C., Weatherston, M., Oakley, B., Grigoriev, M., MacAdams, L. et al., 2016, "Phase II Report for the TRUST Program," PMF-02129 Defense Technical Information Center www.dtic.mil

PERFORMANCE ENHANCEMENT AND PREDICTION OF COMPOSITE MATERIALS WITH EMBEDDED ELECTROSPUN PIEZOELECTRIC SENSORS

Rahman J. Mazed*[1], Muthu Ram Prabhu Elenchezhian[1], Vamsee Vadlamudi[1], Riaz Ahmed[2], Rassel Raihan[1], Kenneth Reifsnider[1]
[1]Institute for Predictive Performance and Methodologies,
The University of Texas at Arlington Research Institute, Fort Worth, TX
[2]Natural and Applied Sciences,
The University of Wisconsin-Green Bay, Green Bay, WI
*Corresponding Author: rahman.mazed@mavs.uta.edu

ABSTRACT

In today's world, continuous fiber reinforced composite materials are extensively used in the aerospace, automotive and other structural industries. Since the applications of such fibers demand for a high safety rating, it is of utmost importance for engineers who design such materials to analyze its safety. When cracks are developed in such materials, the mechanical, electrical, and thermal properties also get altered as a function of the cracks. Previous studies have shown that changes in electrical properties can be directly correlated with the development of cracks in the material which can then be used to predict the remaining life of such materials. In this study we propose to introduce an additional layer of Electrospun Polyvinylidene Fluoride (PVDF) nanofiber in the CFRP matrix. PVDF is a piezoelectric material, meaning it will generate electricity on encountering vibrations. Therefore, when PVDF nanofibers are introduced in the matrix, the stress on the matrix will be translated onto the nanofibers which in turn will output electrical signals. In this study, 3 types of glass fiber composites (embedded with PVDF, embedded with PVDF with Multiwalled Carbon nanotube, and no additional material) are subjected to quasi static tensile tests while the impedance across the material is continuously measured. The change in impedance was then related to the stress the material was subjected to and predict the remaining life of the material.

Keywords: Composites, Electrospinning, PVDF, Impedance Spectroscopy

1. INTRODUCTION

Structural health monitoring systems for composite materials are receiving growing interest in the industry as the use of these high-performance material is increasing in various fields. The aerospace industry has shown interest towards the study as they require high safety factors, minimum downtimes, and reduced structural inspection costs [1]. Structural health monitoring systems must meet certain requirements: they must not cause damage to the composite material, must be in inaccessible areas, and must transmit useful data to investigate state of damage [1,2].

SAMPE Virtual Conference Proceedings, 2020. Society for the Advancement of Material and Process Engineering – North America.

Some of the techniques that have been studied include thermography, acoustic emission, and fiber optics. [1, 2]. Measuring electrical resistance has also proven to detect very low damage levels in terms of fiber breaks. However, these techniques have posed challenges like incompatibility with composite materials or high-level sophisticated equipment needing to be installed in the system to continuously monitor the structure [2].

An alternative method of monitoring structural health is to embed the structures with nanomaterials that can sense crack propagation. Carbon nanotubes (CNT) is an example of such nanoparticles that can change the electrical properties of composite materials [3]. Addition of nanoparticles could induce a positive effect on correlating the change in the resistivity to damage. Composite materials have complex morphology and are inherently dielectric. When they deform, it results in a change in electrical properties that are not significant, which makes it difficult to correlate damage to just small changes in the electrical properties. However, this limitation can be overcome by addition of nanoparticles which enhance the electrical properties and aid the understanding of damage progression in the material.

Introducing nanoparticles into the fiber matrix could pose a challenge. This is where electrospinning technology could play a crucial role in embedding nanoparticles into the fiber matrix. Electrospinning can produce continuous mesh fibers that have diameters in the scale of nanometers. These fibers are referred to as nanofibers and the whole mesh is referred to as a nanomat. The result of continuously producing the fibers reduces the chance of crack propagation, which lead to extraterritorial strength properties because of minimal defects [4]. Another beneficial property is the volume to surface area ratio, this is useful in energy storage. Other properties that nanofibers generally experience include being lightweight, flexible, and highly porous [5]. There are several industries that could benefit from nanomats, such as the medical field or even the military; some of the applications are filtration (facemask), sensor production, protective clothing, tissue engineering, wound dressing, and energy storage [6]. Nanomats are very versatile and like many composite materials, the overall properties can be tailored by modifying fiber orientation. Continuous fibers are classified into three types: solid, core shell, and hollow shell. Solid fibers are made of one material while core shell has a different material on the inside, which allows for further variation of the properties. On the other hand, hollow shell fibers have a cavity running through the fiber which increases surface area and add additional functions [7].

During electrospinning (Figure 1), the high voltage applied to the pumping fluid causes the fluid to become positively charged. Hence, it is attracted towards the grounded collector due to the electrostatic force of attraction between the two. When the electrostatic force of attraction overcomes the surface tension of the fluid, the fluid shoots towards the collector in a whipping motion. The fluid usually consists of a polymer and a volatile solvent; upon contact with the collector, the solvent dries leaving behind the polymer.

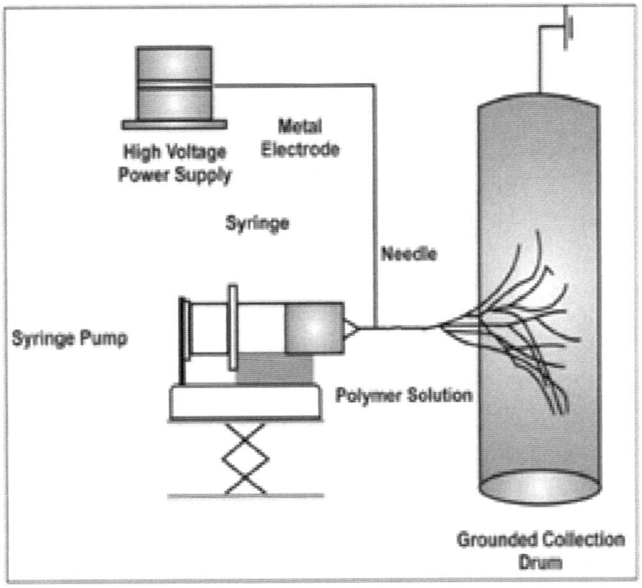

Figure 1: Schematic setup of electrospinning process depicting the polymer collection process triggered by high voltage difference between the syringe and the collector [8]

Polyvinylidene Fluoride (PVDF) is a polymeric material that can be electrospun. It is an important polymeric material that has five crystalline polymorphs: α, β, γ, δ, and ε. The β phase shows excellent piezo, pyro, and ferroelectric properties that can be used in energy storage, spin valve devices, biomedicine, sensors, and smart scaffolds [9]. Electrospinning techniques can be used directly to produce β phase in PVDF by controlling parameters such as temperature, solvent, feeding rate, and distance between tip and collector [10,11]. The technique is very efficient and can be easily scaled. Addition of CNT into the nanofiber has also shown an improvement of β phase in PVDF nanofiber and improved the electrical properties of these fibers. [12]

In this study, all the key elements discussed in the literature have been brought together to make a single case of structural health monitoring of glass fiber composites. Glass fiber composites have been co-cured with electrospun PVDF nanofiber with multiwalled carbon nanotube (MWCNT) and it is subjected to tensile load until failure. Meanwhile, the impedance across the thickness of the laminate is measured by impedance spectroscopy.

2. EXPERIMENTATION

2.1 Material

The nanofiber mats were made from Polyvinylidene fluoride (PVDF) powder (<=100% w/w) purchased from Alfa Aesar, Multiwalled carbon nanotubes (MWCNT) purchased from Sigma Aldrich, and Dimethylformamide (DMF) (>=99% v/v) purchased from Sigma Aldrich. The glass fiber laminate was made from Unidirectional E-Glass fiber with 250 F Epoxy Resin. The tab materials were made using CYCOM 2265-7781 Glass fiber.

2.2 Solution Preparation

For PVDF solution, 20 mL of DMF was mixed with 2g of PVDF in a conical flask. Then the solution was heated at 80 °C while being stirred at 350 rpm for 8 hours. For PVDF with MWCNT solution, 40 mL of DMF was mixed with 0.188g of MWCNT in a conical flask. The solution was then sonified for 5 minutes. After sonification, 3.9 g of PVDF was added to the solution and the solution was heated at 80 °C while being stirred at 350 rpm for 8 hours.

2.3 Electrospinning

2 types of nanomats were produced from the solutions: PVDF and PVDF with Multiwalled Carbon nanotube. For both the solutions the electrospinning process was identical. A 5 mL BD syringe was first filled with the solution. A syringe pump was used to pump the fluid at 1.25 mL/hour. A total volume of 2.5 mL of the solution was dispensed from the syringe. The voltage applied at the tip of the needle of the syringe was set to 20 kV. Aluminum foil was placed over the rotating mandrel to act as the collector. The rotating collector was set to revolve at 3180 rpm. The distance between the tip of the needle and the collector was kept at 0.178 m. Figure 2 shows the electrospinning setup.

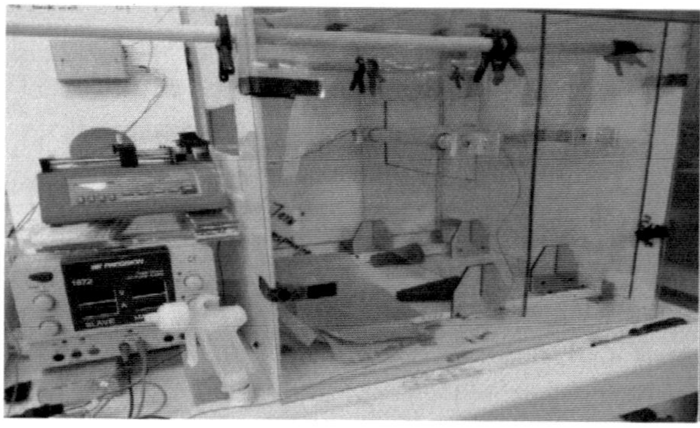

Figure 2: Electrospinning Setup

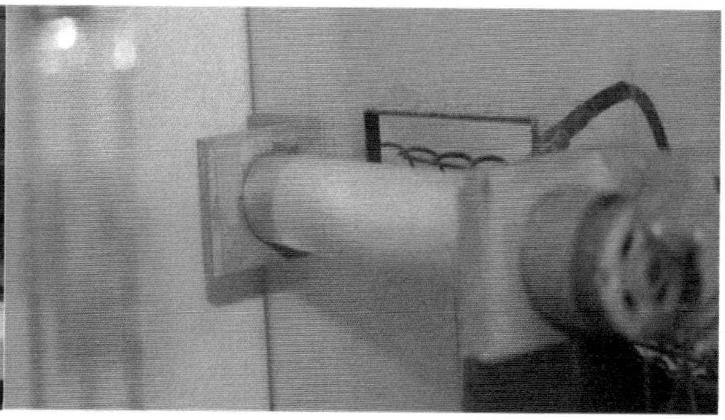

Figure 3: Nanomat Collected on Rotating Mandrel

Figure 4: Nanofiber mat on Aluminum foil

2.4 Lay-up and Molding

The glass fiber was created by cutting 4 plies of Unidirectional E-Glass fiber all set at 0°. The laminae were cut 0.305m×0.305m. The nanofibers were sandwiched between the 4 plies. Figure 5 shows the stacking sequence of the matrix. The dimension of the nanofiber mat was 0.013m×0.064m. The curing cycle for the sample has been shown in Figure 7.

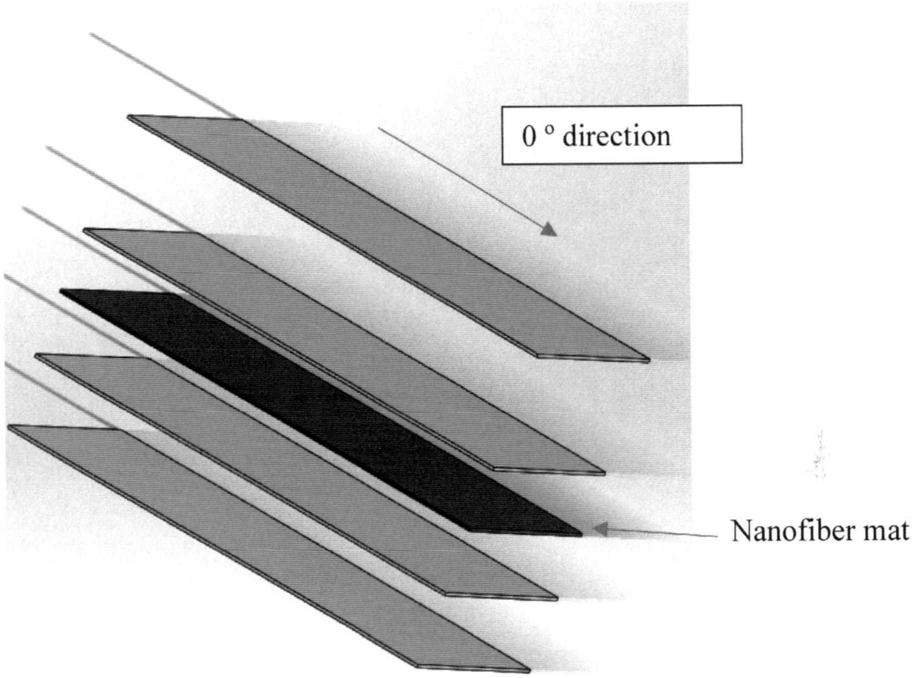

Figure 5: Stacking Sequence of the laminate with nanofiber mat in the mid-plane of the laminate

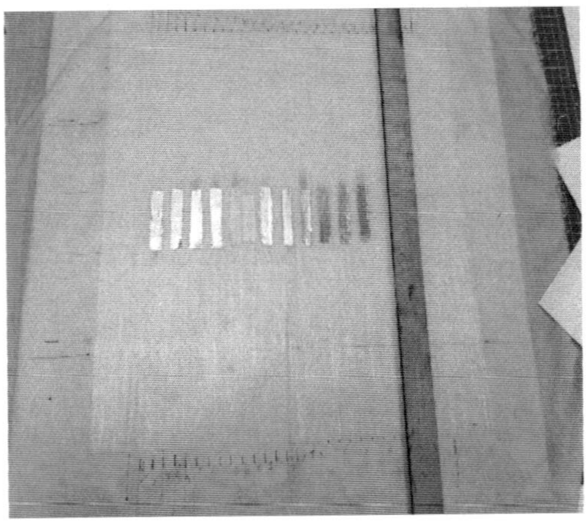

Figure 6: Nanofiber Between Glass Fiber Plies Pre-Curing

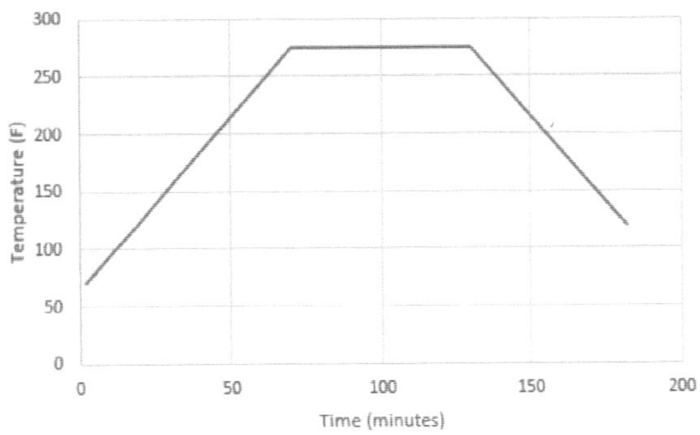

Figure 7: Curing Cycle of the glass fiber prepreg

The aluminum foil that can be seen in Figure 6 was peeled off, leaving behind the nanofibers. The remaining 2 plies were then placed on top. After the lay-up, the laminate was cured using compression molding technique recommended by the manufacturer [13].

After molding, samples were cut into coupons of 0.254m×0.013m as shown in Figure 8. The ends of the coupons were attached to tabs using 0.051m×0.013m tabs made from 8 plies of woven glass fiber. The tab material was manufactured using compression molding technique recommended by the manufacturer [14].

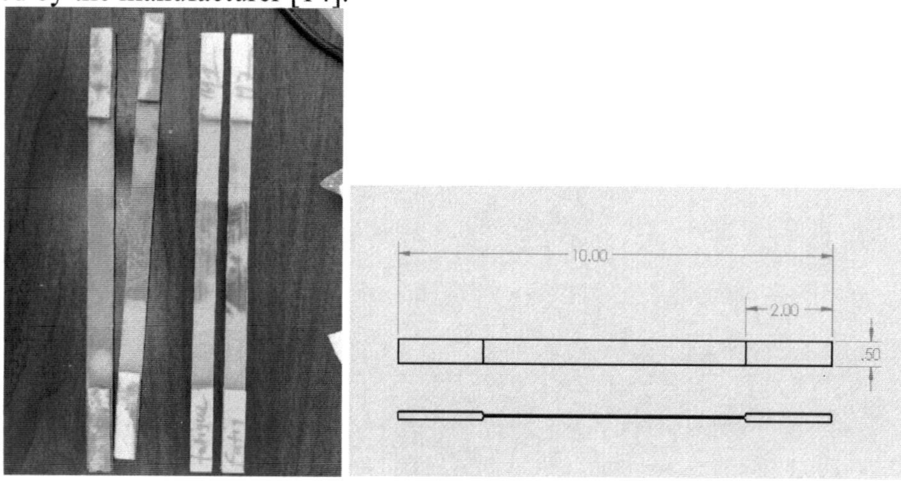

Figure 8: Coupon Sample

2.5 Quasi Static Tensile Test with Impedance Measurement

A total of twelve samples were tested for this experiment, four samples of each kind: virgin samples (without any sensor), PVDF samples, and PVDF with MWCNT samples. The samples were tested using MTS Landmark Servo Hydraulic Test Systems. Quasi Static Tensile Tests were performed on the samples under displacement-controlled configuration. The rate of displacement was set at 0.011 inches/minute. The impedance of the samples was measured using GAMRY Interface 1000A. A Potentiostat test was carried out with the frequency set at 100 Hz and the DC voltage set at 1V rms. The impedance was measured by placing an electrode across the thickness of the sample while the quasi static tensile test was ongoing. Figure 9 shows the in-situ tensile testing setup.

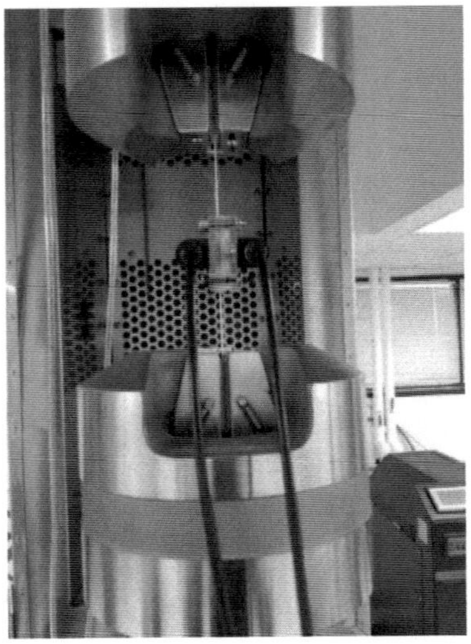

Figure 9: Sample with Electrode

3. RESULTS

The following subsections show the results obtained for each type of sample. The data close to failure was truncated because of the instability in the material behavior.

Virgin Samples

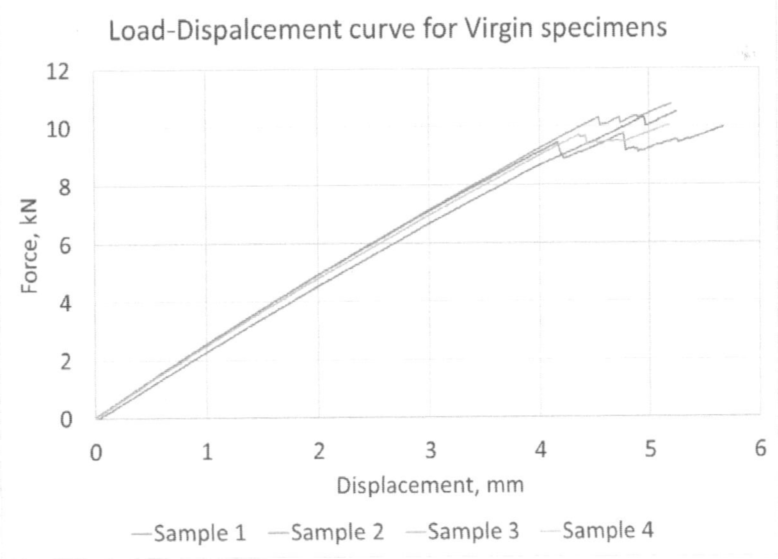

Figure 10: Load displacement curves for virgin samples

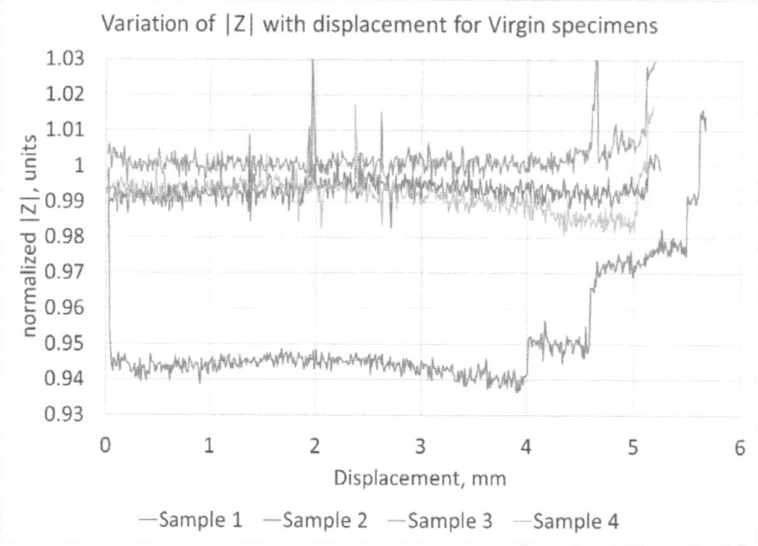

Figure 11: normalized modulus of impedance vs displacement for virgin specimens

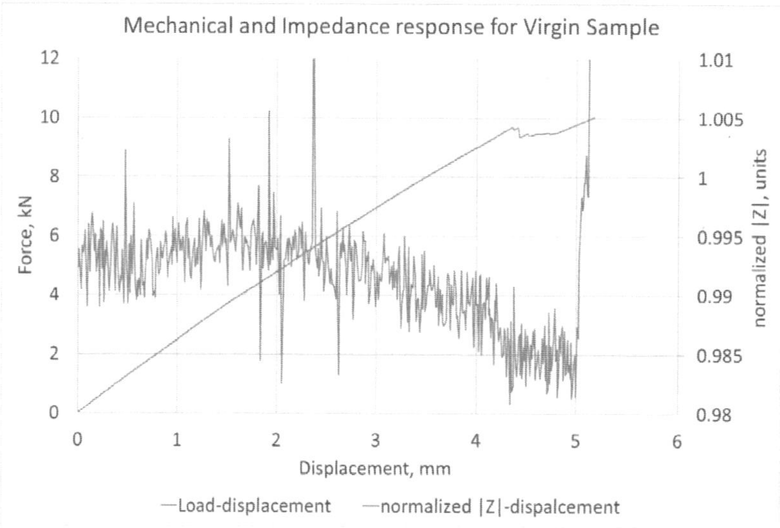

Figure 12: Mechanical and impedance response for a virgin sample

The impedance for each sample is normalized w.r.t. to impedance before initiation of test (t=0) such that the impedance for each specimen starts at 1. Figure 10 shows the load displacement curves for the virgin samples. The average breaking load for this set of specimens was 10.33 kN with a standard deviation of 0.383 kN.

Figure 12 depicts the mechanical and impedance response for one specimen in the set. It can be observed that impedance increases initially, followed by a decrease and sudden jumps close to failure when the fiber splitting starts. This trend is contrary to what was observed in multi directional laminates in [14] where in the impedance initially decreased followed by an increase when delamination initiated. This could be because of the longitudinal matrix cracks that develop in the laminate that lead to debonding (matrix fiber interface) which hinders the charge flow and hence leads to increase in impedance. When the splitting starts, a fracture path is created or a path of least resistance that leads to decrease in impedance as observed in Figure 12. However, detailed microscopic study needs to be performed to better understand the damage development and correlate with variation in impedance.

Figure 11 shows the impedance response for the set of virgin specimens where in it can be observed that all the specimens except sample 3 have a same trend in the impedance change.

Samples with PVDF nanofiber mat

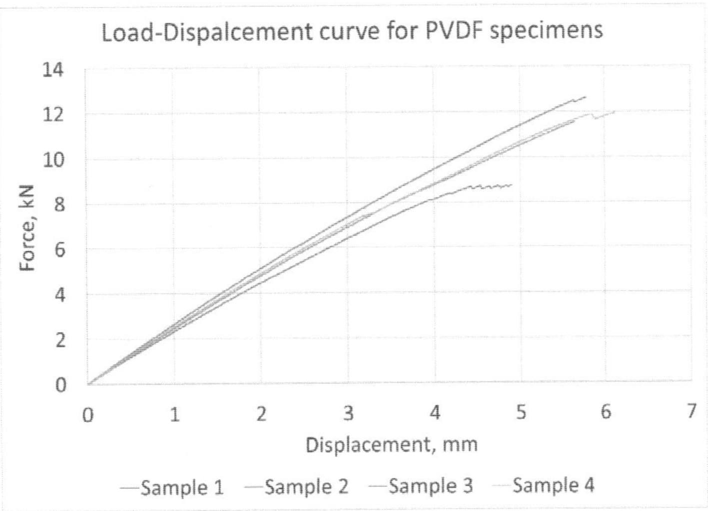

Figure 13: Load displacement curves for PVDF samples

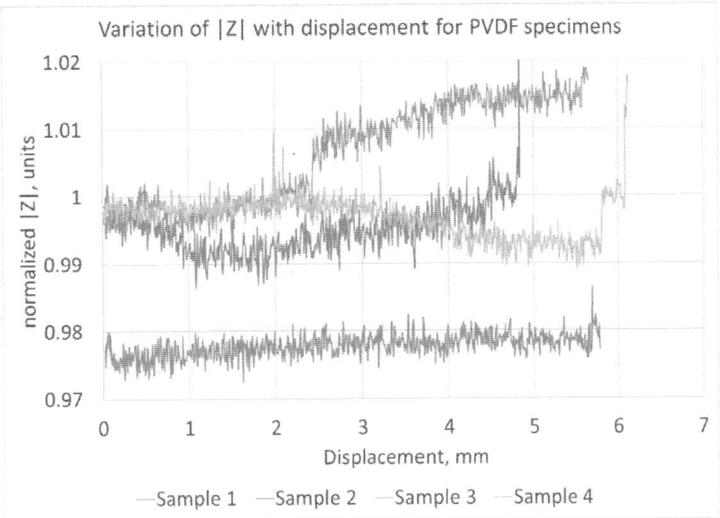

Figure 14: normalized modulus of impedance vs displacement for PVDF specimens

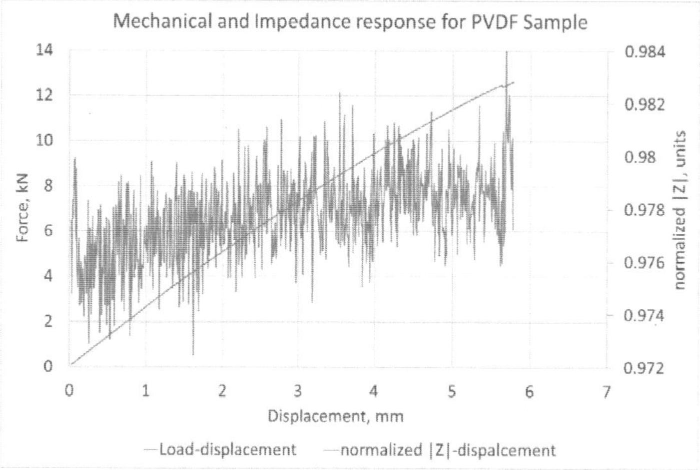

Figure 15: Mechanical and impedance response for a PVDF sample

Figure 13 shows the load displacement curves for the PVDF samples. The average breaking load for this set of specimens was 11.23 kN with a standard deviation of 1.706 kN.

Figure 15 depicts the mechanical and impedance response for one specimen in the set. It can be observed that impedance increases initially, followed by a decrease and sudden jumps close to failure when the fiber splitting starts. Figure 14 shows the impedance response for the PVDF sample set. In this case, it can be observed that the trend is random, i.e. two specimens have the trend in Figure 15, and the remaining two specimens have a contrary trend i.e. an initial decrease followed by an increase. We believe this trend represents the behavior as delamination comes into play (unlike virgin specimens) because of the presence of the nano fiber mat and hence leads to increase in impedance in the end stages as observed in [14]. However, further evidence is required to substantiate these findings and more number of specimens are to be tested to determine the trend followed by the material system.

Samples with PVDF embedded with MWCNT nanofiber mat

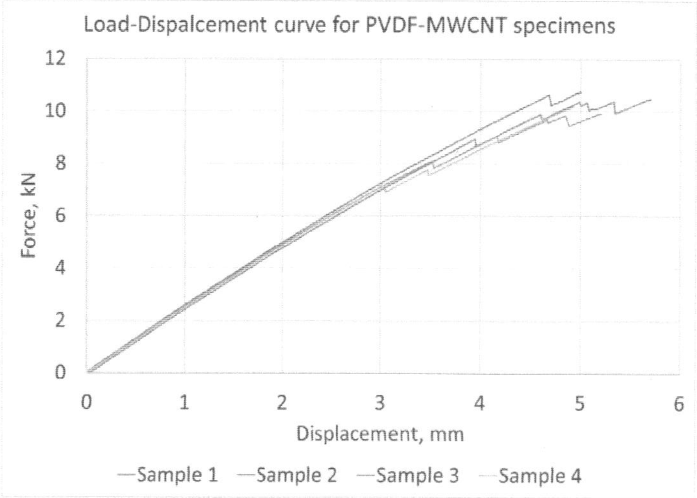

Figure 16: Load displacement curves for PVDF-MWCNT samples

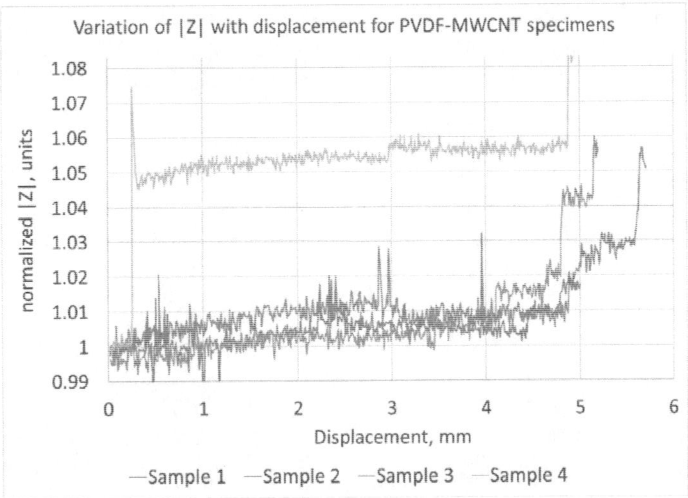

Figure 17: normalized modulus of impedance vs displacement for PVDF-MWCNT specimens

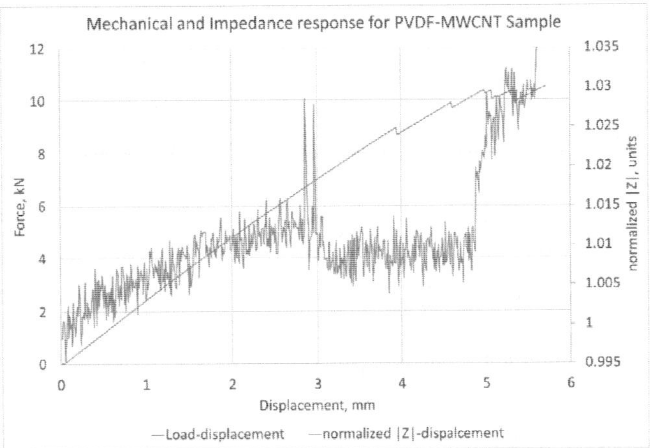

Figure 18: Mechanical and impedance response for a PVDF-MWCNT sample

Figure 16 shows the load displacement curves for the PVDF samples. The average breaking load for this set of specimens was 10.35 kN with a standard deviation of 0.374 kN.
Figure 18 depicts the mechanical and impedance response for one specimen in the set. It can be observed that impedance increases initially, followed by a decrease and sudden jumps close to failure when the fiber splitting starts. Figure 17 shows the impedance response for the PVDF-MWCNT sample set. In this case as well similar to virgin specimens, three specimens have the same trend whereas sample 4 has a continuously increasing trend. However, further studies need to be performed to study the damage development to understand the impedance change.

Figure 19 shows how the individual strands separate once the material starts failing. After the first strand broke, the load applied by the machine and the magnitude of impedance met with a drastic drop for all the samples. When failure started, each individual strand started breaking separately which shows as a load drop and a subsequent increase in impedance.

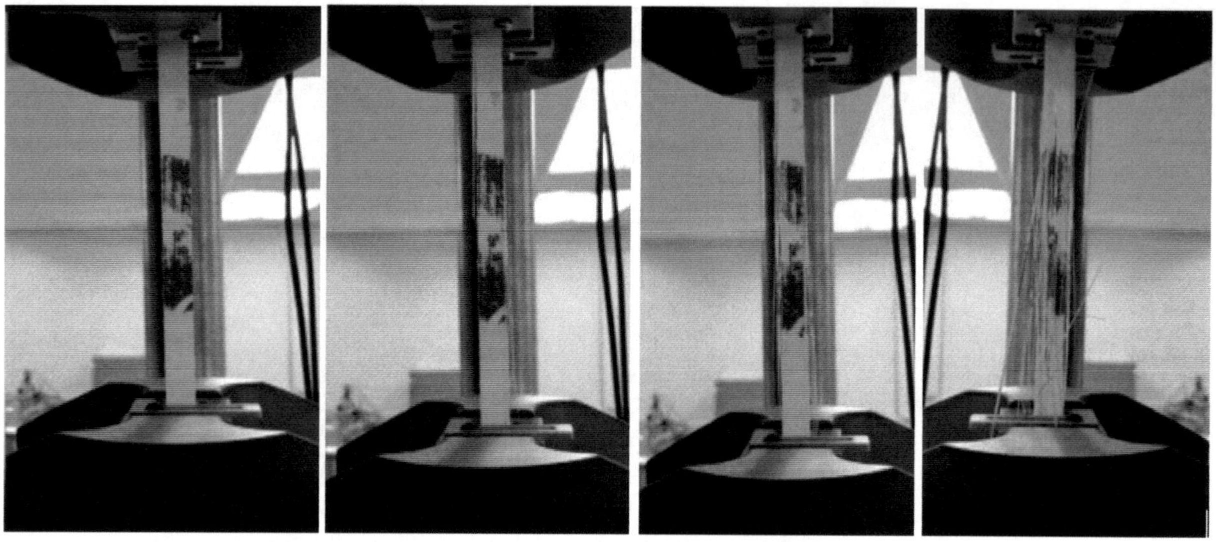

Figure 19: Stages of Failure for Specimen

Figure 20 shows the Breaking Load of the 3 types of samples. It can be observed that with the addition of a nanofiber mat into the fiber matrix, the ultimate tensile stress of the specimen increased. The virgin sample had the lowest breaking load of 10.33 kN with a standard deviation of 0.383 kN, while the samples with PVDF nanomat had the highest breaking load of 11.23 kN with a standard deviation of 1.706 kN. From Figure 20, it can be observed that, the inclusion of PVDF with MWCNT didn't alter the breaking load. Further studies need to be performed to better understand the material behavior and to substantiate these findings and will be reported in subsequent publications.

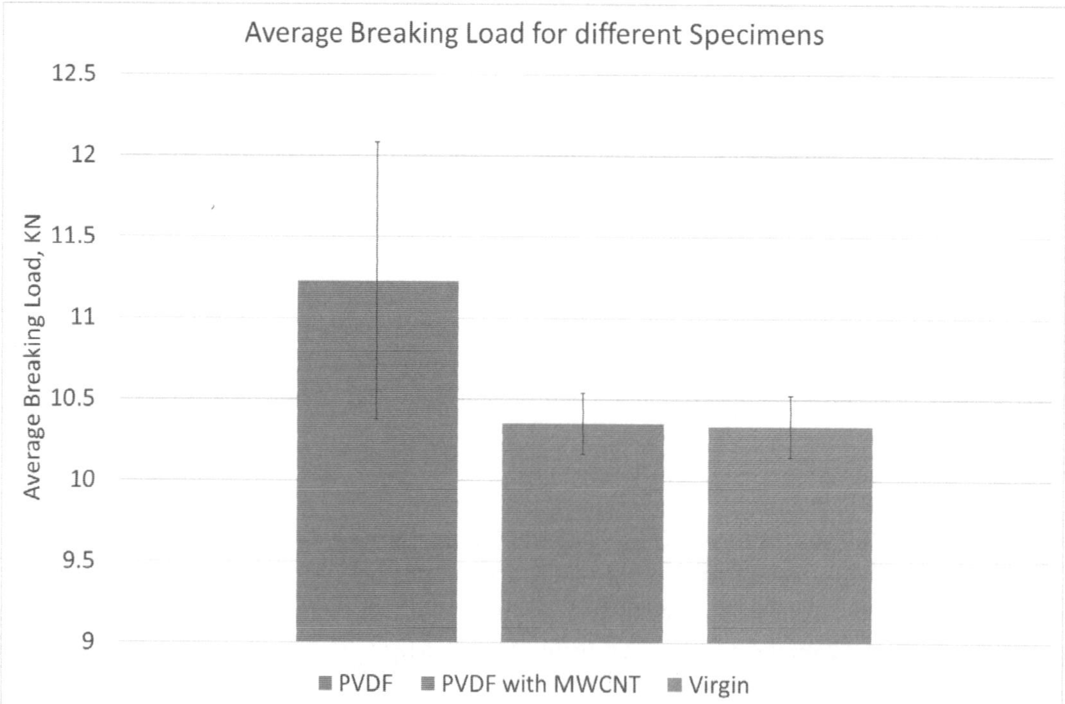

Figure 20: Average breaking load for different specimens

4. CONCLUSIONS

From the experiments conducted during this study, variation of impedance with damage development has been reported. It can be observed that for virgin specimens, an initial increase in impedance followed by a decrease when first strand splitting occurred and an increase when laminate failure started to occur. These observations were contrary to observed variation in literature and hence need to perform detailed microscopic evaluation to understand damage development.

In case of specimens with PVDF, the breaking load increased significantly but the trend in the impedance variation was random and hence inconclusive of how impedance varies with damage development and need to test more samples to better understand material behavior. For specimens with PVDF-MWCNT, the breaking load was similar to virgin specimens and so was the impedance change. To better understand these observations, further study need to be performed and will be reported in subsequent publications.

5. REFERENCE

1. Abry, J. "In Situ Detection of Damage in CFRP Laminates by Electrical Resistance Measurements." *Composites Science and Technology*, vol. 59, no. 6, 1999, pp. 925–935., doi:10.1016/s0266-3538(98)00132-8.
2. Matsuzaki, Ryosuke, et al. "Antenna/Sensor Multifunctional Composites for the Wireless Detection of Damage." *Composites Science and Technology*, vol. 69, no. 15-16, 2009, pp. 2507–2513., doi:10.1016/j.compscitech.2009.07.002.
3. Ferreira, André Duarte B.l., et al. "Multifunctional Material Systems: A State-of-the-Art Review." *Composite Structures*, vol. 151, 2016, pp. 3–35., doi:10.1016/j.compstruct.2016.01.028.
4. Ulubayram, Kezban, et al. "Nanofibers Based Antibacterial Drug Design, Delivery and Applications." *Current Pharmaceutical Design*, vol. 21, no. 15, Oct. 2015, pp. 1930–1943., doi:10.2174/1381612821666150302151804.
5. Li, D., and Y. Xia. "Electrospinning of Nanofibers: Reinventing the Wheel?" *Advanced Materials*, vol. 16, no. 14, 2004, pp. 1151–1170., doi:10.1002/adma.200400719.
6. Murugan, Ramalingam, and Seeram Ramakrishna. "Design Strategies of Tissue Engineering Scaffolds with Controlled Fiber Orientation." *Tissue Engineering*, vol. 13, no. 8, 2007, pp. 1845–1866., doi:10.1089/ten.2006.0078.
7. Li, Dan, et al. "Electrospinning of Polymeric and Ceramic Nanofibers as Uniaxially Aligned Arrays." *Nano Letters*, vol. 3, no. 8, 2003, pp. 1167–1171., doi:10.1021/nl0344256.

8. Zafar, Muhammad, et al. "Potential of Electrospun Nanofibers for Biomedical and Dental Applications." *Materials*, vol. 9, no. 2, 2016, p. 73., doi:10.3390/ma9020073.

9. Ruan, Liuxia et al. "Properties and Applications of the β Phase Poly(vinylidene fluoride)." Polymers vol. 10,3 228. 26 Feb. 2018, doi:10.3390/polym10030228Yee, Wu Aik, et al. "Morphology, Polymorphism Behavior and Molecular Orientation of Electrospun Poly(Vinylidene Fluoride) Fibers." *Polymer*, vol. 48, no. 2, 2007, pp. 512–521., doi:10.1016/j.polymer.2006.11.036.

10. Choi, S.w., et al. "An Electrospun Poly(Vinylidene Fluoride) Nanofibrous Membrane and Its Battery Applications." *Advanced Materials*, vol. 15, no. 23, Mar. 2003, pp. 2027–2032., doi:10.1002/adma.200304617.

11. Liu, Zong-Hsin, et al. "Piezoelectricity of Well-Aligned Electrospun Fiber Composites." *IEEE Sensors Journal*, vol. 13, no. 10, 2013, pp. 4098–4103., doi:10.1109/jsen.2013.2278739.

12. "Prepreg - Fiberglass (E-Glass) - Unidirectional (470 OAW) - 250F Resin - 40' Wide x 0.011' Thick." *Rock West Composites*, www.rockwestcomposites.com/14042-d-group.

13. Boeing. "CYCOM 2265-7781 CYTEC ENGINEERED MATERIALS INC GLASS PHENOLIC PREPREG: Buy Now at Boeing Distribution." *CYCOM 2265-7781 CYTEC ENGINEERED MATERIALS INC GLASS PHENOLIC PREPREG | Buy Now at Boeing Distribution*, www.boeingdistribution.com/product/GLASS PHENOLIC PREPREG/CYCOM 2265-7781/CYC0M2265-7781-60IN.

14. Vadlamudi, Vamsee. "Assessment of Material State in Composites Using Global Dielectric State Variable." PhD diss., 2019. Retrieved from:

https://rc.library.uta.edu/uta-ir/handle/10106/28880

STRUCTURAL HEALTH MONITORING OF COMPOSITES FROM CARBON NANOTUBE (CNT) COATED E-GLASS FIBER

Sidney Wong, Omar Dwidar, Sergio R. Rodriguez Herrera,
Eltahry Elghandour, Ph.D., Amro El Badawy, Ph.D.
California Polytechnic State University, San Luis Obispo
San Luis Obispo, California

ABSTRACT

This research investigated the development of a nanomaterial-based sensor for health monitoring of composite structures. To develop the sensor, carbon nanotube/epoxy mixture (2%wt CNT) was coated on a strand of E-glass fiber to be adhered onto a fiberglass composite specimen. The selection of E-glass fiber and fiberglass plate was largely due to its electrical insulating properties to demonstrate that the carbon nanotube is driving the sensing capabilities through its highly conductive nature. In addition, by adhering the coated E-glass fiber to a fiberglass coupon, the homogeneity and material properties were approximately maintained. Tensile testing of the specimen conducted through a Lloyd LD50 tensile testing machine provided data on the actual strain which was correlated with the experimental differential resistances measured by a multimeter, both at the same specified tensile loading conditions. With two sets of data, the experimental resistance data was calibrated with the actual strain data collected. Ultimately, the experimental sensors created a sample of gauge factors which represents 91.24% probability of replicating the observed range of gauge factors by using the same manufacturing procedures, providing a valid alternative and consistent method to detecting composite damage.

Keywords: Mechanical Engineering, Aerospace Engineering, Structural Engineering, Composites, Strain Gauge, Carbon Nanotubes (CNT), Nanoparticles, Nanotechnology, Epoxy, Tensile Testing, E-Glass Fiber, Health Monitoring System, Mechanics of Materials, Manufacturing

1. INTRODUCTION

The rising popularity of composite materials has influenced the aerospace, automotive, and buildings industries to creating structures with such advanced materials. Although manufacturers are substituting traditional materials with modern composite materials for benefits in high strength to weight ratios and non-corrosive properties, most companies are currently using a dated health monitoring system that consists of utilizing transducers to monitor transmitted waves generated by ultrasonics. This method is time intensive and can be improved with accelerated methods by using an advanced system such as nanotechnology-based sensors to refrain from bonding strain gauges throughout the structure. By exploring methods to detect damage, rather than diagnose damage, in composite structures, critical failure can be efficiently prevented, hence a reduction in time and costs will be achieved while improving the safety of users, occupants, and nearby civilians.

The objectives of this research entail investigations in material selections, manufacturing process, and testing of the sensor. Calibration of the manufactured sensors was performed through comparisons between the experimental sensor data to that from a traditional extensometer adhered to the same sample test subject. The results were verified for accuracy, validity, and the feasibility based on the practicality of the manufacturing process was discussed.

2. BACKGROUND

Recent studies have explored the development of a structural health monitoring system to predict strain, damage, and residual strength of a composite by mixing 2% by weight reduced graphene oxide (rGO) into Araldite LY 556 epoxy and HY 951 hardener with a sonication method and rolling the mixture onto E-glass fibers. By coating the strand of E-glass fiber with conductive rGO, the piezoresistance relating the strain to the force applied was observed when a tensile test was performed [1]. A similar experiment has also been performed with the same methods but instead, using carbon nanotubes (CNT) instead of rGO. Investigators grew CNTs aligned on alumina fiber cloth using a chemical vapor deposition method. The cloth was made into a laminate using West Systems Epoxy Resin 105 and Hardener 206. This sensor was then subjected to an impact test at which a steel ball was dropped at a set height [2]. This type of sensor was also subjected to monatomic tests, cyclic fatigue tests, and load controlled incremental loading/unloading tensile test with use of a tensile testing machine which were compared to a controlled strain gauge data set [3]. Another method of creating such sensor used a Bucky paper film that was rolled onto the surface of the test sample to monitor delamination through a cyclic fatigue test, where a significant difference in electrical resistance between loaded and unloaded conditions was observed as a result of the crack separating and closing [4]. Typically, tensile testing was performed during the initial phases of the study. However, once the tensile tests have proven to be successful, three-point bend tests and vibrational tests have been examined. Once all the tests have been perfected on a coated single strand of fiber, a gridded weave layout would be prototyped as a next step to be able to locate impact [5]. In most tests conducted on the experimental sensors, terminals at each end of the sensors were probed with a multimeter to read the resistance when subjected to a load. Both two- and four-point probe electrical resistance measurements were made possible, but typically a two-point probe was sufficient as the resistance readings noticed were large, in the kilo-ohms range. A four-point probe was only used if small resistances were read and high sensitivity of the multimeter was needed [2][3][6].

The theory behind coating a strand of fiber comes from the way a strain gauge works [7]. A strain gauge simply functions as a wire, where when stretched, or tensioned, the resistance increases, as shown in Equation 1:

$$R = \frac{\rho L}{A} \tag{1}$$

where R is the resistance, ρ is the resistivity, L is the length, and A is the cross-sectional area of the wire.

Creating a sensor that reads resistances from a multimeter, the electrical strain (ε_E) could then be determined by Equation 2 and compared to the mechanical strain (ε_M), from Equation 3, or collected through the data acquisition machine attached to the tensile testing machine. With the electrical strain plotted against the mechanical strain, the gauge factor (GF) could be determined from Equation 4 by taking the slope of the linear plot.

$$\varepsilon_E = \frac{\Delta R}{R} = \frac{R - R_o}{R_o} \tag{2}$$

$$\varepsilon_M = \frac{\Delta L}{L_o} = \frac{L - L_o}{L_o} \tag{3}$$

$$GF = \frac{\varepsilon_E}{\varepsilon_M} \tag{4}$$

The gauge factor is the ratio between the electrical strain to mechanical strain. This ratio represents the sensitivity of the sensor. In addition, a stress-strain curve, generated from data collected from a strain gauge attached on the opposite side of the specimen was plotted on the same figure as the electrical-mechanical strain curve to provide a controlled data set [1][3][7]. By plotting two data sets on a single plot allowed for confirmation that the stress-strain curve and the electric-mechanical strain curve are both linear, as testing was conducted in the elastic region of the test coupon.

3. METHODS

3.1 Materials Selection

The materials selected, based on availability and short lead time, consisted of carbon nanotubes (CNT) as the sensing medium due to its high conductivity, E-Glass Fiber as the base strand coated due to its good electrical insulation properties, as well as its high strength and stiffness and relatively low cost when compared to S-Glass Fibers. West Systems 105 Epoxy Resin and 206 Slow Hardener were chosen for its non-conductive properties, as well as its long cure time of 10-15 hours, providing ample time for coating the fibers. Properties of the multi-walled carbon nanotubes are tabulated in Table 1 [8]. Silver paint was dedicated for creating the two terminals at the end of each sensor with copper foil wrapping around the ends for the probes to attached to when measuring resistance. The NP130HF Thermoset Fiberglass Composite Sheet, donated by Norplex Micarta, was designated as the test sample. Properties of the Norplex Micarta NP130HF is tabulated in Table 2 [9]. An extra benefit of using this Fiberglass composite sheet as the test material is that it maintained the material properties of the strand of E-Glass Fiber selected.

Table 1: Physical Properties of Multi-Walled Carbon Nanotubes [8]

MWNTs -COOH Functionalized 95% Properties	
Property	Specification
Purity	>95%
Content of -COOH	2.0 wt%
Outer Diameter	10-20 nm
Inner Diameter	3-5 nm
Length	5-30 nm
Electrical Conductivity	> 100 s/cm
Bulk Density	0.27 g/cu.cm
True Density	2.1 g/cu.cm
Manufacturing Method	Catalytic CVD

Table 2: Norplex-Micarta NP130HF Fibreglass Coupon Material Properties [9]

Norplex-Micarta NP130HF Material Properties	
Thickness	0.159 cm
Length	22.9 cm
Width	2.54 cm
Elastic Modulus	25.5 MPa

3.2 Manufacturing Processes

The experimental setup began with the synthesis of carbon nanotubes into epoxy resin-hardener. Five parts of West Systems 105 Epoxy Resin and one-part West Systems 206 Slow Hardener by weight was mixed with 2% of carbon nanotubes by total weight. This mixture was then coated onto a strand of E-glass fiber by dipping the stand of fiber into the mixture and placed between vacuum bags. A squeegee method was used to create an even coat of conductive mixture onto the E-glass fiber. Once fully coated, the tacky strands of coated fiber were then dragged along powdered carbon nanotubes to adhere extra nanoparticles on the surface of the sensor. This sensor was then compressed between two flat plates and left for 15 hours to cure. Once cured, the ends of the sensors were painted with silver paint to create the terminals and given time to dry. An initial test to measure resistance by probing both ends with a multimeter proved the conductivity of the sensor. The magnitude of the initial resistance measured were in the order of mega-ohms.

Next the test coupon was trimmed to maximize the 0.159 cm thick, 45.7 cm x 45.7 cm sheet of NP130HF Thermoset Fiberglass Sheet from Norplex Micarta. 2.54 cm x 22.9 cm specimens were cut to make the most out of the resources obtained. One sensor was centered and adhered to each coupon with West Systems 105 Epoxy Resin and 206 Hardener under compression and allowed for curing. Once cured a tab of copper foil was wrapped around the coupon and sensor at the areas where silver paint was applied to allow for the alligator clips from the multimeter to reach the terminals, as shown in Figure 1. The dimensions of the samples created are tabulated in Table 3. It should be noted that Specimen 0 was the first working sensor created during the early manufacturing processes while Specimens 1 through Specimen 3 were created in attempts to replicating the results from Specimen 0.

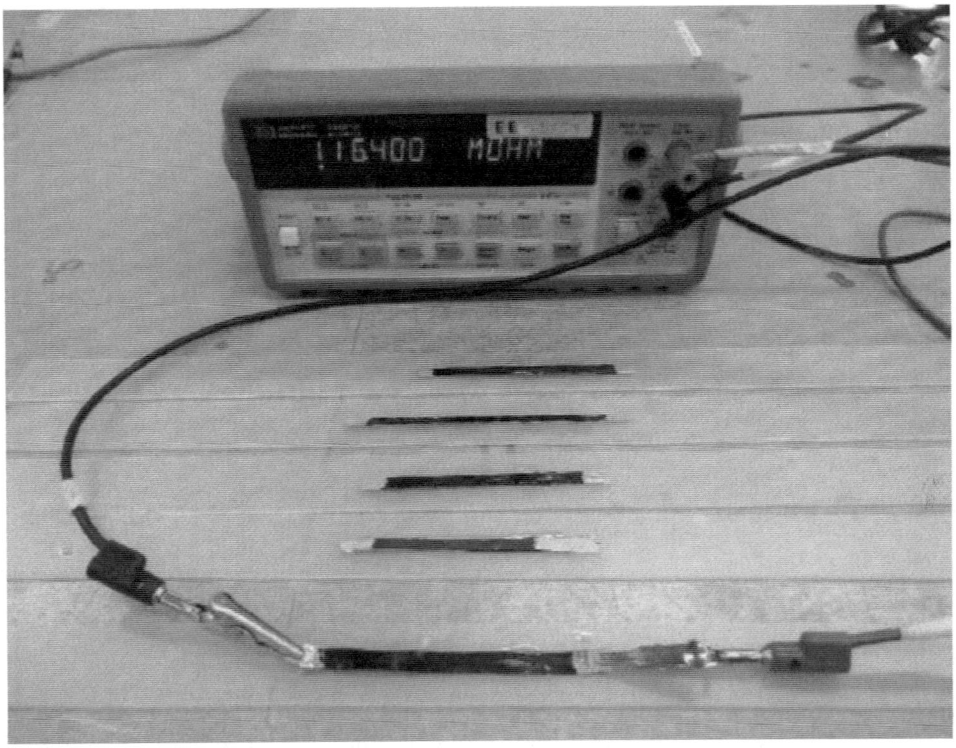

Figure 1: Confirmation of Conductivity in Sensors

Table 3: Specific and Sensor Dimensions

Sample Dimensions (Units in Inches)						
	Coupon			Sensor		
Specimen	Width (cm)	Length (cm)	Thick (cm)	Width (cm)	Length (cm)	Thick (cm)
0	2.60	22.9	.159	0.696	10.21	0.0829
1	2.52			0.668	9.31	0.0737
2	2.58			0.513	9.44	0.0812
3	2.48			0.536	9.65	0.0787

3.3 Testing Procedures and Data Collection

Tensile testing was determined as the first step of testing these sensors, but first a blank specimen without any CNT sensors was tested to 5000 newtons to ensure that the material was still within the elastic range. Since an extensometer was made available, a strain gauge was not needed for data collection. Being within the range of elasticity was confirmed by plotting the stress-strain curve and verifying the linearity of the relationship. Therefore, a maximum load of

5000 newtons was applied on the coupon during tensile testing. Although majority of past experiments used a strain gauge to collect a controlled data set, an extensometer was primarily used to provide the control needed to compare to the experimental CNT sensor created. To collect data from the CNT sensor, the multimeter logged resistances between the two terminals. Each specimen was tested for five trials.

Figure 2: Testing Apparatus

4. RESULTS AND DISCUSSION

4.1 Data Analysis

Once tensile testing was performed, the stress-strain curve of each trial was generated. Before plotting the electrical-mechanical strain curve from the CNT sensor, the electrical strain must be computed using Equation 2 from the resistance data logged. The two curves plotted for each specimen are shown below in Figures 3 through Figure 6. In addition, a correction factor was incorporated into the third data set plotted to correct the electrical strain curve to match the stress-strain curve. The slope of the stress-strain curve is known as the Modulus of Elasticity, while the slope of the electrical-mechanical strain curve is the Gauge Factor. Both, which were determined are tabulated in Table 4 with its mean and standard deviation for each specimen and for all samples and trials. It was noticed that variability of the gauge factor determined for Specimen 0 had the highest variability, as it was an initial experiment. However, the variability was minimized once the manufacturing procedure was solidified. Figure 7 depicts a plot with all the average electrical-mechanical strain curves, centered at the origin for each specimen, with the slope representing the gauge factor.

Table 4: Gauge Factor Determined by

Specimen	Trial	Extensometer (Controlled) Modulus of Elasticity (MPa)	CNT Sensor Gauge Factor
0	1	29483	6.81
	2	27066	2.46
	3	28822	3.43
	4	30291	4.36
	5	30056	4.24
	Mean	29143.6	4.26
	Std. Dev.	1292.7	1.62
1	1	34128	3.64
	2	35303	3.22
	3	30821	3.32
	4	27610	2.81
	5	31946	3.21
	Mean	31961.6	3.24
	Std. Dev.	3004.0	0.30
2	1	31564	3.83
	2	30846	3.34
	3	30759	3.35
	4	30656	3.36
	5	30680	3.29
	Mean	30901	3.43
	Std. Dev.	378.0	0.22
3	1	32164	4.08
	2	32900	3.55
	3	32709	3.54
	4	32723	3.39
	5	32785	3.53
	Mean	32656.2	3.62
	Std. Dev.	285.3	0.27
Total Mean		31165.6	3.64
Total Std. Dev.		2035.8	0.87

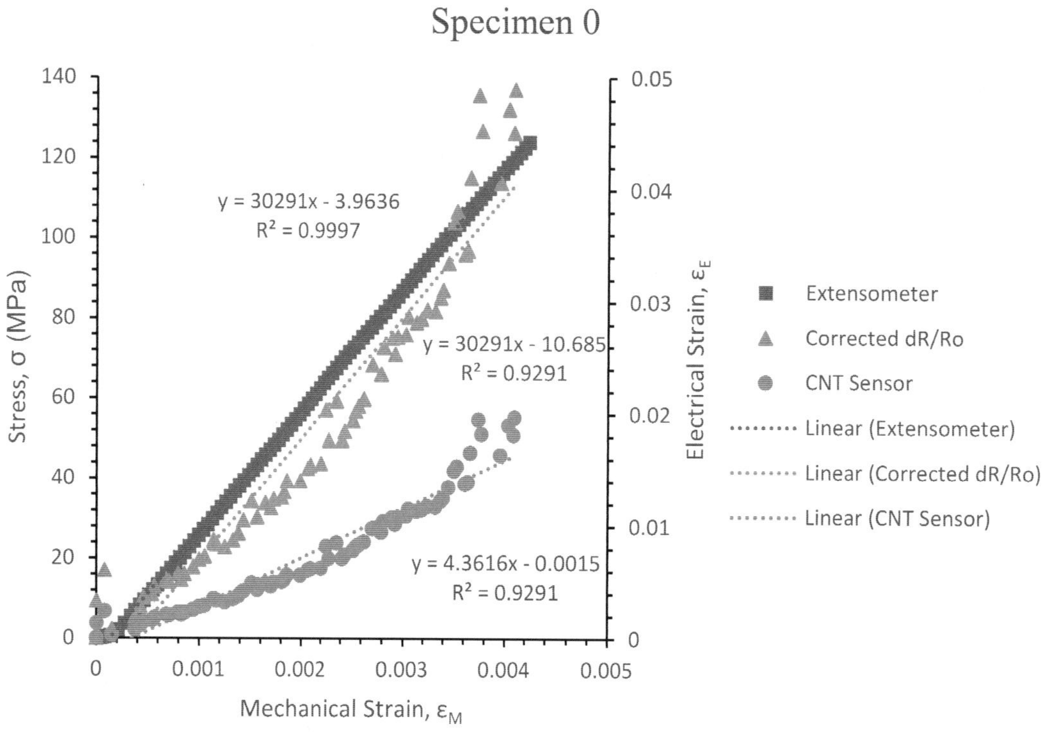

Figure 3: Specimen 0 Data Plot

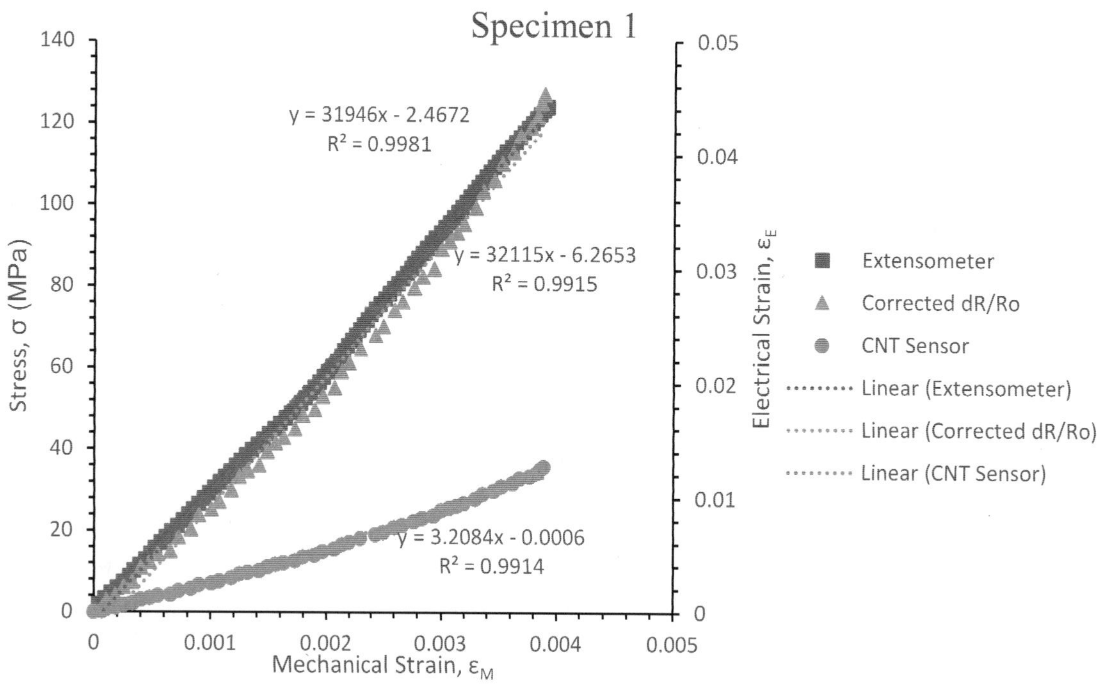

Figure 4: Specimen 1 Data Plot

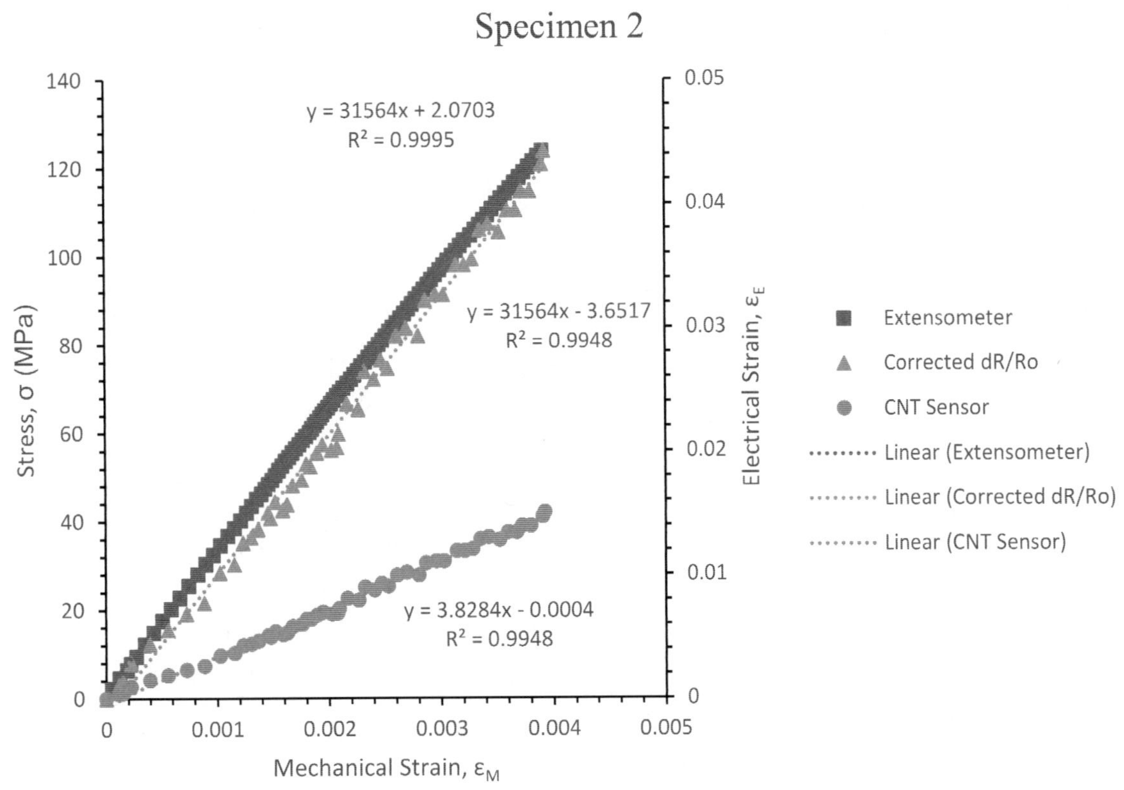

Figure 5: Specimen 2 Data Plot

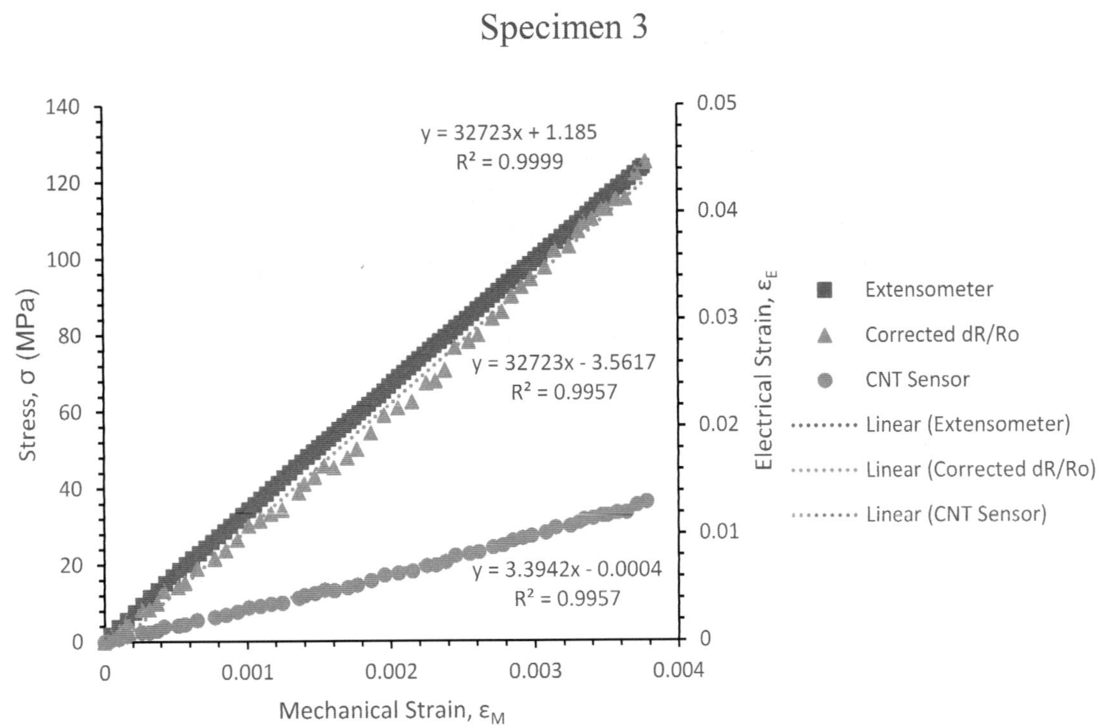

Figure 6: Specimen 3 Data Plot

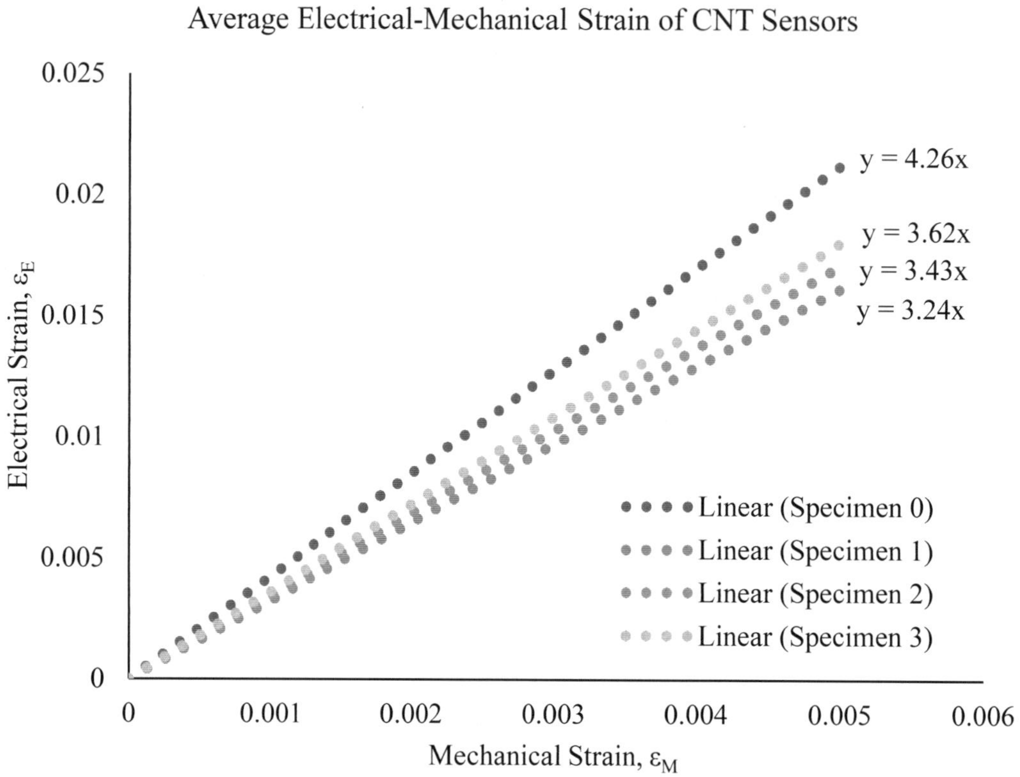

Figure 7: Average Electrical-Mechanical Strain of CNT Sensors with Gauge Factors

4.2 Statistical Analysis

With the raw experimental data and gauge factors tabulated, with the averages and standard deviation computed, a box plot was formed for gauge factor of each sample tested, as shown in Figure 8, respectively. It is visually apparent that the samples created after the initial success of specimen 1 resulted in a smaller spread that is concentrated at its mean modulus of elasticity and gauge factor. The consistency in the span of specimens 1-3 proves the consistency in the testing technique. However, the slight variations in the average values at which the modulus of elasticity and gauge factors were concentrated at shows the inconsistency in manufacturing methods.

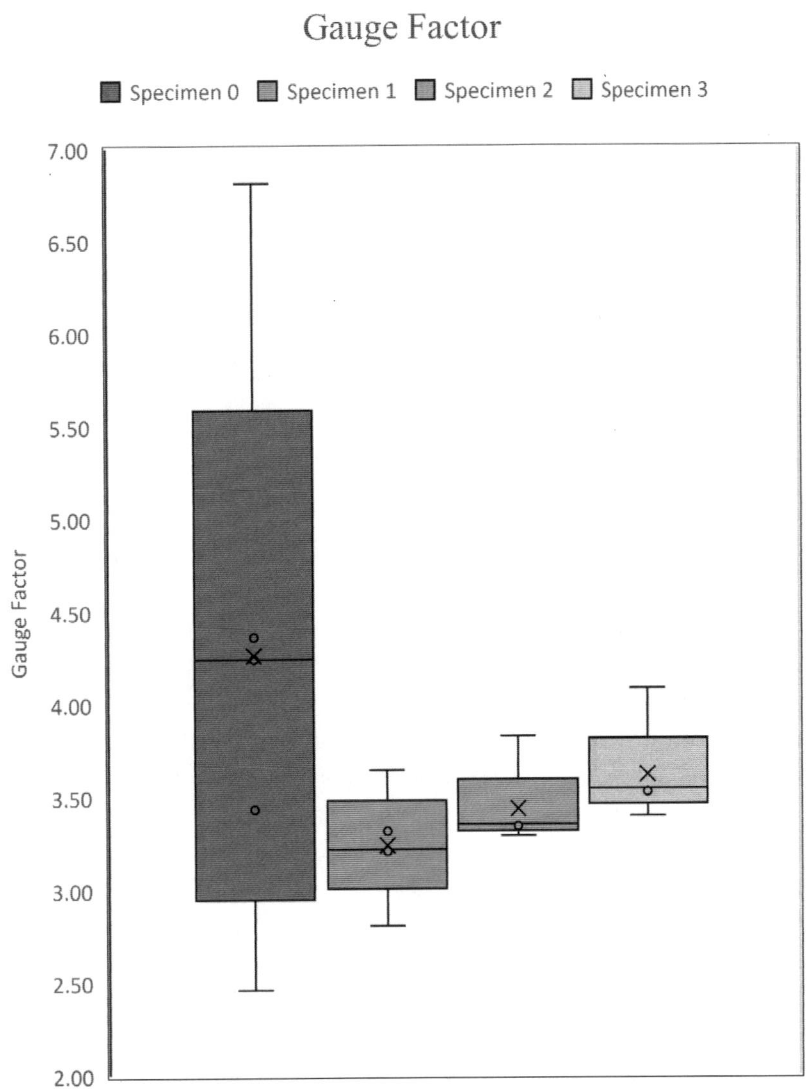

Figure 8: Box Plots of the Gauge Factors Determined for Each Specimen

The gauge factor observed were between the range of 2.46 and 6.81 with an average gauge factor of 3.64, which falls in the range of a typical metal foil strain gauge and aligns with Alexopoulos's results [7]. In comparison, Alexopoulos observed a gauge factor of 3.14 when loading the specimen to 83% of fracture stress, which is approximately a 14.75% difference from the mean of all the samples tested. However, when Alexopoulos loaded the specimen to 100% of fracture stress, a gauge factor of 3.43 was noticed, which resulted in a 5.94% difference from the mean of all the samples tested in this study [7]. With an overall average standard deviation of 0.87, a z-test statistical method for a Gaussian distribution is appropriated by using Equation 9 to find the Z-value at the minima and maximum values of the gauge factors and Equation 10 to find the probability of the gauge factors observed of the population as follows:

$$Mean(GF) = 3.64 \tag{5}$$

$$Std.Dev(GF) = 0.87 \tag{6}$$

$$Z = \frac{X - Mean(GF)}{Std.Dev(GF)} \tag{7}$$

$$P\big(Min(GF) \leq X \leq Max(GF)\big) = P\big(Z(Min(GF)) \leq Z \leq Z(Max(GF))\big) \tag{8}$$

$$Z\big(Min(GF)\big) = \frac{2.46 - 3.64}{0.87} = -1.356 \tag{9}$$

$$Z\big(Max(GF)\big) = \frac{6.81 - 3.64}{0.87} = 3.644 \tag{10}$$

$$P\big(Min(GF) \leq X \leq Max(GF)\big) = P(Z \leq 3.644) - P(Z \leq -1.356) \tag{11}$$

$$P\big(Min(GF) \leq X \leq Max(GF)\big) = 0.9999 - 0.0875 \tag{12}$$

$$P\big(Min(GF) \leq X \leq Max(GF)\big) = 0.9124 = 91.24\% \tag{13}$$

It was determined that the probability that the gauge factor of any test samples will fall within the range observed in this study is 91.24%, assuming a Gaussian distribution.

4.3 Future Recommendations

For future attempts, it is highly recommended to troubleshoot the trial performed by contemplating possible flaws in synthesizing the carbon nanotubes into the resin, as well as the proportions of nanoparticles and the epoxy. Examine the electrode area at the terminals and making sure that a solid connection to the nanotubes exists. Attempt the procedures with different sensing mediums, such as graphene oxide or buckypaper. In addition, if resources are available, try a method in electrophoretic deposition to plate a substrate such as a sheet of tissue paper or fabric to yield an evenly dispersed sensor, similar to buckypaper. If the initial resistance ratings were lower, the sensors would be more sensitive. Study the differences in characteristics when varying the lengths and thicknesses of the sensors. Another method that may yield positive results would incorporate creating a paste with carbon nanotubes infused into a syringe where the sensors are drawn into the material utilizing the sensor.

Once the manufacturing process has been fulfilled and consistent resistance readings observed with calibrations from tensile testing accomplished, it would be a good idea to perform calibrations through bend testing and fatigue testing. Finite element analysis through Solidworks, Ansys, or Abaqus may also be included as an additional form of validation. To take this research further, consider reproducing the same results but with the sensors lined up in a gridded pattern with multiple nodes and be able to locate the point at which a force is acted on the grid. Lastly, repeat the gridded sensor but with this layout embedded within a sandwich of composite layers.

5. CONCLUSION

Creating a manufacturing process for a health monitoring sensor coated with nanoparticles provide a number of challenges to learn from. Performing multiple experiments provided an insight of how the various methods resulted in the creation of a more effective sensor in aerospace, automotive, and building structure applications. The infusion of 2% carbon nanotubes into epoxy by weight and application onto the strand of fibre was proven to be difficult based on the equipment available and lab facilities accessed while finding a data acquisition device compatible with the multimeter used could have allowed for more samples to be tested, which impacted the decisions made and results significantly.

The manufacturing of CNT sensor was proven to be a viable concept. The gauge factor determined were between the range of 2.46 and 6.81, falls between 84.67% of the accepted range of gauge factor 2 to 5 for a typical commercial strain gauge. With an overall average standard deviation of 0.87, the probability that the gauge factor of any test samples will fall within the range observed is 91.24%, assuming a normal Gaussian distribution.

With our method of creating these CNT sensors, it has been shown the consistency of readings through tensile loads. The authors plan to further study the effects of bending, vibrations, and fatigue on the CNT sensors.

6. REFERENCES

[1] Balaji R, Sasikumar M. (2017). Graphene based strain and damage prediction system for polymer composites.

[2] Raghavan, A., Kessler, S.S., Dunn, C.T., Barber, D., Wicks, S., & Wardle, B.L. (2009). Structural Health Monitoring using Carbon Nanotube (CNT) Enhanced Composites.

[3] Aly, K., Li, A., & Bradford, P.D. (2016). Strain sensing in composites using aligned carbon nanotube sheets embedded in the interlaminar region.

[4] Kravchenko, O.G., Pedrazzoli, D., Bonab, V.S., & Manzas-Zloczower, I. (2018). Conductive interlaminar interfaces for structural health monitoring in composite laminates under fatigue loading.

[5] Sasikumar, M., Balaji, R., & Vinothkumar, M. (2018). Nanoparticles-coated glass fibre-based damage localization and monitoring system for polymer composites.

[6] Inam, F., Bhat, B.R., Luhyna, N., & Vo, T. (2013). Comparison of structural health assessment capabilities in epoxy – carbon black and epoxy – carbon nanotube nanocomposites.

[7] Alexopoulos, N.D., Bartholome, C., Poulin, P., & Marioli-Riga, Z. (2010). Structural health monitoring of glass fiber reinforced composites using embedded carbon nanotube (CNT) fibers.

[8] Sky Spring Nanomaterials, Inc. (n.d.). MWNTs 95% 10-20nm. Retrieved from https://www.ssnano.com/inc/sdetail/mwnts-95--10-20-nm/872

[9] Norplex-Micarta. (n.d.). NP130HF – Epoxy Glass Sheet. Retrieved from https://www.norplexmicarta.com/grades/np130hf/

EFFECT OF FAST CURE CYCLES ON MODE I AND MODE II INTERLAMINAR FRACTURE TOUGHNESS FOR INTERLAYER TOUGHENED PREPREG COMPOSITES

Cheng Chen[1], Anoush Poursartip[1], Göran Fernlund[1]
[1]Composites Research Network, Materials Engineering Department, the University of British Columbia,

309-6350 Stores Road

Vancouver, BC Canada V6T 1Z4

ABSTRACT

Faster cure cycles for thermoset prepreg composites are required to achieve higher production rates in the aerospace industry. This work studies the effect of fast cure cycles on the static Mode I (G_{Ic}) and Mode II (G_{IIc}) interlaminar fracture toughness of an interlayer toughened composite material system, T800SC/3900-2B. Composite laminates were processed to the same degree of cure using a manufacturer's recommended cure cycle (MRCC), and three faster cure cycles with high heating rates (>2.8 °C/min) or high curing temperatures (>180 °C). The initiation and propagation G_{Ic} and G_{IIc} were measured using Double Cantilever Beam (DCB) and End-notched Flexure (ENF) tests under static loading, respectively. Results show that curing conditions affect interlayer microstructures and the morphology of toughening particles. However, the initiation and propagation G_{Ic} and G_{IIc} are not sensitive to the applied higher heating rates and curing temperatures. These results suggest that there may be an opportunity to greatly reduce the processing time for T800SC/3900-2B prepregs using faster cure cycles outside of the manufacturer's recommended processing window, yet retaining high interlaminar fracture toughness.

Keywords: Polymer-matrix composites (PMCs), Particle-reinforced composites, Delamination, Fracture toughness
Corresponding author: Göran Fernlund

1. INTRODUCTION

Increasing production rates of carbon fibre structures in commercial airplanes is crucial to meet market needs. Carbon fibre reinforced thermoset composites have been widely used to manufacture primary structures of commercial airplanes such as wings and fuselage. Significant effort has been made to accelerate the manufacturing process for aerospace composite parts using automated composite layup machinery such as automated fibre placement (AFP) and automated tape-laying (ATL). However, the production rate is still limited by the manufacturer's recommended cure cycles (MRCC) for aerospace-grade thermoset prepreg composites. Figure.1

(a) shows that the MRCC usually lasts for 4-16 hours to complete cure and achieve high quality [1–7]. One of the solutions to shorten the cure cycle time is to redesign the resin chemistry. There are many signs of progress in developing and commercializing fast-cure resin systems for high high-volume production in the automotive industry [8,9]. For the aerospace industry, however, due to the high certification cost for a new material system, it is desirable to develop a faster cure cycle for a mature material system that has already been certified for aerospace applications. The material manufacturer usually provides a conservative processing window for its composite material product. Figure.1 (b) shows recommended cure cycle heating rates for thermoset prepreg composite material that are being used for airplane structures. The maximum allowed heating rates for these material systems are usually lower than 5 °C/min. Figure. 2 shows the cure cycle time and the proportion for each stage of the cure cycle using the lowest and highest recommended heating rates for different aerospace thermoset composites. Using the highest heating rates, the cure cycle time is limited by both the heating stage (29-62%) and holding stage (23-53%), as shown in Figure.2 (a-b). Figure.2 (c-d) suggests that the cure cycle time is dominated by the holding stage at 180 °C using the highest recommended heating rates. To shorten the cure cycle time, a faster cure cycle with a higher heating rate and a higher curing temperature may improve the manufacturing efficiency and achieve the same quality without significant change to the existing manufacturing process. However, the question remains whether using fast cure cycles outside of the manufacturer's recommended processing window can retain the final physical and mechanical properties. Commercial interlayer toughened prepreg composites such as Toray's T800SC/3900-2B and Hexcel's HexPly® M21E/IMA were chosen for primary structures of commercial airplanes due to their high delamination and damage resistance [10,11]. Thermoplastic particles in the interlayer and thermoplastic-modified thermoset epoxy introduce more complicated compositions and laminate microstructures for toughened material systems, which may make them more sensitive to curing conditions than non-interlayer toughened material systems. Hunt et al. [12] examined various non-standard cure cycles for HexPly® M21E/IMA and found that the Mode I interlaminar fracture toughness (G_{Ic}) and crack propagation behaviours are dependent on curing conditions. Specimens cured with one-hold cure cycles generally exhibit stick-slip behaviours while specimens cured with two-stage cures cycles show stable crack propagation with a 22% reduction of propagation G_{Ic}. For laminates cured with elevated temperature above 180 °C, the crack tends to migrate from the toughened interlayer to the intralaminar region. Incerti et al. [13] found that the fracture toughness of core-shell rubber (CSR) nano-modified epoxies decreases with increasing curing temperature and that this dependence is more significant for epoxies with higher CSR particle weight fractions, which indicates that the dependence is caused by CSR.

This work studies the effect of fast cure cycles with a high heating rate or a high curing temperature on the static Mode I and Mode II interlaminar fracture toughness of T800SC/3900-2B. Laminates were processed to the same degree of cure (DoC=0.9) using the MRCC and non-standard fast cure cycles. G_{Ic} toughness was measured using Double Cantilever Beam (DCB) tests. Initiation and propagation G_{IIc} toughness were measured using non-precracked (NPC) and precracked (PC) end-notched flexure (ENF) tests, respectively. The laminate microstructure was investigated using an optical microscope. The fracture morphology was examined using scanning electron microscope (SEM). This work suggests that there may be a larger processing window for T800SC/3900-2B than the currently suggested one.

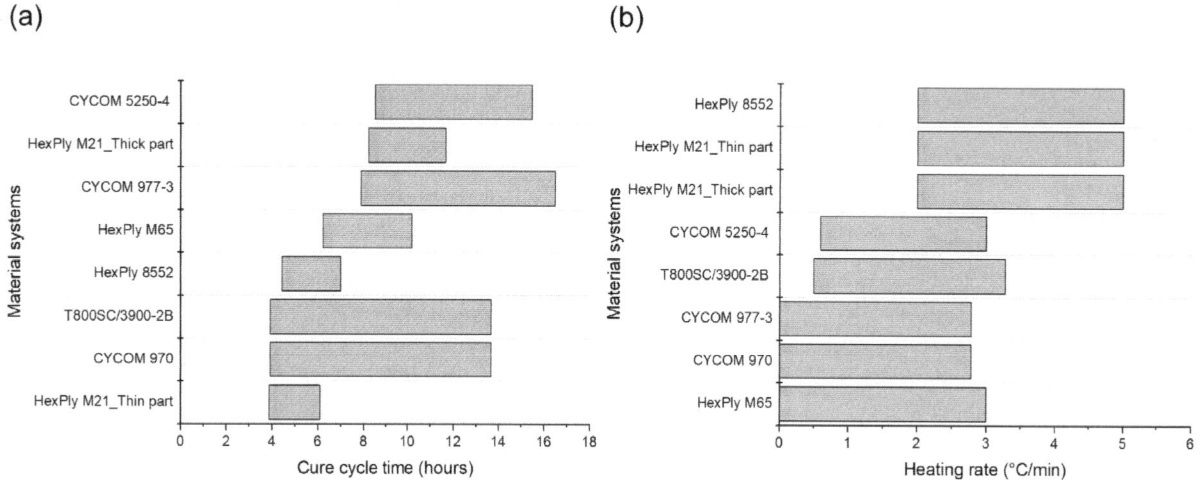

Figureure.1. Cure cycle time and recommended heating rates for different aerospace-grade thermoset prepreg composites [1–7].

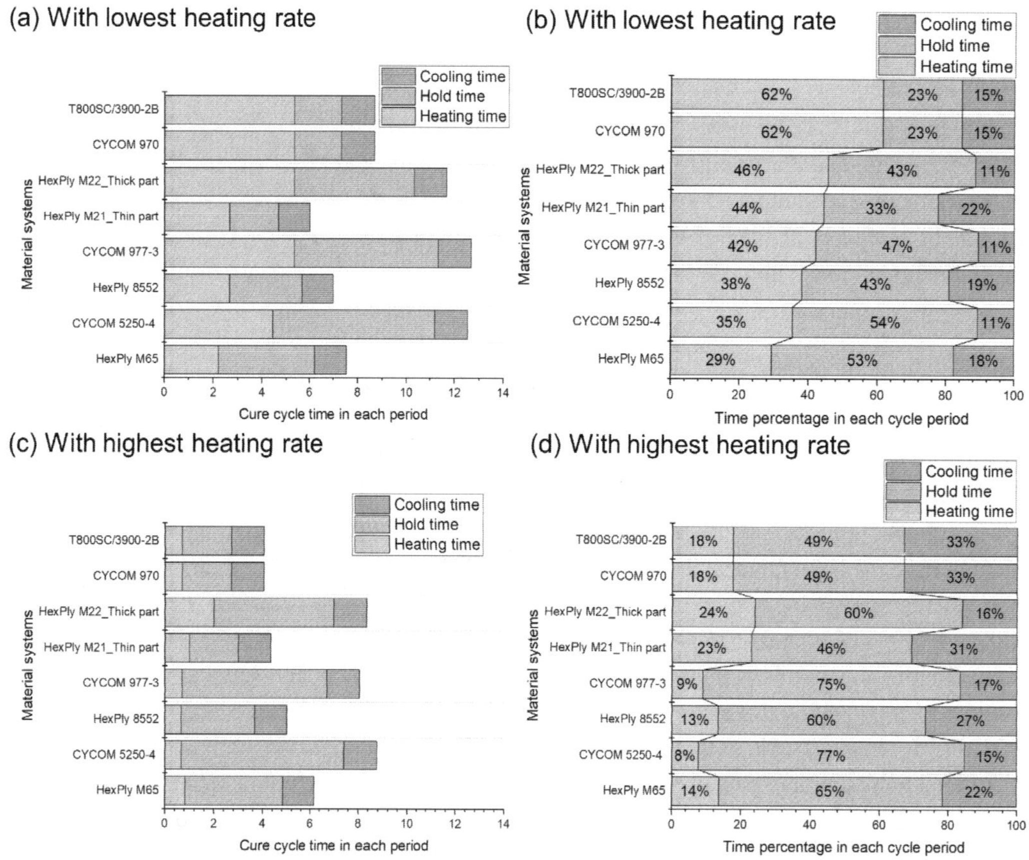

Figureure.2. Processing time and the proportion of time spent at each stage of the cure cycle using lowest and highest recommended heating rates for different aerospace-grade thermoset composites [1–7].

2. EXPERIMENTATION

2.1 Material and sample preparation

An interlayer toughened prepreg, Toray T800SC/3900-2B unidirectional (UD) tape, was used in this study. In this material system, thermoplastic particles are distributed in the interlayer to improve delamination and impact damage resistance. This material system has been widely used in the primary structures of Boeing's commercial airplanes. All laminates with stacking sequence $[0]_{20}$ were manufactured by hand lay-up on a flat aluminum tool with a non-adhesive film at the laminate-tool interface. A 12.5 μm thick FEP film was inserted at the laminate mid-plane during the layup process as a delamination initiator. The initial delamination length, a_0, was approximately 50 mm for DCB specimens and 45 mm for ENF specimens. Rubber dams were placed around laminates to prevent resin bleed. A flat aluminum plate was placed on top of the laminates to obtain a smooth top surface. After debulking under full vacuum at 23 °C for 12 hours, laminates were cured in an autoclave with 85 psi autoclave pressure. Full vacuum was applied under the vacuum bag throughout the cure cycle. Cure cycles with different heating rates and cure temperatures were used to achieve the same final Degree of Cure (DoC=0.90), as shown in Table.1. The final DoC was calculated with an open cure kinetics model [14] exercised using the Raven process simulation software from Convergent Manufacturing Technologies. After cure, DCB specimens with dimensions of 125 mm (length) × 25 mm (width) × 3.86 mm (thickness) and ENF specimens with dimensions of 130 mm (length) × 25 mm (width) × 3.86 mm (thickness) were cut from large laminate panels using a water-lubricated precision diamond saw (Isomet 400). After that, specimens were conditioned in the standard laboratory atmosphere for seven days.

Table 1. Temperature cure cycles

Type of cure cycle	MRCC	High heating rate	High cure temperature	High heating rate and curing temperature
Cure cycle ID	CC1	CC2	CC3	CC4
Heating rate (°C/min)	1.0	7.0	1.0	7.0
Hold temperature (°C)	180	180	200	200
Hold time (min)	110	135	30	40
Cool-down rate (°C/min)	2.78 (same for all)			
Final DoC	0.9 (same for all)			
Total cure cycle time (min)	328	200	275	131
Gel temperature (°C)	173	180	173	200

2.2 Double Cantilever Beam (DCB) test

Double Cantilever Beam test was performed using a universal mechanical testing machine (Instron 5982 Floor Model Testing Systems) following the ASTM 5528-13 standard [15]. Piano hinges were bonded to the specimen using an epoxy adhesive to apply the crack-opening load perpendicular to the debonding plane. The loading rate was 2 mm/min. Five repeats were conducted for each cure cycle. A camera was used to monitor the delamination growth, a. The load (P), load point displacement (δ) and delamination length were recorded simultaneously and used to calculate the strain energy release rate, G_{Ic}. To account for the rotation effect, G_{Ic} was calculated using the Modified Beam Theory (MBT) method by the following equation:

$$G_{Ic} = \frac{3P\delta}{2b(a + |\Delta|)} \quad (1)$$

where Δ is determined experimentally by generating a least square plot of the cube root of compliance, $C^{1/3}$, versus delamination length. b, is the average width of the specimen.

2.3 Quasi-static end-notched flexure (ENF) test

Quasi-static end-notched flexure (ENF) tests were performed using a universal mechanical testing machine (Instron 5982 Floor Model Testing Systems) with a three-point bending fixture according to the ASTM D7905M–14 standard [16], as shown in Figure.1 (a). Five repeats were conducted for each cure cycle. The crosshead loading rate was 0.5 mm/min. Both Non-Precracked (NPC) and Precracked (PC) fracture tests were performed to measure the initiation and propagation G_{IIc}, respectively. NPC tests performed with the 12.5 μm FEP insert as the initial crack. After the crack had grown, a PC test was performed on the same specimen. Two Compliance Calibration (CC) tests were performed at crack lengths of 20 mm and 40 mm, respectively. The CC coefficients, A and m, were determined using a linear least-squares regression analysis of the compliance, C, versus crack length, a, according to the relationship: $\mathrm{C} = \mathrm{A} + \mathrm{m}a^3$. G_{IIc} was then calculated using the following equation:

$$G_{IIc} = \frac{3mP_{Max}^2 a_0^2}{2B} \quad (2)$$

where m is the CC coefficient, P_{Max} is the maximum force from the fracture test, a_0 is the crack length used in the fracture test (30 mm), and B is the specimen width.

2.4 Fractographic analysis

The microstructure of cured laminates was examined from polished 20-ply laminates using an optical microscope (Nikon Eclipse MA200). A scanning electron microscope (FEI Quanta 650, 5 kV) was used to analyze the fracture morphology of the fracture surface of the DCB specimens.

3. RESULTS

3.1 Interlayer microstructure

Thermoplastic particles are introduced to the interlayer of T800SC/3900-2B laminates to improve delamination and impact resistance. The interlayer topology such as particle volume fraction in the interlayer and interlayer thickness impacts the toughening effect. Therefore, it is of great interest to examine whether the interlayer topology is influenced by curing conditions. Figure.3 shows the interlayer microstructure of T800SC/3900-2B laminates processed using a standard cure cycle (CC1) and faster cure cycles (CC2, CC3 and CC4). For the standard cure cycle CC1, particles deform slightly and contact fibres and adjacent particles. The interlayer microstructure for CC3 with a higher curing temperature of 200 °C and a standard heating rate of 1.0 °C/min is similar to that of CC1. Compared to CC1 and CC3, the particle deformation is larger for CC2 with a higher heating rate of 7.0 °C/min and a standard curing temperature of 180 °C. For CC4 with a higher heating rate of 7 °C/min and a higher curing temperature of 200 °C, particles deform significantly and are highly compacted. The interlayer for CC4 is thinner than those for CC1-CC3. Causes of particle deformation and interlayer thinning have been revealed in previous work [17]. Generally, if the gelation temperature of the base resin is lower than the glass transition temperature of thermoplastic particles ($T_g^p = 150$ °C), particles cannot deform due to the constraining effect of gelled resin. On the other hand, If the gelation temperature of the base resin is higher than T_g^p, particles will deform under the fibre bed pressure when the resin is still fluid, which results in interlayer thinning. Table.1 shows that the gel temperature for CC1 and CC3 is the same, which causes similar microstructures for both cure cycles. The gel temperature is 180 °C for CC2 and 200 °C for CC4, which are higher than T_g^p and gel temperatures for CC1 and CC3. Therefore, particle deformation for CC2 and CC4 is larger than CC1 and CC3.

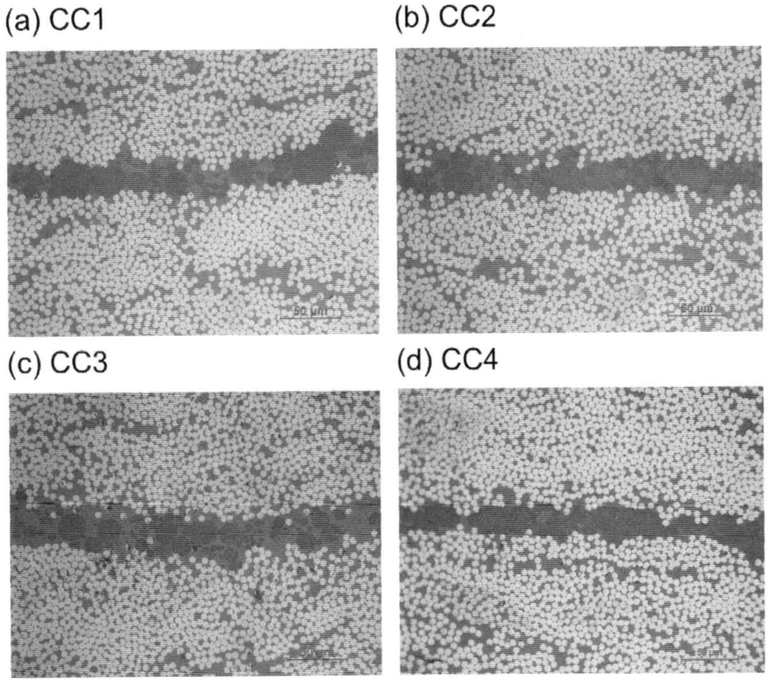

Figure.3. The interlayer microstructure of T800SC/3900-2B processed using different cure cycles

3.2 Mode I Interlaminar fracture toughness

A typical delamination resistance curve (R-curve) for the DCB samples is displayed in Figure.4. Unstable crack growth occurs after the crack initiation from the preimplanted insert and is arrested after. The initial propagation G_{Ic} is lower than initiation G_{Ic} due to loss of the toughening effect. After the crack arrests, propagation G_{Ic} increases as the crack grows due to increasing fibre bridging and finally reaches a plateau. Therefore, the saturated propagation G_{Ic} is higher than the initial propagation G_{Ic}. Figure.5. compares the initial G_{Ic}, initial propagation G_{Ic}, and saturated propagation G_{Ic} for different cure cycles. The initial G_{Ic} for CC2, CC3 and CC4 is slightly higher than that for the standard cure cycle CC1. The initial propagation G_{Ic} and saturated propagation G_{Ic} are not sensitive to curing conditions. The measured initiation G_{Ic} value may be influenced by the shape and quality of crack tip made by the preimplanted insert. Figure.6 shows the fracture morphology for different cure cycles under Mode I loading. The result shows that the crack goes through particles and that particle-resin debonding is not observed, which indicates high particle-resin interfacial strength. Particles processed using CC1 and CC3 have a uniform internal microstructure. However, for CC2 and CC4 with a higher heating rate, particles have a core-shell microstructure. These results show that curing conditions affect the particle morphology and may influence the properties of toughening particles.

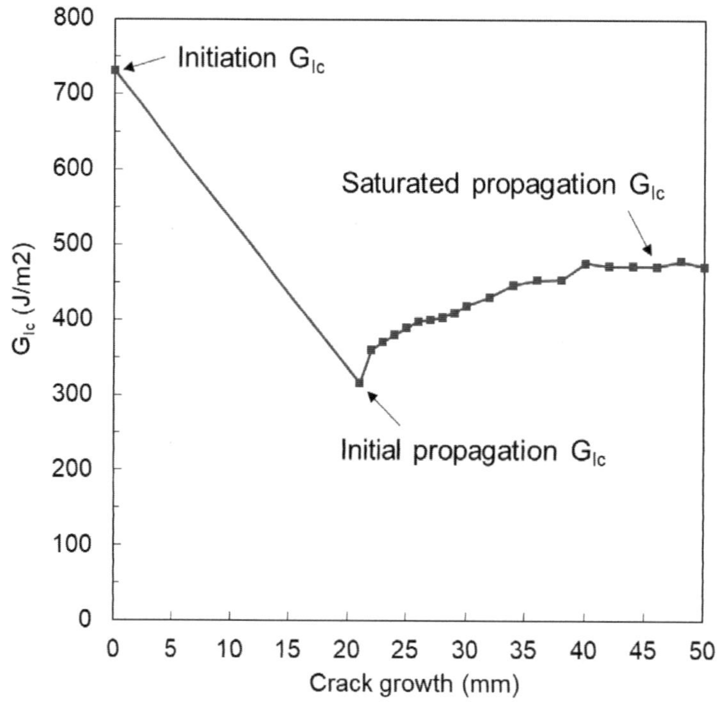

Figure.4. Determination of initial G_{Ic}, initial propagation G_{Ic}, and saturated propagation G_{Ic}.

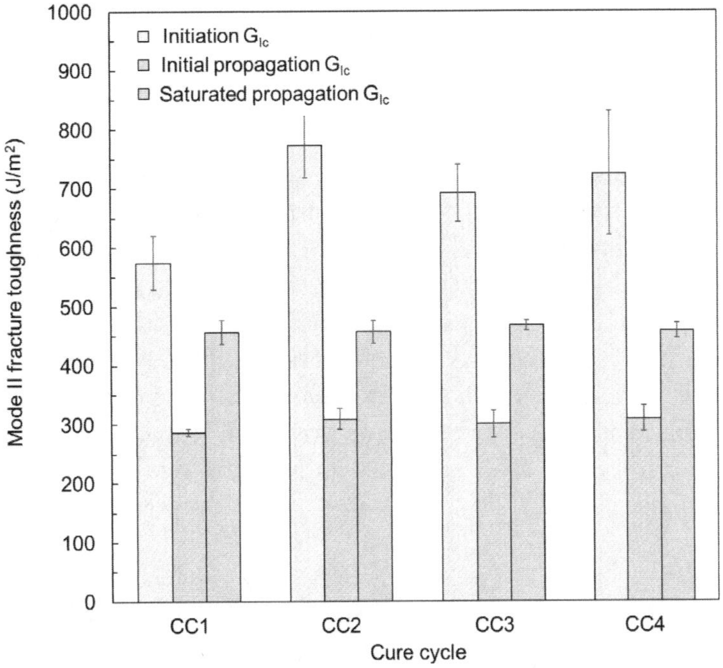

Figure.5. The initial G_{Ic}, initial propagation G_{Ic}, and saturated propagation G_{Ic} at different cure cycles.

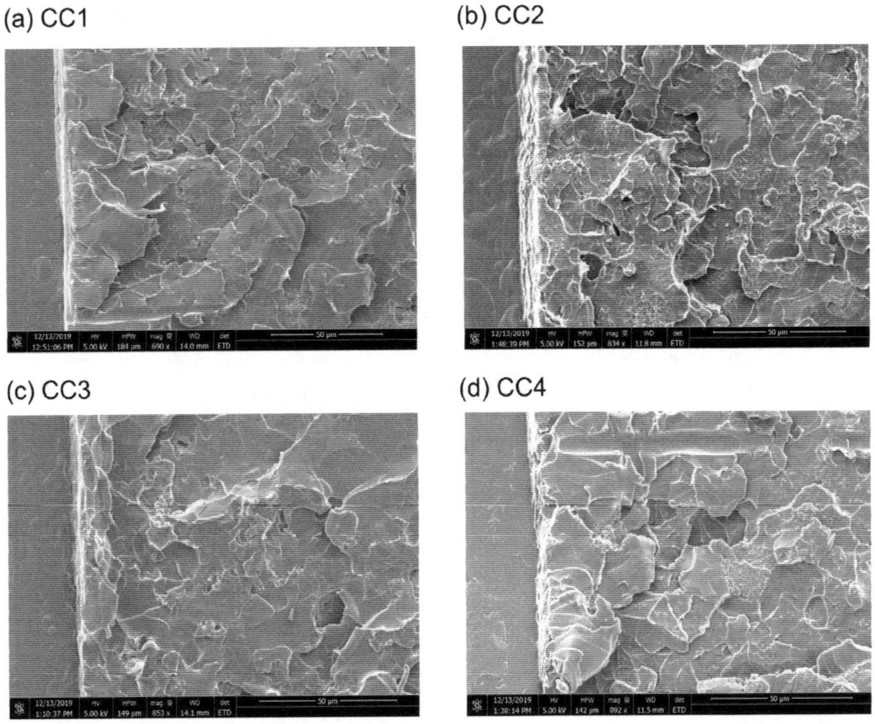

Figure.6. Mode I fracture morphology at different cure cycles.

3.3 Mode II Interlaminar fracture toughness

Figure.7 compares initial and propagation G_{IIc} for different cure cycles. The initiation G_{IIc} was measured from the preimplanted insert, and the first and second propagation G_{IIc} were measured at a distance away from the preimplanted insert. There is a minor difference in initiation and propagation G_{IIc} between these four cure cycles. Interestingly, G_{IIc} decreases slightly as the crack grows, which is due to the crack migration from toughened interlayer to more brittle intralayer, as shown in Figure.8. It has been reported that the rate of crack migration to the intralayer is determined by the curing condition and strain rate [18]. It was found that the rate of crack migration is high for cure cycles with either low heating rate or low curing temperature, which results in a significant reduction of G_{IIc} during crack propagation. For the standard cure cycle and the faster cure cycles used, most of the crack remains in the toughened interlayer during crack propagation. It is desirable to maintain the crack within the interlayer or at the boundary of interlayer to keep the toughening effect such as interlayer plastic deformation. Compared to the standard cure cycle CC1, the cure cycle time for CC2, CC3 and CC4 is reduced significantly while initiation and propagation G_{IIc} are retained.

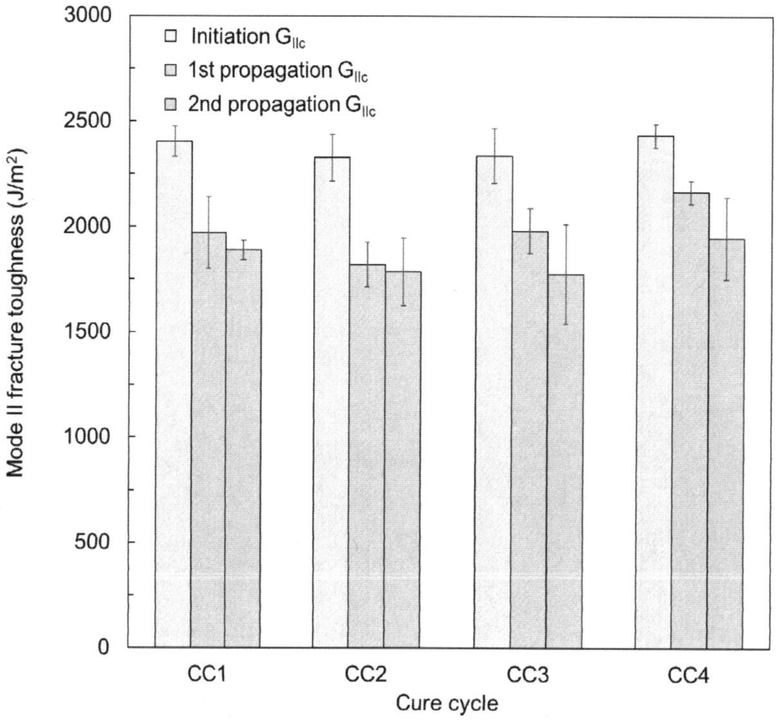

Figure.7. The initial and propagation G_{IIc} at different cure cycles.

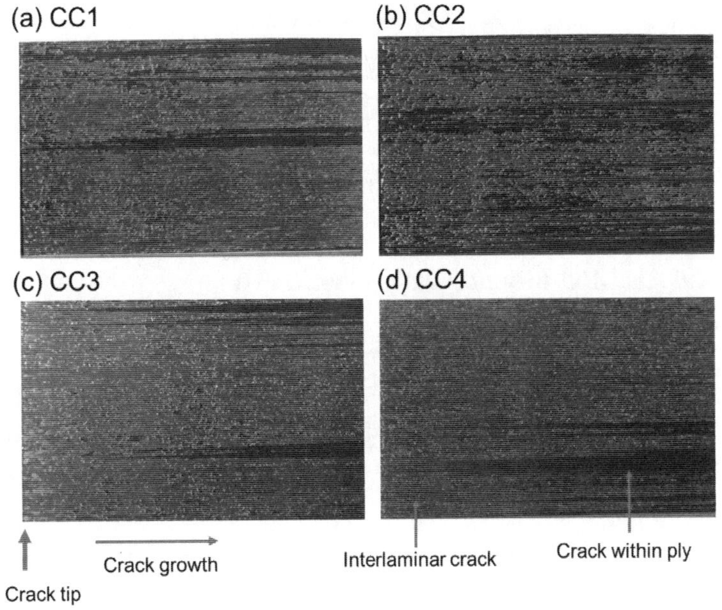

Figure.8. The fracture surface of the bottom side of ENF specimens for at different cure cycles.

4. CONCLUSIONS

This work examined the effect of fast cure cycles on the interlayer microstructure and the quasi-static G_{Ic} and G_{IIc} of T800SC/3900-2B. It was found that not only interlayer microstructures such as interlayer thickness and particle shape but also the particle internal morphology are influenced by curing parameters such as heating rate and curing temperature. Fast cure cycles with higher heating rates or curing temperatures do not affect propagation G_{Ic} but may improve initial G_{Ic}. The results of ENF tests showed that G_{IIc} decreases as the crack grows for all cure cycles, which is a results of local crack migration from the toughened interlayer to the more brittle intralayer. Initiation and propagation G_{IIc} are not sensitive to curing conditions. There have been concerns raised about the negative effect of fast cure cycles on the interlaminar fracture toughness of interlayer toughened composites. This work indicates that higher heating rates and higher curing temperatures outside of the recommended cure cycle do not reduce interlaminar fracture toughness of T800SC/3900-2B. Therefore, there may be a larger processing window for T800SC/3900-2B prepregs than what is currently used. More cure cycles and properties should be exzamined to establish the true processing window for T800SC/3900-2B prepregs in the future.

5. ACKNOWLEDGEMENTS

The authors declare no conflict of interest. This research was financially supported by the Composites Research Network (CRN). We would also like to thank the Boeing Company and Convergent Manufacturing Technologies for their support. The authors acknowledge Dr. Karl M. Nelson and Pete George from the Boeing Company for providing insight and guidance in this research.

6. REFERENCES

[1] Inc. TCMA. 3900 PREPREG SYSTEM Catalogue 2017.

[2] Hexcel. HexPly ® M21 - Product Data Sheet - EU Version. Hexcel 2015.

[3] Cytec Industries. CYCOM ® 5250-4 Technical Data Sheet 2011.

[4] Cytec Industries. 977-3 Epoxy Resin System 2006.

[5] HEXCEL. HexPly ® M65 Product Data Sheet. 2016.

[6] Hexcel. HexPly 8552 Product Data Sheet 2016:1–6.

[7] Cytec Industries Inc. AEROSPACE MATERIALS CYCOM ® 970 Epoxy Resin. 2013.

[8] SLOAN PM& J. Fast and Faster: Rapid-cure resins drive down cycle times. CompositesWorld 2018. https://www.compositesworld.com/articles/fast-and-faster-rapid-cure-epoxies-drive-down-cycle-times.

[9] Nguyen FN, Yoshioka K, Toshiya Kamae IT, Kitano A. Fast-cycle cfrp manufacturing technologies for automobile applications n.d.

[10] Odagiri N, Kishi H, Yamashita M. Development of torayca prepreg p2302 carbon fiber reinforced plastic for aircraft primary structural materials. Adv Compos Mater 1996;5:249–54. doi:10.1163/156855196X00301.

[11] Hexcel. Hexcel ready to fly on the A350 XWB. Reinf Plast 2013;57:25–6. doi:10.1016/s0034-3617(13)70089-4.

[12] Hunt C, Kratz J, Partridge IK. Cure path dependency of mode i fracture toughness in thermoplastic particle interleaf toughened prepreg laminates. Compos Part A Appl Sci Manuf 2016;87:109–14. doi:10.1016/j.compositesa.2016.04.016.

[13] Incerti D, Wang T, Carolan D, Fergusson A. Curing rate effects on the toughness of epoxy polymers. Polymer (Guildf) 2018;159:116–23. doi:10.1016/j.polymer.2018.11.008.

[14] Donna Dykeman. Minimizing uncertainty in cure modeling for composites manufacturing. University of British Columbia, 2008. doi:10.14288/1.0066334.

[15] ASTM D5528-01. Standard test method for mode I interlaminar fracture toughness of unidirectional fiber-reinforced polymer matrix composites. Am Stand Test Methods 2014;03:1–12. doi:10.1520/D5528-13.2.

[16] ASTM D7905. Standard test method for determination of the mode II interlaminar fracture toughness of unidirectional fiber-reinforced polymer matrix composites. Astm 2014:1–18. doi:10.1520/D7905.

[17] Chen C, Poursartip A, Fernlund G. Cure-dependent microstructures and their effect on elastic properties of interlayer toughened thermoset composites 2019. Manuscript submitted for publication.

[18] Chen C, Nesbitt S, Reiner J, Poursartip A, Vaziri R. Cure path dependency of static and dynamic Mode II interlaminar fracture toughness of interlayer toughened composite laminates 2020. Manuscript submitted for publication.

A SUPPLEMENTAL REPORTING TOOL FOR ASSESSING TECHNOLOGY READINESS

Michael P. Thompson
Boeing (Retired)
Tacoma, WA/USA

ABSTRACT

In current practice, technology maturity is generally measured in terms of Technology Readiness Level (TRL). Technology Readiness Levels range from Level 1 (Basic Principles Observed and Reported) up to Level 9 (Actual System Flight Proven). TR Levels are established by answering multiple generic questions to assess technology maturity at each Level. The TR Level definitions are also generic, and the methodology implicitly assumes that levels are worked consecutively (i.e., that all Level 1 work is completed before beginning Level 2, and so on). These characteristics create limitations in terms of fully and accurately illustrating actual technical progress. This paper proposes a supplemental reporting tool for technology maturity assessment, utilizing customizable scaling levels to reflect actual project milestones, along with a simple binary (Met/Not Met) status for each requirement at each scaling level. Full technology development status can thus be summarized with a single-page chart using simple color-coded blocks to show progress at meeting all requirements for all scaling levels up to final production implementation.

1. INTRODUCTION

In any large organization which develops new technologies, there is a need for "snapshot" project status reviews, to enable senior management to quickly assess and compare the maturity of all projects within the organization's portfolio, thus enabling decisions regarding priority, funding, and implementation scheduling for all technologies under development. The most common methodology for these snapshot assessments is the Technology Readiness Assessment concept originally developed by NASA in the 1980's, and since adopted by the Department of Defense, other governmental organizations, and many private companies [1]. Each project is assessed against multiple criteria and its maturity is reported as a "Technology Readiness Level" (TRL), with TRL1 being least mature ("Basic Principles Observed and Reported"), and TRL9 representing a fully mature technology ("Actual System Flight Proven").

The purpose of this paper is to propose a supplemental reporting tool for measuring technology readiness. This new reporting tool is tied to the specific test parts and demonstration articles required to prove out new material and fabrication process technologies and applications. It can be reviewed at a glance to provide substantially more project information than just a numerical TR Level, and reflect actual project development more accurately than the TRA methodology. This new reporting methodology is not a replacement for TRLs – it is intended to be an additional supplement to aid senior management in decision-making.

SAMPE Virtual Conference Proceedings, 2020. Society for the Advancement of Material and Process Engineering – North America.

The scope of this paper is limited, and addresses fabrication of aerospace structural composite parts and assemblies (the author's area of technical experience). However, it is anticipated that many of the principles described herein would also be applicable to metallic structure and components for non-aerospace applications. Much of the literature regarding Technology Readiness Assessment considers development of complete aircraft and space vehicles, which incorporate not only new materials and fabrication processes, but also new mechanical, electrical, avionics, software, control, and integration systems. Projects at that scale are beyond the scope of this paper.

In typical practice, aerospace developmental projects are not targeted at generic technologies (e.g., "prove oven cure works" or "prove this prepreg works" or "prove this forming process works"). Most projects are targeted at specific applications which already have a scheduled production insertion date (e.g., "prove that oven cure can be used to successfully fabricate *this specific new wing spar with all these specific design features,* and be production-ready in two years.") Thus "technology development" (for oven cure in this case) is completely intertwined with application development of the new spar design. It is these types of composite part developmental projects for which this new reporting tool is targeted.

2. CREATING AND EXECUTING THE DEVELOPMENTAL PROGRAM

2.1 The Building Block Approach to Creating the Test/Evaluation Program

The first step in developing a new composite part technology is creation of the test/evaluation program. This test program not only has to generate data which demonstrate acceptable part quality and mechanical strength under all appropriate loading and environmental conditions, its manufacturability evaluations must demonstrate consistency in part quality and compatibility with tooling and other required processes.

The aerospace industry has adopted the "Building Block" approach to designing such test programs for structural part development. Figure 1 illustrates the key features and scaling level descriptions used in the Building Block approach [2].

FAA Advisory Circular AC20-107B [3], dated 9/8/2009, which addresses the subject of showing compliance to FAA regulations for composite aircraft structure, includes a good summary of the rationale and benefits of the Building Block approach: "The strength of the composite structure should be reliably established, incrementally, through a program of analysis and a series of tests conducted using specimens of varying levels of complexity. Often referred to in industry as the 'Building Block' approach, these tests and analyses at the coupon, element, detail, and subcomponent levels can be used to address the issues of variability, environment, structural discontinuity (e.g., joints, cut-outs or other stress risers), damage, manufacturing defects, and design or process-specific details. Typically, testing progresses from simple specimens to more complex elements and details over time. This approach allows the data collected for sufficient analysis correlation and the necessary replicates to quantify variations occurring at the larger structural scales to be economically obtained. The lessons learned from initial tests also help avoid early failures in more complex full scale tests, which are more costly to conduct and often occur later in a certification program schedule."

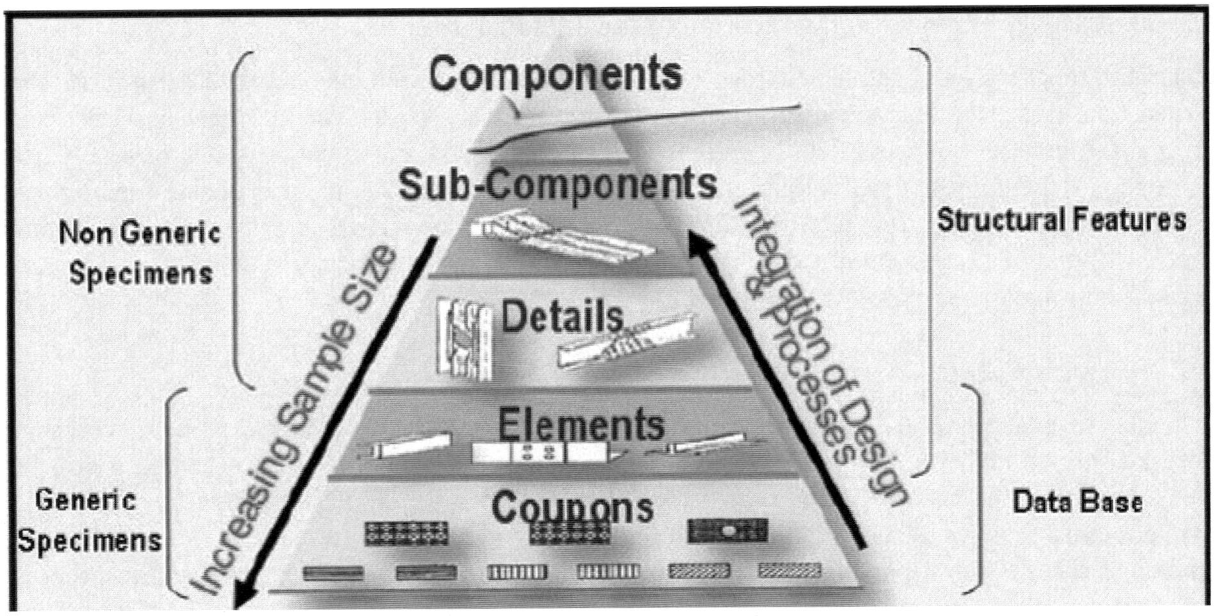

Figure 1. Example Building Block Diagram For Technology Development Projects

Note that the Building Block approach as illustrated in Figure 1 provides only generic scaling levels (Coupon/Element/Detail/Subcomponent/Component) in its guidelines and principles for designing structural test programs. For the project-specific test program, Structures Engineering and other Engineering groups utilize these principles to design the necessary customized hardware articles and test requirements to validate all aspects of the technology development effort. The project-specific test program might include variations on the Building Block scaling levels, or include individual test articles which address more than one of the Building Block levels, and therefore might not neatly fall into one of the five categories shown in Figure 1.

Therefore, for clarity purposes within this paper, the scaling levels associated with the project-specific test parts, demonstration articles, and process compatibility trials shall be collectively designated "Test Article Scaling Levels" (TASLs), to differentiate them from the five Building Block scaling levels and the nine Technology Readiness Assessment scaling levels (TRLs).

Typically, the lowest Test Article Scaling Level (TASL1) comprises one or more flat panel configurations which must meet specific mechanical requirements. The next level up in terms of of size or complexity (TASL2) includes one or more contoured panels which might incorporate convex or concave radii, thick and thin areas, ply drops, and other basic part features. Intermediate TASLs would include demonstration articles to address the specific part features and process needs – perhaps contour joggles or prepreg formability capability or automated tape laydown or inspectability of specific part geometries, and so on. Higher TASLs would address increased part size and thickness and structural integration. The test articles are specifically designed so that collectively, they validate all end item production part requirements. Each test part has its own defined list of requirements which are either Met or Not Met – there is very little subjectivity or ambiguity in these judgments. Success is not achieved until 100% of the defined requirements are Met for all test parts and demonstration articles.

2.2 The Project Engineer and Execution of the Test Program

Once Engineering has established the test program and requirements for each test part, the project engineer takes responsibility for project execution. With even the best-designed test program, there are countless details which must be addressed and process issues to resolve to demonstrate that all test article requirements are met. The project engineer is constrained by test article designs and requirements, but there is considerable latitude for creativity in identifying successful manufacturing processes. The project engineer also must determine how to sequence tasks to optimize overall project progress.

2.2.1 Process Robustness and Specific Part Features

In trying to assess overall technology maturity (or simply asking whether a process works or not), a common reality in composites technology development is that for the vast majority of the part in question, the manufacturing process is fully robust, but the small remaining fraction of the part contains features where all the technology problems occur. Instead of one technology challenge for the entire part, the project becomes divided into several feature-specific process development mini-projects. These problem areas might be as small as a few square inches, but these types of challenges can be intractable and threaten success for the entire project. Such challenges might include exotherms during cure for thick laminate areas, inability to achieve proper consolidation pressure in high contour areas or joggles, excessive entrained air or moisture causing porosity within composite laminate ply drop areas or bondlines, localized ply wrinkling, and so on. It occasionally happens that expensive large-scale process revisions must be developed to accommodate such problem areas. These sorts of issues are a major concern for the project engineer.

2.2.2 Risks of Rigid Adherence to the Implied Building Block Sequence for Development

The Building Block approach as illustrated in Figure 1 carries the implicit assumption that small coupon trials should be completed first to understand small-scale phenomena, before progressing to element trials and then details and so on, up to full-scale components evaluated last. In the field of structural composites, however, strict adherence to this approach may not always be best.

Chemistry and physics obviously do not change from the small-scale to the large-scale. But composites history is littered with costly examples of our imperfect understanding of the physics of composites processing and cure in regard to scaling effects. Such problems typically include resin rich or poor areas, porosity/void formation, delaminations, ply wrinkling, and warpage. A process that consistently produces acceptable laminate for small-scale parts can fail dramatically for large-scale parts. There are currently many limitations to our understanding of what is happening during cure inside a large complex composite part at specific part feature locations regarding resin flow, resin polymerization, resin pressure, and fiber movement. The point here is that for technology upgrade efforts involving new materials, process parameters, or design concepts, no matter how many small parts I have successfully fabricated, even with simulated part features, I may be wrong in assuming that I can now safely scale up to full size. For a new composite material or an innovative cure cycle, we might not know what will happen with a 15-meter (50-foot) configured part until we make one. In such a case, where large-scale processing problems are not identified until late in the project, the cost and effort required to modify process parameters and redo testing required for validation can be exorbitant.

All of this serves as a warning regarding rigid adherence to the sequential nature of the Building Block Approach and the TRA methodology. In structural composites applications, a process engineer who relies exclusively on small test coupons and TRL3-scale "proof of concept" demonstration articles to establish the process parameters for large parts is asking for trouble. The project engineer is well served to fabricate simulated larger parts early in the project, to identify processing issues which might not otherwise appear until prototype part validation (TRL6 or TRL7) is underway. Likewise, demonstration articles with anticipated part features should be fabricated early in the project, to identify which features are likely to cause defects. Final designs for production tooling and parts do not have to be complete to make these demonstration articles. Approximations are usually sufficient, and are certainly better than strict reliance on the process knowledge gleaned from test coupons and small demonstration articles.

In a large complex composites development project, therefore, it is entirely likely that fabrication and testing of both small-scale and selected large-scale demonstration articles will be occurring simultaneously, as will be mid-sized test articles to validate processing for specific part features.

3. ASSESSING TECHNOLOGY READINESS

A large organization may have tens or hundreds of projects in work, and senior management doesn't have time to review voluminous reports on every project when making decisions regarding technology implementation, prioritization, scheduling, planning, and funding. A reporting tool is therefore needed, so that project engineers can provide management with high-level "at-a-glance" status on each technology development effort. Management can then review all these reports to assess the overall portfolio of projects and make appropriate plans for the future.

The generally accepted industry standard for this "snapshot assessment" is the Technology Readiness Assessment (TRA) methodology first developed by NASA in the 1980's. The TRA methodology establishes nine Levels of technology maturity ranked by "Technology Readiness Level" (TRL). The criteria to meet each TR Level are shown in the familiar NASA "Thermometer" diagram shown in Figure 2. Technology maturity at any given time is simply reported to management by Level ranking.

The Technology Readiness Levels are generic. The same nine TR Levels are used for any technology development effort. It doesn't matter if the required technology development involves material upgrades, process improvement, tooling or equipment changes, or new structural designs, it is assumed that the same evaluation methodology can be applied across the board.

The TRA methodology for estimating technology maturity assigns a single Technology Readiness Level to the entire project. There is no consideration that certain aspects of the project might be more advanced than others. For a "TRL4" project, this creates the implicit assumption that every aspect of the project has been successfully completed up to "component validation in a laboratory environment" (per Figure 2), and that work is now progressing to achieve TRL5 ("component validation in a relevant environment"). In more general terms, the TRL approach implies that the project engineer works through a set of TRL3 tasks, then a set of TRL4 tasks, then TRL5, and so on, sequentially progressing through each Level in increasing order.

Engineers and managers understand that workflow is not that orderly, but this is how progress through the TR Levels is communicated.

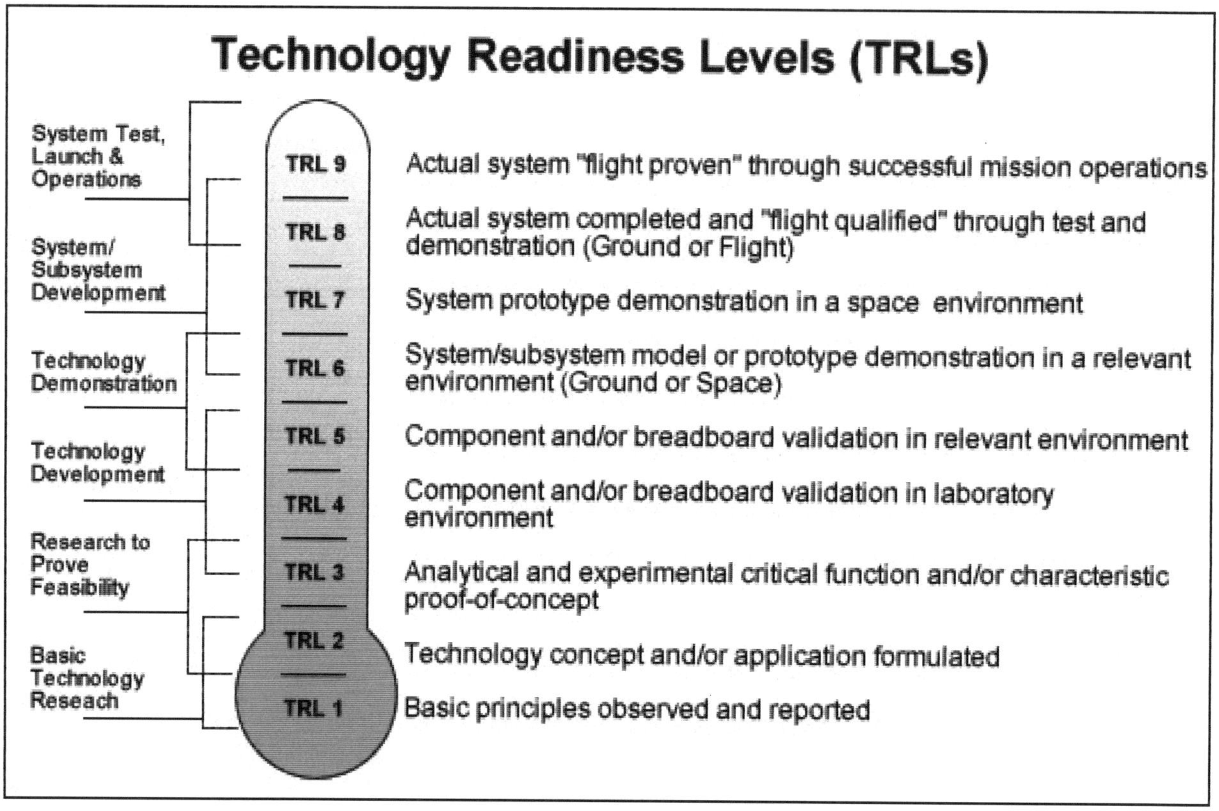

Figure 2. NASA "Thermometer" Diagram Defining the Nine Technology Readiness Levels

3.1 Assessing Maturity and Assigning the Appropriate Technology Readiness Level

As a technology development project progresses, it moves sequentially through the successive TRLs via periodic reviews. The common practice for upgrading TRL status is through answering a series of questions specific to each Technology Readiness Level which interrogate the various aspects of the project for technology maturity. Examples of such questions can be found in the Air Force Research Laboratory (AFRL) "Technology Readiness Calculator" [4], and the Department of Energy's process guide for technology readiness assessment [5]. Other organizations may have their own lists of questions. The questions are structured to have Yes/No answers, but the AFRL methodology also allows for a "percent complete" response to each question. The lists of questions are typically rather lengthy. For example, in transitioning from TRL4 to TRL5, the AFRL Calculator asks 47 different questions, and the Department of Energy process guide asks 34. Ideally, all answers must be "Yes" to advance to the next TR Level.

Similarly to the Technology Readiness Levels themselves, the lists of questions used for level assessment are generic. They are not customized to the project in question. No matter what technology is being developed, the same questions are asked.

3.2 Benefits and Limitations of the TRA Methodology

This generic nature provides the TRA methodology with some attractive features. The simple "TRL4" or "TRL5" status (for example) provides senior managers with an instantly understandable measure of technology maturity. In hearing that one project has reached TRL7 while another is at TRL5, managers have a clear idea of which is closer to production implementation. This also promotes standardization of assessments across the full portfolio of projects. Another positive feature of TRA methodology is that answering all the detail questions forces the project engineer to thoroughly consider all aspects of the development effort.

There are, however, significant limitations to the TRL approach if used alone. Simply stating "this project is at TRL5" provides very limited information. There is no accounting for differences in project complexity or difficulty. There is a certain ambiguity in how to report status for parts where all areas of the part meet requirements except for one or two small design features. Additionally, the TRL status only reflects the Level for which *all* requirements have been met, and does not provide visibility of early work that has been done at higher or more complex developmental levels to mitigate risk.

Furthermore, properly assessing technology maturity via the TRA methodology is not straightforward. Thousands of pages of documents from multiple government agencies provide additional guidance and address challenges with TRA methodology. NASA, in fact, conducted multiple in-depth studies over a recent 5-year period to review TRA methodology and recommend upgrades, culminating in a 2016 report entitled "Final Report of the NASA Technology Readiness Assessment (TRA) Study Team" [6]. In this report, NASA made numerous improvement recommendations, including consolidation of paperwork, clarification of terminology, additional guidance information for TRA evaluations, and increased training.

The NASA report also emphasized the fact that reporting only the TR Level for a given project is not sufficient. The report reinforced the importance of supplementing the TRL assessment by using the "Advancement Degree of Difficulty" (AD^2) methodology for risk assessment of upcoming challenges required to move from the current TR Level to the level required for production implementation [7]. With the AD^2 tool, each aspect of the project is rated in terms of likelihood of success for full production implementation. There are nine AD^2 risk levels, with Risk Level 1 signifying minimal risk and Risk Level 9 signifying the need for completely original research. The AD^2 method also correlates the likelihood of failure with the consequence of failure to assess full perfomance/cost/schedule potential impact to the project. As with TR Levels, assigning the proper AD^2 risk level is not straightforward. One document showing the questions which must be addressed shows 9 questions to be answered for Design and Analysis, 20 questions for Manufacturing, 9 questions for Operations, and 13 questions for Test & Evaluation [8]. Since these questions involve estimation of solution probabilities for as-yet undeveloped technology, they are by nature approximations based on the personal experience of the reviewer(s). Note that for a composites fabrication project that involves a large complex part with multiple specific design features, each of which may require specific process solutions, these questions might have to be asked for every specific design feature.

4. CREATING A MORE SUBSTANTIVE ASSESSMENT REPORT

A more informative TRA supplemental report would incorporate the following features:

- Present overall project status in an easily understood one-page format for quick review
- Reflect the actual customized scaling level test articles to be fabricated
- Denote requirement progress with binary "Met/Not Met" status
- Reflect work being done on multiple scaling levels simultaneously
- Include flexibility for an evolving project to add, modify, or group requirements

Figure 3 illustrates a sample template to illustrate the features of this supplemental "customized binary" technology development snapshot report. This type of report is easily produced using commercially available spreadsheet software.

Test Article Scaling Level 1	Test Article Scaling Level 2	Test Article Scaling Level 3	Test Article Scaling Level 4	Test Article Scaling Level 5	Test Article Scaling Level 6	Test Article Scaling Level 7
Requirement 1	Requirement 1	Requirement 1	Requirement 1	Requirement 1	Requirement 1	Requirement 1
Requirement 2	Requirement 2	Requirement 2	Requirement 2	Requirement 2	Requirement 2	Requirement 2
Requirement 3	Requirement 3	Requirement 3	Requirement 3	Requirement 3	Requirement 3	Requirement 3
Requirement 4	Requirement 4	Requirement 4	Requirement 4	Requirement 4	Requirement 4	Requirement 4
Requirement 5		Requirement 5	Requirement 5	Requirement 5	Requirement 5	Requirement 5
Requirement 6		Requirement 6		Requirement 6	Requirement 6	Requirement 6
Requirement 7				Requirement 7	Requirement 7	Requirement 7
Requirement 8				Requirement 8		Requirement 8

Figure 3. Template Example of New Supplemental Technology Development Reporting Tool

As described previously, the Test Article Scaling Levels (TASLs) are based on the specific physical demonstration articles which must be fabricated, and are therefore customized to the application in question. In an actual report, they can be titled to provide additional descriptive information. The various Engineering groups may assign more or less TASLs than the nine Levels used in the TRA methodology.

Requirements which have been met are shown in green. Requirements not yet tested are shown as blank squares. Failed tests are of particular significance in terms of project risk, and are shown in red to draw management attention. The red squares reflect that some sort of fix is required to achieve project success. There is little to no ambiguity or judgment required regarding test status for any requirement at any Test Article Scaling Level.

The report illustrates work performed to date at all scaling levels. It may be that one or more tests at TASL1 have been delayed and are not yet complete. At the same time, larger test articles may have already been fabricated and some requirements already demonstrated. This report format provides this type of multi-Level visibility.

The report can be modified as the project evolves to add TASLs or modify requirements as needed. As a project progresses, it is very common to find that additional test articles are required to demonstrate full technology maturity due to previously unforeseen issues, or that additional requirements must be created. In addition, as the project moves closer to success, previously met requirements can be grouped for brevity, allowing greater focus on unresolved issues for the benefit of management in decision-making.

4.1 Hypothetical Project for Detailed Illustration

A hypothetical project involving development of a new composite airplane wing spar can be used to illustrate how the new reporting tool functions. This spar will incorporate a new composite material requiring validation of a suitable cure process, and the design will include radius areas requiring automated multi-ply forming processes, severe ply drop regions to transition from thick to thin laminate areas, contour features with joggles and a significant twist section, and attachment areas utilizing hard bagside caul tooling to maintain surface flatness.

The basic technology milestones for such a project would include the following:

- The proposed cure cycle produces acceptable quality laminate (low porosity, voids, etc.)
- Basic composite mechanical properties at coupon level are acceptable
- Properties after impact, fatigue, and environmental exposure are acceptable
- Large-scale mechanical tests are acceptable
- Automated tape laying process is compatible with new material
- Inspection processes are compatible with the new material
- Drill/Trim/Assembly processes are compatible with new material
- Acceptable repair methods are available
- Large-scale parts can be fabricated and meet cure cycle ramp rate requirements
- ALL part features can be fabricated to reliably meet acceptable quality levels:
  - Thick and thin areas and ply drop transitions
  - Joggle areas
  - Convex and concave radii and spar twist sections
  - Attachment areas under hard caul tooling

Figure 4 illustrates the new supplemental technology development status report for this hypothetical spar development project. This illustration assumes roughly a midway point in the overall technology development project (i.e., TRL4 or TRL5).

For this hypothetical project, Engineering has selected seven Test Article Scaling Levels for the test parts required to demonstrate sufficient technology maturity to put this spar into production:

- TASL1 is a set of flat panels of various configurations which must meet certain quality requirements and several mechanical strength requirements.
- TASL2 is a set of small contoured panels to demonstrate capability to fabricate radius areas and ply drop areas, as well as validate that inspection processes are compatible.
- TASL3 comprises larger sub-scale test parts fabricated to separately address specific design features. Level 3A is a test article to evaluate processing for a simulation of features common to the thick end of the part. The Level 3B article addresses the twist section of the spar, and Level 3C addresses attach points with bagside hard cauls.
- TASL4 is a large-sized simulated spar demonstration article. This test article is fabricated early in the development process. In addition to large size, it incorporates expected features in the final part. Although expensive in comparison to the smaller test

articles, this demonstration is used as an "early warning sign" to identify potential scaling issues before final designs, tooling, and process parameters are established.

- TASL5 is a large-scale structural test article.
- TASL6 is a full-scale final configured stand-alone spar.
- TASL7 is the fully assembled spar as mounted on an aircraft wing.

The Figure 4 report illustrates technology development at all scaling levels. In this example, note that the fatigue test under TASL1 is not yet complete, since fatigue cycling may require many months. The project engineer will not stop other work while waiting for the Fatigue test, so multiple tasks for higher TASLs have proceeded and many of the requirements for these more advanced test articles have already been met satisfactorily. However, under strict TRL definitions, one could argue that this project must be rated a TRL2, because until final Fatigue results are available, there is no guarantee that even coupon level performance has been fully validated. By itself, this would be a misleading rating.

The blank squares in the report illustrate the requirements for which testing has not yet been performed, and the red squares indicate the critical issues where demonstration articles failed to meet test requirements.

This one-page report can be reviewed at a glance, another critical characteristic for senior managers who must review and compare dozens of projects for decision-making purposes. With one look at this chart, management can understand what has been done, what has not been done, and what tests have failed, for all demonstration articles and Test Article Scaling Levels. This is substantially more information than is provided simply by stating the TR Level.

Test Article Scaling Level 1	Test Article Scaling Level 2	Sub-Scale Test Articles to Assess Specific Features: Test Article Scaling Level 3a	Test Article Scaling Level 3b	Test Article Scaling Level 3c	Test Article Scaling Level 4	Test Article Scaling Level 5	Test Article Scaling Level 6	Test Article Scaling Level 7
Flat Panels	Contour Panels	Thick End of Part	Twist Section	Attach Points	Large Sim Part	Large Struct Test	Configured Part	Configured Assy
Laminate quality acceptable	Laminate quality acceptable	Laminate quality acceptable	Laminate quality acceptable	Laminate quality acceptable	Laminate quality acceptable	Laminate quality acceptable	Laminate quality acceptable	Laminate quality acceptable
Meets process spec req'ts	Meets process spec req'ts	Meets process spec req'ts	Meets process spec req'ts	Meets process spec req'ts	Meets process spec req'ts	Meets process spec req'ts	Meets process spec req'ts	Meets process spec req'ts
Business case still valid	Business case still valid	Business case still valid	Business case still valid	Business case still valid	Business case still valid	Business case still valid	Business case still valid	Business case still valid
Tension	Prepreg formability	Vacuum bag concept works	Vacuum bag concept works	Attach points flat	Tooling concept works	Acceptable test result	Tooling concept works	Tooling concept works
Compression	Concave/ Convex Radii	Demold/Drill/ Trim	Demold/Drill/ Trim	No ply waves, wrinkles	Bag/Demold/ Drill/Trim		Automated layup	Automated layup
Shear	Thick vs. Thin laminate	Concave/ Convex Radii	Basic Radius	Vacuum bag concept works	Automated Layup		Inspectability	Inspectability
CAI	Internal Ply Drops	Prepreg Formability	Joggles	Drill	Inspectability		Full Structural Test	Final Assembly
Solvent resistance	Inspectability	Internal Ply Drops	Prepreg Formability	Inspectability	Repair Acceptable			
Environmental resistance		Thick vs. Thin laminate	Spar Twist	Assembly Trial				
Fatigue		Automated tape layup	Automated tape layup	Repair Acceptable				
Fracture Toughness		Inspectability	Inspectability					
Drill/Trim Acceptable			Repair Acceptable					
Repair Acceptable								

Evaluation Complete – Meets Req't

Evaluation Not Yet Complete

Failed to Meet Req't – Fix Required

Figure 4. Supplemental Technology Development Status Report For Hypothetical Spar Project

It is recognized that some projects might have a large number of required test articles and/or very long lists of requirements and all this information might not easily fit onto a single-page report. This is especially likely as a project matures and unforeseen challenges occur, requiring additional test demonstration articles and/or new requirements. Adding a second page to the report is one option to address such cases. Another option is to reformat the blocks in the report to group categories for previously met requirements. In terms of management attention, detailed review of previously met requirements is much less important than attention to work not yet completed or especially failed tests. As a project matures, the report can evolve to maintain the emphasis on remaining challenges.

Figure 5 illustrates key columns for the technology development status report for the same spar development project, at a later date than what was shown above in Figure 4. Note that for some columns, previously met requirements are grouped into a single block and described as "All Requirements Met" or "Base Requirements Met." This allows greater emphasis on the cells shown with red text. These are the updated blocks which show that success has been achieved for many of the unmet requirements from Figure 4 (note that most are green), as well as provide visibility of the two new demonstration articles and several new requirements added since the Figure 4 report was generated. The report also highlights a new failed test, representing that manufacturing continues to struggle with achieving properly consolidated wrinkle-free laminate at joggle areas of the twist section of the spar. As a companion to the primary TRL status level, this report provides much greater specificity to management regarding realistic status of the overall project and identification of remaining challenges to production implementation.

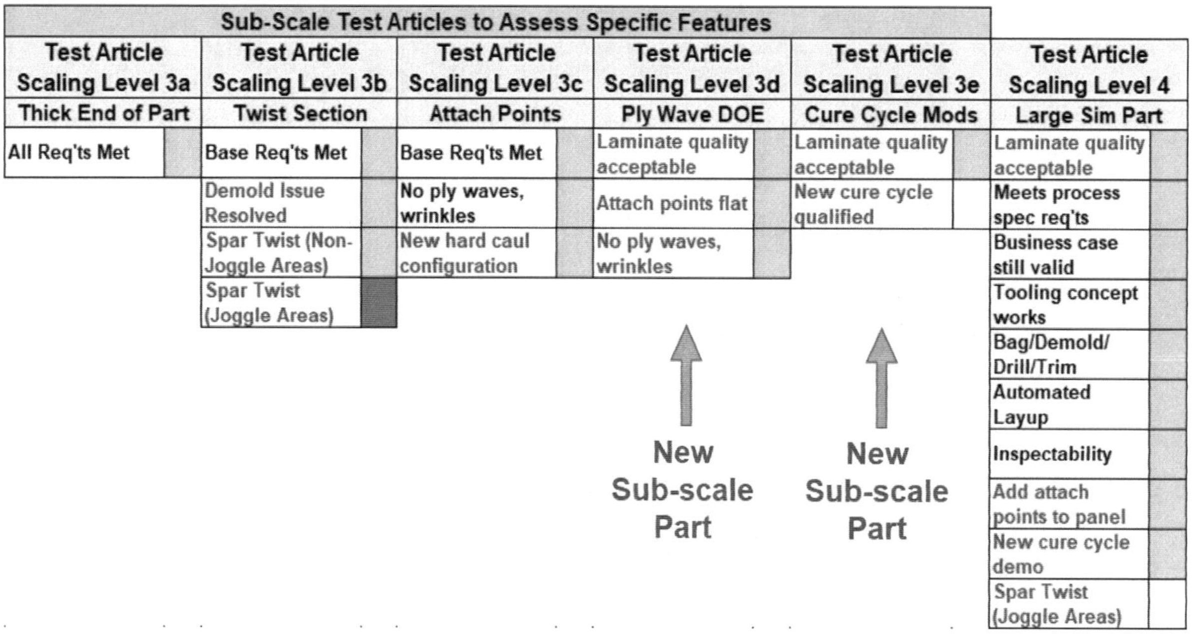

Figure 5. Example Report Illustrating Updated Project Status, and Newly Identified Issues

5. SUMMARY

Measurement of technology maturity via the Technology Readiness Assessment methodology is widely accepted throughout government and industry, and provides a standardized method for technology comparison to support organizational decision-making. The formalized questions used for setting appropriate TRLs are also very helpful in terms of getting project engineers to think through all aspects of their projects.

However, simply reporting technology maturity as a single Technology Readiness Level for an complex developmental project provides management with only a limited understanding of true project status. Reporting by TR Level alone does not provide visibility of the reality that work in complex projects is usually proceeding at multiple scaling levels simultaneously, and does not account for the common situation where technology challenges are confined to small specific part features.

This paper proposes a supplemental reporting vehicle, to be presented to management along with the Technology Readiness Level Assessment. This supplemental report can be easily produced using commercially available spreadsheet software, and shows in table format the "Met/Not Met" status for all requirements at all defined scaling levels for the entire project. This report can evolve as the project progresses, to maintain emphasis on scaling level requirements which remain unmet, especially failed tests where a fix is required.

This new report format communicates a much greater level of technology maturity detail and realism than just the TRL status, while retaining the necessary characteristic that project status can be assessed at a glance.

I wish to gratefully acknowledge the invaluable contributions of Barbara Haynes in organizing and editing this paper and helping me put my ideas together.

6. REFERENCES

1. Towery, N. D., Towery, M., Thomas, A. Report No. FHWA-HRT-17-047. "Technology Readiness Level Guidebook," United States. Department of Transportation. September 2017

2. Abumeri, G., Abdi, F., Raju, K.S., McCloskey, A., Housner, J., Bohner, R. "Cost Effective Computational Approach for Generation of Polymeric Composite Material Allowables for Reduced Testing." *Advances in Composite Materials – Ecodesign and Analysis.* Ed. Brahim Attaf. Rijeka, Croatia: InTech, 2011 DOI (this chapter only): 10.5772/14528

3. Federal Aviation Administration Advisory Circular No. AC20-107B, Section 7b. Subject: Composite Aircraft Structure. September 8, 2009

4. Nolte, W. L., Kennedy, B. C., Dziegiel Jr., R. J. "Technology Readiness Level Calculator." *NDIA Systems Engineering Conference.* October 20, 2003. The AFRL "TRL Calculator" Excel file link can be found at < https://www.dau.edu/cop/stm/lists/tools/allitems.aspx>

5. "Technology Readiness Assessment (TRA)/Technology Maturation Plan (TMP) Process Guide" Table D.6. United States. Department of Energy. March 2008

6. Hirshorn, S. and Jefferies, S. "Final Report of the NASA Technology Readiness Assessment (TRA) Team." United States. National Aeronautics and Space Administration. March 2016.

7. Bilbro, James W. "Using the Advancement Degree of Difficulty (AD^2) as an Input to Risk Management." *Technology Maturity Conference.* Virginia Beach, VA. Sept. 8-12, 2008

8. Bilbro, James W. White Paper – "Systematic Assessment of the Project/Program Impacts of Technological Advancement and Insertion," Revision A. October 2007 < https://www.dau.edu/cop/stm/_layouts/15/WopiFrame.aspx?sourcedoc=/cop/stm/DAU%20Sponsored%20Documents/White%20Paper%20on%20Technology%20Assessment%20Rev%20A.doc&action=default&DefaultItemOpen=1>

DETERMINING MODE I FRACTURE TOUGHNESS OF ADHESIVE COMPOSITE JOINTS: AUTONOMOUS CRACK TRACKING USING MATLAB DEVELOPED PROGRAM

Peter E. Caltagirone[1], Dr. Dylan Cousins[1], David Snowberg[2], Dr. Aaron P. Stebner[1]
Colorado School of Mines[1], National Renewable Energy Lab[2]
Golden, CO

ABSTRACT

Characterizing the fracture toughness of composite material joints is essential to informing design safety factors. ASTM D5528 defines how to calculate the Mode I fracture toughness of an adhesive using a double cantilever beam peel test. The standard specifies visual markers on the sample be inspected during the experiment to measure the crack length, which is needed to calculate the fracture toughness. While this method allows one to determine the fracture toughness, the process is extremely time consuming, tedious, and prone to human error when marking the sample and again when visually inspecting the crack growth relative to those markers. A MATLAB toolbox was written to automatically analyze crack propagation from digital images, create compliance plots, and export the fracture toughness using the modified beam theory method, the compliance calibration method, and the modified compliance calibration method. Rather than using 18 points of data as defined in the traditional method, the program computes the fracture toughness using hundreds of points, reducing noise and error. Different test types are compared i.e. ductile adhesive, brittle adhesive, crack jumping, inhomogeneous surface, etc.; and the results calculated using the ASTM standard method are compared to those exported by the MATLAB program.

Keywords: DCB, Fracture Toughness, MATLAB, ASTM D5528, Crack Propagation, Image Analysis
Corresponding author: Peter E. Caltagirone

SAMPE Virtual Conference Proceedings, 2020. Society for the Advancement of Material and Process Engineering – North America.

1. INTRODUCTION

Adhesive bonds are commonly used to join fiber reinforced polymer composites for the wind and aerospace industries. Understanding of the failure mechanics of adhesive joints enables the design of safe, reliable composite structures. Fracture mechanics are typically characterized using three failure modes: 1) mode I, which describes peeling, 2) mode II, which describes shearing, and 3) mode III, which describes tearing [5]. The most common cause of failure of adhesively bonded fiber reinforced composites is delamination, which is predominantly a mode I failure [1]. This type of failure is governed by the interlaminar fracture toughness of the joint, which is used in delamination failure models including the cohesive zone model (CZM) [1]–[3] and virtual crack closure technique (VCCT) [4]. Without accurate fracture toughness values, models cannot accurately be used to design and predict the failure of adhesively bonded composite structures.

A double cantilever beam (DCB) test is often used to determine the interlaminar (mode I) fracture toughness of adhesive joints as it is relatively easy to fabricate and test such specimens [6], [7]. The fracture toughness of a material is a measure of the critical energy release rate of a propagating crack through the material. A DCB test can be used to determine the fracture toughness of a material as a function of load, displacement, crack length, and specimen dimensions as established through various theories [8]–[14]. The ASTM standard to determine fracture toughness of composite materials (ASTM D5528) [6] suggests the use of three methods; the modified beam theory (MBT) [15], compliance calibration method (CC) [16], and the modified compliance calibration method (MCC) [17] to calculate the mode I fracture toughness.

While the current ASTM method provides a means to determine the mode I fracture toughness of adhesive joints, advances in digital image acquisition and processing that have been made in recent years can improve the use of the standard, especially in terms of improving the certainty of the measurements and simplifying the sample preparation. Specifically, as the test requires measurement of crack length during the delamination failure, the ASTM standard indicates that markings should be made on the sample in 1mm increments, and then visual inspection should be used to note the load and displacement each time the crack advances 1 mm. The accuracy of this visual inspection technique is limited by the accuracy of the marker placement on the sample and the use of only 19 data points. Furthermore, if the crack suddenly advances past more than 1 marker at a time, the data is skewed, and the fracture toughness value will be incorrectly reported. ASTM suggests using a microscope to measure the crack length as it passes the markings and record the force and displacement at that crack length. This process takes 30 minutes to an hour for each side of the sample, thus not only is the testing tedious, but also time consuming.

In this paper, toward addressing these limitations, we present the development of a digital image processing code to automatically measure the crack length and export the fracture toughness values of a DCB test using hundreds-to-thousands of data points in the form of digital images that are continuously acquired during the test. We also provide discussion of the suggested sample preparation method and camera setup. To verify and benchmark the code, multiple cases of crack propagation are demonstrated and a comparison of the code to traditional visual inspection-based calculations is given.

2. MATLAB CODE METHODOLOGY/EXPERIMENT SETUP

2.1 Program Methodology

The purpose of the software is to improve the process used to calculate mode I interlaminar fracture toughness from DCB tests. The program imports the test load and displacement data from a Microsoft Excel file, analyzes a series of images, and locates the edge of the crack in each image. Once the crack length has been determined throughout the test, the program then generates the required compliance plots and calculates the fracture toughness using the modified beam theory as described in ASTM D5528. The fracture toughness is calculated using the three methods found in the standard (MBT, CC, and MCC). The fracture toughness and flexural modulus are exported to the Excel file where the force and displacement data are obtained.

2.1.2 Required Inputs

An Excel file with the axial force and axial displacement as shown in Figure 1 must be placed in the same folder as the images. The force data should start in A4 and the displacement data should start in B4. It is imperative that the data rate be synced for the force, displacement, and images to ensure accurate results (the number of images should match the number of force/displacement data points). Image recording should stop once the crack is at its maximum length.

	A	B
1		
2	Axial Force	Axial Displacement
3	N	mm
4	0.147354126	0.040035695
5	0.372039795	0.083059072
6	0.757476807	0.127678737
7	1.158096313	0.169729814
8	0.908874512	0.211285427
9	1.434158325	0.251406804
10	1.522949219	0.294798985
11	1.560073853	0.338345766
12	2.049987793	0.380361453

Figure 1 - Excel data example

The other required inputs of the program are: initial crack length in mm, specimen width in mm, specimen thickness in mm, binary threshold of the test (recommend 70-95), and whether an export of the crack length video is desired. These inputs are shown in Figure 2.

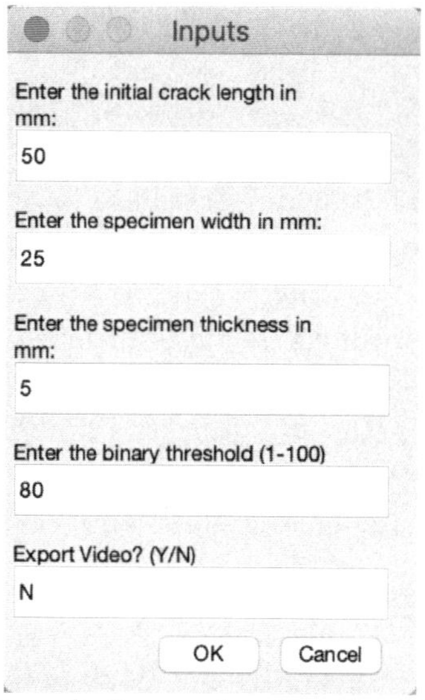

Figure 2 - Program input requirements

The images for one side of the sample should be in order in a single folder along with the data Excel file.

2.1.3 Image Examples

The images should be as clear as possible and the sides of the samples should contain no bubbles or unintended markings. A single black line should be placed on the sample to indicate where the start of the crack is (the end of the Teflon insert). The camera should be setup so that the full sample is shown within the frame, without the hinges in the picture. If desired, a small black marking can be made on the edge of the sample to indicate where to stop the test. An example of a good image frame setup is shown below in Figure 3. It should be noted that for the program to function correctly, the sample must be white with a black background and greyscale images should be used.

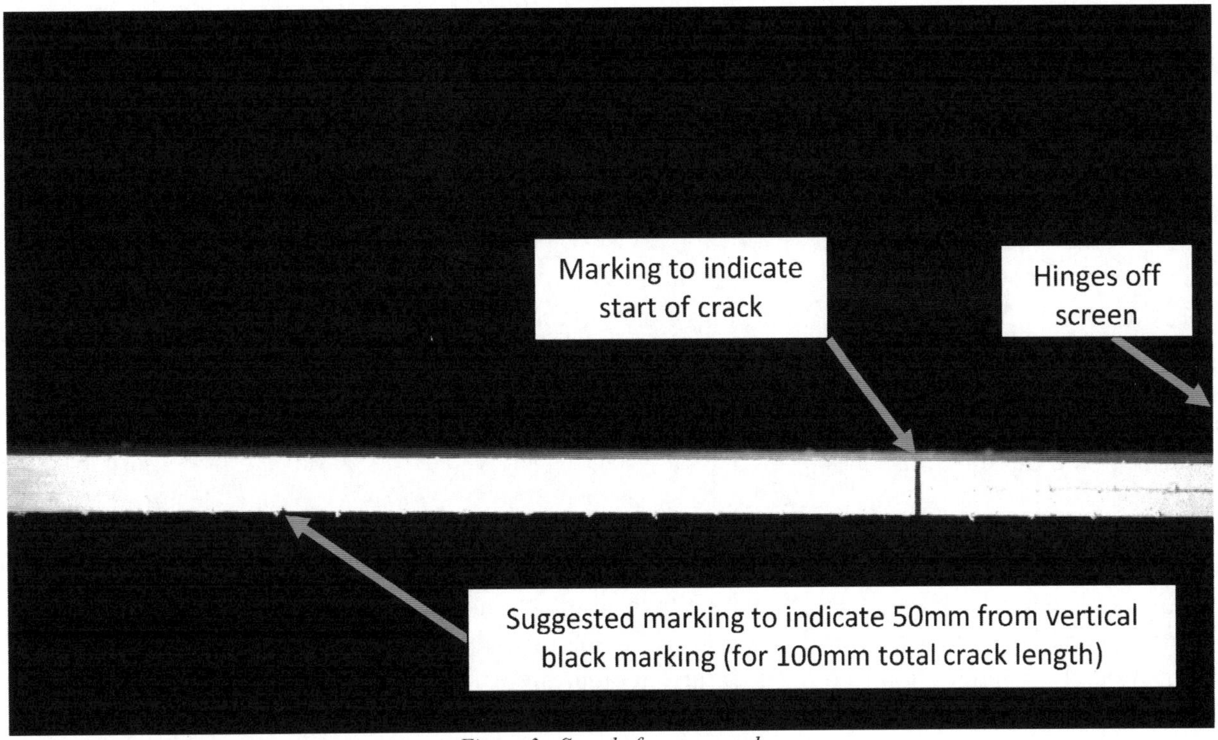

Figure 3 - Sample frame example

As for any testing requiring image acquisition, proper camera setup and lighting is essential to produce good results. For the image above, a FLIR Grasshopper3 camera was used (1920x1200 resolution, 12 megapixels) with a lens length of 45mm. Cameras are placed 0.9m on both side of the sample so crack propagation is obtained from both sides. A schematic of this setup is shown in Figure 4.

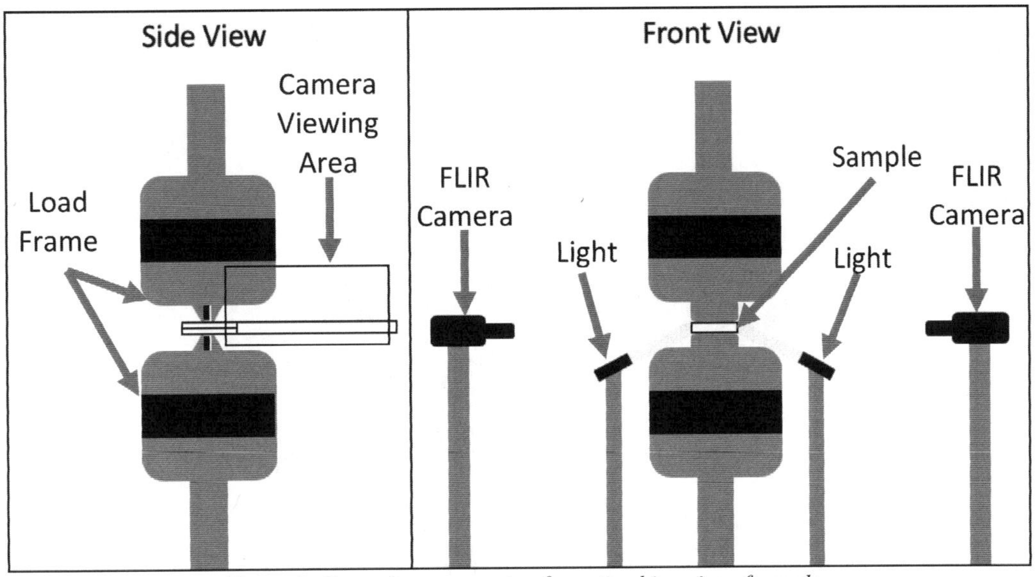

Figure 4 - Example camera setup for optimal imaging of sample

It is possible for the sample to be angled when loaded into the grips if the hinges are not applied perfectly. As the sample is loaded, the tilt is quickly corrected due to the forces acting on the specimen, however this tilt will cause the program to fail. For this reason, a rotation feature was added into the code to correct for any possible tilt occurring after gripping. After finding the edges of the sample, the program uses a least squares line of both sides of the sample to determine the angle and correct for the tilt. Additionally, the program crops each image to the testing area to reduce computational time. To accomplish this, the start of the crack is located in the first image (from the initial black line) and the final edge of the crack is located in the final image. These two points are then used to crop the image to the required testing area.

Once the images are rotated and cropped, the program locates the crack tip in each image. To do this, the top and bottom edges of the sample are first located, then the program searches column by column along the length of the sample for 'dark' pixels. The shade of pixel that determines the crack tip is set using the binary threshold; one of the required user inputs. Once the crack tip is located within each image, the program can use the load, displacement, and crack length to determine the fracture toughness.

As discussed in the introduction, the ASTM standard indicates three methods to calculate the fracture toughness (MBT, CC, MCC). Each method requires a slightly different compliance plot to obtain the equivalent crack length, which takes into account the rotation that occurs at the crack front. For each method, a compliance plot is created, and a LS line is used to determine the equivalent crack length required for accurate calculation of fracture toughness. Once this value is obtained for each of the methods, the program calculates the fracture toughness using the equations given in ASTM D5528.

2.2 Testing Cases

This next section will explore three separate cases that one may encounter when running a DCB test. The first will focus on a perfect sample, one which is parallel to the ground when loaded into the machine and crack propagation occurs evenly without jumping. The second case will focus on a sample that is tilted when loaded into the machine, and the correction and results. The final case will focus on a sample where the crack jumps more than 2mm, which would typically be cause to remove the sample from the batch.

2.2.1 'Perfect' sample

Figure 5 shows the initial image for a 'perfect' test. The sample has no noticeable rotation, the testing area is well lit, and there is good contrast between the sample and the background.

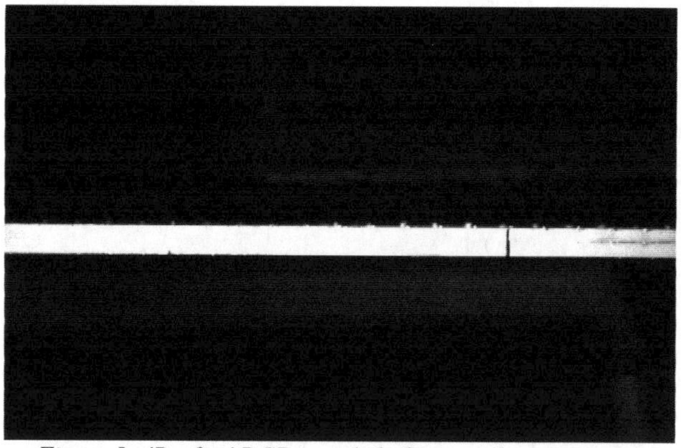

Figure 5 - 'Perfect' DCB sample before crack propagation

Figure 6 shows the result after the cropping procedure has taken place. The first and last images in the sequence are shown as examples. The sample is automatically flipped along the vertical, as analysis occurs from right to left. The cropped images show that only the full testing area of the sample remains after the operation, cutting down the computational time to analyze the crack length. The original images had a resolution of 1200 x 1920 pixels, while the final images a resolution of 645 x 1045 pixels; the required area to analyze is reduced by about 70 %. Additionally, the noise in the background of the initial image was reduced through a contrast operation, leading to more accurate tracking of the crack front.

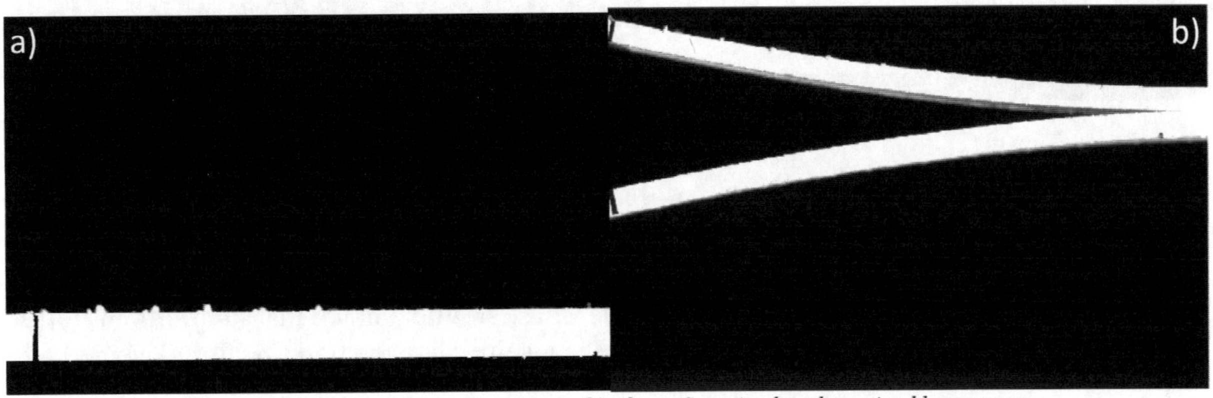

Figure 6 – First image (a) and last image (b) of test. Cropping box determined by program.

2.2.2 Angled sample

As discussed above, sometimes the sample can be angled when gripped in a load frame if the hinges are not adhered properly. This typically resolves itself within the first 10-15 seconds of testing as the even load on each side of the sample causes it to straighten out. If the sample is angled at the start of the test it still creates an issue for the program, thus a rotation correction was implemented into the code. Figure 7 shows an example of this case where the sample is slightly angled within the image (note that the hinges are off screen to the right of the image).

Figure 7 - Rotated DCB sample before crack propagation

Figure 8 shows the results after the rotation and cropping procedure has been completed. Again, the image has been flipped along the vertical to allow for proper analysis. Additionally, there is now no noticeable rotation in either final image. This allows for more accurate analysis of the crack length.

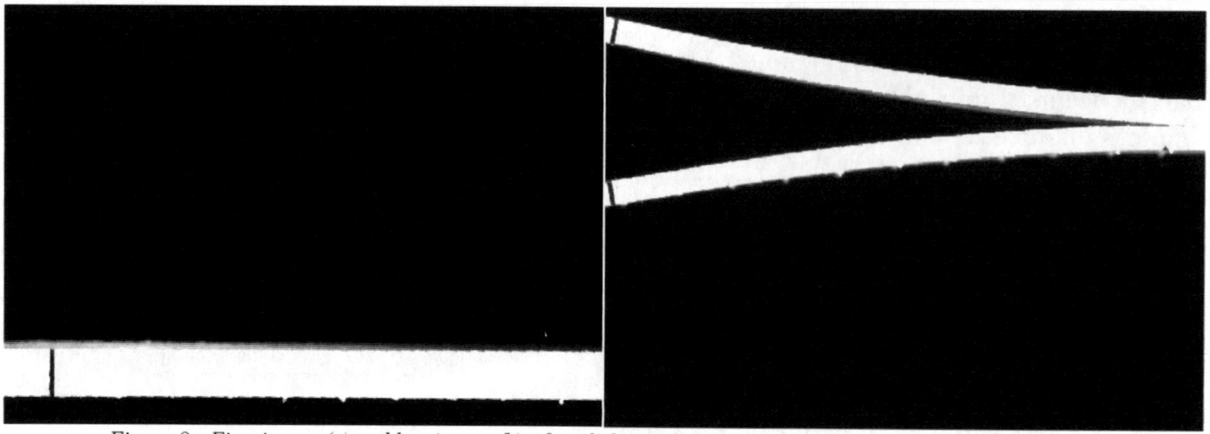

Figure 8 - First image (a) and last image (b) of angled sample test. Cropping box determined by program.

2.2.3 Crack-jump sample

In some cases, a crack will propagate through the sample for a period of time, and then the crack will arrest. Once the required energy is reached to continue crack growth, propagation will continue along the sample. Figure 9 is an example of a test with a crack that jumps 4mm in just over one second, which could be cause to discard the results using the traditional ASTM method.

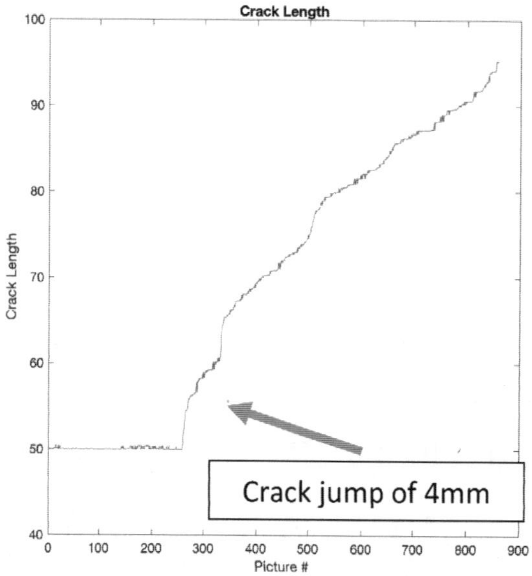

Figure 9 - Crack jump observed during testing

Compliance plots for the test in Figure 9 are shown in Figure 10. The left compliance plot shows a dataset used by the program with over 800 datapoints, whereas the right plot has only 18 datapoints; the number specified by the ASTM D5528 standard. The left plot results in a fracture toughness value of 0.4208 kJ/m^2 and the right plot results in a fracture toughness value of 0.4495 kJ/m^2. Depending on the application, a fracture toughness difference of 6% could make a significant impact on the performance of the system and being able to accurately predict when it will fail.

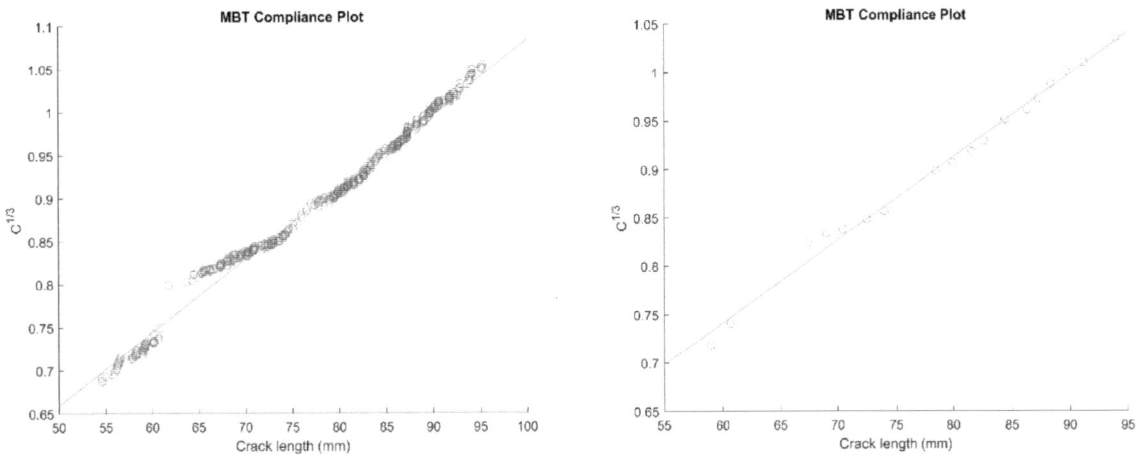

Figure 10 - Compliance plot showing how crack jump can potentially skew data

3. RESULTS

Verification of the code is important to ensure valid data is produced from experiments when using this program. To accomplish this, cracks within images were measured by hand and this data was used to calculate the fracture toughness values. 11 tests were completed and compared to the values output by the program. ImageJ was used to measure 18 data points evenly spaced throughout the tests.

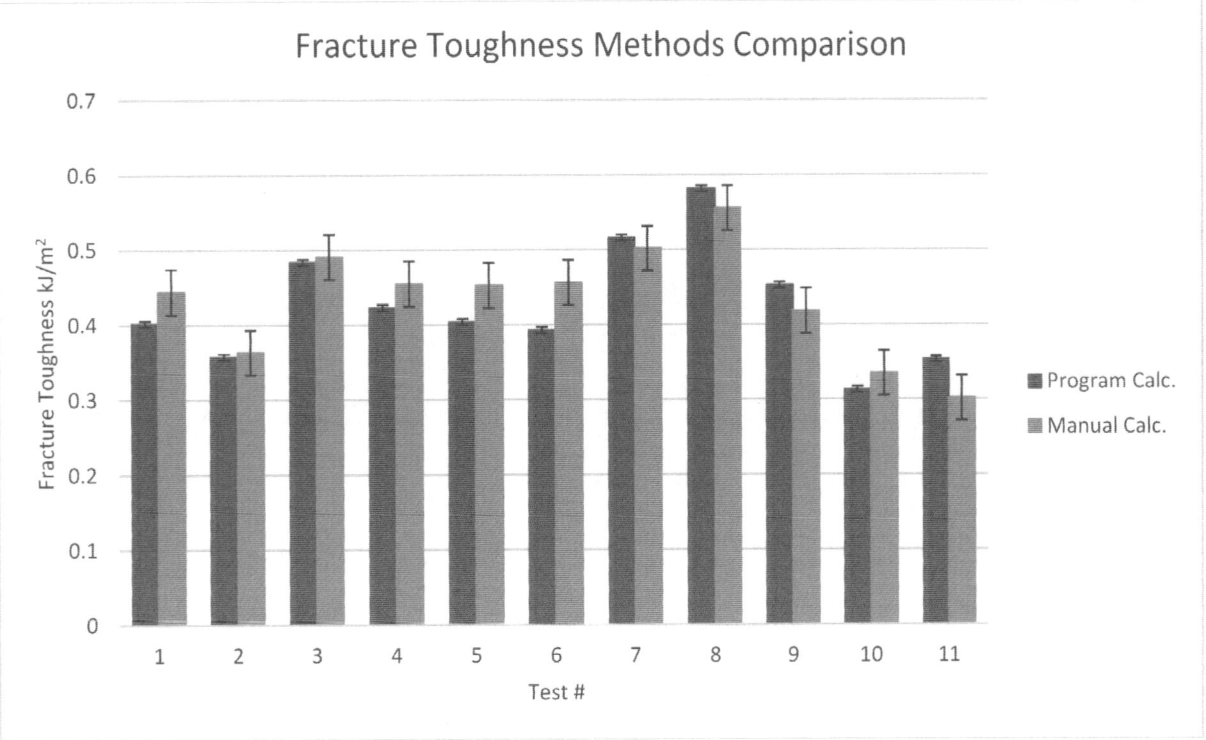

Figure 11 - Comparison of fracture toughness values calculated using the program and by measuring crack length by hand

The error bars on the plot show the 95% confidence interval for the tests. As the program uses close to 1000 datapoints for each calculation, whereas the hand calculation only uses 18, the values calculated by the program are likely more accurate than the ones calculated by hand as any effect the outliers have on the data are reduced by the number of datapoints. Both the program and manual fracture toughness values are very close in each test, varying by less than 0.05 kJ/m2. This indicates the program is working properly in determining the fracture toughness values through autonomous tracking of the crack tip.

4. CONCLUSIONS

As discussed, the ASTM D5528 method for determining the fracture toughness of an adhesive joint between two fiber reinforced composite pieces using a double cantilever beam (DCB) test can be improved through the autonomous crack tracking program written using MATLAB as it allows for easier analysis of the tests and reduces potential errors from sample fabrication or measurement of crack length during the test. Any error from misplaced markings is eliminated when using the program as only one initial marking is required for the location of the start of the crack. Additionally, the number of data points is dependent on the data sampling rate, so thousands of data points could be used to determine the fracture toughness of an adhesive, rather than only 19 per the ASTM standard, which further eliminates errors and also allows for accurate analysis of samples where the crack jumps.

5. ACKNOWLEDGEMENTS

The authors gratefully acknowledge financial support from the State of Colorado Office of Economic Development and International Trade Advanced Industries Program (program manager Katie Woslager) and Colorado Higher Education Competitive Research Authority (CHECRA) through their commitment to the Institute for Advanced Composites Manufacturing and Innovation (IACMI) Wind Energy program. This material is based upon work supported by the U.S. Department of Energy's Office of Energy Efficiency and Renewable Energy (EERE) under the support of the Task 4.2 of Institute for Advanced Composites Manufacturing Innovation (IACMI), Award Number DE-EE006926 managed by John Winkel from DOE and John Unser from IACMI. Academic and National Laboratory partners for this project are: Derek Berry and David Snowberg (National Renewable Energy Lab), Aaron Stebner (Colorado School of Mines), Nathan Sharpe (Purdue), Dayakar Penumadu (University of Tennessee), Douglas Adams (Vanderbilt). Industrial consortium was led by Dana Swan (Arkema), Mingfu Zhang (Johns Manville), and Stephen Nolet (TPI Composites). The views and opinions of authors expressed in this paper or referenced documents do not necessarily state or reflect those of the United States Government or the identified collaborating partners. Authors also acknowledge that important insight and ideas were obtained from academic and industrial collaborators during the project activities who are not being formally acknowledged in this manuscript as co-authors. Materials supplied and manufacturing methods developed by the industrial collaborators are gratefully acknowledged.

6. REFERENCES

[1] G. I. Barenblatt, "The Mathematical Theory of Equilibrium Cracks in Brittle Fracture," *Adv. Appl. Mech.*, 1962.

[2] Y. Mi, M. A. Crisfield, G. A. O. Davies, and H. B. Hellweg, "Progressive delamination using interface elements," *J. Compos. Mater.*, 1998.

[3] W. Xu and A. M. Waas, "Multiple solutions in cohesive zone models of fracture," *Eng. Fract. Mech.*, 2017.

[4] E. F. Rybicki and M. F. Kanninen, "A finite element calculation of stress intensity factors by a modified crack closure integral," *Eng. Fract. Mech.*, 1977.

[5] N. E. Dowling, K. Siva Prasad, and R. Narayanasamy, *Mechanical behavior of materials : engineering methods for deformation, fracture, and fatigue*. Pearson, 2013.

[6] "ASTM D5528-13. Standard test method for Mode I interlaminar fracture toughness of unidirectional fiber-reinforced polymer matrix composites," West Conshohocken, PA, 2013.

[7] A. Brunner, B. Blackman, and P. Davies, "A status report on delamination resistance testing of polymer–matrix composites," *Eng Frac Mech*, vol. 75, no. 9, pp. 2279–2794, 2008.

[8] J. G. Williams, "End corrections for orthotropic DCB specimens," *Compos. Sci. Technol.*, 1989.

[9] P. Qiao and J. Wang, "Novel joint deformation models and their application to delamination fracture analysis," *Compos. Sci. Technol.*, 2005.

[10] M. M. Shokrieh, M. Heidari-Rarani, and M. R. Ayatollahi, "Interlaminar fracture toughness of unidirectional DCB specimens: A novel theoretical approach," *Polym. Test.*, 2012.

[11] S. Wang and C. Harvey, "A theory of one-dimensional fracture," *Compos. Struct.*, 2012.

[12] Z. Suo, G. Bao, B. Fan, and T. C. Wang, "Orthotropy rescaling and implications for fracture in composites," *Int. J. Solids Struct.*, 1991.

[13] M. G. Andrews and R. Massabò, "The effects of shear and near tip deformations on energy release rate and mode mixity of edge-cracked orthotropic layers," *Eng. Fract. Mech.*, 2007.

[14] J. Xie, A. M. Waas, and M. Rassaian, "Closed-form solutions for cohesive zone modeling of delamination toughness tests," *Int. J. Solids Struct.*, 2016.

[15] S. Hashemi, A. J. Kinloch, and J. G. Williams, "Corrections needed in double-cantilever beam tests for assessing the interlaminar failure of fibre-composites," *J. Mater. Sci. Lett.*, 1989.

[16] J. P. Berry, "Determination of fracture surface energies by the cleavage technique," *J. Appl. Phys.*, 1963.

[17] K. Kageyama and M. Hojo, "Proposed methods for interlaminar fracture toughness tests of composite laminates," in *5th U.S./Japan Conference on Composite materials*, 1990, pp. 227–234.

INVESTIGATION OF THE TEMPERATURE INFLUENCE IN THE CONTEXT OF AUTOMATED FIBER PLACEMENT LAYUP ON DOUBLY CURVED TOOLS

Alex Brasington[1], Shingo Miura[2], Tsuyoshi Saotome[2], Ramy Harik[1]

[1]University of South Carolina
McNAIR Center, Department of Mechanical Engineering
1000 Catawba Street
Columbia, SC 29201

[2]Toray Composite Materials America Inc.
19002 50th Avenue East,
Tacoma, WA 98446

ABSTRACT

This study aims to investigate the induced effects on the temperature requirements due to the use of a complex geometry, while also gathering data for two types of materials supplied by Toray Industries. Automated Fiber Placement (AFP) has been sweeping the composites manufacturing industry due to its ability to manufacture parts with high speed, repeatability, and process quality. As a result of the advancement of the machines' capabilities, it is being used to manufacture parts with higher degrees of complexity. The effect of the shape's complexity on process parameters, namely temperature, is not well understood. Experiments consists of using zero- and ninety-degree constant angle layups on a complex tool to thoroughly examine the effect of curvature on the AFP process. Results show that material selection is vital to ensure high quality parts because of the difficulties with adhesion to the complex surface. Repeated layups indicate that an increase in temperature is necessary in concave areas to overcome bridging and steering defects. Heater deviation can also vary and will result in local increases and decreases in temperature. All of these effects must be accounted for when process planning, and should be monitored during AFP manufacturing to ensure high quality parts.

Keywords: Automated Fiber Placement, Process Parameters, Temperature Influence, Complex Tools
Corresponding author: Ramy Harik (harik@cec.sc.edu)

1. INTRODUCTION

Automated Fiber Placement (AFP) is an automated process that consists of a gantry/robotic system with an attached fiber placement head. The AFP head enables multiple strips, or tows, to be laid onto a tool surface to manufacture the laminate. Adhesion between the incoming tows and substrate at specified speeds is ensured by using appropriate process conditions such as heating, compaction, and tensioning systems. A series of tows forms a course, courses are then combined to create a ply, and multiple plies create a laminate. AFP has been consistently progressing the ability to manufacture composite parts with high speed, repeatability, and improved quality. This had led to more industries adopting and researching the technology. Flat, low curvature geometries, and cylinders have been the focus areas of research, but AFP in now being used on parts with increasing complexity. Moreover, AFP is also explored for small and medium parts, versus the traditional large structures. Previously complex parts had to be manufactured through manual lay-up which is a costly and time-consuming process – as well as having accuracy issues – where each composite layer is placed by hand. Although such a process allows for a wider range of applicable surface complexity, it is not feasible for larger structures and the adaptation of the AFP process is beneficial. The AFP machine enables the process to be fully automated with little machine stoppage and manual work. But when AFP defects occur, laying is put on hold and material has to be reworked. The complexity of the part has significant effects on the required AFP process parameters to prevent these defects. There is a difference between 1D, 2D, and 3D layups in terms of process parameters and defect occurrence [1]. 2D steered layups have changes of steering radius within a single ply, but 3D layups also have a change of tool curvature within a single course. This makes it difficult to find good process settings for an entire 3D layup. In order to achieve this, the process parameters must be understood as well as the parameters effect on defect occurrence and their inter-relation. The essential parameter to be considered in this study is temperature, although optimization of compaction, feed rate (head speed), and tow tension are also vital for high quality layups.

Temperature is a crucial factor to ensure proper adhesion of incoming tows to the substrate. Research has shown that proper process temperatures lead to increased adhesion and higher overall part quality [2]. The applied temperature must be in a certain processing window: high enough to ensure adequate material tackiness, but not so high that material degradation occurs. The importance of sufficient temperature is increased when laying up on complex tools with the combination of concave and convex curvatures. The extent of possible curvatures of a tool must be within a certain limit to prevent collisions with the machine being used [3]. Even with this limitation, the curvatures can be severe enough to warrant attention to the associated effects, however published research is limited on this topic. [4] presented a decomposition of the heating elements in AFP to enable better across the roller control of the heat received by the substrate. When placement is over complex parts, heat is received differently by the substrate due to differences in distance across the roller. There are multiple defects that can occur from improper adhesion, however the main defects due to tool curvature are bridging and steering induced defects such as wrinkles, puckers and angle deviation. A bridging defect occurs when a tow does not fully adhere to the concave surface, resulting in a gap between the tow and substrate [1]. Steering is induced by the projection of a constant fiber angle onto a curved surface, producing a tool path with in-plane curvature. Although the radius of the tool path is not significant compared to the minimum steering radius, it produces defects commonly seen in steering experiments. The notable defects of this study are folds, wrinkles, and further bridging. Tow pull up occurs due to the change in length between the inside and outside of the tow. This change in length causes tensile strains leading to the outer part of the tow losing its adherence to the substrate, and possibly bridging [5] [6]. The combination of the tool curvature and induced steering creates a challenging lay-up scenario.

Preventing the aforementioned defects is important to ensure optimal part quality and performance. This study will delve into the required temperatures for proper adhesion along a doubly curved tool. The effects will be quantitatively examined to observe the trend of the necessary temperatures, and a lay-up of all courses will validate experimentation. The knowledge gained will assist in increasing the quality of complex parts produced by AFP machines, therefore expanding its applicability. The remainder of the report will consist of 4 sections. Section 2 will explain the experimental setup and procedures used during experimentation. Section 3 will present the results of trials with single courses and entire ply layups. Section 4 will discuss the findings and their significance. Lastly, Section 5 will consist of a summary of our work and suggestions for future research.

2. EXPERIMENTAL PROCEDURES

2.1 Experimental Setup

These experiments were carried out using an Ingersoll Machine Tools Lynx AFP machine and a doubly curved tool made out of aluminum, both of which can be seen in Figure 1 and Figure 2. This tool allows for experimentation on a doubly curved surface with varying curvature in the zero direction. Heat was applied using the humm3® [7] head attachment seen in Figure 2 (b). This heating source is a flash lamp with rapid heating capabilities along with a small head that makes it beneficial for complex tools. The shape of the crystal allows for focused heating on the substrate and prevents heating of incoming tows. It is known that the temperature is directly related to the power of the heater, and the power is dependent on three variables: voltage, frequency, and pulse time [8]. By setting the frequency and pulse time at 60 Hz and 2000 ms respectively, we can obtain increments of power by controlling the voltage. The lowest voltage used was 100 V while the highest used was 220 V. Head cooling was also used to ensure the tows inside the AFP head remained cool to prevent them from getting tacky and sticking inside the head. The head cooling temperature was set to be 6-8 °C.

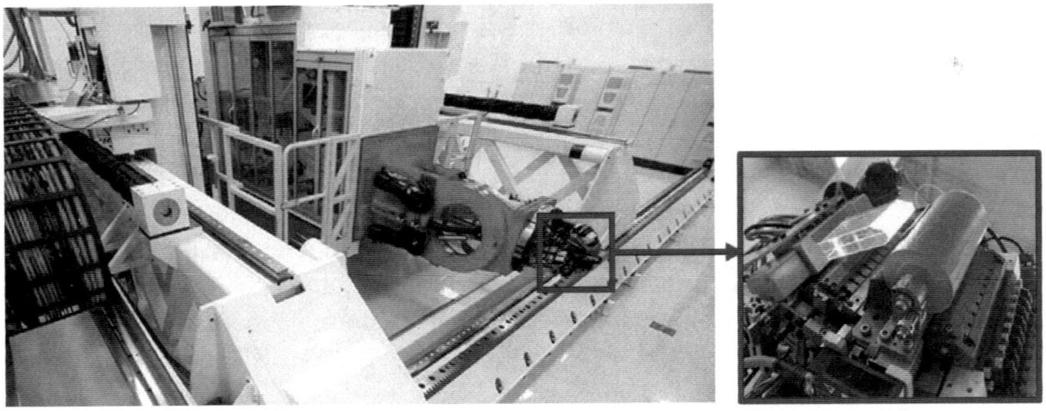

(a) (b)

Figure 1: Image of the AFP machine and fiber placement head

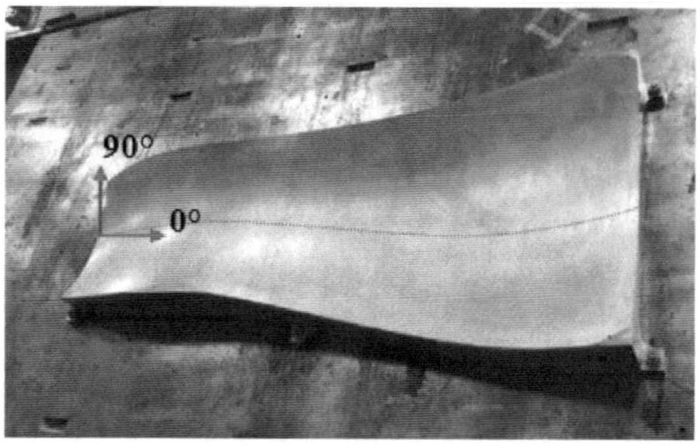

Figure 2: Image of the doubly curved tool used during experimentation

Furthermore, the experimentation used two materials supplied by Toray Industries: material A and material B. Both materials had a tow width of 6.35 mm (0.25 in), and 8 tows were used to create the courses. Different materials were used to ensure that the results were valid between materials with varying properties.

Programming of the AFP machine was performed using the Ingersoll Composite Programming System (iCPS). Careful consideration was taken through simulations and dry runs to ensure that any possible collisions were avoided. After ensuring there were no collisions, courses were ran to determine the highest speed that could be achieved while also remaining constant. This was determined to be 10 m/min. Similar to the speed, the other process parameters remained constant to explore only effects on temperature. The compaction was set at 445 N (100 lbs) and was applied using a roller with a width of 5.5 cm (2.2 in). Lastly, the tow tension was set at 3 N (0.67 lbs).

2.2 Experimental Procedures

A two-stage test plan was used to explore the effects that complex geometries have on required process temperatures. The first stage consisted of experimenting with a single model course in the zero- and ninety-degree directions. The experimental courses are highlighted in red in Figure 3. Four initial plies were laid before experimenting to eliminate variability and heat dissipation into the tool. The zero-degree courses are straight in the AFP roller travel direction, while the curvature of the tool varies significantly as illustrated in the x-z projection. Contrarily, the ninety-degree courses have a varying steering radius as illustrated in y-z projection, while the tool curvature is relatively constant as illustrated in y-x projection. The projections can be seen in Figure 4. This stage allowed for a detailed examination of the experimental courses to analyze any present defects, while also exploring the differing curvatures along the tool. The assessment of the defects was done visually and classified in two parts: (1) how well the incoming tows adhered to the substrate, and (2) how substantial the defects were. Using this assessment, the courses were categorized as either a fully tacked course, or a course that has defects. The fully tacked category includes courses that are fully adhered to the substrate with no large visible defects. The defects category includes courses that are partially or totally untacked from the substrate or have significant defects. Examples of each category can be seen in Figure 5. If a layup resulted in a course with defects, the temperature was incrementally increased until a fully tacked course was achieved. This data provides the lower bound of the materials processing window. The temperature measurements were acquired using a Graphtec data logger (GL980) and thermocouples placed onto the substrate material along the course. This allowed for measurements at various curvature values.

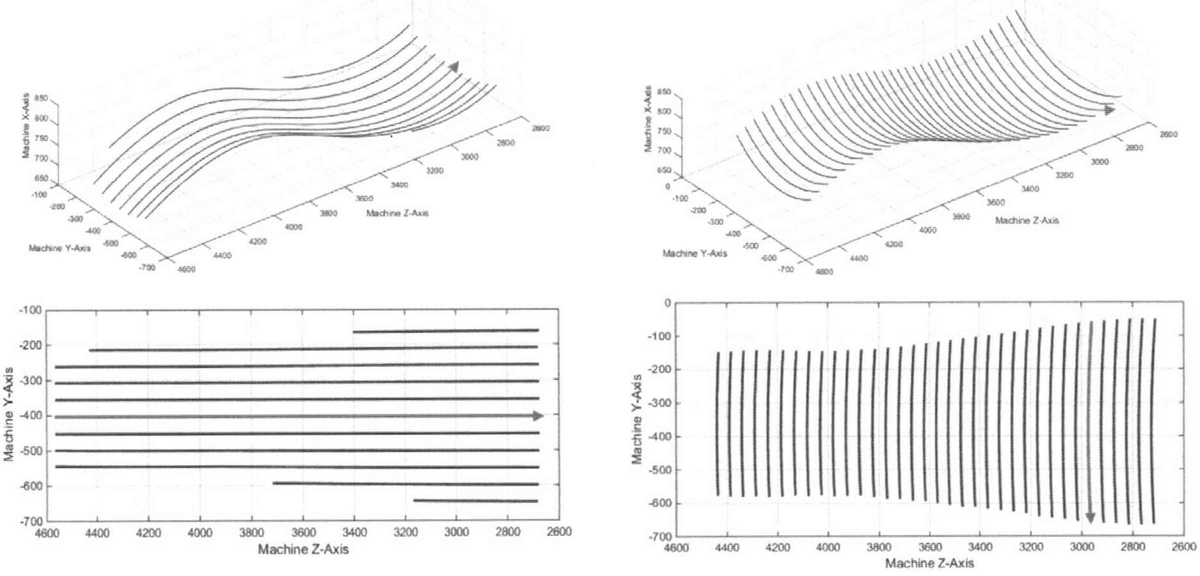

(a) (b)

Figure 3: (a) zero- and (b) ninety-degree courses

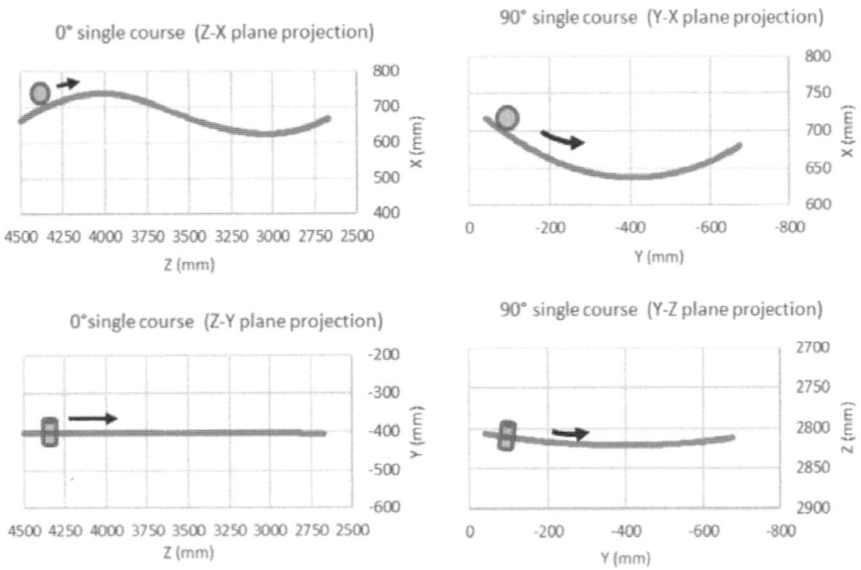

(a) (b)

Figure 4: (a) zero- and (b) ninety-degree courses with reference to roller direction

Stage two of the experiment was used to validate the results from stage one. In this stage, the layup was done for the entire surface to ensure that the required temperatures were valid across the part. Unlike stage one, this stage consisted of little intervention during the manufacturing process. Allowing the manufacturing to be completed ply-by-ply reflects the usual AFP process where each ply is laid and then inspected for defects.

3. RESULTS

3.1 Single Course Layup Trials

Figure 5 shows measured temperature and layup results with material A at multiple points along the zero-degree single course. Each point on the curved dotted line indicates measured substrate temperatures and inspection results at thermocouple locations along the single course. Temperatures are indicated on the y-axis of the graph, and visual inspection results are indicated by green circles (No defects) or red triangles (Defect occurrence). Red painted triangles represent defects such as lifted tows that occurred during layup, while hollow triangles represent defects of untacked tows that occurred at start of the layup. During the layup, machine feed rate (or head speed) and supplied voltage to the humm3® heater was constant. Therefore, the variation of material temperature might be due to the change of the distance between humm3® heater crystal and substrate material. However, in this case, maximum temperature difference was within 5 °C in single layup course.

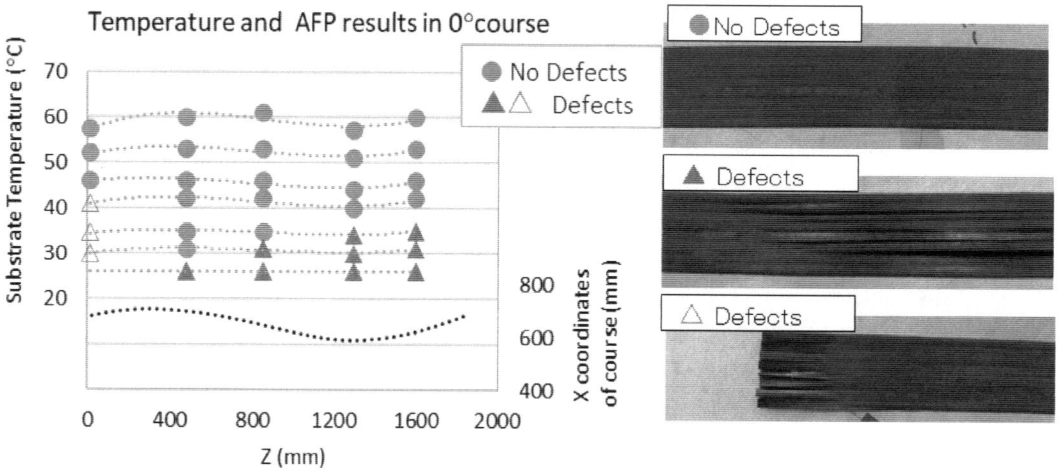

Figure 5: Measured temperature and inspection results for each layup settings with material A

As a whole, when materials were placed with higher temperature, it was more likely to adhere without defects. However, minimum temperature for proper adherence was different depending on the location along the course. The highest required temperature was needed at the beginning of the course, possibly caused by laying tows chilled inside AFP head with chilled roller, or just effect of tow edge's insufficient adhesion. Therefore, these abnormally high temperatures were not considered an effect of tool geometry. It should be noted that minimum temperature for tack was higher at the concave area by more than 5 °C compared to those at the convex area. Though there was not much difference (less than 5 °C) in material temperature throughout a single course, both proper adherence and defects appeared. This suggested that tow defects were caused by course shape.

The same experimentations, measurements and plotting were conducted for single ninety-degree courses with materials A and B. Minimum required temperature for every course and material were recorded and connected respectively to show a clear tendency of required temperature for adherence (Figure 6). Temperatures above the lines in the Figure 6 can be regarded as processing windows for each combination of material and course geometry. In zero-degree single course layup trials, both materials showed that concave areas needed higher temperatures than convex areas for proper adherence. Also, the start of each course required the highest temperature. Meanwhile, ninety-degree courses needed relatively higher temperature at all points along course than the zero-degree layup. These results strongly suggest concave areas generally require higher applied temperatures than convex areas. In addition, unlike the zero-degree results, ninety-degree courses showed almost the same, or slightly higher temperatures at all points along the course.

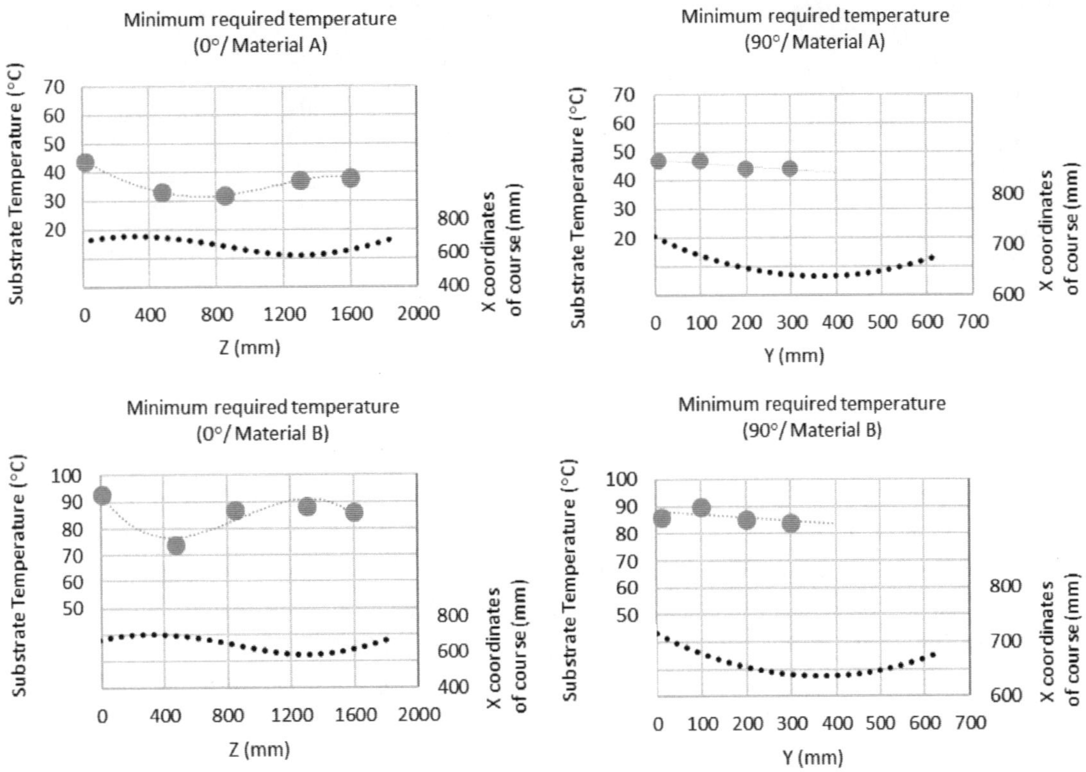

Figure 6: Minimum required temperature trend for adhesion in zero and ninety single course with material A and B

3.2 Surface Layup Trials

The series of results for entire ply layup are shown in Figure 7. There were no bridging defects during the zero-degree layup with material A when tows were laid down with an applied temperature above 50 °C. There were also no significant bridging defects when ninety-degree surface layups were laid with temperatures above 50 °C. Therefore, the AFP process did not have to be paused due to re-work defects during the layup of these plies. On the other hand, ninety-degree surface layups with temperatures above 38 °C brought many defects at the right-most area of the concave portion of the tool. This is in good agreement with the results obtained in first stage of experiments. However, significant defects did not occur after 8th course of the ply. These tendencies were the same when laying up with material B, though the temperature values were over 30 °C higher than those with material A. Therefore, the processing windows for each material obtained from the first stage experiments were well validated, and actual ply manufacturing was successfully demonstrated.

(a) (b) (c)

Figure 7: Entire surface layup with temperatures above and below the processing window for(a) 0 and (b)(c) 90 degree with material A. (Equivalent results were obtained with material B)

4. DISCUSSION

From the experiment, it became clear that concave areas are prone to bridging defects, even though processing temperatures were equivalent to the ones in the convex area. This is assumed to be due to the tow tension and tensile strain based on tow draping. Figure 8 and Figure 9 show schematic images of tows being laid along each course. Figure 8 (a) and (b) represent tows on different area of the zero-degree course. Both tows are exposed to effect of curvature, however the tensile vector component faces toward substrate in Figure 8 (a), but faces the opposite direction in Figure 8 (b). The tensile vector in the direction away from the substrate forces tows to lift up or bridge.

Ninety-degree course experiments brought similar results in terms of required temperature values in the concave area. However, the defects in these courses were characteristic of those shown in Figure 7 (c). The defects of tow lift up occurred at the outside of the course, and particularly only occurred on courses 1-6. Even though the steering radius in any single course was as large as 4500 mm measured from y-z projection of Figure 3, these defects are obviously caused by steering. This partial insufficient adhesion was contributed to tensile strain at the outer edge of the curved tows caused by the change in length between the outside and inside of the course (Figure 9 (b)). These effects should be taken into account to avoid bridging defects when tows are placed onto curved tool surface in both travel and width direction of layup motion.

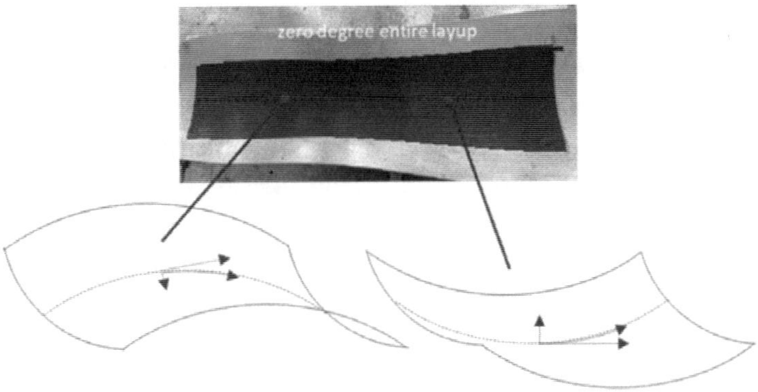

(a) (b)

Figure 8: Schematic image of tow strain caused by curvature and steering: (a) convex and (b) concave for 0-degree course

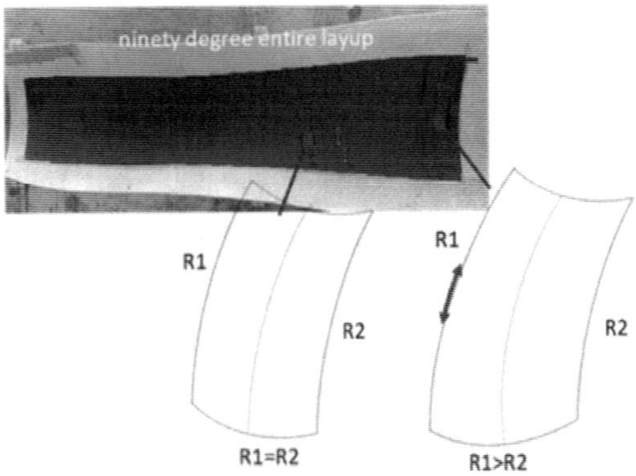

(a) (b)

Figure 9: Schematic image of tow strain caused by curvature and steering: (a) constant and (b) varying steering radius for 90-degree course

5. CONCLUSION

In this work, accurate evaluation of processing temperatures and materials defects through experimental measurement and visual inspection of layups on a double curved tool were achieved. Tool complexity and AFP course shape was analyzed, and its effect on defect occurrence was investigated into detail. The results showed tow-lift and bridging defects on the complex tool occurred not due to heating deviation, but course shape geometry itself. It made it quantitatively clear that concave areas with lower curvature radii need higher temperature for tow placement without defect occurrence. Also, these defects are more likely to occur when tows were placed with steering on curved surfaces. The defects occurrence was contributed to the tensile strain in the tow while being placed on a curved path and a curved tool. This was considered to lead to insufficient adhesion between the tows and the substrate. Additionally, our understanding of the defects and material process parameters was validated through demonstration of entire ply layup with less defect occurrence or machine stoppages. The acquired quantitative information will help to perform 3D fiber placement on complex parts and can give useful insight to solve defects which could occur during 3D layups.

This research can be seen as a basis for future experimentation into how AFP heating can be affected. Firstly, more fiber angles should be explored (±45°), including non-conventional angles (15°, 30°, 60°). These experiments focused on zero- and ninety-degree fiber angles, but structures can have a wider variety than this. Also, more significant steering radii could be explored to examine how the required temperature may change. Applications of AFP layup on complex shapes also need to be explored. This can include things such as doublers, blended wing bodies, as well as smaller complex parts within a structure. Research also needs to consider how the other process parameters can affect the minimum required temperatures, as this will play an important role in determining the overall optimal processing conditions. Furthermore, this study has potential to be expanded to simulate layup on three dimensional tools and predict defects occurrence by combining layup course geometry analysis and actual experimental data. If the simulation is developed and it can predict how to resolve defect occurrence, and it would enable the AFP process to perform better on complex geometries.

6. REFERENCES

[1] R. Harik, C. Saidy, S. Williams, Z. Gurdal and B. Grimsley, "Automated Fiber Placement Defect Identity Cards: Cause, Anticipation, Existence, Significance, and Progression," in *SAMPE 2018 Conference & Exhibition*, Long Beach, CA, 2018.

[2] A. B. Hulcher, J. M. Marchello, J. A. Hinkley, N. J. Johnston and M. A. Lamontia, "Dry Ribbon for Heated Head Automated Fiber Placement," NASA Langley Reserach Center, 2000.

[3] J. Halbritter, R. Harik, A. Zuloaga and M. vanTooren, "Tool Path Generation on Doubly-Curved Free-Form Surfaces," *Computer-Aided Design and Applications,* vol. 14, no. 6, pp. 844-851, 2017. https://doi.org/10.1080/16864360.2017.1287760

[4] K. Xia, R. Harik, J. Herrera, J. Patel and B. Grimsley, "Numerical Simulation of ADP Nip Point Temperature Prediction for Complex Geometries," in *SAMPE 2018 Conference & Exhibition*, Long Beach, California, 2018.

[5] R. Wehbe, "Modeling of Tow Wrinkling In Automated Fiber Placement Based On Geometrical Considerations," University of South Carolina, M.S. thesis, 2017. https://scholarcommons.sc.edu/etd/4449/

[6] G. Rousseau, R. Wehbe, J. Halbritter and R. Harik, "Automated Fiber Placement Path Planning: A state-of-the-art review," *Computer-Aided Design & Applications,* vol. 16, no. 2, pp. 172-203, 2019. https://doi.org/10.14733/cadaps.2019.172-203

[7] Heraus, "humm3 - Intelligent heat for Automated Fibre Placement (AFP)," [Online]. Available: https://www.heraeus.com/media/media/hng/doc_hng/products_and_solutions_1/arc_and_flash_lamps_1/Hume_2019.pdf. [Accessed 22 November 2019].

[8] Baz Radwan, A.(2019). Experimental Analysis of the Automated Fiber Placement Manufacturing Parameters for High and Low Tack Prepreg Material. (Master's thesis). Retrieved from https://scholarcommons.sc.edu/etd/5432

ULTRASONIC CONSOLIDATION OF DRY CARBON FIBER AND POLYPHENYLENE SULFIDE FILM

S. Williams and G. Palardy

Department of Mechanical & Industrial Engineering, Louisiana State University
3304 Patrick F Taylor Hall
Baton Rouge, LA 70803

ABSTRACT

Ultrasonic welding of thermoplastics is a common practice in industry but its use for composite materials is not fully understood yet. Ultrasonic consolidation of carbon fiber/thermoplastic composites uses frictional and viscoelastic heating generated by ultrasonic vibrations to melt the thermoplastic matrix and infuse dry fibers. The advantage of this method is its high speed (<10 s) and low cost compared to traditional techniques such as compression molding. Ultrasonic consolidation was performed on polyphenylene sulfide (PPS) films and woven dry carbon fibers with a Rinco Dynamic 3000 welder. Consolidation pressures ranged from 0.48 MPa to 1.1 MPa. Weld duration was controlled with the welder's travel parameter, setting the distance of compression during the application of vibrations. Viable travel values were determined from fiber compressibility. Void content was obtained with optical microscopy and crystallinity was calculated using differential scanning calorimetry (DSC). Crystallinity was also verified with microhardness measurements. This study demonstrated that ultrasonic consolidation is a viable manufacturing method, producing CF/PPS parts with low void content (< 2 %) and crystallinity values that increased with consolidation pressure.

Corresponding author: Genevieve Palardy (gpalardy@lsu.edu)

1. INTRODUCTION

The need for lightweight structures in the aerospace and automotive industries creates a demand for cost effective manufacturing methods. Thermoplastic composites (TPCs) are well suited for this because of their short manufacturing time and their ability to be re-melted for assembly. Ultrasonic consolidation (USC) of thermoplastic composites is a promising manufacturing technique. This method has the advantages of rapidity, energy efficiency, and ease of automation. Ultrasonic consolidation uses frictional and viscoelastic heating generated by ultrasonic vibrations to melt the thermoplastic matrix and infuse the dry fibers. Ultrasonic welding of thermoplastic parts has been in use since the 1980's [1], but its application to joining thermoplastic composite parts has only been studied more extensively in the past 7 years [2]. There have been very few studies on the ultrasonic-assisted consolidation of composite materials, which is the focus of this study.

Gomer et al. [3] conducted a study on unidirectional dry carbon fibers (CF) and glass fibers (GF) with polyamide 6, polyethylene, and polypropylene films. They successfully demonstrated the feasibility of this method, requiring high tension in the unidirectional dry fibers during consolidation to prevent the fibers from spreading. They reported maximum tensile strengths of 1.2 GPa and a fiber volume fraction of 33 %, but did not investigate the microstructural outcomes. Another similar method was developed by Rizzolo et al. [4], who tested the use of ultrasonic heating for automated fiber placement of unidirectional GF/high-density polyethylene tape and unidirectional CF/polyethylene terephthalate tape. They accomplished this by placing the tapes onto a moving stage under a stationary sonotrode. They demonstrated that the stiffness of these layups decreased by approximately 50 % when the speed of the layup stage during consolidation was tripled.

Crystallinity plays an important role in the mechanical behavior and chemical resistance of a polymer. Higher crystallinity will lead to higher chemical resistance, stiffness, and strength, where an amorphous structure will be less brittle and more impact resistant [5]. Differential scanning calorimetry (DSC) is typically used to calculate the crystallinity of polymers. After ultrasonic heating, the polymer cools down from its meting temperature to room temperature in less than 10 seconds. Cooling rates this rapid are expected to result in a predominantly amorphous thermoplastic matrix [3]. Microhardness tests are a simple method for characterizing the microstructure of a polymer. The microhardness is directly related to the crystallinity of the sample [4] and may be used to compare the crystallinities of samples consolidated under different processing conditions.

During ultrasonic consolidation, pressure and vibrations are applied through a flat sonotrode. In theory, the optimum downward displacement of the sonotrode during consolidation (also called "travel") should be high enough to replace all of the air inside the dry CF plies with melted PPS. This idea is based on Helmus et al. who studied air evacuation from a partially impregnated microstructure [6]. The amount of air inside the CF plies is determined based on the thickness of the dry CF plies, the mass per area of the plies and the volume density of the carbon fibers. The volume fraction V_f of carbon fibers inside the plies is given by Eq. [1]:

$$V_f = \frac{\sigma}{th * \rho_f} \qquad [1]$$

where σ is the surface area density of the plies, *th* is the thickness of the plies, and ρ_f is the volume density of the carbon fibers. The travel T required to impregnate the plies is given by Eq. [2]:

$$T = th * (1 - V_f) \qquad [2]$$

where T is the travel required. Thickness of the plies refers to thickness just before the onset of consolidation once the consolidation pressure has been reached. Compression tests were conducted to determine the relationship between the consolidation pressure and the thickness of the dry CF plies.

This paper is focused on investigating the feasibility of manufacturing CF/PPS composites from woven dry fibers and PPS films using ultrasonic consolidation. This study determined the parameters under which ultrasonic consolidation was successful, and characterized the resulting products. Void content and crystallinity were compared for samples manufactured under different consolidation pressures and travel values. This was done to understand the effects of the consolidation parameters on product characteristics, and to optimize product quality.

2. EXPERIMENTATION

2.1 Materials

PPS films, supplied by Professional Plastics, were 0.08 mm-thick and cut into 50 mm by 50 mm squares. The dry carbon fiber plies, supplied by Composite Envisions, were a 2x2 twill fabric with a surface density of 238 gsm, woven from high density Toray T-300 with a density of 1.76 g/cm^3. The dry fiber fabric was cut into 80 mm by 80 mm squares which were then taped around the edges to minimize fraying. The tape was applied to the CF fabric before the squares were cut out (Figure 1).

Figure 1. Tape was applied to 2x2 twill CF fabric (Toray T-300) before squares were cut out.

2.2 Ultrasonic consolidation

A Rinco Dynamic 3000 ultrasonic welder performed the consolidation. The welder operates at 20 kHz and has a maximum power output of 3000 W. A circular sonotrode with a diameter of 40 mm was used, all tests were conducted with a vibrational amplitude of 38.1 μm. The layup consisted of three plies of dry carbon fiber surrounded by four PPS films, as shown in Figure 2. All CF plies were taped to the jig to prevent motion during consolidation.

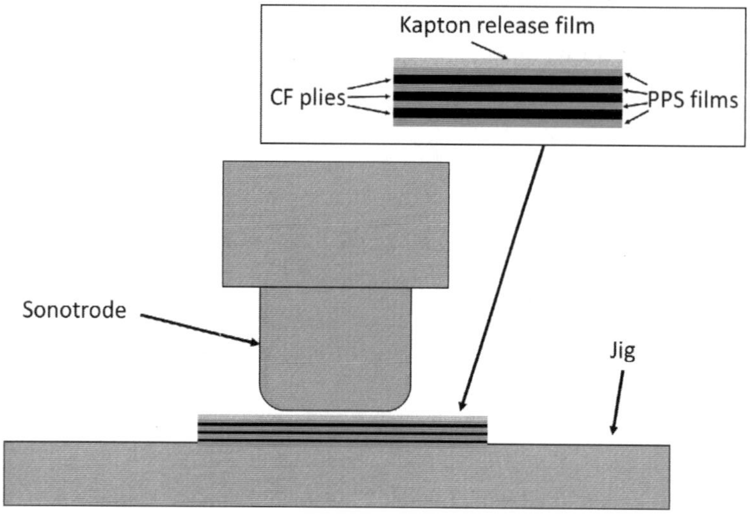

Figure 2. Diagram of ultrasonic consolidation layup.

The process begins when the sonotrode descends onto the materials and presses down with increasing force until the pressure for consolidation is reached. Once that pressure has been reached ultrasonic vibrations are applied. The application of pressure and ultrasonic heating is continued until the sample has been compressed to a prescribed distance since the start of vibrations. This distance is revered to as travel. When the travel distance is met the sonotrode stops transmitting ultrasound and maintains the consolidation pressure for a holding time of 2 seconds. The ultrasonic welder is controlled by a microprocessor referred to as the automated control unit (ACU) which allows the welding parameters such as force, travel, and weld time to be monitored and controlled [7].

Temperature measurements were taken using type K thermocouples from Omega Engineering to record the cooling curves for some of the samples. This was done by placing the thermocouples underneath the layup and in the center of the sonotrode's contact area.

2.3 Dry fiber bed compression

The thickness of the dry CF plies determines the theoretical travel required to fully impregnate the CF plies with PPS. The relationship between the ply thickness and consolidation pressure was determined by placing 10 CF plies on the welding jig and increasing the pressure from 0 MPa to 1.6 MPa. The thickness and pressure were obtained from the sonotrode's displacement and force data generated by the ACU.

2.4 Void content

Cross sectional images of the samples were taken to measure void content. The samples were cut and mounted vertically in epoxy pucks which were then polished using 360, 600, 800, and 1200 grit sand paper, followed by pads sprayed with diamond particle (6 μm, followed by 1 μm). Images of the samples were then taken through a Meiji MT8100F optical microscope at 20x magnification.

The void content was found using ImageJ by selectively picking out voids using brightness thresholds for macro- and micro-voids, as illustrated in Figure 3.

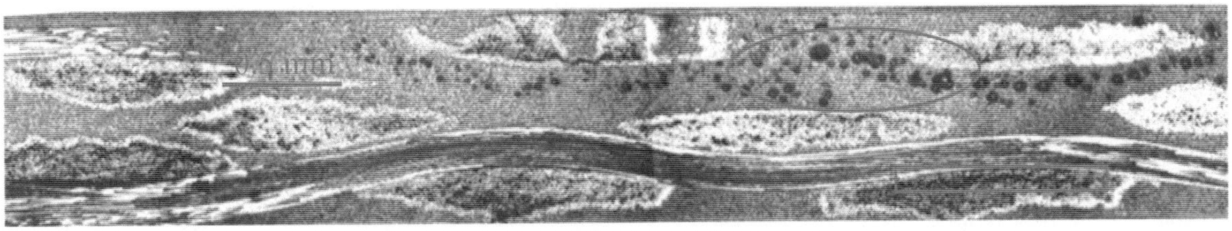

Figure 3. The macro voids are seen as spherical bubbles in the composite's matrix, examples are circled with an oval. Micro-voids were located inside the fiber tows, such as the section contained by the rectangle.

2.5 Crystallinity

A PerkinElmer DSC 4000 was used to determine the crystallinities of the different samples. The samples were cut into pieces between 5 mg and 10 mg, and sealed in a 5 mm diameter aluminum pan. The pan was placed in the DSC with a second empty aluminum pan for reference. The samples were heatup up to 300 °C at 10 °C/min, then remained at 300 °C for 1 minute. It was cooled down to room temperature at 10 °C/min. After 1 minute at room temperature, the cycle was repeated a second time.

Cold crystallization occurs at temperatures above the glass transition temperature (T_g), which is 90 °C for PPS [5]. Cold crystallization is an exothermic process where the polymer chains have gained enough mobility to cross link into semi-crystalline structures [8]. The melting temperature (T_m) of PPS is 285 °C [5], the latent heat of melting is associated with the potential energy of semi-crystalline structures (Figure 4). The degree of crystallinity is calculated using Eq. [3]:

$$X_c = \frac{\Delta H_m - \Delta H_c}{\Delta H_f^0} * 100 \qquad [3]$$

where X_c is the percent crystallinity, ΔH_m is the specific melting enthalpy, ΔH_c is the specific energy of cold crystallization, and ΔH_f^0 is the specific melting enthalpy of an ideal crystal. The values for ΔH_f^0 vary greatly in literature, ranging from 50 J/g to 150.4 J/g [9]. The ΔH_f^0 value used to calculate these crystallinities (Eq. 3) was 100 J/g since that is the median of the range of possible values.

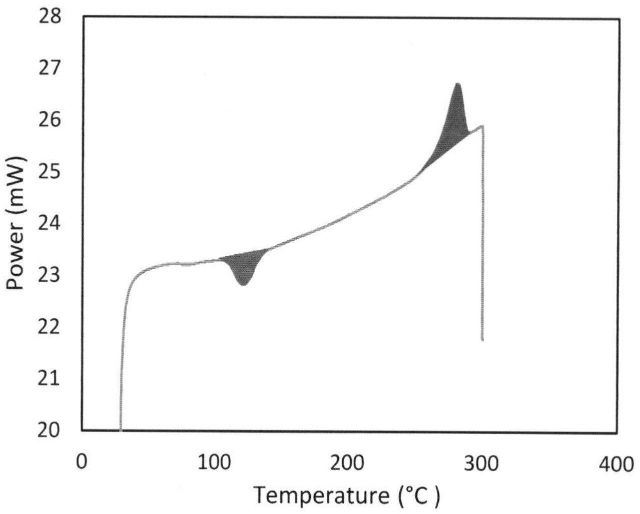

Figure 4. Power transmitted during DSC indicates cold crystallization at 125°C highlighted in blue and melting at 280°C highlighted in red.

ΔH_m and ΔH_c were found by plotting the power with respect to time and integrating the area between the peaks and the power curves that would have existed without the peaks (Figure 5).

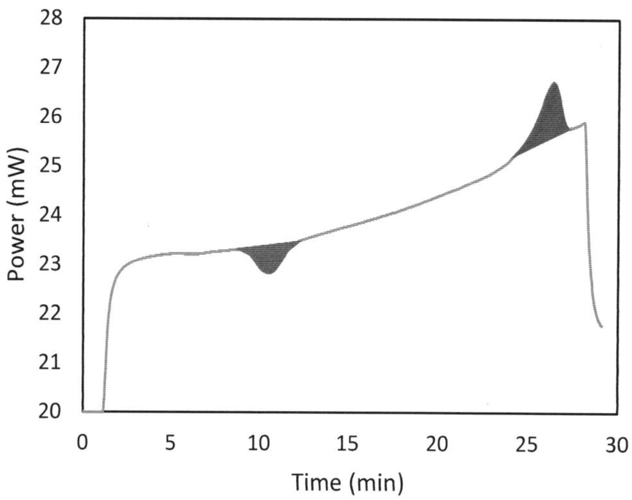

Figure 5. Power peaks are integrated over time for cold crystallization and melting. Highlighted areas are the areas integrated to obtain ΔH_m and ΔH_c.

2.6 Microhardness

A Clark CM-802AT microhardness tester was used to test the microhardness of the samples. Samples were mounted on epoxy casts and fastened onto the stage of the microhardness tester. An area on the sample with all PPS and no CF was found with a 10X objective lens. This area was then compressed for 15 seconds with a square pyramid shaped diamond tip under a load of 500 g. At the end of these 15 seconds the indent left by the tip is square shaped. The diagonals of the square indents were measured through the microscope's 10X objective lens. The resulting microhardness was represented in the following equation:

$$HV = \frac{F}{d1 * d2} \qquad [4]$$

HV is the Vicker's microhardness value, F is the force applied to the sample through the diamond tipped piece, and d1 and d2 are the lengths of the diagonals of the square indentation. Notice the units of HV are in N/mm^2, but it does not represent a pressure.

3. RESULTS

3.1 Compression tests of dry CF plies

The thickness of the dry fibers, the surface area density of the dry fibers (238 gsm), and the volume density of the dry fibers (1.76 g/cm^3) all determine the amount of air remaining in the dry fiber plies. The travel required to remove this air was defined by combining Eq. [1] and Eq. [2]:

$$T = th - \frac{\sigma}{\rho_f} \qquad [5]$$

Compression tests on the dry fibers were carried out with the sonotrode to obtain the thickness of the plies as a function of consolidation pressure, which was then used to generate a relationship between consolidation pressure and optimum travel (Figure 6). The travel represents the downward vertical displacement of the sonotrode during the ultrasonic consolidation process.

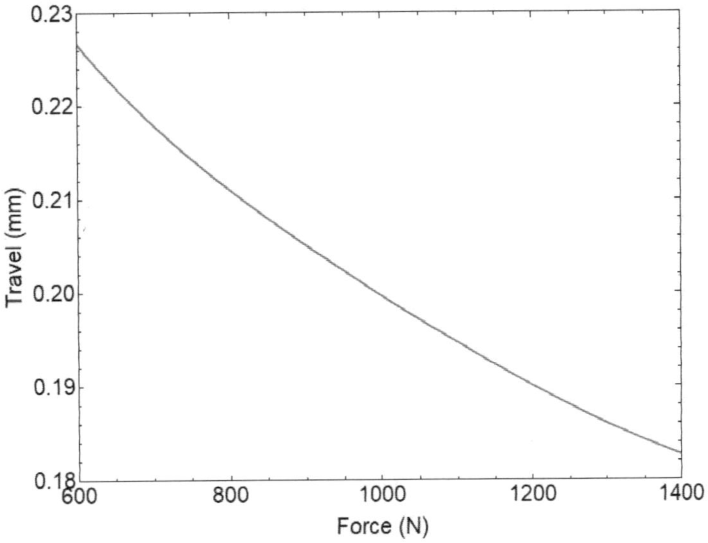

Figure 6. The travel required as a function of Force is found from the compressibility of the dry CF plies.

The resulting optimum travel values for a 3 ply layup at the tested consolidation pressures are shown in Table 1.

Table 1. The optimum travel value is listed for each consolidation pressure. These travel values vary slightly, ranging from 0.18 mm to 0.23 mm.

Consolidation pressure (MPa)	0.48	0.64	0.80	0.95	1.11
Sonotrode Force (N)	600	800	1000	1200	1400
Optimum Travel (mm)	0.23	0.21	0.2	0.19	0.18

3.2 Void content

All samples were consolidated under forces from 600 N to 1200 N in 200 N increments with a 40 mm diameter circular sonotrode, resulting in consolidation pressures of 0.48 MPa to 1.11 MPa. All these samples had travel values of 0.18 mm, 0.20 mm, 0.22mm, or 0.24mm in order to create consistency between different consolidation pressures and to cover the range of values shown in Table 1. Void contents for these samples are shown below (Figure 7).

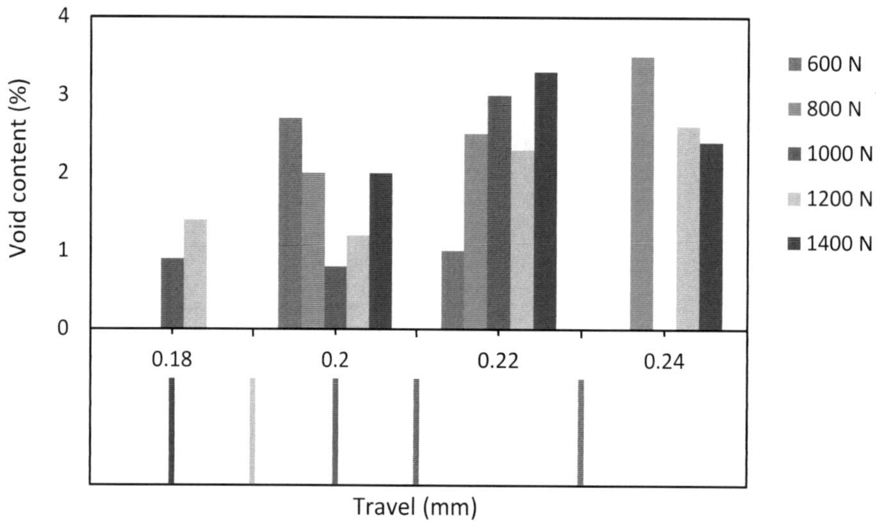

Figure 7. Resulting void contents for samples of varying force and travel values. Theoretical optimum travel values for each force are indicated with a vertical line under the bar graph.

The theoretical travel values for each force value is displayed under the bar graph. The travel values of lowest void content show a general agreement with the theory. The sample with the lowest void content for any given consolidation pressure also has a travel value that is closest to the optimal value. The void contents were averaged for the different forces and compared in Figure 8.

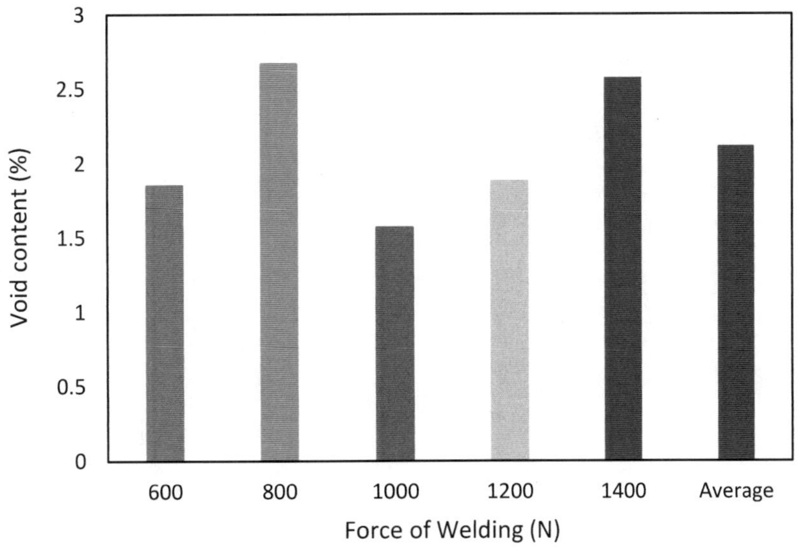

Figure 8. Void contents for each force are compared. The average void content for all samples shown on the right is 2.1 $\pm$ 0.9 %.

A consolidation pressure of 0.95 MPa resulting from a force of 1200 N generated samples with the lowest average void content. Repeated trials are underway to test for repeatability. It should be noted that a void content of 2 % is the quality threshold for the aerospace industry. Most of the samples fell within that threshold, and all samples were manufactured in a short period of time (4 to 8 seconds), making ultrasonic consolidation a high-speed process which does not sacrifice quality.

3.3 Temperature measurements

The temperature profiles for the cooling of the samples immediately after the application of vibrations showed an impressively rapid cooling rate of 125 °C/s between T_m and T_g. This was not surprising because the samples had thicknesses between 0.7 mm and 0.75 mm and remained under consolidation pressure for a holding time of two seconds after the vibrations had stopped. The samples submitted to temperature measurements all had travel values of 0.22 mm and all displayed similar cooling behaviors (Figure 9).

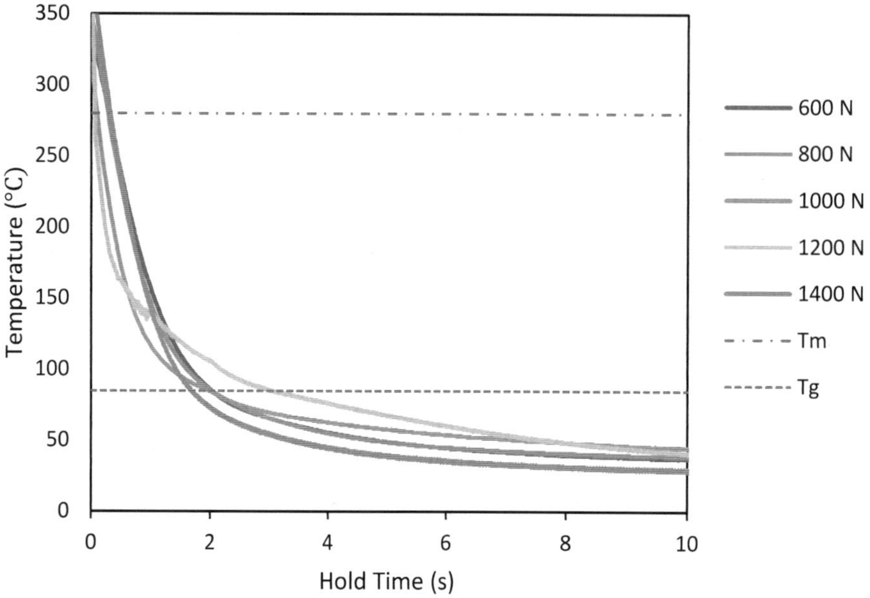

Figure 9. Cooling profiles for consolidated samples immediately after the application of ultrasonic vibrations.

Most samples cooled from T_m to T_g in the first two seconds of hold time with an exception for the 1200 N samples, which took three seconds. They all displayed high cooling rates, but they had no consistent relationship with consolidation force.

3.4 Crystallinity and microhardness

DSC measurements were conducted on the same samples that were subjected to temperature measurements. Cooling rates appeared to be independent of the consolidation force (Figure 9), which suggests that the relationship between consolidation force and crystallinity must be caused by other factors.

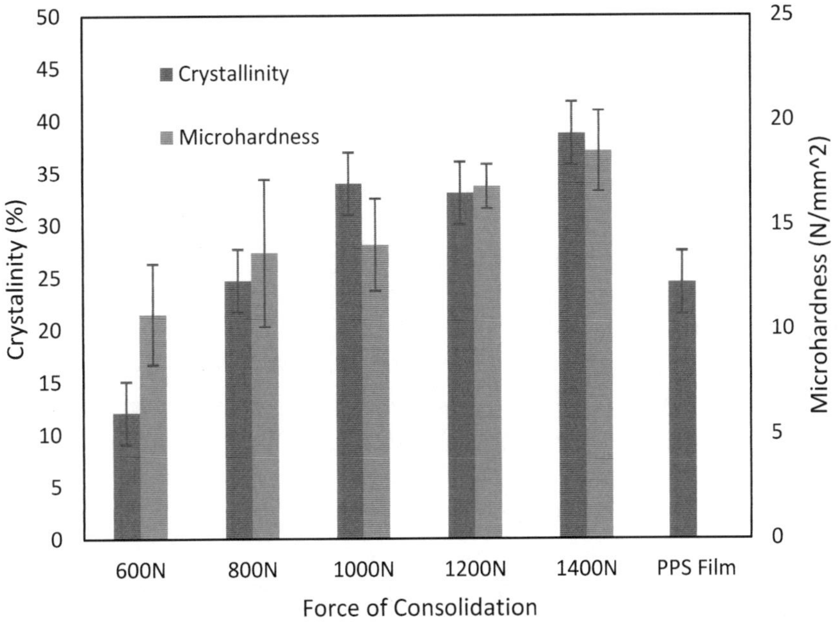

Figure 10. Both crystallinity and microhardness increased with consolidation force. The crystallinity of the unprocessed PPS film is displayed on the right.

The resulting degrees of crystallinity are unexpected considering the rapid cooling rate of these samples. The direct relationship between crystallinity and consolidation force is supported by a similar relationship between microhardness and consolidation force, this is because higher hardness is associated with higher crystallinity [4]. This is likely due to strain induced crystallization [10], which would be greater under higher consolidation forces. Since the ΔH_f^0 values used to calculate crystallinity (Eq. 3) vary greatly in literature [9], further work needs to be carried out to determine this value for the PPS used in these experiments. It should be noted that the trend of increasing crystallinity with increasing consolidation forces remains intact, regardless of the value for ΔH_f^0 used in Eq. 3.

4. CONCLUSIONS

Ultrasonic consolidation of CF/thermoplastic composites is a promising new manufacturing method which has the advantage of its speed and low cost. The samples fabricated for this paper showed low void volume fractions (< 2 %).Void content could be minimized in a predictable fashion that agreed with theory. However, future work is needed to test the repeatability of these void volume fractions.

Increasing consolidation force increased the crystallinity of the PPS matrix and microhardness of the matrix. This relationship between the consolidation force and the matrix crystallinity is a potentially useful tool for controlling the crystallinity of the manufactured product. An accurate value for the specific melting enthalpy of an ideal PPS crystal (ΔH_f^0) should be determined in future work.

Additional future works will include: 1) stiffness and strength tests, under tension and 3-point bending, of ultrasonically consolidated composites in comparison with their compression molded counterparts, 2) further investigation of temperature profiles and crystallinity with respect to ultrasonic consolidation parameters, and 3) real-time flow visualization through an optical welding frame.

5. REFERENCES

1. Tolunay, M.N., P.R. Dawson, and K.K. Wang, *Heating and bonding mechanisms in ultrasonic welding of thermoplastics.* Polymer Engineering & Science, 1983. **23**(13): p. 726-733, DOI 10.1002/pen.760231307.
2. Villegas, I.F., et al., *Process and performance evaluation of ultrasonic, induction and resistance welding of advanced thermoplastic composites.* Journal of Thermoplastic Composite Materials, 2013. **26**(8): p. 1007-1024, DOI 10.1177/0892705712456031 .
3. Gomer, A., et al., *Fabrication of Fiber Reinforced Plastics by Ultrasonic Welding.* Journal of Composites Science, 2018. **2**(3): p. 56.
4. Rizzolo, R.H. and D.F. Walczyk, *Ultrasonic consolidation of thermoplastic composite prepreg for automated fiber placement.* Journal of Thermoplastic Composite Materials, 2016. **29**(11): p. 1480-1497, DOI 10.1177/0892705714565705.
5. Batista, N.L., et al., *Correlation between degree of crystallinity, morphology and mechanical properties of PPS/carbon fiber laminates.* Materials Research-Ibero-American Journal of Materials, 2016. **19**(1): p. 195-201, DOI 10.1590/1980-5373-mr-2015-0453.
6. Helmus, R., et al., *Out-of-autoclave prepreg consolidation: Coupled air evacuation and prepreg impregnation modeling.* Journal of Composite Materials, 2016. **50**(10): p. 1403-1413, DOI 10.1177/0021998315592005 .
7. Villegas, I.F., *In situ monitoring of ultrasonic welding of thermoplastic composites through power and displacement data.* Journal of Thermoplastic Composite Materials, 2015. **28**(1): p. 66-85, DOI 10.1177/0892705712475015.
8. Kenny, J.M. and A. Maffezzoli, *Crystallization kinetics of poly(phenylene sulfide) (PPS) and PPS/carbon fiber composites.* Polymer Engineering and Science, 1991. **31**(8): p. 607-614, DOI 10.1002/pen.760310812.
9. Sacchetti, F., et al., *Effect of cooling rate on the interlaminar fracture toughness of unidirectional Carbon/PPS laminates.* Engineering Fracture Mechanics, 2018. **203**: p. 126-136, DOI 10.1016/j.engfracmech.2018.02.022.
10. Koutras, N., et al., *Characterisation of crystallinity at the interface of ultrasonically welded carbon fibre PPS joints.* Composites Part A: Applied Science and Manufacturing, 2019. **125**: p. 105574, DOI10.1016/j.compositesa.2019.105574 .

STUDY OF VOID EVOLUTION IN EPOXY RESIN DURING CURE

Kaitlin Rigitano[1], Andy George[2], Dan Rapking[1], G.P Tandon[1], Thao Gibson[1]
[1] Structural Materials Division, University of Dayton Research Institute, Dayton, OH
[2] Manufacturing Engineering Department, Brigham Young University, Provo, UT

ABSTRACT

It is well known that voids in composite materials lead to decreased structural performance. It is therefore necessary to understand their behavior in order to minimize the negative effects caused by the presence of these defects. In this work, an experiment was developed to study the evolution of voids in an epoxy material during cure. The experimental apparatus featured transparent glass tooling that allowed voids to be observed using an optical imaging setup under vacuum and elevated temperature conditions. Intentionally induced defects along with naturally evolving porosity were studied in both neat resin films and carbon fiber prepreg materials. Image analysis led to information on the evolution of void content, size, distribution, shape, and movement in an epoxy as a function of the cure cycle, and post-experiment optical microscopy provided insight into the morphology of defects that become trapped within composite materials during cure.

Corresponding author: Andy George, andy_george@byu.edu

1. INTRODUCTION

High performance structural composites have traditionally relied upon autoclave cure of prepreg materials to achieve the highest mechanical properties. Despite their promised advantages, many industry entities have been slow to switch from metals to composites. The major restraining factor has always been the cost and long cycle times to make the highest quality composites in such an autoclave-prepreg process. As the global aerospace market continues to become more competitive, and the automotive industry is constantly pushed to higher fuel economy standards, there is a strong push to develop less costly processes to make composites of similar quality. This is the driver behind out-of-autoclave (OoA) prepreg material systems in use or being developed today [1,2].

An OoA prepreg is designed with lower viscosity and other properties including dry areas to allow gas evacuation [3]. This allows better matrix saturation and degassing of the prepreg without resorting to an autoclave for consolidation pressure. Perhaps the greatest risk to using an OoA prepreg system is the reduced capacity of the manufacturing process to eliminate voids. Voids are any small bubble or pocket of air entrapped in the final cured laminate. These voids can originate in the prepreg manufacture, from handling the prepreg material including placement in the mold and ply stack-up [4], volatiles during cure, and any of several processing faults such as vacuum bag leaks [5].

Voids in composite laminates have a significant negative impact on the mechanical properties of parts [6-8]. For every 1% increase in void content between 0% - 5% yields a decrease in interlaminar shear strength and flexural strength of up to 10% and a decrease in flexural stiffness of up to 5% for every 1% of void content [9].

When the prepreg matrix is not yet gelled, these bubbles may be forced out of the system and to dissolve harmlessly into the resin by pressure. The pressure of an autoclave is usually sufficient to eradicate all voids. But an autoclave imposes significant costs in capital equipment, operation, and cycle time. A significant reduction in part cost can be afforded if the manufacturer can optimize a process enough to not need to resort to autoclave cure. For vacuum-bag-only (VBO) processing of OoA prepreg systems to meet autoclave-cured mechanical standards, the manufacturing process must be well understood in terms of how voids form and dissipate [10,11]. Yet most scientific literature on voids and prepreg discusses the final mechanical effects, as opposed to how to mitigate the voids outside of autoclave cure.

The objective of this work is to understand and capture void development/movement and/or resin movement during debulking/curing through high magnification imaging during the cure process. This was done in an attempt to identify differences that could equate to void evolution over time. Imaging experiments were able to capture visible changes in the voids during the course of the cure cycle.

2. EXPERIMENTATION

Experimentation was based on current literature that describes similar efforts [12-14]. In these experiments, an approximately 152 x 152 mm piece of 5320-1/IM7 prepreg was pressed against a 305 x 305 x 9.5 mm borosilicate glass sheet that had been previously cleaned and prepared with release agent (Figure 1). Some experiments also involved a 15 μm thick film of the same resin system used in the prepreg (Cycom 5320-1 epoxy) [15].

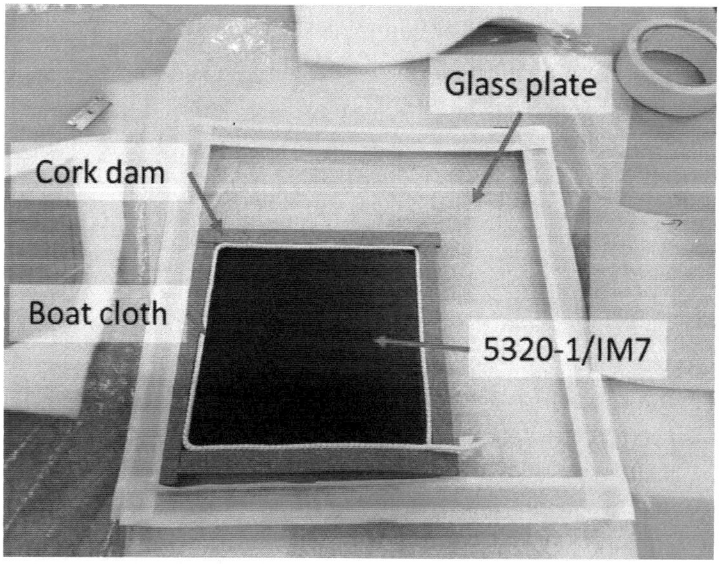

Figure 1. Prepreg on glass plate for void evolution experiments.

The manufacturer-recommended bagging sequence is shown in Figure 2. In this study, a fiberglass breather string and a 12.7 mm wide cork strip were used to replace the sealant tape and fiberglass cloth around the edges of the prepreg. For those experiments involving only resin film, a PTFE semi-permeable membrane was placed above the film, to allow gas evacuation without drawing the resin into the vacuum line.

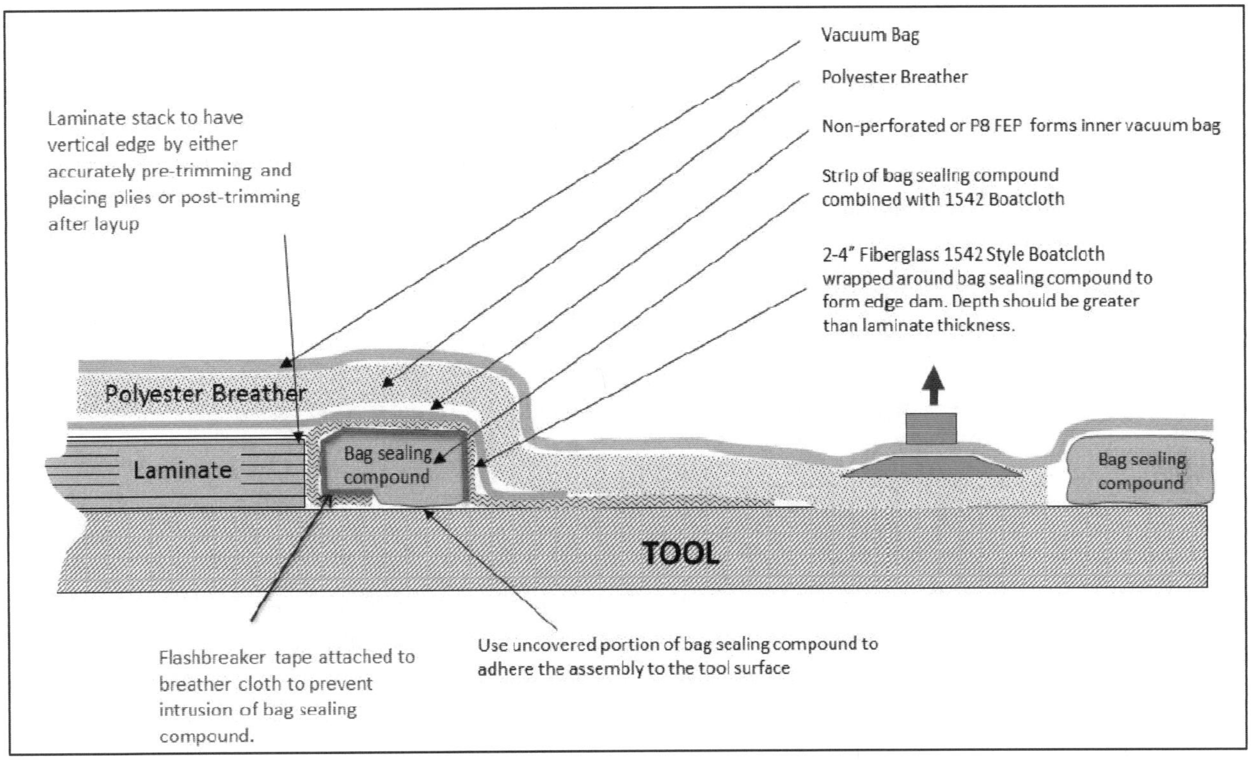

Figure 2. Manufacturer Recommended Bagging

A heating blanket was incorporated into the bagging scheme above the polyester breather layer, and an additional polyester breather layer was placed between the heating blanket and the vacuum bag. The heating blanket was connected to a Cole Palmer Digi-Sense Temperature Controller, and a thermocouple was placed inside the bag and connected to the temperature controller for temperature feedback (Figure 3). The bagged glass plate was elevated (glass plate facing downward) using lab stands, and a camera was placed underneath (Figure 4). The camera was a Point Grey 5-megapixel monochrome camera with a 180 mm focal length lens and a Fostec fiber light source. The camera was set up to image a 12 x 10 mm square in the corner of the prepreg sample, and captured an image every 30 seconds while the experiment was being conducted.

Table 1 lists the seven experimental infusions represented in this study designed to observe how voids evolve throughout a cure cycle. As seen, both prepreg and resin film were used in various combinations, and the film was sometimes punched with a needle to simulate voids and have more control over the initial conditions. For each laminate, full vacuum was applied initially (approximately 960 mbar vacuum pressure), and then the heating was begun. The "Full" cure cycle used for the majority of the experiments was: 1) ramp from ambient (approximately 28°C) to temperature T of 60°C at 1.1 °C/min, then 2) hold at 60°C for two hours, then 3) ramp at 1.1

°C/min to 120°C, then 4) hold at 120°C for 1 hour, before 5) cooling at 2.6 °C/min. These cure cycles are illustrated in Figure 5. Some experiments were conducted with an interrupted cure cycle. Cycle (A) followed the "Full" cycle except the second heating ramp proceeded only until 90°C, and then the heat was turned off and the laminate cooled naturally back to ambient. Cycle (B) only involved the first heating ramp, and only up to 47°C, whereupon the heat was turned off and the resin film cooled naturally back to ambient. These three cure cycle iterations are illustrated in Figure 5, as well as the viscosity profile accompanying the full cure cycle.

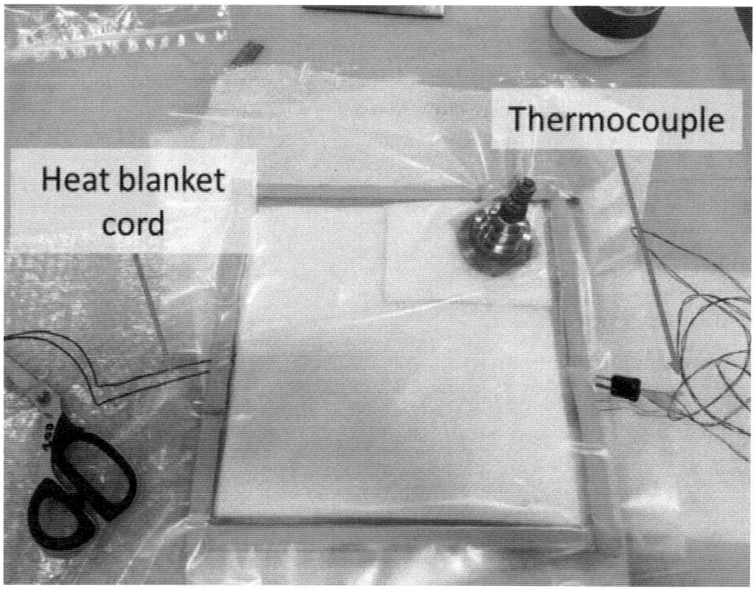

Figure 3. Prepreg was heated with a heat blanket connected to a temperature controller with a thermocouple as a temperature feedback.

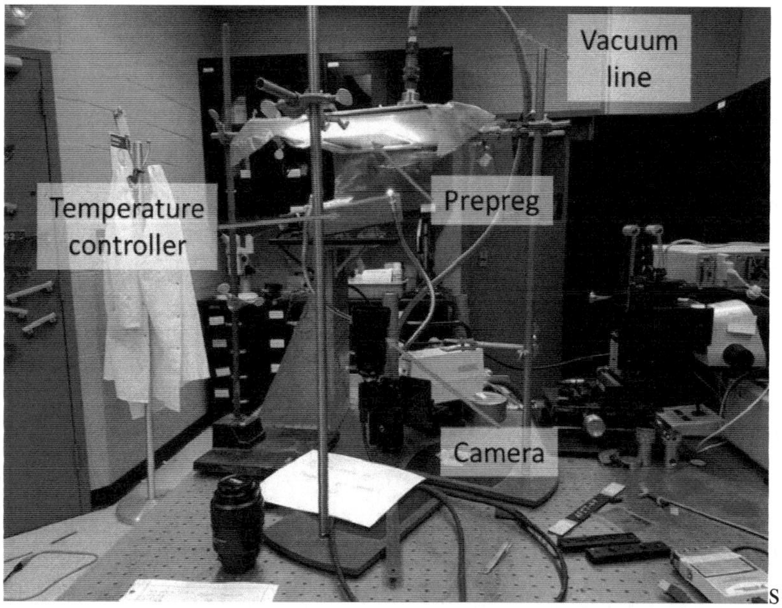

Figure 4. Prepreg/glass plate system was elevated above a camera for imaging while exposed to vacuum and a thermal cycle.

Table 1. Experimental program.

Experiment	Material	Cure
1	Prepreg	Full
2	Prepreg	Full
3	Prepreg	Interrupted (A)
4	Resin film only	Full
5	Resin film only	Full
6	Prepreg behind resin film with punched holes	Full
7	Resin film only, with punched holes	Interrupted (B)

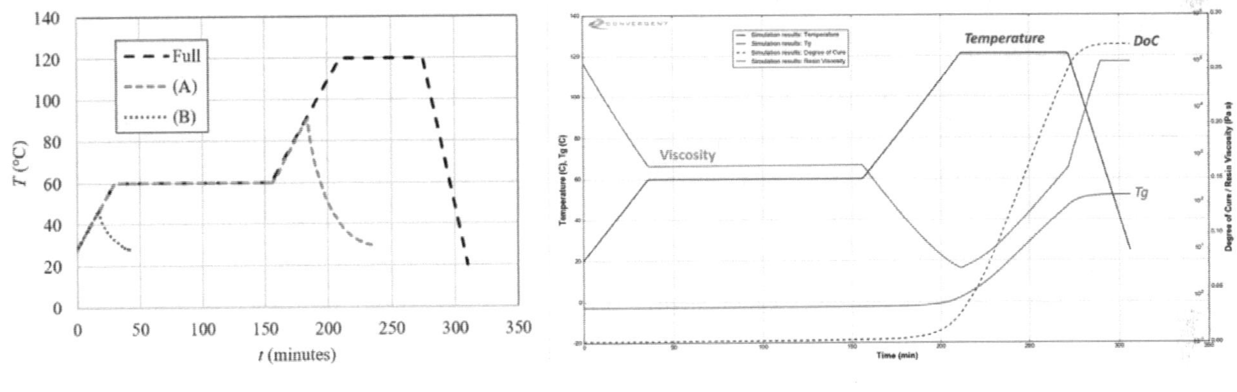

(a) (b)

Figure 5. (a) Temperature cycle used during cure and (b) viscosity profile during cure cycle

3. RESULTS

3.1 Prepreg voids qualitative analysis

Figure 6(a) shows the 12 x 10 mm image area of the prepreg at the start of experiment 1, before any vacuum or heat were applied. The image has a few scratch marks captured from the glass surface which can be seen as white streak lines. The prepreg itself appears rather dark as white light is not able to penetrate the opaque material. However, the faint outlines of individual fibers in the prepreg can be seen. Also seen in the image are white spots of varying shapes and sizes scattered widely across the image foreground.

As the focus of this study is the evolution of bubbles, the prepreg images captured during the cure cycle were modified by subtracting the binary initial image (6(a)) from all subsequent photos. An example image from the same experiment, at the end of the dwell at 60°C, is shown in Figure 6(b), along with the modified image in Figure 6(c), after subtraction of image Figure 6(a). As seen in Figure 6(c), there are still a number of small white spots, which seem to be clustered into streaks which follow the fiber orientation.

To help identify the void features in these images, the last binary image for each experiment (captured after cooling back to ambient) was compared with micrographs of the cured laminate surface after removing it from the mold. Figure 7 shows this micrograph for experiment 1, focused on what looks like a long surface void, or a large area of dry fiber with droplets of resin within. Shown in the inset is the last image taken while still in the mold, where the long white streak is the only prominent feature, and matches in location with the long void from the post-cure micrograph. This suggests that the bright white spots are only reflections, from either small round voids, or resin droplets within a larger void like this streak. There were other, smaller surface voids seen in the post-cure micrograph, that were not immediately visible from the in-mold image, such as the four small white areas seen underneath the long surface void in the micrograph in Figure 7.

(a) (b) (c)

Figure 6. (a) Initial raw prepreg image before vacuum or heating; (b) raw prepreg image during cure experimentation; (c) image-b after subtraction of initial image-a.

Figure 7. Post-cure micrograph of long surface void, with inset from last in-mold image taken for Experiment 1.

Figure 8 shows the last in-mold photograph for experiment 2, along with the entire corresponding 12 x 10 mm field of view in the post-cure micrograph. One long prominent void is visible in the bottom left of the in-mold image; gray-scale thresholding in image analysis, with a very low gray-level threshold makes other voids more clearly visible as well. In experiment 1, the same low gray-level threshold was applied and automatically converted into a binary image for each original experimental image. For the other prepreg experiments, the lighting was non-uniform across the image, thus requiring image analysis filters (size and circularity after dilation and erosion cycles), combined with some manual marking of void features to identify the void features.

An example post-cure micrograph for each of the three prepreg experiments is shown in Figure 9 with a close-up of such surface voids. All of these voids run along the fiber direction and seem to push the fibers up away from the camera, which was placed underneath. The void content v_O was calculated for the full field of view in the final experimental image, as the percentage of the total image area covered in void. The final v_O was 0.6%, 4.0%, and 6.4% respectively for experiment 1, 2, and 3. The interrupted cure would intuitively leave more voids in the laminate than a full cure. But the cause for the discrepancy between the two full cure experiments (1 and 2) remains unknown. From a mechanical point of view, these void features are especially problematic as this concentration of the voids in a few long surface defects would most likely act as a discontinuity and crack initiation site during loading. It would be more preferential to mechanics, given the same v_O, if these voids were smaller and more spread out through the laminate.

(a) (b) (c)

Figure 8. (a) Post-cure micrograph for experiment 2, (b) last in-mold image, (c) mask showing voids.

(a) (b) (c)

Figure 9. Post-cure micrographs of surface voids in prepreg experiments 1 (a), 2 (b), and 3 (c).

Another question is whether these voids result from 1) the as-supplied prepreg manufacture, 2) air entrapped when laying the prepreg on the glass mold, or 3) the cure cycle itself, e.g.

volatilization. Hamill, Centea, and Nutt published a recent study [4] showing the surface porosity in OoA prepreg is primarily from the second choice above, i.e. air trapped at the tool-prepreg interface during layup. Figure 10 presents void mask images for key moments in the experimental duration of each prepreg test, namely when the temperature first hits 60°C, then when it hit 120°C, then the last image i.e. after cool-down. Note that as experiment 3 involved an interrupted cure cycle and never was heated to 120°C, the corresponding image is absent. A significant amount of air was evacuated after the first heating ramp. Little volatilization is expected to occur at temperatures under 60°C, thus the left-hand images in Figure 10 should show nothing formed during processing.

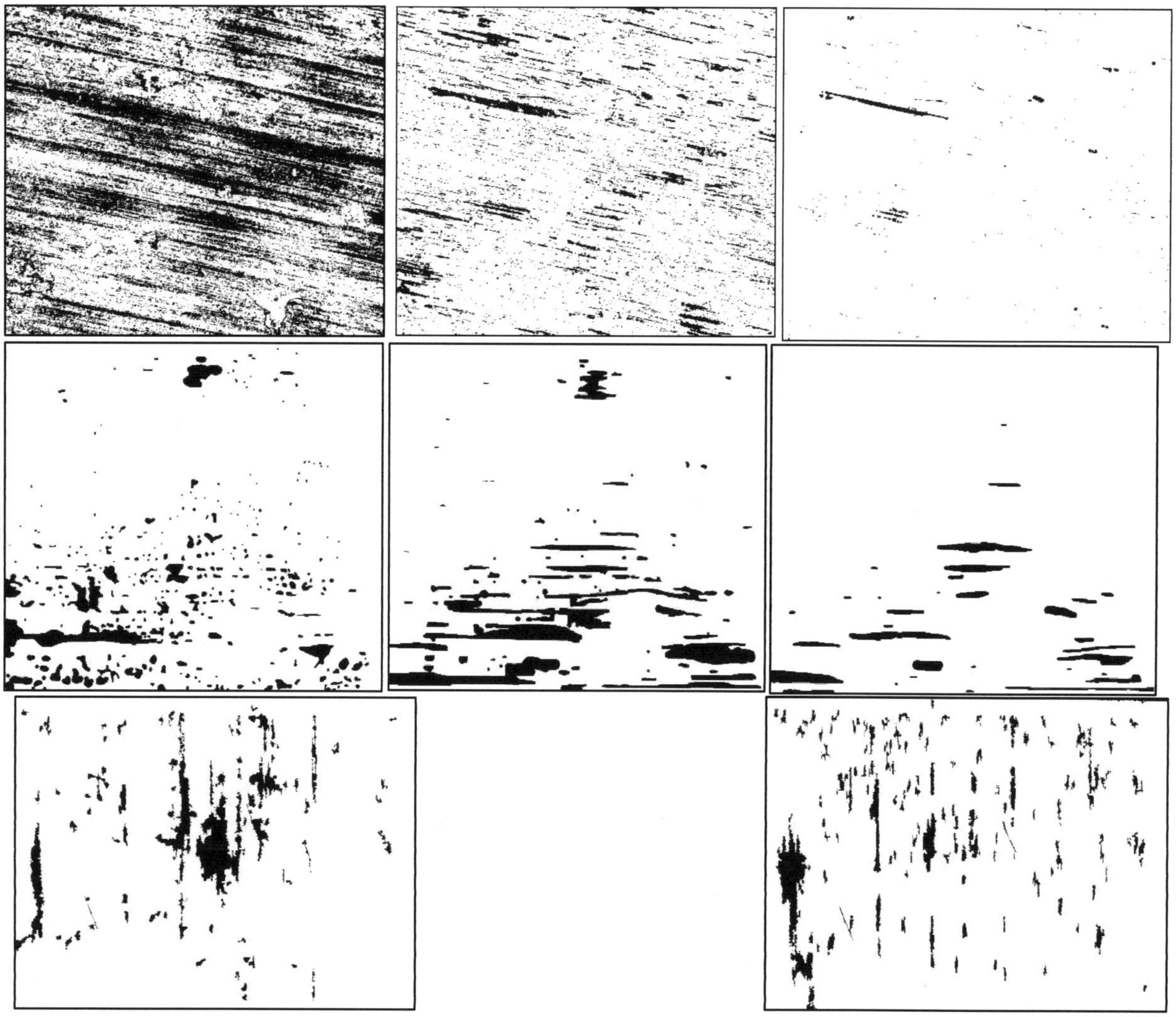

Figure 10. Void mask images (left = reach 60°C, mid = reach 120°C, right = end). Top row: experiment 1, mid row: experiment 2, bottom row: experiment 3 (no photo at 120°C).

In experiment 1, automatic thresholding showed many voids, both streaks and specks. Many of the long voids are gone after heating to 120°C, due to the increasing mobility of the bubbles as the viscosity decreased. Nearly all of the voids are gone by the end, with the main exception being the one long streak. This surviving long void appears to have been present in roughly the same location in the first image. In fact, all voids appear to move little, only to dissipate in place as the air evacuation continues.

As already mentioned, the uneven shading made void detection more difficult. *A posteriori* shading correction attempts [16] were made to no immediate avail. Thus small thin voids as seen in experiment 1 were impossible to detect, since the boundaries of voids were more subjectively chosen during manual painting. The images in experiment 2 show similar trends however, with low- and high-void locations remaining constant throughout, i.e. little bubble mobility. Eventually only a few long streaks remain.

Experiment 3 shows much less void dissipation, most likely due to the interrupted cure cycle never allowing the low viscosities in the other experiments. There is some agreement between initial and final void locations, e.g. the large streak on the left-hand side. For both experiment 2 and 3, the long-skinny bubbles in the initial image seem to survive the cure cycle more than the rounder bubbles in the initial image. The rounder bubbles may be caused by air pockets trapped during placement of the prepreg upon the mold, while the longer bubbles may be from the prepreg manufacture, and somehow be more resistant to gas evacuation.

3.2 Prepreg voids quantitative analysis

Figure 11 shows the average area A, as well as the void content v_0 in select images from each of the prepreg experiments, along with the temperature histories. Note that the irregular shading in experiment 2 made void identification highly time-consuming for each image and therefore less images are recorded here. Both A and v_0 decrease sharply during the initial ramp to 60°C, then level off throughout the hold at 60°C. Experiment 1 showed a further decrease beginning roughly at the middle of the second heating ramp to 120°C, followed by a slight increase in A as most of the small voids disappear only leaving the one long streak (see Figure 10). Experiment 2 showed an increase in A but decrease in v_0 in the middle of the second heating ramp, again from disappearance of the smaller bubbles and leaving the larger bubbles. Experiment 3, with the interrupted cure cycle, begins to cool down at about the same temperature at which the changes occur in the other 2 experiments. It thus stays fairly constant in A and v_0. This data suggests that the initial heating ramp gets rid of most of the initial voids, due to the increased mobility of bubbles from the lower viscosity, although many persistent bubbles remain. Little happens during the two-hour hold at 60°C. A further significant increase in gas evacuation occurs during the second heating ramp from 60°C to 120°C, due to the even greater bubble mobility. The slight increase in v_0 in Experiment 1 during the hold at 60°C, and in experiment 3 near the end of heating are interesting behavior, and may be caused by volatilization before the heating ramp further mobilizes the remaining bubbles.

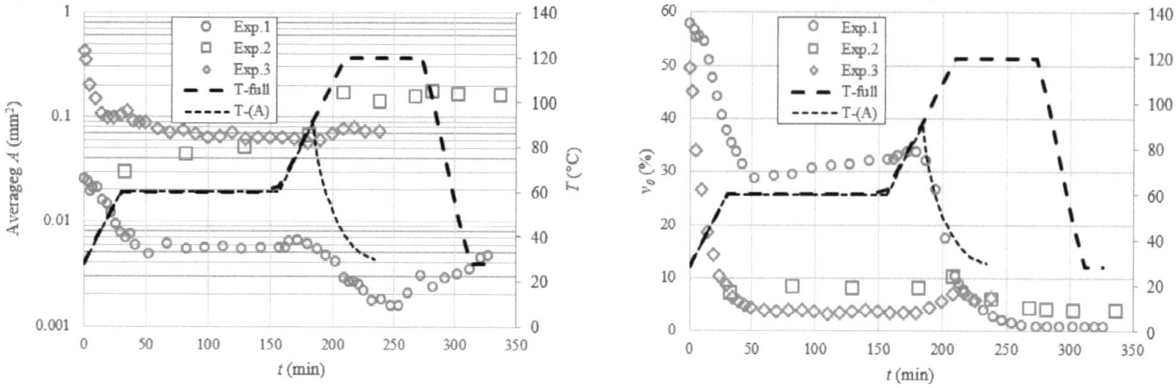

(a) (b)

Figure 11. (a) Average bubble area A and (b) void content v_0

Figure 12 shows the evolution of the aspect ratio φ averaged over all of the voids, calculated by dividing the major axis length by the minor axis length of a bounding ellipse around each void. An increase can be seen during the initial heating, especially in experiment 3, and then another increase during the second heating ramp to 120 °C, especially in experiment 2. This quantifies the observation made earlier that the rounder bubbles are degassed (evacuated) easier than the long skinny voids, causing the aspect ratio of the average bubble to increase.

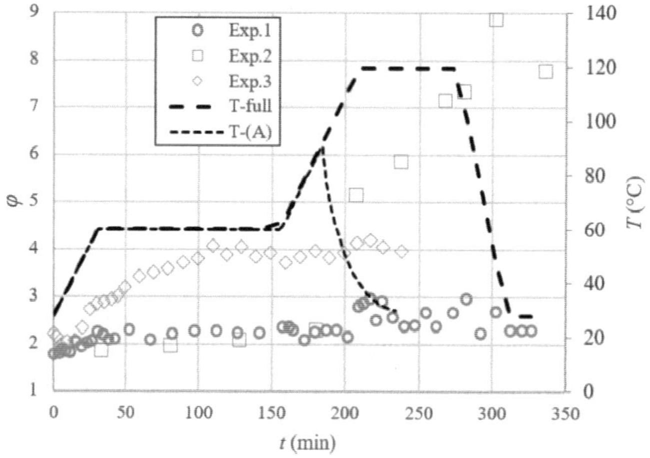

Figure 12. Average aspect ratio φ

Figure 13 illustrates an analysis of the relationship between a void and its neighboring voids, using two-point statistics [17,18]. These are color maps of the clustering of the bubbles, representing the average probability of finding another bubble a particular distance away from any given bubble placed in the middle of the color map image. One can see how initially the bubbles are spread across the image, with bubbles found in any direction. But later on in the experiment the bubbles become more clustered into a streak, where there's little probability of finding a bubble outside of the streak in any direction. These color maps result from experiment 1, but similar clustering behavior was seen in the other two prepreg experiments.

(a) (b) (c)

Figure 13. Two-point statistical maps denoting void clustering for experiment 1: (a) = reach 60°C, (b) = reach 120°C, (c) = end

3.3 Resin film voids qualitative analysis

The fibers in the prepreg cause the voids to be long and stringy, and more difficult to identify and characterize than the usual rounder bubbles in a non-fibrous medium. To better see the effect of the cure cycle on bubbles without the complication of the fibers, various experiments were also performed with thin layers of neat resin film.

Experiment 4 was the first of these, and showed a surface filled with bubbles initially, presumably from all the air entrapped when placing the film on the glass surface. As it heated up, the resin appears to liquefy, leaving bubbles floating as islands in the resin, which appear to be coagulated air in the form of voids. The mold appears to have had enough deviation from flatness, that the bubbles moved with gravity towards a corner of the field of view, thus the bubble distribution is not what it would naturally be. The increasing temperature evacuates the voids more easily than in the prepreg because of the lack of fibers which act as filter impediments to bubble evacuation. The voids are also larger and rounder than in the prepreg experiments, again due to the lack of fibers to constrain the bubbles to conform to the thin inter-fiber gaps.

Experiment 5 was performed similarly, but the experiment setup was leveled to help prevent resin flow due to gravity. Hundreds of bubbles are in in the initial images, barely visible against the cross-hatch pattern of the PTFE membrane (Figure 14). Although more flat, with less obvious directionality to the movement, these bubbles are more mobile than in the prepreg experiments, and they are evacuated quickly through the porous membrane, with no bubbles visible after reaching 120°C.

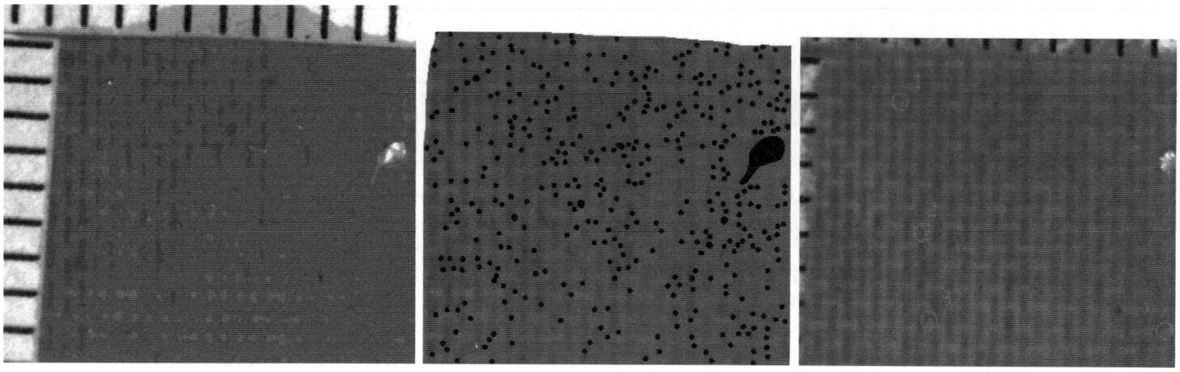

(a) (b) (c)

Figure 14. (a) Experiment 5 image at beginning of hold at 60°C, (b) with marked locations of voids, and (c) image at end of hold at 60°C. Black tick marks near the image edges represent mm spacing.

To evaluate the evolution of these bubbles, an image at the end of the hold at 60°C was selected, and four of the most prominent bubbles were identified (Figure 14 (c)). These holes were then identified in images before and after. At the beginning of the experiment, there was a tiny mark for each of these four bubbles, and all four increase in size during the heating ramp to 60°C when the original image was taken. This increase in size is possibly from air coming out of the liquefied resin with the vacuum application, but the cause is unknown. Little change occurs throughout the 60°C hold. During the second ramp, these four along with all other bubbles disappear. Some voids move to the right out of the field of view, but most simply shrink to nothing, presumably by gas evacuation through the membrane.

Large voids were purposefully introduced into the other resin film experiments, to allow larger and more-pronounced features to follow during void evolution through the cure cycle. This was done with by punching 250 to 950 μm diameter holes in the film with various sized needles. Experiment 6 was the first with punched holes, and was performed with a layer of prepreg behind the film. This served two purposes: 1) to give a dark background against which voids were easier to automatically identify than against the membrane cross-hatch pattern, and 2) so that the gas evacuation was forced through prepreg instead of simply a PTFE membrane and would thus be possible more similar to the prepreg experiments. But the opaque prepreg layer also darkened the images which again required low gray-level thresholding to see any of the voids, and made it impossible to see any of the faint small bubbles seen in the experiments 4 and 5.

Figure 15 illustrates the punched hole pattern for experiment 6, and their location in the initial mask image showing the punched holes. With vacuum and heat, the voids begin to evolve in both size, shape, and location (Figure 16). The ramp to 60°C again causes most of the air to be degassed, followed by stagnant behavior again in the hold at 60°C. The ramp to 120°C causes further degassing and alignment of the bubbles with the direction of the fibers in the adjacent prepreg layer, presumably as these large degas-resistant bubbles are drawn through the prepreg for evacuation. By the end of the dwell at 120°C, most of the air in these elongated bubbles has been pulled through, and at the end of the experiment, there is little left of only one of the original 12 tracked voids.

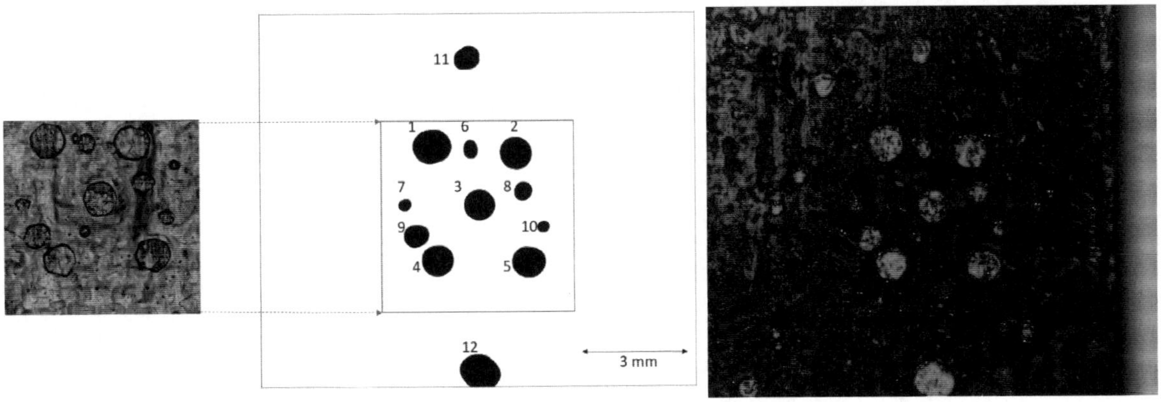

(a) (b)

Figure 15. (a) Experiment 6 punched hole pattern, and (b) their identification by gray-level thresholding in the initial image.

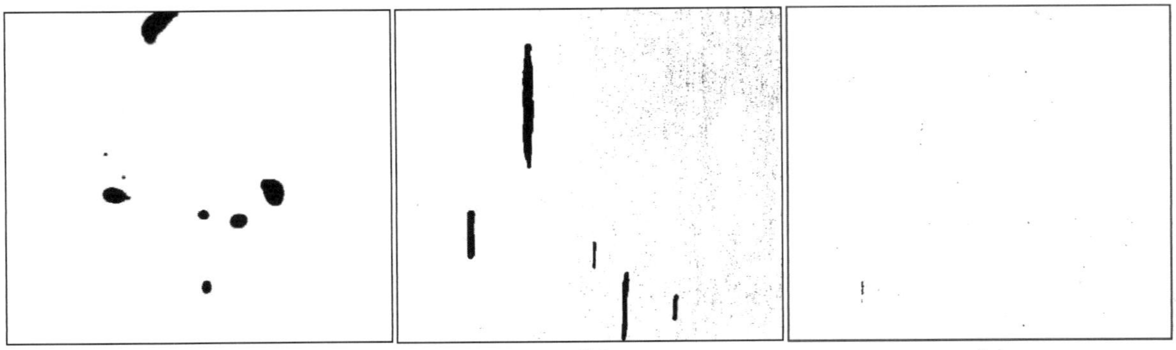

(a) (b) (c)

Figure 16. Experiment 6 void mask images: (a) = reach 60°C, (b) = reach 120°C, (c) = end

In the last experiment (7), five holes were punched in the film but no prepreg was used. The cure cycle was interrupted with very little heating, only ramping to 47°C before allowing to cool naturally. The limitation on temperature and thus high viscosity limit degassing, and many voids remain in the image at the end of the experiment, although in much smaller sizes. The punched holes and five more prominent voids from the initial image were again tracked through the experiment, and the resulting masks of the first and last image are shown in Figure 17. The voids in the last image are difficult to see due to their small size. They have not moved in location much, but have greatly decreased in size through the heating ramp, as short as it was. This again shows how much of the void evacuation is accomplished by the first heating ramp.

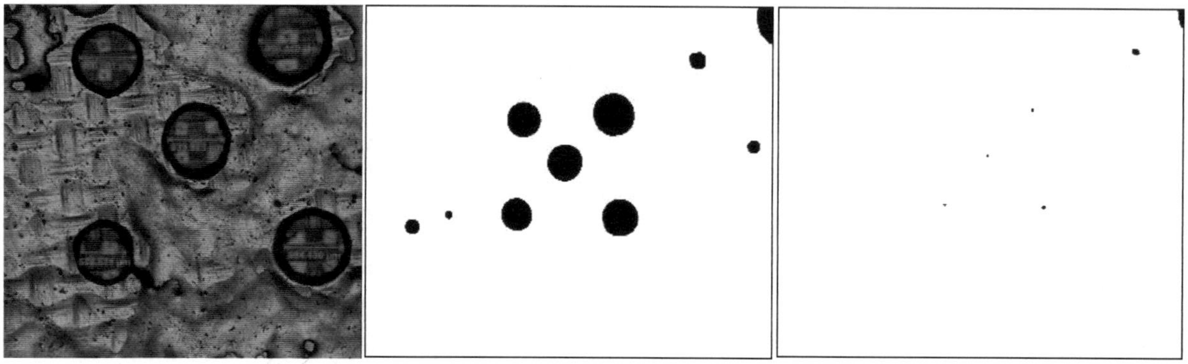

(a) (b) (c)

Figure 17. (a) Experiment 7 punched hole pattern, and (b) first and (c) last mask images showing the five punched holes in addition to five prominent voids.

3.4 Resin film voids quantitative analysis

Quantitative analysis of the film experiments was complicated by the image background. The voids in experiment 4 were hand-painted due to the difficulties with automated void detection arising from the membrane's cross-hatch pattern (Figure 14). The voids in experiment 5 were more numerous, and had fainter outlines exacerbating the challenges. Without clear outlines, the shape of the bubbles was also impossible to characterize, except for one particularly large void, visible on the far right side of Figure 14(a). Thus each void in select images was simply marked for distribution and total bubble count studies (Figure 14). Experiment 6 proved to be too dark to allow for any kind of detection of the many small bubbles in these images, and thus no quantitative results are available. Figure 18 plots the total bubble count for experiments 4 and 5, showing a roughly linear decrease for experiment 5, down to essentially 0 by the end of the ramp to 120°C.

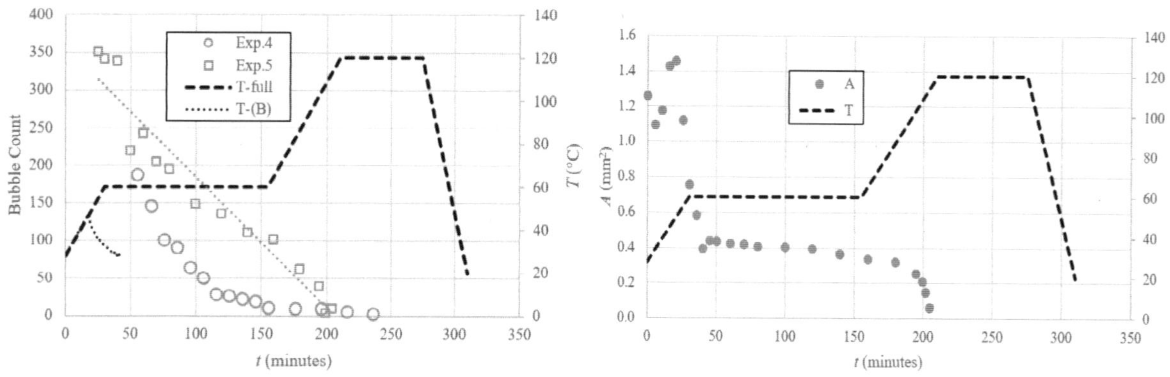

(a) (b)

Figure 18. (a) Total bubble count for experiment 4 and 5 and (b) largest bubble area in experiment 5

The area of the one large irregularly shaped bubble in experiment 5 was measured and the results are also shown in Figure 18. The area of the bubble drops rapidly during the ramp to 60°C, then stays more constant (with a steady slight decrease) throughout the dwell at 60 °C. The area decreases faster and faster once the second temperature ramp from 60°C to 120°C begins, and the bubble is gone by the time the temperature reaches 120°C.

Experiment 7, although challenged by the membrane cross-hatch background (Figure 19), had the best focus of the film experiments, and thus was targeted for a machine learning (ML) based attempt at automated void detection of the hundreds of bubbles in each image. This was accomplished using Mask R-CNN [19] as implemented in "detectron2" an open source object detection library developed by Facebook AI Research [20]. Mask R-CNN is a deep neural network that identifies objects of interest and produces corresponding segmentation masks [19]. A Python script was built to generate masks as well as train/retrain the model to fit the data sets. Figure 19 shows an example image overlaid with the ML results for the first image, in which the five punched holes are clearly visible. Figure 19 also shows two resulting mask images, the one corresponding to the same initial image, and the one from the very last image, showing a scattered distribution of small bubbles.

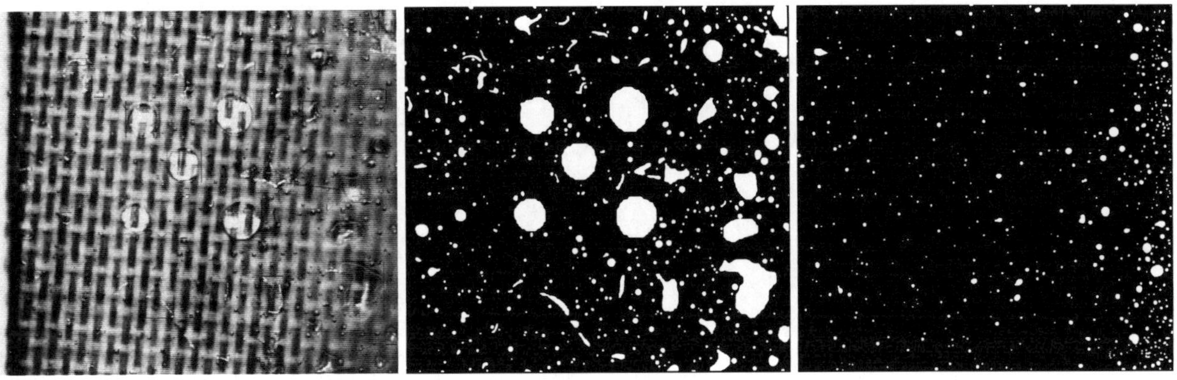

(a) (b) (c)

Figure 19. (a) Machine learning void identification overlay and (b) resulting mask image for experiment 7's initial image, and (c) resultant final mask image.

Figure 20 shows the two-point statistics for Experiment 7: the initial image, when the heat is turned off at 47°C, and the final image after reaching ambient temperature. The clustering initially shows the cross-pattern distribution of the five punched holes as they are by far the largest voids. These punched holes decrease in size and during the heating ramp, and the distribution ends up as scattering of many very small voids at the end.

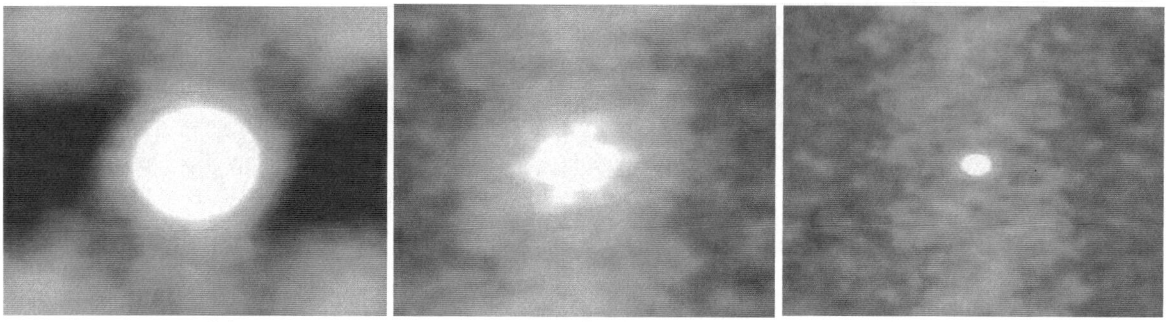

(a) (b) (c)

Figure 20. Two-point statistical maps denoting void clustering for experiment 7: (a) = initial, (b) = start cooling, (c) = end

Figure 21 plots the average area, void content, bubble count, and aspect ratio throughout the short cure cycle. The temperature profile is overlaid on these graphs without a scale, only to show the time-position of the end of heating. As seen, average A and v_0 follow an identical trend, as the number of bubbles stays around 530. Both decrease quickly, then level off as the heating ends and the sample cools down, slowing gas evacuation due to rising viscosity. The void content drops from 12% initially to approximately 2% in this short and interrupted cure cycle. The aspect ratio drops slightly, meaning the bubbles become rounder until the heating ends and then they level off.

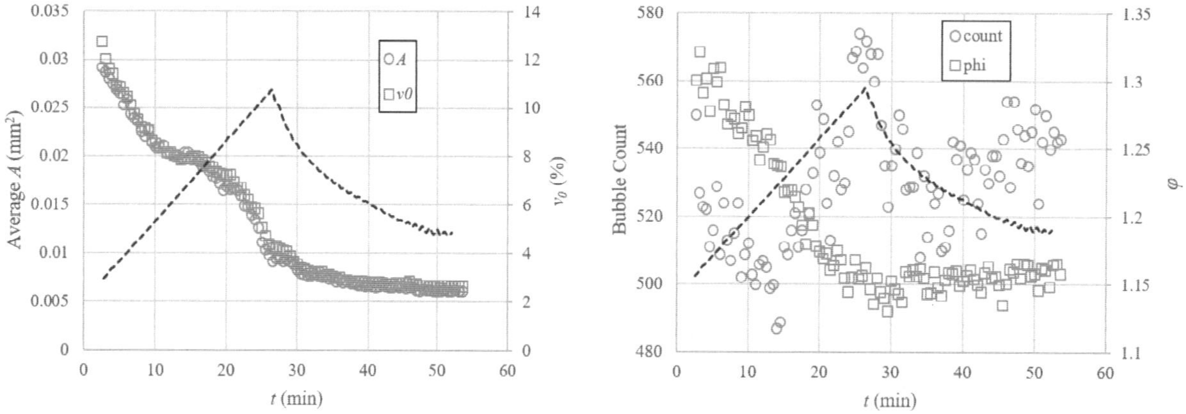

(a) (b)

Figure 21. (a) Average area, void content, (b) bubble count, and aspect ratio

4. CONCLUSIONS

The cure of both prepreg and epoxy film was monitored by imaging the evolution of voids throughout both full and interrupted cure cycles. In prepreg, there is a significant amount of trapped air at the beginning of the cure cycles. Through an initial ramp to 60°C, most of this air is evacuated through the breather cloth to the vacuum. Little degassing occurs over the hold time at 60°C. A further, smaller reduction in void content occurs during the second ramp from 60 to 120°C. Long skinny voids are common, as constrained by the fibers, some of which remain through the entire cure cycle and seem resistant to degassing. In films, a similar pattern is observed, where most of the degassing occurs in the first heating ramp, and little beyond that until the second heating ramp.

There are unique challenges to identifying voids in each of the experimental iterations. Manual identification, automated grey-level thresholding, various image analysis tools, and a machine learning script were all used to assist in gathering the data for this analysis.

5. ACKNOLWEDGEMENTS

The authors are grateful to Floid Gilbert for performing the machine learning bubble detection scripts used in the film-only experiment. Additionally, the authors would like to acknowledge our sponsor, Advanced Computational Technology, LLC, for this work performed under contract 16-ACT-2018-001-UDRI, and AFRL SBIR contract FA8650-18-C-5004, program manager Dr. Rick Hall of AFRL/RXCC. The information in this paper was approved for public release, distribution unlimited by 88ABW-2020-1145.

6. REFERENCES

1. Gardiner, G. “Out-of-autoclave prepregs: hype or revolution?” High Perform Compos (2011).
2. Centea, Timotei, Lessa K. Grunenfelder, and Steven R. Nutt. "A review of out-of-autoclave prepregs–Material properties, process phenomena, and manufacturing considerations." Composites Part A: Applied Science and Manufacturing 70 (2015): 132-154.
3. Ridgard, C. “Out of autoclave composite technology for aerospace, defense and space structures.” Proc SAMPE 2009 conf. Baltimore, MD, 2009.
4. Hamill, L, Centea, T., and Nutt, S. "Surface porosity during vacuum bag-only prepreg processing: Causes and mitigation strategies." Composites Part A: Applied Science and Manufacturing 75 (2015): 1-10.
5. Centea, T., and Hubert, P. "Out-of-autoclave prepreg consolidation under deficient pressure conditions." Journal of Composite Materials 48.16 (2014): 2033-2045.
6. Olivier, P., Cottu, J., and Ferret, B. "Effects of cure cycle pressure and voids on some mechanical properties of carbon/epoxy laminates." Composites 26.7 (1995): 509-515.
7. "Composite Materials Handbook," MIL-HDBK-17-3F. Department of Defense, 3(5) (2002)
8. L. Ling, B. Zhang, D. Wang and Z. Wu. "Effects of Cure Cycle on Void Content and Mechanical Properties of Composite Laminates." Composite Structures 73 (2006): 303-309.
9. Ghiorse, S. R. "Effects of Void Content on the Mechanical Properties of Carbon/Epoxy Laminates." SAMPE Quarterly 24(2) (1993): 54-59.
10. Lucas S, Howard S, Senger J. Vacuum bag only processing: improving prepreg out-time and porosity for large composite structure. Proc SAMPE 2010 conf. Seattle, WA, 2010.

11. Sutter JK, Kenner WS, Pelham L, Miller SG, Polis DL, Hou T, et al. Comparison of autoclave and out-of-autoclave composites. Proc 42nd int SAMPE tech conf. Salt Lake City, UT, 2010.

12. Helmus, R. et al. “An experimental technique to characterize interplay void formation in unidirectional prepregs.” *Journal of Composite Materials* 51(5) (2017): 579-591.

13. Hu, W. et al. “In situ monitoring and analysis of void evolution in unidirectional prepreg.” *Journal of Composite Materials* 52(21) (2018): 2847-2858.

14. Gangloff, J. et al. “Entrapment and venting of bubbles during vacuum bag prepreg processing.” *Journal of Composite Materials* 51(18): (2017), 2757-2768.

15. https://www.solvay.com/en/product/cycom-5320-1#product-documents (2020)

16. Bonnet, N. “Shading correction (a posteriori).” UMRS-INSERM (Reims, France) https://imagej.nih.gov/ij/plugins/inserm514/Documentation/A_posteriori_shading_correction_514_v3/A_posteriori_shading_correction_514_v3.html, (2020).

17. Fullwood, D., D. Gerrard, A. George, & D. Halverson. “Dispersion metrics for composites – a machine learning based analysis.” Proceedings of SAMPE International Conference, Long Beach, USA, 2013.

18. Niezgoda, S., Fullwood, D., and Kalidindi, S. “Delineation of the space of 2-point correlations in a composite material system.” *Acta Mater* 56 (2008): 5285–5292.

19. He, Kaiming, et al. "Mask r-cnn. corr abs/1703.06870 (2017)." arXiv preprint arXiv:1703.06870, https://dblp.org/rec/bib/journals/corr/HeGDG17, (2017).

20. Wu, Y. et al. “Detectron2.” https://github.com/facebookresearch/detectron2, (2019).

MODIFIED EPOXY MATRIX RESINS FOR REDUCED DEPENDENCE ON REDUNDANT FASTENERS IN SECONDARY-BONDED COMPOSITE STRUCTURES

Frank L. Palmieri[1], Tyler B. Hudson[1], Austin J. Smith[1], Roberto J. Cano[1], Jin Ho Kang[2], Yi Lin[2], Kathryn M.Crossett[3], Bryson Clifford[3], Isaac J. Barnett[3], and John W. Connell[1]
[1]NASA Langley Research Center, Hampton, VA, 23681, USA
[2]National Institute of Aerospace, Hampton, VA, 23666, USA
[3]NASA Internships Fellowships and Scholarships Program, Hampton, VA, 23681, USA

ABSTRACT

In structural bonds, the interface between adherend and adhesive is nearly two-dimensional making it susceptible to minute quantities of contamination that can lead to weak bonds. No methods currently exist to measure the strength of an adhesive bond nondestructively. For these reasons, regulating organizations such as the Federal Aviation Administration (FAA) often require redundant load paths in secondary-bonded, primary-structures to achieve certification for civil transport aircraft. To meet this requirement, manufacturers commonly install redundant fasteners (~10^5 fasteners in a twin-aisle airframe with composite fuselage and wings), which add tremendous complexity, time and cost, and significantly reduce performance of the airframe. Under NASA's Convergent Aeronautics Solutions (CAS) Program, the Adhesive Free Bonding of Composites (AERoBOND) project is investigating reformulated aerospace epoxy matrix resins to enable reflow and diffusion of the resin at the interface during a secondary bonding/cure process. The mixing process eliminates the material discontinuity at the interface, thereby removing the dependence of bond performance on adhesion across a nearly two-dimensional boundary. The AERoBOND interface is similar to the interlaminar region formed during a co-cure process, so joint performance depends on the cohesive properties of the matrix resin, which can be measured, inspected non-destructively, and certified for flight with fewer redundant fasteners. This presentation describes the ongoing development of stoichiometrically-offset epoxy resins to control the degree of cure in composite joints and enable interfacial mixing during secondary bonding/curing. Details on the resin formulation, laminate fabrication, and mechanical testing will also be presented. The precracked mode-II fracture toughness measured from preliminary AERoBOND joints was similar for joints fabricated with conventional epoxy matrix materials.

Keywords: thermoset, hardener, prepreg, assembly
Corresponding author: Frank Palmieri – frank.l.palmieri@nasa.gov

SAMPE Virtual Conference Proceedings, 2020. Society for the Advancement of Material and Process Engineering – North America.

1. INTRODUCTION

Polymer matrix composites are used in high performance aerospace structures because of their excellent specific strength, toughness and stiffness along the fiber. To realize the full performance advantages of composites, complex, built-up structures must be assembled using adhesives, but uncertainty in bond strength often requires manufacturers to install bolts or other crack-arrest features to ensure safety in critical applications and achieve certification [1]. The inherent uncertainty in adhesive bonds stems from the material discontinuity at the composite-to-adhesive interfaces, which are susceptible to contamination [2]. In contrast, co-cured composites, although limited in size and complexity, result in predictable structures that are certifiable for commercial aviation with reduced dependence on redundant load paths [1].

1.1 Example Assembly Process

The proposed technology uses a stoichiometric offset of the hardener-to-epoxy ratio on the faying surfaces of laminates. Assembly of the components in a subsequent "secondary-co-cure" process results in a joint with no material discontinuities (Figure 1).

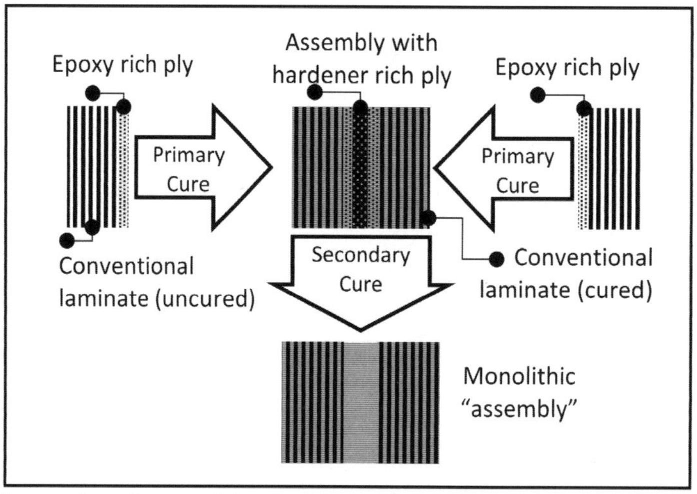

Figure 1. Schematic of assembly process using offset resin and reinforcing fiber.

In one embodiment of this technique, composite components are prepared with surfaces that are stoichiometrically rich with epoxy functional groups (Figure 1). During the primary cure, the epoxy rich (ER) resin mixes with the conventional resin, but the offset stoichiometry in the ER surfaces limits the advancement of molecular weight, and the resin on the faying surfaces remains flowable at elevated temperature with intact reactive groups even after the primary cure. In step 2, the composite panels are assembled with a ply of hardener rich (HR) material between the ER surfaces. During secondary cure, the ER and HR plies intermix and cure to form a composite assembly with no discernable interface, analogous to a conventional laminate. During both cure processes, increased temperature reduces the viscosity of the uncured resin allowing flow, diffusion, and consolidation. During the secondary cure step, intermixing of the HR and ER resins occurs, which eliminates material discontinuity at the joint. By combining the HR and ER resins, stoichiometric equivalence is achieved, and the molecular weight of the resin can advance until vitrification occurs.

1.2 Critical Material and Process Parameters

The successful fabrication of an HR/ER joint depends on multiple, interdependent material and process parameters including precursor selection, stoichiometric offset, initial degree of cure, fiber volume fraction, ply thickness, number of plies, and cure cycle temperature profile. Previous work using rheology and calorimetry indicated an ER r-value ≤ 0.15 prevented gelation during primary cure where r is defined in equation 1 as:

Equation 1: $$r = \frac{eq_H}{eq_E} \equiv \frac{molar\ equivalents\ of\ hardener}{molar\ equivalents\ of\ epoxy}$$

In conventional resin formulations, the r-value is typically 0.8 in order to obtain complete cure.[3]

This report describes a sub-set of over 20 experiments conducted to optimize materials and processes to maximize mechanical properties of the assembled joint. Infrared (IR) spectroscopy was used to measure the chemical state of the ER surface after primary cure. Laminates were fabricated with various HR and ER layer thicknesses, inspected using ultrasonic testing, and mechanically tested to measure fracture toughness.

1.3 Considerations for Aerospace Implementation

The goal for joints assembled using offset resin chemistry should be similar in mechanical performance of co-cured assemblies, not necessarily bonded joints, for implementation in future aircraft structures. The inherent reliability and predictability of co-cured and analogously offset resin assembly is the critical factor in moving from bonded joint with redundant fasteners to monolithic laminate structures with minimal fastener requirement. Therefore, the goal of this research is not to exceed the mechanical performance of joints structures, but rather to work toward similar performance and reliability of co-cured joints.

2. EXPERIMENTATION

2.1 Materials and Methods

Epoxy resins were formulated from two components: API-60® (part A) epoxy resin supplied by Kaneka North America with an epoxy equivalent weight of 131 g/mol and diethyltoluenediamine (DETDA, part B) hardener supplied by Alpha Chemistry as shown in Figure 2.

Resins were formulated from parts A and B in a resin kettle in batches of 0.25-1.2 kg by heating to 100 °C and agitating with an overhead mechanical stirrer for 60-90 min under a nitrogen atmosphere. To fabricate mechanical test specimens, prepreg was prepared from HexTow® IM7G 12K carbon fiber from Hexcel Corporation® and offset resins with r-values of about 0.15 and 2.5 for the ER and HR plies, respectively. Methyl ethyl ketone (MEK), obtained from Sigma Aldrich, was used to dilute the resin for prepreg preparation. Hexply® IM7/8552, 35%, 190 gsm unidirectional tape was obtained from Hexcel Corporation® and used as backing for the mechanical test specimens.

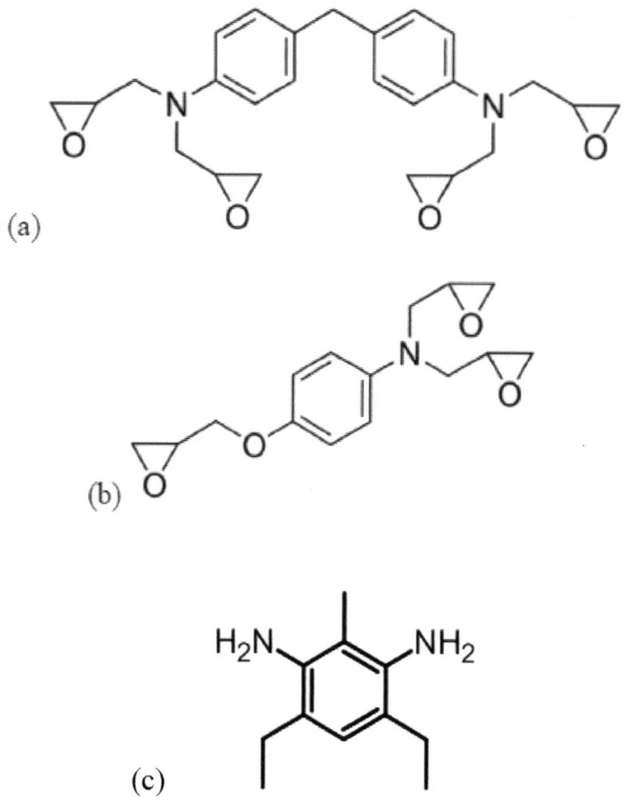

Figure 2. Structures of API-60® epoxy resin including the tetrafunctional epoxy, 4,4′-methylenebis(N,N-diglycidylaniline) (a) and the trifunctional epoxy, N,N-diglycidyl-4-glycidyloxyaniline (b), and the tetrafunctional hardener diethytoluenediamine (DETDA) (c). Mole ratio of the monomers (a) to (b) was known as 3:1.

2.2 Characterization and Analysis

Differential scanning calorimetry (DSC) was conducted on offset resins using a TA Instruments Q20 modulated DSC (MDSC®) with a heating rate of 3 °C/min. Samples of approximately 3 mg were hermetically sealed in aluminum pans and cured at 180 °C for 2 h before cooling to -40 °C and ramping to 280 °C to measure the glass transition temperature (T_g) and residual heat of reaction.

Unidirectional prepreg tape was prepared using a custom prepregger from a resin solution of 70-85 wt.% solids and the compliment of MEK. In addition to prepregging, HR material was also prepared by hand painting the MEK/HR resin solution (30% to 50% solids in MEK) onto a plain weave, e-glass fiber fabric. Twenty-three-ply composite panels were prepared by laying up the Hexcel® Hexply® 8552 prepreg, ER prepreg, and HR material in a 30 cm by 15 cm format according to [$Hexcel_9$,ER_2,HR,ER_2,$Hexcel_9$]. Each panel was cured in an autoclave using the two-step process in Figure 1. Primary cure produced two "half-panels" each with 11 plies [$Hexcel_9$,ER_2]. The half-panels were then assembled with a ply of HR material in contact with the ER surfaces and returned to the autoclave for secondary cure. All cure cycles had a final hold of at least 1 h at 178 °C. Various intermediate hold conditions (time and temperature) were tested to optimize primary and secondary cure processes.

Resin chemistry was characterized by IR spectroscopy to determine the *r*-value at the surface after primary cure. The *r*-values were calculated based on the relative peak heights at 907, 1514, and 2961 cm^{-1} using an empirical calibration curve developed from a series of resin samples of known *r*-value and degree of cure. Ultrasonic inspection in pulse-echo mode was conducted on a MISTRAS® UPK-T60-HS high speed C-scan system fitted with a NDT Automation® 10.0 MHz/13 mm immersion transducer (IU10G1). End-notched flexure (ENF) testing was conducted according to ASTM D7905-14 to measure mode II fracture toughness using six replicate specimens [4]. Figure 3 shows the test configuration using a 3-point bend configuration to load the specimen. The sides of the specimen are painted white for visualization of crack extension.

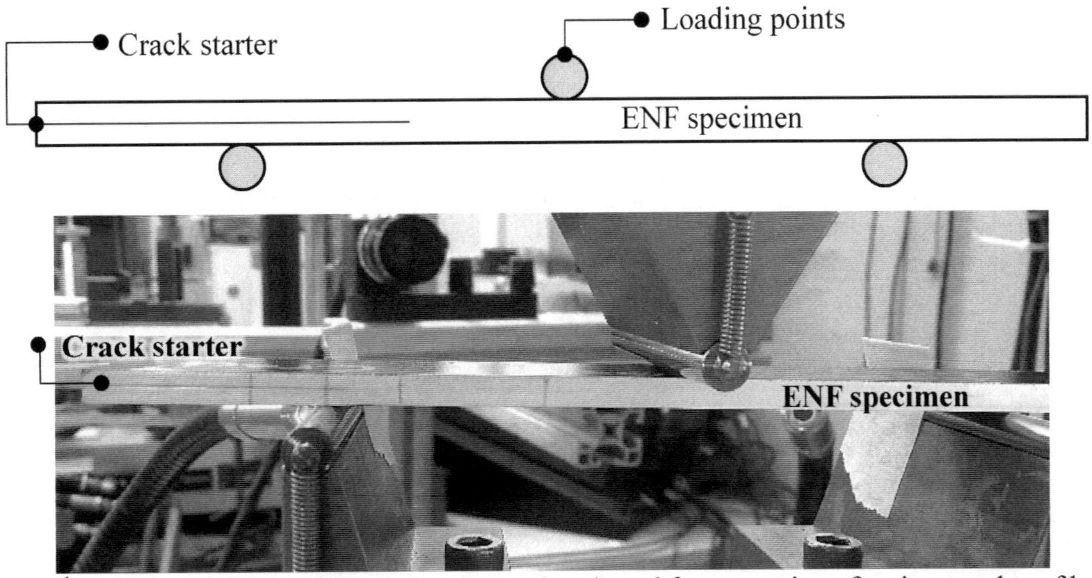

Figure 3. Specimen schematic and a photograph of test setup for the ENF test.

3. RESULTS

Process development for fabricating joints from offset resins followed an iterative experimental method. Critical findings from several preliminary experiments are summarized in section 3.1. Detailed results from four experiments are presented in section 3.2 where *r*-values for the ER and HR resins were 0.15 and 2.5, respectively.

3.1 Preliminary Process Development

Preliminary experiments indicated that extremely high hardener concentrations (r > 5) and crystalline solid hardeners (DDS) hampered diffusion and bonding in the secondary cure process. During the primary cure, the ER resin at the surface must not advance past the gel point (~45% conversion for this system) to maintain the mobility needed to affect mixing during the secondary cure. The *r*-value of the ER surface, inferred from IR spectral analysis (Table 1), indicates the relative advancement of the surface polymer due to ingress of hardener from the bulk of the laminate or homopolymerization of the epoxy. ER surfaces with *r*-values less than ~0.3 resulted in measurable joint properties whereas *r*-values greater than ~0.3 (data not included here) formed no joints. Early testing indicated that the ER layer required a minimum fiber areal weight (>150 g/m^2, FAW) and resin content (~55% RC) to remain sufficiently flowable and reactive after the primary cure. Further, the final hold time at 178 °C was limited to 1 h to prevent excessive etherfication of the ER matrix.

3.2 Detailed Experimental Data

Table 1 shows results for experimental joints. In all cases, the primary cure cycle included a 3 h hold at 107 °C before the final hold for 1 h at 178 °C. The secondary cure had a 3 h hold at 178 °C to ensure complete cure of the DETA hardened system. The resin areal weight (RAW) was varied for the ER and HR plies used in the four experiments according to Table 1. Additionally, two carrier materials were tested in the HR ply. From Table 1, it appears that higher ER and HR RAW improves fracture toughness of the laminate joint.

Table 1. Test parameters and results used in experiments 1-4 (E1-E4). Scrim denotes the plain weave glass fiber scrim cloth and CFT denotes a unidirectional carbon fiber tape used as the HR resin carrier. The baseline fracture toughness used to calculate % of baseline values was 739 J/m^2 [5]. ER *r*-values were measured after primary cure. *Fracture toughness measured from samples with no precrack for E1 whereas all other values had a precrack.

Experimental #	1	2	3	4
Parameter Set	18	20	21	15
ER-RAW (g/m^2)	81	222	222	200
HR-RAW (g/m^2)	190	190	117	240
HR carrier	Scrim	Scrim	CFT	Scrim
ER *r*-value	0.128	0.290	0.175	0.238
$G_{II\text{-}PC}$ (J/m^2)	88±33*	372±61	280±34	616±127
% of Baseline	12	51	38	84

3.2.1 Non-destructive Inspection

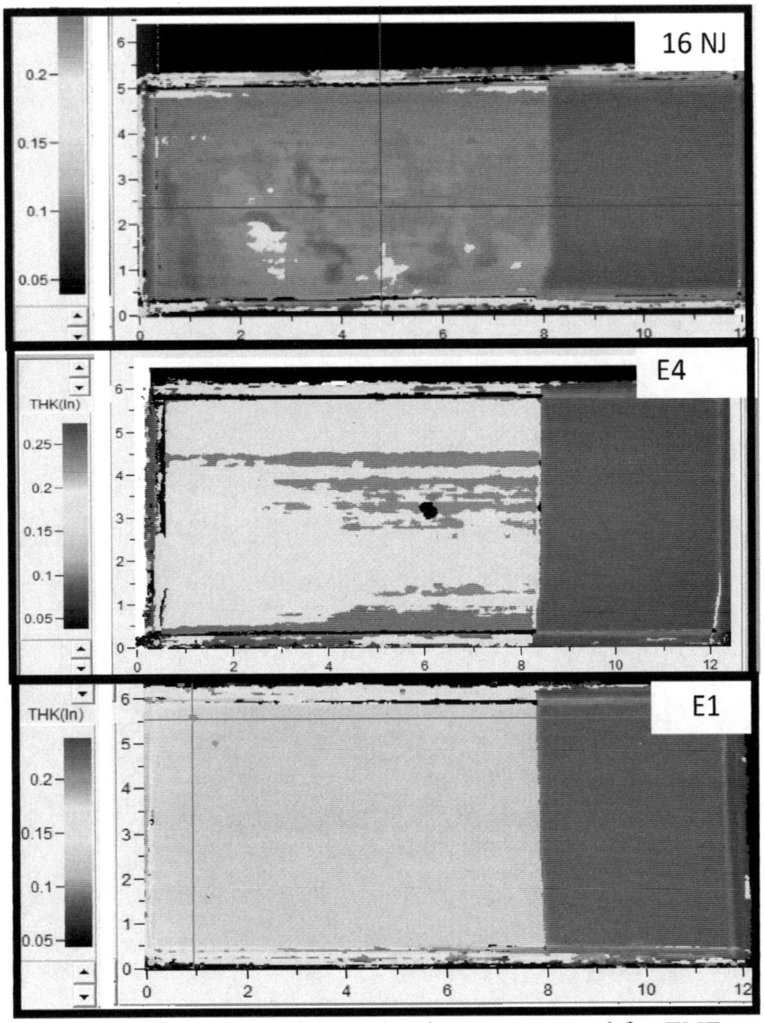

Figure 4. Time of flight C-scan images of three laminates prepared for ENF testing. NJ indicates an experiment that formed no joint and E4 and E1 coorespond to the experiment numbers in Table 1. All scales are in inches.

Figure 4 shows time of flight C-scan results for three experiments with vastly differing joint properties. A blue pixel indicates the largest reflection occurred at the mid-plane and a yellow pixel indicates the largest reflection occurred at the bottom surface. In each image, the dark blue field on the right ¼ of the image is due to the crack starter film, which creates an intentional disbond in the joint needed for ENF testing. The image labeled NJ formed no joint and the disbond in the C-scan image is readily apparent as the light blue field in the left ¾ of the image. Image E4 shows a large region of yellow color indicating most of the panel has uniform properties through the thickness with small blue areas that have measurable different acoustic properties. Finally, E1 indicates a uniform laminate with no significant variations in properties through the thickness. The C-scan results were successful at finding joined and delaminated panels prior to mechanical testing, but the C-scan data was not a good predictor of joint mechanical performance, but more an indicator of joint homogeneity. Contrary to the images, fracture toughness of E4 greatly exceeded those of E1 (Table 2).

3.2.2 Microscopy

In Figure 5, a microsection taken from E4 shows the resin rich interlaminar region between the ER and HR plies. The bright circles are the polished cross sections of carbon fibers, approximately 6 microns in diameter. The microsection shows the successful diffusion and mixing of the ER and HR resins in E4 such that no interface or inhomogeneity is visible in the joint.

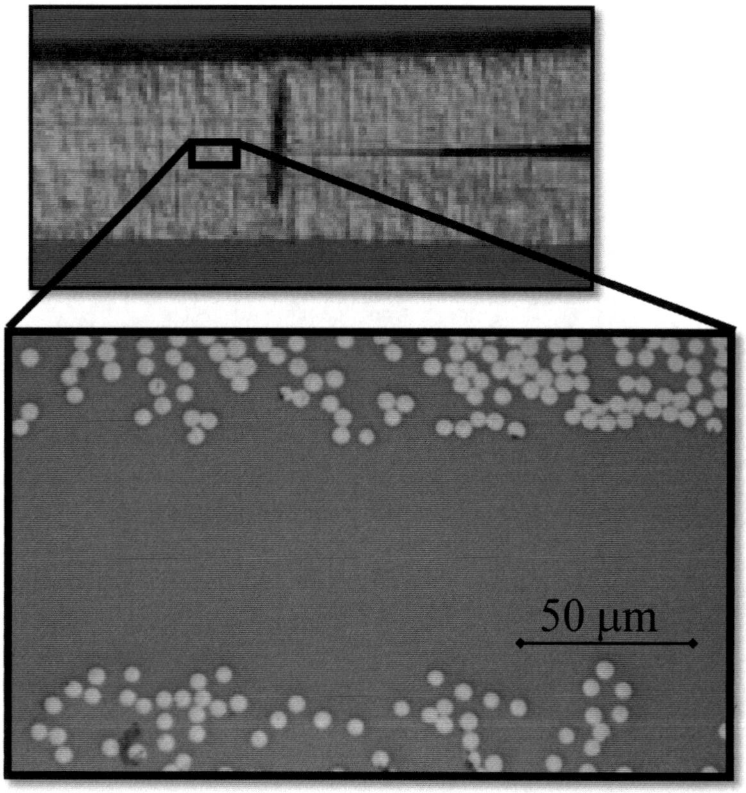

Figure 5. Images of E4 composite ER/HR joint showing the entire side of the specimen (top) and a polished cross-section of the combined and cured ER and HR layers (bottom).

3.2.3 Mechanical Performance

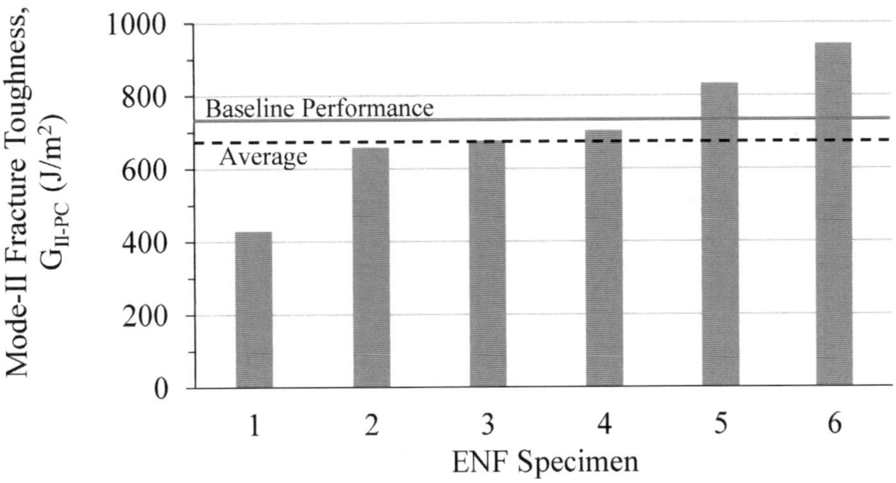

Figure 6: Precracked mode-II fracture toughness results per specimen for E4. Blue bars are tape-activated cracks and green bars are for scrim-activated cracks.

The precracked (PC) mode-II fracture toughness of individual specimens from E4 is shown in Figure 6. The blue bars were obtained using a tape-activated test scheme where the crack was driven into the ER tape ply. The green bars were measured using a scrim-activated setup meaning the crack front was driven into the woven glass carrier of the HR ply. The scrim-activated tests exceeded the baseline laminate fracture toughness because the fabric architecture creates a more tortuous path for the crack as it extends, thereby increasing the resistance to crack growth at that interface.

4. CONCLUSIONS

Stoichiometrically offset epoxy resins were used to assemble composite laminate joints with no discernable bondline in optical micrographs after reflow and mixing of resins at the interface during cure. Laminates with sufficient RAW of ER material on the surface after primary cure maintained a surface r-value below 0.3 and retained sufficient mobility to intermix and polymerize during the secondary cure. Increasing the RAW of both the HR and ER led to improved mode-II facture toughness, which reached 84% of baseline fracture toughness measured from conventional laminates.

5. REFERENCES

1. Kruse, T., Fuertes, T. A. S., Koerwien, T. and Geistbeck, M. "Bonding of CFRP primary aerospace structures - boundary conditions for certification in relation with new design and technology developments," *Proceedings of the 2015 International SAMPE Tech. Conf.* Seattle, WA, 2014 Society for the Advancement of Materials and Process Engineering. CD-ROM.

2. Palmieri, F., Ledesma, R., Fulton, T., Arthur, A., Eldridge, K., Thibeault, S., Lin, Y., Wohl, C. J. and Connell, J. W. "Picosecond pulsed laser ablation for the surface preparation of epoxy composites," *Proceedings of the 2017 International SAMPE Tech. Conf.* Seattle, WA, 2017 Society for the Advancement of Materials and Process Engineering. CD-ROM.

3. Palmieri, F. L., Hudson, T. B., Cano, R. J., Tastepe, E., Rufeisen, D., Ahmed, L., Wohl, C. J. and Connell, J. W. "Adhesive Joining of Composite Laminates Using Epoxy Resins with Stoichiometric Offset," Hilton Head, SC, 2/18/19; 2019 Annual Meeting of the Adhesion Society.

4. Standard Test Method for Determination of the Mode II Interlaminar Fracture Toughness of Unidirectional Fiber-Reinforced Polymer Matrix Composites, 2014, ASTM International, West Conshohocken, PA, 2014, www.astm.org.

5. O'Brien, T. K., Johnston, W. M. and Toland, G. J., Mode II Interlaminar Fracture Toughness and Fatigue Characterization of Graphite Epoxy Composite Material, NASA/TM-2010-216838 NASA Langley Research Center. Hampton, VA, 2010.

INVESTIGATION OF THE EFFECT OF THICKNESS AND SURFACE ROUGHNESS ON ADHESIVE JOINT STRENGTH

Başar ALTINTAŞ[1,2], Tuncay KARAÇAY[1], Bülent ACAR[2]
[1]Gazi University, Maltepe, Ankara 06570, Turkey
[2]Roketsan Inc. Elmadağ, Ankara 06780, Turkey

ABSTRACT

Adhesives are widely used in many different areas such as aerospace, automotive and marine applications. Due to their low costs and mechanical advantages, adhesive joints have become preferable solutions. However, strength of these type of joints may vary with manufacturing process. In this study, adhesive thickness and surface roughness which are major factors that affect joint strengths were investigated. Uniaxial tensile and shear tests were performed for determining failure loads. Both tests were carried out in accordance with ASTM standards. Tensile tests have been performed for three different surface roughness and three different adhesive thickness. For tensile test specimens, a modification was done to curing mold in order to eliminate alignment problem between clamping jaws of test machine. In that way, scatter in test results were reduced. Standard thick adherend shear test (TAST) specimens were used for shear tests and only surface roughness effect was investigated. Test results showed the relation between joint strength and mentioned factors.

Keywords: adhesive, surface roughness, thickness, tensile test, strength
Corresponding author: Başar ALTINTAŞ

1. INTRODUCTION

Development in adhesive materials and bonding techniques made adhesively bonded joints to better solutions for many applications. They have many advantages in mechanical and economical aspects. The main advantages of bonded joints over mechanically fastened joints such as bolted or riveted joints are their uniform stress distribution. Fatigue resistance of bonded joints are better because there is no hole or assembly detail that makes stress concentration effect on the structure. Additionally, lack of these holes, assembly details etc., means less production process and lower costs. On the other hand, joint strength is affected by some parameters that related to adhesive application and adherend manufacturing. Surface roughness and bond line thickness are the major factors.

Roughness values are highly dependent to surface treatment methods which are done during manufacturing. Many researchers have studied to effect of surface roughness and treatment on adhesive bond strength. Hirulkar et al. [1] investigated the influence of mechanical surface treatment methods including sanding, grinding, polishing and lapping with aluminum SLJ specimens. There are also many researches focused on chemical surface treatment methods. Correia et al. [2] worked on effect of surface treatment on adhesive bond strength. They used different acid anodizing methods for surface treatment. Xu et al. [3] also used acid anodizing on their work and surface morphologies with different water contact angles were shown. Zhan et al. [4] made similar study with different grades of silicon carbide abrasive papers and acidic solution. Plasma surface treatment is another way to change roughness and adhesive properties. Leahy et al. [5] employed plasma surface treatment to improve the bond strength of an adhesive joint of a glass-fiber-reinforced and commercially pure titanium. Contact angles for oxygen and argon plasma treated specimens were shown. Effect of surface roughness on fatigue behaviour is also investigated. Studies can be found in [6-7].

Another major factor mentioned before is adhesive bond line thickness. Akhavan-Safar [8] et al. performed series of shear tests and FE analysis with different bond line thicknesses. They used chemical treated aluminum SLJ specimens for tests. Castagnetti et al. [9] performed experimental campaign, which examined two structural adhesives in two different joints. For the TBJ, four different values of adhesive thickness were examined under torsion load. On the other hand, three different thickness were examined for SLJ. Lißner et al. [10] made a comprehensive study on adhesive interfaces. They investigated influence of influence bond line thickness, loading rate, and deformation mode. Void distribution of the investigated bond line thicknesses represented with X-ray tomography. They also determined parameters for CZM and simulate mix-mode deformation. Fractured surface details were showed with SEM. Shokrian et al. [11] investigated effects of both surface treatment and adhesive thickness on shear strength. They used AFM on aluminum substrates and 3D topology results of different surface treatments were shown. They also measured different shapes of microcapsule sizes with SEM. Finally, studies which are dealt with effect of bond line thickness on interfacial strength can be found in [12-13].

In this work, effects of surface roughness and bond line thickness on adhesive joint strength are investigated. Mechanical surface treatments were done and DP490 epoxy based structural adhesive was used for tests. Tensile tests were performed for three different bond line thickness and surface roughness. Test results showed that rupture load is very sensitive to pin hole alignment. Tensile tests were continued after some modifications done on curing mold. For shear tests, standard thick adherend shear test (TAST) specimens were used. Only two different surface roughness were tested since bond line thickness is constant for this type of specimen.

2. TEST

2.1 Test Procedure

Both tensile and shear test specimens manufactured of aluminum alloy 2024-T3. 3M Scotch-Welt DP490 was used as adhesive. DP490 is a black, thixotropic, gap filling two-component epoxy adhesive with particularly good application characteristics. It is designed for use where toughness and high strength are required and shows special benefits in the construction of composite assemblies [14]. Mechanical treatment methods were used for roughening the aluminum surface. Different grades of sandpapers were used, and sand blasting process were done in order to determine surface roughness magnitudes. Dummy aluminum plates were grinded in one direction with P120, P220 and P360 grit sized sandpapers also sand blasting process was taken. Surfaces were grinded in only one direction. Then, surface roughness measured with Mitutoyo SJ-201 roughness tester (Figure 1). Roughness values are shown in Table 1.

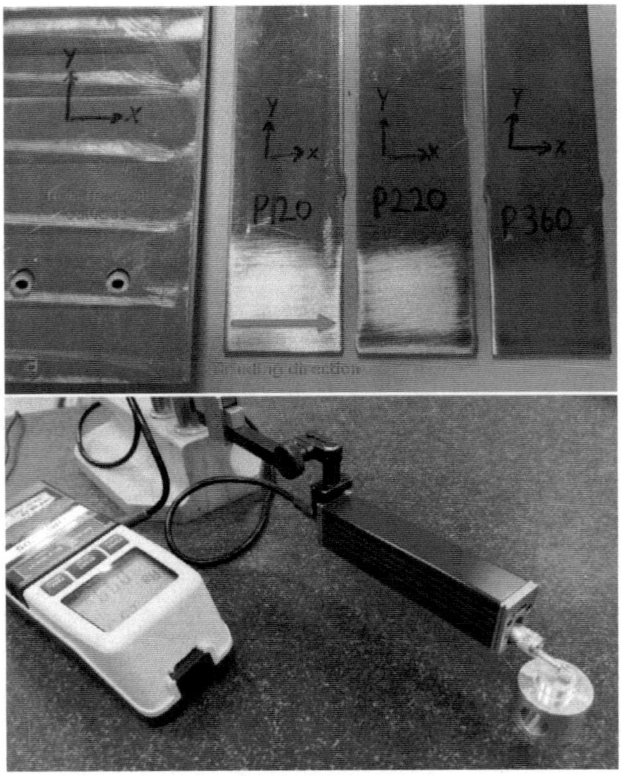

Figure 1. Dummy plates with different surface roughness and grinding direction (a). Roughness measurement of specimen grinded randomly with P220 sandpaper (b).

Table 1. Surface roughness values of measured surfaces.

Surface	**Direction**	**Ra (µm)**	**Rz (µm)**
Non-treated	(Almost Equal)	0.14	1.03
Grinded (P120 Sandpaper)	X	1.08	10.12
	Y	1.32	9.82
Grinded (P220 Sandpaper)	X	0.69	4.43
	Y	0.79	7.47
Grinded (P360 Sandpaper)	X	0.37	4.64
	Y	0.55	3.98
Sand Blasting (with fine sand)	(Almost Equal)	2.27	16.37
Grinded randomly (P220 Sandpaper)	(Almost Equal)	0.56	5.47

After the roughness magnitudes were determined P220 grit sized sandpaper and sand blasting process were selected for tests. Before adhesive application, bond surfaces were grinded randomly and surface roughness were measured again. After adhesive application, mold assemblies were cured at 80°C in temperature-controlled oven for one hour. Finally, all tensile and shear test were conducted at room temperature under 1.3 mm/min speed.

2.2 Tensile Test Specimen

For tensile tests, specimens were designed simply so that no extra parts or fixtures needed for tensile test machine. However, on this type of specimens bending may occur if pin holes are misaligned or contact point of pin is undetermined. It is an unwanted side effect and its level cannot be determined directly. In order to eliminate these problems, specimen rotation in curing mold was constrained with pins. Additionally, a notched pin was used so contact between specimen and pin is always clear. Test specimen dimensions, curing mold assembly and notched pin are shown in Figure 2.

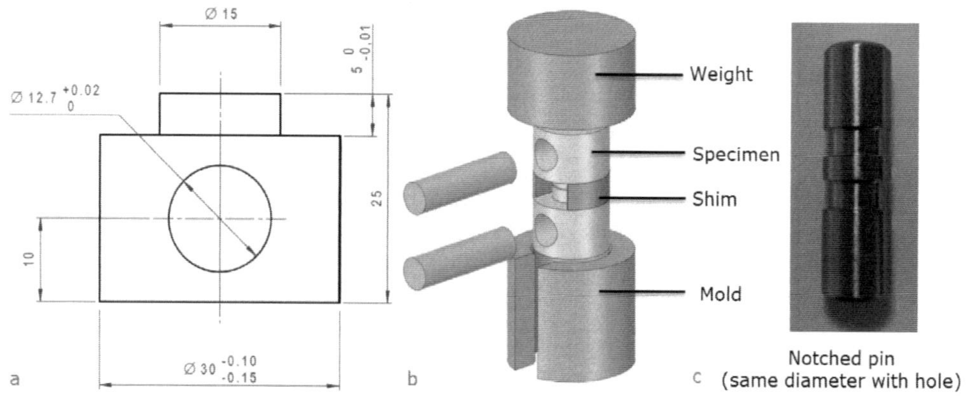

Figure 2. Specimen dimensions (a), curing assembly (b) and notched pin used in tests (c).

Tests were conducted for three different surfaces and three different adhesive thicknesses. These surfaces are non-treated, grinded with P220 sandpaper, and sand blasted surfaces. Thickness was adjusted with precision manufactured shims. Tested adhesive thicknesses are 0.3 mm, 0.5 mm and 1 mm.

2.3 Shear Test Specimen

Thick adherend shear test (TAST) specimens specified in IS0 11003-2 [15] standard was used in shear tests. Specimen dimensions and curing assembly are shown in Figure 3.

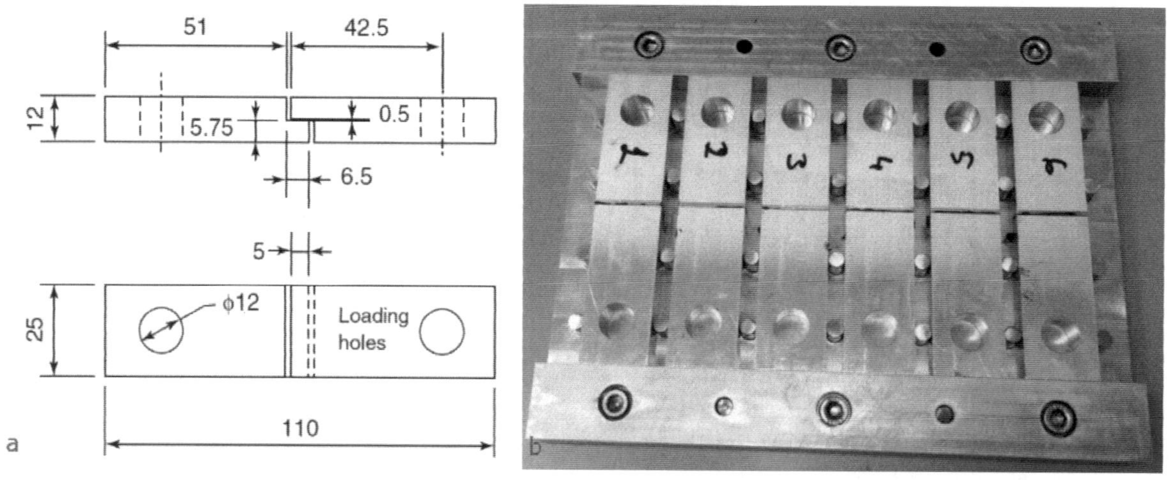

Figure 3. TAST specimen dimensions taken from [17] (a) and curing assembly with 1.5 mm thick shims (b).

There are two 1.5 mm steel shims for each test specimen. To maintain geometric dimensions while curing, shims must be inserted between each gap. Mold release spray was applied to all in contact surfaces with shim before applying adhesive, because shim should not be bonded to specimen. Notched pin was used for test to clarify contact point. All surface treatment and curing processes are same as tensile tests but shear tests were conducted for three different surface roughness and only 0.5 mm adhesive thickness.

3. RESULTS

As shown in Figure 4, for non-treated surfaces, tensile and shear strengths are lower than treated ones. However, there is not significant change between grinded and sand blasted results. It is seen that tensile properties are increasing with surface roughness until a limit value.

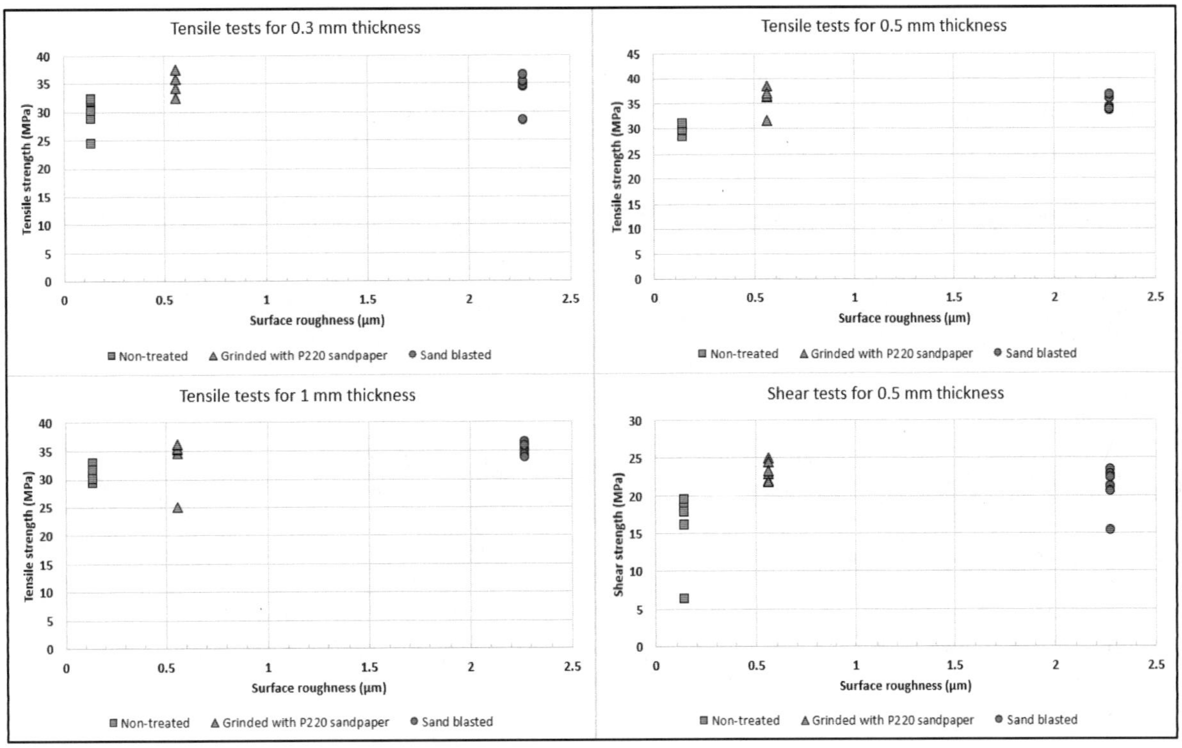

Figure 4. Tensile and shear test results.

Results showed that, for DP490 adhesive, there is no significant effect of bond line thickness on strength. However, it is seen that failure type related to thickness. Adhesive-cohesive failure was seen on 0.3 mm tensile test specimens while fully cohesive failure was observed on 0.5 mm and 1 mm test specimens. During the tests, it was noticed that some of non-treated test specimens have bonding defects. These specimens were ruptured well before expected. The reason of this problem thought to be low roughness values. Failure surfaces is shown in Figure 5.

Figure 5. Adhesive-cohesive failure (a), Cohesive faiulure(b) and bonding defect (c).

As shown in graphics, some test results are lower than its mean value. Fracture surfaces of these specimens were examined with optical microscope and big cavities were seen (Figure 6). These cavities, which are larger than others, cause a decrease in strength.

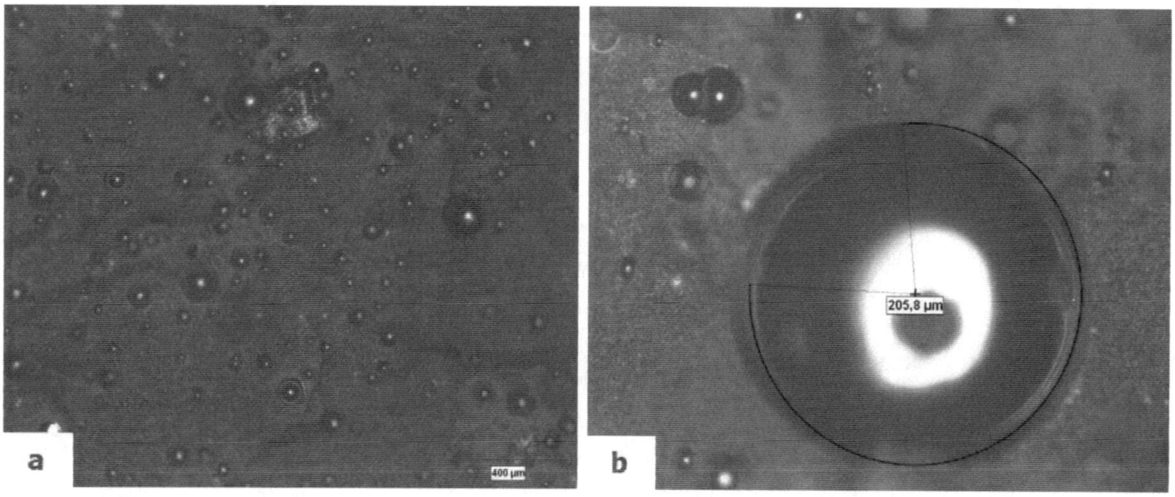

Figure 6. Normal size cavities (a) and large cavities. (b).

4. CONCLUSIONS

Effect of thickness and surface roughness on adhesively bonded joints is investigated. For this purpose, appropriate test specimens were manufactured. Tensile tests were conducted for different roughness and bond line thicknesses. On the other hand, shear tests were only dealt with surface roughness. Test results showed that surface roughness affect tensile and shear strength until a limit value. After that, there is not any significant change on strength. It was seen that there Some bonding defects were seen on non-treated surfaces. For tested adhesive, there is no significant effect of bond line thickness on strength, but failure type is changing. Finally, fractured surfaces were examined, and it was seen that specimens with larger cavities ruptured under lower loads.

5. REFERENCES

[1] Nitesh. S. Hirulkar, Pankaj.R. Jaiswal, Pirondi Alessandro, Paulo Reis, *Influence of Mechanical surface treatment on the strength of mixed adhesive joint*, Materials Today: Proceedings 5 (2018) 18776–18788, DOI: 10.1016/j.matpr.2018.06.224

[2] Sergio Correia, Vitor Anes, Luis Reis, *Effect of surface treatment on adhesively bonded aluminium-aluminium joints regarding aeronautical structures*, Engineering Failure Analysis 84 (2018) 34–45, DOI: 10.1016/j.engfailanal.2017.10.010

[3] Yiwei Xu, Huaguan Li, Yizhou Shen, Senyun Liu, Wentao Wang and Jie Tao, *Improvement of adhesion performance between aluminum alloy sheet and epoxy based on anodizing technique*, International Journal of Adhesion and Adhesives, DOI: 10.1016/j.ijadhadh.2016.05.007

[4] Xiaohong Zhan, Jicheng Chen, Cheng Gu, Qingyu Peng, Jie Chen & Yanhong Wei (2016): *Study on effects of pre-treatment and surface roughness on tensile-shear strength of 2060 Al-Li alloy adhesive joints*, The Journal of Adhesion, DOI: 10.1080/00218464.2015.1124765

[5] W. Leahy, V. Barron, M. Buggy, T. Young, A. Mas, F. Schue, T. McCabe & M. Bridge, (2001) *Plasma Surface Treatment of Aerospace Materials for Enhanced Adhesive Bonding*, The Journal of Adhesion, 77:3, 215-249, DOI: 10.1080/00218460108030739

[6] S. Azari, M. Papini & J. K. Spelt (2010) *Effect of Surface Roughness on the Performance of Adhesive Joints Under Static and Cyclic Loading*, The Journal of Adhesion, 86:7, 742-764, DOI: 10.1080/00218464.2010.482430, DOI: 10.1080/00218464.2010.482430

[7] Y. Boutar, S. Naïmi, S. Mezlini, R.J.C. Carbas, L.F.M. da Silva and M. Ben Sik Ali, *Fatigue resistance of an aluminium one-component polyurethane adhesive joint for the automotive industry: Effect of surface roughness and adhesive thickness*, International Journal of Adhesion and Adhesives, DOI: 10.1016/j.ijadhadh.2018.02.012

[8] A. Akhavan-Safar, M.R. Ayatollahi, L.F.M. da Silva, *Strength prediction of adhesively bonded single lap joints with different bondline thicknesses: A critical longitudinal strain approach*, International Journal of Solids and Structures (2017), DOI: 10.1016/j.ijsolstr.2017.01.022

[9] D. Castagnetti, A. Spaggiari & E. Dragoni (2011): *Effect of Bondline Thickness on the Static Strength of Structural Adhesives Under Nearly-Homogeneous Shear Stresses*, The Journal of Adhesion, 87:7-8, 780-803, DOI: 10.1080/00218464.2011.597309

[10] M. Lißner, E. Alabort, H. Cui, R. Rito, B.R.K. Blackman, N. Petrinic, *Experimental characterisation and numerical modelling of the influence of bondline thickness, loading rate, and deformation mode on the response of ductile adhesive interfaces*, Journal of the Mechanics and Physics of Solids 130 (2019) 349–369, DOI: 10.1016/j.jmps.2019.06.011

[11] Minoo D. Shokrian, Karim Shelesh-Nezhad, Reza Najjar, *The effects of Al surface treatment, adhesive thickness and microcapsule inclusion on the shear strength of bonded joints*, International Journal of Adhesion and Adhesives 89 (2019) 139–147, DOI: 10.1016/j.ijadhadh.2019.01.001

[12] Gefu Ji, Zhenyu Ouyang, Guoqiang Li, Samuel Ibekwe, Su-Seng Pang, *Effects of adhesive thickness on global and local Mode-I interfacial fracture of bonded joints*, International Journal of Solids and Structures 47 (2010) 2445–2458, DOI: 10.1016/j.ijsolstr.2010.05.006

[13] Wei Xu, Yueguang Wei, *Influence of adhesive thickness on local interface fracture and overall strength of metallic adhesive bonding structures*, International Journal of Adhesion & Adhesives 40(2013)158–167, DOI: 10.1016/j.ijadhadh.2012.07.012

[14] 3M™ Scotch-Weld™ DP490 – Product Datasheet, 1996

[15] ASTM D897-08(2016), Standard Test Method for Tensile Properties of Adhesive Bonds, ASTM International, West Conshohocken, PA, 2016, www.astm.org

[16] ISO 11003-2:2019Adhesives — Determination of shear behaviour of structural adhesives — Part 2: Tensile test method using thick adherends

[17] Lucas F.M. da Silva, David A. Dillard, Bamber Blackman, Robert D. Adams, *Testing Adhesive Joints: Best Practices*, John Wiley & Sons, 2012

COMPRESSION MOLDING OF COMPLEX THERMOSET LAMINATES

Travis R. Adams, Timothy J. Luchini, Jared B. Hughes, Steven M. Shewchuk, Adam Martinez, Gail Hahn

The Boeing Company
Saint Louis, MO 63134

ABSTRACT

Traditionally, high-rate small parts have been fabricated out of metal due to low cost, and high producibility. Following after trends in the automotive sector, certain aerospace-grade resins have shown processing flexibility while maintaining high mechanical and physical performance. The ability to fabricate complex composite laminates with a compression molding process using aerospace resins is essential for achieving low cost, structurally efficient components. The reduction in composites processing time through compression molding also enables favorable economic trades against metallic baselines, while providing increased component performance and providing weight savings. Challenges exist as the laminate thickness increases, tight radius of curvature are incorporated, and unidirectional tape is utilized. An undulating part was fabricated using aerospace prepregs and compression molding techniques. This shows that unidirectional tape can be used to fabricate constant cross section laminates with thick complex features in an automated process with direct forming. The demonstration article was fabricated using CYCOM® 5320-1 carbon fiber tape and fabric prepregs utilizing an on tool cycle time of 30 minutes to enable high throughput. The fabrication of this article will lead to further development of structure that requires quasi-isotropic properties coupled with complex, stiffening geometry. In addition, the ability to compression mold this geometry opens the option of geometrically stiffened structure where previously cost prohibitive.

Corresponding author: Gail Hahn

1. INTRODUCTION

Recently, there has been expanded interest in developing processes that enable the fabrication of high-rate (>1000 parts per year), high quality, thermoset components, [1] and [2]. Thermosets traditionally have a wide variety of product forms that lend themselves to complex forming and isothermal processing. Combined with these product forms, technology has been developed that allows the compression molding of aerospace grade thermoset materials [3].

Compression molding of thermosets offers an out-of-autoclave solution to traditional, autoclave baselines. Utilizing a closed mold tooling concept [3] and isothermal temperatures, on-tool times of 30 minutes or less can be achieved. When paired with a free-standing post-cure, mechanical and physical properties can be attained that compare well with those from an autoclave, or vacuum bag processes.

As the utilization of composites continues to grow outside of traditional aerospace fixed wing applications, the need for rate capable processes increases. New applications are continually being explored in commercial applications like urban air mobility platforms and defense applications like high performance weapons. These applications require the performance characteristics and weight savings offered by carbon fiber composites with the rapid cycle times traditionally targeted by industrial applications. Aerospace incumbent materials have shown promise in this area as well as developmental materials being researched by the traditional aerospace supply base.

1.1 Component Definition

A multifunctional team including Design, Materials and Process, and Strength Engineers developed an experimental laminate, referred to as TS-RAPM-012. This laminate was created to meet a set of mechanical performance requirements and hit a set of processing milestones in the RAPM program [3-8], in conjunction with the Tailorable Feedstock and Forming (TFF) initiative [1]. This part has several characteristics that are of interest when a high stiffness, buckle resistant laminate is desired. Of note, it has a nominal thickness of 6.3 mm, tight 12.7 mm radii which are formed in a corrugated pattern. Figure 1 shows a modeled representation of this component. In its untrimmed state (not shown), 44mm of excess material extends past the outer curves, and beyond its length.

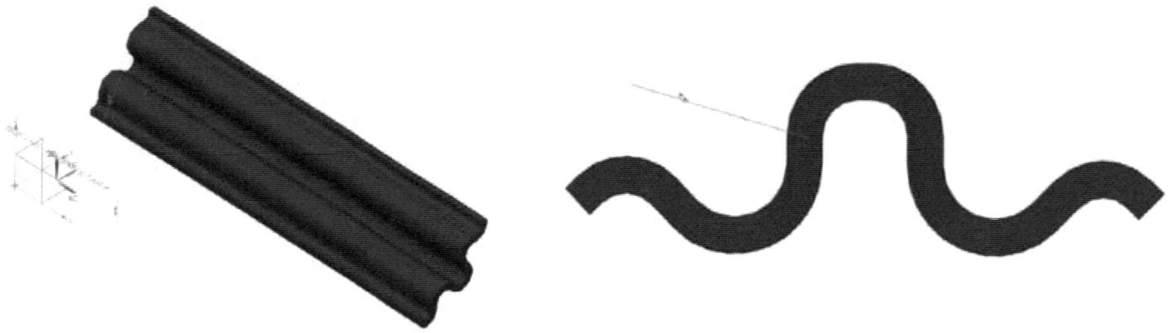

Figure 1 - TS-RAPM-012

To meet the performance requirements, a blend of multiple textile variants (Table 1) was used. A quasi-isotropic layup was used utilizing primarily intermediate modulus, sandwiched between fabric and glass plies. TS-RAPM-012 utilizes the CYCOM® 5320-1 resin system, which has shown the flexibility required for shorter length, isothermal cure cycles [3].

Table 1 - TS-RAPM-012 Material Composition

Material Designation	Resin System	Fiber Type	Weave
1	CYCOM® 5320-1	Glass	Style 108
2	CYCOM® 5320-1	T650	8 Harness Satin
3	CYCOM® 5320-1	IM7	Unidirectional

1.2 Compression Molding of Complex Geometry

Compression molding of thermosets is the process of placing a planar, collated blank of material into a two part mold installed in a hydraulic, heated press. Under hydraulic pressure and temperature, the liquid thermoset material is allowed to cure. This process is capable of high-volume fabrication of high-quality laminates [3]. Utilizing prior development in the RAPM program [3], it has been shown that a 30 minute, isothermal cure cycle at 177 °C, followed by a free standing 177 °C post cure for 2 hours provides adequate mechanical and physical properties for the CYCOM® 5320-1 resin system. The short on-tool cure cycle provides a laminate enough green strength for extraction from the tool in order to be subsequently batch processed in an oven at a later time. In addition to these thermal parameters, several other decisions must be made which are dependent on part geometry and tooling used. These parameters include the need to perform a pre-consolidation, and if a pre-heat processes is needed. Pre-consolidation is the process of using heat and vacuum to remove bulk from a collated ply-stack; this is essentially a hot-debulk. Pre-heating a blank provides improved formability of the blank by quickly reducing resin viscosity. An example of the planned TS-RAPM-012 process flow is shown in Figure 2.

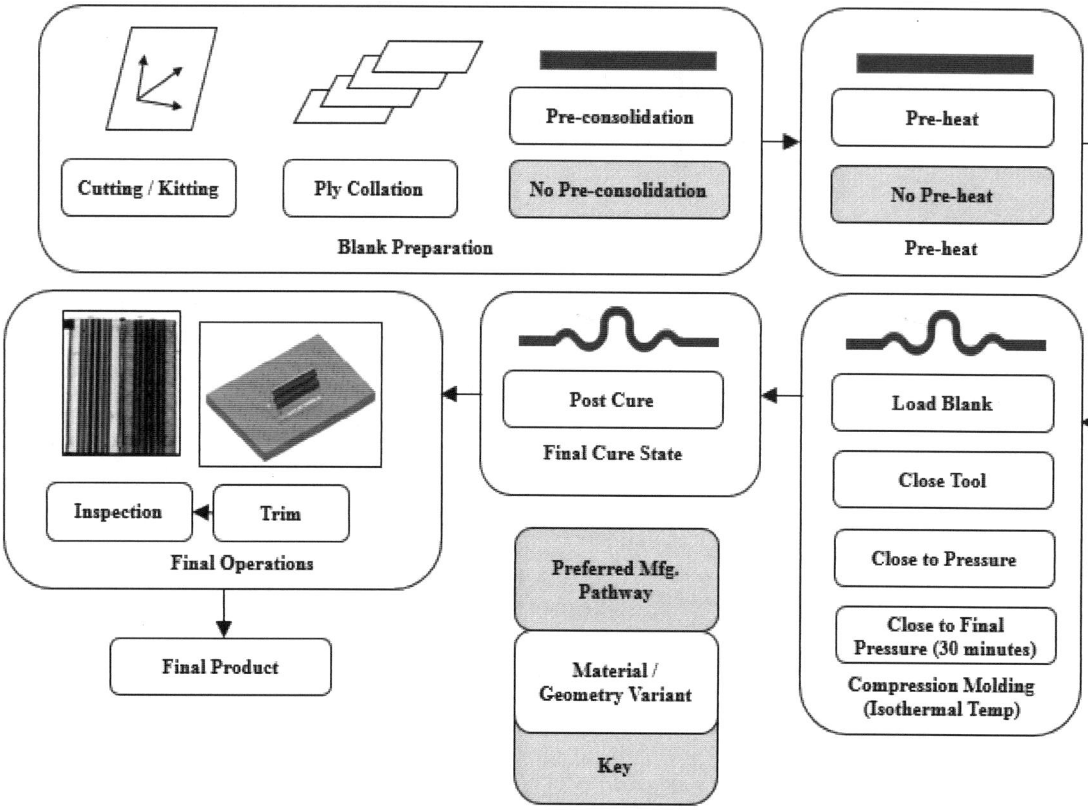

Figure 2 - Compression Molding Process for TS-RAPM-012

2. EXPERIMENTATION

2.1 Tooling

2.1.1 Clamp Fixture

A Surface Generation clamp fixture with Performance to Functional Specification (PtFS™) heating/cooling control technology was selected for the fabrication of TS-RAPM-012. The purpose of this advanced press is to tightly control the thermal input into the laminate using a pixelated tooling approach [4]. Being able to control individual pixels allows for consistent and uniform heat flow into the laminate. In addition, this press offers the flexibility to control consolidation pressure as a desired set point, or by controlling the mold gap by adjusting the platen offset, effectively creating a desired gap between mold faces. This press has 150 tonne clamping capacity, with a temperature ceiling of over 400°C. The design and build of this manufacturing cell is shown in Figure 3.

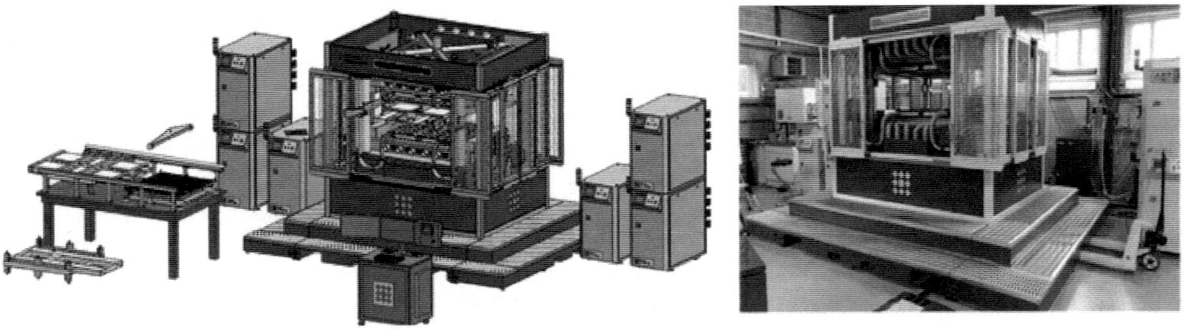

Figure 3 – CF-OaO-RAPM PtFS Manufacturing Cell

2.1.2 Laminate Mold

The mold for this laminate, designated CMD-TS-RAPM-012-501, was specifically designed for use on CF-OaO-RAPM with PtFS [4]. Fabricated from P20 steel, the mold utilizes both integrated resin and vacuum seals. Given the overall concave shape of the female-side mold, the vacuum seal allows the tool cavity to be evacuated during consolidation. The resin seal is designed to keep the resin from migrating out of the molding cavity, allowing the laminate to retain internal hydrostatic pressure, a key component to the fabrication of a quality compression molded laminate. As a process mitigation, the tool cavity was purposely fabricated thinner than nominal laminate thickness. The purpose of this was account for statistical variation in the resin content of the material. If resin content is below nominal, this "thinned out" cavity will still allow the laminate to maintain internal hydrostatic pressure. A representation of the mold faces is shown in Figure 4.

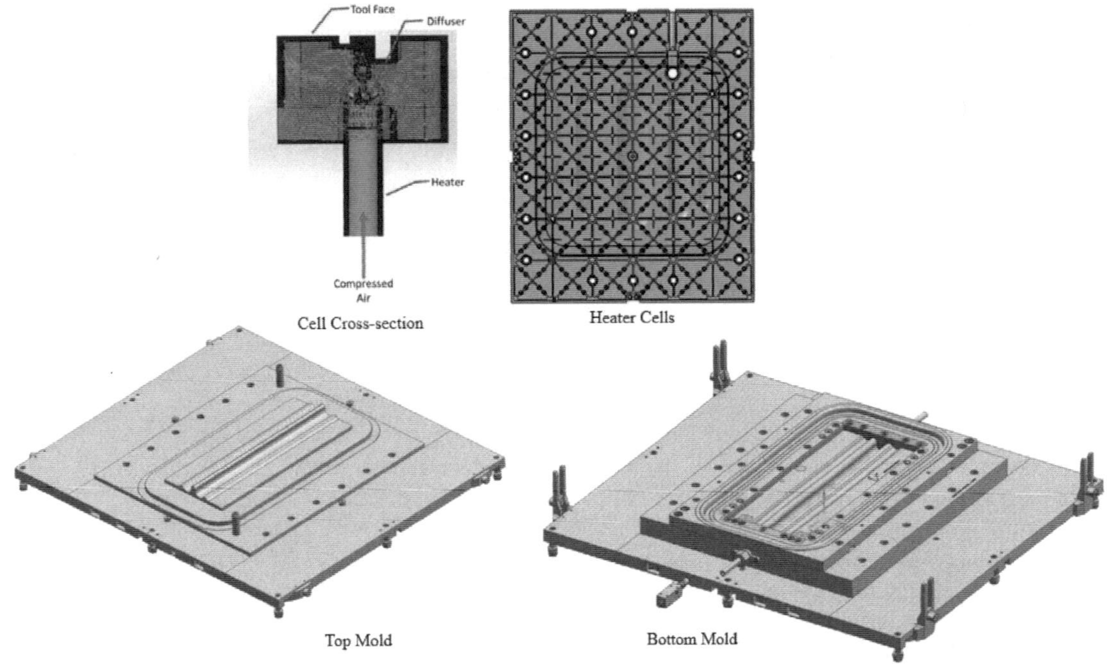

Figure 4 - CMD-TS-RAPM-012-501 Mold Configuration

A tool thermal uniformity survey was conducted on these mold faces to properly tune the thermal input into each of these tooling cells, or pixels. The hot and cold areas were also important to identify as these were locations that would be tested for degree of cure and glass transition temperature.

2.1.3 Supplemental tooling

During the development of TS-RAPM-012 it was determined that additional tools outside of the mold and clamping fixture would be required. The first was an additively manufactured trim fixture. This tool is designed with integral vacuums to hold the part in place. It was machined using a 5-axis Hurco. In addition, while the final CMD-TS-RAPM-012-501 mold was being manufactured, an autoclave mold was machined to create risk reduction articles in an autoclave. The purpose was to develop a baseline for part quality as well as understand the labor benefits achieved using a compression molding process against a traditional hand lay-up. These tools are shown in Figure 5.

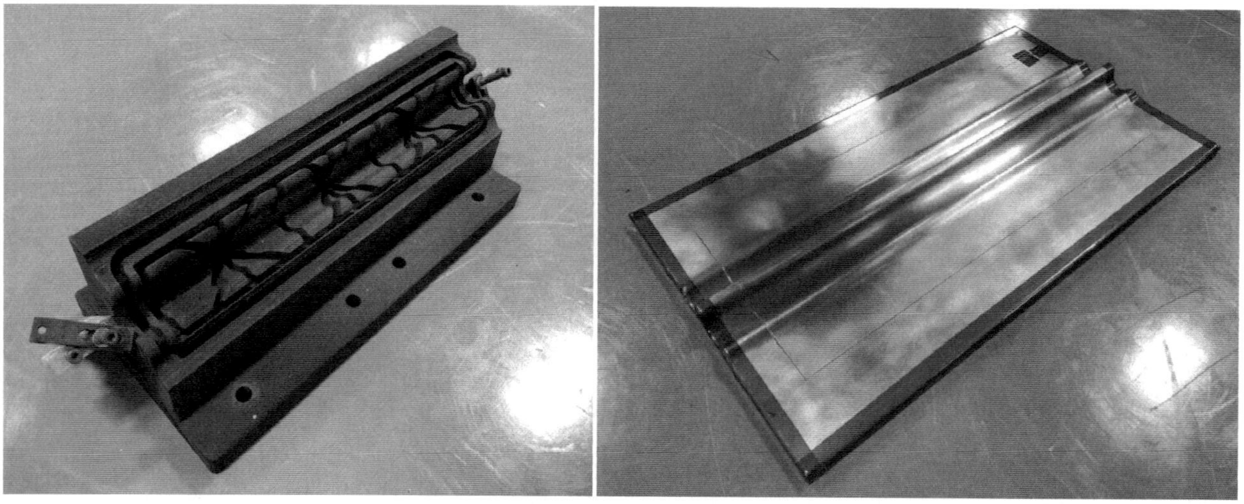

Additive Trim Fixture | Autoclave Mold

Figure 5 - Supplemental Tooling

2.2 Part Trials

2.2.1 Autoclave Baseline

To understand what traditional hand lay-up laminate would look like, and to compare labor differences to a compression molding process, several laminates were fabricated as risk reduction articles. Important information regarding time to fabricate was collected.

First, because of the deep, sudden changes in the geometry of this part, getting acceptable bag side quality was difficult. Multiple bagging schemes were utilized in order to achieve a smooth bag-side surface, and wrinkle-free laminates. Woven and nonwoven breather, as well as a flexible rubber caul were used. Because this was a trial tool, a caul sheet was not fabricated. The best part was fabricated using non-woven breather during processing. However, this laminate took three shifts to complete the hand lay-up operation, meaning it is not a feasible solution to high rate applications. This information is valuable when deciding if a compression molding process is an appropriate for a laminate of this complexity.

2.2.2 Forming Trials

An appreciable amount of work on the compression molding of continuous fiber thermosets had been accomplished prior to the fabrication of this component. Using previous process development and lessons learned, a solid baseline process was already in place as a starting point [3]. The difference from previous experiences are the aggressive geometry, and the use of a quasi-isotropic lay-up. The CF-OAO-RAPM PtFS controller was programmed to apply a minimal amount of pressure on the laminate until a proper viscosity was achieved. After this initial step, final pressure was applied to finish out the cure of the laminate. Only one trial was post-cured in a batch oven, and the rest post-cured in the clamping fixture to reduce time between trials. Very little was changed in regards to the press programming between trials. Several trials were completed building off of what was observed from visual, dimensional, and ultrasonic C-Scan inspections. Select laminates were subsequently dissected and chosen to undergo additional tests including photo-microscopy, degree of cure, and glass transition temperature. Table 2 outlines the trials that were completed in this exercise. After an acceptable laminate was fabricated, four additional laminates were fabricated using the same process. These additional laminates prove the consistency of the process parameters.

Table 2 - TS-RAPM-012 Trial Run Parameters

Trial	Cold Form	Pre-Consolidation (min / °C)	Pre-Heat	Cure Time (min)	Run Notes
0	Yes	No	No	30	First run - Cork Dam used; no release film
1	Yes	No	No	30	Release film used to aid in extraction: (2) layers on bottom, (1) on top; No cork dam
2	Yes	No	No	30	No release film; no cork dam
3	Yes	Attempted	No	N/A	Part destroyed during run
4	No	No	No	30	First part without cold form; no release film; no cork dam
5	No	40 / 116	3 / 179	30	Blank pre-consolidated flat; new release film scheme; cork dams used
6	Yes	No	No	N/A	Test part for new release film scheme. Part not cured
7	No	No	No	30	New release film scheme; cork dams were used
8	No	40 / 116	3 / 179	30	Blank pre-consolidated flat; cork dams used; no release film

In regards to the process parameters noted, several warrant discussion. First, the idea of cold-forming a previously collated ply-stack was derived from a desire to see if the plies would suffer damage during the actual pressing operation. After this was done the first time (Trial 0), it was thought that cold-forming could provide some benefit in order to pre-shape the charges. This was abandoned after Trial 3 due to being too cumbersome, time consuming, and the blank relaxing back to a flat shape. The other reason this was being done was to fill the mold cavity. The ply kits were designed to completely fill the mold cavity; trying to press a full sized flat charge into a fully assembled tool resulted in the tool shear wall clamping down on the ply stack, destroying it. After this was discovered, the side walls were removed in order to cold-form a whole stack of plies. When the cold-forming was abandoned, the blanks were trimmed down to just fit within the periphery of the assembled tool. Consequently, the blank would stop short of completely filling the mold, causing resin loss. Cork dams were inserted into the press to provide mass to fill volume.

Next, pre-consolidation (or hot-debulk), is a process that has been utilized with thick laminates to increase part quality. The process is performed in an oven, under full vacuum (>20inHg). The laminate is placed on a plate and bagged with woven fiberglass breather, and pre-consolidated in an oven for a determined amount of time. For this part, the parameters are shown in Table 2. It is preferred to not perform this step in order to reduce processing time, and keep a low thermal history on the resin. It must be observed that a pre-heat process is highly recommended when a blank is pre-consolidated. If placed in a press cold, the materials ability to form complex geometry is hampered by the high resin viscosity of the staged part. For these trials, the blank was simply placed into the hot mold, and closed as close as possible to the flat blank thickness without touching. After 3 minutes, the press operation proceeded as normal. It is recommended that an automated infrared pre-heat be used in a production scenario to reduce variability and part handling [3].

Last, in order to aid in laminate removal after cure or improve hydrostatic pressure, two configurations of release film was used. The first configuration used in Trial 1, was simply sandwiching the blank in 3 layers of film; 2 on the female side, 1 on male. The thought was the side with 2 layers would create a slip plane to help with forming. This was ultimately abandoned because of surface wrinkles in the cured laminate. The second configuration stretched two layers taut across the mold, with the blank in-between. This provided good results, but caused an unacceptably thin laminate.

2.2.3 Trial Evaluation

Two criteria were used during the part forming trials to evaluate acceptability. All parts were submitted for ultrasonic inspection utilizing a 1MHz and 5MHz scan, while several parts were chosen for dissection. This second group of parts were evaluated for laminate thickness. Figure 6 shows both the 1Mhz and 5Mhz scan. It is important to note that only the 5MHz would be an acceptable frequency to correlate an attenuation level to porosity. However, the 1MHz scan was able to more reliable penetrate the aggressive curvature. This was used to verify that the parts were consistent visually across most geometry of the laminate.

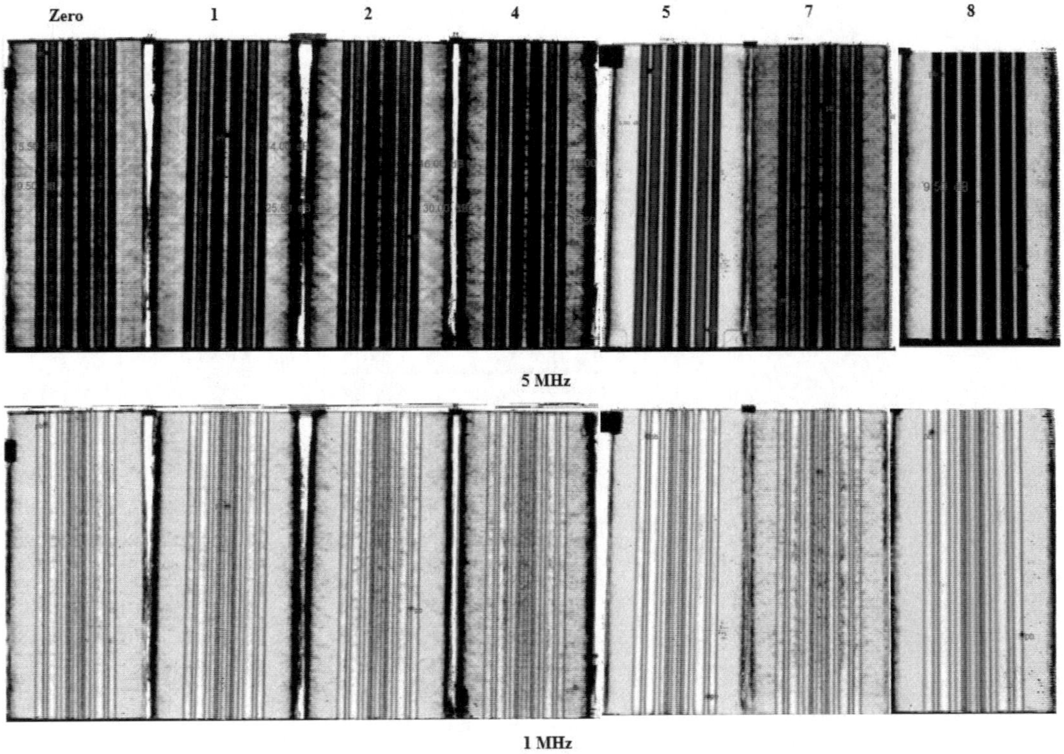

Figure 6 - TS-RAPM-012 Trial NDT Inspection Results

It is important to note that the dark regions around the radii will show increased attenuation. This is due to the ultrasonic system used only scanning on a 2D plane. However, a few observations should be made from evaluating these images. First, pre-consolidation has a large impact on perceived part quality. The drop in attenuation levels in Trials 5 and 8 are evidence of this. Second, the cold forming appears to have correlation to lower attenuation, meaning that the shaping of the plies prior to cure may have reduced the laminate porosity.

In addition to ultrasonic inspection, thickness measurements were taken on several parts, and is shown in Table 3.

Table 3 - TS-RAPM-012 Thickness Results

Trial	Thickness min/max (mm)[1]
0	5.8/6.4
1	N/A
2	N/A
3	N/A
4	5.8/6.4
5	5.6/6.3
6	
7	5.6/6.0
8	5.8/6.0

Notes:

1 - Nominal thickness is 6.3mm; permissible minimum is 5.7mm

The key take-away from the thickness data is the impact of both the pre-consolidation and the "new release film scheme" had on the overall part thickness. Both appeared to have created a thinner part, although the pre-consolidation appears to have abated this problem when combined with the release film. Trial 8 was an effort to fine tune the process, and was successful. A part with low attenuation levels, and desirable thickness was manufactured.

3. RESULTS

3.1 Non-Destructive Evaluation

In addition to the trial parts detailed above, four additional parts were successfully fabricated using an identical process to Trial 8. The NDT results are shown in Figure 7.

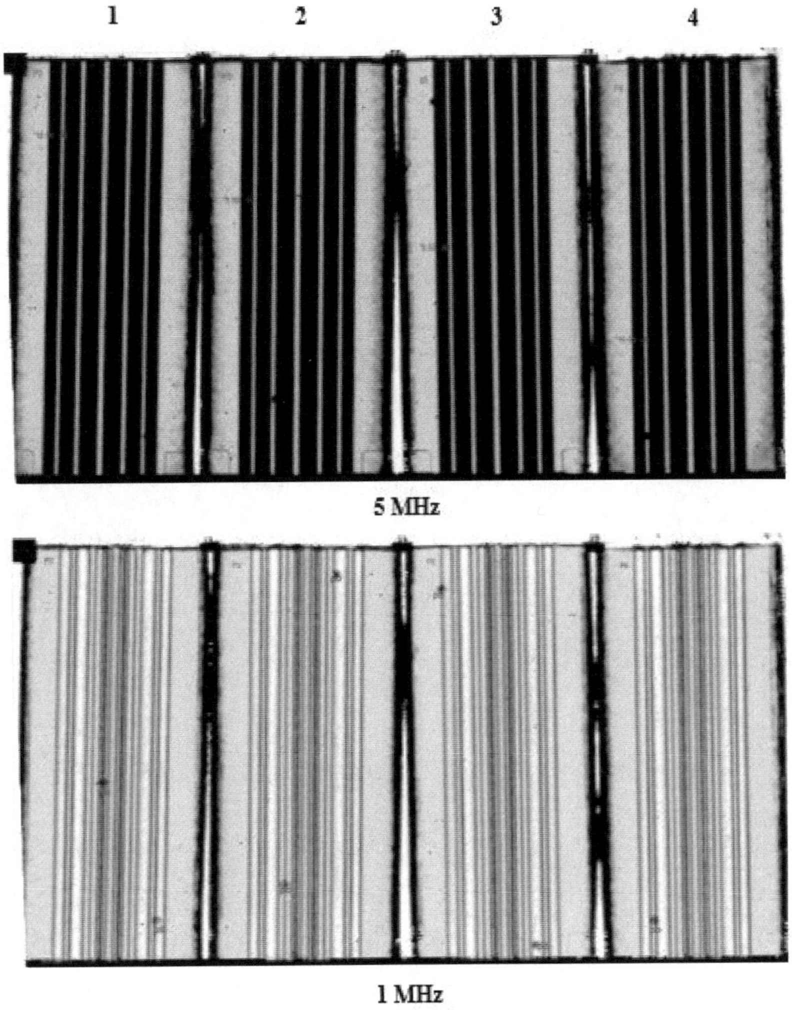

Figure 7 - TS-RAPM-012 Final NDT Results

All subsequent parts show excellent consistency with Trial 8 and each other; each part ranges between 9.5 and 10 dB with the 5MHz scan. Visually, the 1 MHz scans are nearly identical. These scans were later correlated with photo-microscopy and void content analysis to ensure the reliability of the scan method. It is important to note that the flat excess material was left on for inspection purposes. It was later trimmed off during the final trim operation. A complete laminate is shown in Figure 8.

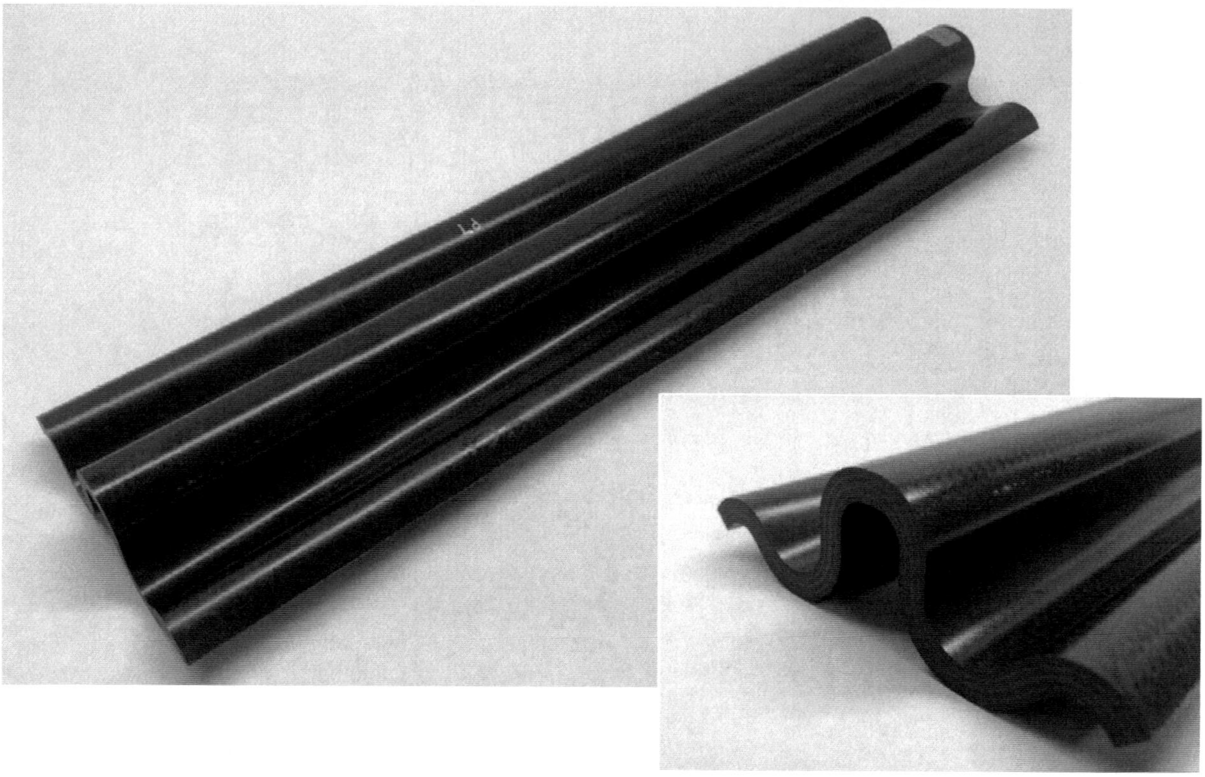

Figure 8 - Completed TS-RAPM-012 Laminate

3.2 Photo-Microscopy

In addition to the ultrasonic inspection, the quality of the laminate internally was also of great interest. Particularly, how much the fibers were disturbed during the forming from a flat blank, and any other anomalies. A dissection map is shown in Figure 9.

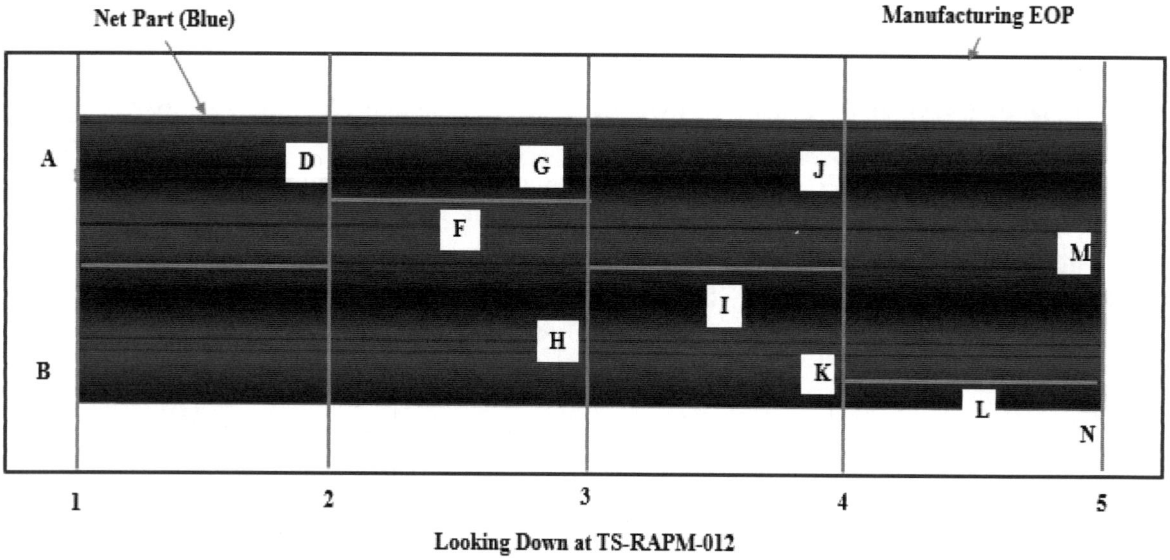

Figure 9 - TS-RAPM-012 Dissection Locations

Four trials were completely dissected, and Trial 8 was partially dissected as to retain the laminate whole. Figure 10 shows a sampling of the images taken of Trial 8, as it represents the final process used.

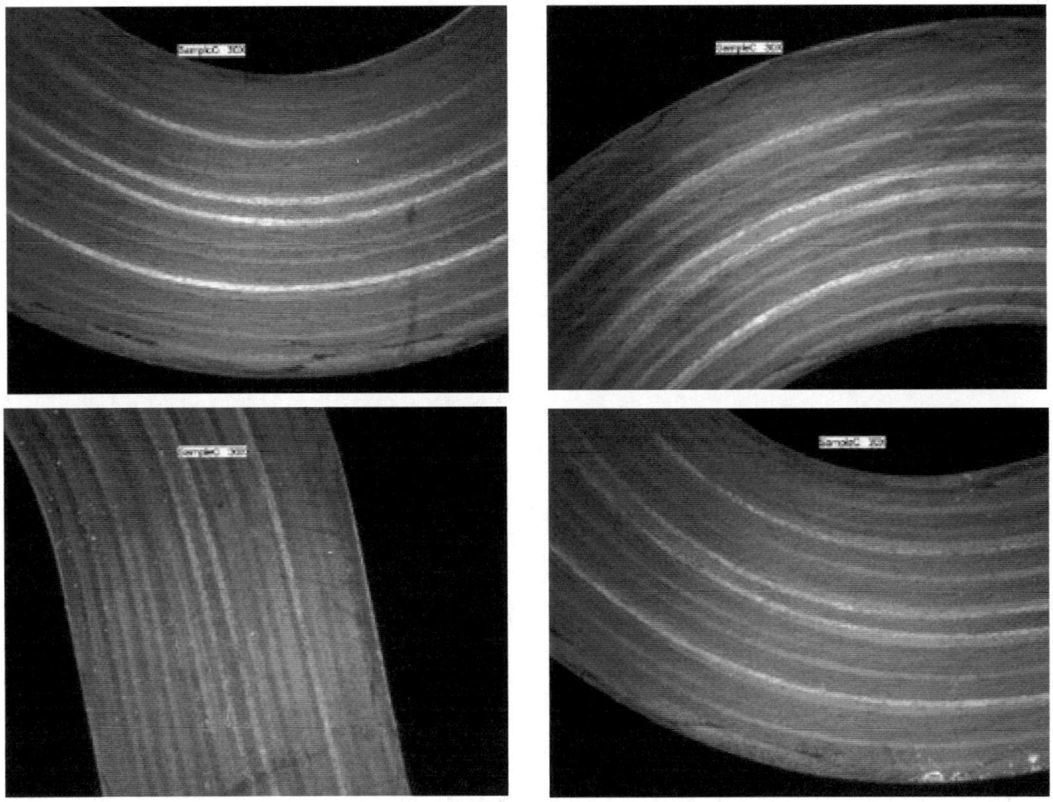

Figure 10 - TS-RAPM-012 Curved Geometry

Overall, the quality of laminate is acceptable. For the majority of sections observed, the fibers remain true and without disruption. Occasionally, small amounts of porosity were observed but as documented later, only in low levels.

3.3 Physical Testing

The last verification of the process was to verify degree of cure, glass transition temperature, void content, and fiber volume fraction.

Table 4 - TS-RAPM-012 Degree of Cure Results

Sample	Mass	Tg half height (2nd Heat)	Residual exotherm	Degree of cure	Exotherm Peak
	(mg)	(C)	(J/g)	(%)	(C)
Prepreg	21.41	214.0	143.7		233.39
T5 DOC 1	13.5	215.9	6.8	95.3%	243.01
T5 DOC 2	11.02	217.1	7.3	94.9%	243.27
T5 DOC 3	11.79	215.4	8.3	94.2%	245.69
T7 DOC 1	12.74	221.2	5.7	96.0%	244.35
T7 DOC 2	13.27	220.9	4.7	96.7%	243.45
T7 DOC 3	11.52	218.6	5.0	96.5%	241.46

Using Differential Scanning Calorimetry (DSC), the degree of cure was determined for six samples; three from Trial 5 and three from Trial 7. Trial 5 was used as a proxy for Trial 8 due to it having identical thermal history. Trial 7 was chosen to see how much difference there would be between a laminate that was pre-consolidated and one that was not. As shown in Table 4, there is very little difference, and the degree of cure is within the expected range of this resin system.

Table 5 - TS-RAPM-012 Glass Transition Temperature Results

Sample	Tg
	(C)
T5-1	207
T5-2	203
T5-3	205
T7-1	205
T7-2	202
T7-3	203

The dry glass transition temperature was determined using Dynamic Mechanical Analysis (DMA), and the results are shown in Table 5. The rational for using Trial's 5 and 7 remain the same. The glass transition temperature is within the expected range of this resin system.

Table 6 - TS-RAPM-012 Fiber Volume and Void Content

Sample	Resin Volume Content	Fiber Volume Content	Void Volume Content
T5-C-1	41.81%	58.40%	-0.21%
T5-C-2	40.63%	59.32%	0.06%
T5-C-3	38.81%	60.79%	0.39%
T5-C-4	38.69%	60.85%	0.46%
T5-C-5	37.96%	61.56%	0.48%
T5-C-6	37.29%	62.04%	0.67%
T5-C-7	38.32%	61.31%	0.36%
T5-C-8	37.90%	61.56%	0.54%
T8-1	38.64%	61.06%	0.31%
T8-2	37.71%	61.89%	0.40%
T8-3	37.27%	62.37%	0.36%
T8-4	37.53%	61.87%	0.60%
T8-5	39.38%	60.80%	-0.18%
T8-6	41.18%	59.11%	-0.29%
T8-7	39.06%	60.84%	0.10%
T8-8	37.11%	62.58%	0.31%

Several samples were taken from Trials 5 and 8 to evaluate fiber volume and void content, with results shown in Table 6. Trial 5 originally had sixteen samples taken from the laminate, but due to specimen error, eight samples were discarded. Of the remaining sixteen samples, three samples show negative values. It is suspected that specimen preparation was to blame as the outer fiberglass ply needed to be mechanically removed. Overall, fiber volume content and void content are within acceptable ranges.

4. CONCLUSIONS

The results indicate that after a rapid sprint controlling the relevant processing variables, a high quality, direct stamped, thermoset composite prepreg laminate with complex curvature can be manufactured that enables new high part volume applications. The results show that this is a viable process meeting standard quality requirements and helped to highlight new tooling and equipment approaches. These findings are consistent with the findings completed in other RAPM [3-8] technology tracks and show novel approaches for structural composite fabrication.

5. ACKNOWLEDGEMENTS

This effort was jointly accomplished by a Boeing-led team and the United States Government (Defense Advanced Research Projects Agency, DARPA) under the guidance of the United States Army Research Office, ARO. The authors would like to acknowledge the guidance and support of Dr. Jan Vandenbrande of DARPA/DSO and Dr. David Stepp of Army Research Office (ARO). The information in this paper was approved for public release by 19-162554-ETT and DISTAR 32485. Research was sponsored by the Defense Advanced Research Projects Agency (DARPA) and was accomplished under Cooperative Agreement Number W911NF-16-2-0087. The views and conclusions contained in this document are those of the authors and should not be interpreted as representing the official policies, either expressed or implied, of the Defense Advanced Research Projects Agency or the U.S. Government. The U.S. Government is authorized to reproduce and distribute reprints for Government purposes notwithstanding any copyright notation herein.

6. REFERENCES

1. Vandenbrande, J. "Tailorable Feedstock and Forming (TFF)" https://www.darpa.mil/program/tailorable-feedstock-and-forming
2. Koon, R., Rademacher, M., and Getty, M. "Challenges and Opportunities to Transitioning Automotive Composites to Aerospace." CAMX Conference Proceedings. Orlando, FL, December 12-14, 2017.
3. Luchini, T.J., Rodriguez, A.J., Rogers, S.A. Bras, A., Whysall, A., Russell, R., Lucas, S., and Hahn, G.L. "Spring Frame Press Fabrication of Aerospace Production Components." SAMPE 2019 Conference Proceedings. Charlotte, NC, May 20-23, 2019.
4. Shewchuk, S.M., Halford, B., Matlack, M.P., Sharpe, A. and Massey, P. "Development of Scalable Dynamic Control Architectures for Flexible Composites Manufacturing Work Cells." SAMPE 2019 Conference Proceedings. Charlotte, NC, May 20-23, 2019.
5. Hahn, G.L., Tsotsis, T.K., "Rapid High-Performance Molding (RAPM) of Small Parts," SAMPE 2019 Conference Proceedings. Charlotte, NC, May 20-23, 2019.
6. Charles, D.A., Matlack, M.P., Hahn, G.L. "Agile Analysis of Manufacturing Cell Costs for Molding of Small Composite Parts." SAMPE 2019 Conference Proceedings. Charlotte, NC, May 20-23, 2019.
7. Tsotsis, T.K., Cespedes-Gonzales, G., Wiener, M., Cohen, L. J., Calamito, D., Costantino, S., and Klunker, F. "Fabrication of a Complex Part with Deep-Draw Sections by Resin Transfer Molding." SAMPE 2019 Conference Proceedings. Charlotte, NC, May 20-23, 2019.
8. Schell, J.J., Shewchuk, S. M., Leach, D., McCrea, T., Abrams, N., Postera, R., and Hahn, G. L. "Rapid High Performance Molding of Structural Thermoplastic Composite Parts." SAMPE 2019 Conference Proceedings. Charlotte, NC, May 20-23, 2019.

NON-LINEAR BEHAVIOUR OF TOWPREG TENSILE MODULUS FOR AUTOMATED FIBER PLACEMENT (AFP)

Massimo Carboni and Mehdi Hojjati
Concordia Center for Composites, Concordia University
Montreal, Quebec, Canada

ABSTRACT

Modern composite manufacturing techniques using Automated Fiber Placement (AFP) technology incorporate simulations and Finite Element Analysis (FEA) prior to manufacturing of the part. In order to have acceptable results from the simulation, the material properties of uncured towpreg is required. One of those properties is tensile modulus. This study investigates the linear and non-linear behavior of the tensile modulus of a single uncured carbon fiber towpreg. Tensile tests were performed, and the force and displacement were monitored. Stress-strain curves were generated, and the modulus was calculated. It was observed that the modulus at the beginning of the load is nonlinear and as the load increases, it becomes constant and closer to the cured laminate. Additionally, investigation was made to determine the cause for the non-linear modulus through studying the fiber waviness of the uncured prepreg by microscopic observations. Comparisons were made between the waviness of untested and uncured carbon fiber tows with the tested uncured carbon fiber tows.

1. INTRODUCTION

In the domain of composite manufacturing, Automated Fiber Placement (AFP) and Automated Tape Laying (ATL) are new processes that are replacing traditional techniques such as hand lay-up. ATL and AFP are similar process which are faster, cost-efficient and provide a viable alternative in manufacturing large complex structures [1-3]. Both manufacturing techniques rely on process modeling simulations to predict the path of the fibers and to control the process parameters in order to avoid the formation of different types of defects. This process model can be used to control tape overlap/ gap as well as tow buckling and wrinkle formation. One of the limiting factors in process modeling for AFP & ATL is the reliance on using the cured composite material properties for process modeling when the material properties for the uncured state should be used instead [1]. Variability exists in the material properties between both states, with the properties of the uncured state being significantly lower [2]. This study will address the linear and non-linear behavior of the tensile modulus of uncured thermoset carbon fiber tows used in AFP and ATL simulations.

A few studies have been done to investigate the material properties of uncured carbon fiber prepreg. The variability of uncured carbon fiber epoxy prepreg were investigated where the tensile modulus was discussed [2]. While this study focused on all types of variability in prepreg including mass properties and waviness, measurements on the tensile response of uncured prepreg was also touched upon. They performed a uniaxial tensile test on a stack of 13 tows of unidirectional carbon fiber/ epoxy prepreg along the fiber direction and measured the stress/strain response using a 25 kN Instron test machine. The study attributed the tensile response to fiber misalignment in the prepreg and while a stress/strain plot was presented and characterized a “lead-in” area on non-linearity followed by a linear region. The lead-in area is attributed to the fiber waviness [4].

Researchers at North Western University, supported by a subcontract from the Ford Motor Company, published their findings on the mechanical response of woven composite prepreg during forming processes [3]. Focusing on the mechanical response and properties of the prepreg system, the uniaxial tension test was performed on uncured prepreg to obtain the tensile modulus. Similar to Ref [4], the stress/ strain plots of the tensile response, over varying temperatures, all exhibited a region of non-linearity followed by a linear region. This report studied the effects of temperature on the non-linear section of the plot and determined that as temperature increases, the area of non-linearity also increases. However, again the phenomenon and the reasoning behind the non-linear region were not addressed, and the stiffness represented by the linear region of the plot was selected to describe the response of the uncured prepreg [5].

This report will concentrate on studying the tensile modulus of unidirectional (UD) thermoset, carbon fiber/epoxy resin systems, as it is one of the primary material properties required for AFP & ATL process modeling [1,5]. To measure the tensile modulus, uniaxial tensile testing was conducted and the engineering stress strain curve was derived from the force and displacement data acquired during the testing. This study will focus on classifying the non-linear region on the stress/strain plot of the tensile response, as this is the region the material will be exhibiting during AFP & ATL manufacturing. As seen in literature, in the uncured state, the material properties of carbon fiber prepreg are observed to be temperature dependent and, like cured composites, the fiber waviness also affects the material properties in the uncured state [4-8]. Testing temperatures should remain below 90 °C to prevent the excess curing during the lay-up. For our experiments, testing was conducted at room temperature 25 °C. Additionally, fiber waviness were measured though microscopic analysis before and after testing to study any variations in the waviness and its effect on the tensile modulus was discussed.

2. EXPERIMENTATION

2.1 Material and Equipment

The AFP prepreg used for experimentation in this study was manufactured by Solvay and consists of 12KHTS carbon fiber continuous filaments and a CYCOM 977-2 epoxy resin system. The prepreg used in AFP and ATL manufacturing are different than traditional prepreg as they are prepared as a several meter long and 6 mm wide continuous strip of material wound around a spool. This spool of prepreg is then able to be feed through the delivery head of an AFP or ATL machine and deposited onto a tool. For experimental tensile testing, samples of this prepreg with dimensions of 127 mm long, 6 mm wide and 0.127 mm thick were used. Samples were tested in a HOSKIN machine with a 5 kN load cell.

2.2 Tensile Behavior of Uncured Prepreg

To characterize the tensile behavior of uncured carbon fiber prepreg, the tensile force and the displacement at the head of the machine were recorded to calculate the engineering stress and strain. Tensile force was applied along the fiber direction of the material (direction 1) and the machine was set to a displacement of 2 mm/min. Prior to testing, samples were removed from the freezer and left to sit at room temperature for 1 hour. The towpreg was wrpped around a steel block to be used inside the tensile machine. It was reported that the main source of failure during tensile testing of uncured prepregs was cause due to the material slippage in the grips. The viscous nature and the low friction coefficient of the uncured epoxy can cause the material to slip out of the grips, resulting in an inaccurate reading of displacement/ strain in the results [4,5,10]. To prevent slippage from occurring, four short steel blocks were used in conjunction with the grips of the machine. The steel blocks were made from standard keystock of cross section 0.5 cm by 0.5 cm and a length of 4 cm. One side of the uncured prepreg was wound around a steel block, as shown in Figure 1a, and then compressed by the other steel block in the grips of the HOSKIN machine (Figure 1b). Between 4 cm to 5 cm of prepreg was used to wound around the block. The samples were first set up and clamped in the lower grip, then they were aligned and clamped in the upper grip. This was to insure no misalignment in the sample setup and to ensure the tensile load was applied in the fiber direction. This type of setup within the grips was able to prevent slipping as the grips viscous epoxy was no longer in contact with the grips, instead the force holding the sample in place was acting directly onto the steel blocks.

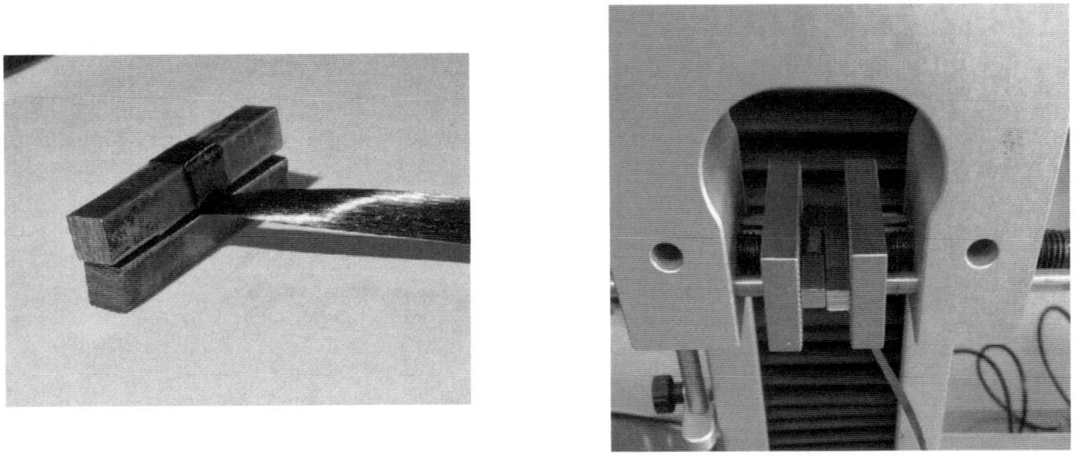

(a) (b)

Figure 1. Test sample for uniaxial tensile testing wrapped around steel blocks to prevent slippage.

The gauge length of each specimen tested was kept at 127 mm, while the actual length of each sample was a maximum of 200 mm as the extra length of material was required to wound around the steel block. One ply of material was used, for a total thickness of approximately 0.127 mm and a width of 6 mm. In total, 24 samples were tested in this configuration. Engineering stress was calculated using the recorded reaction force in the load cell (F_1) and the dimensions of the sample, width (w), and thickness (t) as:

$$\sigma_1 = {F_1}/{(w * t)} \qquad [1]$$

Engineering strain was calculated using the original gauge length of the sample (L_g) with the displacement measured in the head of the machine (ΔL).

$$\varepsilon_1 = {\Delta L}/{L} \qquad [2]$$

2.3 Microscopic Imaging

To investigate and observe the fiber waviness and fiber misalignment, microscopic images and analysis were performed on the prepreg samples prior to and post tensile testing. Samples were prepared by embedding them in an epoxy resin puck, and then polishing them until the sample was exposed on one end. Once the samples were polished, microscopic images were taken at 10x magnification. A MATLAB code developed at the Chalmers University of Technology was used for characterizing the degree of fiber misalignment in microscopic images [4]. Using this software, the change in misalignment and in turn the change in waviness before and after testing can be studied.

3. RESULTS & DISCUSSION

3.1 Uniaxial Tensile Testing

3.1.1 Evaluation of Test Method

As there are no standards for testing uncured prepreg, the accuracy and reliability of our test method was evaluated. To conduct our evaluation, test samples were setup using the steel blocks and were subjected to an initial load of 100 N. Samples were then released and the tensile force was returned to a predefined load, where it held the samples for a short time before they were tensile tested once more. The return loads chosen for each sample were 40 N, 30 N, 20 N and 10 N respectively. Figure 2 shows the resultant plot for all 4 samples.

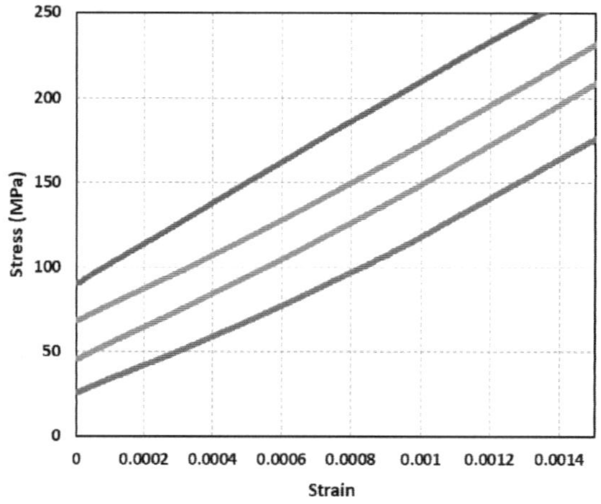

Figure 2. Experimental results for identification of slippage in the grips

Table 1. Maximum Young's modulus recorded in linear section of stress vs. strain plot

Sample #	Young's Modulus (GPa)	Sample #	Young's Modulus (GPa)
1	113.78	3	113.69
2	115.54	4	112.99

As shown in Ffigure 2 and Table 1, all samples tested exhibits a linear modulus with similar slope of approximatly 114 GPa. The similarities in the slope indicate good repeatability of our test method. It was determined that samples should be loaded under a pre-lod prior to testing. For this experimentation, the pre-load was selected to be approximatly 100 N. The purpose of the pre-load is to ensure that the clamping force is sufficient, prevent any slippage in the system and to ensure that the same is properly aligned in the testing fixture.

3.1.2 Uniaxial Tensile Test Results

Eight specimens from two distinct batches of material were tested. The results indicate that the tensile response of an uncured carbon fiber can be characterized as a non-linear behavior at the beginning of the test followed by a linear region. The same behavior has been reported in the literature where the non-linear region described as a lead-in region of the reduced modulus [2]. From experimental observations, the lead-in (non-linear) region of the stress strain curve for our prepreg material was observed up until 0.2% strain where it then followed by a linear trend. Due to the observed failure in the prepreg at a force of over 500 N, in the form of fiber splitting and tearing of the fibers, the tensile response was plotted up to 0.5-0.6% strain. Figure 3 shows the average stress strain plot generated from all 8 samples.

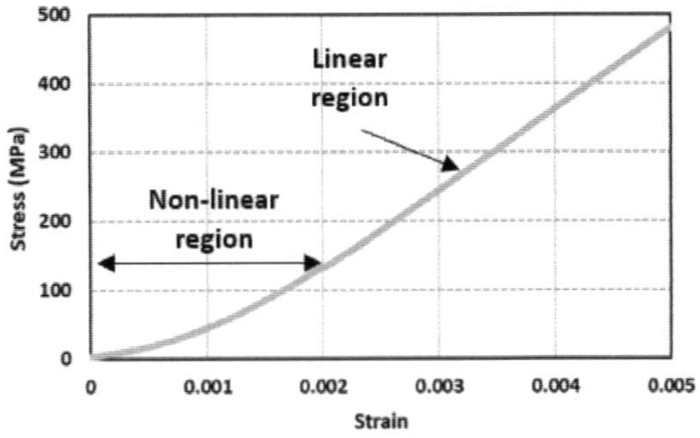

Figure 3. Experimental Results of Tensile Behavior of Uncured Prepreg

It is important to note that Figure 3 represents an average of all sample and when comparing the stress strain curve over individual samples and even over the two different batches of identical material, there is a large standard deviation between results. Figure 4 shows the tensile response between two batches of identical Solvay prepreg. While both plots exhibit the same characteristic lead-in area followed by an area of linear response, the two plots have their own distinct behaviour.

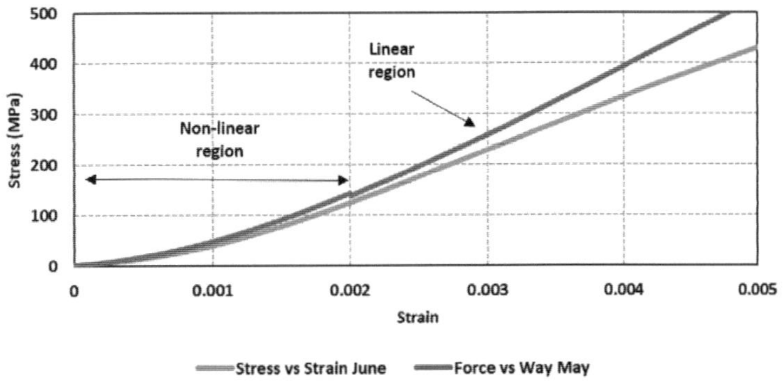

Figure 4. Difference in Tensile Response Between two different spools of towpreg materials.

It is possible that the variations in tensile modulus are attributed to the differences in waviness between the samples. As the waviness in uncured prepreg can change due to the ease of fiber movement within the uncured viscous epoxy, the angle of misalignment will then vary per sample, and in turn introduces a source of variability in values for the tensile response. This can be further supported when comparing the tensile response of each sample individually. The tensile response between samples, even those within the same batch of material, have a high standard deviation between each other. Each sample was cut from the same roll of AFP prepreg and theoretically should have the exact same properties. However as recorded experimentally, each sample exhibits a different tensile response due to the varying waviness and misalignment of the fibers between each sample caused by the movement of the fibers in the viscous epoxy matrix. One aspect of the tensile response curve that stays constant between all samples are the lead-in and linear regions caused by the re-aligning of the misaligned fibers. As all samples experience a certain degree of misalignment, they in turn all display a non-linear gain in modulus as the fibers align themselves along the direction of the applied load.

Studying the tensile response of all samples as an average, shown in Figure 3, the lead-in region of non-linear modulus lies between 0%-0.2% strain and 0-100 MPa of stress. Correspondingly, this relates to a maximum of 50 N of applied uniaxial tensile force. In AFP & ATL manufacturing, tensile loads are expected to be close to 25 N, however it is difficult to measure exactly what the applied loads are during this process. Nonetheless, during AFP & ATL manufacturing, the tensile modulus of the uncured prepreg is expected to be in the non-linear region of its response and thus modeling of the manufacturing process must consider the waviness, misalignment and the changing modulus in this region. Microscopic analysis provides clear observations of what is happening at the micro level of the prepreg and is a means of developing a microscale model to support the experimental results.

3.2 Microscopic Observations

To capture the effect of fiber waviness and misalignment in the prepreg, microscopic analysis was done on the prepreg samples prior to and post tensile testing. Figure 5 represents microscopic images of one sample at 10x magnification. Visually, the waviness is shown to have changed after the sample was tested in uniaxial tension. The waviness appears to have diminished along the fiber direction and the overall misalignment of the fibers has been reduced. To further support the visual observations, the photos were processed using the MATLAB software developed by D. Whilhelmsson as mentioned in the previous section [4].

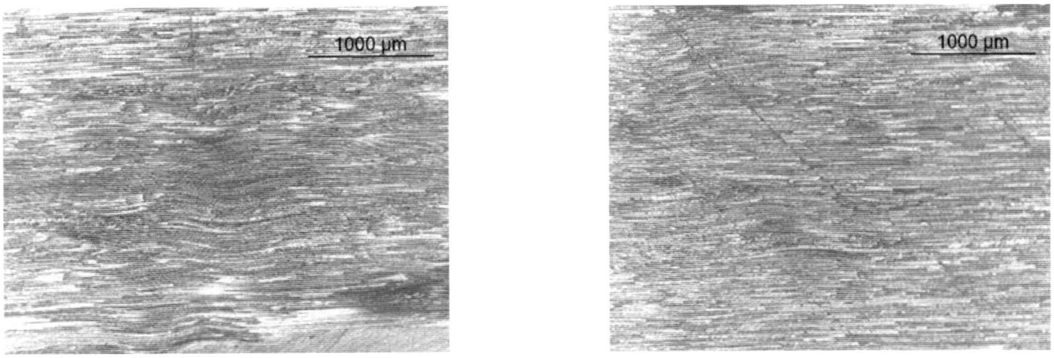

(a) (b)

Figure 5. Microscopic images of an uncured carbon fiber towpreg a) before and b)after uniaxial tensile testing

The results of the MATLAB analysis shown in Table 2 support the visual findings of the microscopic images. The overall mean misalignment angle, the maximum misalignment angle and the standard deviation of angles have all been reduced after uniaxial tensile testing. Following classical lamination theory, this means that the theoretical tensile modulus of the same prepreg sample after testing is higher than if no uniaxial load was applied. Additionally, the Figure 4 justifies the assumption that there is an event of fiber re-alignment during the tensile testing experiment and that the lead-in area of the tensile response plot is attributed to this movement of fibers.

Table 2. Results of Fiber Misalignment Within Carbon Fiber Prepreg Before and After Uniaxial Tensile Testing

	Sample Prior to Testing	**Sample After Testing**
Successfully Measured Area	89.1%	93.4%
Mean Angle (Deg)	1.750	1.170
Mean Misalignment Angle (Deg)	3.722	2.773
Maximum Misalignment Angle (Deg)	15.500	11.904
Standard Deviation of Angles (Deg)	4.696	3.575

3.3 Mathematical Model

Using the experimental data obtaining through uniaxial tensile testing, a mathematical equation was derived to predict the modulus of the uncured prepreg at different load levels. The model is based on the derivative of a second order polynomial fitted to the average stress vs. strain plot of all the samples. Figure 6 represents the average stress vs. strain plot over all samples along with the second order polynomial fitted to the non-linear region of the graph. The trend-line was fitted at a starting Strain value of 0.0002, as to remove any initial noise or possible testing error that can occur at the start of the experimentation. The trendline and resulting equations will be able to predict the initial stress & strain points and the expected modulus of the uncured prepreg at 0 N.

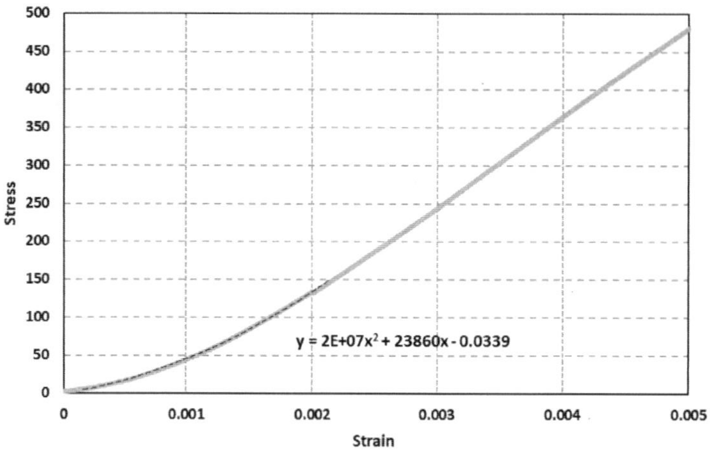

Figure 6. Experimental Results of Tensile Behavior of Uncured Prepreg with trendline

$$y = ax^2 + bx + c \quad [3]$$

$$\frac{dy}{dx} = 2ax + b \quad [4]$$

$$\left({}^{1}/_{a} * f\right) = ax^2 + bx + c \quad [5]$$

$$x = \frac{-b \pm \sqrt{b^2 - 4a\left[c - \left({}^{1}/_{a} * f\right)\right]}}{2a} \quad [5]$$

$$\frac{dy}{dx} = 2a\left(\frac{-b \pm \sqrt{b^2 - 4ad}}{2a}\right) + b \quad [6]$$

$$E^{eff} = \left|\sqrt{23860^2 - 4(2 * 10^7)(-0.0339 - 2.22f\,)}\right| \text{ (MPa)} \quad [7]$$

Equation 7 represents a mathematical solution for calculating the effecting Young's Modulus of the uncured prepreg system in terms of force f in Newtons. This equation is valid only between strain values of 0-0.2% and is only applicable for the material used in our experimentation. Applying Equation 7 to AFP or ATL manufacturing, where the expected applied tensile load is approximately 25 N, the resultant Young's modulus is calculated to be approximately 70 GPa. At 0 N, the modulus is approximately 24 GPa.

Studying the geometry of a single fiber, a mathematical model was also developed to predict the force based on the amplitude and period of the waveform. The fiber was modelled with the assumption that waviness follows a sinusoidal function with a period (k), and amplitude (a) as seen in Figure 7.

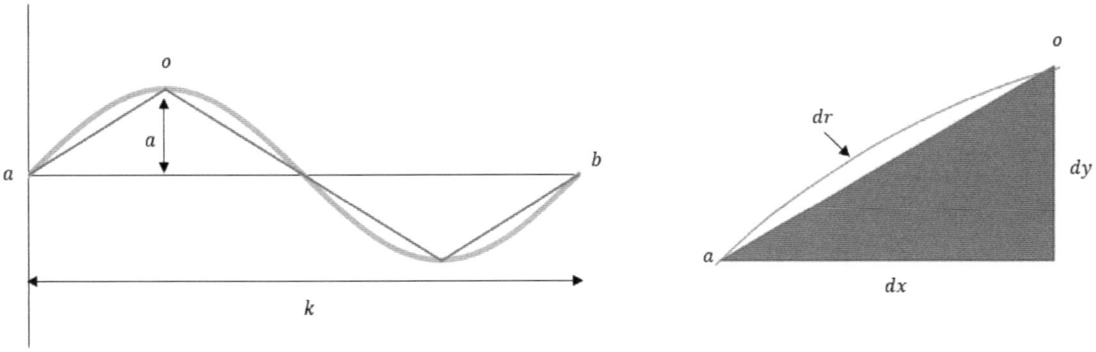

Figure 7. Generalized Sinusoidal Waveform of a Single Fiber.

$$L = \int_a^b \sqrt{1 + \left(\frac{dy}{dx}\right)^2}\, dx \quad [8]$$

$$y = a * sin[k(x)] \quad [9]$$

$$\left(\frac{dy}{dx}\right) = k * a * \cos[k(x)] \quad [10]$$

$$L = \int_a^b \sqrt{1 + (k * a * \cos[k(x)])^2} dx \quad [11]$$

$$\int_0^{950} \sqrt{1 + \left(\left({}^{2\pi}/_{b}\right) * 36.8871 * \cos\left[\left({}^{2\pi}/_{b}\right)(x)\right]\right)^2}\, dx = 963.982\ \mu m \quad [12]$$

$$L = \int_a^{b+\Delta b} \sqrt{1 + \left((k + \Delta k) * (a + \Delta a) * \cos[(k + \Delta)(x)]\right)^2} dx \quad [13]$$

Using Equation 11 to calculate the true length of the sinusoidal graph, the rate of change of the amplitude *a* as the length *b* can be numerically determined using Equation 13. MATLAB was used to numerically solve for the change in amplitude by equating the integral in Equation 13 to the arc length solve for in Equation 12. Figure 8 represents the resultant values.

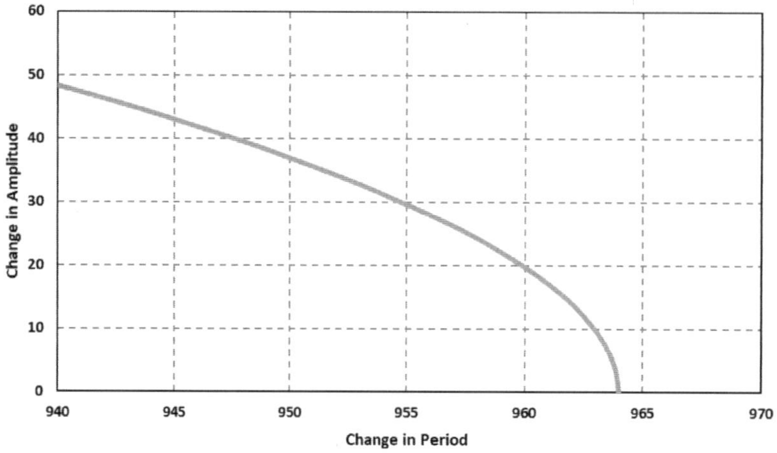

Figure 8. Ratio Between the Change in Amplitude & the Change in Period

Based on the graphical results in Figure 8, the ratio between the change in amplitude vs wavelength follows a non-linear trend. The sharp drop in amplitude as the period approaches the arc length is attributed to the values approaching the node of the periodic function. Uses these values, and the stiffness equations developed by M. Geers et al, an equation based on the geometric wave of a single fiber can be developed to solve for the axial force and the bending moment exerted on the system [5].

$$\left\{ \begin{matrix} N \\ M_2 \end{matrix} \right\} = \begin{bmatrix} a_{11}^{hom} & a_{13}^{hom} \\ a_{31}^{hom} & a_{33}^{hom} \end{bmatrix} * \left\{ \begin{matrix} E^E \\ E^{C_1} \end{matrix} \right\} \qquad [14]$$

In Equation 14, N represents the axial force while M_2 represents the bending moment.

$$\begin{cases} a_{11}^{hom} = \dfrac{E cos\theta}{\dfrac{cos^2\theta}{A} + \dfrac{x_1^2}{I}} \\ a_{33}^{hom} = EI cos\theta \\ a_{13}^{hom} = a_{31}^{hom} = 0 \end{cases} \qquad [15]$$

In Equation 15, L represents the period length, I, the moment of inertia, E, the Young's Modulus, and θ, the cross-section inclination angle as shown in Figure 7. Using Equation 14 [5], and the results in Figure 8, the axial force and the moment in a single fiber can be determined based on the geometric properties of the sine waveform. Furthermore, with initial microscopy of a sample and knowing the relationship between the change in amplitude and the change in wavelength, the average axial force and bending moment in the prepreg can be predicted based on the initial average wavelength on the fibers. Using these resultant calculated force and moment, the effective Young's Modulus of the uncured prepreg system can be determined by outside of experimental testing. Future work is still in progress to further develop the equations for calculating force in a fiber following a periodic functions based on the research of E. Syerko et al. and P. Cartraud et al. [6], [7].

4. CONCLUSIONS

Studying the mechanical response of uncured carbon fiber prepreg systems is important for AFP and ATL manufacturing, as these properties will be used to simulate the process and predict the overall quality of the part. Inaccurate simulations can lead to defects in the automated manufacturing process which results in a negative time and cost incurrence. The behavior of uncured towpreg under tensile testing was investigated. It was shown how at the beginning of its tensile response, the fibers in the uncured prepreg system exhibit a moment of re-alignment and movement in which the waviness and the angle of misalignment within the system is corrected. This movement within the uncured and viscous epoxy system results in a non-linear tensile response, until a certain point where the majority of the fibers are aligned to the direction of the force, and the tensile modulus turns linear. This changed in modulus due, to the change in angle of the fibers, can be mathematically solved using classical lamination theory and was recorded in experimentation over 24 samples from 2 batches of material. Microscopic imaging was performed to capture this change in fiber angle in the towpreg. Using images of an uncured prepreg sample at 10x magnification taken prior to, and after uniaxial tensile testing shows a clear change in the angle of misalignment and the waviness of the carbon fibers. A mathematical equation was proposed to predict the towpreg modulus at different load level.

Acknowledgments

The authors would like to acknowledge the financial support from the Natural Sciences and Engineering Research Council of Canada (NSERC). Material was provided by Bombardier Aerospace.

5. REFERENCES

[1] D. Leutz, M. Vermilyea, S. Bel, and R. Hinterhölzl, "Forming Simulation of Thick AFP Laminates and Comparison with Live CT Imaging," *Appl. Compos. Mater.*, vol. 23, no. 4, pp. 583–600, Aug. 2016, DOI: 10.1007/s10443-016-9475-6

[2] K. Potter, C. Langer, B. Hodgkiss, and S. Lamb, "Sources of variability in uncured aerospace grade unidirectional carbon fibre epoxy preimpregnate," *Compos. Part A Appl. Sci. Manuf.*, vol. 38, no. 3, pp. 905–916, Mar. 2007, DOI 10.1016/j.compositesa.2006.07.010

[3] W. Zhang *et al.*, "Experimental Methods to Characterize the Woven Composite Prepreg Behavior During the Preforming Process," in *Proceedings of the American Society for Composites--Thirty-First Technical Conference*, 2016.

[4] D. Wilhelmsson and L. E. Asp, "A high resolution method for characterisation of fibre misalignment angles in composites," *Compos. Sci. Technol.*, vol. 165, pp. 214–221, 2018, DOI: https://doi.org/10.1016/j.compscitech.2018.07.002

[5] M. Geers *et al.*, "Computational homogenization of structures and materials To cite this version : HAL Id : hal-01419916 Computational homogenization of structures and materials," pp. 0–12, 2016.

[6] E. Syerko, A. A. Diskovsky, I. V. Andrianov, S. Comas-Cardona, and C. Binetruy, "Corrugated beams mechanical behavior modeling by the homogenization method," *Int. J. Solids Struct.*, vol. 50, no. 6, pp. 928–936, 2013, DOI: 10.1016/j.ijsolstr.2012.11.013

[7] P. Cartraud and T. Messager, "Computational homogenization of periodic beam-like structures," *Int. J. Solids Struct.*, vol. 43, no. 3–4, pp. 686–696, 2006, DOI: 10.1016/j.ijsolstr.2005.03.063

A HOLISTIC APPROACH TO COMPOSITES QUALITY USING ARTIFICIAL INTELLIGENCE-ENABLED AUTOMATIC INSPECTION

Scott Blake
Aligned Vision
27 Industrial Avenue
Chelmsford, Massachusetts 01824 USA

ABSTRACT

The advent of automatic inspection in composites fabrication has led to great progress in inspection speed and quality, and generates a large quantity of inspection data, which can serve as a rich data source for closed-loop manufacturing. These advances represent only the first wave of innovation made feasible by this technology. Not only is automatic inspection capable of quickly verifying key features of each ply in a large component, it also can be leveraged to inspect a multitude of material, process and component attributes not included in current quality programs. By viewing inspection as part of a holistic quality solution, fabricators can transition to an approach in which they monitor in-process key performance indicators (KPIs) and adjust process parameters in real time to reduce non-conformance events. Artificial intelligence is a critical enabling technology, speeding the development of analysis algorithms for each attribute, and also employing inspection data for Deep Learning and continuous process improvement. This paper explores the range of attributes that automatic inspection may monitor, the method for incorporating this new level of monitoring into a holistic quality approach, and the ways in which a holistic quality program might advance the state of the art in composites fabrication.

Corresponding author: Scott Blake

1. INTRODUCTION

In a paper on its automated inspection system used in large-scale automated fiber placement (AFP) cells, Electroimpact reports that inspection data can approach 1 TB (terabyte) per part for large complex composite parts [1]. In Electroimpact's initial application of AFP in-process automatic inspection, this raw inspection data is processed on the fly to verify overlap and gap width as well as ply boundary location, and to flag non-conformances and suspect locations.

The value of the large "data lake" produced by automatic inspection extends well beyond in-process inspection. It supports a holistic approach to quality in which the data is aggregated and contextualized as manufacturing intelligence. Continuous process and product improvement on an unprecedented scale, relying on this rich manufacturing intelligence, is in the offing. The capabilities of in-process automatic inspection technology, along with the untapped value of generated data lakes, may be leveraged to advance composites fabrication quality efforts in two ways, both of which rely on artificial intelligence (AI) for implementation. First, automatic inspection may be applied to a multitude of material, process and component attributes not included in current composites quality programs. AI-enabled application development opens this door and allows fabricators to monitor numerous in-process key performance indicators (KPIs) and adjust process parameters in real time to reduce non-conformance events. Second, AI Deep Learning mechanisms will employ the manufacturing intelligence aggregated from automatic inspection data to implement closed-loop continuous improvement, in which manufacturing insights fed back to design and engineering are used to improve product design and fabrication processes.

1.1 Artificial Intelligence and Automatic Inspection

AI is defined as the ability of a device to perceive its environment and determine the best course of action to achieve intended goals [2]. In industrial settings, two sub-categories of AI predominate: Machine Learning, which is the ability of a computer to use data to improve performance of a particular task without being explicitly programmed to do so; and Deep Learning, which also uses data to improve performance, but typically on larger data sets and more complex logic networks [3]. In relation to automatic inspection, Machine Learning is the AI mechanism used for application development while Deep Learning enables closed-loop product and process improvement.

1.2 Automatic Inspection Technology

Illustrative of the technologies available for in-process automatic inspection, the Electroimpact automated inspection system includes two elements. First, a laser profilometer is mounted to the AFP deposition head. From this position, it collects data in parallel with material lay down. By projecting a laser line and measuring the distance to points along that laser line, the profilometer creates a profile of the surface that is analyzed to identify and measure features, in this case, overlaps and gaps between tow lanes. Because the profilometer travels with the head, it is able to detect tow position relative to other tows, making it an effective means of verifying that the widths of overlaps and gaps between AFP tow lanes are within tolerance.

Second, to verify absolute position of tow ends — that is, their position relative not just to neighboring tow lanes but to the component's overall layout — a large-field automatic inspection system is mounted to a second gantry (Figure 1) [4]. This system is capable of inspecting in detail any area on the surface, independent of the deposition head. It consists of laser projection and machine vision hardware, and laser templating and automatic inspection software. The system (Figure 2) is designed to inspect detailed regions of interest anywhere on a large complex surface (typically 5m by 5m). Each unit of the system projects its laser and captures images within a 60° (±30°) angle. Meeting Boeing D6-55902 requirements, the unit achieves an accuracy of 0.38 mm (0.015 in.) in a standard 3m by 3m (10 ft by 10 ft) field at a distance of 3m (10 ft) from the unit.

Figure 1. Large-field automatic inspection system mounted to a second gantry over a large AFP cell. Two inspection units are circled in red.

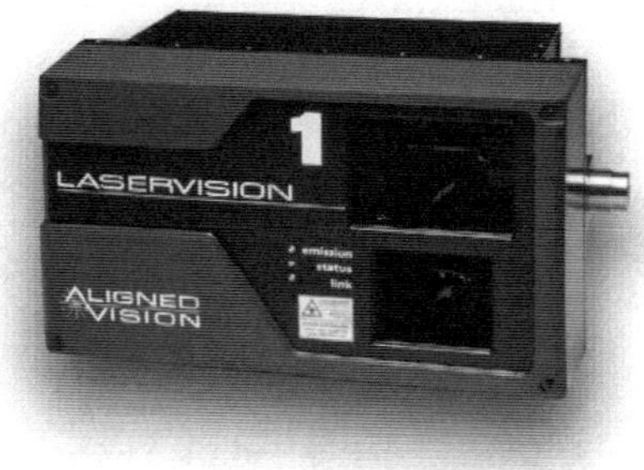

Figure 2. Large-field automatic inspection unit consists of laser projector with integrated machine vision system.

To perform automatic inspection, the vision component of the system directly accesses and applies manufacturing data to aim a high-resolution, high-magnification camera, which captures "calibrated images" under data control. "Calibrated imaging" refers to the images being captured along with a photogrammetric transform that defines the relationship between the camera and the feature being imaged. This enables measurements in the image that correspond accurately to features on the surface of the part. That is, the transform enables each pixel in the 2D image to be dimensioned relative to the 3D surface and in the coordinate system of the part. The analyzed images support a "gauging" function, in which the surface containing the features being imaged is assumed to be correct and the location of features on that surface are assessed for translations and rotations. Any features found to be out of tolerance are flagged and highlighted on the surface by the laser projector (Figure 3), enabling near-real-time assessment and repair before any flaw is covered by subsequent plies.

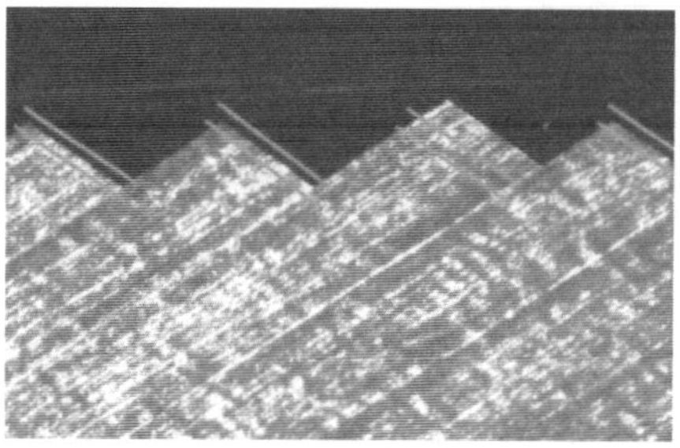

Figure 3. Analysis enabled by transform. Out-of-tolerance tow placement identified.

1.3 Currently Inspected Attributes

Tow end and edge positions define ply boundaries in AFP layup. Automatic inspection technology is also able to measure ply boundaries created through other layup methods (manual, Automated Tape Placement, etc.). As with many new technologies, the first attributes to which automatic inspection has been applied are those commonly inspected by the previous state of the art, in this case, manual inspection. In addition to gap and overlap widths, attributes to which laser profilometers have been applied include pits, bulges and distortions in filament-wound cylinders [5]; and automated dry material placement [6]. Large-field automatic inspection has been used to verify (or flag) material type (unidirectional versus woven plies, for example), fiber orientation (Figure 4), and foreign objects and debris (FOD).

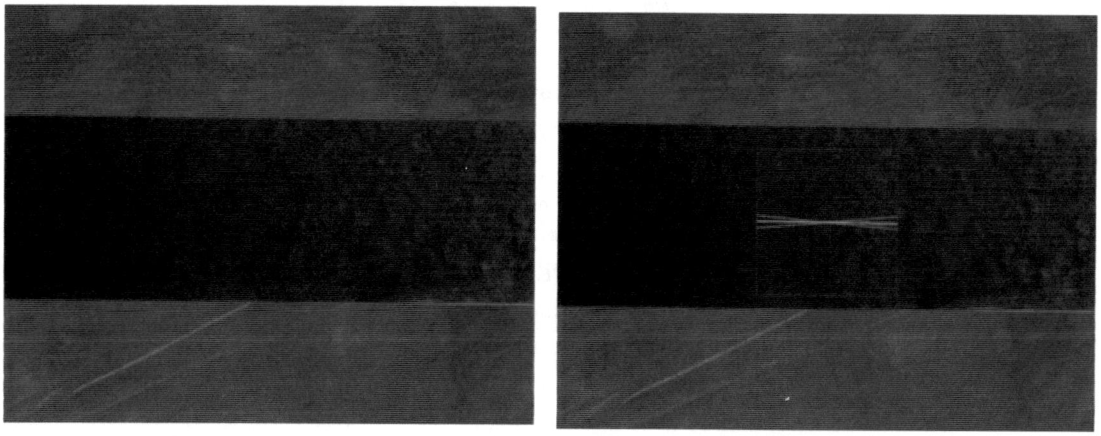

```
visionuni version 20180105

All read in: 0.547000
 Calcs done: 1.687000
angle expected -89.727594
angle found -89.283803
angle diff 0.443792  tol 5.000000
REFERENCE_ANGLE -89.727594
IMAGE_RESULT?ORIENT+ANGLE+-89.283803+PASS
```

Figure 4. Raw (left) and analyzed (right) image from large-field automatic inspection system verifying that fiber orientation is within tolerance.

To detect FOD, the automatic inspection system's analysis algorithm identifies any abrupt changes in optical surface characteristics where none is expected. Such changes prompt the system to flag the location, and the laser projector draws a boundary around the FOD on the surface (Figure 5). FOD detection extends to specific kinds of FOD and application areas. For example, detecting peel ply in critical bonding applications requires the development of algorithms specific to the environment in which bonding will take place, the kinds of composite surfaces to be bonded, and the type of peel ply being used. More about algorithm development is discussed under Section 2, Quality Approaches, below.

Figure 5. Detected Foreign Objects and Debris (FOD) detected by automatic inspection are pinpointed on the surface by laser outline.

In addition to extending the specific attributes to which automatic inspection is applied (e.g. type of FOD), developments also continue regarding the mechanisms by which automatic inspection occurs; and this too increases the demands on algorithm development. For example, new camera technology improves the ability to recognize edges, layers, fiber orientation and FOD before processing an image with analysis algorithms (Figure 6).

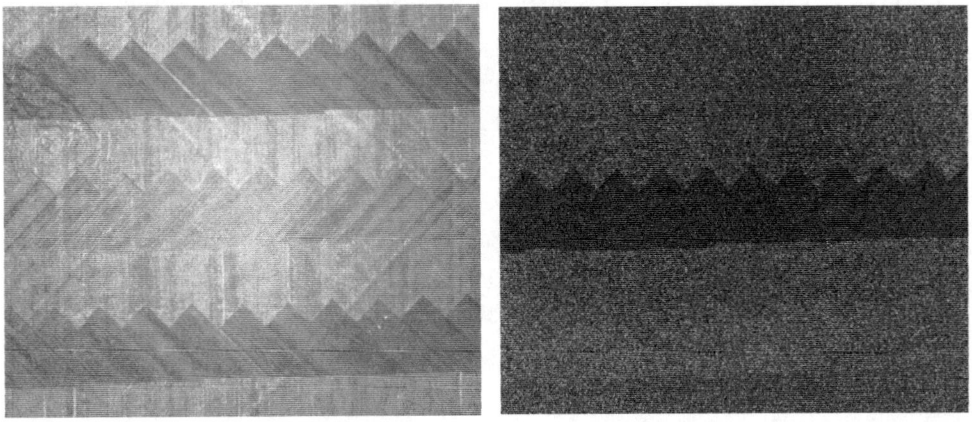

Figure 6. Camera advancement improves image capture and also places new demands on application development.

1.4 Additional Attributes to Inspect

Automatic inspection systems are capable of verifying other important attributes of composites fabrication, attributes that are critical to part performance and that are extremely difficult to detect through human visual inspection. Prior to automatic inspection, the only alternative has been metrology, which is infeasible for most applications because it is so labor- and time-intensive. (Additionally, some metrology systems provide a costly and unnecessary level of precision.) To measure some of these attributes with metrology, a "point cloud" comprised of hundreds of thousands or millions of points has to be captured and converted into a surface. The modeled surface must then be aligned with a nominal surface and the two compared. In contrast, the large-field automatic inspection system uses much less data-intensive expectation-driven analysis: it generates an acceptable pattern and its tolerances directly from the CAD model. The system's machine vision captures an as-built laser pattern from the actual surface being inspected. Automatic image analysis then compares the CAD-generated and as-built patterns, producing a go/no-go metric.

One significant area for which automatic inspection has the potential to markedly improve quality is raw material inspection. In high performance composites applications, flaws in raw materials may diminish critical attributes. Yet currently, common practice is to inspect raw material manually or not at all. Manual inspection here faces the same difficulties as in the fabrication cell: human limitations such as distractedness, fatigue and verification bias decrease inspection effectiveness. Moreover, manual inspection is very time consuming. In the holistic quality approach outlined below, automatic inspection and material tracking systems could work together on the same software platform, making raw material inspection an easily streamlined and beneficial process.

Within the composites fabrication cell, attributes to which automatic inspection may be applied include wrinkles, bridging and shear. Inspection of these attributes has been assessed at the Advanced Composites Centre for Innovation and Science (ACCIS), University of Bristol, U.K. [7]. Wrinkles occur when material does not conform completely to the tool surface. To detect wrinkles, the automatic inspection system's analysis algorithm identifies discontinuities in a laser reference pattern that would not be expected when viewing an unwrinkled surface. The approximate size of a detected wrinkle is determined through the defined geometric relationships among the projection source and pattern, the camera, and each surface (Figure 7).

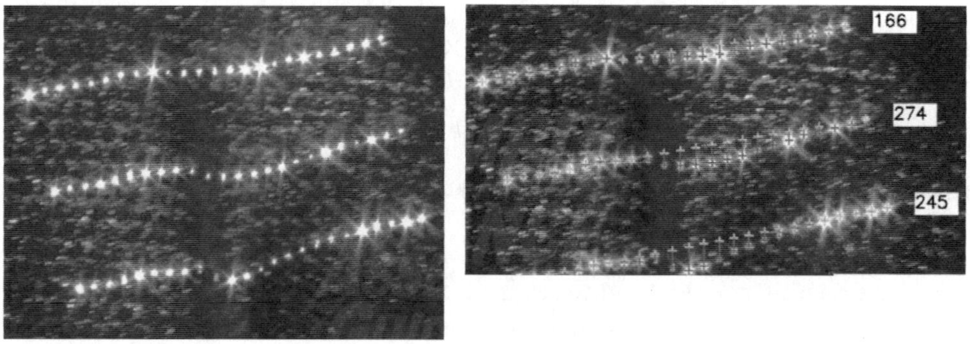

Figure 7. Automatic detection of 0.8mm wrinkle in robotic layup system

Bridging occurs when material does not conform fully to the tool surface in a concave region. With secondary bridging, material that initially conforms is pulled out of conformity as material is draped or creased elsewhere. The automated inspection system may be programmed not only to detect bridging but also to automatically re-inspect areas that are susceptible to secondary bridging (Figure 8).

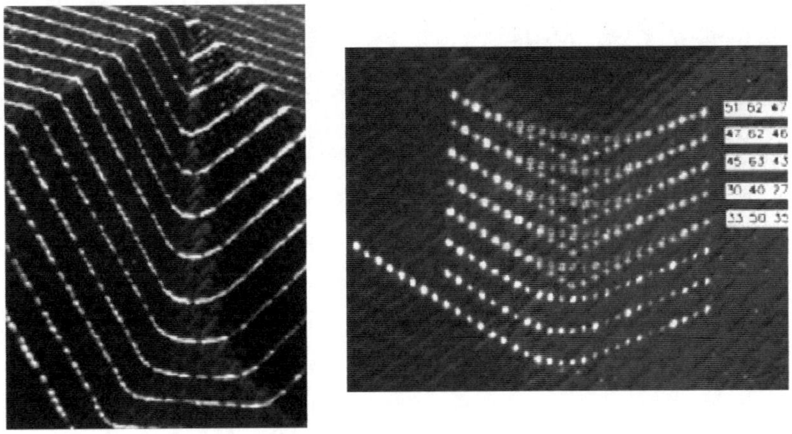

Figure 8. Difference between expected and actual location of projected reference points indicates proportion of bridge displacement.

Shear — the distortion of the material's fiber orientation as it is deformed in three dimensions — often occurs by design in composite fabrication. Excessive shear, however, may weaken the finished product. To detect shear, the automatic inspection system illuminates the surface to maximize the visibility of each tow. Analysis algorithms track the appearance of each tow in images captured by the system's aimed camera and calculate the angular change in direction along the length of the tow. Maximum positive and negative angular changes are automatically calculated (Figure 9).

Figure 9. Automatic discovery of fiber paths in material sheared in press

Beyond composites fabrication, automatic inspection technologies are applicable to post-processing steps like machining, as well as assembly and other manufacturing processes. The large-field automatic inspection system described here has the capability to inspect for any visible feature's presence or absence, as well as its location on a surface. For example, it may be used to verify not only the position of fasteners in an assembly, but also the type of fasteners being used (Figure 10).

Figure 10. Automatic inspection detects fastener position and type. The captured image (left, from 10 ft away) is high resolution and shows excellent detail under zoom (center and right).

2. QUALITY APPROACHES

The automation of inspection tasks is one of numerous areas in which digitalization is driving fabrication quality and speed to new levels. Digitalization should not be confused with digitization: the latter simply takes analog information and converts it to digital form (e.g. scanning a physical surface to create a digital model) while the former generates, analyzes and manages data. Ultimately, digitalization leads to the complete Digital Factory, also known as Smart Manufacturing or Industry 4.0. Digitalization typically is implemented in a step-by-step fashion. Major phases of this implementation include digital point solutions, integrated manufacturing operations management (MOM) solutions, and closed-loop manufacturing operations. As a key element of quality efforts, automatic inspection is following this stepwise path toward a holistic quality approach, sometimes referred to as Quality 4.0.

2.1 Automatic Inspection as a Digital Point Solution

Automatic inspection technology is often introduced to a fabrication floor as a stand-alone technology. The system imports needed data from CAD and composites-specific CAM files (such as Fibersim or CPD). In the case of the large-field automatic inspection system discussed here, these files direct the system to aim the laser projector/machine vision unit and capture the desired images. The system's computer then performs image analysis and provides confirmation of analyzed features, or sends an alert regarding a non-conformance (Figure 11).

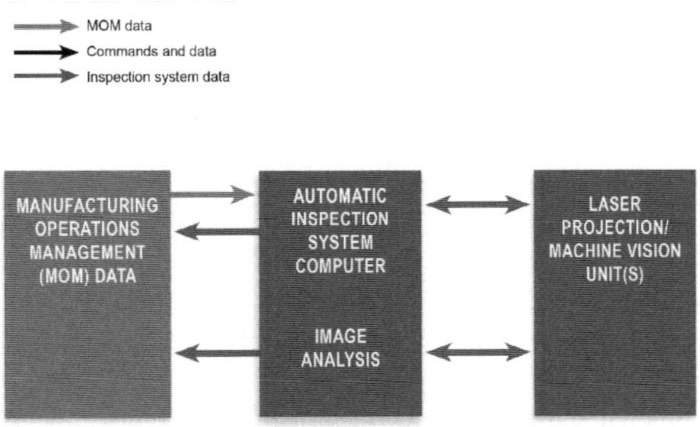

Figure 11. As a digital point solution, an automatic inspection system operates as a stand-alone, importing design and control data from the manufacturing operations management (MOM) system and exporting as-built data back to the MOM system.

Application of automatic inspection to each specific fabrication effort is unique. As with virtually all automated technologies and contrary to a common misconception, application is not as simple as setting up the equipment and switching it on. Because of the relatively high number of variables associated with fabrication of composite materials — fiber and resin type, fiber form, fiber orientation, ply schedules, resin infusion or prepregging method, layup mechanism and cure mechanism, to name a few — application development for automated manufacturing processes has been a tedious and time-consuming process. This is no less true of automatic inspection, for which the analysis algorithms used to assess digital images collected via machine vision have been hand-engineered.

With the high number of composites fabrication variables, plus additional factors like lighting within the work cell, developing an analysis algorithm for a particular application via hand-engineering typically takes several weeks of dedicated effort. To accelerate implementation of automatic inspection to each application, recent experimentation suggests that Machine Learning is an effective and accurate approach to application development [8]. In a matter of one to two days, a Machine Learning program may be trained on a set of images to detect a particular attribute.

Machine Learning for automatic inspection applications can be conducted away from manufacturing floor, which means that manufacturers may be able to transition to automatic inspection quickly and with little to no disruption of shop floor activities. Manufacturers employing laser templating could replace a standard unit with a laser projection/machine vision unit and continue to conduct laser templating operations while Machine Learning uses harvested images to develop the specific application. The only additional requirement would be the deliberate creation of flaws to provide a full data set for the Machine Learning system.

To create an algorithm that is ready to detect all potential flaws, the fabricator would create artifacts containing those flaws, as well as correctly fabricated regions. Then in an environment similar to the manufacturing floor, the machine vision system would capture images that would be used to train the Machine Learning system.

2.2 Integration of Automatic Inspection into Manufacturing Operations Management

Using a software development kit or other integration process, automatic inspection can be integrated into a manufacturing operations management (MOM) system (Figure 12), and more specifically, into the quality management functionality of a MOM system. On an integrated MOM platform, automatic inspection becomes a critical function within the quality ecosystem, maximizing a fabricator's ability to respond to real-time manufacturing quality events. To illustrate, an integrated MOM system performs machine and work cell monitoring alongside in-process inspection of components as they are built. When a non-conformance event is detected by an automatic inspection system, a machine operational parameter related to that non-conformance may be adjusted by an operator, or even by the system itself. For example, a gap between tow lanes that is outside of tolerance may point to a dropped tow, which can be corrected in real time on the AFP.

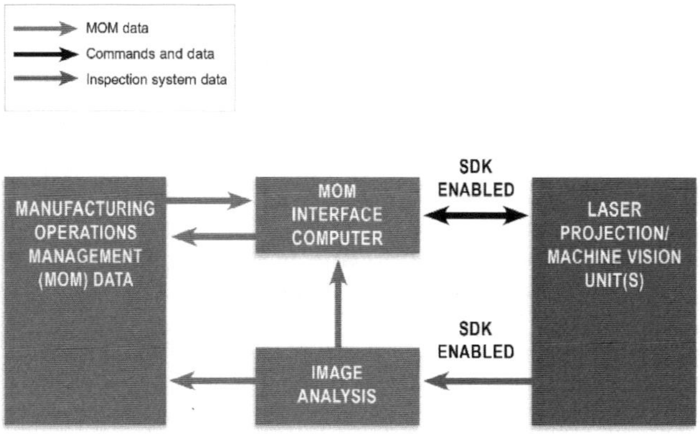

Figure 12. Integrated into a manufacturing operations management (MOM) system, automatic inspection becomes one of the critical functions within a quality ecosystem.

2.3 Manufacturing Intelligence and Closed-Loop Manufacturing

Automatic inspection systems generate detailed, as-built data for each part — an as-built Digital Twin, or virtual representation of each individual part's specific attributes. The Digital Twin includes quantitative data as well as the image set captured by automatic inspection. While the automatic inspection system captures detailed images of small regions within the surface of a large component, it has the capacity to combine these images and produce a single image of a complete ply surface (Figure 13). As a result, the as-built Digital Twin provides both visual and quantitative traceability for each individual part — documentation of critical attributes that can accompany the part throughout its operational life.

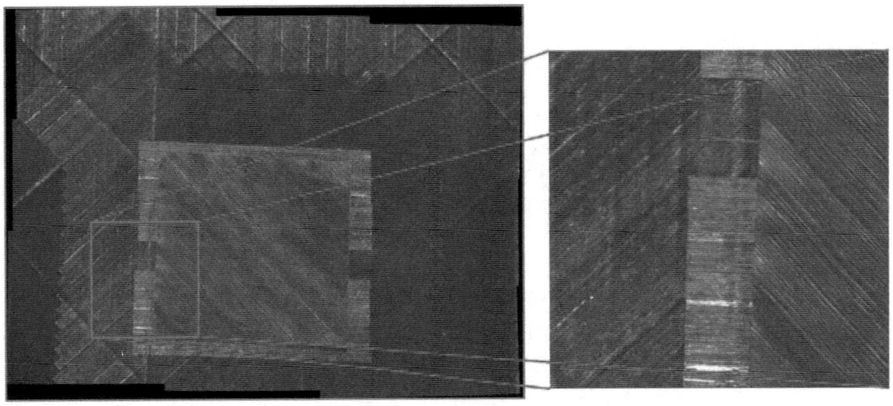

Figure 13. Nine captured images combined into one (left). Algorithm in the automatic inspection system aligns images seamlessly.

Comprehensive quality documentation is one outcome of automatic inspection. In a fully integrated system and with the availability of Deep Learning processes, this detailed, as-built data also feeds predictive and prescriptive analysis, as well as root cause analysis. Applying Deep Learning to a large database of inspection data and feeding the insights gained back to product design and process engineering will lead to improvements in simulation, modeling, product quality and process efficiency and efficacy.

3. CONCLUSIONS

In composites manufacturing, a holistic quality program places all quality-related data — from material lot and out-time to post-processing component features and attributes — on the same software platform. Data generated by in-process automatic inspection on this platform enables real-time corrective actions that extend beyond rework of a detected non-conformance to adjustment of operational parameters related to that non-conformance. An immediate increase in part quality and decrease in scrap rate are the first outcomes of this integrated approach. Beyond improved quality of individual components, closed-loop manufacturing utilizes the manufacturing intelligence aggregated from multiple component data and contextualized within the full set of operational parameters. This manufacturing intelligence feeds Deep Learning efforts, which in turn enable continuous improvements to product design, process engineering, operational conditions and settings, and resulting product quality.

4. REFERENCES

1. Cemenska, J., Rudberg, T., Henscheid, M., Lauletta, A., et al., "AFP Automated Inspection System Performance and Expectations," SAE Technical Paper 2017-01-2150, 2017, DOI:10.4271/2017-01-2150.

2. Hahn, T., "Artificial Intelligence: Optimizing Industrial Operations." The European Files 54 (2018): 30.

3. "Artificial Intelligence (AI) vs. Machine Learning vs Deep Learning." 25 Jan 2019: https://skymind.ai/wiki/ai-vs-machine-learning-vs-deep-learning.

4. Black, Sara, "Improving Composites Processing with Automated Inspection." *CompositesWorld* 4:2 (February 2018).

5. Saulsberry, R., "Development of a Composite Pressure Vessel Laser Profilometry and Eddy Current Scanning System to Meet Manufacturing and Analytical Needs." NASA Engineering & Safety Center news release, 30 November 2016.

6. Black, Sara, "Improving Composites Processing with Automated Inspection, Part II." *CompositesWorld* 4:6 (June 2018).

7. Potter, K., Ward, C., Crowley, D., Elkington, M., Blake, S., Whittington, D., "Real Time Gauging of Wrinkle, Shear and Bridge Formation in an Automated Composite Manufacturing Process," presented at the Composites and Advanced Materials Expo, September 27-29, 2016

8. Blake, Scott, "Recent Advances in Artificial Intelligence Applications to Composites Fabrication," presented at CAMX, 26 September 2019.

THE EFFECT OF MATERIAL VARIABILITY ON THE PRE-GELATION BEHAVIOUR OF PREPREG COMPOSITES

Caitlin Duffner[1], Navid Zobeiry[2], Anoush Poursartip[1]
1. The University of British Columbia, Vancouver, BC, Canada
2. University of Washington, Seattle, WA, USA

ABSTRACT

Processes such as forming and AFP have become increasingly popular, and these manufacturing techniques require manipulation of the uncured pre-impregnated thermoset matrix composite material (prepreg). Interest in the initial state or pre-gelation behaviour of prepregs has grown due to these manufacturing trends as well as the increasing recent research connecting this regime to processing defects such as wrinkling and porosity. Historically, the pre-gelation behaviour of prepreg has been relatively unexplored due to the challenges that traditional testing methods have in handling the soft raw prepreg. It is shown here that Digital Image Correlation (DIC) can capture the prepreg behaviour through the entirety of a cure cycle on both a local and global level. This work presents the global response of the prepreg, showing the true pre-gelation response, which differs markedly from what has traditionally been presented as the expected behaviour. Additionally, the variability in local prepreg response can be linked back to local strain variations observed using DIC. With the insight into the variability of a ply and the pre-gelation strain response, the findings in this work emphasize the importance of understanding and accurately capturing the initial prepreg condition.

Corresponding author: Anoush Poursartip

1. INTRODUCTION

The composites industry has seen a shift in recent years, increasingly incorporating newer manufacturing methods such as AFP and forming. These newer methods involve handling raw material in different ways and therefore, a better understanding of the mechanisms that drive the raw material behaviour is needed. Recent research has focussed on the impact of the initial condition of the material on the development of defects such as wrinkling and porosity, which develop early during the process cycle [1–3]. However, it appears that no one has directly characterized the pre-gelation free-strain response of prepreg, and the literature typically assumes

SAMPE Virtual Conference Proceedings, 2020. Society for the Advancement of Material and Process Engineering – North America.

that the pre-gelation free strain response should follow a path similar to that depicted in Figure 1 [4].

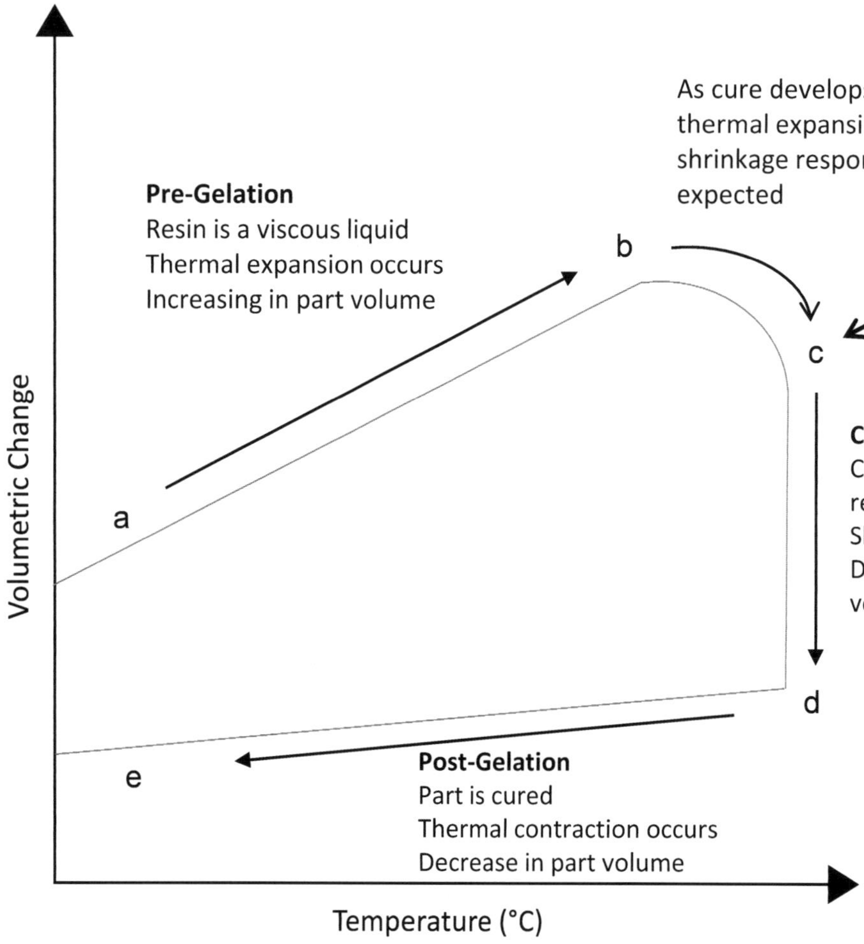

Figure 1. Schematic of the volumetric changes that a thermoset based prepreg system is traditionally expected to undergo during processing, as adapted from literature [4]. When heating initially, the resin in the ply is a liquid and is expected to expand (a-b). At some point during the heating cycle, the cure reaction will begin and will contribute to volumetric shrinkage (b-c and c-d). Finally, when the prepreg is cooled thermal contraction will shrink the prepreg (d-e)

It is typically believed that when heated, the uncured prepreg will expand through the thickness of the ply and in the direction perpendicular to the fibres. As resin cure starts, it is expected that this will lead to a corresponding shrinkage in the prepreg, with a linear relationship between strain and the degree of cure (DoC) [4]. In essence, the literature considers the resin in the pre-gelation prepreg to be acting as a very low modulus solid. This work evaluates this assumption, by directly observing the pre-gelation behaviour of prepreg using digital image correlation (DIC), a non-contact measurement technique. The results presented here are part of a more extensive study into pre-gelation prepreg behaviour [5]. Throughout this study, the importance of the variability in the material and strain response during the early stages of processing is highlighted.

2. EXPERIMENTATION

2.1 Material

This study used an autoclave thermoset system composite material, AS4/8552 unidirectional (UD) prepreg (HexPly AS4 12K/8552; 35 %; 190 AW) [6]. Previous work has created process models for the resin cure kinetics and viscosity development [7,8]. These process models have been implemented in RAVEN simulation software [8], which has been used in this study to obtain cure and viscosity profiles, as well as the gelation point, for the different test temperature cycles.

2.2 Digital Image Correlation (DIC)

DIC systems use image analysis to observe the development of strain in a material. In the past, this technique had been used to study resin flow, crack development, and stress development in composite materials [9–15]. The use of a non-contact measurement technique in this study allowed the strain development of a prepreg sample through the complete process cycle to be captured. The DIC set-up captured the in-plane strains (ε_{11} and ε_{22}) of the UD prepreg sample.

2.2.1 DIC Equipment Set-Up

This study used a Correlated Solutions VIC-3D system. Two digital cameras were positioned approximately 100 cm above the sample position. The cameras were pointed at a 30° angle towards the sample. The cameras were connected to a computer and controlled by VIC 3D image acquisition software. LED lights placed on either side of the sample were used to illuminate the sample allowing a high contrast image to be captured.
The system was calibrated using a 9 x 4-inch calibration panel placed on the heading plate where the sample sits during testing. Several images of the calibration panels at varying positions were used to establish the height and angle of the cameras by tracking target spots on the panel.

2.2.2 Heating System

The samples were heated using a Peltier heating system controlled by an Anton Paar Rheometer. The Peltier heating system has a temperature range between -20 °C to 200 °C and was selected for its fast and accurate temperature control. Teflon tape was placed on the surface of the plate to reduce interaction between the sample and the substrate.

The temperature cycle selected for this experiment was designed to limit cure advancement during the temperature ramp segment of the cycle, isolating the thermal and chemical free strains. Figure 2 shows the cure cycle.

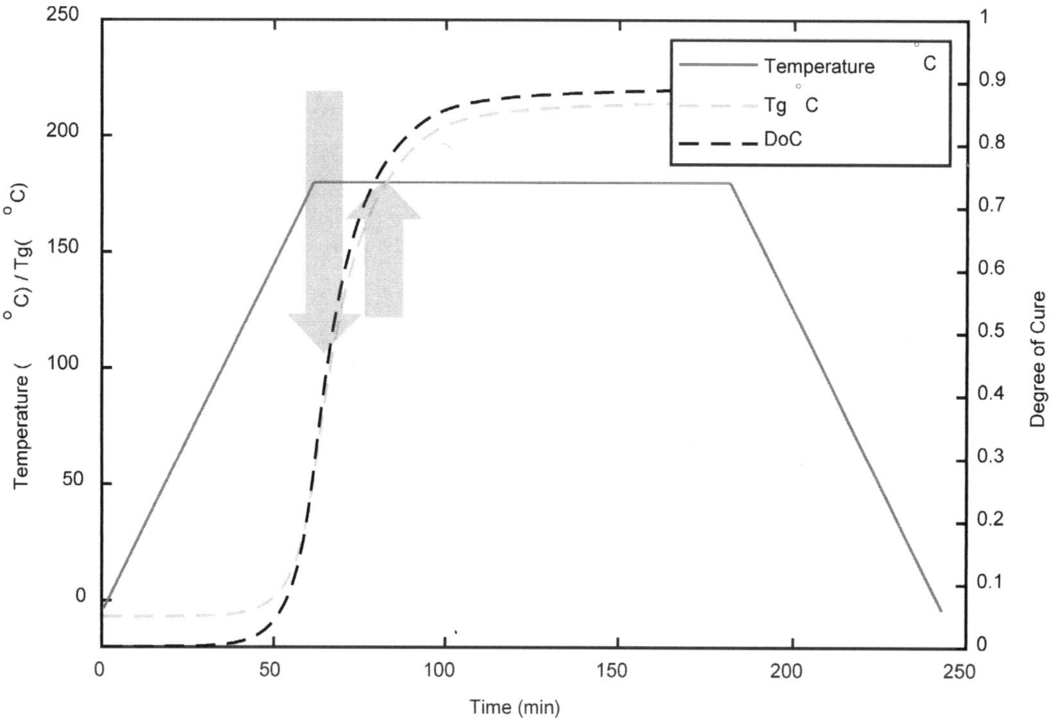

Figure 2. Temperature Cycle 1 used for DIC experiments. One hold cycle- ramp from -5 °C to 180 °C at a rate of 3 °C /min. Hold isothermally at 180 °C for 120 min. Ramp down to -5 °C at 3 °C /min.

2.2.3 Sample Preparation

AS4/8552 unidirectional prepreg was cut into squares approximately 25 mm by 25 mm. Samples of either 1 or 4 plies were made using a unidirectional lay up and de-bulked at room temperature for 10 minutes. After being laid up, samples were painted using a high-temperature Rust-Oleum white [16], and then a speckled layer of KRYLON black spray paint with a matt finish [17] was applied resulting in a random, high contrast speckle pattern.

2.2.4 Image Acquisition and Processing

After the sample was prepared and placed on the heating plate, the temperature cycle and the image acquisition were started in unison. Ever 30 seconds, an image was taken until the completion of the temperature cycle. Temperature data was recorded every second.

To process the images, the Correlated Solutions software was used. The sample was selected, and the software tracked the displacement of the speckles on the sample resulting in the local strain data. The average strain at each time step can be found for the sample during post-processing of the data. To extract the average strain data the extraction tool in the software was used. Lines running perpendicular to the fibre direction were drawn across the sample were selected using the data extraction tool. Once marked the average strain data were extracted for the selected line. The average of these results is taken, resulting in the average global strain in the transverse direction (ε_{22}) to the fibres.

3. RESULTS

3.1 Global Transverse Strain Behaviour

DIC results of the average transverse strain behaviour shows the pre-gelation behaviour response differs markedly from the traditionally assumed behaviour in the literature, as depicted in Figure 1. As opposed to observing a thermal expansion on the initial heating ramp, the results, as shown in Figure 3, showed only a small initial expansion at the start of the heating ramp, followed by a significant drop in strain.

Figure 3 shows the results for one DIC test performed on a 4 ply UD sample processed with the temperature cycle described in Figure 2. The results seen in the test presented here are consistent with, and representative of nine other tests with one hold temperature cycles and four tests with two hold cycles [5].

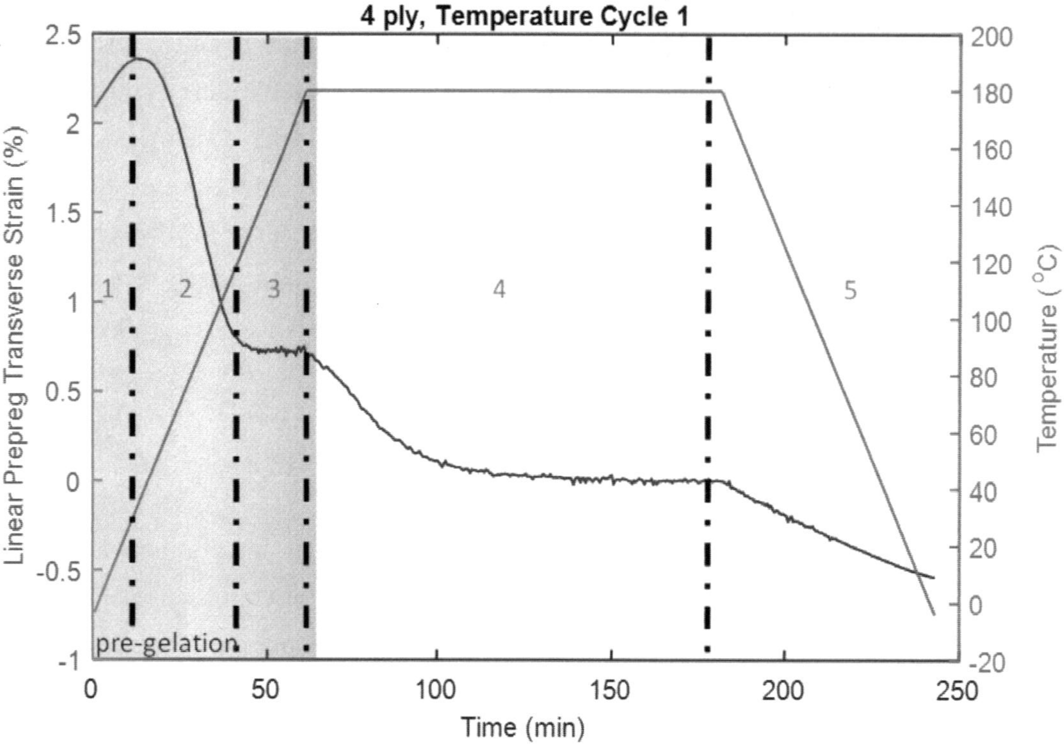

Figure 3. Measured temperature and linear prepreg transverse strain from DIC as a function of time. Zones delineate shifts in strain behaviour: Zones 1-3 identify different behaviour during the heating ramp, Zone 4 is the isothermal hold section of the test where cure advances, and Zone 5 is the cooldown ramp.

3.1.1 Observed behaviour in different zones

The average linear prepreg transverse strain response is presented in Figure 3 and Figure 4. The results have been broken down into zones based on shifts in the strain response. By examining each of the zones, a better understanding of the cause for the observed behaviour can be gained. As the test starts, the sample is heated at a rate of 3 °C/min. Zone 1 focuses on the strain response during the start of this heating ramp, between -5 °C and approximately 30 ± 5 °C. During this time, the sample expands thermally. The observed transverse coefficient of thermal expansion (CTE) is 89×10^{-6} C^{o-1}. For AS4/8552 prepreg or equivalent calculated linear prepreg values based on 8552 neat resin measurements, the literature reports CTE liquid values ranges from 74×10^{-6} °C^{-1} to 373×10^{-6} °C^{-1} [18–20]. The resin viscosity in Zone 1 is initially relatively high (above 600 KPa.s) but drops to about 9 KPa.s at 30 °C at as the prepreg is heated. As the viscosity drops further during the heating ramp, the material expands less, until it begins to shrink in Zone 2. The initial CTE response of the prepreg as captured in Zone 1 is consistent with the range reported previously.

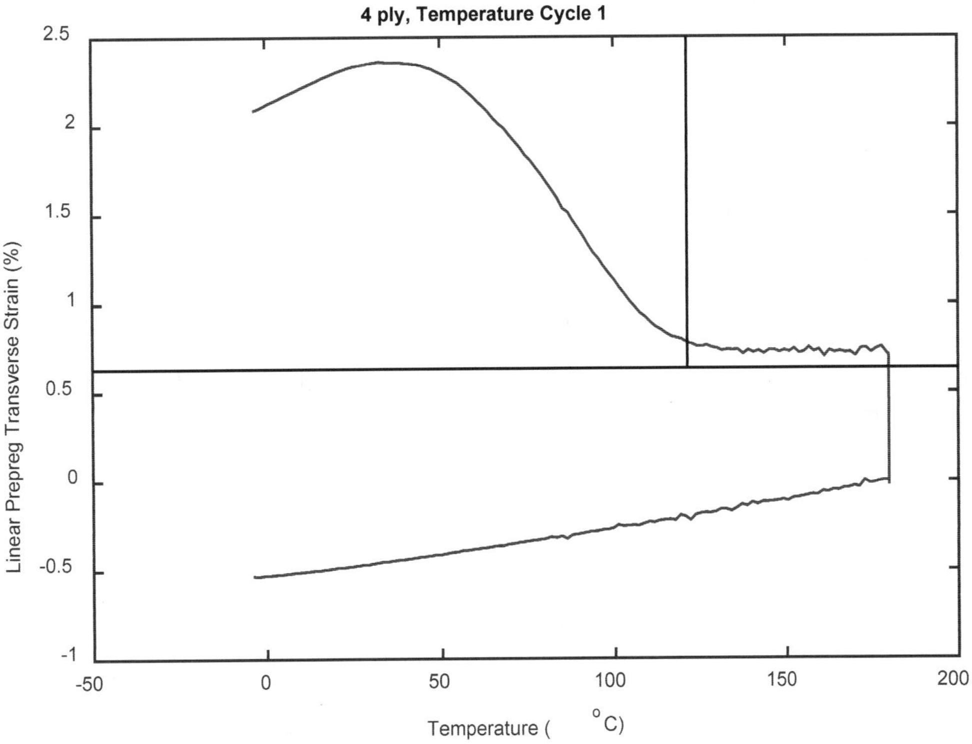

Figure 4. Measured linear prepreg transverse strain as a function of applied temperature, with Zones 1-5 from Figure 3 identified. This observed behaviour is markedly different from the traditionally expected behaviour as shown in Figure 1.

Zone 2 shows unexpected behaviour, in that the prepreg is shrinking in the transverse direction even as it is being heated. Previous work has highlighted that the prepreg is not fully saturated, and there is a variable resin distribution within the fibre bed of AS4/8552 UD prepreg. This leads to self-consolidation within the ply when heated [5,21,22]. The driving force leading to this ply saturation is the surface tension driven capillary action which occurs between the resin and the fibre bed within the core of a ply. The strain drop observed in this work can be linked back to this surface tension effect [22]. For the test presented in this paper, the strain response showed a decrease of -1.34 % between 40 °C and 105 °C. In this temperature range, the resin viscosity is low enough to flow (~3300 Pa.s down to ~20 Pa.s) and as the resin flows to fill voids within the prepreg, the ply consolidates.

In Zone 3, the material shows little change in strain. Although the sample is still being heated, cure starts in earnest (see Figure 2) and therefore in this zone, conflicting contributions of thermal expansion and cure induced shrinkage are expected. One might still expect the effect of thermal expansion to dominate initially, although at a decreasing rate, as in the trend depicted between points b and c in Figure 1. The absence of a change in strain observed in Zone 3 could be a result of continuing resin flow at the microscopic scale, or perhaps shear flow of the resin relative to the fibre bed.

Zone 4 is during the isothermal hold and is therefore the isolated effect of cure shrinkage. Cure shrinkage of the neat resin is believed to be linear with cure [19]; however, the DIC results for the prepreg show a transition between pre- and post- gelation response. Pre-gelation, the linear coefficient of cure shrinkage (CCS), was found to be -0.37 %, and post gelation the CCS was found to be -2.18 %. The prepreg post-gelation response is consistent with the neat resin values reported in the literature [19]. The low CCS of the prepreg in the pre-gelation regime suggests that the resin cannot fully overcome the fibre-bed resistance and thus partially flows or shears relative to the fibre bed. Once the resin gels, the resin can overcome the fibre-bed resistance, and the prepreg CCS is consistent with the resin CCS.

Zone 5 represents the cool-down portion of the test. The prepreg in this zone is post-gelation, and as it cools down acts as a glassy solid. The thermal contraction yields a glassy CTE of 29.56×10^{-6} °C^{-1} which is constant with the values of $33.3 \pm 0.07 \times 10^{-6}$ C$^{\circ-1}$ reported in the literature [23].

3.2 Local Strain Variation Pre-gelation

The strain maps for each DIC image taken during the test can be used to visualize the local strain variations across a sample. 3D plots allow the visualization of the surface profile of the sample with the strain variable overlaid. Examination of these strain maps shows the non-uniformity of strain with time, particularly in the pre-gelation regime. Snapshots of the sample profile at key moments during the heating ramp are shown in Figure 5.

At the start of the heating cycle, the 3D DIC surface plot is uniform in strain (all orange), but as the sample is heated, bands of different strain magnitude appear. This variation is consistent with the change in the global strain response between Zone 1 and Zone 2. It appears that in the areas where surface-tension-driven resin flow has led to ply saturation, the strain maps show variability, with some regions showing negative strain responses double other regions. This would imply that within the ply, some regions see a larger consolidation than others. This is consistent with other studies that have shown the variability in a ply of AS4/8552 [5,24].

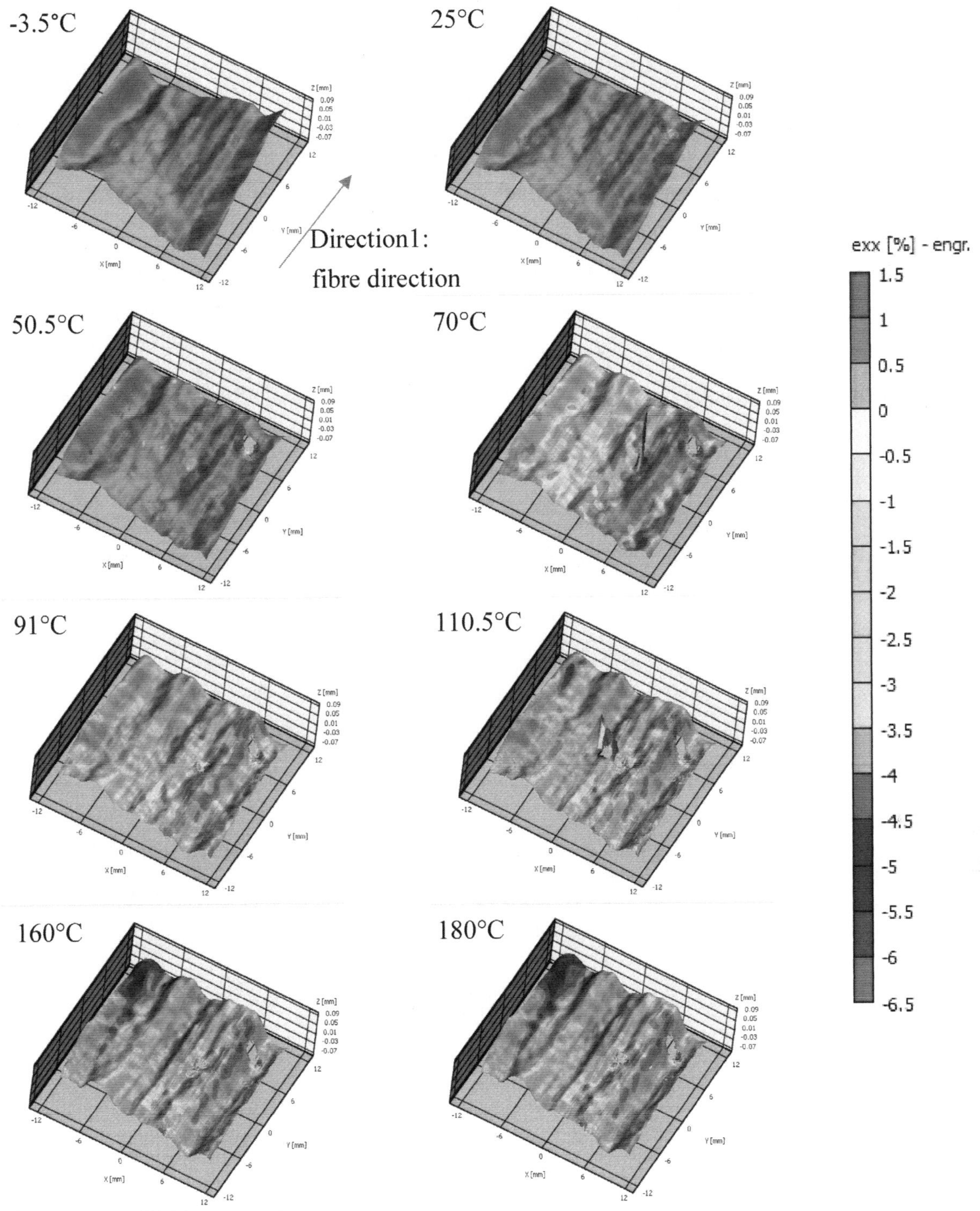

Figure 5. Strain maps for a 4-ply test of AS4/8552 processed using temperate cycle 1. Key points during the temperature ramp from -5 °C to 180 °C at a rate of 3 °C /min are shown. After 50.5 °C, the strain pattern shifts from a uniform strain (orange in colour) to a more random pattern, with some areas showing localized strain values double the average strain.

4. SUMMARY AND CONCLUSIONS

DIC image analysis results show that in the pre-gelation state, prepreg free strain behaviour is markedly different from what is typically expected, and solid-mechanics-based descriptions are not applicable. In particular, it appears that resin flow in and around the fibre-bed has a dominant impact. Only after gelation, when the modulus of resin increases, does the prepreg increasingly follow the expected behaviour. Additionally, the results introduced in this paper highlight the variability in prepreg response, particularly once the resin viscosity drops significantly during heating. The local strain patterns are consistent with the global strain trends although further study is needed to understand the details of the observed local and global response.

As manufacturing shifts towards automation using AFP and forming, understanding and control of the condition of the incoming prepreg and its subsequent handling should help the quest for defect reduction. The work presented in this paper not only gives insight into the underlying mechanisms that determine pre-gelation prepreg response but suggests a pathway to the interpretation of non-destructive imaging results gathered during automated manufacturing processes.

5. ACKNOWLEDGMENTS

The authors would like to thank the industrial members of the Composites Research Network (The Boeing Company, Convergent Manufacturing Technologies Inc., Toray Americas, Avcorp Industries) for their financial support as well as the Natural Sciences and Engineering Research Council of Canada (NSERC). We gratefully acknowledge our colleagues for their technical insight and recommendations on this project: Professor Goran Fernlund, Dr. Christophe Mobuchon, and Dr. Casey Keulen.

6. REFERENCES

1. Stewart AL, Poursartip A. 2018."Characterization of fibre alignment in as-received aerospace grade unidirectional prepreg". *Compos Part A Appl Sci Manuf* 2018;112:239–49. doi:10.1016/j.compositesa.2018.04.018.
2. Mohseni SM, Fernlund G, Lane M. 2018."Cure cycle design to suppress moisture-driven bubble growth in polymer composites". *J Compos Mater* 2018;52:1821–32. doi:10.1177/0021998317734393.
3. Stewart AL. 2018.Misalignment defects in unidirectional composite materials. The University of British Columbia, 2018. doi:10.14288/1.0375794.
4. Garstka T, Ersoy N, Potter KDD, Wisnom MRR. 2007."In situ measurements of through-the-thickness strains during processing of AS4/8552 composite". *Compos Part A Appl Sci Manuf* 2007;38:2517–26. doi:10.1016/j.compositesa.2007.07.018.
5. Duffner C. 2019."Experimental Study of the Pre-gelation Behaviour of Composite

Prepreg" 2019. doi:10.14288/1.0387323.
6. HEXCEL. 2016."HexPly® 8552 Product Data Sheet". *Hexcel* 2016:1–6.
7. Van Ee D, Poursartip A. 2009."NCAMP Hexply Material Properties Database for use with COMPRO CCA and Raven". *Natl Cent Adv Mater Perform* 2009:141.
8. Convergent Manufacturing. 2017."RAVEN simulation software" 2017. http://www.convergent.ca/products/raven-simulation-software.
9. Mehdikhani M, Aravand M, Sabuncuoglu B, Callens MG, Lomov S V., Gorbatikh L. 2016."Full-field strain measurements at the micro-scale in fiber-reinforced composites using digital image correlation". *Compos Struct* 2016;140:192–201. doi:10.1016/j.compstruct.2015.12.020.
10. Knowles KR, Tu J, Wiggins JS. 2017."Thermal and volumetric property analysis of polymer networks and composites using elevated temperature digital image correlation". *Polym Test* 2017;58:48–53. doi:10.1016/j.polymertesting.2016.12.013.
11. Heinz SR, Wiggins JS. 2010."Uniaxial compression analysis of glassy polymer networks using digital image correlation". *Polym Test* 2010;29:925–32. doi:10.1016/j.polymertesting.2010.08.001.
12. Bing P, Hui-min X, Tao H, Asundi A. 2009."Measurement of coefficient of thermal expansion of films using digital image correlation method". *Polym Test* 2009;28:75–83. doi:10.1016/j.polymertesting.2008.11.004.
13. Jerabek M, Major Z, Lang RW. 2010."Strain determination of polymeric materials using digital image correlation". *Polym Test* 2010;29:407–16. doi:10.1016/j.polymertesting.2010.01.005.
14. Jackson MB, Heinz SR, Wiggins JS. 2012."Fluid ingress strain analysis of glassy polymer networks using digital image correlation". *Polym Test* 2012;31:1131–9. doi:10.1016/j.polymertesting.2012.08.007.
15. Zobeiry N, Vaziri R, Poursartip A. 2015."Characterization of strain-softening behaviour and failure mechanisms of composites under tension and compression". *Compos Part A Appl Sci Manuf* 2015:29–41.
16. Rust-Oleum Consumer Brands Canada. 2015.Rust-Oleum White Safety Data Sheet.
17. Products Krylon. 2019.KRYLON® Camouflage Paint Black-Safety data sheet.
18. Mobuchon C, Poursartip A. 2019."Linear dependences of the cure shrinkage and thermal expansion of thermoset polymers on degree of cure", unpublished work.
19. Mobuchon C, Zobeiry N, Duffner C, Poursartip A. 2018."A Multifaceted Approach for Process Characterization of Polymer- matrix Composites". 12th Canada-Japan Work. Compos., Takayama: , p. 1–8.
20. Chatterjee S. 2014."Thermo-mechanical characterization of 8552 resin and AS4/8552 prepreg", unpublished work.
21. Zobeiry N, Duffner C. 2018."Measuring the negative pressure during processing of advanced composites". *Compos Struct* 2018;203:11–7. doi:10.1016/j.compstruct.2018.06.123.
22. DUFFNER C, ZOBEIRY N, POURSARTIP A. 2018."Examination of Pre-gelation Behavior in AS4/8552 Prepreg Composites". Am. Soc. Compos. 2018, . doi:10.12783/asc33/25974.
23. Ersoy N, Garstka T, Potter K, Wisnom MR, Porter D, Clegg M, et al. 2010."Development of the properties of a carbon fibre reinforced thermosetting composite through cure".

Compos Part A Appl Sci Manuf 2010;41:401–9. doi:10.1016/j.compositesa.2009.11.007.
24. Duffner, C, Courteau-Godmaire H, Zobeiry N, Poursartip A. 2019."Characterization of Prepreg Initial Morphology". 11th Can. Conf. Compos, Kelowna, July 2019.

Effect of Gaps on the Damage Initiation and Failure of Thin Composite Laminates Manufactured by AFP Under Out-of-Plane Loading

Mohammadhossein Ghayour[1*], Mehdi Hojjati[1], Rajamohan Ganesan[1]
Concordia Center for Composites, Department of Mechanical, Industrial and Aerospace Engineering
Concordia University
Montreal, Quebec, Canada

ABSTRACT

Advanced robotics as one of the elements of the smart factory has attracted interests in the composite industry through the last decade, and Automated Fiber Placement has found its way to be one of the well-established composite fabrication techniques in both aerospace and automated industries. However, because of using the fiber tows instead of composite prepreg sheets, new types of defects may be formed during the fiber deposition by a robot arm. These defects can induce interlaminar/intralaminar damages under the service loading, which reduces the mechanical performance of the final products. Although the effect of these defects on the performance of the composite structure can be found in the literature, there is still a lack of knowledge in the mechanism of interaction between the manufacturing defects and damages. In the present study, the effect of automated manufacturing gaps on the damage initiation and propagation in Carbon/Epoxy composite beams under out-of-plane loading is investigated. Short beam shear and three-point bending tests are carried out on the samples with defects. Flexural and interlaminar shear strengths of the samples are measured and compared with the samples with no defects (baseline sample). Furthermore, a series of microscopic observations are performed, and the effect of the manufacturing defects on the delamination and matrix damages initiation and also on matrix cracks propagation is investigated.

1. INTRODUCTION

Automated Fiber Placement (AFP) is an advanced robot technology for the manufacturing of large composite structures, especially in the aircraft industry. Despite many benefits of using this technology, such as rapid manufacturing, structures manufactured by AFP techniques are prone to have induced defects during the fiber deposition process. These defects can be categorized as gap, overlap, missing tow, tow wrinkling, upfolding, twisted tow, bridging, and crowning (Figure 1). However, gap is the most probable type of defects that may occur between two adjacent fiber tows during the automated fiber deposition. This defect may bring local material and geometrical discontinuities that affect the mechanical performance of the composite structures, which cannot be neglected [1]. It can also accelerate the process of catastrophic failure under different loading conditions, initiate the internal damages and decrease the durability of the structures. These defects can even affect the flexural and in-plane stiffness of the structures, which might be an issue for the structures in which the deformation is one of the main design criteria.

SAMPE Virtual Conference Proceedings, 2020. Society for the Advancement of Material and Process Engineering – North America.

Experimental studies show that the effect of gaps on the tensile strength is negligible [2, 3]. However, a reduction of compressive strength between 5-27% can be expected for the composite plate under compressive loading [3–6]. The wide range of the effectiveness comes from the fact that different scenarios of the embedding the gaps into the structures can be defined. For instance, an experimental study on the quasi-isotropic laminates with 2.0 mm intentional gap and overlap between the tapes has been performed in Ref. [7] and it was shown that compressive strengths of the composite beams could be reduced by around 7% and 15%, respectively. However, the existence of the stress concentration due to having an open hole makes this effect negligible to consider [8]. An experimental in-plane fatigue test has also been performed on a unidirectional laminate with a single intentional gap with the dimensions of 0.125×0.25 square inch in Ref. [9]. Results reveal that the effect of gaps becomes more severe with increasing the maximum applied stress. However, this effect is negligible for the low level of fatigue stress.

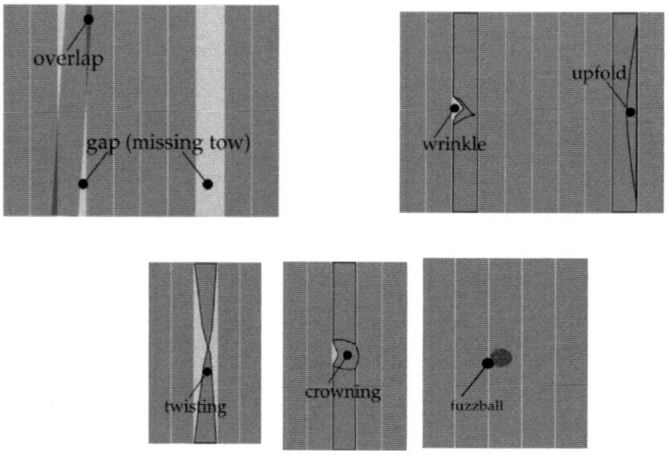

Figure 1. Different types of AFP manufacturing defect (reproduced from ref. [10])

It is also important to note that, although the induced gaps are the result of machine malfunction for regular composite laminates, it is impossible to avoid these defects in the new concept of composite manufacturing with fiber steering. Designing the structure with the curvilinear fiber path changes the in-plane stress distribution and causes a considerable increase in the buckling capacity of the composite structures [11–17]. However, the existence of the induced gaps is still a significant issue for these types of structures.

Although the effect of fiber tow gaps on the ultimate strengths of the composite plates under in-plane loading can be found in literature, there is still a lack of knowledge on the effect of induced gaps for the structures under out-of-plane loading. Furthermore, the interaction of this defect with the interlaminar and intralaminar damages needs to be studied.

The main objective of the present paper is to study the effect of fiber two gaps induced in the thin quasi-isotropic Carbon/Epoxy plates on the ultimate strength and also on the damage propagation under transverse loading. For this purpose, standard flexural and short-beam shear tests are carried out to observe the effect of the manufacturing defects on the flexural behavior and interlaminar shear strength of the composite beams. Furthermore, microscopic observation was performed on the damaged sample to investigate the interaction of the defects with delamination initiation and matrix cracks propagation.

2. EXPERIMENTAL PROCEDURE

In this study, the hand-layup (HL) technique was chosen for fabricating the composite plate with no defects. These samples, which are made by using prepreg sheets, are considered as baseline samples. A ZX130L Kawasaki 6-axis articulated robot arm was used to fabricate the samples with the periodical gaps between fiber tows (AFP-G). The process of embedding the gaps between each fiber course was proceeded by adjusting the AFP software to consider the fiber tow's width, 2.0 mm larger than the real width, which is 6.35 mm (Figure 2). These gaps cause 8% of the free spaces (Gap percentage) in the whole laminate during the fiber deposition.

Two composite plates with the dimensions of 19×17 inches were fabricated with both HL and AFP techniques and have been vacuumed and cured in an autoclave for six hours under the constant pressure of 90 psi and curing temperature of 180° C. An Aluminum hard tool (Caul plate) was also used to reduce the out-of-plane fiber waviness caused by free spaces between the fiber tows. The plates are then cut into the desired samples for both flexural and short beam shear tests. Table 1 shows the properties of the composite beams prepared for this study.

Table 1. Geometrical properties of the experimental samples

Type of the test	Sample Dimensions	Stacking Sequence	Material	Curing Type
Three-point bending	100 ×13 mm^2	$[0_2/45_2/90_2/-45_2]_s$	Cycom 977-2 HTS 40	Autoclave
		$[90_2/-45_2/0_2/45_2]_s$		
Short beam shear	15 ×5 mm^2	$[0_2/45_2/90_2/-45_2]_s$		

Figure 2. Automated Fiber Placement process (a) Fiber deposition by a robot arm (b) Fiber tow gap between the fiber courses

The effect of free spaces between the fiber tows on the quality of the cured plates was investigated by measuring the thickness and also by the microscopic observation of the cross-section of the intact samples. Figure 3.a. shows that embedding the 8% of the gap during fiber deposition may decrease the final thickness of the plates by 8%. This change has a significant effect on the flexural behavior of the samples. For example, it can reduce the second moment of area (I_{xx}) of the composite beam by 25% (Figure 3.b.). Figure 4 shows the cross-section view of the AFP-G samples at gap area. It can be seen that the thickness of the -45° plies has been reduced at the gap area, which has been filled with resin.

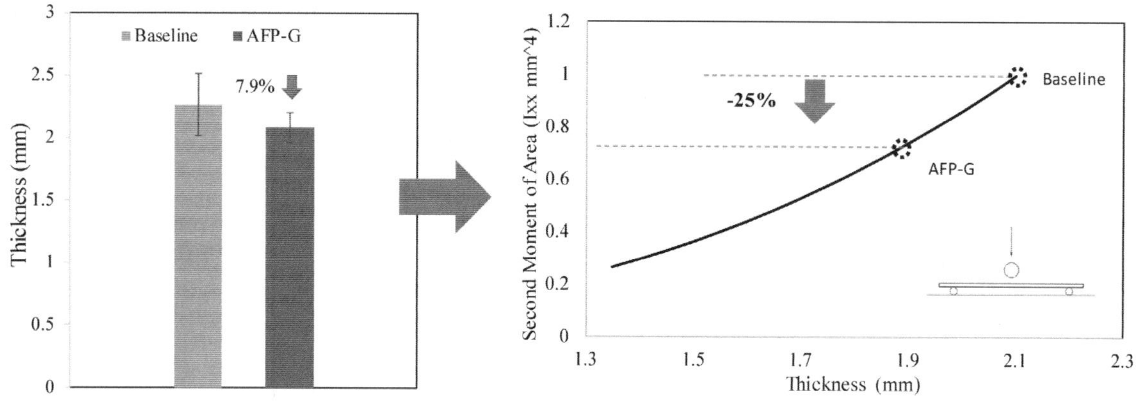

(a) (b)

Figure 3. Effect of fiber tow gaps on (a) thickness reduction (b) second moment of area

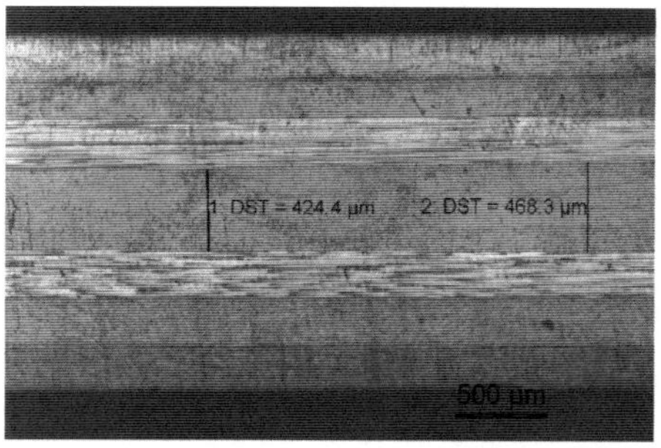

Figure 4. Local thickness reduction at the gap area

The other effect of the existence of the pre-set gaps on the quality of the composite samples can be found in Figure 5. The preset gap affects the uniformity of the Fiber Volume Fraction (FVF) in the gap area. For instance, FVF drops by 50% at the -45°lamina and reaches 27% at the center of the gap area. It is also worth to note that the width of the resin-rich area (pre-set gap area) is reduced due to fibers movement during the curing process. However, the domain of the FVF reduction seems to remain around 2.0 mm, which is equal to the width of the intentional gap.

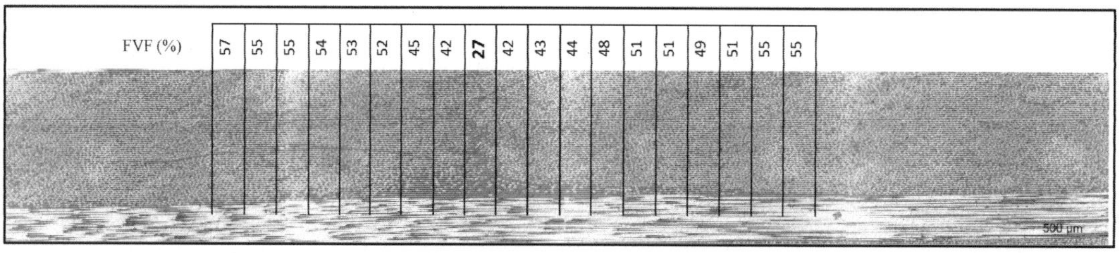

Figure 5. Distribution of the Fiber Volume Fractions at the gap zone at 45° lamina

The effect of the distributed gaps on the out-of-plane behavior and interlaminar behavior of the composite beams was investigated by performing a series of three-point bending and short-beam shear (SBS) tests. Figure 6 shows the dimensions and applied boundary conditions for flexural and SBS tests setup, which are based on ASTM D7264 and ASTM D2344 standards, respectively. It can be seen that the specimens for the flexural tests have the dimensions of 100×13 mm with a span length of 80 mm (Figure 6.a.). Besides, two 5.0 mm diameter rollers simulate the simply-supported boundary conditions, and the out-of-plane load is applied by a 10 mm diameter roller. The same boundary condition but with a crosshead roller with diameter of 5.0 mm is used for SBS tests (Figure 6.b.). A universal testing machine with a maximum load capacity of 5 kN was used for measuring the flexural strength and Interlaminar Shear Strength (ILSS) of the composite beams. The machine can record 50 data per second with the displacement and load precision of 0.01 mm and 0.1 N, respectively. In order to achieve consistency in the results, four samples per each category were subjected to loading with the loading rate of 1.0 mm/min.

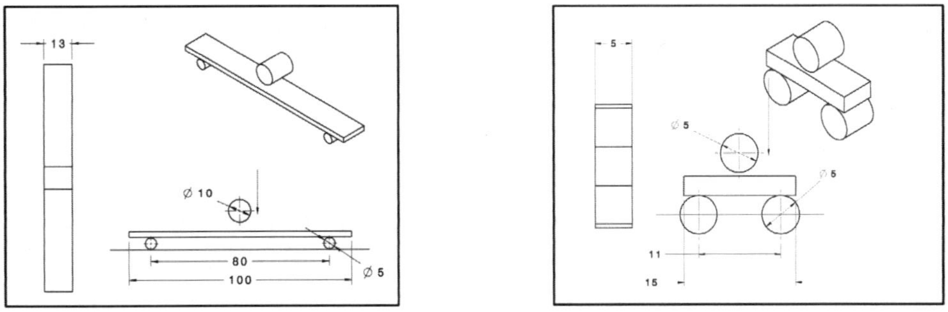

(a) (b)

Figure 6. Dimensions and boundary condition of the specimens for the (a) three-point flexural test and (b) Short-beam shear test (All dimensions are in mm)

3. RESULTS AND DISCUSSION

It is recommended by ASTM standard to find the flexural stiffness and strength and also ILSS of the beams by the following formulae:

Flexural strength
$$\sigma_m = \frac{3P_m L}{2bh^2} \quad [1]$$

Flexural Stiffness
$$E_F^{Secant} = \frac{L^3 m}{4bh^3} \quad [2]$$

Interlaminar Shear Strength
$$F^{sbs} = 0.75\frac{P_m}{bh} \quad [3]$$

In the above, P_m is the maximum applied load, L is the span length, b, and h are the width and thickness of the specimens, respectively, and m is the slope of the secant of the force-displacement curve. Note that a reference value of 2.1 mm (baseline thickness) has been used for the thickness of all specimens.

Table 2 shows a summary of the results for both flexural and short beam shear tests. It shows that the effect of fiber tow gaps on the flexural stiffness and strength, and interlaminar shear strength is significant. While the flexural stiffness of the baseline beam is around 75 GPa, there is 35% reduction for the samples with periodic gaps. Almost the same reduction can be expected for the flexural strength. In this case, flexural stiffnesses are about 1200 MPa and 840 MPa for the baseline and AFP-G samples, respectively. It is important to note that part of this reduction comes from the assumption of using the reference thickness for both baseline and AFP-G samples. For example, applying the actual thickness for AFP-G samples in the formula (2), the flexural stiffness of the samples would be around 65 MPa. It means only 12% reduction in flexural stiffness is the result of the thickness reduction and material properties reduction due to the existence of resin-rich area. The reason for using the reference thickness instead of the actual one is that widespread gaps can take place at a partial zone of a structure in the actual practice. It is more practical to consider the effect of the manufacturing gap as local material degradation rather than local thickness reduction. The same explanation can be said for the reduction of the flexural strength.

Furthermore, there is no difference in the failure mode of the quasi-isotropic beams under flexural loading. For both series of the beams, ultimate failure is the fiber breakage under compression loading at zero layers and the delamination between the zero and 45° laminae (Figure 7).

Table 2. A comparison of the flexural behavior and ILSS of the composite beams with gaps and baseline samples

	Baseline	AFP-G	Difference (%)
Flexural Stiffness (GPa)	74.08 Std. Dev. = 1.7	47.98 Std. Dev. = 1.88	35
Flexural Strength (MPa)	1232 Std. Dev. = 20	836.6 Std. Dev. = 26	32
Interlaminar Shear Strength (MPa)	94.86 Std. Dev. = 2.7	81.6 Std. Dev. = 4.8	13

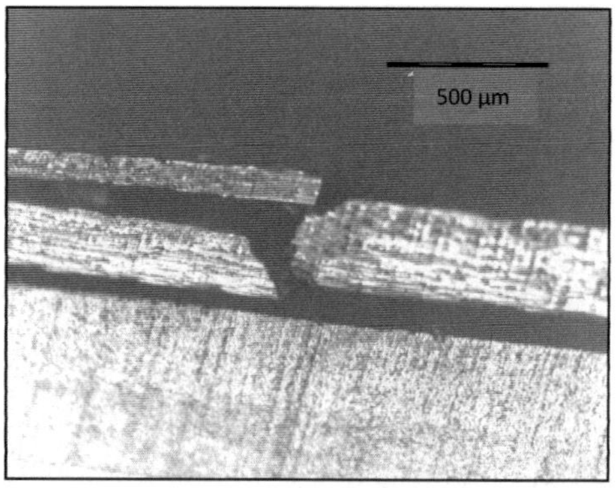

Figure 7. Failure mode of the composite beams under flexural loading

Fiber tow gaps also reduce the interlaminar shear strength of the composite laminates. It was shown in Table 2 that this defect can decrease the ILSS by around 13%. Note that because of the small dimensions of the short beams, there is a possibility of not having gap area in the tested samples, which means that all of the studied samples are not necessarily the representative of the samples with the gap. In order to overcome this issue, more than 15 samples were tested experimentally, and then by using microscopic observation, the average value has been calculated from the samples with the gap. Looking at the zone of delamination initiation for the AFP-G samples, it was realized that the resin pockets can be the sources of delamination initiation for the beams under short beam shear tests. Figure 8 shows two different short beams that have been failed under transverse loading. It is clear that the delamination, which is the primary mechanism of the failure, initiates from the resin-rich areas which have been created due to the intentional gaps.

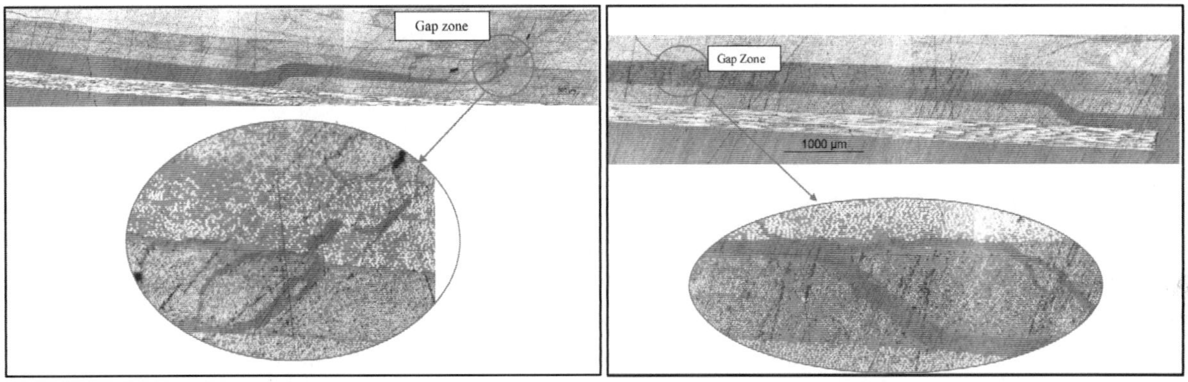

Figure 8. Delamination initiation from the resin-rich area induced by fiber tow gaps

The effect of the intentional gaps on the Matrix Crack (MC) propagation in the beams under transverse loading was investigated by three-point bending tests on the composite beams with the stacking sequence of $[90_2/-45_2/0_2/45_2]_s$.

For this purpose, the crack density of the bottom layers (90° plies) at different crosshead loads was measured. Figure 9.a. shows the load versus displacement curve for both baseline and AFP-G samples. It shows that the intentional gaps have significant effect on the loading capacity of the composite beams with low flexural rigidity. Manufacturing gaps can reduce the loading capacity of the beams with the stacking sequence of $[90_2/-45_2/0_2/45_2]_s$ by about 50%. Although matrix cracking is not a catastrophic failure, it reduces the stiffness of the materials and also might induce the delamination. MC also reduces the level of the stress due to free surface area and causes stress perturbation at MC zone [18, 19]. In this study, the crack density of the 90° plies at the bottom was measured experimentally at different loads. These layers are under tension load perpendicular to the fiber directions. Thus, matrix cracks propagate from the spot with the highest level of the stress, which in this case it places at the middle of the beam. Figure 9.b shows the graph of the crack density versus the longitudinal strain of the bottom surface at the center of the beams. Note that, the longitudinal strain (ϵ_{xx}) was calculated based on the Euler-Bernoulli beam theory. It is evident that the MC initiates at lower strain for AFP-G sample. However, there is no major difference between the MC propagation for both samples. That probably shows that gap defects do not have any significant effect of the matrix crack propagation. Furthermore, no evidence on the interaction of MC and gap defects was also found by microscopic observation. However, more experimental studies are required to confirm this statement.

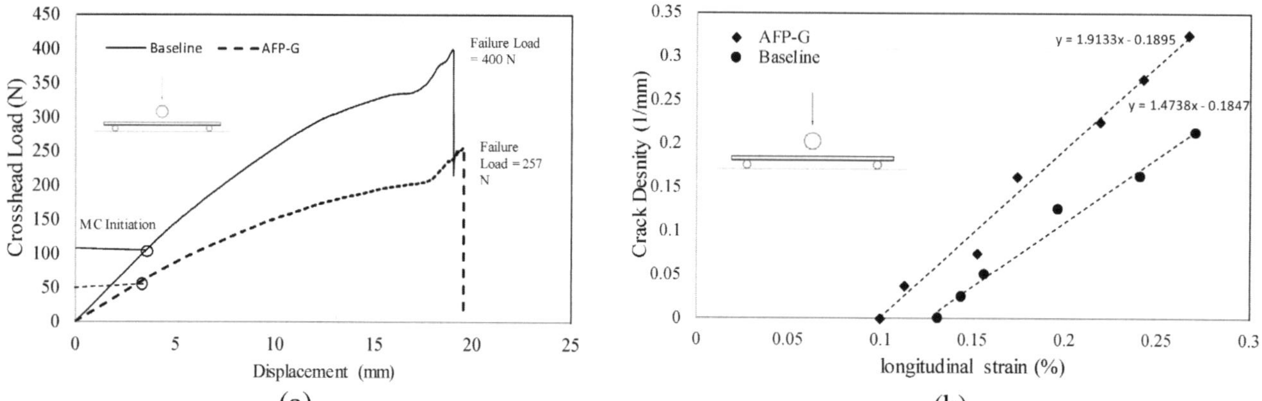

Figure 9. Comparison of Crosshead load versus (a) transverse displacement (b) crack density of 90° plies for AFP-G and baseline samples with the stacking sequence of $[90_2/-45_2/0_2/45_2]_s$ under flexural loading

4. CONCLUSION

In this paper, the effect of fiber tow gaps on the interlaminar and intralaminar damage initiation and propagation and also on the failure of the thin composite beams under transverse loading was investigated experimentally. For this purpose, both three-point bending and short beam shear tests were selected to measure the flexural stiffness and strength and also the interlaminar shear strength of the beams in the presence of the gaps. A quasi-static composite plate with periodic gaps between the fiber tows was fabricated by Automated Fiber Placement technique. This plate is the representative of the samples with the manufacturing gap and the results were compared with that of the samples with no defects (baseline samples).

Results show that the flexural stiffness of the beams are reduced by around 35% due to thickness reduction and material degradation caused by fiber tow gaps. The same percentage of reduction can also be expected for the flexural stiffness of the quasi-isotropic beams. Interlaminar shear strength is also decreased by about 13%. Microscopic observation shows that the delamination was initiated at the gap areas for the short beam samples with defects. It emphasizes that the resin-rich areas caused by gaps are prone to be the sources of the delamination initiation.

The effect of the fiber tow gaps on the matrix cracks propagation under bending load was studied by performing three-point bending test on the beams with the stacking sequence of $[90_2/-45_2/0_2/45_2]_s$ in which the matrix cracks propagate at 90° layers under tensile loads. No significant difference between the results was found for both the samples with the gaps and the baseline. It probably shows that the effect of the gaps on the matrix crack initiation and propagation can be neglected. However, more studies are required to confirm this statement.

5. ACKNOWLEDGMENT

The authors would like to acknowledge the support provided by the Concordia University, Concordia Center for Composites (CONCOM), Natural Sciences and Engineering Research Council of Canada (NSERC), Bombardier Aerospace and also FRQNT Doctoral Research Scholarship (BX2).

6. References

[1] Lukaszewicz DH-JA, Ward C, Potter KD. "The engineering aspects of automated prepreg layup: History, present and future". *Compos Part B Eng* 2012; 43: 997–1009.

[2] Lan M, Cartié D, Davies P, Baley C. "Influence of embedded gap and overlap fiber placement defects on the microstructure and shear and compression properties of carbon–epoxy laminates". *Compos Part A Appl Sci Manuf* 2016; 82: 198–207.

[3] Croft K, Lessard L, Pasini D, Hojjati M, Chen J. "Experimental study of the effect of automated fiber placement induced defects on performance of composite laminates". *Manuf* 2011; 42: 484–491.

[4] Guin WE, Jackson JR, Bosley CM. "Effects of tow-to-tow gaps in composite laminates fabricated via automated fiber placement". *Compos Part A Appl Sci Manuf*; 115. Epub ahead of print 2018. DOI: 10.1016/j.compositesa.2018.09.014.

[5] Sawicki AJ, Minguet PJ. "Effect of intraply overlaps and gaps upon the compression strength of composite laminates". In: *Collection of Technical Papers - AIAA/ASME/ASCE/AHS/ASC Structures, Structural Dynamics and Materials Conference*. AIAA, 1998, pp. 744–754.

[6] Nimbal SS, Banker MM, Roopa A, Varughese B, Sundaram R. "Effect of Gap Induced Waviness on Compressive Strength of Laminated Composites". *Mater Today Proc* 2017; 4: 8355–8369.

[7] Woigk W, Hallett SR, Jones MI, Kuhtz M, Hornig A, Gude M. "Experimental investigation of the effect of defects in Automated Fibre Placement produced composite laminates". *Compos Struct* 2018; 201: 1004–1017.

[8] Falcó O, Mayugo JA, Lopes CS, Gascons N, Costa J. "Variable-stiffness composite panels: Defect tolerance under in-plane tensile loading". *Compos Part A Appl Sci Manuf* 2014; 63: 21–31.

[9] Elsherbini YM, Hoa S V. "Experimental and numerical investigation of the effect of gaps on fatigue behavior of unidirectional carbon/epoxy automated fiber placement laminates". *J Compos Mater* 2016; 51: 759–772.

[10] Heinecke F, Willberg C. "Manufacturing-Induced Imperfections in Composite Parts Manufactured via Automated Fiber Placement". *Journal of Composites Science*; 3. Epub ahead of print 2019. DOI: 10.3390/jcs3020056.

[11] Hyer MW, Lee HH. "The use of curvilinear fiber format to improve buckling resistance of composite plates with central circular holes". *Compos Struct* 1991; 18: 239–261.

[12] Marouene A, Boukhili R, Chen J, Yousefpour A. "Buckling behavior of variable-stiffness composite laminates manufactured by the tow-drop method". *Compos Struct* 2016; 139: 243–253.

[13] Z. Gürdal, B. F. Tatting, K. C. Wu, J. H. Starnes J. "Variable Stiffness Panels, PART 1: Effects of Stiffness Variation on In-Plane and Bending Responses". *Compos Part A Appl Sci Manuf* 2005; 39: 911–922.

[14] Lopes CS, Gürdal Z, Camanho PP. "Variable-stiffness composite panels: Buckling and first-ply failure improvements over straight-fibre laminates". *Comput Struct* 2008; 86: 897–907.

[15] Setoodeh S, Abdalla MM, IJsselmuiden ST, Gürdal Z. "Design of variable-stiffness composite panels for maximum buckling load". *Compos Struct* 2009; 87: 109–117.

[16] Rouhi M, Ghayoor H, Fortin-Simpson J, Zacchia TT, Hoa S V., Hojjati M. "Design, manufacturing, and testing of a variable stiffness composite cylinder". *Compos Struct* 2018; 184: 146–152.

[17] Fayazbakhsh K, Arian Nik M, Pasini D, Lessard L. "Defect layer method to capture effect of gaps and overlaps in variable stiffness laminates made by Automated Fiber Placement". *Compos Struct* 2013; 97: 245–251.

[18] Ghayour M, Hosseini-Toudeshky H, Jalalvand M, Barbero EJ. "Micro/macro approach for prediction of matrix cracking evolution in laminated composites". *J Compos Mater* 2016; 50: 2647–2659.

[19] Ghayour M, Chitsaz N, Hosseini-Toudeshky H, Barbero EJ. "Enhanced variational approach for damage analysis of laminated composite". *Mech Adv Mater Struct* 2019; 1–11.

IN-SITU CONSOLIDATION OF COMPLEX CURVATURE THERMOPLASTIC COMPOSITE PART MANUFACTURED BY AUTOMATED FIBER PLACEMENT

Dan Ursenbach[+], Justin Merotte[++], Noushin Bahramshahi[+]
[+] Collins Aerospace, Chula Vista, CA
[++] Coriolis Composites, Queven, France

ABSTRACT

This paper presents a project at Collins Aerospace on the manufacturing of complex curvature thermoplastic composite parts with laser assisted fiber placement (LAFP) in-situ consolidation. To date, most of the work done in in-situ consolidated thermoplastic AFP parts have been limited to simple curvature parts, or body of revolution parts where the AFP process has been used much like a Filament Winding process. This work investigates the processing of a complex 3D part with double curvature surface and confined space constraints via LAFP in-situ consolidation. A true 3D complex curvature part inevitably contains fiber convergence zones, and the gaps associated with such zones. These gaps present a significant challenge to achieving good laminate properties, and a subsequent design optimization was performed to minimize them. Other challenges characteristic of in-situ and thermoplastic processing included print-through of tool release film and splices, which require industry solutions. Finally, a set of validation articles were fabricated by applying the appropriate processing settings. While the finished validation articles had limited defect areas, they were deemed acceptable for postlamination operations like bonding or induction welding of thermoplastic stiffeners.

Corresponding author: Dan Ursenbach, Daniel.Ursenbach@collins.com

1. INTRODUCTION

Carbon fiber reinforced thermoplastic composite materials have been available for many years. Although the theoretical promise of high volume, low cost thermoplastic composite parts has been discussed and explored, there has yet to be a major commercial aerospace manufacturing program that combines the large scale, complex integration, and competitive cost required by modern reinforced skin fuselage or nacelle aerostructures. Thanks to advances in automation, such as laser assisted fiber placement (LAFP), and production thermoplastic welding processes, the aerospace industry is on the cusp of achieving the necessary processes and fundamental understanding to bring this to reality.

To achieve the ultimate low cost reinforced nacelle or fuselage skin assembly and eliminating subsequent joining process, one can use in-situ only processing, where aircraft skin is fused via the laser assisted AFP process over pre-consolidated frames and stringers. This in-situ melt consolidation process was first explored at a large scale by the Sikorsky SARAP demonstrator in 2006-2007. [1]

Laminates made with in-situ only processing and PAEK family of polymer prepregs have yet to achieve anything much more than 75% of autoclave or vacuum bag properties, even when using low melt temperature PEEK (e.g., Toray's TC1225 or Victrex's AE-250). Additionally, for a fuselage and nacelle skin melt consolidation process, it is required that the skin is placed on a convex tool, as shown in Figure 1. This processing method requires the final ply of the AFP sequence to be the aerodynamic surface. As such, all the defects occurring during the layup tend to print through to the surface, and therefore the process is very sensitive to small variations.

Conversely, when a concave tool is used, the tool side surface becomes the aerodynamic surface, and the process is subsequently less sensitive to material and process variation. The skin can even be post-consolidated before stiffener joining. However, this process choice does not allow the option of melt consolidation of the skins directly to the frames and stiffeners, and therefore results in a higher process cost.

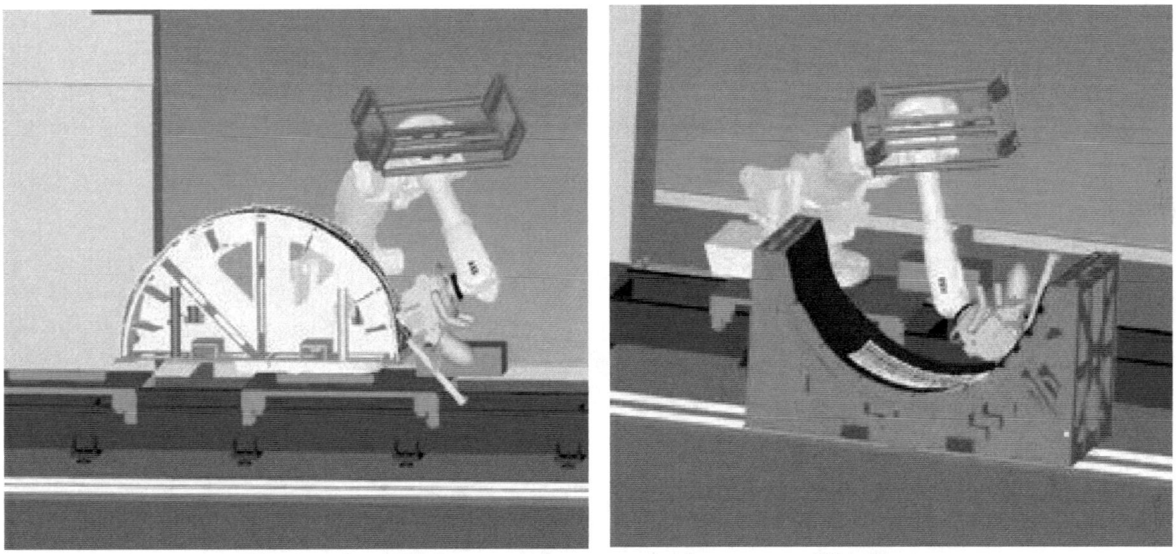

Figure 1: AFP lay up of convex and concave 3D shapes

This paper investigates both approaches to evaluate the limitations and state of the art. Included in the outcomes of the investigation are a list of needed material and process improvements that are required before in-situ melt bonding can become a viable aerostructure manufacturing process.

2. EXPERIMENTATION

2.1 Approach

Before beginning the layup of the 3D parts, various material and process parameters were explored via flat panel fabrication and inspection. These included variations in speed, temperature, and materials. These evaluations were done both to provide needed inputs to the 3D part layup, and to determine mechanical performance as a function of material and process parameters. This flat panel investigation is reported in a companion paper in this SAMPE conference [2].

2.2 Materials

Two commercially available slit tapes from Toray were used in this study, a traditional PEEK prepreg material, TC-1200, along with a newer "low melt" or LM-PEEK resin system, TC-1225. Both prepreg systems used Hexcel IM7 fiber. Only the LM-PEEK system was used on the 3D fiber placed part.

2.3 Machine details

The machine used to manufacture the panels and 3D part was a CORIOLIS® C1, with 8 tows x 6.35mm (1/4 inch) width per tow head mounted on an ABB IRB6650 robot, itself mounted on a GUDEL TM90 linear axis. This machine can layup simultaneously up to 8 continuous tows at speeds up to 1000mm/s. This machine is the property of the ComposiTIC Research Centre based in Ploemeur, Brittany, France.

In order to heat the material to the required processing temperature, a Laserline LDF6000-100 6kW Infrared laser source was used. A laser homogenizer optic (Figure 2.a) mounted on the fiber placement head created a rectangular laser beam with a cross section of 8mm height and 56mm width for flat panels and 8x40mm² for 3D complex shape. An optical fiber guided the laser energy from laser source to the fiber placement head (Figure 2).

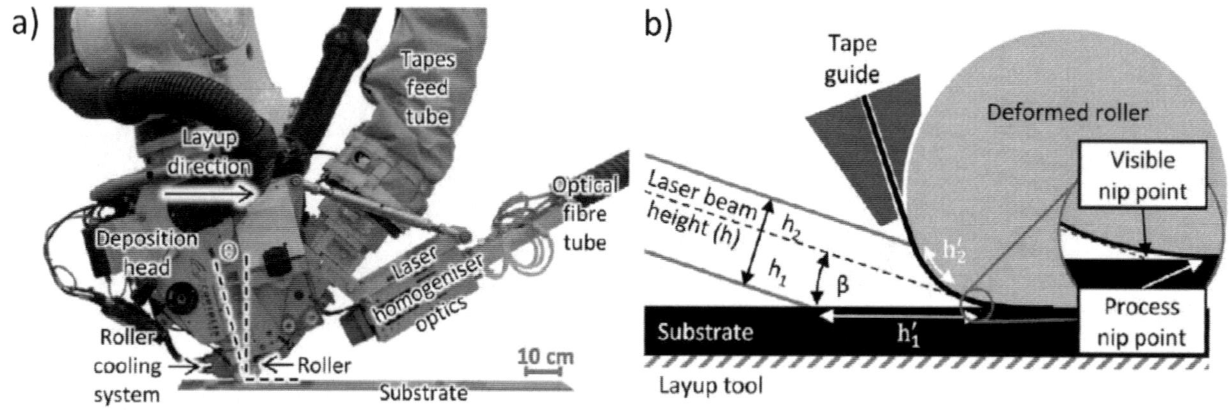

Figure 2 a) CORIOLIS Automated Fiber Placement (AFP) robotic machine in thermoplastic configuration b) Definition of layup parameters [3]

2.4 Molds and 3D part

For the flat panel investigation, both controlled heated and room temperature tools were used to evaluate the influence of heated substrate. For complex 3D shape manufacturing, two demonstrator molds, one convex and one concave, showing typical topology singularities were used. Neither of these tools was heated during AFP processing. As discussed, the convex tool demonstrator represents the ultimate low cost process choice, where aircraft skin can be melt consolidated over pre-consolidated frames and stringers during in-situ fiber placement.

The second 3D complex layup was done on a concave tool to achieve high quality outer surface. This configuration also allowed us to evaluate the ability of the AFP head to fit into confined complex spaces (Figure 1).

The part lay up in both 3D scenarios was a simple quasi isotropic laminate built with 8 full plies $[45/90/-45/0]_s$. The Toray TC1225 LM-PEEK was used due to its better laminate properties and larger processing window.

2.5 Programming software

CADFiber® and CATFiber® for CATIA® (Dassault Systemes) were used to program and analyze AFP machine trajectories. The mold surface was approximated using an adapted mesh and implemented into the software. Layup strategies were determined using the simulation tools developed in the software (roller crush, laser beam visualization, fiber deviation angle, etc.).

2.6 Layup and characterization

The different laminates were laid up using a polyimide film as a first ply deposition solution. Prior to layup on complex surfaces, dry runs were conducted in order to validate programming strategies. The layup was then performed with ply inspection between each ply.

Process temperatures during layup were measured using an infra-red camera Optris PI4000 with an acquisition frequency of 120 Hz. The measured temperature (Figure 3) was an average of the temperature values located at the nip point, defined here as the first point of contact between the fed fibers and the substrate detected by the IR camera. Material emissivity is supposed to be equal to 1 as it is assumed that all the wave reflections are absorbed by the material [4]. This measurement technique is very sensitive to surface defects and material quality (recorded temperature variations can reach up to 80°C at a constant laser power). Since the variation due to surface defects is much larger than that of the potential variation in material emissivity, an assumed material emissivity equal to 1 is reasonable.

The recorded temperature is the average of the values measured on a distance smaller than deposition width and located at the nip point. Considering the limitation given above, the layup temperature is given with an error equal to ±25°C, between minimum and maximum recorded values.

Temperature is a function of laser power which is adjusted to reach the desired processing temperature.

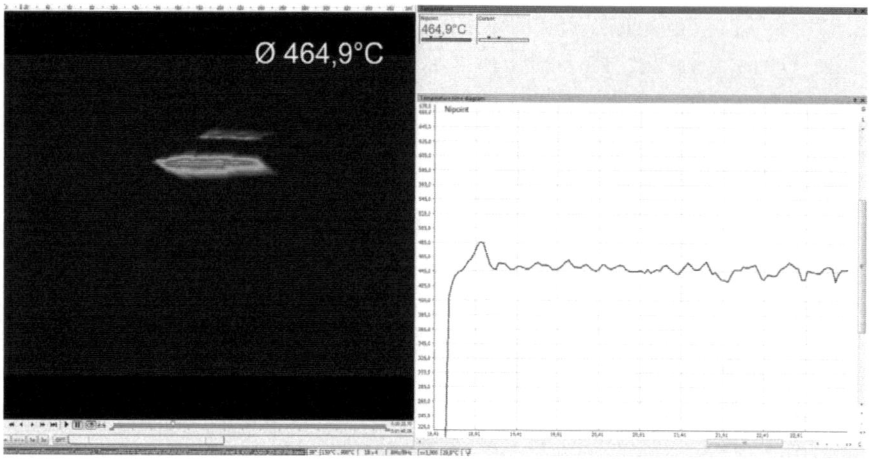

Figure 3 Temperature measurement using IR camera: recording during in-situ layup

From previous studies [5-8], it has been showed that optimum in-situ consolidation was obtained when applying measured temperature of around Tm+100°C. Hence, PEEK materials are processed at 450°C while LM-PEEK materials are processed at 410°C.

Roller hardness is a key point while talking about LAFP. Indeed, the best properties on flat panels have found to be obtained with unconformable metallic rollers. On this study, due to the high complexity of the demonstrator part, conformable rollers must be used. Hence, two different roller hardnesses and widths (leading to different compaction pressure) were tested and their impact on surface quality and porosity content evaluated. This work was conducted on PEEK material laid up on flat panels only. Porosity content on each panel was evaluated using cross section microscopy and density measurement.

Even though better sample microstructure was obtained on heated tooling, the 3D complex part was laid up on room temperature tool. However, the remaining layup settings, such as layup temperature, speed and roller hardness derived from the flat panel study, were applied.

2.7 3D Part Programming

Once the tooling surface was discretized into a mesh, the part ply contours were applied. Due to the double curvature of the surface, it was necessary to sectorize the surface to keep the fiber deviation as low as possible from the optimum direction. The fiber deviation analysis and the resulting contour sectorization is given for one ply in Figure 4.

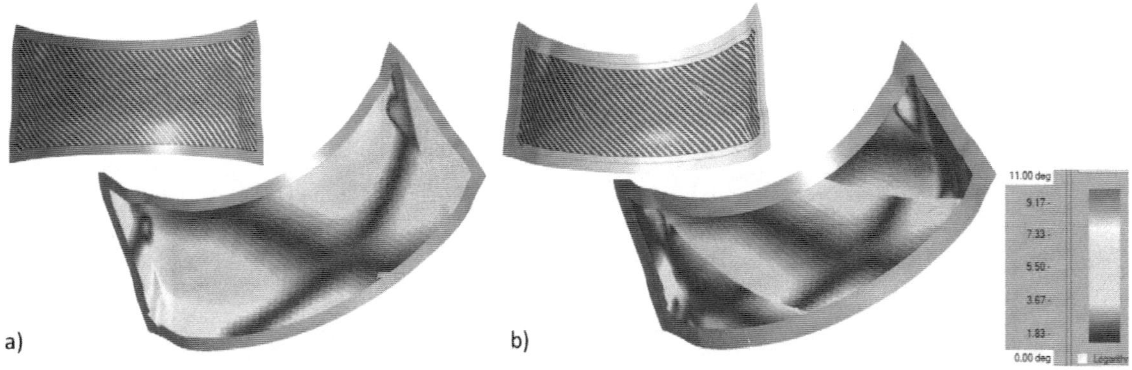

Figure 4 Fiber orientation analysis results for two different strategies: a) one single sector (max fiber deviation angle=10.92 deg), b) three separate sectors (max fiber deviation angle=5.02 deg)

Triangular gaps and overlaps are the inconvenient consequence of contour sectorization (Figure 5). When no post consolidation is applied, these gaps can be transformed into porosities if the matrix flow is insufficient. In order to minimize the impact of such gaps, those defects were shifted in space from one ply to another.

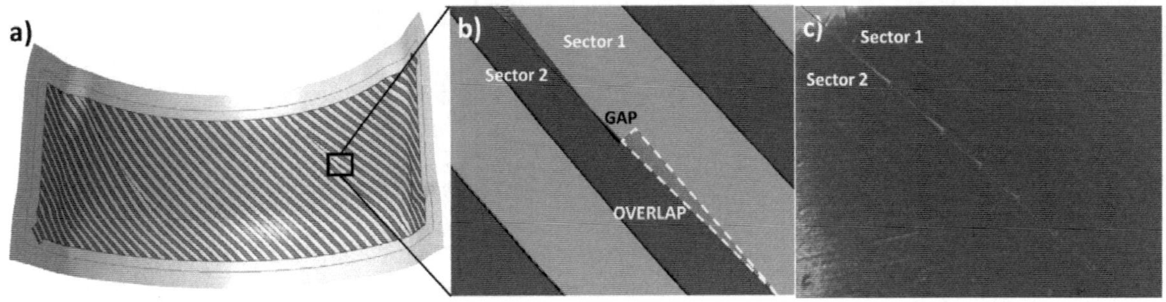

Figure 5 Creation of gap/overlap at the border of each sector (a,b: software illustration, c: triangular gaps on real part)

Once sectorization was complete, a second check was performed using the roller/surface contact analysis tools. This software tool helps to estimate the required roller deformation in order to ensure full contact between the roller and the surface. From this analysis (Figure 6), it was determined that no additional contour sectorization was necessary. The roller deformation properties were adequate to apply constant pressure over the contact area.

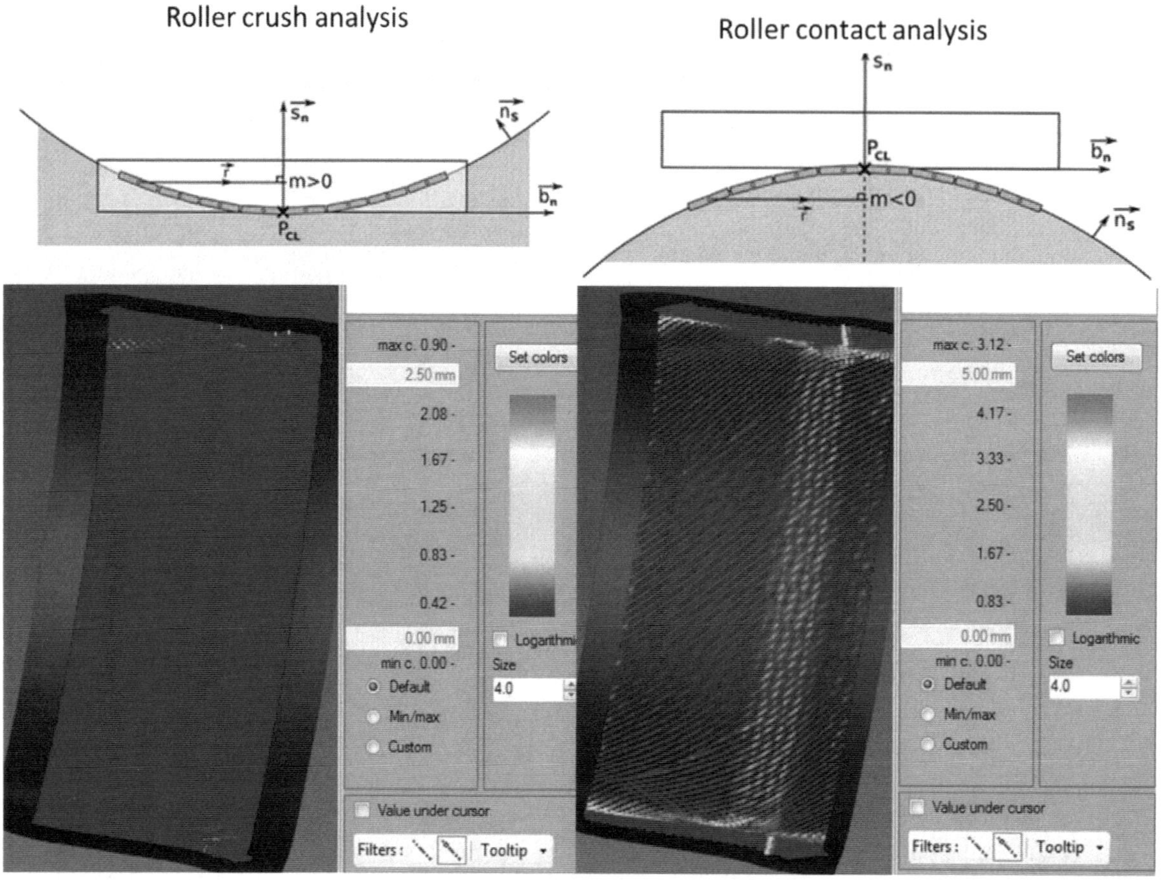

Figure 6 Analysis values for deposition roller deformation and roller/surface contact

The final optimization performed for the part is called laser heating feasibility. To ensure proper in-situ consolidation, laser angle must remain below a given tilt angle limit. Moreover, when the tooling surface contains local convex areas, the laser beam can be partially blocked by the substrate (i.e., shadow effect) before it reaches the nip-point (Figure 7.a). Local tilting values must then be adjusted to prevent inadequate heating (Figure 7.b).

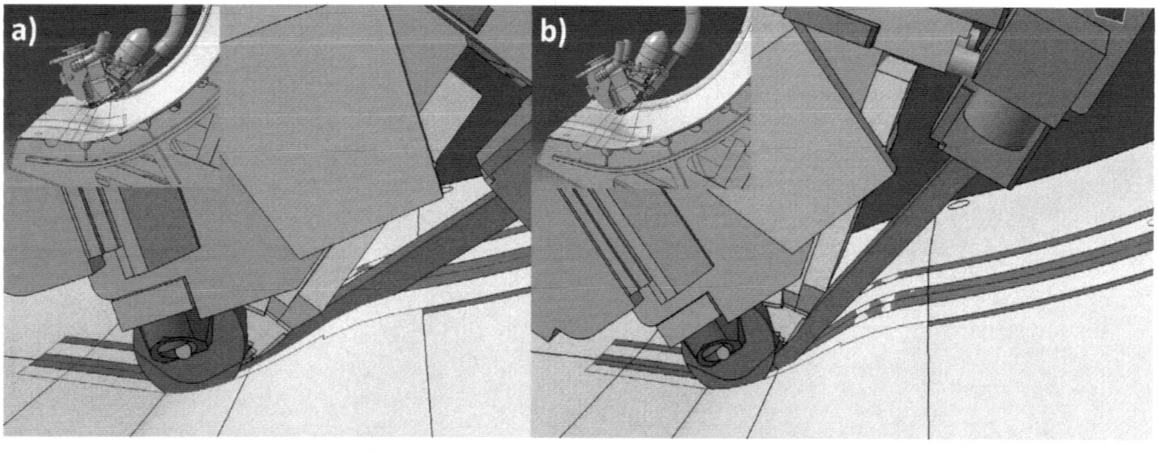

Figure 7 Laser beam visualization a) without sufficient head tilt angle (shadow) b) with corrected tilt angle

Finally, before generating the program that will be read by the machine, a tool path simulation is conducted in order to detect any trajectory deviation or collision between the machine and the tooling (Figure 8.a).

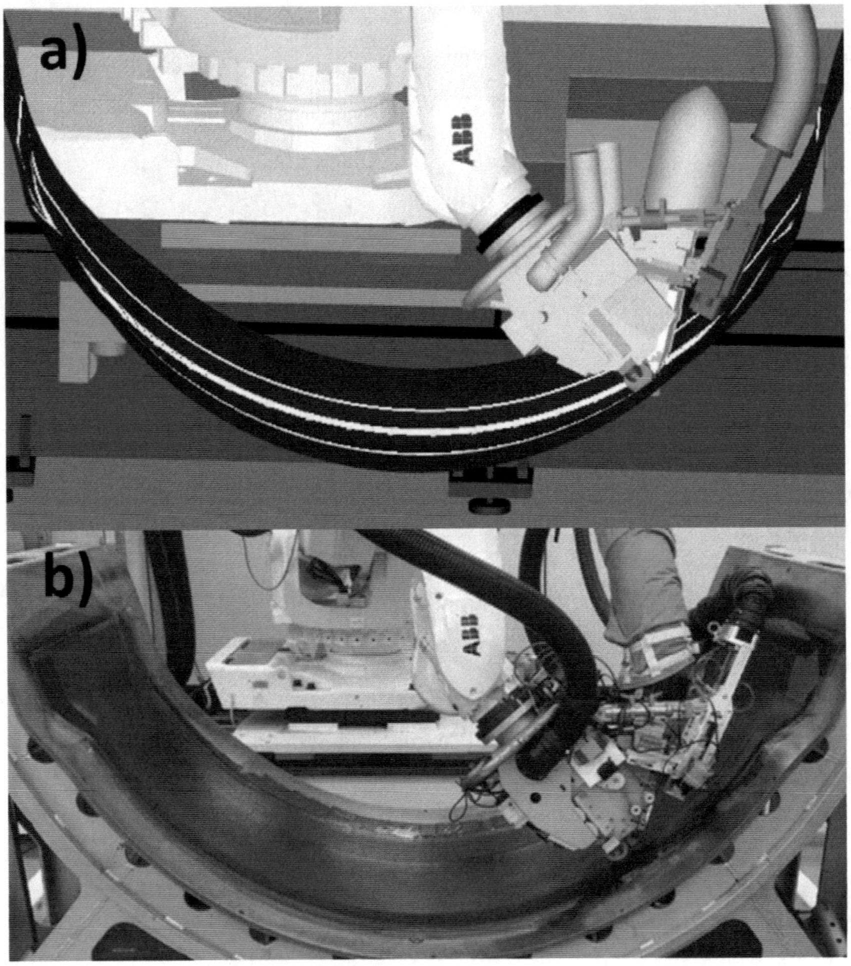

Figure 8 a) Layup simulation with CADFiber software, b) Corresponding real layup

3. RESULTS

3.1 3D Part Manufacturing

Tool preparation before layup required installing a release film. At this point, the only option to create the necessary balance of tackiness and release was polyimide release film. However, the double curvature surface required discretization of the release film. Each section was linked to the other by adhesive tape. As a result, the overlaps created by the tapes printed on the first ply of the part (Figure 9), and also can result in print through to final layer, which is discussed later.

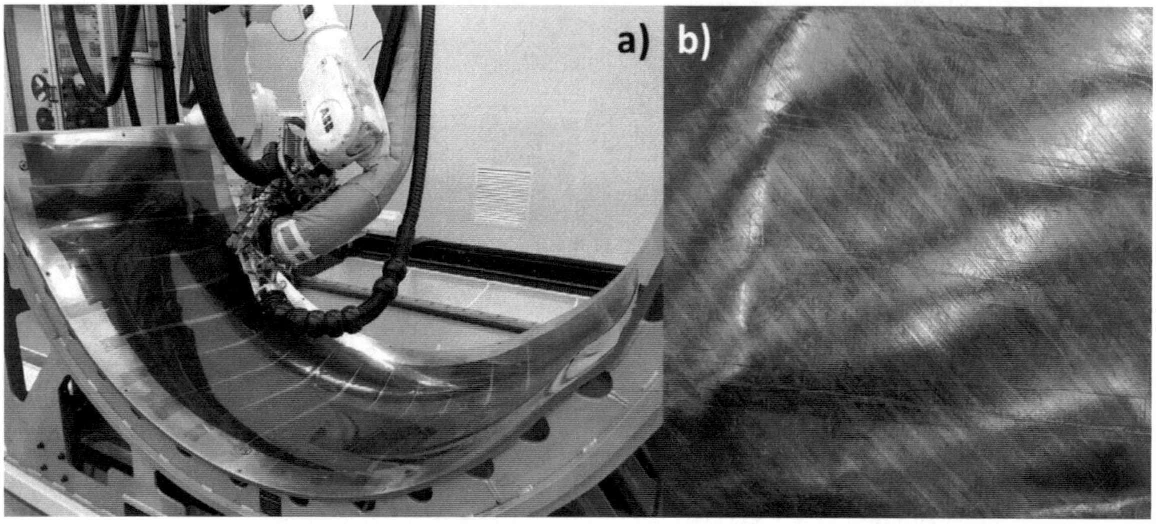

Figure 9 a) Polyimide film strips positioning on the concave tool, b) Print through of the tape on the first ply of the part

Once the program was generated by the CADFiber software and the layup parameters applied, layup was performed on the complex 3D shape. The entire required surface was deposited by the AFP machine without any egregious defects: no fiber bridging, or untacked tows.

For the convex tool layup, outside surface condition was critical since, in production, this would be the aero surface layer. As can be seen in Figure 10, print-through from both polyimide release film, tape and overlaps/gaps were present on the surface. It should be noted that for the layup shown, no overlaps were permitted in the programming, and the resulting gaps were visually significant. Later iterations switched to a split, where the convergence of zones were shared as 50% overlap and 50% gap area, and resulting surface defects were found to be visually less severe.

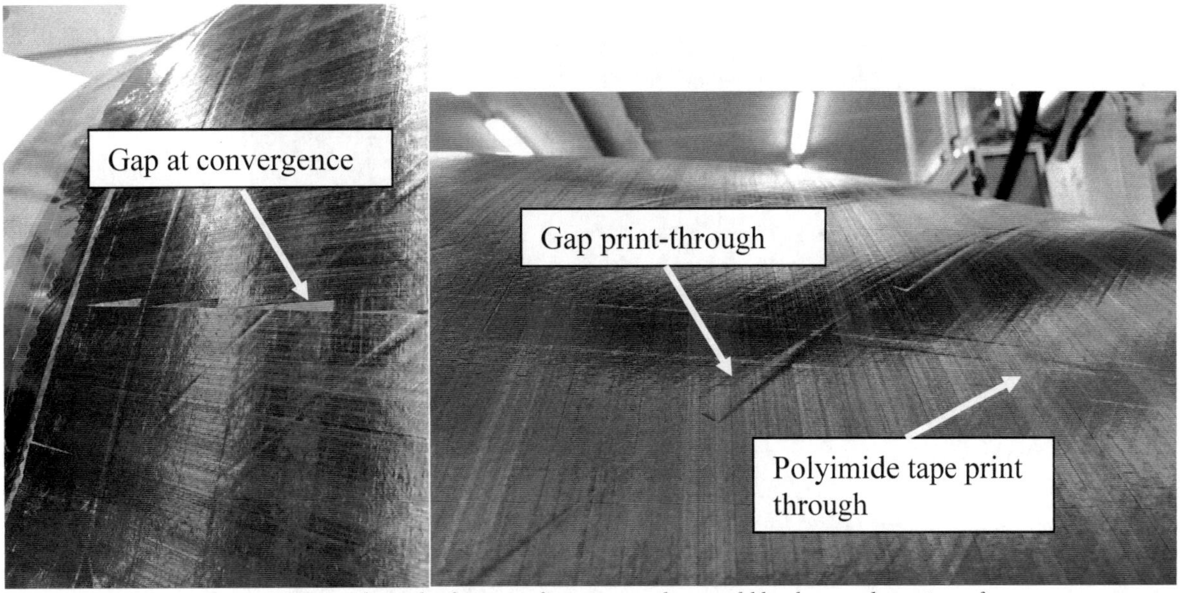

Figure 10 Print through of process features on what would be the aerodynamic surface

In terms of in-process defects, unlike thermoset AFP, the deposited tows are immediately fused to the underlying layer, hence it is very difficult to repair defects created during AFP. When the AFP parameters are properly set, defects occurring are mostly due to the raw material: tape splices and localized lack of impregnation. When properly marked, tape splices can be detected by the machine and removed before being deposited. However, such operation increases significantly the layup time and the scrap rate. Impregnation flaws create gaps that are difficult to fill, especially if viscous matrices are used.

Figure 11.a shows the final part in concave mold. One can notice the overall good quality of the in-situ manufactured part. Moreover, due to its good surface aspect, the part was considered acceptable for post lamination operations. Hence, Figure 11.b shows the finished skin with a stamped stiffener adhesively bonded to the inside surface.

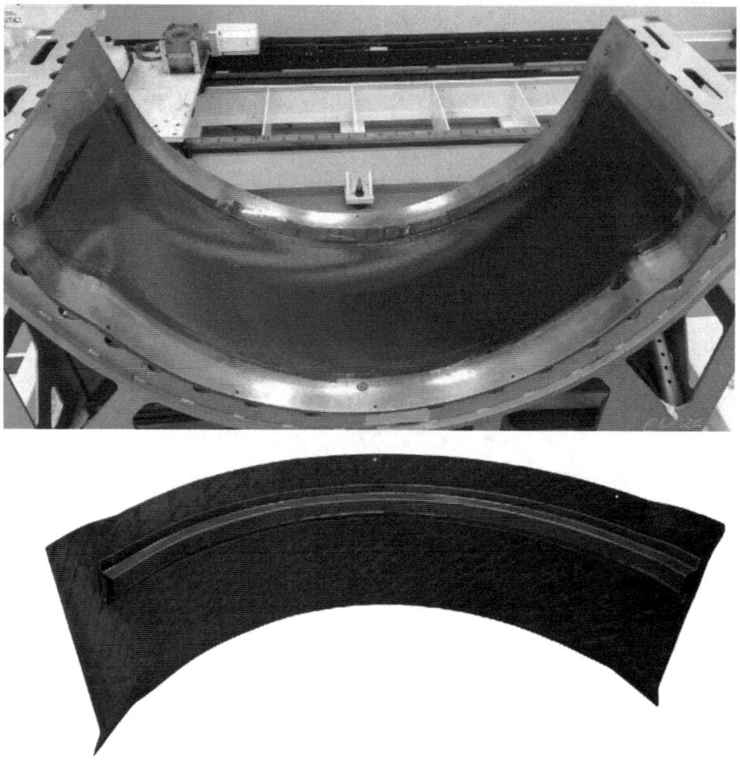

Figure 11 a) Final manufactured part, b) with bonded stiffener

4. CONCLUSIONS

Even though this work clearly shows the applicability of LAFP for complex 3D shaped parts, and the potential for in-situ only fiber placement, several issues need to be addressed before production can be realized. While the issues are fewer for concave tooled skin layups, this option requires settling for a less cost-efficient manufacturing solution.

The primary issues with the convex-tooled in-situ melt consolidated stiffened structure, not in any particular order, include: 1) print through of the release film borders/tape joints on the part, 2) print through of the overlaps/gaps required when converging zones on double-curvature parts, 3) print through due to variation in stamped part (i.e., frame flange) thicknesses that are melt consolidated, and 4) print through of tow splices. Solutions to each of these challenges are necessary across the industry. Indeed, we suggest that collaborative endeavors among the raw material, consumable and processing companies and institutions be encouraged to solve these universal problems in the thermoplastic composite supply chain.

5. REFERENCES

[1] Cartensen T. A., Townsend W., and Goodworth A., "Development and Validation of a Virtual Prototype Airframe Design as Part of the Survivable Affordable Repairable Airframe Program," Proceedings of the 64th American Helicopter Society Forum, Montreal, Canada, April 29-May 1, 2008.

[2] Salah L, Bahramshahi N, Ursenbach D, Merotte J, Mechanical Properties of Carbon-Fiber Thermoplastic Composites Manufactured by Laser Assisted Fiber Placement (LAFP). SAMPE Proceedings, Seattle, WA, May 2020.

[3] Di Francesco M, Veldenz L, Dell'Anno G, Potter K. Heater power control for multi-material, variable speed Automated Fibre Placement. Compos Part Appl Sci Manuf 2017;101:408-21. https://doi.org/10.1016/j.compositesa.2017.06.015.

[4] Stokes-Griffin CM, Compston P. A combined optical-thermal model for near-infrared laser heating of thermoplastic composites in an automated tape placement process. Compos Part Appl Sci Manuf 2015;75:104-15. https://doi.org/10.1016/j.compositesa.2014.08.006.

[5] Cartié D. Automated fiber placement of thermoplastic materials: effects of process parameters on the mechanical properties of the laminate, Long Beach: 2018.

[6] Merotte J, Green S, Padey D, Larroque G. PAEK composite prepreg: fast and efficient part manufacturing. JEC Compos Mag 2019.

[7] Merotte J, Green S, Padey D, Larroque G. Unique Polyaryletherketone (PAEK) Prepreg Allows high versatility in Composite Parts Manufacturing using Automated Layup, Anaheim: 2019.

[8] Merotte J, Cartié D, Henry K, Grohens Y, Hamlyn A. automated fiber placement of thermoplastic materials: pursuit of low porosity without the autoclave, Charlotte: 2019.

THE FUNDAMENTAL REQUIREMENTS TO ACHIEVE INDUSTRY 4.0

Paul M. Oberle
3D Infotech, Inc.
Irvine, CA

ABSTRACT

Sources of error (uncertainty) in any manufacturing or dimensional inspection process should be minimized to ensure reliable, consistent, and safe operation. The many possible sources of uncertainty [1] include environmental, mechanical, and human errors. There is a broadening opportunity to reduce or eliminate these uncertainties and engage people on the production floor to make significant productivity improvements that the entire enterprise can appreciate. Any manufacturing process (subtractive/additive/hybrid) benefits and we look at how Universal Metrology Automation® streamlines the way leaders implement an intelligent and rapid quality control process which scales with production need. The challenges related to developing integrated solutions and the methods/platforms available that reduce or eliminate risk will be discussed. There are conventional advantages available immediately to innovators. Those who have already begun pursuing Industry 4.0 [2] understand why metrology is the key to the "Smart Factory". Wherein automated metrology delivers objective data, instantly highlights trends and outliers, delivers intelligent guidance and decisions, and allows technicians to focus critical thinking and problem solving on meaningful events in the day-to-day production environment.

Some plant managers might see the metrology lab as the quality assurance cost center, but metrology is critical to production and product innovation. True it can be labor intensive but when paired with automation it optimizes how we leverage human talent and problem solving. Wasting talent on repetitive tasks is no longer a requirement, nor do labor-intensive activities generate a competitive edge. Leaders see this as an opportunity to empower skilled individuals towards activities that are value added to the organization and promote the digital intelligence that is your competitive advantage and achieves Industry 4.0 today. Can we call it Just in time [3] Quality?

Keywords: Automation, Metrology, Robot, Quality, Industry 4.0
Corresponding author: Paul Oberle, paul@3dinfotech.com

1. INTRODUCTION

Manufacturers make products; the goal is to deliver something that someone wants because that product will improve an aspect of life in some way. Anything a manufacturer can do to ensure customer satisfaction with adequate yields to support demand at the lowest cost of production is a competitive advantage. This may be obvious to many people.

Automation and robotics have been used by manufacturers for a very long time to address the market's demand for "better, faster, less expensive" – in 1961 General Motors became the first major manufacturer to use robotics in a production environment [4]. However, markets demand ever more complex parts in greater volume. Wouldn't it be nice if our machines could sense when they need to be re-calibrated and maintained? Industry 4.0 is focused on the optimization and maintenance of automated processes using sensing, analytics, and inter-machine data exchange to generate real-world improvement [5].

Limited only by the amount and quality of collected data, 3D measurement is the key to understanding how well something was made and how to correct the inevitable mechanical drift that occurs in any manufacturing line [6]. "Sensing" is the fundamental requirement to achieve Industry 4.0, more specifically 3D sensing. This is evidenced by the impact conventional quality control has in many stages of a product's lifecycle. If we could gather data near or at the same rate as production and if this high-fidelity data was distilled into a format that could be easily communicated with peers, or leveraged immediately within our CNC processes, the ability to implement optimal production becomes a reality and we have a smart factory.

This paper examines how Universal Metrology Automation® (UMA) streamlines the way manufacturing leaders may fully integrate dimensional inspection to realize the vision of Industry 4.0 and how it can scale with changing requirements. The challenges related to developing an integrated solution will be discussed along with the methods and platforms available that ensure success by reducing risk, complexity, and deployment time.

In any pursuit, having a map of where you are and where you want to be is the difference between reacting to, or controlling your destiny. The more detailed the map the better you can see opportunities or recognize if you are straying off-course before it's too late.

2. BACKGROUND

We will assume that you have some knowledge of measurement [7], its origins, standards, economic impact [8], and some of the most commonly used measurement tools available today. With the introduction of the Kings Foot in the late 1700's and then the Metric System [9], technological development of metrology was relatively sparse until development of the Michelson Interferometer in 1887. Roughly 70-years later in the 1960's we get the Coordinate Measurement Machine (CMM) and 30 years later CMMs persist but are being rapidly overtaken by portable, non-contact 2D and 3D vision systems ("3D scanning") [10]. Collectively anything that has to do with measurement including the Kings Foot is metrology equipment and in 2020 we have more metrology equipment options to choose from than ever before. The one common trend has been to capture more data, more frequently, and more precisely – 3D vision leads this trend [11].

Regarding non-contact 3D vision systems specifically, there are many viable 3D scanners from well-known companies like API, Creaform, FARO, Hexagon, Nikon, and ZEISS. Meanwhile, smaller companies from all over the world continue to challenge the status quo with 3D vision systems with ever improving performance/price ratios. "3D scanning" is the most rapidly growing segment in metrology and produces orders of magnitude higher resolution data than tactile alternatives.

3. METROLOGY EQUIPMENT

The term metrology has been used to describe the science of measuring and inspecting parts and assemblies. Typically, measurements can be taken by contact and/or non-contact methods. Contact methods can vary significantly, ranging from calipers and other hand-held gages that allow an operator to manually take discrete measurements through the process of observation to Coordinate Measurement Machines (CMMs) (Figure 1) which allow contact probes to locate X,Y,Z coordinates in 3D space manually or via Computer Numerical Control (CNC) devices, using software to record points. With hand-held manual gauges, recording measurements is typically done with the operator taking note of the measurements observed. Software driven metrology accomplishes this without the risk of error caused by manual entry. Software driven metrology also analyzes the data and present reports for part or process review, helping make decisions about the manufacturing process.

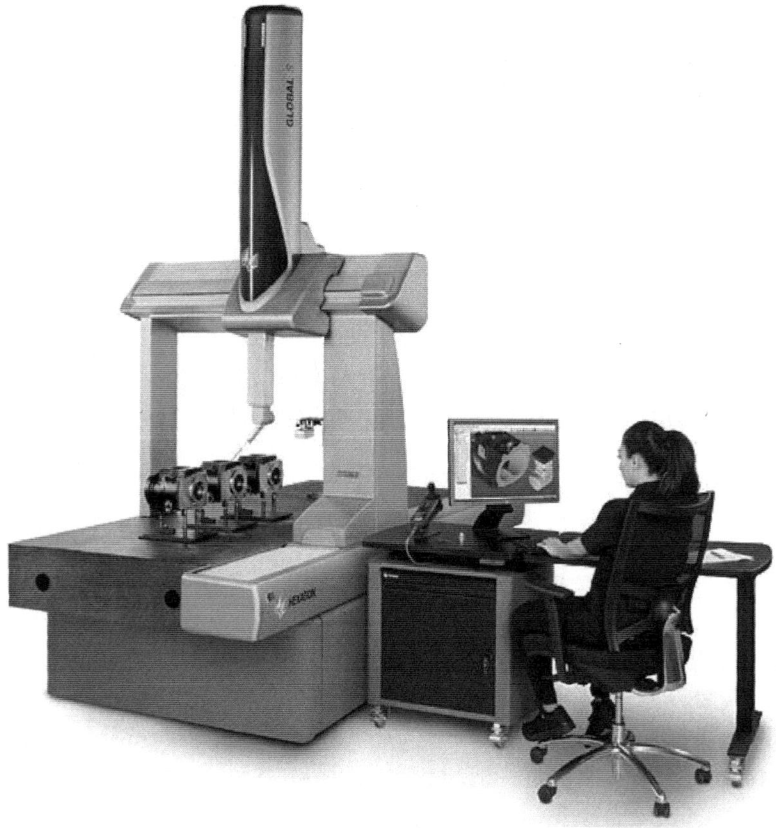

Figure 1 - Contact measurement with a CMM

As a natural phenomenon of technology evolution, metrology gauges have now become faster and more portable. By taking advantage of advances in laser and camera technology various commercially available devices have been released for industrial use. One type of laser scanner uses a single laser point that is projected onto a surface and observed by a camera at a known angle and base line distance from the laser source. This can be used to isolate the point of contact as an X, Y, Z coordinate. The same can be achieved by projecting a laser line (Figure 2) or a series of laser lines to gain more points over a greater surface area in a short time span. While another common tool called structured light (Figure 3) uses the combination of a camera and a projected pattern of light to triangulate 3D coordinates. This leap in data collection speed has helped reduce the time required to complete a measurement cycle and the complexity of integrating multiple laser scanners to compensate for an individual sensor's speed deficiency. Modern laser scanners allow an operator to simply wave the laser scanner over the part (Figure 4), as if they were "painting" the part, while visualizing the measurement points on a computer screen to see if they have achieved adequate coverage of the workpiece [12]. The generation of millions of data points per second has put pressure on downstream software used to process the data. Such software needs to be able to provide real time visualization of the data, which requires rapid polygonization and shading of the data using sophisticated algorithms that have been optimized using both the Central Processing Unit (CPU) and the Graphical Processing Unit (GPU) of the computer.

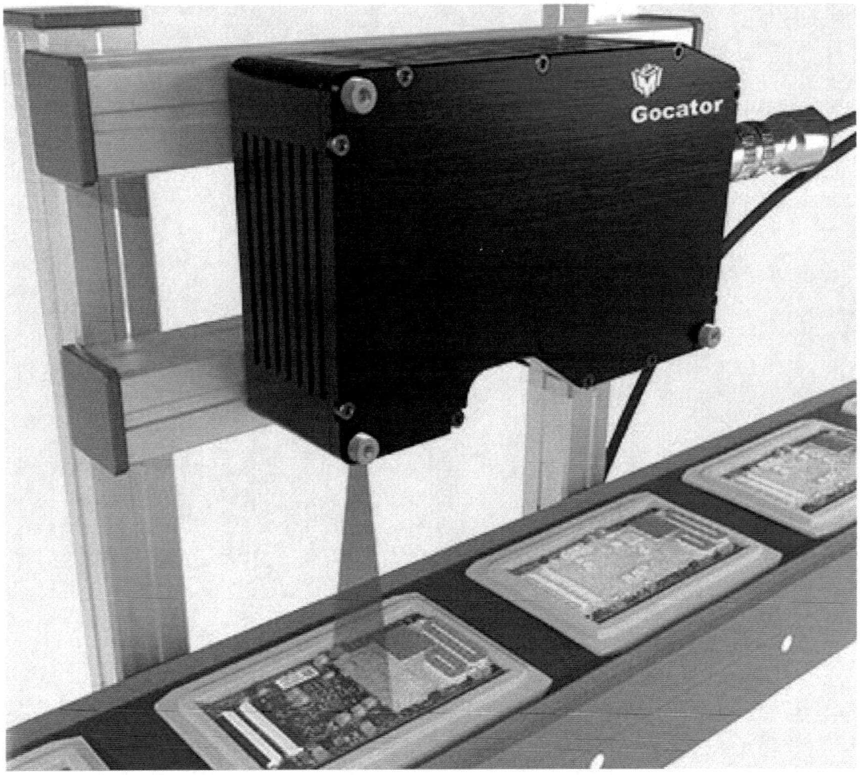

Figure 2 - Laser profile scanner used on a printed circuit board inspection application

Figure 3 - Structured light 3D scanner on a tripod measuring a compressor wheel

Figure 4 - Handheld 3D laser scanner inspecting an aerospace assembly

Traditionally laser technology has demonstrated some disadvantages that have impacted device selection for some applications. Lasers tend to flare as they bounce off the surface that is being measured. This results in noise which needs to be removed using noise-reduction algorithms. The remaining points are averaged to generate a single point, which is part of a series of points sampled in the laser profile. This results in the resolution or point spacing of the measuring device needing to be limited. In contrast, camera-based devices can be affected by highly reflective or matte black surfaces. A thin film or powder is applied to the surface temporarily, in some cases, to combat this issue (while "blue light" tends to perform better on shiny surfaces than "white light", filtering only for the blue color spectrum of light does not eliminate the problem). Camera based 3D scanners use pixels to evaluate the surface being measured. The higher the resolution of the camera, and smaller the area of measurement or Field of View (FOV), the smaller the point spacing that can be achieved. Such systems require both the part and the camera to be perfectly stationary while the image is being captured. The time to capture an image can range from milliseconds to 5 seconds per measurement. The length of time for exposure varies based on the luster or reflectivity of the surface. There continues to be innovation in the metrology sensor market yielding hundreds of potential options to choose from.

4. METROLOGY SOFTWARE

The existence of large amounts of data has resulted in the development of niche software for processing the data. Conventional Computer Aided Design (CAD) software is not capable of performing the tasks related to capture, visualize, and analyze such vast data sets. Some metrology equipment manufacturers develop their own proprietary software that is used with the hardware they supply (Figure 5). This software can have a cost advantage, however due to the rapid innovation in laser and camera technology proprietary metrology software could potentially lock a user into a product eco-system that does not always allow the user to choose the best tool for a specific application.

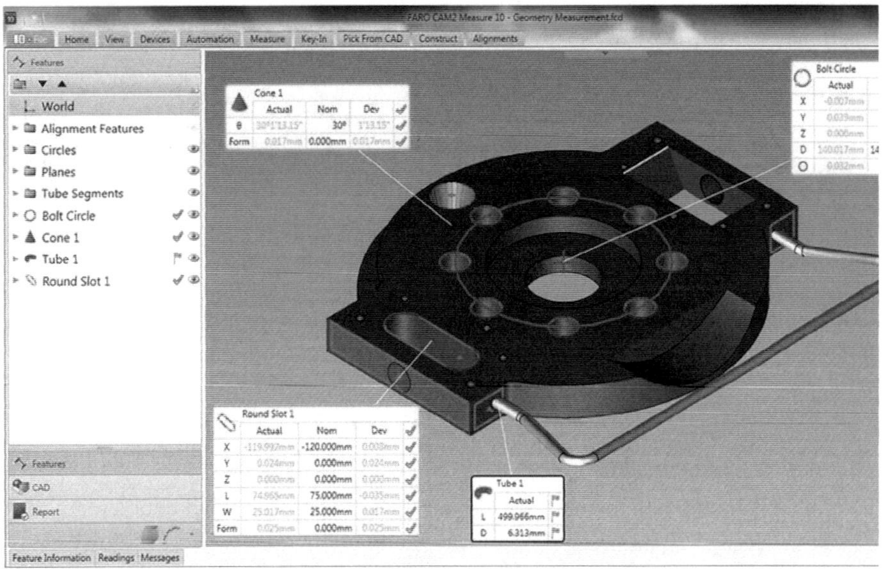

Figure 5 - Proprietary metrology software

PolyWorks® (Figure 6) has emerged as the metrology software of choice among leading aerospace and automotive companies for its breadth and depth of functionality related to metrology and quality control applications. This versatility includes the ability to work with native CAD geometry with Geometric Dimension and Tolerance (GD&T) information, perform a variety of data alignments, work with datum reference schemes, fit feature primitives, execute large data analysis, generate profiles and cross sections, work with 2D and 3D calipers, display thickness values, and calculate Statistical Process Control (SPC) to name a few.

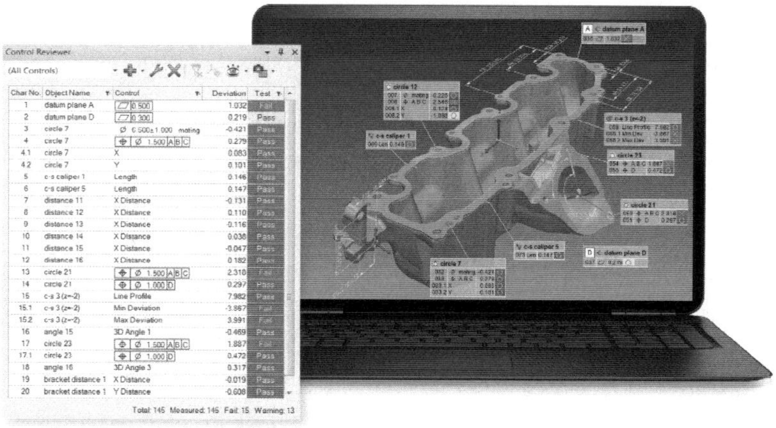

Figure 6 - 3D inspection in PolyWorks with feature and smartGD&T controls

PolyWorks software also has a plug-in architecture allowing third party software development to write macro scripts to automate existing functions or to use its native Software Development Kit (SDK) to write advanced customized routines that are typically not available in alternative metrology software. PolyWorks' versatility and power are ideal for deployment in hand-held laser scanning applications. Camera based scanners can also take advantage of PolyWorks, as an offline data analysis engine. The universal metrology platform of PolyWorks enables operators to be device independent and work with a wide variety of measurement devices [13].

5. SMART CAMERA

The term smart camera refers to a machine vision system (laser or camera based) used as part of a more complex automated system [14]. Smart cameras are used to capture images, extract application-specific information, process the information, and make decisions based on the results. Smart cameras are usually placed on a production line and compare real time images of products coming down the line to specific inspection criteria.

Machine vision systems for inspection can gauge hole diameters and feature distances, detect the presence or absence of key components, verify product alignment or orientation, and inspect for damages such as bends, cracks, and holes. The ability to complete the inspection tasks can be augmented with the addition of a robotic arm that can traverse around the part being inspected. Just as the robot assists the camera in doing its job, the camera also assists the robot, providing guidance for pick and place operations such as removing parts from bins, arranging parts on pallets, and finding and aligning assembly components.

6. METROLOGY AUTOMATION

The objective of automating metrology is to reduce or remove the bottleneck posed by manual (one-by-one) inspection of parts to check for defects. Besides providing a productivity gain through consistent results, metrology automation also helps mitigate some of the sources of uncertainty from production and inspection process. Reducing scrap and eliminating tedious, labor-intensive, repetitive inspection tasks are two of the obvious profit and safety improving benefits achieved with automated metrology.

Metrology automation addresses the low productivity hurdles many factories currently face in their quality control operations. By eliminating human error, mistakes, and mishandling of products, managers can have a higher level of confidence in production quality reporting [15]. This can help prioritize any issues that need to be investigated in the production process, provide faster response to customer complaints, reduce "red-tagging" of products, and improve shipment timelines.

7. 2D VS. 3D

2D automated systems can be extremely fast, accurate, cost effective, robust, and reliable. Machine vision can perform hundreds or even thousands of inspections per minute. The images captured have such a high resolution that details too small to see with the human eye can be easily evaluated. Certain measurements, such as distances and diameters, can be performed more quickly and accurately than a 3D system. Using machine vision for robot guidance also reduces the need for expensive hard tooling to maintain part position during inspection, enabling manufacturers to build multiple products on the same production line. Additionally, smart cameras can read barcodes, data matrices, and other alphanumeric strings. These identification capabilities provide traceability and lead to tighter process control. Presence or absence detection is performed much more quickly and reliably with machine vision than with a human.

3D measurement devices make up for the shortcomings of 2D devices. Metrology software allows for comparing the 3D data back to its original CAD model. This model can contain valuable product manufacturing information (PMI) as part of a collaborative product development system. A GD&T engine that strictly adheres to ASME Y14.5-2009 and ISO 1101 is an indispensable feature for 3D metrology software [16]. Advanced alignment capabilities allow the part to be viewed in various 3D orientations. These alignments can be used to mimic a virtual assembly between multiple components, mating them virtually to check for future interference issues or to analyze flush and gap deviations. Another fundamental tool for analyzing 3D data is the deviation map, which provides an easy to understand visual representation of how the part compares back to the original design. All these tools work together to make root causes in complex manufacturing processes much easier to identify.

8. AUTOMATION

Automation has been used to reduce human involvement in manufacturing processes. It has evolved through the office and factory alike with computer systems, software, CNC machines, motion control system, robots, and so forth [17]. Industrial robots (Figure 7) have emerged in the form of articulating arms that can be driven to any position within their reach to perform a variety of tasks. Material handling, painting, polishing, applying adhesive, welding, part assembly, and drilling are some of the most common applications robots have been deployed for in manufacturing. While mobile industrial robots are seeing rapid growth and have become practical contributors to material handling applications, in this section we will focus on articulating robotic arms.

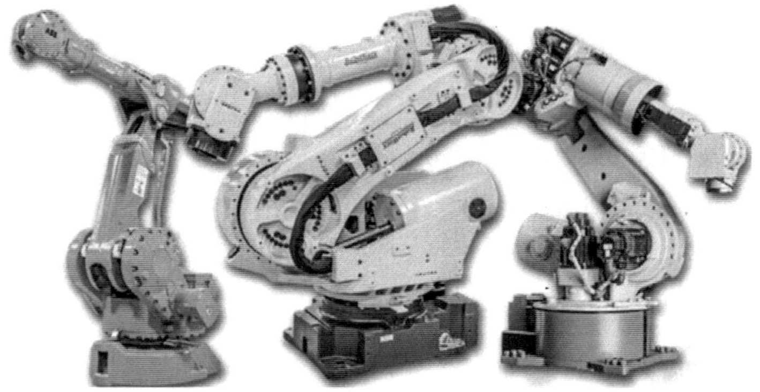

Figure 7 - A trio of industrial robots from different manufacturers

The two primary types of articulating robots are industrial and collaborative. Conventional industrial robots are typically programmed using a teach pendant (Figure 8) or a remote computer and deployed in a factory for heavy duty and high-speed applications. Such robots need to work in a protected area where operators are not allowed to enter, unless all systems are switched off. If an un-intended intrusion in the working area of the robot occurs, there is a possibility that the robot may collide with the operator or a workpiece. Since industrial robots are not inherently equipped with sensing capabilities accidents can be sudden and costly, in the worst case fatal.

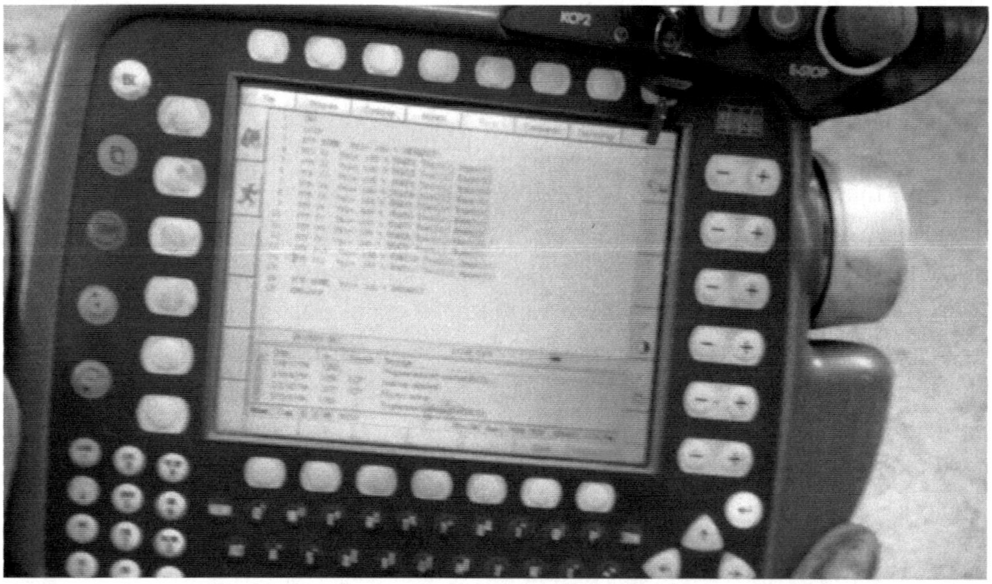

Figure 8 - Industrial robot teach pendant with programmed sequence on screen

The need for a more flexible and human friendly robot has given rise to collaborative robots (also referred to as "cobots") (Figure 9) that have certain safety features and sensing capabilities built in [18]. Each joint has sensors and brakes, so that if a collision occurs, the robot immediately stops. The sensitivity can be set in such a way that neither the operator, nor the part would be harmed.

Figure 9 - Collaborative robots being programmed by two colleagues in a factory

Both types of robots typically have six joints and operate much like a human arm, from shoulder to wrist, with 360 degrees of motion in each joint. By attaching an End of Arm Tool (EOAT), the robot can act as a human with a hand, holding a tool that an person might hold when performing a manual operation. There are many types of EOATs available and new ones can be designed, depending on the intended use of the robot. As a matter of fact, the advent of 3D printing has made design, fabrication, and testing of new EOATs fast and easy.

For metrology, the EOAT consists of a scanner and a scanner mounting bracket attached at the end of the robot arm. Each robot comes equipped with a controller with its own Operating System (OS). There is support for software programming and communication with other devices using Modbus or TCP/IP protocols. If the foundational software on the robot controller is not enough for a task, supplementary communication protocols can be provided by PC based software connected via an ethernet cable.

9. AUTOMATION SOFTWARE

Traditionally, robots have been controlled using the teach pendant and manually programmed to go to a series of locations and execute commands. The teach pendant is connected to a robot controller capable of sending the robot to any position in 3D space within its functional reach. In order to calculate the position of the robot, a series of joints work in conjunction using Inverse Kinematics (IK) to map the location of the robot end point. This calculation is done behind the scenes, inside the controller, without involvement by the operator.

Sequential programming with the robot controller is effective for discrete applications with limited variables such as picking and placing parts, drilling holes, or dispensing adhesives in the same place each time. However, for applications that include more variables (like environmental variables which cannot be simulated fully), teach pendant programming is time consuming, cumbersome, and in some cases entirely in-effective.

Metrology automation is one such application where sequential programming with the teach pendant and robot controller leaves much to be desired. Here use of a PC workstation connected to the robot controller provides an inherent advantage with superior computing power over the controller. The application software running on the PC must be capable of performing the tasks of the teach pendant, sensing, synchronizing the joint positions from the robot, and maintaining the kinematic chain of all the joints in order to ensure precision in the robotic measurement process.

Custom software has been used in the past for individual projects. The advantage of a custom software is that when designed by process experts it can be an effective way of both accomplishing a discrete task as well as locking down the interactive components in an automated metrology application to perform only the task its deployed to do without deviation. The problem is that within the product lifecycle of most industries today revisions will be made to improve the product, fix an issue that is impacting the product's function, or an entirely new product line might be introduced. It is profitable for the developer of these custom systems when changes occur because this typically creates additional on-site work that can demand a significant amount of billable time and resources.

Streamline© automation software (Figure 10) has been developed to offer flexibility, broader functionality, and ease-of-use for the end-user [19]. All common industrial operator/administrator paradigms are included with permission control, whereby an advanced user can be authorized to create new inspection routines while plant floor operators may be restricted to only run existing routines. The touch-screen interface makes it easy to navigate through a host of part numbers and "play" the inspection routine or execution of the inspection routine can be selected and driven by barcode. Streamline is versatile in that it supports a multitude of robots and metrology devices, both 2D and 3D. A typical workflow for an operator is shown in Figure 11 below.

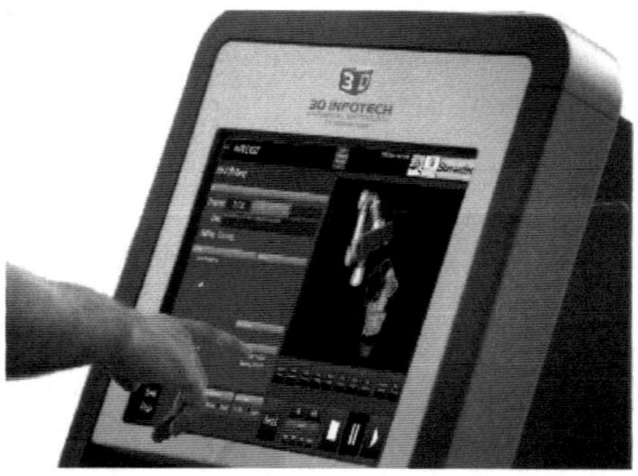

Figure 10 - Touchscreen interface of Streamline automation software

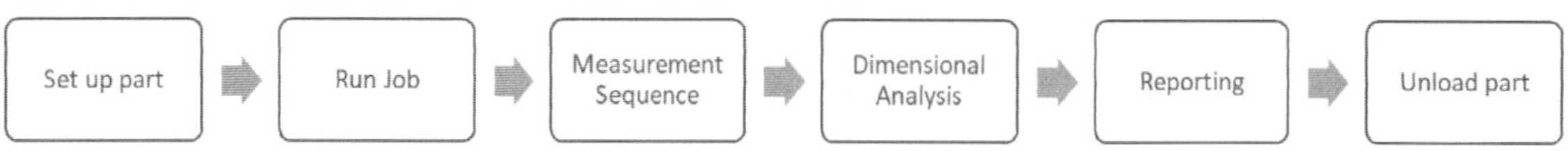

Figure 11 - Automated quality control process workflow

All this is done seamlessly without the need for operator knowledge of any programming or technical inspection coding. The operator "plays" the job in Streamline, which then automatically triggers the robot to perform the measurement sequence. Once measurement is complete, Streamline automatically triggers PolyWorks® to perform the dimensional analysis and reporting. Upon completion, the part and its critical features are validated, and the operator is prompted to unload the part.

Streamline supports additional axis/robot(s) for part manipulation, including motion control via a CNC controlled rail or rotary stage. Figure 12 below shows an example of a system that has a rotary stage and an inverted rail that provides operators with 8-axis of motion to measure any product that can fit within the measurement volume and be supported on the rotary stage, including the complex part shown.

Figure 12 - Operator loading a workpiece for automated inspection

In addition to keeping track of inspection routine progress/status, Streamline maintains a history of completed inspections by piece per part number scanned. This offers traceability and statistical process control over a batch of parts. This can be easily and automatically exported to MS Excel, a third-party data analysis software, or even a database on the network. Maintaining a digital thread on the network is key to performance evaluation and control, and highly encouraged in every deployment of automated metrology.

10. OPERATOR GUIDANCE/MACHINE VISION FOR PART PLACEMENT

With automation in general the spatial coordinates of the target object/workpiece relative to the robot is critical. In metrology automation this remains true, however there is more flexibility on part placement compared to a machining operation, for instance. Depending on the type of 2D/3D vision system used, this flexibility in workpiece placement (Figure 13) can be range between an inch (sometimes more) to a few millimeters based on the 3D vision system's field of view. It is probable that the operator may not load a workpiece in the same exact position every time (as a robot would) before running the routine to inspect it. This variation in the location of the workpiece can result in a lack of data collected in specific areas. Strategies can be used to reduce the risk of placement variation. One method is to use hard fixturing that ensures repeatable workpiece placement, another is to use visual guides that the operator uses for workpiece positioning.

Streamline has this functionality built-in (Figure 14). By using a series of images and integrated messages, the operator is instructed to place the part consistently where it belongs.

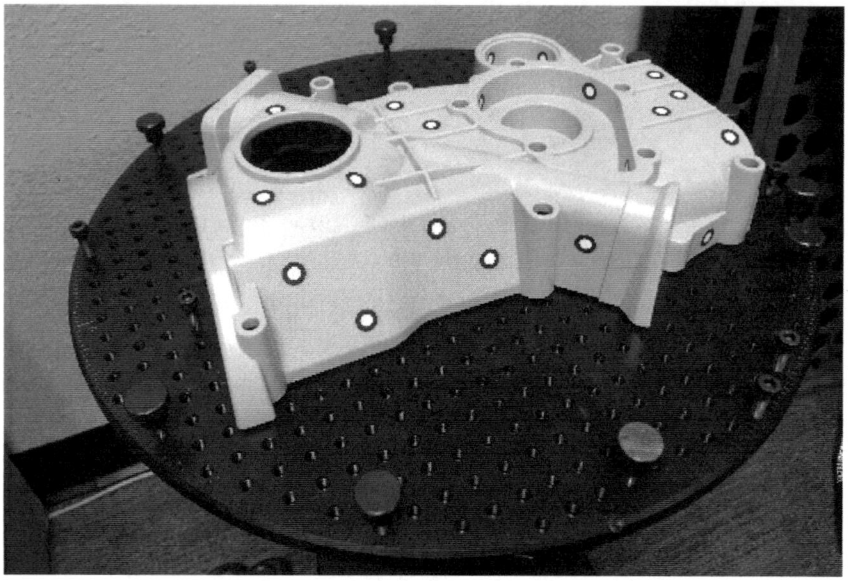

Figure 13 - Workpiece placement in on a rotary stage

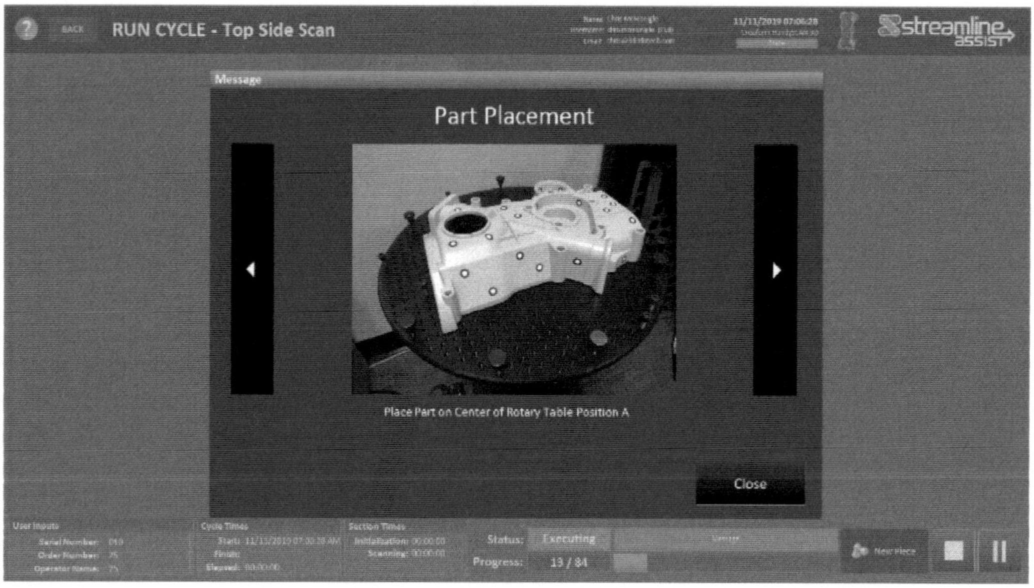

Figure 14 - Visual workpiece placement definition and guidance

Although helpful, these methods may not be enough because there is no absolute requirement that the part be precisely in the desired location prior to scanning. Rather than being limited by any given sensor technologies, the part placement module of Streamline makes precise placement easy to train and implement at will. An external camera evaluates the position of the workpiece and presents a green light to the operator once the part is in the correct position/orientation (Figure 15). During the workpiece loading process, the operator is given quantitative and qualitative visual information to inform how much and in what direction (if at all) the workpiece needs to be repositioned (Figure 16). Camera selection is important for the accuracy of the workpiece placement algorithm. While a web camera or equivalent is supported, using an industrial machine vision camera with superior pixel level accuracy provides significantly better capability to confirm a repeatable part position.

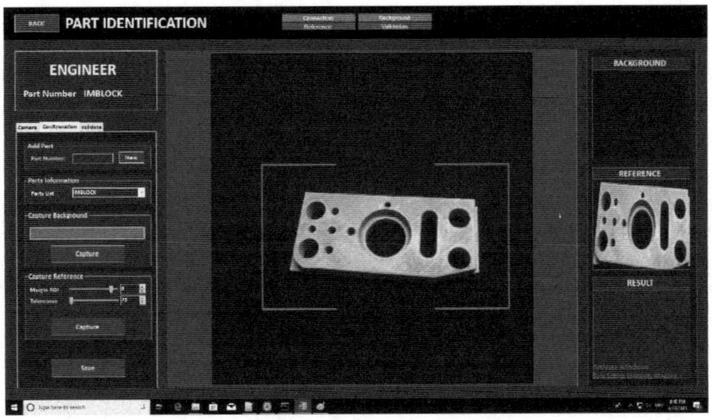

Figure 15 - Teaching proper workpiece position, camera learns position and automatically detects edges

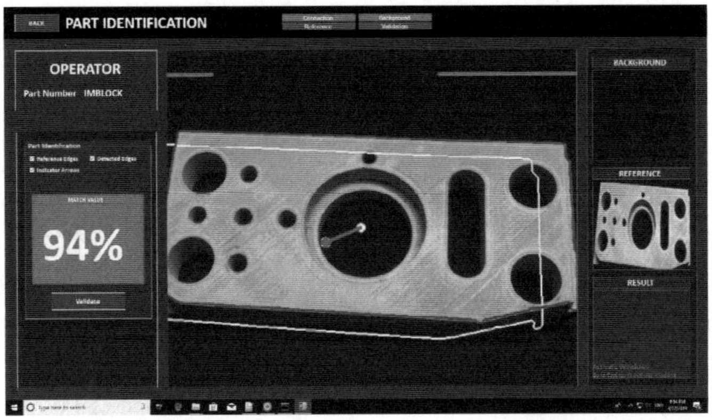

Figure 16 - Dynamic visual workpiece guidance with camera-based edge detection and user defined acceptable tolerance

The images above illustrate how the software is taught the ideal position for workpiece placement, the very first time. The image captured becomes a reference image. Various setting parameters can be adjusted to ensure background noise is eliminated and ambient lighting conditions are managed accordingly. When a new part is placed near the same location, the camera captures the image at a rate of 60Hz providing real-time feedback to the operator regarding the workpiece positioning with respect to the reference location recorded. Edge detection algorithms display the outline of the critical features of the part, while a guidance arrow directs the operator to move the part in a certain direction and amplitude. In the image above, the green 94% confidence indicates that the part is placed within the acceptable postional tolerance. This tolerance is customizable for different types of setups.

11. PROJECTION OF RESULTS

Certain applications require (or benefit from) marking for re-work. It is also helpful to be able to communicate the inspection results directly on the part (Figure 17). This can be accomplished by using augmented reality software to project the graphical and numerical information directly on the part. By measuring the workpiece and comparing it to the CAD nominal data, the deviation values can be calculated and then projected back onto the workpiece [20]. The images that follow show an example of a metal product that has some defects (Figure 18).

Figure 17 - Features with numerical results projected

Figure 18 - Graphical color map projection

The projection system can be integrated directly with the scanning system as a second EOAT on the robot. The image below shows one example how the scanner and the projector are on the same robot, in Figure 19 below the scanner and projector are mounted at the same time but at 90 degrees relative to each other. Once the measurement and inspection is completed, the robot goes back over the workpiece and projects the results directly to the areas of concern (Figure 20).

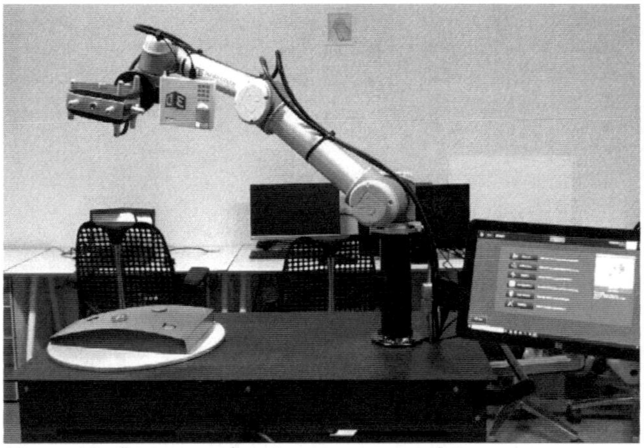

Figure 19 - Scanner and projector on same robot

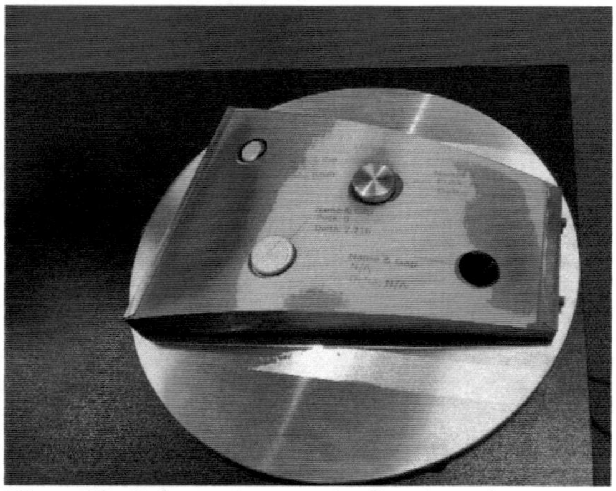

Figure 20 - Robot projects results directly on the part

12. ROBOT GUIDANCE FOR MEASUREMENT PATH CREATION

Current robotic measurement systems rely on an existing robot path to perform the scanning operation. This requires an advanced user to program the robot, typically using numerically indexed joint positions, and set-up programs based on unique part dimensions and characteristics. This activity can also be accomplished using the workpiece itself, the operator may interactively move the robot over the workpiece within the field of view of the attached 3D vision system on the end of the robot and record the robotic motion to be played back for the inspection/measurement task (Figure 21).

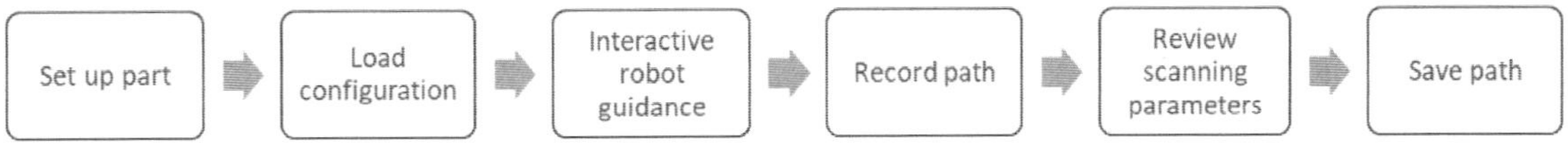

Figure 21 - Workflow for interactive robot teaching of measurement path

Alternatively, the robotic scanning path can be generated off-line using a CAD model of the part if a robust kinematic chain and 3D model of the overall robotic cell is available (Figure 22). This requires the availability of the workpiece CAD model.

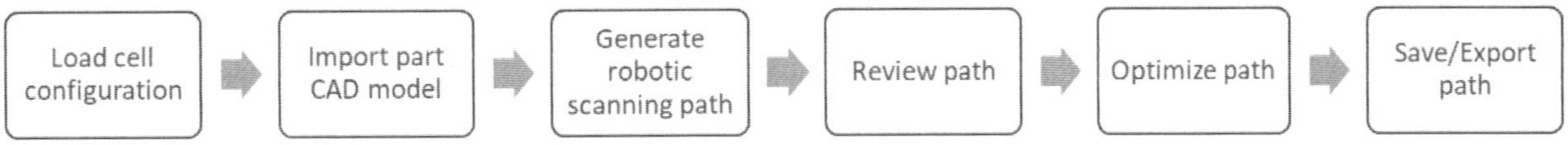

Figure 22 - Workflow for CAD based simulation of measurement path

This approach allows the complete simulation of the inspection job prior to measuring a real workpiece. Simulation is an effective method of creating inspection programs. However, a virtually taught robot program should be tested for safety and efficacy. For instance, a 3D vision system's field of view can be described numerically, however, the way in which the laser or structured light camera responds to the angle of orientation of the EOAT and the surface properties of a workpiece cannot be simulated completely. At the very least, it is considered best practice for an experienced operator to carefully observe and be prepared to make optimizations to a robotic path that has been simulated, "Trust but verify" is the operative principle.

While all these approaches are proven and allow seamless execution of an inspection program, there are possibilities to further automate and improve the process. Let's consider a scenario where a workpiece has been placed in the work area and the robot has been taught to measure the part (Figure 23) When the second piece is placed, it is possible that the position is not in the exact same location (Figure 24).

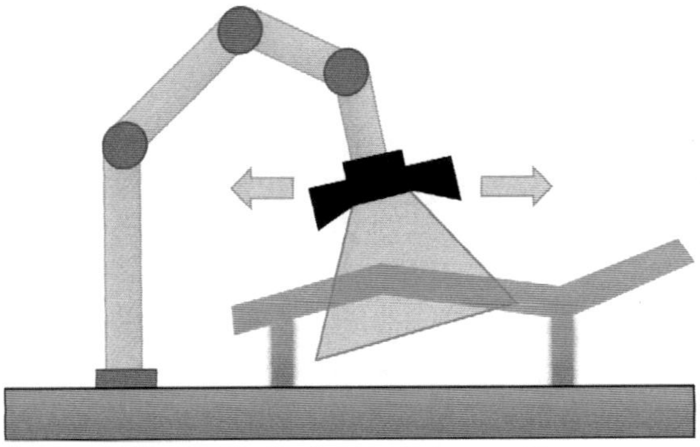

Figure 23 - Perfect workpiece position

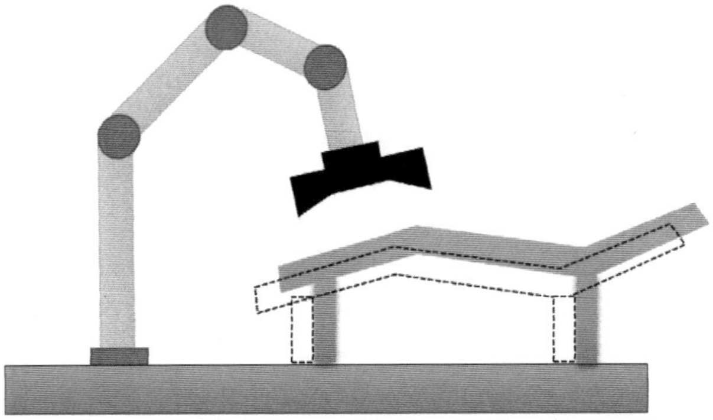

Figure 24 - Workpiece position not ideal or displaced

Ideally, the robot is capable of "self-adjusting" to the change in workpiece position. That way the operator should not have to worry as much about accurate part placement, as long as the part is in the working volume of the robotic scanning system. By using additional vision cameras in the inspection cell, the robot path can be adjusted dynamically to adapt to a shift in the workpiece that has been placed in the work area (Figure 25). If a path for the workpiece being inspected already exists, then the robot path can be adapted based upon auxiliary vision or sensor feedback to compensate for the shift in location of the workpiece.

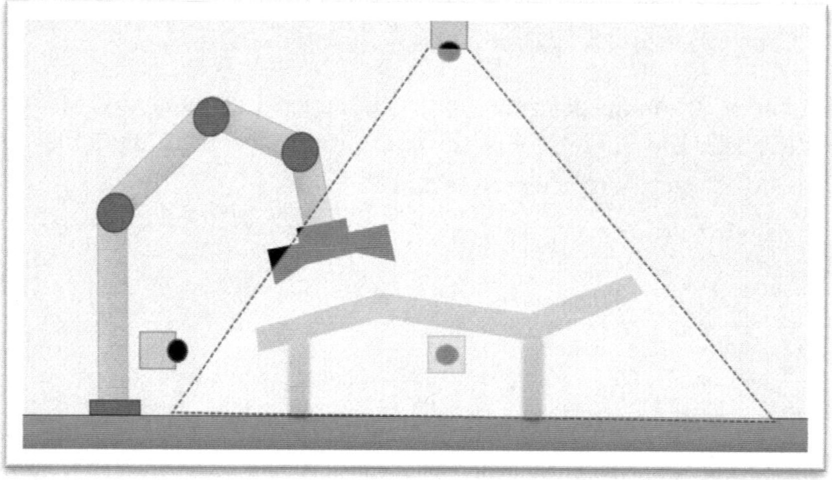

Figure 25 - Robotic inspection cell with vision system to analyze the workpiece position in measurement volume

The algorithm that needs to be developed in order to accomplish this uses silhouette lines of the part from each view and combines them into a single 3D polygonal mesh. Such an algorithm can be accurate enough to generate a robotic scanning path, but insufficient to be used for effective inspection of the part.

The 3D vision system is still required to ensure accurate measurement data is collected for inspection of parts. Simply using three images may not be enough to capture the details on every type of part. Therefore, it is important to place the part on a rotary stage that allows three images to be captured at regularly spaced rotary angles. For example, a user might setup a sequence for images to be collected every 10 degrees. This will result in 36 sets of images, or a total of 108 images that need to be merged into a single polygonal model. This is possible based on the tests that have been conducted by 3D Infotech in its research and development department.

The machine vision cameras used for the image capture do not need to be extremely expensive, but rather, just need to have accurate pixilation. Depending on the level of detail in the family of parts, 4MP cameras (for example) may be sufficient to calculate the 3D polygonal mesh.

Once the mesh has been calculated, it can be used to offset an existing robotic scanning path or generate a new one. For new parts, this will be extremely helpful because the new method will employ the best of both the current approaches that are available today. Since an actual part is being used, versus a CAD model, there will be a better correlation of the path to the real world, eliminating the need to verify the simulated robotic scan path. Additionally, since there is a digital replica of the part as a polygonal mesh, there won't be any further need to interactively teach the robot any new scanning paths. This sophisticated level of automation is driven primarily through software development, testing, and optimization of algorithms. The hardware components are required primarily in the testing phase.

13. MACHINE LEARNING

The image data being collected for robotic guidance can be used for building a data model for machine learning. This can be particularly useful for visual inspection, as opposed to dimensional inspection. Machine learning is the scientific study of algorithms and statistical models that computer systems use to perform a specific task without using explicit instructions, relying on patterns and inference instead. It is seen as a subset of artificial intelligence [21]. Machine learning algorithms build a mathematical model based on sample data, known as "training data", in order to make predictions or decisions without being explicitly programmed to perform the task. Machine learning algorithms are used in a wide variety of applications, such as computer vision, where it is difficult or infeasible to develop a conventional algorithm for effectively performing the task.

Machine learning is closely related to computational statistics, which focuses on making predictions using computers. The inspection data that is collected by the automated metrology system can also be used to focus on exploratory data analysis through unsupervised learning. This can be used for predictive analytics of parts that are being manufactured.

14. CONCLUSION

Metrology automation offers significant advantages over conventional methods of inspection. By using a robotic system, human operators are freed up from tasks subject to human error and repetitive process fatigue, allowing them to pursue other tasks such as investigative work, trouble shooting, communication, and product development. Allowing the robot to take over mundane, repetitive tasks will create an inspection process that fosters increased productivity and eliminates bottlenecking in quality control, while providing highly repeatable, high-quality, results.

Metrology automation can be further improved by implementing procedures such as operator guided part placement and robot guided measurement paths. The process of evaluating defects or performing rework can be communicated directly on the part with augmented reality. These technologies, as discussed in this paper, make it easier for operators to do their work and reduce the risk of mistakes. Assuming the volume of parts being measured daily is significant, smarter sensors, both 2D and 3D, will generate large volumes of actionable data that is essential for machine learning and to capitalize on the full value of metrology at all stages of the process intelligence pyramid (Figure 26). In addition, machine learning is the foundation for autonomously guided robots, which can be crucial in finding process level problems. With machine learning algorithms, it is possible to gain increased control over the manufacturing process, accelerate and improve manufacturing decision-making, and strategically re-deploy precious human resources to the tasks they excel at.

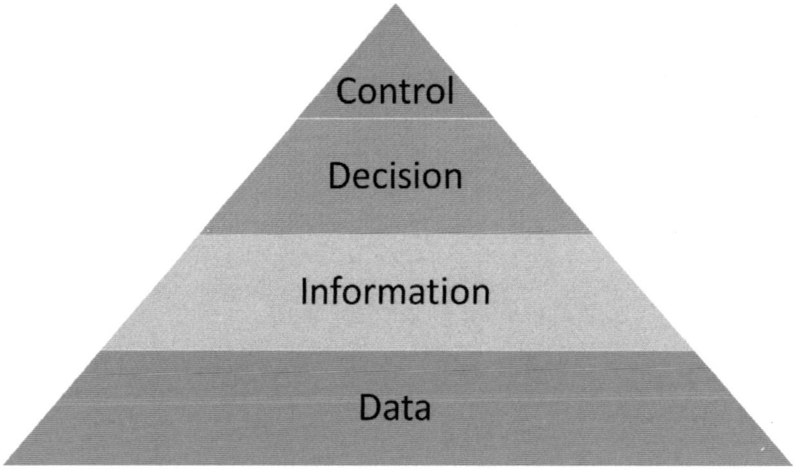

Figure 26 - Process intelligence pyramid

15.REFERENCES

[1] JCGM, "Guide to the Expression of Uncertainty in Measurement (GUM)," September 1995 - 2012. [Online]. Available: https://www.bipm.org/en/publications/guides/gum.html.

[2] B. S. Mark Cotteleer, "Forces of change: Industry 4.0," 18 December 2007. [Online]. Available: https://www2.deloitte.com/us/en/insights/focus/industry-4-0/overview.html.

[3] D. K. N. B. H. F. T. Akbar Javadian Kootanaee, "Just-in-Time Manufacturing System: From Introduction to Implementation," vol. 1, no. 2, 2013.

[4] "History of Industrial Robots," , . [Online]. Available: https://www.robots.com/education/industrial-history. [Accessed 19 2 2020].

[5] H. . Lasi, P. . Fettke, H.-G. . Kemper, T. . Feld and M. B. Hoffmann, "Industry 4.0," *Business & Information Systems Engineering,* vol. 6, no. 4, pp. 239-242, 2014.

[6] K. G. Harding, "Challenges and opportunities for 3D optical metrology: what is needed today from an industry perspective," *Proceedings of SPIE,* vol. 7066, no. , p. , 2008.

[7] P. . Howarth and F. . Redgrave, Metrology - in short, ed., vol. , , : Danish Fundamental Metrology & National Physical Laboratory, , p. .

[8] H. G. Semerjian and R. L. Watters, "Impact of measurement and standards infrastucture on the national economy and international trade," *Measurement,* vol. 27, no. 3, pp. 179-196, 2000.

[9] "History of Metrology," , . [Online]. Available: http://www.msc-conf.com/history-of-metrology/. [Accessed 19 2 2020].

[10] Creaform, "The history of metrology from Galileo to optical systems," 25 October 2016. [Online]. Available: https://www.creaform3d.com/blog/the-history-of-metrology-from-galileo-to-optical-systems/.

[11] B. . Breuckmann, "25 Years of High Definition 3D Scanning: History, State of the Art, Outlook," , 2014. [Online]. Available: https://dl.acm.org/citation.cfm?id=2737773. [Accessed 19 2 2020].

[12] M. . Givi, L. . Cournoyer, G. . Reain and B. J. Eves, "Performance evaluation of a portable 3D imaging system," *Precision Engineering-journal of The International Societies for Precision Engineering and Nanotechnology,* vol. 59, no. , pp. 156-165, 2019.

[13] M. . Buttigieg, "PolyWorks at NASA Armstrong Flight Research Center," , 2016. [Online]. Available: https://ntrs.nasa.gov/search.jsp?r=20160010347. [Accessed 19 2 2020].

[14] Wikipedia, "Wikipedia," 9 October 2019. [Online]. Available: https://en.wikipedia.org/wiki/Smart_camera.

[15] J. Martin, "Metrology Automation is Here and Now," 1 April 2014. [Online]. Available: https://www.qualitymag.com/articles/91770-metrology-automation-is-here-and-now.

[16] G. . Ameta, R. R. Lipman, S. P. Moylan and P. . Witherell, "Investigating the Role of Geometric Dimensioning and Tolerancing in Additive Manufacturing," *Journal of Mechanical Design,* vol. 137, no. 11, p. 111401, 2015.

[17] S. . Shipp, N. . Gupta, B. . Lal, J. A. Scott, C. L. Weber, M. S. Finnin, M. . Blake, S. . Newsome and S. . Thomas, "Emerging Global Trends in Advanced Manufacturing," , 2012. [Online]. Available: https://wilsoncenter.org/sites/default/files/emerging_global_trends_in_advanced_manufacturing.pdf. [Accessed 19 2 2020].

[18] K. Pittman, "A History of Collaborative Robots: From Intelligent Lift Assists to Cobots," 28 October 2016. [Online]. Available: https://www.engineering.com/AdvancedManufacturing/ArticleID/13540/A-History-of-Collaborative-Robots-From-Intelligent-Lift-Assists-to-Cobots.aspx.

[19] Design News, "Golden Mousetrap Awards," 2019. [Online]. Available: https://goldenmousetraps.designnews.com/2019-winners.

[20] K. Mills, "PolyWorks Plug-in Provides AR Results Projection," 18 September 2018. [Online]. Available: https://metrology.news/polyworks-plug-in-provides-ar-results-projection/.

[21] J. . Bloom, "Industrial Machine Learning," , 2017. [Online]. Available: http://dblp.uni-trier.de/db/conf/kdd/kdd2017.html. [Accessed 19 2 2020].

[22] United Nations Industrial Development Organization, "Emerging trends in global manufacturing industries," United Nations Industrial Development Organization, Vienna, 2013.

[23] K. Mills, "Automated Metrology Software - Market Report Available," 12 December 2016. [Online]. Available: https://metrology.news/metrology-software-market-report-issued/.

[24] Markets and Markets, "3D Metrology Market worth $16.2 billion by 2024," Markets and Markets, Northbrook, 2019.

[25] O. Fräfel, "Automation: what it means for companies," 20 April 2018. [Online]. Available: https://www.credit-suisse.com/ch/en/articles/private-banking/automatisierung-was-sie-fur-unternehmen-bedeutet-201804.html.

[26] "World Robotics 2016; Executive Summary World Robotics 2016 Industrial Robots," , . [Online]. Available: https://ifr.org/img/uploads/Executive_Summary_WR_Industrial_Robots_20161.pdf. [Accessed 19 2 2020].

Bonding to Bulk Metallic Glass using Aerospace-Grade Structural Adhesive

Rose Roberts[1], Elizabeth Kidd[1], Evan Hilgemann[2], and Giles Dillingham[1]

[1]BTG Labs

Cincinnati, OH

[2]Jet Propulsion Lab, California Institute of Technology
Pasadena, CA

ABSTRACT

Bulk metallic glasses, also known as amorphous metals (AMs) are non-crystalline metals with growing applications in aerospace. Due to their amorphous nature, AMs generally exhibit stronger mechanical properties than most crystalline alloys and are more corrosion resistant. The limited reactivity of a corrosion resistant surface may explain the difficulty in establishing robust bonds to certain adhesives, especially commonly used structural adhesives for aerospace applications. Here, an amine cured aerospace adhesive was used for bonding an iron-nickel based AM satellite component. Various types of processes were explored to avoid adhesive failure and increase bond strength. It was found that the AM required specific chemical derivatization of the metal surface in order to achieve cohesive failure using the epoxy adhesive. Surface analysis was used to determine the mechanisms of adhesion enhancement. This work helps explain the difficulties in bonding AM materials and provides strategies for design of future bonded AM components.

Corresponding author: Giles Dillingham

1. INTRODUCTION

The ability to directly image earth-like planets in neighboring solar systems is limited by glare of stars washing out the signal from orbiting planets. This glare must be suppressed in order to image planets capable of supporting life. NASA Jet Propulsion Lab's Starshade project aims to use a large (20-30 m diameter) shade to block out the light of a star and enable an orbiting space telescope to image surrounding planets. The shape of Starshade is precisely controlled to minimize glare caused by both diffraction of starlight as well as scatter of solar light along its perimeter [1].

The edges of the Starshade must also be robust to fracture and plastic deformation, and be able to withstand cold temperatures, radiation, and other conditions experienced in space. Amorphous metals are highly resistant to corrosion and have mechanical properties fitting for this application [2,3]. Due to the lack of grain boundaries inherent in amorphous metals, the material can be precisely chemically etched and meet strict micron level precision requirements [4].

However, bondline failures have been observed when adhering the amorphous metal foil to a carbon fiber substrate [5]. Poor adhesion is defined as interfacial failure between the substrate and the adhesive, also called adhesive failure. Good adhesion is defined as cohesive failure, or failure within the adhesive so that adhesive can be observed on both sides of the failed bond. There are several aerospace-grade structural adhesive epoxies commonly used for similar applications, such as LOCTITE® EA 9394, that exhibit good adhesion for most substrates. However, it has been found that EA 9394 exhibits poor adhesion to amorphous metal. Cleanliness, surface treatments, roughness, metallurgical/surface composition, crystallinity, silane primer coatings, and an alternative adhesive were used to investigate the cause of poor adhesion between EA 9394 and JPL's amorphous metal used for the Starshade project.

2. EXPERIMENTATION

2.1 Materials

Amorphous metal (AM) foil was purchased from Metglas (ferrous-based, zirconium-based, and titanium-based; 0.0381 mm thick) and spring steel (0.0508 mm thick) was purchased from McMaster Carr. Bonds to aluminum coupons were made using LOCTITE® EA 9394 epoxy adhesive and ZAP brand Zap-A-Gap PT-02 cyanoacrylate glue. Silane primers included 3-aminopropyltriethoxsilane and *n*-2-aminotheyl-3-aminopropyltrimethoxysilane, purchased from Gelest. Water (HPLC grade), methanol (≥99.6%), acetone (ACS grade, ≥99.5%), isopropanol (ACS grade, ≥99.5%), and ethanol (ACS grade, ≥99.5%) were purchased from Sigma Aldrich.

2.2 Methods

2.2.1 Amorphous Metal and Spring Steel Foils Surface Preparation

Several methods were pursued to prepare the AM and spring steel foils for bonding. AM foils were treated using solvent cleaning, plasma treatment, grit blasting, silane primer, and/or crystalline-inducing temperatures. Spring steel was treated only with silane primer after solvent cleaning. All surfaces were characterized using water contact angle (WCA) before treatment to track homogeneity of surface cleanliness.

2.2.2 Bonding Peel Test Specimens

Peel test specimens were assembled by first preparing the foils with the appropriate treatment (described in Section 2.2.1 above and Table 1). Aluminum coupons were cleaned with 4 unidirectional wipes of acetone followed by 4 uniaxial wipes of ethanol or isopropanol, with a final abrasion step using stearate-free 180-grit sandpaper on a pneumatic hand sander. Aluminum coupons were monitored with WCA to ensure a strong adhesive-aluminum bond. WCA was 25° or lower on coupons before bonding. Adhesive was then applied to the aluminum coupon. EA 9394 was spread using a wooden stir stick while the cyanoacrylate adhesive wet out the surface of the coupon without any additional mechanical spreading techniques. The aluminum coupon was then pressed onto the prepared AM or spring steel surface until squeeze-out (adhesive) appeared on all sides. Squeeze-out was removed using the edge of a razor blade. Approximately 100 grams of mass was applied to the specimens to prevent void formation during curing. Specimens bonded using EA 9394 were cured at 45 °C for 2 hours and specimens bonded with cyanoacrylate adhesive were cured at room temperature overnight.

Table 1. Treatment and Adhesive Systems for Bonding Peel Specimens

Foil Material	Adhesive	Treatment
Fe-based AM	EA 9394 epoxy	Solvent cleaning (acetone and isopropanol)
		Plasma treatment
		Grit-blasting
		Induced crystallinity (annealed between 400-800 °C)
		Silane (primary amine and secondary amine types)
	Cyanoacrylate	Solvent cleaning (acetone and ethanol)
		Plasma treatment
Zr- and Ti-based AM	EA 9394	Solvent Cleaning
		Plasma treatment
Spring Steel	EA 9394	Plasma treatment
		Silane (secondary amine type only)

2.2.3 Qualitative Peel Tests

Peel tests were performed qualitatively since orders of magnitude differences were necessary for the application and the importance of failure mode is greater than overall peel strength. Differences between small and large amounts of force required was recorded along with failure mode. A cohesive failure mode (failure within the adhesive rather than the substrate-adhesive interface) was considered ideal, as this indicates that the weakest part of the bond lies within the adhesive rather than the interface.

2.2.4 Equipment

Water contact angle was measured using a Surface Analyst 3001. Fourier Transform Infrared Spectroscopy was conducted using a Thermoscientific Nicolet with a grazing angle reflection accessory (Harrick Scientific). Plasma treatment was performed using a PlasmaTreat Openair-Plasma system with a rotating nozzle. X-ray photoelectron spectroscopy (XPS) was performed using Service Physics SSX-100 ESCA with an aluminum anode and 800 μm spot size. Grit-blasting was performed using filtered compressed air and 220 μm aluminum oxide media.

3. RESULTS

3.1 Adhesive Performance of Solvent Cleaned, Plasma Treated, and Grit-Blasted Fe-AM Foil

Because poor or inconsistent adhesion is frequently due to surface contaminants, the cleanliness of the Fe-AM foil was investigated first. WCA, FTIR-RAIRS, and XPS were used to determine whether any chemical species known to interfere with adhesion could be identified on the surface.

3.1.1 Water contact angle

Because water contact angle is sensitive to monomolecular levels of contaminants, it is a convenient indicator of surface cleanliness. WCA was measured on as received samples, after solvent cleaning, after plasma treatment, and after grit-blasting. Plasma treatment will both clean the surface of small amounts of residual organics and dehydrate the oxide. In addition to removal of organic contaminants, grit blasting will remove pre-existing oxide, providing a thin, rough, native oxide for bonding.

Figure 1 shows the water contact angle response of the surfaces to treatment. Solvent cleaning with IPA and acetone did not significantly change the WCA, indicating that the as received surfaces were relatively clean, and that the contaminants were not readily removable by solvent wiping. Plasma treatment decreased the WCA to about 10° and grit blasting decreased the WCA to 5°, indicating an increase in the number of active polar groups, most likely -OH.

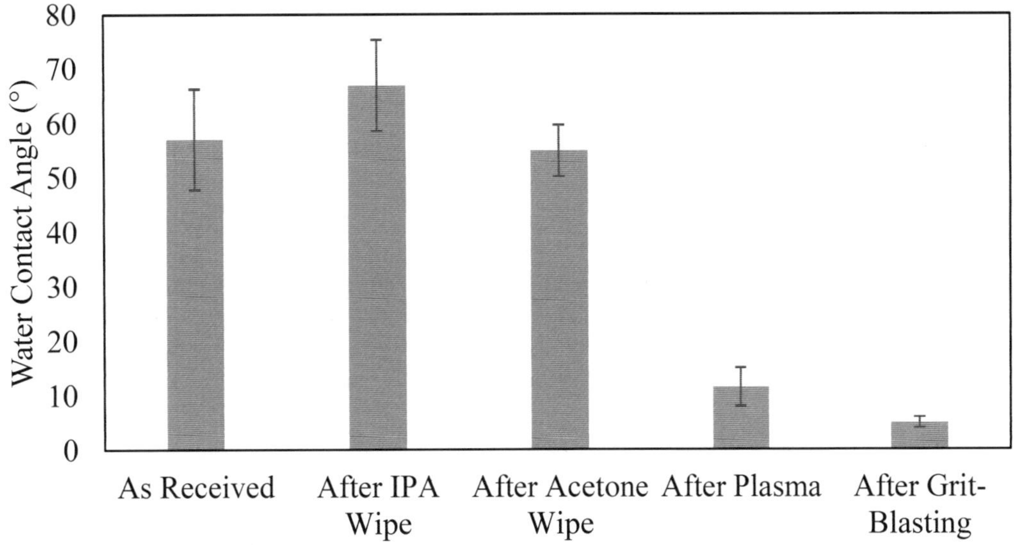

Figure 1. Water contact angle of Fe-AM as received, after solvent cleaning, after plasma treatment, and after grit-blasting. Error bars represent standard deviation.

3.1.2 FTIR-RAIRS Analysis

Fourier transform reflection-absorption infrared spectroscopy (FTIR-RAIRS) is sensitive to monolayer levels of organic molecules on metal surfaces. Spectra (Figure 2) of as received Fe-AM foils show no measurable amount of contamination on the surface of the foils unless intentionally contaminated, for example by a fingerprint. In this case, small peaks associated with hydrocarbons appear near 2900 cm^{-1}.

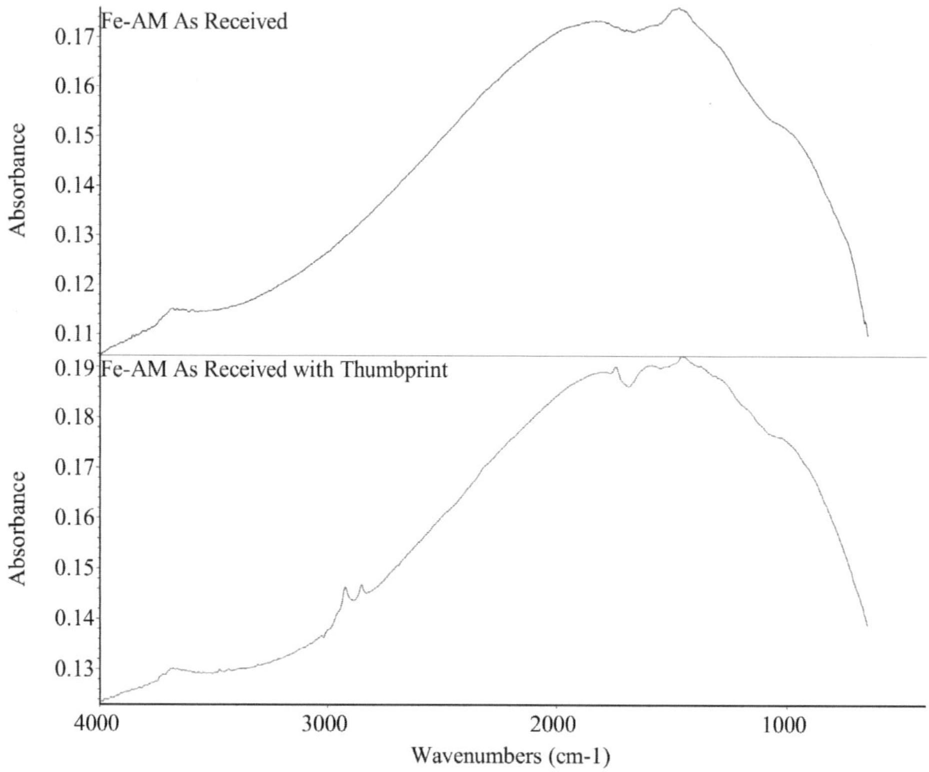

Figure 2. FTIR-RAIRS spectra of Fe-AM.

3.1.3 Adhesive Performance of Clean, Plasma Treated, and Grit-Blasted Fe-AM Foils

After cleaning, plasma treatment, or grit blasting, Fe-AM foils were bonded using EA 9394, cured, and peeled. All samples exhibited complete interfacial failure at the foil-epoxy bond, meaning no epoxy remained on the foil surface after peeling. Figure 3 shows a representative image of the failed samples as well as a grit-blasted surface. It can be observed that the texture of the foil has been imprinted on the epoxy, thus the epoxy wetted to the surface of the foil, however a strong bond failed to form.

3.1.4 XPS Analysis of Failed Surfaces

In order to better understand the cause of poor failure mode of the epoxy-AM bond, as received and failed foils were analyzed in XPS to determine whether any contamination was present that was not measurable in FTIR and whether a non-visible layer of epoxy remained on the surface of a failed foil. Resulting atomic compositions are given in Figure 4.

Carbon and oxygen were present in amounts consistent with a metal surface exposed to atmosphere, for both as received and failed foil surfaces. Other elements besides nitrogen are present in amounts similar to values expected based on the composition of the foil given by the manufacturer. Nitrogen should not be present on the surface of a clean foil. However, a relatively large amount of nitrogen (8% of the surface elements) is present on the foil side of the failure surface. This indicates that a residue from the epoxy remains on the foil after being bonded and peeled.

Figure 3. Representative image of A) an interfacially failed peel test using EA 9394 and cleaned Fe-AM foil. A grit-blasted Fe-AM foil (B) and associated peel test (C) are also shown.

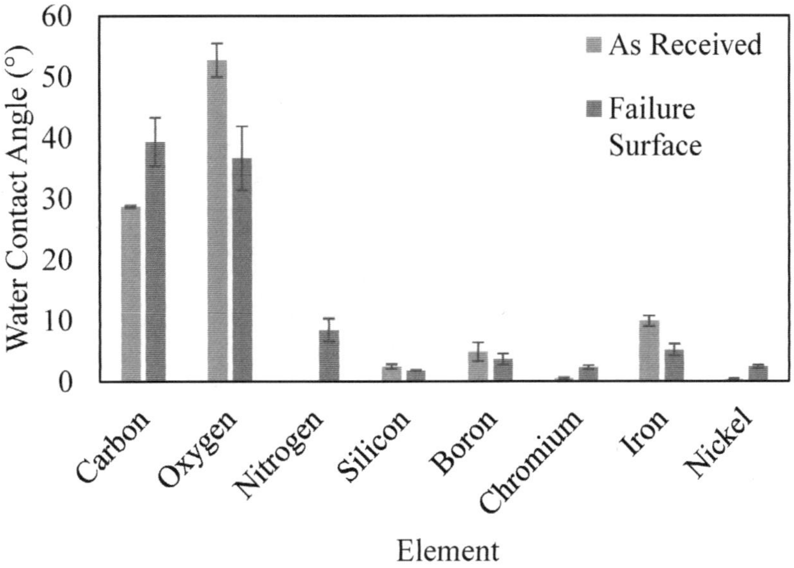

Figure 4. Atomic composition of the surface of as received Fe-AM and the Fe-AM failure surface via XPS analysis. Each bar represents the average of three locations, error bars representing the standard deviation.

Like many other structural epoxy adhesives, EA 9394 uses an amine driven mechanism to cure the epoxy [6,7]. Based on an estimate of the stoichiometry of DGEBA/amine reactions, the expected atomic % of nitrogen in the cured epoxy is approximately 2%. A large excess of nitrogen on the surface indicates that uncured amines have segregated to the foil surface. As boron is known to complex with amines, a theory is that the nitrogen interacts with boron groups on the surface causing it to preferentially adsorb to the surface. This would remove curing agent from the near-surface regions of the adhesive and prevent full curing. To test whether boron was interfering with the near-surface epoxy curing reaction, foils that did not contain boron were explored.

3.2 Adhesive Performance using Zr- and Ti-based AM Foils

Zirconium- and titanium-based foils used did not contain boron, according to the manufacturer. Foil samples were bonded as received, after solvent wiping, and after plasma treating. Fe-AM and food-grade aluminum foil were used to compare failure modes and peel strength.

Images of the failure mode are given in Figure 5. Failure mode for all samples appeared to be interfacial and peel strength was similarly low compared with the Fe-based foil. Aluminum foil also failed interfacially, however a little more force was needed for peeling compared with the AM foils. This suggests that boron at the surface is not interfering with epoxy curing and it is not the composition of the amorphous metal that is causing poor adhesion to EA 9394. The composition of the adhesive was explored next.

3.3 Adhesive Performance of Cyanoacrylate Adhesive

Cyanoacrylate adhesives can be strong, however they do not use amine chemistry to cure. While the application of the AM prevents the final use of cyanoacrylates in the field, this proof-of-concept experiment was meant to determine if other families of adhesives are able to bond to the AM foils.

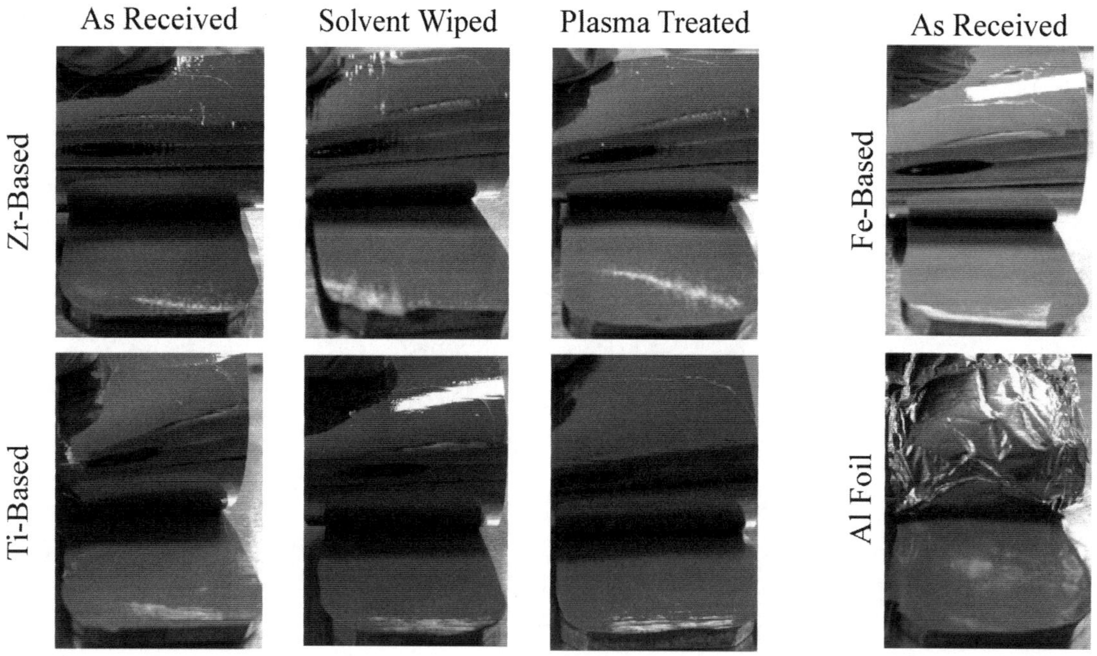

Figure 5. Peel tests of as received, solvent wiped, or plasma treated Zr- and Ti-based amorphous metal foils. Fe-based AM and aluminum foil also represented as control samples.

Cyanoacrylate was used to bond clean and plasma treated Fe-AM foil. Images of failure mode are provided in Figure 6. Cohesive failure was achieved, with adhesive remaining on both the foil and aluminum substrate. A significant amount of force was required to peel the foil from the coupon compared with EA 9394. Samples that were not plasma treated showed better adhesive performance than plasma treated samples. It is likely that plasma dehydrated the surface, and cyanoacrylates require some humidity to cure.

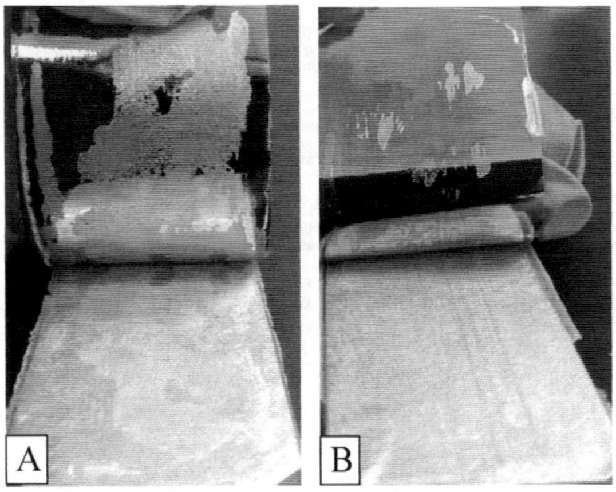

Figure 6. Peel test of A) solvent cleaned and B) plasma treated Fe-AM foil using cyanoacrylate adhesive.

3.4 Adhesive Performance of Crystallinity-induced Fe-AM Foil

Surface cleanliness and metallurgical composition of the surface do not increase adhesion performance. It has been shown that crystallographic orientation can change wettability of liquids on a surface [8]. This can be attributed to active groups available on a surface based on the crystal orientation. However, it was postulated that because there are no crystalline grain boundaries for the epoxy to bind to, perhaps the lack of crystallinity is causing too few attachment points to achieve a good bond. It was attempted to induce crystallinity in the foils to determine whether this was the cause of poor adhesion. Foils were annealed at 400 (10 or 30 min), 600 (10 or 30 min), or 800 °C (3 hrs, then left in oven as it cooled overnight).

Figure 7 shows surfaces of the foils after recrystallization. No visual changes to the surface were observed for foils annealed at 400 °C. A color change on samples annealing at 600 °C indicate that an oxide layer may have formed. The foil annealed at 800 °C showed heavy oxidation (rust), of which some was able to be removed with careful abrasion before bonding.

Bonded samples behaved differently depending on the temperature of annealing. Samples annealed at 400 °C peeled easily and interfacially, in a manner similar to untreated AM. Samples annealed at 600 °C took some force to peel away from the epoxy, however became quite brittle and the foil would break apart in small sections. The colored coating that had formed on the foil was removed with the epoxy, indicating a weakly bound oxide layer had formed during annealing that did not improve adhesion. The sample annealed at 800 °C showed improved adhesion, including some cohesive failure and additional strength to peel. However, the physical changes to the foil are not conducive for the optical properties needed for the Starshade application.

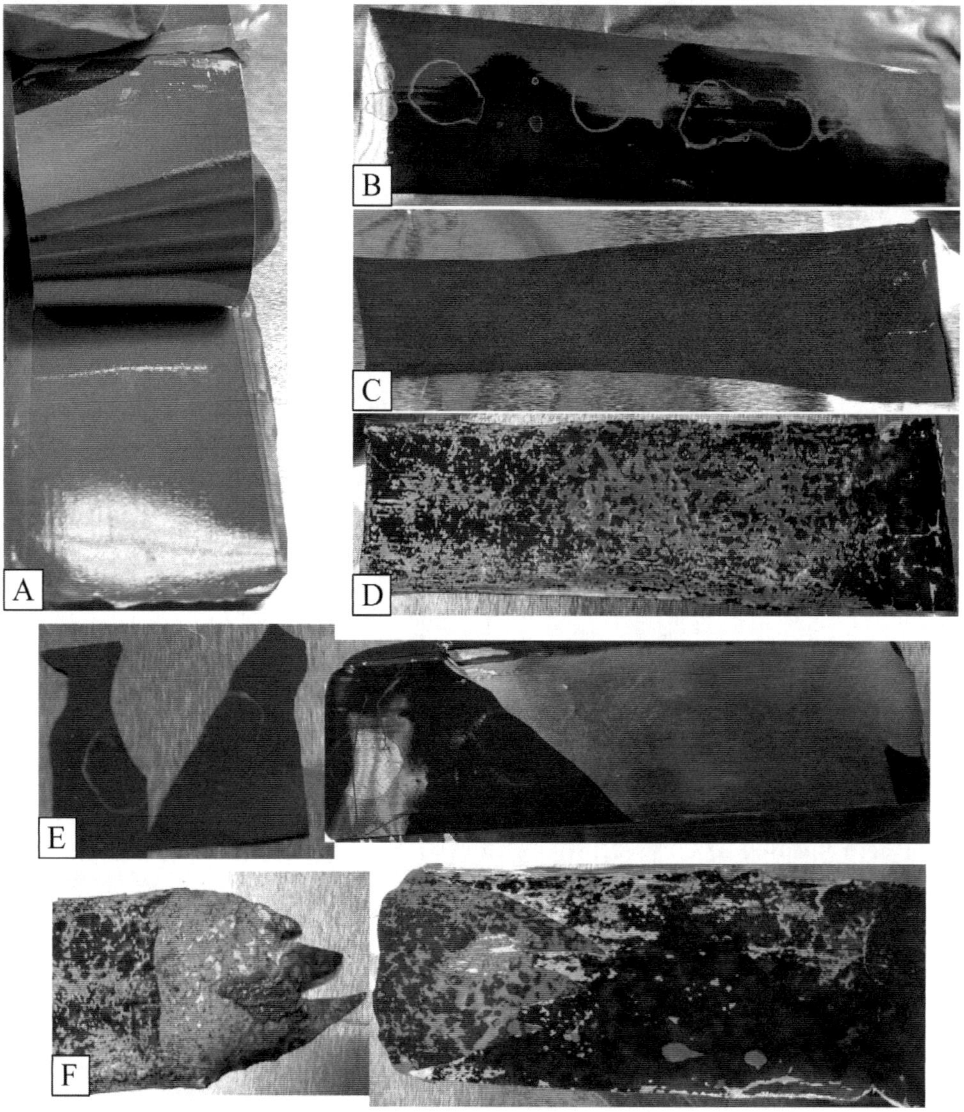

Figure 7. A) 400 °C for 10 min; no visual changes in surface, no increased adhesion, B) 600 °C, 30 min; spots are from water contact angle (removed color), C) and D) 800 °C, 3 hrs, before and after rust removal, E) 600 °C, 30 min; brittle failure of the foil, oxide remains on epoxy, F) 800 °C, 3 hrs; failure of the foil, oxide remains on epoxy, some epoxy remains on foil.

3.5 Adhesive Performance of Silane-primed Foils

Using a conversion coating on the surface to increase chemical bonds to the adhesive were explored using silane primers. 3-aminopropyltriethoxysilane (contains a primary amine) and *n*-2-aminotheyl-3-aminopropyltrimethoxysilane (contains both primary and secondary amines) were used to explore changes in adhesion. Figure 8 shows silane coatings on foils.

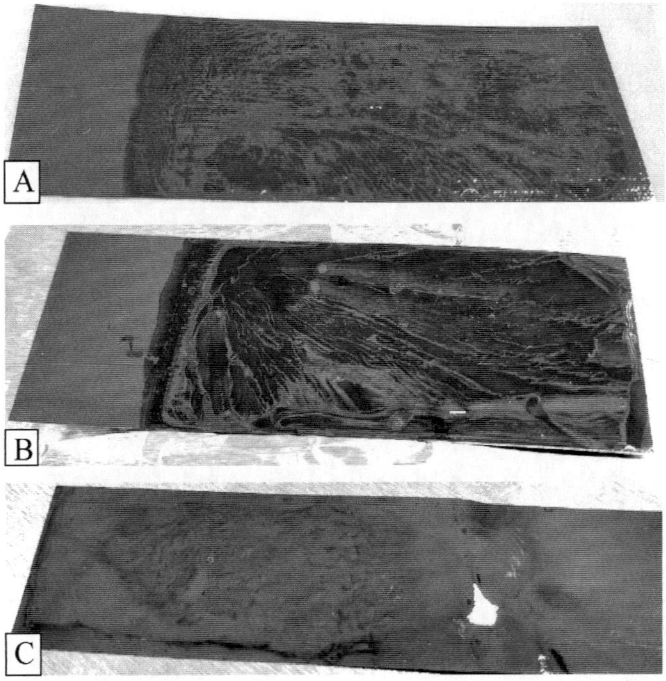

Figure 8. Silane coatings on Fe-AM foils that produced the best adhesion performance. A) air-cured diamine silane coating, B) oven-cured diamine silane coating, and C) air-cured primary amine silane coating. Note: the white spot on the air-cured primary amine sample is a glare, not due to the silane coating.

Various conditions were used to achieve increased crosslinking and thickness of the silane layer. FTIR-RAIRS was used to track amount of crosslinking, based on Si-O-Si peak frequency shift, and thickness of silane layer, based on area under the peak. Figure 9 shows some of this data, and marks which conditions (amount of crosslinking and layer thickness) produced the best adhesion performance. Adhesion performance ranged from no change in adhesion performance to large amounts of cohesive failure and increased strength (Figure 10) immediately after normal curing procedures. It is interesting to note that several weeks after the bond was initially peeled, the cohesive failure mode changed into adhesive failure mode. Figure 10 also shows a sample that was partially peeled immediately after bonding, then four weeks later was peeled fully. This implies that the cohesive failure mode may have been caused by the silane layer interfering with the ability of the epoxy to cure at the surface, increasing the time to achieve a full cure.

With some conditions of creating the silane layer, it was observed that a layer of tacky material was present between the epoxy and the foil. When peeled, this layer could be heard audibly de-sticking to the surface. When pressed back to the surface, the tackiness remained and the foil regained some adhesion to the epoxy surface. Figure 11 provides images of the tacky layer, seen moderately on the Fe-AM foil, and a much thicker layer on spring steel. Some tackiness remained at least 10 days after initial cure, although strength may have decreased slightly.

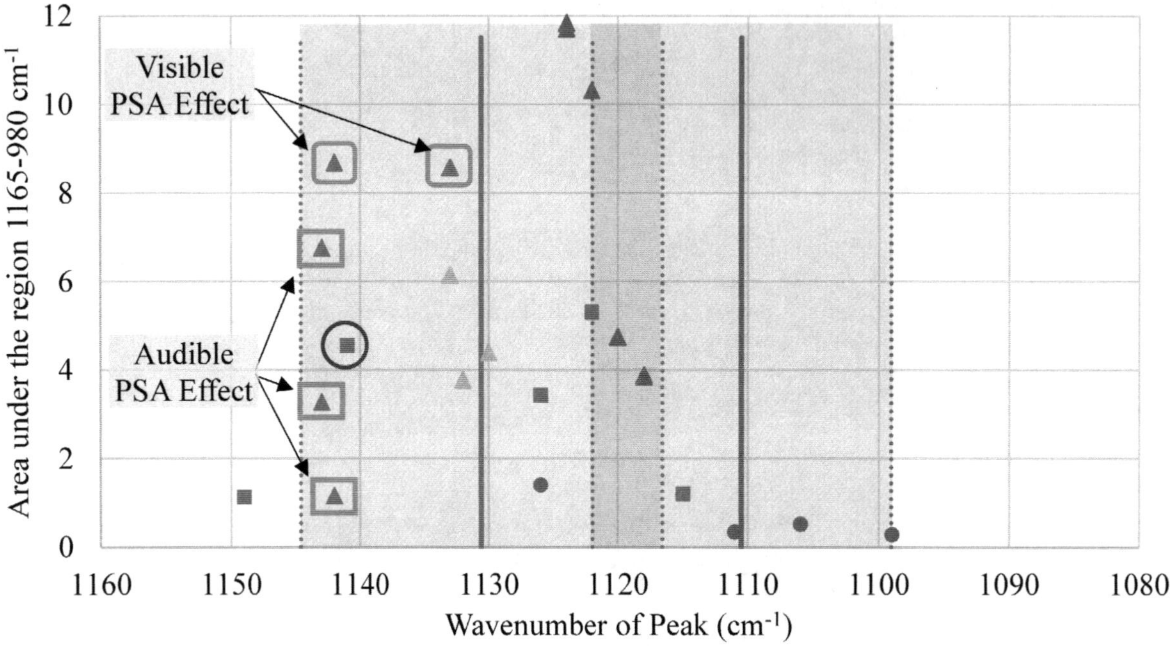

Figure 9. Peak frequency compared with area of the peak of cured silane layer on foils using FTIR-RAIRS. Solid vertical lines represent average peak frequency of samples giving best (green) and worst (blue) adhesion performance. Shaded areas, bound by dotted lines, represent the standard deviation around the average. The circled green square is the sample that showed the most cohesive failure and required the most strength to peel.

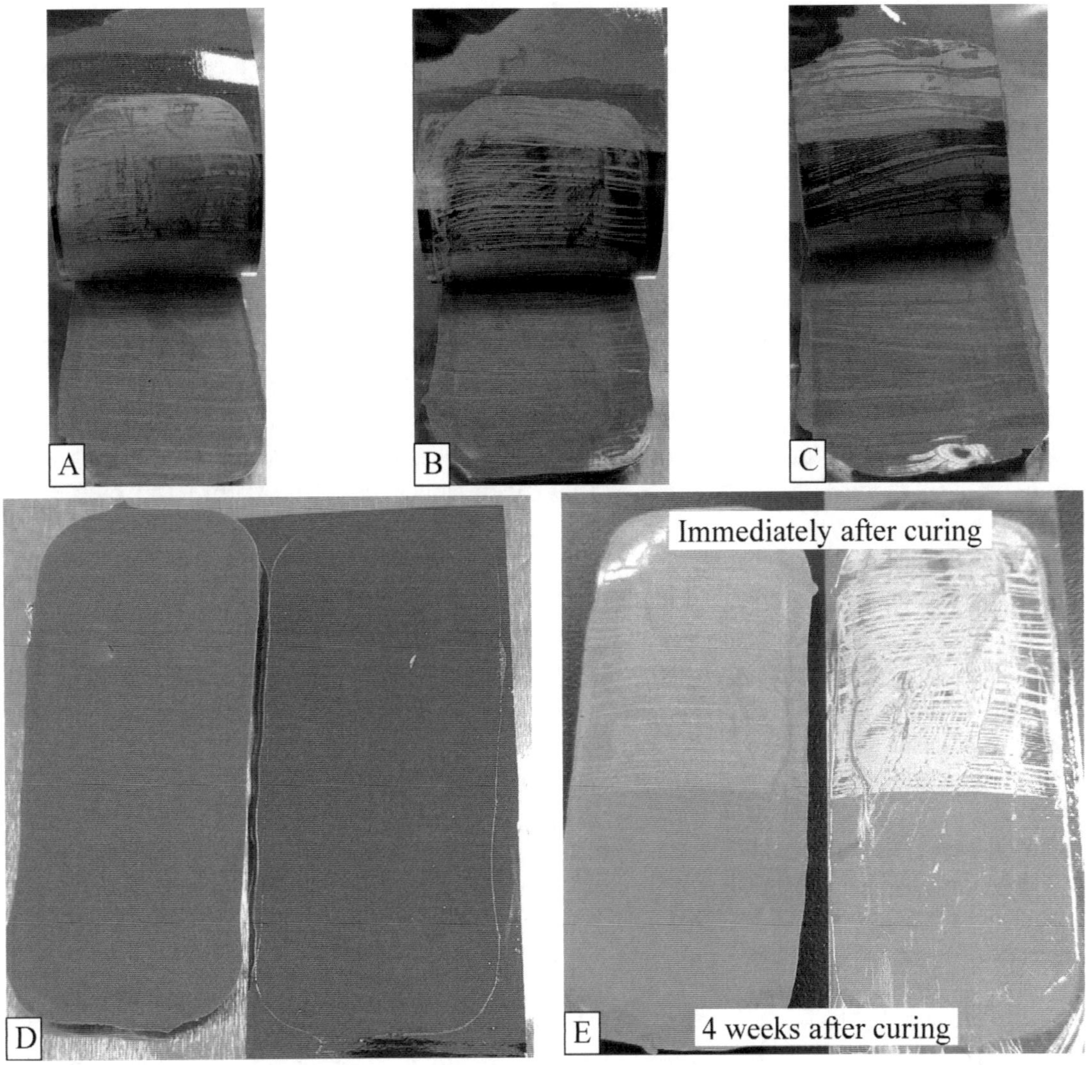

Figure 10. A) Air-cured diamine silane coating, B) oven-cured diamine silane coating, C) air-cured primary amine silane coating, D) as received AM failure surfaces, E) oven-cured diamine silane coating failure surface immediately after curing and 4 weeks after curing.

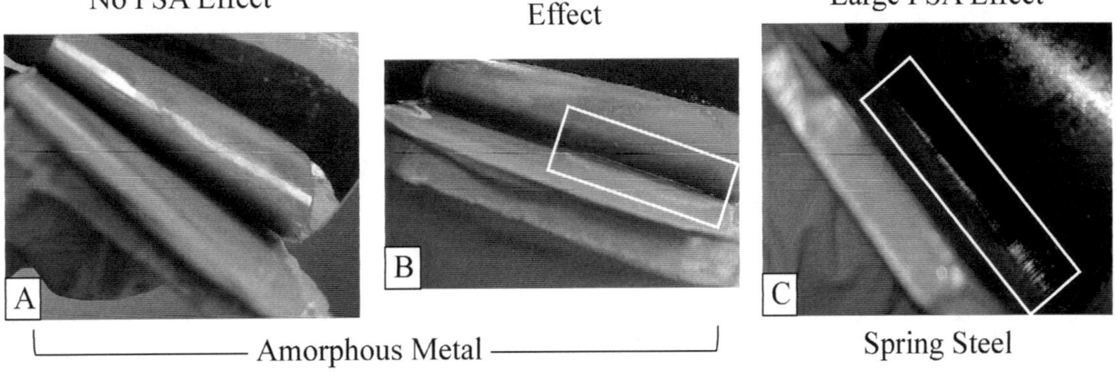

Figure 11. Pressure sensitive adhesive effect can be visualized by tacky layer between epoxy and foil surfaces, observed as polymeric bridges between the surfaces. A) No PSA effect can be visualized on amorphous metal with no silane coating. B) A moderate PSA effect can be visualized by a thin tacky layer when a moderate amount of silane is used on amorphous metal. C) A large PSA effect can be seen with a thick tacky layer when a large amount of silane is used on spring steel.

4. DISCUSSION

Amorphous metal foil is to be used for the Jet Propulsion Lab's Starshade project to provide an optically precise edge that will be resistant to fracture, plastic deformation, and corrosion in space applications. The foil is bonded to support structures using EA 9394 epoxy, a well-characterized structural adhesive commonly used in aerospace applications. However, only poor adhesion performance of this bond during peel tests has been achieved. To explore the cause of poor adhesion, several factors were explored.

Cleaning, plasma treatment, and grit-blasting a surface are often used to create chemically active surfaces that adhere readily and achieve good adhesive performance, which is characterized by a cohesive failure mode (adhesive presence homogenously found on both failure surfaces, high strength required for failure). However, these factors did not improve adhesive performance on the amorphous metal foil, characterized by adhesive failure (failure at the interface, adhesive on only one of the failure surfaces, low strength required for failure).

Metallurgical composition of the foils was explored by choosing foil that did not contain boron, which was though to be interfering with a full epoxy cure. Foil crystallinity was also considered. Heating the foils to high temperatures to induce crystallinity created weak oxide layers, embrittled the foils, and did not improve adhesion performance.

Converting the surface chemistry of the foils by using silane primers improve adhesion initially after curing the epoxy, however this improved performance is lost over several weeks. These silane films were very thick (as evidenced by their visual appearance). Interpenetration of the epoxy with these films would change the stoichiometry of the adhesive in the near-interface regions, potentially interfering with epoxy curing. This undercured zone remains tacky after the standard curing cycle, and shows long-range deformation during peel, similar to a pressure sensitive adhesive. After aging, curing progresses to the point where the ability of this interpenetrated silane-epoxy zone no longer shows long range deformation, and any adhesion improvement due to this effect is lost. The cyanoacrylate adhesive that was investigated showed significantly improved adhesion over the epoxy, and this bond strength remains after at least 4 weeks.

Cleanliness, hydroxyl groups on the surface, roughness, metallurgy, and surface chemistry do not appear to improve adhesion after the epoxy has fully cured. In instances where adhesion was improved, the adhesive at the interface did not provide the stiff, high-yield strength structural adhesive properties normally attributed to EA 9394, due to either not being fully cured at the interface or because of a different adhesive with a lower yield strength being used. Improved adhesion was present when the adhesive could plastically deform. This implies that poor adhesion between the amorphous metal and EA 9394 is more likely to be caused by mechanical stresses at the failure point rather than being due to chemical adhesion at the interface.

Adhesion between two surfaces is governed by Equation 1, where G is the toughness of an adhesive joint (energy/unit area), W_A is the work of adhesion (related to surface energy), and $\varphi(v,T)$ is the viscoelastic deformation energy, which is dependent on the viscoelasticity of the material at a specific temperature [9].

$$G = W_A(1 + \varphi(v,T)) \qquad [1]$$

PSAs exhibit very large viscoelastic energy dissipation compared with structural adhesives. In this case, $\varphi(v,T)$ can be thousands of times greater than in structural adhesives, and W_A becomes less important. This explains why PSA tape can stick to many surfaces, including "non-stick" materials such as polytetrafluoroethylene, without heavy cleaning or surface treatment, since surface cleanliness and treatments are less important. The viscosity of the PSA enables the force from the peel test to be distributed across a large volume of adhesive, thus increasing the force required to fail the bond [9].

Structural adhesives require high yield strengths and high moduli. These requirements severely limit $\varphi(v,T)$. For these systems, adhesion depends much more on W_A than $\varphi(v,T)$. W_A is highly dependent on clean surfaces and surface treatments to ensure a high surface energy [9].

Peel tests concentrate applied load into a very small area. If the adhesive is incapable of long-range plastic deformation and the substrate is a high-yield material, strength will be low. If the substrate being peeled has a low yield strength, the energy distribution from the peel force can be dispersed into the substrate rather than the bond, and the peel strength will increase [10]. However, if the substrate being peeled has a high yield strength, all the energy from the peel test goes into breaking the bond.

The behavior exhibited by the heavy silane treatment is similar to that observed by a pressure sensitive adhesive (PSA). While the application of bonding epoxy to AM for space applications may not warrant using a viscosity- and temperature-dependent PSA effect, it provides a deeper understanding into why an epoxy does not bond well to an AM.

5. CONCLUSIONS

Amorphous metal foil used for the JPL Starshade project does not bond well to the aerospace epoxy EA 9394. Cleanliness, surface treatment, roughness, metallurgical composition, crystallinity, silane primers, and cyanoacrylate adhesives were explored to increase adhesive performance. It was found that increased adhesion occurred when the adhesive yield strength was lower than fully cured EA 9394, indicating that mechanical stresses rather than chemical composition of the surface is the most likely cause to poor adhesion.

6. ACKNOWLEDGMENTS

Portions of this research were funded through a subcontract with the Jet Propulsion Laboratory, California Institute of Technology, under a contract with the National Aeronautics and Space Administration (80NM0018D004).

7. REFERENCES

1. "Developing Starshade Technology to Image Earthsized Exoplanets around Neighboring Stars." United States. National Aeronautics and Space Administration. *2017 Science Mission Directorate: Technology Highlights* Washing, D.C.: GPO, 2017.

2. Diegle, Ronald B. "Chemical Properties of Metallic Glasses." *Journal of Non-Crystalline Solids* 61&62 (1984): 601-612.

3. Jafary-Zadeh, Mehdi; Kumar, Gideon P.; Branicio, Paulo S.; Seifi, Mohsen; Lewandowski, John J.; and Cui, Fangsen. "A Critical Review on Metallic Glasses as Structural Materials for Cardiovascular Stent Applications." *Journal of Functional Biomaterials* 9(19) (2018). DOI:10.3390/jfb9010019

4. Steeves, John; Martin, Stefan; Webb, David; Lisman, Douglas; Shaklan, Stuart. "Precision Optical Edges for a Starshade External Occulter." *Advances in Optical and Mechanical Technologies for Telescopes and Instrumentation II*. Edinburgh, UK, 22 July 2016. Proc. SPIE 9912. Paper. DOI: 10.1117/12.2233409

5. Hilgemann, Evan; McKeithen, Dylan; Saltarelli, Nicholas; Ferguson, Maxwell; Steeves, John; Webb, David; Shaklan, Stuart; Martin, Stefan; and Lisman, Douglas. "Advancements in precision edges for a starshade external occulter." *Optical Engineering + Applications.* San Diego, California, United States, 9 September 2019. Proc. SPIE 11117. Paper and presentation. DOI: 10.1117/12.2530160

6. "Safety Data Sheet: LOCTITE EA 9394 Aero Part A." Henkel (2018).

7. "Safety Data Sheet: LOCTITE EA 9394 Aero Part B." Henkel (2018).

8. Shen, P.; Fujii, H.; Nogi, K. "Effect of substrate crystallographic orientation on wettability and adhesion in several representative systems." *Journal of Materials Processing Technology* 155-156 (2004): 1256-1260. DOI:10.1016/j.jmatprotec.2004.04.212

9. Abbott, Steven. Adhesion Science: Principles and Practice. Lancaster, Pennsylvania: DEStech Publications, Inc., 2015.

10. Li, Qian; Batra, Romesh C.; Graham, Ian; Dillard, David A. "Examining T-peel specimen bond length effects: Experimental and numerical explorations of transitions to steady-state debonding." *International Journal of Solids and Structures* 180-181 (2019): 72-83. DOI: 10.1016/j.ijsolstr.2019.07.012

DEVELOPMENT OF INTERMITTENT LAYER HEIGHT CONTROL USING LINE SCANNING FOR SUCCESSIVE TOOLPATH ADAPTION IN LASER MATERIAL DEPOSITION

Jan Bremer[1], Tim Fleischhauer[1], Andres Gasser[1], Johannes Henrich Schleifenbaum[1,2]

1: Fraunhofer Institute for Laser Technology
2: RWTH Aachen University – Digital Additive Production
Aachen, Germany

ABSTRACT

Laser Material Deposition (LMD), also known as Directed Energy Deposition (DED), is an additive manufacturing (AM) technology used in repair and manufacturing applications of high-value components. It is commonly used in industries such as aerospace, tooling and turbomachinery. The ability to produce near-net-shape parts results in low material losses and can lead to reduction of production costs and lead times compared to conventional machining processes. In hybrid manufacturing, LMD is used together with conventional processes to exploit both technologies' benefits for production of large parts with complex structures. Increasing stability and geometric accuracy of LMD processes is key enabler for further expansion of applications of this manufacturing technology. Current approaches focus on parameter studies, development of predictive modeling as well as closed-loop control. This work proposes a novel approach for component geometry control in LMD. This approach is based on an intermittent geometry control. LMD builds are split into separate sections. After depositing a section or block, a laser line scanner is utilized to measure the detected block geometry in the form of a point cloud. An algorithm is developed that calculates local height deviations from target geometry and adjusts the feed rate along the tool center point (TCP) path. Experiments are conducted in order to establish control parameters such as reference distance between adjacent support points along the TCP path (minimum and maximum feed rate increment) and functional relation between layer height and feed rate. Thus, dependence of layer height with respect to absolute part height is investigated.

Keywords: Laser Material Deposition, Directed Energy Deposition, laser line scanner, adaptive processing, layer-height-control, point cloud manipulation
Corresponding author: Jan Bremer

1. INTRODUCTION

Production of high performance and lightweight components is a crucial task in industries such as aerospace or automotive. They play a key role in terms of sustainability and performance enhancement. In manufacturing of metal components in aerospace applications, state of the art conventional machining processes come with high buy-to-fly ratios, i.e. huge ratios between the weights of raw material used to produce components compared to the weights of the final parts. Up to 90% of the raw material is wasted [1]. Correspondingly, high production costs occur, resulting from the usage of high-value materials such as titanium alloys or Ni-based superalloys. [2, 3]

Combining conventional manufacturing processes with additive manufacturing (AM), so-called hybrid manufacturing, yields considerable improvements compared to sole conventional manufacturing. One of the commonly used AM technologies is Laser Material Deposition (LMD). It offers the potential to manufacture complex forms and near-net-shape parts, resulting in significantly reduced production costs and lead times, as well as energy and raw material savings. [3, 4]

LMD is an AM technology where energy emitted from a laser source is utilized to locally melt a substrate. Filler material in the form of powder or wire is transported into the melt pool. Powder is carried by an inert gas jet. The melt pool is shielded by an additional inert gas jet. After solidification, the filler material forms a track on the substrate. Layers of material are formed by the deposition of multiple tracks. After depositing a layer, the relative distance between cladding head and substrate is increased by an estimated layer height. [5]

If the estimated layer height differs from the deposited layer height beyond certain limits, the process becomes unstable and must be aborted. Investigations into control strategies of the deposited layer height are conducted to increase process stability and geometrical accuracy.

To control the deposited layer height in a wire LMD (WLMD) process, *Garmendia et al.* [6] propose to use the feed rate as the actuating variable. A stationary structured light geometry scanner is utilized to measure each deposited layer. The deposition process is interrupted for measuring. Subsequently, the feed rate of the next layer is adjusted locally to deviations between the estimated and the deposited layer height along the tool center point (TCP) path. A functional relation between feed rate and layer height is established experimentally for this purpose. [6]

Depositing a certain number of layers blockwise and iteratively recalculating the number of layers to be deposited in a subsequent block is another approach to layer height control. Block height is measured with a stationary structured light sensor after deposition. Recalculation is then based on differences between measured block height and estimated block height. [7]

Another approach to control layer height is to adjust the average layer height of the following block based on the measurement of the previously deposited block. Again, a structured light sensor is used. [8]

Further investigations in controlling the deposited height include adjusting laser power [9] or feed rate [10] based on charge couple device (CCD) cameras. A laser line scanner (LLS) is utilized by *Heralić et al.* [11] to measure layer height and subsequently adjust wire feed rate in a WLMD process. Powder feed rate has a considerable influence on track height [12]. However, it is rarely considered in height control because of its comparatively long response time [12, 13]. Several guidelines for stable line scanning results, such as the use of LLS with a wavelength ranging from 405 nm to 520 nm or the sandblasting of substrates, are recommended. [14]

The algorithm presented in this paper is based on the work of *Garmendia et al.* [6]. Novel is the approach to use a LLS to measure the height of deposited blocks. The control algorithm is improved in terms of selecting measurement points to calculate local height deviations and experimentally determining boundary conditions for the feed rate adjustments based on the capabilities of the used LMD system.

2. EXPERIMENTAL SETUP

2.1 System setup

Experiments are conducted on a LMD system consisting of a 90kg payload industrial robot and a variable beam geometry powder processing optic. A continuous coaxial powder delivery nozzle is used. Beam source is a diode laser with maximum power output of 5000W and a wavelength of 940 – 1060 ± 10 nm. A line scanner with a nominal measuring line width of 100 mm and a wavelength of 405 nm is used. The system setup is depicted in Figure 1.

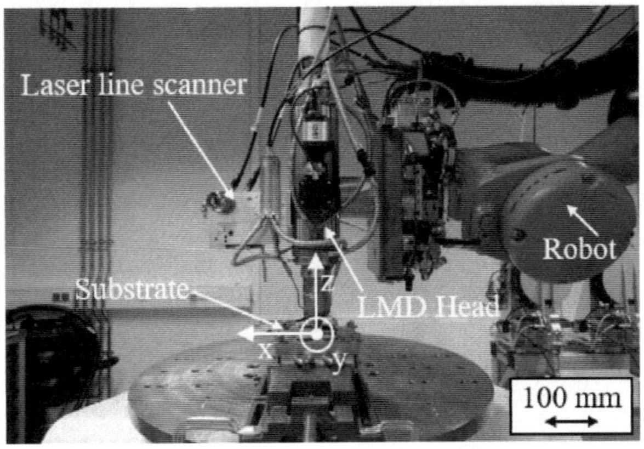

Figure 1. System setup

Plasma atomized Inconel 718 powder with a grain size distribution ranging from 15 to 45 µm is used. Chemical composition is depicted in table 1.

Table 1. Chemical Composition of Inconel 718

Element	Cr	Co	Fe	Mo	Ni	Nb	Ta	Ti
Mass fraction [%]	17 - 21	0 - 1	13 - 25	2.8 - 3.3	50 - 55	4.75 - 5.5	0 - 0.05	0.65 - 1.15

2.2 Process parameters

All measurements are conducted using the process parameters listed in table 2. Nominal feed rate is 33.3 mm/s. Feed rate variations in experiments and within the height control strategy are always denoted as deviation percentages from the nominal feed rate.

Table 2. Process parameters

Parameter	**Value**
Feed rate	33.3 mm/s
Laser power	1000 W
Beam diameter	1.6 mm
Powder feed rate	9.5 g/min
Powder delivery gas	3.5 NL/min

Parameter	**Value**
Powder focus offset	-1 mm
Shielding gas flow rate	10.7 NL/min
Track Overlap	50%
Estimated layer height	0.65 mm

2.3 Measurement technologies

A LLS measures two-dimensional height profiles by utilizing the light section principle. Three-dimensional scanning of an object is achieved by concatenating height profiles obtained at a fixed

frequency whilst scanner and object are moved relative to each other. In this work, height profiles are measured at a frequency of 83.3 Hz. Resolution of the scan line is 1280 points which yields a distance of 78 μm between adjacent points on a nominal scan line of 100 mm. Smallest height difference between adjacent points is 12 μm for the given configuration. A line scanning software is used to acquire and synchronize line scanning and machine data, resulting in output of a point cloud of the scanned geometry in the working coordinate system.
To evaluate feed rate limits considered for the height control strategy, a laser tracker is used. The laser tracker measures positions of a retroreflector attached to the processing optics at a fixed frequency of 1000 Hz. Measurement of positions is afflicted with an error of maximum ±27 μm for the given system setup.

2.4 Height control strategy

The approach investigated in this work is to deposit a fixed number of layers within a single block. After depositing a block, block height is measured with the LLS and local deviations from the estimated height are calculated. The TCP path is split into segments to adjust the feed rate locally with respect to local height deviations. Each segment is assigned a local feed rate. If local height deviations Δh_{local} stay within desired stability limits, it is assumed that adjusting the feed rate is sufficient to compensate them. Stability limits are set according to equation 1. The next block is deposited with locally adjusted feed rates. If local height deviations exceed stability limits, the current LMD process is aborted. Layer heights are recalculated based on the current height and a new LMD process is started, following the same control strategy. The principle of the height control strategy is depicted in Figure 2.

$$-1\text{ mm} \leq \Delta h_{local} = h_{set} - h_{local} \leq 3\text{ mm} \tag{1}$$

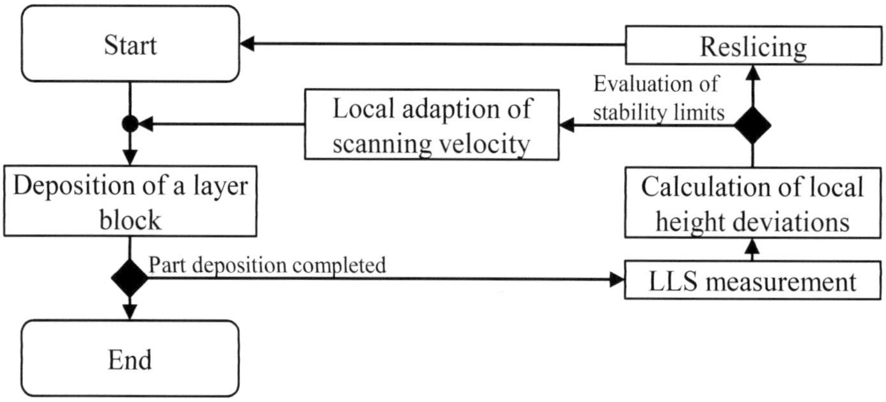

Figure 2. Principle of the height control strategy

Initially within the height control strategy each section of the path, where material is deposited, is split into segments. These sections are framed by discrete points $P_S = (x_S, y_S, z_S)^T$ and $P_E = (x_E, y_E, z_E)^T$, as depicted in Figure 3a. In this work, curved trajectories are approximated by linear sections.

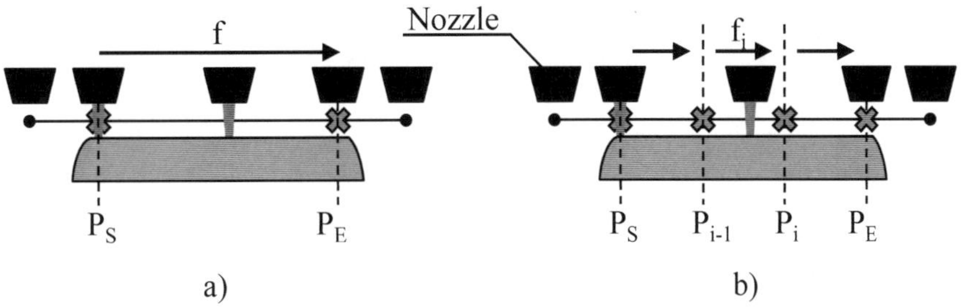

Figure 3. Preparation of a TCP path for application of the height control strategy
a) Given TCP path
b) TCP path with support points

To interpolate the path from P_s to P_E with support points P_i as depicted in Figure 3b, a normalized direction vector n_{line} is calculated with equation 2.

$$n_{line} = \begin{pmatrix} x_E - x_S \\ y_E - y_S \\ z_E - z_S \end{pmatrix} \cdot \frac{1}{\sqrt{(x_E - x_S)^2 + (y_E - y_S)^2 + (z_E - z_S)^2}} \tag{2}$$

A reference distance Δs, used to calculate the number of support points to be placed between P_S and P_E, must be established experimentally. The number of support points can then be calculated with equation 3, where $d(P_S, P_E)$ is the euclidean distance between two points and $\lfloor x \rfloor$ is the rounding off function for a value x.

$$n = \left\lfloor \frac{d(P_S, P_E)}{\Delta s} \right\rfloor \tag{3}$$

Since the number of support points is rounded, the effective distance Δs_{eff} between support points is subsequently calculated with equation 4. By rounding off it is assured that the effective distance Δs_{eff} between two support points is always larger than the reference distance Δs.

$$\Delta s_{eff} = \frac{d(P_S, P_E)}{n} \tag{4}$$

The distance between P_S and P_E is filled with support points according to equation 5.

$$P_i = P_S + n_{line} \cdot \Delta s_{eff} \cdot i \qquad i = 1 \ldots n_{line} \tag{5}$$

The LLS measurement of the deposited surface yields data in the form of a three-dimensional point cloud C^{LLS}. For each support point P_i, a set C_i^{LLS} of point cloud points $P_k^{LLS} = \left(x_k^{LLS}, y_k^{LLS}, z_k^{LLS}\right)^T$ is determined that lies within a certain influence area. The influence area is defined as a rectangle ranging from P_{i-1} to P_i, thus having an edge length in the direction of the normalized direction vector n_{line} of Δs_{eff}. Perpendicular to n_{line}, the edge length is set to the beam diameter d. To determine, whether a point P_k^{LLS} lies within the influence area of a support point P_i, distance vector r_{i-1} in the x-y-plane is calculated with equation 6.

$$r_{i-1} = \begin{pmatrix} x_k^{LLS} \\ y_k^{LLS} \end{pmatrix} - \begin{pmatrix} x_{i-1} \\ y_{i-1} \end{pmatrix} \quad (6)$$

Transformation from the global x-y-coordinate system representation r_{i-1} into a local $\tilde{x}$-$\tilde{y}$-coordinate system representation $\tilde{r}_{i-1} = (\tilde{x}_{i-1}, \tilde{y}_{i-1})^T$ is conducted with equation 7 for simplifying the determination.

$$\tilde{r}_{i-1} = \begin{bmatrix} -\sin(\alpha) & -\cos(\alpha) \\ -\cos(\alpha) & \sin(\alpha) \end{bmatrix} \cdot r_{i-1} \quad (7)$$

Orientation angle α is calculated with equation 8.

$$\alpha = \cos^{-1}\left(n_{line} \cdot \begin{pmatrix} 0 \\ 1 \\ 0 \end{pmatrix} \right) \quad (8)$$

An exemplary influence area is given in Figure 4.

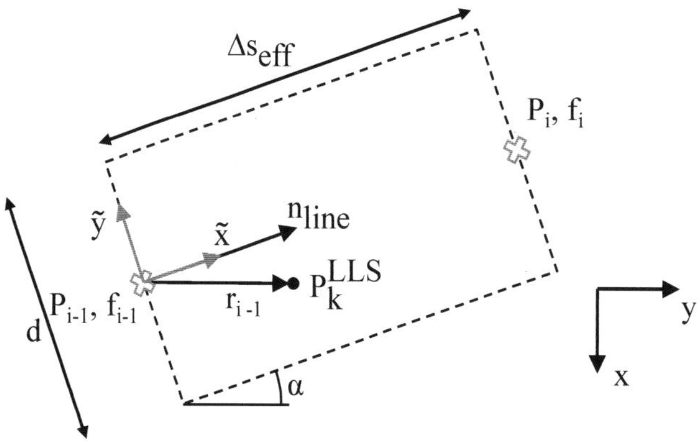

Figure 4. Influence area for a support point P_i

Determination of the set C_i^{LLS} of all point cloud points that belong to the influence area of a given support point P_i leads to evaluating equation 9 for each point of the point cloud.

$$C_i^{LLS} = \left\{ P_k^{LLS} \in C^{LLS}: 0 \leq \tilde{x}_{i-1} \leq \Delta s_{eff,i} \wedge -\frac{d}{2} \leq \tilde{y}_{i-1} \leq \frac{d}{2} \right\} \quad k = 0 \ldots \left| C^{LLS} \right| \quad (9)$$

Local height at a support point P_i is calculated with equation 10.

$$h_{local,i} = \frac{1}{\left| C_i^{LLS} \right|} \cdot \sum_{P_k^{LLS} \in C_i^{LLS}} z_k^{LLS} \quad (10)$$

From the difference $\Delta h_{local,i} = h_{set} - h_{local,i}$ between the estimated height h_{set} and the deposited height $h_{local,i}$ a local adaption of the feed rate f_i is conducted based on a functional relation

$f_i = f(\Delta h_{local,i})$. The functional relation is determined experimentally in this work. A visualization of the above described interaction of TCP path and point cloud data is depicted in Figure 5.

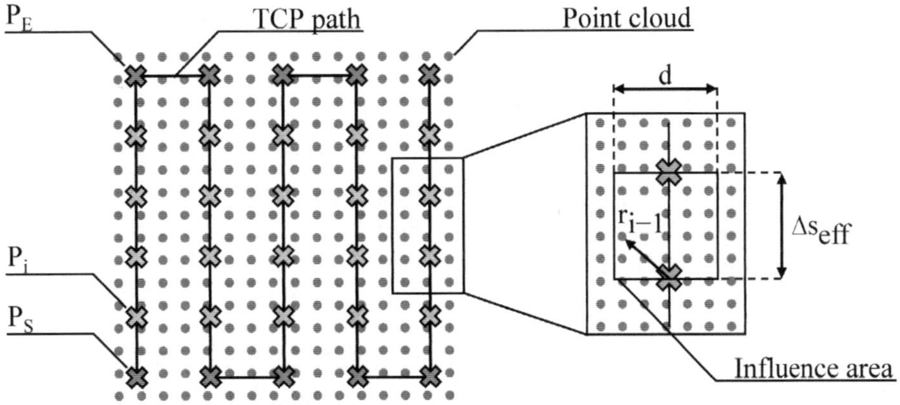

Figure 5. Overlapping of TCP path and point cloud data

Due to the nature of the experimental setup, the alignment of point cloud measurements with the nominal CAD model used for comparison as well as the LLS measurements can contain errors. Reasons can be, among others, geometrical tolerances in the mounting of the LLS, latency in triggering of LLS measurements in the control system as well as errors in LLS tool data in the control system and misalignment of the specimens due to mounting errors.

A Random Sample Consensus (RANSAC) algorithm is used to compensate tilting of the point cloud. RANSAC is a non-deterministic, iterative algorithm, used to fit a model to data that is heavily disturbed by outliers. [15]

In this work, the RANSAC algorithm is implemented to estimate the substrate plane in the given point cloud data. Height of a point is then calculated as projected distance from the estimated plane, resulting in a height profile of a deposited object. In Figure 6 an exemplary height profile calculation is depicted. Deposited object is a cuboid with an edge length of 25 mm and an estimated height of 15 mm.

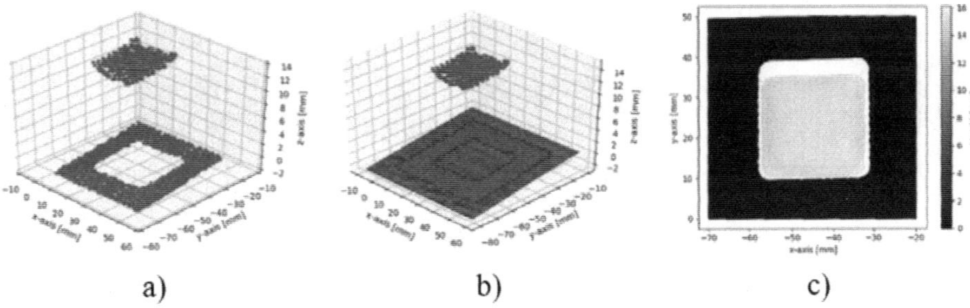

Figure 6. Application of the RANSAC algorithm
a) Raw point cloud
b) Estimated substrate plane
c) Height profile

After tilting compensation with the RANSAC algorithm, absolute position errors are compensated. The center point of all height profile points whose height exceeds a tolerance band around the

value zero is calculated. Each point of the point cloud is shifted by the distance between the calculated center point and the geometrical center point, obtained from the Computer Aided Design (CAD) data of the deposited object.
The height control strategy is implemented as a Python script, utilizing the NumPy module.

2.5 Experimental investigations

To implement the height control strategy on the LMD system, several parameters and relations must be determined experimentally. Four experiments are conducted to establish a reference distance Δs as well as a maximum and minimum feed rate increment between adjacent support points and a functional relation between layer height and feed rate.

Experiment 1
Linear TCP paths of 25 mm length are examined at different feed rates to investigate, whether actual feed rates can meet required set feed rates within an interval of ±50% around the nominal feed rate. Smallest feed rate increments investigated are 1% steps within an interval of ±5% around the nominal feed rate. Each measurement is repeated three times. Measurements are conducted in the x-direction as well as the y-direction of the base coordinate system, depicted in Figure 1. TCP positions are measured at distinct points in time using the laser tracker system. Feed rate is calculated from obtained position data.

Experiment 2
Reference distance Δs between two adjacent support points with respect to a maximum feed rate increment is established by examining a TCP path as depicted in Figure 7. A linear TCP path is filled with support points at a fixed distance Δs. Distinct distance values 0.5 mm, 1 mm, 2 mm, 3 mm, 5 mm and 10 mm are investigated. Feed rate is varied stepwise in increments ranging from $f_{Nom} \cdot (1 - 3\Delta f)$ to $f_{Nom} \cdot (1 + 3\Delta f)$, with Δf describing fixed percentages 1%, 5% and 10%, respectively. The feed rate trajectory contains two maxima, resulting in 26 support points on the TCP path. A full factorial investigation of Δf and Δs is conducted. Each combination is repeated three times. TCP positions are measured at distinct points in time using the laser tracker system. Feed rate is calculated from obtained position data.

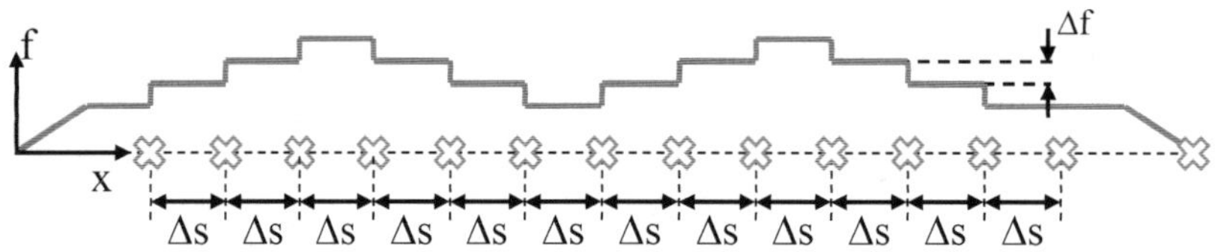

Figure 7. Experiment to establish a reference distance Δs

Experiment 3
Twelve cuboids, consisting of three blocks with eight layers each, are deposited at nominal feed rate. Block height is measured with the LLS to investigate nominal layer height, deposited with the process parameter set. Influence of the distance to the substrate plane on the deposited layer height is investigated. Build-up strategy consists of contour trajectories and hatch trajectories in the form of meanders, fly-in and fly-out paths are integrated.

Experiment 4

To determine a functional relation between layer height and feed rate, cuboids, consisting of three blocks with four layers each, are deposited at various feed rates. Block height is measured with the LLS. Layer height is calculated from block height measurements. Feed rate is varied within an interval of ±50% around nominal feed rate at increments of 10%. For each feed rate, three cuboids are deposited.

Experiment 5
Six cuboids are deposited in two steps for evaluation of the experimentally determined control parameters and functional relations implemented in the layer height control strategy. In the first step, tilted planes are deposited by decreasing the feed rate from the maximum feed rate limit to the minimum feed rate limit when moving the TCP in positive y-direction. Feed rate is decreased by the maximum feed rate increment Δf on segments with the length of the reference support point distance Δs. Tilted planes consist of four deposited layers each. In the second step, a block of four layers is deposited on top of each tilted plane. Layer height control is applied to the deposition process of three blocks. The remaining blocks are deposited without layer height control strategy as reference.

3. RESULTS

The capability of the LMD system to meet set feed rates is examined in experiment 1. Differences between measured actual feed rates f_{act} and set feed rates f_{set} are depicted in Figure 8. Set feed rates are denoted with either a letter M or P, indicating a negative or a positive deviation from the nominal feed rate, followed by a percentage value. For example, feed rate FM50 refers to a feed rate $f = f_{nom} \cdot (1 - 50/100)$.

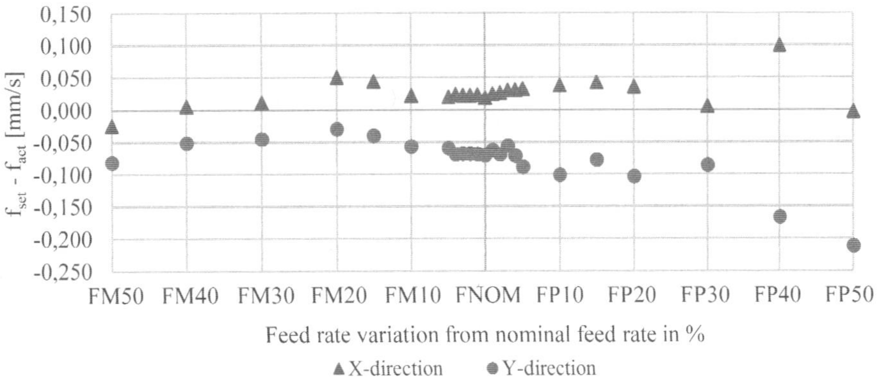

Figure 8. LMD system capabilities to meet set feed rates

Over the whole range of investigated feed rates, deviations from set feed rate lie within a limit of 0.21 mm/s. On average, deviations amount to 0.09% of the set feed rate in x-direction and 0.23% in y-direction. Maximum amount in x-direction is 0.21% at a feed rate variation of +40%, in y-direction 0.49% at a feed rate variation of -50%.
For implementation of the height control strategy it is hence assumed that the LMD system is capable of meeting required set feed rate. Even small feed rate variations of 1% can be distinguished. Therefore, no minimum feed rate increments are included in the height control strategy. Since deviations in x- and y-direction are of similar magnitude, possible influence of

direction is neglected in the following experiments. Isotropic behavior of the LMD system is assumed within the working space.
Stepwise variation of feed rate is applied on a TCP path in experiment 2 to investigate the influence of support point distance Δs in correlation with feed rate increment Δf on the feed rate accuracy. As depicted in Figure 9a, the LMD system cannot follow a stepwise variation of the set feed rate when the support point distance is set to 0.5 mm. Applying a support point distance of 10 mm yields an accurate actual feed rate profile, as depicted in Figure 9b. In the given example, a feed rate increment of 10% is used.

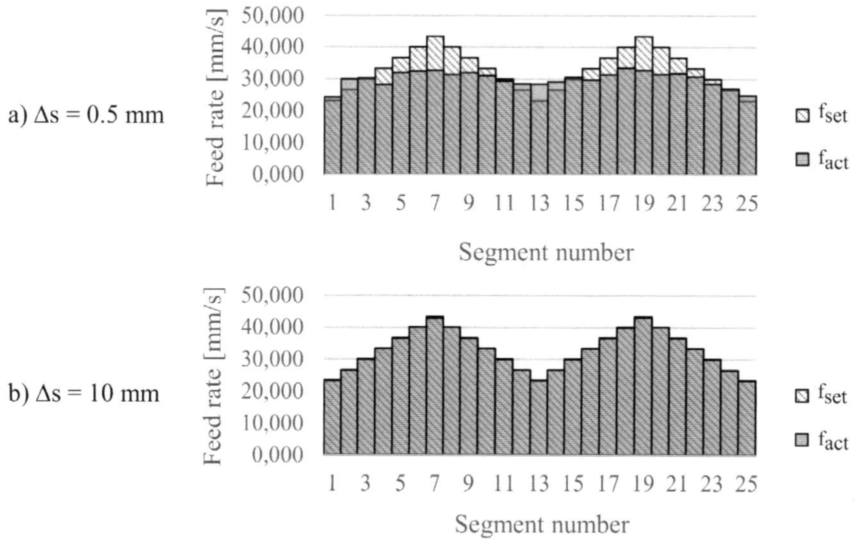

Figure 9. Comparison of the influence of support point distances on the feed rate accuracy

For quantification purposes, feed rate error $E_{\Delta s}^{\Delta f}$ is defined in equation 11. The feed rate error describes the ratio between the sum of the absolute values of all feed rate deviations, occurring on each line segment enclosed by two adjacent support points and the sum of the corresponding set fee rates.

$$E_{\Delta s}^{\Delta f} = \frac{\sum_{Seg} \left| f_{set,\,Seg} - f_{act,\,Seg} \right|}{\sum_{Seg} f_{set,\,Seg}} \cdot 100 \qquad (11)$$

Calculated feed rate error is depicted in Figure 10 for all investigated combinations of support point distance and feed rate increment. It decreases with increasing support point distance as well as with decreasing feed rate increment. Support point distances of more than 2 mm yield feed rate errors of less than 2% for all investigated feed rate increments.

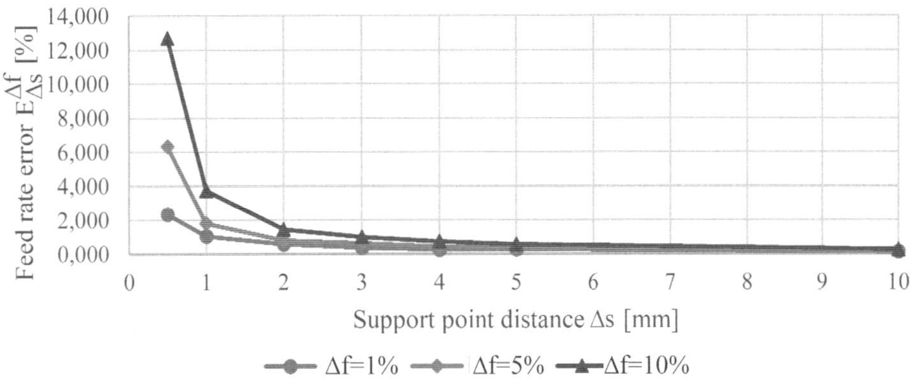

Figure 10. Evaluation of the feed rate error

For implementation in the height control strategy, a reference support point distance of 2 mm is chosen. A maximum feed rate increment for adjacent support points of 10% is implemented in the height control strategy.

In experiment 3, cuboids are deposited using the process parameter set and the nominal feed rate. Each cuboid consists of three blocks. After depositing a block, block height is measured with the LLS. Overall 36 blocks are deposited. An average block height of 4.97 mm is calculated with a standard deviation of 0.084 mm, resulting in a layer height of 0.621 mm which is slightly smaller than the 0.65 mm estimated layer height process parameter.

To investigate the influence of the distance from the substrate on the deposited layer height, deviation of the measured block height from the average block height is depicted as a box plot in Figure 11. Block height of the first block, deposited directly on the substrate, is, on average, 98% of the average block height. The third block, which is deposited at the largest distance from the substrate, amounts to 102% of the average block height. Standard deviation of the first block deposited is about one magnitude larger than of the remaining ones.

Since the block height deviates about only 4% within an absolute distance of 15 mm to the substrate, the influence of the distance from the substrate on the deposited layer height is neglected in the height control strategy.

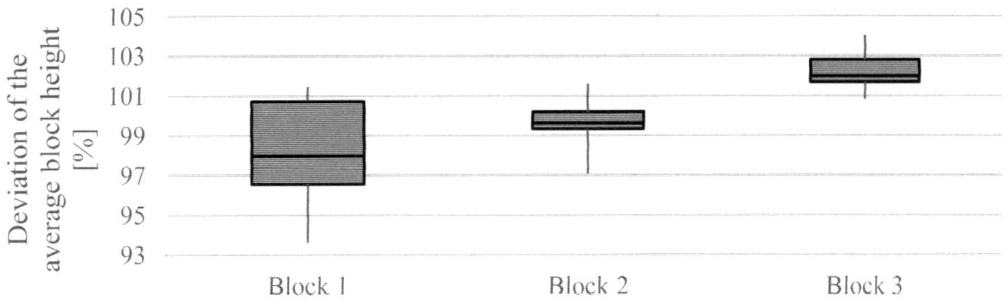

Figure 11. Box plot representation of deviations from the average block height

Functional relation between feed rate and layer height is investigated in experiment 4. Three cuboids, consisting of three blocks each, are deposited for each investigated feed rate, resulting in nine blocks to evaluate for each feed rate. A box plot representation of the block height distribution with respect to the feed rate is depicted in Figure 12.

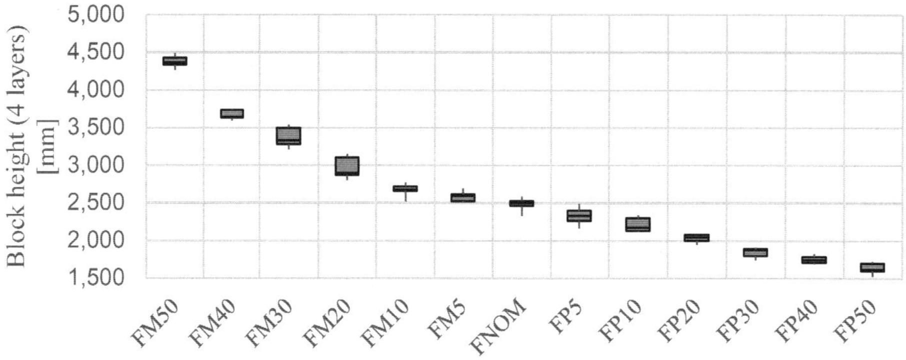

Figure 12. Box plot representation of the block height distribution with respect to the feed rate

Block height increases more with decreasing feed rates than it decreases with increasing feed rates. Linear regression is applied to the measurements in an area of ± 30% around the nominal feed rate, yielding a model equation (equation 12) with a R^2-value of 0.977.

$$h\,[mm] = -\,0.0061 \cdot f\,[\%] + 0.6237\,[mm] \tag{12}$$

For implementation in the height control strategy, equation 13 is derived from equation 12. It describes the relation between feed rate and local height deviations. Local height deviation is divided by four, because the LLS measurements of the deposited height are conducted every four layers within the first implementation of the height control strategy.

$$f_i \left[\frac{m}{s}\right] = f_{Nom} \left[\frac{m}{s}\right] - 0.0546 \left[\frac{\frac{m}{s}}{mm}\right] \cdot \frac{\Delta h_{local}}{4}\,[mm] \tag{13}$$

In experiment 5, tilted planes are deposited by reducing the feed rate from FP30 to FM30 in feed rate increments of 10% when moving the TCP in positive y-direction. Each track consists of seven feed rate segments with a length of $\Delta s = 2$ mm each. This yields, together with the contour tracks, an edge length of the deposited cuboids of 18 mm. Expected geometric features minimum height h_{min}, maximum height h_{max} and tilting angle α of the tilted plane are calculated in table 3.

Table 3. Expected geometric quantities of the tilted plane

Geometric quantity	Equation	Expected value
h_{max}	$[-0.0061 \cdot (-30) + 0.06237] \cdot 4$	3.23 mm
h_{min}	$[-0.0061 \cdot 30 + 0.06237] \cdot 4$	1.76 mm
α	$\tan^{-1}([h_{max} - h_{min}] / 18)$	4.67°

Evaluation of the measured point cloud data is conducted by evaluating each height profile obtained by the LLS on its own. All points that belong to the tilted plane are determined for each laser line. Linear regression is applied to these points to obtain the tilting angle α. Minimum and maximum height are obtained as the maximum and minimum height value of the points. An exemplary point cloud of a tilted plane is depicted in Figure 13. Points marked in orange color are used for evaluation.

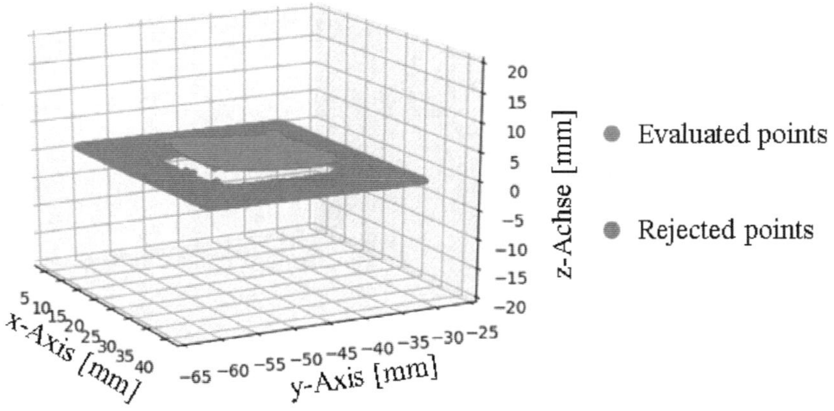

Figure 13. Evaluation of a tilted plane

Calculating the average values of each LLS measurement and each tilted plane yields an average tilting angle $\alpha = 4.428°$, an average maximum height $h_{max} = 2.876$ mm, an average minimum height $h_{min} = 1.651$ mm and an average R^2-Value of 0.926. Deviation between expected and measured tilting angle is 0.242°. Geometric features of the deposited tilted planes are in good accordance with expected values.

Further deposition is conducted on top of the tilted planes to further evaluate the ability of the algorithm to compensate uneven surfaces. Single LLS line measurements of the deposition result without layer height control (Figure 14a) and of the deposition with applied layer height control (Figure 14b) are depicted in Figure 14. Set height of the deposited block is 5 mm.

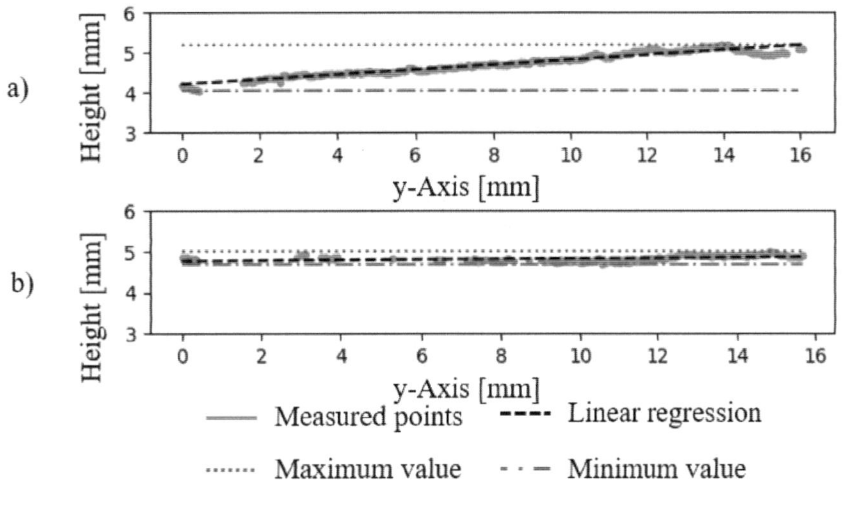

Figure 14. Comparison of controlled and uncontrolled deposition on top of the tilted plane
a) Uncontrolled deposition
b) Controlled deposition

Application of the layer height control strategy results in the tilting of the plane being almost completely leveled. On average, a resulting tilting angle of -0.389° remains. The uncontrolled deposition of another block yields a resulting tilting angle of 3.293°. The tilting of the deposited cuboid is decreased through self-stabilization of the process by 1.135° without the control algorithm. Cuboids deposited with and without layer height control strategy are shown in Figure 15.

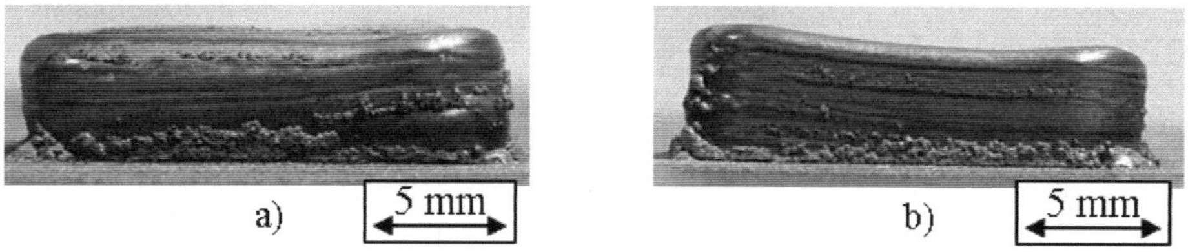

Figure 15. Cuboids deposited on tilted initial build
a) With applied layer height control strategy
b) Without applied layer height control strategy

4. CONCLUSIONS

This work presents a novel approach on process control in laser material deposition to increase stability and geometric accuracy. Previous work was extended by utilizing a laser line scanner to measure the height of deposited LMD layers. Point cloud data was processed to calculate local height deviations from a required set geometry, utilizing the random sample consensus algorithm for evaluation. Feed rate was varied along the TCP path to compensate local height deviations.
Experiments were conducted to investigate the control parameters needed for implementation of the height control strategy. Reference distance between adjacent support points on the TCP path was set to 2 mm. Maximum feed rate increment was 10%, implementation of a minimum feed rate increment was not necessary. Feed rate was varied within a ±30% interval around the nominal process feed rate. Adequate correlation between layer height and feed rate was found within this interval. Deposition experiments of cuboid specimens revealed no relevant dependency of the layer height with respect to the absolute part height.
Evaluation of the layer height control strategy was conducted by depositing tilted planes and subsequently applying the layer height control strategy to build leveled cuboids on top of them. Uncontrolled deposition on top of the tilted planes was conducted as reference. Tilted planes with a tilting angle of 4.428° were deposited, which was a deviation of only 0.242° from the expected tilting angle. Without layer height control, further deposition resulted in a final tilting angle of 3.293°. Applying the layer height control strategy yielded a remaining tilting angle of -0.389°, showing the capability to compensate uneven substrates or previously unstable process conditions.
Further work will be conducted to evaluate the performance of a layer-height controlled LMD process.. Metallurgical properties of produced parts will be investigated. Evaluation of the developed height control strategy will be conducted through experiments on specimen geometries as well as parts from industrial applications.

5. REFERENCES

[1] D. Konitzer, S. Duclos, and T. Rockstroh, "Materials for sustainable turbine engine development," *MRS Bull.*, vol. 37, no. 4, pp. 383–387, 2012.

[2] R. Dehoff *et al.*, "Case Study: Additive Manufacturing of Aerospace Brackets," *Advanced Materials and Processes*, vol. 171, 2013.

[3] J. M. Flynn, A. Shokrani, S. T. Newman, and V. Dhokia, "Hybrid additive and subtractive machine tools – Research and industrial developments," *International Journal of Machine Tools and Manufacture*, vol. 101, pp. 79–101, 2016.

[4] J. I. Arrizubieta, M. Cortina, J. E. Ruiz, and A. Lamikiz, "Combination of Laser Material Deposition and Laser Surface Processes for the Holistic Manufacture of Inconel 718 Components," (eng), *Materials (Basel, Switzerland)*, vol. 11, no. 7, 2018.

[5] F. Klocke, *Fertigungsverfahren 5*. Berlin, Heidelberg: Springer Berlin Heidelberg, 2015.

[6] I. Garmendia, J. Pujana, A. Lamikiz, J. Flores, and M. Madarieta, "Development of an Intra-Layer Adaptive Toolpath Generation Control Procedure in the Laser Metal Wire Deposition Process," (eng), *Materials (Basel, Switzerland)*, vol. 12, no. 3, 2019.

[7] I. Garmendia, J. Leunda, J. Pujana, and A. Lamikiz, "In-process height control during laser metal deposition based on structured light 3D scanning," *Procedia CIRP*, vol. 68, pp. 375–380, 2018.

[8] I. Garmendia, J. Pujana, A. Lamikiz, M. Madarieta, and J. Leunda, "Structured light-based height control for laser metal deposition," *Journal of Manufacturing Processes*, vol. 42, pp. 20–27, 2019.

[9] L. Song, V. Bagavath-Singh, B. Dutta, and J. Mazumder, "Control of melt pool temperature and deposition height during direct metal deposition process," *Int J Adv Manuf Technol*, vol. 58, no. 1-4, pp. 247–256, 2012.

[10] M. Zeinali and A. Khajepour, "Height Control in Laser Cladding Using Adaptive Sliding Mode Technique: Theory and Experiment," *Int J Adv Manuf Technol*, vol. 132, no. 4, p. 501, 2010.

[11] A. Heralić, A.-K. Christiansson, and B. Lennartson, "Height control of laser metal-wire deposition based on iterative learning control and 3D scanning," *Optics and Lasers in Engineering*, vol. 50, no. 9, pp. 1230–1241, 2012.

[12] H. Qi, M. Azer, and P. Singh, "Adaptive toolpath deposition method for laser net shape manufacturing and repair of turbine compressor airfoils," *Int J Adv Manuf Technol*, vol. 48, no. 1-4, pp. 121–131, 2010.

[13] D. Hu and R. Kovacevic, "Sensing, modeling and control for laser-based additive manufacturing," *International Journal of Machine Tools and Manufacture*, vol. 43, no. 1, pp. 51–60, 2003.

[14] M. Buhr, J. Weber, J.-P. Wenzl, M. Möller, and C. Emmelmann, "Influences of process conditions on stability of sensor controlled robot-based laser metal deposition," *Procedia CIRP*, vol. 74, pp. 149–153, 2018.

[15] M. A. Fischler and R. C. Bolles, "Random Sample Consensus: A Paradigm for Model Fitting with Applications to Image Analysis and Automated Cartography," *Communications of the ACM*, vol. 24, no. 6, 381-395, 1981.

TAILORING EPOXY FOAM PROPERTIES WITH AN INNOVATIVE PROCESS APPROACH USING CO_2 AS BLOWING AGENT

Simon T. Kaysser[1]*, Christian A. Keun[1], Christian Bethke[2], Volker Altstädt[2,3]

[1] CompriseTec GmbH
Hamburg, Hamburg/Germany

[2] Department of Polymer Engineering, University of Bayreuth
Bayreuth, Bavaria/Germany

[3] Bavarian Polymer Institute and Bayreuth Institute of Macromolecular Research, University of Bayreuth
Bayreuth, Bavaria/Germany

ABSTRACT

For modern applications in aviation and electric mobility, foamed polymers are in strong demand due to their low density, tailorable mechanical properties and thermal insulation. The properties of polymer foams are derived from their morphology and matrix. More demanding applications in aviation, for example sandwich structures with SMC outer layers and foamed epoxy core, request specific foam systems to meet flame retardance regulations as well as thermo-mechanical process requirements. However, the production of epoxy foam systems, especially without using hazardous chemical blowing agents, for high temperature applications is a challenge still to be resolved.

CompriseTec GmbH and the Department of Polymer Engineering (University of Bayreuth) are developing tailored epoxy foam systems using an innovative approach. Therefore, CO_2 blocked amine hardeners in the form of carbamates combined with adaptable pre-curing systems and additives are used to tailor foam properties. A tailorable morphology, high mechanical and temperature properties (Tg = 150 – 200 °C) can be achieved in combination with further features, such as flame retardance. In the EUROSTARS funded project "DiaPore", epoxy carbamate foam systems were developed showing the potential of tailoring specific properties, focusing on low to high density and medium to ultra-fine closed cell morphology. First flame-retardant modified systems were successfully evaluated.

*Corresponding author: Simon T. Kaysser (kaysser@comprisetec.de)
** Presented by: Dr. Christian A. Keun

1. INTRODUCTION

Because of the growing importance of lightweight design for different applications, especially in the automotive and aviation sector, lightweight material solutions such as polymer foams and

sandwich structures are in high demand. This report focuses on an alternative approach to produce rigid thermoset foam systems, tailorable to specific requirements.
Most aircraft interior components, such as panels, hat racks, galleys etc. are based on glass- or carbon-fiber filled phenolic resin, because of their strict FST (Fire, Smoke, Toxicity) regulations according to FAR 25.853. Flame-retardant modified epoxy resin systems, using particle modification, e.g. aluminum hydroxide (ATH) or ammonium polyphosphate (APP) or reactive phosphorus compounds such as dihydrooxaphosphaphenanthrene (DOPO) are state-of-the-art and a promising alternative to phenolic resins [1]. The FST testing is conducted mainly with Ohio State University (OSU) apparatus or cone calorimeters or open flame tests [2-4]. A promising approach is the usage of Sheet Molding Compound (SMC) technology due to its potential for producing complex geometries, short cycle times and net shape curing for the production of sandwich panels. Fette et al. showed the feasibility of the integration of rigid foam cores between SMC layers during the curing step in a hot-pressing process [5,6]. For such demanding applications, where high mechanical requirements, especially compressive strength and stiffness at high temperature, meet flame retardants regulations, new polymer foam systems have to be developed in order to further increase the applicability of modern lightweight technologies.
Epoxy foams have outstanding properties regarding mechanical, thermal and chemical resistance as well as electrical properties and moisture resistance [7-9] and are used since the early 1950's, mainly in electronics, marine, aeronautics and space applications [8, 10, 11-13]. However, complex processing and hazardous chemical blowing agents prevent the large-scale application of epoxy foams in a lot of application fields. Common regulations, for example SNAP or REACH, list many chemical blowing agents used for epoxy foam systems in the past and focus others still used today, as many of them and/or their obtained byproducts are toxic, harmful or explosive. Thus, a foaming process, using only CO_2 as harmless blowing agent, gets more attractive. The main challenge of epoxy foaming is the timing of hardening and cell growth while foaming. Because of the step-polymerization process from liquid monomer to solid three-dimensional molecular network, the viscosity of the foam system drops dramatically while foaming and curing, which makes the application of physical blowing agents, such as supercritical CO_2, challenging. Chemical blowing agents can be used, but its decomposition has to be precisely adapted to the specific foam resin and hardener system in order to achieve fine foam morphology, resulting in good mechanical properties [9, 14]. One promising approach for CO_2 as blowing agent is the use of carbonates (f.e. sodium bicarbonate, $NaHCO_3$) or carbamate salts. Carbamates are products of CO_2 reacting with certain amine hardeners. They are already known from applications in lightweight adhesives for hollow chamber structures. [12, 13]. The CO_2-blocking of the amine curing agents (exothermal) takes place at room temperature in a simple 3-phase-stirrer reactor with primary and secondary amines. For completion, it has to be noted that certain tertiary amines can undergo a carbonate formation only.
Both reaction products, the carbamate salts and the carbonates, can act as latent curing and blowing agents. The use of carbamate salts is more suitable as no byproducts (H_2O) are involved. Recent research on carbamates showed different suitable common amine hardeners such as N-Aminoethylpiperazine (AEP), m-Xylenediamine (mXDA) and 4,4'Diaminodicyclohexyl-methane (DDCM) for the formation of carbamate salts [9, 19, 20]. The curing and foaming reaction is controlled by the specific decomposition temperature and rate of the carbamate salt. Therefore, the viscosity of the epoxy resin system can be adapted, resulting in tailorable homogeneous foam morphologies. There are multiple approaches to further optimize the morphology and thereby the properties of epoxy foams. Takiguchi [8] and Lyu et al. [14] showed, that the viscosity during the

foaming step can be adjusted by pre-curing, resulting in a more controllable cell growth. Besides the pre-curing approach, the addition of particles with a nucleating effect can be beneficial, as well as the preferred limited-foaming process in contrast to free-foaming [15]. Epoxy foams can also be produced using hollow particles (f.e. polymeric microspheres or glass hollow spheres) to obtain a syntactic foam. However, the increased viscosity of the system, resulting from the very high filler degree of hollow-spheres, can be a challenge while processing, especially in combination with additional fillers, f.e. flame retardants [16-18]. One main factor for the compressive behavior of porous structures is its morphology. A finer foam morphology (lower average cell diameter) leads to higher compressive strength and stiffness. As shown by recent studies, the foam morphology can be controlled using a pre-curing approach by mixing blocked and unblocked amine curing agents [20]. The thermal and mechanical properties are important factors for the applicability of a foam in sandwich-SMC-processes. Trials showed necessary loads of approx. 2.5 – 3.0 MPa and higher at 135 - 155 °C, depended on part geometry and material composition. A foam compressive modulus higher than 150 MPa is suitable to prevent foam collapse.
In the project, presented in this paper, alternative hardeners were evaluated regarding their ability to form carbamate salts. Isophorone Diamine (IPDA) was found as suitable new hardener system for an increased Tg. Diglycidylether of Bisphenole A (DGEBA) and Epoxy Novolac (EN) were used as epoxy resin systems. For flame retardance, different commercially available flame retardants have been used and evaluated. In this paper, it is shown, that the properties of resulting epoxy-based foams can be tailored in a wide range of properties, for example density, morphology, thermal and mechanical properties as well as burning behavior, in order to fit to the requirements for specific applications in the aviation and other applications for rigid foams.

2. EXPERIMENTATION

2.1 Materials

As resin system, an epoxy based novolac (EN, DEN431, Olin, Missouri, United States) with an epoxy equivalent weight (EEW) of 173.5 g/mol, and a pre-determined viscosity of 19 Pa s (25 °C) was used. For further trials, Diglycidylether of Bisphenole A (DGEBA, DER331, Olin, Missouri, United States) with a molecular weight (Mw) of 340.4 g/mol, an EWW of 187 g/mol and a pre-determined viscosity of 13 Pa s (25 °C) was used. As amine hardener system, Isophorone Diamine (IPDA) with a Mw of 170.3 g/mol and an amino hydrogen equivalent weight (AHEW) of 42.6 g/mol, (Aradur 22962, Huntsman, Texas, United States) was used as received. As flame retardants aluminum hydroxide (ATH) is used as particle (medium grade Reflamal S2 and fine grade S30, Dadco Alumina and Chemicals Limited, St Peter Port, Channel Islands, United Kingdom, supplied by WTH Walter Thieme Handel GmbH, Stade, Germany) with 30 and 40 wt%. Furthermore, reactive phosphorus compounds based on dihydrooxaphosphaphenanthrene (DOPO, Struktol Polyphlox 3760, Schill + Seilacher "Struktol" GmbH, Hamburg, Germany, EEW 340 g/mol) are used with a total amount of 6.0 % P in the total system. Additionally, a phosphorus based (aluminium diethylphosphinate, further named exo) flame retardant (Exolit OP935, Clariant, Muttenz, Switzerland) is used with a corresponding P content of 2 %, to the total resin mixture. Furthermore, a methyl tallow bis(2-hydroxyethyl) ammonium montmorillonite type nano clay (Cloisite 30B, BYK Additives & Instruments, Wesel, Germany) is used with an amount of 20 wt%. For foaming of specimen for compression and thermal analysis as well as morphology, an aluminum mold with a cavity of 30 x 30 x 10 mm^3 and a bulk volume of 9 cm^3 was used. For burning tests, an aluminum mold with a cavity of 80 x 80 x 2 mm^3 and a bulk volume of 12.8 cm^3

was used. As mold release agent, Loctite Frekote 770-NC from Henkel (Duesseldorf, Germany) was used.

2.2 Procedures

2.2.1 CO_2 blocking reaction and carbamate salt preparation

The synthesis of the CO_2-blocked amine curing agent (carbamate salt) was conducted according to Ren et al. and Bethke et al. [19,20].

2.2.2 Dispersion

For the dispersion of the dried and milled carbamate salt powder in epoxy resin, a three-roll mill process was used, using a 120 EH – 450 three-roll mill (EXAKT Advanced Technologies GmbH, Norderstedt, Germany). The carbamate and epoxy resin were pre-mixed by hand in a stoichiometric ratio and then milled at 50 °C with 200 1/min and gap sizes of 15 and 5 µm for two runs. Afterwards, the dispersion was then stored refrigerated. Additional flame-retardant particles were dispersed using an overhead stirrer.

2.2.3 Foaming

Before foaming, the mold was set at room temperature. The resin-carbamate dispersion was processed at ambient conditions. First, the mold was treated with release agent. Afterwards, a specific amount of resin–hardener mixture and particles were prepared and dispersed using an overhead stirrer. For pre-cured systems, a ratio of 30 % unblocked IPDA with 70 % B-IPDA was used. The mixture was filled in the prepared cavity of the aluminum mold (30 x 30 x 10 mm^3) and transferred into a laboratory oven at 160 °C for 90 min for foaming and curing. For the pre-cured specimens, except *D-n-500-p*, a second aluminum tool with a cavity of 80 x 80 x 2 mm^3 was used for cone calorimeter specimens. After curing, the mold was cooled down to room temperature prior to demolding. Table 1 summarizes all specimen used in this research. Resin and harder in a stoichiometric ratio were used for all samples.

2.2.4 Characterization

Differential scanning calorimetry (DSC) was conducted with a DSC1 Stare System (Mettler Toledo, Columbus, OH, USA). All measurements were carried out in the range of -30 to 300 °C with N_2 at 50 ml/min. Thermogravimetric analysis (TGA) was conducted with a TGA/SDTA851e system (Mettler Toledo, Columbus, OH, USA) in order to investigate the decomposition behavior in the range of 30 – 350 °C. The CO_2 content was calculated theoretically by the amount of available amino groups per molecule and the results gained from TGA.
In order to determine the viscosity profile during curing and the gel point, rheological investigations were conducted with MCR301 rheometer (Anton Paar, Graz, Austria) equipped with a 25-mm plate. Dynamic measurements were carried out from 25 °C to 150 °C at a heating rate of 2 °C/min with $\omega = 1$ rad/s and $\gamma = 10\%$. Isothermal measurements were carried out with $\omega = 1$ rad/s and $\gamma = 10\%$, and temperatures were chosen depending on the system.

Table 1: Epoxy foam specimens, evaluated in this report

Sample	Resin	Hardener	Additive	Filler degree [wt.%]	Curing t [min]	Curing T [°C]	Pre-curing t [min]	Pre-curing T [°C]	Density [g/cm³]
D-n-p	DGEBA	IPDA/ B-IPDA	-		60	180	300	60	0.796
D-DEPAl-p	DGEBA	IPDA/ B-IPDA	exo	8	60	180	300	60	0.677
D-Cloiste-p	DGEBA	IPDA/ B-IPDA	Clay	20	60	180	300	60	0.856
D-ATHS2-40-p	DGEBA	IPDA/ B-IPDA	ATH S2	40	60	180	300	60	0.942
EN-n-p	EN	IPDA/ B-IPDA	-	-	60	180	60	75	0.284
EN-ATHS30-30-p	EN	IPDA/ B-IPDA	ATH S30	30	60	180	60	75	0.275
EN-DOPO-p	EN	IPDA/ B-IPDA	Polyphlox 3760	40 (resin)	60	180	60	75	0.283
EN-DOPO-ATHS30-30-p	EN	IPDA/ B-IPDA	Polyphlox 3760/ ATH S30	25 (resin) / 30 ATH	60	180	60	75	0.289
EN-n-300	EN	B-IPDA	-	-	90	160	-	-	0.292
EN-n-500	EN	B-IPDA	-	-	90	160	-	-	0.475
EN-ATHS30-30	EN	B-IPDA	ATH S30	30	60	160	-	-	0.405
EN-ATHS2-30	EN	B-IPDA	ATH S2	30	90	160	-	-	0.430
EN-ATHS2-40	EN	B-IPDA	ATH S2	40	90	160	-	-	0.506
EN-ATHS2-50	EN	B-IPDA	ATH S2	50	90	160	-	-	0.605
EN-DOPO	EN	B-IPDA	Polyphlox 3760	40 (resin)	60	160	-	-	0.274
EN-DOPO-ATHS30-30	EN	B-IPDA	Polyphlox 3760 / ATH S30	25 (resin) / 30 ATH	60	160	-	-	0.436
D-n-200	DGEBA	B-IPDA	-	-	90	140	-	-	0.216
D-n-300	DGEBA	B-IPDA	-	-	90	140	-	-	0.293
D-n-500	DGEBA	B-IPDA	-	-	90	140	-	-	0.484
D-n-500-p	DGEBA	IPDA/ B-IPDA	-	-	60	180	300	60	0.489

The glass transition temperature (T_g) was determined by dynamic mechanical analysis (DMA) with a Rheometric Scientific RDA III (Rheoservice GmbH & Co. KG, Reichelsheim, Germany) device, analyzing the samples in the range of 30 – 200 °C. The optical characterization of the cell structure was carried out via SEM analysis with a JSM-6510 (Jeol, Akishima, Tokyo, Japan). Samples were sputtered with a 13-nm gold layer (99.9 % purity) by using a Sputter Coater 108auto (Cressington, Waltford, England). The density of the specimen was measured via weighing cut out specimen, skin layer removed, with a laboratory scale by the Archimedes principle, referring to DIN EN ISO 845.

Compression tests were carried out using a 20 kN universal test machine ZMART PRO 1455 (ZwickRoell, Ulm, Germany), referring to DIN EN ISO 844:2014-11 with a preload of 1 N and a test speed of 1 mm/min up to a deformation of 70%. Initial burning tests were carried out based on the horizontal burn test of UL 94. Cone calorimeter tests were conducted with an i-Cone Cone Calorimeter from Fire Testing Technology (East Grinstead, United Kingdom) according to ISO 5660 with a heat flux of 35 kW m^{-2} and 25 mm gap size.

3. RESULTS

3.1 Decomposition of CO_2-blocked amine hardener

Since carbamate salts function as latent blowing agents, it is important to understand the unblocking behavior and specific temperatures in order to choose suitable resin systems and process temperatures and thereby obtain application-specific tailored epoxy foams. Therefore, TGA measurements of the blocked (carbamate salt; *B-IPDA*) and unblocked IPDA (*IPDA*) were carried out and compared in Figure 1a.

As mentioned before for carbamate systems, based on CO_2-blocked AEP (B-AEP) by Ren et al. [19] and Bethke et al. [20], the decomposition of a CO_2-blocked amine curing agent shows three main steps. The first decomposition step with an onset temperature 82 °C (1 wt.% weight loss), a peak temperature of 130 °C and 29 wt.% loss indicates the main release of CO_2. Due do vaporization effects during the measurement, a certain amount of IPDA is also expected to evaporate in this range, as a weight loss of 9 wt.% was detected in this range during neat IPDA measurement. Since, CO_2 is released, the decomposition onset temperature indicates also the curing onset via unblocked IPDA. The second decomposition step with a peak temperature at 148 °C indicates the release of additional CO_2 and more IPDA. The third decomposition step up to 164 °C is related to the total vaporization of the remaining IPDA.

3.2 Reactivity of carbamate epoxy system

In order to tailor the foam properties, the knowledge of certain process temperatures, is necessary. Therefore, Figures 1b-c show the DSC curve of B-IPDA as well as the curing behavior of EN with IPDA and B-IPDA. It is shown, that B-IPDA is stable until an onset temperature of 82 °C. Because the peak decomposition (Figure 1a) temperature of B-IPDA (147 °C) is higher than the reaction peak of EN-IPDA (110 °C), the reaction of EN with B-IPDA is slightly shifted to higher temperatures compared to the unblocked IPDA. Furthermore, the degradation of carbamate salts is an endothermic process, leading to a draw of energy from the curing reaction. In order to control the foam morphology of epoxy foams, while using carbamate salts as blowing agents, the resin viscosity is one main influence parameter as shown in the literature. A higher viscosity of the resin system leads to more homogeneous morphology with a smaller cell size distribution. [20]

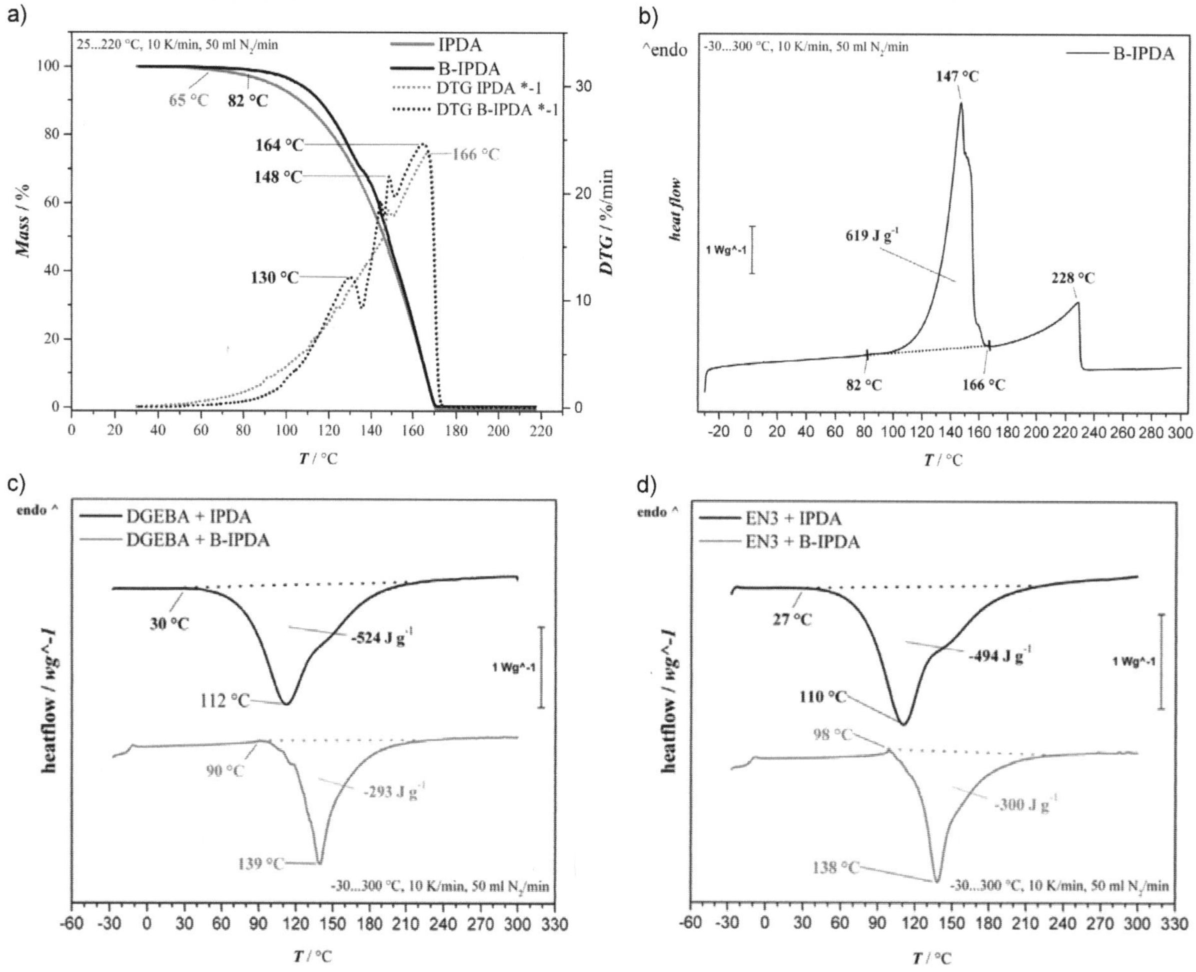

Figure 1a): Comparison of degradation behavior of IPDA and B-IPDA in the TGA. The 1st derivative is multiplied by -1 for proper illustration,
Figure 1b): Characteristic decomposition temperatures of B-IPDA as observed from DSC,
Figure 1c): Curing behavior of DGEBA with B-IPDA as observed from DSC,
Figure 1d): Curing behavior of EN with B-IPDA as observed from DSC.

As stated in section 2, DGEBA and EN are used as epoxy resin systems in this project. The viscosity over temperature of EN with IPDA and B-IPDA are summarized in Figure 2.
At room temperature EN shows an initial viscosity of 19 Pa s which drops down to 0.14 Pa s at 82 °C, the initial decomposition temperature of B-IPDA, and further to 0.04 Pa s at the temperature of the first decomposition peak of B-IPDA (130 °C). DGEBA has an initial viscosity of 13 Pa s. At 82 °C, the viscosity dropped to 0.07 Pa s and further to 0.02 Pa s at the temperature of the first decomposition peak of B-IPDA (130 °C). The addition of 100 % stoichiometric IPDA at DGEBA and EN leads to a temperature induced drop down of the viscosity up to 60°C. At this point the curing reaction dominates leading to a change in the slope. The curing reaction leads to a total gelation of the systems around 102 °C for EN and 104 °C for DGEBA. This behavior can be changed when less than 100 % stoichiometric amounts of IPDA are added. Several pre-trials lead to the conclusion that an amount of 30 % is suitable, as both resin systems reach a plateau region in the dynamic measurement, where the viscosity remains nearly constant over longer temperature ranges. In case of EN around 0.66 and for DGEBA at 0.33 Pa s.

Thus, for the pre-curing of DGEBA and EN a mixture of 30 wt.% unblocked and 70 wt.% blocked IPDA was chosen. However, for comparison samples containing B-IPDA only where investigated, too. Both systems were measured with a 120 min precuring at 60 °C. It has to be noted in advance, that the released CO_2 is disturbing the measurement, as material can be pushed out of the plate and bubbles entrapped. The observed change in viscosity can be observed in Figure 2.

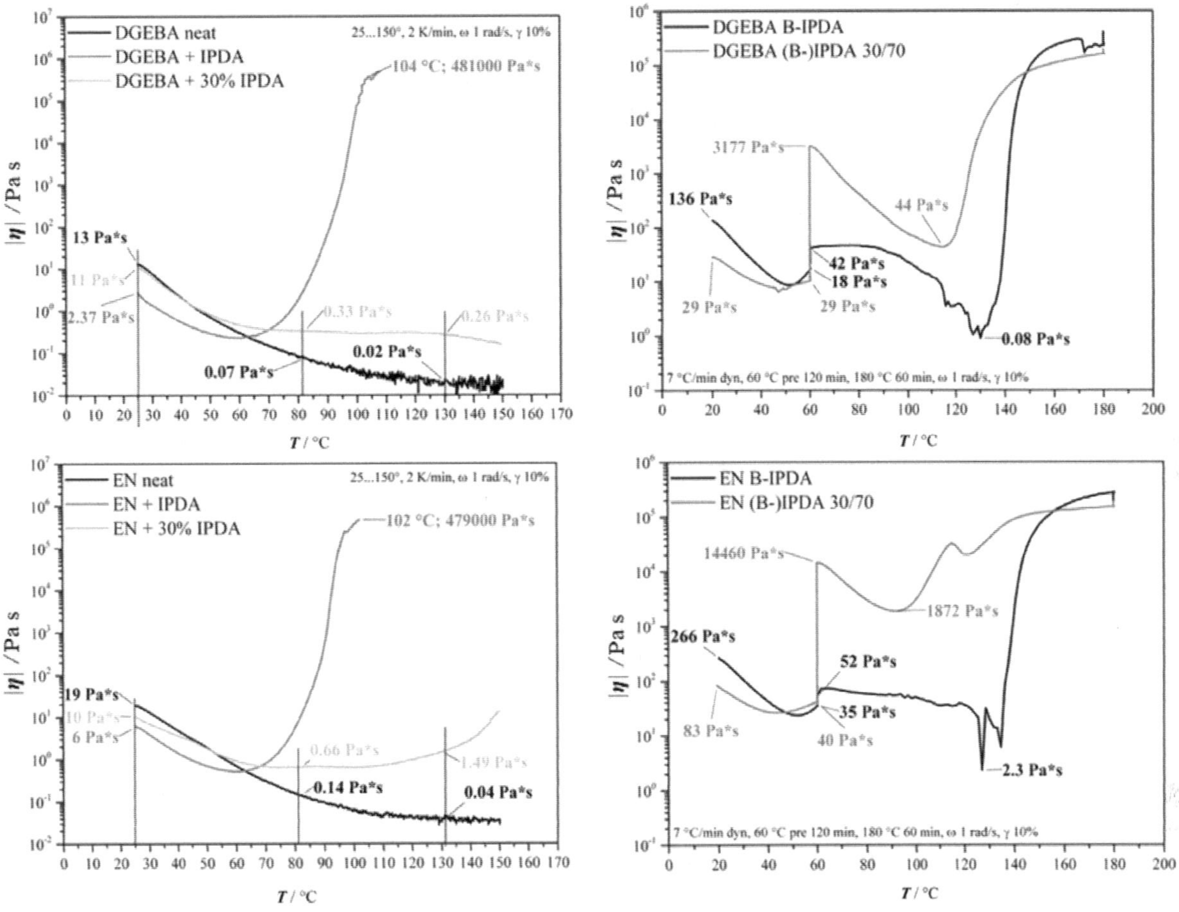

Figure 2. Viscosity over temperature of DGEBA and EN with IPDA and B-IPDA systems

As it can be observed from Figure 2, that the blocked hardener shows only slight increase of viscosity during pre-curing, namely 17 Pa s in case of EN and 24 Pa s in case of DGEBA. These effects may result from small amounts of unblocking B-IPDA during this time caused by the shearing effects during the measurement and solubility effects in the resin, which effects along the whole measurement, especially at elevated temperatures. When 30 wt.% of the B-IPDA are replaced by unblocked IPDA, a pre-curing effect can be observed. Within the 120 min of pre-curing, the viscosity of the EN system is increasing from 40 to 14460, while at DGEBA it is increasing from 29 to 3177 Pa s. When the temperature is increased further to 180 °C, a thermally induced drop-down in viscosity can be observed, supported by a plasticizing effect of released CO_2 from B-IPDA decomposition. In case of EN with 30 % IPDA a minimum of 1872 Pa s is reached, before the network-growth overcome these effects. This can be seen as minimum-viscosity before bubble stabilization takes place. Thus, the viscosity before foaming can be adjusted. A slight second drop down around 125 °C can be observed, resulting in further CO_2 release as observed from TGA. The EN B-IPDA sample shows neatly a plateau until around 123 °C, where the CO_2 release is very high and a minimum viscosity of 2.3 Pa s is reached. Same effects can be observed at DGEBA. However, the lower network strength due to the two functional DGEBA is resulting in a weak plateau region with an early drop of viscosity and significant drop down to 0.08 Pa s before foam stabilization. In conclusion, the higher functionality of the EN (2.7) in comparison to DGEBA (2) allows a higher viscosity level before foaming in addition to a faster and more stable network growth. These results lead to the expectation of a better cell morphology and shorter cycle times compared to DGEBA. This is in accordance to the findings of Bethke et. al. [20], where higher functionality of resins and higher resin viscosity was found to be beneficial for morphology control. The contribution of B-IPDA to the precuring can be seen as neglectable, approaching the idea of Ren et. al. of a latent carbamate curing system under ambient conditions [19].

3.3 Resulting foam density and morphology

For the application in sandwich structures, a low density and high mechanical performance, are key factors. State of the art flame retardant epoxy foam systems are in the medium-density spectrum, f.e. Sicomin PB 410 with a standard density of 0.5 g/cm³ and a measured average cell diameter of approx. 79.61 µm. In order to gain a low density while maintaining attractive mechanical properties, a homogeneous foam morphology is crucial. In Table 1, the density of the foams, developed in the framework of this project, are listed. It can be summarized, that in this foaming approach, using only CO_2 blocked amine curing agents in the form of carbamate salts as blowing agents, medium to high foam densities can be achieved. When using EN as base resin system, without any pre-curing, more homogeneous foam structures can be achieved compared to foams using DGEBA. All foams were produced with sealed molds using a limited foaming approach. The density of unmodified foams can be tailored in the range of approximately 0.2 g/cm³ to 0.5 g/cm³ and above, while maintaining a homogeneous medium fine to fine morphology. With a pre-curing approach, using a mixture of 30 wt.% unblocked and 70 wt.% CO_2-blocked IPDA, the density of the foams increases to high density foams (0.5 g/cm³ and higher), due to a higher matrix viscosity and lower CO_2-content of the foamed system. Using high filler degrees, as ATH in this application, leads to an increase of density, as the foamable matrix content is reduced and replaced with higher density (2.4 g/cm³) particles. Thus, the introduction of a pre-curing approach increases the density of the foams, it is also an effective way of tailoring foam cell size and distribution as shown in former studies [14, 20]. When using DGEBA and B-IPDA for example, the average cell size decreases from 166.22 µm (standard deviation 190.5 µm) to 54.19 µm (standard deviation: 25.83 µm). Through the introduction of 40 wt.% ATH, the average cell size decreases further to 42.21 µm (standard deviation: 23.97 µm). Figure 3 shows a comparison between the foam morphology of carbamate (B-IPDA) foamed DGEBA with and without pre-curing. As it can be seen, pre-curing, using as well carbamate salt as unblocked amine curing agents, is an effective way to tailor the foam morphology and density. At higher pre-curing temperatures (70 °C and above), a highly bimodal cell morphology results in the final foam, as carbamates already start to decompose. Thus, lower pre-curing temperatures have to be preferred in order to achieve a fine and homogeneous foam morphology. Figure 3 shows SEM images of the core region of foamed specimen, using DGEBA/B-IPDA/IPDA with a pre-curing approach and different flame retardants. All of these specimens show a very fine foam morphology with a maximum cell size in the range of 100 – 150 µm in diameter. Neat foam systems show the higher homogeneity in cell size, while the addition of high filler degrees of particles – ATH or clay – lead not only to a significant increase in density, but also to a bimodal foam morphology.

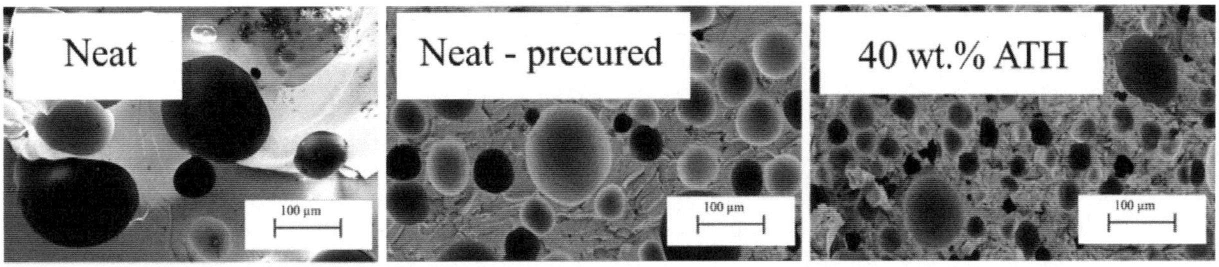

Figure 3: SEM images of DGEBA/B-IPDA foams, neat and precured with ATH

It can be concluded, that the pre-curing approach is suitable to achieve fine and very fine foam morphologies. However, long cycle-times (about 4 h at 60 °C) are disadvantageous for an economic foam production. Therefore, other approaches like the introduction of a higher viscosity resin system or filler particles, ATH in this case, can be beneficial. As described in section 3.2 the viscosity at foaming temperature of EN is not sufficiently higher than DGEBA to achieve a tolerable morphology. In Figure 4, the resulting morphology of EN foams, foamed with B-IPDA and IPDA, modified with ATH and/or Polyphlox 3760 (*DOPO*) are shown.

Figure 4: top: Comparison of cross-cut 30 x 30 x 10 mm³ EN/B-IPDA specimens with and without flame retardants; bottom: corresponding SEM images

As it can be seen, the modification of EN foam systems with particle additives lead to a much finer foam morphology with a bimodular cell size distribution (average cell size for EN-DOPO-ATHS30-30 without precuring: 113.37 µm, standard deviation: 123.28 µm).

3.4 Thermal properties

Advanced foam applications, especially in the field of sandwich panels and the aviation sector, require a high thermal performance. For the production of sandwich-SMC panels for example, process temperatures of up to 150 °C at the cavity can occur. Therefore, a high glass transition temperature is a relevant requirement. The glass transition temperatures of the specimen are listed in Table 2. It is shown, that the glass transition temperature (*Peak tan_delta*) of foam systems, using DGEBA are slightly higher in comparison with EN based foams. It can be also concluded, that the addition of filler particles, as well as the modification of density, does not have a significant influence of EN based foams. All of those specimens exhibit a very high glass transition temperature of approx. 158 °C, higher than the Sandwich-SMC process temperatures (approx. 150 °C). The introduction of DOPO causes a decrease in glass transition temperatures due to a lower interconnectivity in the molecular network of the cured polymer.

3.5 Mechanical compression properties

Besides the thermal, also mechanical properties are relevant requirements for foam systems for lightweight applications. For sandwich structures, and especially Sandwich-SMC panels due to the high pressure at the production process (approx. 2.5 MPa), a high compressive strength and stiffness are key features. Figure 5 shows exemplary compressive stress-strain-diagrams of particle-filled and unfilled epoxy foams systems, foamed with CO_2-blocked amine curing agents. Table 2 shows the summarized results of the compression tests of the developed foams.

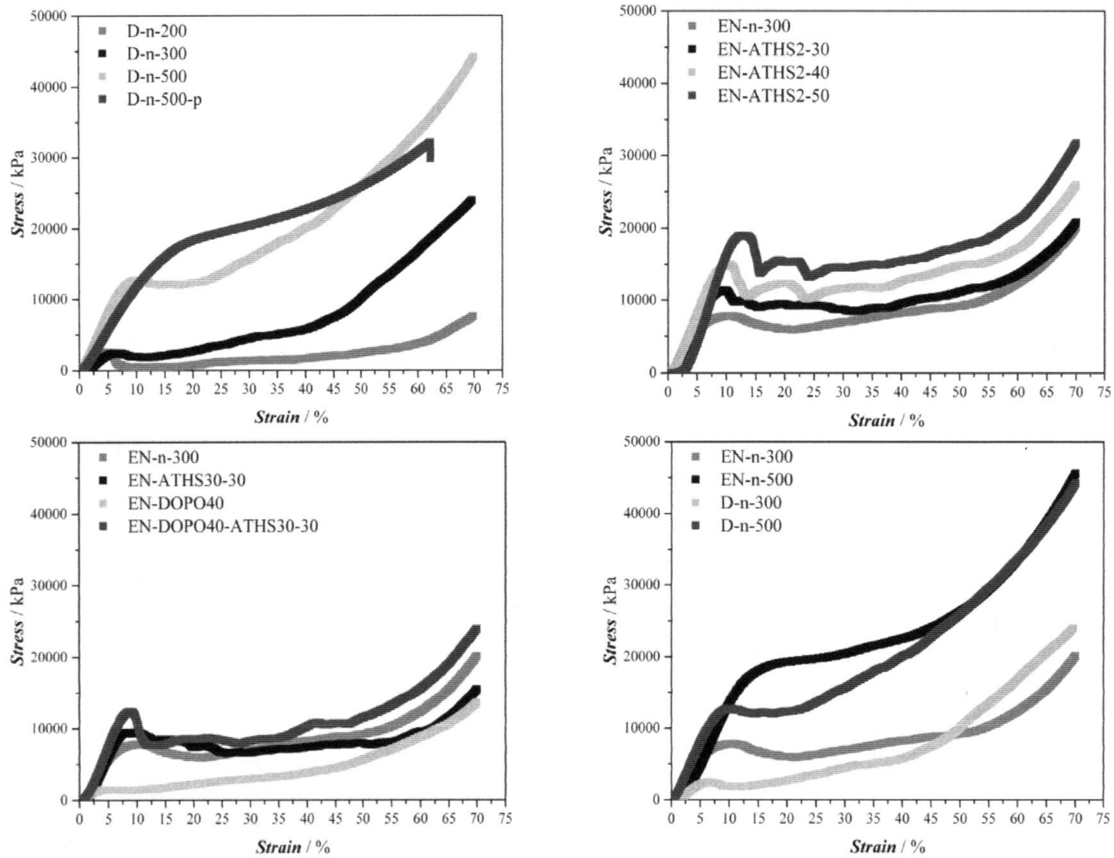

Figure 5: Exemplary compressive stress-strain-diagrams of modified epoxy foams systems

As it can be seen and expected, a higher foam density leads to a higher compressive strength and compressive modulus. For example, the EN/B-IPDA system without pre-curing exhibits an increase in compressive modulus of 130.667 MPa and a stress at 10 % strain of 6.9 MPa at a foam density of approx. 0.3 g/cm³ to 195 MPa compressive modulus and 13.41 MPa stress at 10 % strain at a foam density of approx. 0.5 g/cm³. Thus, as it can be seen in Table 2 by comparing the results for DGEBA/B-IPDA/IPDA with a foam density of approx. 0.5 g/cm³, an increased homogeneity of the foam morphology, achieved by using pre-curing with, does not lead to higher compressive mechanical properties.

Table 2: Density, glass transition temperature (derived from DMA) and compressive properties

Sample	Density [g/cm³]	*Tg* [°C] (Peak tan_delta)	Compressive Young's modulus [MPa]	Stress at 10 % strain [MPa]	Stress at break [MPa]	Strain at break [%]
EN-n-300	0.292	158	130.67	6.9	6.91	8.83
EN-n-500	0.475	158	195	13.41	n/a	n/a
EN-ATHS30-30	0.405	157	184.33	11.08	11.17	8.17
EN-ATHS2-30	0.430	158	163	10.9	11.03	9
EN-ATHS2-40	0.506	158	199.33	14.33	14.43	9.07
EN-ATHS2-50	0.605	n/a	250	17.05	19	9.5
EN-DOPO	0.274	113	53.85	1.11	1.47	4.2
EN-DOPO-ATHS30-30	0.436	130	207.67	11.56	12.47	7.6
D-n-200	0.216	159	86.5	0.42	2.53	4.1
D-n-300	0.293	165	132.5	1.9	2.27	3.8
D-n-500	0.484	160	166	12.73	12.8	9.5
D-n-500-p	0.489	162	142	11.67	n/a	n/a

In summary, unmodified EN-based epoxy foams exhibit superior mechanical properties in compression, compared to DGEBA-based foam systems of the same density. The introduction of stiff filler particles, ATH in this case, lead to a variation of the compressive behavior. As Table 2 shows, by comparing the results for the neat EN-based foam system "EN-n-300" with ATH-modified variations, an increase in compressive young's modulus as well as in stress at 10 % strain and stress at break occurs with a higher filler degree of ATH Reflamal S2. Also, a slight increase of strain at break occurs with an increasing filler degree. As shown before, the modification with finer particles leads to an increased nucleating effect and a finer foam morphology, resulting in an increase in mechanical properties. In comparison, compressive young's modulus and strength are slightly superior, when using a finer grade of ATH with the same filler degree (Table 2: EN-ATHS30-30, EN-ATHS2-30) with a slight decrease in strain at break. The use of Polyphlox 3760 (*DOPO*) in EN/B-IPDA (25 wt.% Polyphlox, 75 wt.% EN) in combination with ATH (S30, total filler degree: 30 wt.%) has only a slight influence on the mechanical properties in compression. If DOPO is used without ATH with a higher content (40 wt.% Polyphlox, 60 wt.% EN), the compression properties of the foam are significantly lowered. It was shown, that the mechanical compression properties of epoxy foams, foamed with a carbamate salt approach, can be tailored to meet application specific properties via the variation of density, morphology (via pre-curing) or the addition of filler particles with a nucleating effect.

3.6 Burning behavior

For a first test series, the burning behavior of DGEBA foams, using a carbamate foaming approach and modified with different commercially available flame retardants were evaluated by Cone calorimeter tests according to ISO 5660 (heat flux 35 kW/m^{-2}; gap size 25 mm). Figure 6 summarizes the resulting heat release rate over time for these specimens as well as the relative values for time to ignition (TTI), the peak heat release rate (pHRR), total heat release (THR) and total smoke released (TSR).

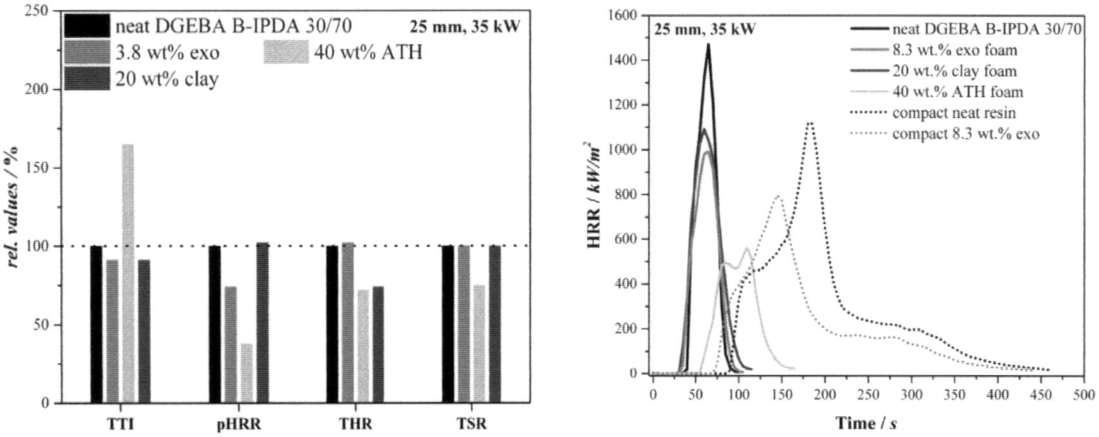

Figure 6: Cone calorimeter test results of modified DGEBA/B-IPDA foams

In comparison, the specimen, modified with 40 wt.% ATH, performs best. The specimen, modified with 20 wt.% clay and 5.22 wt.% exo showed similar performance, inferior to the ATH modified specimen. However, the addition of 40 wt.% of ATH also leads to a significant increase in density. Therefore, further trials were carried out, using a lower ATH filler degree and an EN-based foam matrix system. The results are shown in Figure 7.

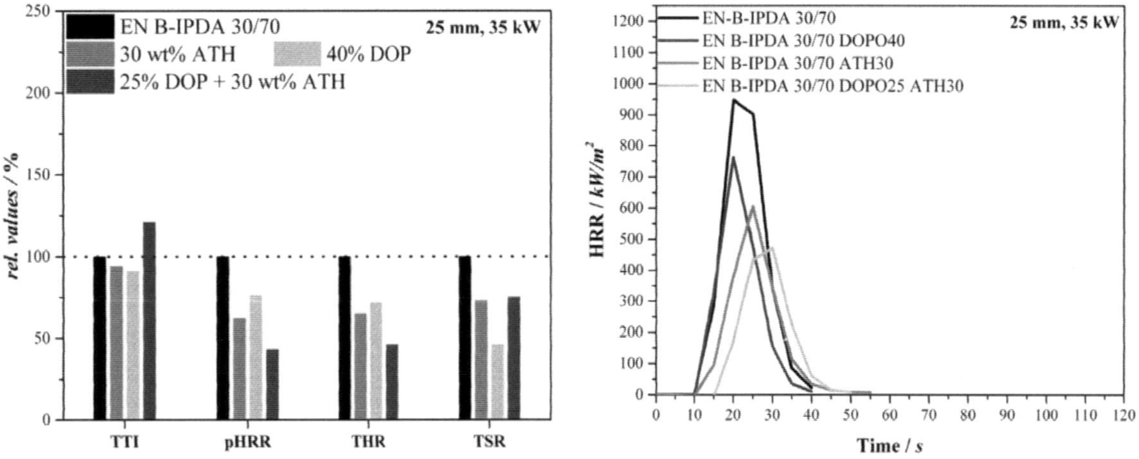

Figure 7: Cone calorimeter test results of modified EN/B-IPDA foams

In comparison the specimen, modified with a combination of DOPO (25 % resin content) and ATH (30 wt.%) showed the best results with a significant decrease in peak heat release rate (pHRR) and total heat release (THR) compared to the unmodified foam. Because of the superior performance of the EN-based foams, only these were further tested in modified horizontal burning tests, according to UL94. Due to the produced sample size, only shortened specimen could be tested, only characterizing the burning behavior to the 25 mm mark. While the unmodified EN-n-300 specimen fails the criteria of the UL94 HB rating, all specimen, modified with flame retardants pass this test – see Table 3 for further details. All specimen, modified with ATH did not exhibit a flame spread to the 25 mm mark.

Table 3: Results of modified horizontal burn tests (according to UL94)

Sample	Time to 25mm mark [s]	Time to extinction [s]	Dripping [Y/N]	Pass/Fail
EN-n-300	15.9	59.35	Y	Fail
EN-ATHS30-30	-	45.64	N	Pass
EN-ATHS2-30	-	45.6	N	Pass
EN-ATHS2-40	-	33.96	N	Pass
EN-ATHS2-50	-	2	N	Pass
EN-DOPO	17.19	24.54	N	Pass
EN-DOPO-ATHS30-30	-	15.54	N	Pass

Following results can be concluded from the horizontal burn tests: A higher filler degree in ATH leads to a shorter time to extinction. Dripping can be prevented with the modification of ATH and/or DOPO. The modification with DOPO and ATH shows a very good performance while maintaining a relatively low density, compared to the 50 wt.% ATH filled specimens. The ATH particle size does not have a significant influence on the horizontal burn results.

4. CONCLUSIONS

It can be concluded, that the innovative carbamate foaming approach is suitable to produce technical epoxy foams with tailorable thermo-mechanical properties and density as well as morphology. This foaming approach shows an environmentally friendly alternative to established foaming techniques of thermosets, that can be tailored to a wide range of applications in automotive and aviation industry (f.e. Sandwich-SMC aircraft interior panels).

Due to the latent properties of the carbamates at low temperatures, the storability of the systems as One-Pack-Formulations at room temperature can be significantly elongated, providing new ways for processing approaches. It was shown, that the compressive properties can be modified using a higher foam density, a modified foam morphology via a pre-curing approach and the modification with filler particles (f.e. ATH). Different flame retardants were evaluated. The EN/B-IPDA composition, modified with a combination of DOPO and ATH shows the best compromise between burning behavior and low density. In future work, the FST performance of the developed foams, according to JAR- 25, §25.853 (c) and Appendix F, Part IV, Change 15 has to be evaluated, using an Ohio State University (OSU) rate of heat release apparatus, enlarging the field of possible applications.

5. FUNDING

The research and development project "DiaPore" is funded by European Union and the German Federal Ministry of Education and Research in the funding program EUROSTARS.

6. ACKNOWLEDGMENTS

The authors kindly thank WTH Stade, Schill & Seilacher, Huntsman, and Olin Blue Cube Germany for material samples used in this research, as well as Exakt Advanced Technologies for supporting this work. The Department of Inorganic Chemistry II at the University of Bayreuth for the ability and assistance to proceed the cone measurements, and the Bavarian Polymer Institute (BPI) for the ability to work in their labs and equipment.

7. REFERENCES

[1]: Weil, E.D. & Levchik, S., "A Review of Current Flame Retardant Systems for Epoxy Resins." *JOURNAL OF FIRE SCIENCES,* 22 (2004): 25-40. doi: 10.1177/0734904104038107.

[2] Asaro, R.J., Lattimer, B. & Ramroth, W., "Structural response of FRP composites during fire." *Composite Structures* 87(4) (2009): 382-393. doi: 10.1016/j.compstruct.2008.02.018.

[3] Bai, Y. & Keller, T., "Time Dependence of Material Properties of FRP Composites in Fire." *Journal of Composite Materials* 43(21) (2009): 2469-2484. doi: 10.1177/0021998309344641.

[4] Gibson, A.G., "The Integrity of Polymer Composites during and after Fire." *Journal of Composite Materials* 38(15) (2004): 1283-1307. doi: 10.1177/0021998304042733.

[5] Fette, M., Reiss, M., Hentschel, M. & Stöß, N., „Technologie zur Herstellung von komplexen und funktionalen SMC-Schaum-Sandwich-Strukturen für Luftfahrt- und Automobilindustrie." *LightweightDesign* 3 (2016): 52-55.

[6] Fette, M., Reiss, M., Stöß, N. & Hentschel, M., "New Approach for the Efficient Manufacturing of Sandwich Structures Based on Sheet Moulding Compounds." *Advanced Materials Research* 1140 (2016): 264-271. doi: 10.4028/www.scientific.net/amr.1140.264.

[7]: Pascalut, J.-P. & Williams, R.J.J., "Epoxy Polymers, New Materials and Innovations." Weinheim: Wiley-VCH Verlag GmbH & Co. KGaA, 2010.

[8]: Takiguchi, O., Ishikawa, D. & Sugimoto, M., "Effect of Rheological Behavior of Epoxy during Precuring on Foaming." *J. Appl. Polym. Sci.* 110 (2008): 657-662. doi: 10.1002/app28727.

[9]: Ren, Q. & Zhu, S., "One-Pack Epoxy Foaming with CO2 as Latent Blowing Agent." *ACS Macro Lett.* 4 (2015): 693-697. doi: 10.1021/acsmacrolett.5b00320.

[10]: Kurek, K. & Bledzki, A., "Fatigue Behavior of Composites with Foamed Matrix." *J. Reinf. Plast. Compos.* 13 (1994): 1116-1134. doi: 10.1177/073168449401301204.

[11]: Kurek, K. & Bledzki, A., "Mechanical Behavior of Polyurethane- And Epoxy Foams and Their Glass Fiber Composites." *Mech. Compos. Mater.* 30 (1994): 105-109. doi: 10.1007/BF00635839.

[12]: Hajimichael, M., Lewis, A. & Scholey, D., "Investigation and Development of Epoxy Foams." *Br. Polym. J.* 18 (1986): 307-311. doi: 10.1002/pi.4980180506.

[13]: Kühlkamp, A., Mauz, O. & S. Göwecke, "Process for Making Epoxy Resin Foam Plastics." *U.S. Patent* US3406131, 15 October 1968.

[14]: Lyu, J., Liu, T. & Xi, Z., "Effect of pre-curing process on epoxy resin foaming using carbon dioxide as blowing agent." *J. of cell. Plast.* 53 (2017): 181-197. doi: 10.1177/0021955X16639235

[15]: Wang, L., Zhang, C., Gong, W., Ji, Y., Qin, S. & He, L., "Preparation of Microcellular Epoxy Foams through a Limited-Foaming Process: A Contradiction with the Time-Temperature-Transformation Cure Diagram." *Adv. Mater.* 30 (2018): 1703992. doi: 10.1002/adma.201703992

[16]: Yu, Q., Zhao, Y., Dong, A. & Y. Li, "Mechanical properties of EPS filled syntactic foams prepared by VARTM." *Composites Part B* 136 (2018): 126-134. doi: 10.1016/j.compositesb.2017.07.053

[17]: Gazzani, S.E., Nassiet, V. & Habas, J., "High Temperature Epoxy Foam: Optimization of Process Parameters." *Polymers* 8 (2016): 215. doi: 10.3390/polym8060215.

[18]: Gupta, N., "A Functionally Graded Syntactic Foam Material for High Energy Absorption under Compression." *Mater. Lett.* 61 (2007): 979-982. doi: 10.1016/j.matlet.2006.06.033.

[19]: Ren, Q. & Zhu, S., "Development of Epoxy Foaming with CO2 as Latent Blowing Agent and Principle in Selection of Amine Curing Agent." *Ind. Eng. Chem. Res.* 54 (2015): 11056-11064. doi: 10.1021/acs.iecr.5b03069.

[20]: Bethke, C., Sanchez-Vazquez, S. A., Raps, D., Bakis, G., Bard, S., Ngoc, D. & Altstädt, V., "Effect of Resin and Blocked/Unblocked Hardener Mixture on the Production of Epoxy Foams with CO2 Blocked Hardener in Batch Foaming Process." *Polymers* 11 (2019): 793. doi: 10.3390/polym11050793.

ENABLING RESPONSIVE REAL-TIME INSPECTION OF THE AUTOMATED FIBER PLACEMENT PROCESS

Maxime Rivard[1], Marc Palardy-Sim[1], Guy Lamouche[1], Steven Roy[1], Christian Padioleau[1], André Beauchesne[1], Daniel Levesque[1], Francis Boismenu[1], Louis-Guy Dicaire[1], Jonathan Boisvert[1], Shawn Peters[1], Jihua Chen[1], Marc-André Octeau[1], Julieta Barroeta Robles[1], François Ferland[1], Martin Tanguay[1], Jay Hissett[2], David Swope[2], Stephen Albers[2], Robert Harper[2], Ken Wright[2], Brad Buhrkuhl[3], Marcus Klakken[3], Gil Lund[3], and Ali Yousefpour[1]

[1] National Research Council Canada
1200 Montreal Road
Ottawa, Ontario, Canada, K1K 2E1

[2] Fives Cincinnati
2200 Litton Lane
Hebron, KY, 41048

[3] Fives Lund LLC
13536 Beacon Coal Mine Rd S
Seattle, WA, 98178

ABSTRACT

Automated Fiber Placement (AFP) is used to manufacture large and complex parts in the aerospace industry. A time-consuming portion of this fabrication process remains the inspection and quality control, which are largely performed visually after each deposited layer. This conference proceeding showcases a disruptive, responsive, and reliable solution based on the Fives In-Process Inspection system enabled by the National Research Council of Canada (NRC) Optical Coherence Tomography (OCT) sensor to perform in-process defect monitoring of the fiber layup. Measurements are taken close to the material deposition location without inhibiting the optimal machine path or slowing down the layup process. Assessment of the quality of a deposition takes place concurrently while the layup head applies material. This responsive feedback loop enables adaptive control of the AFP fabrication process. Technical details on how the OCT based inspection system's data flow has been integrated within the manufacturing process of a Fives Viper AFP machine are provided to explain the system's real-time response and high-resolution measurements. Results obtained in an industrial setting using the sensor installed on a production AFP machine are presented.

Corresponding author: Maxime.Rivard@nrc-cnrc.gc.ca

1. INTRODUCTION

An AFP machine fabricates a part with layers of composite material ply surfaces. Each ply consists of machine courses, which are composed of a multiplicity of tows layed down in a single pass of the layup head. For example, a machine may lay up to 32 tows of 3.175 mm width corresponding to a maximum course width of 101.6 mm. A set of adjacent courses with a common orientation constitute a ply, and a set of plies constitute a part. To ensure the finished part does not contain any critical defects that may be introduced by this additive manufacturing process, each deposited ply must be inspected and reworked, if necessary, before the next ply is layed.

The implementation of a high-fidelity AFP inspection system is a sought-after solution for machine manufacturers and end-users alike. Traditional ply-by-ply manual and visual inspection methods can consume as much as 30-60% of the total AFP production time. In addition to significantly reducing production time and cost, in-process inspection provides greater repeatability, removing operator bias and enabling adaptive process control. Furthermore, it allows part manufacturers to quantitatively demonstrate how the “as-built” part truly deviates from the “as-designed” model, an information that is critical for the definition of acceptable manufacturing tolerances.

The implementation of a real-time AFP inspection solution that does not impede the process, but rather becomes an integral part of the equipment, represents quite a challenge and has only achieved partial success in past implementations. An inspection system using a laser analog distance sensor was patented as early as 1996 by Kitson et al. The system could be mounted on the head of an AFP machine to perform real-time detection of predictable geometry-induced end of band gaps and overlaps as well as unpredictable manufacturing-induced gaps and overlaps along the length of a course [1]. Since then, a number of systems based on laser line triangulation sensors have been investigated to great depth as an inspection technique capable of defect detection [2-4]. Finally, thermographic sensing techniques have been used to demonstrate the feasibility of detecting defects by their unique temperature signature [5, 6]. While these systems have demonstrated some success in a few applications, a number of material and integration challenges limit their widespread application.

In previous papers, an automated fiber placement inspection system based on Optical Coherence Tomography (OCT) was presented [7, 8]. This sensing technique is employed to address the key challenges that must be overcome for an AFP inspection solution to be effective. This modular solution allows for a compact optical scanner design which is well integrated on the AFP head. The measurement rate of OCT also enables inspection without slowing down the fabrication process. The robust sensitivity of OCT is crucial to perform accurate measurements on any composite material, which often returns low optical power.

This paper presents some background on the OCT technology along with its key benefits applicable to in-process, real-time AFP inspection. The process of developing a 3D “as-built” dataset concurrent to part fabrication is described. The part manufacturing workflow, coordinate system management, and inspection data processing pipeline explain the integration of the inspection system within the AFP machine. Finally, measurement results from the inspection system integrated with a Fives Viper AFP machine are shown.

2. OPTICAL COHERENCE TOMOGRAPHY IMAGING

Optical Coherence Tomography (OCT) is an imaging technique initially proposed in 1991 that is similar to ultrasound imaging; however, with OCT, optical waves are used to probe the sample instead of mechanical waves [9, 10]. OCT is primarily used in the biomedical field to image transparent and semi-transparent tissue such as the retina and artery walls since it can penetrate up to a depth of a few mm under the tissue's surface [10]. Today, many commercial OCT systems are available, and the technique is used clinically in medical applications to diagnose patients. Outside of the medical world, OCT has also found a niche application in real-time welding inspection [11].

A simplified representation of the OCT measurement system employed in this work is depicted in Figure 1. It is based on the swept-source OCT technology [10]. A laser emits light, continuously sweeping over a broad wavelength spectrum, in an interferometer that divides the light between a reference arm and a sample arm. The reference arm has a mirror at the end of its optical path while the sample arm directs the light towards the object to be inspected. The light returned by the reference mirror and the sample is recombined and sent to a detector. The interference between the light returned from each arm contains information, encoded in the different wavelengths, on the mismatch in optical length between the path going through the reference arm and the path going through the sample arm. The location of any reflecting structures in the sample arm is recovered from a Fourier transform of the interference signal. An important feature of this technology is that the scanning system in the sample arm can be tailored to any size and shape required to accommodate a wide variety of applications and available space.

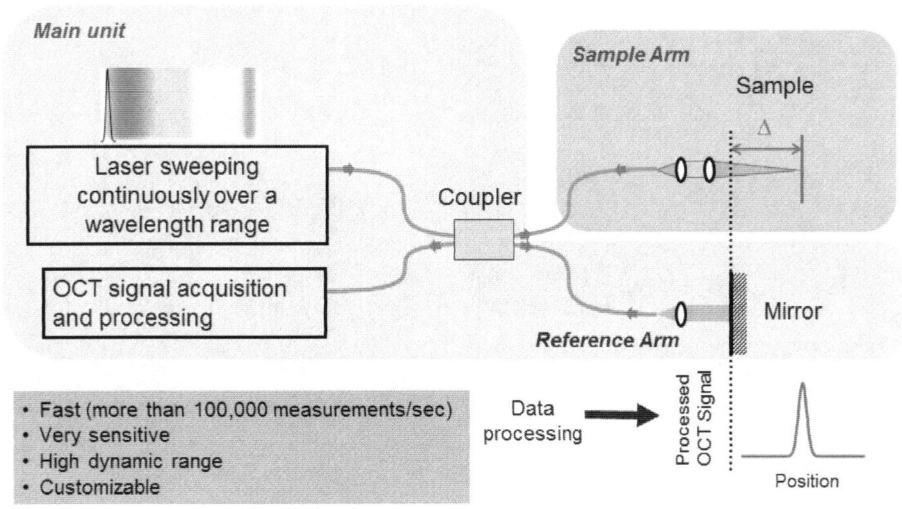

Figure 1: A simplified representation of an Optical Coherence Tomography system.

The carbon fiber based prepreg materials widely used for AFP are opaque to light, thus preventing penetrative measurements within the material. Nonetheless, the technique is a highly effective implementation of surface profilometry. It is capable of accurately imaging carbon fibers because of its high resolution and robust imaging capabilities. Previous publications have discussed in detail, the beneficial characteristics of OCT based measurement systems [7, 8]. They will be summarized again below as background.

The benefits of using OCT in a real-time sensing application are fivefold. Firstly, the scanning-head, or sample arm depicted in Figure 1, can be separated from the rest of the inspection system, facilitating low footprint sensor integration within the AFP head, which has limited space and inertial mass available for additional hardware. Secondly, the optical path is confocal (light is sent and received from the same pinhole point), thus, facilitating miniaturization and optimal mounting configuration of the scanning-head. Furthermore, because of the confocal optical configuration, light that does not originate from the focal point on the sample surface is rejected by the light collection optics of the sensor scanning head, giving OCT based measurement systems greater immunity to ambient light noise [11]. Thirdly, the measured signal varies proportionally to the square root of the intensity of the signal collected from the illumination of the sample surface. This dramatically increases the dynamic range of light intensity that can be handled by the system. In other words, OCT based measurement systems can perform measurement on surfaces returning high or low quantity of light without the need to change the sensor configuration to accommodate these variations in the signal. Fourthly, the system can detect weak signals from a sample surface because the signal returned from the reference arm remains high which helps maintain a measurable interference pattern. Since carbon fiber based prepreg materials return less light when inspected at higher incidence angle, this feature contributes to maintain accurate surface measurements with OCT systems even at high incidence angles [8]. Finally, the measurement range of OCT systems can be adjusted dynamically to keep track of surfaces with complex geometries. Modification of the length of the reference arm with a delay line and displacement of the focal point enables on-the-fly adjustments to maintain the inspected surface within the sensor's measurement range even if it moves closer to or away from the sensor head.

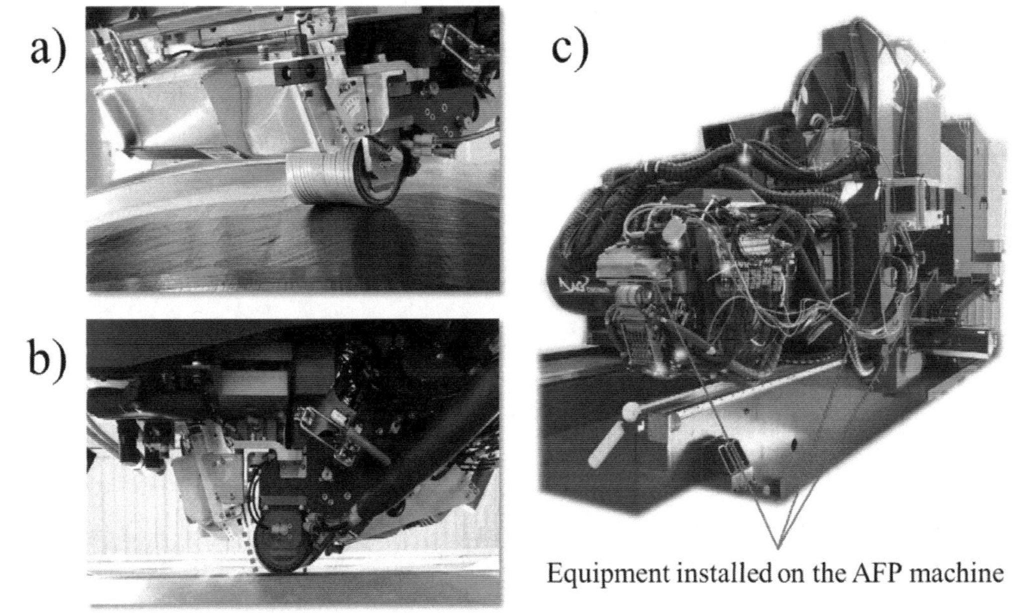

Figure 2: (a-b) OCT scanning head installed on an AFP head at an inclined angle. (c) All measurement system components installed on the Viper 4000 AFP machine: scanning head and the two enclosures (gray and black).

Figure 2 shows the OCT hardware integrated within NRC's Fives Viper 4000 AFP machine located at Stelia Aerospace in Mirabel, Quebec. Figure 2 a) and b) show the sensor scanning head installed on the AFP head. As the course is layed, the sensor scans a line perpendicular to the layup direction in the part's surface plane, after the compaction roller. The scanning head is inclined with respect to the compaction direction to obtain measurements as close as possible to the compaction roller nip point. This angle is roughly 20° and can be seen in Figure 2 b) where the optical path is depicted by the red dotted line. As shown in Figure 2 c), all the remaining hardware components are installed in two enclosures mounted on the side of the machine where they do not impact the high accuracy and acceleration of the layup head. The interferometer's sample arm, referenced earlier, extends from the black main enclosure in Figure 2 c) to the scanning head.

3. PART DEVELOPMENT AND MANUFACTURING WORKFLOW

In order to determine whether located features are classified as defects, the "as-built" sensor measurement scans must be compared to the "as-designed" digital part model. To do this, multiple data sources must be processed and combined correctly.

The process begins with the part design data. This data starts as a 3D solid model and is converted into composite ply surfaces via a CAM tool like Siemens FiberSim or CATIA Composite Workbench. These surfaces along with the fiber orientation of each ply (a.k.a. the part ply book) are the basic building blocks needed to generate an AFP machine program and represent a digital version of the perfect part. Next, the machine programming software tool is used to convert the ply surface definitions into an AFP machine program. The Fives Advanced Composites Environment Suite (ACES) offline programming software was used for this work. ACES imports the ply data from a CAM tool and generates a machine program, ply boundary definitions, and the inspection assets required to transform the measurements from the inspection sensor reference to an absolute, located, 3D point cloud dataset, namely, the "as-built" part. It accounts for the relationship of the tool surface to the layup head during the generation of this machine program to avoid collisions while maintaining contact. After the part program is generated, it is loaded into the AFP machine control system and used to manufacture the part. Machine alignment is performed prior to the layup to determine the relationship of the program to the tool surface. During layup, the In-Process Inspection system processes data simultaneously from both the AFP machine (head locations) and the inspection sensor (surface profiles measurements) to create a surface map of the ply. A profile is a collection of relative elevations or heights along a scanned width of the part surface. This custom Simultaneous Location and Mapping (SLAM) software module is a key ingredient to a responsive measurement system. The last step in generating the "as-built" part is the real-time alignment and comparison of the 3D point cloud of sensor measurements to the original CAD ply surfaces. The differences between these two datasets are candidates for deviations and defects. Those falling outside of manufacturing tolerances must be reworked before continuing layup of the next ply. With the outline of the process summarized above, the hardware alignment and required dataset reference frame transformations must be well defined.

4. COORDINATE REFERENCE FRAME MANAGEMENT

In order to correctly align the datasets in the process described above, the coordinate reference frames of each source must be carefully defined and managed. The following list describes these reference frames and Figure 3 demonstrates their relation to one another in the part workflow.

1. Design Reference Frame (DRF) – The part coordinate system which shows its location within the parent assembly. This definition originates from the customer's design or a larger assembly within which the part is integrated (e.g.: airplane coordinate system).

2. Program Reference Frame (PRF) – Once the specific AFP machine model and layup tool is defined, the program reference frame represents the initially aligned coordinate reference used in the machine program during New Production Introduction (NPI). This program is "posted" from the offline programming tool to the AFP control system.

3. Machine Reference Frame (MRF) – When the program is setup on the machine, the specific location of the layup tool in relation to the AFP head is determined via a probing process. The offset parameters from the ideal program is calculated and stored in the AFP control system. This frame is also sometimes referred to as world-coordinate frame since it describes the coordinates commonly used by the machine operators.

4. Sensor Reference Frame (SRF) – The inspection system provides a 2D profile for each scan line. Currently, this reference frame originates in the optical scanner of the OCT system. The origin of this reference frame moves with the AFP head relatively to the tool during the fabrication process which makes it a complex operation to accurately convert coordinates from the SRF to the DRF.

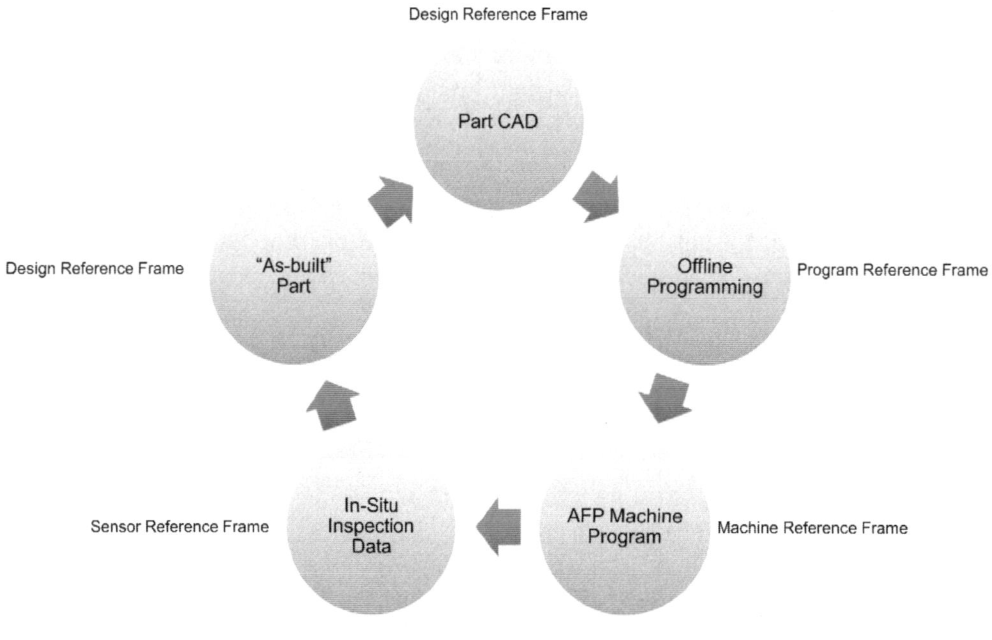

Figure 3: Relationship between reference frames used in a part workflow.

5. INSPECTION DATA PROCESSING PIPELINE

The data processing required to generate the "as-built" component involves two major activities to go from offline programming to processed data. First is the New Product Introduction (NPI) activity where the offline programming application provides the part program to the AFP machine and the part plan to the In-Process Inspection system. This step is a one-time process and is completed before layup. The second activity is the layup of the part and simultaneous streaming of inspection sensor data through the real-time processing pipeline. This step, outlined in Figure 4 a), involves first processing high bandwidth raw datagrams generated by the scanning head transducer into surface profiles during the scan conversion step. Next, these surface profiles scans are synchronized with the machine position stream before being fused together using the Simultaneous Location and Mapping (SLAM) module of the inspection system. The refined data obtained at this stage can then be stored and used to identify features, determine if they are defects (out of tolerance features) providing immediate process control information to the AFP operator.

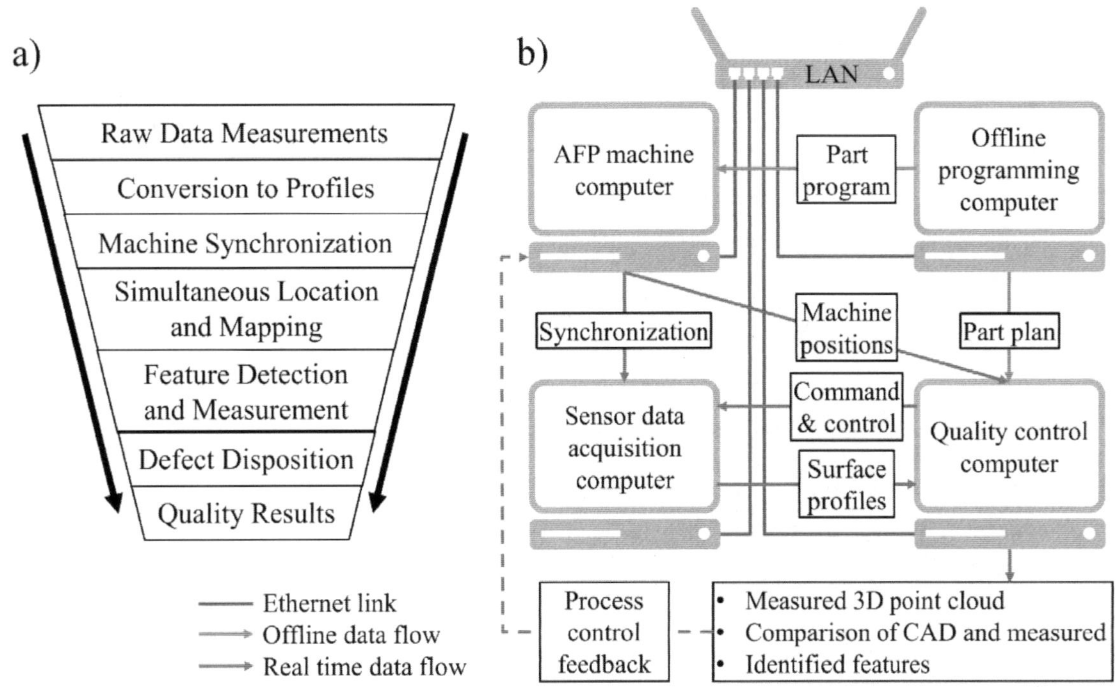

Figure 4: Overview of the data pipeline of the inspection system integrated to an AFP machine. (a) Contextual view of the real-time streaming pipeline leading to quality results. (b) Links between elements of the system.

Figure 4 b) shows the basic data flow between the offline programming computer running ACES, the AFP machine, and the In-Process Inspection system. This diagram shows how data is collected and processed into a real-time "as-built" part dataset. Prior to layup, the offline ACES programming software is used to generate the part program which is sent to the AFP machine computer as well as the part plan which is sent to the defect analysis computer (green arrows). This step is the one-time process completed before layup. Once layup starts, there is an active synchronization process between the In-Process Inspection system measurements and the AFP head's position and orientation across the tool surface. Both of these data streams are sent to the quality control computer where the surface profile data is processed into a located 3D point cloud ("as-built"), compared to the layup plan ("as-designed"), and analyzed for out of tolerance defects. These actions, shown as red lines in Figure 4 b), are performed concurrent to the AFP layup with results available immediately after a completed course to facilitate prompt feedback and determine if the process is in control and can continue layup.

IN-PROCESS INSPECTION RESULTS

To demonstrate the capabilities of the technology, measurements were collected in a production-like environment with the OCT based In-Process Inspection system mounted on the Viper 4000 AFP machine. The OCT based inspection system has the ability to detect a variety of defects. These defects are grouped into three types: Gaps/overlaps, placement errors (missing tow, tow end), and topical defects (splice, wrinkles, twists). Figure 5 shows a single course containing many tow starts and cuts layed up on a flat tool. A perpendicular full course, also layed up by AFP formed the underlying layer. This trial demonstrates the ability of the system to accurately detect both gaps and overlaps between adjacent tows as well as tow end placement.

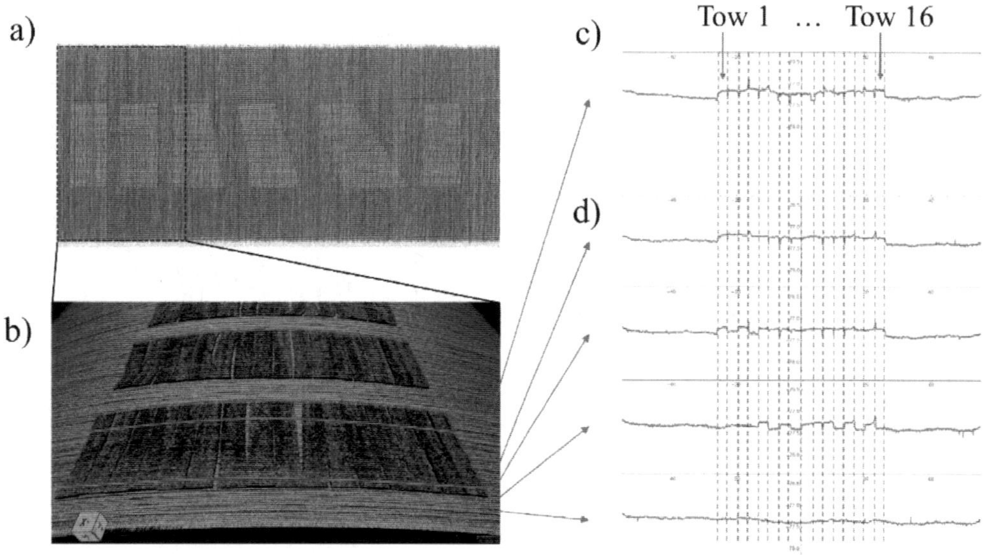

Figure 5: Trial measurement. (a) Image formed of consecutive profiles in the sensor reference frame. The image dimension is approximately 1500 mm in length by 100 mm in width. (b) Raw tomography volume data from which profiles are extracted by performing a peak detection. (c) Visible gaps and overlaps in a profile. (d) Four consecutive profiles that cover a start of band. The profile width covers 100 mm and the depth scale covers 2mm.

The intermediate measurement results from this trial are shown in Figure 5 a), where consecutive profiles from the sensor are combined to form a heat map image. Each pixel represents a single depth measurement and the heat map scale covers approximately 2 mm of depth in Figure 5 a). Figure 5 b) is a 3D image, with transparency, of the raw data showing the whole tomographic volume measured by the sensor. The points located at the surface provide the strongest signal and are the most apparent in the rendering. These points are isolated and located with a peak detection algorithm to provide depth measurements. The collection of depth measurements along the scan line are then regrouped to form surface profiles like those shown in Figure 5 c) and d). The profile shown in Figure 5 c) was taken in the middle of the deposited course. Based on the known nominal tow width and number of tows in a course, regions of interests can be identified in the profile. In this case 16 tows were layed up and identified. At the junction of adjacent tows, the presence and width of gaps or overlaps are measured. Figure 5 d) shows four consecutive profiles taken at the beginning of the course. Each individual tow's start location can be identified as it appears.

Figure 6 shows an example of splice, wrinkle, twist, and missing tow defects that were manually added to an existing course, with an 8-ply quasi-isotropic underlying laminate. Each pixel in the heat map images shown in Figure 6 b) to f) represents a single depth measurement. As seen in Figure 6 b) to f), all defects can be readily identified and classified despite the build-up of underlying layers containing numerous uncorrected gaps and overlaps, resulting in the topology visible in the figure. This topology is enhanced by the heat map scale that covers approximately 1 mm of depth.

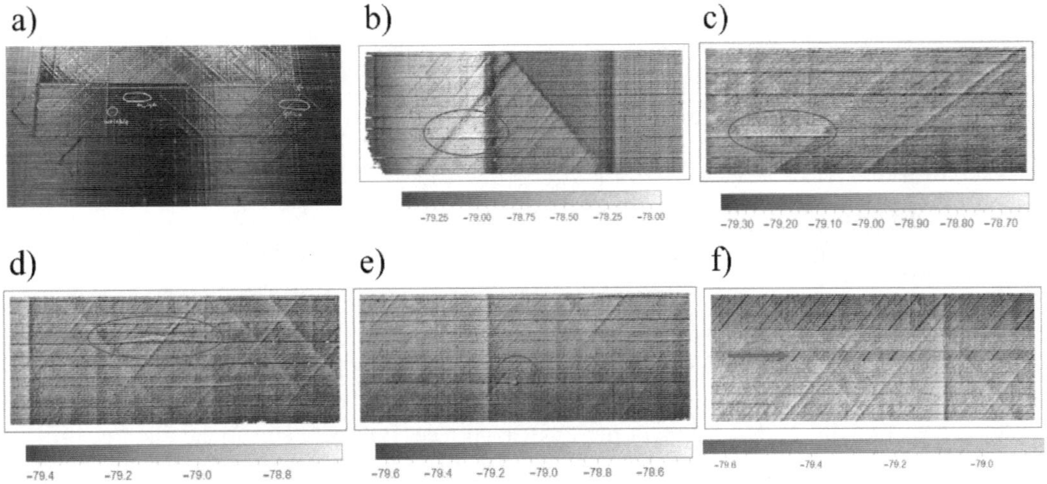

Figure 6: Flat surface with various defects. (a) Picture of the surface. (b-c) Tow splices. (d) Tow twist. (e) Tow wrinkle. (f) Missing tow. (b-f) The images dimensions are approximately 100 mm in length by 40 mm in width.

The intermediate results described above only show one basic level of feature detection offered by this technology. The data is presented in a relative reference (Sensor Reference Frame), and thus, has limited application on more complex part geometry. In order to demonstrate the ability to accurately locate and correct points in space, a single course layed up over the bevel section of complex shaped tool was measured. The course is highlighted in green in the ACES offline programming application in Figure 7 a). When the layup surface contains complex curvatures, the AFP head reorientation and change of speed during layup cause apparent image distortions from the true ply surface, as shown in Figure 7 b). These are evidenced by the discoloration (large blue triangular shadow followed by large bright yellow section) and apparent irregularities in the orientation of the tows in the underlying ply. Note that these discoloration represent a large topographic feature since the heat map scale covers 20 mm of depth in Figure 7 b). To correct for this artifact, the profiles measured by the sensor are transformed from the relative Sensor Reference Frame to an absolute 3D point cloud in the Design Reference Frame.

The procedure for creating a 3D point cloud described in this paper is applied and the results are shown in Figure 7 c). After converting the measured data to a 3D point cloud, the artifacts from machine reorientation and change of speed during layup are gone and the surface no longer appears distorted. At this stage, the “as-built” part can be directly compared to the “as-designed” CAD model.

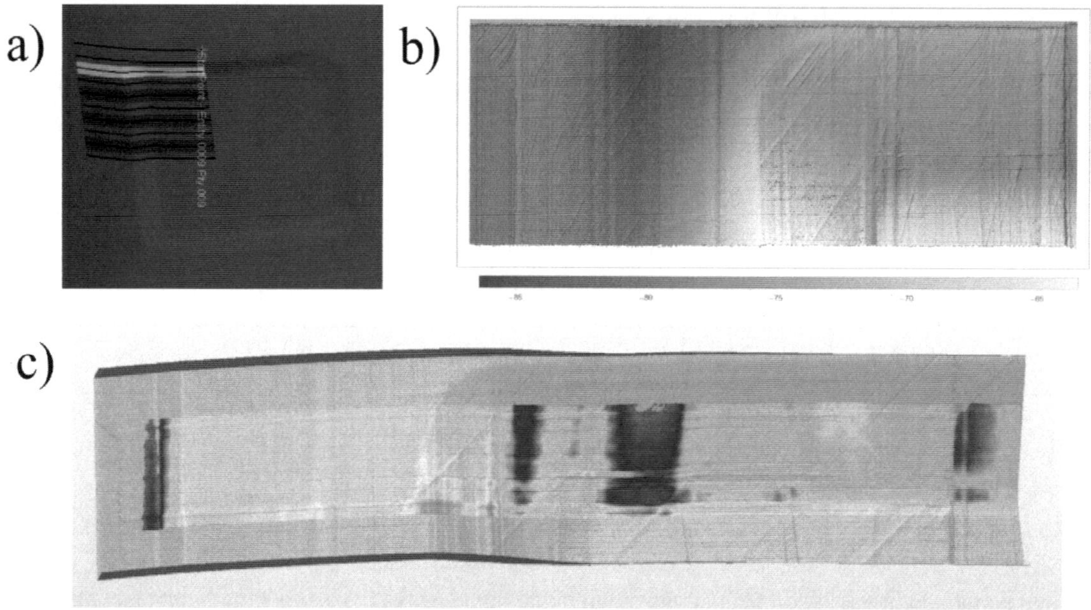

Figure 7: Measurement on a curved surface. (a) View of the programmed course in ACES (in green). (b) Consecutive profiles in the Sensor Reference Frame. The scan is approximately 100 mm wide, but dimensions along length are distorted. (c) 3D point cloud of the "as-built" part, located on the tool CAD model.

6. CONCLUSION

Gathering comprehensive surface information of a composite part with an in-process inspection system during its real-time AFP fabrication is a key enabler for comparing an "as-built" part to its "as-designed" specification. This paper describes how the NRC OCT sensor paired with the Fives In-Process Inspection system has overcome these challenges to generate accurate 3D point cloud datasets which can be used for inspection, process development, and real-time process control. The seamless integration of this measurement platform on a production AFP machine was outlined and results were shown to demonstrate the implementation of the solution.

7. REFERENCES

1. Kitson, L. E., Rock, D. K., & Eder, J. E., "Composite material laser flaw detection", *United States Patent*, #5562788, 1996.
2. Cemenska, J., et al., "AFP Automated Inspection System Performance and Expectations," *SAE Technical Paper* 2017-01-2150, 2017.

   [https://doi.org/10.4271/2017-01-2150]
3. Schmitt, R., et al., "In-process 3D laser measurement to control the fiber tape-laying for composite production," *SPIE Photonics Europe*, Brussels, 2010.

   [https://doi.org/10.1117/12.853882]
4. Weimer, C., et al., "Increasing the Productivity of CFRP Production Processes by Robustness and Reliability Enhancement," *CAMX*, Anaheim, 2016.

5. Juarez, P.D., et al., "Advances in in-situ inspection of automated fiber placement systems," *SPIE*, Baltimore, 2016.

[https://doi.org/10.1117/12.2223028]

6. Schmidt, C., et al., "Thermal Imaging as a Solution for Reliable Monitoring of AFP Processes," *3rd International Symposium on Automated Composites Manufacturing*, Montreal, 2017.

7. Palardy-Sim, M., et al., "Next Generation Inspection Solution for Automated Fibre Placement," *4th International Symposium on Automated Composites Manufacturing*, Montreal, 2019.

8. Palardy-Sim, M., et al., "Advances in a Next Generation Measurement & Inspection System for Automated Fibre Placement," *CAMX*, Anaheim, 2019.

9. Huang, D., et al., "Optical coherence tomography," *Science*, 254, 1178-1181, 1991.

[DOI: 10.1126/science.1957169]

10. Drexler, W., & Fujimoto, J. G., "Optical Coherence Tomography," *Springer International Publishing*, 2nd ed., 2015.

[ISBN: 978-3-319-06419-2]

11. Webster, P. J. L., et al., "Automatic laser welding and milling with in situ inline coherent imaging," *OPTICS LETTERS*, 39, 21, 6217-6220, 2014.

[https://doi.org/10.1364/OL.39.006217]

CHALLENGES OF AEROSPACE STRUCTURAL PART GEOMETRIES FOR HIGH-RATE COMPRESSION MOLDING

Aurele Bras[1], Alejandro J. Rodriguez[2], Richard Russell[1], Timothy J. Luchini[3], Travis Adams[3], Adam Whysall[1], Scott A. Rogers[2], Scott Lucas[2], Gail L. Hahn[3]

[1]Solvay Composite Materials, Heanor, UK DE75 7SP
[2]Solvay Composite Materials, Anaheim, CA 92806
[3]The Boeing Company, Saint Louis, MO 63134

ABSTRACT

Thermoset matrices continuous carbon fiber composites show opportunities for high-rate production of aerospace small structural parts when processed in compression molding. Previous studies have presented the material development and automated process development for press forming of carbon fiber composites. Takt times under 30 minutes were demonstrated for primary and secondary structures. This was achieved by building a flat blank, which was then formed into final shape and cured without pre-forming step.

A set of generic parts and associated steel tooling were generated. They include a range of geometries which present challenges for sub-30 minutes forming of a flat blank. The parts manufactured achieved aerospace structural performance. The paper presents how some of the main challenges were overcome, through both process and material development. The parts were designed by Boeing; pressed in Solvay's research and innovation facilities in Heanor, UK; and analyzed at Solvay research center in Anaheim, CA and at Boeing Saint Louis, MO USA.

Corresponding author: Gail Hahn

1. INTRODUCTION

In order to support the growth of new forms of high rate aerospace products [1], such as single aisle aircraft, drones, and flying taxis, a joint effort has been carried out by Solvay and Boeing in developing novel material and compression molding process [2].

The compression molding process developed has enabled takt times down to 20 minutes, thanks to both material chemistry, advanced automation, and the ability for a 2D blank to be formed isothermally into its final shape without pre-forming [3].

The constraint of forming parts in one step is particularly challenging when considering aerospace primary and secondary structures, presenting complex geometries, among which include vertical edges, thickness variations, beads, deep draws, double curvatures, etc.

The geometrical challenges have been investigated and solved through the design and manufacture of 4 geometries, one after the other and with increasing levels of complexity. The main prepreg materials considered are based on CYCOM® 5320-1 and CYCOM® EP2750 resin systems combined with woven T650 3k 8HS fibers. Other resin systems have also been considered but quickly discarded due to not meeting performance or manufacturability targets. The following Table 1 presents the main differences between CYCOM® 5320-1 and CYCOM® EP2750. The parts were designed at Boeing, the tools were made at C-Con GmbH, parts were manufactured at Solvay's Applications Centre in Heanor, UK, and tested at Solvay Anaheim, CA and Boeing St. Louis, MO USA.

Table 1. Comparison between CYCOM® 5320-1 and CYCOM® EP2750 pre-pregs [3]

	CYCOM® 5320-1	**CYCOM® EP2750**
Fibre	T650	T650
Weave	8 Harness satin	8 Harness satin
Resin content	36% wt.	40% wt.
Toughening	Medium	Medium
Impregnation	Medium to low	High
Bulk	High	Low
Tack	High	Tailored

The design of the three geometries studied is based on the anticipated challenges of molding at high-rate for aerospace applications as well as the complexity of an actual primary structure part. Features Include:

- Deep draw sections
- Thicknesses variations from 3.8 mm to 8.9 mm (0.15 inches to 0.3 inches)
- Double curvatures
- Vertical flanges

In chronological order, the geometries developed were:

<u>Beaded panel:</u>

- Beads of various geometry parameters
- Joggles
- Pad-up local thickness variations

- An edge going from 70° to 90° draft angle

Rib part:

- Centered pad-up thickness variation
- 4-sides vertical flanges
- Trimmed holes in the thickest area

C-Channel part:

- A deep draw U-section
- Vertical flanges
- Outer radius different from inner radius, resulting in a change of section shape
- Change in thickness along with the change in section shape

The compression molding process included both double diaphragm forming (DDF) and spring-frame in the first parts, then focuses exclusively on spring-frame. Both DDF and spring-frame use the principle of holding a flat blank and carrying it through an infrared (IR) preheat stage before shuttling it on top of the mold. The steel mold then closes following the press automated program, before applying full pressure to the prepreg for the rest of the on-tool time. Figure 1 below shows the overall process.

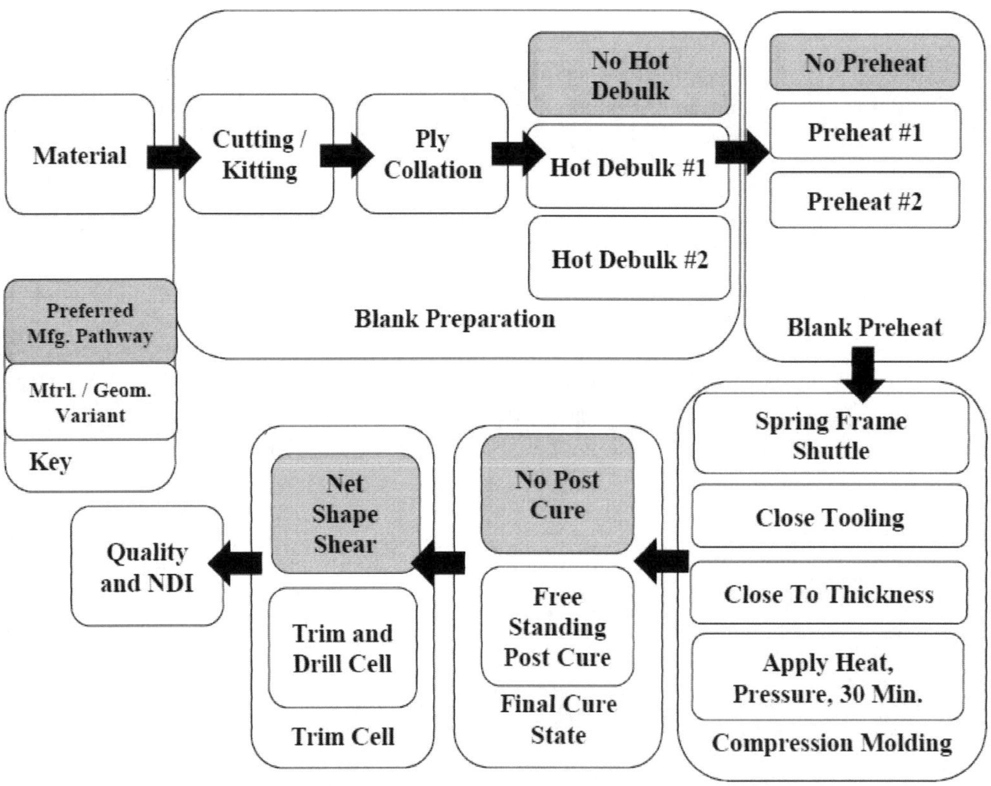

Figure 1. Flowchart of the automated process considered [4]

The main quality outcomes assessed are:

- Void content, using Non-destructive Test NDT (C-scan) analysis and photomicroscopy
- Resin content, using acid digestion method
- Cured ply thickness
- Degree of cure through the measurement of Tg on Differential Scanning Calorimetry (DSC) and Thermo-Mechanical Analysis (TMA)
- Mechanical performance, mainly through Short Beam Shear (SBS) test

2. EXPERIMENTATION AND RESULTS

2.1 TS-RAPM-009 Beaded Panel

Figure 2 illustrates the TS-RAPM-009 beaded panel. It is a 600 mm long flat part with various geometries parallel beads. The panel thickness builds up from 3 mm to 6 mm in one corner (pad-up). Another corner presents a joggle at constant thickness. The opposite edge presents a vertical flange with a draft angle varying linearly from 0° to 20°.

<u>Objectives:</u>

- Narrowing down material and process selection:
    - Plain weave (PW) is compared to 8-harness (8HS) for both CYCOM® 5320-1 and CYCOM® EP2750 systems
    - DDF and spring-frame blank carrying methods are compared

- Building correlation with process simulation models developed in house [5]

- Determining spring-in (reduction of enclosed angle after cure) for the lay-up considered

- Assessing compatibility of aerospace quality criteria with:
    - Bead geometries
    - Changes of thickness

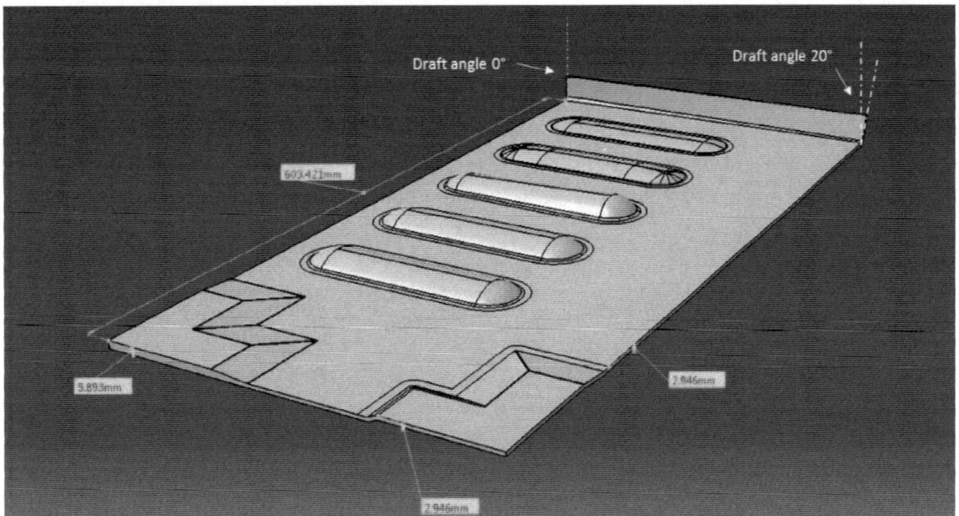

Figure 2. TS-RAPM-009 Beaded panel geometry.

The beaded panel study was carried out in Solvay's applications facility in the UK. The press equipment is capable of automatically shuttling the flat blanks into infrared preheating, then into the mold tool (Figure 3).

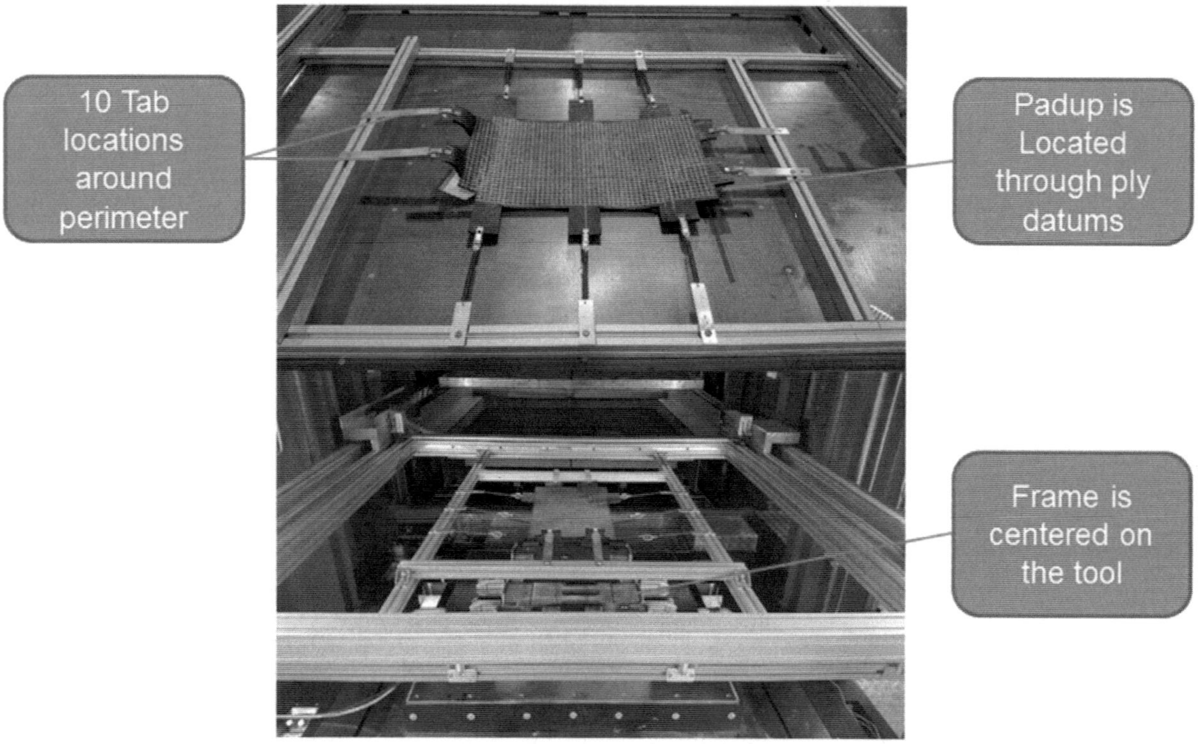

Figure 3. Beaded panel spring-frame configuration

The final part can be visualized on Figure 4. The dwell time has an impact on resin overflow: the image on the right presents a part with greater overflow than on the left, due to using 8 minutes dwell rather than 10 minutes.

Figure 4. TS-RAPM-009 beaded panel parts after spring-frame molding. Both parts are made using CYCOM® EP2750. On the left, a dwell time of 10 minutes was used. On the right, a dwell time of 8 minutes was used

A higher flow of resin has a positive effect on fiber wetting and pressure build-up in the tool cavity. The two above parts are compared on the below C-scan, which shows that the 8 minutes dwell time led to a lower void content. This is particularly true in the pad-up area. The sensitivity of this area to void content is due to its proximity to the edge of the tool cavity. There, two factors combine in providing a negative effect on porosity content:

1- The tool cavity does not present a shear edge or a reduction in section on the surrounding of the part. Therefore the pressure in the cavity decreases in the vicinity of the part edge, especially in thicker regions, where this section is actually bigger.
2- The beaded panel presents a relatively small, off-center pad-up area. This makes the pad-up plies subject to potential slippage through the process, due to spring tension, resin flow, and absence of back pressure from the outer edge of the part.

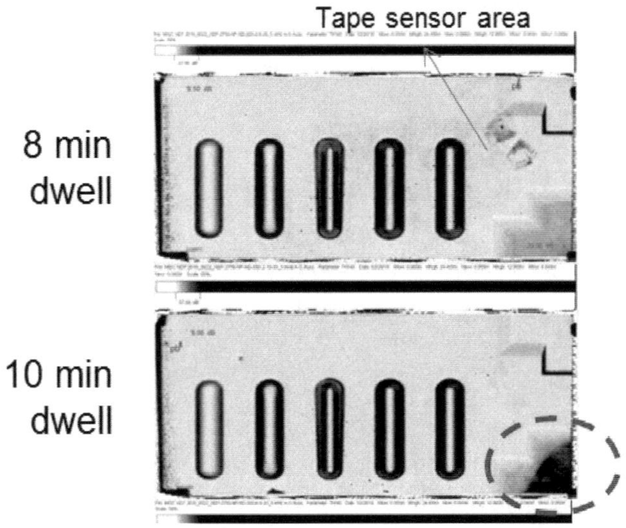

Figure 5. C-scan analysis of TS-RAPM-009 beaded panels shown in Figure 4, presenting respectively 8 minutes dwell time and 10 minutes dwell time

The pad-up zone for the above two parts is observed in photo-microscopy, visible in Figure 6. Although 8 minute dwell does not remove all porosity from the part, there is significant improvement compared to the 10 minute dwell time part. The void content goes from 5.5% to 0.4% on the padup area.

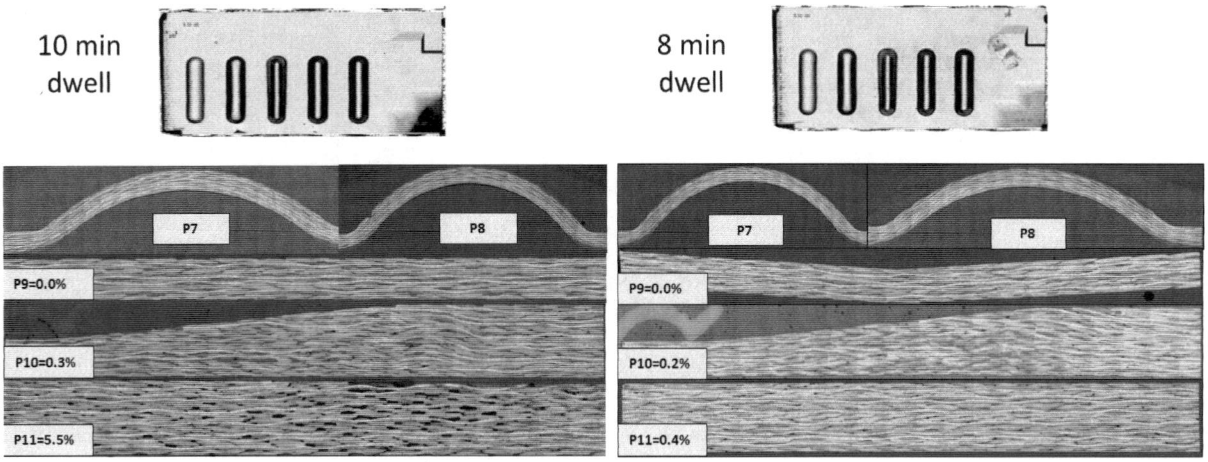

Figure 6. Photo-microscopy comparison of the beaded panel parts shown in Figure 4 and Figure 5. The section observed are extracted from the thin, tapered, thick, and beaded areas of the panels.

Unlike autoclave processing, compression molding takes advantage of forced resin flow for increasing part quality due to higher pressures applied. The process repeatability and manufacturing window are smaller for a material which is close to the final part's fiber volume fraction (Vf). Figure 7 shows that parts made with CYCOM® 5320-1 end up on the lower resin content limit due to the application of the necessary resin flow during the process to achieve sufficient part quality.

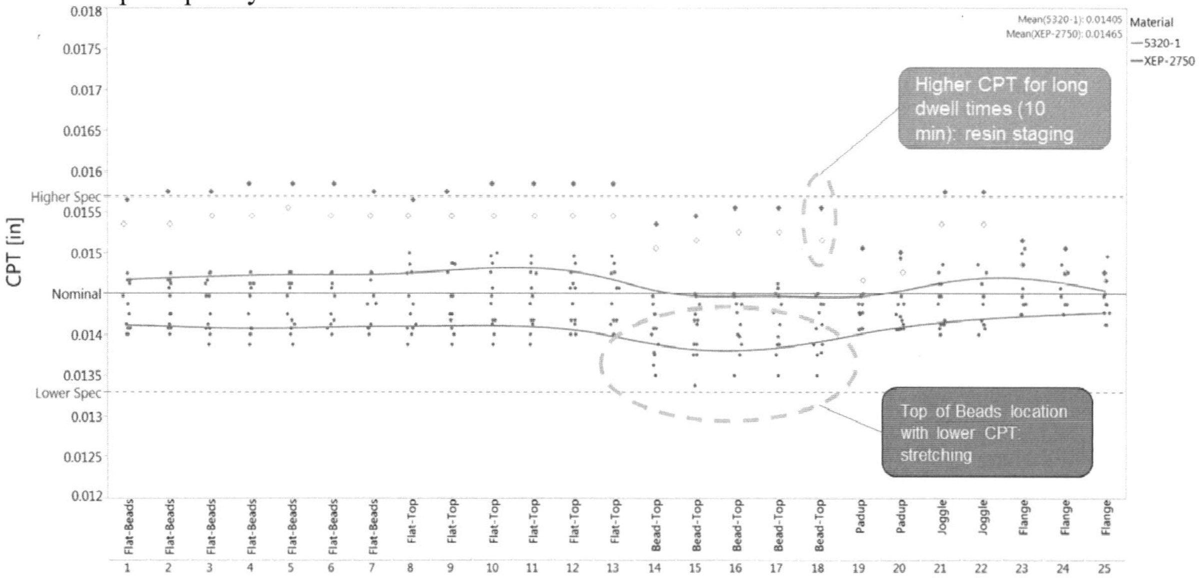

Figure 7. Cured Ply Thickness (CPT) of TS-RAPM-009 panels at 25 locations

Another key observation from Figure 7 is the reduction in CPT for the top of the beads, which can be verified by microscopy, see Figure 6. The fibers are locally put in tension and will create a Vf gradient from the bead's outer radius to the bead's inner radius. This local tension also propagates into fiber disturbance on the flat zone of the panel. This has been captured and integrated by the process simulation model developed for compression molding at Solvay [5], see Figure 8.

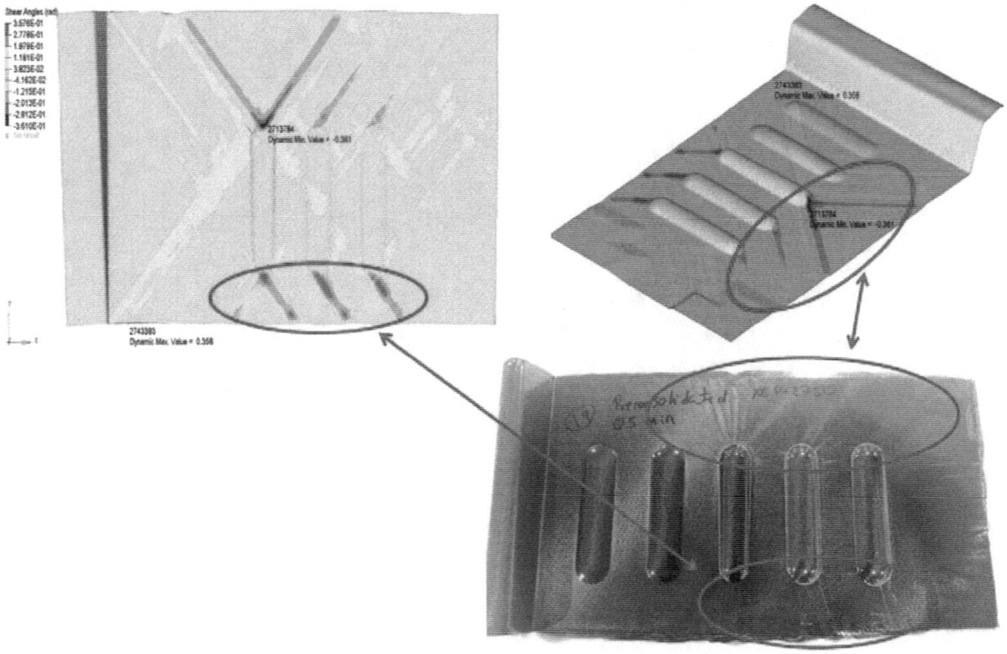

Figure 8. Visual correlation between process simulation and manufactured panel

2.2 TS-RAPM-002 Rib Part

Figure 9 illustrates the TS-RAPM-002 rib configuration. The thickness builds up in the center of the part, from 3.3 mm to 5.2 mm. The longer edge of the part is 540 mm long and the depth of the draw is 42 mm on the final trimmed part. The 4 walls are vertical. The spring-in compensation angle determined during the beaded panel studies has been applied to the tool geometry

Objectives:

- Integration of Transformer Film (patent pending) for increasing hydrostatic pressure for low resin content CYCOM® 5320-1 prepreg

- Assessment of surrounding vertical edge quality and finish

- Development of automated blank preparation in order to mitigate the hand layup uncertainty brought by positioning accurately a set of pad-ups which have no common edge with the rest of the part.

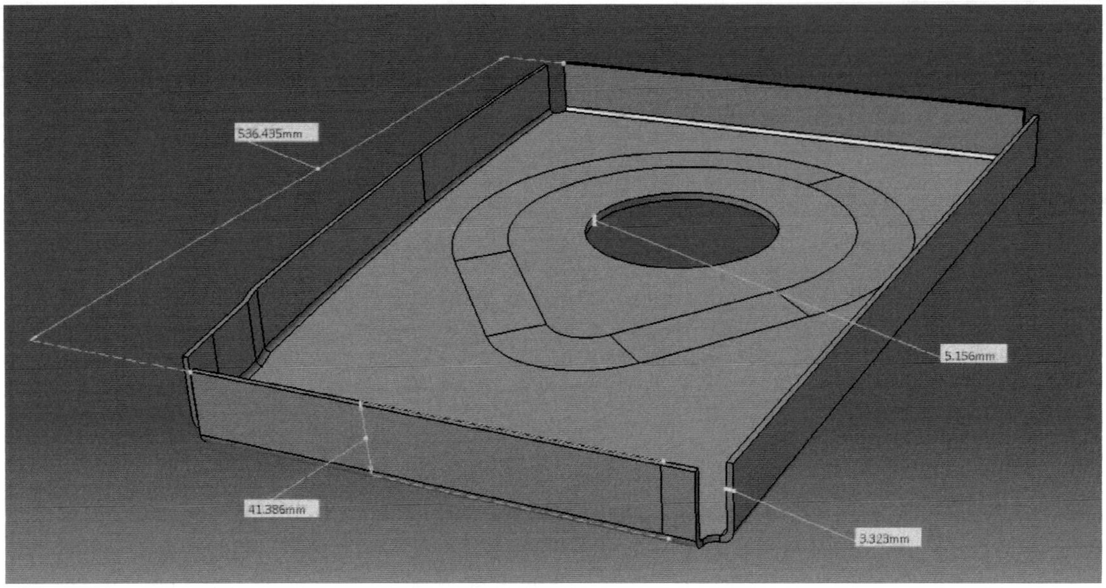

Figure 9. TS-RAPM-002 rib part geometry.

Observations made on the beaded panels led to higher tension capability and spring location modularity on the spring-frame. Figure 10 highlights loading and release of the parts for both spring-frame and Double Diaphragm Forming (DDF) method. Due to the deep draw of the TS-RAPM-002 rib, material excess is required in order to attach the spring clamps, whereas material excess can be more limited when using DDF. This effect is further increased here by tool design which does not allow necessary room for the spring clamps to be in close vicinity with the tool cavity.

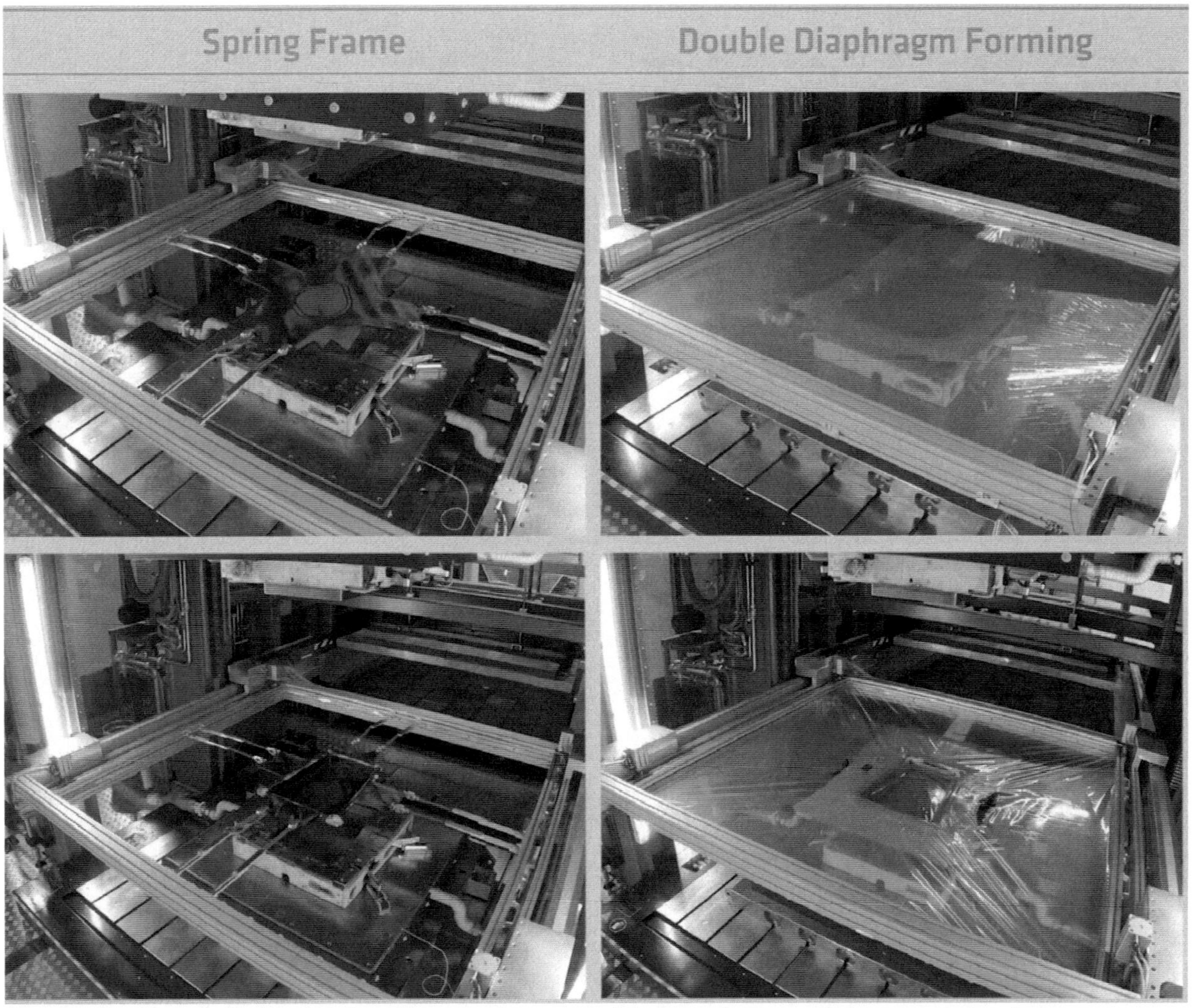

Figure 10. On top: blank shuttling from preheat to tool before closing. On bottom: part release

With regards to the TS-RAPM-009 beaded panel, Figure 7, the same observations can be made regarding the cured ply thickness (CPT) of CYCOM® 5320-1 TS-RAPM-002 rib, which is on average lower than the one of CYCOM® EP2750, Figure 11. The CPT is higher on the flat parts than on the vertical edges, regardless of the material used. The hypothesis is that this is due to flow restriction induced by the geometry of the part. The resin is kept within the central flat area of the part, as the vertical flanges act as a resin seal. This is helping compaction and part finish in the central flat area. However, the vertical flanges suffer from strong shear during tool closing, bringing surface roughness (but no porosity, see Figure 13). This is mitigated by the use of a lower bulk material, such as CYCOM® EP2750. Pre-consolidation of CYCOM® 5320-1 is a viable way to reduce bulk factor.

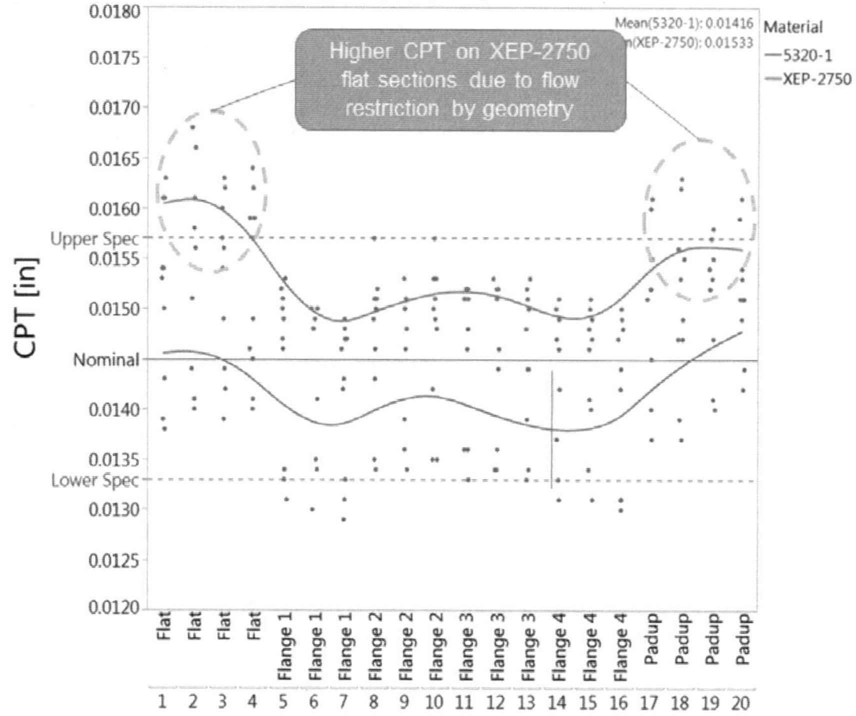

Figure 11. rib part CPT by location

In order to increase resin content from nominal quantity for CYCOM® 5320-1, a Transformer Film (patent pending) was applied ahead of the compression process. As a result, the average CPT of a low porosity CYCOM® 5320-1 part has been brought up to nominal (Figure 12). This has the effect of being able to move away from pre-consolidation, which is a time consuming extra step to the industrial automated manufacturing scenario. In summary, the use of Transformer Film allows the manufacturing of high-quality part with net resin content.

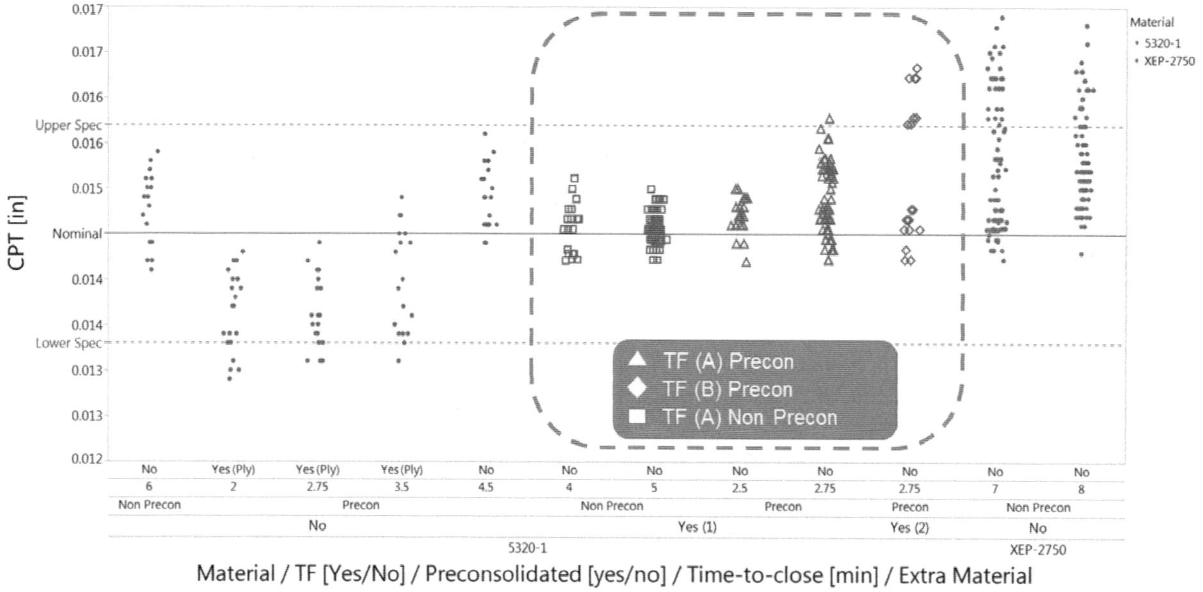

Figure 12. CPT increase from CYCOM® 5320-1 (blue dots), to CYCOM® 5320-1 with "TF" liquid shims (blue polygons), to CYCOM® EP2750 (red dots)

Figure 13 provides photo-micrographs of cross sections demonstrating the quality of the TS-RAPM-002 part discussed.

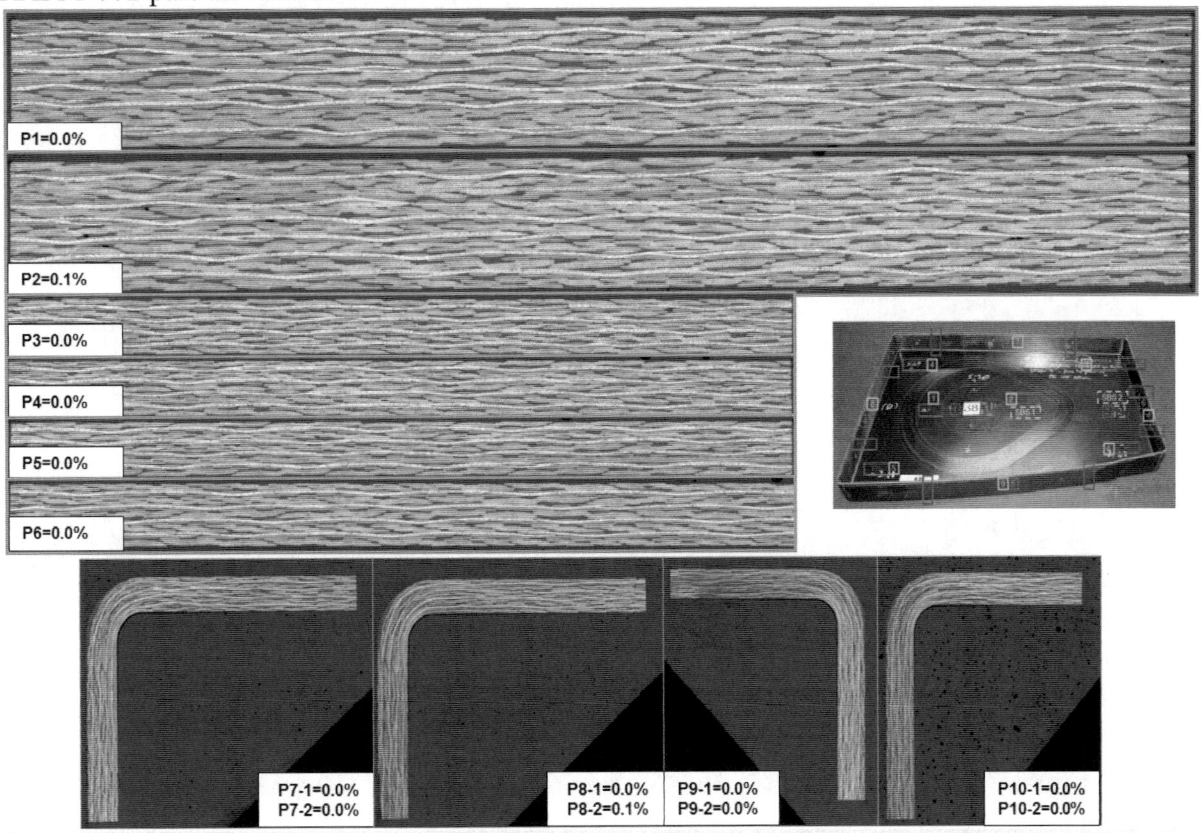

Figure 13. TS-RAPM-002 rib part photo-microscopy. The section are taken from the red areas on the rib part photograph.

2.3 TS-RAPM-003 C-Channel Part

Figure 14 illustrates the TS-RAPM-003 Curved Channel. The part is 640 mm long and presents a U-section 40 mm deep on the final trimmed piece. The inner radius of the section is greater than the outer radius. The thickness transitions from 6.6 mm in the narrow section of the part to 8.8 mm in the thick section of the part.

Objectives:

- Observation of the effects of a higher thicknesses on part quality and manufacturability

- Observation of the effects of an asymmetrical change in section on part quality and manufacturability

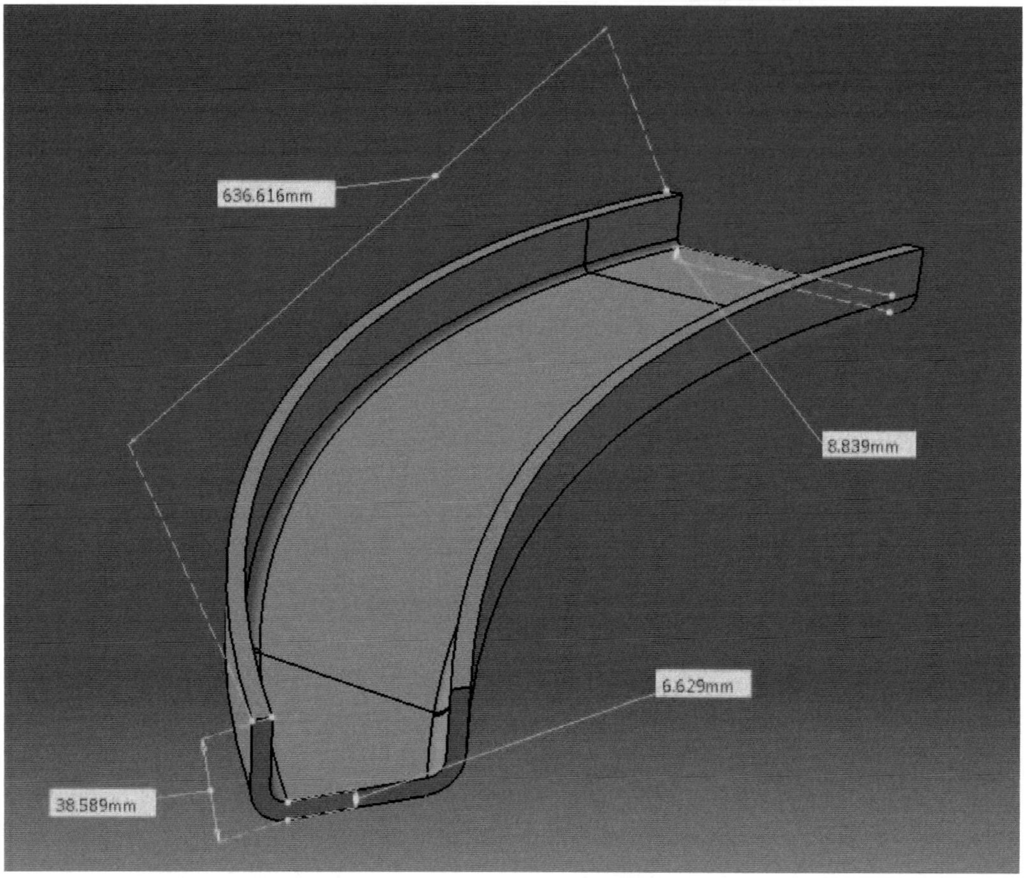

Figure 14. TS-RAPM-003 C-Channel part geometry.

Due to its thickness, the C-channel part is substantially heavier than the previous part configurations discussed. Therefore, the spring-frame has more attachment points than with previous shapes, as shown on Figure 15. The study of the beaded panel presented in the section above led to predicting wrinkle propagation through the part when the 2D blank is formed into multiple curvatures – this, however, is very dependent on fabric and viscosity profile of the part during forming. In the case of the C-channel geometry, the U-section evolves along the part in thickness, and the outer curvature differs from the inner curvature. This translates into potential fiber distortion through the part, which is visible on C-scans for the initial manufacturing trials (Figure 16).

Figure 15. C-channel spring-frame configuration.

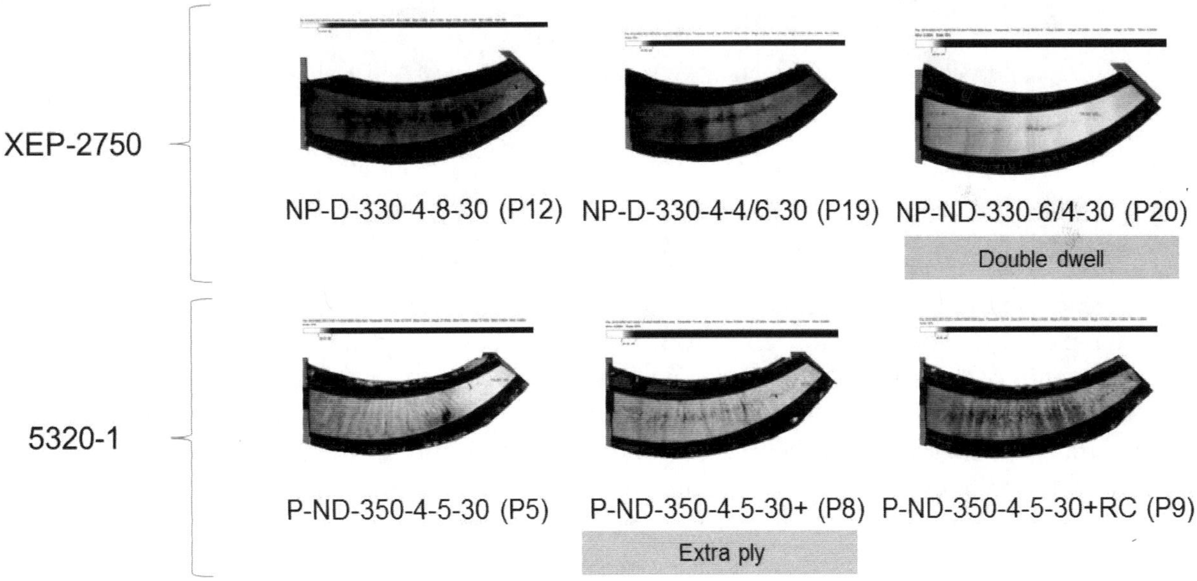

Figure 16. C-channel parts C-scan analysis. Initial manufacturing trials parts manufactured using CYCOM® EP2750 (top) and preconsolidated CYCOM® 5320-1 (bottom).

The C-scans shown on Figure 16 above highlight the presence of wrinkle propagating from the inner radius to the outer radius of the parts for parts using preconsolidated blanks. The grey scale indicates the wrinkles also carry porosity. For CYCOM® EP2750, the "double dwell" improved porosity and removed wrinkles from the parts. This consists of applying a certain amount of pressure during the dwell, in a sequence which follows the viscosity build-up of the material. For CYCOM® 5320-1, the addition of an extra ply reduced porosity and wrinkle magnitude but did not eliminate them at first. This is explained by the increased viscosity of the CYCOM® 5320-1 blanks advanced during the preconsolidation process – higher viscosity increases the potential of getting wrinkles on the parts.

In order to resolve the wrinkling and porosity of parts manufactured on the initial trials, the Transformer Film was used to increase the resin content on the material prior to manufacturing. This increases in resin content eliminates the need of preconsolidating and also guarantees a low porosity part but enabling the hydrostatic pressure on the mold cavity. Figure 17 shows the same part manufactured without preconsolidation and by addition of the Transformer Film. As shown in the C-Scans, wrinkles are significantly reduced and the parts show low porosity.

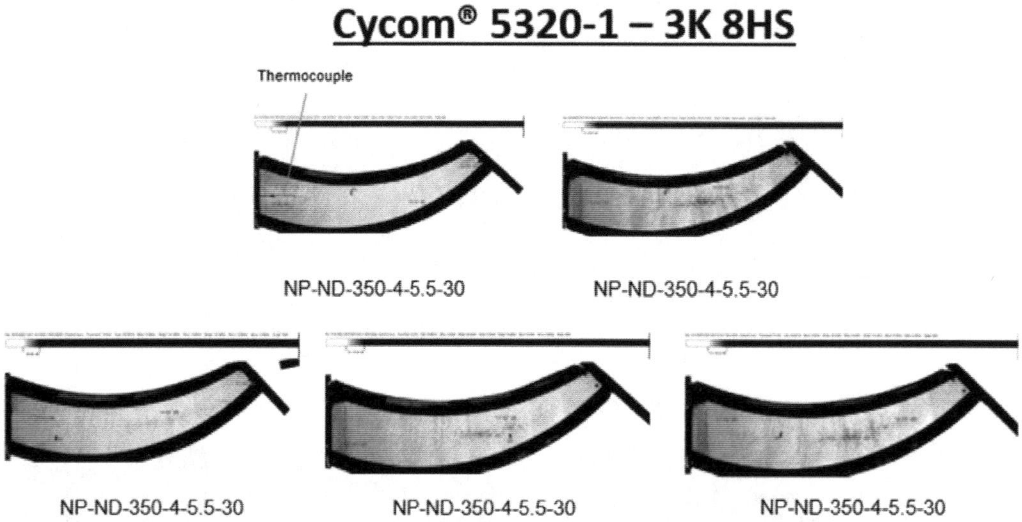

Figure 17. Manufacturing of C-Channel parts through compression molding using CYCOM® 5320-1 3K 8HS and Transformer Film. Repeated manufacturing of parts using same materials and processes show robustness of this high-rate manufacturing process

Figure 18 shows the effect of double dwell on part thickness. It is possible to retain more resin within the part while having satisfactory wetting of the fibers. Therefore the thickness is enabled to build up and reach nominal whilst keeping low porosity levels and low fiber distortion levels. The early pressure application of the double dwell leads some of the resin contained in the blank to flow from the thick to the thin area of the part. This results in thicknesses slightly beyond nominal on the thin side of the part (Figure 18) due to higher resin content. This also explains brighter C-scan reading in the thin area of the part (Figure 16).

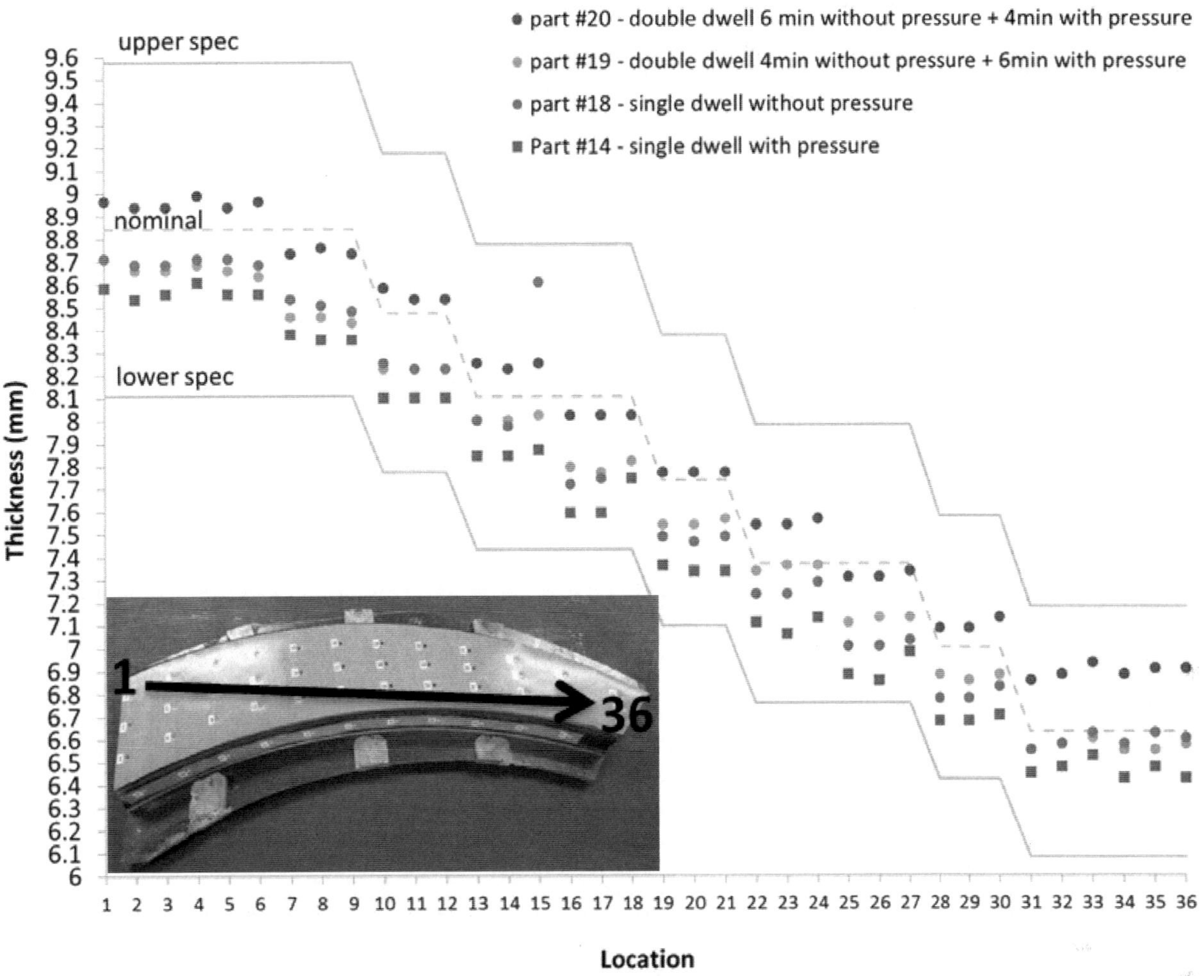

Figure 18. C-channel thickness measurement vs location. 36 samples have been extracted and measured on microscope. Those points go from the thicker area of the part to the thinner one.

3. CONCLUSIONS

The design and manufacture of a set of challenging part geometries has enabled generating knowledge towards the industrialization of high-rate manufacturing of aerospace structural parts by compression molding. Key pieces of learning are summarized by category.

Material design and formulation:

- Low bulk is beneficial for part quality, specifically to contain porosity levels and increase surface finish in vertical edges. This is achievable through high levels of impregnation during prepreg manufacture

- Higher resin content allows a wider manufacturing parameter window, as well as increased process repeatability

Tooling design:

- A reduction in section in the surrounding of the tool cavity will help pressure build up, particularly for parts with varying thicknesses

Equipment and machinery:

- Ideally, the pressing equipment should include:

  - IR preheat station
  - Automated carrier frame shuttling
  - Tight parallelism tolerances between platens
  - The ability to control and monitor position of platens to a tight tolerance
  - Ideally, the ability to press to low forces

Process cycle:

- "Double dwell" is beneficial for thick parts, but can induce CPT unevenness in parts of varying thicknesses

- A shorter dwell will increase fiber wetting but can lead to resin starvation, insufficient pressure build-up, or insufficient cured ply thicknesses

- Pre-consolidation allows reducing fiber disturbance in vertical walls, and contributes to reducing overall porosity levels

-Use of Transformer Film increases the process robustness by guaranteeing hydrostatic pressure in the mold cavity during consolidation

4. ACKNOWLEDGMENTS

This effort was jointly accomplished by a Boeing-led team and the United States Government (Defense Advanced Research Projects Agency, DARPA) under the guidance of the United States Army Research Office, ARO. The authors would like to acknowledge the guidance and support of Dr. Jan Vandenbrande of DARPA/DSO and Dr. David Stepp of Army Research Office (ARO). The information in this paper was approved for public release by 19-162544 and DISTAR 32485. Research was sponsored by the Defense Advanced Research Projects Agency (DARPA) and was accomplished under Cooperative Agreement Number W911NF-16-2-0087. The views and conclusions contained in this document are those of the authors and should not be interpreted as representing the official policies, either expressed or implied, of the Defense Advanced Research Projects Agency or the U.S. Government. The U.S. Government is authorized to reproduce and distribute reprints for Government purposes notwithstanding any copyright notation herein.

5. REFERENCES

1. J. Vandenbrande, "Tailorable Feedstock and Forming (TFF)," DARPA, [Online]. Available: https://www.darpa.mil/program/tailorable-feedstock-and-forming. [Accessed 04th December 2019].
2. G. L. Hahn and T. K. Tsotsis, "Rapid High-Performance Molding (RAPM) of Small Parts," in SAMPE 2019 Conference Proceedings, Charlotte, 2019.
3. A. Bras, A. J. Rodriguez, S. A. Rogers, A. Whysall, R. Russell, J. Brudenell, S. Lucas, T. Wybrow and S. Hill, "High-rate Manufacturing of Aerospace Structural Parts through Compression Moulding," in SAMPE Europe Conference Proceedings, Nantes, 2019.
4. T. J. Luchini, A. J. Rodriguez, S. A. Rogers, A. Bras, A. Whysall, R. Russell, S. Lucas and G. L. Hahn, "Spring Frame Press Fabrication of Aerospace Production Components," in SAMPE America conference, Charlotte, NC, 2019
5. F. Bressan, "Methodology for Press Forming Simulation of a Continuous Fiber-Reinforced Polymer Composite," in SAMPE America Conference Proceedings, Seattle, 2020.

METHODOLOGY FOR PRESS FORMING SIMULATION OF A CONTINUOUS FIBER-REINFORCED POLYMER COMPOSITE

Bressan, Fabio[1]; Jovas, Bob[2]; Luchini, Timothy[3]; Rodriguez, Alejandro[1]; Rogers, Scott[1]; Russell, Richard[2]; Hahn Gail[3]

[1]Solvay Composite Materials, Anaheim, CA 92806
[2]Solvay Composite Materials, Heanor, UK
[3]The Boeing Company, Saint Louis, MO 63134

ABSTRACT

Numerical simulation is a powerful tool in manufacturing since it provides a lower cost and faster analysis than actual trial-and-error testing. Press forming manufacturing simulation can be realized with many different system configurations including different boundary conditions and material set up. This work presents numerical analysis, advanced material card development and validation of a real component manufactured through press forming using thermoset prepreg material. The simulation was performed with an explicit numerical code and is visually correlated with a final experimental results.

Corresponding author: Gail Hahn

1. INTRODUCTION

Press forming of CFRP parts provides high rate, repeatability and cost reduction when coupled with mass production techniques in industrial and aerospace applications. Simulation of the part-forming process using an FEM numerical approach is an effective method to predict defects and facilitate the design for manufacturing [1-11].

The aim of this project is to evaluate the state of the art of simulation with traditional general purpose and specialized software present in the market. The software analysis requires the properties of the material to be represented by a software material card.

The high deformability of the thermoset composite pre-preg material at expected molding temperature was characterized by Purdue University laboratories and the material card was developed within the Solvay Virtual Engineering Team.

A manufacturing simulation analysis was performed for two different part geometries: a beaded panel representative of a geometrically stiffened structures and a C-frame part representative of a structural geometry. In particular for the beaded panel, the prediction capability was evaluated with a visual correlation to the real experimental manufacturing process.

The limits and the advantages of using a numerical method to predict some of the defects in the experimental test was evaluated with a preliminary experimental correlation analysis.

1.1 Press prepreg manufacturing process

Press manufacturing of parts from continuous fiber-reinforced thermoset composites has been specifically developed for high volume aero and automotive applications. Fast cure cycles and compatibility with automated processes (robots) are important to the press forming process.

Figure 1 shows the process flow for a part manufactured through compression molding. The first step of the process is to prepare the 2D flat blank; this step encompasses the cutting and kitting of the plies, ply collation by pick & place or hand layup followed by a RT debulk. After blank preparation the blank is loaded on the spring frame and loaded onto the press for processing. The blank is then shuttled to the IR preheat station where it resides for a short period of time before being moved into the tool on the press where it is formed and consolidated. Further details are shown in [12].

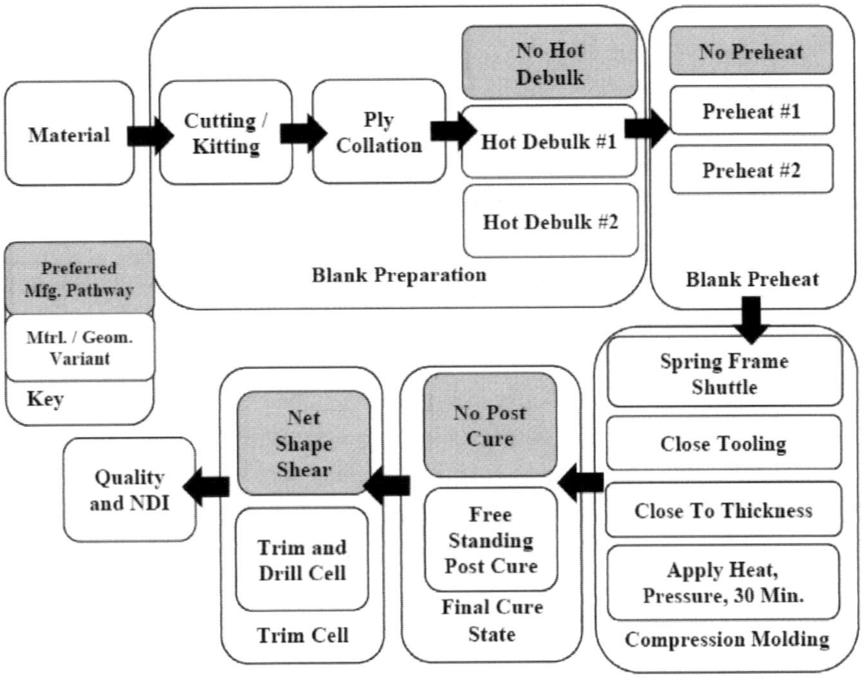

Figure 1. Simplified spring frame compression molding process for aerospace capable composite parts [12]

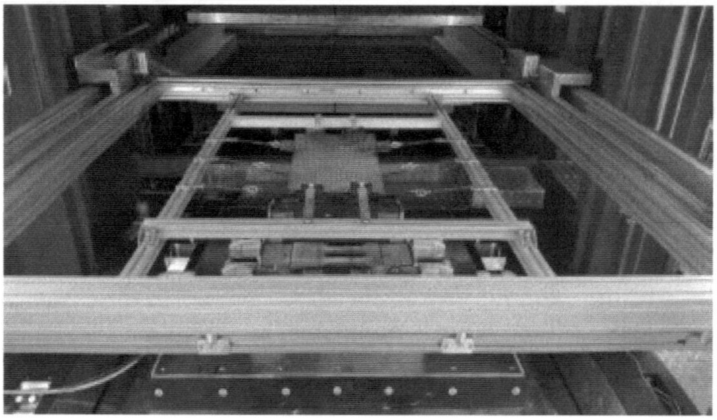

Figure 2. Beaded panel blank loaded into the spring frame ready to be shuttled into the preheat station [12]

Press tools are more expensive than a typical hand layup approach and simulation of the forming process before tool machining verifies both the design parameters and allows exploration of the optimum layup for a new component.

1.2 State of the Art of Press Forming Simulation

During the preliminary stage of this work, different commercial numerical codes were evaluated. The main evaluation criteria considered were primarily the accuracy, secondarily the speed of the codes, and finally the quality of the material models implemented (material cards definition).

Most of the general purpose finite elements commercial codes are well established in the metals industry. The software solvers are calibrated and predictive with material cards for homogenous materials. The automotive industry uses forming simulation mainly for sheet metal for external panels. Forming simulation for metal can predict the residual stress in the part which allows the designer to optimize thickness, reduce waste and improve the overall process for manufacturing. Forming simulation for complex shapes is currently solved with the implementation of explicit commercial computer codes. This numerical solution is fast and accurate in predicting wrinkling and improving the design for manufacturing of new geometries.

The three major challenges for any numerical simulation for manufacturing are:

1. Thermoset materials requires a special material card that need to take into account the interdependence between viscosity and temperature. Different than Thermoplastic, Thermoset materials are not reversible [7].

2. Material card is mesh dependent.

3. The prediction of the wrinkling and fiber direction for complex geometries requires an explicit code. Implicit solvers are not fast enough with very fine mesh. Implicit solvers can't solve multiple contacts between layers.

For this specific project the solution proposed was performed by Ls-dyna explicit code because it was the technology, at that time, which provided an excellent material card and a good compromise of speed and accuracy. Hypermesh and Lsprepost were used for pre and post definition and J-composites was used for auto/semi-auto processing material card form the experimental tests.

Additional tools were integrated in the process of simulation in order to develop an accurate material card from the experimental test log files. In particular J-sol form J-composite. It was also possible to develop specific TCL (Altair) scripts to accelerate the time consuming preprocessor procedure. LaminateTools (from Anaglyph) was the tools used for generating flat pattern of each single layers.

2. EXPERIMENTATION

2.1 Solvay Material Characterization of CYCOM® 5320-1 prepreg

Solvay CYCOM® 5320-1 prepreg was characterized at forming temperature of 370 F after being pre-consolidated. The textile used was a T650-35 3K 8HS at 370 gsm for the generation of the material cards using LS-DYNA solver.

2.2 Cure Cycle

All material samples were characterized and tested in accordance with the temperature cycle specified in Figure 3. Temperature was maintained as close as possible to the profile with the test beginning at the region labeled as "Forming 1". Test speeds were chosen to make sure tests finished in less than 30 seconds to ensure that there wasn't a significant property change developed during the test. The environmental chamber used to conduct the high temperature tests is shown in Figure 4. The actual temperature data vs. the target temperature data is shown in Figure 5. As the figure shows, the oven was able to track the temperature closely except at the very end where it was not quite able to keep up with the desired heating rate. Some tests were started at around 125 F (as shown) due a lack of oven cooling and the delay in cooling to room temperature [2].

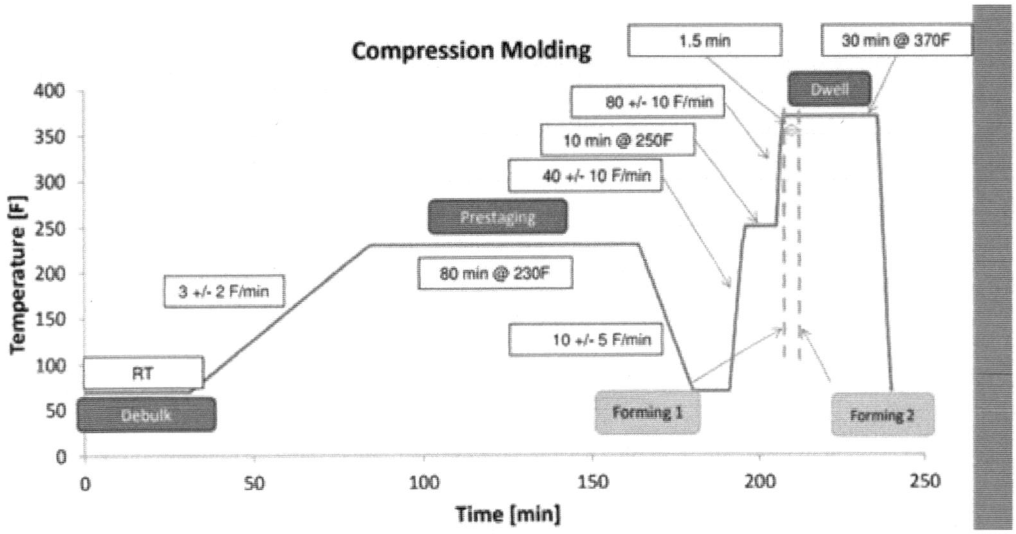

Figure 3. Temperature cycle used for characterization of the mechanical properties of CYCOM® 5320-1 3K 8HS prepreg.

Figure 4. Temperature chamber used to characterize the prepreg material at forming temperature.

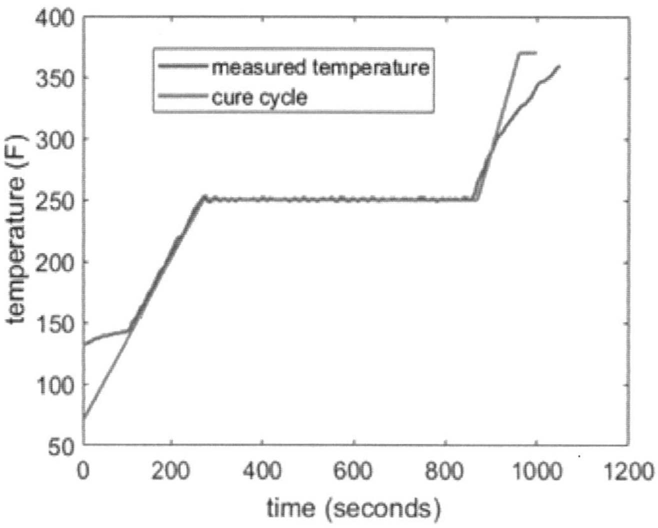

Figure 5. Target temperature vs. logged temperature during the characterization of the prepreg material.

Tensile Modulus, Shear Modulus, Bending Stiffness and Friction Coefficient

In order to generate the material card for any commercial solvers, it is necessary to perform a minimum of four fundamental tests as shown in Figure 6.

Figure 6. Tests performed for tensile modulus, shear modulus, bending stiffness, and friction coefficient tests.

2.2.1 Tensile Modulus

The tensile modulus measurements were attempted at high temperature. However, in order to accurately measure modulus there must be a way to verify strain in the sample. Typically this is done either by attaching a strain gauge to the sample or by using digital image correlation (DIC). Unfortunately, because neither strain gauging nor DIC was able to be successfully performed, the room temperature tensile modulus was assumed to be roughly equal to the high temperature tensile modulus. Tensile modulus is dominated by the fiber modulus and the assumption was that the final property of 3.1 GPa tested for room temperature was an accurate estimation for the numerical simulation.

2.2.2 Bending Stiffness (Flexural Modulus)

The flexural modulus was measured according to the standard ASTM D7264 in the 3 point bend configuration. Figure 7 shows the test setup for the flexural modulus results. The span length of the sample was 78 mm. The width was 25.4 mm and the thickness was 2.6 mm. The machine head rate was set at 10 mm per minute. The force displacement curves and the stress strain curves were then generated [4]. The modulus was estimated using the data up to 1% strain. The flexural modulus was calculated to be 9.7 MPa.

Figure 7. 3-point bending test set-up.

2.2.3 In plane Shear Modulus

The in-plane shear modulus of the prepreg was measured using a picture frame test [1]. Figure 8 shows the picture frame test setup before beginning the test and as it nears completion. The force and displacement are measured by the load frame and are converted into shear force and shear strain to get the shear modulus of the material. The machine was pulled at 3 mm per second and the length of each side of the picture frame fixture is 190 mm. Figure 8 (right) shows the force-shear angle curves.

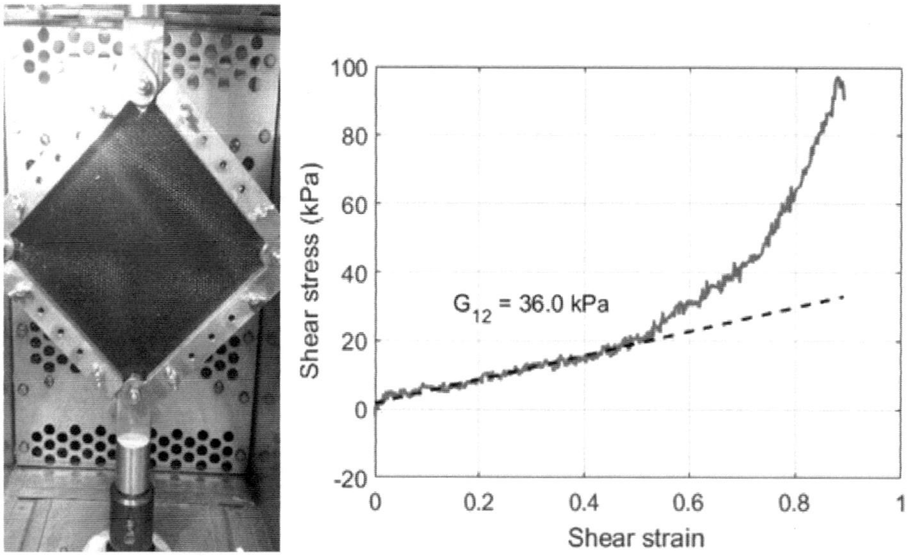

Figure 8. Left: Picture frame test setup. Right: Shear modulus estimation based on data log.

The shear modulus is calculated using the initial flat slope of the shear stress-shear strain curve Figure 7 shows an example of the shear modulus being estimated using linear regression in the linear region with an estimate of 36 kPa.

2.2.4 *Friction Coefficient*

Friction tests were performed at room temperature because of concerns that the high temperature values would be extremely sensitive to degree of cure. The reported room temperature friction values were 0.19 for the tool-ply friction and 0.31 for the ply-ply friction.

2.2.5 *Summary of Mechanical Solvay CYCOM® 5320-1 prepreg Properties*

Property	Value
Tensile Modulus*	3.1 GPa
Flexural Modulus	9.7 MPa
In-Plane Shear Modulus	36 kPa
Tool-Ply Friction Coefficient*	0.19
Ply-Ply Friction Coefficient*	0.31
*Values were measured at room temperature	

2.2.5.1 *LS-DYNA material card calibration*

Material cards must be identified using an inverse analysis loop where the characterization test is performed virtually and the material parameters are adjusted according to some objective function, typically the sum of the squares of the distances between the model and experimental curves [10]. In this project, the J-composites preprocessor software was used to derive the material model.

Experimental conditions were designed in order to collect the most appropriate results for the material characterization in its 'forming conditions'. By the time of the study, J-composites software (Figure 9) was the most efficient solution to automate the process of LS-Dyna material card model identification.

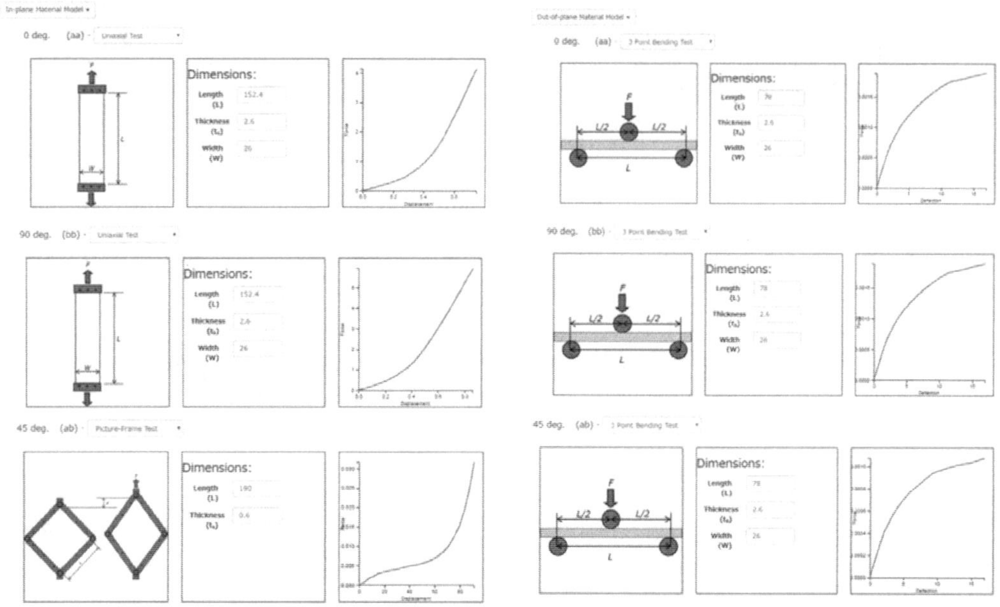

Figure 9. J-Composite preprocessing of experimental results for: Tensile (0, 90), Picture frame, 3 point bending (0,90, 45).

The picture frame was used for numerical correlation and calibration of the mesh dependency. Figure 10 represents a numerical and experimental correlation for Picture Frame test. The correlation shows a high level of accuracy and prediction for a flat simulation until the locking angle [5] [8] and it shows an accurate Material Card Model Characterization [3]. A similar approach was used for the bias extension test.

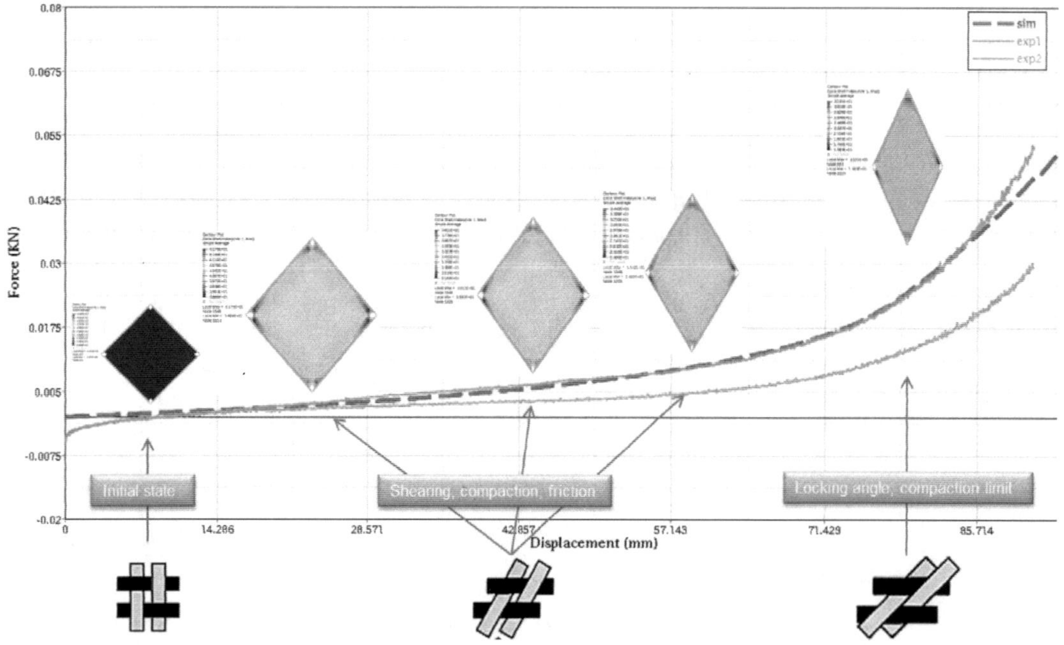

Figure 10. Picture frame test correlation between data and model used for the material cards

Ls-Dyna MAT278 (Figure 11) was used for the press forming simulation presented in this article. This material model was developed for draping and curing analysis of pre-pregged woven carbon fibers; the model is a mixture of MAT234 & MAT277, where MAT234 provides the reorientation & locking phenomenon of fibers, while MAT277 provides the viscoelastic behavior of epoxy resin. The matrix resin along with the fiber orientation during deformation contribute to the overall stress state.

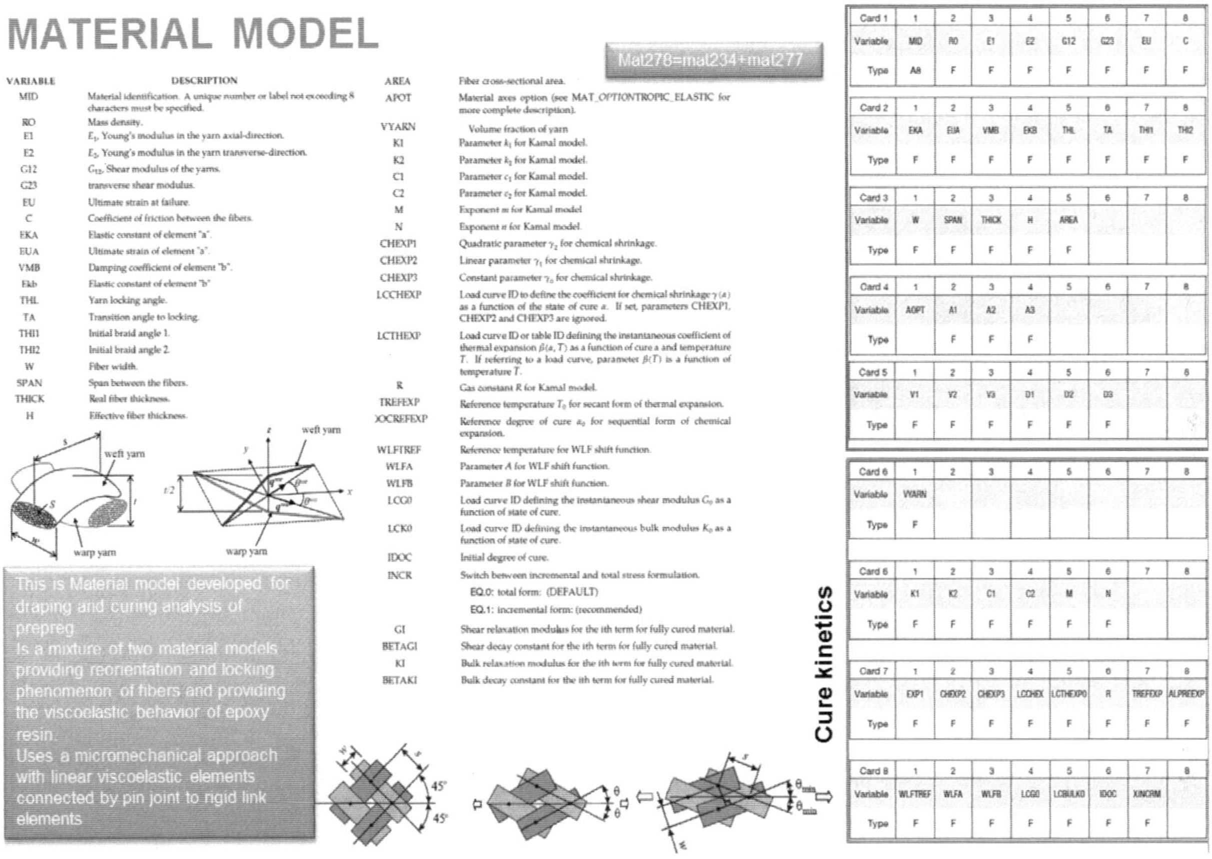

Figure 11. LS-Dyna Material Card Definition.

3. RESULTS

The beaded panel and c-frame parts were simulated and performed in sequence one after the other. The results from the simulation are showed in the following sections.

3.1 Beaded Panel Geometry

The overall manufacturing simulation of the beaded panel is show in Figure 12. In the simulation, a blank is created and placed between two solid surfaces that represent the tool hard surfaces for matched metal tooling.

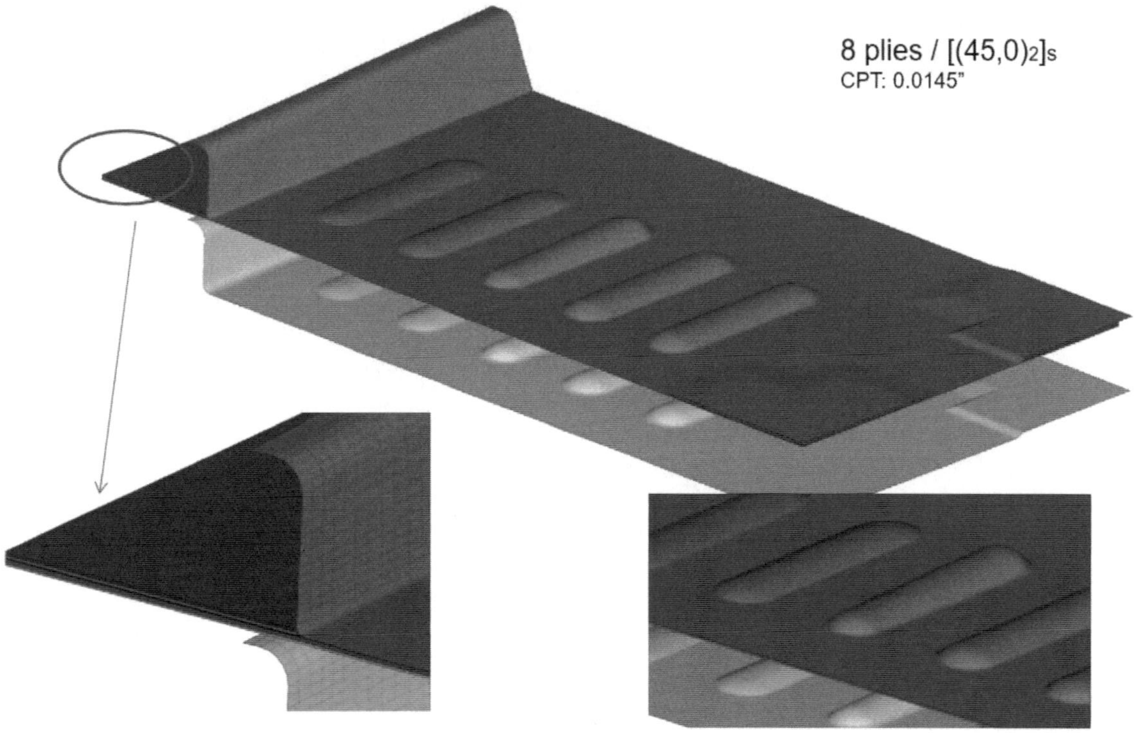

Figure 12. Beaded panel geometrical details.

Full laminate layup is simulated with multiple layers of shell elements built independently. Each mesh is aligned with a fiber direction, Figure 13. The Finite Element Model (FEM) included a mesh of 1.7 million elements and was simulated in 3h and 30 min using a computational system including 64 cores of 3.47GHz Intel Xeon processors [6] [9].

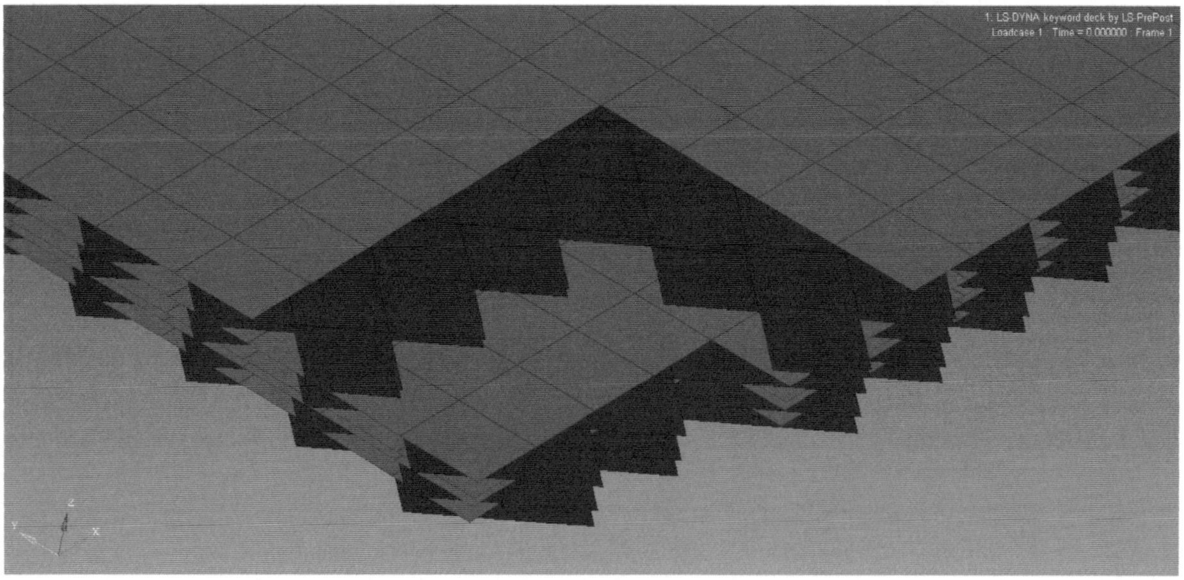

Figure 13. Multiple layers mesh dependency.

The post-processing result of the external layers is shown in the following Figures 14 and 15. The shear angle contour of ply 8 (a 45 degree ply) shows distortion lines at an angle deviation of 20.6 deg.

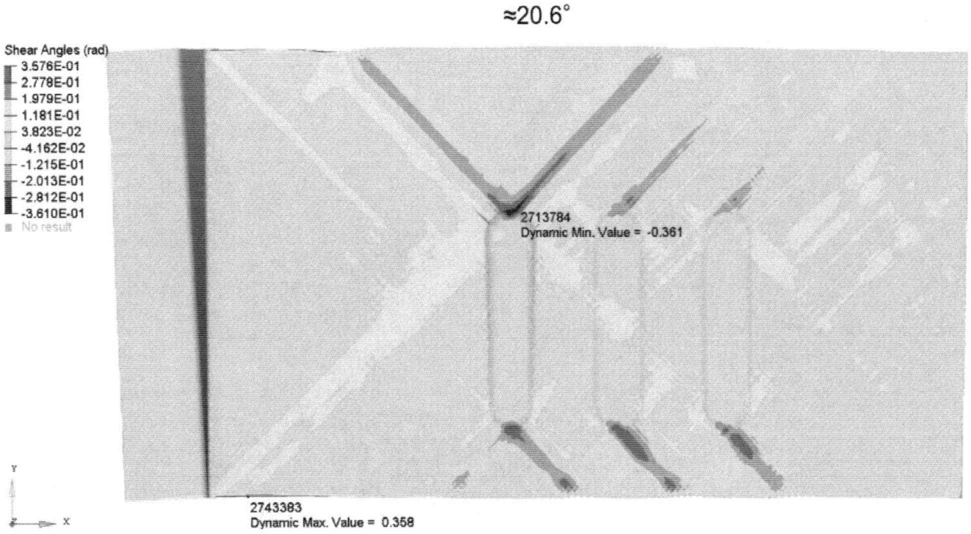

Figure 14. Shear angle deviation at ply 8 for the beaded panel geometry.

Figure 15 on the 0/90 degree ply 2 shows the global angle deviation is between 16.6 degrees to 18.9 degrees.

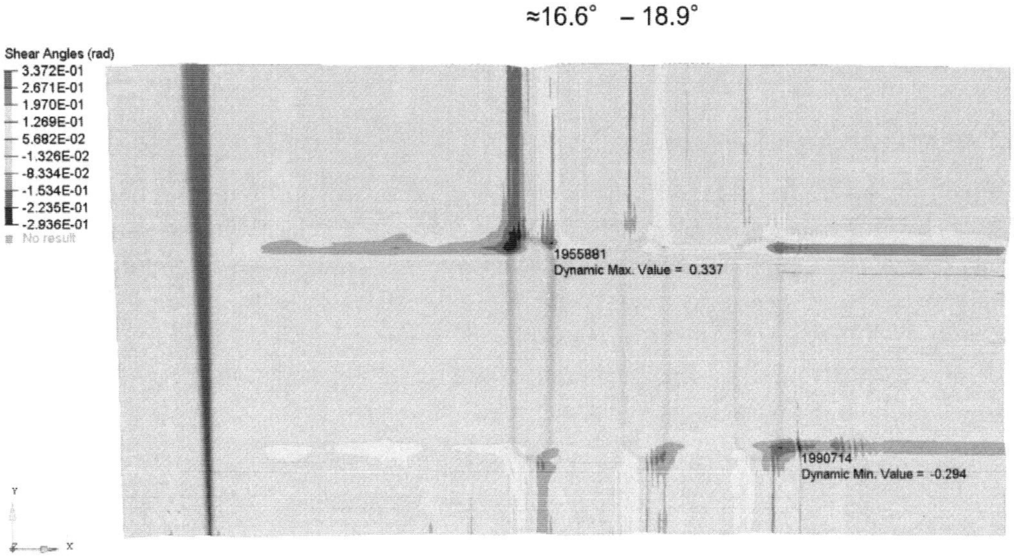

Figure 15. Shear angle fiber deviation at ply 2 for the beaded panel geometry.

In terms of correlation, a quick visual inspection was performed where a good correlation is found between the simulated and experimental part, as shown in Figure 16. As seen on the pictures, the largest fiber deviation occurs on the center bead which corresponds to the larger out-of-plane geometrical feature (note: the center bead is intentionally higher on one end to test drape).

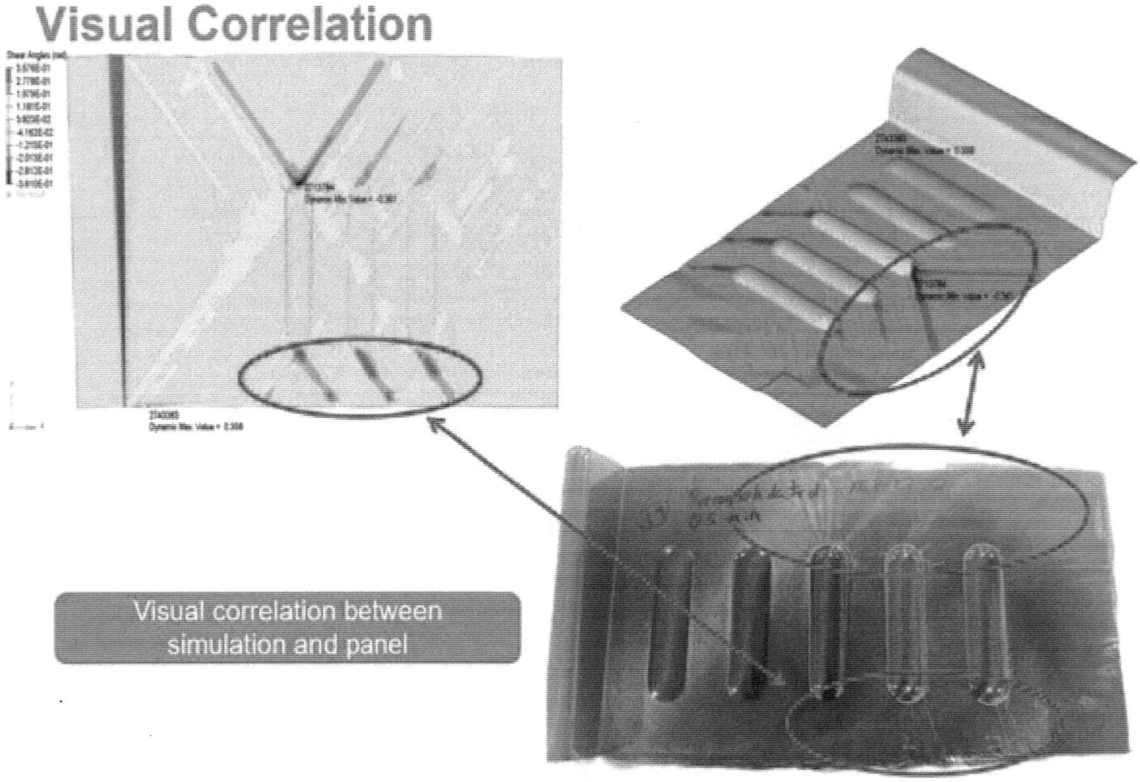

Figure 16. Visual and numerical correlation of fiber deviation between simulated and experimental results.

3.2 C-frame Geometry

The overall manufacturing set up for the C-channel component is shown in Figure 17. The model consisted in 33 plies, 34 contacts between plies, and 560,000 deformable elements. The model was run on 64 cores, 3.47 GHz CPU which yielded a simulation time of 7 hours. Just as in the previous geometry, the model consisted of two solid (metal) surfaces representing the tool with the 2D flat blank representation sandwiched between the surfaces at the beginning of the process. The flat blank is held by 10 springs distributed on the perimeter of the parts, 3 springs on each of the long sections and 2 springs on the narrow sections.

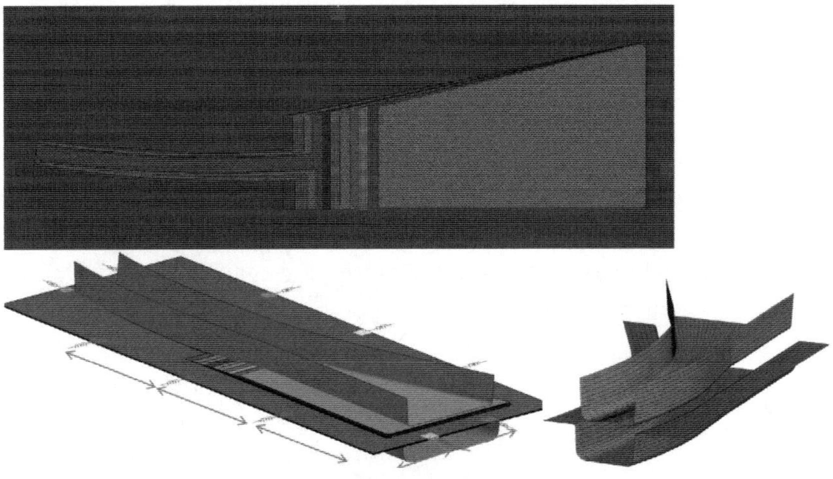

Figure 17. C-Frame simulation setup.

Figure 18 shows the simulation result after the forming process under specific spring boundary conditions. As seen in the Figure, the blank deforms to the desired tool shape with areas in the narrow section showing higher probability of wrinkling in the simulated setup. Estimated maximum fiber deviation for this geometry is 8 degrees.

Figure 18. Results of the manufacturing simulation of the C-frame geometry.

For this second geometry, the C-frame part, there are three important learnings and challenges:

1. Large & variable thicknesses are much more difficult to simulate and predict.

2. In the forming of multi-layered composites, out-of-plane wrinkling occurs due to the combination of intra-ply material deformations and inter-ply slippage.

3. The identification of the flat pattern of local reinforcements for large thickness was a complex procedure with an interactive trial and error process.

Table 1 shows the representation of different cross-sections of the part in different areas of the design. In a few sections of that part there are visible defects like wrinkling or compaction issues. Section (a) of the Table shows the area where high pressure in the internal curvature may generate potential thickness variations. In sections (b) and (c) show where potential for fiber deviation, wrinkling, and porosity increase due to the ply distortion as a result of manufacturing. Correlation between part manufacturing and numerical results were not available and is subject of future work.

Table 1.Representation of different cross-sections of the part and potential for fiber deviation based on the current manufacturing simulation scenario.

a) High pressure in internal curvature Non-uniform thickness	
b) Fiber waviness Compaction issues / voids	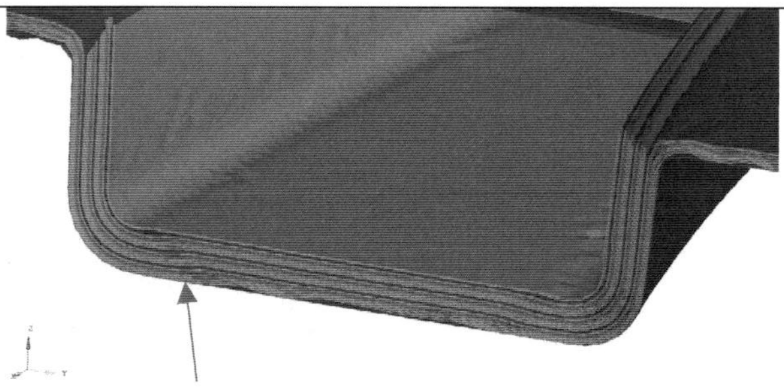
c) Fiber waviness Compaction issues / voids	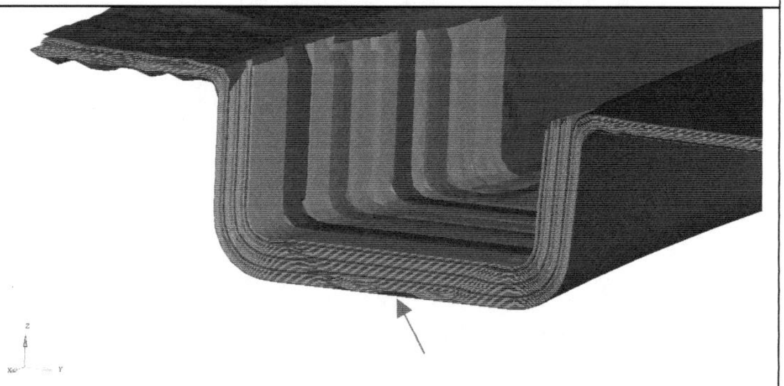

4. CONCLUSIONS

The proposed methodology of material modeling with the application of Ls-Dyna explicit numerical code is able to simulate the behavior of the fabric material. The simulations performed accounts for the fiber reorientation based on the material card developed. The fabric prepreg was characterized using a picture frame distortional test until the fiber locking angle was reached. The correlation was in good agreement with laboratory experiments.

The study of the beaded panel presented a good visual correlation between predicted fiber distortion in the forming simulation and the real composite part. This first simulation and correlation was used to identify the appropriate mesh density and the accuracy of the material card. Using the methodology developed for the first geometry, a second geometry was simulated and analyzed for fiber deviation. Both case scenarios show viability of the proposed approach by establishing the global simulation methodology. A more detailed correlation between numerical and experimental results is needed; this will be the focus of future work.

Simulation modeling of thermoset material is a great tool with a good visual correlation to actual parts. More development need to be done on correlation technologies like tomography and other scan methods. Another important future activity is developing a material card including methods to address viscosity and temperature dependency, especially for thermosets.

In conclusion, this work presents an efficient tool and methodology for predicting fiber direction, highlights potential fabric wrinkling, reduces tooling risk, and helps identify factors related to the press forming process for continuous fiber composite structures.

5. ACKNOWLEDGMENTS

This effort was jointly accomplished by a Boeing-led team and the United States Government (Defense Advanced Research Projects Agency, DARPA) under the guidance of the United States Army Research Office, ARO. The authors would like to acknowledge the guidance and support of Dr. Jan Vandenbrande of DARPA/DSO and Dr. David Stepp of Army Research Office (ARO). The information in this paper was approved for public release by 19-162544-ETT and DISTAR 32485. Research was sponsored by the Defense Advanced Research Projects Agency (DARPA) and was accomplished under Cooperative Agreement Number W911NF-16-2-0087. The views and conclusions contained in this document are those of the authors and should not be interpreted as representing the official policies, either expressed or implied, of the Defense Advanced Research Projects Agency or the U.S. Government. The U.S. Government is authorized to reproduce and distribute reprints for Government purposes notwithstanding any copyright notation herein.

6. REFERENCES

[1] Harrison, P., M. J. Clifford, and A. C. Long. "Shear characterization of viscous woven textile composites: a comparison between picture frame and bias extension experiments." Composites Science and Technology 64.10 (2004): 1453-1465.

[2] Kenny, J. M., A. Apicella, and L. Nicolais. "A model for the thermal and chemo-rheological behavior of thermosets. I: Processing of epoxy-based composites." Polymer Engineering & Science 29.15 (1989): 973-983.

[3] Zhang, Weizhao, Huaqing Ren, Biao Liang, Danielle Zeng, Xuming Su, Jeffrey Dahl, Mansour Mirdamadi, Qiangsheng Zhao, and Jian Cao. "A non-orthogonal material model of woven composites in the preforming process." CIRP Annals 66, no. 1 (2017): 257-260.

[4] Liang, Biao, Nahiene Hamila, Mickaël Peillon, and Philippe Boisse. "Analysis of thermoplastic prepreg bending stiffness during manufacturing and of its influence on wrinkling simulations." Composites Part A: Applied Science and Manufacturing 67 (2014): 111-122.

[5] Hamila, Nahiene, and Philippe Boisse. "Locking in simulation of composite reinforcement deformations. Analysis and treatment." Composites Part A: Applied Science and Manufacturing 53 (2013): 109-117.

[6] Guzman-Maldonado, E., N. Hamila, P. Boisse, and J. Bikard. "Thermomechanical analysis, modelling and simulation of the forming of pre-impregnated thermoplastics composites." Composites Part A: Applied Science and Manufacturing 78 (2015): 211-222.

[7] Guzman-Maldonado, E., N. Hamila, N. Naouar, G. Moulin, and P. Boisse. "Simulation of thermoplastic prepreg thermoforming based on a visco-hyperelastic model and a thermal homogenization." Materials & Design 93 (2016): 431-442.

[8] Hamila, Nahiène, and Philippe Boisse. "Tension locking in finite-element analyses of textile composite reinforcement deformation." Comptes Rendus Mécanique 341, no. 6 (2013): 508-519.

[9] Boisse, Philippe, Nahiene Hamila, Emmanuelle Vidal-Sallé, and François Dumont. "Simulation of wrinkling during textile composite reinforcement forming. Influence of tensile, in-plane shear and bending stiffnesses." Composites Science and Technology 71, no. 5 (2011): 683-692.

[10] Nishi, Masato, and Tei Hirashima. "Forming simulation of textile composites using LS-DYNA." In Proceedings of the 10th European LS-DYNA Conference, Würzburg, Germany. 2015.

[11] Schommer, Dominic, Miro Duhovic, and Joachim Hausmann. "Modeling of non-isothermal thermoforming of fabric reinforced thermoplastic composites." In Proceedings 10th European LS-Dyna Conference. 2015.

[12] Luchini, T.J., Rodriguez, A.J., Rogers, S.A. Bras, A., Whysall, A., Russell, R., Lucas, S., and Hahn, G.L. “Spring Frame Press Fabrication of Aerospace Production Components.” SAMPE 2019 Conference Proceedings. Charlotte, NC, May 20-23, 2019

THERMOPLASTIC COMPOSITE PROCESSING USING SUPERPLASTIC FORMED MAGNESIUM BLADDERS

Steven M. Shewchuk, Jared B Hughes, Corey M. Vasel, Gail L. Hahn
The Boeing Company, Saint Louis, MO 63134

ABSTRACT

This paper discusses a method of manufacturing thermoplastic composite parts using superplastic formed AZ31 magnesium bladders. This method superplastically forms magnesium against the composite part at temperatures lower than traditional superplastic forming temperatures and above thermoplastic composite melting temperatures to consolidate the thermoplastic composite. Historically, superplastic forming has been completed with pressurized argon gas. An investigation into the use of nitrogen gas was completed and the resulting parts were evaluated. The part configurations are evaluated for attenuation, surface roughness, and ply locations on a gentle contoured single sided tool surface with single use bladders. It was found that the use of nitrogen as compared to argon as a bladder pressurization gas had a negative impact on the part quality.

Corresponding author: Hahn, Gail

1. INTRODUCTION

The RApid High Performance Molding (RAPM) for Small Parts program has investigated and demonstrated the ability to fabricate PEEK thermoplastic preforms by applying compaction pressure with a superplastic formed magnesium bladder [4] [5] [6]. Previously, these parts were used in a secondary stamp forming processing step. The part trials completed for this paper consolidated PEKK thermoplastic composite material from loose plies to a net shape composite part using an AZ31 Magnesium bladder on a low curvature single sided tool. With this process change, nitrogen was used as a pressurization gas to improve compatibility with automation equipment. The part surface quality, ply locations, and consolidation quality were evaluated for the parts.

1.1 Audience

This paper was written for those experienced in thermoplastic composite part fabrication. A new processing method for parts fabricated in a press is discussed. The paper discusses process modification and parameters of importance and evaluates the resulting parts. For those familiar with thermoplastic composite processing this paper will provide an overview of the superplastic formed AZ31 magnesium bladder processing of parts.

1.2 Background on Magnesium Bladder Processing

The RAPM program discussed an alternative to match metal tooling for thermoplastic part fabrication. This single sided tooling approach has lower tooling requirements. Historically, bladder materials have been unable to withstand the processing temperatures with the elongation needed for processing. As a result alternative bladder materials were evaluated. Superplastic forming of AZ31 magnesium has been used in automotive industries, the material is relatively low cost, and the supply base exists. For this reason it was chosen as a bladder material candidate earlier in the program. At the beginning of the program the bladders were processed with argon with successful preform fabrication. To be more compatible with current automation equipment design for use with nitrogen, the processing gas was changed. This also had the additional benefit of nitrogen gas being lower cost.

1.3 Tooling and Part Geometry

The parts were fabricated from a PEKK thermoplastic polymer matrix with unidirectional carbon fiber reinforcement. The parts had a balanced and symmetric layup. The tooling is shown in Figure 1. The tooling was manufactured from Invar 42 to best match the thermal expansion of the composite material. The tool sat with the tool CMD-TP-RAPM-003-501 tool inside of a pixilated active heating and cooling clamping fixture capable of applying 150 ton clamping force and heating to over 400°C fabricated by Surface Generation Ltd [4] (CF-OaO-RAPM-001). The tool has a gentle contour to manufacture a part 40.6 cm x 40.6 cm (16 inches x 16 inches) and 1.3cm (0.5 inches) in height.

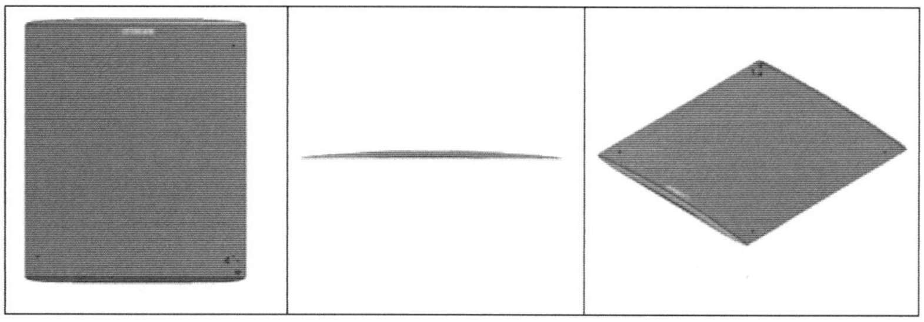

Figure 1 - CMD-TP-RAPM-017-501 top view (left), front view (middle), isometric view (right)

Three part geometries were evaluated with varying thickness and ply drops. The parts are shown in Figure 2. The three different Geometries (flat, ply drop, and picture frame) have 32 plies in their thick regions and 16 plies in their thin regions. Between the two regions are a 20:1 ply drop ratio.

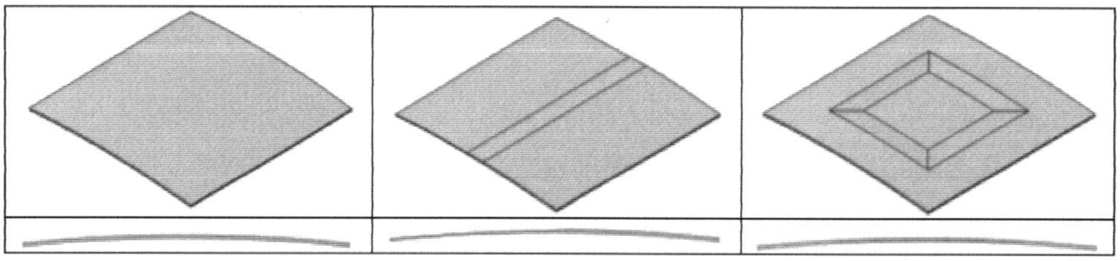

Figure 2 - Part geometry 1 (left), geometry 2 (middle), and geometry 3 (right)

In this paper the part name will reference the part geometry (G), and Serial Number (N). The full part name would be TP-RAPM-017-G-N but the part names will be referenced as 017-G-N as shorthand where G is the geometry and N is the serial number for the specific geometry.

1.4 Manufacturing Equipment and Method

The parts were manufactured on the CF-OaO-RAPM-001 tool clamping fixture with and without the accompanying Pressclave pressure controller. The tooling was placed inside the CMD-TP-RAPM-003 picture frame underneath an AZ31 magnesium bladder. The CF-OaO-RAPM-001 clamping fixture allows for the rapid heating and cooling of the tool face using 180 independent heating and cooling zones. The CMD-TP-RAPM-003-501 picture frame tool allows for the pressurization of the magnesium bladder at elevated temperatures against the thermoplastic part using pressurized gas to superplastic form the bladder against the composite at melt temperatures. This superplastic forming of the bladder is what provides the necessary compaction force against the thermoplastic part.

Traditionally superplastic forming of AZ31 Magnesium is completed at temperatures above 400°C however, at lower temperatures, magnesium will form at low strain rates. This manufacturing method takes advantage of the low temperature forming to debulk, compact and consolidate the thermoplastic composite. The bladders apply compaction pressure to the thermoplastic composite during the heat up. As the composite material reached its melt temperature pressure was increased against the composite. During the cool down, pressure was maintained to ensure consolidation.

1.5 Superplastic Forming Simulation

Prior to manufacturing of the composite parts the following finite element analysis was used to predict material thinning and pressure requirements for the superplastic forming (SPF) of an AZ31 magnesium sheet. The SPF process makes use of extensive microstructural deformation mechanisms. The complexity of these mechanisms makes development of accurate constitutive material models difficult. LS-Dyna is frequently used to perform analysis of SPF processes. LS-Dyna is an advanced general-purpose finite element program capable of simulating complex real world problems. LS-Dyna code models a highly nonlinear, transient dynamic finite element analysis using explicit time integration [2]. The large deformations and material non-linearity of the SPF process requires a solver that it set up for such mechanisms.

2. EXPERIMENTATION

2.1 Magnesium Super Plastic Forming Simulation

The geometry of the model is shown in Figure 3. This is a model of the picture frame tool that the consolidation tool sits inside of. The CMD-TP-RAPM-017-501 tool was located inside the boundary of the picture frame beneath the magnesium sheet. The consolidation tool was removed for a conservative value which would require the magnesium to elongate more in the simulation than in application. The model consists of two components, a stainless steel tool and magnesium sheet. Each component was defined with its own set of properties and material models. The surfaces of each component were used to generate a mesh using uniform thickness shell elements of the type ELFORM 16. ELFORM 16 assigns a fully integrated shell element based on Reissner-Mindlin kinematic assumption. [1] This element type was selected due to its balance in compatibility with implicit simulations without being relatively computationally expensive.

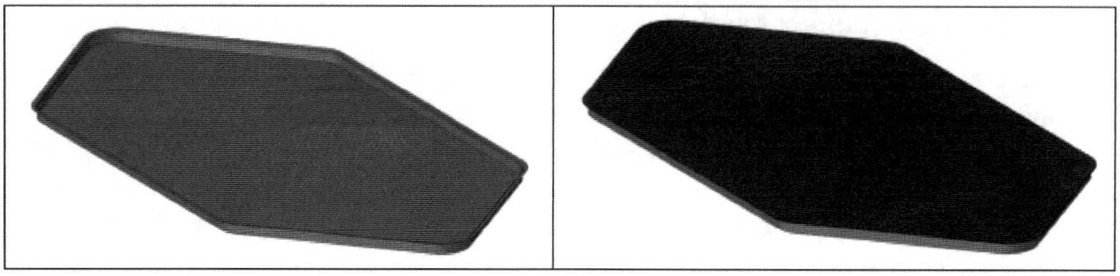

Figure 3 - Forming tool for the superplastic forming simulation (left), forming simulation with the magnesium sheet (right)

The material model for the magnesium sheet was based on constitutive equations developed by Taleff et al [3]. The paper derives strain rate equation and temperature dependent modulus the for Mg AZ31 forming at 450˚C. Equation 1 shows the strain rate and Equation 2 shows the dynamic unrelaxed temperature dependent Young's modulus.

$$\dot{\varepsilon} = 7.494 \times 10^{8} \left(\frac{\sigma}{E}\right)^{1.57} \exp\left(\frac{-Q_1}{RT}\right) + 2.517 \times 10^{24} \left(\frac{\sigma}{E}\right)^{5.1} \exp\left(\frac{-Q_2}{RT}\right) \quad [1]$$

$$E(T) = 48{,}700 - 8.59T - 0.0195T^2 \quad [2]$$

Where:

$$Q_1 = 91\ \frac{kJ}{mole} \qquad Q_2 = 138\ \frac{kJ}{mole} \qquad \dot{\varepsilon} = Strain\ Rate$$

$$E = Youngs\ Modulus \quad \sigma = Flow\ \text{Stress} \quad R = Ideal\ Gas\ Constant\ for\ Air$$

$$T = Temperature\ (K)$$

MATL 64 in LS-Dyna is a material model that implements properties for a strain rate sensitive elasto-plastic material with power law hardening [3]. The constitutive relationship is of the form is shown in Equation 3.

$$\sigma = k\varepsilon^{m}\dot{\varepsilon}^{n} \quad [3]$$

Where:

$$\sigma = Flow\ Stress \quad \varepsilon = Strain \quad \dot{\varepsilon} = Strain\ Rate \quad k = Strength\ Coefficient$$

$$m = Strain\ Rate\ Sensitivity\ Exponent \quad n = Strain\ Hardening\ Exponent$$

Solving Equations 1 & 2 for a given temperature and strain rate yields the corresponding flow stress. Once a flow stress is determined, the corresponding k, m, and n coefficients were estimated by fitting the curve to the true stress vs true strain curve [3]. Using these two data sets k, m, and n coefficients were selected to best match the curve for a given flow stress/strain rate pair. By selecting coefficients of: k=85, m=0.225, and n=0.005, the following curve fits were generated for 400˚C.

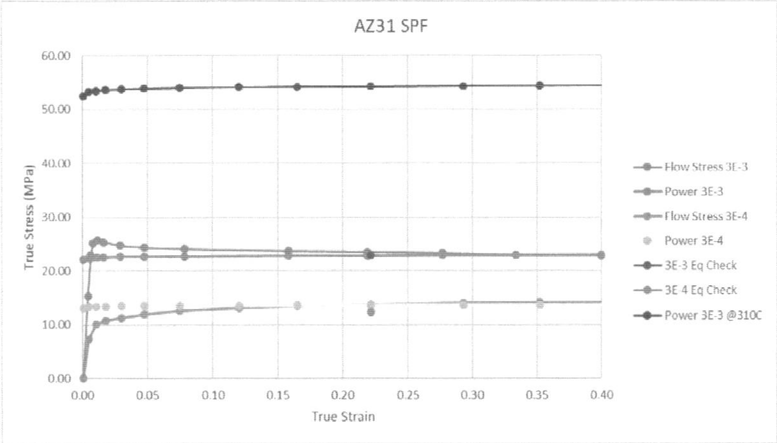

Figure 4 - Power law curve fits vs test data from material models for simulation of superplastic Mg alloy sheet forming [3]

Figure 4 shows that the curve fits for 400 °C were a sufficient representation of the AZ31 alloy up to 0.4 true strain. Unfortunately, no data was available for the parameters of interest for our FEA. The temperature of interest was 310 °C, and the fully formed geometry would likely require true strains greater than 0.5. This material model was unverified in this temperature range but was used in testing prior to part fabrication as the best available approximation. Since the material model is unverified one of the recommendations to gain further confidence in these results would be to perform ASTM E2448 superplastic tensile testing to verify the constitutive equations used here are representative. The tool properties (density, modulus, Poisson's, etc.) were taken from design handbooks, and assigned MATL 20 [2]. However they were of little relevance as the part was assigned to be perfectly rigid and fixed in space. The remainder of the model was created using the following parameters. All nodes on the perimeter of the sheet were fixed from displacement and rotations. All surfaces on the sheet were applied a tabular pressure loading vs time as seen in Table 1.

Table 1 - Pressure schedule for the forming simulation

Time (s)	Pressure bar (psi)
0	0 (0)
5	2.07 (30)
30	9.65 (140)
300	9.65 (140)
330	0 (0)

The sheet was assigned contact properties that included: Static friction coefficient = 0.4, Dynamic friction coefficient = 0.2, Contact stiffness = 0.1, Viscous damping coefficient = 10%.

2.2 Part Processing

Each part was fabricated with a new 1.6 mm (0.063 inch) thick AZ31 Magnesium bladder. The ply kits were cut using a drag blade Gerber cutter and ultrasonically welded together following collation. These ply kits were then placed within the scribe line of the cure tool inside the clamping fixture. The ply kit prior to consolidation is shown in Figure 5.

Figure 5 - Ply kit on the consolidation tool prior to fabrication

The tool was closed and the part was heated to 385 ± 15 °C. During the heating of the part various pressurization cycles were tested. Following the heating the part was cooled below 220 °C, pressure was released, and the part was extracted from the clamping fixture. To validate the part temperatures, a thermal profile was completed by measure the tool face in every independent heating channel and adjusting set points until all the tool face channels were within the 385 ± 15 °C temperature band. The thermal profile was validated by imbedding thermocouples into the thermoplastic ply kit. The thermocouples were imbedded in the locations shown and produced the thermal validation plot in Figure 6.

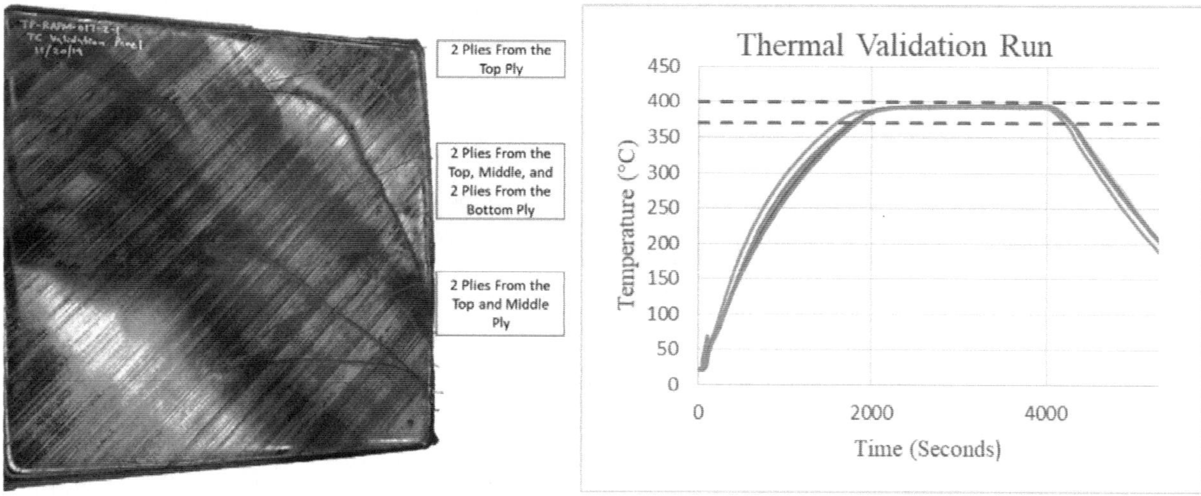

Figure 6 - Thermal profile article thermocouple locations (left), Imbedded part thermocouples in the thermal validation run (right)

2.3 Pressurization

It has been demonstrated that using pressurized argon inside the magnesium bladder will provide a satisfactory surface to the resulting composite parts. There was interest in using a lower cost nitrogen gas as a pressurization method. As of December 2019 nitrogen costs approximately 2/3 the cost of argon gas. Additionally, argon gas reduces the lifespan of the seals used inside of the automated pressurization equipment (Pressclave), which was designed for use with nitrogen gas. The following pressure cycles were tested with either nitrogen or argon. The control was part TP-RAPM-017-1-1. This part used argon for pressurization, applied an initial pressure of 1.38 bar (20 psi) at room temperature and ramped pressure to 9.65 bar (140 psi) after the lagging part thermocouple reached the melt temperature. The part was then cooled at 10 °C/minute under pressure until the part thermocouple was below 220 °C, then the part was extracted from the clamping fixture.

While evaluating the use of nitrogen gas as the pressurization method, processing steps were altered based on part quality observations. Iterations to the pressurization schedule as well as timing of pressurization were conducted. Initial pressure was changed from 1.38 bar (20 psi) to 2.07 bar (30 psi), 3.45 bar (50 psi), and 4.83 bar (70 psi) to determine the optimal compaction pressure with nitrogen. Pressurization schedules were tuned from 0.70 bar (10 psi) per minute to 0.70 bar per 15 seconds when ramping to the maximum compaction pressure of 9.65 bar (140 psi). The timing of maximum pressurization was evaluated to investigate the part quality from the application of pressure just prior to the polymer melt and just after the polymer melt. All parts were cooled down at the same rate of 10 °C/minute to below 220 °C before removal. A total of 10 parts were fabrication in 3 different geometries throughout the process optimization process. These parts were evaluated with combinations of NDI, surface roughness, dimensional tolerance and destructive inspection.

3. RESULTS

3.1 Simulation Results

The first checks of the FEM output was that the material was flowing at speeds appropriate to the constitutive models. The strain rate target for the process was 3.0E-3, by sampling the largest element deformations over time was produced. Figure 7 shows that all of the elements sampled move at strain rates appropriate to the process over the first 20 seconds. After 20 seconds some of the elements contact the tool and cease to displace further. The pressure ramp rate could potentially be increased without causing the sheet to form too quickly.

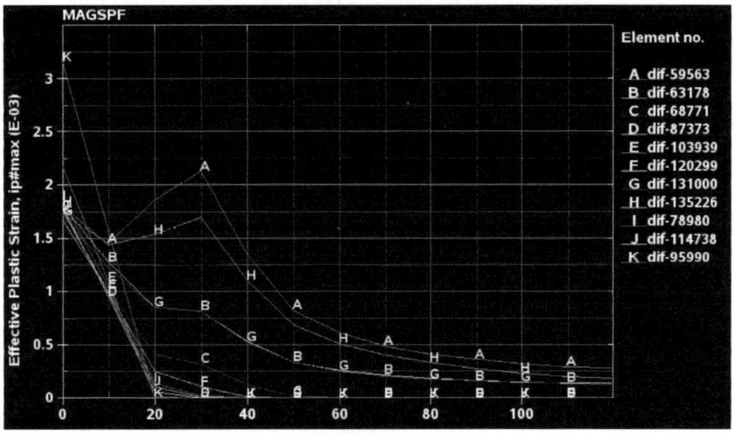

Figure 7 - Elemental strain rate vs time

One of the primary concerns with the lower temperature on forming the AZ31 sheet was the need to apply compaction pressure to the thermoplastic part before the melt temperature of the composite at 310 °C. It was expected that the part was considered to be under full pressure when the magnesium bladder was in contact with the bottom of the picture frame. This ignores the actual consolidation tool and provides a conservative estimate. The part being processed is approximately 18" across. Figure 8 shows the approximate time to contact an 18" span on the tool. Based on this analysis it was shown that the magnesium would contact an 18" wide span on the tool after 30 seconds if pressure was introduced when the tool hit 310 °C. This full coverage time can potentially be reduced by applying pressure earlier in the processing cycle as the part is heated to its melt temperature.

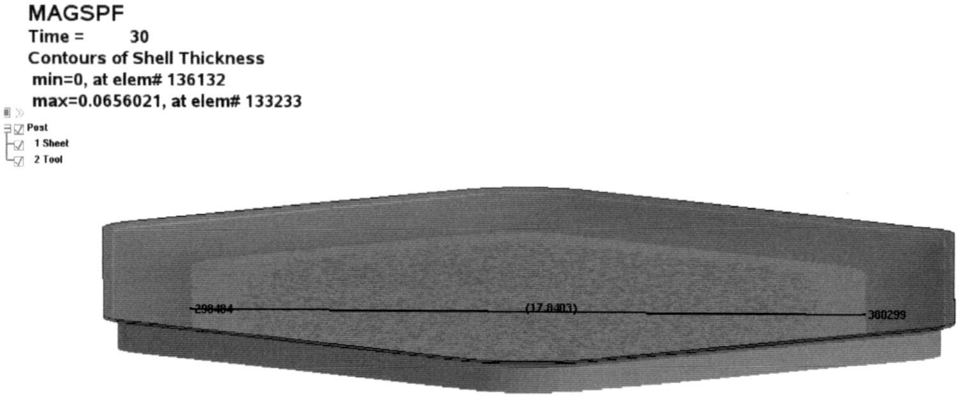

Figure 8 - Sheet contact with the tool at 30 seconds

Standard checks of the total strain (i.e. resulting shell thickness) and stresses were monitored throughout the process to look for any indications of a material failure. Results are shown in Figure 9. Over the full pressure schedule sampled the material only undergoes a strain of approximately 0.15 in the largest forming regions.

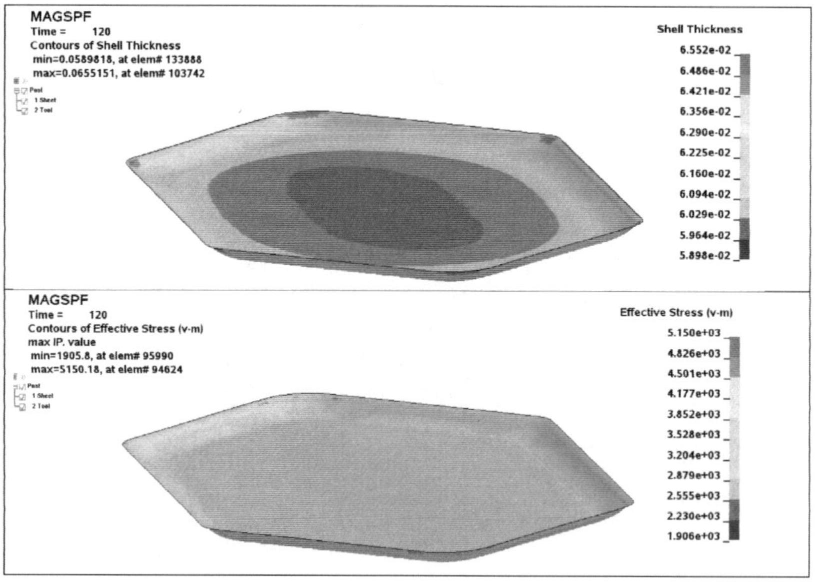

Figure 9 - Sheet thickness at steady state (top), sheet stresses as steady state (sottom)

3.2 NDI and Surface Roughness Results

The following non-destructive inspection (NDI) scans were completed with through transmission C-scans and a surface profilometer. The surface roughness was measured in 9 locations shown in Figure 10. Each location had the roughness measured in the 0°, +45°, -45°, and 90° directions. These values were reported in Table 2.

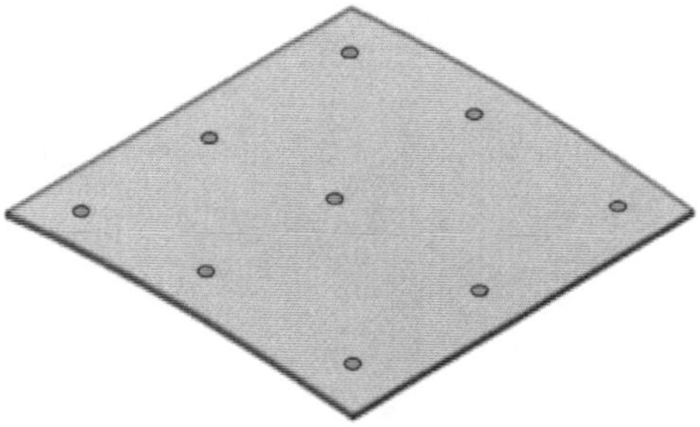

Figure 10 - Locations of surface roughness measurements

Table 2 - Surface roughness measurements

Surface Roughness (µin)	017-1-1	017-2-3	017-2-4	017-3-1	017-3-2	017-3-4
Min	24	15	23	37	39	22
Max	67	101	128	115	104	70
Standard Dev	10.23	21.51	27.97	17.80	18.36	12.89
Average	44.89	53.58	59.94	65.78	71.44	49.16

NDI scans can be seen for each of the three manufactured geometries in Figure 11, Figure 12, and Figure 13. These images were taken under 5 MHz by C-scan. Differences in the surface quality can be seen for each geometry with 017-1-1 being the argon gas control part Figure 11 (a).

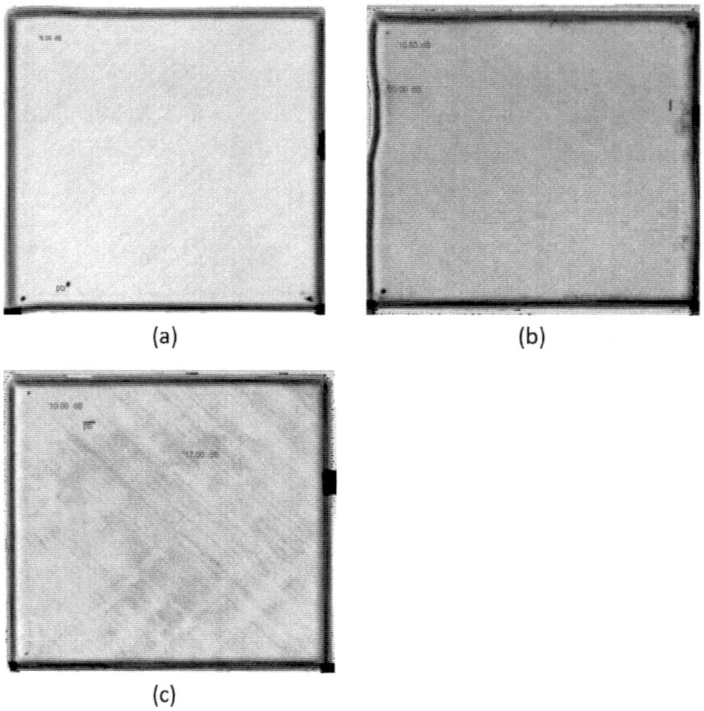

Figure 11 - NDI of flat panel geometry (a) 017-1-1, (b) 017-1-2, (c) 017-1-3

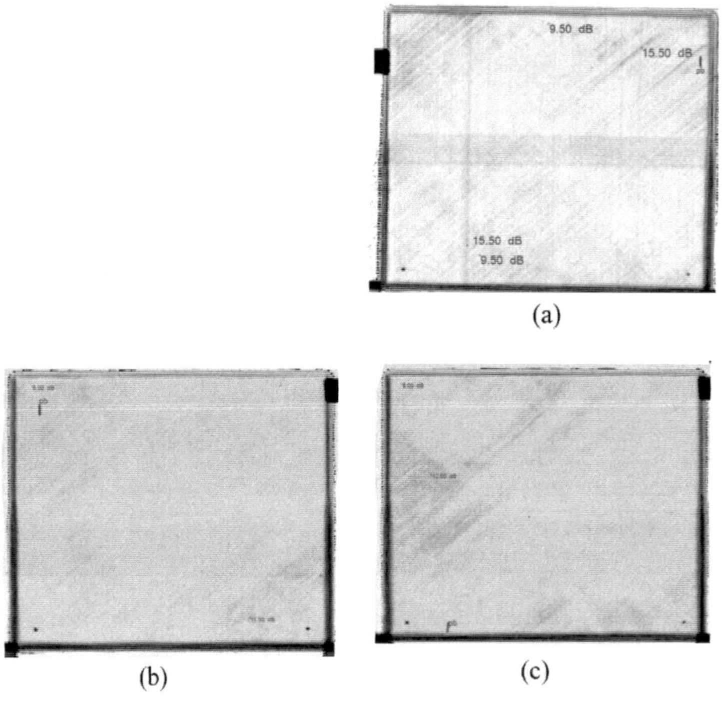

Figure 12 - NDI of ply drop panels (a) 017-2-2, (b) 017-2-3, (c) 017-2-4

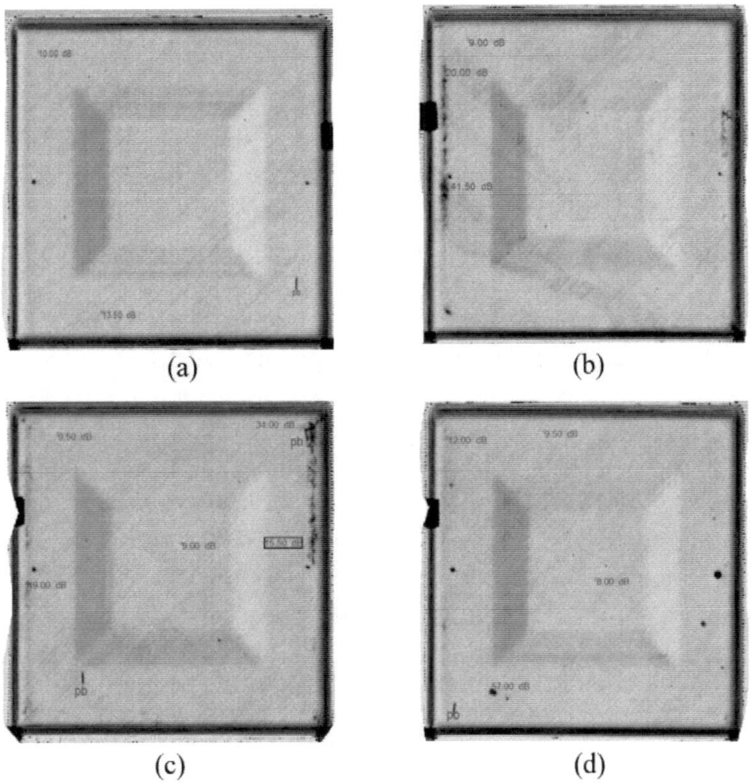

Figure 13 - NDI of picture frame panel (a) 017-3-1, (b) 017-3-2, (c) 017-3-3, (d) 017-3-4

It is expected that panel regions of 32 plies have an attenuation level of less than 20.09 dB and regions with 16 plies have an attenuation level of 14.85 dB. All of the panels have high attenuation in the two locations where the ultrasonic tack weld took place. These tack welds were excluded from evaluation. Future panels would have these welds located outside of the part net trim line. Panels 017-3-4 has good consolidation and a surface roughness that was comparable to the argon panel 017-1-1, however there are 2 dime sized delamination away from the edge. As a result the part would need to be remade.

The panels 017-1-1, 017-3-4, 017-3-3 and 017-3-2 were processed using the same temperature and pressure cycle. These panel 017-3-3 and 017-3-2 had regions of poor consolidation mainly near the edges of the plies. Trimming 30.5mm (1.2 in) from the manufacturing edge of part removed these edge effects from the panel resulting in an acceptable part. These regions are expected to be the last regions of contact for the magnesium bladders. As a method of reducing the manufacturing excess the initial compaction pressure was increased from 1.37 bar (20 psi) to 2.07 bar (30 psi) prior to heat-up. This additional compaction force resulted in a more uniform part consolidation, especially near the edges. The increase in pressure had a negative effect on the surface roughness on the bladder side of the part. The baseline surface roughness form the argon produced panel 44.88 Ra with a max of 67 Ra. The additional pressure panel had an average Ra of 65.78 Ra and a max of 115 Ra. The surface of the resulting panel no longer had a glossy surface and was matte in appearance.

All the parts that were manufactured with nitrogen bladder pressure were loosely adhered to the magnesium bladder following consolidation. Two methods were used to attempt to improve consolidation, improve surface finish and reduce this adhesion. The first method was to sand the bladder smooth with 150 grit sand paper. Comparing 017-1-2 and 017-1-3, where 017-1-3 was fabricated with a sanded bladder, shows that the bladder surface roughness did not have an effect on the part quality. The second method was to apply pressure earlier in the processing cycle prior to the part melt. A thermal profile identified the time when the max part temperature reached the onset of melt for the PEKK material. Parts 017-2-2, 017-2-3 and 017-2-4 used this method with varying room temperature pressure application. The initial room temperature pressures that were used were 2.07 bar, 3.45 bar, and 4.83 bar (30 psi, 50 psi, and 70 psi).

Part 017-2-2 used the initial 2.07 bar (30 psi) application. The resulting part attenuation in the 16 ply region exceeded the attenuation limit. This part would have been needed to be remade. The part 017-2-3 and 017-2-4 used 3.45 bar (50 psi) and 4.83 bar (70 psi) respectively. Both panels had acceptable levels of attenuation. Part 017-2-4 had higher levels of attenuation and a rougher surface finish. It also appears that a wrinkle had formed in the ply drops down the center of the part. The higher initial pressure likely prevented the plies from sliding past each other during processing thereby trapping the wrinkle in the part.

3.3 Ply Placement and Thickness Measurements

The ply drops within the parts geometry 2 and geometry 3. The cut locations are shown in Figure 14. Geometry 2 had ply drop positions measured in sample 2-2-A while Geometry 3 had ply drops measured in location 3-3-A and 3-3-C.

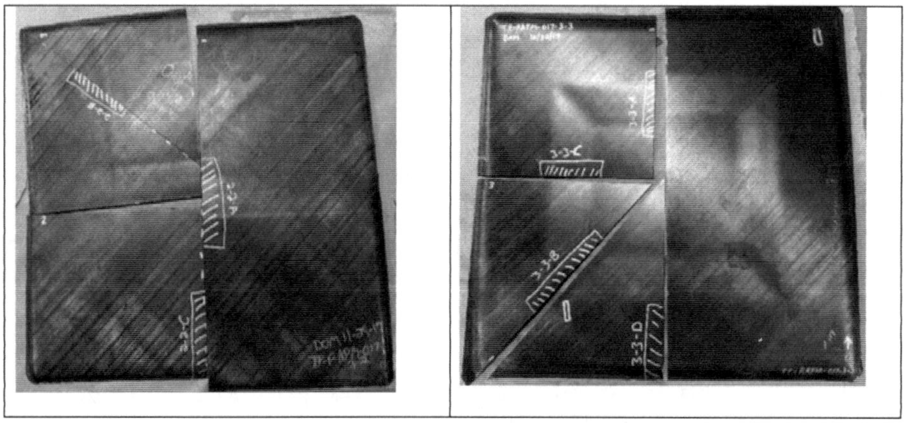

Figure 14- Section cut locations for geometry 2 (left) and geometry 3 (right)

Expected distances between various plies and the actual distance is shown in Table 3. The ply drop ratio that was used was 20:1. A ply was expected to be dropped every 2.76 mm (0.109 in). These ply drops were identified from cross sections shown in Figure 15.

Table 3 Ply Drop Locations in Geometry 2 and 3

2-2-A					3-3-A					3-3-C			
Measure #	Ply Drop	Expected (mm)	Actual (mm)		Measure #	Ply Drop	Expected (mm)	Actual (mm)		Measure #	Ply Drop	Expected (mm)	Actual (mm)
1	1	2.794	1.778		1	1	2.794	2.570		1	1	2.794	1.790
3	2	2.794	2.743		4	2	2.794	2.362		2	2	2.794	2.827
4	3	2.794	3.683		5	3	2.794	1.918		3	3	2.794	4.524
5	4	2.794	3.200		6	4	2.794	1.107		4	4	2.794	3.526
6	5	2.794	2.769		7	5	2.794	2.253		5	5	2.794	3.028
7	6	2.794	3.277		8	6	2.794	2.982		6	6	2.794	2.764
8	7	2.794	0.584			7	ND	ND		7	7	2.794	3.086
	8	ND	ND			8	ND	ND		8	8	2.794	2.482
		Average	2.576				Average	2.199				Average	3.003
		Stdev	1.063				Stdev	0.641				Stdev	0.795

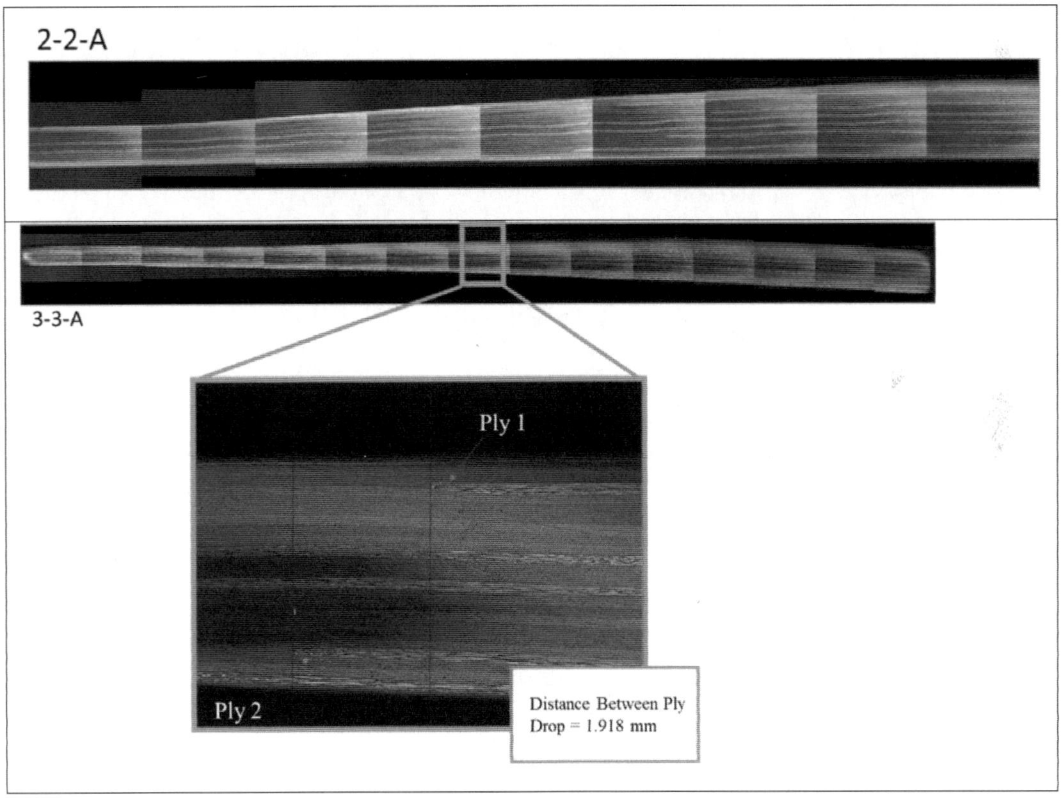

Figure 15 - Cross section of ply drops from 2-2-A (top) ply drop measurement from 3-3-A (bottom)

Bulk thickness measurements were verified in the parts 017-3-1 through 017-3-4. The average, standard deviation, minimum and maximum values are shown in Table 4. The panels have a thickness increase near the edge of the part just prior to the thinning of the laminate at the end. This thickening is occur with a 22 mm (0.87 inch) distance from the manufacturing edge of part. A cross section of thickness is shown in Figure 16.

Table 4 - Bulk thickness values and limits (mm)

	017-3-1		017-3-2		017-3-3		017-3-4		Limit	
	Outside	Inside	Outside	Inside	Outside	Inside	Outside	Inside	Outside	Inside
Avg	4.59	2.30	4.54	2.27	4.58	2.29	4.52	2.26	4.41	2.21
Stdev	0.04	0.02	0.04	0.03	0.03	0.23	0.04	0.02		
Min	4.47	2.25	4.41	2.19	4.47	2.22	4.44	2.19	4.11	2.05
Max	4.79	2.40	4.73	2.45	4.67	2.37	4.70	2.34	4.72	2.36

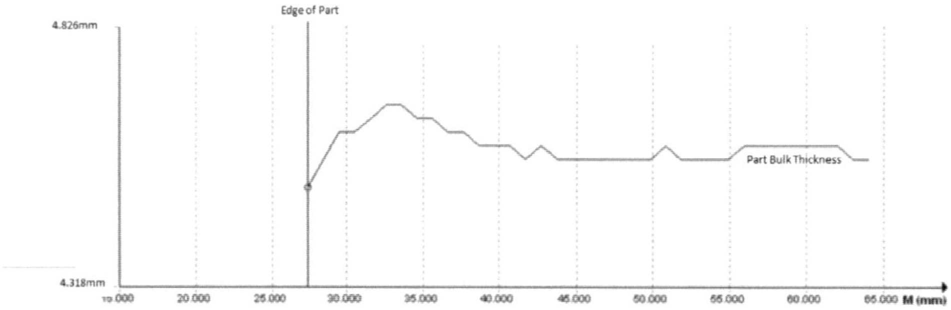

Figure 16 Laminate thickness swell near the edge of part

4. CONCLUSIONS

Panels fabricated using the argon gas show lower attenuation in NDI. The 6 dB attenuation for 32 plies of composite is lower than any other panel fabricated with the nitrogen bladder pressure. The argon bladder pressure part had lower attenuation than parts with regions of only 16 plies as well. In addition the parts fabricated with nitrogen gas have a surface roughness and a matte finish that varied more than the argon panels. Continued development of AZ31 magnesium bladder processing parameters of thermoplastic composites with argon processing gas will need to be completed. Argon processed parts produced higher quality laminates.

High attenuation levels near the edge of laminates due to low initial processing pressure typically occur in the outer 30.5 mm (1.2 in) from manufacturing edge of part. With processing parameters adjustments this edge trim could be reduced to a standard 25.4mm (1 inch) of trim to conservatively account for the edge of part thickening that occurs from the bladder use.

The pressure and temperature cycles used in part 017-1-1 and part 017-2-3 produced the parts with the best results. Both cycles applied sufficient compaction pressure to consolidate the plies without preventing the plies from being able to slide past each other preventing wrinkling.

5. ACKNOWLEDGEMENTS

This effort was jointly accomplished by a Boeing-led team and the United States Government (Defense Advanced Research Projects Agency, DARPA) under the guidance of the United States

Army Research Office, ARO. The authors would like to acknowledge the guidance and support of Dr. Jan Vandenbrande of DARPA/DSO and Dr. David Stepp of Army Research Office (ARO). The information in this paper was approved for public release by 19-162544 and DISTAR 32485. Research was sponsored by the Defense Advanced Research Projects Agency (DARPA) and was accomplished under Cooperative Agreement Number W911NF-16-2-0087. The views and conclusions contained in this document are those of the authors and should not be interpreted as representing the official policies, either expressed or implied, of the Defense Advanced Research Projects Agency or the U.S. Government. The U.S. Government is authorized to reproduce and distribute reprints for Government purposes notwithstanding any copyright notation herein.

The authors thank Ravi Verma for use of data [3] for Figure 4.

6. REFERENCES

[1] A. Haufe, K. Schweizerhof, and P. Dubois, Properties and Limits: Review of Shell Element Formulations. Developer Forum 2013. Filderstadt, Germany. 24 September 2013.

[2] LS-Dyna Keyword User's Manual. Volume II, Material Models. LS-Dyna R8.0, 18 March 2015

[3] Taleff, E.M., Hector, L.G., Verma, R. et al. J. of Materi Eng and Perform (2010) 19: 488. https://doi.org/10.1007/s11665-010-9612-6

[4] Hahn, G.L., Tsotsis, T.K., "Rapid High-Performance Molding (RAPM) of Small Parts," SAMPE 2019 Conference Proceedings. Charlotte, NC, May 20-23, 2019.

[5] Shewchuk, S.M., Halford, B., Matlack, M.P., Sharpe, A. and Massey, P. "Development of Scalable Dynamic Control Architectures for Flexible Composites Manufacturing Work Cells." SAMPE 2019 Conference Proceedings. Charlotte, NC, May 20-23, 2019.

[6] Schell, J.J., Shewchuk, S. M., Leach, D., McCrea, T., Abrams, N., Postera, R., and Hahn, G. L. "Rapid High Performance Molding of Structural Thermoplastic Composite Parts." SAMPE 2019 Conference Proceedings. Charlotte, NC, May 20-23, 2019.

A SMART ROBOTIC CELL FOR AUTOMATING COMPOSITE PREPREG LAYUP

Rishi K. Malhan[1], Aniruddha V. Shembekar[1], Rex Jomy Joseph[1], Ariyan M. Kabir[1], Prahar M. Bhatt[1], Brual Shah[1], Wenping Zhao[2], Steven Nutt[3], and Satyandra K. Gupta[1]

[1]Center for Advanced Manufacturing, University of Southern California
[2]Raytheon Technologies Research Center (legacy- United Technologies Research Center)
[3]M.C. Gill Composites Center, University of Southern California

ABSTRACT

Hand layup is a commonly used process for making composite structures from several plies of carbon-fiber prepreg. The process involves multiple human operators manipulating and conforming layers of prepreg to a mold. The manual layup process is ergonomically challenging, tedious, and limits throughput. Moreover, different operators may perform the process differently and hence introduce inconsistency. We have developed a smart robotic cell to automate the prepreg sheet layup process. The cell uses multiple robots to manipulate and drape sheets over a tool. A human expert provides a sequence to conform the ply and types of end-effectors to be used as input to the system. The system automatically generates trajectories for the robots that can achieve the specified layup. This paper describes a systematic approach for selecting and placing robots in the cell and description of tools and end effectors needed for utilizing the cell. We also describe planning algorithms for (a) generating plans to grasp and manipulate the ply and (b) generating feasible robot trajectories. Our system can generate plans in a computationally efficient manner for complex parts. We demonstrate the automated layup by conducting physical experiments on an industry-inspired mold using the generated plans. Our system can perform sheet layup at speed comparable to human operators.

1. INTRODUCTION

Composites are engineered materials most often consisting of continuous aligned carbon fibers embedded within an engineered polymer matrix. High-performance composites are frequently produced from prepregs, or fiber beds pre-impregnated with a catalyzed but uncured polymer resin [1]. Composites enable production of lightweight structures with high specific mechanical properties, tailored anisotropic structural response, and high durability. For these reasons, they are enabling technologies for multiple sectors of national importance, including aerospace, defense, clean energy, and ground transportation. The U.S. composites industry is growing at an annual rate of 5% and was valued at $7.5 billion in 2015, with 62% of this value located in manufacturing-intensive sectors. The Federal Government has identified lightweight, high-strength, and high-stiffness composites as a critical technology for the Department of Defense, as well as an essential cross-cutting technology for U.S. manufacturing, with applications in transportation (aerospace, automotive), efficient power generation, and infrastructure.

High-performance composites are often produced by laying down prepreg material on a contoured tool in a specific orientation. The layup (or material placement) step is critical to both part quality and productivity [3]. After a layup, the resulting laminate is cured using heat and pressure to develop the desired material properties. Successful layup is critical because the ply stacking sequence governs mechanical response and because defects caused by improper layup (e.g., wrinkles, ply bridging at corners, air gaps, porosity [2]) can lead to part rejection or rework.

Prepregs are available in narrow forms (fiber tows or tapes measuring 3 - 300 mm wide) or wide woven roll sheet stock, for example, 1200 mm wide. Tows and tapes are unidirectional, lightweight, compliant, and compatible with automated fiber placement (AFP) or automated tape layup (ATL), processes that consist of laying down material using a placement head attached to a robot (gantry or arm) [4-5]. However, AFP and ATL are incompatible with small or medium-sized parts whose complex features or production characteristics pose capability, accessibility, size, or cost challenges. Recently, the use of composites in the automotive industry has led to the development of high-rate production methods for vehicle parts using liquid molding. However, such methods require substantial infrastructure investments. For these reasons, most small and medium-sized complex-shaped parts are currently produced by hand layup.

Hand layup is time-consuming and slow, expensive (due to labor costs), and can result in part-to-part variability because of human factors that are difficult to control (e.g., operator experience and dexterity). Thus, there are substantial technical and economic incentives for replacing hand layup with an automated process that can (1) improve productivity and reduce the risk of operator injury due to repetitive motion injuries, and other ergonomically challenging tasks, (2) provide consistent layup quality through a formalized and optimized process, reducing the risk of human-induced defects, and (3) increase throughput and reduce overall costs. The hand layup process was studied by Elkington et al. [6,7] and Prabhu et al. [8] to identify methods used for deforming and draping prepreg to a complex-shaped tool surface. Such et al. [9] investigated the use of tracking and projection technologies to assist human operators during layup. Ward et al. [10] developed a system for automated removal of backing paper, applicable to both hand layup and future automated solutions. Buckingham et al. and Molfino et al. [11] proposed approaches for automating the manufacturing of composite broad goods. However, the systems prioritized sheet translation rather than accurate draping (deformation and conformation) to the tool. Elkington et al. [12], building on prior work at the University of Bristol, demonstrated an automated prepreg sheet layup process, which our work leverages and expands upon. The approach utilizes a six-axis industrial robot and tailored end effectors to conform prepreg sheets to a tool. Recently, Elkington et al. [13] reviewed prior literature on automated prepreg draping. Furthermore, human-robot collaboration techniques have also been developed for the manipulation of flexible materials [14]. Here, we build upon prior work by developing a conceptual framework that leverages advances in prepreg processing and collaborative robotics [14-16], and focuses on draping without pre-shearing. Das and Nilankan [17] designed an optimization-based motion planner to solve the problem of deformable object manipulation by utilizing a controller that accounts for modeling uncertainty associated with shape changes.

Our goal is to develop a robotic cell using robots to reduce manual labor in the composite layup. The human-robot interaction is necessary for risk-averse sectors such as aerospace, where efficient production is critical, but full automation is unlikely if the risk of process-induced defects exists. This paper describes a systematic approach for selecting and placing robots in the cell and description of tools and end effectors needed for utilizing the cell. We also describe planning algorithms for (a) generating plans to grasp and manipulate the ply and (b) generating feasible robot trajectories. Our system can generate plans in a computationally efficient manner for complex parts. We demonstrate the automated layup by conducting physical experiments on an industry-inspired mold using the generated plans.

2. SYSTEM DESIGN

Recent advances in robotics are attractive for composite manufacturing. Modern robots are highly repeatable and can potentially be used for the layup on complex tools. On-line planning capabilities are providing foundations for automatically generating plans and eliminating the need for manually writing programs. There are new on-line sensing modalities such as 3D vision that can be used for real-time process monitoring, defect detection, and repair plan generation. The concept of a hybrid robotic cell enables accepting appropriate interventions from human operators when robots cannot resolve problems on their own.

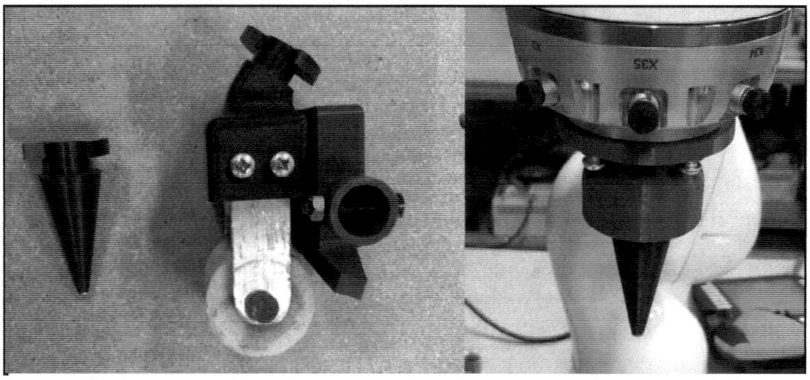

Figure 1: Tool Changing Mechanism for End Effector

We have created a robotic cell that supports collaboration among three robots. Two or more robots collaborate to manipulate the material and place it on the tool. This requires performing sheet stretching, shearing, and compaction operations. One robot manipulates the tool to ensure that other agents can access necessary regions. A human operator (with assistance from robots) performs inspection, repair, and material placement tasks deemed too difficult for robots. The human operator also troubleshoots problems and modifies automatically generated plans. Finally, the cell includes a centralized planner to coordinate the motions of all the robots.

We have conducted preliminary tests by 3D printing molds that have concave and convex regions. Based on these preliminary tests, we have developed requirements for the automated robot instruction generation system [18-19]. The automated robotic instruction generation system will need to compute both draping and grasping robot trajectories over the mold. Following are the key instruction generation system requirements:

1. Users should be able to guide draping paths based on this previous experience and draping simulation. This requires development of a user-friendly GUI.

2. When human experts perform the layup process, they know exactly where to start the layup process and how layup should progress so that there will be no wrinkle formation. For automated robot instruction generation, the planner will identify the priority sequence of layup regions for the mold and appropriately generate the robot motion plans.

3. Draping robot motions need to be computed based on the user-defined path. A planner will be required to generate robot joint angles for tool motion that will take into account the process and robot constraints.

4. A grasp planner will be needed to compute the grasping locations of robots, which will enable robots to hold the sheet and exert the necessary force to stretch the sheet, so that wrinkle formation is avoided.

5. When robot trajectories are discontinuous, point to point trajectory planner will be required to generate collision-free repositioning robot motions.

6. Robot trajectories will be generated in the form of a series of locations for robot end effectors to travel. For accurate layup, robots need to go to the exact location specified by the planner, for which a higher degree of accuracy in part placement is needed. Hence an accurate part registration planner is required.

7. The motion between the draping robot and the grasping robots will need to be synchronized to avoid any possibility of collision.

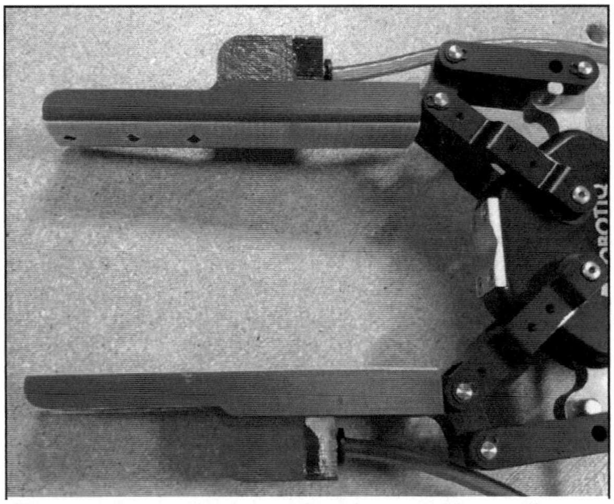

Figure 2: Preliminary Design of Sheet Grasping Fingers

8. A simple tool changing mechanism will need to be incorporated to incorporate different sized rollers and dibbers. Figure 1 shows the 3-D printed tool holder mounted on the KUKA iiwa LBR 7 robotic manipulator and the roller that can be attached to the holder by simply pushing and rotating. The planner will need to specify a tool change command to utilize the appropriate tool.

9. The grasping robots will need to frequently change their grasping locations and continuously need to handle the sticky composite sheet. To effectively release the sheet without it being stuck to the gripper finger, a new grasping end effector will be needed. For this, we will need to make use of the custom grippers with the air pressure to release the sheet. Figure 2 shows a preliminary prototype of custom gripper fingers that perform satisfactorily during the preliminary test. The grasp planner needs to compute a set of feasible grasping location so that grasp-ing locations can be changed to avoid defect formation.

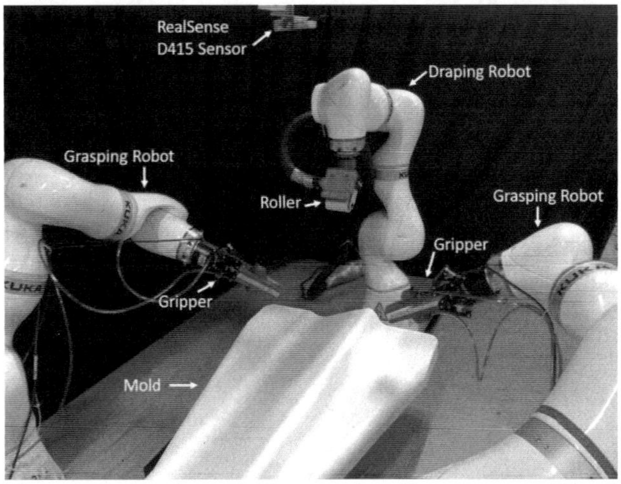

Figure 3: Designed Composite Sheet Layup Work Cell

The draping robots need to conform to the surface of the mold during the layup as well as apply enough force to adhere to the prepregs on mold surface. Thus, we require a robot that operates under force control and provide enough compliance. The grasping robots are required to maintain a certain amount of tension while grasping the prepreg sheet to avoid dropping. Moreover, the grasping robots also needs to be compliant to draping robot's actions and provide enough slack during the layup operation. Thus, the robots selected will need to have force control. Considering the reachability requirements for the test molds, we have selected one KUKA LBR iiwa 14 R820 robotic manipulators for draping and two KUKA LBR iiwa 7 800 robotic manipulators for grasping the sheet. These manipulators have in-built force sensing and force control technology that will provide aid to the layup process.

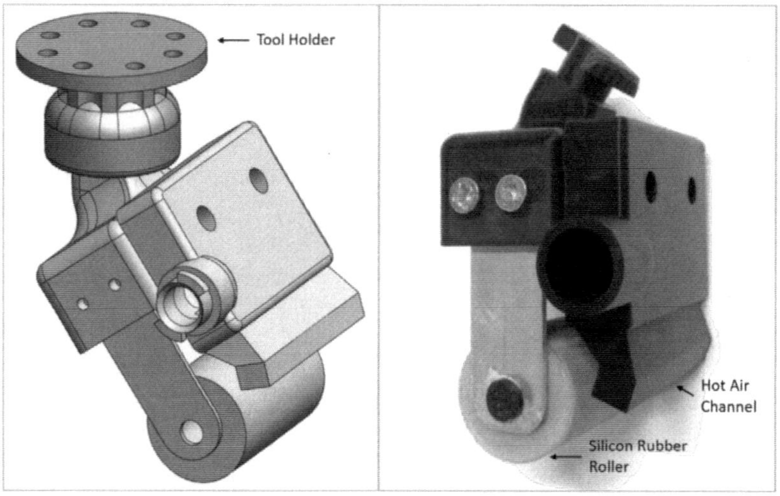

Figure 4: Flat Roller End Effector

Robotic manipulators should be placed in the work cell in such a way that their configuration space overlap should be as minimum as possible to avoid collisions, but on the other hand, they should cover as much mold region as possible to ensure smooth trajectories and avoid singularities. The robot-tool layout was determined using a combination of sampling, optimization, and search-based algorithms [20-21], utilizing knowledge of kinematic constraints to minimize the chance of collisions [22]. The mold placement planner generates the position and orientation for the mold to be placed in the robotic work cell, where it ensures the maximum reachability of the draping robot over the mold surface. Figure 3 shows the 3-robots work cell for the automated sheet layup process.

For the process monitoring and intervention control, perception technology is needed. We have added the RealSense D-415 RGB-D sensor in the work cell, which will continuously monitor the composite sheet being draped over the mold. This sensor will allow the intervention controller to track the sheet in real-time and provide necessary actions to robots' controllers in case of any issues.

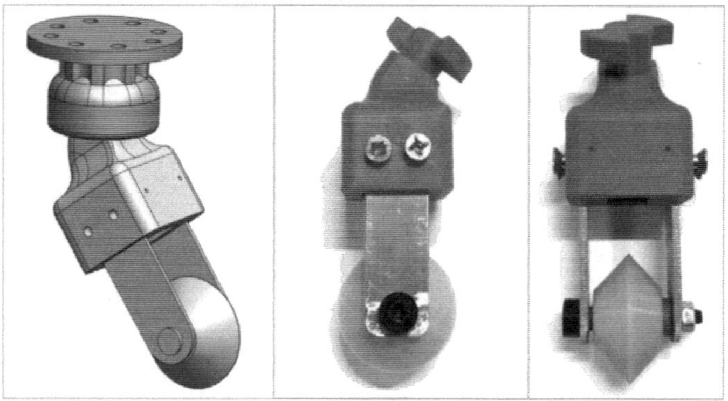

Figure 5: Conical Roller End Effector

2.1 Draping robot end effectors

The primary requirements for draping robot end-effector are to maintain a draping force, provide a compliant contact with the sheet, and avoid damage during layup. Considering the above constraints, we have designed the different sized roller end effectors for draping robots. Figure 4 shows the CAD model and the physical roller end effector. The head of the end effector is designed in such a way that it can be easily inserted into or removed from the tool holder (which is fixed to the robot wrist) without any mechanism.

The main body which holds the rubber roller and the air channel is 3D printed with ABS to reduce the overall weight of the end effector. The metal plates are used to connect the rubber roller to the body to support the force exerted by the robot. The silicon rubber is made of Smooth-On Mold Max with a shore hardness of A40. This rubber roller provides the necessary cushioning between robot and sheet to prevent any possible damage in the layup process. The sheet is slightly heated with hot air passing through channels which activates the resin and improves the adhesion to maintain the tackiness of the resin on the composite sheet.

In some mold surface geometries, sharp corners are present in the mold. If the corner radius is too small, roller cannot access the surface near a corner, and ply bridging can occur. To avoid that, we have designed the conical roller. Figure 5 shows the CAD model and the physical design of the conical roller. With such shape of the roller, ply is draped appropriately over sharp corners of the mold. The main body and the hot air channel of the conical roller are the same as the flat cylindrical roller. For small concave/convex corners where rolling motion is difficult to perform, specialized dibbers can be used.

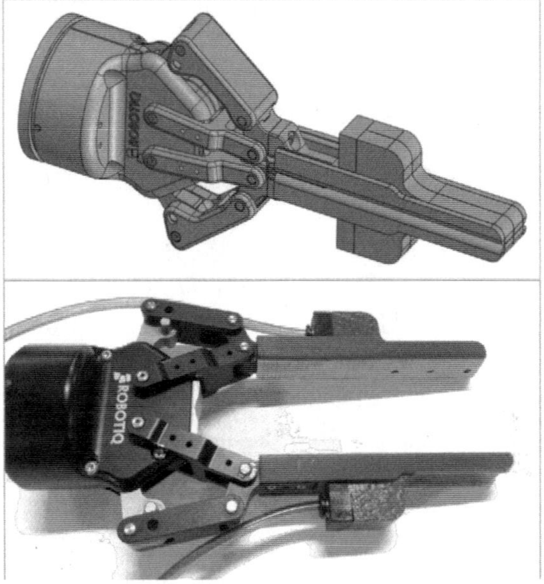

Figure 6: Grasping Robot End Effector

2.2 Grasping robot end effectors

To provide the necessary force and for the ease of gripping, we have used the Robotiq 2F-85 gripper. This gripper allows us to control the gripping force and easily manipulate the sheet. As the gripper is holding the sheet with a certain force, opening the gripper does not release the sheet completely due to the stickiness of resin. Hence, we have designed the custom gripping fingers which have PTFE tape on the contact side of the fingers. PTFE is not affected by the stickiness of the resin and releases the sheet as soon as the gripper is opened. We also have added the pneumatic control for pressurized air release from the inner side of the finger. This will ensure the sheet release completely when the gripper is opened. Figure 6 shows the grasping robot end effector for the layup process.

2.3 User interface

The main requirements for the user interface are to:

1. Select the mold and the tool for draping
2. Generate the tool path over the mold for draping robot using geodesic curves.
3. Generate automated instructions for robot motion
4. Control robot motions
5. Simulate the layup process

We have developed the user interface for the composite sheet layup process using the ROS 3D visualizer package RVIZ. RVIZ lets you add, and control visualize mesh models, point clouds, etc. We have also added the control panel buttons, which will enable the user to handle the complete layup process flexibly. Figure 7 shows the user interface for the layup process.

There are seven main components of the user interface:

a. Notification Bar: Notification bar shows the information regarding process steps to be followed by the user, the existing process going on in the simulator, as well as errors and warnings whenever required.

b. Mold and Tool Selection: The first thing for the user to do is to select the appropriate mold for the layup from the drop-down menu. Users can select one or many tools from the drop-down menu to generate the appropriate draping trajectories.

c. Toolpath Generation: Once the draping sequence generation button is clicked, a new window will pop up with the only mold where the user can generate the toolpath by drawing trajectories for a layup. Users will also have the option of selecting the existing toolpaths from the system. Once the toolpath is generated, the user can either determine the mold placement location or choose to select the existing mold location for a layup.

d. Mold registration: Mold localization with respect to the reference frame (draping robot base frame in this case) is performed in two stages - (i) vision-based registration and (ii) contact-based registration. If the user chooses to select the existing mold location, then the vision-based registration step will be eliminated, and the user will perform the contact-based registration. In contact-based registration, another window will pop up, which will prompt the user to generate the probing trajectories for contact-based registration. Users can simulate the probing trajectories to ensure correct robot motion over the mold. Once contact-based registration is performed, the part will be localized with respect to the reference frame with submillimeter accuracy.

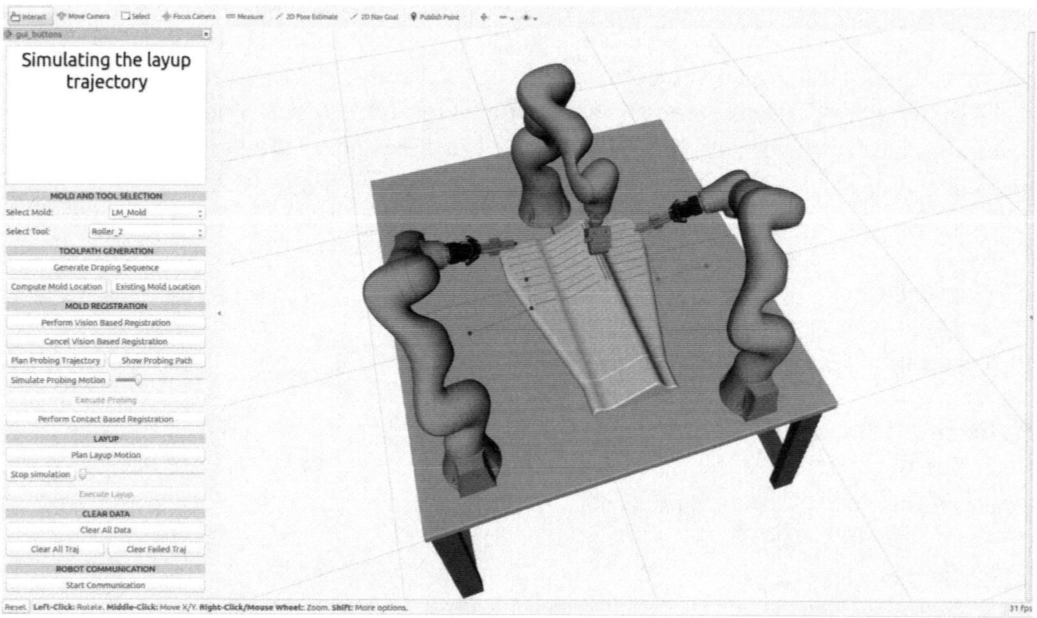

Figure 7: User Interface for Robotic Layup Process

e. Layup process: Once the user presses the plan layup button, the robot motion sequence is generated. Both draping and grasping robot trajectories will appear on the screen. Successful draping trajectories will appear as green, while failed trajectories will appear as red. If there are any failed trajectories, the user can regenerate the draping sequence and re-localize the mold. Users can simulate the robot motion to ensure correctness.

f. Clear data: Users can clear all data and start the complete process again with either the same or the different mold. Users can also clear the failed trajectories from the window to visualize the feasible robot motion for the layup

g. Robot Communication: Users can control the actual robot motion for the layup from this inter-face. Once the communication is established with robots, the user can track the real-time motion of the robots, and, also start or stop the layup motion.

3. ROBOT INSTRUCTION GENERATION

3.1.1 Draping robot motion

Depending upon the mold shape and layup area, the skilled operator generates the toolpaths for the draping robot motion using GUI. Each toolpath is a series of waypoints over which robot end effector needs to travel. We can always make use of these waypoints directly to generate the robot motion, but the path executed by the robot may not satisfy motion constraints, like continuity constraint, velocity constraint, collision constraint, etc. For reliable layup, all the motion constraints need to be evaluated during the trajectory computation phase. Hence, a low-level constrained trajectory generator is used to compute the robot joint angles for tool motion. The input to this draping trajectory generator module is sequence waypoints with respect to the robot frame and the robot tool center point (TCP). The output of this planner, set of joint angles for motion, is sent to the robot controller for motion execution.

3.1.2 Grasping robot motion

Based on the toolpaths for layup, mold geometry, and the mold location, grasp locations in the cartesian frame are computed using a grasp planner. These grasp locations are then converted into configuration space using grasp robot path generation module and sent to the respective grasping robots' controllers for motion execution. If a modification to the grasp plan is required by the process, we use machine learning techniques to refine the plan online [23].

3.1.3 Point-to-point trajectory motion

In the case of a draping robot path, a motion plan is generated only for robot travel along toolpath. This trajectory generator module does not consider the motion between the two successive toolpaths. If the toolpaths are not continuous, then the draping robot needs to readjust its position and orientation for the start of every new path. The same problem appears in the grasp plan generation as it is not concerned with the robot motion during the regrasp. Point-to-point (PTP) trajectory planner is used in these cases for generation motion execution instructions doing the intermediate phases. This PTP planner takes the start and end joint angles configurations, and locations and sizes of environmental objects as inputs and compute the path for robot motion, which is collision-free. Therefore, for the layup execution, constrained trajectory motion, combined with PTP motion instructions, are sent to robot controllers.

3.1.4 Intervention controller robot instructions:

The intervention controller motion execution plans are generated during the online process monitoring when the actual layup is started. Based on the regrasping positions to avoid any drooping or to eliminate wrinkles, robot motion is computed in joint configuration space and sent to the robot controller as an interruption signal.

3.1.5 Robots communication and synchronization

The robot instructions are transferred to robot controllers using TCP/IP protocols. Low-level socket programming is implemented for a secure connection and data transfer between multiple robots with PC. Each robot, as a server, has a one-to-one communication line with PC, which acts as the client. The robots' synchronization is handled by the PC, which sequentially sends the motion instructions to robot controllers for layup execution. Robot controllers also have separate communication lines for handling interrupt signals given by the intervention controller. These intervention signals are given priority, and regrasp motion is executed first before resuming the layup.

3.2.1 Path Constrained Trajectory Planner

Finding an exact solution involves solving a non-linear optimization problem in the configuration space for every waypoint in a path. We propose a path-constrained trajectory planner that will evaluate reachability over the paths by enforcing the constraints. All the waypoints are transformed with respect to the base of the robot. An optimization problem is solved to find the joint configuration. Constraints involved in this optimization routine are:

- Configuration Space Constraints: The joint angles, joint velocities, and joint torques of the robot are limited within the upper and lower bounds.

- Instantaneous Velocity Constraint: We need to ensure that the robot must meet the minimum desired velocity while executing the paths. For a waypoint, we can apply a rigid body transformation to find the corresponding robot flange pose. An instantaneous velocity vector can be defined. This vector is comprised of position and orientation velocities. A Jacobian matrix J is used to express the differential kinematics of the robot.

- Transition Velocity Constraint: Instantaneous velocity constraint does not ensure if the joint velocity is sufficient to transition. We introduce a transition constraint for this purpose.

- Continuity Constraint: We need to avoid any large joint angle changes while executing the paths. The Jacobian matrix can be used for two close configurations are similar. We use Pearson's correlation coefficient between the Jacobian matrices at consecutive configurations to enforce the continuity constraint.

- Force Constraint: The robot must apply the desired force on the path which is being executed. The Jacobian matrix establishes the relationship between the joint torque vector and the force moment vector at the end-effector. Assuming the inertial effects to be negligible under constant velocity, we use the Jacobian matrix to enforce the constraint on the joint configuration of the robot.

- Collision Constraint: We model the robot and the roller tool using a set of spheres. We compute a (Euclidean Distance Transform) EDT map for the environment for faster collision detection.

- Singularity: Manipulability index is used to keep the robot away from singularities. We can analytically compute this index at any given configuration.

A solution, if exists, may not be found by the optimization algorithm if all the constraints are enforced at once. In our approach, we apply the position, orientation, and continuity constraint first to guide the end-effector in the region where position and orientation constraint is guaranteed to be satisfied. Continuity constraint is more relaxed as it directs the algorithm to find a solution in position and orientation regions, which is closer to the previous solution. Singularity constraint, on the other hand, will truncate the region to avoid singular solutions. So, continuity and singularity constraints are applied along with position and orientation. We then apply velocity and force constraints. The constraint which imposes the most restrictions is the collision constraint.

Due to complex geometries, the tool is likely to collide with the mold. So, we enforce the collision constraint in the end, which will have the smallest feasible region. However, even after using a successive application of constraints strategy, it is not guaranteed to find a solution if it exists. It is highly dependent on the initial guess used in the optimization algorithm. If the initial guess is closer to a region where all the constraints except one are being satisfied, the algorithm can get stuck in the local minima.

3.2.2 Grasp Planner

Our algorithm is an any-time state-space search algorithm. It generates a grasp plan and then updates it further to reduce the cost. The algorithm can be terminated if a time limit is reached, the cost cannot be decreased any further, or all the nodes have been evaluated. Our goal is to use domain-dependent heuristics to generate a low-cost initial plan and obtain an optimal resolution solution in a computationally tractable time. We use an ordering heuristic that prioritizes the nodes in every stage. The ordering heuristic first explores the state space to prioritize the nodes (grasp points), which are promising. We begin by exploring a small number of grasp locations for every node. We evaluate two grasp locations, which are equally spaced between the mold surface and a maximum height above the mold. We pick the location with a lower cost and assign its grasp location to the corresponding grasp point. All the grasp points in a stage are then ordered based on the respective assigned costs. We then use the generated lower cost grasp locations as an initial guess for the gradient descent step used in local assignment search. The heuristic significantly reduces the computational cost. It allows us to generate good initial guess for the gradient descent search during the location assignment. A minimum cost solution is then found with fewer iterations as opposed to exploring the entire search space for the optimal solution. Additionally, prioritizing nodes helps us evaluate the grasp points, which can have a lower cost at the beginning of the algorithm. Therefore, our algorithm generates grasp plans with a low initial effort com-pared to randomly selecting nodes.

This algorithm is used to find solutions for the grasp plans in each of the two-state spaces corresponding to each robot separately in our case. Consider the mold A shown in Figure 8 as an illustration. The two independent solutions are found for robot-1 (RB1) and robot-2 (RB2). The first two images show that only the unconformed particles in a robot's virtual partition are considered during evaluation. The unconformed particles in the other robot's partition are considered to be fixed at their draped position during the simulation. The third image in the sequence shows the combined solution, which is generated from the nominal grasp locations. We can see that the ply collides with the mold. We use a state-space search to reposition the grippers of both the robots in order to minimize the cost. We use gradient descent to adjust for the grasp locations, which gives us a solution shown in the fourth image.

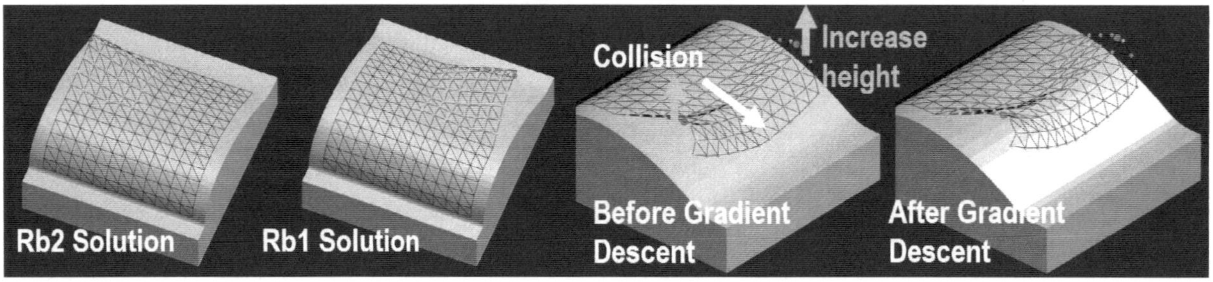

Figure 8: (Left to Right) Independent solutions obtained for robot-1 (RB1) and robot-2 (RB2), ply state simulated for combined solution obtained from the nominal grasp locations, solution improved using a state space search using the nominal grasp locations

3.3.1 Intervention Control

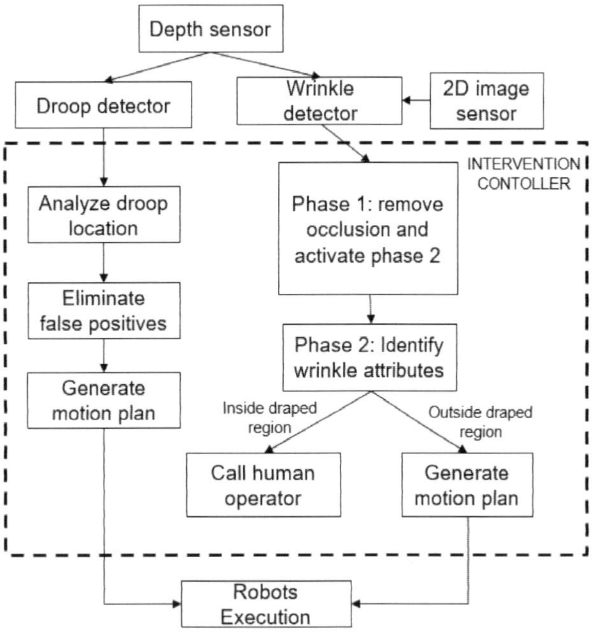

Figure 9: Intervention Control Architecture

This controller takes direct control of robotic arms to prevent collisions. Figure 9 illustrates the defect analysis and corrective action flow of the intervention controller. The intervention controller simultaneously processes the signals received from both the droop detection module and the wrinkle detection module. Appropriate grasp plans are computed based on the drooping or wrinkles attributes and sent to the robot controller for motion execution.

The online monitoring of ply (sheet) requires us to constantly monitor the ply state during the motion execution. If there is only one camera in the cell, then in some cases, either draping or grasping robots can cast occlusions to the ply view. Therefore, we have implemented the intervention control in 2-stages with four cameras. The first stage of intervention control includes the detection of ply using depth sensors from different angles, while the second stage includes ply capture and analysis from the more accurate 2D camera.

3.3.2 Droop Avoidance

Drooping location signals are analyzed based on their distance from the draped region. The undraped region surrounding the draped region is not categorized as drooping, as ply in this region will be closer to the mold. For generating the motion plan, higher weight is given to the drooping locations which are away from the draped region (see Figure 10).

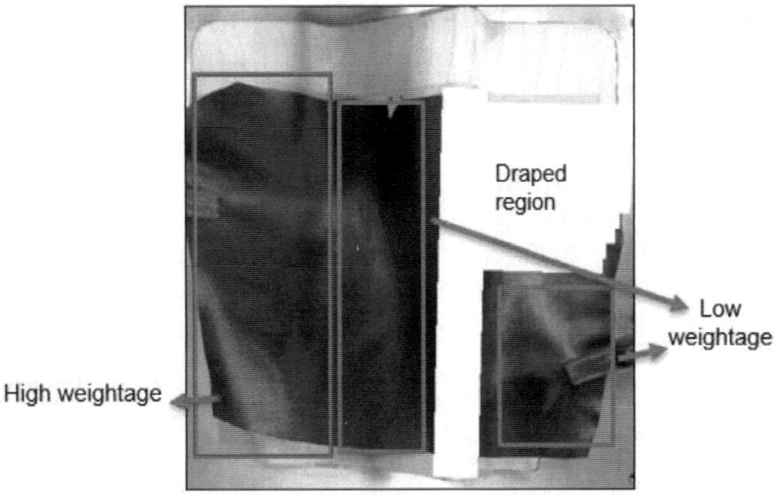

Figure 10: Assigning weightage for undraped region

Using the droop location and the weight metric, the nearest grasping robot is identified. The current grasp pose of that robot is altered in both translation and orientation. The orientation component is important to avoid any possible bending of the ply at the grasping point when the sheet is lifted. The new robot pose is obtained by moving along the curve whose center lies on the nearest point in the draped region.

3.3.3 Wrinkle Elimination

Phase 1: getting wrinkle locations using depth sensors - Due to the inaccuracy of depth sensors, data may contain both false positives (not detecting wrinkles even when it exists) and false negatives (detecting wrinkles even if there isn't any). Hence, when the callback for phase 1 is activated, the draping robot halts the motion and goes to the home position. This eliminates ply occlusion due to the robot and activates the more-accurate image-based wrinkle detection module.

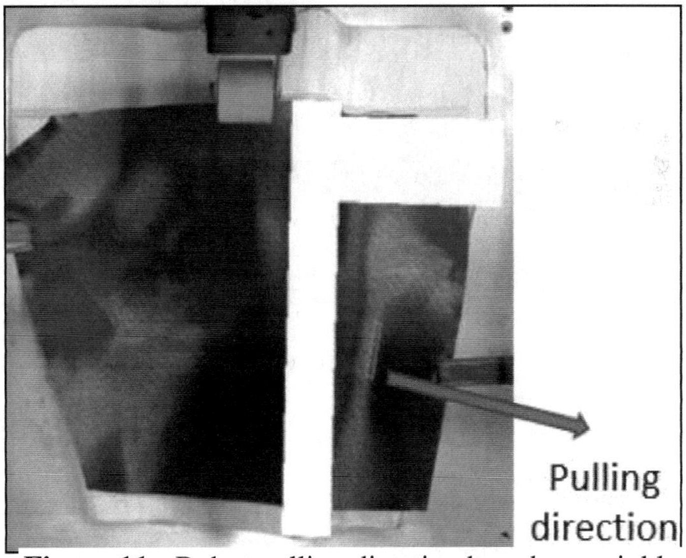

Figure 11: Robot pulling direction based on wrinkle orientation

Phase 2: analyzing wrinkle attributes to generate a corrective motion plan - Image-based wrinkle detection correctly identifies and eliminates false negatives and false positives. From the remaining wrinkles, preference is given to the wrinkles identified inside the draped region as they cannot be easily rectified by robots. Hence, in that case, the system will call the operator to come and rectify the wrinkle. Other wrinkles are sorted based on their sizes and priority given to more prominent wrinkles. Wrinkles appearing inside already fixed regions are ignored. Using the wrinkle attributes (wrinkle centroid, bounding box encircling the wrinkle, bounding box orientation, etc.), the direction of wrinkle is analyzed, and the grasp plan is generated to traverse along the perpendicular direction of wrinkle. Figure 11 shows the direction of grasping robot motion to pull the wrinkle out.

4. Results

We tested the system with a complex industry part. Figure 12 shows steps of the robotic layup process for this part. The mold dimensions are 711 x 558 x 76 mm. It took 7 minutes to layup one sheet using the robots. The ply tolerance was within +/- 2.54 mm and +/- 10°. Process is repeatable within 500 microns through thickness. Conformity and compaction were achieved within tolerances. No visible damage such as wrinkles, voids, or bridging was observed. Quality was comparable to hand layup operation. 92% reduction in human labor was achieved.

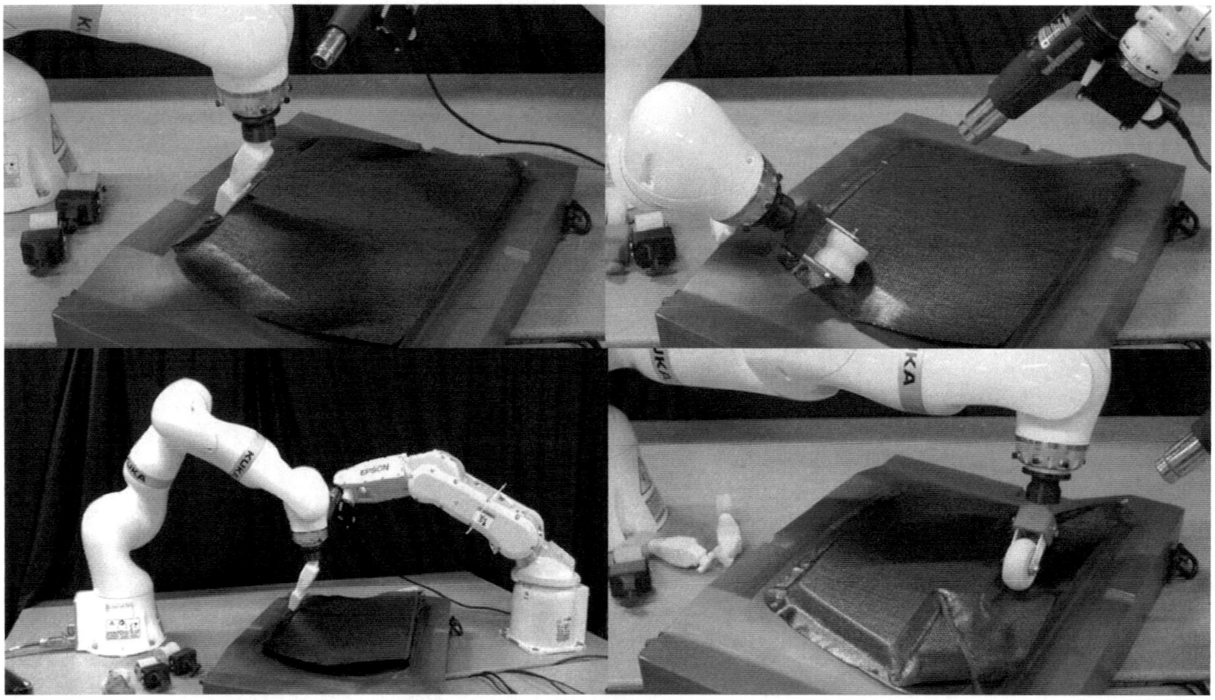

Figure 12: Layup over an industrial test case

5. Conclusions and Future Works

This paper describes an approach for automated prepreg sheet layup using a robotic cell. Conceptually, the method relies on development of a manufacturing cell comprising multiple robots, whose task execution is based on automated instruction generation and aided by process simulations and input from human experts at key stages. Aspects of the approach were implemented within a lab-scale test cell, and an experimental investigation was performed to assess viability, evaluate technical challenges, and identify areas where R&D efforts should be focused. Results show that automated prepreg sheet layup is viable, and potentially competitive with hand layup even for complex-shaped parts but may require human input at key stages. This work also showed that Robot instructions can be generated automatically without any need for robot programming. We also illustrated that layup process can be monitored, and interventions can be performed to prevent defect formation. The future directions for work include the following:

- We need to standardize the robot end-effectors' design.
- For appropriate adhesion between the mold and the ply, controlling ambient temperature is critical. We need to integrate a thermal management system with the cell.
- Significant variability in sheet behavior was observed. This may due to different types of woven carbon fiber, change in ambient temperatures, different refrigeration conditions, etc. This will require on-line adaptation of plans based on observed performance.
- Fast automated tool change is needed, as multiple tools are swapped during the process.

6. Acknowledgements

This work was supported in part by Subaward No. ARM-17-01-F-20 from the Advanced Robotics for Manufacturing ("ARM") under Agreement Number W911NF-17-3-0004 sponsored by the Office of the Secretary of Defense. ARM Project Management was provided by Matthew S. Fischer. The views and conclusions contained in this document are those of the authors and should not be interpreted as representing the official policies, either expressed or implied, of either ARM or the Office of the Secretary of Defense of the U.S. Government. The U.S. Government is authorized to reproduce and distribute reprints for Government purposes notwithstanding any copyright notation herein.

REFERENCES

[1] Campbell, F.C. "Chapter 1 – Introduction to Composite Materials and Processes: Unique Materials that Require Unique Processes". *Manufacturing Processes for Advanced Composites*. Elsevier, 2004.
[2] Campbell, F.C., Mallow, A.R., and Browning, C.E. "Porosity in Carbon Fiber Composites: An Overview of Causes." *Journal of Advanced Materials* 26(4) (1995): 18–33.
[3] Campbell, F.C. "Chapter 5 – Ply Collation: A Major Cost Driver". *Manufacturing Processes for Advanced Composites*. Elsevier, 2004.
[4] Manson, J.-A.E., Wakeman, M.D., and Bernet, N. "Composite Processing and Manufacturing - An Overview." In A. Kelly and C. Zweben, eds., *Comprehensive Composite Materials*. Elsevier Science Ltd., 2000.
[5] Lukaszewicz, D.H.J.A., Ward, C., and Potter, K.D. "The engineering aspects of automated prepreg layup: History, present and future." *Composites Part B: Engineering* 43(3) (2012): 997–1009.
[6] Elkington, M., Bloom, D., Ward, C., Chatzimichali, A., and Potter, K. "Hand layup: understanding the manual process." *Advanced Manufacturing: Polymer & Composites Science* 1(3) (2015): 138–151.
[7] Elkington, M., Ward, C., Chatzimichali, A., and Potter, K. "Studying effects of preshearing on hand layup." *Advanced Manufacturing: Polymer & Composites Science* 1(2) (2015): 80–93.
[8] Prabhu, V.A., Elkington, M., Crowley, D., Tiwari, A., and Ward, C. "Digitisation of manual composite layup task knowledge using gaming technology." *Composites Part B: Engineering* 112 (2017): 314–326.
[9] Such, M., Ward, C., Hutabarat, W., and Tiwari, A. "Intelligent composite layup by the application of low cost tracking and Projection technologies." *Procedia CIRP* 25(C) (2014):

122–131.
[10] Ward, C., Bhatnagar, V., and Potter, K. "Developing an Automated System for the Removal of Protective Films From Pre-Preg Material, To Remove a Manufacturing Bottleneck in Terms of Pick and Place Automation." *Proceedings of the SAMPE SETEC 13 Conference*, 2013.
[11] Molfino, R., Zoppi, M., Cepolina, F., Yousef, J., and Cepolina, E.E. "Design of a Hyper-flexible cell for handling 3D Carbon fiber fabric." *Recent Advances in Mechanical Engineering and Mechanics* (2014): 165–170.
[12] Elkington, M., Ward, C., and Potter, K. "Automated layup of sheet prepregs on complex moulds." *Journal of Advanced Materials* (2016): 70–84.
[13] Elkington, M., Sarkytbayev, A., and Ward. "Automated Composite Draping: A Review." *Proceedings of the SAMPE 2017 Technical Conference*. Society for the Advancement of Material and Process Engineering (2017).
[14] Kruse D., Radke R., and Wen J. "Collaborative human-robot manipulation of highly deformable materials." *Proceedings of the 2015 IEEE International Conference on Robotics and Automation (ICRA)*. Institute of Electrical and Electronics Engineers. (2015): 3782-3787.
[15] Peng. Z and Li Y. "Position/force control of two manipulators handling a flexible payload based on finite-element model." *Proceedings of the 2007 IEEE International Conference on Robotics and Biomimetics (ROBIO)*. Institute of Electrical and Electronics Engineers. (2007): 2178-2182.
[16] Zhang P., and Li Y. "Simulations and trajectory tracking of two manipulators manipulating a flexible payload." *Proceedings of the 2008 IEEE Conference on Robotics, Automation and Mechatronics*. Institute of Electrical and Electronics Engineers. (2008): 72-77.
[17] Das J., and Nilanjan S. "Autonomous shape control of a deformable object by multiple manipulators." *Journal of Intelligent & Robotic Systems* 62(1) (2011): 3-27.
[18] Malhan R.K., Kabir A.M., Shembekar A.V., Shah B., Centea T., and Gupta S.K. "Hybrid Cells for Multi-Layer Prepreg Composite Sheet Layup". *IEEE International Conference on Automation Science and Engineering (CASE)*, Munich, Germany, August 2018,pp:1466-1472.
[19] Malhan R.K., Kabir A.M., Shah B., Centea T., and Gupta S.K. "Automated prepreg sheet placement using collaborative robotics". *In: Society for the Advancement of Material and Process Engineering (SAMPE) Technical Conference and Exhibition*, 2018.
[20] Malhan R.K., Kabir A.M., Shah B., and Gupta S.K. "Identifying Feasible Workpiece Placement with Respect to Redundant Manipulator for Complex Manufacturing Tasks". *IEEE International Conference on Robotics and Automation (ICRA 2019)*, Montreal, Canada, May 20 – May 24, 2019.
[21] Malhan R.K., Kabir A.M., Shah B., Centea T., and Gupta S. K. "Determining Feasible Robot Placements in Robotic Cells for Composite Prepreg Sheet Layup". *ASME Manufacturing Science and Engineering Conference*, Erie, PA, June 2019.
[22] Pan J., Chitta S., and Manocha D. "FCL: A general purpose library for collision and proximity queries". *Proceedings of the 2012 IEEE International Conference on Robotics and Automation*. (2012): 3859–3866.
[23] Malhan R.K., Joseph R.J., Shembekar A.V., Kabir A.M., Bhatt P.M., and Gupta S.K. "Online Grasp Plan Refinement for Reducing Defects During Robotic Layup of Composite Prepreg Sheets." *IEEE International Conference on Robotics and Automation (ICRA)*, Paris, France, May 31-June 4, 2020.

ASSESSMENT OF EFFECTS OF ALUMINA NANOFIBERS ON WETTABILITY, VISCOSITY, AND MECHANICAL PROPERTIES OF UNSATURATED POLYESTER

Harish S. P. Kallagunta[1,2], Subash Panta[2], Aziza Nahar[2], and Jitendra S. Tate[1,2,*]
Texas State University
601 University Drive, San Marcos, TX, 78666
[1]Materials Science Engineering and Commercialization Program, Texas State University
[2]Ingram School of Engineering, Texas State University
JT31@txstate.edu; 512-245-1826

ABSTRACT

Increase in viscosity, poor wettability and agglomeration are the major issues when using nanoparticles to enhance the properties of polymeric materials. Processability of composites becomes challenging with the increase in the viscosity. A study has been made on the effects of alumina nanofiber modification on the viscosity and mechanical strength of unsaturated polyester resin. Additionally, wettability studies of the alumina nanofiber with unsaturated polyester resin using contact angle measurements were compared. Tension, Flexure and Izod impact tests were performed on 0.25wt%, 0.5wt%, 0.75wt% and 1wt% loading levels of alumina nanofibers in unsaturated polyester matrix and were compared to neat composites. Alumina nanofibers were dispersed in unsaturated polyester resin using three roll mills. Nanomodified and neat polyester coupons were cast in silicon molds according to the standards. TEM Images revealed good dispersion. Silane treatment proved to improve the homogeneity in dispersion and positively affect the viscosity. In terms of mechanical properties, silane treatment proved to be effective at lower loading levels. However, at higher loading levels, silane treatment was outperformed by untreated alumina nanofibers.

1. INTRODUCTION

Due to the covalently bonded network with high crosslinking density and localized plastic deformation in front of the crack-tip under mechanical loading leads to catastrophic failure in thermoset resins. Polyester resin among the thermoset resins have highest applications in the industry despite their low mechanical properties, due to their affordability and ease of processing. Nanoparticle modification of polymeric materials has gained prominence in the recent decade in the development of composite materials to make them multifunctional. Nanoparticles are more effective because of the high surface area to volume ratio. Hence very low filler content is required to achieve desired properties. However, the incorporation nanoparticles into the polymer matrix is challenging because of complications such as increase in viscosity, agglomeration and poor exfoliation [1-3].

Diverse varieties of nanoparticles are studied to enhance the performance of the unsaturated polyester resin. Carbon nanotubes[4-6], Nanosilica[7, 8], Fumed silica [9-11], nanoclay [12], halloysite nanotubes[13, 14] and elastomers[15] modification of unsaturated polyester is studied by various researchers in understanding its effect on the static, dynamic, thermal and electrical properties. Yu et al [16]studied graphene oxide as the coupling agent for graphene nanoplatelets/epoxy composites and found improvement in toughness by 110% at 0.86wt% loading. Lavoratti et al[17] has studied the effects of methoxy and ethoxy silane on the graphene nanoplatelets. The study has shown improvement in tensile strength by 21.7% and glass transition varied by 8% at 0.25wt% loading level of (3-glycidyloxypropyl) Trimethoxy silane treated graphene nanoplatelets. Tee et al [18], has demonstrated that the silane treated silver nanoparticles in epoxy composites further enhances the multifunctionality of the composites. The silane treatment has also proven effective to increase dispersibility of silver nanoparticles. Fracture toughness K_{IC} and G_{IC} improvement with the use of silane functionalization of graphene oxide for epoxy composites by 39% and 83% is observed in studies by Wan et al[19]. Improvement in thermal stability at 0.5wt% and improvement of tensile, flexure properties with silane functionalization was observed. Amide and imide based functionalization of silica nanoparticles for polyester nanocomposites is studied by Ghosh et al and the group observed increased thermal stability in the nanocomposites[20]. Alumina nanofibers, which are relatively new form of nanomaterials, are studied on their effectiveness in strengthening the polymer matrix for fiber reinforced composite materials. Mechanical, thermal stability and flammability characteristics of PA6 thermoplastics modified with alumina nanofibers by Wu et al [21] reported a significant effect of alumina nanofiber modification on the mechanical properties although there is no change in thermal and flammability properties. Raikitin and Grodensky [22], reported an increase in mechanical property by 72% by adding alumina nanofibers to the epoxy resin.

This research study is aimed at understanding the effects of alumina nanofibers on the physical, morphological and mechanical properties when incorporated into the polyester matrix. The effect of surface treatment on the alumina nanofibers on the above properties at four different loading levels was compared with the untreated alumina nanofibers modified polyester.

2. MATERIALS AND MANUFACTURING

2.1 Material System:

The material system used for this study comprises of unsaturated polyester resin SIL47DA-2949 with a viscosity of 396 cP from Interplastic Corporation. Alumina nanofibers with minimal aspect ratio 18-20 under trade name Nafen™ from ANF technologies were used as nanofillers. Benzoyl Peroxide was used as the curing agent with heat as an external stimulator. One set of polymer samples are modified with untreated alumina nanofibers and the second set were modified with vinyl silane coupling agent treated alumina nanofibers. Loading levels of 0.25wt%, 0.5wt%, 0.75wt% and 1wt% were used for both polymers.

2.2 Sample fabrication

2.2.1 Dispersion using calendaring approach

Three-roll milling was used for dispersing Nafen™ into the polyester resin. In every pass, by adjusting the gap setting between the rollers a high shear force was achieved to reduce the clumps and clusters formed. The gap between the rollers was set at 25μm, 15μm, and 7.5μm respectively for three passes. Vinyl silane coupling agent treated alumina nanofibers were also dispersed using the same approach.

Figure 1: Three roll mill and illustration of dispersion using calendaring approach (Courtesy: www.exaktusa.com)

2.2.2 Fabrication of polymer test coupons

The untreated alumina nanofiber (UT-ANF) modified polyester resin and vinyl silane treated alumina nanofiber (VT-ANF) modified polyester resin were cast into silicon molds made as per the ASTM standards and were cured in the oven at 80C as per the specifications of personnel at

Interplastic corporation. The cured modified polyester coupons were then post-cured for three hours at 110 °C in the programmable oven.

3. EXPERIMENTATION

3.1 TEM Imaging

In order to evaluate the dispersion state of alumina nanofibers in the polyester, Transmission Electron Microscopy was performed on the polymer coupons. Copper grids were used on which thin sections of polymer modified with alumina nanofiber obtained using ultra microtome are placed. The grids were then viewed in a high-resolution transmission electron microscope.

3.2 Refractive Index

The refractive index measurements were calculated using Bausch & Lomb Abbe-3L Refractometer. The sample is then carefully placed on the prism to form a thin film across the whole prism and the prism was closed. Adjusting the lamp so the light shines on the prism and looking through the eyepiece, the index of refraction is recorded.

3.3 Viscosity

The effects of adding UT-ANF and VT-ANF on the viscosity of the unsaturated polyester resin were studied using the Thermo Scientific Haake Viscotester. Along with the effects of surface treatment, the loading weight percentage of 0.25wt%, 0.5wt%, 0.75wt% and 1wt% were also studied and reported as compared to viscosity of neat unsaturated polyester resin. 15mL of the sample was placed inside the immersion tube and a rotor (coaxial cylinder) was used to measure the shear rate and viscosity. All the samples are tested at room temperature. The shear rate vs shear stress plots were graphed, and viscosity is measured as the slope of the curve.

3.4 Wettability

Rame-hart contact angle goniometer is used to measure the contact angle to get and understanding of the wettability. The UT-ANF and VT-ANF modified polyester resin droplets are carefully dropped on to the GPC substrate and with the aid of a telescopic eyepiece the angle of the resin on the substrate is measured.

3.5 Mechanical testing

The static testing on the untreated and vinyl treated Alumina nanofiber (ANF) modified polyester coupons is performed using UTS electromechanical system. Tensile, Flexure and Izod impact tests performed on the polyester coupons (UPE) modified with untreated and surface treated

Alumina nanofibers (ANF). The tensile tests were conducted according to ASTM D638 at a loading rate of 1mm/min. The ultimate tensile strength (UTS), modulus and percentage elongation were obtained from this test. Flexure testing was conducted according to the ASTM D790 standard at a loading rate of 0.10mm/min. Izod impact testing was done according to the ASTM D256 standard and were subjected to a constant energy of 9J, to obtain the impact strength of the polymer coupons.

4. RESULTS AND DISCUSSION

4.1 TEM Analysis

The TEM images as shown in Fig 2, revealed agglomeration of ANF nanoparticles in UPE at higher loading levels, whereas a decent quality of dispersion of ANF nanoparticles in lower loadings is observed. It was observed that ANF nanoparticles are retaining their fiber like morphology even after subjecting to high shear forces via three roll milling. Although individual ANF nanoparticles are visible, it is observed that the fibers are not completely de-bundled and distributed. The lack of de-bundling might be because of absence of any surface treatment. It would be interesting to see the effect of surface treatment of ANF on the de-bundling and distribution aspect of the nanoparticles during dispersion.

Figure 2: TEM images showing dispersion of Untreated ANF in UPE at loading levels of (a) 0.25wt%, (b) 0.5wt%, (c) 0.75wt% and (d) 1wt%

4.2 Refractive index

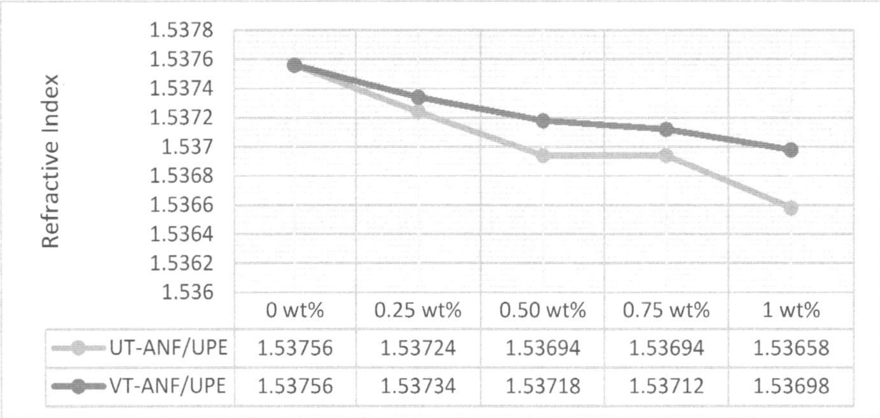

Figure 3: A summary of comparison of refractive index of UT-ANF/UPE and VT-ANF/UPE at loading levels of (a) 0.25wt%, (b) 0.5wt%, (c) 0.75wt% and (d) 1wt%

The refractive index measurement is ratio of velocity of light in vacuum to the velocity of light in the polyester resin medium. The refractive index values showed consistency for respective loading wt% as shown in Fig 3. The VT-ANF/UPE as compared to UT-ANF/UPE showed good refractive index values meaning that the silane treatment of alumina nanofibers provided a good measure for homogenous dispersion. The refractive index values as compared to the control are very less indicating the presence of alumina nanofibers, the decrease in the values with increased loading wt% is due to the light encountering alumina nanofibers and deflecting the light reducing the index of refraction.

4.3 Viscosity

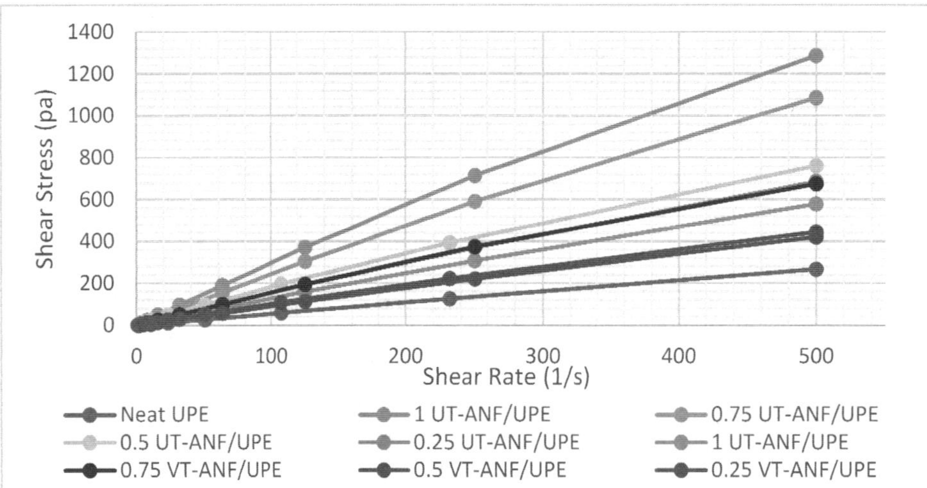

Figure 4: A summary of viscosity change of UT-ANF/UPE and VT-ANF/UPE at loading levels of (a) 0.25wt%, (b) 0.5wt%, (c) 0.75wt% and (d) 1wt%

Table 1: Viscosity values of UT-ANF/UPE and VT-ANF/UPE

Loading wt%	Viscosity (cP)	
	UT-ANF/UPE	VT-ANF/UPE
0.00	534.3	
0.25	1380.6	844.2
0.50	1528.8	890.9
0.75	2186.9	1364.0
1.00	2601.0	1162.3

The shear rate vs shear stress plots are given in figure 4. The graphs of shear stress vs shear rate showed a linear relation, which means the non-modified resin, shows a Newtonian behavior. A look at the plots gives the understanding of change in viscosity with increase in the loading wt% of alumina nanofibers in the polyester. The surface treatment of alumina nanofibers helped in reducing the viscosity of the polyester resin with respect to the as received alumina nanofiber modified polyester. It is evident from this plot that with surface modification, the viscosity of nanomodified polyester can be achieved and thus processing challenges can be addressed. Table 1 gives an understanding of the viscosity changes with addition of nanoparticles and the effect of surface treatment on the viscosity property. It is quite evident that, surface treatment of alumina nanofibers has a positive effect on the change in viscosity. The addition of surface functional groups reduces the attraction among the nanoparticles thus increasing homogenous dispersion. For affordable processes such as vacuum assisted resin transfer molding (VARTM) or layup, the ideal viscosity range is 100-3000cP. When using vacuum assisted resin transfer molding, high viscosity of the resin poses a challenge of increased time of resin infusion thus increasing the fabrication time. In case of hand layup, high viscosity of the resin increases the air pockets during the wetting of the fabrics. From the above result, a surface treatment of the nanoparticles can help reduce the processing challenges associated with high viscosity.

4.4 Wettability

The degree to which liquids wet a fiber is a measure of the bonding between the fiber and the resin, which results in efficient stress transfer between matrix and reinforcement. The contact angles measured provide information about wettability, essential in selecting or adjusting the modifications on nanomaterial or fiber surfaces. The nanoparticle modification of polyester resin changes the variation of the surface energy of the resin droplets when added to the substrate thus varying the wettability of the glass fibers. Surface treated alumina nanofibers during the dispersion process broke down into smaller nanofibers providing continuous surface reducing the surface energy and thus increasing the wettability. The agglomerated lumps during the higher concentration levels, creates a continuous rough surface thus by increasing the wetting angle. However, the results in this study suggest there is not big influence in loading percentage of alumina nanofiber except for the silane modification.

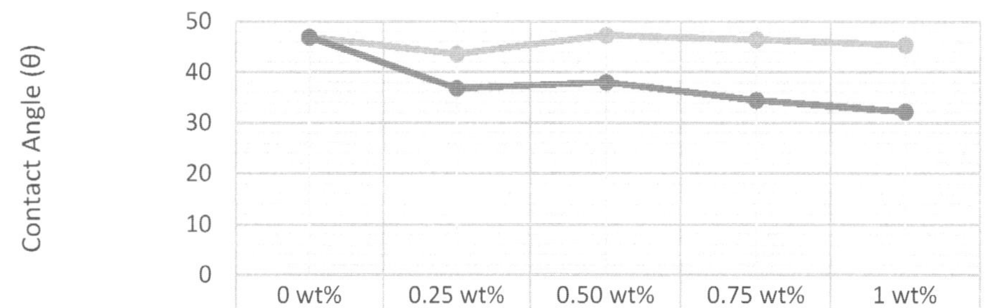

	0 wt%	0.25 wt%	0.50 wt%	0.75 wt%	1 wt%
UT-ANF/UPE	46.93	43.6	47.27	46.45	45.38
VT-ANF/UPE	46.93	36.8	37.97	34.43	32.18

Figure 5: A summary of contact angle measurements of UT-ANF/UPE and VT-ANF/UPE at loading levels of (a) 0.25wt%, (b) 0.5wt%, (c) 0.75wt% and (d) 1wt%

4.5 Mechanical properties

Tensile Properties

As observed in Fig 6, no significant improvement in the tensile strength of the ANF/UPE was observed, although an increase of tensile modulus at 0.25wt% loading level is observed. The tensile properties when used with both UT-ANF and VT-ANF, initially declined, but regained at a loading of 1wt%. The tensile strength of the VT-ANF/UPE as compared to the UT-ANF/UPE is a result of the homogenous dispersion, which is facilitated through the surface treatment. The tensile modulus of UT-ANF/UPE is high as compared to the VT-ANF/UPE. Equilibrium toughness as shown in Fig 7, has seen a consistent improvement with silane treated alumina nanofiber modified polyester coupons. This behavior shows that, the surface treatment enhanced the bonding between the matrix and nanofiber, thus increasing the work done on the sample. The three-roll mill process induces high shearing forces as the distance between the rollers is decreased, which possibly dealt with efficient de-bundling of alumina nanofibers during each pass. However, it was observed from TEM images at higher loading levels, the alumina nanofibers tend to agglomerate which is possibly due to the high surface areas of nanofibers, meaning that even the aspect ratio is maintained, they tend to act as stress concentrators.

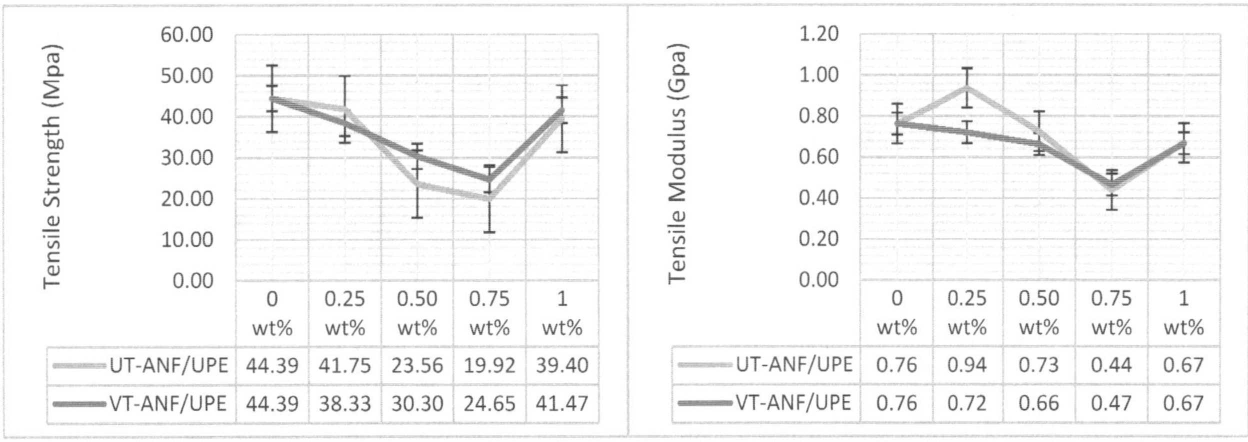

Figure 6: Tensile strength and tensile modulus comparison of UT-ANF/UPE and VT-ANF/UPE at loading levels of (a) 0.25wt%, (b) 0.5wt%, (c) 0.75wt% and (d) 1wt%

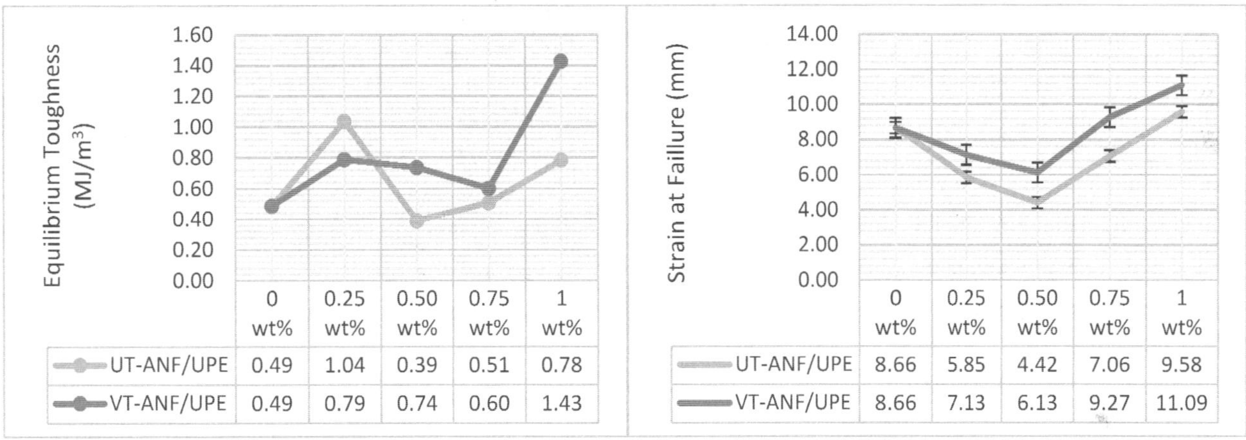

Figure 7: Equilibrium toughness and strain at failure of UT-ANF/UPE and VT-ANF/UPE at loading levels of (a) 0.25wt%, (b) 0.5wt%, (c) 0.75wt% and (d) 1wt%

Flexure Properties

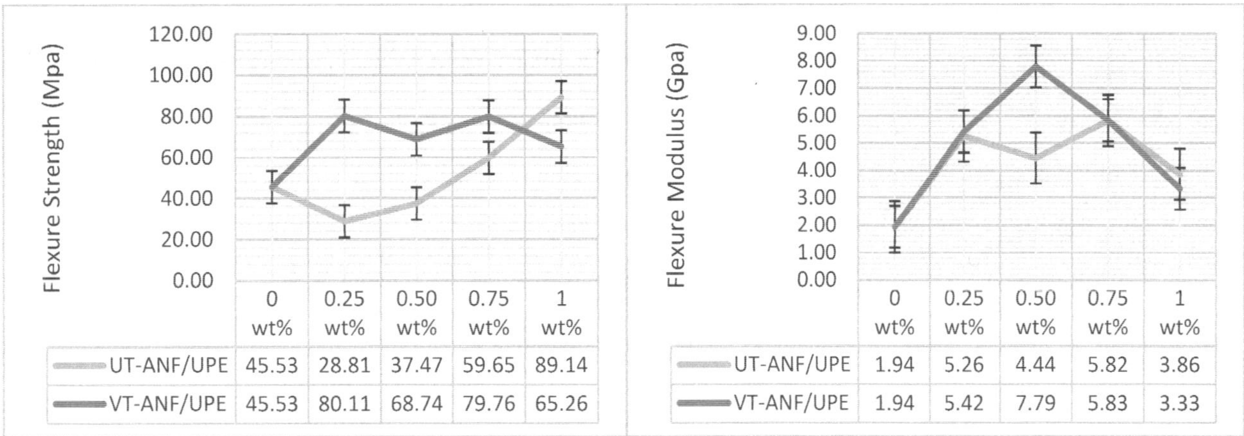

Figure 8: A summary of flexure properties of UT-ANF/UPE and VT-ANF/UPE at loading levels of (a) 0.25wt%, (b) 0.5wt%, (c) 0.75wt% and (d) 1wt%

The flexure properties as shown in fig 8, shows the effect of nanoparticles in the polyester resin. Due to the unique fiber like morphology, the alumina nanofibers could deflect the force, thus causing an increase in the properties. The silane treatment of the alumina nanofibers has proven to be effective in increasing in the flexural strength and flexural modulus of the polyester resin coupons. The as received alumina nanofibers modified polyester composites initially has seen a decrease in their strength values but regained their performance after 0.50wt%. It is well known that increase in flexure strength and flexure modulus would result in higher toughness values, which is a result of the difficulty of crack initiation and propagation within matrix.

Izod Impact

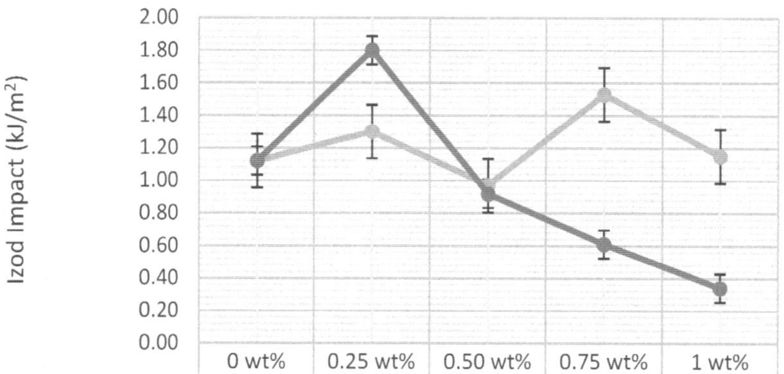

	0 wt%	0.25 wt%	0.50 wt%	0.75 wt%	1 wt%
UT-ANF/UPE	1.12	1.30	0.97	1.53	1.15
VT-ANF/UPE	1.12	1.80	0.92	0.61	0.34

Figure 9: A summary of Izod impact of UT-ANF/UPE and VT-ANF/UPE at loading levels of (a) 0.25wt%, (b) 0.5wt%, (c) 0.75wt% and (d) 1wt%

The Izod impact strength of the nanomodified polyester coupons is reported in fig 9. The impact results show a dependence of the loading percentage on the impact energy absorbed by the polymer coupons. The VT-ANF/UPE initially has seen an increase in the energy absorbed at 0.25wt%, but further the energy absorbed drastically decreased. The UT-ANF/UPE has seen mixed results. At 0.25wt% loading level of VT-ANF/UPE the impact toughness of the polymer increased by 60% and the impact strength of UT-ANF/UPE polymer composite is increased by 36.6%.

CONCLUSION

Transmission electron microscopy, viscosity, wettability, refractive index, tensile, flexure and Izod impact property of polyester coupons modified with untreated as received alumina nanofibers and silane treated alumina nanofibers is studied and compared. TEM Images showed agglomeration but decent dispersion at higher loading levels. As expected, viscosity increased with loading level of alumina nanofibers. Silane treatment has proven effective to impart mechanical properties at reduced viscosity. The refractive index also proves the efficiency of silane treatment of alumina nanofibers in promoting homogenous dispersion into polyester resin.

1. The viscosity rise observed was not substantial in polyester modified with silane treated alumina nanofibers. At 0.25wt% loading level, the viscosity rises in VT-ANF/UPE was only 58% as opposed to 158% at same loading level of UT-ANF in polyester.
2. Regardless of the silane treatment of alumina nanofibers, major of the mechanical properties showed good improvement at 0.25wt% loading level.
3. The equilibrium toughness, which is an important parameter against deformation, has seen an improvement of 112.2% at 0.25wt% loading level of UT-ANF in polyester. With silane treatment at 1wt% loading level has seen a maximum improvement of 191.8%. The silane treatment of alumina nanofibers.
4. Flexural properties have improved with addition of alumina nanofibers. However, with silane treatment it was observed that 0.25wt% loading level the strength improved by 43.3%.
5. Izod impact property of the polyester modified with VT-ANF at 0.25wt% loading level has seen maximum improvement by 60.7%.
6. It can be concluded form this study that the inclusion of alumina nanofibers has shown high potential of improving the mechanical properties. It is validated that surface treatment of alumina nanofibers has the potential to improve the properties significantly at lower loading levels. Selection of low loading levels is important in order to tackle processing challenges and less defects into the composites.
7. It is imperative to consider using different silane coupling agents compatible with polyester and alumina nanofibers to evaluate their effects on mechanical properties at reduced viscosity and homogenous dispersion.

ACKNOWLEDGMENTS

The authors are grateful to Mr. Tim Ferland of ANF Technology and Mr. David Gerald of Interplastic Corporation for their kind support in this research. The authors highly appreciate the help of Dr. Ray Cook, Jason Wagner, and Ruben Villarreal for waterjet cutting of test samples. The authors would also like to thank the Ms. Mariana Ocampo and Ms. Alissa Savage of ARSC at Texas State University for the training and support for TEM and SEM Imaging. The authors are thankful to Arigbabowo Oluwasola and Elizabeth Alvizo of Advanced Composites Laboratory for their support and help at various stages of testing.

References

1. Crawford, R. and J. Throne, *ROTATIONAL MOLDING POLYMERS.* 2002. p. 19-68.
2. Chu, F., et al., *Hierarchical Structure: An effective Strategy to Enhance the Mechanical Performance and Fire Safety of Unsaturated Polyester Resin.* ACS Applied Materials & Interfaces, 2019. **11**(32): p. 29436-29447.
3. *H. V. Boenig, Unsaturated Polyesters, Structure and Properties. Elsevier Publishing Company, Amsterdam 1964. 222 S., DIN A 5, Preis DM 31.—.* Fette, Seifen, Anstrichmittel, 1966. **68**(1): p. 30-30.
4. Wang, S. and J. Qiu, *Enhancing thermal conductivity of glass fiber/polymer composites through carbon nanotubes incorporation.* Composites Part B: Engineering, 2010. **41**(7): p. 533-536.
5. Seyhan, A.T., et al., *Critical aspects related to processing of carbon nanotube/unsaturated thermoset polyester nanocomposites.* European Polymer Journal, 2007. **43**(2): p. 374-379.
6. Agnihotri, P., S. Basu, and K.K. Kar, *Effect of carbon nanotube length and density on the properties of carbon nanotube-coated carbon fiber/polyester composites.* Carbon, 2011. **49**(9): p. 3098-3106.
7. Jaya Vinse Ruban, Y., S. Ginil Mon, and D. Vetha Roy, *Mechanical and thermal studies of unsaturated polyester-toughened epoxy composites filled with amine-functionalized nanosilica.* Applied Nanoscience, 2013. **3**(1): p. 7-12.
8. Zhou, S.-X., et al., *Effect of nanosilica on the properties of polyester-based polyurethane.* Journal of Applied Polymer Science, 2003. **88**(1): p. 189-193.
9. Morote-Martínez, V., V. Pascual-Sánchez, and J.M. Martín-Martínez, *Improvement in mechanical and structural integrity of natural stone by applying unsaturated polyester resin-nanosilica hybrid thin coating.* European Polymer Journal, 2008. **44**(10): p. 3146-3155.
10. Marinkovic, A., et al., *Mechanical properties of composites based on unsaturated polyester resins obtained by chemical recycling of poly(ethylene terephthalate).* Hemijska industrija, 2013. **67**: p. 913-922.
11. Mirabedini, S.M. and A. Kiamanesh, *The effect of micro and nano-sized particles on mechanical and adhesion properties of a clear polyester powder coating.* Progress in Organic Coatings, 2013. **76**(11): p. 1625-1632.
12. Albdiry, M.T. and B.F. Yousif, *Toughening of brittle polyester with functionalized halloysite nanocomposites.* Composites Part B: Engineering, 2019. **160**: p. 94-109.
13. Albdiry, M.T., H. Ku, and B.F. Yousif, *Impact fracture behaviour of silane-treated halloysite nanotubes-reinforced unsaturated polyester.* Engineering Failure Analysis, 2013. **35**: p. 718-725.
14. Saharudin, S.M., R. Atif, and F. Inam, *Effect of Short-Term Water Exposure on the Mechanical Properties of Halloysite Nanotube-Multi Layer Graphene Reinforced Polyester Nanocomposites.* Polymers, 2017. **9**(1).

15. Benny Cherian, A. and E.T. Thachil, *Toughening Studies of an Unsaturated Polyester Resin Using Maleated Elastomers.* Progress in Rubber and Plastics Technology, 2001. **17**(4): p. 205-224.
16. Yu, Z. and L.T. Drzal, *Functionalized graphene oxide as coupling agent for graphene nanoplatelet/epoxy composites.* Polymer Composites, 2019. **n/a**(n/a).
17. Lavoratti, A., A.J. Zattera, and S.C. Amico, *Mechanical and dynamic-mechanical properties of silane-treated graphite nanoplatelet/epoxy composites.* Journal of Applied Polymer Science, 2018. **135**(45): p. 46724.
18. Tee, D.I., et al., *Effect of silane-based coupling agent on the properties of silver nanoparticles filled epoxy composites.* Composites Science and Technology, 2007. **67**(11): p. 2584-2591.
19. Wan, Y.-J., et al., *Mechanical properties of epoxy composites filled with silane-functionalized graphene oxide.* Composites Part A: Applied Science and Manufacturing, 2014. **64**: p. 79-89.
20. Ghosh, S., S.K. Goswami, and L.J. Mathias, *Surface modification of nano-silica with amides and imides for use in polyester nanocomposites.* Journal of Materials Chemistry A, 2013. **1**(19): p. 6073-6080.
21. Hao Wu, M.K.a.J.H.K. *Functionalized Nafen™ Alumina Nanofiber (ANF) Reinforced Polyamide 6 Nanocomposites: Mechanical, Thermal And Flame Retardant Properties.* in *Society for the Advancement of Material and Process Engineering*. 2015. Baltimore, MD.
22. I. Grodensky, A.R. *POLYMER NANOCOMPOSITES BASED ON NAFEN™ – UNIQUE ALUMINA NANOFIBERS.* in *Society for the Advancement of Material and Process Engineering* 2014. Seattle, WA.

THERMAL AND ELECTROCHEMICAL PROPERTIES OF POLYVINYLIDENE FLOURIDE, FUMED ALUMINUM OXIDE, & SINGLE WALLED CARBON NANOTUBE NANOCOMPOSITES FOR SUPERCAPACITOR APPLICATIONS

Ruchinda S. Gooneratne, Jude Iroh
Department of Mechanical and Materials Engineering, University of Cincinnati
Cincinnati, OH

ABSTRACT

The effect of dispersing single walled carbon nanotubes and fumed alumina in a polyvinylidene fluoride matrix was carried out to identify the thermal and electrical properties for flexible supercapacitor electrode applications. Addition of CNTs helped improve the polymer's conductive and charge storage capabilities. Characterization was carried out to understand the thermal characteristics of the nanocomposites using TGA and DSC which indicated a lowering in thermal stability upon addition of the nanofillers. Electrochemical tests were also conducted in the form of cyclic voltammetry and electrochemical impedance spectroscopy using 1M H_2SO_4 electrolyte, yielding specific gravimetric capacitance values close to 72.88 F/g. Results also identified the improvement in specific capacitance of the composite upon addition of very low (1wt%) of fumed alumina.

Keywords: Supercapacitors, polyvinylidene fluoride, fumed aluminum oxide, carbon nanotube, energy storage

1. INTRODUCTION

Flexible electronic storage devices are fast becoming the path forward in a world governed by energy drawing mobile devices which constantly are in need of replenishment. With battery technology being pushed to its limits, another candidate for energy storage arises in supercapacitors. Supercapacitors (SCs) have progressed over the recent past to being researched and developed to replace Li-ion batteries due to their incredibly high power-densities, charge cyclability, lifespan, and overall ease, as well as speed at which they can be charged over a wider range of temperatures. However, the limitations currently faced include low storage capacities and fast discharging rates. To address these shortcomings, a vast amount of research and funding have been channeled towards developing new materials and methods.

The effect of dispersing fumed alumina (high surface area aluminum oxide) as well as single walled carbon nanotubes with polyvinylidene fluoride was carried out in this study. The aim was to produce a conductive polymer composite for the purpose of charge storage capabilities for the development of supercapacitor electrode material. Improving/synthesizing SC material with high surface to volume ratio, porosity, and electrical conductivity are key to overcoming the limitations currently faced. Graphene sheets are widely popular in most literature with this regard as it satisfies most of the required criteria [1].

The use of PVDF as a binder has been proven to be an effective polymer to suspend activated carbon with low internal resistance. As a result, for the purpose of this study, PVDF was used as the matrix for the SWCNTs, which are themselves highly conductive and possess charge holding capacity [2, 3].

2. EXPERIMENTATION

2.1 Materials

Polyvinylidene fluoride (PVDF) and single wall carbon nanotubes (SWCNTs) were provided by ATEN Industries LLC. Highly dispersed fumed aluminum oxide (Al_2O_3) under the product name AEROXIDE ® Alu C was provided by Evonik Industries. n,n-Dimethylformamide (DMF), ACS Reagent Grade was purchased from Right Price Chemicals and used as the solvent for dispersions carried out in this system.

2.2 Method

For the synthesis of the polymer composites, all solute masses were kept at 2g in total. Neat PVDF films with 0 wt% of nanofillers were initially created by dispersing the polymer powder in DMF via mechanical solution mixing in stepwise additions at room temperature. Varying wt% of fumed alumina and SWCNTs (1%, 5%, 10%, 50% respectively) were systematically varied alone with PVDF, as well as in combination with each other in the PVDF matrix. These combinations were carried out to study the effect of the nanofillers on the polymer. Each nanofiller (according to the respective wt%) was dispersed individually in DMF solvent under probe sonication and mechanical stirring at 5°C for around 45 min.

Once a uniform dispersion was visually obtained under the time frame, the contents were carefully poured into a premixed solution of PVDF/DMF in a beaker under slow stirring to blend the polymer and nanofillers together for around 8 hours until a viscous slurry was formed. Thin films of each slurry were coated using an MTI vacuum assisted film coater (coated at 0.5mm wet thickness) over a Teflon sheet and cured using the heater attachment set at around 45°C overnight until all visible solvent had completely dried out. The cured polymer composite films with thicknesses of around 0.03mm were simply peeled off the Teflon to be tested.

2.3 Characterization

The effect of the addition of fumed alumina and/or SWCNTs on the thermal characteristics of the polymer such as degradation and melting temperatures, enthalpy of melting, and degree of crystallinity were studied via thermogravimetric analysis and differential scanning calorimetry using TA Instruments' Q50 and Q20 respectively. TGA and DSC were carried out on each sample with a ramping rate of 5°C/min starting from room temperature to 600°C and 450°C respectively under an inert nitrogen atmosphere. The melting peaks were observed for shifts and used to calculate the degree of crystallinity (χ_c) using 100% crystalline PVDF heat of fusion (ΔH_o) of 105J/g according to the formula [6]:

$$\chi_c = \frac{\Delta H_m}{99\% \times \Delta H_o} \times 100\% \quad [1]$$

Electrochemical testing for charge storage in terms of calculating capacitance was carried out by a series of cyclic voltammetry, ranging from scan rates of 5, 10, 25, 50, & 100mV/s in 1mol H_2SO_4 as the electrolyte from 0V to 1V. Electro chemical impedance was also measured for the samples from a frequency of 1MHz to 0.01Hz against an open circuit potential of 1V. These were all carried out on a Gamry Reference 3000 potentiostat using a Ag/AgCl reference electrode.

3. RESULTS & DISCUSSION

Thermogravimetric analysis of the samples indicated, that upon increasing the wt% addition of either fumed alumina or SWCNTs alone, the temperature at which the degradation of the polymer composite began was reduced.

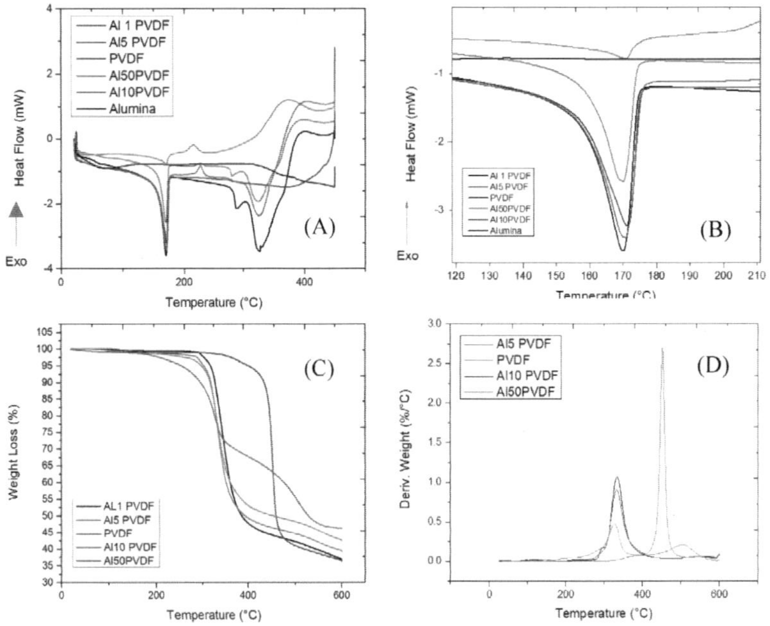

Figure 1. (A) DSC curve of PVDF with varying wt% of aluminum oxide carried out from room temperature to 450°C under nitrogen atmosphere. (B) Zoomed in portion of the DSC melting peaks. (C) TGA curve run from room temperature to 600°C under nitrogen atmosphere. (D) DTG curve.

This depreciation in thermal stability was also observed in the combinatorial wt% addition of both SWCNTs and fumed alumina, thus indicating the lowering of the thermal stability of the system upon addition of these nanofillers.

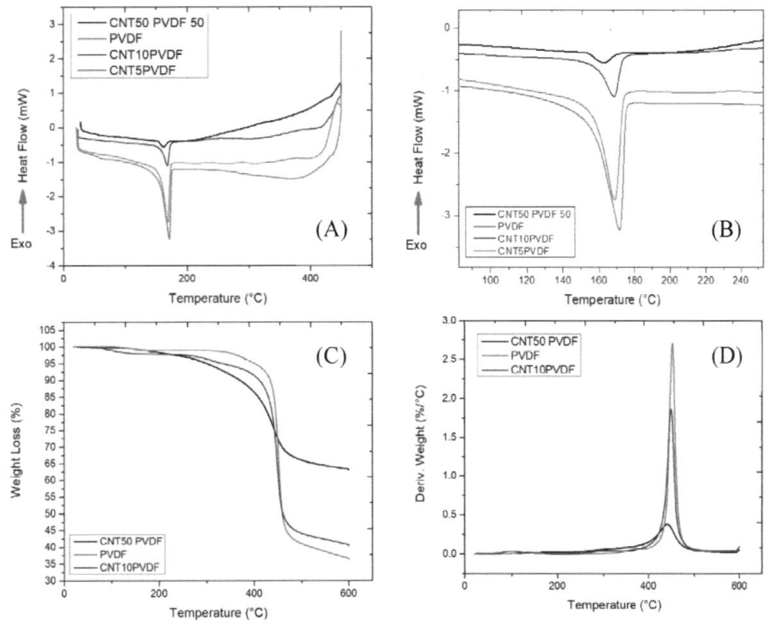

Figure 2. (A) DSC curve of PVDF with varying wt% of single wall carbon nanotubes carried out from room temperature to 450°C under nitrogen atmosphere. (B) Zoomed in portion of the DSC melting peaks. (C) TGA curve run from room temperature to 600°C under nitrogen atmosphere. (D) DTG curve

The TGA graphs in figure1C indicate that neat PVDF begins to degrade close to 450°C. The derivative weight loss curve in figure 2D shows an interesting trend with the addition of only SWCNTs in PVDF, shifting the neat PVDF peak at 452°C to as much as 440°C at 50 wt% loading. However, this shift (figure 1D) is much more prominent in the fumed alumina PVDF nanocomposite, reducing from 452°C to 327°C at 50 wt% loading.

As can be seen from the results in table 1, the addition of fumed alumina doesn't have a considerable decrease in the melting point temperature of the PVDF system. Nevertheless, its melting point temperature is inversely proportional upon the addition of alumina. The enthalpy of melting is also seen to decrease upon added loading of alumina into the system, with the lowest enthalpy of melting observed at highest loading.

Addition of SWCNTs also decrease the melting temperature of the neat PVDF system, but more considerably - close to a decrease of almost 9°C with a 50 wt% loading. This also resulted in the drop of enthalpy of melting by about 86.0%.

An interesting exothermic peak is observed in alumina based PVDF nanocomposites in figure 1A and figure 3A after the melting peaks between 200°C to 240°C. The shift in the peaks of combined alumina SWCNT PVDF nanocomposites are also observed, but less shifted towards lower temperatures than variants of the polymer with only alumina nanofiller present. The DTG curves in figure 1D indicate the most apparent weight losses occurring at relatively lower temperatures in the composites containing only alumina (around 340°C) as compared to those containing only SWCNT (around 450°C). The higher the content of alumina the lower the DTG peak height and temperature at which it was present.

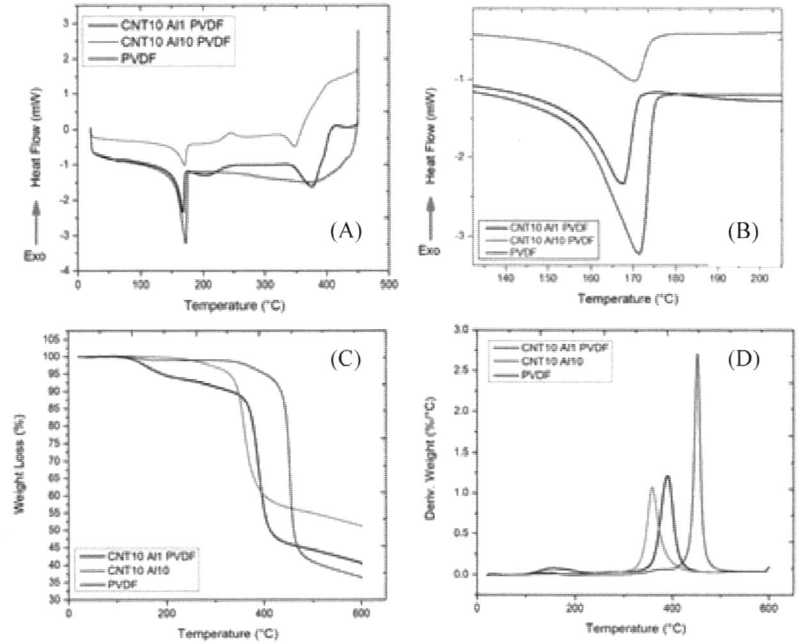

Figure 3. (A) DSC curve of PVDF with varying wt% of single wall carbon nanotubes and fumed alumina carried out from room temperature to 450°C under nitrogen atmosphere. (B) Zoomed in portion of the DSC melting peaks. (C) TGA curve run from room temperature to 600°C under nitrogen atmosphere. (D) DTG curve.

Using equation [1] the degrees of crystallization are displayed in Table 1.0.

Table 1. Enthalpy of melting ΔH_m and degree of crystallinity χ_C of nanocomposite samples.

Composite	**Tm (°C)**	**ΔH_0 (heat of fusion of 100% crystalline PVDF) (J/g)**	**ΔH_m (J/g)**	**χ_C (%)**
Neat PVDF	171.6	105	53.2	51.3
5 wt% SWCNT	168.9	105	56.4	54.3
10 wt% SWCNT	168.6	105	54.7	52.7

50 wt% SWCNT	162.7	105	7.4	7.1
1 wt% Al_2O_3	170.2	105	57.0	54.9
5 wt% Al_2O_3	170.0	105	38.0	36.6
10 wt% Al_2O_3	170.8	105	52.1	50.1
50 wt% Al_2O_3	171.0	105	21.5	20.7
1 wt% SWCNT, 1 wt% Al_2O_3	172.6	105	64.6	62.2
5 wt% SWCNT, 1 wt% Al_2O_3	169.7	105	47.2	45.4
1 wt% SWCNT, 5 wt% Al_2O_3	169.5	105	54.2	52.2
10 wt% SWCNT, 1 wt% Al_2O_3	167.7	105	40.7	39.2
1 wt% SWCNT, 10 wt% Al_2O_3	170.3	105	50.4	48.5
5 wt% SWCNT, 5 wt% Al_2O_3	168.3	105	45.5	43.8
10 wt% SWCNT, 10 wt% Al_2O_3	170.4	105	37.4	36.0

Table 1 shows an overall decrease in degree of crystallinity upon addition of the fillers, except for a marked increase in the sample that contains 1 wt% CNT, 1 wt% Al_2O_3 (χ_C =62.2%). The lowest degree of crystallinity is observed at 50 wt% CNT. This lowering of the polymer composite's thermal stability can be related to a study done by Ma, J., Haque, R. I., & Larsen [6] who proposed that the addition of the nano fillers also bring with it contaminants of metal catalyst by-products, as well as the presence of defects in the SWCNT which may account for this decrease in thermal stability as compared to neat PVDF. Furthermore, according to the same study, the impact of nano fibers was mentioned to have improved the degree of crystallization, where in this case the presence of very low wt% of fumed alumina or SWCNT, either together or separately have contributed to do so.

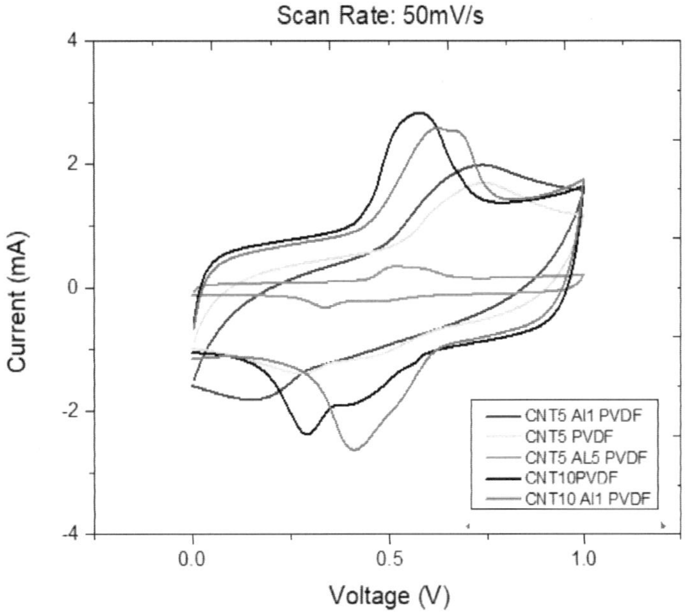

Figure 4. Cyclic voltammogram of nanocomposite samples carried out at 50mV/s scan rate in 1M H_2SO_4 at room temperature using a Ag/AgCl reference electrode.

The improvement of the capacitance as shown in the voltammogram in figure 4 is due to the highly conductive nature of carbon nanotubes and the porosity of the structure formed [3]. Higher loading of SWCNTs increases the conductive nature of the polymer, resulting in a higher capacitance value.

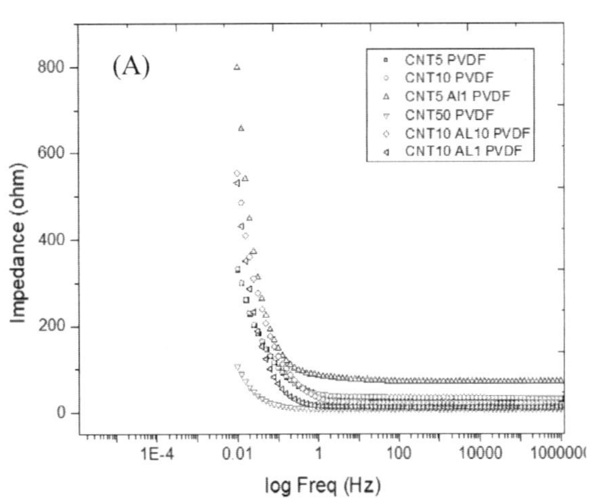

Figure 5. Electrochemical impedance spectroscopy showing (A) Bode plot and (B) Nyquist plot.

The results of gravimetric capacitance calculated using CV for a scan rate of 50mV/s are shown tabulated in Table 2 below.

Table 2. Gravimetric capacitance measured using cyclic voltammetry.

Composite	Specific Gravimetric Capacitance (F/g)
5 wt% SWCNT	6.43
10 wt% SWCNT	9.24
50 wt% SWCNT	72.88
5 wt% SWCNT, 1 wt% Al_2O_3	8.8
5 wt% SWCNT, 5 wt% Al_2O_3	1.69
10 wt% SWCNT, 1 wt% Al_2O_3	12.04
10 wt% SWCNT, 10 wt% Al_2O_3	3.72

The CV curve in figure 4 shows a very visible increase in charge storage as the SWCNT loading is increased. The addition of alumina in the composite decreases the gravimetric capacitance but only at higher weight percentages. As observed, a 1 wt% addition of alumina in a SWCNT/PVDF composite improves the charge storage capability per gram compared with a PVDF composite having the same wt% of SWCNT alone.

The EIS curves of the Nyquist plot (figure 5B) show that increasing wt% of the SWCNTs results in a reduction of the resistance of the polymer with the electrolyte interface. The plot also gives a good demonstration of the pore resistance that impacts the diffusion resistance within the bulk of the material by running it at frequency ranges of 0.01Hz to 1MHz. At higher alumina loading it can be observed that a higher resistance is introduced, shifting the plot to the right. [5]

4. CONCLUSIONS

The addition of SWCNT into the PVDF polymer matrix was observed to improve the electrical conductivity of the composite systematically with increasing wt%. The gravimetric capacitance was improved significantly (upto 12 times) from a 6.4F/g at 5 wt% to 72.9F/g at 50 wt% loading of SWCNT. The fumed alumina was not observed to improve the charge storage capabilities of the polymer composite but did show a slight increase in the gravimetric capacitance for only a very small loading percentage (5wt% SWCNT 1%Al_2O_3) as compared to the polymer composite with the same loading of SWCNT (5 wt% CNT) – an improvement in gravimetric capacitance by 2.37F/g.

Addition of both the nano fillers did not improve the thermal stability of PVDF but reduced the onset of degradation temperature, with fumed alumina reducing it more than SWCNT of the same weight loading. In terms of a flexible and chemically resistant composite supercapacitor electrode material, this study shows that fumed alumina, SWCNT, PVDF can be used for charge storage applications. Effective results of these composite electrodes indicate them to be composed of a higher loading percentage of SWCNT and with very low percentage of alumina.

5. REFERENCES

1. Nomura, K. 4.4 V supercapacitors based on super-stable mesoporous carbon sheet made of edge-free graphene walls. *The Royal Society of Chemistry* . (2019).
2. P, Samata; S, Shahbaz; R.M, Holmukhe. Study of PVDF based electrode structure in supercapacitors., Volume 7, Issue 4. International Journal of Engineering and Technology: (UAE), 2018
3. Noked, M., Okashy, S., Zimrin, T. and Aurbach, D. (2012), Composite Carbon Nanotube/Carbon Electrodes for Electrical Double-Layer Super Capacitors. Angew. Chem. Int. Ed., 51: 1568-1571. doi:10.1002/anie.201104334
4. Evgeny Senokos, Y. O. Energy storage in structural composites by introducing CNT fiber/polymer electrolyte interleaves. *Scientific Reports*. (2018).

5. Okafor, Patricia. "Processing and Characterization of Graphene/Polyimide-Nickel Oxide Hybrid Nanocomposites for Advanced Energy Storage in Supercapacitor Applications." Electronic Thesis or Dissertation. University of Cincinnati, 2016
6. Ma, J., Haque, R. I., & Larsen, R. M. Crystallization and mechanical properties of functionalized single-walled carbon nanotubes/polyvinylidene fluoride composites. Journal of Reinforced Plastics and Composites, 31(21), 1417–1425. 2016. https://doi.org/10.1177/0731684412456747

EMBODIED ENERGY OF PYROLYSIS AND SOLVOLYSIS PROCESSES FOR RECYCLING CARBON FIBER REINFORCED POLYMER WASTE

Komal Kooduvalli[1,2,3], Samantha Ford[1,2,4], Sabrinna Romero[1,2,5], Isaac Sloan[1,2,6], Hayley Coughlin[1,2], John Unser[2], Uday Vaidya[1,2,3], Soydan Ozcan[2,3]

[1]Fibers and Composites Manufacturing Facility, The University of Tennessee, 1321 White Ave, Knoxville, TN 37996, USA
[2]Institute for Advanced Composites Manufacturing Innovation (IACMI), 2360 Cherahala Blvd, Knoxville, TN 37932
[3]Energy and Transportation Sciences Division, Oak Ridge National Laboratory, 1 Bethel Valley Rd, Oak Ridge, TN 37831, USA
[4]Chatham University, 1 Woodland Rd, Pittsburgh, PA 15232
[5]Humboldt State University, 1 Harpst St, Arcata, CA 95521
[6]Colorado State University, Fort Collins, CO 80523

ABSTRACT

Composites are being increasingly used in aerospace, automotive, energy, gas storage, marine, infrastructure, sporting goods, and other secondary industries. The key drivers for composites is light weight, high specific strength, and durability. The composites market is projected to grow from $72.58 billion to $115.43 billion by 2022 (CAGR of 8.13% between 2017-2022) with carbon fiber demand alone projected to soar to 150,000 metric tons by 2020 globally, implying that much of this will reach their End-of-Life (EOL) stage in the coming years. Due to rising disposal cost of landfills and lack of space, it is important to diversify composite waste streams in order to address feasible options related to recovery, reuse, and remanufacture. The Institute for Advanced Composites Manufacturing Innovation (IACMI) has partnered with several industry collaborators to address the issue of carbon fiber reinforced polymer (CFRP) waste. This paper highlights the significance, complexities, application, and the embodied energy (EE) associated with two types of recycling processes for EOL CFRP. The cumulative energy demand (CED) method was applied to analyze the pyrolysis and solvolysis processes using the life cycle assessment (LCA) software SimaPro v.9.0.0.33 and the FRPC Energy Use Estimation Tool developed by the Oak Ridge National Laboratory. Data was sourced to model the amount of fiber, resin, and embodied energy that may be recovered from each recycling system considering 1 kg of carbon fiber-epoxy laminate as benchmark. It was found that a continuous natural gas furnace based pyrolysis system consumes a total of 52 MJ/kg by default and 42 MJ/kg while reusing syngas generated within the system as avoided energy. The supercritical solvolysis process considered for analysis presented a total embodied energy of 257 MJ/kg.

Corresponding author: Komal Kooduvalli kkooduva@utk.edu

1. INTRODUCTION

Carbon fiber (CF) composites have the capability of being used in ways that propel positive sustainability implications while considering manufacturing through the environmental, business, and humanitarian lens. For example, the Boeing 787 aircraft weight was significantly reduced by using CF composites, enabling fuel economy and cutting company costs in addition to reduced environmental emissions, ultimately translating to a cleaner supply chain. However, the end of life for these aerospace products still needs to be addressed for downstream processes. One aircraft contains an estimated 20 tons of carbon fiber composites [1]. Previous literature has suggested that landfilling CF composites is an environmentally friendly solution as these materials remain inert for long periods of time [2]. Despite this, landfilling may become an unreliable EOL option for CFRP waste due to several limiting factors due to limited space and rising costs. Current recycling methods available for CFRP available are categorized as: mechanical, pyrolysis, fluidized bed, and chemical [3]. In this study, the pyrolysis (thermal) and solvolysis (chemical) options are investigated.

1.1 Pyrolysis

Recycling CF composites comes with unique challenges when paired with thermoset resins due to their cross-linked molecular structure. Thermosets typically are not able to be reformed for second generation products and are not naturally biodegradable [4]. However, recycling via pyrolysis has emerged to address these challenges. Pyrolysis is the most widely adopted method in the industry for recycling CF composites, and therefore, one of the focus areas of this study. Figure 1 shows the general flow of the pyrolysis process when applied to EOL CFRP composites. Pyrolysis involves heating composites in an environment void of oxygen under an inert medium such as nitrogen at a set temperature range that decomposes the composite matrix into three components: gas, oil, and fibers [5]. A comprehensive LCA is needed to analyze the environmental impacts that rCF can have when replacing vCF. LCA was chosen to create a detailed understanding of the pyrolysis process that will account for the environmental burdens of the process along with implications of energy recovery.

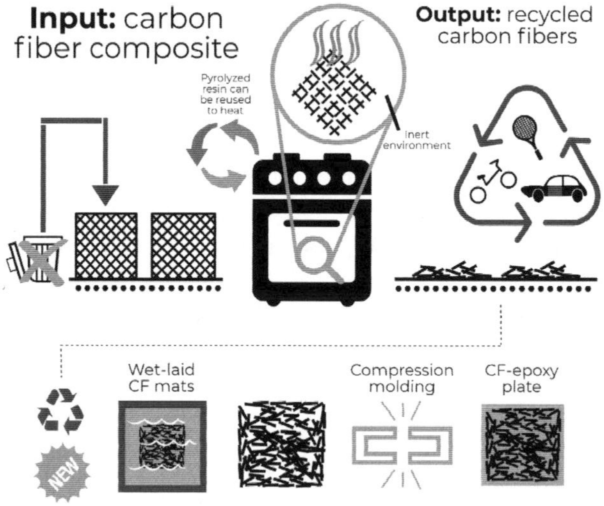

Figure 1. The pyrolysis process when applied to CFRP to promote circularity

1.2 Solvolysis

Solvolysis is a chemical process that uses a solvent to separate the resin and fibers of CFRP waste. Employing sustainable business practices while reducing cost applies to industries using CFRP. An analysis to quantify the embodied energy of 1 kg of CFRP undergoing solvolysis was also conducted. Solvolysis is a chemical process that can either be conducted at low or high temperatures. High temperature solvolysis is also referred to as supercritical solvolysis. This process uses a solvent at its supercritical temperature to separate resin and fiber from CFRP waste [6]. Addition of a catalyst or multiple solvents may be able to decrease operation conditions required [6]. The range of temperatures that supercritical solvolysis is conducted at is 375-650 °C and pressure ranges from 22-35 MPa [7]. Figure 2 displays the supercritical solvolysis route for the given CFRP part. The benefit of using supercritical solvolysis compared to the other methods is that the recovered fibers have similar properties to those of the virgin material [8]. Through the study Knight conducted, it was found that there was 100% retention of the single filament tensile strength and modulus [9]. The purpose of this study is to analyze the embodied energy of a supercritical solvolysis system using the life cycle assessment software SimaPro (© Copyright PRé 2020) to model this process.

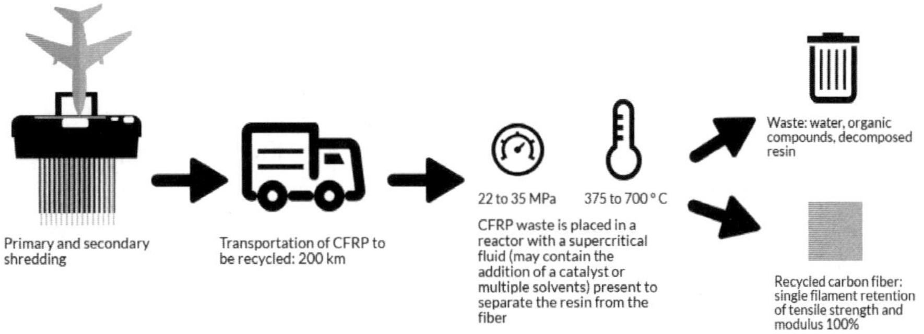

Figure 2. Supercritical solvolysis process considering EOL CFRP parts

2. EXPERIMENTATION

2.1 Materials and Methods

Information was obtained from a combination of scientific literature, databases within SimaPro (Ecoinvent 3 – allocation, cut-off by classification – unit; ELCD; Industry Data 2.0, US-EI 2.2; and USLCI) [10-12], and the FRPC Energy Estimator Tool [13] complemented by extensive industry interviews (company technicians, managers, and researchers) involved in pyrolysis and solvolysis composites recycling industry.

2.1.1 Pyrolysis machinery specification and processing information

Figure 3 represents the culmination of information gathered to create a realistic industry schematic of the pyrolysis furnace starting with the sorting process. Transportation, size reduction of the composite part, and post-processing steps were added to create a complete pyrolysis recycling system. This process was then used as the baseline for SimaPro calculations. Machinery specifications were obtained from scientific articles or directly from the manufacturers. Steps considered in the LCA are as follows; primary shredding, secondary shredding, sorting, furnace, scrubber, hydrocarbon cracker, combing, precision cutting, and transportation.

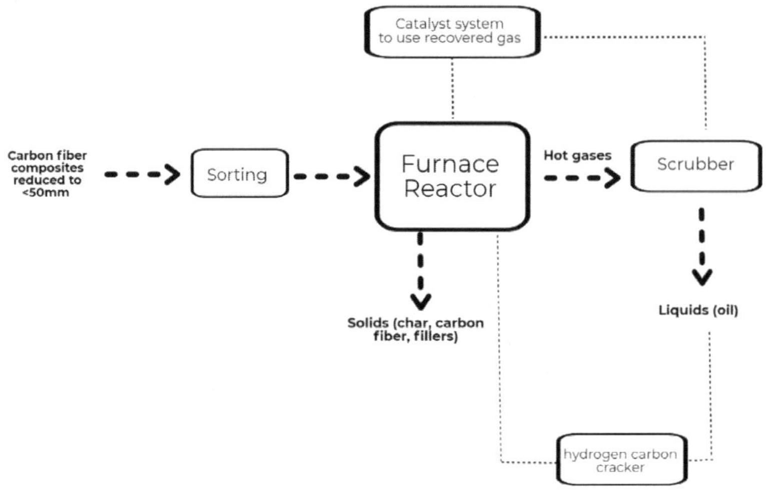

Figure 3. Pyrolysis furnace system boundaries

A 50:50 CF-epoxy composite was taken as the representative specimen for the pyrolysis process. The processes used to make the virgin CF-epoxy composite include total transportation after raw material extraction (as transportation for this life stage is accounted for in the material's embodied energy values), which was assumed to be 200 km, and manufacturing process involving wet-layup followed by compression molding.

2.1.2 Pyrolysis process estimations and assumptions

A number of general assumptions were made due to the proprietary nature of some information relating to the pyrolysis furnace design and company innovations.

1. The furnace was assumed to be a continuous design capable of processing 3800 lb every 1.25 hours, run at a frequency that recycles 1200 metric tons of CF per year.
2. Epoxy breaks down into oil and gas in the furnace, with exact ratios taken from a laboratory study that used a small static bed reactor running at 600 °C. These were then extrapolated to industry scale [14].
3. Environmental impacts of individual gases formed from the epoxy breakdown were not considered due to their containment and reuse within the system.
4. Regarding post-processing of recycled CF, combing and carding were calculated based on industry information considering a throughput of 3000 lb/hr.
5. Scrubber values were based on a standard cyclone design recommended by the Environmental Protection Agency [15, 16].
6. For furnace systems that had the ability to reuse oil, data from hydrocrackers were taken, however any catalytic system with the ability to break apart oil molecules could theoretically be used.
7. In order to obtain syngas densities, temperatures taken for individual chemical densities ranged from 500-600 °C.

It was understood that the pyrolysis process energy would consume a substantial amount of energy as a result of being a large continuous process. However, the syngas that is created within the furnace may be reused within the same process in the place of natural gas as fuel to run the same system as avoided energy. These scenarios are found to be most prevalent within the industry currently and have been considered for analysis.

2.1.3 Solvolysis solvent and equipment selection

A variety of solvents may be used during solvolysis of CFRP. However, the scenario chosen for analysis follows the work of Knight where 150 kg of CFRP was analyzed with 700 kg of water as the solvent [9]. This scenario represents a functional unit of 1 kg CFRP with a 50:50 ratio of carbon fiber to epoxy. Throughout literature review, power ratings of solvolysis boilers and reactors were not provided for the desired industrial scale. Therefore, a high pressure reactor that could maintain the required temperature and pressure of supercritical water and a volume of 2 m^3 was chosen [17]. These properties are based on the work of Knight where an industrial sized reactor was defined as 1.93 m^3 [9]. An average distance of 200 km was considered for transportation of the CFRP waste. Some studies state that the solvent after solvolysis may be reused [18]. However, was not considered in this study as the pathway of such processes were not clearly specified.

3. RESULTS AND DISCUSSION

3.1 Life Cycle Assessment

Embodied energy is the total amount of energy that is found within the supply chain of a given product, process or service [19, 20]. A Life Cycle Assessment (LCA) is the concept that caters to all of the environmental impacts associated with a certain product or process [21, 22]. The Cumulative Energy Demand (CED) method [23, 24] was the method that was used to calculate the embodied energy of both pyrolysis and solvolysis processes for a given functional unit of 1 kg CFRP scrap adhering to the ISO 14040 and 14044 standards [25, 26]. Embodied energy was calculated using the LCA software SimaPro [10] with some values referenced from the FRPC Energy Estimator Tool, developed by the Oak Ridge National Laboratory [13]. The goal of the study was to perform a gate-to-gate embodied energy analysis considering both the pyrolysis and solvolysis recycling technologies. Medium voltage US electricity grid dataset from the SimaPro US-EI library DATASMART, provided by Long Trail Sustainability, 2015 [11] was considered for modeling.

3.1.1 Pyrolysis process impacts

In regards to the pyrolysis model, the most energy intensive step in the pyrolysis process was found to be that of the furnace, with transportation, size reduction, and post-processing steps accounting for a total of only 1.168 MJ/kg. To calculate the EE of the pyrolysis furnace system, values for sorting, furnace, and scrubber, were combined from Table 1. A sensitivity analysis was run based on the ability of the furnace to reuse gas to offset energy needed to heat the system. The most advanced pyrolysis process has the ability to avoid energy consumption by using the syngas *and oil* produced to power the same system. However, this case is not considered for the study. The most energy intensive scenario was the default scenario for the pyrolysis furnace where neither gas nor oil were reused to power the process with a furnace EE of 51.02 MJ/kg (Table 1). The most common system in industry today is a pyrolysis furnace with the capacity to only capture and reuse the syngas (excluding oil) created to power the same furnace. This type of system considering an energy offset (as avoided electricity) resulted in furnace EE of 41.32 MJ/kg (Table 2).

Table 1. Embodied energy per process step within a natural gas based furnace pyrolysis process. Default scenario (without avoided energy consideration) results reflect a total EE of 52 MJ/kg

Impact category	Unit	Step 1: Primary Shredding	Step 2: Secondary Shredding	Step 3: Sorting	Step 4B: Furnace Continuous-Natural Gas	Step 5: Scrubber w/o avoided product	Step 6: Hydrocracker w/o avoided product	Step 7: Combing	Step 8: Precision cutting	Transport
Total	MJ	1.65E-01	4.36E-01	2.69E-01	5.06E+01	7.01E-07	1.92E-01	6.48E-05	4.35E-05	5.67E-01
Non renewable, fossil	MJ	1.14E-01	3.02E-01	1.87E-01	4.73E+01	4.85E-07	1.33E-01	4.48E-05	3.01E-05	5.63E-01
Non-renewable, nuclear	MJ	4.44E-02	1.17E-01	7.25E-02	2.84E+00	1.89E-07	5.16E-02	1.74E-05	1.17E-05	3.60E-03
Non-renewable, biomass	MJ	1.26E-09	3.34E-09	2.07E-09	9.00E-08	5.38E-15	1.47E-09	4.96E-13	3.33E-13	1.30E-07
Renewable, biomass	MJ	1.83E-04	4.83E-04	2.98E-04	7.17E-02	7.76E-10	2.12E-04	7.17E-08	4.81E-08	9.29E-05
Renewable, wind, solar, geothermal	MJ	2.90E-03	7.68E-03	4.74E-03	9.79E-02	1.23E-08	3.38E-03	1.14E-06	7.65E-07	1.24E-04
Renewable, water	MJ	3.32E-03	8.78E-03	5.42E-03	2.51E-01	1.41E-08	3.86E-03	1.30E-06	8.75E-07	3.19E-04
		Size reduction				Pyrolysis		Post-processing		Transportation
		0.601				51.019		0.0001		0.567

Table 2. Embodied energy per process step within a natural gas based furnace pyrolysis process. When considering avoided electricity from syngas reuse, the total EE is 42 MJ/kg

Impact category	Unit	Step 1: Primary Shredding	Step 2: Secondary Shredding	Step 3: Sorting	Step 4B: Furnace Continuous-Natural Gas	Step 5: Scrubber	Step 6: Hydrocracker w/o avoided product	Step 7: Combing	Step 8: Precision cutting	Transport
Total	MJ	1.65E-01	4.36E-01	2.69E-01	5.06E+01	-9.70E+00	1.92E-01	6.48E-05	4.35E-05	5.67E-01
Non renewable, fossil	MJ	1.14E-01	3.02E-01	1.87E-01	4.73E+01	-6.71E+00	1.33E-01	4.48E-05	3.01E-05	5.63E-01
Non-renewable, nuclear	MJ	4.44E-02	1.17E-01	7.25E-02	2.84E+00	-2.61E+00	5.16E-02	1.74E-05	1.17E-05	3.60E-03
Non-renewable, biomass	MJ	1.26E-09	3.34E-09	2.07E-09	9.00E-08	-7.43E-08	1.47E-09	4.96E-13	3.33E-13	1.30E-07
Renewable, biomass	MJ	1.83E-04	4.83E-04	2.98E-04	7.17E-02	-1.07E-02	2.12E-04	7.17E-08	4.81E-08	9.29E-05
Renewable, wind, solar, geothermal	MJ	2.90E-03	7.68E-03	4.74E-03	9.79E-02	-1.71E-01	3.38E-03	1.14E-06	7.65E-07	1.24E-04
Renewable, water	MJ	3.32E-03	8.78E-03	5.42E-03	2.51E-01	-1.95E-01	3.86E-03	1.30E-06	8.75E-07	3.19E-04
		Size reduction				Pyrolysis		Post-processing		Transportation
		0.601				41.323		0.0001		0.567

3.1.2 Solvolysis process impacts

Table 3 outlines the embodied energy during every step of the solvolysis process including the energy required to achieve supercritical condition, to power the reactor and amount of deionized water that is used within the system. The values for size reduction, solvolysis, post-processing and transportation values were kept constant as per the pyrolysis process. Values for reuse were not modeled as this is not clear from published literature. It is seen that within the solvolysis process, the maximum amount of energy is consumed at the stage where supercritical water needs to be reached, with EE of 255.16 MJ/kg.

Table 3. Embodied energy results of the supercritical solvolysis process. Total EE associated with the solvolysis process of 1 kg CFRP using supercritical water is 257 MJ/kg

Impact category	Unit	Step 1 -Primary Shredding	Step 2 - Secondary Shredding	Step 3 - Achieving supercritical condition	Step 4 - Reactor	Step 5D - Water (H20)	Step 6 - Combing	Step 7 - Precision cutting	Transportation
Total	MJ	0.16	0.44	255.16	1.04	0.02	0.00	0.00	0.57
Non renewable, fossil	MJ	1.14E-01	3.02E-01	2.29E+02	7.22E-01	1.32E-02	4.48E-05	3.01E-05	5.63E-01
Non-renewable, nuclear	MJ	4.44E-02	1.17E-01	2.31E+01	2.81E-01	5.13E-03	1.74E-05	1.17E-05	3.60E-03
Non-renewable, biomass	MJ	1.26E-09	3.34E-09	6.68E-07	8.00E-09	4.51E-10	4.96E-13	3.33E-13	1.30E-07
Renewable, biomass	MJ	1.83E-04	4.83E-04	1.09E-01	1.15E-03	4.40E-04	7.17E-08	4.81E-08	9.29E-05
Renewable, wind, solar, geothermal	MJ	2.90E-03	7.68E-03	1.49E+00	1.84E-02	1.77E-04	1.14E-06	7.65E-07	1.24E-04
Renewable, water	MJ	3.32E-03	8.78E-03	1.73E+00	2.10E-02	4.55E-04	1.30E-06	8.75E-07	3.19E-04
		Size Reduction			Solvolysis		Post-processing		Transportation
		0.601			256.227		0.0001		0.567

4. CONCLUSIONS

Results indicate that in both recycling technology cases, transportation resulted in lower embodied energy (0.567 MJ/kg) compared to both size reduction (0.601 MJ/kg) and post processing (0.0001 MJ/kg) steps (Tables 1, 2, and 3).

4.1 Sustainability of pyrolyzed CF-epoxy composites

Industry is primarily concerned with the amount of energy and cost associated with recycling CF-epoxy composites. The current analysis shows that the most basic pyrolysis system with no reuse capability to offset energy will still have lower EE than that of vCF which has an EE range of 183-286 MJ/kg [27] and more recently some citing 1155-1166 MJ/kg [13]. This difference is even more significant while considering a pyrolysis process utilizing a natural gas furnace with the ability to offset energy demands. According to these results, reusing the same syngas produced during the process has potential to bring the overall EE of the process down by 19%. While considering pricing, the cost of vCF ranges from \$33-66/kg depending on varying grades where rCF can be sold for a fraction of that cost based on the scale of the recycling facility ranging from anywhere between \$5-15/kg [28]. This fact, along with the energy savings that is seen between vCF and rCF makes for a strong economic and environmental case in terms of recycling CF-epoxy composites via pyrolysis. Due to the extremely high EE of CF as a raw material, the energy required for the pyrolysis process itself is relatively low, resulting in reduced total EE for the recycled product as virgin manufacturing and procurement burdens are removed. Furthermore, if the pyrolysis process is optimized further by reusing both the syngas and oil, the EE of the recycled product could be further reduced.

4.2 Sustainability of CF-epoxy composites undergoing solvolysis

The embodied energy for achieving supercritical conditions is the source of the energy peak with respect to this process. Therefore, ways to improve the efficiency of this reaction or cutting down on the dwell time with the use of catalysts, and reusing solvents that are used for separation may alleviate this stress. Knight found that this process removed up to 99% of a highly cross linked resin with 100% retention of single filament tensile strength and modulus [9]. Supercritical solvolysis as a method for recycling CFRP is still a novel technology. Therefore, more research is required to analyze other environmental impacts associated with the process and how resins and solvents may be reused.

4.3 Closing Remarks

Pyrolysis has proven to be a sustainable pathway for CFRP EOL scrap. Although the process itself has scope for improvement, industry interviews suggest that pyrolysis optimization methods (syngas reuse, oil use, etc.) may become the norm in the coming years. In addition, supporting this sector of the composites industry has the potential to create supply for the ever-increasing demand of carbon fiber. In regards to supercritical solvolysis, using water as a solvent would help recover fibers that possess almost identical properties to those of virgin material. More research is needed to understand other environmental impacts such as greenhouse gas emission, acidification potential, respiratory effects, among others while considering the process, route for resin reuse, and applications for recycled carbon fiber. Future research publishing the effect of catalyst and solvent use in reducing process dwell time would substantially assist in understanding the energy consumption related to this significant reduction in cycle times.

5. ACKNOWLEDGEMENT

We would like to thank Ryan Ginder, Philip Barnett, and Ryan Spencer from The University of Tennessee, Chuck Ludwig from CHZ Technologies, Tim Spahn from Carbon Fiber Recycling, Inc., Mark Mauhar from Carbon Conversions, Mathilde Poulet and Alasdair Gledhill from ELG Carbon Fibre, Andrew Maxey from Vartega Inc., and Adherent Technologies, Inc. for their generous guidance on the currently existing recycling technologies and valuable data contribution presented in this paper.

6. FUNDING

The authors would like to acknowledge the support of the Institute for Advanced Composites Manufacturing Innovation (IACMI)—The Composites Institute. The information, data, and work presented herein was funded in part by the US Department of Energy, Office of Energy Efficiency and Renewable Energy, under Award Number DE-EE0006926.

7. REFERENCES

[1] S. M. C. Alves, F. S. da Silva, M. V. Donadon, R. R. Garcia, and E. J. Corat, "Process and characterization of reclaimed carbon fiber composites by pyrolysis and oxidation, assisted by thermal plasma to avoid pollutants emissions," *Journal of Composite Materials,* vol. 52, no. 10, pp. 1379-1398, 2018.

[2] Y. Khalil, "Comparative environmental and human health evaluations of thermolysis and solvolysis recycling technologies of carbon fiber reinforced polymer waste," *Waste Management,* vol. 76, pp. 767-778, 2018.

[3] S. Pimenta and S. T. Pinho, "Recycling carbon fibre reinforced polymers for structural applications: Technology review and market outlook," *Waste management,* vol. 31, no. 2, pp. 378-392, 2011.

[4] K. Wong, C. Rudd, S. Pickering, and X. Liu, "Composites recycling solutions for the aviation industry," *Science China Technological Sciences,* vol. 60, no. 9, pp. 1291-1300, 2017.

[5] S. Naqvi, H. M. Prabhakara, E. Bramer, W. Dierkes, R. Akkerman, and G. Brem, "A critical review on recycling of end-of-life carbon fibre/glass fibre reinforced composites waste using pyrolysis towards a circular economy," *Resources, conservation and recycling,* vol. 136, pp. 118-129, 2018.

[6] G. Oliveux, L. O. Dandy, and G. A. Leeke, "Current status of recycling of fibre reinforced polymers: Review of technologies, reuse and resulting properties," *Progress in Materials Science,* vol. 72, pp. 61-99, 2015.

[7] C. Morin, A. Loppinet-Serani, F. Cansell, and C. Aymonier, "Near-and supercritical solvolysis of carbon fibre reinforced polymers (CFRPs) for recycling carbon fibers as a valuable resource: State of the art," *The journal of supercritical fluids,* vol. 66, pp. 232-240, 2012.

[8] M. Prinçaud, C. Aymonier, A. Loppinet-Serani, N. Perry, and G. Sonnemann, "Environmental feasibility of the recycling of carbon fibers from CFRPs by solvolysis

using supercritical water," *ACS Sustainable Chemistry & Engineering,* vol. 2, no. 6, pp. 1498-1502, 2014.
[9] C. C. Knight, "Recycling high-performance carbon fiber reinforced polymer composites using sub-critical and supercritical water," 2013.
[10] "SimaPro," 9.0.0.33 ed: PRé Consultants, 2018.
[11] "DATASMART LCI Package," ed: Long Trail Sustainability, 2015.
[12] G. Wernet, C. Bauer, B. Steubing, J. Reinhard, E. Moreno-Ruiz, and B. Weidema, "The ecoinvent database version 3 (part I): overview and methodology," *The International Journal of Life Cycle Assessment,* vol. 21, no. 9, pp. 1218-1230, 2016.
[13] S. Das and K. Armstrong. FRPC Energy Use Estimation Tool [Online]. Available: http://energytoolestimator.com/
[14] A. Cunliffe, N. Jones, and P. Williams, "Pyrolysis of composite plastic waste," *Environmental technology,* vol. 24, no. 5, pp. 653-663, 2003.
[15] *Monitoring by Control Technique - Wet Scrubber For Particulate Matter*. Available: https://www.epa.gov/air-emissions-monitoring-knowledge-base/monitoring-control-technique-wet-scrubber-particulate-matter
[16] P. H. Daniel Mussatti, "Air Pollution Control Cost Manual," in "Section 6: Particulate Matter Controls; Chapter 2: Wet Scrubbers for Particulate Matter," U.S. EPAJuly 15, 2002 2002, Available: https://www.epa.gov/air-emissions-monitoring-knowledge-base/monitoring-control-technique-wet-scrubber-particulate-matter, Accessed on: June 20, 2019.
[17] White Coffee. (March 12). *"Organic Breakfast Blend"* Available: http://www.whitecoffee.com/store/organic-breakfast-blend/227fdc28-e64b-4673-a499-f7a7fbad0087?category=single%20serve
[18] H. U. Sokoli, "Chemical Solvolysis as an Approach to Recycle Fibre Reinforced Thermoset Polymer Composites and Close the End-of the Life Cycle."
[19] A. A. Morini, M. J. Ribeiro, and D. Hotza, "Early-stage materials selection based on embodied energy and carbon footprint," *Materials & Design,* vol. 178, p. 107861, 2019.
[20] "Embodied energy," vol. 2017, ed: Composites UK, 2018.
[21] G. Rebitzer *et al.*, "Life cycle assessment: Part 1: Framework, goal and scope definition, inventory analysis, and applications," *Environment international,* vol. 30, no. 5, pp. 701-720, 2004.
[22] M. Duda and J. S. Shaw, "Life cycle assessment," *Society,* journal article vol. 35, no. 1, pp. 38-43, November 01 1997.
[23] D. Gürzenich, J. Mathur, N. K. Bansal, and H.-J. Wagner, "Cumulative energy demand for selected renewable energy technologies," *The International Journal of Life Cycle Assessment,* vol. 4, no. 3, pp. 143-149, 1999.
[24] S. Kara and S. Manmek, "Composites: Calculating their Embodied Energy," *The University of New South Wales,* 2009.
[25] *Environmental management-life cycleassessment-principles and framework*, 2006.
[26] *Environmental management-life cycle assessment-requirements and guidelines*, 2006.
[27] T. Suzuki and J. Takahashi, "Prediction of energy intensity of carbon fiber reinforced plastics for mass-produced passenger cars," in *Proceedings of 9th Japan International SAMPE Symposium*, 2005, pp. 14-19.

[28] F. Meng, J. McKechnie, and S. J. Pickering, "An assessment of financial viability of recycled carbon fibre in automotive applications," *Composites Part A: Applied Science and Manufacturing,* vol. 109, pp. 207-220, 2018.

EXAMINATION OF THE SIZE EFFECT ON STRENGTH IN A HEMP FABRIC COMPOSITE

Leigh N. Gaither[a*], *Christopher Reidel, PhD.*[a], *Yawen Li, PhD.*[a]
[a]*Lawrence Technical University*
Southfield, MI USA

*leigh_gaither@yahoo.com

ABSTRACT

The variation of strength in tension and flexural test specimens for composites containing glass or carbon fiber is well documented in the literature [1]. While holding the fraction of fiber content constant and scaling up the test specimen size, research has indicated a significant diminishing in tensile and flexural strength in unidirectional fibrous composites [2]. The size effect states that for fibrous composites there is distinct reduction of tensile strength as the net cross-section of a beam in 4-point bending is increasing. Size effects are important from a fundamental point of view in understanding failure. They are also important in a practical sense since tests to establish values of material strength are normally performed on relatively small specimens. As very little is known concerning the size effect for composites bearing the natural fibers this research will show a correlation in data to a discernible reduction in tensile strength for a given size in a structural member. The multifunctional attributes are show in the mechanical connection of the large samples where steel imbedded in the material does not increase its sensitivity to strain.

Corresponding author: leigh_gaither@yahoo.com

1. INTRODUCTION

At the advent of World War II, Henry Ford working from the Ford Nutritional Laboratory in Dearborn Michigan, for the development of agriculturally derived products, brought to the automotive industry, with the assistance of George Washington Carver, a method to turn soybeans into plastics and combined with Biobased filaments such as hemp fibers, produced the first composite vehicle of north America. Working together, Ford and Carver in a shared commitment to the development of advanced materials for the use in building cars, while deriving the fuels to power them, with all the materials adapted from the agricultural domain.

*Figure 1: Henry Ford & George Washington Carver Collaborating on Soybean Plastics***Invalid source specified.**.

The experiments resulted in the creation of a soybean "plastic-bodied car." Ford eventually unveiled the "Soybean Car" on August 13, 1941 at Dearborn Days, an annual community festival.

Figure 2: Soy Bean Lab [2].

The outbreak of World War II and the suspension of automobile production forced Ford to abandon his efforts to mass-produce plastic car bodies. Until 1943, however, he maintained that he would build them as soon as the war was over, even though engineers had run into problems; the plastic ended up costing much more than steel because the panels took a long time to cure in their molds.

Ford died in 1946 and, according to the Ford Motor Company, the idea just seems to have gotten lost in the files (*Atlanta Journal and Constitution* 1973) [3].

The research reflected in this paper builds on the pioneering work of these dedicated men to the discovery of products made in sustainable materials. This effort will show that a composite derived from the combination of woven hemp fabric casted in a matrix of epoxy resin will show strength variation which is dependent on the size of the sample tested. The data collected during this experiment in tensile strength of sample composites will be evaluated using the Weibull distribution method.

2. EXPERIMENTATION

2.1 Materials & Testing Methods

The woven hemp fabric was supplied by the Rawganique Company as a product that was not produced in any sweatshop facility from USA, Canada and Europe. Furthermore there was no animal testing in connection of this material and it is free of GMO, PVC, dioxin, formaldehyde, acids, and other industrial chemicals commonly used in manufacturing. The fabric is referred to as HWC2, a canvas like consistence and weight.

Figure 3: What is this?

Figure 4: Is the Ruler Backward?

The epoxy used in this assembly is the West system with 105 resin and the 205 hardener produced by the Gougeon Brother, Inc. of Bay City, Michigan.

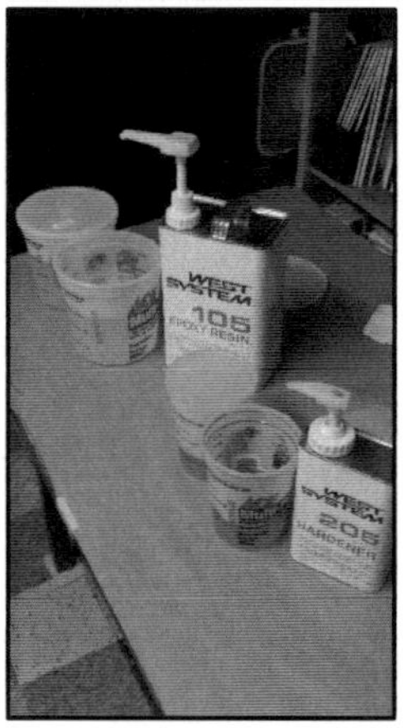

Figure 5: Epoxy Components.

Every sample was constructed so that the warp of the fabric will see the tensile load during testing and each layer will be parallel with the others. No pretreatment was done to the fabric prior to layup. All work in preparation of the samples was performed at Tulane Engineering consultants in Taylor, Michigan. The shape for each sample was hand shaped with the aid of a template that could be resized for each sample required. The cutting was with carbide tools only and no refinishing of cut surface was done.

Figure 6: Carbide cutting tool.

The volume for each layer count was measured by sandwiching the number of required layers together under the weight of an aluminum platen then this size was reproduced as a regulator frame during vacuum forming.

Figure 7: Regulator Frame with Escape channels.

The individual layers were saturated prior to lay-up than the excess amount of epoxy was drawn from the clamping fixture under a vacuum force of 10 inches of mercury.

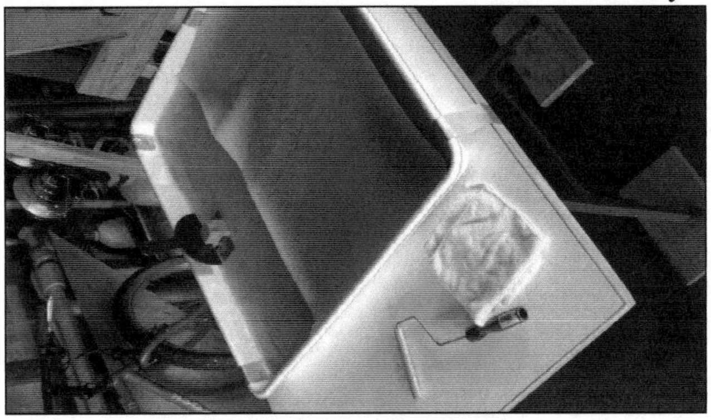

Figure 8: Layer Saturation.

Figure 9: Stacked Layers of Saturated Fabric.

After stacking the required number of layers the aluminum plate was positioned over the pile then the rubber sheet drawn over all the included elements.

Figure 10: Aluminum Plate with distributed weight.

To aid the compression phase during evacuation, settling weights are added over the rubber sheet. The aluminum sheet must be thick enough to minimize deflection under these weight there by making an irregular thickness composite sheet.

A small amount of epoxy was poured out onto a flat surface to cure to be an indicator of the setting rate of this current sheet.

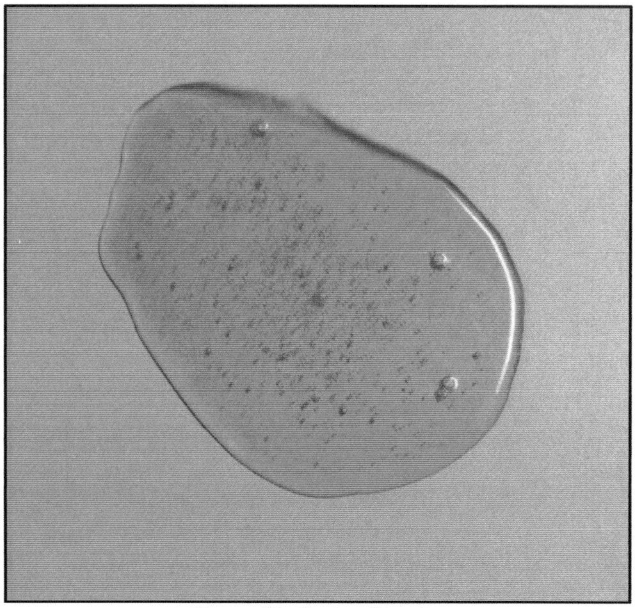

Figure 11: Curing Sample.

After the epoxy sample has become sufficiently ridged to resist bending, the vacuum is discontinued and left to normal air pressure for the remainder of 24 hours. After which the fixture can be stripped from the vacuum table and trimmed for sawing.

Figure 12: Sample Post Epoxy, excess Epoxy drawn out through escape ports.

An important consideration to casting with epoxy resins is not to let large amounts of mixture stand in the mixing pail, epoxy under goes an exothermic reaction and tremendous heat can be generated within a deep volume of this material. To counteract this tendency each layer is saturated individually with enough mix then it is placed flat on a vinyl plastic sheet till needed for the final stack up. This way a single layer can have 45 minutes of open time.

The image of the cured sheet shows how the still fluid epoxy is drawn from the casting cavity through the corrals surrounding the layup. There are narrow vents milled into the regulator frame to allow the emergence of epoxy from around the curing stack. Also notice that large amounts of air are being drawn from the fabric and replaced by epoxy during this setting phase.

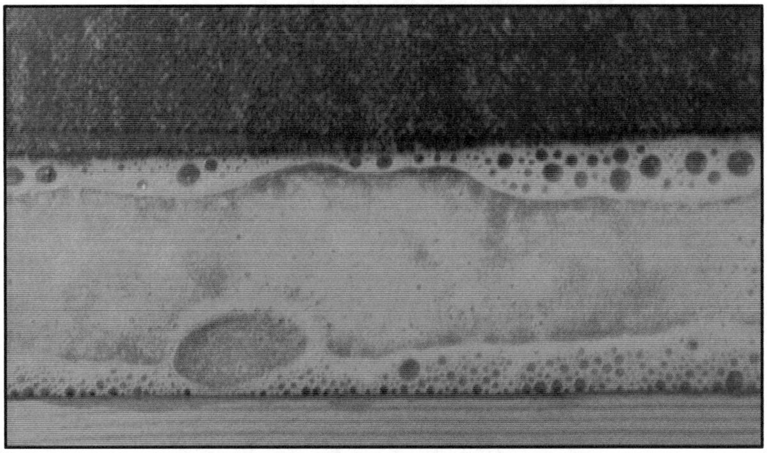

Figure 13: Epoxy with escaped air from fabric.

This is very important to the complete merging of epoxy to the fabric, without the outward bound motion of fluid epoxy with air entrainment vast regions of none bounding layers will occur.

The entire cured sheet can be pried from the flat table and made ready for sawing to the tab width required by first trimming the frame work away.

Figure 14: Table saw used to cut specimens.

The table saw is fitted with a 40 carbide tooth 12” blade and slowed to 40Hz with VFD unit. After squaring the sheet the actual sample blanks are sawn.

Figure 15: Blanks being cut to size.

These blanks are then prepared for mounting into the shaping template by drilling locator holes through the material at precise locations.

Figure 16: Drilled holes for specimen testing rig.

After adjustments to the convertible templet are made to conform to the require cross section width the shaping to size is performed.

Figure 17: Specimens are shaped using a template.

This procedure is repeated approximately 30 times for each of the four thickness used and of the two widths for each thickness. The first attempt to have the samples brake at a particular spot following ASTM periodical led to repeated failures.

Figure 18: Parallel specimens continued to break in the incorrect location.

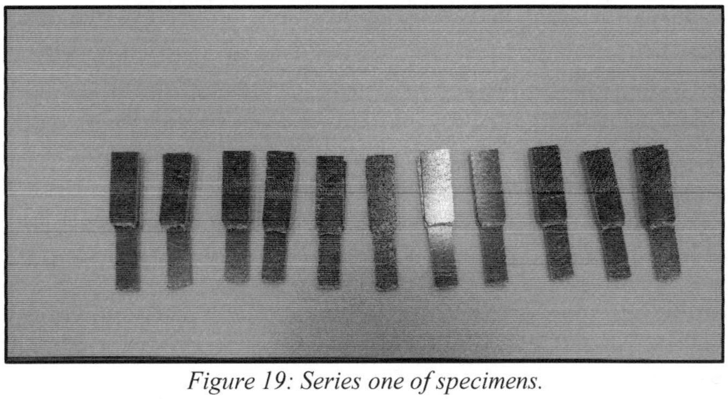

Figure 19: Series one of specimens.

The final sample shape took two years to discover both for the proper brake location and gripping the oversize tabs.

Figure 20: Custom adjustable gripping rig to maintain consistent hold on specimens.

This image shows how the griping solution was solved by using steel inserts within the tab region and a through pin to give the needed holding power to rupture these large samples. The steel was 16 gage 4130a sheet metal casted into the sheets during layup. The sawing of these samples was accomplished using a 200 tooth HSS hollow ground cold saw blade mounted in the table saw and run at 9 Hz as adjusted with the VFD unit. A word of caution: do not run a cold saw blade above the recommended RPM's as defined by the manufacture. This was a one of a kind sawing technique that gave remarkable results to size and finish without having to alter the sample material in the process.

2.2 Tensile Testing the Samples

Tensile test were conducted for each size family of samples using the MTS Q100 test machine without measuring the absolute strain only the force and specimen cross section were recorded.

Figure 21: Specimen shape and gripping rig allows the breaks to occur where intended.

Figure 22: Discoloration of Specimen at breaking point.

Typically as the sample was being subjected to tensile force a characteristic color change would occur in the narrow of the sample at half the rupture strength then closer to rupture force multiple snapping noises were audible and finally a load discharge upon rupturing. The rupture face was consolidated without loose particles ejected from the surface.

Full size specimen with modified grip and pin arrangement. The gradated strain field is shown here surrounding the rupture face. With these large samples a grip tightening routine had to be developed to ensure consistent holding results. The square head bolts seen here were cross tightened to full allowable torque for these harden ½-13 set screw bolts. The bolts bare onto the grip blocks which have serrated texture scribed into their surfaces.

3. RESULTS

The fracture of the tension specimens occurred throughout the population within the middle third of the test region and no failures occurred within the clamping section. Two typical views of broken specimens are shown in Figs. 22 and 21. For all tests the maximum load was recorded and the strength was calculated on the bases of the individually measured minimal cross-sections.

GROUP	1	2	3	4	5	6	7	8
Average Area (mm^2)	564	389	191	136	39	87	276	94
Average Stress (MPa)	79	83	88	90	62	71	81	69
Deviation Area ± (mm^2)	33	16	13	13	4	8	5	7
Deviation Stress ± (MPa)	4	5	7	4	7	7	4	8
LAYERS	10		5		1	3	7	3

Table 1: Each layer construction group and corresponding physical statistics.

The cumulative frequency for those groups that had two cross-section areas constructed for a given thickness are shown here.

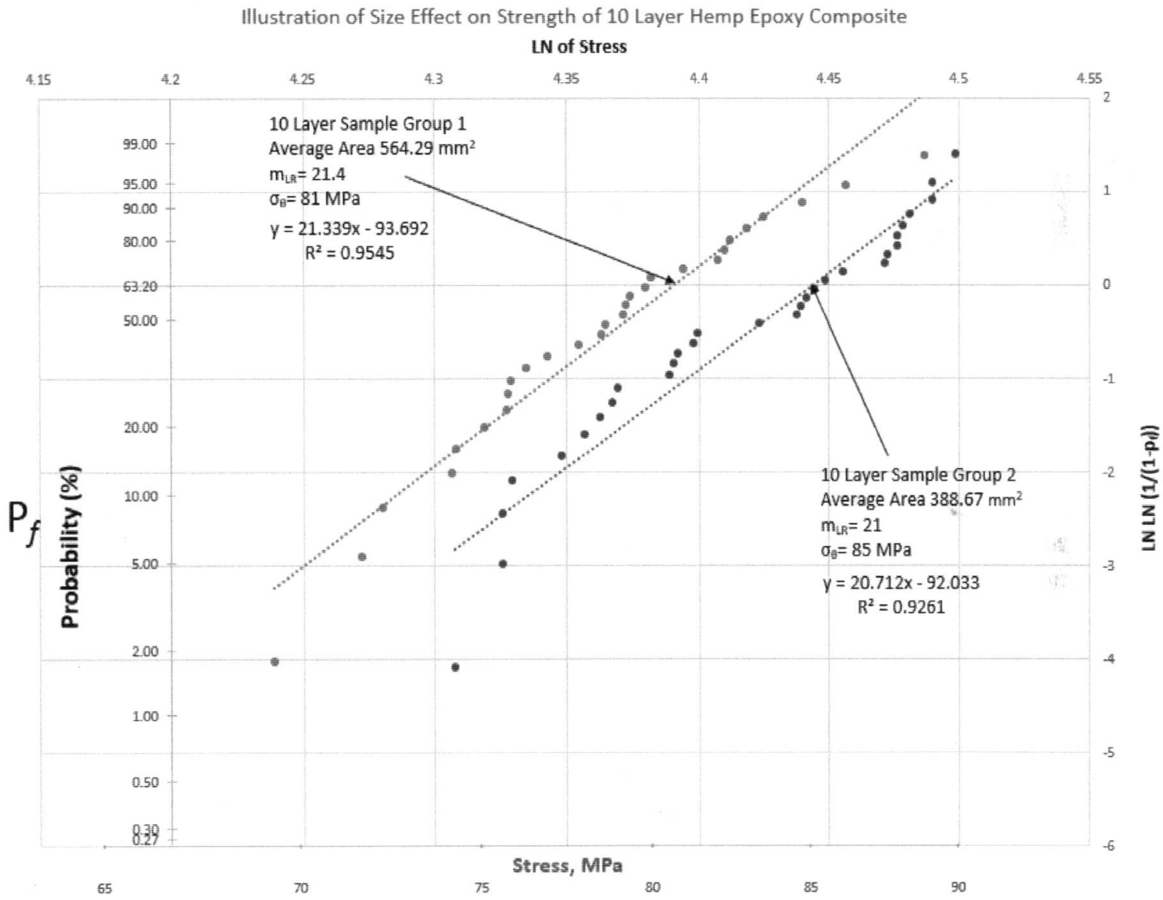

Figure 23: 10 Layer Hemp Epoxy Composite.

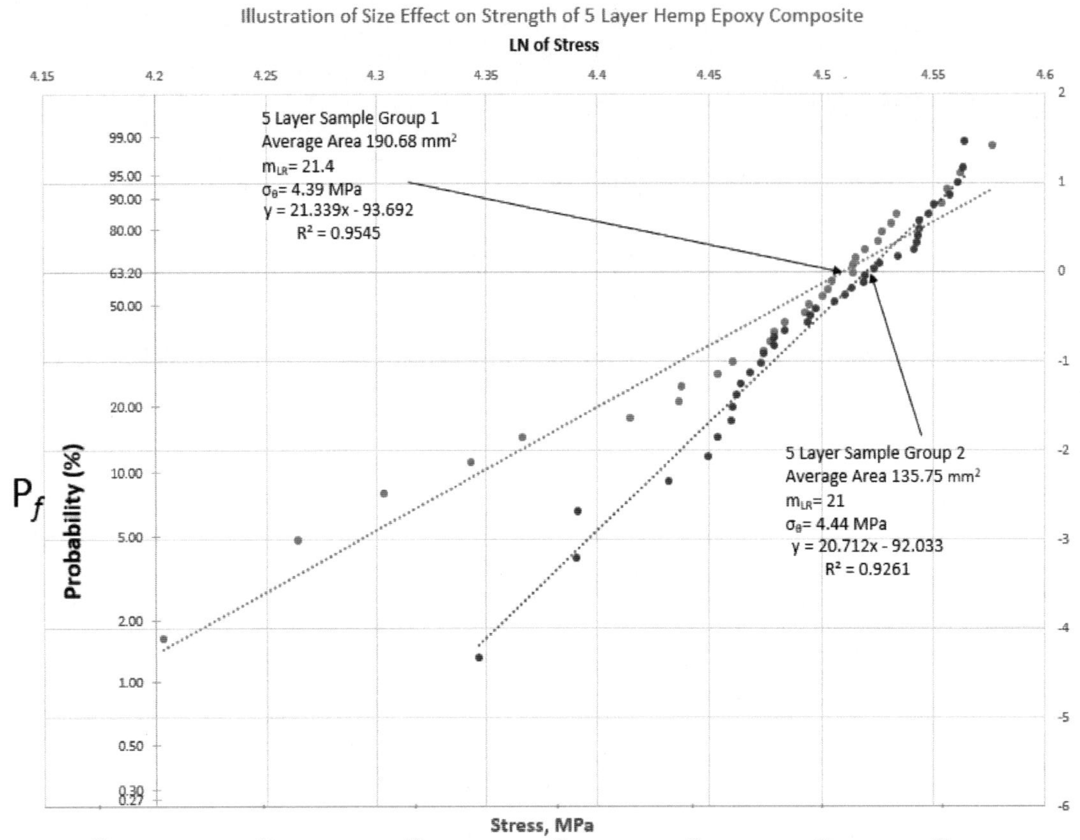

Figure 24: 5 Layer Hemp Composite.

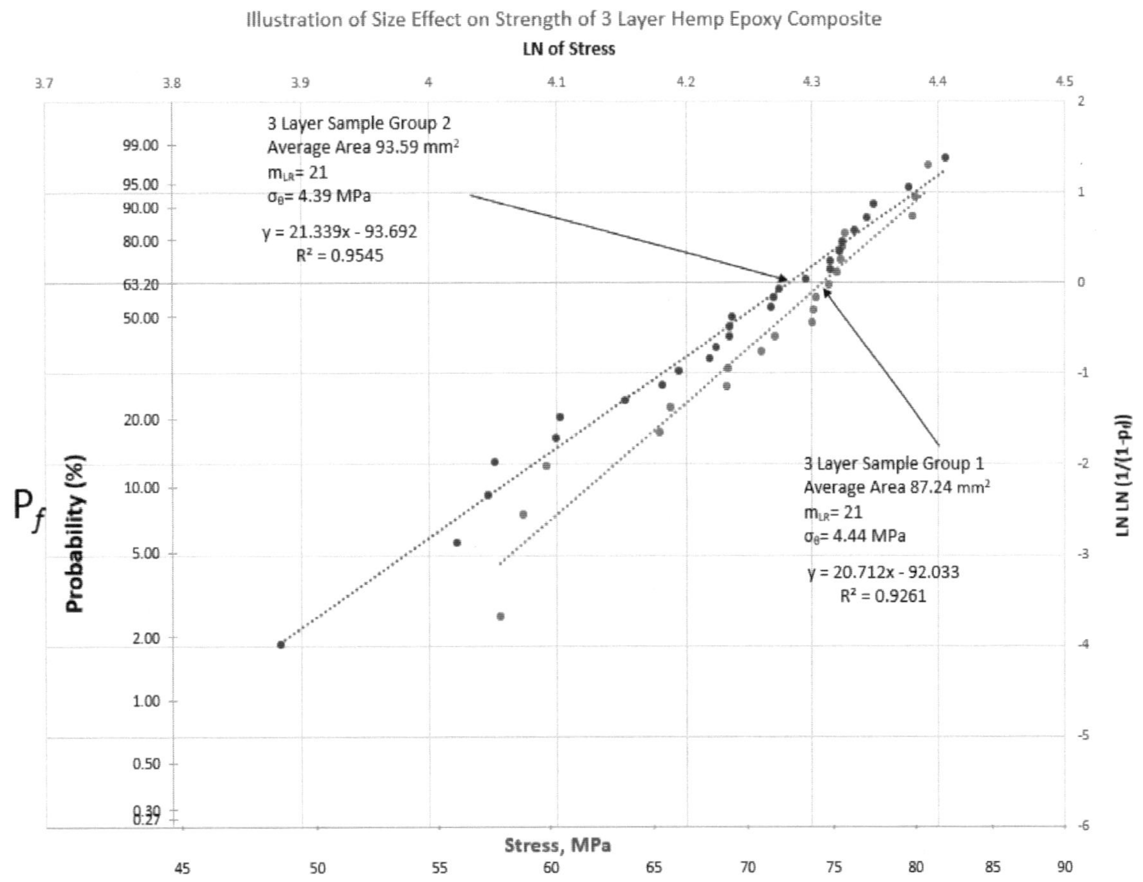

Figure 25: 3 Layer Hemp Composite.

In each case the right shift towards higher strength is associated with a decrease in cross-section area.

As pointed out by Quinn et al, that it can be seen from the charts that the slopes (Weibull moduli) indicate the strength widths. The similar slopes suggest that the same flaw types were active in similarly constructed samples and this verified by fractography analysis. See photos of SEM. The wiggles in the curves are not unusual and are common in small size sample sets. A high modulus, or steep slope, is associated with a narrow strength distribution. The characteristic strength, or Weibull scale parameter sigma sub theta, indicates the distribution location along the abscissa (x) axis, and is expected to move according to test specimen size or the amount of material that is highly stressed [4].

$$P_f = 1 - \exp\left(-(\sigma/\sigma_\theta)^m\right)$$
$$\ln(1 - P_f) = -(\sigma/\ \sigma_\theta)^m$$
$$ln\left[\ln\left(1/(1 - P_f)\right)\right] = m \ln\sigma - m \ln\sigma_\theta$$

1: The two-parameter Weibull distribution used for the test data analysis.

3.1 Fractography Analysis of Specimens

Samples were withdrawn from the specimen population two from each of the 8 groups which represent the highest and lowest braking stress from the tensile test. Each of these had a cut-away part of 1 cm square by the thickness of that group prepared for examination in the Quanta FEG 450 scanning electron microscope. The objective of this investigation is to find evidence of rupture initiation and how the fracture progressed. Pictured in the Figure is a representative finding of each sample.

Figure 26: Specimen 128, Low Breaking Stress.

The 'fish eye' object in the cross section of fracture zone which measures approximately 100 micrometers across, has emanating from its boarders the characteristic heckles of rupture which point to the start of fracture. The object is an air bubble frozen into the epoxy while setting. Surrounding the hemisphere are the strands of hemp fabric which were pulled from the opposite half of the broke specimen. Finally are the broken fibers themselves which nearly square across the brake. The fiber appears to have greater ductility when subjected to tensile forces then the bonding epoxy matrix.

The sequence of events within the fracture event are as follows(1) the load is accommodated equally across the narrow width of the sample, (2)as load increases the bubbles are first to brake as evidenced by auditable cracking and a visible color change in the epoxy, (3)The epoxy face ruptures and the entire load transfer to the fibers directed parallel to the load direction, (4) there is pull out of some fibers with subsequent separation of the fibers from each other which accounts for the long fiber bundles near the 'fish eye', (5) when enough of the fracture has taken place so as to diminish the effective cross section supporting the load, a final rupture occurs at the matrix and the still bonded fiber brakes across the remaining final section.

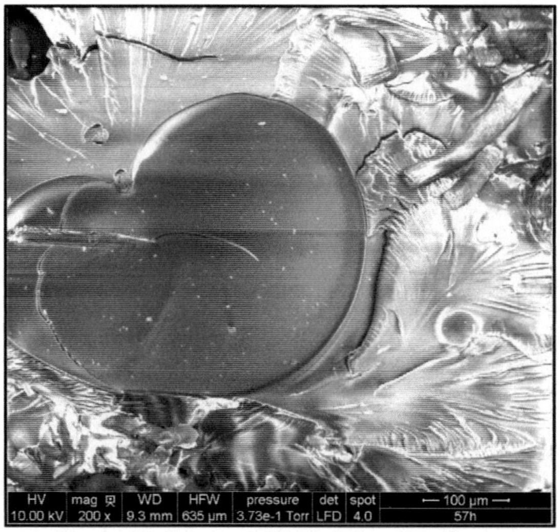

Figure 27: Specimen 57, High Breaking Stress.

The fiber bundle theory holds that load shifts from broke fibers to the local surrounding fibers until enough weak fibers are broke to cause a final rupture of the section. This disrobes well the action of a fiber bundle such as rope or braid but with a matrix present engulfing the fibers a multi-layer process is occurring during failure.

This graph shows the rise in strength with the increase number of layers of fabric until a maximum is reached then additional layers decrease the strength.

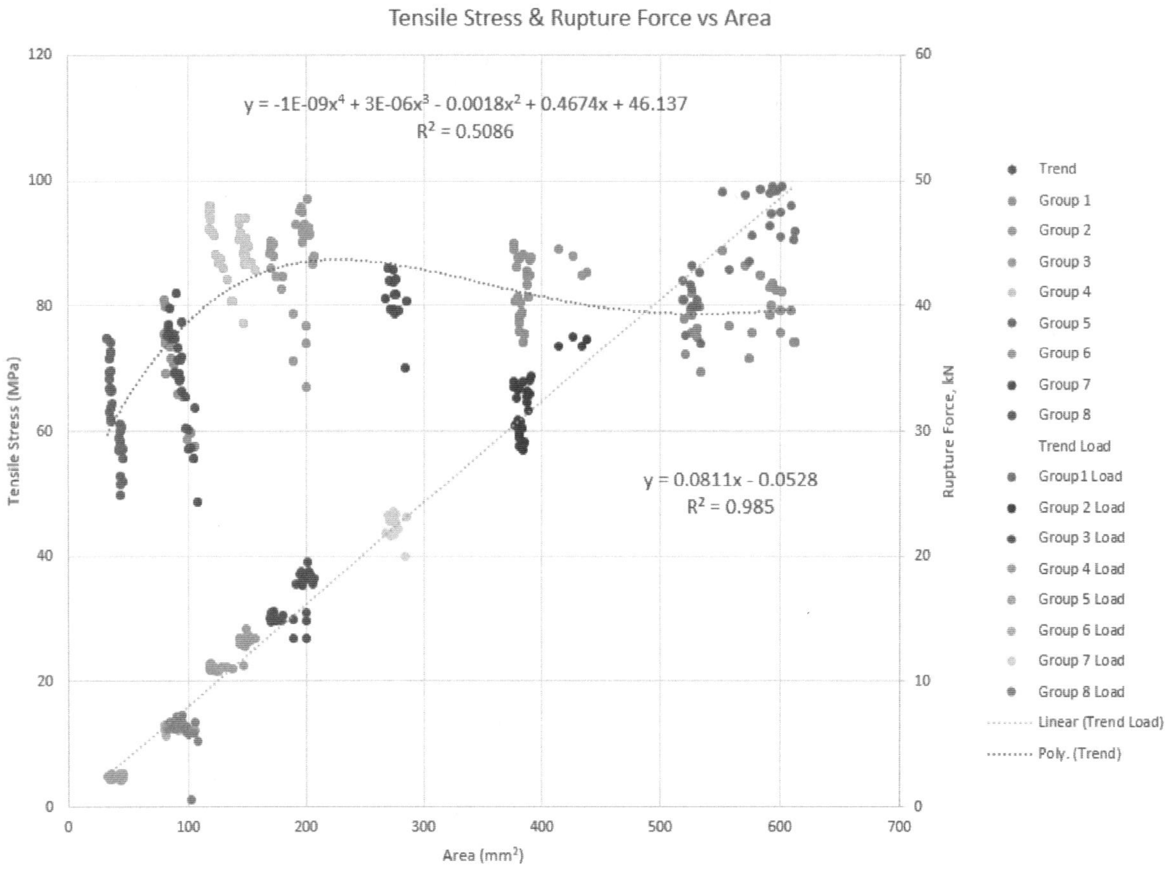

Figure 28: The straight line plot is the terminal load vs. the increase in cross section area. The load carrying capacity is increasing with size whereas the ultimate stress in decreasing.

4. CONCLUSIONS

We have shown the complete assembly process of hemp composites with the material selection and casting form design through to the sample cutting and sizing of each specimen to demonstrate the size effect on the tensile strength of woven fabric hemp composites. A method of specimen preparation was developed after preliminary trial runs so a consistent casted sheet resulted leading to uniform samples. The specimens grips were developed by responding to the holding dynamics afforded the composite samples, slipping out of the standard grips proved unsatisfactory to this experiment. The sample size and profile as a result of the adjustable templet for shaping each specimen was a project unto itself but the consistence of sample size was proof to the extra effort. A technique to manage the flow of epoxy into and out of the regulator frame used during casting of each sample sheet was brought about by observations and alteration to subsequent preparation sections. At the time of this testing there were no provisions to measure the strain associated with each tensile test and cross head motion in the test machine was not a reliable measurement as some slippage in the grips was present and the gage length would be too large for precise strain capture.

The data collected was subjected to plotting in a bar chart to reveal if a normal distribution was present. Our results shown that a shift left was present but this was within agreeable range to use the Weibull plotting methods for data characteristic demonstration. The graphics illustrate the presence of a size effect on the tensile strength for the three layer count groups produced in this experiment. A secondary graph was devised to show a correspondence between the number of layers of fabric and the relative cross section size of the sample. This graph with all data shown and the best fit polynomial line, helps in the understanding the influence of layer count as being greater than the cross section size in each sample.

Fractography evidence suggest that during extension of the specimens a complete picture of the rupture zone in each sample is a collection of events coupled at first in parallel, the fabric and matrices, than a transition to failures in series prior to a rapid cascading and final separation of the test sample. A crack initiation region is clearly seen commencing from an air bubble and progressing into nearby fabric. This causing a loosening of the bond that surrounds the fibers and the subsequent pullout of those fiber bundles. The witnessing of audible cracking noise prior to sample fracture with discoloration on the sample surface lends credence the actual sequence of events leading to rupture in each specimen. The epoxy matrix is primarily susceptible to tensile loading as first seen at the bubble flaw. The loss of bonding to the fibers is hastened through a detachment of matrix from the fiber by the mechanical enlargement of the matrix surrounding the fiber. Each sample investigated by electron microscope shows a similar sequence to rupture regardless of how many layers are used in its construction.

5. REFERENCES

[1] K. E. Jackson, "Workshop on Scaling Effects in Composite Materials and Structures," in *NASA Conference Publication 3271*, Virginia, 1994.

[2] M. R. Wisnom, "Size Effects in the Testing of Fibre-Composite Materials," *Composites Science and Technology,* vol. 59, pp. 1937-1957, 1999.

[3] W. Shurtleff and A. Aoyagi, "Henry Ford and His Employees: Work with Soy," SOYINFO CENTER, 2004. [Online]. Available: https://www.soyinfocenter.com/HSS/henry_ford_and_employees.php. [Accessed 4 January 2020].

[4] G. D. Quinn and J. B. Quinn, "A practical and systematic review of Weibull statistics forreporting strengths of dental materials," *Dent Mater,* vol. 26, no. 2, pp. 135-147, 2010.

PRODUCTION AND TENSILE TESTING OF ROTATIONALLY MOLDED HYBRID COMPOSITE TIE RODS

Jonas Nieschlag, Sven Coutandin, Jürgen Fleischer
Karlsruhe Institute of Technology (KIT), wbk Institute of Production Science,
Karlsruhe, Germany

ABSTRACT

An innovative production processes for manufacturing rotationally symmetric FRP-metal components, such as drive shafts or tie rods, is the rotational molding process. In the course of this process, a dry fiber preform and metallic load-introduction elements are inserted into a two-piece mold and subsequently clamped into a spindle. The matrix is injected directly into the rotating mold. Due to the arising centrifugal forces, the preform is impregnated and the component cures under rotation. In comparison to conventional joining processes, such as adhesive bonding or bolt connections, the metallic components as well as the FRP part are intrinsically joined during the forming process. A downstream joining process is not required. The joint is based either on the adhesive property of the matrix system or on a form-fit geometry with undercuts. The paper addresses the production and tensile testing of tie rods. Different rod geometries and different surface treatments, including sandblasting, knurling, and arc spraying, are compared and evaluated.

1. INTRODUCTION

1.1 Motivation

The world today is much more aware of the finite nature of energy resources, and in response, environmental regulations are becoming ever more stringent. Nevertheless, mobility is more and more increasing. One possibility to face this challenge is to rely on lightweight constructions to reduce the energy consumption and contribute to sustainable development. For components that need to meet high mechanical requirements at a low mass, the use of fiber reinforced plastics (FRP) is appropriative because of their excellent mechanical properties. Lightweight constructions with FRP reduce the energy consumption of moving systems during their operating phase or increase the payload of different applications.

Due to its force flow, the tension rod constitutes a highly suitable component for the use of these anisotropic FRP. Tie rods could be used for aircraft and spacecraft applications, such as actuators for landing flaps, lightweight hydraulic actuators or construction elements for solar panels [7, 8, 20, 21]. The loads are usually introduced into the composite tie rod with metallic load introduction elements, which are attached to the rest of the system. These metallic load introduction elements are joined to the FRP structure by using either friction locking, a form-fit, or substance bond. Although the component consists of FRP and metal, this fact should not be confused with multifunctionality. The joining technologies required to obtain these hybrid components can generally be divided into two groups: intrinsic and extrinsic hybridization methods. An intrinsic hybrid component is an integral component for which the connection of the different materials is carried out by primary shaping or forming the metallic or fibre-reinforced component. Hence, no extrinsic joining process, such as riveting, screwing or adhesive bonding is needed [3]. There are many different production technologies for producing hybrid composite tie rods. Selecting a process depends on the target cost, number and requirements of the product. This paper presents a new manufacturing process for the production of composite tie rods by rotational moulding, which is based on the principle of centrifugal force. In order to evaluate the mechanical strength, a large test row of composite tie rods with different geometries and surface treatments is manufactured and tested.

1.2 State of the Art

According to the current state of the art, hollow structures made of FRP for drive shafts or tie rods are produced by winding [16], pultrusion [11] or blow molding [1]. One variant of the winding process is wet filament winding where a roving is drawn through a thermoplastic matrix bath and wound onto a heated core. Another possibility is prepreg winding, in which both thermoset and thermoplastic prepregs are used. The pultrusion process also uses either thermoset or thermoplastic materials. In addition, pull braiding pultrusion provides the option of implementing reinforcements transverse to the production direction [12]. In general, the pultrusion process shows the highest productivity compared to other production technologies. A less common method to manufacture hollow FRP structures is the blow molding process, where a blow hose is pressurized so that the dry fiber preform is pressed against the tool form. The matrix is then injected into the cavity between blow hose and tool form. If the mold is heated, blow molding can also be applied for thermoplastic materials [1]. Impregnation processes such as resin transfer molding (RTM) or vacuum assisted transfer molding (VARTM) can likewise be used. Similar to blow molding with thermosets, a dry fiber preform is needed. This preform is either manufactured through braiding or winding of fabrics. Additionally, a core must be used in order to avoid accumulation of excess matrix in the cavity. The core must then be removed in a downstream process step. In order to attach the hollow FRP structures to the remaining system or install bearings and gearwheels, metallic elements are necessary. Established technologies for joining metallic load introduction elements with hollow FRP structures are interference fit [18], adhesive bonding [9], bolt connections, pin connections [19], form fit [1, 17, 21] or joints created by friction locking [6].

An innovative process for the intrinsic production of hollow structures made of FRP and metal is the rotational molding process [2, 4]. A dry and hollow fiber preform is inserted into a mold, together with the metallic load introduction elements. The mold is sealed and then clamped in a spindle. Afterwards, the thermoset matrix system is injected under rotation. The dry fiber preform is infiltrated due to the generated centrifugal forces and hardens under rotation. Hence, the metallic load introduction elements can be intrinsically joined by co-cured bonding [15], a form fit or a combination of the two. Compared to established technologies, no additional process step is necessary. Furthermore, expenses for rotational molding are rather low, as only a simple mold along with a spindle are required.

1.3 Objective and Approach

First studies of the rotational molding process have already been conducted and a rough feasibility of the process could be demonstrated as both fiber volume content and geometric fidelity are satisfactory [4, 14]. Therefore, the objective of this paper is to evaluate the mechanical strength and the energy absorption during failure of the intrinsically manufactured FRP-metal joint. As a test component, a rotationally molded hybrid composite tie rod is selected. This hybrid composite tie rod is manufactured in varying ways. The first variant is a tie rod with a cylindrical co-cured bonded lap joint (see Figure 1a). The second variant has a form-fit geometry (see Figure 1b). Due to the adhesive properties of the epoxy matrix used, the second variant has a joint which is a combination of a form fit and a co-cured bonding. In the present state of research, various approaches exist to increase the mechanical strength of hybrid FRP-metal joints [1, 5, 6, 13, 22]. Therefore, different surface treatments for the metallic load introduction elements are chosen, such as grit blasting, knurling, and arc spraying. Some treatments clearly increase the manufacturing costs. Nevertheless, all treatments must be compared to one another in order to allow a final assessment. A large series of samples is manufactured and quasistatic tension tests are conducted to evaluate the strength and the energy absorption of the components. The failure of the specimens is analyzed with microsection.

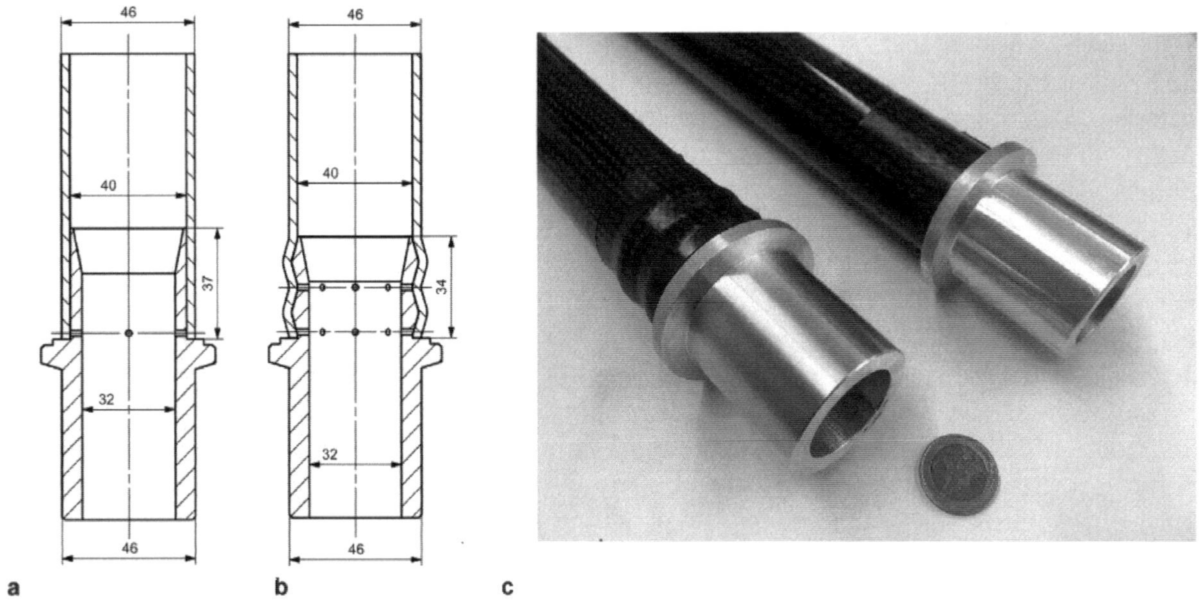

Figure 1. Overview of the different variants: **(a)** tie rod with a cylindrical co-cured bonded lap joint, **(b)** tie rod with a combination of a form fit and a co-cured bonding, **(c)** manufactured components.

1.4 Overview of Contents

Section 2 describes the manufacturing of the rotationally molded hybrid composite tie rods and the different treatments in detail. Furthermore, the test setup for the tensile tests is explained. In Section 3, the results of the tensile tests and the microsections are presented. The different treatments are compared to one another, and the achieved joint strength is discussed. Conclusions are drawn in Section 4 and finally, a short outlook is provided.

2. EXPERIMENTATION

2.1 Manufacturing of Test Specimens

First, the load introduction elements are turned out of steel. As described in Figure 1, there are two different component variants. In addition, different coatings and treatments are carried out (see

Table 1). In order to have an appropriate reference, the first treatment is an eight minutes cleaning carried out in an ultrasonic bath with acetone. The load introduction element is later joined to the FRP hollow structure by co-cured bonding using rotational molding. The second reference is a downstream adhesive bonded variant 1 with the structural adhesive Permabond ET5428. The cylindrical carbon fiber reinforced plastic (CFRP) structure is rotationally molded and later adhesively bonded to the metallic load introduction element. The aim of producing this reference is to compare the intrinsic approach with the common extrinsic production.

Table 1. Coatings and surface treatments of the metallic load introduction elements

Coatings and treatments	**Variant 1: (co-cured bonded lap joint)**	**Variant 2: (combination of form fit and co-cured bonding)**	**Example image of variant 2**
Reference 1: No treatment, only cleaning	x	x	
Reference 2: Adhesive bonding after production	x		
Grit blasting	x	x	
Arc spraying	x		
Knurling	x	x	
Coating with structural adhesive	x	x	

All other test specimens are mechanically treated or coated and produced by rotational molding. Mechanical adhesion is improved by increasing the surface roughness of the load inducing elements. A simple and cost-effective process to increase roughness is grit blasting. The test specimens are therefore grit blasted with corundum and 8 bar so that a roughness of Ra 2.7 µm is achieved. Another option to increase the surface roughness is the arc spraying process, in which wolfram carbide is melted and sprayed onto the workpiece surface at a very high speed. The gained surface roughness for the given load introduction elements is Ra 9.8 µm. Knurling deforms the load introduction element on the mesoscale. For variant 1, a RGE 30° knurling profile is chosen with two different profile depths of 0.4 mm and 0.7 mm. Variant 2 is structured with a similar profile and a depth of 0.7 mm by milling. The last treatment is a coating performed before rotational molding with the structural adhesive Permabond ET5428.

The processing of the load introduction elements is followed by the production of the dry fiber preform. Three layers of carbon fiber braided sleeves from Siltex are used for this (see Table 2). Components with a knurled surface get a braided sleeve with similar fiber orientation as first layer but a lower amount of filaments is used because this sleeve fits better into the knurled structures. The braided sleeve is then drawn layer by layer onto a core and applied together with the binder powder EPIKOTE™ 05311. After that, the binder powder is activated layer-wise by heat to obtain a dimensionally stable preform, which can be placed into the two-piece mold together with the load introduction elements. Subsequently, the mold is sealed and preheated for half an hour at 60 °C in a heating oven. By preheating the mold, the viscosity of the matrix system can be influenced, and thus improve infiltration and accelerate the curing cycle.

Table 2. Manufacturing parameters

Parameter	Value
Ply stack	$\pm 32°_1 / \pm 22°_1 / \pm 32°_1$ (Siltex 6430, 6K/ Siltex 4803, 6K/ Siltex 6430, 6K)
Ply stack for components with knurled surface	$\pm 30°_2 / \pm 22°_1 / \pm 32°_1$ (Siltex 3770, 3K/ Siltex 4803, 6K/ Siltex 6430, 6K)
Preheating	30 minutes at 60 °C
Resin	NEUKADUR EP 986
Hardener	NEUKADUR Hardener 269
Binder powder	EPIKOTE™ 05311
Centrifugal speed	3600 rpm
Rotation time	2.5 hours
Tempering	18 hours according to instructions

The tool is then clamped in a turning machine and set in rotation with 3600 rpm. A unit injects the epoxy matrix NEUKADUR EP 986 directly into the mold [10]. Due to centrifugal forces, the dry fibers are impregnated (see Figure 2). The radial impregnation takes place before the axial overlap area is impregnated [14]. To obtain a pore-free FRP structure, ventilation holes are drilled into the metallic load introduction elements as shown in Figure 1 and Table 1. After 2.5 hours, the rotation is stopped, and the mold is put into a heating oven for tempering according to the manufacturer's instructions. Subsequently, the hybrid composite tie rods are demolded.

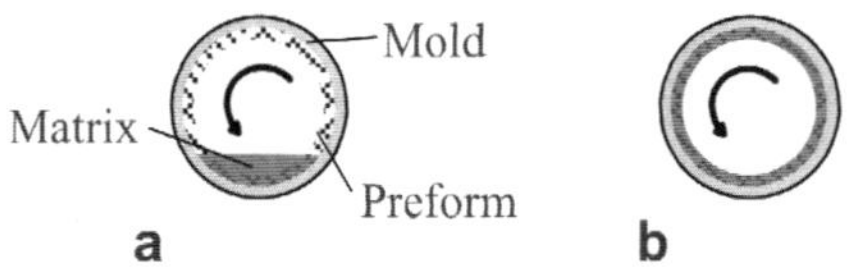

Figure 2. Process sequence: (**a**) injection of matrix system during rotation, (**b**) completely infiltrated component.

2.2 Tension Tests of Test Specimens

With the objective to increase the number of test specimen, the tie rod is cut in the middle, to obtain two test specimen, each consisting of one hollow CFRP structure with the corresponding load introduction element. One test specimen comes from the right and the other from the left mold half. Hence, four test specimen are available per variant. Using appropriative collets and a pressure of 140 bar, the test specimen are attached to the testing machine (see Figure 3). The specimens are then quasistatically tested at a speed of 2 mm per minute.

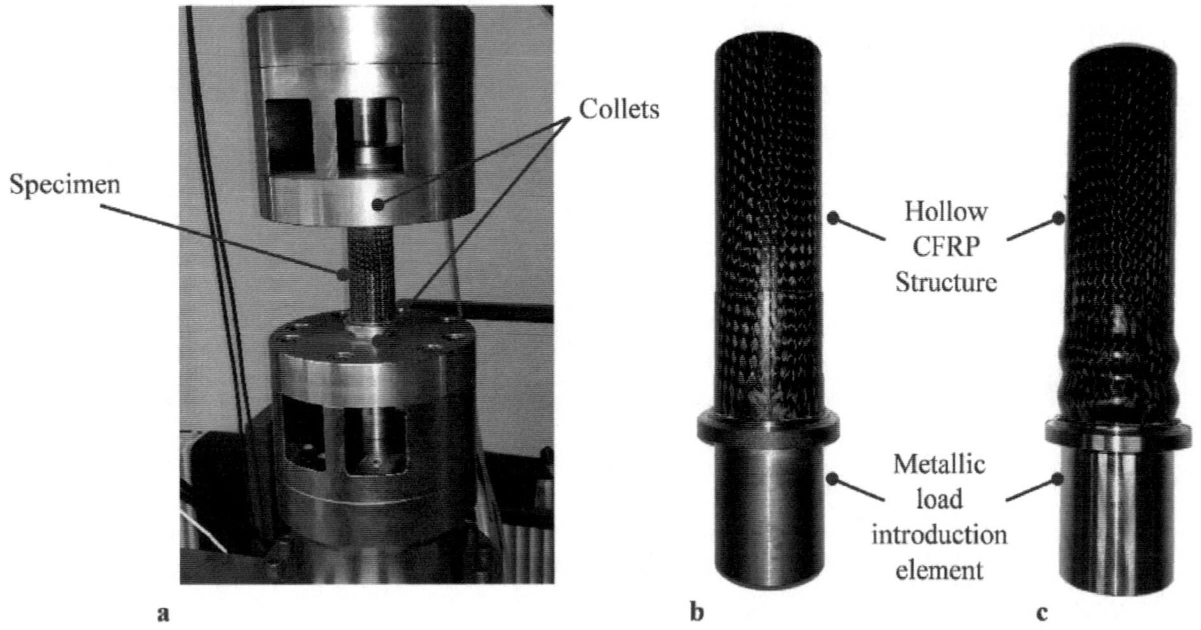

Figure 3. Overview tensile tests: **(a)** test setup, **(b)** variant 1, test specimen with co-cured bonding c variant 2, test specimen with combination of form fit and co-cured bonding.

3. RESULTS

3.1 Mechanical Joint-Strength and Force-Displacement Analysis

The results indicate that the test specimens with untreated load introduction elements feature significantly lower strengths compared to mechanically treated or coated specimens (see Figure 4.) Simply cleaning the specimens with acetone (untreated reference) does thus not seem to be sufficient in order to transfer high forces between the CRFP-metal joint. Furthermore, it can be seen that all test specimens from the left half of the mold achieve very poorer results compared to the right half (see Figure 5). It can be concluded, that the left half of the mold produced by milling is not sufficiently accurate, resulting in a parallel and angular offset between the load introduction element and the CFRP structure, which negatively effects the strength. Therefore, the components from the left half of the mold are not considered in Figure 6. Some of the specimens slip out of the collets and could not be tested to failure (see Figure 5, grit-blasted specimen 2 on the right side of the mold). Particularly high joint strengths for variant 1 are achieved by coating with structural adhesive, grit blasting and arc spraying. The strength can be increased both by coating and by mechanical treatment. Comparing the results of the arc spray treatment with Ra 9.8 μm to those of the grit blasting treatment with Ra 2.7 μm indicates, that a higher surface roughness positively influences the joint strength. Compared to reference 2, which is adhesively bonded after production with Permabond ET5428, the intrinsically rotation molded specimens demonstrate increased strength. Due to the centrifugal force, the load introduction element is concentrically integrated into the CRFP structure. The reference 2 specimens may have been less accurately bonded, which could explain the difference in joint strength.

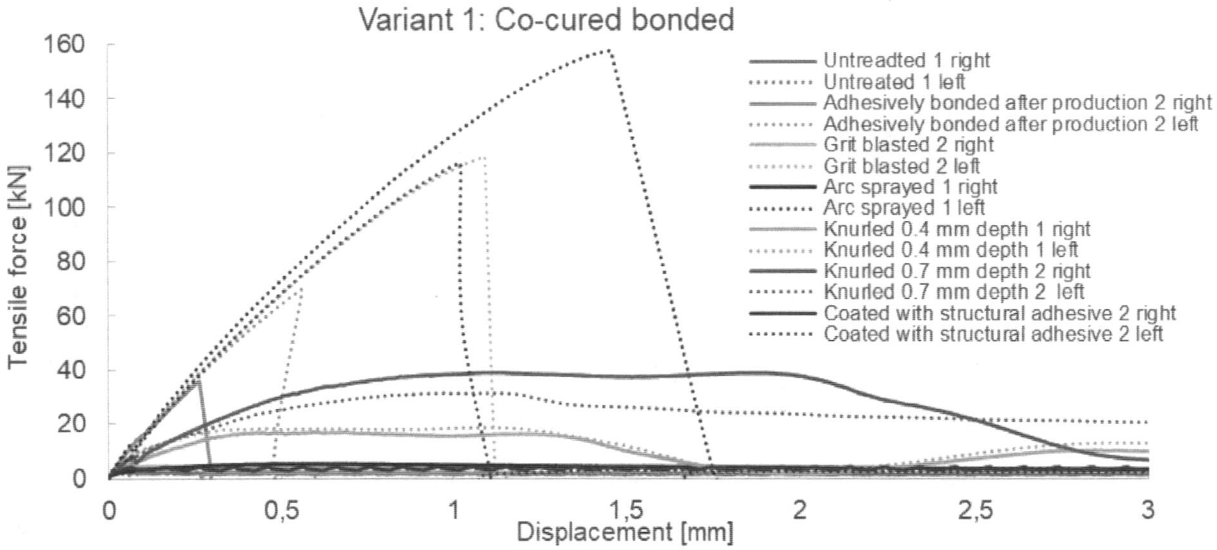

Figure 4. Force-displacement diagram of variant 1 with co-cured bonding specimens. The data displays a selection of the specimens and is classified into first and second specimen as well as left and right mold side.

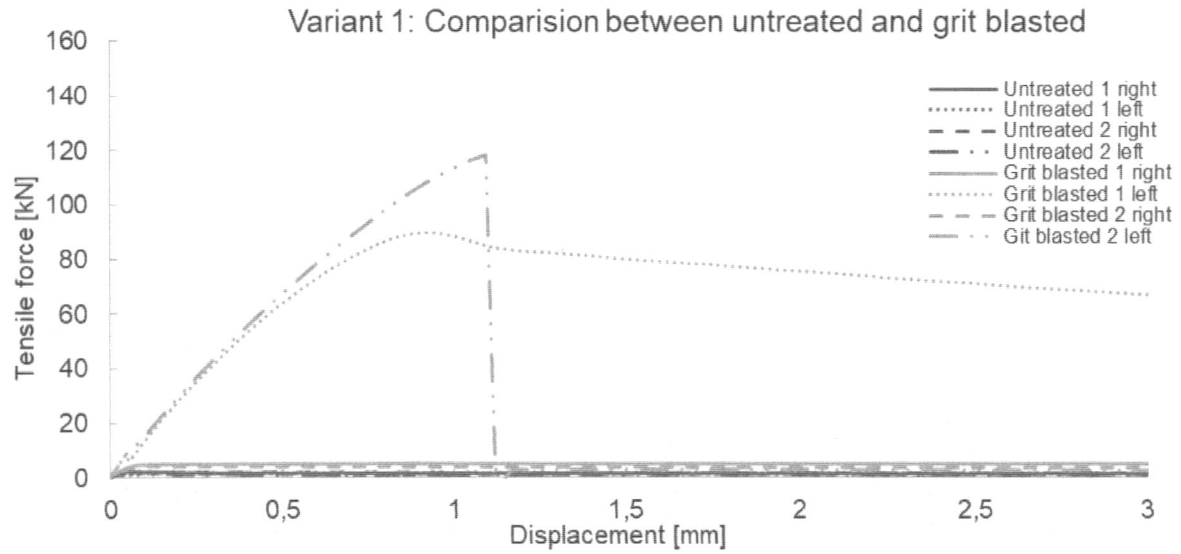

Figure 5. Force-displacement comparison between untreated and grit-blasted specimens for all tested parts

The knurled specimens are of moderate strength, whereby a positive influence can be observed with increasing knurling depth. As opposed to the other components, no abrupt failure occurs but a successive failure with a high energy absorption. By means of knurling, a meso-form fit is created, which is responsible for the energy absorption. Hence, the knurled components are suitable for use in crash-relevant structures. Figure 6 summarizes the maximum tensile forces with the corresponding joint strength and the standard deviation. The joint strength is the shear stress which is related to the surface area of the metallic load introduction elements, taking into account the ventilation holes. The best specimen with an arc spray treatment fails at a force of 157 kN, resulting in a maximum joint strength of approximately 34 MPa.

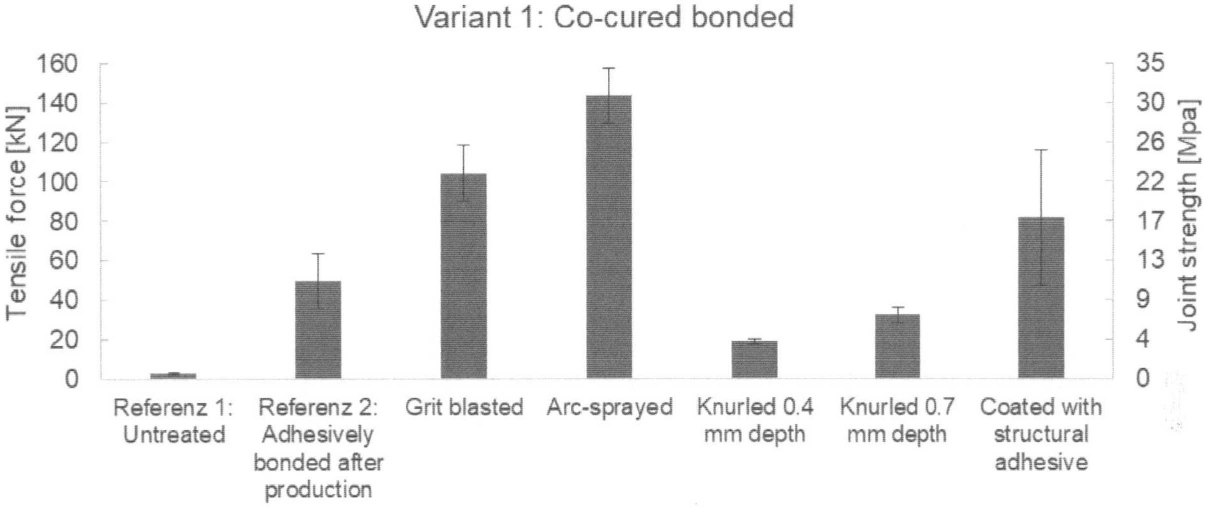

Figure 6. Bar chart with maximum tensile force and joint strength of variant 1 specimens with co-cured bonding

In contrast to variant 1 with a co-cured bonded joint, the variant 2 specimens show a combination of a form fit and co-cured bonding. Figure 7 displays the quasistatic tensile test results of this variant. In general, the strength values are lower than the ones of variant 1.

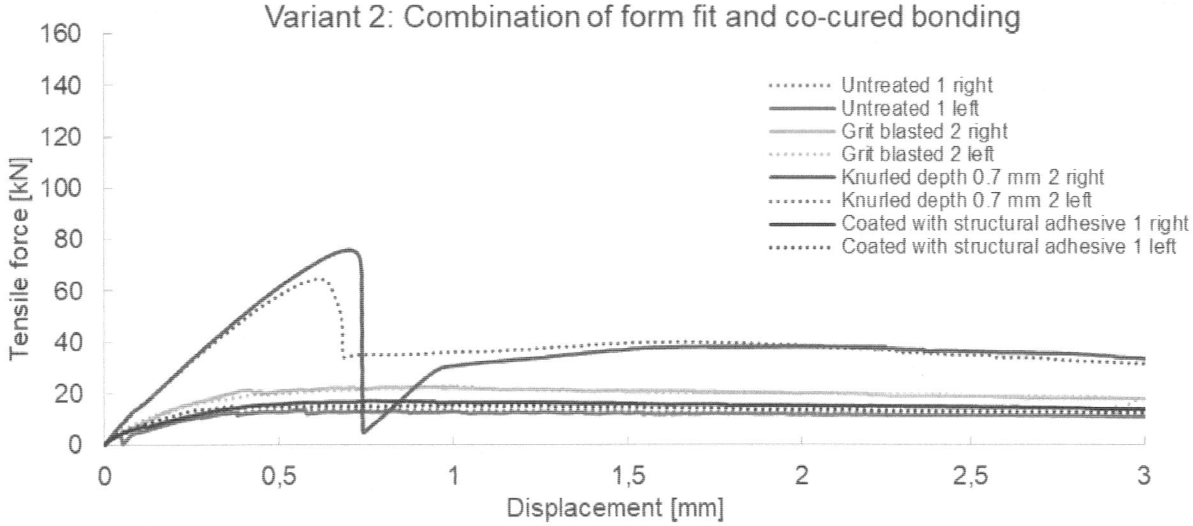

Figure 7. Force-displacement diagram of variant 2 with combination of form fit and co-cured bonding

Even mechanical treatments such as grit blasting or coating with structural adhesive do not generate any significant added value. A clear advantage is only achieved with knurling on the mesoscale. Similar to the knurled specimens of variant 1, all specimens of variant 2 contain energy absorption in the failure process, which is much higher than the energy absorption of variant 1. This energy absorption can be attributed to the form fit. The comparison of variants 1 and 2 reveals that high joint strengths can be achieved by co-cured bonding with appropriate treatment, whereas the use of form-fit geometries can lead to great energy absorption in the event of failure, which can be relevant for crash applications.

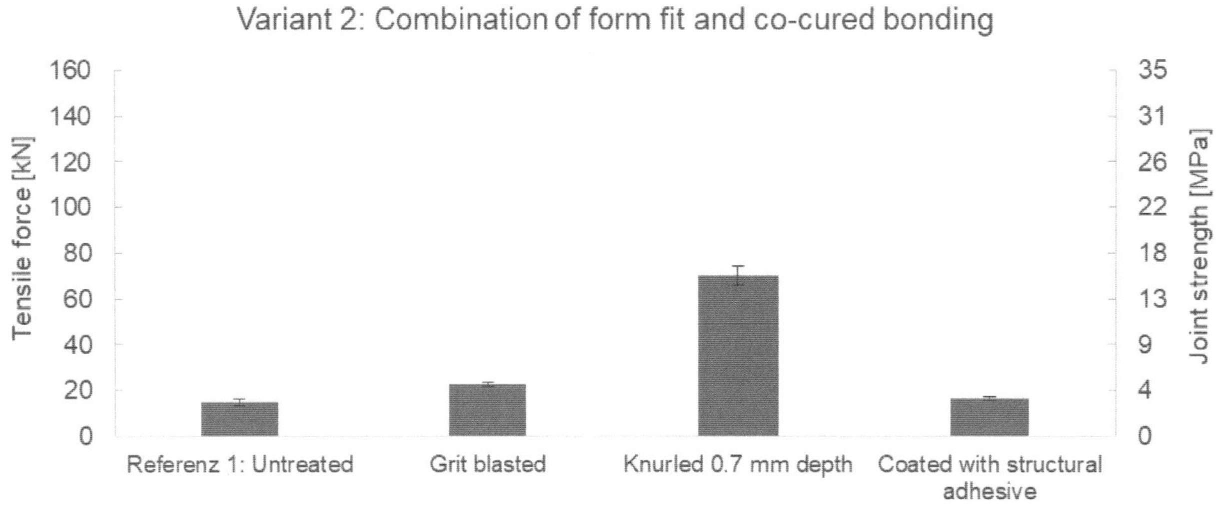

Figure 8. Bar chart with maximum tensile force and joint strength of variant 2 specimens

3.2 Microsections of the Interface

Microscopic analyses are carried out in order to analyze the knurled interface between the CFRP and the metal more precisely (see Figure 9). It can be concluded that the individual fiber bundles do not completely fit into the knurled structure. Furthermore, a pure matrix area between the laminate and the metal is visible. In addition, a wrinkle of the carbon fiber preform is formed in the mold parting line, as shown in the picture on the left. In future experiments, a preform with smaller fiber bundles should be used to allow them to be fully embedded in the knurled structure. In addition, more attention must be paid to the preform's production to avoid wrinkles when closing the two halves of the mold.

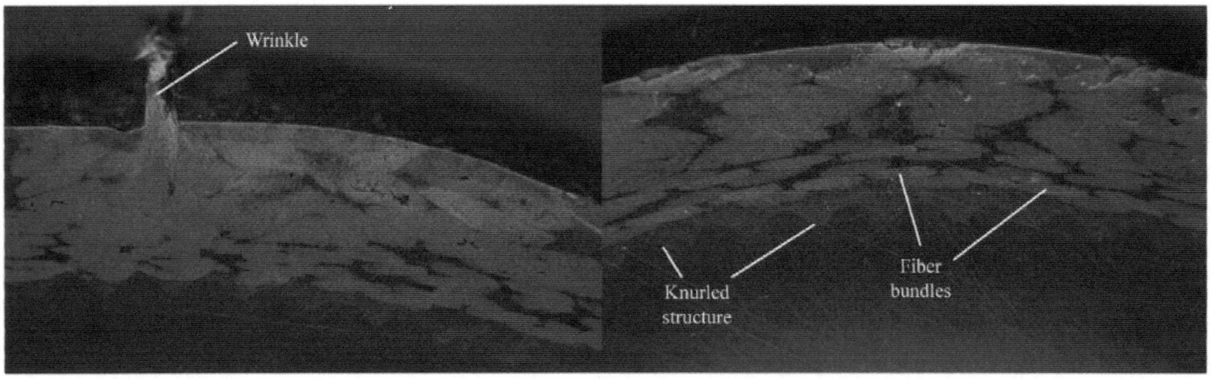

Figure 9. Microsections of the knurled interface between CFRP and metal

4. CONCLUSIONS AND OUTLOOK

In this paper, the intrinsic manufacturing of a hybrid composite tie rod with rotational molding has been presented. Different mechanical surface treatments and coatings have been compared with one other on the basis of different specimen geometries. In general, the results have shown that higher joint strengths can be achieved by appropriate treatments. For components with co-cured bonding, mechanical treatments that increase the surface roughness, such as arc spraying, turn out to be suitable. The failure of these specimens occurs abruptly without any energy absorption after the first failure. In order to avoid parallel and angular offsets, the mold for this intrinsic technology must be produced highly accurately. In contrast to co-cured bonded test specimens, samples with a combination of a form fit and co-cured bonding show a moderate joint strength but good energy absorption. A surface treatment with a structure similar to a knurled interface is particularly well suited for these types of components. The energy absorption behavior during failure qualifies these components for crash applications. However, microsections reveal a pure matrix area between the CRFP and the metal.

The achieved results show a reasonable joint strength for the chosen treatments. In future works, numerical simulations have to be conducted to optimize the joint geometry. In addition, the precision of the mold shape must be increased, and the filament bundles should integrate more into the knurled surface to further increase joint strength. Finally, fast curing matrix systems with heat supply should be used in order to significantly reduce production times.

5. ACKNOWLEDGEMENTS

This paper is based on investigations of the collaborative research program "Schwerpunktprogramm 1712" which has been kindly supported by the German Research Foundation (DFG – Deutsche Forschungsgemeinschaft) and the project "FaMeZug" which has been funded by the Federal Ministry of Economic Affairs and Energy. The authors wish to thank Philipp Eyer and Pauline Pagès for their support during production and test execution. Furthermore, the authors wish to express their gratitude to Siltex Flecht- und Isoliertechnologie Holzmüller GmbH & Co.KG for providing the braided sleeves and the Institute for Applied Materials (KIT) for granting the opportunity to use their testing machines.

6. REFERENCES

1. Barfuss D, Grützner R, Hirsch F et al. "Multi-scale structuring for thermoplastic-metal contour joints of hollow profiles." *Prod. Eng. Res. Devel.* 12(2) (2018): 229–238. doi: 10.1007/s11740-018-0800-9
2. Ehleben M, Schürmann H "Manufacturing of centrifuged continuous fibre-reinforced precision pipes with thermoplastic matrix." *Composites Science and Technology* 66(15) (2006): 2601–2609. doi: 10.1016/j.compscitech.2006.03.015
3. Fleischer J, Ochs A, Dosch S "The future of lightweight manufacturing—production-related challenges when hybridizing metals and continuous fiber-reinforced plastics. In: Proceedings of the International Conference on New Developments in Sheet Metal Forming, Stuttgart, Germany, pp 22–23
4. Fleischer J, Koch S-F, Coutandin S "Manufacturing of polygon fiber reinforced plastic profiles by rotational molding and intrinsic hybridization." *Prod. Eng. Res. Devel.* 9(3) (2015): 317–328. doi: 10.1007/s11740-015-0620-0

5. Gebhardt J, Fleischer J "Experimental investigation and performance enhancement of inserts in composite parts." *Procedia CIRP* 23 (2014): 7–12. doi: 10.1016/j.procir.2014.10.084
6. Gerhard T, Friedrich C "Mechanical fastening of carbon composite tubes, numerical calculation of axial loading capacity and experimental verification." *Composites Part B: Engineering* 67 (2014): 391–399. doi: 10.1016/j.compositesb.2014.07.024
7. Hufenbach W, Gude M, Helms O et al. "Novel load-introduction systems for lightweight aircraft components in composite-metal-hybrid design (2008)
8. Jegley DC, Wu KC, Phelps JE et al. "Structural efficiency of composite struts for aerospace applications." *Journal of Spacecraft and Rockets* 49(5) (2012): 915–924. doi: 10.2514/1.60053
9. Kim KS, Kim WT, Jun EJ "Optimal tubular adhesive-bonded lap joint of the carbon fiber epoxy composite shaft." *Composite structures* 21(3) (1992): 163–176. doi: 10.1016/0263-8223(92)90016-6
10. Koch S-F "Fügen von Metall-Faserverbund-Hybridwelle im Schleuderverfahren: ein Beitrag zur fertigungsgerechten intrinsischen Hybridisierung". Shaker Verlag (2017)
11. Meyer R "Handbook of pultrusion technology". Springer Science & Business Media (2012)
12. Michaeli W, Jürss D "Thermoplastic pull-braiding: pultrusion of profiles with braided fibre lay-up and thermoplastic matrix system (PP)." *Composites Part A: Applied Science and Manufacturing* 27(1) (1996): 3–7. doi: 10.1016/1359-835X(95)00004-L
13. Molitor P, Barron V, Young T "Surface treatment of titanium for adhesive bonding to polymer composites: a review." *International Journal of Adhesion and Adhesives* 21(2) (2001): 129–136. doi: 10.1016/S0143-7496(00)00044-0
14. Nieschlag J, Ruhland P, Coutandin S et al. "Rotational Molding for the Production of Hybrid FRP Metal Tension and Compression Rods with Form Fit. In: Production at the leading edge of technology. Springer (2019), pp 131–138
15. Park SW, Kim HS "Optimum design of the co-cured double lap joint composed of aluminum and carbon epoxy composite." *Composite structures* 75(1-4) (2006): 289–297. doi: 10.1016/j.compstruct.2006.04.031
16. Peters ST "Composite filament winding". ASM International (2011)
17. Schütze R "Lightweight carbon fibre rods and truss structures." *Materials & design* 18(4-6) (1997): 231–238. doi: 10.1016/S0261-3069(97)00056-3
18. T Fischer HS (ed) "Der Pressverband als Krafteinleitungs-Element in Faser-Kunststoff-Verbund-Wellen. 11 (2005)
19. Ucsnik S, Scheerer M, Zaremba S et al. "Experimental investigation of a novel hybrid metal–composite joining technology." *Composites Part A: Applied Science and Manufacturing* 41(3) (2010): 369–374. doi: 10.1016/j.compositesa.2009.11.003
20. W Brewster J "Design and Manufacture of Structurally Efficient Tapered Struts (2010)
21. Welsch M, Funck R "Development of Lightweight–Control Rod made of Fibre Composite Material for Aviation"
22. Zinn C, Bobbert M, Dammann C et al. "Shear strength and failure behaviour of laser nano-structured and conventionally pre-treated interfaces in intrinsically manufactured CFRP-steel hybrids." *Composites Part B: Engineering* 151 (2018): 173–185. doi: 10.1016/j.compositesb.2018.05.030

EFFECTS OF SURFACE TREATMENT OF ALUMINA NANOFIBERS ON MECHANICAL AND THERMAL PROPERTIES OF POLYESTER E-GLASS COMPOSITES

Harish S.P. Kallagunta[1,2], Jitendra S. Tate[1,2,*], Gary W. Beall[1,3], Subash Panta[2], Maria Beldeque[2], Muzaffer Akanda[2], Rafiqul Islam[2]

Texas State University 601 University Drive, San Marcos, TX, 78666
[1]Materials Science, Engineering and Commercialization Program, Texas State University
[2]Ingram School of Engineering, Texas State University
[3]Department of Chemistry and Biochemistry, Texas State University
JT31@txstate.edu; 512-245-1826

ABSTRACT:

Dispersibility of the nanoparticles in the matrix is one of the critical factors influencing various properties of nanocomposites. To address the issue of agglomeration in nanocomposites and negate the effects on properties, surface treatment of nanoparticles is seen as the viable solution. This study is focused on the influence of alumina nanofiber (Nafen™) loadings and surface treatment of the alumina nanofibers (Nafen™) on thermal and mechanical properties of polyester/E-Glass composites. Loading levels of 0.25wt%, 0.5wt%,0.75wt% and 1wt% for both untreated alumina nanofibers and silane surface treated alumina nanofibers were selected and are dispersed into polyester resin using three-roll mill. Transmission Electron Microscopy (TEM) revealed that at higher loading levels, the alumina nanofibers tend to agglomerate. Thermogravimetric analysis was performed on the unreinforced nanocomposite coupons and the increase in thermal degradation behavior with addition of nanoparticles. Silane treatment of alumina nanofibers have seemed to improve the lower onset degradation temperature. Static mechanical studies were performed. The mechanical behavior improved with addition of alumina nanofibers. Major improvement in properties is observed at 0.25wt% and 1wt% silane treated alumina nanofiber modified glass polymer composites.

1. INTRODUCTION

Glass fiber composites have become cost-effective solution in the industrial composite market against high cost carbon fiber composites which are limited to aerospace applications. Unsaturated polyester resin infused glass fiber composite due to its good mechanical performance and low cost have gained prominence. Addition of nanoparticles to these composites have proven to be more effective in enhancing the mechanical performance without drastically increasing the cost. It was noticed that ceramic nanoparticle inclusion into thermosetting polymers improves the resistance of composites to wear, self-cleaning, self-healing and good adhesive property [1, 2]. Ionic and covalent bonding characteristic of ceramic nanoparticles offers good thermal conductivity, electrical resistivity and low dielectric constant, which are major concerns for miniaturization and high computing speed [3-5]. Nanoscale precipitation and point defects introduction through incorporation of nanoparticles can provide excellent thermal insulation properties which find applications in subsea, construction and piping industry[6, 7]. The nano-porosity and electrostatic attraction offered by alumina nanofibers with high surface areas is seen as a great potential for water treatment in order to reduce the contaminants through adsorption. However, much research is needed in order to assess its impact on environment and human health [8, 9]. Alumina nanofibers can find applications where high thermal stability is desired. The high thermal stability often can lead to improved fire resistance and char formation offering thermal protection to composites at affordable cost.

Glass fiber reinforced polymer nanocomposites have been researched for a variety of applications that require impact resistance, flame retardancy, and improved fatigue performance. Research utilizing different nanoparticles to enhance the mechanical properties have been of focus in designing for improved fracture toughness. Silica nanoparticles are studied for their effectiveness in increasing the mechanical properties when added to the glass fiber reinforced epoxy composites and are relatively inexpensive, nontoxic, biocompatible, highly thermal resistant and effective as mechanical reinforcements[10]. In another study Tate et al [11] investigated the influence of 6, 7 and 8wt% nanosilica on glass fiber reinforced epoxy composites in static mechanical and fatigue loading. It was observed that 6wt% nanosilica loading has greater effect on tensile strength, percentage elongation and interlaminar shear strength. 6wt% loaded nanosilica epoxy glass fiber composites showed a modulus loss of 45% as compared to 65% modulus loss of neat specimen. In another study by [12],[13] 20wt% and 25wt% loading of nanosilica, the Youngs modulus showed an increase of 23% and 78% respectively. Rostamiyan's research group in their study compared the effects of Nano-clay and Nano-silica on flexural properties. The group found that the hybridization of glass fiber with epoxy, nanosilica and nanoclay showed promising results with little variance [14]. Addition of other micro particles along with the nanomaterials has also been studied to evaluate their effect on the impact strength and interfacial adhesion of the nanocomposites. Tate et al, conducted experiments on glass fiber epoxy composites modified with 6.5wt% rubber microparticles in combination with 8.1wt% Nano-silica and reported an effective impact energy dissipation to surrounding area [15]. Nano alumina with particle morphology are also studied by various researchers. Zheng et al [16] reported that enhancement in impact strength by 84% is observed with addition of 20wt% alumina nanoparticle loading. Afrouzian et al [17] studied the effects of nanosilica on the tensile, flexural and ballistic impact properties of glass fiber epoxy composite laminates and found significant effect on mechanical behavior of the composite laminates under study. The virgin specimens and composites modified with loadings of 0.5wt%, 1wt% and 3wt% of nanosilica are compared to each other and lowest loading of nanosilica is found to outperform the remaining by 47%. Alumina nanofibers which are relatively new form of nanomaterials has very limited studies performed on their effectiveness in strengthening the polymer matrix for fiber reinforced composite materials. Some of the studies conducted on alumina nanofibers are: Mechanical, thermal stability and flammability characteristics of PA6 thermoplastics modified with alumina nanofibers by Wu et al.,[18] reported a significant effect of alumina nanofiber modification on the mechanical properties although there is no change in thermal and flammability properties. Rakitin.A and Grodensky [19], reported an increase in mechanical property by 72% by adding alumina nanofibers to the epoxy resin. Previous studies have indicated positive effect of alumina nanofiber in glass fiber epoxy composites under low impact loading at 2.5wt%[20]. Silane surface treatment using chloro-silane proved ineffective and inhibited ballistic impact property at 0.25wt% and 0.5wt% of silane treated alumina nanofiber loading[21].

The objective of the present study is to evaluate the mechanical performance of alumina nanofiber modification of unsaturated polyester reinforced with woven E-glass fibers. In addition to low density and fiber morphology which offers the benefits of tailoring the aspect ratio, it has been considered economical than other forms of ceramic nanoparticles in terms of manufacturability and monetary cost. Alumina nanofiber at four different loading levels (0.25wt%, 0.5wt%, 0.75wt%

and 1wt%) is studied to evaluate the variation in static properties. The effect of surface treatment of alumina nanofibers on the thermal stability and dispersion is studied along with its effect on the fracture behavior of the glass polyester composite material system.

2. MATERIALS AND MANUFACTURING

2.1 Material System

The material system used for this study comprises of 0/90 woven E-Glass fabric from Owens corning having GSM 888.0 as the reinforcement and unsaturated polyester resin SIL47DA-2949 with a viscosity of 396 cP from Interplastic Corporation as the matrix. Alumina nanofibers with minimal aspect ratio 18-20 under trade name Nafen^{TM} from ANF Technology are used as nanofillers. We use BPO as the curing agent with heat as an external stimulator. One set of composites were modified with untreated alumina nanofibers and the second set were modified with vinyl silane coupling agent treated alumina nanofibers. Loading levels of 0.25wt%, 0.5wt%, 0.75wt% and 1wt% were used for both sets of composites laminates. The glass polyester composites modified with untreated alumina nanofibers are here from referred to as UT-ANF/GPC and the composites modified with vinyl silane treated alumina nanofibers as VT-ANF/GPC.

2.2 Fabrication of Laminates

2.2.1 Dispersion using calendaring approach

Three roll milling was used for dispersing Nafen^{TM} into the polyester resin. In every pass, by adjusting the gap setting between the rollers a high shear force is achieved to reduce the clumps and clusters formed. The gap between the rollers is set at 25μm, 15μm, and 7.5μm respectively for three passes. Vinyl silane coupling agent treated alumina nanofibers are also dispersed using the same approach.

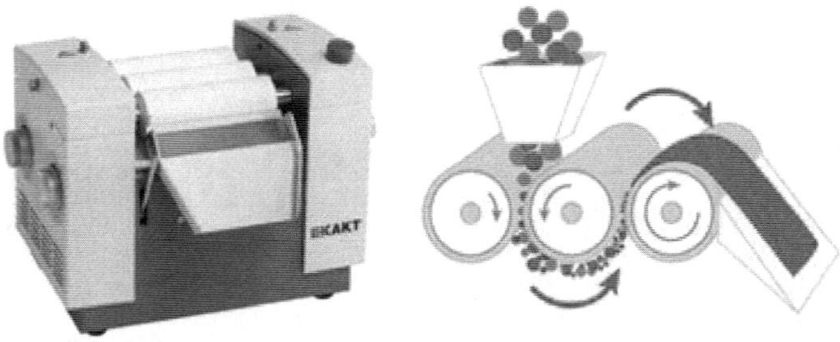

Figure 1: Three roll mill and illustration of dispersion using calendaring approach (Courtesy: www.exaktusa.com)

2.2.2 Fabrication of composite laminates and sample extraction

The E-glass fabrics were passed through a resin bath which allowed uniform wetting of the fabric. Fourteen layers were stacked and consolidated by using a pressure of 1.43 MPa and a platen heat of 82°C in the hydraulic compression press. The laminate was then post-cured for three hours at 110 °C in the programmable oven. Samples are cut according to the ASTM standards using a water jet to prevent edge delamination.

3. EXPERIMENTATION

3.1 TEM Imaging

In order to evaluate the dispersion state of alumina nanofibers in the polyester, Transmission Electron Microscopy was performed on the polymer coupons. The thin sections of polymer modified with alumina nanofiber were obtained using ultramicrotome which were placed on copper grids. The grids were then viewed in a high-resolution transmission electron microscope.

3.2 Thermogravimetric (TGA) Analysis

TGA analysis was performed to at 10°C/min ramp rate from room temperature up to 600°C in nitrogen atmosphere using a Discovery SDT650 from TA instruments. An air flow rate of 50mL/min was used.

3.3 Static Mechanical Testing

The static testing on the untreated and vinyl treated Alumina nanofiber (ANF) modified Glass Polyester Composites (GPC) is performed using MTS 810 Servo Hydraulic System. Table 1 gives an illustration of the static tests performed on the Glass Polyester Composites (GPC) modified with untreated and surface treated Alumina nanofibers (ANF).

Table 1: Static test standards and parameters performed on ANF-GPC

ASTM Standard	Property	Sample Dimensions	Loading Rate
D 3039	Tension	10in x 1in x 't'	2 mm/min
D 7264	Flexure	5.5in x 0.5in x 't'	1 mm/min
D 2344	Short Beam Shear Test	6t x 2t x t	1mm/min

3.4 SEM Imaging

SEM imaging was performed using Helios Nanolab 400 to analyze the failure modes at 10.0kV and a current range between 0.69amp and 2.7amp. The samples were previously treated with EMS150T ES sputter coater to add a layer of carbon of at least 20nm.

4. RESULTS AND DISCUSSION

4.1 TEM Analysis

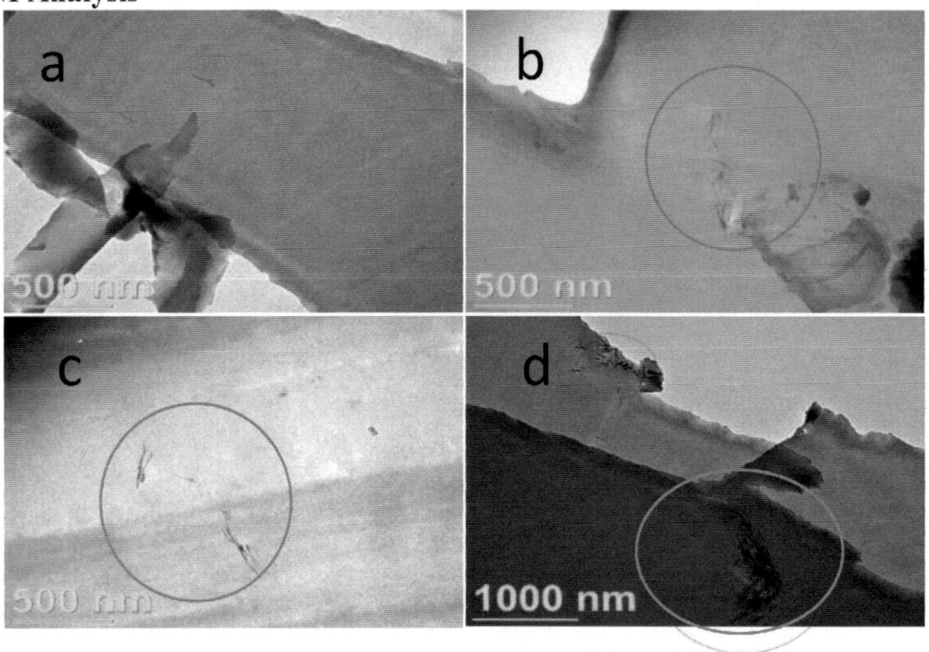

Figure 2: TEM images showing dispersion of Untreated ANF in UPE at loading levels of (a) 0.25wt%, (b) 0.5wt%, (c) 0.75wt% and (d) 1wt%

The TEM images revealed agglomeration of ANF nanoparticles in UPE at higher loading levels, whereas a decent quality of dispersion of ANF nanoparticles in lower loadings is observed. It was observed that ANF nanoparticles retained their fiber like morphology even after subjecting to high shear forces via three roll milling. Although individual ANF nanoparticles were visible, it was observed that the fibers are not completely de-bundled and distributed. The lack of de-bundling might be because of absence of any surface treatment. It would be interesting to see the effect of surface treatment of ANF on the de-bundling and distribution aspect of the nanoparticles during dispersion.

4.2 Thermogravimetric (TGA) Analysis
The TGA results on the formulations are reported in fig 3 and the results are as expected from addition of ANF. As compared to neat polyester, ANF modified polyester showed increase in thermal stability. The weight loss from the TGA curves of the vinyl silane surface treated is more than the untreated silane surface treated, as this is due to the combustion of the silane agent. The decomposition temperatures calculated from the thermographs are summarized in table 2. With the increase of loading content, the peak mass loss temperature increased. The lower onset degradation temperature was enhanced with surface treatment of alumina nanofiber due to promoted bonding between polyester resin and VT-ANF. However, the degradation temperature at higher temperatures as indicated by 50wt% mass loss temperature is common in polymer systems.

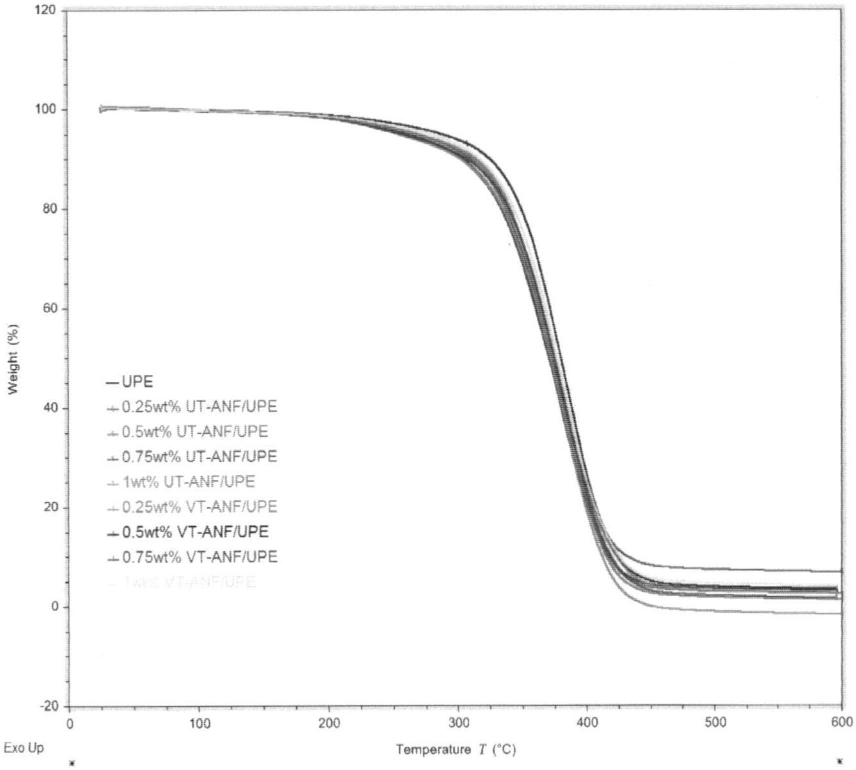

Figure 3: TGA thermographs showing dispersion of Untreated ANF in UPE at loading levels of (a) 0.25wt%, (b) 0.5wt%, (c) 0.75wt% and (d) 1wt%

Table 2: Summary of decomposition temperatures and residual mass

Sample	T_{dec} 10wt%		T_{dec} 50wt%		Peak mass loss temperature	
	°C		°C		°C	
	UT-ANF/UPE	VT-ANF/UPE	UT-ANF/UPE	VT-ANF/UPE	UT-ANF/UPE	VT-ANF/UPE
0wt%	305.30		370.24		374.65	
0.25wt%	309.94	315.22	372.35	371.92	382.03	373.05
0.50wt%	313.37	327.12	374.09	380.05	384.71	393.15
0.75wt%	303.32	308.61	372.11	374.76	381.37	393.84
1.00wt%	307.29	319.20	370.91	376.83	385.44	392.28

4.3 Static Mechanical Properties

A summary of mechanical test results on the glass polyester composites modified with untreated and vinyl silane treated ANF tabulated in table 3 and the percentage improvement in properties is shown in table 4. The ANF modification proven to enhance the mechanical properties even if it is untreated or vinyl silane surface treated. It seemed the addition of silane treated alumina nanofibers to the glass polyester composites had shown no greater change and the improvement is consistent at all the loading levels. However, with the untreated alumina nanofiber modified composites there seem to be a trend. The tensile properties improved at 0.25wt%, seen a decrement at 0.5wt% and 0.75wt% and then improved at 1wt%. Similar behavior is observed in flexure properties.

As observed from the percentage improvement in properties as summarized in table 4, the maximum of improvement in properties is seen in vinyl silane treated alumina nanofibers modified glass polyester composites, with all the properties peaking at 0.25wt% and 1wt% loading level of VT-ANF/GPC.

Table 3: Results of static mechanical test of ANF/GPC (Standard deviations are in parenthesis)

Tests	Control	0.25 wt%		0.5wt%		0.75wt%		1wt%	
		UT	VT	UT	VT	UT	VT	UT	VT
Tensile Strength (MPa)	410.82 (28.51)	521.32 (19.11)	507.45 (11.88)	506.25 (14.88)	508.23 (8.18)	453.18 (16.02)	520.35 (0.223)	524.60 (5.56)	520.44 (7.26)
Tensile Modulus (GPa)	18.62 (2.86)	20.64 (0.54)	22.69 (1.51)	17.52 (0.64)	25.25 (2.29)	18.87 (1.41)	23.24 (1.92)	21.78 (0.32)	24.88 (1.57)
% Elongation	6.72 (0.50)	4.96 (0.41)	3.03 (0.79)	4.09 (0.18)	2.90 (1.08)	4.63 (0.22)	1.86 (0.14)	5.43 (0.73)	5.50 (1.98)
Equilibrium Toughness (MJ/m^3)	7.08	10.29	10.22	10.02	10.07	8.70	10.44	10.24	9.95
Flexure Strength (MPa)	112.52 (17.14)	117.11 (14.91)	126.18 (13.05)	93.67 (3.44)	119.45 (3.10)	115.51 (12.86)	121.60 (12.25)	133.53 (14.87)	129.69 (3.47)

Flexure Modulus (GPa)	22.82 (7.59)	24.46 (2.36)	28.02 (4.98)	21.25 (1.05)	26.20 (1.88)	23.21 (1.51)	27.98 (4.05)	29.24 (1.56)	30.98 (2.79)
Short Beam Strength (MPa)	12.61 (0.85)	13.40 (1.29)	14.93 (2.84)	10.45 (0.85)	11.94 (1.08)	12.69 (2.16)	10.72 (1.85)	10.66 (1.37)	13.68 (2.63)

Table 4: Percentage improvement in static mechanical properties of ANF/GPC

Tests	Control	0.25 wt%		0.5wt%		0.75wt%		1wt%	
		UT	VT	UT	VT	UT	VT	UT	VT
Tensile Strength	-	26.90	23.52	23.23	23.71	10.31	26.66	27.70	26.88
Tensile Modulus	-	10.85	21.86	-5.91	35.61	1.34	24.81	16.97	33.62
% Elongation	-	-26.19	-54.91	-39.14	-56.85	-31.10	-72.32	-19.20	-18.15
Equilibrium Toughness	-	45.34	44.35	41.53	42.23	22.88	47.76	44.63	40.54
Flexure Strength	-	4.08	12.14	-16.5	6.16	2.66	8.07	18.67	15.26
Flexure Modulus	-	7.19	22.79	-6.88	14.81	1.71	22.61	28.13	35.76
ILSS	-	6.26	18.40	-17.13	-5.31	0.63	-14.99	-15.46	8.49

The tensile strength properties of both sets of composites as compared to control composites showed an improvement of approximately 20%. However, the tensile modulus of VT-ANF/GPC set of composites are significantly higher than the UT-ANF/GPC. The equilibrium toughness, which is a measure of the work done on the sample before failure is at an average of 43% for both set of composites. The major change associated with the silane treatment of the alumina nanofibers is the interlaminar shear strength, which peaked at 0.25wt% loading level. It was observed that the VT-ANF/GPC due to the presence silane-coupling agent contributed to higher bonding strengths and improved ILSS property by 18.40%.

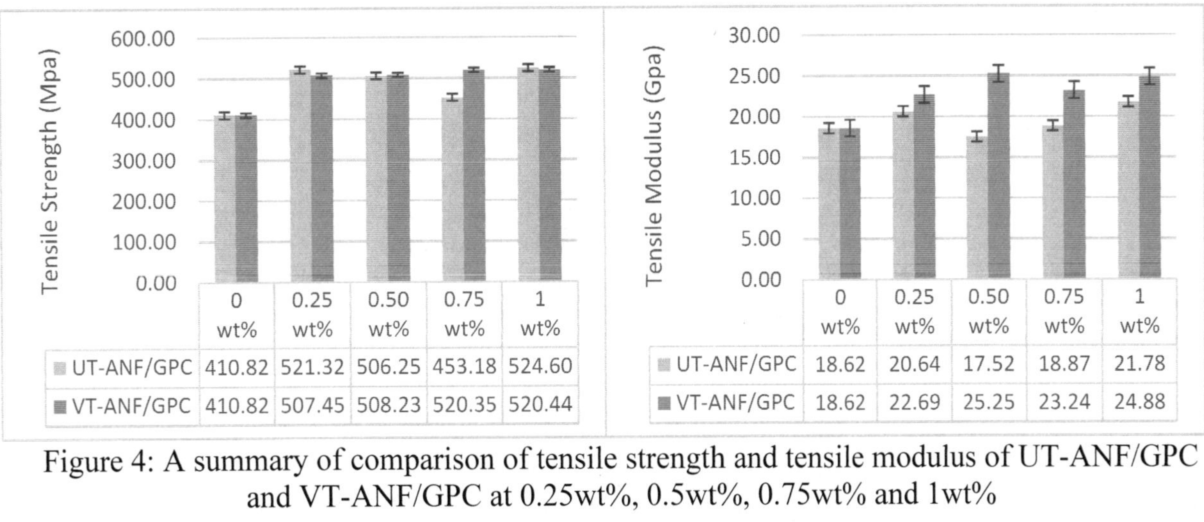

Figure 4: A summary of comparison of tensile strength and tensile modulus of UT-ANF/GPC and VT-ANF/GPC at 0.25wt%, 0.5wt%, 0.75wt% and 1wt%

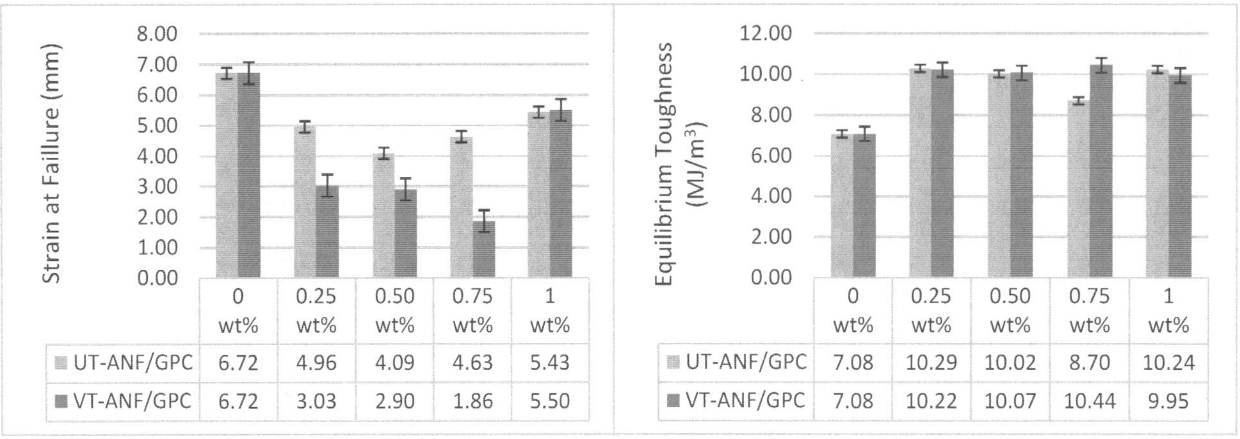

Figure 5: A summary of comparison of strain at failure and equilibrium toughness of UT-ANF/GPC and VT-ANF/GPC at 0.25wt%, 0.5wt%, 0.75wt% and 1wt%

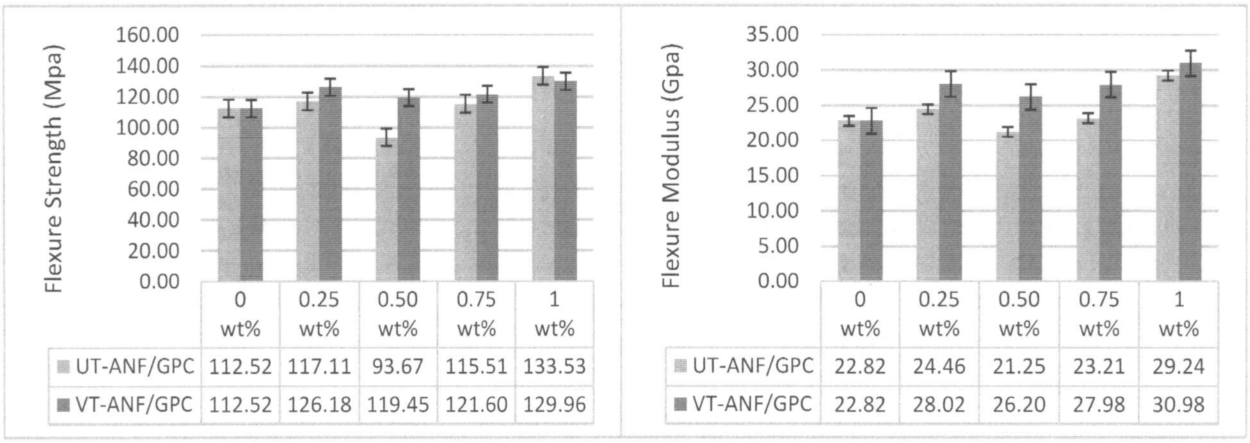

Figure 6: A summary of comparison of Flexure Strength and Flexure Modulus of UT-ANF/GPC and VT-ANF/GPC at 0.25wt%, 0.5wt%, 0.75wt% and 1wt%

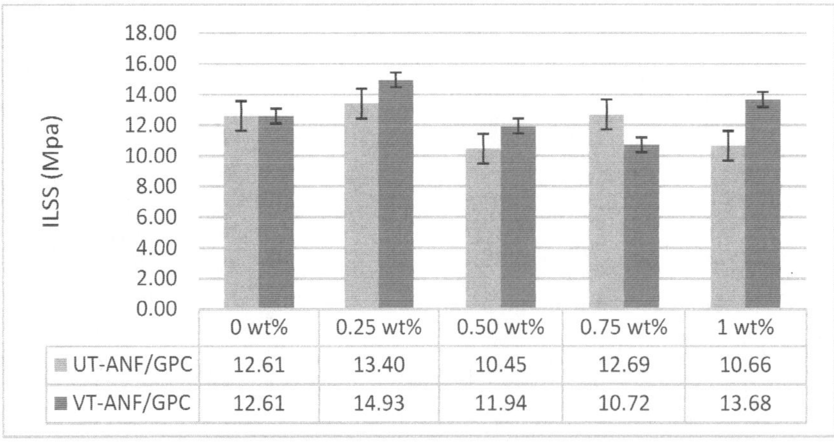

Figure 7: A comparison of ILSS of UT-ANF/GPC and VT-ANF/GPC at 0.25wt%, 0.5wt%, 0.75wt% and 1wt%

4.4 SEM Analysis

Figure 8 and Figure 9 shows the cross-sectional SEM images of samples after testing to understand the damage induced. The UT-ANF/GPC composite SEM images are shown in Figure 7, and they reveal fiber matrix debonding, fiber breakage, delamination, and cracks which are quite expected as it is a complete failure. Although the failure and damage are clearly visible, the interaction of UT-ANF modified polyester with the glass fibers cannot be clearly established. In case of VT-ANF/GPC, more of fiber pull out is prevalent and the fiber matrix debonding is less as compared to UT-ANF/GPC. AS for the short beam specimens considered here, the degree of bonding in each sample varied.

The morphology of fracture surfaces as seen in the figures 8 and 9 revealed that the composites are behaving more brittle than ductile which is expected from the glass reinforced polyester composites which is as seen from the morphologies of the fiber breakage. The surface topography of the silane treated alumina nanofiber filled composites showed smooth surfaces, with little fiber breakages and cleavages showing fiber pullout. The amount of resin still bonded to the fibers as seen in figure 9, reveals the enhancement in fiber adhesion to the matrix. In case of both sets of composites, the fiber is visible after the failure, meaning that there is uneven coating of fibers during hand layup. The SEM images reveal that failure at the interface is predominant than the rupture of the fibers.

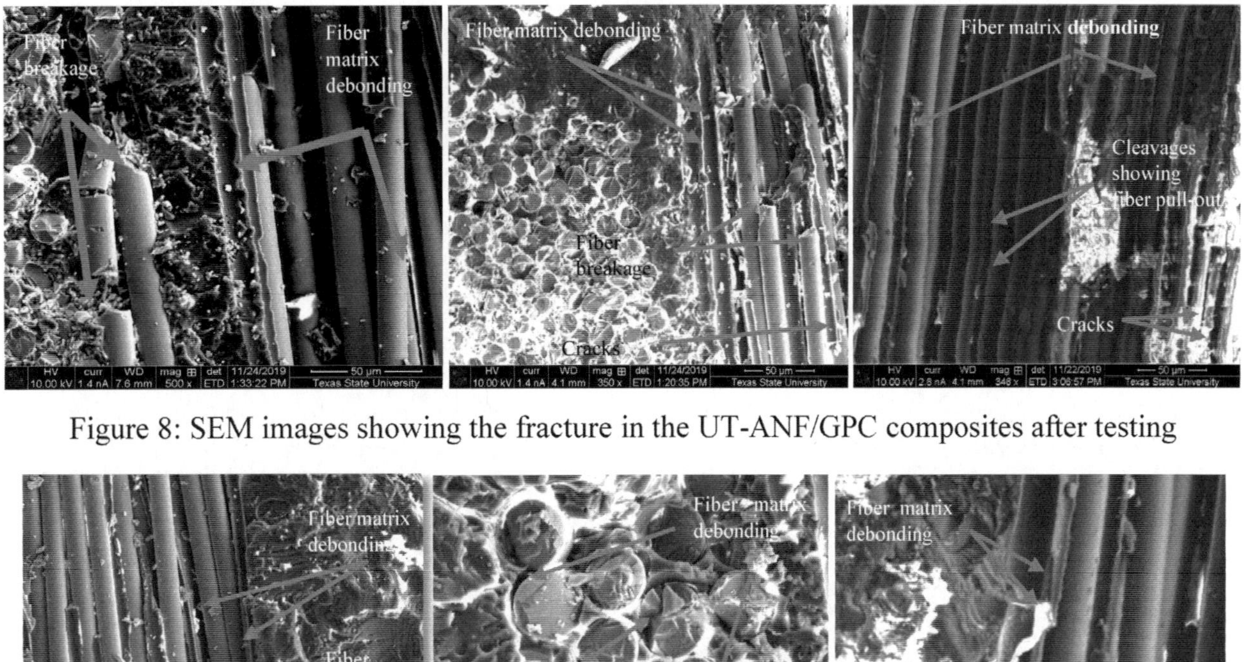

Figure 8: SEM images showing the fracture in the UT-ANF/GPC composites after testing

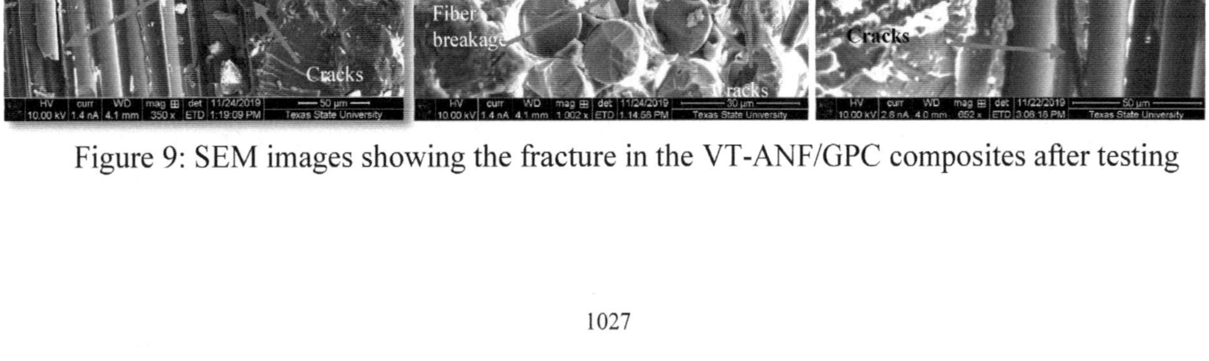

Figure 9: SEM images showing the fracture in the VT-ANF/GPC composites after testing

CONCLUSION

Transmission electron microscopy, thermo gravimetry analysis, tensile, flexure, interlaminar shear stress and scanning electron microscopy on the glass epoxy composites modified with untreated as received alumina nanofibers and silane treated alumina nanofibers is studied and compared. Observations in terms of dispersion, thermal stability, mechanical performance and fracture analysis on tested samples are reported in this article. TEM Images showed agglomeration at higher loadings but decent dispersion at lower loading. TGA thermograms revealed the improvement in thermal stability with increased loading level as measured by the peak mass loss. Silane treated alumina nanofiber peak mass loss temperature is at the maximum. The property improvement is obvious in alumina nanofibers. However, the major of the property values enhanced at 0.25wt% and 1wt% silane treated alumina nanofiber modified GPC. Interlaminar shear stress which is an important measure for strength against delamination has seen an improvement by 18.40% at 0.25wt% loading level.

ACKNOWLEDGMENTS

The authors are grateful to Mr. Tim Ferland of ANF Technology, Mr. David Gerald of Interplastic Corporation of Hexion, Inc., and Mr. Luis Arroys of Owens Corning, for their kind support in this research. The authors highly appreciate the help of Dr Ray Cook, Jason Wagner, and Ruben Villarreal for waterjet cutting of test samples. The authors would also like to thank the ARSC at Texas State University for the training and support for TEM and SEM Imaging. The authors are thankful to members of Advanced Composites Laboratory for their support and help at various stages of testing.

REFERENCES

1. Askar, K.A. and K. Song, *Chapter 4 - Epoxy-Based Multifunctional Nanocomposites*, in *Polymer-Based Multifunctional Nanocomposites and Their Applications*, K. Song, C. Liu, and J.Z. Guo, Editors. 2019, Elsevier. p. 111-135.
2. Gilbert, E.N., B.S. Hayes, and J.C. Seferis, *Nano-Alumina Modified Epoxy Based Film Adhesives.* Polymer Engineering and Science, 2003. **43**(5): p. 1096-1104.
3. Ting, W., F. Tseng, and K. Chiou. *Advanced materials with low dielectric properties and highly thermal conductivity.* in *2017 International Conference on Electronics Packaging (ICEP)*. 2017.
4. Mallik, S., et al., *Investigation of thermal management materials for automotive electronic control units.* Applied Thermal Engineering, 2011. **31**(2-3): p. 355-362.
5. Li, J., et al., *Extremely Low Thermal Conductivity in Thermoelectric Ge0.55Pb0.45Te Solid Solutions via Se Substitution.* Chemistry of Materials, 2016. **28**(17): p. 6367-6373.
6. Papadopoulos, A.M., *State of the art in thermal insulation materials and aims for future developments.* Energy and Buildings, 2005. **37**(1): p. 77-86.
7. Wang, F., J.M. Zhang, and Q.G. Tang, *Research Progress on Thermal Insulation Materials.* Advanced Materials Research, 2012. **427**: p. 157-162.
8. Brown, J. and M.D. Sobsey, *Ceramic media amended with metal oxide for the capture of viruses in drinking water.* Environmental Technology, 2009. **30**(4): p. 379-391.
9. Lau, B., et al., *Removal of nano and microparticles by granular filter media coated with nanoporous aluminium oxide.* Water science and technology : a journal of the International Association on Water Pollution Research, 2004. **50**: p. 223-8.

10. Landowski, M., et al., *Impact damage in SiO2 nanoparticle enhanced epoxy – Carbon fibre composites.* Composites Part B: Engineering, 2017. **113**: p. 91-99.
11. Tate, J.S., et al., *Tension-tension fatigue performance and stiffness degradation of nanosilica-modified glass fiber-reinforced composites.* Journal of Composite Materials, 2018. **52**(6): p. 823-34.
12. Carolan, D., et al., *Toughening of epoxy-based hybrid nanocomposites.* Polymer, 2016. **97**: p. 179-190.
13. Kothmann, M.H., et al., *Fatigue crack propagation behaviour of epoxy resins modified with silica-nanoparticles.* Polymer, 2015. **60**: p. 157-163.
14. Rostamiyan, Y., et al., *Experimental and optimizing flexural strength of epoxy-based nanocomposite: Effect of using nano silica and nano clay by using response surface design methodology.* Materials & Design, 2015. **69**: p. 96-104.
15. J.S. Tate, E.T., Swapnil Gaikwad, Stephen Sprener, Isidro Rosas, and Matt Andrews, *Low-velocity impact studies on glass reinforced composites using rubber micro-particles and silica nano-particles modified epoxy resin.* Journal of Nanoscience, Nanoengineering, and Applications, 2013. **3**(1).
16. Zhang, J.-X., et al., *The Influence of High Content Nano-Al2O3 on the Properties of Epoxy Resin Composites AU - Zheng, Ya-Ping.* Polymer-Plastics Technology and Engineering, 2009. **48**(4): p. 384-388.
17. Afrouzian, A., et al., *Effect of nano-particles on the tensile, flexural and perforation properties of the glass/epoxy composites.* Journal of Reinforced Plastics and Composites, 2017. **36**(12): p. 900-16.
18. Hao Wu, M.K.a.J.H.K. *Functionalized Nafen™ Alumina Nanofiber (ANF) Reinforced Polyamide 6 Nanocomposites: Mechanical, Thermal And Flame Retardant Properties*. in *Society for the Advancement of Material and Process Engineering*. 2015. Baltimore, MD.
19. I. Grodensky, A.R. *POLYMER NANOCOMPOSITES BASED ON NAFEN™ – UNIQUE ALUMINA NANOFIBERS*. in *Society for the Advancement of Material and Process Engineering* 2014. Seattle, WA.
20. Kallagunta, H. and J. Tate, *Low-velocity impact behavior of glass fiber epoxy composites modified with nanoceramic particles.* Journal of Composite Materials, 2019: p. 002199831989343.
21. Kallagunta, H. and J. Tate, *Ballistic Performance of E-Glass Polyester Composites Modified with Silane Treated Alumina Nanofibers*. 2019.

CARBON NANOMATERIAL BASED MULTIFUNCTIONAL FABRICS FOR CHARACTERIZING HUMAN JOINT MOTION

Amit Chaudhari[1,2] Abhishek Jerome[1,3] Sagar M Doshi[1,2] Michael Keefe[1,2] Erik Thostenson[1,2,3]

1- Center for Composite Materials, 2- Department of Mechanical Engineering,
3- Department of Materials Science and Engineering
University of Delaware, Newark, DE 19716

ABSTRACT

Scalable processes such as electrophoretic deposition and dip coating are used to deposit carbon nanotubes (CNTs) on non-conductive fabrics such as cotton, nylon, wool, and aramid to develop multifunctional, smart garments. Functionalized CNTs create a flexible and electrically conductive nanocomposite film on the surface of the fibers. The CNTs also impart piezoresistive properties to the fabrics making them an ideal candidate for sensor-related applications, where the electrical resistance of the fabric changes with applied pressure. This property can be used to create functional garments that can be used to measure/characterize human joint motion.

The ease of use for assistive devices, particularly in physical-therapy applications would be improved if the devices are comfortable to wear and more helpful to standard ordinary exercises. In such wearable assistive devices, real-time sensing and feedback are critical for seamless movements and spontaneous motion support. In this research, we are depositing functionalized CNTs onto a non-woven aramid fabric to create flexible pressure sensors that are used to measure the applied pressure by a Kevlar sleeve on a manikin arm during flexion and extension. When the manikin arm wrapped with Kevlar sleeve is flexed, the CNT coated sensor underneath the sleeve is compressed due to the applied pressure, which causes a decrease in the electrical resistance. In this paper, we discuss the experimental characterization and applications of these sensors and their potential applicability to provide real-time feedback in assistive devices.

Keywords: Carbon Nanotubes, Multi Scale Composites, Human-Fabric Interaction, Smart Fabric
Corresponding author: Amit Chaudhari, amitc@udel.edu

1. INTRODUCTION

The world health organization recently projected that the worldwide number of people over 60 years of age would double by 2050 [1]. Touch sensitive assistive devices embraced the new idea of developing equipment to optimize human interaction and support. A soft-type wearable robots are preferred over hard robots due to discomfort associated with higher weight and bulkiness of the latter [2]. Despite the latest technical advancements, the development of assistive devices is facing various challenges such as designing and placing actuators, sensing external disturbances, and controlling the structure of these systems.

SAMPE Virtual Conference Proceedings, 2020. Society for the Advancement of Material and Process Engineering – North America.

Human-machine interaction is important for communication with assistive devices. Van *et al.* [3] have developed a multi-layered (two dielectrics and one conductive layer) robotic sensing skin for proximity and pressure sensing towards the development of the soft human-robot interface. Gerhardt and co-workers [4] developed a textile friction analyzer for the tribological assessment of skin–fabric interactions and measured human-fabric contact-pressure using pressure-sensitive film (Tekscan Inc). Gerratt *et al.* [5] also reported an e-skin designed to be worn over the hand, monitor live finger movement, and register distributed pressure along the entire length of the finger. Tekscan I-SCAN50 and Fujifilm pressure sensing skin are commercially available tactile sensing systems. These sensors have challenges in real-time repeatable sensing and are not easily integrable in garments as a result of which, they are not comfortable to wear for long times. In human-machine interaction, it is also important to understand the influence of various parameters, such as fabric properties, sleeve-fitting, and sleeve friction, affecting the human-garment interface. Erol *et al.* [6] demonstrated a direct connection between a key element related to tactile comfort and task performance: the friction between the representative arm and the sleeve. He developed an experimentally validated finite element model to parametrically study the effect of friction.

CNT nanocomposites have emerged as potential materials for sensor applications due to their piezoresistive mechanical-electrical coupling properties. Multiple flexible strain sensors using CNTs were proposed by researchers [7- 9] and found applications in human joint characterization. Ali *et al.* [10] developed CNT based composite element and tested it as a pressure sensor. Doshi and Thostenson [11] developed a non-woven aramid textile-based CNT pressure sensor with an ultra-wide sensing range (0.0025 MPa – 40 MPa). These sensors, combined with portable electronics, can be used for real-time sensing applications [12].

In the present study, a CNT-based non-woven aramid pressure sensor is prepared and tested to measure human-fabric interaction close to the elbow joint. Mechanical testing, using a manikin arm, is performed to measure human-fabric interaction. Finite element analysis (FEA) was used to model the human-fabric interaction and validated with experiments to study the behavior of forces experienced by manikin arm near the elbow due to human fabric interaction.

2. EXPERIMENT AND SIMULATION

Sensors were manufactured using a dip-coating process as described in our prior work [9, 13]. Non-woven aramid fabric (20601, Technical Fiber Products, Inc.) with an areal weight of 50 g/m^2 was coated using a commercially available aqueous dispersion of multi-walled CNTs. To lower the viscosity of the dispersion, 1 part of the dispersion was mixed with 2 parts of ultrapure water by weight. Uniform mixing was ensured by processing diluted dispersion using a centrifugal mixer (THINKY® ARM-310) at 2000 rpm for 120 s and then 30 minutes of sonication in an ultrasonic bath (Branson® 1510). Due to the hydrophobic nature of aramid fabric, the fabric was dipped in ultrapure water until the fabric was completely wet. To coat the fabric, the dispersion was poured in a flat-bottomed container, and a piece of fabric was dipped in the diluted dispersion for 10 minutes on each side. The fabric was then dried in a convection oven for 60 minutes at 150°C.

Figure 1: Photograph of sensor pre-loading and calibration.

Sensors were cut to a size of 100 mm in length and 25.4 mm wide. Electrodes and lead wires were attached to the specimen using conductive silver paint (Flash Dry, SPI Supplies,) and a 2 part conductive silver epoxy resin (EPOXIES® 40-3900, Epoxies, Etc.). The electrodes were attached 90 mm apart. Prior to testing on manikin arm, the sensor was pre-loaded up to the maximum expected force of 1200 N using a universal testing machine (Instron 5567), Figure 1 for improved and repeatable response followed by cyclic loading, as discussed by Doshi and Thostenson [11]. The sensor was electrically insulated to avoid interference with resistance measurement. Resistance was recorded using a voltage-current meter (Keithley 6430 sub-femtoamp remote sourcemeter). The current was measured across the electrodes on the application of a constant source voltage, and resistance and force measurements were synchronized using a customized LabVIEW program.

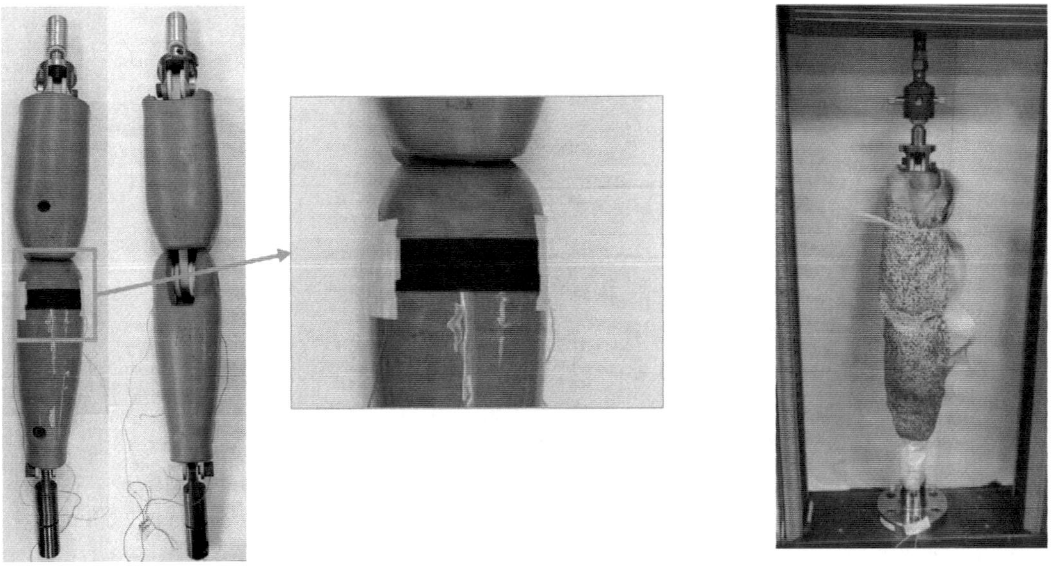

(a) (b)

Figure 2: (a) Sensor location on a Hybrid-III manikin arm and (b) manikin arm test setup with Kevlar sleeve.

A Hybrid III manikin-dummy right arm was used for mechanical testing. A cylindrically shaped fabric sleeve, diameter 100 mm, was sewn from hydrophobic Style 706 plain-woven Kevlar fabric (600 denier KM2 woven, weight 184 g/m^2, 0.23 mm thick). This fabric is well characterized and can be used in passive actuation in assistive devices. CNT coated aramid fabric pressure sensor is attached to the manikin arm near the elbow joint, Figure 2a. The Kevlar sleeve was mounted directly on the manikin arm, Figure 2b. The manikin arm was tested using a screw-driven mechanical testing load frame. Prior to loading, the arm is bent at a small angle (50 mm initial downward displacement from full extension) at the initiation of the experiments to avoid locking of the joint when crosshead of the Instron moves downward, causing the bending of the elbow joint. The prescribed motion corresponds to a combination of elbow flexion motion and shoulder adduction on the coronal plane keeping the wrist joint and shoulder joint in one line.

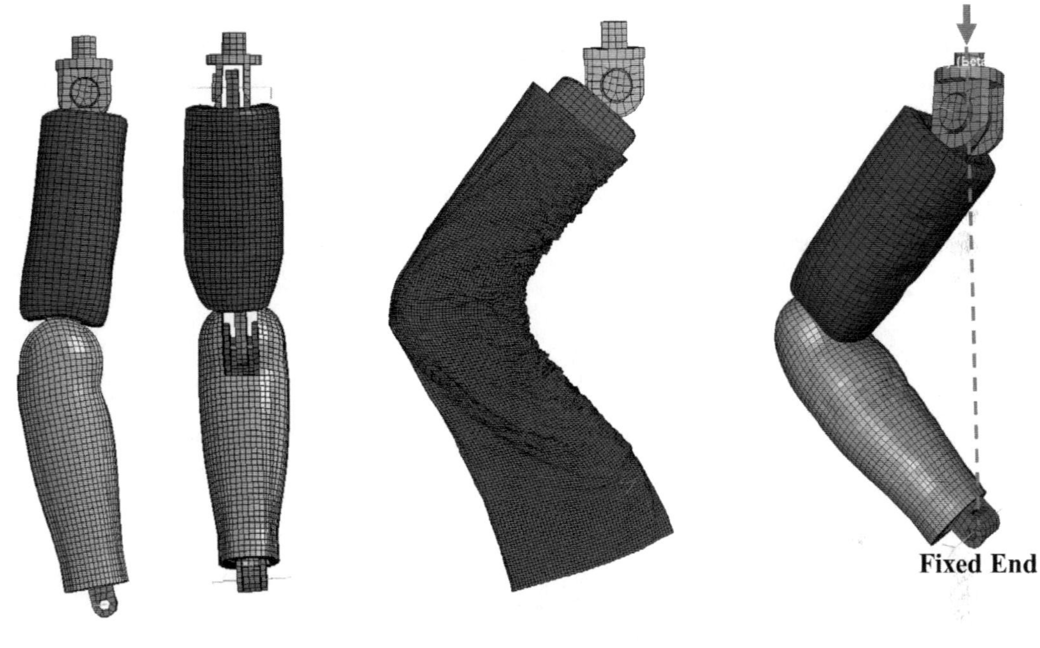

(a) (b) (c)

Figure3: (a) FE model of manikin arm, (b) manikin arm model with sleeve and (c) loading direction of the manikin arm

The finite element model developed in a previous study [6] was utilized for modeling the sleeve/arm interaction. 'Detailed hybrid III 50th percentile dummy model for LS Dyna' was downloaded and rotated to keep in the vertical position of the arm similar to the experimental configuration, Figure 3a. The FEA model has five joints defined and modeled as elastic-plastic stiffness and frictional moment as inputs. These parameters of joints were fine-tuned to achieve the force values similar to the experiment. The Kevlar sleeve was modeled as a fully integrated shell element of LS-DYNA and a refined mesh is used to capture the wrinkling of fabric, Figure 3b. Fabric material was defined as linear orthotropic material for the initial study. To simulate the motion similar to experiment, the shoulder joint was accelerated and deaccelerated along a line of motion, Figure 3c, keeping the wrist pin fixed.

3. RESULTS AND DISCUSSION

The sensor was pre-loaded to a pressure of 0.52 N/mm^2 (1.5 times of the expected pressure) to ensure the repeatability and consistency in the response. The sensor was loaded and the crosshead was held for approximately 20 seconds in the same position. It can be seen in Figure 4a, that there is close to 15% resistance change recorded and it was stable during the hold time. Repeated loading was also tested and verified for repeatability.

To examine the sleeve/arm interaction, the manikin arm was tested with and without the sleeve. The force values for both cases are shown in Figure 4b. The test with the sleeve requires a higher force for bending the manikin arm. Without the sleeve, the applied force is only dependent on the torque required to overcome joint friction. Due to the frictional interactions between the Kevlar sleeve and the arm, the compressive force required is higher with the sleeve. The additional force needed to bend the manikin arm with the sleeve can be calculated by these two forces i.e., subtracting the force required to bend the manikin arm from the force required to bend the arm with a sleeve.

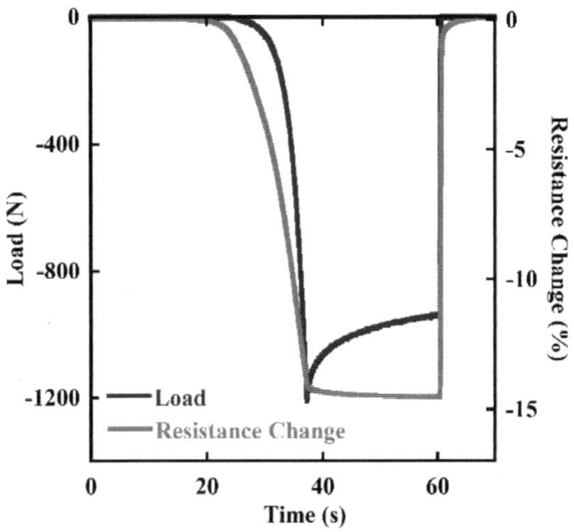

(a)

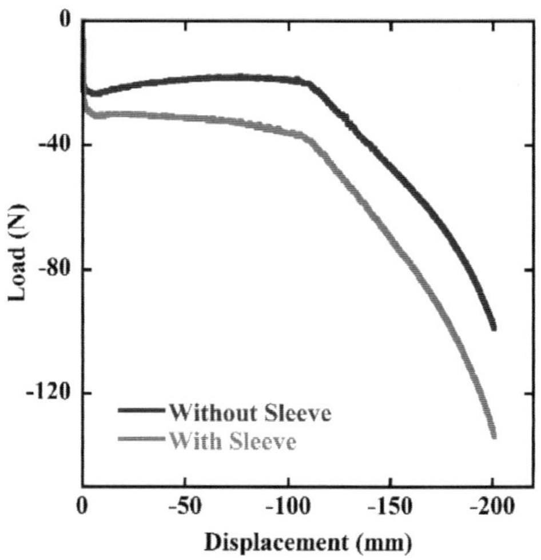

(b)

Figure 4: (a) Sensor response during pretesting and (b) force required to bend the arm with and without the Kevlar sleeve.

Resistance data from the sensor was measured for vertical downward displacement of the crosshead. Initially the crosshead was kept 50 mm downward from the full straight arm position. Resistance at this position was recorded as initial resistance for calculation of the resistance change. The sleeve is in contact with the manikin arm near sensor. During flexion motion, pressure is applied to the sensor by the fabric and resistance decreases. Under extension the resistance comes back to its original value. A slight overshoot of resistance during extension is likely because the sensor sticks slightly to fabric while sleeve gets separated from the sensor gradually. Similar behavior was also observed in previous study [11]. Multiple cycles of 120 mm displacement were repeated, and about 2.5 % resistance change was recorded (Figure 5). The response is consistent and repeatable.

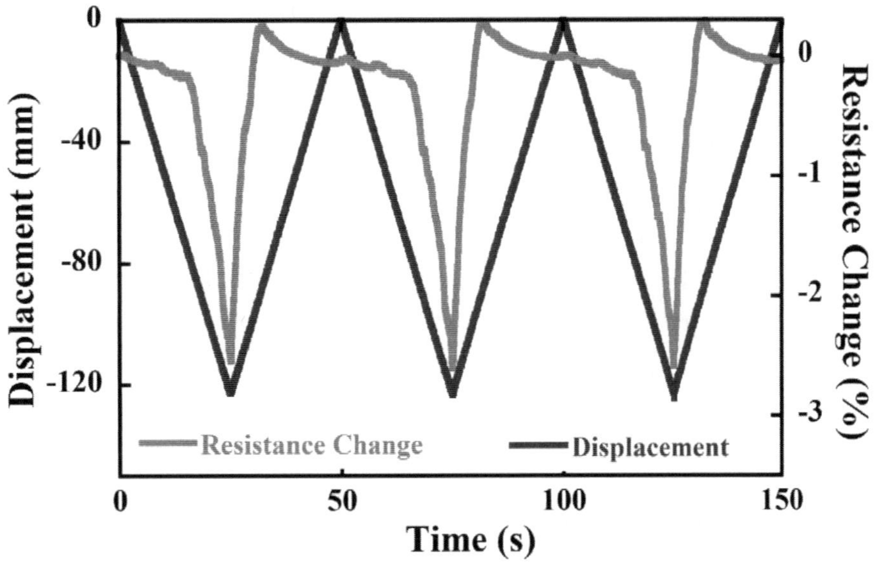

Figure 5: CNT sensor response in multiple displacement cycles.

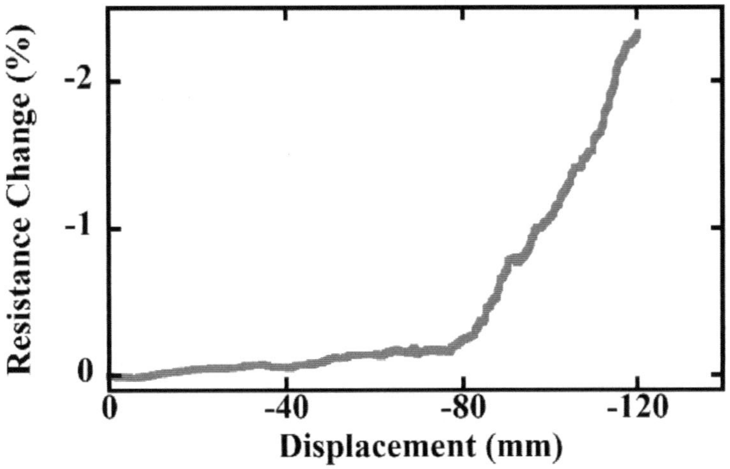

Figure 6: Sensor response in a single flexion motion.

Figure 6 shows resistance change with displacement for an isolated flexion cycle. It is observed that during flexion, the sleeve experiences tensile forces (sensor side). In the first 80 mm of crosshead displacement, no significant pressure was applied by the sleeve on the CNT pressure sensor. In the remaining cycle, pressure was applied gradually on the sensor, and a linear decrease in resistance change was recorded.

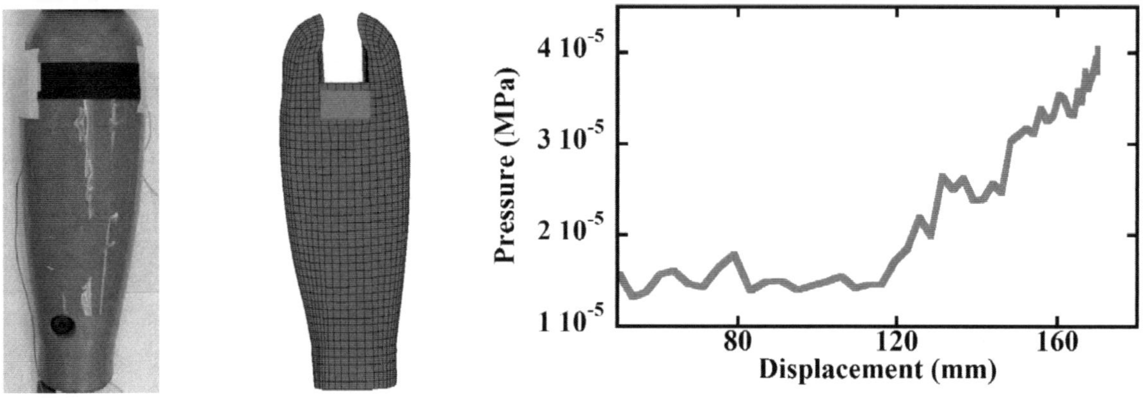

(a) (b) (c)

Figure 7: (a) Sensor location, (b) element location for average pressure measurement in the finite element model, and (c) average pressure during flexion at location shown in (b).

FEA was performed for a single cycle with a displacement rate similar to realistic human motion, flexion and extension in less than one second. No significant dynamic effects were expected, as established in a previous study [6]. FEA results show that the sleeve is loading the arm symmetrically across the centerline. Pressure was applied gradually with flexion with time due to the contact between sleeve and arm close to sensor location (Figure 7a). The average pressure due to contact between sleeve and arm was calculated for the location shown in Figure 7b. Finite element analysis was performed for 170 mm downward displacement of shoulder joint. Figure 7c shows the average calculated pressure applied by the fabric over the sensor area. Data is shown starting at an initial displacement of 50 mm to be consistent with the initial displacement from mechanical testing. Pressure variation calculated in FEA matches closely with the experimental resistance response recorded by CNT sensor.

Fujifilm Prescale® tactile pressure indicating sensor film provide by Sensor Products Inc. was used to measure the pressure applied by fabric sleeve on the manikin arm at the sensor location, Figure 8a. Extreme low, 0.05 N/mm^2 to 0.2 N/mm^2 (7.2 psi – 28 psi) pressure range, type film was used and tested for one cycle. Transparent thin film changes its color, Figure 8a, based on the applied pressure by the Kevlar sleeve on the manikin arm. Figure 8b shows the pressure distribution after analysis of the film using a Topaq® Pressure Analysis System. The pressure distribution indicates that on flexion motion, maximum pressure was observed near centerline with an average value of pressure 0.08 N/mm^2 (12 psi). It also supports that 2.5 % resistance change in CNT pressure sensor corresponds to 0.08 N/mm^2 pressure applied by fabric on the centerline of the pressure sensor. Discontinuation in the centerline pressure profile on moving towards forearm is likely because the thin pressure film is not conformal to manikin arm surface and unable to capture the interaction between manikin arm and fabric sleeve. Another factor contributing to the missing pressure data is the limitation of thin film pressure sensor to record pressure values in the range of 0.05 N/mm^2 to 0.2 N/mm^2 (7.2 – 28 psi). Pressure value below 0.05 $N.mm^2$ (7.2 psi) will not be recorded. A different pressure film, ultra extreme low (Fujifilm Prescale® Tactile Pressure Indicating Sensor Film) type, is required to measure pressure below 0.05 $N.mm^2$ (7.2 psi). Pressure variation is according to the behavior observed during testing that fabric is loading manikin arm gradually with an increase in flexion.

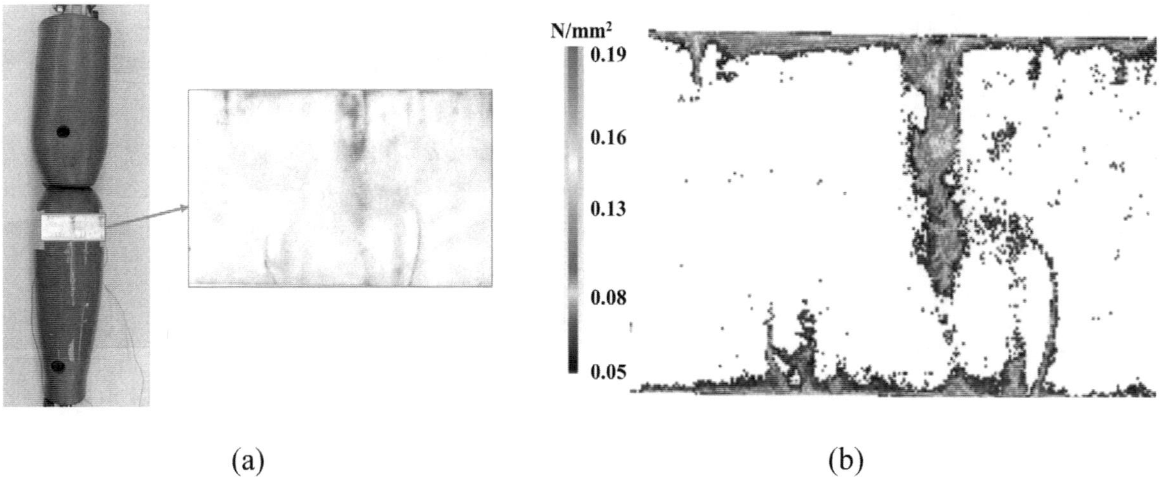

Figure 8: (a) Tactile pressure indicating film placement on manikin arm and response after one cycle and (b) pseudocolor pressure representation of the film.

4. CONCLUSIONS

A non-woven Aramid CNT-based pressure sensor is capable of recording fabric-manikin arm interaction based on resistance change. The fabric based CNT sensor can be integrated with garments to provide real time sensing to assistive devices. Once verified and validated for different material sleeves, finite element model developed can be used to do parametric studies such as examining sleeve fitting, sleeve diameter, and friction between manikin arm and sleeve.

Future work will focus on modeling natural flexion motion, keeping the arm horizontal. In such a setup, the joint angle can be characterized with pressure sensor. After the calibration of CNT pressure sensor with commercially available thin-film tactile pressure sensors, the sensors will be characterized for spatial mapping of pressure response using electrical impedance tomography as discussed by Dai and Thostenson [14].

5. ACKNOWLEDGMENTS

The authors gratefully acknowledge the funding support from the UNIDEL foundation. The authors also acknowledge Sensor Products Inc. for providing Fujifilm tactile pressure sensor and its analysis. The authors thank Mr. John Thiravong for assistance with mechanical testing.

6. REFERENCES

[1] WHO (World Health Organization), Executive Boardroom , WHO Headquarters, pp. 1–3, 2017.

[2] S. J. Park and C. H. Park, *Suit-type Wearable Robot Powered by Shape-memory-alloy-based Fabric Muscle*, Sci. Rep., vol. 9, no. 1, pp. 1–8, 2019.

https://doi.org/10.1038/s41598-019-45722-x

[3] V. A. Ho, S. Hirai, and K. Naraki, *Fabric interface with proximity and tactile sensation for human-robot interaction*, IEEE Int. Conf. Intell. Robot. Syst., vol. 2016-Novem, pp. 238–245, 2016. https://doi.org/10.1109/IROS.2016.7759061

[4] L. C. Gerhardt, N. Mattle, G. U. Schrade, N. D. Spencer, and S. Derler, *Study of skin-fabric interactions of relevance to decubitus: Friction and contact-pressure measurements*, Ski. Res. Technol., vol. 14, no. 1, pp. 77–88, 2008. https://doi.org/10.1111/j.1600-0846.2007.00264.x

[5] A. P. Gerratt, H. O. Michaud, and S. P. Lacour, *Elastomeric electronic skin for prosthetic tactile sensation*, Adv. Funct. Mater., vol. 25, no. 15, pp. 2287–2295, 2015. https://doi.org/10.1002/adfm.201404365

[6] O. Erol, E. D. Wetzel, and M. Keefe, *Simulation of a textile sleeve on a manikin arm undergoing elbow flexion: effect of arm-sleeve friction*, J. Text. Inst., vol. 106, no. 10, pp. 1135–1146, 2015. https://doi.org/10.1080/00405000.2014.977523

[7] T. Yamada et al., *A stretchable carbon nanotube strain sensor for human-motion detection*, Nat. Nanotechnol., vol. 6, no. 5, pp. 296–301, 2011. https://doi.org/10.1038/nnano.2011.36

[8] X. Wang, J. Li, H. Song, H. Huang, and J. Gou, *Highly Stretchable and Wearable Strain Sensor Based on Printable Carbon Nanotube Layers/Polydimethylsiloxane Composites with Adjustable Sensitivity*, ACS Appl. Mater. Interfaces, vol. 10, no. 8, pp. 7371–7380, 2018. https://doi.org/10.1021/acsami.7b17766

[9] A. Chaudhari, S. M. Doshi, C. Murray, and E. T. Thostenson, *Highly-sensitive carbon nanotube based sensors using everyday fabrics for human motion analysis*, Int. SAMPE Tech. Conf., vol. 2019-May, pp. 1–7, 2019. https://doi.org/10.33599/nasampe/s.19.1412

[10] A. Ali, A. Khan, K. S. Karimov, A. Ali, and A. Daud Khan, *Pressure Sensitive Sensors Based on Carbon Nanotubes, Graphene, and Its Composites*, J. Nanomater., vol. 2018, 2018. https://doi.org/10.1155/2018/9592610

[11] S. M. Doshi and E. T. Thostenson, *Thin and Flexible Carbon Nanotube-Based Pressure Sensors with Ultrawide Sensing Range*, ACS Sensors, vol. 3, no. 7, pp. 1276–1282, 2018. https://doi.org/10.1021/acssensors.8b00378

[12] S. M. Doshi, A. Chaudhari, C. Murray, J. Higginson, and E. T. Thostenson, *Carbon nanotube coated fabric-based thin and flexible pressure sensors with ultra-wide sensing range*, Int. SAMPE Tech. Conf., vol. 2019-May, 2019. https://doi.org/10.33599/nasampe/s.19.1411

[13] S. Ahmed, S. Doshi, T. Schumacher, E. T. Thostenson, and J. McConnell, *Development of a Novel Integrated Strengthening and Sensing Methodology for Steel Structures Using CNT-Based Composites*, J. Struct. Eng. (United States), vol. 143, no. 4, pp. 1–10, 2017. https://doi.org/10.1061/(ASCE)ST.1943-541X.0001697

[14] H. Dai and E. T. Thostenson, *Large-Area Carbon Nanotube-Based Flexible Composites for Ultra-Wide Range Pressure Sensing and Spatial Pressure Mapping*, ACS Appl. Mater. Interfaces, 2019. https://doi.org/10.1021/acsami.9b17100

CURE BEHAVIOR OF A FURAN-BASED EPOXY-AMINE THERMOSETTING SYSTEM FOR LIQUID MOLDING APPLICATIONS

Xi Chu, Zachary Hinton, Dr. John J. La Scala*, Dr. Giuseppe R. Palmese

Drexel University

3141 Chestnut Street
Philadelphia, PA USA

grp27@drexel.edu

*U.S. Army Research Laboratory
Aberdeen Proving Ground MD 21005

ABSTRACT

This study investigates the cure behavior of a fully furan-based epoxy-amine network in an effort to help guide the process of preparing composites using vacuum assisted resin transfer molding process (VARTM). The resin system was cured stoichiometrically at room temperature, 60 ℃ and 80 ℃. Gelation and vitrification times were obtained using NIR spectroscopy and oscillatory isothermal rheology measurements. A Time-Temperature-Transformation isothermal cure diagram was constructed to summarize these data. The furan-based system possesses a favorable processing window for VARTM. In particular, the viscosity of the system remains below 1000 cp for the first 2.4 h of isothermal room temperature cure, and gelation does not appear to occur at this temperature because it is preceded by vitrification.

1. INTRODUCTION

This work investigates the chemorheological behavior of a novel furan-based epoxy-amine network in an effort to help guide the process of preparing composites using the vacuum assisted resin transfer molding (VARTM) process. Ongoing research shows that furan-based epoxy-amine systems possess excellent mechanical and fire resistance characteristic [1,2,3]. These systems are therefore good candidates for use as matrix materials in fiber reinforced composites. Common epoxy composite resin systems are based on diglycidyl ether of bisphenol A (DGEBA) cured with aromatic or cycloaliphatic diamines. Human health and environmental concerns regarding bisphenol A (BPA) has generated significant interest in the development of BPA free epoxy alternatives[4,5,6]. There is a great diversity of biomass that can serve as feedstock for the generation of alternative epoxy monomers and curing agents with resulting thermoset characteristics suitable for composite applications. Difuran-diamine (DFDA) is a bio-based monomer of interest that has been found to impart good thermomechanical characteristics when used as curing agent in epoxy formulations [1]. In this communication, the reaction behavior of a new furan-based diepoxy (FDE), cured with DFDA was investigated. The reaction behavior was evaluated using near infrared spectroscopy (NIR) at isothermal cure temperatures including room temperature, 60°C and 80°C. Rheology measurements were conducted under the same isothermal reaction temperatures to assess viscosity and gelation behavior. An preliminary isothermal Time-Temperature-Transformation (TTT) cure diagram was generated to aid the selection of VARTM processing conditions for the FDE-DFDA system.

2. EXPERIMENTATION

2.1 Materials

Chemical structures of the bio-based epoxy resin FDE and amine curing agent DFDA used in this study are shown in **Figure 1** Both furan-based resins were synthesized and purified in the laboratory, characterized by H^1NMR. The EEW value of FDE was determined by following ASTM epoxide titration technique, D 1652-97. The titrated value, 105 g/eq, agrees very well with the theoretical value. The DFDA amine curing agent has a theoretical AHEW value of 51.6 g/eq. The FDE/DFDA system was evaluated at stoichiometry using the EEW and AHEW given above.

2.2 Fourier transform infrared (FTIR) spectroscopy

Cure kinetics studies of FDE/DFDA were conducted using A Nexus 6700 FTIR spectrometer (Thermo-Nicolet Corp.) with a CaF2 beam splitter and a DTGS detector. Experiments were conducted in the near infrared (N-IR, 4000-8000 cm-1) at room temperature, 60 °C and 80 °C. Spectra were obtained at regular time intervals using 4 cm^{-1} resolution with 32 scans per spectrum. The epoxy peak of FDE at 4529 cm^{-1} was used to track the epoxy concentration [E] as a function of time.

2.3 Reactivity Study Experimental Setup

Low hydroxyl silica glass tubes with an inside diameter of 1.60 ± 0.05 mm and 65 mm length were employed as sample holders. Samples were prepared by mixing 1.000 gram of FDE and a stoichiometric amount of DFDA. The mixtures were then degassed and injected into the glass tube. A temperature-controlled aluminum sample holder shown schematically in **Figure 2** was used for the N-IR experiments. This device allowed for control of temperature within ± 0.5 °C of the set point temperature. A background scan was conducted before each experiment using the glass tube before filling [10].

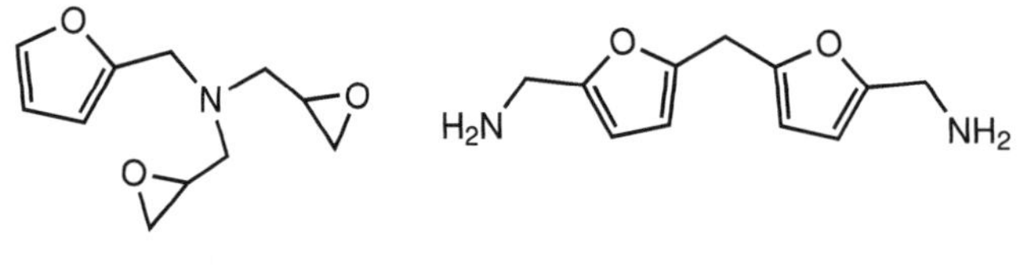

a. b.

Figure 1. Chemical structures of (a) FDE, (b)DFDA.

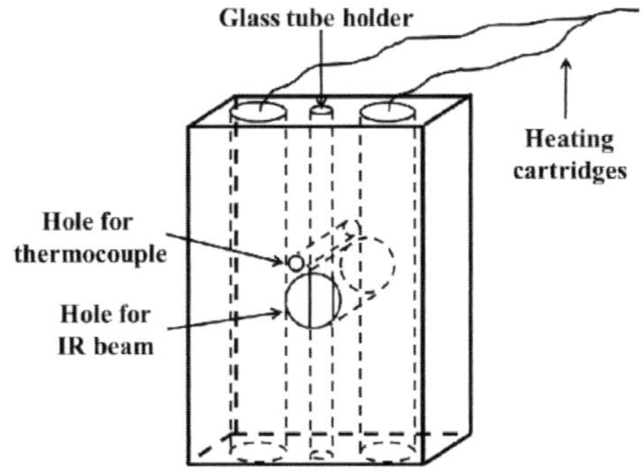

Figure 2. Schematic diagram heated sample holder used for NIR experiments[10].

2.4 Rheological study on curing behaviors at different temperature

Rheology of the curing resin was measured using continuous small amplitude oscillatory shear (SAOS) with a DHR-3 rheometer (TA Instruments, New Castle, DE). Liquid resins were loaded between disposable, aluminum parallel plates with a diameter of 25 mm and gap height of 700 mm. For temperatures other than room-temperature, the plates were heated using Peltier heating to the desired temperature and equilibrated. The sample was then loaded and equilibrated. During curing, a continuous frequency of 1 rad/s and strain amplitude of 1% was applied while maintaining a constant gap. Moduli were measured close to vitrification for most samples. Note that the magnitudes of moduli are artificially high beyond gelation as the sample dimensions are reduced to the point of underfilling the rheometer. At short times, the modulus of the liquid resin is below machine limits.

2.5 Differential Scanning Calorimetry (DSC)

A TA Instruments, Q2000 Differential Scanning Calorimeter (DSC) was used to measure the heat of reaction of FDE/DFDA cured stoichiometrically. FDE/DFDA was heated under inert atmosphere from room temperature to 200 °C at a heating rate of 1 °C/min. The heat of reaction for this system cured stoichiometrically was found to be 560.9 J/g.

3. RESULTS

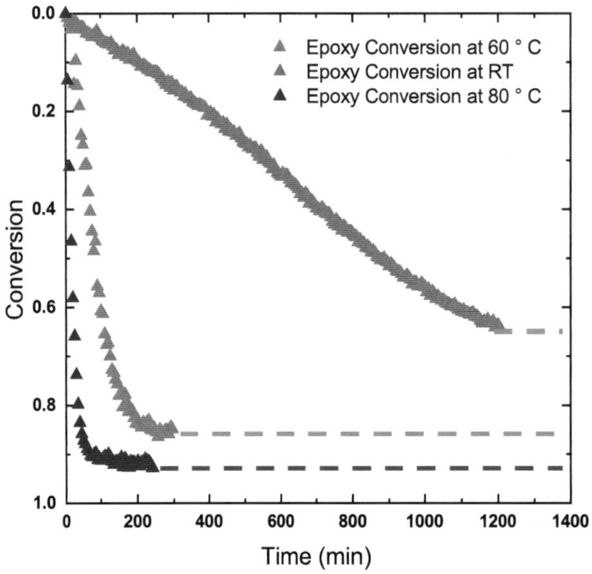

Figure 3. Epoxy conversion vs. time of FDE/DFDA system at room temperature, 60 °C and 80 °C.

Plots of epoxy conversion versus time for the FDE/DFDA cured isothermally at room temperature, 60°C and 80°C are given in **Figure 3**. The plots show typical cure behavior of aliphatic epoxy-amine systems. The reaction proceeds slowly at room temperature and increases significantly as the isothermal cure temperature is increased. The conversion profiles in the early stages of cure exhibit autocatalytic behavior. Moreover, for all temperatures evaluated, the latter stages of cure highlight diffusion limited reaction resulting from vitrification that limits final conversion. The T_g at full conversion of this system is 94°C, as obtained from the loss modulus peak of a DMA (Dynamic mechanical analysis) plot. For the purpose of our analysis, we estimate the time to vitrification by the time point at which conversion ceases to change noticeably. This conversion is also associated with the T_g of the system reaching the isothermal cure temperature ($T_g = T_{cure}$), Times required for the system to vitrify based on isothermal reaction behavior are summarized in **Table 1**. At room temperature, more than 20 hours are required to reach vitrification. Vitrification occurs more quickly at 60 °C and 80 °C (5 hours and 2.76 hours, respectively).

Understanding the chemorheolgical behavior of a resin system is important in order to select appropriate processing conditions. Oscillatory rheology experiments were conducted using the FDE/DFDA system at the same isothermal temperatures employed for the cure studies. **Figure 4** contains data from these experiments in the form of storage modulus and loss modulus plotted as a function of time for the three temperature conditions. The time to gelation is important for defining the processing window of thermosetting resin systems. Once the system gels it cannot be made to flow. For epoxy-amine systems of as given formulation, gelation occurs at a specific value of conversion. The crossover point of loss and storage modulus traces can be used to define the gel point. Crossover points are clearly identified in the 60°C and 80°C cases. Crossover occurs at 110 min for 60 °C and 36 min for 80°C cure. The room temperature isothermal cure experiment shows the value of the loss modulus grows close to that of the storage modulus but remains lower with the two curves becoming parallel at long times. This behavior suggests that gelation might not occur before vitrification in this case. The conversion versus time data given in **Figure 3** was used to determine gel point conversion associated with the gel times obtained from the rheology experiments. These values are reported in **Table 1** and are 68% and 73% conversion for 60°C and 80°C cure respectively.

Using classical Flory-Stockmayer theory for gelation, the following equation predicts critical conversion at gelation for a system at stoichiometry where $\boldsymbol{f_E}$ is the epoxy molecule functionality and $\boldsymbol{f_A}$ is the amine curing agent functionality.

$$\alpha_{gel} = \frac{1}{[(f_E - 1)(f_A - 1)]^{1/2}}$$

For the system being considered $\boldsymbol{f_E} = 2$ and $\boldsymbol{f_A} = 4$ so that the predicted $\boldsymbol{\alpha_{gel}}$ is 57.7%. The predicted value is significantly lower than the measured values. The theory assumes uniform reactivity (i.e. no difference between primary and secondary amine) and no cyclization. Higher reactivity of primary versus secondary amine and cyclization both increase the conversion needed for gelation. In previous work the reactivity ratio of furan systems was found to favor primary epoxy-amine reactions [1,2,10) and it is known that glycidyl amine systems form networks with significant cyclization [12] ; therefore, the higher measured gel conversion is reasonable.

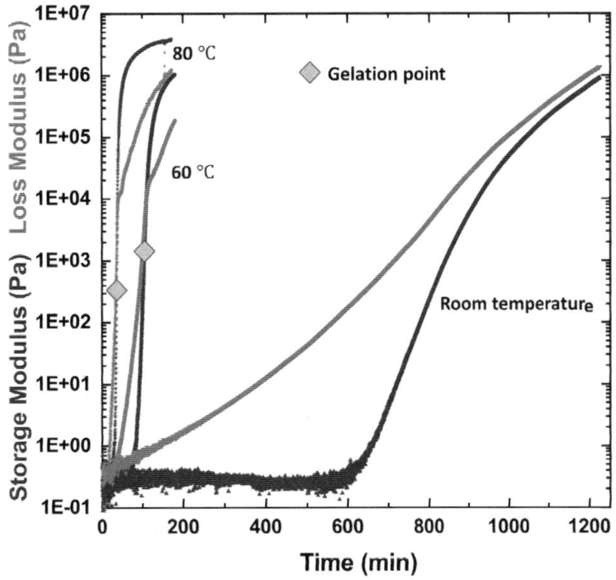

Figure 4. Storage modulus and loss modulus of FDE/DFDA system during curing at room temperature, 60 °C and 80 °C.

Table 1. Summary of gelation time (t_g), gelation conversion (α_g), vitrification time (t_v) and vitrification conversion (α_v) with respect to time for isothermal cure at room temperature, 60 °C and 80 °C.

	t_{gel} (mins)	α_{gel} (%)	t_v (mins)	α_v(%)
Room Temperature	--	-	~1200	64
60 °C	110	68	300	84
80 °C	36	73	166	93

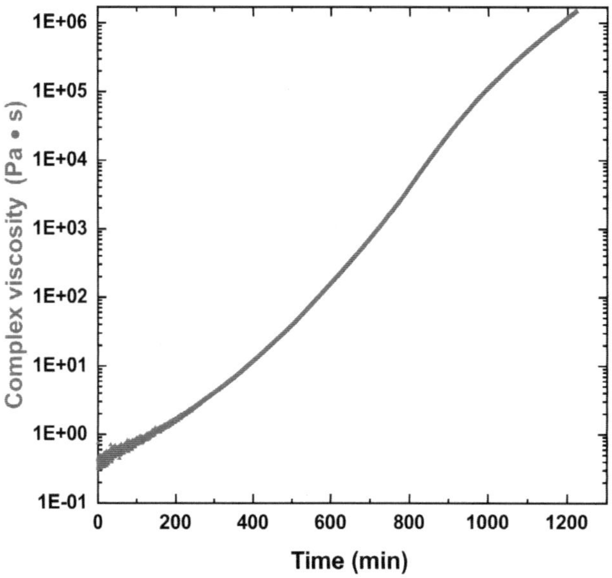

Figure 5. Complex viscosity vs. Time of FDE/DFDA network at room temperature. curing.

The complex viscosity with respect to time at room temperature that was obtained from the rheological study is given in **Figure 5** VARTM requires low viscosity resin systems matrix (100-1000 cp) at the processing temperature. At room temperature the viscosity of FDE/DFDA system reaches 1000 cp 2.4 hours after mixing which provided a usefully long processing window for composite preparation.

4. CONCLUSION

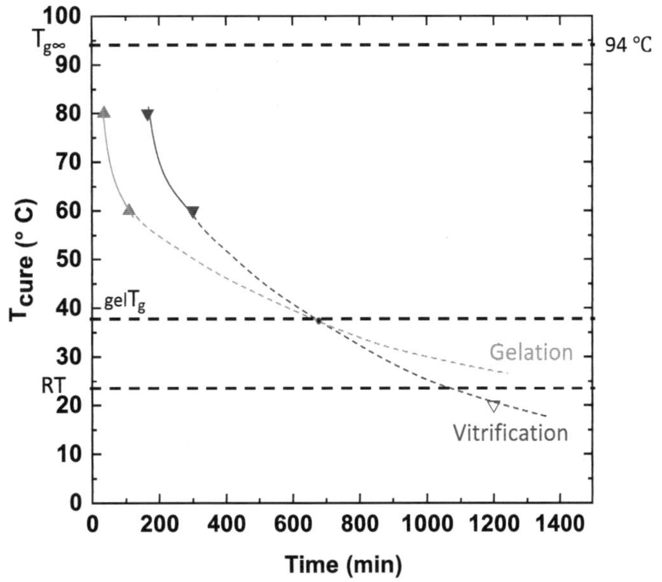

Figure 6. TTT cure diagram of FDE/DFDA system showing gelation (green) and vitrification curves (blue).

Times to vitrification (t_v) determined from isothermal NIR cure experiments at three temperatures as well as times to gelation (t_{gel}) determined from isothermal rheology experiments for the same temperatures are plotted in **Figure 6** to obtain a Time-Temperature-Transformation (TTT) isothermal cure diagram for the FDE/DFDA system. In addition to the gelation and vitrification curves, full cure T_g ($T_{g\infty}$ = 94°C) and an estimate of the isothermal cure temperature at which gelation and vitrification simultaneously ($_{gel}Tg$ ~ 37°C) are shown, $_{gel}Tg$ greater than room temperature indicates a favorable room temperature processing window because gelation will not occur at this cure temperature. This coupled with low room temperature viscosity (< 1000 cp for the first 2.4 h of cure) make the FDE/DFDA system a good choice for VARTM processing.

5. ACKNOWLEDGEMENT

Research was sponsored by the Army Research Laboratory and was accomplished under Cooperative Agreement W911NF-16-2-0225. The views and conclusions contained in this document are those of the authors and should not be interpreted as representing the official policies, either expressed or implied, of the Army Research Laboratory or the U.S. Government. The U.S.

Government is authorized to reproduce and distribute reprints for Government purposes notwithstanding any copyright notation herein.

6. REFERENCES

1. Hu, Fengshuo.; La Scala, John. J.; Sadler, Joshua. M.; Palmese, Giuseppe. R., "Preparation and Characterization of Fully Furan-Based Renewable Thermosetting Epoxy-Amine Systems." Macromol. Chem. Phys (2015), 216, 1441-1446. DOI: 10.1002/macp.201500142.
2. Hu, F., Yadav, S. K., La Scala, J. J., Sadler, J. M., & Palmese, G. R., "Preparation and Characterization of Fully Furan-Based Renewable Thermosetting Epoxy-Amine Systems." Macromolecular Chemistry and Physics(2015), 216(13), 1441–1446. https://doi.org/10.1002/macp.201500142.
3. Meng, J., Zeng, Y., Chen, P., Zhang, J., Yao, C., Fang, Z., & Guo, K., "New ultrastiff bio-furan epoxy networks with high T g : Facile synthesis to excellent properties." European Polymer Journal(2019), 121(October), 109292. https://doi.org/10.1016/j.eurpolymj.2019.109292.
4. R. C. Thompson, C. J. Moore, F. S. vom Saal, S. H. Swan, Philos. Trans. R. Soc. London, B 2009, 364, 2153.
5. Phenol, The National Institute for Occupational Safety and Health, Centers for Disease Control and Prevention, Atlanta, Georgia 2015.
6. Bisphenol A, The National Institute of Environmental Health Sciences, National Institute of Health, Research Triangle Park, North Carolina 2010.
7. McCrum, N. G.; Buckley, C. P.; Bucknall, C. B., Principles of polymer engineering. Oxford University Press.: New York, 1997.
8. Raquez, J. M.; Deleglise, M.; Lacrampe, M. F.; Krawczak, P., Progress in Polymer Science 2010, 35 (4), 487-509.
9. Enns, J. B., & Gillham, J. K., "Effect of the extent of cure on the modulus, glass transition, water absorptio, and density of an amine-cured epoxy." Journal of Applied Polymer Science(1983), 28(9), 2831–2846. https://doi.org/10.1002/app.1983.070280914, Princeton University, Princeton, New Jersey.
10. Hu, F. (2016). Structure - Property Relationships of Furanyl Thermosetting Polymer Materials Derived from Biobased Feedstocks. Drexel.
11. Peng, X., & Gillham, J. K., "Time-Temperature-Transformation (TTT) Cure Diagrams : Relationship Between Tg and the Temperature and Time of Cure for Epoxy Systems." Journal of Applied Polymer Science(1985), *30*, 4685–4696, Princeton University, Princeton, New Jersey.
12. Matějka, L., & Dušek, K., "Mechanism and Kinetics of Curing of Epoxides Based on Diglycidylamine with Aromatic Amines. 1. The Reaction of Diglycidylaniline with Secondary Amines." Macromolecules(1989), 22(7), 2902–2910. https://doi.org/10.1021/ma00197a005

BIO-BASED BENZOXAZINE MONOMER AND POLYMER BASED ON VANILLIN AND DIFURFURYL DIAMINE

Mengwen Yu, Santosh K. Yadav, John J. La Scala, Giuseppe R. Palmese
Drexel University
Philadelphia, PA/USA

ABSTRACT

A type of bio-based benzoxazine monomer based on difurfuryl diamine (DFDA) obtained by the coupling of furfuryl amine using formaldehyde will be discussed. The benzoxazine monomer was synthesized by reacting DFDA with renewable phenolic vanillin and formaldehyde. This benzoxazine system was purified by precipitation to obtain solid powders that melt at 55 °C to form low viscosity liquids that can be readily used to impregnate fiber reinforcements. Onset cure temperature is 163 °C and the autocatalytic reaction behavior was observed. The resulting polybenzoxazine has about 244 °C after post curing. It was found that the polybenzoxazine network containing high content of furan rings in the backbone provides the improvement of 800 °C char yield compared with Bisphenol A (BPA) based polybenzoxaizines greater than 50% measured by TGA temperature ramps in an inert environment. This paper will discuss processing and thermal characteristics of this highly renewable material, which are useful for high temperature applications.

Keywords: Bio-based, Polybenzoxazine, vanillin, furfurylamine
Corresponding author: Giuseppe R. Palmese, grp27@drexel.edu

1. INTRODUCTION

Benzoxazines belong to heterocyclic compounds based on an aromatic ring connected with a six-membered heterocycle containing one oxygen and one nitrogen atom. These monomers are usually synthesized by the Mannich reaction of phenols, formaldehyde, and amines.[1,2] Polybenzoxazines possess good performance characteristics, especially thermal performance, such as high char yield, high glass transition temperature, and fire resistance performance. They also possess advantages including good electric properties, low melt viscosity of the precursors, curing without a strong catalyst, and near-zero volume change upon cure.[3,4] Therefore, they can be used for electronics applications and as matrix materials for fiber-reinforced composite materials that require high temperature capabilities.

An increasing number of researchers have focused on bio-based materials as replacements for petroleum-based materials to mitigate environmental concerns and limited resources. For polybenzoxazines, a number of bio-based building blocks have been used to synthesize benzoxazine monomers. Lochab and Varma synthesized difunctional benzoxazines derived from bisphenol-A and cardanol.[5] Sini and coworkers synthesized furfurylamine- and vanillin-based benzoxazines and their curing behavior.[6] Shen and his group synthesized DFDA- and phenol-based polybenzoxazines with good thermal stability and adhesion properties.[7] To our knowledge there is no work available combining bio-based phenols and DFDA to synthesize benzoxazine monomers with higher bio-based content. In this investigation, DFDA- and vanillin-based benzoxazine was synthesized and characterized. Cure kinetics were evaluated using FTIR spectroscopy. Additionally, thermomechanical behavior of the system was assessed by dynamic mechanical analysis (DMA) using carbon fiber fabric composites impregnated with the benzoxazine monomer powder.

2. EXPERIMENTATION

2.1 Materials

Vanillin, furfurylamine, paraformaldehyde, chloroform, sodium hydroxide, diethyl ether, magnesium sulfate anhydrous, formaldehyde solution (37 %), and hydrochloric acid solution (37 %) were purchased from Sigma Aldrich and used as received.

2.2 Synthesis of (5,5'- methylenebis(furan-5,2-diyl))dimethanamine (DFDA) and benzoxazine monomer

DFDA was synthesized by following the method provided by Holfinger et al. and resulted in a brown viscous liquid.[8] DFDA- and Vanillin-based benzoxazine (BZ-DFDA-Vanillin) was synthesized by Mannich reaction in the presence of solvent. Chloroform, paraformaldehyde and DFDA were introduced into the round bottomed flask and mixed at room temperature for 30 min. After adding vanillin, the mixture was stirred at 70 °C and refluxed for 20 h. The liquid of mixture was washed with sodium hydroxide solution and distilled water, and last, a crystalized yellow powder was obtained by crystallization (Yield: 65 %, Purity: 95 %). ^{1}H-NMR of BZ-DFDA-Vanillin (500 MHz, $CDCl_3$, ppm): δ 9.82 (s, 2H); 7.45 (s, 2H); 7.15 (s, 2H); 6.19 (d, 2H); 6.03 (d, 2H); 5.03 (s, 4H); 4.10 (s, 6H); 3.98 (s, 4H); 3.94 (s, 4H); 3.89 (s, 2H).

2.3 Preparation of polybenzoxazine-carbon fiber fabric composites

PolyBZ-DFDA-Vanillin/CF composites were prepared by powder impregnation and compression molding. Carbon fiber (AS4 G', 5HS 12K) was cut to 110 mm × 110 mm and placed in a steel mold with the cavity size 110 mm × 110 mm× 1.5 mm. Six alternating layers of carbon fiber and benzoxazine powder were pressed into the cavity, and the mold was preheated first to 100 °C for 1h and then 130 °C 1h to melt powder in the press at 3 MPa pressure. Then the composite was cured under 150 °C 1 h, 180 °C 2 h, 200 °C 1 h. The fiber volume fraction of the resulting composite was about 45%.

3. RESULTS

3.1 Structure characterization of BZ-DFDA-Vanillin

The structure of the new benzoxazine monomer was confirmed by ^{1}H NMR spectra, shown in Fig. 1A). The protons of oxazine rings -Ar-CH_2-N- and -OCH_2-N- give rise to characteristics peaks at 4.10 and 5.03 ppm, respectively. The resonances at 6.03 and 6.09 ppm are assigned to furan ring protons. The spectrum also displays two resonances centered at 4.10 and 9.82 ppm that correspond to the formyl and methyl groups consistent with the structure of vanillin. In addition, the theoretical values calculated for the molecular structure of this novel benzoxazine correlates well with the ^{1}H NMR elemental analysis of BZ-DFDA-Vanillin shown in the experiment section. The Mid-FTIR spectra shown in Fig. 1B) is also consistent with the structure of BZ-DFDA-Vanillin. Characteristic absorption peaks for the furan rings are observed at 1584 and 785 cm^{-1}. The presence of the formyl group of vanillin is confirmed by the peak absorption at 1686 cm^{-1}. Moreover, the absorption peaks at 1020 cm^{-1} (asymmetric stretching of C-N-C) and 1220 cm^{-1} (Ar-C-O stretch), indicate that an oxazine ring fuses to the benzene ring of vanillin.[9]

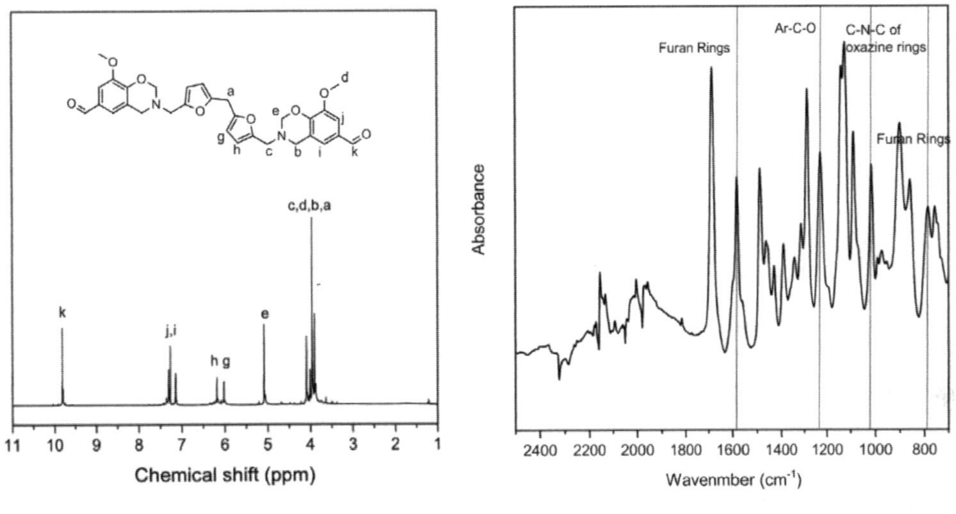

Fig. 1 A) ^{1}H-NMR spectra of BZ-DFDA-Vanillin; B) FTIR spectra of BZ-DFDA-Vanillin.

3.2 Curing behavior of BZ-DFDA-Vanillin

The thermal behavior of BZ-DFDA-Vanillin was monitored using Differential Scanning Calorimetry (DSC) as shown in Fig. 2A). The BZ-DFDA-Vanillin benzoxazine monomer exhibited a melting transition at 55 °C that is 45 °C lower than the 100 °C melting point of di-benzene-based benzoxazines, indicating the furan rings have an impact on reducing melting temperature. In addition to the melting endotherm, BZ-DFDA-Vanillin exhibits an exothermic peak that starts around 160 °C and ends around 200 °C. From the previous work of other groups, it is known that the ring-opening curing process of benzoxazines typically show exothermic peaks between 200-280 °C.[10] This novel difuran based benzoxazine exhibited exothermic behavior in a range 50°C lower than incumbent systems. Additionally, the processing window of BZ-DFDA-Vanillin, as defined by the temperature difference between the onset curing temperature and the melting point, is wider than that of benzene containing systems. It was found that the BZ-DFDA-Vanillin followed a ring-opening polymerization mechanism with the elimination of carboxylic groups of vanillin, which was observed by Sini's group who studied furfurylamine- and vanillin-based benzoxazine.[6,11] More details regarding the curing mechanisms of BZ-DFDA-Vanillin will be discussed in a future paper from our group.

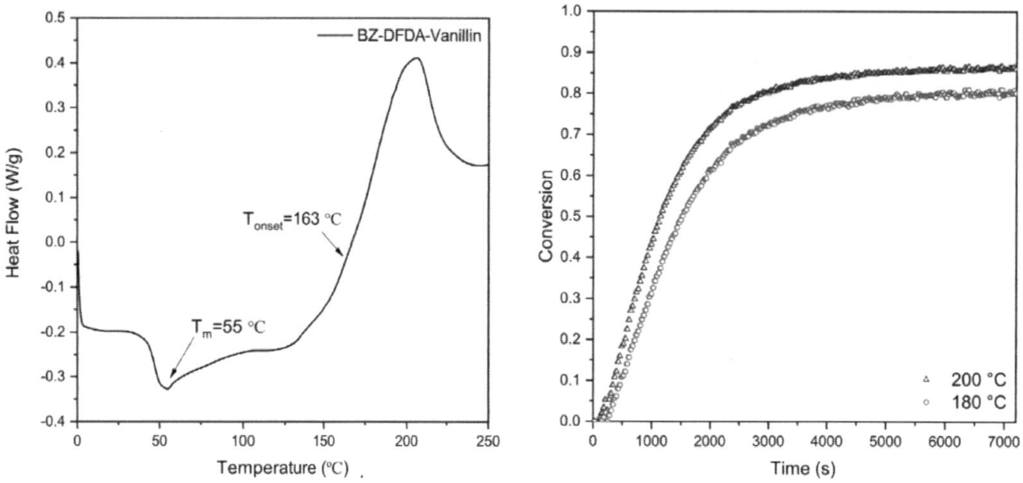

Fig. 2 A) DSC thermogram of BZ-DFDA-Vanillin showing melt transition and peak exotherm; B) conversion as a function of time of BZ-DFDA-Vanillin at 180° and 200 °C isothermal cure conditions.

The curing kinetics of benzoxazine monomers were characterized by FTIR isothermal experiments. The conversion of benzoxazines was calculated by quantifying the height change of absorption bands of oxazine rings (1220 cm^{-1}). Fig. 2B) shows cure profiles at 180° and 200 °C. The shape of these curves is consistent with autocatalytic reaction behavior. The polymerization of BZ-DFDA-Vanillin reached a steady state of conversion within about 1 hour without the use of a catalyst. At 200 °C, the material achieved about 90 % final conversion, which is consistent with the exothermic behavior from the DSC thermogram.

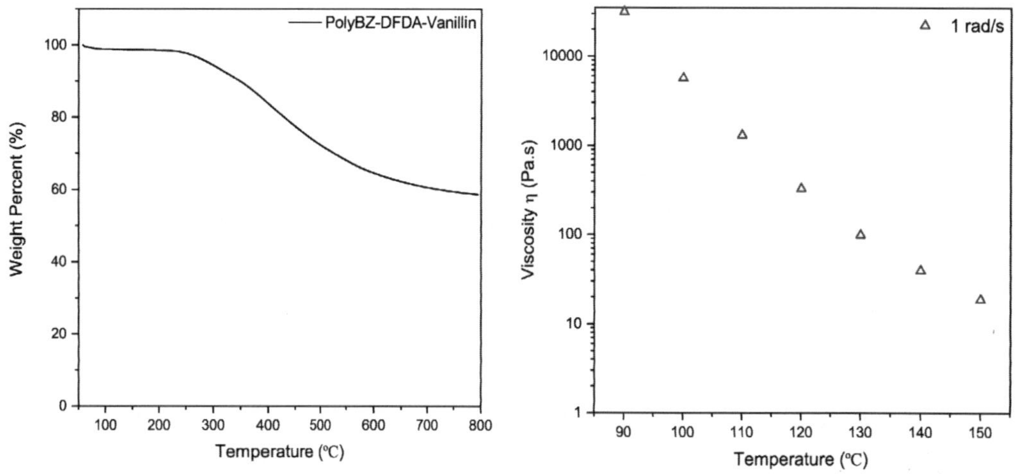

Fig. 3 A) TGA thermogram of PolyBZ-DFDA-Vanillin; B) viscosity of melted BZ-DFDA-Vanillin versus temperature.

3.3 Thermal stability of the bio-based polybenzoxazine

The thermal stability of PolyBZ-DFDA-Vanillin was observed by Thermogravimetric Analysis (TGA) in argon atmosphere. Fig. 3A) shows the thermogram obtained at 10 °C/minute and Table

1 includes the char yield at 800°C from the thermogram (Row 7). Also included for comparison in Table 1 are char yield literature values for other systems. The TGA curve in Fig. 3A) indicates that difuran-containing polybenzoxazine has 61% char yield at 800 °C. It should be noted that BZ-DFDA-Vanillin lost about 10% weight due to the elimination of carboxylic groups during curing process, so PolyBZ-DFDA-Vanillin possesses around 50 % char yield based on initial weight of precursors. Compared with 62% char yield of DFDA- and phenol-based polybenzoxazine PolyBZ-F that was studied by Shen's group (Row 6), the DFDA-vanillin based polybenzoxazine is similar but contains significantly higher renewable content. Furthermore, comparison with typical dibenzene-based polybenzoxazine, such as Bisphenol A-based polybenzoxazine (Rows 1–5 in Table 1), show that the furan-containing network has a positive impact on enriching the char formation of polybenzoxazines.[10]

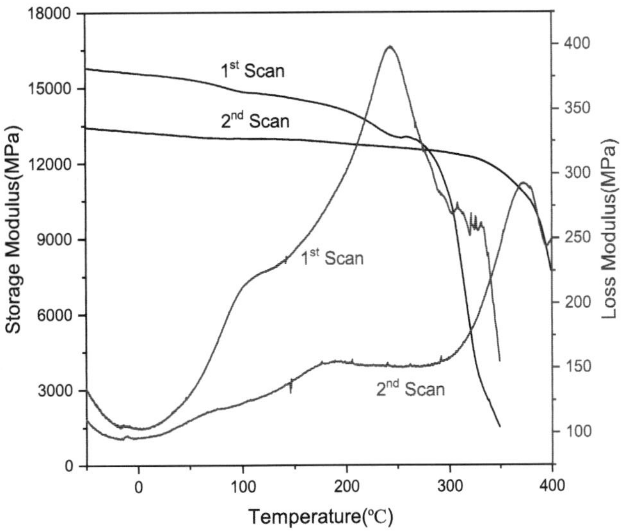

Fig. 4 DMA curves of PolyBZ-DFDA-Vanillin/ CF composites.

3.4 Dynamic mechanical properties of PolyBZ-DFDA-Vanillin/Carbon Fiber composites

Prior to the preparation of polybenzoxazine carbon fiber composites, the viscosity of the monomer was measured using TA Instruments DHR-3 determine if the melted precursor is able to infiltrate into the carbon fiber fabric. The viscosity as a function of temperature of BZ-DFDA-Vanillin is shown in Fig. 3B). The viscosity decreases steadily after the melt transition and is 1000 Pa·s at 100 °C, which is low enough to infiltrate the fabric. This value was selected as the starting temperature for compression processing of the composite bar for DMA testing. DMA was used to investigate the dynamic mechanical properties of bio-based polybenzoxazine carbon fiber composites. Fig. 4 presents the storage modulus and loss modulus curves of this cured material and were obtained using a ramp rate of 2 °C/min from -50 °C to 350 °C followed by a second scan. The storage modulus of the composite is well above 10 GPa at room temperature as is expected because of the 3 layers carbon fiber reinforcement. The glass transition temperature (T_g) of the material shows significant difference between the two scans. The first run following a 200 °C post curing temperature resulted in T_g of 244 °C, however, the T_g increased to 372 °C on the second scan because the material was heated to 350 °C during the first run. The curing kinetics study of the BZ-DFDA-Vanillin showed 90% benzoxazine conversion at a 200 °C isothermal cure temperature. The 244°C T_g could be a result of the first run temperature ramp causing continued benzoxazine conversion. This would not explain the much higher T_g (372 °C) obtained on the second run. It was found that PolyBZ-DFDA-Vanillin starts to decompose and form char beginning at 280 °C from TGA curves, so the heating to 350°C would certainly affect the network structure via reactions other than benzoxazine ring opening polymerization.

Table 1 Thermal and mechanical properties of polybenzoxazines derived therefrom.

Abbr.	**Phenol**	**Amine**	**Char Yield at 800°C (%)**	**T_g (°C)**
BA-a	BPA	Aniline	30[4]	170[4]
BA-ot	BPA	o-Toluidine	32[4]	114[4]
BA-mt	BPA	m-Toluidine	31[4]	210[4]
BA-pt	BPA	p-Toluidine	32[4]	158[4]
BA-35x	BPA	3,5-Xylidine	28[4]	245[4]
BZ-F	Phenol	DFDA	62[7]	270[7]
BZ-DFDA-Vanillin	Vanillin	DFDA	61	244

Table 1 also contains T_g data for a number of BZ resins. Of the BPA-based polybenzoxazines, only BZ-35x has a T_g close to that of the BZ-DFDA-Vanillin system. But this system has only 28% char yield, The T_g's of the rest of the BPA-based polybenzoxazines are significantly lower. It is clear that the DFDA- and vanillin-based polybenzoxazines performed better both in terms of T_g and char yield.[12] PolyBZ-F has a 270 °C T_g but these systems were post cured at 220 °C. Thus, the performance of the fully bio-based polybenzoxazine has similar performance characteristics to those of the Phenol-DFDA system, suggesting that the furan building block is responsible for the improved performance characteristics.

4. CONCLUSIONS

A novel bio-based benzoxazine monomer derived from furfurylamine and vanillin was synthesized with high purity. ^{1}H NMR and FTIR characterization of this material was used to confirm the structure of BZ-DFDA-Vanillin. BZ-DFDA-Vanillin exhibited autocatalytic ring-opening curing kinetics as observed by Mid-IR spectroscopy. The PolyBZ-DFDA-Vanillin system showed better thermal stability and higher T_g than BPA-based polybenzoxazines. PolyBZ-DFDA-Vanillin has higher renewable content and similar performance characteristics when compared to DFDA-phenol-based polybenzoxazine reported in the literature.

5. ACKNOWLEDGMENTS

We acknowledge financial support from CCDC-Army Research Laboratory under cooperative agreement W911NF-16-2-0225. We thank the Department of Chemistry in Drexel University for instruments use.

6. REFERENCES

1. Lligadas G, Tüzün A, Ronda JC, Galià M, Cádiz V. Polybenzoxazines: New players in the bio-based polymer arena. *Polym Chem.* 2014;5(23):6636-6644. doi:10.1039/c4py00914b

2. Ghosh NN, Kiskan B, Yagci Y. Polybenzoxazines-New high performance thermosetting resins: Synthesis and properties. *Prog Polym Sci.* 2007;32(11):1344-1391. doi:10.1016/j.progpolymsci.2007.07.002

3. Takeichi T, Kawauchi T, Agag T. High Performance Polybenzoxazines as a Novel Type of Phenolic Resin. *Polym J.* 2008;40(12):1121-1131. doi:10.1295/polymj.pj2008072

4. Ishida H, Agag T. Handbook of Benzoxazine Resins. *Handb Benzoxazine Resins.* 2011:740480. doi:10.1016/C2010-0-66598-9

5. Lochab B, Varma IK, Bijwe J. Thermal behaviour of cardanol-based benzoxazines: Monomers and polymers. *J Therm Anal Calorim.* 2010;102(2):769-774. doi:10.1007/s10973-010-0736-6

6. Sini NK, Bijwe J, Varma IK. Renewable benzoxazine monomer from Vanillin: Synthesis, characterization, and studies on curing behavior. *J Polym Sci Part A Polym Chem.* 2014;52(1):7-11. doi:10.1002/pola.26981

7. Shen X, Dai J, Liu Y, Liu X, Zhu J. Synthesis of high performance polybenzoxazine networks from bio-based furfurylamine: Furan vs benzene ring. *Polymer (Guildf).* 2017;122(March 2018):258-269. doi:10.1016/j.polymer.2017.06.075

8. Holfinger MS, Conner AH, Holm DR, Hill CG. Synthesis of Difurfuryl Diamines by the Acidic Condensation of Furfurylamine with Aldehydes and Their Mechanism of Formation. *J Org Chem.* 1995;60(6):1595-1598. doi:10.1021/jo00111a017

9. Wang C, Sun J, Liu X, Sudo A, Endo T. Synthesis and copolymerization of fully bio-based benzoxazines from guaiacol, furfurylamine and stearylamine. *Green Chem.* 2012;14(10):2799-2806. doi:10.1039/c2gc35796h

10. Ishida H, Sanders DP. Regioselectivity and network structure of difunctional alkyl-substituted aromatic amine-based polybenzoxazines. *Macromolecules.* 2000;33(22):8149-8157. doi:10.1021/ma991836t

11. Sini NK, Bijwe J, Varma IK. Thermal behaviour of bis-benzoxazines derived from renewable feed stock "vanillin." *Polym Degrad Stab.* 2014;109:270-277. doi:10.1016/j.polymdegradstab.2014.07.015

12. Ishida H, Allen DJ. Physical and mechanical characterization of near-zero shrinkage polybenzoxazines. *J Polym Sci Part B Polym Phys.* 1996;34(6):1019-1030. doi:10.1002/(SICI)1099-0488(19960430)34:6<1019::AID-POLB1>3.0.CO;2-T

DEVELOPMENT AND CHARACTERIZATION OF GRAPHENE NANOPLATELETS FILLED CARBON FIBER/BENZOXAZINE/EPOXY COMPOSITES

Poom Narongdej, Steven J. Nava, Jack T. Denk, Ehsan Barjasteh
California State University, Long Beach (CSULB)
Long Beach, California/U.S.A.

ABSTRACT

In this study, we report an advancement in electrical, mechanical, and thermo-mechanical properties of benzoxazine-epoxy copolymer and its resultant composites by the addition of Graphene Nanoplatelets (GNPs). The GNPs were incorporated at different loadings in a copolymer consisting of bisphenol-A based benzoxazine (BZ) and bi-functional cycloaliphatic epoxy resins (CER). A three-roll mill (3RM) calendering device was utilized to disperse the GNPs in the copolymer matrix. The composites were characterized for their electrical, mechanical, and thermomechanical properties. The addition of GNPs showed an improvement of greater than 35 % in both modulus and strength with the addition of GNPs at 0.9 wt.%. A similar trend was observed in the storage moduli, where the maximum value was obtained from the sample at the same nanoparticle loading. The electrical conductivity was achieved at the electrical percolation threshold between 0.6 and 0.9 wt.% of the GNP loadings. The BZ/CER/GNPs composites were used as a matrix to impregnate carbon fibers through the vacuum-assisted resin transfer molding (VARTM) process. The resultant composite laminates were then tested for their mechanical and electrical properties.

Corresponding author: Dr. Ehsan Barjasteh

1. INTRODUCTION

Benzoxazine resin is a new class of thermoset resin that is suitable for high-performance applications due to its outstanding mechanical and thermal properties [1-4], good flame retardance [1,4], high char yield [1,4], low water absorption [1,3,4], excellent adhesive property [2], and electrical insulation [5,6]. It is also suitable for carbon-fiber composite manufacturing using out-of-autoclave (OOA) processes, such as vacuum-assisted resin transfer molding (VARTM). This is owed to the BZ resin's advantages, including no catalyst or coupling agent required for curing [1,4], no by-product formed in a cured sample [1,4,7], and near-zero volumetric shrinkage or expansion upon curing [1,3,4,8]. Because the OOA technique allows composite parts to be fabricated at an increasing volume with lower cost, the OOA-processable resin with developed performances is highly demanded in multiple industries.

An increase in the applications of benzoxazine and OOA manufacturing leads to multiple modifications of the resin's properties and process. Several works have reported a great compatibility and property enhancement by alloying benzoxazine with various types of epoxy resins [9,10,11]. For instance, our previous study showed that blending the bisphenol-A based benzoxazine with cycloaliphatic epoxy resin with a concentration greater than 25 wt.% exhibited a significant enhancement in a glass transition temperature (T_g) and crosslinking density. Furthermore, the resultant resin provided a suitable viscosity that eased the VARTM process [12]. However, the incorporation of epoxy led to a decrease in both tensile strength and modulus [12]. The modification of the BZ/CER system thus was required to enhance the diminished properties.

Graphene is a promising candidate as a nanofiller for enhancing the mechanical properties of the developed BZ/CER resin due to its outstanding mechanical modulus and strength of approximately 1 TPa and 130 GPa, respectively [13,14]. A single two-dimensional sheet of graphene can be obtained from overcoming the interaction force between layers of graphite. This can be achieved by multiple approaches such as ultrasonication, ball milling, and mechanical calendering [15]. In order to obtain the full advantages of utilizing graphene as a nanofiller, it is necessary for the nanoparticles to be homogenously dispersed throughout the entire matrix [16,17].

One of the commonly used methods to disperse the nanoparticles is ultrasonication. This approach requires a low-viscosity medium to assist the dispersion of the nanoparticles; therefore, solvents such as chloroform [18], acetone [19], ethanol [20], and tetrahydrofuran (THF) [21,22] are commonly used in several studies. However, these solvents are required to be removed from the system prior to the curing process to prevent negative effects on thermal and mechanical properties processed by any residual solvent [23,24]. This adds a complexity and extra cost to the fabrication process. For this reason, a solvent-free nanocomposite manufacturing approach becomes a key research interest. In our previous study [12], both ultrasonication and mechanical calendering using the 3RM were implemented to disperse GNPs in the BZ/CER matrix. Both mechanical and electrical performances of the 3RM samples were found to outperform the ultrasonication samples. Furthermore, this solvent-free mechanical calendering approach was suitable for high-volume nanocomposite manufacturing due to its simplicity, shorter manufacturing time, and uniform dispersion of nanoparticles in a host matrix.

In this study, GNPs at various ratios were incorporated in the BZ/CER resin using the 3RM calendering technique. The cured samples were characterized for their mechanical, thermomechanical, and electrical properties. The resultant nanocomposite with the most favorable ratio of GNPs was chosen to reinforce carbon fibers using the VARTM method. The cured composite was tested for its mechanical and electrical performances.

2. EXPERIMENTATION

2.1 Materials

Bisphenol A-based benzoxazine resin (BZ) was supplied from Huntsman Corporation with a density of 1.22 g/cm^3. Cycloaliphatic epoxy resin (CER), 3,4-epoxycyclohexylmethy 3,4-epoxycyclohexane carboxylate, was obtained from Lindau Chemicals Inc. with a density of 1.165 g/cm^3 under the chemical name of Lindoxy 190 (131-143 g/equiv.). Unidirectional carbon fiber fabrics were received from Textile Products Inc. with an areal weight of 335 gsm. Graphite nanoparticles were purchased from Ashbury Graphite Mills with a surface area of 350 m^2/g, under its commercial name of Nano 24.

2.2 Nanocomposite Fabrication

The ratio between BZ and CER (62.5:37.5) was chosen from our previous research [12], which provided a suitable viscosity for the VARTM process as well as improved thermal properties. Therefore, the same mixing ratio was investigated further with the incorporation of GNPs in this study.

The reference BZ/CER sample was mixed in a temperature-controlled oil bath at 110 °C with a magnetic stirrer rotating at 375 rpm until a homogenous mixture was achieved. The BZ/CER/GNP nanocomposite was fabricated by first adding the GNPs in CER at different ratios (0.6, 0.9, 1.8, and 2.4 wt.%). The GNP/CER mixture was then milled nine cycles in total until a uniform dispersion was obtained. The mixture was then transferred to a 20 ml vial and was immersed in the temperature-controlled oil bath at 110 °C. The BZ powder was gradually added to the GNP/CER mixture under a control shear force generated by a magnetic stirrer rotating at 375 rpm. Both reference and BZ/CER/GNP mixtures were placed between aluminum plates covered with the non-porous Teflon-coated fiberglass sheet to manufacture a thin film. The rests were casted in the aluminum mold for a surface resistivity test. The samples were cured at 180 °C for 2 hours, 200 °C for 2 hours, and 220 °C for 2 hours.

2.3 Composite Fabrication

The reference carbon-fiber composite laminate was manufactured without any GNP incorporation. The resin was prepared by mixing CER and BZ using a high-shear mechanical mixer in a temperature-controlled oil bath. The temperature was controlled at 110 °C, and the stirring speed was maintained at 800 rpm.

A specific ratio of GNP was incorporated in the CER. The mixture was then poured into a 3RM calendaring device (Torrey Hills Technologies) for nine cycles to ensure a uniform dispersion of GNPs. The BZ was then added gradually to the GNP/CER mixture to prevent a rapid increase in viscosity using the same mixing condition as the reference panel.

Twelve plies of dry carbon fiber with a dimension of 30.5 cm × 30.5 cm were stacked in a unidirectional sequence and were fabricated using the VARTM process. Both reference and GNP-filled composite laminates were infused at 110 °C and were cured at 180 °C for 2 hours, 200 °C for 2 hours, and 220 °C for 2 hours.

2.4 Characterization

A tensile test for thin-film samples was conducted at room temperature, in accordance with ASTM D1708, using the ADMET eXpert 4200 micro-tensile tester with a 445 N load cell. The average thickness and width of the specimens were 0.1 and 5 mm, respectively. The gauge length was initially set at 22 mm with the movement speed of 1 mm/min. Five specimens were tested for each GNP loading, and stress-strain curves were plotted to calculate the modulus of elasticity, tensile strength, and percent elongation.

A dynamic mechanical analysis of the nanocomposite was conducted using a DMA instrument (Q800, TA Instruments) equipped with a tensile clamp, in accordance with ASTM D4065. The temperature was changed from 30 °C to 350 °C with a heating rate of 5 °C/min.

An electrical property of the nanocomposite was determined based on surface resistivity of the casted samples using the four-point probe (Signatone Pro4) connected with the Keithley 2461 sourcemeter. The distance between measuring pins was fixed at 1 mm. The resistivity of the carbon fiber composites was measured by a two-point measurement with electrodes attached on the specimens' surfaces by conductive carbon tapes. The resistivity of the samples was measured on a rectangular bar with average dimensions of 8.9 cm × 3.2 cm × 0.39 cm.

A short beam shear test was determined using a universal mechanical test frame (Shimadzu) equipped with a 10 kN load cell, in accordance with ASTM D2344. The interlaminar shear strength (ILSS) of the carbon fiber composite was determined from samples that were tested at the room temperature with a span-to-thickness ratio of 4:1 and a crosshead speed of 1.27 mm/min. At least five samples were tested, and the average values were reported.

A flexural test was performed on the carbon fiber composite in order to measure the flexural modulus and strength. The test was conducted at the room temperature using a universal mechanical test frame (Shimadzu) mounted with a 10 kN load cell. A support span-to-thickness ratio of 32:1 and crosshead speed of 1 mm/min were applied in accordance with ASTM D7264.

3. RESULTS AND DISCUSSION

3.1 Dispersion of Graphene Nanoplatelets in Epoxy Matrix

A use of solvent is essential in easing the dispersion and preventing sedimentation of GNPs in a host matrix. In order to prevent any possible negative effects caused by remaining solvent prior to the curing process and to minimize the manufacturing steps, CER was chosen as a dispersing medium in this study to alleviate the distribution of GNPs due to its low viscosity.

It is important to ensure a uniform dispersion of GNPs in the matrix and maintain homogeneity without any sign of re-agglomeration. A small ratio of GNPs (i.e., 0.01 wt.%) was initially added and dispersed in the CER under nine cycles with the 3RM until no obvious cluster of the nanoparticles was observed in the mixture. A reference sample was prepared by physically mixing 0.01 wt.% GNPs with CER for a comparison as shown in **Figure 1**. The reference sample was observed to have several large agglomerations in the mixture and sediments at the bottom of the container. On the other hand, the mixture dispersed by the 3RM showed a homogenous mixture with a uniform dispersion of GNPs in the CER matrix.

After a three-day storage, the sample dispersed by the 3RM showed no signs of re-agglomeration among GNP particles and the mixture remained completely homogeneous. In contrast, the sample mixed by hand showed discontinuity in particle distribution and sediment at the bottom of the container. Therefore, it could be concluded that CER is a suitable candidate to replace typical solvents used for dispersion and the 3RM method is sufficient for dispersion of GNPs.

Figure 1. Three-day storage of the CER/GNPs mixture at room temperature for the agglomeration observation.

3.2 Graphene Nanoplatelets Filled Benzoxazine-Epoxy Nanocomposite

3.2.1 Tensile Properties

A reference specimen was manufactured without an addition of GNPs represented as 0 wt.% in **Figure 2**, followed by samples with increasing GNP contents. The tensile test was performed to determine the effect of GNP addition on the specimen's stiffness, strength, and elongation.

Figure 2a showed an improvement in the modulus of elasticity with an increasing GNP content up to 0.9 wt.%. At this ratio, the modulus reached 4.3 GPa, which was equivalent to a 53.6 % increase from the reference. This significant enhancement of the tensile properties was due to the outstanding stiffness of graphene (~1 TPa [13,14]) that constricted the mobility of polymer chains and shared a portion of load from the host matrix [17,24,25,26,27]. However, a further addition of GNPs beyond this ratio resulted in a decline of sample's stiffness. This was a result of GNP agglomerations at high nanofiller contents that minimized the effectiveness of the GNP's aspect ratio, which reduced the stress transfer from the matrix [24].

The trend in tensile strength was similar to the modulus of elasticity with a maximum increase of 36.9 % at 0.9 wt.%. This enhancement in the tensile strength was the result of a uniform dispersion of GNPs in the BZ/CER matrix as well as the graphene's exceptional tensile strength of nearly 130 GPa [13,14]. A further increase in the GNP content resulted in a reduction of the tensile strength as it had a high tendency to re-agglomerate. The agglomerate acted as a defect center and stress-concentration point when a load was applied [22,23,25].

The elongation of the GNP-filled nanocomposite was found to decrease with an increased GNP concentration, as shown in **Figure 2c**. At higher GNP contents, the particles began to contact each other and restricted the degree of chain formation under loading [21,22,28]. The resultant nanocomposite therefore became more brittle.

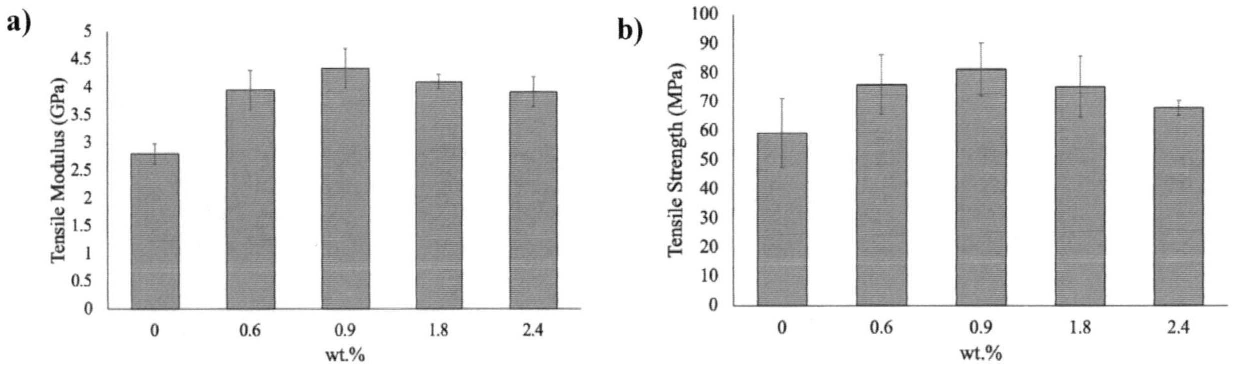

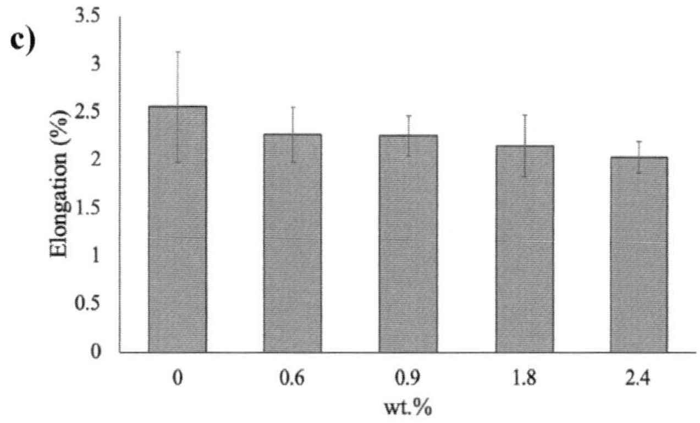

Figure 2. Tensile moduli (a), tensile strength (b), and percent elongation (c) of nanocomposites with different GNP contents dispersed by 3RM approach.

3.2.2 Thermomechanical Properties

The effect of GNP addition on the thermo-mechanical properties was investigated. **Figure 3** shows the storage modulus and loss factor (Tan δ) values with respect to temperature. It was observed that the initial storage modulus values at 35 °C increased with the increasing GNP concentration up to the maximum at 0.9 wt.%. With a further inclusion of GNPs beyond this content, a decrease in the storage modulus value was observed. The increase in storage modulus at 35 °C corresponded with the results obtained in the static tensile test, indicating a stronger stiffness due to a greater dispersion of GNPs in the specimens. The T_g values of all samples were determined and compared at the peak of the Tan δ curves. No significant change in the T_g was observed with the incorporation of GNPs.

Figure 3. Storage moduli (left) and Tan δ (right) curves of different GNP contents by 3RM approach.

3.2.3 Electrical Properties

In general, neat epoxy and benzoxazine resins exhibit an insulative behavior [5,6,17,22,29]. However, electrically conductive composites are highly demanded in the electronic and aerospace industries. One approach to overcome this limitation was to incorporate conductive nanofillers such as GNPs into the BZ/CER resin. In this study, the resistivity of the GNP-filled nanocomposites was found to reduce by several orders of magnitude due to a continuous formation of conductive networks of the graphene sheets.

Various GNP contents were incorporated to the BZ/CER resin up to the point where its resistivity decreased dramatically. This point, known as a percolation threshold, indicated a pathway formation for electron transfer in the nanocomposite [22]. The governing factors that determined the percolation threshold included a specific surface area, aspect ratio, dispersion uniformity, and interparticle distance of GNPs [17,22,30].

The electrical resistivity for each concentration of GNP-filled BZ/CER nanocomposite can be seen in **Figure 4**. A significant drop in electrical resistance was observed between 0.6 and 0.9 wt.% GNPs. This sudden drop in resistivity suggested a concentration where the percolation threshold could be found. The insulative behavior at low GNP contents was due to a large distance between GNPs that resulted in a discontinuous electrical path for electron mobility. With further addition of GNPs, the interparticle distance became small enough to promote the electron transfer [17,21].

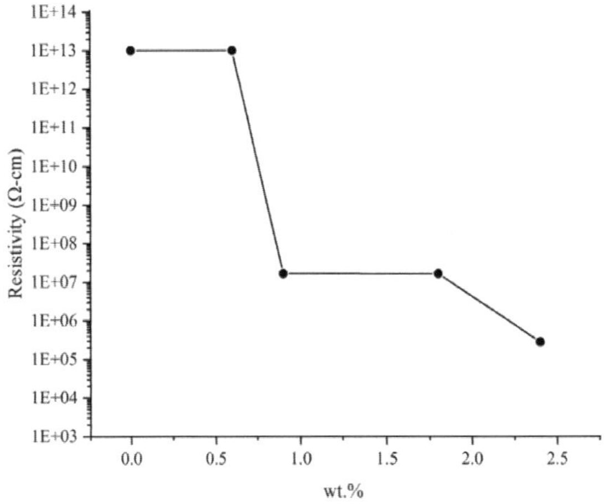

Figure 4. Electrical resistivity at different GNP contents dispersed by 3RM approach.

3.3 Composites

3.3.1 Mechanical Properties

Based on the results from different GNP contents in the nanocomposite, 0.9wt.% showed the optimum performances in all mechanical, thermo-mechanical, and electrical properties. A further study aimed to incorporate the GNP-filled nanocomposite in a carbon-fiber laminate. Short beam shear and flexural tests were then conducted to determine the mechanical properties of the resultant laminate.

The ILSS values were determined from the short beam shear test, which directly linked with the interfacial adhesion between fiber and matrix [31]. The reference laminate showed an ILSS value at 76.6 MPa. The incorporation of GNPs increased the value up to 80 MPa, which was a 4.5 % increase from the reference. The addition of GNPs could have influenced the mechanical interlocking between the fiber and matrix leading to an increase in shear modulus and reduced the interlaminar stress concentrations at the interphase [31].

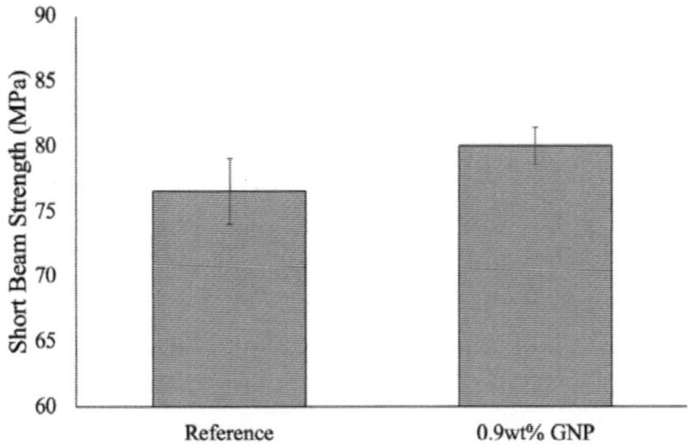

Figure 5. Interlaminar shear strength of composite laminates with and without GNPs.

Flexural properties of both the reference and GNP-filled composites were determined by conducting a flexural test. **Figure 6** compares the flexural modulus between the two specimens, with an increase from 101.03 GPa up to 111.82 GPa. This was equivalent to 10.68 % increase from the reference. Similarly, the flexural strength showed an increase of 4.8 %, from 1311.74 MPa to 1374.97 MPa. A small increase in both flexural modulus and strength values was also reported by Qin et al. [31] determining that the flexural properties in the 0° direction were primarily dominant by the properties of fibers.

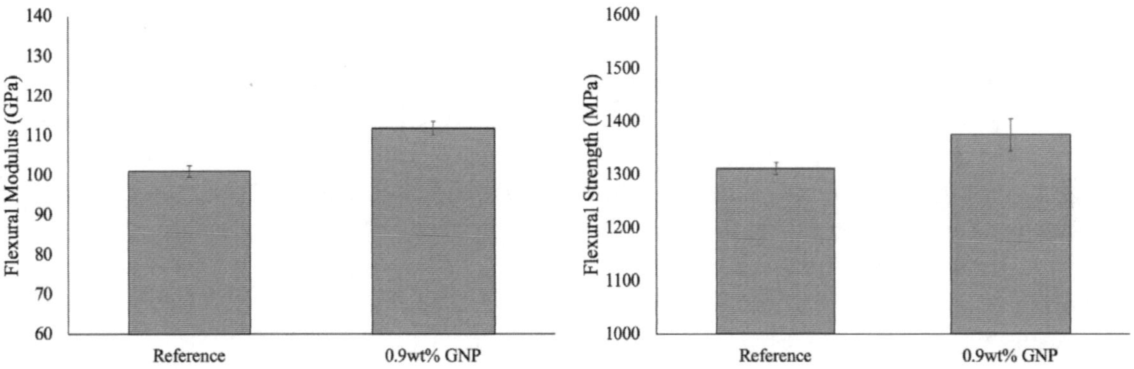

Figure 6. Flexural properties of composite laminates with and without GNPs: flexural modulus (left), flexural strength (right).

3.3.2 Electrical Properties

In general, dry carbon fibers are electrically conductive, but both BZ and CER resins possess an insulative behavior [5,6,17,22,29]. When the BZ/CER copolymer was infused into the laminate, it blocked the conductive path between the ply layers as well as surfaces. As mentioned in the previous section, the electrical percolation threshold was found at 0.9 wt.% GNPs. The same GNP content was then added to the BZ/CER resin to investigate the effect of conductive nanofillers on electrical resistivity.

The electrical resistance of both reference and GNP-filed composites were measured by a two-point probe method. Both of the reference and GNP samples were cut into a rectangular shape.

The length of each specimen bar was parallel to the direction of fibers, and the width was perpendicular to the fiber direction.

Table 1 shows the resistance results of each laminate. The incorporation of GNPs led to a reduction in electrical resistance in all directions. The reference laminate was observed to have an average volume resistivity of 2.44 Ω-cm, while the addition of 0.9 wt.% GNPs resulted in a 41.5 % reduction of the composite's volume resistivity to 1.42 Ω-cm. The electrical resistance was influenced by the GNPs within the composite panel, which may have assisted in creating more electrical networks for the electron transfer [31]. A conductive path between the probe points was created on both sides of the composites' surfaces to determine the surface resistivity in each coordinate direction and volume resistivity of the entire panel.

Table 1. Electrical properties of composite laminates with and without GNPs.

Resistance	Reference laminate	GNP-filled laminate
Average 0°-direction resistance (Ω)	15.13	11.76
Average 90°-direction resistance (Ω)	2.47	1.36
Average through thickness resistance (Ω)	2.58	1.28
Volume resistivity (ρ) (Ω-cm)	2.44	1.42

3.3.3 Analysis and Optimization of Manufacturing Process

It was observed from the resultant panel that it was difficult to obtain a uniform dispersion of the GNP-filled laminate. Although a uniform dispersion of GNPs in the matrix could be obtained by the use of 3RM at 0.9 wt.%, during the resin transfer infusion, the GNPs did not flow uniformly. This resulted in a high concentration of GNPs at the inlet and both edges of the panel, as shown by the arrows in Figure 7. Therefore, the manufacturing process of GNP-filled composites should be further investigated and optimized.

In order to optimize the manufacturing process, it is beneficial to initially visualize the flow of GNPs during the infusion process. This can be conducted by infusing the GNPs at a low weight fraction (e.g., 0.01 wt.%) into a lighter color material, such as fiberglass. Once the flow of GNP is observed, there are several methods can be further implemented to improve the dispersion quality. One method is to reduce the infusion pressure in order to ensure that the GNPs can more easily flow in between dry fiber plies. Another approach is to replace the infusion step with a B-staging technique by placing resin films in between carbon fiber plies. The layup can then be cured with a suitable pressure and temperature using a heat press [32].

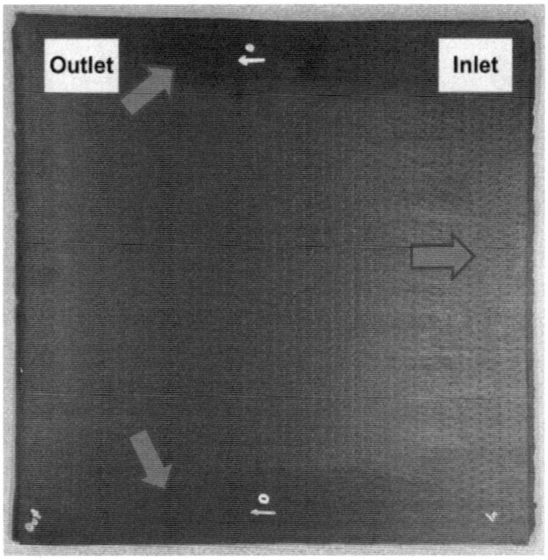

Figure 7. Bottom side of the resultant GNP-filled composite laminate.

4. CONCLUSIONS

The incorporation of GNPs in the BZ/CER copolymer was investigated in this study. It was found that increasing GNP contents up to 0.9 wt.% enhanced both mechanical and thermo-mechanical properties. A further nanoparticle addition beyond this content resulted in a decrease in both tensile performances and storage moduli. No significant change in the glass transition temperature was observed with the incorporation of GNPs. The electrical properties were determined based on the resistivity of the nanocomposite, and the percolation threshold was found between 0.6 and 0.9 wt.%.

The GNP content of 0.9 wt.% was determined to provide the optimum properties; therefore, the same ratio was used during composite fabrication. The GNP-filled composite showed improvements in both flexural properties and interlaminar shear strength. The electrical resistance decreased in the composite containing GNPs. Although improvements in mechanical and electrical properties could be obtained with the addition of GNPs, the GNP content was found to be ununiform throughout the entire laminate via the VARTM process. Therefore, an optimization of manufacturing processes should be further investigated.

5. REFERENCES

1. Rimdusit, S., Jubsilp, C., and Tiptipakorn, S. "Alloys and Composites of Polybenzoxazines." *Singapore: Springer* (2013): 29–46.

2. Zou, X., Yang, X., Xu, M., Jia, K., and Liu, X. "Curing Behaviors and Properties of Allyl- and Benzoxazine-Functional Phthalonitrile with Improved Processability." *Journal of Polymer Research* 23(1) (2016): 2.

3. Zhang, S., Yang, P., Bai, Y., Zhou, T., Zhu, R., and Gu, Y. "Polybenzoxazines: Thermal Responsiveness of Hydrogen Bonds and Application as Latent Curing Agents for Thermosetting Resins." *ACS Omega* 2(4) (2017): 1529–1534.

4. Ke, W., Rumin, W., and Jinfang, Z. "Curing Kinetics of Hybrid Networks Composed of Benzoxazine and Multifunctional Novolac Epoxy." *International Journal of Polymer Science* (2015): 1–9.

5. Zúñiga, C., Bonnaud, L., Lligadas, G., Ronda, J. C., Galià, M., Cádiz, V., and Dubois, P. "Convenient and Solventless Preparation of Pure Carbon Nanotube/ Polybenzoxazine Nanocomposites with Low Percolation Threshold and Improved Thermal and Fire Properties." *Journal of Materials Chemistry A* 2(19) (2014): 6814-6822.

6. Dumas, L., Bonnaud, L., Olivier, M., Poorteman, M., and Dubois, P. "High Performance Benzoxazine/CNT Nanohybrid Network - An Easy and Scalable Way to Combine Attractive Properties." *European Polymer Journal* (2014): 218-225.

7. Uchida, S., Kawauchi, T., Furukawa, N., and Takeichi, T. "Polymer Alloys of High-Molecular-Weight Benzoxazine and Epoxy Resin." *High Performance Polymer* 26(7) (2014): 846–855.

8. Ghosh, N. N., Kiskan, B., and Yagci, Y. "Polybenzoxazines-New High Performance Thermosetting Resins: Synthesis and Properties." *Progress inPolymer Science* 32(11) (2007): 1344–1391.

9. Ishida, H., and Allen, D. J. "Mechanical Characterization of Copolymers Based on Benzoxazine and Epoxy." *Polymer* 37(20) (1996): 4487–4495.

10. Rao, B. S., Reddy, K. R., Pathak, S. K., and Pasala, A. R. "Benzoxazine-Epoxy Copolymers: Effect of Molecular Weight and Crosslinking on Thermal and Viscoelastic Properties." *Polymer International* 54(10) (2005): 1371-1376.

11. Zhao, P., Zhou, Q., Deng, Y., Zhu, R., and Gu, Y. "A Novel Benzoxazine/Epoxy Blend with Multiphase Structure." *RSC Advanse* 4(1) (2014): 238-242.

12. Barjasteh, E., Gouni, S., Sutanto, C., and Narongdej, P. "Bisphenol-A Benzoxazine and Cycloaliphatic Epoxy Copolymer for Composite Processing by Resin Infusion." *Journal of Composite Materials* 53(13) (2018): 1777-1790.

13. Liu, S., Chevali, V. S., Xu, Z., Hui, D., and Wang, H. "A Review of Extending Performance of Epoxy Resins Using Carbon Nanomaterials." *Composites Part B: Engineering* 136 (2018): 197-214.

14. Young, R. J., Kinloch, I. A., Gong, L., and Novoselov, K. S. "The Mechanics of Graphene Nanocomposites: A Review." *Compososite Science and Technology* 72(12) (2012): 1459–1476.

15. Yi, Min, and Zhigang Shen. "A Review on Mechanical Exfoliation for the Scalable Production of Graphene." *Journal of Materials Chemistry A* 3(22) (2015): 11700-11715.

16. Domun, N., Hadavinia, H., Zhang, T., Sainsbury, T., Liaghat, G. H., and Vahid, S. "Improving the Fracture Toughness and the Strength of Epoxy Using Nanomaterials – a Review of the Current Status." *Nanoscale* 7(23) (2015): 10294–10329.

17. Ahmadi-Moghadam, B., and Taheri, F. "Effect of Processing Parameters on the Structure and Multi-Functional Performance of Epoxy/GNP-Nanocomposites." *Journal of Materials Science* 49(18) (2014): 6180-6190.

18. Ghaleb, Z. A., Mariatti, M., and Ariff, Z. M. "Graphene Nanoparticle Dispersion in Epoxy Thin Film Composites for Electronic Applications: Effect on Tensile, Electrical and Thermal Properties." *Journal of Materials Science: Materials in Electronics* 28(1) (2017): 808-817.

19. Shen, X. J., Liu, Y., Xiao, H. M., Feng, Q. P., Yu, Z. Z., and Fu, S. Y. "The Reinforcing Effect of Graphene Nanosheets on the Cryogenic Mechanical Properties of Epoxy Resins." *Composite Science and Technology* 72(13) (2012): 1581-1587.

20. Tang, L. C., Wan, Y. J., Yan, D., Pei, Y. B., Zhao, L., Li, Y. B., Wu, L. Bin, Jiang, J. X., and Lai, G. Q. "The Effect of Graphene Dispersion on the Mechanical Properties of Graphene/Epoxy Composites." *Carbon* 60 (2013): 16-27.

21. Araby, S., Meng, Q., Zhang, L., Kang, H., Majewski, P., Tang, Y., and Ma, J. "Electrically and Thermally Conductive Elastomer/Graphene Nanocomposites by Solution Mixing." *Polymers* 55(1) (2014): 201-210.

22. Ma, J., Meng, Q., Zaman, I., Zhu, S., Michelmore, A., Kawashima, N., Wang, C. H., and Kuan, H. C. "Development of Polymer Composites Using Modified, High-Structural Integrity Graphene Platelets." *Composites Science and Technology* 91 (2014): 82-90.

23. Wei, J., Atif, R., Vo, T., and Inam, F. "Graphene Nanoplatelets in Epoxy System: Dispersion, Reaggregation, and Mechanical Properties of Nanocomposites." *Journal of Nanomaterials* 16(1) (2015): 374.

24. Chong, H. M., Hinder, S. J., and Taylor, A. C. "Graphene Nanoplatelet-Modified Epoxy: Effect of Aspect Ratio and Surface Functionality on Mechanical Properties and Toughening Mechanisms." *Journal of Materials Science* 51(19) (2016): 8764-8790.

25. Zaman, I., Manshoor, B., Khalid, A., Meng, Q., and Araby, S. "Interface Modification of Clay and Graphene Platelets Reinforced Epoxy Nanocomposites: A Comparative Study." *Journal of Materials Science* 49(17) (2014): 5856-5865.

26. Zaman, I., Kuan, H. C., Dai, J., Kawashima, N., Michelmore, A., Sovi, A., Dong, S., Luong, L., and Ma, J. "From Carbon Nanotubes and Silicate Layers to Graphene Platelets for Polymer Nanocomposites." *Nanoscales* 4(15) (2012): 4578-4586.

27. Zaman, I., Nor, F. M., Manshoor, B., Khalid, A., and Araby, S. "Influence of Interface on Epoxy/Clay Nanocomposites: 2. Mechanical and Thermal Dynamic Properties." *Procedia Manufacturing* 2 (2015): 23-27.

28. Prolongo, M. G., Salom, C., Arribas, C., Sánchez-Cabezudo, M., Masegosa, R. M., and Prolongo, S. G. "Influence of Graphene Nanoplatelets on Curing and Mechanical Properties of Graphene/Epoxy Nanocomposites." *Journal of Thermal Analysis and Calorimetry* 125(2) (2016): 629-636.

29. Ghaleb, Z. A., Mariatti, M., and Ariff, Z. M. "Properties of Graphene Nanopowder and Multi-Walled Carbon Nanotube-Filled Epoxy Thin-Film Nanocomposites for Electronic Applications: The Effect of Sonication Time and Filler Loading." *Composite Part A: Applied Science and Manufacturing* 58 (2014): 77-83.

30. Prolongo, S. G., Moriche, R., Jiménez-Suárez, A., Sánchez, M., and Ureña, A. "Advantages and Disadvantages of the Addition of Graphene Nanoplatelets to Epoxy Resins." *European Polymer Journal* 61 (2014): 206-214.

31. Qin, W., Vautard, F., Drzal, L. T., and Yu, J. "Mechanical and Electrical Properties of Carbon Fiber Composites with Incorporation of Graphene Nanoplatelets at the Fiber-Matrix Interphase." *Composites Part B: Engineering* 69 (2015): 335-341.

32. Kim, C. U., Kim, S. J., Park, J. C., and Song, J. I. "Fabrication and Evaluation of Mechanical Properties of CF/GNP Composites." *Procedia Manufacturing* 2 (2015): 368-373.

THE Z-DIRECTIONAL DC ELECTRICAL CONDUCTIVITY OF CARBON NANOFIBER Z-THREADED POROUS CFRP COMPOSITES MANUFACTURED BY RADIAL FLOW ALIGNMENT TECHNIQUE

Bikash Ranabhat[1,2], Sebastian Kirmse[1,2], Michael Johnson [1], Kuang-Ting Hsiao[1,*]

University of South Alabama
Department of Mechanical Engineering, 150 Jaguar Drive, Shelby Hall, Suite 3130
Mobile, AL, 36608
1.Department of Mechanical Engineering
2.Systems Engineering Program
*Corresponding Author: Email: kthsiao@southalabama.edu, Phone: +1 (251) 460-7889

ABSTRACT

The non-conductive polymer matrix impedes carbon fiber's excellent electrical conductivity in the carbon fiber reinforced polymer (CFRP) composites. Due to its low z-directional DC electrical conductivity, CFRP composites has limited application capabilities in lightning strike protection (LSP) and electromagnetic shielding. This research demonstrates the ability to manufacture lightweight and electrically conductive (i.e., z-direction) CNFs z-threaded porous CFRP composite utilizing just 50 % of epoxy used in the traditional composites. In this study, the radial flow alignment technique was used to z-thread carbon nanofibers (CNFs) into the carbon fiber fabric. Experimental results showed the z-directional DC electrical conductivity in the 0.85 wt% CNF z-threaded porous (P-ZT-CFRP) was 21.06 S/m, which is an increase of 778 % and 107 % compared to the traditional composite sample and P-CFRP (without nano-reinforcement) laminate, respectively.

Keywords: Z-threaded CFRP, Electrical conductivity, Carbon nanofiber, Radial flow alignment.

1. INTRODUCTION

Carbon fiber reinforced polymer (CFRP) composites have a high strength-to-weight, better anti-fatigue, anti-aging features, high in-plane electrical conductivity, and corrosion-resistant features. With the characteristics of lightweight and high performance, CFRP composites are highly desired materials for aerospace, automotive, portable electronics, medical devices, and marine instruments. However, the out-of-plane direction (i.e., z-direction) electrical conductivity is minimal compared to the in-plane direction. The minimal z-direction electrical conductivity has limited its applications in aircraft, wind turbine blades, electromagnetic shielding, lightning strike protection (LSP), damage detecting, strain/stress sensing, chemical, and thermal sensing [1,2]. The low conductive nature of CFRP composite is mainly due to the insulating matrix. Though, the polymer matrix reinforces the carbon fiber, it is also the weakest point of the material system. It makes the composite vulnerable to matrix-sensitive damage (e.g., compressive failure, delamination, and shear failure). The polymer matrix used, significantly reduces the electrical and thermal conductivity, especially in the through-thickness direction. Merely removing the polymer matrix out of the composite is not a feasible solution to the previously mentioned problems because the carbon fiber requires reinforcement to form a lightweight and robust material. Researchers all over the world have conducted studies attempting to overcome the weakness of the polymer matrix in the composite and reported the use of nanofillers (carbon nanotubes and carbon nanofibers) as a potential solution. Moreover, researchers have aligned the nanofillers in the host matrix to attain desired CFRP composite properties using different alignment techniques, e.g., mechanical force [3,4], magnetic field [5,6], electric field [7,8], shear flows [9], electrospinning [10,11], radial flow alignment (RFA) [12], and aligned resin film transfer method [13].

1.1 Related work

Based on the previously conducted research, CNFs z-threaded CFRP laminates have been found useful in enhancing the thermal, electrical, and mechanical properties of CFRP laminates, which is presented in Table 1. The purpose of this paper is: (1) to prepare a polymer matrix containing 50 % of the epoxy used in the traditional composites, (2) use the radial flow alignment technique (RFA) in the new polymer matrix developed, (3) to develop ultralight porous CNF z-threaded nanocomposite, and (4) to measure through-thickness DC electrical conductivity (DC EC) of the porous CNFs z-threaded nanocomposite. The radial flow alignment technique is useful to produce CNFs z-threaded prepreg tape of width around 1 inch wide, which makes RFA a suitable candidate for the prepreg tape production methodology for additive manufacturing. The other advantages of RFA include the cost and time to set up this technology is less expensive compared to resin flow transfer (sponge method) [15].

Table 1. Z-threaded CFRP laminates thermal, electrical, and mechanical properties.

Material Property	**CNF wt%**	**CNF Type**	**CF Type**	**Resin Type**	**Mfg. Method**	**Control CFRP**	**ZT-CFRP**	**Improvement**
Mode -I Delamination [14]	0.1-0.3	PR-24-XT-PS	T300 plain weave	EPON 862/Epikure-W	Resin-flow transfer (sponge)	305 J/m^2	397 J/m^2	29 %
Through-thickness DC EC [12]	1	PR-24-XT-PS	AS4 UD	EPON 862/Epikure-W	Radial flow alignment	0.161 S/m	16.24 S/m	9987 %
Through-thickness DC EC [15]	0.5, 1, 2	PR-24-XT-PS	T700 UD	EPON 862/Epikure-W	Radial flow alignment	2.4 S/m	8.56, 13.46, 16.9 S/m resp.	257 %, 461 %, 604 %, resp.
Through-thickness DC EC [18]	2	PR-24-LD-HHT	AS4 UD	EPON 862/Epikure-W	Resin-flow transfer (sponge)	0.33 S/m	9.81 S/m	2971 %
Longitudinal Compressive Strength [16]	1	PR-24-XT-HHT	AS4 UD	EPON 862/Epikure-W	Resin-flow transfer (sponge)	673.85 MPa	773.76 MPa	14.83 %
CF: Carbon fiber; UD: Unidirectional; PS: pyrolytically stripped carbon nanofibers; HHT: High-heat treated (up to 3000 °C) carbon nanofibers; XT: Debulked form.								

2. EXPERIMENTATION

2.1 Materials

The materials that were used to manufacture the porous nanocomposites along with their suppliers are presented in Table 2.

Table 2. List of materials and its suppliers

Materials	Suppliers
Carbon fiber: Toray T-700 unidirectional (680 g/m^2 areal weight, 1.8 g/cm^3 fiber density, 12k tow size).	Toray
Polymer matrix: Epon 862 & Epikure-W catalyst (100:26.5)	Miller-Stephenson Chemical Co
Surfactants:BYK-9076 & BYK-9077	BYK USA, Inc
CNFs: PR-24-XT-HHT	Applied Science, Inc
Acetone	Fisher Scientific International, Inc

Two different samples were manufactured; a porous CFRP nanocomposite containing 0.85 wt% of CNF z-threads (i.e., 0.85 wt% P-ZT-CFRP) and porous CFRP composite without CNFs (i.e., P-CFRP) to study the z-directional (i.e., through-thickness) DC electrical conductivity.

2.2 Manufacturing of Porous Nanocomposites

The manufacturing of the porous nanocomposites was carried out through the following process: (1) preparation of matrix system, (2) radial flow alignment (RFA) for z-threading of CNFs into the fabric, and (3) Out of Autoclave-Vacuum Bag Only (OOA-VBO) curing process.

The matrix system was prepared by mixing two mixtures. The first mixture (M1) was a mixture of acetone/surfactants/CNFs in the ratio of 50, 1, and 0.85 % by weight, respectively. The second mixture (M2) was the mixture of Epon 862 and Epikure-W catalyst in the ratio of 100:26.5.

2.2.1 Preparation of matrix system

The M1 and M2 mixtures were prepared simultaneously. The M1 mixture was subjected to high shear mixing for a total of 40 minutes with 20 minutes in a clockwise direction and 20 minutes in a counterclockwise direction. While at the same time, M2 was heated in the oven at 120 °C for 50 minutes with continuous stirring at an interval of every 10 minutes. Once the M1 mixture came out of high shear mixing, it was subjected to sonication for 40 minutes. The M2 mixture was taken out of the oven and allowed to cool to 40 °C. The M1 mixture, once out of sonication, was gradually mixed with the M2 mixture with continuous mechanical stirring and subjected to high shear mixing for 10 minutes to make the mixture M.

2.2.2 Radial flow alignment for P-ZT-CFRP

The z-aligned porous prepregs (i.e., CNFs z-threaded porous CFRP prepreg) were prepared using a radial flow alignment technique to z-thread CNFs into the T-700 fabric as described in [17]. The carbon fiber was conformally rolled on the curve surface of the perforated hollow roller. The open end of the perforated roller was connected to the vacuum pump while the other end was closed. A perforated hollow roller was employed to draw the mixture M towards the center of the roller. The converging flow of the viscous mixture M causes the CNFs being stretched, aligned and forced to thread through the fabric. Thus, produced z-threaded porous CFRP prepreg was transferred to nylon bagging and stored in the freezer at -18 °C for later use.

2.2.3 Hand layup for P-CFRP

The Hand layup technique was used to prepare the P-CFRP prepreg plies. The mixture of Epon 862 and Epikure-W in the ratio 100:26.5 was B-staged in the oven at 120 °C for 50 minutes and allowed to cool to 40 °C. Then 50 wt% acetone was gradually mixed through mechanical mixing. The prepared final mixture was then poured on top of the fabric and rolled with a roller for uniform impregnation of resin. Then, the P-CFRP prepreg was transferred to nylon bagging and stored in the freezer for later use.

2.2.4 Out of Autoclave-Vacuum Bag Only (OOA-VBO) Curing process.

Five of each P-CFRP and P-ZT-CFRP prepreg plies were stacked and completely cured to form P-CFRP laminates and P-ZT-CFRP laminates via OOA-VBO curing process. The layup scheme used in the OOA-VBO process is shown in Figure 1, while the curing cycle is presented in Table 3. More details about the layup scheme are described in [12]. Once the samples were cured, a computer numerical control (CNC) router (CNC Shark Pro HD from Next Wave Automation) was used to cut the samples into the dimension of 80 mm length and 13 mm width. The samples were then polished on both the top and bottom surfaces and cleaned with acetone before testing the samples for z-directional DC electrical conductivity. The overall sample manufacturing and testing are shown in Figure 2.

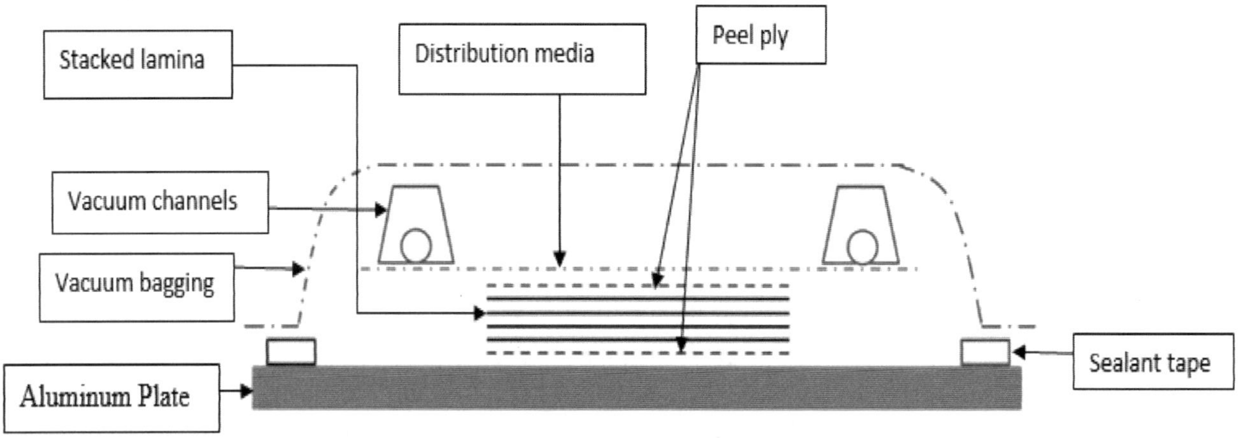

Figure 1. OOA-VBO layup scheme used for the preparation of samples [12]

Table 3. OOA-VBO curing cycle used for samples [12]

Time (min)	Temperature (°C)	Vacuum Pump
60	Room temperature (23)	On
30	45	On
120	120	On
240	180	Off

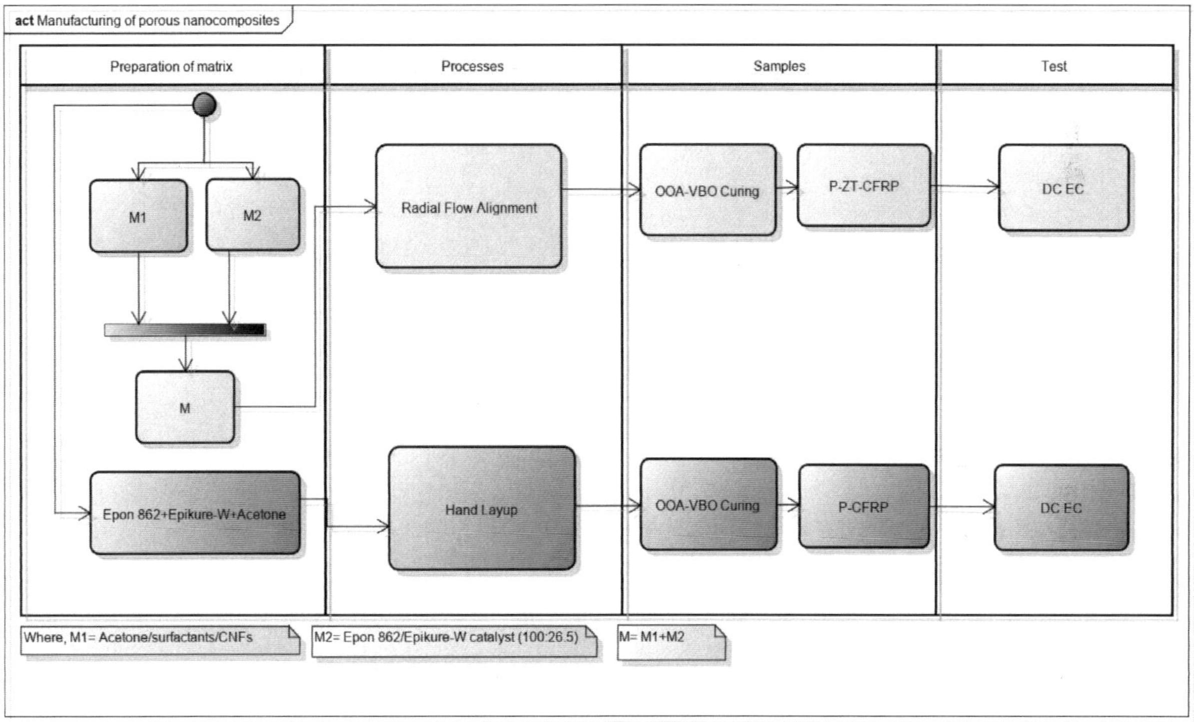

Figure 2. Activity diagram for samples preparation and testing.

2.3 Z-directional DC electrical conductivity test

A four-probe technique was employed to measure the electrical resistance of the samples. The testing procedure, as described in the publication by Bikash et al. [12], was followed to measure electrical resistance R (Ω) of the P-ZT-CFRP and P-CFRP samples. Then Equation 1 was used to calculate the through-thickness DC electrical conductivity σ (S/m). The z-directional DC electrical conductivity testing apparatus is shown in Figure 3, and more details about its construction are published in [12].

$$\sigma = \frac{L}{R * A} \qquad (1)$$

Where, σ = specific electrical conductivity of the CFRP laminate sample (S/m).

L = Z-direction thickness of CFRP laminate sample (m)

A = Area of the CFRP laminate sample (m^2)

R = Electrical resistance measured (ohm)

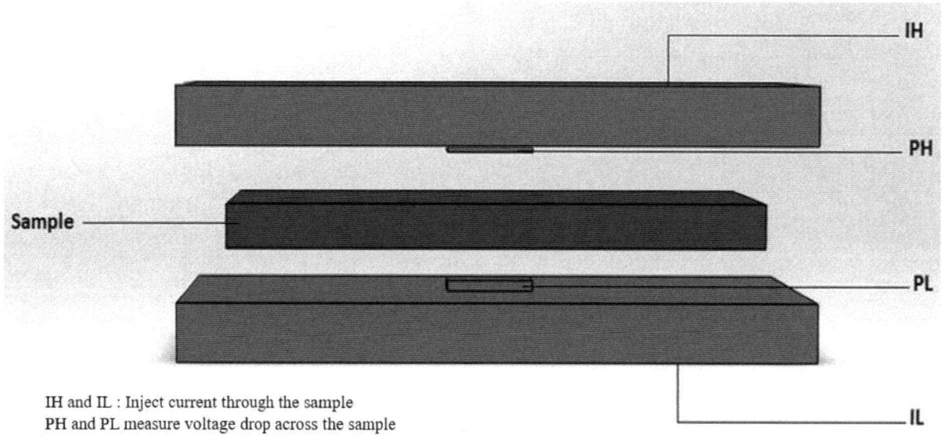

Figure 3. Schematic of the four-probe z-directional DC electrical conductivity testing apparatus [12].

3. RESULTS

The radial flow alignment technique [17] was verified and validated by the authors [12,15] as an effective method to align and z-thread CNFs into the CFRP prepreg through SEM images and through-thickness DC electrical conductivity values. The through-thickness DC electrical conductivity results are summarized in Table 4, together with the results for traditional CFRP, 0.5, 1, and 2 wt% CNFs z-threaded CFRP laminate from a previous study [15].

Table 4. Statistical analysis of z-directional DC electrical conductivities measured by four-probe test for traditional CFRP, 0.5, 1, and 2 wt% CNF z-threaded CFRP laminates [15], porous CFRP laminate, and 0.85 wt% CNF z-threaded P-CFRP laminate.

Samples	Mean DC z-electrical conductivity (S/m)	Standard deviations (S/m)	Coefficient of variation (%)	% Increase	Normalized based on control CFRP
Traditional CFRP laminate [15]	2.40	0.126	5.30	0	1 X
0.5 wt% CNFs z-threaded CFRP laminate [15]	8.56	0.769	9	257	3.57 X
1 wt% CNFs z-threaded CFRP laminate [15]	13.46	1.460	10.80	461	5.61 X
2 wt% CNFs z-threaded CFRP laminate [15]	16.90	1.436	8.50	604	7.04 X
Porous CFRP laminate	10.19	0.730	7.17	325	4.25 X
0.85 wt% CNFs z-threaded P-CFRP laminate	21.06	1.118	5.31	778	8.78 X

The results show that even without the addition of any nanofillers, z-direction DC electrical conductivity of the P-CFRP laminate is 325 % and 19 % higher compared to the traditional CFRP and 0.5 wt% CNF z-threaded CFRP, respectively. The increase in the z-directional DC electrical conductivity of the P-CFRP laminate is due to the use of a reduced amount of low-conductive/insulating resin material during the composite manufacturing process. The reduction of resin amount in the CFRP composite was achieved by adding acetone in the matrix and removed acetone during the curing process, creating a porous and conductive composite. The z-directional DC electrical conductivity observed in the 0.85 wt% CNFs z-threaded P-CFRP was 21.06 S/m, which is a 778 % increase compared to the control sample (i.e., traditional CFRP) and 107 % increase in comparison with the P-CFRP laminate. The z-direction DC electrical conductivity for the P-CFRP was 10.19 S/m, which means it can satisfy and exceed the electromagnetic shielding's electrical conductivity requirement of 10 S/m [19]. At the same time, 0.85 wt% CNFs z-threaded P-CFRP comfortably surpass the electrical conductivity requirement for electromagnetic shielding applications. Normalizing all the conductivity values based on the conductivity of the traditional CFRP sample, it was reported that the 0.85 wt% P-ZT-CFRP laminate was approximately 8.78 times as conductive as the traditional CFRP laminate and 2.07 times as conductive as the P-CFRP.

4. CONCLUSIONS

In this study, a porous composite (P-CFRP) was manufactured using hand layup while 0.85 wt% CNFs z-threaded porous nanocomposite (i.e., P-ZT-CFRP) was fabricated using a radial flow alignment process. The porous structure in the composite was created in the sample through the vaporization of the acetone during the curing process. The electrically conductive pathway formed by the alignment of carbon nanofibers in the though-thickness direction using a radial flow alignment technique and the use of a reduced amount of resin to manufacture the porous nanocomposite significantly enhanced the z-direction DC electrical conductivity of the sample. The DC electrical conductivity recorded for the porous composite (P-CFRP) and 0.85 wt% CNF z-threaded porous nanocomposite (P-ZT-CFRP) was 10.19 S/m and 21.06 S/m, respectively. The DC electrical conductivity value of 0.85 wt% CNF z-threaded porous nanocomposite was the highest value obtained for the z-threaded CFRP laminate [12,15,18].

In conclusion, the z-directional DC electrical conductivity value for the P-CFRP and 0.85 wt% CNF z-threaded P-CFRP satisfied the electromagnetic shielding applications requirement (i.e. 10 S/m) [19]. In the future, more tests like interlaminar shear strength and tension tests will be performed on the CNFs z-threaded P-CFRP nanocomposite to make it available for a broader range of applications. The future goal of the study is to make the prepreg tape suitable for magnetic compaction force assisted additive manufacturing [20].

5. ACKNOWLEDGMENTS

The authors would like to acknowledge the financial support from the Alabama Department of Commerce through the Alabama Innovation Fund (Award number: 150436) and the Graduate Assistantship from the Systems Engineering Program at the University of South Alabama. The authors are grateful for the carbon fiber materials provided by Toray, the surfactants provided by BYK USA, Inc., and the carbon nanofiber material provided by Applied Science, Inc.

6. REFERENCES

1. Abry, J. C., Y. K. Choi, A. Chateauminois, B. Dalloz, G. Giraud, and M. Salvia. "In-situ monitoring of damage in CFRP laminates by means of AC and DC measurements." *Composites Science and Technology* 61, no. 6 (2001): 855-864.

2. Yokozeki, Tomohiro, Teruya Goto, Tatsuhiro Takahashi, Danna Qian, Shouji Itou, Yoshiyasu Hirano, Yuichi Ishida, Masaru Ishibashi, and Toshio Ogasawara. "Development and characterization of CFRP using a polyaniline-based conductive thermoset matrix." *Composites Science and Technology* 117 (2015): 277-281.

3. Jin, L., C. Bower, and O. Zhou. "Alignment of carbon nanotubes in a polymer matrix by mechanical stretching." Applied physics letters 73, no. 9 (1998): 1197-1199.

4. Najeeb, C. K., Jingbo Chang, Jae-Hyeok Lee, and Jae-Ho Kim. "Fabrication of aligned ultra-thin transparent conductive films of single-walled carbon nanotubes by a compression/sliding method." *Scripta Materialia* 64, no. 2 (2011): 126-129.

5. Smith, B. W., Z. Benes, D. E. Luzzi, J. E. Fischer, D. A. Walters, M. J. Casavant, J. Schmidt, and R. E. Smalley. "Structural anisotropy of magnetically aligned single wall carbon nanotube films." Applied Physics Letters 77, no. 5 (2000): 663-665.

6. Choi, E. S., J. S. Brooks, D. L. Eaton, M. S. Al-Haik, M. Y. Hussaini, H. Garmestani, D. Li, and K. Dahmen. "Enhancement of thermal and electrical properties of carbon nanotube polymer composites by magnetic field processing." *Journal of Applied physics* 94, no. 9 (2003): 6034-6039.

7. Krupke R, Hennrich F, kappes MM, Lohneysen HV." Surface conductance induced dielectrophoresis of semiconducting single-walled carbon nanotubes." Nano Lett 2004:4:1395-9.

8. Zhu, Yue-Feng, Chen Ma, Wei Zhang, Ren-Ping Zhang, Nikhil Koratkar, and Ji Liang. "Alignment of multiwalled carbon nanotubes in bulk epoxy composites via electric field." Journal of Applied Physics 105, no. 5 (2009): 054319.

9. Vigolo, Brigitte, Alain Penicaud, Claude Coulon, Cédric Sauder, René Pailler, Catherine Journet, Patrick Bernier, and Philippe Poulin. "Macroscopic fibers and ribbons of oriented carbon nanotubes." *Science* 290, no. 5495 (2000): 1331-1334.

10. Ko, Frank, Yury Gogotsi, Ashraf Ali, Nevin Naguib, Haihui Ye, G. L. Yang, Christopher Li, and Peter Willis. "Electrospinning of continuous carbon nanotube-filled nanofiber yarns." Advanced materials 15, no. 14 (2003): 1161-1165.

11. Sen, Rahul, Bin Zhao, Daniel Perea, Mikhail E. Itkis, Hui Hu, James Love, Elena Bekyarova, and Robert C. Haddon. "Preparation of single-walled carbon nanotube reinforced polystyrene and polyurethane nanofibers and membranes by electrospinning." Nano letters 4, no. 3 (2004): 459-464.

12. Ranabhat, B., and Hsiao, K. "Improve the Through-Thickness Electrical Conductivity of CFRP Improve the Through-Thickness Electrical Conductivity of CFRP Laminate Using Flow-Aligned Carbon Nanofiber Z-Threads," in *Proceedings of SAMPE 2018 (Society for the Advancement of Material and Process Engineering), Long Beach, CA, May, 21-24*, 2018, SE18--1100.

13. K. Hsiao and G. Hickman, "Method for Manufacturing Nano-Structurally Aligned Multi-Scale Composites," WO/2015/017321, 06-Feb-2015.

14. Hsiao, K.T., Scruggs, A.M., Brewer, J.S., Hickman, G.J.S., McDonald, E.E., and Henderson, K. "Effect of carbon nanofiber z-threads on mode-I delamination toughness of carbon fiber reinforced plastic laminates," Composites Part A: Applied Science and Manufacturing 91 (2016): 324–335. DOI: 10.1016/j.compositesa.2016.10.022.

15. Ranabhat, Bikash, Sebastian Kirmse, and Kuang-Ting Hsiao."Roles of carbon nanofiber alignment and concentration for the improvement of z-directional DC electrical conductivity of CNF z-threaded CFRP laminates manufactured using radial flow alignment technique," in *Proceedings of CAMX 2019, Anaheim, CA, 2019.*

16. Kirmse, S., Kim, K., Ranabhat, B., and Hsiao, K.-T. "Effects of Carbon Nanofiber Z-Threads on the Longitudinal Compressive Strength of Unidirectional CFRP Laminates," in Proceedings of SAMPE 2019 - Charlotte, NC, May 20-23, 2019, TP19--1531. DOI: 10.33599/nasampe/s.19.1531.

17. Hsiao, Kuang-Ting." Apparatus and method for directional alignment of nanofibers in a porous medium" US Provisional Serial NO. 62/003,982; PCT/US2015/033000; Pub.No.: WO/2015/184151 A1 (2015).

18. Kirmse, S., Ranabhat, B., and Hsiao, K.-T. "A Preliminary study of the electrical and interlaminar shear properties of a porous CFRP composite laminate containing carbon nanofiber z-threads," in *Proceedings of CAMX 2019 - Anaheim, CA, 2019.*

19. Moniruzzaman, Mohammad, and Karen I. Winey. "Polymer nanocomposites containing carbon nanotubes." Macromolecules 39, no. 16 (2006): 5194-5205.

20. Ranabhat, B., Kirmse, S., and Hsiao, K. "Feasibility Study of Novel Magnetic Compaction Force Assisted Additive Manufacturing (MCFA-AM) Methodology for Continuous Carbon Fiber Reinforced Polymer (C-CFRP) Composites," in *SAMPE 2019 - Charlotte, NC*, 2019, TP19--1535. DOI: 10.33599/nasampe/s.19.1535

CHARACTERIZATION AND SUSTAINABLE RETTING OF RAGWEED FIBERS

Elizabeth N. Alvizo[1], Jitendra S. Tate[1,2*], Serra E. Holthaus[3], and Jennifer A. Irvin[2,3]

Texas State University
San Marcos, TX, 78666
[1]Ingram School of Engineering, Texas State University
[2]Materials Science, Engineering and Commercialization Program, Texas State University
[3]Department of Chemistry and Biochemistry, Texas State University
[*]JT31@txstate.edu; 512-245-1826

ABSTRACT

This work aims to characterize the North American native plant, *Ambrosia trifida,* commonly known as ragweed. The purpose of this research is to build a cost-effective manufacturing process that will produce quality microfibers to be used as a reinforcement in polymer matrix composite (PMC) applications. Chemical retting methods were employed and evaluated based on cellulose content extracted. Hydrogen peroxide (H_2O_2) was used to expedite the retting process as previous experiments have shown it to produce cellulose with less impurities. Water retting was conducted in parallel with H_2O_2. Distilled, tap, and river water were compared based on impurities and time allotted. Sodium hydroxide chemical treatment was used to cleanse the fiber from non-cellulose material. The overall mechanical properties of natural fiber reinforced PMC are highly dependent on morphology, aspect ratio, hydrophilic tendency and dimensional stability of the fibers. Scanning Electron Microscopy (SEM) was used for morphology studies of the fiber. Mechanical properties were identified using MTS Exceed. Tensile strength, tensile modulus, and elongation at break mechanical properties were compared to commercially available natural fibers, such as jute, kenaf, hemp, and flax. After retting, the ragweed fibers were ground up and converted to cellulose acetate through esterification. The soluble cellulose acetate was characterized using Fourier Transform Infrared Spectroscopy (FT-IR) and titrated to determine the degree of esterification. Properties of ragweed based cellulose acetate were compared to commercially sourced cellulose acetate. The resultant cellulose acetate was a soluble polymer that could be electrospun to form a nanofiber.

KEYWORDS: *Fiber-reinforced composites, natural fiber, electrospinning, surface treatments, cellulose, cellulose acetate*

INTRODUCTION

A renewed interest for natural fibers is expected as the global economy strives for sustainable industrial practices. Natural fiber advantages are low specific weight resulting in high specific strength and stiffness; sustainability in that production requires little energy and CO_2 is used while oxygen is given back to the environment; desirable for low-wage countries due to low investment at low cost for materials; friendly processing regarding tool wear; possibility for thermal recycling; and good thermal and acoustic insulating properties. (van Rijswijk, Brouwer & Beukers, 2001) However moderate mechanical properties restrain the fibers from high-tech applications resulting in products restricted to interior and non-structural components.

Cellulose nanocomposites are not only limited to their low-weight, environmentally friendly processing, and flexibility. Cellulose nanocomposites find application in many areas such as packaging, optoelectric, barrier, electronics, biomedical, paper, textiles, military, etc. (Feldman, 2015) Thus, there is a need to improve the compatibility between cellulose nanoparticles and hydrophobic polymer matrices through understanding chemical modifications.

The research of this paper serves to employ Giant ragweed, or *Ambrosia trifida*, as a competitor in the market for natural fiber production. *A. trifida* is native to the U.S. and Canada and is seen as a pest prolific crop due to its resilience and pollen production; it is known to be a hardy plant and a discomfort to allergic persons. It does especially well in disturbed soils. Considering many agricultural fields are disturbed an ideal environment is created for *A. trifida* and it becomes a common and prolific competitive weed. More importantly, A. trifida has now become glyphosate (the main ingredient in RoundUp©) tolerant in crop fields and is now a difficult weed to manage (Vink et al. 2012; Westhoven et al. 2009; Powles 2008) Ragweed may grow up to 3.6-m (12-ft) and it is characterized by small, greenish flowers borne on upright terminal spikes and deep-green lobed or dissected leaves. Ragweed is a type of Bast fiber that generally has a wood core surrounded by a stem. Within the stems are fibers bundles, each containing individual fiber cells, or filaments. Figure 1 shows the cross section of different sizes of Ragweed fiber. The layering is noticeably distinct due to age and diameter of the crop. The filaments are the cellulose and hemicellulose that are held together by a pectin and lignin matrix. Retting, or degumming, is a process of removing non-cellulosic material attached to fibers, such as, hemicellulose, lignin, and pectin. The percentage of each non-cellulose material depends on the type of Bast fiber. Table 1 shows the cellulose content in commercially available natural fibers.

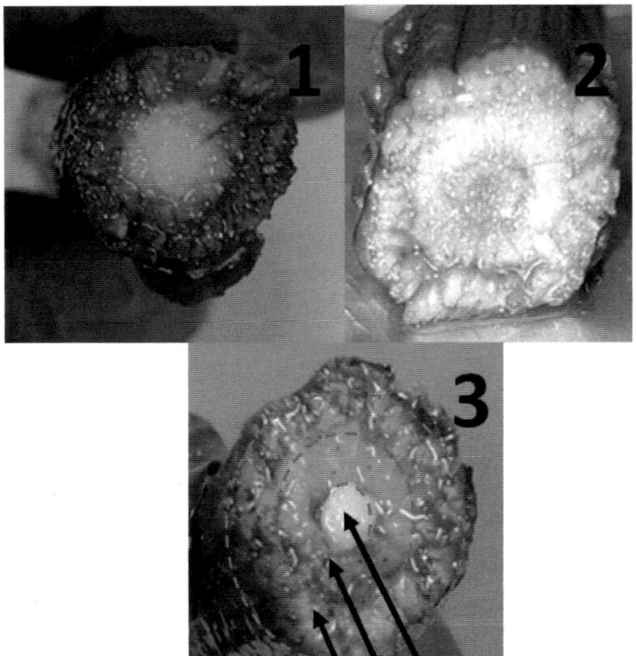

Figure 1. (1) Fiber diameter less than 10-mm (2) Fiber diameter greater than 10-mm (3) Breakdown of layering in fiber

Table 1. Approximate chemical composition (%) of cellulose fibers

Fiber	Cellulose	Hemicellulose	Pectin	Lignin	Fat/wax
Flax	62 – 71	16 – 18	1.8 - 2.0	2.0 -2.5	1.5
Hemp	67 –75	16 – 18	0.8	2.9 - 3.3	0.7
Jute	59 – 71	12 – 13	0.2 - 4.4	11.8 - 12.9	0.5
Source: Kraessog et al., 1996; Lewin and Pearce, 1998, in Cavaco-Paulo,A. And Gubitz, G.M., Texile					

processing with enzymes, Woodhead Publishing Ltd, UK, 2003.

Traditional retting methods for separating the Bast fiber from the inner core are done by dew or water retting. Both methods may require 14 to 28 days to degrade non-cellulose material. Alternative retting methods such as mechanical decortication, chemical, heat, and enzymatic treatments have been reported with mixed findings. (Tahir et. Al. 2011) Balance of pH for water, temperature and stem thickness are known to play a factor in the quality and efficiency of retting as well. Post-retting alkali (NaOH) solution treatment were employed to aid in binding of fiber and matrix composite material. Literature has shown that concentration and temperature of the solution is vital in exploring improvements in mechanical properties. (Ray et. Al. 2001) Table 2 show properties of natural fibers comparing Flax, Hemp, and Jute without alkali treatment.

Table 2. Properties of natural fibers

Properties	**Fiber**		
	Flax	Hemp	Jute
Density g/cm3	1.4	1.48	1.46
Tensile strength** 10E6 N/m2	800 -1500	550 - 900	400 - 800
E – modulus Gpa	60 - 80	70	10 - 30
Elongation at failure (%)	1.2 - 1.6	1.6	1.8
Moisture absorption (%)	7	8	12
Price/Kg ($), raw (mat/fabric)	1.3 (1.7/3.8)	0.6 - 1.8 (2/4)	0.35 (.5/0.9 - 2)

**tensile strength depends strongly on type of fiber being in a bundle or a single filament

Source: http://www.fao.org/3/Y1873E/y1873e0a.htm#bm10

Electrical charge can be applied to a polymer solution and destabilizes the droplet to form a conical shape known as a Taylor Cone. When Coulombic repulsion exceeds the surface tension, the Taylor cone extends, and the polymer solution travels straight and then bends and loops. During this process, the polymer solution dries and solidifies onto the collector, forming fine nanofibers. Electrospinning has multiple parameters that can affect the quality and properties of the nanofibers, such as the polymer's molecular weight, flow rate, diameter of the needle tip, electric potential, and polymer concentration.

MATERIALS AND METHODS

Ragweed fibers were harvested in June 2019. Hydrogen peroxide retting was conducted in parallel with three different water types. The three different water were tap, distilled and river. River water was collected from the San Marcos river. After the specified retting time, the hydrogen peroxide fibers were rinsed with DI water and then left to dry under a vacuum. Water retted fibers were simply left to dry under the vacuum after removing from retting bin. Fibers were randomly selected to bath in NaOH for 1 hour. The Sodium hydroxide treatment was done using 15% concentration sodium hydroxide. Figure 2 shows a flowchart of ragweed processing. Mechanical testing was done using ASTM C1557 –14 and a 1-kN load cell was used to allow for precise measurements. Table 3 shows how long each fiber was spent retting. The JEOL JSM-6010 PLUS/LA Scanning Electron Microscope (SEM) was used to record SEM images and the Quorum Technologies EMS150T ES was used to prep the samples with a layer of carbon. The samples were nonconductive and were prone to charging under the microscope, so it was important to sputter coat the images for best quality.

The Bruker Tensor II FTIR was equipped with OPUS 7.5 software and a Harrick SplitPea ATR accessory. The instrument was used to acquire FTR-ATR spectra of several types of cellulose (commercially sourced, as well as the outer and inner cores of ragweed-sourced) and of cellulose acetate. The sample was loaded onto the ATR accessory plate and pressed against the crystal. The spectrm was recorded, and the ATR accessory plate was cleaned with DI water so the process could be repeated for the other samples.

Electrospinning was attempted using a Spraybase electrospinning instrument, Gamma high voltage box, New Era Pump Systems, Inc. syringe pump, and Vexta brushless DC motor. Based on the method reported by Konwarh et al., (2013), the cellulose acetate fibers were dissolved in a 24% (w/v) solution of chloroform and dimethylacetamide (1:1 ratio). 5 mL of the solution was loaded into a 10 mL syringe and set on the syringe pump to release at 0.05 mL/min. The voltage was steadily increased until it reached 15 kV.

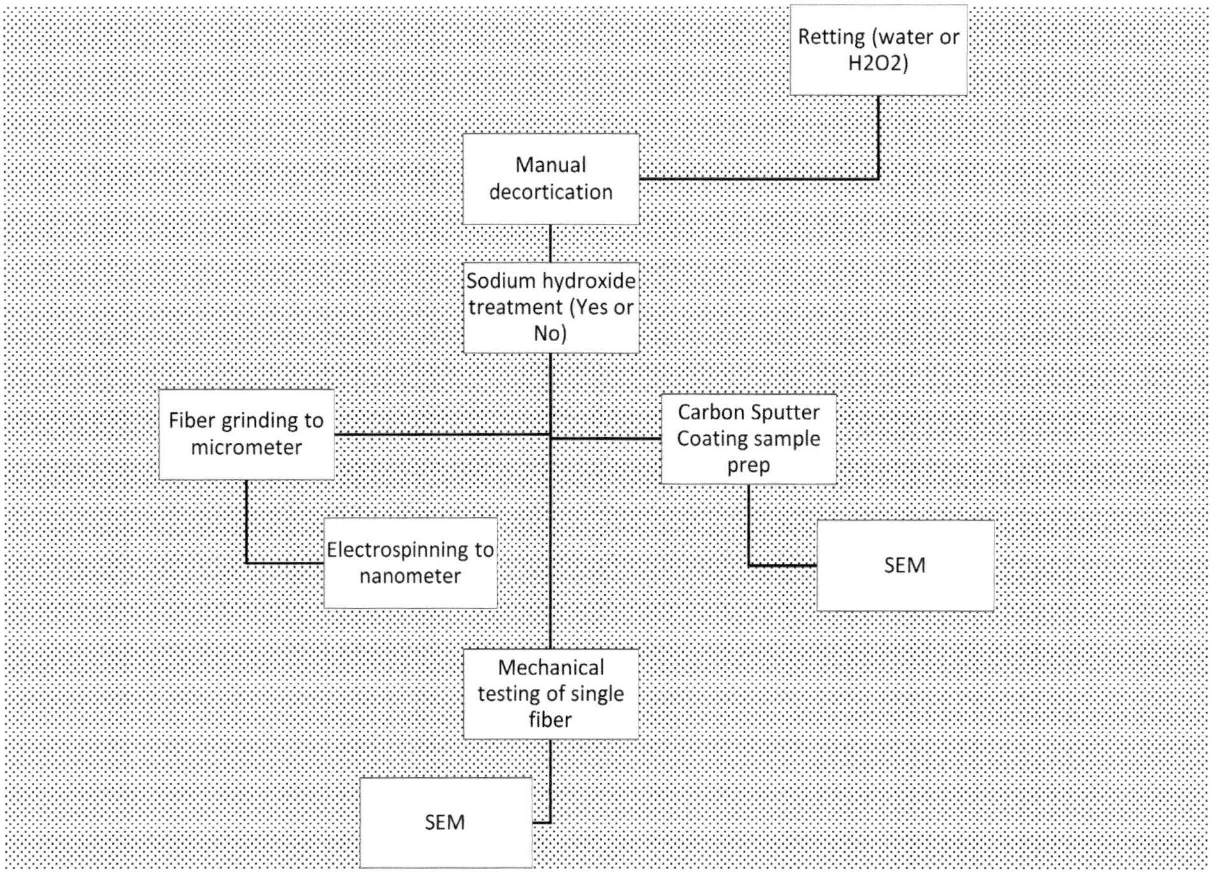

Figure 2. Diagram of ragweed processing outline

Table 3. Retting type and duration inputs and outputs	
Retting methods	Days spent retting
River water sf	14
Distilled water sf	14

Tap water sf	14
Hydrogen peroxide sf	7
Sf =short fiber	

Lab grade cellulose from Ward's science was esterified through the following process, modified from two literature procedures (Rosenthal (1967) and Meng et al., (2017)). Cellulose (0.8g) was suspended in glacial acetic acid (50 mL) at room temperature for 15 minutes. Then, concentrated sulfuric acid (0.14 mL) was added to catalyze the reaction, which was stirred at room temperature for 5 minutes. Finally, acetic anhydride (28 mL) was added, and the reaction was left to stir at 45 °C for 24 hours in air. The resulting cellulose acetate was added dropwise to de-ionized (DI) water that was stirred at 900 rpm. The resultant precipitate was collected by filtration through paper dried under vacuum at room temperature overnight. For further purification, the cellulose acetate was dissolved in chloroform and added dropwise into rapidly stirring methanol, filtered, and vacuum dried under the same conditions.

RESULTS

The table below shows tensile test results from previous experiments using the specified fiber types. Side view of the fiber show microscopic differences in the fibers but how that affects mechanical properties is not fully understood. More tests need to be done in order to form a reputable assumption of the mechanical properties. Mechanical testing was tricky in that fibers were very sensitive when securing in the grips. Attempts were made to test the most recent fibers however the data was not ideal considering issues with failing at the grips of the test specimen. Mechanical testing shows that fibers are nowhere near superior compared to hemp, flax, and jute fibers however when fibers were bathed in NaOH it shows that properties were significantly improved. This supports previous literature in that alkali treatments are important to understand and implement.

Table 4. Tensile Mechanical Properties of Ragweed Fibers

Fiber Type Tested	Number of Tests	Average Peak Load (N)	Average Area (mm2)	Average Peak Stress (MPa)	Average Strain at Break (mm/mm)	Modulus (Total Secant, MPa)	Average Strain at Yield (mm/mm)
H2O 7Days 20mm	3	10.93909	0.357578	30.59217849	0.0166	1842.902319	0.00906
H2O 7Days 40mm	3	5.877736667	0.357578	16.43763505	0.0161	1020.971121	0.0116
NaOH 20mm	3	26.57434667	0.36312	73.18337372	0.0244	2999.318595	0.0244
NaOH 40mm	3	25.15211333	0.36312	69.26667034	0.016	4329.166896	0.0159

DISCUSSION AND CONCLUSIONS

Fibers that appeared to produce more mucous were easier to decorticate and it was important to decorticate right after retting. Allowing the fiber to dry made it more difficult to remove the inner core and allowed increased risk of possible damage to the fiber. It appears that 100% removal of core was impossible considering that small fragments would remain on some fibers, especially with larger fiber diameters. All water retted fibers had a strong unpleasant odor, especially the river retted water. Hydrogen peroxide fibers had less mucous but finer fibers as the result. Photographs of retted fibers can be seen in Figure 3.

Figure 3. (top) Water retting fibers showing mucous after retting (bottom) hydrogen peroxide fibers with less mucous

SEM imaging of the fibers (Figures 4 and 5) showed the fibers to be around 300 micrometers in diameter. NaOH fibers produced concentrated areas of unknown white dots. Hydrogen peroxide interestingly did not produce a clean even fiber. However, on a macroscopic scale the hydrogen peroxide retted fibers appeared to be finer in nature.

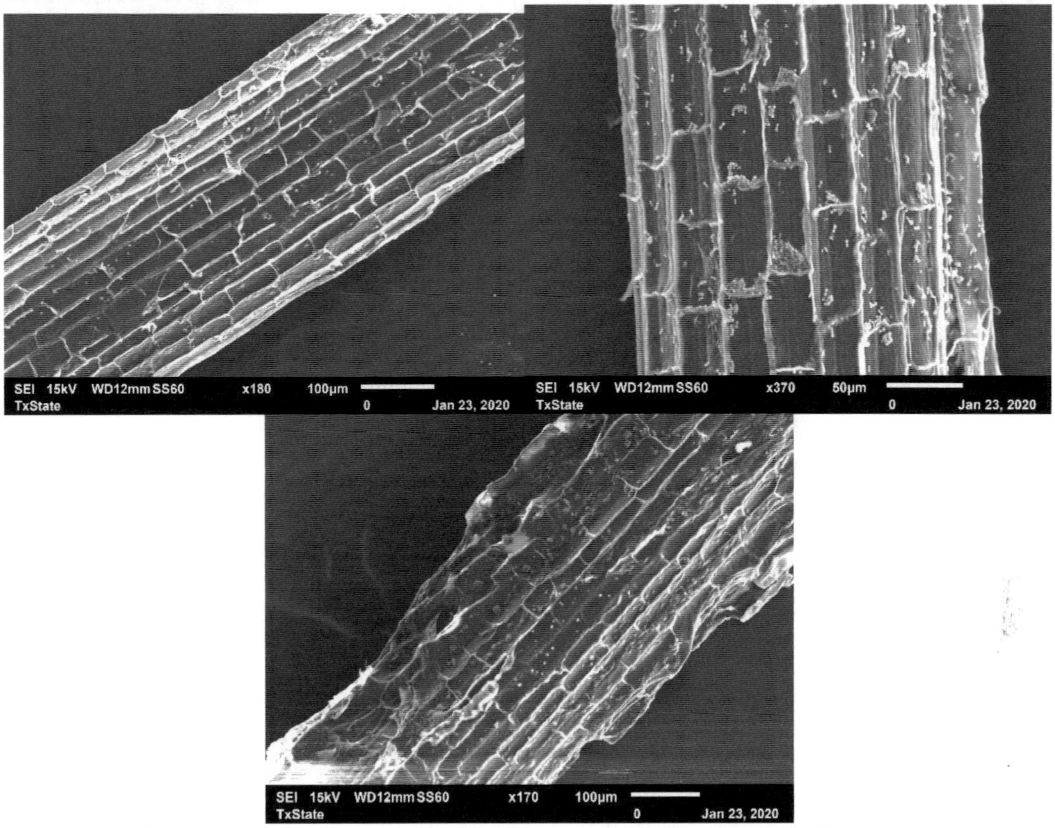

Figure 4. (top left) Regular tap water fiber retting (top right) Fiber after NaOH bath (bottom) Hydrogen peroxide fiber retting

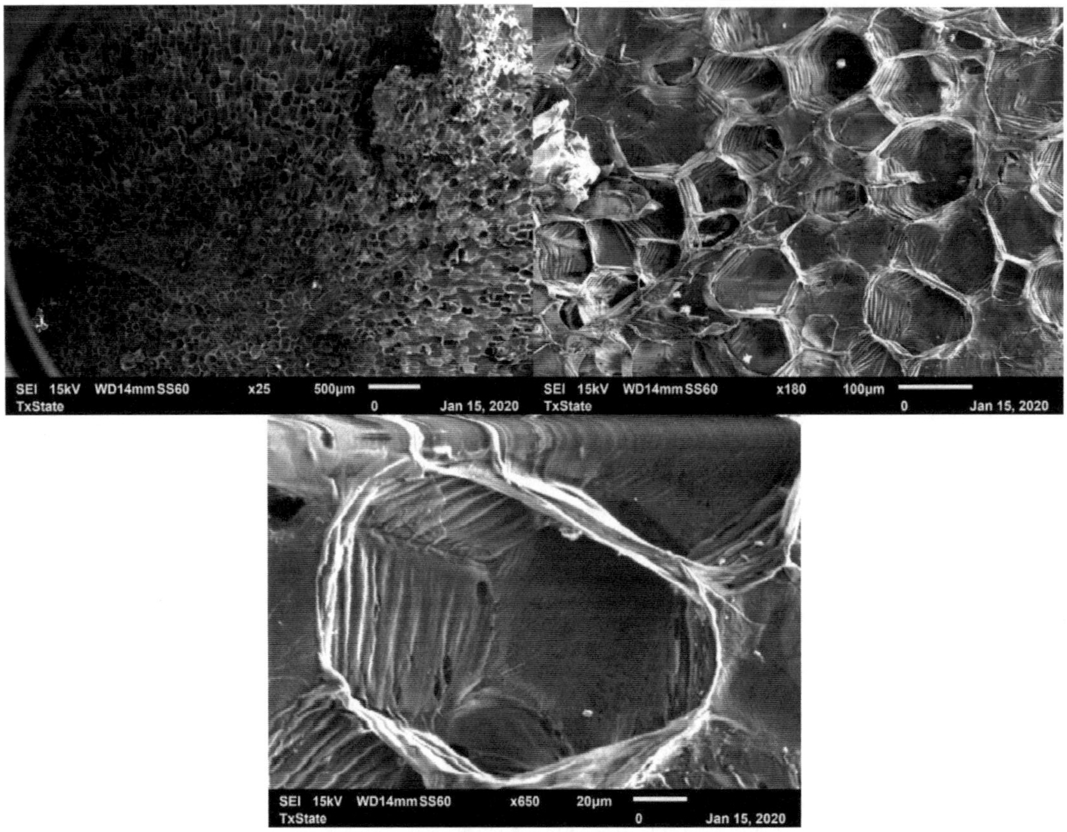

Figure 5. Bast fiber cross section

Commercial cellulose was used to evaluate esterification (Scheme 1) protocols. Spectroscopic evidence for successful esterification of commercial cellulose and successful isolation of cellulose from the inner core of ragweed can be seen in Figure 6. Samples of inner core cellulose (gray line in Figure 6) and commercial cellulose (blue) both reveal similar FTIR spectra that are consistent with values found in the literature (Shoba et al., 2018). The spectrum of outer core cellulose (orange) is somewhat similar to that of commercial cellulose and inner core cellulose, although outer core cellulose exhibits an additional peak at 1652 cm^{-1} that is characteristic of carbonyl groups. This may be residual organic species such as chlorophyll that were not decomposed under the retting conditions used for that sample (DI water only, no NaOH or H_2O_2). Characteristic cellulose absorptions include C-O-C (1057 and 1108 cm^{-1}), CH_2/CH_3 (2900 cm^{-1}), and O-H (3344 cm^{-1}). Additional peaks consistent with cellulose acetate (green) include ester C-O-C (1239 cm^{-1}) and C=O (1749 cm^{-1}). Differences in peak intensities between samples are due to sample-to-sample inconsistencies on the ATR crystal compounded by differences in sample morphologies.The absorption an 2359 cm^{-1} in all samples is an artifact from background atmospheric gases. Next steps include esterification of ragweed-sourced cellulose and gel permeation chromatography to determine molecular weight of cellulose acetate.

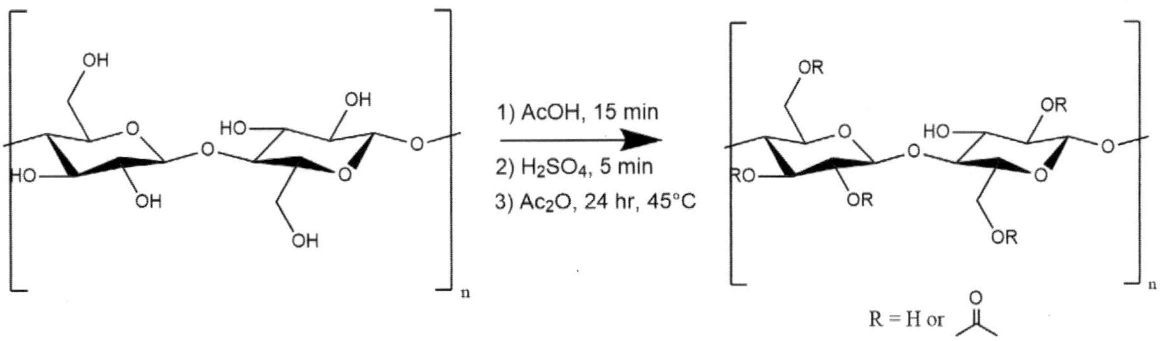

Scheme 1. Esterification of cellulose to produce cellulose acetate.

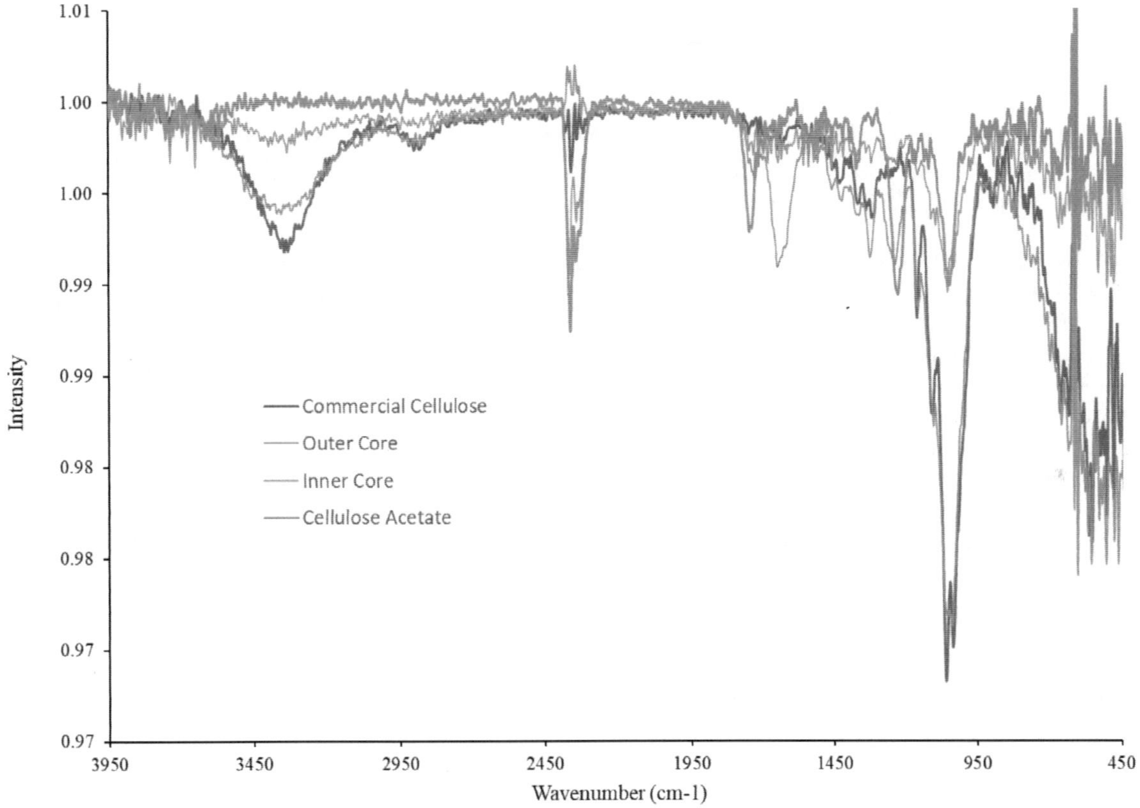

Figure 6. FTIR spectra of commercial cellulose (blue), outer core of ragweed-sourced cellulose (orange), inner core of ragweed-sourced cellulose (gray), and cellulose acetate prepared by esterification of commercial cellulose (green).

Multiple attempts to prepare electrospun nanofibers of cellulose acetate were unsuccessful, producing powders rather than nanofibers. Powders are often formed when viscosity and/or concentration are too low, but the sample preparation and electrospinning parameters used duplicated those reported by Konwarh et al. (2013). We suspect that the molecular weight of our cellulose might be lower than that used by Konwarh. Furthermore, while electrospinning, the polymer would start to precipitate on the needle before it could reach the collector, presumably because the acetone was evaporating. Next steps include molecular weight determination, determination of degree of esterification, and electrospinning in alternative solvent systems as well as with esterified ragweed-sourced cellulose.

REFERENCES

Van Rijswijk, K., Brouwer, W., & Beukers, A. (2001). Application of Natural Fibre Composites in the Development of Rural Societies. Retrieved 19 January 2020, from http://www.fao.org/tempref/GI/Reserved/FTP_FaoRne/morelinks/Publications/English/Natural-Fibre-Composites.pdf

Vink, J. P., Soltani, N., Robinson, D. E., Tardif, F. J., Lawton, M. B., & Sikkema, P. H. (2012). Occurrence and distribution of glyphosate-resistant giant ragweed (Ambrosia trifida L.) in southwestern Ontario. *Canadian Journal of Plant Science*, *92*(3), 533-539.

Powles, S. B. (2008). Evolved glyphosate-resistant weeds around the world: lessons to be learnt. *Pest management science*, *64*(4), 360-365.

Andrew M. Westhoven, Vince M. Davis, Kevin D. Gibson, Stephen C. Weller, and William G. Johnson (*2008*) Field Presence of Glyphosate-Resistant Horseweed (Conyza Canadensis), Common Lambsquarters (Chenopodium Album), and Giant Ragweed (Ambrosia Trifida) Biotypes with Elevated Tolerance to Glyphosate. Weed Technology: July 2008, Vol. 22, No. 3, pp. 544- 548.

M. Tahir, Paridah & Ahmed, Amel & SaifulAzry, Syeed & Ahmed, Zakiah. (2011). Retting process of some bast plant fibres and its effect on fibre quality: A review. BioResources. 6. 5260-5281.
Ray, D., Sarkar, B. K., Rana, A. K., & Bose, N. R. (2001). Effect of alkali treated jute fibres on composite properties. *Bulletin of materials science*, *24*(2), 129-135.

Konwarh, R., Karak, N., & Misra, M. (2013). Electrospun cellulose acetate nanofibers: The present status and gamut of biotechnological applications. *Biotechnology Advances*, *31*(4), 421–437. https://doi.org/10.1016/j.biotechadv.2013.01.002

Meng, C., Cao, G.-P., Yan, Y.-Z., Zhao, E.-Y., Hou, L.-Y., & Shi, H.-Y. (2017). Synthesis of cellulose acetate propionate with controllable contents and distributions of acetyl and propionyl in the C2, C3 and C6 positions. *Reaction Kinetics, Mechanisms and Catalysis*, *122*(2), 1031–1047. https://doi.org/10.1007/s11144-017-1260-5

Rosenthal, A. J. (1967). The role of acid catalysts in the manufacture of cellulose acetate. *Pure and Applied Chemistry*, *14*, 535–546. https://doi.org/10.1351/pac196714030535

Shoba, B., Jeyanthi, J., & Vairam, S. (2018). Synthesis, characterization of cellulose acetate membrane and application for the treatment of oily wastewater. *Environmental Technology,* https://doi.org/10.1080/09593330.2018.1543353.

Feldman, D. (2015). Cellulose Nanocomposites. *Journal of Macromolecular Science, Part A: Pure and Applied Chemistry*, 52(4), pp.322-329.

SCALABILITY POTENTIAL OF NANOCELLULOSE GLASS FIBER COATINGS FOR LIGHTWEIGHT COMPOSITES

Ejaz Haque, Tae Joong Jeong, Shadi Shariatnia, Dorrin Jarrahbashi, Amir Asadi, Tequila Harris, Robert J. Moon, Kyriaki Kalaitzidou

Georgia Institute of Technology
813 Ferst Dr. NW
Atlanta, GA 30332

ABSTRACT

High levels of greenhouse gas emissions in the transportation sector have sparked a rise in demand for lightweight materials in automotive manufacturing. Developing a material with the appropriate balance of performance, cost, and weight to displace existing structural technologies has proven challenging, however. Some promise has been shown in the use of nanocellulose as a secondary reinforcement phase in fiber-reinforced composites, but process scalability remains a significant barrier to their incorporation at the manufacturer level. Coating the fiber surface with nanocellulose is a conceptually scalable approach to enhancing composite properties through toughening of the fiber-matrix interphase. However, while improvements in mechanical properties have been observed via this approach, the relationship between processing methodology and surface coating properties has yet to be thoroughly studied. This work aims to demonstrate the scalability of nanocellulose fiber coatings and their associated application space using different coating processes. Two scalable thin film deposition methods, slot die and spray coating, will be evaluated. The characteristics of fiber coatings achieved using these methods will be compared using SEM, XPS, and TGA to establish specific advantages of each process. Composite interfacial mechanical properties will also be compared against each other and against lab scale results.

Corresponding author: Ejaz Haque

1. INTRODUCTION

Nanomaterial reinforcement is a popular method in the literature for improving the mechanical performance of fiber-reinforced composites. Studies have demonstrated the potential of various nanomaterials such as nylon to increase mechanical properties without significantly impacting composite mass, making this approach ideal for lightweight applications [1]. However, many of the promising results in the literature to date come with the caveat that they have been achieved using lab-scale methods. Due to the nature of the application space for parts made using this technology, scalability is paramount and is a major barrier to commercial implementation.

Cellulose nanocrystals (CNC) present a unique opportunity to achieve this milestone. Not only have they been shown to be effective interface modifiers on the lab scale, but market projections show that their cost, an important external factor with regard to scalability, is expected to drop in the near future [2]. Addition of CNC onto the surfaces of glass fibers used in the production of

SAMPE Virtual Conference Proceedings, 2020. Society for the Advancement of Material and Process Engineering – North America.

fiberglass has already been shown to enhance the fiber-matrix interface in the literature [3]. If automated coating technology could be used to achieve similar improvements, the scalability barrier in the production of lightweight CNC-enhanced composites could be overcome.

This work focuses on the application of CNC-based coatings to glass fibers using a slot die system. Through mechanical testing and comparison to literature results, scalability of the coating can be demonstrated should similar performance be measured. This comparison will be shown alongside observations from microscopic and spectroscopic techniques to better illustrate the impact of the slot die technique on the glass fiber surface.

2. EXPERIMENTATION

2.1 Materials

The glass fibers coated in this series of experiments were ME1510 multiend rovings (TEX 4800, 10 μm diameter) purchased from Owens Corning (Oak Brook, IL, USA). The CNC used in these coatings were a product of the Forest Products Laboratory (Madison, WI, USA) modified with methyl(triphenyl) phosphonium counterions by the laboratory of Douglas M. Fox of American University (Washington, D.C., USA). The CNC prior to modification were reported to have a width between 5-20 nm and a length between 150-200 nm, and the ion exchange modification was not found to significantly affect these dimensions [4].

The epoxy system used for fabrication of tensile coupons consisted of US Composites (West Palm Beach, FL, USA) 635 thin epoxy and 556 slow amine hardener. This epoxy was additionally used in the formation of an emulsion used for fiber coating along with Pluronic P123, a poloxamer of molar mass approximately 5800 g/mol purchased from Sigma Aldrich (St. Louis, MO, USA).

2.2 Solution Preparation

The epoxy/poloxamer emulsion used for fiber coating was selected as the CNC dispersing agent of interest based on previously reported performance metrics [3]. It was prepared by mixing the poloxamer, CNC, and water magnetically for an hour in an ice bath. The epoxy was then added to the solution and mixing continued for another 30 minutes. Sonication was subsequently used to break up potentially agglomerated CNC particles. Epoxy and poloxamer constituted 5 wt% of all solutions (in the form of a 4:1 epoxy:poloxamer ratio), and CNC constituted an additional 1 wt% in the solutions in which it was incorporated.

Table 1. Coating formulations and their respective shorthand notation.

Coating Label	Coating Method	Coating Composition
GF0	N/A (as-received)	N/A
DC1	Dip Coat [3]	5 wt% Emulsion
SD1	Slot Die	5 wt% Emulsion
DC2	Dip Coat	5 wt% Emulsion + 1 wt% CNC
SD2	Slot Die	5 wt% Emulsion + 1 wt% CNC

2.3 Slot Die Fiber Coating

A stainless steel slot die mounted on a Roll-to-Roll (R2R) system equipped with a velocity-controlled carrier web and a volumetric flow pump with syringe was used to coat the CNC-epoxy solutions. The slot die was 160 mm wide, with a coating gap of 0.1 mm. The processing conditions include: a slot gap of 300 μm, the flow rate of 1 mL/min, and carrier web velocity of 0.5 mm/s.

The glass fibers were adhered to a sacrificial acrylic plate using tape at the two extreme ends of the rovings. 10 mm were left between each roving to ensure no overlap. The acrylic plate containing the glass fibers was placed on the carrier web just below the slot die. After coating the glass fibers, the samples were left to dry at room temperature for 24 hours. Slot die allowed for uniform coating of the CNC-epoxy solution on the glass fiber mat.

2.4 Surface Property Assessment

Morphological assessment of coated fiber surfaces was performed using a Zeiss Ultra-60 FESEM at an accelerating voltage of 5 kV. After coating, fibers were dried overnight under ambient conditions and sputter coated in gold for 30 seconds. Both the coated rovings as well as single fibers pulled from said rovings were sputter coated so as to understand the effects of fiber removal, a necessary step in the fabrication of single fiber tensile coupons.

Surface elemental analysis was performed on the coated rovings using a Thermo Scientific K-Alpha spectrometer equipped with a 1.486 KeV aluminum K-alpha x-ray source. Characteristic signals from carbon, oxygen, nitrogen, silicon, sulfur, and phosphorous were used to confirm the presence of the coatings and qualitatively assess their thicknesses.

Mass of the coating adhered to the fiber surface was determined using a TA Instruments Q50 TGA. 10-20 mg samples of coated rovings were heated to 600°C at 10°C/min to determine total mass lost. Results for this process are reported as an average of 3 samples.

2.5 Mechanical Property Assessment

Interfacial mechanical properties were determined using the single fiber fragmentation test (SFFT) [5]. Epoxy dogbone coupons each containing a single glass fiber were cured then tested in tension according to ASTM D638. A tensile load was applied in the axial direction of the fiber with a crosshead displacement rate of 1 mm/min for 3 minutes to ensure saturation of fiber fracture phenomena within each sample. Following tensile testing, fiber fracture behavior was analyzed using a Leica DM2500 polarized light optical microscope. Due to the direct relationship between fracture quantity and interfacial shear strength, fracture count is used as a representative surrogate for mechanical properties in this work [3]. Results are compared to literature data that was collected using a lab-scale dip coating procedure and the same formulations described in this work to assess scalability of the coating technology.

The coupons were made by mounting coated, dried glass fibers on a mold using heat-resistant tape such that the fibers maintained tension throughout the curing process. The resin, a 2:1 mixture of epoxy to hardener that was stirred magnetically and degassed under vacuum, was poured over the fibers until the mold cavities were filled. The filled mold was placed in an oven at a temperature cycle of 100°C for 1 hour followed by 120°C for 2 hours. The mold was then removed from the oven and allowed to cool overnight under ambient conditions prior to removal of the cured samples.

3. RESULTS

The following results are representative of phenomena observed for glass fibers coated using a slot die process. While a separate set of fibers has been coated using a spray process, analysis of those samples is ongoing and may be presented alongside the current data at a later time.

3.1 SEM Morphology

Differences in the appearance of the fiber surface as a result of slot die coating can be seen in Figure 1. Uncoated, as-received fibers were found to be nearly uniform, with patches of pre-existing sizing on the surface and continuous layers of the same sizing between fibers to hold them together within the roving. These features were mostly obscured upon application of SD1, which resulted in a coating so thick that even the curvature of the individual fibers was difficult to distinguish. Interestingly, the same was not observed in the case of SD2; while visual confirmation of the presence of coating was certainly possible, it was sufficiently thin that characteristics of the underlying fiber were occasionally still visible.

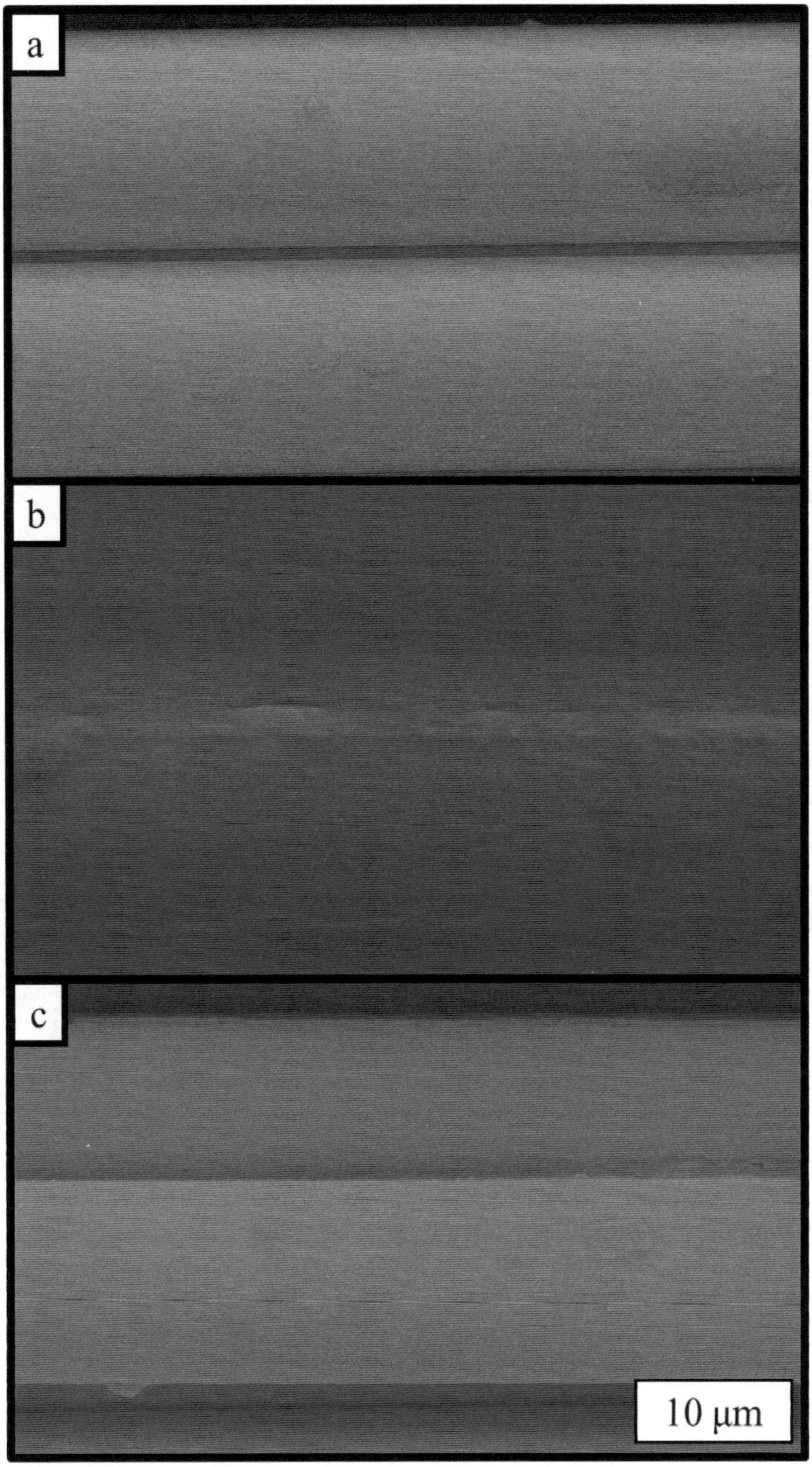

Figure 1. SEM images of fiber rovings a) GF0 (uncoated), b) SD1 (5 wt% emulsion), and c) SD2 (5 wt% emulsion + 1 wt% CNC).

Given the identical slot die parameters used for both coatings, the inconsistent thicknesses between SD1 and SD2 are most likely a consequence of the difference in their formulations. Based on previously reported results for the same formulations, it is possible that the presence of CNC may enhance the wetting potential of the emulsion on the glass fiber surface [6]. This is further supported by the observation in this work that the pure emulsion dewetted from small areas of the bare rovings, resulting in shrunken, highly concentrated pools of coating in those areas, whereas the emulsion with CNC impregnated the rovings without issue.

In addition to coated rovings, fibers removed from coated rovings were imaged to understand the morphology of the single fibers used in SFFT. Tearing of the coating layer was observed, leaving some areas uncoated and others with excess coating stripped from adjacent fibers. In the case of SD2, the CNC may have acted as "defects" in the tearing process, resulting in more inhomogeneities as compared to the clean tears observed in Figure 2a.

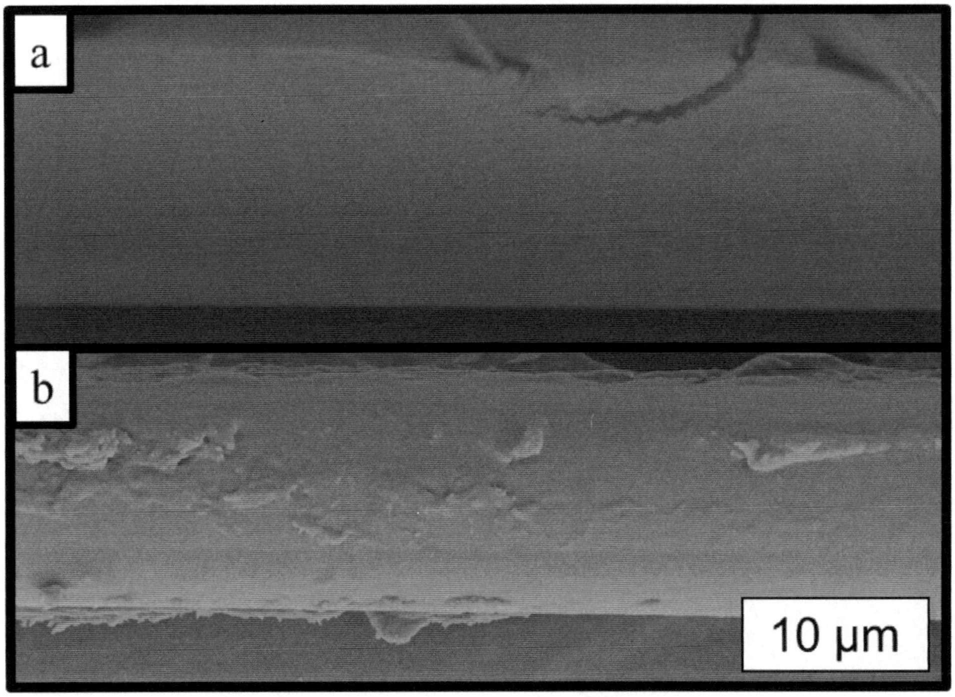

Figure 2. Broken coating observed on fibers a) SD1 (5 wt% emulsion) and b) SD2 (5 wt% emulsion + 1 wt% CNC). Tearing and remnants of coating torn from adjacent fibers can be seen on the single fibers imaged.

3.2 Interfacial Mechanical Enhancement

SFFT results from literature are compared to results for slot die coated fibers in Figure 3.

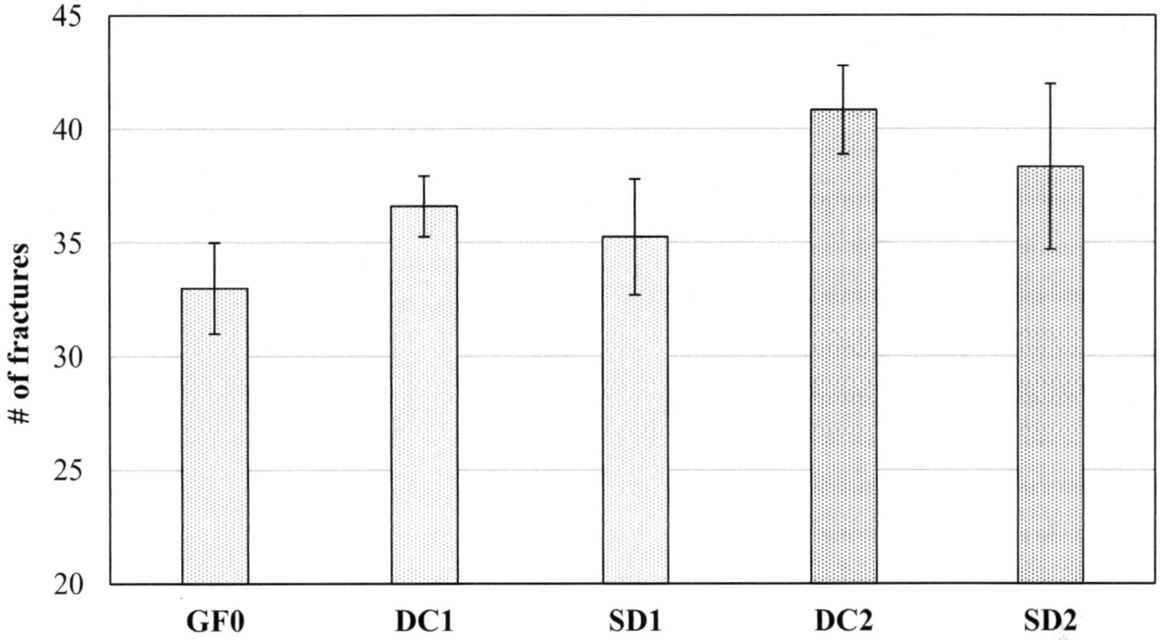

Figure 3. Post-SFFT fractures observed for as-received fibers (GF0), fibers coated in emulsion (DC1 and SD1), and fibers coated in a mixture of emulsion and CNC (DC2 and SD2). Error bars are one standard deviation.

A comparison of average values reveals that slot die coating improves interfacial shear strength in the tested composites, and CNC enhances this effect. However, these averages were found to be slightly lower than their literature counterparts and calculated standard deviations were higher. This is likely explained by the phenomena observed in Figure 2; the manual separation of fibers results in tearing of the coating. In the literature experiments, single fibers were separated from their rovings prior to coating, meaning no damage to the coating was caused by the fabrication of SFFT samples. As a result, slot die coating yielded the same best-case scenario as the literature (fibers with uniform coating coverage) but suffered a dramatically weaker worst case (fibers with sections of coating stripped away), resulting in a wider range of fracture counts. This worst-case scenario is likely irrelevant to practical application due to the use of chopped rovings rather than single fibers in sheet molding compound fabrication. However, accounting for it lends important context to the trends described in Figure 3, revealing very comparable results to the lab-scale and thus suggesting that CNC coatings can offer scalable interfacial improvements.

3.3 Fracture Geometry and Features

Polarized light optical microscopy was used to count fractures after tensile testing, and images taken of said fractures are shown in Figure 4 below.

Figure 4. Polarized light microscope images of single fiber fracture behavior for a) GF0 (uncoated), b) SD1 (5 wt% emulsion), c) SD2 (5 wt% emulsion + 1 wt% CNC).

The fracture behavior observed in these samples largely mirrors what can be found in the literature. Uncoated fibers (Figure 4a) typically displayed weak local interfaces with long, black matrix debonding regions surrounded by dull birefringence. Coating the fibers resulted in generally stronger interfaces that did not de-bond upon application of tensile load. These fractures took on a cone-shaped geometry with bright birefringence patterns as seen in Figures 4b and 4c. Weak fracture patterns were occasionally observed for SD1 and SD2 as well, another likely result of the torn coating phenomenon that results in locally weak interfaces.

3.4 Coating Surface Characteristics

Observations regarding the thickness of applied coatings was corroborated using TGA. As expected from Figure 1, the results in Table 2 show that the highest mass loading was observed in the case of SD1. Additionally, the highest standard deviation was observed in this case, likely due to the dewetting phenomenon that resulted in areas of the fiber with higher concentrations of coating than others. The smooth application of SD2 resulted in a comparatively more homogenous set of coatings, thus the lower average loading and standard deviation.

Table 2. Mass of coating adhered to fiber roving surfaces as determined by TGA.

Coating	**Mass Loss (%)**
GF0	0.82 ± 0.09
SD1	7.23 ± 0.60
SD2	4.77 ± 0.03

These results were further investigated through the use of XPS. Differences in the carbon signals between coated and uncoated fibers indicate the successful application of coatings via the slot die mechanism. The C1s peak at approximately 287 eV, which is not observed in Figure 5a, is representative of a C-C bond present in the triblock copolymer used in the SD1 and SD2 emulsion.

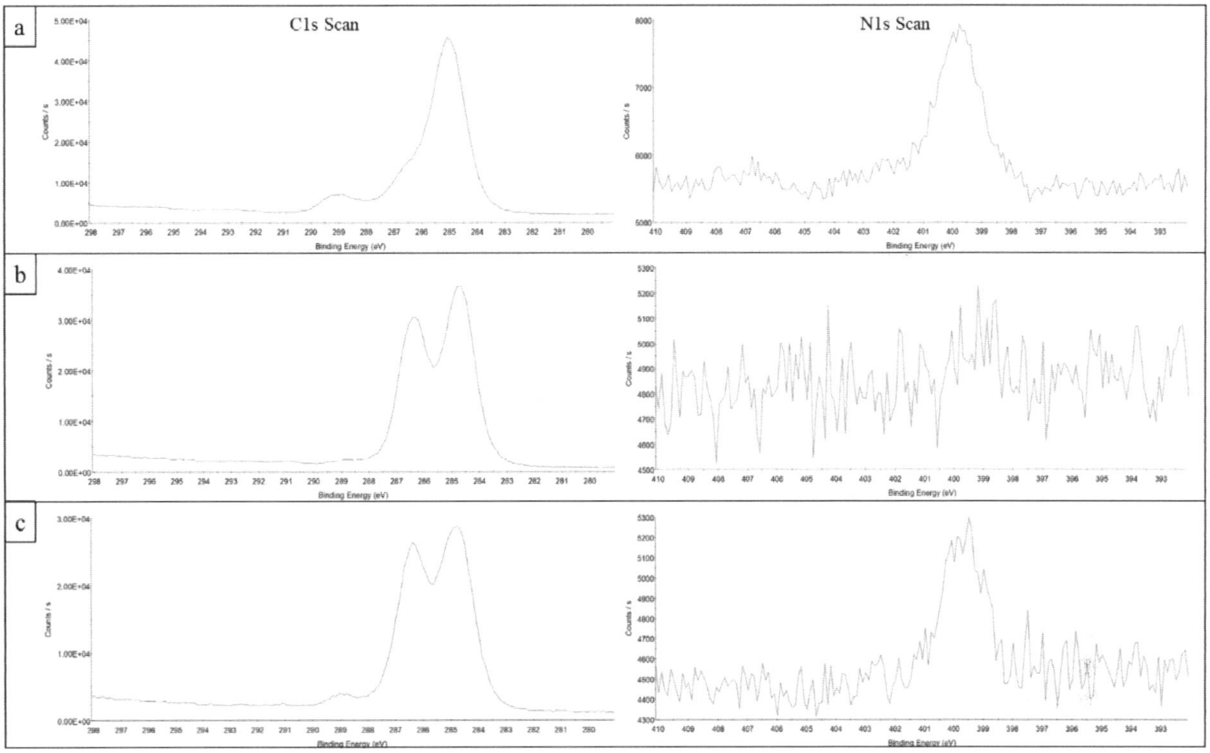

Figure 5. XPS C1s and N1s spectra for a) GF0 (uncoated), b) SD1 (5 wt% emulsion), c) SD2 (5 wt% emulsion + 1 wt% CNC).

The decrease in N1s signals is also a consequence of coating, as nitrogen is only present in the aminosilane-based sizing applied to the fiber surface by the manufacturer. Due to the limited penetration depth of the x-ray beam, no distinct N1s signal is seen in Figure 5b, the thickest coating applied. Similarly, the C1s signal from the bare fiber surface at approximately 289 eV shrinks considerably upon coating, particularly in the case of SD1.

4. CONCLUSIONS

A slot die system was used to apply CNC-based coatings to glass fiber rovings. Single fibers were removed from these coated rovings and cured in tensile coupons to assess interfacial shear strength. The data showed an enhancement in mechanical properties comparable to the improvements seen in the literature for lab-scale processes. SEM, optical microscopy, TGA, and XPS all reflected improved wetting characteristics for formulations containing CNC, resulting in slightly thinner and more uniform coatings. These results suggest that the addition of CNC to glass fiber coatings for the purpose of composite lightweighting can be viably performed in scalable fashion.

The authors would like to thank P^3 Nano and the U.S. Endowment for Forestry and Communities for funding this work, as well as Owens Corning for providing glass fibers.

5. REFERENCES

1. Garces, J.M., et al., *Polymeric nanocomposites for automotive applications.* Advanced Materials, 2000. **12**(23): p. 1835-1839. doi: 10.1002/1521-4095(200012)12:23<1835::AID-ADMA1835>3.0.CO;2-T
2. de Assis, C.A., et al., *Conversion Economics of Forest Biomaterials: Risk and Financial Analysis of CNC Manufacturing.* Biofuels, Bioproducts and Biorefining, 2017. doi: 10.1002/bbb.1782
3. Goswami, J., et al., *The Effect of Cellulose Nanocrystal Coatings on the Glass Fiber–Epoxy Interphase.* Materials, 2019. **12**(12): p. 1951. doi: 10.3390/ma12121951
4. Fox, D.M., et al., *Simultaneously tailoring surface energies and thermal stabilities of cellulose nanocrystals using ion exchange: effects on polymer composite properties for transportation, infrastructure, and renewable energy applications.* ACS applied materials & interfaces, 2016. **8**(40): p. 27270-27281. doi: 10.1021/acsami.6b06083
5. Kelly, A. and a.W. Tyson, *Tensile properties of fibre-reinforced metals: copper/tungsten and copper/molybdenum.* Journal of the Mechanics and Physics of Solids, 1965. **13**(6): p. 329-350. doi: 10.1016/0022-5096(65)90035-9
6. Haque, E., et al. *Nanocellulose Fiber Sizing for Fiberglass Composites.* in *Proceedings of the American Society for Composites—Thirty-fourth Technical Conference.* 2019. doi: 10.12783/asc34/31326

CARBON-INFILTRATED CARBON NANOTUBE COATING APPLICATIONS: SCALPEL BLADES AND EXTERNAL FIXATORS

Stephanie R. Morco, Jaclyn M. Larsen, Brian D. Jensen, Anton E. Bowden
Department of Mechanical Engineering, Brigham Young University
Provo, UT

ABSTRACT

A carbon-infiltrated carbon nanotube (CICNT) coating grown on stainless steel has a broad range of possible applications due to its non-pharmaceutical biofilm resistance [1]. This property may provide a solution to surgical site infection (SSI), so scalpel blades and external fixator pins were coated with CICNT and subjected to mechanical testing and MRSA biofilm testing. This scalpel blade coating is one of the first reports of CNT fabrication using 400-series SS as a substrate and a catalyst [2]. The CICNT coating did not prove to have superior anti-bacterial performance to a bare scalpel blade, but this work provides greater insight on CICNT properties. Hardness that was lost during the CICNT manufacturing process was recoverable, and the coating did not delaminate under the force of a diamond scribe. Although the CICNT on these scalpel blades did not exhibit MRSA biofilm resistance, new theories were offered on what factors are critical for the CICNT nanostructure to resist biofilm. The external fixator pin application proved to have significantly more promise than the scalpel blades. The bare stainless steel pin was completely covered in MRSA biofilm, while both the vertically-aligned and randomly-oriented CICNT-coated pins each allowed only a few cells to attach. Although there is more testing to be done, a CICNT coating may be an exceptional way to prevent infection while reducing antibiotic use.

Corresponding author: Anton E. Bowden

1. INTRODUCTION

1.1 The Need for an Antibacterial Coating

Surgical site infection (SSI) continues to affect hundreds of thousands of people in the US each year [1]. With a 2-5% incidence rate and \$3.5-\$10 billion annual US cost, these infections are now the most common and costly type of hospital-acquired infection [1, 2]. For orthopaedic joint replacements, approximately half of SSIs are periprosthetic joint infections (PJIs) [3], which may be considered the most devastating complication of these surgeries [4]. The bacteria species that cause most SSI and PJI are part of the normal skin flora [5, 6], so although the evidence is limited, the scalpel may potentially cause some of these infections [7]. Therefore, coating a scalpel with an antibacterial coating may prove beneficial.

Along with addressing the scalpel as an infection source, external fixator pins have a remarkably high infection rate. The pin tract infection rate is 21% to 42% of cases [8]. Yet despite this risk inherent with the device's transdermal pathway, they are still used because they play an important role in cases that need reduced soft tissue damage, avoidance of any infected areas, and adjustability [9]. Therefore, coating an external fixator pin with an antibacterial coating may also prove beneficial.

1.2 Challenges with Current Solutions

There are a variety of efforts aimed at addressing the challenge of SSI for external fixator pin infection. Coating pins with silver, iodine, or gentamicin-coated sleeves has shown evidence of reducing infection [10]. However, all of these options are based on their chemistry, so there is a concern of cytotoxic effects [11]. Additionally, antibiotic-based coatings are limited by drug-resistant bacteria strains, such as methicillin-resistant S. aureus (MRSA), and biofilm formation that makes systemic antibiotics ineffective [11].

1.3 Benefits of a CICNT Coating

A carbon-infiltrated carbon nanotube (CICNT) coating may offer an improved solution. Recent in vitro testing on CICNT coatings provides evidence that this coating resists MRSA cell attachment and biofilm formation due to its nanostructure rather than its chemistry [12]. Although additional testing is required, the results are promising, especially because it can be easily coated directly onto stainless steel (used in both scalpel blades and external fixation pins) and on different geometries. Other advantages of a CICNT coating include that it is low cost [13], heat stable, and would have a theoretically infinite shelf life. Because of this, potential applications are being investigated. The objective of this research is to test the feasibility and effectiveness of coating scalpel blades and/or external fixator pins with CICNT to resist bacterial infection.

2. EXPERIMENTATION

2.1 Scalpel Blade Fabrication and Testing

CICNT was coated directly onto size 10 scalpel blades with a process optimized for this 400-series stainless steel (SS). The process sequence was a heat treatment, CNT growth, then carbon infiltration, all in a CVD tube furnace. The heat treatment was at 800°C in argon (Ar, flowing at 250 sccm) for 30 minutes, the CNT growth was at 700°C in hydrogen (H2, flowing at 230 sccm) and ethylene (C2H4, flowing at 250 sccm) for 15 minutes, and the carbon infiltration was at 850°C in Ar and C2H4 for 25 minutes. Controls were uncoated size 10 scalpel blades.

The CICNT coatings on scalpel blades then underwent various mechanical tests: cutting performance, hardness changes, and scratch durability. Cutting performance was tested by an experienced dissection technician cutting raw chicken with the skin intact using two coated and two uncoated blades. Four types of cuts were performed: plunge cut, 4-6 inch longitudinal cut, skin and tissue separation cut, and a cut that strikes the bone. This allowed for a subjective feedback on cutting feel and sharpness. Hardness change was tested on a Rockwell C scale in five locations (two on the body and three along the edge of the blade) on two of each type of treatment: (1) uncoated scalpels, (2) heat treated scalpels that underwent the time-temperatures protocol for growing CICNT, but without the carbon source, (3) CICNT-coated scalpels, (4) heat treated scalpels quenched in oil directly following the "infiltration" step, and (5) CICNT-coated scalpels quenched in oil directly following infiltration. The CICNT coating's scratch durability was tested with a diamond scribe scratching a coated scalpel blade nine times: three different pressures (light, medium, strong) in three locations. These were then SEM imaged for visual results.

Finally, the scalpel blade coating was tested for MRSA biofilm resistance. Three coated and uncoated scalpel blades were placed in a CDC biofilm reactor with MRSA-inoculated tryptic soy broth (TSB). The reactor was placed on a hot plate set to 34°C and 130 rpm for 24 hours. To prepare for SEM imaging, samples were rinsed in phosphate-buffered saline (PBS), fixed with glutaraldehyde, dehydrated with ethanol, and coated with gold [14]. SEM images were taken at 5000x magnification in five locations: three along the sharpened edge and two on the body of the blade. This magnification was used to view the morphology and spread of MRSA cells.

2.2 External Fixator Pin Fabrication and Testing

External fixator pins were coated with CI-CNT using two CNT catalyst methods: layered and direct. The pins were 316L SS rods (McMaster 89325K91) with a 0.25-inch diameter and approximately 1-inch length. An uncoated SS rod was used as a control.

The layered method produced a CNT forest by a using a traditional, externally-applied catalyst [15]. First, a diffusion barrier, an 80-nm thick layer of alumina, was deposited onto the pins by rotating the pins during an electron-beam deposition [15]. The pins were also rotated during the thermal evaporation of the CNT catalyst, a 4 nm thick layer of iron. A CNT forest was grown then infiltrated in a CVD tube furnace. The growth occurred at 750°C with hydrogen (flowing at 230 sccm) and ethylene (flowing at 250 sccm) for 1 minute, and the infiltration occurred at 900°C under the same gases and flow rates for 22 minutes.

The direct method produced a CNT forest by modifying the SS substrate to be a catalyst – similar to the scalpel blade coating fabrication. This was performed in the CVD furnace in continuity: heat treatment, CNT growth, then CNT infiltration. The heat treatment occurred at 800°C with air (flowing at approximately 200 sccm) for 5 minutes, the CNT growth occurred at 700°C with hydrogen (flowing at 230 sccm) and ethylene (flowing at 250 sccm) for 15 minutes, and the CNT infiltration occurred at 850°C with argon (flowing at 250 sccm) and ethylene (flowing at 250 sccm) for 25 minutes.

Three external fixator pins were tested for MRSA biofilm resistance – one of each type of pin. A pin that was coated by the layered method, one that was coated by the direct method, and one uncoated pin as a control were fixed to coupons using paraffin wax then placed in a CDC biofilm reactor under the same conditions as the scalpel blades for the first 24 hours. For the subsequent 24 hours, the nutrients in the broth were replenished by flowing fresh TSB diluted to 10% strength (~6.9 mL/min). After 48 hours, the pins were prepared for SEM imaging using the same process as the scalpels above. SEM images were consistently taken at the apex of the pins at 1,000x and 10,000x to view the morphology and spread of MRSA cells.

3. RESULTS AND DISCUSSION

3.1 Scalpel Blades

3.1.1 CICNT Coating Morphology

The CICNT coating on the scalpel blades was randomly-oriented and mostly uniform (Figure 1). There were no distinct patches of overgrown CICNT that are sometimes called bush-like growths [17]. Another observation was that the body of the blades had longer CICNT and more CICNT height variation than the edge of the blades. A randomly-oriented forest was expected because most methods for growing CNT using SS as the substrate and catalyst do not produce vertically-aligned forests [18]. Because the surface structure is still similar to vertically-aligned CICNT on the nanoscale, a randomly-oriented forest has previously shown evidence of biofilm resistance properties [1].

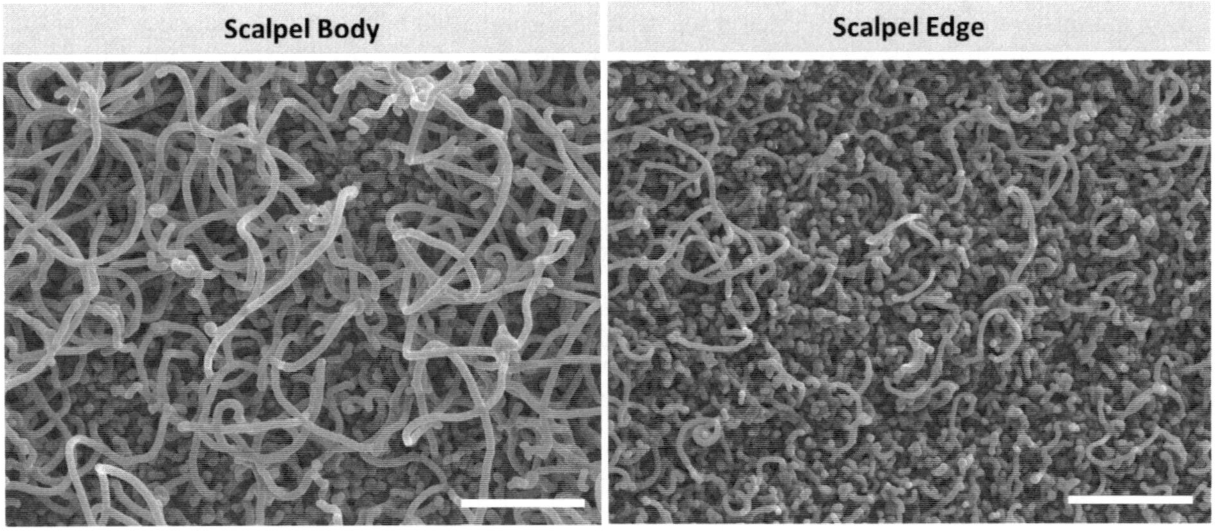

Figure 1: Top view of the body (left) and blade edge (right) of a CICNT-coated scalpel blade. Scale bars are 5 μm (10,000x magnification).

3.1.2 Cutting Performance

An experienced dissection technician performed the various cuts through raw chicken using the CICNT-coated scalpel blades to test their function. The coated and uncoated blades successfully dissected the chicken using a plunge cut, a 4-6 inch longitudinal cut, a skin and tissue separation cut, and a cut that strikes the bone. There were no black spots indicating detachment of CICNT in the chicken; however, the technician feedback indicated that the coated blades were less sharp than the uncoated blades. Since a blade's sharpness over prolonged use is directly correlated to its hardness, this prompted a look into a hardness recovery process for the CICNT-coated blades. This is detailed in the next section.

3.1.3 Scalpel Hardness

The CICNT fabrication process may affect the scalpel's hardness, and because a material's hardness determines how sharp a blade stays, this property was investigated. The hardness of a blade with no treatment ("Uncoated") was compared to four other treatments. Two treatments were treated and cooled in the furnace ("Heat Treat" and "CICNT"), and two treatments were quenched in oil after the 800°C infiltration in order to harden the blades ("Heat Treat & Quench" and "CICNT & Quench"). The oil quenched blades showed increased hardness (Figure 2). The blades that went through the CICNT process without carbon gas ("Heat Treat" and "Heat Treat & Quench") each show decreased hardness compared to their CICNT counterparts (Figure 2).

The results indicated that the temperatures required to coat scalpels with CICNT significantly reduce the hardness of the blade, but this hardness can be almost completely recovered. The heat treatment without CICNT reduced the hardness from an average of 53 HRC to an average of 31

HRC. The carburizing from the ethylene during the growth and infiltration of CICNT caused the hardness to increase to an average of 41 HRC. Approximately the same amount of hardness could be recovered by an oil quench directly following the "infiltration" step during the heat treatment (average of 43 HRC). The hardness was nearly recovered by combining the carburizing from the CICNT growth and infiltration with the oil quench directly following the infiltration step (average of 48 HRC). This shows that a CICNT-coated scalpel blade could be nearly as capable at maintaining a sharp edge.

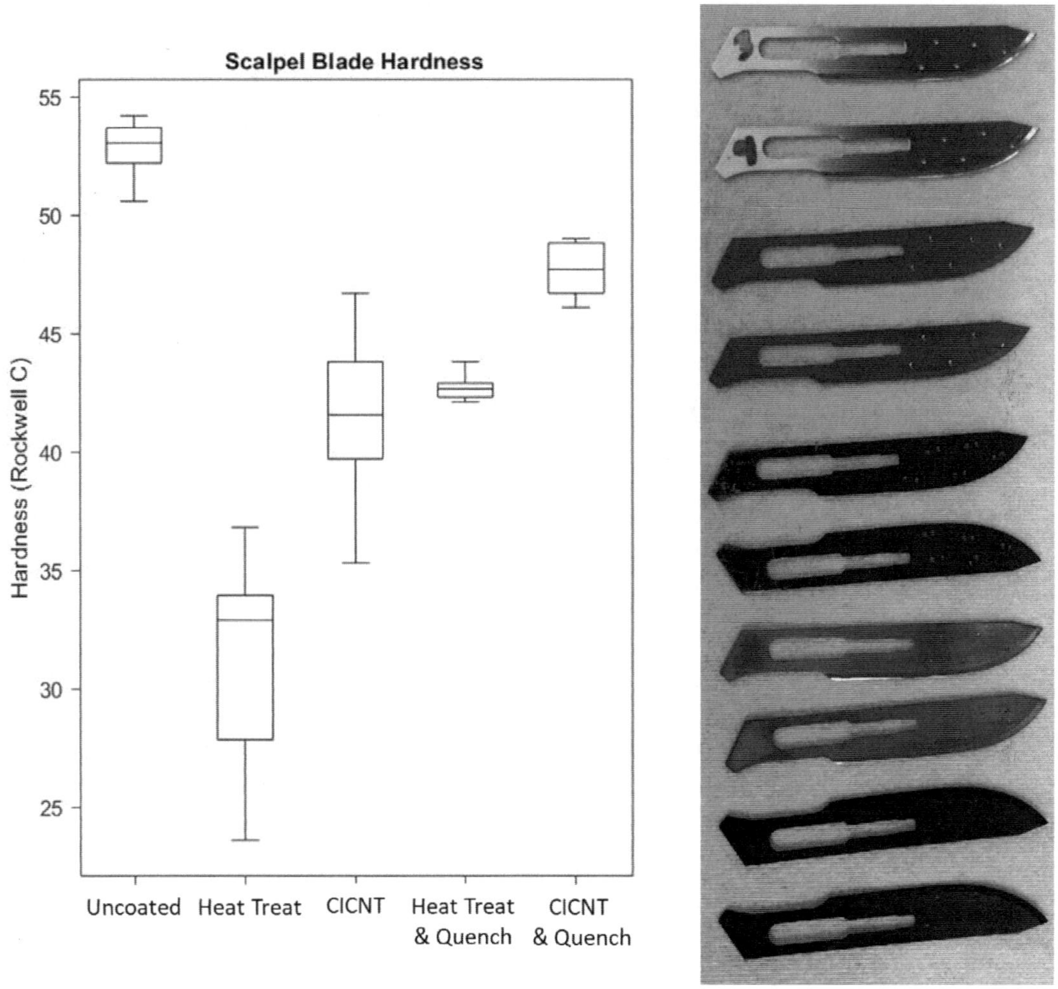

Figure 2: Scalpel hardness measurements for various treatments (left). Photo of the scalpel blades measured (right). Indentations show the measurement locations. Treatment pairs from top to bottom: uncoated, heat treated, CICNT-coated, heat treated and quenched, CICNT-coated and quenched.

3.1.4 Scratch Durability

CICNT-coated scalpel blades were able to withstand the hardness of a diamond scribe scratch with only a local disturbance rather than delamination. SEM images of the scratches are shown in Figure 3. As the force increased, both surfaces naturally exhibited a deeper and wider disturbance. Compared to bare scalpel, the disturbance was wider on the CICNT-coated blade than the uncoated blade. As measured in the previous section, the hardness of the CICNT-coated blades is reduced without an oil quench, so the wider disturbance may be due to the 12 HRC reduction in hardness.

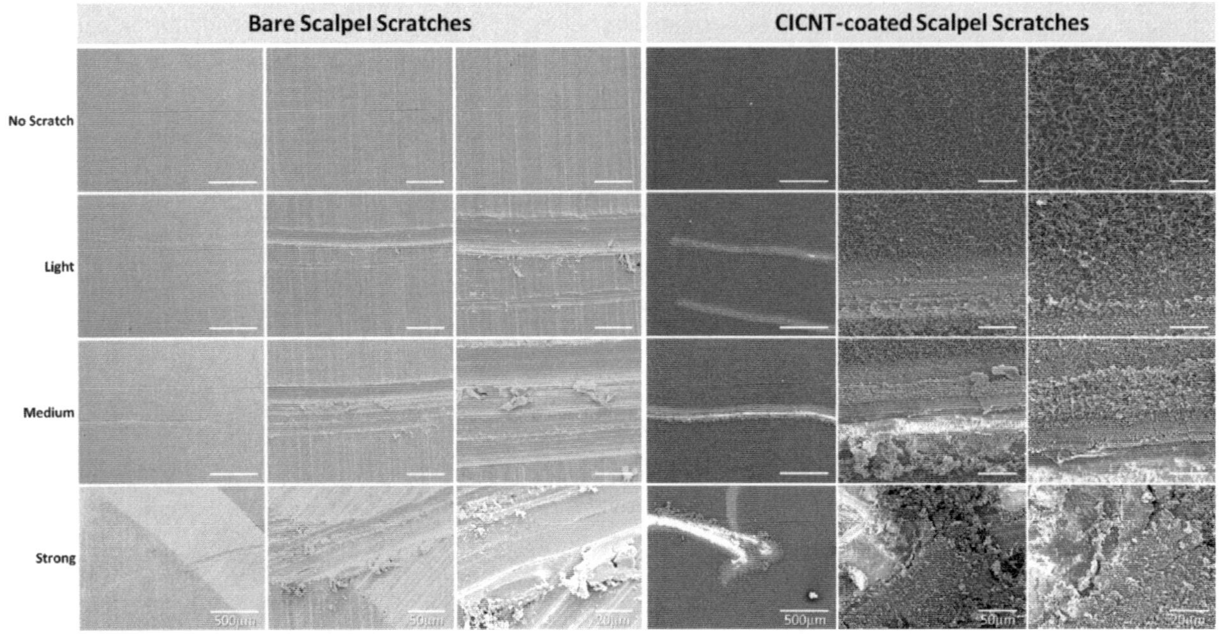

Figure 3; SEM images of scratched scalpels. Bare scalpel scratches (left set of images) and CICNT-coated scalpel scratches (right set of images) with increasing magnifications left to right (scale bars: 500 μm, 50 μm, and 20 μm for each set) and increasing pressures top to bottom (no scratch, light pressure, medium pressure, and strong pressure) are shown.

3.1.5 MRSA Biofilm Resistance

Following 24 hours in the MRSA biofilm reactor, the CICNT-coated scalpels allowed more biofilm attachment than the uncoated scalpels (Figure 4); however, comparing these results to a CICNT coating that resists biofilm expands our understanding on critical dimensions needed for a nanostructure to resist MRSA biofilm. The main difference between the scalpel coating and other successful CICNT coatings is that the CICNT heights on the scalpels were significantly more variable. This variation allowed the longer nanotubes to spread out throughout the surface, and this may affect the results in a variety of ways.

Because biofilm formation is directly related to the shear stress caused by the fluid dynamics from the rotating paddle [19], the flow dynamics at the surface may be different on this CICNT-coating than on the CICNT-coated fixator pins (Section 4.2) or on the coatings reported in previous experiments [1]. The additional roughness caused by the extra-long CICNT may have altered the shear stress to a more optimal level for biofilm formation. The individual CICNT may have caused additional flow disturbances on the nanoscale as the broth flowed over all the nanotubes.

Because biofilm formation is also directly related to the stiffness of its substrate [20], the unusual result could be caused by the lower stiffness in the longer nanotubes. There is an optimal stiffness for S. aureus biofilm formation [20], and MRSA most likely has the same response. Modeling the extra-long nanotubes as beams, their stiffness would be significantly reduced because the length of the beam, or nanotube, variable is to the fourth power. Therefore, although CICNT has a high modulus of elasticity [21], the stiffness of the scalpel blade coating may have changed enough to become optimal for MRSA biofilm formation.

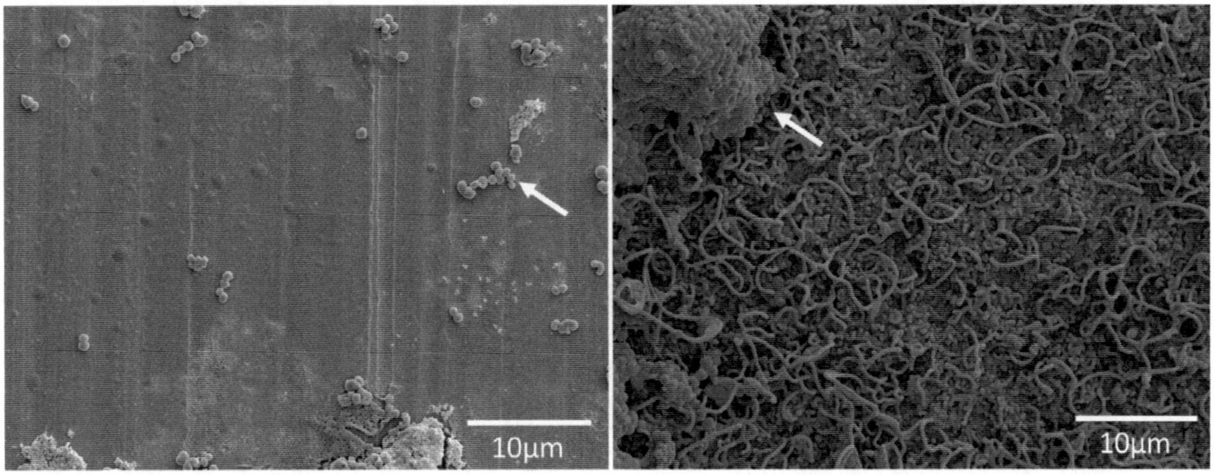

Figure 4: SEM images taken on the body of the uncoated (left) and CICNT-coated (right) scalpels after 24 hours in a MRSA biofilm reactor. The largest groups of MRSA cells on each image are indicated by arrows. The CICNT-coated blade had more biofilm formation.

3.2 External Fixator Pins

3.2.1 CICNT Coating Morphology

The CICNT fabrication method using a layered catalyst produced a vertically-aligned forest, and the method using a direct catalyst produced a randomly-oriented forest. SEM images of these are shown in the left columns of Figure 5. The CICNT forest from the layered method has relatively smooth and uniform surface, whereas the CICNT forest from the direct method shows bush-like growths.

3.2.2 MRSA Biofilm Resistance

Following 48 hours in a MRSA biofilm reactor, SEM images of the fixator pins showed significantly less biofilm formation for both types of CICNT coatings compared to the bare SS pin (Figure 5). MRSA cells were identified by their spherical shape, smooth edges, and roughly 0.8μm diameter, and then marked in Figure 6. The bare SS pin was entirely covered with MRSA cells, while there are only a handful of cells on the pins with a CICNT coating. The cell aggregation on the bare SS pin indicated mature biofilm formation on the surface, and the lack of cells attached to both types of CICNT coating indicated biofilm resistance. This biofilm resistance has an advantage over other anti-bacterial coatings, such as pharmaceuticals or silver, because its mechanism does not kill the bacteria.

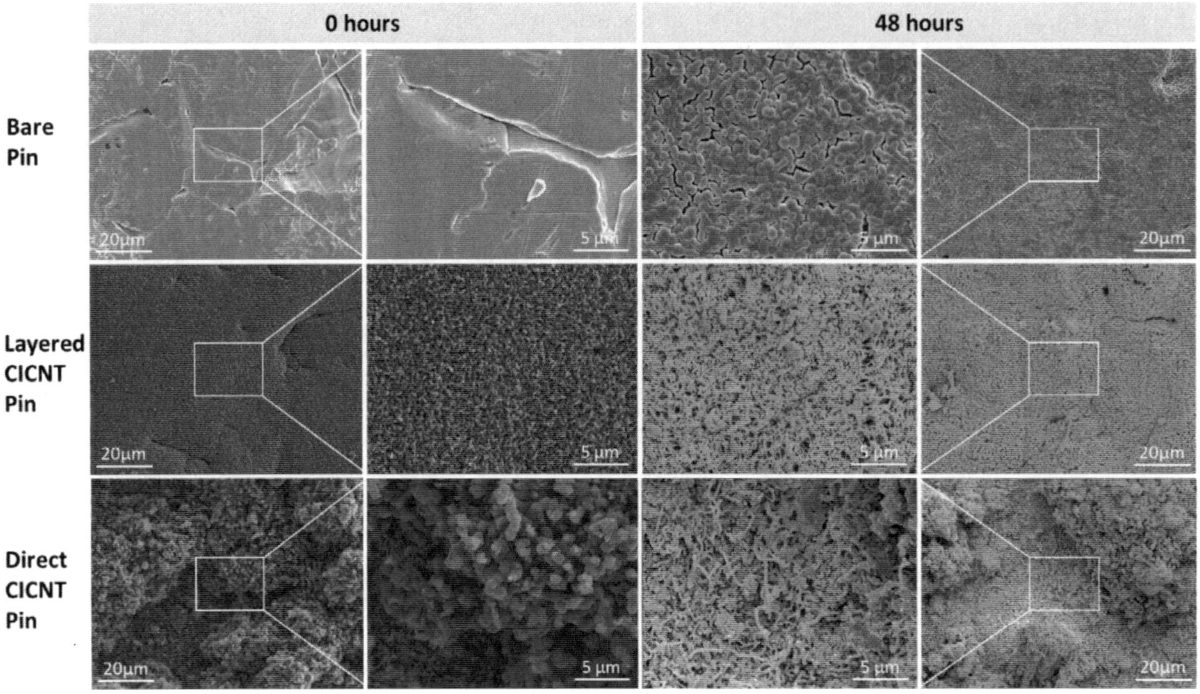

Figure 5: SEM images of each material before (left) and after (right) the experiment. After 48 hours, MRSA cells were identified by their spherical shape, smooth edges, and roughly 0.8µm diameter. Magnifications are 1,000x and 10,000x.

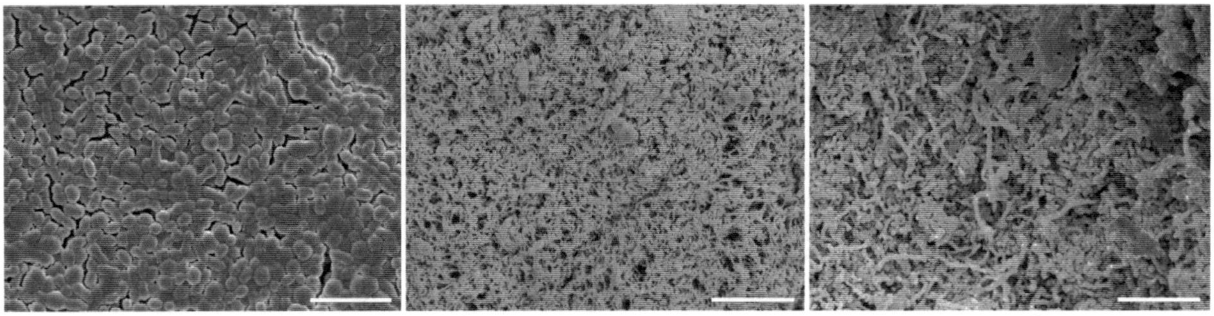

Figure 6: MRSA cells marked in red for each material: bare SS, layered CICNT pin, and direct CICNT pin. Image locations are the same as those in Figure 5. Scale bars are 5 µm.

3.3 Limitations and Future Work

As with any study of a new coating, there are limitations that provide more areas to research. More extensive testing must be done to further understand and verify the MRSA biofilm resistance seen from CICNT coatings. In general, these tests would include additional variables such as physical variations of CICNT, higher sample sizes, additional bacterial strains, and multiple experiment lengths. Specifically, since this study showed exceptional MRSA biofilm resistance on external fixator pins coated with CICNT, more testing must be done to ensure that the CICNT coatings promote skin-implant integration because infection is regularly the result of this lack of integration [22].

The disturbance of the CICNT coating caused by the diamond scribe during the scratch durability testing likely removed some CICNT from the substrate. This may be a problem in vivo if the infiltration is removed from the CICNT to expose the CNT to the body. Although CICNT have shown biocompatibility, it is questionable if bare CNT is biocompatible because the evidence is inconclusive and there are multiple factors that affect its toxicity [23, 24]. Additional testing would need to be done.

4. CONCLUSIONS

Because CICNT-coating that has been shown to resist biofilm formation based on its nanostructure rather than its chemistry [1], a solution for SSI was proposed by coating scalpel blades and external fixator pins with CICNT and subjecting them to mechanical and MRSA biofilm tests. On the scalpel blades, the CICNT coating did not exhibit superior performance to a bare scalpel blade, but this work provides greater insight on CICNT properties. The hardness that was lost during the CICNT manufacturing process was recoverable, and the coating did not delaminate under the force of a diamond scribe. Although the CICNT-coating on scalpel blades did not show MRSA biofilm resistance, the variables critical for the CICNT nanostructure to resist biofilm are better understood. The external fixator pin application proved to have significantly more promise than the scalpel blades. The bare SS pin was completely covered in MRSA biofilm, while both the vertically-aligned and randomly-oriented CICNT-coated pins each allowed only a few cells to attach. Although there is more testing to be done, we are optimistic that a CICNT coating will prove to be an exceptional way to prevent infection while reducing antibiotic use.

5. REFERENCES

1. Morco, S., et al., Structural Antibacterial Properties of Carbon-Infiltrated Carbon Nanotube Coatings, in BMES. 2016.
2. Hu, J.J., et al., Tribological behavior and graphitization of carbon nanotubes grown on 440C stainless steel. Tribology Letters, 2005. DOI: 10.1007/s11249-005-5091-6.
3. Ban, K.A., et al., American College of Surgeons and Surgical Infection Society: Surgical Site Infection Guidelines, 2016 Update. J Am Coll Surg, 2017. DOI: 10.1016/j.jamcollsurg.2016.10.029.
4. Anderson, D.J., et al., Strategies to prevent surgical site infections in acute care hospitals: 2014 update. Infect Control Hosp Epidemiol, 2014. DOI: 10.1086/676022.
5. Dryden, M., Prosthetic joint infection: managing infection in a bionic era. J Antimicrob Chemother, 2014. DOI: 10.1093/jac/dku246.
6. Bjerke-Kroll, B.T., et al., Periprosthetic joint infections treated with two-stage revision over 14 years: an evolving microbiology profile. J Arthroplasty, 2014. DOI: 10.1016/j.arth.2013.09.053.
7. Tande, A.J. and R. Patel, Prosthetic joint infection. Clin Microbiol Rev, 2014. DOI: 10.1128/CMR.00111-13.
8. Calina, D., et al., Antimicrobial resistance development following surgical site infections. Mol Med Rep, 2017. DOI: 10.3892/mmr.2016.6034.

9. Alaee, F., et al., General Assembly, Prevention, Operating Room - Surgical Technique: Proceedings of International Consensus on Orthopedic Infections. J Arthroplasty, 2019. DOI: 10.1016/j.arth.2018.09.064.

10. Guerado, E., J.R. Cano, and F. Fernandez-Sanchez, Pin tract infection prophylaxis and treatment. Injury, 2019. DOI: 10.1016/j.injury.2019.03.044.

11. Bliven, E.K., et al., External fixation of the lower extremities: Biomechanical perspective and recent innovations. Injury, 2019. DOI: 10.1016/j.injury.2019.03.041.

12. Shirai, T., et al., Prevention of pin tract infection with iodine-supported titanium pins. J Orthop Sci, 2014. DOI: 10.1007/s00776-014-0561-z.

13. Jennison, T., M. McNally, and H. Pandit, Prevention of infection in external fixator pin sites. Acta Biomater, 2014. DOI: 10.1016/j.actbio.2013.09.019.

14. Baddour, C.E., et al., A simple thermal CVD method for carbon nanotube synthesis on stainless steel 304 without the addition of an external catalyst. Carbon, 2009. DOI: https://doi.org/10.1016/j.carbon.2008.10.038.

15. Williams, D.L. and R.D. Bloebaum, Observing the Biofilm Matrix of Staphylococcus epidermidis ATCC 35984 Grown Using the CDC Biofilm Reactor. Microscopy and Microanalysis, 2010. DOI: 10.1017/S143192760999136x.

16. Robison, W., B. Jensen, and A. Bowden, Patterned Carbon Nanotube Growth on Convex Cylindrical Stainless Steel Substrates for the Production of Coronary Stents. 2016. DOI: 10.1115/DETC2016-60243.

17. Prawitt, D.G., D. Porter, and B.D. Jensen, Patterned Carbon Nanotube Growth on Stainless Steel in International Design Engineering Technical Conferences and Computers and Information in Engineering Conference. ASME: Quebec City, Quebec, Canada. 2018 DOI: https://doi.org/10.1115/DETC2018-86130.

18. Zhuo, C., et al., Oxidative heat treatment of 316L stainless steel for effective catalytic growth of carbon nanotubes. Applied Surface Science, 2014. DOI: 10.1016/j.apsusc.2014.05.189.

19. Williams, D.L., et al., Growth substrate may influence biofilm susceptibility to antibiotics. PloS one, 2019. DOI: 10.1371/journal.pone.0206774.

20. Yu, C., et al., Nanoprobe-based force spectroscopy as a versatile platform for probing the mechanical adhesion of bacteria. Nanoscale, 2019. DOI: 10.1039/c8nr10338k.

21. Hanna, B.H., et al., Mechanical Property Measurement of Carbon Infiltrated Carbon Nanotube Structures for Compliant Micromechanisms. Journal of Microelectromechanical Systems, 2014. DOI: 10.1109/jmems.2014.2312847.

22. Peramo, A. and C.L. Marcelo, Bioengineering the skin-implant interface: the use of regenerative therapies in implanted devices. Ann Biomed Eng, 2010. DOI: 10.1007/s10439-010-9937-1.

23. Mohanta, D., et al., Carbon nanotubes: Evaluation of toxicity at biointerfaces. Journal of Pharmaceutical Analysis, 2019. DOI: https://doi.org/10.1016/j.jpha.2019.04.003.

24. Ignat, S.R., et al., Versatile Biomaterial Platform Enriched with Graphene Oxide and Carbon Nanotubes for Multiple Tissue Engineering Applications. Int J Mol Sci, 2019. DOI: 10.3390/ijms20163868.

A FACTORY-CENTRIC WORKFORCE DEVELOPMENT APPROACH FOR AEROSPACE INDUSTRY

Navid Zobeiry[1], Charles Seaton[2], Marco Salviato[3], Xu Chen[4], Ashis Banerjee[4, 5], Santosh Devasia[4], Jihui Yang[1], Agnes Blom-Schieber[6], James Buttrick[6], Samuel Pedigo[6]

[1] Materials Science & Engineering, University of Washington, Seattle, WA

[2] Consultant and Instructor for FAA on Composite Materials Education, Seattle, WA

[3] Aeronautics & Astronautics, University of Washington, Seattle, WA

[4] Mechanical Engineering, University of Washington, Seattle, WA

[5] Industrial & Systems Engineering, University of Washington, Seattle, WA

[6] The Boeing Company

ABSTRACT

Given the rapid transformation of the aerospace sector in the last decade, teaching practices should prepare engineers to face fast-paced industries that are dealing with exceedingly complex problems. Now more than ever we need engineers who are capable of working and communicating effectively within large and multi-disciplinary groups, considering the introduction of new material systems such as advanced composites in primary structural elements, development of automated processing methods such as Automated Fiber Placement (AFP), and the transition to interconnectivity among production systems, workers, products and customers. Exceedingly we need to train well-rounded and practical-minded engineers and scientists. At the University of Washington, we are dedicated to support the aerospace industry by training the next generation of engineers. A new 16,000 sq. ft. facility, Advanced Composites Center (ACC), will be dedicated for manufacturing of aerospace composite parts. Taking advantage of automation and AFP processing, sensor technologies, autoclaves and other manufacturing equipment, we aim to replicate a small factory within the university environment at the ACC. Partnering with industry, this facility will enable students to gain practical experience working on industry relevant problems within a factory setting. In addition, while working on industrial projects, students will gain experience on business, intellectual property (IP) and project management aspects.

Keywords: Workforce development, Aerospace, Composites Manufacturing, Industry 4.0
Corresponding author: Navid Zobeiry
Boeing release-of-information record: #20-162593-BCA

1. INTRODUCTION

Training of aerospace engineers currently relies on established teaching practices at universities to deliver fundamental concepts of materials science, data science, manufacturing, mechanics and structural engineering in classrooms followed by limited hands-on experience in laboratories. In recent years, more emphasis is placed on project-based teaching approaches and flipped classroom ideas where students first learn fundamentals on their own using pre-recorded online materials, and then gain practical experience while working on group projects in classrooms/laboratories.

For composites, typical courses at universities cover topics such as introduction to raw materials (resin, fiber, fabric architecture, prepreg, sandwich core…), processing methods (prepreg hand lay-up, Resin Transfer Molding…), and heavy emphasis on mechanics (Laminated Plate Theory, failure theories, sandwich theory…). Hands-on experience may include introduction to fabrication methods such as prepreg hand lay-up, bagging and autoclave curing [1]. In some cases, emphasis may be placed on more advanced topics such as thermal management during manufacturing, Non-Destructive Inspection (NDI), repair, and safety considerations in design, manufacturing and maintenance. Toward completion of the degree, capstone projects defined in collaboration with industry, give students a rare opportunity to gain practical experience while working on industry-relevant topics.

However, considering specific needs of the aerospace industry, topics such as building-block approach, certification and substantiation of composite aircraft, maintenance, and size and production scaling considerations in manufacturing are largely neglected in typical university course syllabi. Our teaching practices must evolve to train well-rounded and practical-minded engineers equipped with necessary skill sets to address industry needs. The needs of the aerospace industry have changed in the past two decades, with the implementation of automated and semi-automated fabrication methods such as Automated Fiber Placement (AFP), the formulation and application of science-based process simulation tools [2], and more recently, the embracement of industry 4.0 concepts to use factory data for machine learning and data-driven modeling [3].

The Federal Aviation Administration (FAA) has focused on creating industry curriculum standards for composite materials on three main functional disciplines: Manufacturing, Maintenance and Structural engineering [4-6]. Participation is mostly online by practicing engineers and non-degree seeking students through Continuing Education Programs. Here are some of the offered courses by FAA:

- *Composites Manufacturing Technology Safety Awareness* (CMgfT) [7], which covers topics such as introduction to raw materials, processing, machining, inspection and bonding.

- *Composite Structural Engineering Technology* (CSET) [8], which focuses on certification of composite civil aircraft structure and cover topics such as engineering, manufacturing, maintenance, certification and structural substantiation. Additional topics such as flutter, crashworthiness, fire safety and lightning protection are also covered. An optional hands-on lab [9] introduces students to basic prepreg cutting, hand layup, bagging, NDI and scarf repair techniques.

By comparison of the above courses with typical course syllabus at universities, the lack of focus at the university level on relevant topics required for practicing aerospace engineers is highlighted. Moreover, it shows that there is room to enhance the learning experience in current workforce development programs by introducing factory-centric hands-on experience, relevant case studies, and by putting more emphasis on project-based teaching. Critical to this learning experience is the question of how to define an effective workforce development program within the university environment which accelerates the acquisition of necessary skills by students to be successful in the aerospace industry.

At the University of Washington, we are setting up a unique facility focusing on advanced composites which incorporates engineering, manufacturing and maintenance disciplines in an integrated product team environment. This new 16,000 sq. ft. facility, the Advanced Composites Center (ACC), will replicate a small-scale factory equipped with necessary equipment, including an AFP machine, autoclaves/ovens, hot press, automatic ply cutter, NDI and mechanical testing equipment. This new facility will provide a unique platform to create a factory-centric workforce development program to train next generation aerospace engineers by linking academic knowledge of composites with the needs of industry and practicing engineers, mechanics and factory workers. The ACC provides opportunities for students and engineers in training to gain practical experience on manufacturing, maintenance and engineering practices, and can be used to educate university students and serve as a platform for the continuing education of practicing engineers. We will build upon current FAA established courses by offering fundamental courses in a variety of teaching venues, including classroom and asynchronous/synchronous remote and online methodologies to provide accessibility to a wide audience. Relevant case studies which apply learning content to real-world situations will be delivered in advanced classrooms for online participation and discussions by remote subject matter experts to enhance the learning experience. Future research projects at ACC will be used to further develop case studies to enhance the learning experience of students [10]. The educational content and venue for the proposed workforce development program are discussed next along with an introduction to the new ACC center at the University of Washington.

2. EDUCATION CONTENT AND VENUE

Building upon current FAA courses [7-9], we propose a new workforce development program with the following four main components:

- **Fundamental courses** to introduce students to topics such as raw materials, processing methods, building block approach, certification and substantiation, maintenance and repair, and design. These courses will be given in advanced classrooms equipped for online participation.

- **Case studies** to enhance the learning experience of participants and to introduce important lessons learned in the aerospace industry. These will be combined with the fundamentals in advanced classrooms.

- **Hands-on labs** at ACC to gain experience in manufacturing, machining, NDI, maintenance and repair. ACC will be equipped for online participation and virtual tours.

- **Industrial/research projects** for students and engineers-in-training to gain practical experience while working on industry-relevant topics in multi-disciplinary groups.

The university must involve industry to define research projects and case studies for the workforce program, while soliciting industry feedback for continuous improvement to make this a sustainable program. Fundamental courses will be an overview of the practice and challenges of certifying commercial composite aircraft using composite materials, and would include the subjects of maintenance, manufacturing and structural engineering. Following established FAA course content curriculum standards, we propose the following two courses for fundamentals:

- Course I: Composite Materials Maintenance and Manufacturing
  - Essential properties and characteristics of composite materials
  - Manufacturing processes (raw materials, transport and storage, curing and solidification, machining, NDI, bonding and assembly)
  - Maintenance (teamwork, safety, maintenance and repair, source documentation, damage detection, characterization and repair substantiation, repair processes, regulations overview)
  - Engineering design for manufacturing and maintenance (design considerations, thermal management, manufacturing implementation, maintenance implementation)
- Course II: Composite Materials Structural Engineering
  - Composite applications (challenges, integrated product teams)
  - Material, processing, and fabrication development (material and process control, defects and damage, protection of structure)
  - Design development (requirements, criteria and objectives, lamination theory and analysis methods, material allowables, design values and knockdown factors, structural bonding, structural bolted joints)
  - Structural substantiation (regulations and guidance, certification approaches, damage and defects, building block testing and analysis, large scale testing and modeling)
  - Flutter, crashworthiness, fire safety and lightning protection
  - Engineering special topics (maintenance: structural repair development and substantiation, manufacturing: certification conformity process)

The main components of the proposed workforce development program and venues are schematically shown in Figure 1 below.

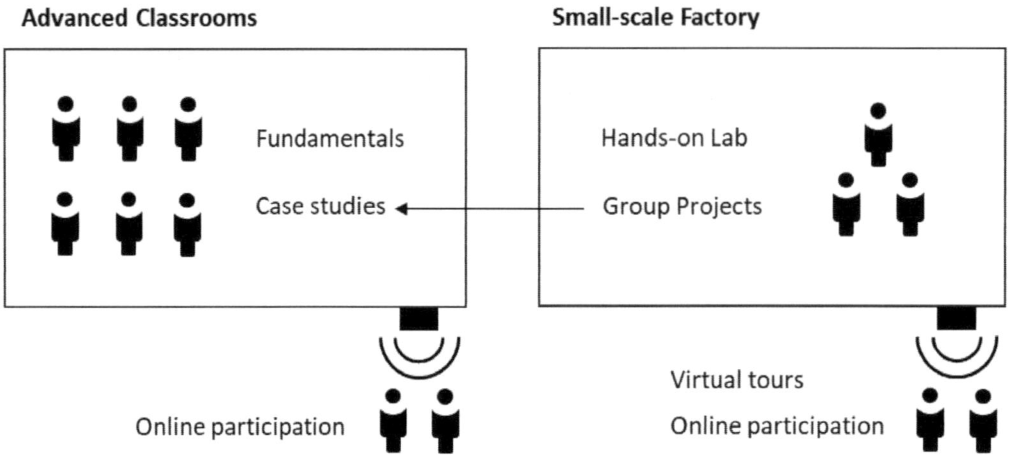

Figure 1: Schematic of the main components of the proposed workforce development program.

3. WORKFORCE TRAINING

Manufacturing and rework processes through factory Material Review Boards (MRB) will be replicated in the Advanced Composites Center as shown schematically in Figure 2. The center will be used as a testbed for the workforce development program described in the previous section. Given the recent transformations in manufacturing industry and also rise of machine learning approaches, it is essential for such a center to be equipped with sensors (e.g. infrared and digital cameras, thermocouples, pressure sensors…) for data mining. We propose that the manufacturing data to be used in the workforce program to support case studies while introducing students and engineers to data science and machine learning techniques.

Students and engineers in training can gain practical experience by participating in hands-on lab or industrial/research projects in such a center. Aside from research projects defined by industrial members, capstone projects for students can also be defined within the factory setting at such a center. Results may be used to develop case studies for fundamental courses where appropriate, depending on the intellectual property agreements of a project. We also propose that such a center to be equipped for virtual tours and online participation, either asynchronously through a Learning Management System (LMS) such as Canvas or Blackboard, or synchronously utilizing advanced classroom technologies.

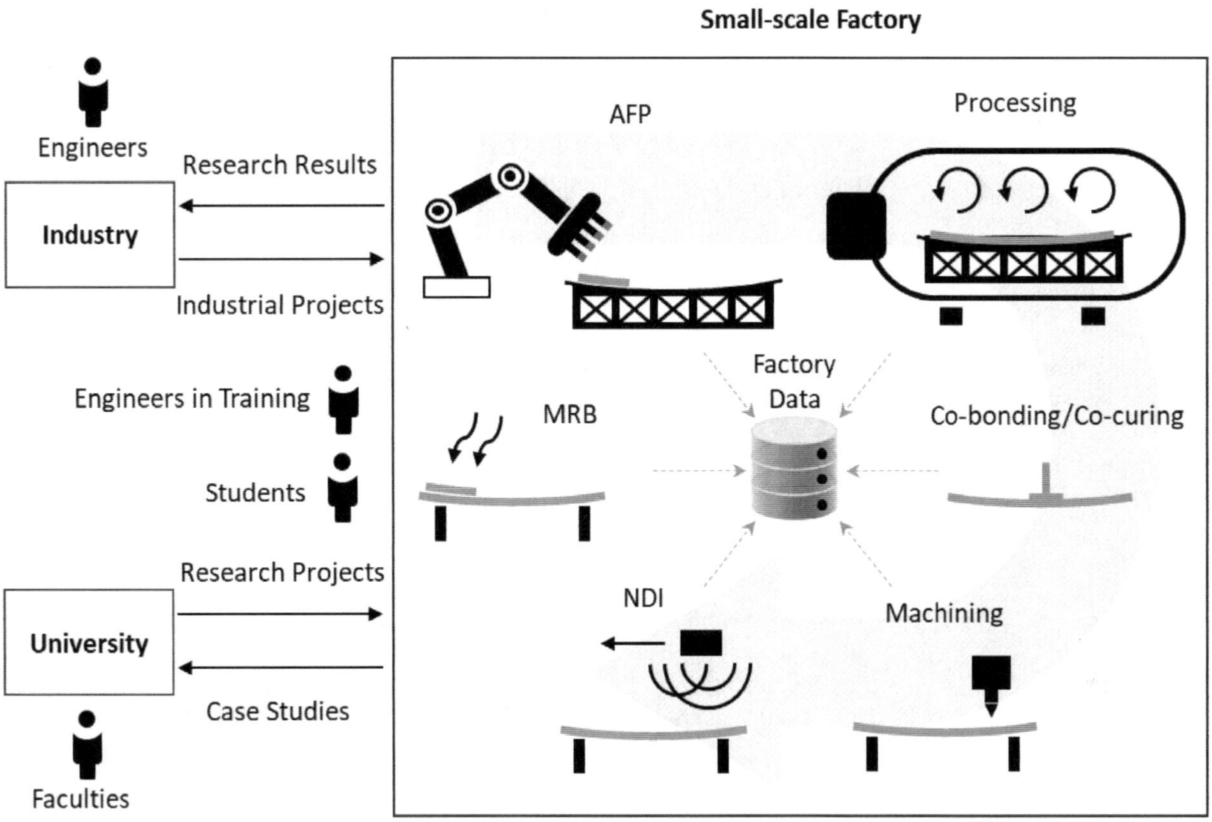

Figure 2: Workforce development program at the ACC.

4. RESEARCH DIRECTIONS

The mission of a similar center described in the previous section should be to provide a robust innovation ecosystem for industry and academia to advance the field of digital manufacturing and maintenance of advanced composites. In terms of fabrication processes, the focus should be on AFP processing and autoclave curing, but other fabrication methods such as hot drape forming, hot press processing and out-of-autoclave curing. Research projects should be defined in collaboration with industry partners or through government funding programs. Some of the main current research topics are listed here:

- Optimization of existing manufacturing processes using a combination of material/process characterization, process simulation, failure simulation, building block mechanical testing, and machine learning. Factory data will be used for characterization and optimization using machine learning. This is schematically shown in Figure 3

- Remote repair of composite parts in the service environment

- Research on different aspects of bonding including repair and quality control, and challenges of achieving thermal uniformity

- Investigating effects of defects by establishing process-microstructure-performance correlation using a combination of simulation, testing, and machine learning

- Developing new material systems/processes

- Investigating material aging and recycling methods

- Leveraging the acquired knowledge to develop certification guidelines for advanced composites

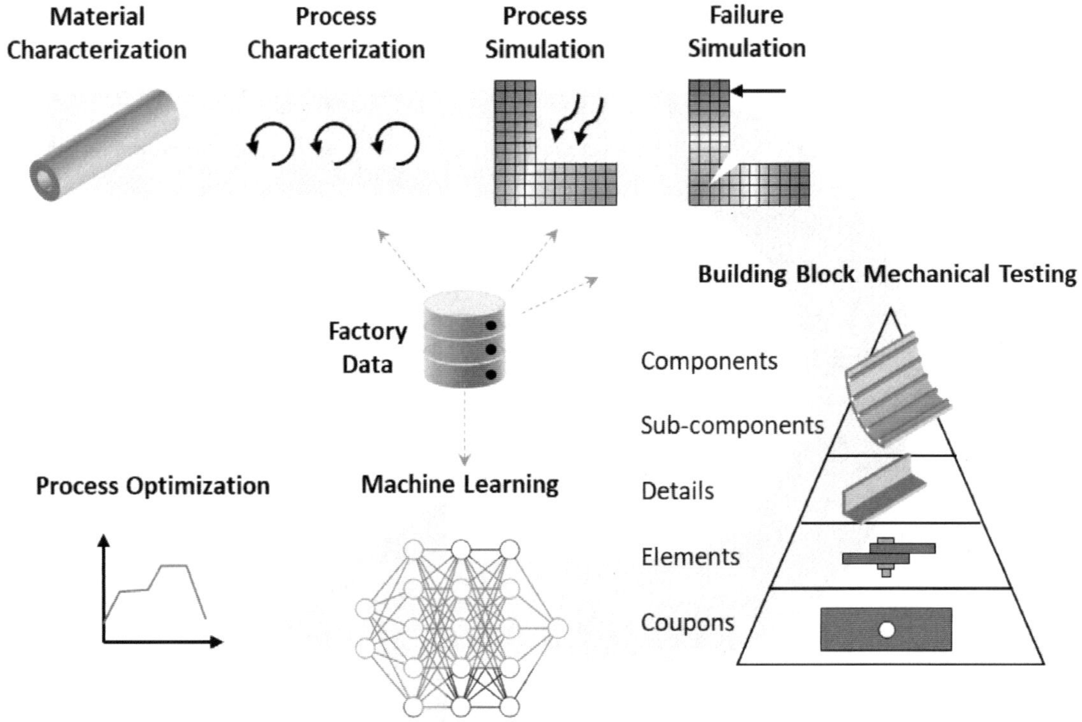

Figure 3: Research on process optimization at the Advanced Composites Center at the University of Washington.

5. SUMMARY

This paper proposes a new factory-centric workforce development program for aerospace engineers. The main components of the program are fundamental courses, case studies, hands-on labs and industrial projects in a variety of learning environments and venues. Fundamental courses will be developed based on established FAA courses and will be enhanced with relevant industrial case studies. While fundamental courses and case studies will be delivered in advanced classrooms and through remote learning management systems for online participation, hands-on labs and research projects will be conducted in Advanced Composites Center (ACC), a new research facility at the University of Washington. This 16,000 sq. ft. center will replicate a small-scale factory equipped with AFP and autoclave processing, and equipment for machining, NDI, and repair.

6. REFERENCES

1. G. Fernlund, C. Mobuchon and N. Zobeiry, *2.3 Autoclave Processing*, In: Beaumont P.W.R. and Zweben C.H. (eds.), Comprehensive Composite Materials II, 2:42-62, Oxford: Academic Press, 2018.
2. N. Zobeiry, A. Forghani, C. Li, R. Thorpe, K. Gordnian, R. Vaziri, G. Fernlund and A. Poursartip, *Multi-scale characterization and representation of composite materials during processing*, Philosophical Transactions of the Royal Society of London A: Mathematical, Physical and Engineering Sciences, DOI: 10.1098/rsta.2015.0278, 374:20150278, 2016.
3. N. Zobeiry and A. Poursartip, *Theory-guided machine learning for processing of advanced composites*, Advanced Materials, adma.201906492, 2020.
4. L. Ilcewicz, L. Cheng, J. Hafenricher and C. Seaton, *Guidelines for the development of a critical composite maintenance and repair issues awareness course*, FAA Technical Center Report (DOT/FAA/AR-08/54), 2009.
5. C. Seaton, D. Casteller, G. Derheim and H. Offermann, *Online course and regional laboratory development in composite maintenance education, including training report manual*, FAA Technical Center Report (DOT/FAA/AR-09/3), 2009.
6. C. Seaton, L. Ilcewicz, *FAA Composite Structural Engineering Technology Safety Awareness Course, based on work administered through FAA and Wichita State University*, FAA Technical Center Report (DOT/FAA/AR-TBD), 2020
7. FAA22005001, *FAA Composite Manufacturing Technology Safety Awareness course (CMfgT)*, based on work administered through FAA and facilitated by Convergent Manufacturing Technologies, AIR-520, 2018.
8. FAA27200078, *FAA Composite Structural Engineering Technology safety awareness course (CSET)*, based on work administered through the FAA and Wichita State University, facilitated by Charles Seaton, Consultant, AIR-520, 2018.
9. FAA27200080, *FAA Composite Structural Engineering Technology (CSET) Hands-on Lab*, AIR-520, 2018.
10. C. Seaton, S. Richter, *Nonconforming Composite Repairs: Case Study Analysis*, FAA Technical Center Report (DOT/FAA/TC-14/20), 2015

AUTHOR INDEX